THE
MATHEMATICAL PAPERS OF
ISAAC NEWTON
VOLUME VIII
1697-1722

The marked-up first page of the printer's copy for the 1704 'De Quadratura Curvarum' (1, 2, §3.1).

THE
MATHEMATICAL PAPERS OF
ISAAC NEWTON

VOLUME VIII
1697-1722

EDITED BY
D. T. WHITESIDE

WITH THE ASSISTANCE IN PUBLICATION OF
A. PRAG

CAMBRIDGE UNIVERSITY PRESS

CAMBRIDGE
LONDON   NEW YORK   NEW ROCHELLE
MELBOURNE   SYDNEY

CAMBRIDGE UNIVERSITY PRESS
Cambridge, New York, Melbourne, Madrid, Cape Town, Singapore, São Paulo

Cambridge University Press
The Edinburgh Building, Cambridge CB2 8RU, UK

Published in the United States of America by Cambridge University Press, New York

www.cambridge.org
Information on this title: www.cambridge.org/9780521201032

Notes, commentary and transcriptions
© Cambridge University Press 1981

First published 1981
This digitally printed version 2008

A catalogue record for this publication is available from the British Library

ISBN 978-0-521-20103-2 hardback
ISBN 978-0-521-04591-9 paperback
ISBN 978-0-521-72054-0 paperback set (8 volumes)

TO SIMON AND PHILIPPA

PREFACE

This, the disconnected mathematical chronicle of a great man's decline from vigorous middle age into an increasingly frail senescence rent by prolonged dispute with Leibniz and his adherents over discoverer's 'right' (so it was then understood) to the calculus, must inevitably be a volume tinged with some sadness. Apart from a general sorrow that we must all of us die (and grow old only to see our powers decline with the passing years) I myself feel a particular regret that Newton, for all his long life, never found anyone in his own image to whom he might properly pass on the many fruits of his mathematical genius, but was instead put barrenly to making hollow noises against the innovator of differential calculus and his lieutenant, Johann Bernoulli. It has certainly been my editor's good fortune to find, three centuries on, so many of his autograph writings which have survived intact yet still unpublished; but it is also a mark of Newton's own extensive failure to impress the fact— as distinct from the hearsay—of his mathematical achievement upon his contemporaries and successors that these papers have never before been printed, or even widely known to exist (let alone have their location specified).

When he left Cambridge for London and office at the Mint in April 1696 Newton was only a little over three months past his fifty-third birthday, with three decades of life yet ahead of him, but resolved henceforth to devote his attention to the 'Kings business', inevitable interruptions aside. As Eliot well knew, however, a world in decline ends not with a bang but a whimper. So in this eighth and last volume of Newton's collected mathematical writings and scribbles over the quarter-century from January 1697, when he responded successfully to Johann Bernoulli's challenge to identify the brachistochrone on his return home tired from a day's work, up to early 1722, when he corrected a small slip in the second edition of his *Arithmetica Universalis* even as its pages stood waiting in the printer's shop, there can be few surprises. Perhaps the only piece which reveals the same sustained fire and insight as those of Newton at the height of his maturity is the complicated tangle of texts wherein he in autumn 1712 located and corrected the previous error in Proposition X of Book 2 of the *Principia* at Bernoulli's behest. What other items are here gathered will, I hope, seem interesting and important enough to repay, partially at least, their present elaborate edition and reissue. Every-where, the effort has been made adequately but concisely to sketch in the back-grounds to the individual papers, though many times with the more isolated and fragmentary ones a 'tutored guess' (as they say) is really all I have been able to hazard as to their precise date and purpose. The scholarly world will in years to come judge with due severity the success or failure of my efforts.

As always, for liberty to print the greatest portion of the texts here set out I stand indebted to the Syndics of Cambridge University Library. One or two minor gaps in the coverage of this main holding of the scientific portion of the 'Portsmouth Collection' (as, unified but unexplored, it was till a century ago) are filled by other papers, deriving from William Jones, now in private possession and by related letters and documents in the Libraries of the Royal Society, London, and King's College, Cambridge: to their owners and custodians I likewise extend my thanks for permission here to print such texts of theirs as I have thought fit. For many years past the University of Cambridge has taken me under its wing, but I must also mention those other institutions, the Sloan Foundation of New York and the Leverhulme Trust, and also the Master and Fellows of Trinity College, Cambridge, who once joined so generously to give me financial support.

It is wholly just that my old friend and colleague Adolf Prag endures to share the title-page of this final volume of Newton's mathematical papers with me. In his seventies he remains the ever-willing, near-omniscient helper that he has always been, and without his furnishing and correction of a wide spectrum of matters literary, technical and historical this edition would have been much the poorer in its detail. Nor let me forget to repeat from the previous volume my dedicatory tribute to my erstwhile other mentor and friend Michael Hoskin for his unstinted warmth of encouragement over so many past years.

To the Syndics of the Cambridge University Press my thanks for supporting the vision of an untried young man of seeing into print a complete edition of Newton's mathematical papers: one which could so easily have rested a pipe-dream. To the numerous staff of the Press's Publishing and Printing Divisions who have in the two decades since had anything—and everything— to do with transforming my handwritten script to be the beautiful *objet d'art et de vertu* which it is here become, my special word of appreciation. Above all to my sub-editor these last sixteen years, Judith Butcher, my gratitude for so ably personifying in her care and intelligence and good sense all that is for me best in the traditions of the Press. But, in so saying, not to lose sight of the many others—draughtsmen, typesetters, copyreaders, all (too few of whom I have ever met face to face)—who have joined in putting their professional skills at the service of this edition.

In dedicating this volume, lastly, I fulfil a standing promise to my children. To my wife, Ruth, who has tolerated so much in me I owe everything.

D.T.W.
9 July 1980

EDITORIAL NOTE

This final volume closely follows its seven predecessors in its style and presentation, and we need not dwell overlong upon its few points of idiosyncrasy. Our basic purpose remains, as ever, to set out the text of Newton's far from always carefully written originals as faithfully as the linearities and other restrictions of the squared-off printed page allow; but where the original is deficient, physically or in logical sense, we have (when this has seemed not to betray the essential significance of Newton's intended meaning) not been shy in filling it out with verbal interpolations, and a few wider extrapolations, put within square brackets. So it is that, other things being equal, we have here preserved all of Newton's own contractions, superscripts and archaic, inconsistent spellings and capitalizations along with his—often total lack of—punctuation, mostly keeping our editorial reorganizations and restructurings to a minimum unless the jejuneness and illogicality of the autograph text presses us to make appropriate intercalation and transposition or other adjustment of its phrases. The English versions of Newton's main Latin pieces which are here set facing them (on right-hand pages) may, we hope, be read with profit in their own right, but their principal *raison d'être* is to be read in conjunction with their Latin parents (on the opposite preceding pages), and as such they have been kept deliberately literal. As before, however, we have not seen point in translating into English many of the minor Latin pieces which add only minimally to other, more elaborate texts which are here treated in full; nor are any of the secondary Latin writings cited in excerpt also rendered *Anglice* except in special circumstances. In both Latin and parallel English versions a twin vertical rule in the margin remains our editorial symbol that the passage so singled out has been cancelled by Newton in the original. Our standardizations of Newton's own symbols and mathematical notations mirror those employed in previous volumes, and we need not here tarry over them. Forwards and backwards reference within this volume, we would lastly remind, is sometimes made according to the convention '**1**, 3, § 2' (by which understand '[Part] **1**, [Section] 3, [Subsection] 2'), while we cite from earlier volumes by the code 'vII: 177–80' (which pinpoints '[Volume] vII: [pages] 177–80').

A special word about indexes. Here, as in previous volumes, we attach in preliminary (pages xxxi–lv below) an 'Analytical Table of Contents', and also append at the end of the book a short alphabetical 'Index of Names'. In introduction to the first volume (I: xiv) the reader was also promised in this final one 'an elaborate, systematically cross-listed general index' which would be 'a sufficient pointer to the macrocosm of individuals who, along with their works, have entered these pages'. Over the years, however, friends and colleagues have dissuaded us from having a single omnium-gatherum terminal index, convincing us of the over-riding advantages of breaking this into a variety of subordinate ones, and of placing these in a separate 'Index Volume'. And so (with the ready agreement of the Cambridge Press) it will now be. Here will be collected alphabetical name- and subject-catalogues in fairly conventional form, along with others listing *inter alia* the primary Newton autographs printed in the edition's eight volumes (ordered by location and manuscript sequence) and other primary and secondary sources there cited. Opportunity will also be taken to correct such editorial howlers—not many as yet!—as have surfaced in the seven earlier volumes, and of course to print any mathematical autographs of Newton's which may yet come to light.

GENERAL INTRODUCTION

In each of the previous volumes it has been our practice to introduce the portion of Newton's mathematical papers which we go on to edit by presenting an overview of the pattern of his day-to-day activities during the period, along with some brief comment upon his other intellectual interests at the time. During the more than three decades in which he lived at Cambridge, growing from immature young undergraduate to seasoned middle-aged professor in a tight academic community effectively shut off from the outside workaday world, such things seemed to us to be pertinent. Our running commentaries there, we felt, helped to build up a coherent picture of the character and style of the man who, in greatest part himself unaided except by the books which he read,[1] made the mathematical discoveries which we proceeded to chronicle in full (and maybe fulsome) detail by reproducing the texts of the autograph sheets on which he wrote them down. After April 1696, when Newton (then in his 54th year) exchanged the quiet isolation of his Fenland university for the busy, commercial whirl of the metropolis (as, first, Warden of the London Mint and soon, from 1699 onwards, its Master), he gave up forever an existence in which matters of the intellect were of central importance. There seems no point here in tracing even a broad outline of his new professional career as administrator of the nation's coinage. Suffice it to remind that his position gave him temporal prestige and a substantial income, while, after 1700 at least, allowing him free time enough to make use of both, notably in injecting new vigour (if a somewhat imperious one) into the then ailing Royal Society: of which he was elected President at the Anniversary meeting in 1703, and remained so till his death more than twenty years later on 20 March 1726/7.[2] For details of how he passed the last three decades of his life—in London and its environs, other than for a brief return to Cambridge in April 1705 to be knighted by Queen Anne[3] and also a first(!) outing to

---

(1) A *catalogue raisonné* of the more than two thousand books in Newton's possession at his death—founded on the eighteenth-century 'Huggins' and Musgrave listings of these (see I: 9, note (23)), but with the valuable addition of their present locations where these are known—is now available on pages 82–265 of John Harrison's *The Library of Isaac Newton* (Cambridge, 1978), there complemented (*ibid.*: 1–27 and 58–78 respectively) by perceptive essays on Newton's habits as a book-user and on the composition of his library by subject.

(2) He had earlier been a member of the Society's Council in 1697–8 and again in 1699–1700. On Newton's Presidency see pages 117–51 of Henry Lyons, *The Royal Society, 1660–1940. A History of its Administration under its Charters* (Cambridge, 1944).

(3) Followed within weeks (see v: 9, note (33)) by his rejection at the polls in Cambridge as a Whig candidate when the political tide was running strongly for the Tories. We may well choose to see political undertones in both Newton and the Queen journeying to Cambridge

Oxford in his seventy-fifth year[4]—we may refer the reader to the existing standard biographies of Newton,[5] fully adequate though none of these is. Where external event does impinge in the background to the documents which we present in following pages, it is of course duly and amply recorded in editorial preface and in relevant footnote.

The collective bulk of these scattered pieces (amassed here, it is true, from a whole quarter of a century of his life) will surely surprise the many who still cleave to the longstanding tradition[6] that Newton's old age was mathematically barren. Their general quality falls, of course, well below the level of his best writings of former days. Only for the rarest of moments did he in penning them experience the white heat of creative insight and discovery as

---

for the knighthood ceremony when it could so much more easily have been performed in London. 'It was talked among us', wrote Stukeley (then an undergraduate at Christ's) after describing the royal progress, 'that one purpose of the Queens, was to recommend...Newton to the choice of the University by her powerful progress; and chiefly projected by the Earl of Hallifax, minded that great man should receive from, and give honor to the world, in a more publick life' (*Memoirs of Sir Isaac Newton's Life by William Stukeley,...1752*, ed. A. Hastings White (London, 1936): 9–10).

(4) 'In August [1720] Sir Isaac went to Oxford in company of Dr John Kiel; he having not been there before' (Stukeley, *Memoirs*: 13). So far as we know, Newton's impressions of his sister Oxbridge city on this (only?) visit are not recorded.

(5) For all their deadweight of obsolete Victorian platitude the pertinent chapters of the second volume of David Brewster's *Memoirs of the Life, Writings and Discoveries of Sir Isaac Newton* (Edinburgh, 1855) are still not superseded; while the notes to the 'Synoptical View of Newton's Life' which Joseph Edleston set in introduction (pages xxi–lxxxi) to his edition of the *Correspondence of Sir Isaac Newton and Professor Cotes* (London, 1850) remain ever fresh in their precision of documentation. Of the two more recent full-scale attempts to comprehend Newton's life and his contemporary impact the relevant later chapters of L. T. More's *Isaac Newton: A Biography* (New York, 1934) are untrustworthy in their detail, superficial in their appreciation of scientific issues, and cloying in their frequent moralizings; the third part (pages 229–392) of F. E .Manuel's *A Portrait of Isaac Newton* (Cambridge, Massachusetts, 1968) is a vivid set of snapshots of its subject's activities 'In London Town' which adduces much new and interesting source material, both manuscript and printed, but many readers will rapidly become wearied by the unproven 'psycho-historical' (and especially sexual) posturings in which its accurate portrayal of the historical Newton is relentlessly dressed up. In *Sir Isaac Newton* (London, 1960)—a far more modest work—H. D. Anthony has some interesting things to say about his London residences and his work at the Mint; the latter is studied in much fuller detail by John Craig in his *Newton at the Mint* (Cambridge, 1946) and subsequent related essays. (For full details see P. and R. Wallis, *Newton and Newtoniana, 1672–1975* (London, 1977): 209. The wealth of letters, to, from and regarding Newton during the last thirty years of his life is now at last gathered in its essential entirety in the four concluding volumes (ed. J. F. Scott and A. R. Hall/L. Tilling) of *The Correspondence of Isaac Newton* (Cambridge, 1967, 1975/1976/1978). Just as we write this, news comes that R. S. Westfall has at last put to press his long-awaited 'intellectual' biography of Newton, but this will concentrate rather upon his scientific thought than on his day-to-day contacts with his contemporaries.

(6) Nurtured by Newton himself, to be sure; see below.

he had so many times previously known it. But he retained till his last years enough of the afterglow of that fire to be able at will competently to reshape unpublished earlier papers for the press—those, notably, 'On the quadrature of curves'[7] and on 'The method of [finite] differences',[8] but also (so it was his intention) his youthful 'Problems for construing æquations'[9]—as well as in a peerless way on occasion to locate and rectify a subtle slip in texts already allowed by him into print.[10] His readiness to make public response, anonymously as that may be,[11] to two widely separated challenges from Johann

---

(7) The 1704 *editio princeps* of this 'Tractatus de Quadratura Curvarum' joined to his 1693 'Geometriæ Liber secundus' (VII: 507–53) a newly composed prefatory 'Introductio' and terminal 'Scholium'; see **1, 2** below. This he had it in mind to publish in the spring of 1712 with the further propositions on methods of expansion into infinite series whose text is set out in **1, 5** following, but nothing came of the intention.

(8) See **1, 4, §2**. The text of this 'Methodus Differentialis', which he allowed William Jones to put into print for him in 1711, was in its greatest part a word-for-word reissue of the scheme of interpolation through central finite differences (IV: 52–68) which he had set down in his Waste Book around the autumn of 1676; the two propositions and final scholium newly appended (their preliminary drafts are given in **1, 4, §1**) add very little.

(9) See II: 450–516. This Newton reshaped and shortened, around the autumn of 1706, to be the tract 'De constructione Problematum Geometricorum' which is printed for the first time in **1, 3, §3** below.

(10) Notably in Proposition 10 of the *Principia*'s second book in its 1687 *editio princeps*, and in the concluding 'Æquationum Constructio linearis' of his old Lucasian lectures on algebra (= V: 54–490) which William Whiston brought out in 1707—to Newton's considerable early apprehension (see V: 9–11)—as his *Arithmetica Universalis; sive de Compositione et Resolutione Arithmetica Liber*. In **1, 6** below we set out the essential portion of the papers in which Newton, having been warned by Johann Bernoulli that there must be some error in it since he could show a particular instance of its result to be false (without being able properly to pinpoint wherein the proposition itself was mistaken), came over several days of intensive probing in late September and early October of 1712 to identify where he had gone wrong in the general reasoning of his *Principia* proposition, and then adroitly to correct it (multiply so, in fact, though in the second edition which appeared the next summer room could be found for only one of the alternative arguments). The considerably less subtle slip made by him in his early 1680's algebraic lecture in constructing the roots of a cubic equation through a wrongly described conchoidal mechanism (see V: 466–9) was seemingly caught by no one else during the fifteen years after the 1707 *Arithmetica* appeared, and was retrieved by Newton himself at a late stage in the passage through press of the second edition of 1722 (his preliminary computations are reproduced in **1, 9, §2** below) on a reprinted leaf (pages 315/316) pasted onto the stub of the clipped-out faulty one. This act furnishes us with the terminal date which in this edition of his papers we put to Newton's almost sixty-year-long career as a professional mathematician.

(11) No doubt this self-effacing keeping back of his author's name in publishing these replies had roots deep in his childhood (in the Puritanical doctrines of self-denial then ground into him) and in his natural reserve and withdrawn personality, but it also cloaked a more commonplace fear of leaving himself vulnerable to counter-attack if he showed himself more openly. We have already (see VII: xxv, note (62)) dwelt at some length on the blustering broadside which he fired at John Flamsteed on 6 January 1698/9 'upon hearing occasionally that you had sent a letter to Dr Wallis...&...mentioned me therein with respect to y^e

Bernoulli—first, in late January 1697, to define the curved path of fall under gravity in least time,[12] and also to solve a complicated problem in Cartesian geometry;[13] the second, commanded by Leibniz nineteen years afterwards in December 1715 as a test of the 'pulse of the English analysts', to determine the set of orthogonals to a given family of curves[14]—is evidence that he was never more than outwardly reluctant to apply his wise old master's hand to things mathematical. It is, sadly, an indication of the gradual decline of his power and grasp with the passing years that, whereas he was (in his middle 50's) entirely successful in resolving the first pair of problems, he found it impossible (as he neared his middle 70's) narrowly to come to grips with the latter 'défi'.[15] The deterioration in his mathematical sharpness was already apparent in between times when Newton twice failed to appreciate the same subtle paralogism in would-be solutions by Abraham de Moivre and John Craige (separately submitted to him for his private comment) to a further longstanding problem of Bernoulli's to 'transform a geometrical [*sc.* algebraic] curve into an infinity of others, likewise geometrical but of different species, which are each of the same length as the one proposed'; indeed, in jotting down his own skeleton of an equivalent 'solution' on Craige's paper he fell unwittingly into their common error of constructing but a linear transformation of the referent Cartesian axes which leaves the given curve itself

---

Theory of yᵉ Moon': 'I was', Newton went on, 'concerned to be publickly brought upon yᵉ stage about what perhaps will never be fitted for yᵉ publick....I do not love to be printed upon every occasion,[,] much less to be dunned & teezed by foreigners about Mathematical things or to be thought by our own people to be trifling away my time about them when I should be about yᵉ Kings business' (*Correspondence of Newton*, 4, 1967: 296). But it was the palpable inadequacies of his lunar theory which really bit at him there, leading him to blaze away at a passing, entirely innocent reference of Flamsteed's to it.

(12) See pages 72–4 below. Precisely how Newton arrived in 1697 at his solution that this *curva brachystochrona* is a vertical cycloid is not known (see our comment in note (8) thereto), but we append on pages 86–90 the fleshed-out bones of a variant demonstration of the result— a *purée* of a clumsily presented mode of proof published by Nicolas Fatio de Duillier the year before—which he sent to David Gregory three years afterwards in 1700.

(13) See pages 76–8 in sequel. Again, Newton's exact derivation of his undemonstrated 'Solutio' is not known, but in it he made evident appeal to his longstanding knowledge of the expression of the sums of powers of the roots of an equation in terms of the symmetrical functions of these which are the equation's coefficients (see our notes (11) and (12) thereto).

(14) See page 62.

(15) Newton in fact attempted a 'General Solution' of Leibniz' Bernoullian problem (the several successive drafts of this are reproduced in 1, 7 below) but could only give a broad and none-too-adequate argument justifying the presumption that a (well-behaved) family of curves does possess a set of curves which meet its members everywhere at right angles; compare page 65 below. He made no attempt, so far as we know, to solve a yet more complicated problem of constructing orthogonals (see 7: note (25)) which Leibniz sent—but Bernoulli once more had contrived—further to test the 'pulse' of the 'English analysts' some months afterwards in the early spring of 1716.

untransmuted.[16] He knew forty years earlier[17] as much as Bernoulli in 1705 about the effects of such axis-transformation, and would in his prime have been equally quick to detect the fallacy here. Only the stiffness and forgetfulness of old age now betrayed him, blinkering his eyes.

(16) Craige's holograph sheet 'De Linearum Curvarum Longitudine. Authore Jo: Craig'—differing in only a couple of words from the article by him of the same title later printed in *Philosophical Transactions*, **26**, No. 314 (for March/April 1707/8): §IV: 64–5 [= *Acta Eruditorum* (August 1710): 352, whence it was further reprinted by Gabriel Cramer in his edition of *Johannis Bernoulli Opera Omnia*, **1** (Lausanne/Geneva, 1742): 449–50]—is preserved as f. 79 of ULC. Add. 3971.4. The texts of Bernoulli's original 'Probléme à Resoudre. Une Courbe algébraïque (vulgairement appellée géométrique) étant donnée, la transformer en une infinité d'autres aussi géométriques, mais d'especes differentes, lesquelles soient chacune de même longeur que la proposée' (*Journal des Sçavans* for 12 February 1703 (N.S.): 112), of the special case of Craige's would-be solution which appeared in *Philosophical Transactions*, **24**, No. 289 (for January/February 1703/4): 1527–8, and of Bernoulli's own previously published *genuina Solutio* by 'reptory motion' in *Acta Eruditorum* (August 1705): 347–59 are to be found conveniently collected in Bernoulli's *Opera*, **1**: 406–24. In essence Craige employed Diophantus' parametrization of the Pythagorean triad $a^2 + b^2 = c^2$ by $a = 2mn$, $b = m^2 - n^2$, $c = m^2 + n^2$ to transform the given curve defined by perpendicular Cartesian coordinates s, z into that of manifestly equal arc-length defined by the perpendicular coordinates $x = (a/c)s + (b/c)z$, $y = -(b/c)s + (a/c)z$, not realizing that this is but a rotation of the axes—namely, anticlockwise through the angle $\cot^{-1}(a/b)$—which leaves the curve itself unchanged. Newton's sole annotation upon Craige's manuscript, 'Eodem recidit si sit $x = \dfrac{ms + nz}{\sqrt{mm + nn}}$ et $y = \dfrac{mz - ns}{\sqrt{mm + nn}}$' (at the foot of its recto side), is the minimal improvement(?) whereby the axes are rotated through $\cot^{-1}(m/n)$. Having been made brusquely aware of his howler by Bernoulli (*via* a letter to a common acquaintance, William Burnet, the pertinent portion of which Bernoulli soon after made public in *Miscellanea Berolinensia*, **1**: 184 [= *Opera*, **1**: 451–2]), Craige made frank admission of his previous failure to observe the identity of the given and 'transmuted' curves in a paragraph appended to his 'Logarithmotechnia Generalis' (*Philosophical Transactions*, **27**, No. 328 (for October 1710): 191–5); this was reprinted as a separate 'Johannis Craigii Additio ad Schediasma de Linearum Curvarum Longitudine' in *Acta Eruditorum* (July 1713): 315 [= *Opera*, **1**: 452].

De Moivre had independently put what was basically Craige's 'solution' privately to Bernoulli in a letter of 27 July 1705, there adjoining that 'j'avois fait voir mon calcul à M. Newton et...il avoit témoigné en être extremement satisfait. Quinze jours après le lui avoir montré, il dit...qu'il avoit travaillé à mon probleme depuis qu'il ne m'avoit vu et qu'il étoit arrivé à mon expression sans calcul'; to which Bernoulli replied the next February that he had 'y...découvert tout aussitôt le paralogisme que vous commettez, parce que je me souvenois d'avoir eu de pareilles pensées il y a bien 8 ans, mais dont je reconnus l'insuffisance peu après....Vous vous pouvez consoler...que M. Newton lui-même...n'a pourtant pas pu s'en apercevoir.' (See K. Wollenschläger, 'Der mathematische Briefwechsel zwischen Johann I Bernoulli und Abraham de Moivre' [= *Verhandlungen der Naturforschenden Gesellschaft in Basel*, **43**, 1933: 151–37]: 210–11, 222–3.) Evidently de Moivre did not in 1706 pass on to Newton the news that they had been so deeply misled, or the latter would surely have informed Craige of his equivalent paralogism a year afterwards.

(17) See I: 155–212, *passim*.

After the death of John Wallis in October 1703—and with his own elevation the next month to the Presidency of the Royal Society—Newton became the respected doyen of British mathematics: a senior whose experienced judgement was widely appealed to, even if it was realized that his seal of approval did not always guarantee the correctness or quality of what it was set upon. David Gregory (till his death in October 1708)[18] and de Moivre (during the decade from April 1704 in which he conducted an extensive correspondence with Johann Bernoulli)[18] allow us a succession of vivid glimpses of his mathematical activities behind scenes in his early London years, as each was privileged to observe them and take part in them. Of the two, de Moivre was by far the deeper, more original intellect, and what he has to say must carry more weight than Gregory's mostly superficial memoranda, though the latter's day-to-day jottings down of remembered conversations and current gossip have their importance in recording for posterity a mass of not wholly trivial ephemera which would otherwise have vanished without trace. As we would expect, each wished to shine in Newton's eyes and was not chary of upstaging the other when it suited him. Gregory, in particular, was not a little jealous of the slightly younger but abler de Moivre, and naturally resentful when the latter pointed to his shortcomings. How Newton himself reacted with diplomatic tact when in 1704 de Moivre found an absurdity in Gregory's discussion of the Cassinoid curve as a viable planetary path two years previously in his *Astronomia*—gently firm in his counsel to the one that he should publish his detection of Gregory's 'exceedingly grave fallacy' in the *Philosophical Transactions*, while subsequently advising the other quietly to make public amends 'by the by' in the midst of a separate article there on 'Cassini's *Orbita*'[19]—is a caution to us not too rigidly to damn him with the image of

(18) See the editions respectively by W. G. Hiscock of 'Extracts from David Gregory's Memoranda, 1677–1708' (as *David Gregory, Isaac Newton and their Circle*, Oxford, 1937); and by Karl Wollenschläger of 'Der mathematische Briefwechsel [1704–14] zwischen Johann I Bernoulli und Abraham de Moivre' (note (16) above). Many other, chiefly technical memoranda of Gregory's survive in Oxford (Christ Church), London (at the Royal Society) and Edinburgh (University Library), and we ourselves have drawn freely upon these yet unpublished diaries and papers. De Moivre's relations with Newton after 1714 are but shadowily known, but he clearly became one of his staunchest aides, particularly in his contacts with the Continentals. (See the indexes to the sixth and seventh volumes of *The Correspondence of Isaac Newton* (Cambridge, 1976/1978), *s.v.* 'De Moivre, Abraham'.)

(19) Gregory had in Lib. III, Prop. VIII of his *Astronomiæ Physicæ & Geometricæ Elementa* (Oxford, 1702): 217–18 'proved' that the Cassinoid—the quartic curve the product of the instantaneous distances of a point of which from two fixed poles is constant—has, where it is conceived to be traversed by a body centrally attracted to one focus (and so round which the *radius vector* traverses focal areas proportional to the orbital time), an exact equant at the other (round which, that is, the body moves at an angular rate proportional to the time); as to be sure the inventor, G. D. Cassini, of this 'nouvelle espece d'Ellipse' had asserted in presenting it to the world a dozen years earlier. (See Jacques Ozanam's *Dictionaire Mathematique, ou*

'autocrat' which certainly at other times he could readily project to his contemporaries. Nor, of course, to suppose that all upon which Newton was asked to exert his mathematical judgement had so academic or so highly serious a context. He was, for instance, asked at some yet undetermined time by a certain John Lacy, 'one of his Majesties surveyors', to settle a coffee-house argument as to whether or not a hilly region has a greater surface-area than the plane contained within the same boundaries; and he took not a little trouble in drafting his reply.[20] The late J. M. Keynes has ably analysed the

Idée Generale des Mathematiques (Paris/Amsterdam, 1691): 435–53, where elaborate exposition is given of 'l'Hypothese de Monsieur *Cassini*...telle que je l'ay apprise dans sa conversation'. Cassini himself wrote much more briefly of his 'Ellipse' two years later in his tract *De l'Origine et du Progrès de l'Astronomie* (Paris, 1693). That this involved a 'paralogisme fort grossier' de Moivre was quick to underline by deducing a subtangential 'expression...admirable par sa simplicité' for the true orbit of a body possessed of an exact 'upper-focus' equant (see Wollenschläger, 'Briefwechsel': 193): one with which Newton, when shown it, 'fut extreme-ment satisfait, et m'exhorta à la donner dans les *Transact. philos.* ajoutant qu'elle plairoit aux geometres', or so de Moivre reported to Johann Bernoulli two years afterwards on 3 March 1704/5 (*ibid.*: 194). Upon, he there continued, speaking to Newton of 'la peine que je me faisois de contredire un homme qui avoit quelque reputation', he had been reassured that 'à l'egard de M. Gregory...je pouvois me dispenser de parler de lui', but 'je ne l'ai pourtant point imprimée' in order not to cause Gregory 'le moindre chagrin'. Instead he informed Gregory of the serious error he had committed, so that the latter might publicly retrieve it in his own way. Gregory's considerable discomfiture at being thus made aware of his clumsy mistake is evident from his private note at this time that in his 'paper (to be insert in the *transact.* si res sit integra) I would speak fully of the Orb. Cassin. & change prop. 8. lib. 3 *Astr.* inserting all that Cassini & Ozanam (I suppose in his dictionary) has said of it, referring the rest to an answer. Si non, I must doe my best by annoying my enemys' (Hiscock, *David Gregory*...: 19); but in a calmer mood on 22 October 1704 he wrote down that 'Mʳ Newton... would have me bring in my Apology by the by in the middle of the paper about Cassini's Orbita, and not at the beginning or ending of it' (*ibid.*: 20). And so it was done in the published article 'De Orbita Cassiniana' (*Philosophical Transactions*, 24, No. 293 [for September/October 1704]: 1704–6), towards whose close Gregory retracted with a minimum of fuss his mistaken notion about the curve's equant 'uti perperam nuper sentiebam'—'Mais il n'a pas touché à mon probleme' de Moivre coldly commented to Bernoulli in his March 1705 letter ('Briefwechsel': 194). (The latter's response the following July was typically to state (*ibid.*: 202): 'Je ne m'étonne pas que M. Gregory ait bronché, car ce n'est pas la premiere fois. Quel misérable, quel grossier paralogisme n'a-t-il pas commis dans sa prétendue solution de la chainette inserée dans les *Transactions* de 1697 pag. 637...'; on which see v: 522, note (7).)

(20) Reproduced (not everywhere adequately) in *Correspondence*, 7, 1978: 361–2 from Newton's heavily corrected original. In his lengthy reply to Lacy's query—given 'etsi ego non is sum cujus authoritati in dijudicandis Mathematicorum controversijs tantum deferas'—Newton appealed to the general inequality of the surface-areas of two convex bodies, one contained entirely within the other, which 'Archimedes alijque demonstraverunt' (see Lambanomena 3 and 4 to 'On the sphere and cylinder' [= E. J. Dijksterhuis, *Archimedes* (Copenhagen, 1956): 145]) to support his main point, and then (as we fill in his illegible words) went on: 'Mensurantur autem loca montosa, ut nosti, distinguendo in plura quorum-libet solidorum segmenta quorum superficies Geometris notæ sunt. Sed hæc methodus non nisi a Geometriæ peritissimis idꝗ cum labore maximo adhiberi potest, & veritatem satis

content of a stray document 'The Case of Trin. Coll.' (then newly bought by him at the 1936 Sotheby's sale) wherein, having been approached in 1708 by the Master, Richard Bentley, to intercede in a dispute between him and the Fellows over how the College revenues ought to be divided, Newton appealed simultaneously to the Old Statutes and a parabolic approximation in contriving a 'fair' apportionment[21]—or rather one just to Bentley's inflated notion of what his proper lion's share of Trinity's income ought to be. (Even so, the latter was insufficiently impressed by the cogencies of Newton's memorandum—if indeed, we must add, he ever saw any version of it—to put it formally forward to the College in attempted arbitration.) Whether, like so many modern economic theoreticians, Newton truly expected the realities of human intercourse neatly and rationally to obey such simple laws we do not know, but his sound and practical business sense is elsewhere (except, perhaps, for his failure in 1720 to sell stock he owned in the South Sea Company when at its peak before the 'Bubble' burst[22]) amply witnessed.

---

appropinquabit si montis superficies in plura momenta quamproxime plana resolvatur, nisi ubi prior methodus in aliqua parte montis adversarij [adhiberi posset in qua non] commode obvenerit [posterior], ubi spero post [debitas] æquationes præstari non posteriorem ubi error omnis quamvis [magnus] est etiamnum in defectu versatur.'

(21) Keynes' letter of 22 April 1939 to the Master of Trinity (then G. M. Trevelyan) making gift of the manuscript (now Trinity. R.4.48^a) to the College—he had bought it at the Sotheby sale three years earlier of the Newton papers still remaining in the Portsmouth family's possession, where (together with its rough first draft) it formed Lot 310—is printed in *Isis*, **49**, 1958: 174–6 under the title 'A Mathematical Analysis by Newton of a Problem in College Administration'. By the Old Statutes of the College an M.A. was assigned 5 nobles (1 noble = £$\frac{1}{3}$) for livery and 8 for his stipend, while the corresponding sums for a B.D. were 6 and 12 nobles, and for a D.D. 7 and 15 respectively, all in addition to rooms and commons, which Newton somewhat arbitrarily estimated as worth 13 nobles *per annum*, and to dividends on surplus revenue, whose amount varied with seniority; while the Master received 12 nobles for his livery and an undivided 300 for stipend and commons jointly (and a College dividend three times that of the Senior Fellow). Computing that, where l is the livery and s the stipend, there is *quamproxime* in the case of M.A.s, B.D.s and D.D.s $s^2 \approx 81l - 342$, Newton then set $l = 12$ to infer that the Master ought proportionately to receive a stipend of ($\sqrt{630} \approx$)25 nobles, and so a commons of (300 − 25 or) 275 nobles; and then suggested that the annual dividends should be paid out to M.A.s, B.D.s, D.D.s and the Master respectively in the same proportion as their total effective incomes of 26, 31, 35 and 312 nobles according to this interpretation of the Statutes. (He quietly ignored the alternative parabolic formula $s = \frac{1}{2}l(19 - l) - 27$, mathematically precise though this is, doubtless because it would have given the Master exactly the same stipend as the D.D.s under his rule....) Keynes makes the good point that Bentley's final proposal to the Fellows, even though it increased the Master's allowances in kind (unchanged by Newton's plan), yielded him a less favourable percentage in total income.

(22) See R. de Villamil, *Newton: the man* (London, 1931): 19–27. Newton bought £5000 of stock through the Bank of England on 27 July 1720, at 180 per cent in a rapidly rising market, and never sold it after the 'crash': at his death in 1727, when (see I: xviii) his estate was frozen by the Prerogative Court of Canterbury before eventually being made over to be shared among his eight surviving step-nephews and step-nieces, South Sea stock was still worth 104 per cent of its nominal value, so that he lost a little over £2000 on his original investment

One thing is not to be denied. As he passed into his 70's, with the 'editio secunda auctior et emendatior' of his *Principia* appeared, and so much of his mental energy increasingly given over to prosecuting his case for outright priority in the discovery of calculus against the Leibnizian 'second inventors',[23] the love of pure mathematics which had captivated him in his youth, held him engrossed through many of his mature years, and yet flickered on into late middle age, at last deserted Newton. 'Now an old man,' he wrote to Johann Bernoulli in September 1719, 'I take no least delight in mathematical studies'.[24] And the same refrain is heard constantly in his correspondence after 1713. Everywhere he gave it as his excuse for not entering into such matters that he had given up their active pursuit when he retired from academic life. 'Its now above eighteen years', he wrote in autumn 1714 to Remond de Montmort, thanking him for the gift of the newly published second edition of his *Essay d'Analyse sur les Jeux de Hasard*,[25] 'since I left off the study of Mathematicks & the disuse of thinking upon these things makes it difficult for me to take them into consideration.'[26] And something more than a year afterwards he similarly replied to an unknown correspondent who had earlier requested him to look over an 'instrument for performing plane

---

(ignoring annual dividends and inflation). The original entries documenting the 1720 and 1727 transactions may still be consulted in the Bank's ledgers of the period; see also the section of the 'inventory' of Newton's 'Goods, Chattels and Credits', made for the Prerogative Court on 5 May 1727, which itemizes his 'Government securities, &c' (de Villamil: 55).

(23) On this squabble, actively pursued by Newton—if mostly behind scenes—throughout the decade 1712–22 with few intermissions, see our introduction to Part 2 below.

(24) 'Studijs Mathematicis jam senex minime delector' (see *Correspondence*, **7**: 69); his similar, slightly qualified plea to Pierre Varignon the year before to be excused active scientific interchange had been: 'Jam quadraginta sunt anni et amplius ex quo commercium per literas circa res Philosophicas & mathematicas habere desij....Nam et Senio confectus a rebus Philosophicis et mathematicis quantum per negotia Societatis Regalis licet, me abstineo' (*ibid.*: 2). This fits well with Whiston's subsequent observation that 'I have heard Sir *Isaac Newton* say, that no old Men (excepting Dr. *Wallis* [!]) love Mathematicks' (*Memoirs of the Life and Writings of Mr. William Whiston...Written by himself*, London, ₁1749: 315–16).

(25) Montmort had also earlier presented him a few months after its publication with the first 1708 edition of his *Essay* (his accompanying letter is given in *Correspondence*, **4**, 1967: 533–4), but without recorded reaction to his invitation to be told of 'les defauts que vous y trouverez'. The gift copy of the second edition (now Trinity. NQ.10.26) is, a few dog-earings of the page corners—one pointing to a mention of Newton's name in the text, and the flyleaf inscription 'Pour Monsieur Newton Par son tres humble serviteur Remond de Monmort' apart—unmarked; compare Harrison's *Library of Newton* (note (1) above): 139.

(26) From the English draft—no Latin letter itself is known—on ULC. Add. 3968.41: 99ᵛ, printed in full by Harrison in his *Library of Newton*: 14, note 1. (On the same manuscript sheet, for what it interests, are found first versions of pages 180–2 of the 'Account' of his 1712 *Commercium Epistolicum D. Johannis Collins et Aliorum de Analysi promota* which Newton published a few months afterwards in the *Philosophical Transactions*; see **2**, Appendix 2 below.) 'But however', he went on, 'I have run my eyes over your performan[ce] & cannot but applaud it much. I wish heartily that France may always flourish with men who improve these sciences.'

Trigonometry at Sea' with the response that it was 'now above 20 years since I left of Mathematicks', adding the further self-exculpation for not examining it in detail that 'I never applied my self much to practical mathematicks, [&] they are now much out of my mind'.[27] Let us look tolerantly upon such evasive phrases from one who in his prime had possessed a supreme capacity for sustained concentration on the ordered and structured, the precious gift of 'industry & a patient thought' as he had described it to Bentley a quarter of a century before.[28] Take it as a sad *memento mori* that he could no longer, in slow decline into senility, feel except as a remembrance the lively allure of what once had excited him and given him pleasure.

Since Newton's death in 1727 there have been many and varied attempts to present and to analyse his mathematical achievement. Of his own younger contemporaries the man, William Jones, whom he allowed to bring out the gathering of his shorter pieces, two previously published and two hitherto unprinted, which appeared in 1711 under the collective title—making no mention of their author's name—of 'Analysis through series of quantities, their fluxions and their differences; together with an enumeration of lines of third order'[29] was the one who possessed the fullest acquaintance with the

(27) See the last paragraph of Newton's draft letter of 22 March (1716?) printed in *Correspondence*, 6: 211–12 (where the date '1715' suggested for its composition must be at least a year too early); he had used virtually the same phrase of 'having left off Mathematicks 20 years ago' in a cancelled sentence of one draft of his response to Conti on 26 February (see *ibid.*: 285). In vindication, nonetheless, of an earlier 'Judgement' of his upon the instrument— what he there said we do not know—he now firmly adjoined that it was 'safer for the Kings fleet to keep to the Trigonometrical operations by Tables which are sufficiently expedite then in stead of them to use instruments wch are less exact & tend only to make dispatch by slubbering over the operations' (*ibid.*: 212).

(28) In his letter of 10 December 1692; see *Correspondence*, 3: 233 (and compare our own I: 9, note (24)). We also have Gregory's word, in a jotting of around 1705 (Hiscock, *David Gregory...* (note (18) above: 25), that 'The best way of overcoming a difficult Probleme is to solve it in some particular easy cases. This gives much light into the general solution. By this way Sir Isaac Newton says he overcame the most difficult things.'

(29) *Analysis per Quantitatum Series, Fluxiones, ac Differentias: cum Enumeratione Linearum Tertii Ordinis* (London, 1711). Let us remind that, along with minimally corrected reprintings—from their 1704 *editio princeps* by Newton himself in appendix to his *Opticks* (see page 21 below and VII: 565)—of the texts of (pages 37–66) his 1693 [+1703] 'De Quadratura Curvarum' and (pages 67–92) his 1695 'Enumeratio' of cubic curves, first printings were here given of (pages 1–38) his 1669 'De Analysi' and (pages 93–101, 'Coronidis loco' in Jones' phrase) the 'Methodus Differentialis' newly composed in augmentation of its 1676 original (see note (8) preceding). Of the book's origin Jones stated in his introductory 'Præfatio Editoris' (signature a1^r) that 'Haec Editio casui, sed tamen non sine ipsius [Newtoni] consensu prius impetrato, ortum acceptum refert', passing to explain in sequel that, having come two years earlier into possession of Collins' transcript of the 'De Analysi' (on this see II: 168–9 and 207, note (2)) and subsequently approached its author for his permission to publish it, Newton 'licentiam...non solum summa cum humanitate concessit; sed & insuper veniam dedit

preserved corpus of Newton's mathematical writings. Over and above what he there edited Jones also came, as we have noticed a number of times before,[30] to have access to the more important of Newton's earliest papers, in their original autographs in some part but mostly in transcripts (then newly come into Jones' possession and still privately owned) which John Collins had made soon after his first meeting their author in late 1669,[31] and also to the mass of Newton's verbal outpourings during the decade from 1712 in defence of his priority of discovery in calculus.[32] We wish we could say that Jones greatly profited either himself or those in contact with him by giving any illuminating discussion of the quality of Newton's mathematical mind or what it accomplished. True it is that, when his friend Thomas Birch began in the early 1730's to gather material for a first comprehensive biographical account of Newton's life and career[33] and approached him for an expert survey of its mathematical aspects, Jones prepared a 24-page narrative stretching from 'the year 1663 [*sic*], a little before Christmas' when 'M.ʳ Newton, being then about 21 years old, bought Schooten's *Miscellanies* and Carte's *Geometry*, (as is

reliqua ipsius colligendi, quae [jam in lucem prodierunt et] ad idem argumentum spectabant' (a1ʳ).

(30) See the indexes to Volumes I–IV, VI and VII, *s.v.* 'Jones, William'.

(31) See II: x, 280–1.

(32) To Jones' jackdaw gatherings—and, let us be fair, to his trained connoisseur's eye for the collectable—we owe the preservation not least of Newton's autograph draft of the report in April 1712 of the 'Royal Society Committee', confirming his absolute priority over Leibniz, which is reproduced in **2**, Appendix 1.4 below.

(33) That which subsequently appeared over his identifying initial 'T' (for 'T[homas Birch]') in *A General Dictionary, Historical and Critical: In which a New and Accurate Translation of that of the Celebrated Mr. Bayle...is included; and interspersed with several thousand Lives never before published*, **7** (London, 1738): 776–802. (A French version of the *Dictionary* appeared at Amsterdam/The Hague in the 1750's as a *Nouveau Dictionnaire Historique et Critique, Pour server de Supplement ou de Continuation au Dictionnaire...de M.ʳ Pierre Bayle* under the direction of— more strictly 'par'—Jacques George de Chaufepié. The translation of Birch's article appears on pages 35–64 of its 'Tome Troisieme', but its authorship is no longer recorded.) This is the basis of the much more widely known unsigned article on 'NEWTON (Sir *Isaac*)' in *Biographia Britannica: or, The Lives of the most eminent Persons who have flourished in Great Britain and Ireland, from the earliest Ages, down to the present Times; collected from the best Authorities, both Printed and Manuscript, and digested in the Manner of Mr. Bayle's Historical and Critical Dictionary*, **5** (London, 1760): 3210–44. Here, let us remark, was born a myth that endured for the next two hundred years. Whereas Birch in his original article had very properly given the source of his mathematical information as 'a very learned friend of our author, in some observations communicated to us' with a marginal note that this was 'William Jones Esq; F.R.S.' (*General Dictionary*, **7**: 777), the unknown hack who rewrote it—evidently aware that Jones had since died, and that his library and papers remained at Shirburn Castle—omitted mention of Jones and in his note [A] (*Biographia Britannica*, **5**: 3210) he recast Birch's phrase to give as the basis of what was said regarding Newton's introduction to mathematics through Oughtred's *Clavis* and Descartes' *Geometrie* (now 'in the beginning[!] of the year 1663...') 'some original papers, now in the hands of the Earl of Macclesfield' which the innocent reader naturally (if erroneously) took to be Newton's own. See also note (36).

found by the account of his expences in those days;) which *Geometry* and
Oughtred's *Clavis* he had read about [*sic*] half a year before...' down to 'the
years 1684, 1685, and...1686' when 'upon the request of the Royal Society,
and Dᵣ Halley's importunities...he wrote his book of *Principles*'⁽³⁴⁾—merely,
however, to have Birch in his published article wholly ignore the 'paper
dated 13ᵗʰ November 1665' and the 'small Tract wrote in October 1666',
lengthy *verbatim* extracts from both of which fill Jones' central pages,⁽³⁵⁾ and
elsewhere make no mention of the 'larger Tract...wrote...in the year
1671'.⁽³⁶⁾ A pity that Birch denied his eighteenth-century readers any fuller

(34) We quote the opening and closing phrases of Jones' untitled autograph, now preserved
in Cambridge University Library (as Add. 3960.2) along with Newton's parent writings on
fluxions from the Portsmouth Collection. How it came to be there included we do not know,
but presume that John Conduitt acquired it from Jones soon after its immediate purpose was
fulfilled.

(35) Excerpts from this tract (Proposition 7 and Problems 1, 2, 3 and 9 [= 1: 414, 416,
419–22, 424] make up half of Jones' piece (pages 8–19); a quotation *in extenso* of the first
part—less Examples 2 and 3—of the paper of 13 November 1665 immediately precedes
(pages 5–7 [= 1: 383–4, 385–6]). These, together with brief mention of the two other Waste
Book papers of 20 May 1665 (see 1: 272–80) where 'the same method is set down in other
words, and fluxions applied to their fluents are there represented by dotted letters' and
'another leaf dated 16 May 1666' (see 1: 392–9) where 'a general method of resolving
Problems by motion is set down, in seven Propositions; the last of which, is the same with that
contained in the paper of 13 November 1665, tho' exprest in other words', support Jones'
main conclusion (page 19) that 'Mᵣ Newton in the Year 1665 invented the method of
Fluxions;...the letters *p*, *q*, *r* being...put for the fluxions of the fluents *x*, *y*, *z*. And...in the
year 1666 he improved this method and extended it to second Fluxions, using sometimes
letters without dots, and sometimes letters with one or two dots for quantities involving first
& second Fluxions'.

(36) It is true that Jones in his account merely lists (page 22) the headings of the twelve
problems embraced in the main section (see III: 74–328) of Newton's treatise, with little more
comment than that he 'illustrates each with great variety of Examples: and here'—*vide*
III: 244–54 (and compare pages 132–46 below)—'gives the Theorems for comparing other
Curves with the Conic sections, reduced into a Catalogue, which are published in his book of
Quadratures'; and that he was scarcely more forthcoming in his preceding brief mention
(pages 20–1) of the 1669 'De Analysi'. He clearly expected, however, that Birch would print
the greater part of his account *verbatim*, for at its end he wrote (page 24) that 'This Transcript
has been made and afterwards revised with the greatest care; but notwithstanding in case of
its being printed, the press ought to be examined...in some places perhaps from Sᵣ Isaac
Newton's own papers....If the figures are engraved, they ought to be done with more
exactness than they could be done here', and he further adjoined the 'Q[uery]? if it would not
be acceptable and usefull to this design if the Tract [*sc.* of October 1666] beginning page 11...
should be given whole & entire'. In the event Birch found room (*General Dictionary*, 7: 777,
note [A]) only to quote the greater part of Jones' first four pages—treating of Newton's
Wallisian induction of the general binomial theorem in the winter of 1664/5 by 'interpoling'
its expansions for positive integral indices 'and thereby very much enlarged the bounds of
analysis, and laid the foundations of making it universal' (compare 1: 104–11)—'communicated to us by the learned William Jones Esq; F.R.S. in whose hands those papers of Sir
Isaac Newton [*viz.* in Collins' transcription of ULC. Add. 4000.15ʳ–17ᵛ] are'. (On which
potentially misleading attribution see our previous remark in note (33) above.)

glimpse of what Newton had attained in his early calculus researches; yet we cannot shed many tears on Jones' behalf. His was no more than a loose compilation of half-understood excerpts from Newton's own writings on and about his early discoveries, put together without any attempt on his own part critically to appreciate their context or analyse their technical content, and leaving even the most evident inconsistencies in Newton's shaky imposed chronology uncorrected. It is not unfitting that so unscholarly a piece has remained unpublished to the present day, when it is now a mere historical curiosity.

The copies themselves of Newton's minor papers which Jones either acquired or himself made from their original did, however, during their author's last years and for some decades after his death come to have a limited circulation in and around London which we should not overlook. The ageing third-rate mathematician John Colson in the 1730's gained an undeserved, belated reputation—and Newton's Lucasian Professorship at Cambridge—by rendering Jones' transcript of the 1671 tract into English.[37] And when Samuel Horsley published the original Latin text of this 'Geometria Ana-

We could, let us add, cite the Newtonian originals for virtually every sentence in Jones' parroted pastiche, but it seems to us not worth the while to do so. Jones' sole merit is to have ordered his textual clippings into some sort of reasonable chronological order, and occasionally to have smoothed the transition from one to the next with an appropriate coupling phrase. But not to say that all Jones' donkey-work went completely for nothing. As we briefly mentioned in the first volume (i: xxix, note (39)), when David Brewster came to prepare his elaborate *Memoirs...of Sir Isaac Newton* (see note (5) above) in the 1830's, he—incapable of comprehending the technical content of Newton's papers himself (and perhaps even without access to many of them; see i: xxix–xxx)—was happy to make considerable use of Jones' account in writing up his own scarcely more perceptive summaries there of his hero's mathematical achievements: what little, in particular, he has to say (*Memoirs*, **2**: 12–14) regarding Newton's fundamental paper of 13 November 1665 and the tract of October the next year is a straightforward (and unacknowledged) crib of what Jones set out about each of these. (To Jones' manuscript itself Brewster at this time contributed the immensely perceptive judgement, scrawled down the right-hand side of its first page, that it is a 'Paper relating to the Method of Fluxions'.)

(37) And publishing it, together with his own pedestrian attempt at a commentary, as *The Method of Fluxions and Infinite Series; with its Application to the Geometry of Curve-lines. By the Inventor Sir Isaac Newton.... Translated from the Author's Latin Original not yet made publick. To which is subjoin'd a A Perpetual Comment upon the whole Work, Consisting of Annotations, Illustrations, and Supplements, In order to make this Treatise A compleat Institution for the use of Learners* (London, 1736); compare iii: notes (32) and (33). This 'Master of Sir Joseph Williamson's free Mathematical-School at Rochester [in Kent]', as the book's subtitle placed him, was already in his 60th year when—supported by Trinity's Master, Robert Smith, against an even more elderly de Moivre 'almost in his dotage' (W. W. Rouse Ball, *A History of the Study of Mathematics at Cambridge* (London, 1889): 101)—he was in 1739 elected successor to the blind Nicholas Saunderson in the Lucasian Chair: an appointment more and more widely admitted to have been a mistake as this 'plain honest man of great industry and assiduity' (William Cole in his 'Collections for an *Athenæ Cantabrigienses*': C 200) but without the least spark of talent lingered on in the position for twenty-one years to become finally a senile semi-recluse.

lytica' for the first time in the opening volume of his collected works of Newton in 1779 he used, by his own statement, not only the autograph manuscript (probably by then lacking its first leaf) but also again the Jones transcript (whence he drew his own title) and a secondary copy made from this long before by James Wilson.[38] Already on 15 December 1720 Wilson could write to Newton himself, informing him that he had seen 'the other day, in the hands of a certain person'—Jones, we presume[39]—'several Mathematical Papers...transcribed from your Manuscripts [which] chiefly related to the Doctrine of Series and Fluxions, and seemed to be taken out of the Treatises you wrote on those Subjects in the years 1666 and 1671';[40] and five weeks later he left with Newton, 'so that you may compare them with the Originals, to see after what manner they have been copied', his own secondary transcriptions of 'the 2ᵈ, 3ᵈ and 4ᵗʰ [Problems] of your Treatise wrote in 1671' and 'a paper in english containing 5 Problems [*sc.* of curves]', further adding that the other papers he had been shown—which 'seemed to have not been so well copied as these, so I did not write them out'—included 'several Problems [*viz.* from the October 1666 tract] as to find the Curvature and Areas of Curves, and to compare Curves together, &c' and also 'a paper [*sc.* in appendix to the 1671 tract] containing 6 Examples, shewing how to deduce the Areas of Curves from the Tables in your Quadratures, with Constructions and Synthetick Demonstrations...by the means of Moments, [in] Analogy to what the Ancients have done on the like occasions'.[41] All innocently rather than

---

(38) 'In hoc libro edendo tribus usi sumus MSS; in quibus et Auctoris Autographus erat. Alium, ignoti cujusdam scribae manu exaratum, nobiscum communicavit...Dominus Carolus Cavendish; cum ipse eum olim a Jonesio acceperat. Tertium de Cavendisiano, ut videtur, propriâ manu exscripserat...Jacobus Wilson....Hunc nobis tradidit Joannes Nourse, Bibliopola Regius...' (*Isaaci Newtoni Opera quæ exstant omnia*, **1** (London, 1779): 390. It seems, however, much more likely that Wilson's secondary copy of the 1671 tract was transcribed directly from Jones' primary one; compare our earlier remark in III: 32, note (2).

(39) Or a pupil of his, at least. Inside the front leaf of the rearranged transcript in his hand of Newton's October 1666 tract which is now ULC. Add. 3960.1: 1–50 Wilson later wrote: 'The Transcriber has here put $\dot{x}$, $\dot{y}$ and $\dot{z}$ for p, q and r of the Original, as is also done'—see III: 73, note (87)—'in the English Translation of Sʳ Isaac's Book of Fluxions wrote in 1671 and printed by Mʳ Colson in 1736. Here seems to be some transpositions and interpolations, as Mʳ Jones was wont to make in those papers of Sʳ Isaac Newton, which he distributed to his Scholars'—on his ensuing page 50 he names a 'Mʳ Watkins' as one such—'that none might make a perfect book out of them. Thus I have seen papers from him containing parts of the above mentioned tract wrote in 1671, and of another book of Sʳ Isaac's intituled *Constructiones Geometricæ Æquationum*'—his 1672 'Problems for construing æquations' (see note (44) below)—'which were some in the original Latin and others translated into English'.

(40) See *Correspondence of Newton*, **7**: 107.

(41) *Correspondence*, **7**: 125. The originals of the texts whose copies Wilson passed back to Newton for checking are reproduced on III: 82–150 and II: 172–84 respectively; that with variant synthetic demonstration of examples of quadrature by the tables in Problem 9 of the

cheekily done, but, since Jones had never, after copying it, himself returned the original of the latter of the two texts with whose transcripts he was now furnished, we will readily appreciate why Newton ignored Wilson's request to him in postscript that 'when you have perused these papers, you would be pleased to seal them up, and to leave them with your servants, that I may have them again upon calling for them some time or other'.[42] We have in earlier volumes noticed a number of times[43] that Wilson came, through Jones, to know of still others of Newton's papers whose originals have become publicly accessible to scholars—where yet they are—only since 1888 when Lord Portsmouth made over his family's holding of their scientific portion to Cambridge University.[44] For more than a century before that, all the outside world knew of such pieces as the 'Problems for construing æquations' passed by Newton to Collins in August 1672 were the informative digests of these given by Wilson in his account of Newton's mathematical discoveries appended to his 1761 edition of the works of Benjamin Robins.[45] Except for the few lines of general description of it inserted by Newton himself in his 'Observations' the next year upon Leibniz' letter to Conti of 9 April 1716 (N.S.),[46] the only portions of the crucially important October 1666 tract which were available in print to interested students until 1962, when its full text was published for the first time, were the *verbatim* extracts and summaries which Wilson also

---

1671 tract (see III: 266–84) is presumably the one whose text we appended thereto on III: 328–52.

(42) *Correspondence*, **7**: 127. The transcripts by Wilson of the late 1660's 'Problems of Curves' (the original remains with Jones' papers in private possession) and of Problems 2–4 of the 1671 tract are still with the corpus of Newton's own writings in Cambridge University Library, catalogued as Add. 3958.5: 75ʳ–76ᵛ (compare II: 175, note (2)) and Add. 3960.1 respectively.

(43) See for instance I: 400, note (1); III: 11, note (28) and 13, note (31).

(44) On the drawn-out negotiations to this end over the previous decade and a half see I: xxx–xxxiii.

(45) *Mathematical Tracts of the late Benjamin Robins, Esq;* ... **2** (London, 1761): Appendix: 340–61. Wilson's outline of the contents of Newton's piece (*ibid.*: 346–7)—he knew it under the title 'Constructiones Geometricæ Æquationum' which Jones set on his secondary transcript of Collins' contemporary copy of the 'Problems' (still remaining with his papers in private possession); compare note (39) above—has been cited *in extenso* on II: 293–4. *Inter alia* Wilson was also familiar (see *Mathematical Tracts*, **2**: 351; his transcript of the paper is preserved on ULC. Add. 3960.1: 48–50) with the 'Notæ in [Leibnitii Tentamen de motuum cœlestium causis] *Acta Eruditorum* an. [16]89 p. 84 & sequ.' which Newton wrote in 1714 in attempted refutation of Leibniz' celestial mechanics and whose original (acquired yet again by Jones and once more remaining with his papers) was till very recently unpublished. (It will now be found in *Correspondence*, **6**: 116–17.)

(46) In appendix, namely, to his augmented reissue of Joseph Raphson's *History of Fluxions* [both English and Latin versions] (London, ₂1717): 111–19, especially 116. The pertinent passage is quoted virtually in full on pages 514–15 below.

there included.[47] The main text of the preceding paper of 13 November 1665[48] whose bare, undemonstrated 'Resolution' Newton had similarly cited in his 1717 remarks[49] to uphold the early date claimed by him for his 'invention' of fluxions was at length made public in 1838 by Stephen Rigaud, but this from a secondary *copia vera* of the original penned by Newton at the time of the dispute over calculus priority (and now in private possession) which the jackdawish William Jones had yet again been quick to acquire.[50]

We traced in detail in the first volume[51] the tangled story of what happened to Newton's own papers after he died, stressing how little use was made of them even by those editors and scholars, like Horsley and Brewster, who had relatively free access to them when they were still united at Hurstbourne Park. That we need not here repeat. For a century and a half after his death Newton's private writings were to all intents and purposes consigned to a limbo from which their scientific portion emerged in the 1870's, through the fifth Earl's generosity, as the 'Portsmouth Collection' in Cambridge University Library,[52] while their remainder were less nobly dispersed worldwide by public sale in 1936, a number once more to disappear into the un-

(47) *Mathematical Tracts of...Robins*, **2**: 351–6. The complete text of the tract (= 1: 400–48) was first printed, from Newton's original (now ULC. Add. 3958.3: 48ᵛ–63ᵛ), in A. R. and M. B. Hall's *Unpublished Scientific Papers of Isaac Newton* (Cambridge, 1962), 15–64.

(48) See 1: 382–4 (and compare note (36) above).

(49) See page 514 below.

(50) S. P. Rigaud, *Historical Essay on the First Publication of Sir Isaac Newton's Principia* (Oxford, 1838): Appendix No. II, 'Early Notice of Fluxions': 20–3. Newton here 'copied the words' only of the first part of his 1665 'discourse'—omitting its 'Example 3ᵈ' [= 1: 384] —and then noted that 'In the same discourse there follows'—see 1: 386–9 for the full text— 'the application of this method to the drawing of tangents, by finding the determination of the motion of any point which describes the curve: and also to the finding the radius of curvity of any curve at any point, by making the perpendicular to the curve move upon it at right angles, and finding that point of the perpendicular which is in least motion...'. The whole piece is, in fact, a redraft of his 1717 'Observations' upon Leibniz' letter to Conti which, cast into Latin, he afterwards set as an intended 'Enarratio plenior' of the Scholium to Lemma II of Book 2 of his *Principia* in its coming third edition (see **2**, Appendix 5 below). Much as there, he here too went on (*Essay*: Appendix: 23–4) briefly to epitomize the content of his two other Waste Book papers of 20 May 1665 and 16 May 1666, and of his 'small tract, dated in October 1666' where 'the same method is again set down without pricked letters, and... applied...to the solving of problems concerning tangents, curvatures of curves, the greatest or least curvatures, the squaring of curvilinear figures, comparing their areas with the areas of simpler curves, finding the lengths of curves, finding such curves whose lengths may be defined by equations or by the lengths of other curves, and about the centres of gravity of figures'. (The untutored nineteenth-century reader, however, could have made little sense of Rigaud's transcription of Newton's final two sentences where, before abruptly breaking off, he added: 'And for finding the curvature of curves without calculation the following rule is set down. Let X'—that is, ↄ [!]—'signify the given equation   '.)

(51) See 1: xviii–xxx.

(52) Again see 1: xxx–xxxiii.

known of individual possession, but the majority, happily, to find a more permanent public resting-place.[53] Of the mathematical treasures which were newly opened up to scholarly investigation in 1888 at Cambridge some few were studied in immediately following years by Rouse Ball in a trio of pioneering essays,[54] while one or two other items were more briefly explored in ensuing decades by Alexander Witting[55] and Duncan Fraser.[56] But these were isolated voices crying in the wilderness. Right into the middle 50's of the present century the received view of Newton's mathematical insight, power and innovatory achievement—as perceptively analysed and assessed above all by J. E. Hofmann[57] and H. W. Turnbull[58]—has been founded in its essentials upon his principal published works: in calculus, his 1669 'De Analysi', his 1671 treatise on infinite series and fluxions, and his 1690's 'De Quadratura Curvarum'; elsewhere, his 1670's and early 1680's lectures on algebraic 'Arithmetica Universalis', his 1695 'Enumeratio' of cubic curves into species, his 1710 'Methodus' of finite differences; not to forget the first two books 'De motu Corporum' of his magisterial *Principia*.

And there of course, with very few exceptions, has been his historical impact. We cannot and do not expect that all the wealth of unprinted text and related ancillary information which we present here and in the seven previous volumes will do other than deepen and round out the basic realities of that accepted image, without seriously controverting its substance. We do hope that, as the years pass, our edition of Newton's mathematical papers may drive

(53) See I: xxxv. P. E. Spargo is currently engaged in preparing a revised reprint of John Taylor's Sotheby sale *Catalogue of the Newton Papers Sold by Order of The Viscount Lymington* [on 13/14 July] *1936* which will incorporate a finding list of the present locations of the individual items auctioned, in so far as these are known. Through the efforts of J. M. Keynes (and of his brother Geoffrey) about a third of these are now in King's College, Cambridge.

(54) See I: xxxiv, notes (49), (50) and (51) for their titles (and the indexes to Volumes II, IV, VI and VII, *s.v.* 'Ball, W. W. R.' for the details of how well we have profited from them ourselves).

(55) See I: xxxiv, note (53) and 322, note (1).

(56) See I: xxxiv–xxxv, note (54), and also IV: 22–3, note (2) and 36, note (2).

(57) 'Studien zur Vorgeschichte des Prioritätstreites zwischen Leibniz und Newton um die Entdeckung der höheren Analysis. I: Materialien zur ersten mathematischen Schaffensperiode Newtons (1665–1675)' (*Abhandlungen der Preussischen Akademie der Wissenschaften, Jahrgang 1943. Math.-naturw. Klasse Nr. 2*). It is a deep regret to us that our own early volumes should have so soon rendered obsolete this magnificently documented study of Newton's early mathematical development: one produced in tercentenary commemoration of the birth of the scientific hero of a country with which his own was (as the propagandistic phrase then put it) in 'total' war, at a time when Britain could only most scrappily honour his memory.

(58) *The Mathematical Discoveries of Newton* (London, 1945). Despite its many inaccuracies of detail—and, inevitably, its near-total ignorance of his early, formative papers—this affords a readable, well-rounded discussion of Newton's achievement which is broadly accurate in its stresses.

out the entrenched myths about their provenance and character which ought to be no longer tenable, and strangle at mis-birth the yet more fanciful notions which float in the train of currently trundling band-wagons. In any context where proper respect is had for documented evidence it ought henceforth to be impossible aggressively to assert the too simple Marxist *idée fixe* that 'Newton was primarily a physicist....Mathematics was for him [only] a mathematical aid for the formulation and above all the explicit application of his laws of motion';[59] or wishfully to imagine that his unpublished papers bear out, in any sense of 'fact', the 'startling revelation...that he devoted more time and energy to the Hermetic pursuit of alchemy than he did to his mathematical studies'.[60] Newton has left on record his own scorn for such vague and (at best) partial truths in a draft of the preface to his 1704 *Opticks* where he wrote:[61]

'But if...you feign Hypotheses & think by them to explain all...you may make a plausible systeme of Philosophy for getting your self a name, but your systeme will be little better then a Romance....Tis much better to do a little with certainty & leave the rest for others that come after you then to explain all things by conjecture without making sure of any thing.'

Such has been our own editorial philosophy these past twenty and more years. May we hope that Newton himself, had he lived, would have looked not

---

(59) Our translation of H. J. Treder's 'Newton war primär Physiker....Die Mathematik war für [ihn] methodisches Hilfsmittel für die Formulierung und vor allem für die explizite Anwendung seiner Bewegungsgesetze' (two representative phrases on page 7 of his 'Isaac Newton und die Begründung der mathematischen Prinzipien der Naturphilosophie. Zu Newtons 250. Todestag am 21.3.1977' [=*Sitzungsberichte der Akademie der Wissenschaften der DDR/Mathematik-Naturwissenschaften-Technik Jahrgang 1977, Nr. 12/N*, Berlin, 1977]). We would make it clear that we do not reject all Marxist—or, indeed, all 'Hermetical' or 'psycho-historical'—interpretations of Newton's personality and thought outright, but only where these clash with the past truth as we have come to know it through the documents by which all such conjectured suppositions and would-be explanations must stand or fall. It is not *historically* true that Newton was 'primarily a physicist'. There is on the other hand, for all its patent absurdities and the much in it that is now badly outdated, a hard core in Boris Hessen's celebrated essay on 'The Social and Economic Roots of Newton's *Principia*' (*Science at the Cross Roads*, London, 1931: 149–212) which we should at least consider with respect, even though we may not accept its past reality.

(60) So Frances Yates in 'Did Newton connect his maths and alchemy?' (*Times Higher Education Supplement*, No. 282 [for 18 March 1977]: 13), perhaps here drawing on the yet more absurd declaration by P. J. French that 'It is well known...that, in terms of volume, Newton's mystical writings far exceed his practical scientific ones' (*John Dee*, London, 1972: 161, note 3), maybe in turn based upon Christopher Hill's highly dubious statement that 'both Napier and Newton attached more importance to their researches into the *Apocalypse* than to logarithms or the law of gravitation' (*Intellectual Origins of the English Revolution*, Oxford, 1965: 7)....And so the myth unravels into thin air and speculation.

(61) ULC Add. 3970.3: 480ᵛ (→ 479ʳ). The full text of this draft preface is printed by J. E. McGuire in *British Journal for the History of Science*, **5**, 1970: 185 (→ 183).

unkindly on our present efforts to raise a worthy and comprehensive monument to his mathematical achievement. It has been our continuing privilege daily to have had an intimate contact, through fading ink on ageing paper though it be, with one who (as we put it when we began[62]) took a giant's leap forward in so many areas of science and intellectual thought, and of whose attainment so much endures undyingly, ever freshly on. *Newtone, te salutamus.*

---

(62) See 1: xxxvi. Neil Armstrong all but quoted us two years afterwards when he set foot on the Moon's surface!

ANALYTICAL TABLE
OF CONTENTS

PART 1
SOLUTIONS TO CHALLENGE-PROBLEMS, REVISIONS OF EARLIER RESEARCHES, AND GENERAL RETROSPECTIONS
(1697–1722)

The twin problems of Johann Bernoulli's 'Programma' of New Year's Day, solved by Newton on 29(–30) January 'before he went to bed', 3. The first problem of the brachistochrone: earlier attempt by Galileo to identify the curve of fall in least time, 4. Leibniz' determination of its first-order defining differential equation (but not that the curve itself is therefore a cycloid), 5. The problem already privately (*via* an unnamed Swiss) put to David Gregory in December 1696: his confusion as to how to reduce the brachistochrone condition, 6. The second, geometrical problem needing 'only half an hour's deep concentration' to master it, 7. The quality of the snap solutions given to both by Newton (rusty after a year's break with academic life), 9. Bernoulli's comment of '*Ex ungue Leonem*...': the context in which this aphorism was cited, 10. Other solutions over 1697–9 to the brachistochrone problem by Leibniz, Jakob Bernoulli and Sault, 11. Yet others by Craige, Gregory and Fatio de Duillier, 12. Johann Bernoulli's own elegant model of the least-time curve as a (Fermatian) optical ray-path: Machin's ineffective attempt (1718) to extend this to a general central-force field, 13. Plans by Newton *c.* 1702 to print his (yet unpublished) tracts on 'Quadratures' and 'Curves of the 2^d Genre' in annexe to his forthcoming *Opticks*, 14. Accelerated by his anger that Cheyne would not withdraw his own coming 'Inverse Method of Fluxions' from the press, 15. Newton's overhasty inference that Cheyne in his *Fluxionum Methodus Inversa* (1703) had plagiarized his 'De Quadratura Curvarum' lacks foundation, 18. The 1693 manuscript revised for publication: initial drafts (late 1703) hint at a much expanded treatise, with wide application of the new quadrature techniques, 20. Revisions of the tract's 'Introductio' more modestly outline the possibilities of a developed calculus of fluxions (through the Taylorian expansion of incremented ordinates): a related new terminal 'Scholium Generale' reaffirms that hereby 'the path is laid to greater things', 22. The *De Quadratura Curvarum* receives a cool initial reception upon publication: in England its niceties are not fully comprehended even by de Moivre, 24. Leibniz' hostile (anonymous) review in the *Acta Eruditorum* (1705): 'In place of Leibnizian differences Mr Newton employs fluxions...in the manner that Fabry substituted advances of motion for Cavalierian indivisibles', 25. John Keill's counter (1711) that it was rather Leibniz who had published Newton's prior discovery of an arithmetic of fluxions: this was was to trigger off a decade of squabble over calculus priorities, in which the literal truth of Leibniz' standpoint was soon lost from sight, 28. The text of the *De Quadratura* reissued without change in 1706, and with some light revision in 1711 (in Jones' compendium of Newtonian *Analysis*), 30. Cotes' interest in its 'second Table' of quadratures (effectible through the areas of ellipses/hyperbolas): his drawing in 1709 of Newton's attention to two errors in it, 32. His notion of a general 'harmony' of such 'measures of angles and [logarithmic] ratios' (circular/hyperbolic functions), 33. His removal of a limitation to the sixth 'Form' (independently attained by Newton himself?), 34. Robins' demonstration (1727) of the *De Quadratura*'s previously unproven Proposition XI, 34. Later editions and translations; eighteenth-century commentaries, 35.

Geometrical jottings by Newton in the early 1700's: other stray numerical (optical, astronomical, day-to-day arithmetical) computations of the period, 36. More sustained reworkings by him of previous expositions of geometrical analysis: the revamped 1672 'Problems for construing æquations' (*viz.* by conchoidal neusis and intersecting conics), now renamed the 'De constructione Problematum Geometricorum', and augmented with a brand-new 'verging' construction of the roots of a general quartic equation, 37. This intended to replace the corresponding terminal section on the 'linear' construction of equations in the 1707 *Arithmetica Universalis?*, 38. The restyled 'Method of Differences': its main Propositions I–IV are unchanged repeats of 1676 equivalents in the Waste Book, 38. To these in late 1710 were adjoined a brief introduction (explaining the notion of divided differences) and a final scholium (appending simple rules for approximate intercalation), to form the 1711 *Methodus Differentialis*, 39. Cotes' 1707 Cambridge lectures on finite differences (independently of Newton): his later elegant extension of Newton's schemes for quadrature through equidistant ordinates, 40. Lack of interest in the topic on the Continent, 42. Later editions of the *Methodus*, 43.

Cotes' seeing of the revised edition of the *Principia* ($_2$1713) through press, 45. Newton's thoughts (Easter 1712) of implementing its appeals, specific and general, to be 'granted the quadrature of curvilinear figures' by reshaping the 1704 *De Quadratura* to be an amplified 'Analysis per quantitates fluentes et earum momenta', 45. Of its major additions Propositions XII/XV (applying the binomial theorem to derive the basic circular and logarithmic series) are founded on Newton's 1676 *epistola posterior* to Leibniz; Propositions XIII/XIV (treating the series solution of given algebraic/fluxional equations) adapt *mutatis mutandis* a basic algorithm of the 1691 'De quadratura' *via* Wallis' 1692 exposition of this in his *Opera*, 47. New reflections on Proposition X of the *Principia*'s second book: the deep-seated error in the 1687 *editio princeps* in its assignment here of the ratio of resistance to (constant downwards) gravity yielding motion in a given (vertical) path, 48. First caught by Johann Bernoulli (early 1710) in the instance of a semicircle: the necessity of increasing Newton's stated ratio by a factor of 3/2 announced by him to the Paris *Académie* in 1711, along with his would-be 'explanation' of Newton's slip (as deriving from a deeper failure to understand the true import of the successive coefficients of a Taylor expansion), 49. Newton hears of his error only in autumn 1712 (from Bernoulli's nephew, Niklaus, then visiting London) when the pertinent pages of the *editio secunda*, passed unchanged by Cotes, have been printed off, 52. The failure of his initial attempts to adjust his 1687 argument, without any major reconstruction of its detail, 53. An alternative approach, comparing increments in the projectile's motion at the same ends of 'successive' infinitesimal arcs, at length succeeds: Newton dashes off a rough note to Niklaus about 'ye Problem set right', 53. To reassure himself, he proceeds to work the equivalent argument in which the 'curving' arc of flight is represented by a polygonal chain of infinitesimal links, each rectilinear; and then reformulates his curvilinear approach by now taking the new base variable of the body to be the horizontal advance (and not the time of passage), 54. This last deftly mortised by Cotes into the waiting *Principia* (January 1713) with minor excisions, 54. Bernoulli accepts the validity of Newton's revised argument, but not that he had himself earlier slipped in mistakenly thinking that Newton had confused second and third derivatives with like coefficients of series in the 1687 original, 55. Confirming 'evidence that the calculus of fluxions was not begotten before the calculus of differences', sent privately by him to Leibniz in June 1713, at once printed in a *Charta volans* as 'the judgement of an eminent mathematician', 57. Bernoulli's stubborn persistence in his opinion till his death, 58. Lagrange's definitive examination of the technical issues (in his 1797 *Théorie des Fonctions Analytiques*), 60. Edleston's minor historical glosses, 61.

Leibniz' request to Bernoulli (November 1715) to be 'reminded' of a problem requiring a descent to 'differentio-differentials' which might be put to the 'English' to test their mathematical 'pulse': Bernoulli responds with that of constructing orthogonals to given classes of curves, 62. An easy hyperbolic instance is put, through Conti, to British mathematicians at large (January 1716): particular solutions are readily forthcoming from Stirling, Machin, Pemberton and Taylor, 63. Newton attempts a 'General Solution' (printed anonymously in the *Phil. Trans.*); but he is beaten by the subtleties of defining the condition of orthogonality between families of curves, 63. Bernoulli thinks that its 'witless' author cannot be Newton (and guesses it to be Taylor), 66. Newton does not attempt the harder Bernoullian problem of orthogonals sent over by Leibniz in April 1716: its tardy fluxions solution (using infinite series) by Taylor in 1717, 66.

Drafts by Newton of a preface to an intended third edition of the *Principia* in the late 1710's, expounding his notions of analysis and synthesis, and asserting how he had employed these in writing his book: his difficulty, however, in there pinpointing instances of where he had employed a prior analysis, afterwards composed in classical style, 68. His final spurt of mathematical activity in the run-up to bringing out the second edition of his *Arithmetica Universalis*: roughings out of his method of 'exterminating' a common variable between two simultaneous equations (quadratics, cubics, quartics), 70. His last known piece of original mathematics: the calculations whereby he caught his error, apparently yet undetected by anyone else, in his description of the pivoted-*cum*-sliding 'set-square' for describing a general conchoidal cubic, 71.

The autograph draft of Newton's published solution (Royal Society. N. 1.61b). 'Problem I. To investigate the curve in which a heavy body falling under the force of its own weight shall descend most swiftly from any given point to any other given point', 72. Newton's undemonstrated construction of the required curve, in the supposition that this is a (vertical) cycloid, 74. 'Problem II. To find a curve such that, if from any point a straight line be drawn to intercept it (in two points), the sum of any given power of the segments cut off be constant': Newton constructs the instances where the powers are successively linear, square, cube, '&c' by stating limitations (derived from his study in 1665 of such symmetrical functions of its roots) upon the coefficients of the curve's defining equation, 74. Analysis of this solution (broadly equivalent to Bernoulli's, but deeper in its insight?), 77. 'In much the same way curves can be found which shall cut off in a rotating straight line three or more segments having like properties', 77.

APPENDIX 1. The full Latin text of Bernoulli's printed 'Programma' of New Year's Day 1697 (from Newton's own copy, now Royal Society. Classified Papers I, 30). Bernoulli's claimed justification for issuing such public mathematical challenges (as 'touchstones' upon which 'the most outstanding analysts' may 'test their methods and stretch their powers'), 80. The 'mechanico-geometrical' problem of the curve of swiftest descent (already posed six months before in the *Acta Eruditorum* without receiving reply other than a communication from Leibniz that he had solved it), 82. The second 'purely geometrical' problem of sums of powers of segments cut off by a rotating transversal (now added 'in New Year's Gift' to a reissue of the first), 84.

---

* N.B. Unless otherwise specified, citations here and below are of manuscripts in the University Library, Cambridge.

the 'dual method of the mathematical sciences': algebra (symbolic arithmetic) uses analysis, 'whence it also is called analysis', but 'I shall develop it conjointly with ordinary arithmetic as a single, more general one', 182. [4] (Add. 3963.14: 156^r/156^v). Classical geometrical analysis: 'plane' problems are to be resolved by Euclid's *Data*, 'solid' ones by ascertaining points of intersection between 'solid' (conic) loci, 184. The Apollonian 'determinate section' cut out in a line by a conic, shown to be 'solid' by way of Apollonius' work 'On cutting off a [given] ratio', 186. Apollonius' related work 'On cutting off a [given] space (product)', 186. 'By this method we have, following in the track of the ancients, constructed all [solid] problems', 188.

APPENDIX 1. Prior versions of Newton's reflections on the nature of arithmetic and geometry, and on the resolution and composition of problems. [1] (Add. 3963.15: 171^v). A first draft of his observations regarding arithmetic (common and 'specious') and geometry, 190. [2] (*ibid.*: 174^v). A tentative recasting, quickly abandoned, 191. [3] (*ibid.*: 173^r). A first version of §2.2 (immediately preceding it on the manuscript page), 191. [4] (*ibid.*: 173^r/174^r). A draft bridging §2.2 and §2.3 (of little interest in itself), 192. [5] (*ibid.*: 175^r/176^r). A cancelled final portion of the manuscript reproduced in §2.3: following the model of those of Euclid for geometry (in Barrow's 1655 Latin edition of the *Elements*) Newton lists comparable definitions, axioms and postulates for a 'general arithmetic' (embracing common arithmetic and algebra), 195. [6] (Add. 3963.14: 156^r). A first version of §2.4 (immediately before it in the manuscript), 197.

APPENDIX 2 (Add. 3963.14: 156^v). Derivation of the equation of a circle in oblique Cartesian coordinates, 199.

§3 (Add. 3963.11: 113^r–114^v/117^r/118^r). 'On the construction of geometrical problems' (by neusis): preparatory definitions and postulates, 200. Problems 1/ and 2: 'To draw a straight line/ and to extend it', 200. Problem 3. 'To draw a straight line through two given points', and (in extension) Problem 4: 'To describe a plane through three given points', 202. Problem 5: 'To draw a circle with given centre and radius', 204. Problems 6/7: 'From a given point on/outside it to set a given length along/to a given line', 204. 'And on these principles all Euclidean problems can be constructed', 204. Problem 8: 'Between two lines to set a given length inclined through a given point', 204. Problem 9: 'To cut three lines by a line so that the intercepts shall be of given length', 206. Further (cancelled) problems on describing a sphere, cylinder and cone, and on cutting the last by a given plane, 206. The 'most ancient' geometers were interested in the constructs of solid geometry only in themselves, for their own properties, 206. 'About Plato's time' geometers began to see the advantage in employing the plane sections of a cone in other constructions; but 'to cut a cone by a plane is a difficult operation', and for that reason Archimedes preferred equivalent neusis construction of 'solid' problems, 208. Continual bisection/trisection of an angle by such 'verging' (taken over unchanged, like most of what follows, from the parent 1672 'Problems'), 208. 'Seeing that today's mathematicians love to reduce everything to equations, it will not be a bore to give such [neusis] construction of cubics and quartics', 210. Problem 1: 'Verging' construction of the general quadratic $x^2 = px + q$, 212. Problem 2: The reduced cubic $x^3 = qx + r$ solved by a linear neusis, 212. And 'another way' by a circular one, 214. Problem 3: The full cubic resolved by attaching a given root, and then (by Prob. 4 following) constructing the ensuing quartic, 216. Problem 4: The (brand new) determination by linear neusis of the (real) roots of the general quartic $x^4 = px^3 + qx^2 + rx + s$, distinguished into cases according as q is negative or positive, 216.

Application to extracting fluents in series (by term-by-term integration of the corresponding series expansion of the fluxion), 301. Examples: $\int 1/(1+x^2)\,.dx$ and $\int \sqrt{(a^2-x^2)}\,.dx$, 301. [4] (Add. 3960.6: 94/92). A minimally revised account of the binomial expansion, again solidly based on (with massive direct citation from) the 1676 *epistola prior*, 302. Outlines for added corollaries 'on areas and fluents' and 'on universal analysis' (meaning what?), with one of a scholium explicitly identifying the precise relationship of the successive coefficients ('first, second, third, ... moments') in the series expansion of an incremented fluent, 303. Draft of the opening to Prop. XIII of the 'Analysis', 304. With outlines of adjoined corollaries 'on finding area' and 'on transmutation of series', and ensuing enunciations (only) of Prop. XIV and its generalization to embrace second-order fluxions, 305. [5] (*ibid.*: 91/92). A minimal recasting of [4]: the general truth of the binomial expansion is now justified by appeal to a Wallisian principle of continuity, 306. Corollary 1 opens the way to computing areas of curves and lengths of arcs by means of series, 307. Corollary 2: 'Hence also analysis is rendered far more universal', 307. In scholium Newton again touches on the fluxional interpretation of the coefficients in a Taylorian expansion, 308. [6] (*ibid.*: 92/93). Corollary 1 now reworked to be a separate 'Prop. XIII', requiring 'to find a fluent quantity, some fluxion of which is a binomial power of a second uniform fluent, in an infinite series', 308. 'Solution' and 'Explanation through examples', 308. Scholium: 'by series of this sort analysis is much broadened', 311. [7] (Add. 3965.12: 218^r). Newton's rough checking computation (up to fifth powers of z) of the accuracy of the instance worked in Proposition XIV of the 'Analysis', 311.

6. PROPOSITION X OF THE PRINCIPIA'S SECOND BOOK REWORKED (AUTUMN 1712)   312

The historical context summarized, 312.

§1. Tottering steps towards achieving a valid resolution of this basic problem in the resisted motion of bodies under constant downwards gravity (late September 1712). [1] (Add. 3965.12: 196^r). A desultory opening calculation, at once abandoned, 313 (footnote). Lightly recasting the defective 1687 argument yields a new general measure of resistance to gravity, 314. Applied to the prime example of trajectory in a semicircle this gives a correct result; but through compensating errors, 314. [2] (*ibid.*: 197^v/197^r). A fuller remodelling of the 1687 argument falls into its same subtle error, 316. As becomes broadly apparent to Newton when he applies his new measure of resistance to gravity in the prime *exemplum* of the semicircle, 318. [3] (*ibid.*: 196^v). A more radical recasting; but Newton now commits two new independent errors which only partially cancel each other out, 320. Whence in the case of a semicircular trajectory his adjusted measure produces once more his impossible 1687 result (*viz.* that the decelerating resistance to motion along the curve is equal to the accelerating component of gravity the opposite way), 322. [4] (*ibid.*: 200^r). In a fresh approach Newton seeks to construct the equation of motion (not vertically but) in the momentary tangential direction, 324. Thence to err carelessly and in mathematical detail, 326. Unconcluded calculations applying the ensuing 'corrected' measure to the instance of the semicircle, 326. If carried through, these would determine the resistance to be twice gravity, and hence (impossibly) constant, 328. [5] (*ibid.*: 201^r). In a minimal reshaping, Newton impercipiently repeats both his previous slips, 328. [6] (*ibid.*: 219^r [+220^r]/219^v). He proceeds radically to alter the basis of his 1687 argument, now taking the horizontal component of the resisted motion to be the base fluent, 330. And is rewarded at last with an accurate (if complicated) expression of the ratio of resistance to gravity, 334. Thus obtaining, when he goes on to apply it to the prime instance of semicircular motion, Johann Bernoulli's (independently derived) value of the ratio there, 334. At once Newton dashes off a brief covering letter to Niklaus Bernoulli, desiring him to 'shew [this solution] to your Unkle' (but was it sent?), 337.

§2. Refining the corrected argument, contriving a variant mode of proof, and tailoring the refashioned solution to fit the space available in the waiting second edition of the *Principia* (October 1712). [1] (Add. 3965.12: 190^r–191^r). A carefully penned fleshing-out of the bare bones of the newly corrected solution (intended to be passed to Bernoulli early in October?). The minimally ameliorated main text, 338. Its application to 'Exempl. 1' of resisted motion under gravity in a semicircle, 342. Examples 2–4 (of the unresisted Galileian parabola and of the conic/higher-order hyperbolas) are understood to be similarly treated (with the ratio of resistance to gravity in them likewise 'merely' increased by a factor of 3/2), 345. [2] (*ibid.*: 198^r–199^r). The equivalent alternative argument where the forces of resistance and gravity are conceived to act, over infinitesimal 'moments' of time, not continuously but *simul et semel* at the beginning of each 'successive' instant, 346. The measure of the ratio of resistance to gravity thereby resulting, 348. This applied to all four examples in the 1687 *Principia*, 350. The more detailed computations which survive (on Add. 3968.41: 119^v) for 'Exempl. 3' of the conic hyperbola, 352 (footnote). [3] (*ibid.*: 192^r–193^v). The definitive 'tenth Proposition of the second Book corrected' as sent (with some slight verbal polishing and a couple of terminal additions here included) to Cotes at the beginning of January 1713. The lightly touched-up main text, 354. Johann Bernoulli's article in the *Acta Eruditorum* 'Leibnizianly' establishing essentially the same result (then in press, but not to appear till February/March): Newton's reaction when shown a 'preprint' by de Moivre, 361 (footnote). Application to the primary example of the semicircle, 362. 'Exempl. 2[–4]' are understood to follow *mutatis mutandis* as in the 1687 *Principia*, 366. Scholium: through lack of space in the waiting *editio secunda* of the *Principia* the alternative argument (presupposing resistance and force to act intermittently in impulses at 'successive' instants) is compressed into an opening 5-line paragraph, 366. The third-order fluxional equation of motion obtained (in heavily disguised geometrical form) when the resistance is supposed to vary as some *n*th power of the instantaneous speed, 368. Bernoulli's continuing blindness to the depth and accuracy of Newton's result: Newton's own seeming belief as early as 1694 that the general solution of the 'physically true' case $n = 2$ was 'in his power' (David Gregory), 369. Bernoulli's publishing in 1719 of the Leibnizian equivalent of Newton's fluxional equation as (did he genuinely believe so?) his own 'invention', 371.

Appendix 1 (*Principia*, ₁1687: 260–74). The text of Proposition X as initially printed. The general problem enunciated (in the confusing 'particular' instance where resistance is taken to be as the density of the medium and the square of the projectile's speed, but the relationship of density to gravity is not specified, 373. Ensuing deduction of the ratio of resistance to gravity (wrong by a 'simple' factor of 3/2), 374. The crucial *faux pas* in Newton's reasoning identified, 376 (footnote). Bernoulli's erroneous attempt (1711) to locate the error in the general argument (as arising from Newton's 'ignorance' of the meaning of the coefficients in a Taylor expansion) after he had faulted its application to the instance of semicircular motion (and no other), 377. 'Exempl. 1' of the semicircle as elaborated in the 1687 *Principia*: extracting the (Taylorian) expansions of the incremented/decremented ordinates, with a digression to urge that there is 'a use not to be despised of these series in the solution of problems depending on tangents and the curvature of curves', 378. The faulty general measure of the ratio of resistance to gravity thereto applied to yield the conclusion (impossible, but not appreciated to be so for a quarter of a century till Bernoulli) that in the semicircle the (decelerating) resistance to motion is exactly equal to the oppositely directed (accelerating) component of gravity along the curve, 380. Example 2: the upright 'Galileian' parabola (resistance to motion in it is accurately determined to be nil), 381. Example 3: a conic hyperbola with one asymptote vertical, 381. Example 4: a like hyperbola of higher order, 384. Scholium: 'it is clear that a resisted projectile path more nearly approaches a hyperbola than a parabola, and can in practice not disadvantageously be replaced by one of these', 385. Eight rules for applying the general hyperbolas of

Example 4 to observed 'reality', 386. Including (Rule 7) the rule of false position employed graphically to construct the meet of the horizontal with the projectile path, given a number of points upon the latter *ex phænomenis*, 387. 'What has been said is readily applied to parabolas': an instance, 389.

APPENDIX 2. Initial attempts by Newton (late September 1712) to locate the defect in his 1687 reasoning. [1] (Add. 3965.10: 109^r/103^r). A first check by Newton on the accuracy of his computation in the crucial 'Exempl. 1' of the semicircle: presupposing the truth of his measure of the ratio of resistance to gravity in the general case, he faultlessly again deduces in this instance the impossible particular result (deficient, as Bernoulli had just shown him, by a factor of 3/2) refuting the general result; but where is its error?, 391. A second, variant calculation produces the same erroneous result, 393. In a third computation Newton for the first time considers tangential motions in equal successive 'instants' the same way, but cannot shake off his *idée fixe* that the corresponding deviations therefrom must (to sufficient accuracy) be equal: the impossible 1687 result again follows, 394. [2] (private). Using this same new approach, he attacks the general problem; but he likewise omits to take account of the accelerative component of downwards gravity acting instantaneously along the curve, 395. [3] (Add. 3968.18: 261^r). A minor recasting of this falls into the same oversight, 397. [4] (Add. 3965.12: 219^v/220^r). Having (in §1.6) at last achieved his goal of a correct general measure of the ratio of resistance to gravity, Newton drafts two corollaries simplifying his complicated expression of it (but fails herein now to notice two mutually cancelling terms,) 398. In a third corollary he sketches, much as in the 1687 *Principia*, the gains in expanding an incremented ordinate as a 'converging series', listing the geometrical denotations and significances of the successive coefficients (in determining tangent-slope, curvature and the higher-order 'variations' of this), 399. An ensuing recomputation of resistance to gravity in the prime instance of the semicircle is broken off unfinished (but, had Newton carried it through, the impossible 1687 ratio would once more have been yielded), 401. [5] (Add. 3968.18: 261^r). A first try at deriving the general measure of the ratio of resistance to gravity, according to the alternative notion that these forces of deceleration and acceleration shall act 'once and instantaneously' at the beginning of each succeeding moment of time; inevitably, when Newton omits to take a necessary square root at a crucial point a false result is obtained, 402. [6] (Add. 3965.12: 198^r). A corrected light remodelling (in immediate prelude to §2.2 above): the correct Bernoullian ratio of resistance to gravity at last ensues in the instance of the semicircle, 403. [7] (Add. 3965.10: 136^v). The general measure expressed in terms of the coefficients of the Taylorian expansion at the point, 404. [8] (*ibid.*: 135^v–136^v). A preliminary recasting of §2.1 to be §2.3: the clarification is introduced of giving explicit denotations to the (unequal) infinitesimal times in which the resisted motion takes place along successive arcs (in the presupposition now that the horizontal component of the motion is uniform), 405. [9] (*ibid.*: 136^v). Newton adumbrates the further intended example of a hyperbolico-logarithmic curve, 409. But breaks off when the calculations become too complicated for him to control, 410. [10] (Add. 3968.37: 540^v). Computation of the variation of density when the resistance is as differing powers of the speed, 411. [11] (*ibid.*:). Rough calculations testing whether two other types of logarithmic curve can be traversed under a resistance proportional to the square of the speed (in a medium of uniform density), 412. Newton's curious attempts at 'integration' of the ensuing equations, 413.

APPENDIX 3. Three drafts of a last, seemingly finally successful, endeavour to salvage the essence of the 1687 mode of argument (late 1712). [1] (Add. 3965.10: 135^r). A first attempt is thoroughly spoilt by lapses both dynamical and mathematical, 415. [2] (*ibid.*: 136^v). A mathematical slip aborts an unfinished recasting, 417. [3] (*ibid.*: 135^r/135^v). Further reworking at length yields a revised argument spoilt only by a crucial numerical slip, 417.

Proof (from the scrap now Add. 3968.41: 132ᵛ) that Newton here afterwards made good his lapse, 420.

reason we synthetically demonstrated propositions [in the *Principia*] after finding them through our analysis', 444. Except for three cited instances, 444–6. The analytical discovery of the rest was by way of the 'method of fluxions and moments' whose elements are demonstrated synthetically in Book 2, Lemma II, 446. 'And he who knows to compose propositions found analytically will not be ignorant of how to resolve them when synthetically demonstrated', 448. [2] (Add. 3968.9: 107ʳ). A minimal reworking in which Pappus is cited as authority for our modern knowledge of the twin method of analysis/synthesis employed by the 'ancients': the superiority of geometrical composition is that it lets propositions 'most impress the mind with clarity', 448. 'And so I synthetically demonstrated in the *Principia* propositions which I found through analysis', though 'certainly I could have written them out analytically with less trouble', 450. [3] (private). A further light recasting, 450. Errors are easily introduced in 'specious' (algebraic) analysis, and only geometrical synthesis can be trusted to yield certainty, 452. [4] (Add. 3968.9: 109ʳ/109ᵛ). The most finished text. The 'ancients' synthetically proved what they had discovered by analysis in order to achieve that certainty attainable only in 'splendidly composed demonstrations', 452. So it is that 'the synthetic method of fluxions and moments occurs widespread in the following treatise', 454. Rare specimens of the use of the analytical method in the *Principia*, 454. That cited in Book 2, Prop. XIV is a case of Newton misunderstanding his own earlier (suppressed) analysis: his confusion explained, 456. 'My purview in the *Principia* was not elaborately to explain its mathematical methods, but merely to treat physics and, principally, the motion of the heavens', 456. The concluding paragraphs in the manuscript (of essentially no mathematical import), 457.

PART 2

NEWTON'S VARIED EFFORTS TO SUBSTANTIATE HIS CLAIMS TO CALCULUS PRIORITY
(1712–1722)

public his early mathematical papers, content in England (as Leibniz in Germany) to be regarded as founder of the 'new' mathematics, 470. Fatio de Duillier's claim in 1699, after having had a sight of some of these papers, to priority of calculus discovery on his behalf, 471. Leibniz' fury at being dubbed 'second inventor' by Fatio (with the hint that he might have 'borrowed' something of his *calculus differentialis* from Newton), 471. Leibniz' (anonymous) 1705 *Acta* review of Newton's *De Quadratura*, comparing the latter's method of fluxions *vis-à-vis* his own (by implication prior) one of differences to Fabry's earlier 'substitution' of 'progressions of motion' for Cavalieri's indivisibles, 472. Keill's 1708 rebuff that it was rather Leibniz who had 'with changes in name and notation' published Newton's previous discovery of an 'arithmetic of fluxions', 473. Leibniz' anger when (February 1711) he read Keill's 'insinuation': his complaint to the Royal Society demanding public apology for being wronged, 474. Newton's delicate position, as President called upon by two Fellows of the Society to take sides in a clash of opinion over his first 'invention': he is at length (April 1711) persuaded to back Keill after being shown Leibniz' 1705 *Acta* review, 474. Keill's reply, having been 'looked over' by Newton, is sent (late May 1711) with the semi-official backing of the Society: the basic documentary claim for Newton's priority is founded on his exposition of his rule for tangents in 1672 ('found among Collins' papers'), and on his '1676' theorems for quadrature by series (as afterwards set out in the *De Quadratura*), upon both of which Leibniz 'must have' drawn, 477. In riposte (December 1711) Leibniz calls upon the Society to 'curb' such 'empty and unjust bawlings' of one with 'too little experience of times past', 480. Newton forthwith addresses the assembled Society, and a Committee is at once set up (March 1712) to deliver its report on the matter, vindicating a month afterwards Newton's 'rights' as first discoverer, 480. The underlying *Realpolitik*: the 'genuine and authentic' evidence considered by the Committee was selected by Newton, and its returned Report drafted by him, 481. The consequent decision of the Society to publish the Committee's findings: Newton himself in fact makes the selection of related letters and documents, writing all the attached footnotes, connecting 'editorial' statements and the introductory 'Ad Lectorem', 483. The (privately printed) *Commercium Epistolicum* finally appears (January 1713), and is distributed to interested individuals and institutions, 485. The second edition of the *Commercium* (1722): its variorum text (1856) by Lefort, 486. The inadequacies of Newton's 'watertight' claim therein to be 'first inventor', 487. Leibniz' *Acta* review (February 1712) of Jones' 1711 *Analysis*: 'first and last ratios differ from differentials merely in the manner of speech and rigour of demonstration', 488. In counter (spring? 1713) Newton drafts a set of 'Observations' (intended for publication in the *Acta* or *Phil. Trans.*?): in particular, he makes basic distinction between the 'fluxion' (instantaneous finite speed) of a 'flowing quantity', and its Leibnizian (infinitesimal) momentaneous increment/decrement, 488. Initial reactions to the *Commercium*: Leibniz (away in Vienna without immediate access to a copy) accepts at second hand Johann Bernoulli's judgement (7 June) that in it 'there is no slightest trace of the [pointed] letters which Newton now employs in place of differentials', while his error in *Principia* 2, X makes it 'patent that he had not yet [in 1687] come to know the true method of differentiating differentials', 490. And, yet sight unseen of the *Commercium*, sets this in pride of place in an anonymous fly-sheet (29 July 1713) as the verdict of 'an eminent mathematician', 492. Keill's 'Response' (much worked over by Newton beforehand) in the *Journal Literaire* (summer 1714): *inter alia* Bernoulli's spurious assignation of the error in Book 2, Proposition X of the 1687 *Principia* is refuted, and counter-charge made of an analogous lapse by Leibniz in his own 1689 'Tentamen' on celestial motion, 494. A further article by Keill (again spoon-fed by Newton) in the 1716 *Journal*: Bernoulli's ensuing lambaste of him in the *Acta*, and the increasingly arid sequels to this side-dispute, 495. Newton's separate intention of more broadly 'himself' replying (autumn 1713) to the Leibniz–Bernoulli *charta volans* in a series of annotations on the *Commercium*, 496. The fourteen queries he then set himself, upon which his refutation ought to be based, 497. In his roughed-out answers he allows Leibniz independent 'second invention' of 'universal analysis', but effectively charges him with plagiarism from the *Principia* in his 1689 *Acta* essays, 499. Attempts by Chamberlayne and 'sGravesande at reconciliation (early 1714) come to nothing, 499. Instead, Newton begins (summer 1714) to draft parallel accounts of 'the Method of Fluxions until 1676' and 'the Differential Method from 1677': Leibniz 'began where Dr Barrow left off' in adopting 'a proper calculation to it', 500. These are combined (autumn 1714) in an 'Account' of the *Commercium* published soon after (January/February 1715) anonymously in the *Phil. Trans.* [= Appendix 2 below], 501. Summary of its 'proofs' of Newton's priority and its adjoined suggestions of Leibniz' possible plagiarism, 501. The identity of its author an open secret in England, 503. Keill's introduction to the French version of the 'Account' in the *Journal Literaire* (aimed deliberately to confuse Continental conjecture about its author?), 503. Bernoulli's 1697 challenge-problem on constructing orthogonals to a family of curves resurrected by Leibniz (November 1715) to 'test the pulse of our English analysts': a particular hyperbolic instance is appended by him (early December) to his letter to Conti in mounting a broader attack upon Newton's 'somewhat strange philosophy', 504. Conti circulates the letter and its *Apostille* at the English Court (midwinter 1715/16): royal pressure is put upon Newton to make proper reply, 506. His answer (late February 1716): Leibniz must accept the documents in the *Commercium* as fact, and renounce the judgement of the 'pretended Mathematician' in the 1713 fly-sheet, 506. Leibniz returns a long reply (March) with an opening stab against Newton's 'forlorn hope', Keill and Cotes, 507. Huygens (not Barrow or Newton) had been his mathematical mentor: he defends the essential truth of the charge in the anonymous 1705 *Acta* review (which he is still unwilling to own as his) that Newton had indeed 'substituted' fluxions for differentials, though now claiming that no charge of plagiarism was implied, 508. Newton seethes with rage in a first intended direct reply (late spring 1716) *via* Conti: when, thirty years before, Craige (who 'still remembers this') had shown him Leibniz' 1684 *Acta* article expounding its elements, his reaction had been that the *calculus differentialis* was 'mine in a new dress', 509. This intended castigation of Leibniz' 'tricking, double-faced, maskerading management' is given over (early summer 1716) in favour of an equivalent set of 'Observations' on Leibniz' March letter to be issued in appendix to a reissue of (the deceased) Raphson's *History of Fluxions*, 510. 'I still have in my custody the original Manuscripts of several mathematical papers written in 1664, 1665 & 1666': in first version extensive extracts are cited from those of 13 November 1665 and 16 May 1666, and from the October 1666 tract, 513. But only short digests appear in the published 'Observations' (even so, no fuller account was given to the world by Newton of his earliest calculus discoveries), 514. Leibniz dies suddenly (November 1716): Newton hastens his 'Observations' into print, but otherwise for a while all is quiet, 517. Leibniz' 'elogies' in the Leipzig *Acta* and Paris *Histoire*: Newton (in private) sternly itemizes the faults in both (1717/1718), being especially harsh about the 'defamatory Libell' of the 1713 fly-sheet, 518. Desmaizeaux' project (1718) of gathering Leibniz' letters on natural philosophy in two volumes, the second to print *inter alia* those dealing with Leibniz' claims to calculus priority: Newton furnishes (summer 1718) a made-to-measure French translation (by de Moivre?) of his 'Observations', previously printed (for insertion in Continental copies of Raphson's *History*?) but never distributed, 520. On receiving proof-sheets of Desmaizeaux' *Recueil*, Vol. 2, Newton's immediate reaction is to reorder its letters into a better reading sequence, and make other small changes (all but one accepted), 521. And then to draft more elaborate *corrigenda*, one a page-long addition (not used), 522. The succeeding versions of his covering letter to Desmaizeaux parade a familiar variety of evidence to 'confirm' his priority of discovery in calculus, 523. At length Newton puts his letter aside, to commence (late summer 1718) an elaborate new 'History of the Differential Method' in 'Supplement' to his 'Remarks' intended for the *Recueil* (compare Appendix 3 below): one of the drafts of this is the celebrated 'Portsmouth memorandum' where Newton looks back to the 'prime [in the middle 1660's] of my age for invention', 524. Desmaizeaux treads softly in casting (1719) his 'simple historical narrative' of the dispute in preface to the *Recueil*, 525. When it becomes clear to Newton that his own preferred 'true' account will not there be published, he sets himself variously to propagate it in a planned (abortive) re-edition of his *Principia* with the *De Quadratura*, in introductory prefaces and adjoined historical scholia (see Appendixes 4–8), 527. These are significant in giving Newton's ultimate opinions upon his 'first invention' in their fullest form, 528.

Varignon's efforts (from late 1718 onwards) to make peace between Newton and Leibniz' surviving lieutenant Johann Bernoulli: the insurmountable gulf of the latter's *idée fixe* that fluxions are 'inferior' to differentials, 528. Newton takes an appeasing letter from Bernoulli (July 1719) as categorical denial of being the 'eminent mathematician' who passed judgement upon him in the 1713 fly-sheet, 529. And grows angry at Bernoulli's response (December 1719) that he could not, 'because I do not keep copies of all my letters', be 'entirely' sure that he was not: straightaway he drafts a set of extensive annotations upon the 1713 *Judicium* (see Appendix 9), 530. An uneasy truce between the two ensues, to be broken (early 1721) when for the first time Bernoulli reads Newton's 'Observations' upon Leibniz' 1716 letter (as Frenchified in Desmaizeaux' *Recueil*), 531. Several times over the next months Varignon (through de Moivre) tries to persuade Newton to retract his most sarcastic appellations therein of 'pretended Mathematician' and 'Knight errant': to have him call Bernoulli instead a 'Don Quixot in Mathematicks' (a phrase withdrawn, however, from the finished letter passed to Varignon in late July 1721), 532. In a subsequent concession (September 1721) Leibniz alone is blamed by Newton as 'the knight errant who provoked me to this duel for the favour of the infinitesimal method', 534. Bernoulli's continuing intransigence spurs Newton (autumn 1721) to dust off his annotations of two years before on the 1713 fly-sheet (see Appendix 9), appending these to a reprinting of the *Commercium Epistolicum* along with a slightly altered Latin version of his 1715 *Phil. Trans.* 'Account' of it (see Appendix 2), 534. His care in polishing his (anonymous) introductory address 'To the Reader' (see Appendix 10 for its fullest version, longer than that printed): his initial intention of there citing his 1672 rule for tangents in an anachronistically remoulded form, 535. The *Commercium...iterum impressum* finally published (1722): Newton heeds a last-minute plea from Varignon to remove a phrase offensive to Bernoulli, 537. The wisdom of hindsight: our fuller present-day awareness of the subtlety and complication of the paths which may lead to 'discovery' divests the old notion of 'first inventor' of its mystique of moral primacy, 538.

APPENDIX 1. PAVING THE WAY TO THE 'COMMERCIUM EPISTOLICUM' (1712)   539

[1] (Add. 3965.8: 79ᵛ). Newton's address to the Royal Society (February 1712) giving his 'considerations' of Leibniz' recent letter of complaint regarding Keill. The background epitomized (see also pages 469–80 preceding), 539. 'I did not see the papers in the *Acta* till last summer & therefore had no hand in beginning this controversy': the fluxional and differential methods are 'one & yᵉ same variously explained', and Leibniz '& his friends allow that I was the inventor', 540. 'I applied it to abstracted æquations before [1669] & thereby made it general....I heard nothing of [Leibniz] having the method before 1677', 541. [2] (Add. 3968.19: 289ʳ). Newton's preface to his planned 'Extracts of yᵉ MS Papers of Mʳ John Collins concerning some late improvements of Algebra' (March 1712). 'Archimedes drew tangents to Spirals & Mʳ Fermat applied the method to Equations', 541. '& thereby drew tangents to curve lines....Mʳ Newton met wᵗʰ this method upon reading Descartes's *Geometry* wᵗʰ the Commentators thereon', 542. Fermat's method improved by James Gregory and Barrow (late 1660's): Newton's *De Analysi* (1669) containing his own more general methods of series and momentaneous increase 'together', 542. [3] (Add. 3968.19: 289ʳ). A more contracted introduction to the planned 'Extracts', 544. [4] (private) Newton's autograph draft (April 1712) of the report to be delivered by the Royal Society Committee on the dispute (whose members seem meekly to have gone along with it with minimal exceptions), 545. The four listed 'reasons for wᶜʰ we reccon Mʳ Newton

the first inventor': above all, Leibniz could have obtained the essence of his variant 'Differential method' ('one & the same wth the method of Fluxions, excepting the name & mode of notation') between 1673 and 1676 from Collins (who 'was free in communicating what he had received from Mr N.'), 546. [5] (Add. 3968.37: 551^r–553^v + 554^v). The fullest English draft of Newton's 'Ad Lectorem' introducing the collection of Collins' letters and papers, now (*c.* mid-summer 1712) fast growing. Breakdown by page of the Collinsian *Commercium Epistolicum* as eventually printed, 547 (footnote). 'The occasion of publishing this Collection of Letters' was 'Mr Leibnitz taking offence at a passage in a discourse of Mr Keill in *Phil. Trans.* A.C. 1708': being 'twice pressed' by Leibniz, and 'having no means of enquiring into this matter by [independent] living evidence', the Royal Society appointed a Committee to 'search out & examin' the pertinent papers and so establish 'who was the first inventor of the method', 547. The 'Letters & Papers till the year 1676 inclusively' show that Newton had already by then a 'general method of solving Problemes by reducing them to equations finite or infinite...& by deducing fluents & their moments from one another by means of those equations': the evidence of the 1669 *De Analysi* and the 1671 tract, 550. From 1673 onwards Leibniz was 'upon another method', but 'there appears nothing of his knowing the Differential method before 1677', 550. Hence Newton is 'first inventor', and whether Leibniz discovered the method afterwards by himself is 'a question of no consequence', 551. By his 'general method' Newton understands his 'method of series & fluents taken together': his method of series he was free in communicating to Leibniz in the middle 1670's, while Leibniz also had then 'a general description' of his other method 'with examples in drawing of tangents, squaring of curves & solving of inverse problemes of Tangents' (in the December 1672 letter to Collins, and the 1676 *epistolæ prior/posterior*), 551. When Leibniz in 1677 sent his own method to Newton he 'forgot to acknowledge that he had but newly found it'—as he had earlier 'forgot' the receipt of eight series from Collins, later communicating to 'his friends at Paris an Opusculum upon one of them [Gregory's for the inverse-tangent] as his own'—and represented it as a discovery 'to be recconed among the principal inventions of Analysis', 553. In 1689 (and more fully in 1693) Leibniz 'began to pretend', moreover, to Newton's method of assuming the terms of a series and determining these 'by the conditions of the Problem', 555. The mistaken 'Proposition' 19 of Leibniz' 1689 *Tentamen* on celestial motions, 'wch is the chief of Mr Newton's relating to Planets': Leibniz is blamed 'not for committing some errors but for adapting a calculation to another man's Proposition with a designe to make himself the inventor', 556. Proposition I of the *De Quadratura* was 'set down' in the 1676 *epistola posterior* as 'the foundation of the method of fluxions', and the eight following Propositions were also 'found before': yet Leibniz 'continues to justify the Account that he has given of the book', 557. The Continental 'Gentlemen' who have used the differential method, particularly L'Hospital, Varignon and the Bernoulli brothers, were 'strangers' to this correspondence ('it was before their time'), 558. [6] (Add. 3968.37: 549^r/549^v). A shorter recasting of [5], afterwards to be rendered word for word into Latin as the printed 'Ad Lectorem', 558. 'Some Notes are added to ye Letters to enable such Readers as want leasure to...see the sense of them', 560.

APPENDIX 2. NEWTON'S ANONYMOUS ACCOUNT (LATE 1714) OF THE 'COMMERCIUM' 561

Extracts from the original printed version (*Phil. Trans.*, **29**, No. 242 [for January/February 1714/15]: 173–208 (+208–24), adapted to Newton's autograph drafts in Add. 3968: *passim*. (For a detailed listing see note (3) on page 562.) The preliminary 'Accounts' in which, from autumn 1713 onwards, he began to marshal 'proofs' of his priority in discovery of 'universal' analysis (by series and fluxional increments), along with ungenerous estimate of Leibniz' 'second invention' of his own variant *calculus differentialis* 'from the year 1677', *vide* 561, note (1). Previous reviews of the *Commercium* 'published abroad' are 'all very imperfect', 562. 'We will first give an account of that part of the method wch consists in resolving finite æquations into infinite [series] & squaring curvilinear areas thereby': Wallis (1657) on so resolving fractions 'by perpetual division', 563. Brouncker's squaring of the hyperbola thereby, and Mercator's demonstration of this by division *à la* Wallis (1668): and 'soon after' James Gregory's 'Geometrical Demonstration thereof', 564. Newton's *De Analysi* (1669) with its 'general method' of doing the like 'in all figures', 564. Collins' wide notice of Newton's tract in the early 1670's to mathematicians in England and abroad, citing 'several sentences out of it' (though its full text was not to be printed for forty years, *viz.* by Jones in 1711), 565. The three rules upon which it 'founds the method of Quadratures': that of Wallis for squaring any rational 'dignity' (power), 568. That for compounding these with + and − signs: the third (to be used 'when the Quadrature does not otherwise succeed') whereby reduction is made to an infinite series whose terms are separately to be squared, 569. Newton's communications of the essential part of this to Leibniz in his *epistolæ prior/posterior* of June/October 1676 (having been desired to 'explain the original' of it), 569. In the *De Analysi*, further, 'from the moments of time he gave the names of moments to the momentaneous increases generated', representing the 'uniform fluxion of time, or of any exponent of time, by an unit', 570. And performing the ensuing calculation 'by the Geometry of the Ancients without any approximation' till the very end, when he 'supposes that the moment *o* [of time] decreases *in infinitum* & vanishes' (though sometimes 'when only investigating a Proposition, for making dispatch he supposes the moment to be infinitely little [from the first] & uses all manner of approximations wch he conceives will produce no error in the conclusion'), 571. In the *Principia* 'velocities are the first fluxions & their increase the second': and 'in Lib. II. Prop. XIV he calls the second difference [!] the difference of moments', 572. Leibniz' first steps in summing simple infinite series: 'See the mystery!', 574. His receipt (*via* Collins) in April 1675 of eight of Newton's series, and Gregory's for the inverse tangent, 595. 'Recompensing' Newton in August 1676 by sending back the last as his own (independent) invention, 577. Other claims of his to 'coinvention' of series (e.g. Newton's for the exponential) 'though the method of finding them [by reversion] was sent him at his own request & he did not yet understand it', 577. Leibniz' response to Newton's *epistola prior* of June 1676: 'its certain that he had not then found out the reduction of problems either to differential equations or to converging series', 578. 'He found this new Analysis therefore...not before 1677' (further would-be supporting evidence is given), 579. Newton communicated his tangent method to Collins in December 1672, asserting that 'This is one particular, or rather a Corollary of a Generall Method [of analysis]...interwoven wth working in æquations by reducing them to infinite series', 580. The basic rules of Sluse's 1673 tangent method (applied to 'differentiating' an arbitrary *n*th power, *n* = 3): Leibniz in November 1676 is 'thinking upon' its 'improvement into a general method', 581. Newton's *epistola posterior* to Leibniz in October 1676: its 'sentence exprest enigmatically' (when—and if—unravelled) 'puts it past all dispute that he had invented the method of fluxions before that time', 582. Leibniz 'began first to propose his differential method' only in his reply (June 1677), 583. L'Hospital in 'his' preface to his 1696 *Analyse* (not knowing Newton's *De Analysi* or his 1676 letters) 'recconed that Mr Leibnitz began where Mr Barrow left off', 583. The thesis that Leibniz owed an unacknowledged basic debt to Barrow urged yet more strongly by Jakob Bernoulli in the *Acta* (January 1691), 584. Barrow's threefold rule for tangents (1670): 'exactly ye same' as Leibniz', 586. The 'Elements' of Leibniz' differential method as published by him in the *Acta* (October 1684) and 'illustrated wth Examples': this 'conteins nothing more then what Mr Newton had affirmed of his general method', 587. Leibniz' preference for *dx* as a 'modification of *x*' (*Acta*, 1686): he 'knew he might have used [other] letters with Dr Barrow', 588. Newton's *Principia* (1687) is 'almost wholly of this calculus' according to L'Hospital: in Lemma II of its second book 'the Elements of this calculus are demonstrated synthetically', while a scholium at its end gives Leibniz due credit for his variant, independent invention of it, 588. Excerpts from Newton's letters of June/October 1676 were first published by Wallis in his *Algebra* (printed 'in English 1683 [!] & in Latin 1693'), 589. In 1695, 'soon after he had intimation from Holland [that] Mr Newtons notions of fluxions passed there wth applause by the name of the Differential Method of Mr Leibnitz', Wallis set the record publicly straight in preface to his *Opera*: 'Newton contrived his method of fluxions ten years and more before Leibniz produced his equivalent differential calculus, using variant forms of expression as it may be', 590. Leibniz in ensuing private correspondence with Wallis 'pretended not that he himself had the method before 1677,...allowed that the methods agreed in the main [but] challenged to himself... differential equations', 591. Even after Fatio 'suggested' (in 1699) that Leibniz, the 'second inventor of this calculus', might have borrowed 'something' from Newton, Leibniz 'did not deny that Mr Newton was the oldest inventor by many years nor asserted any thing more to himself then that he had found the method apart', 592. But began to make such a claim only after the death in 1703 of Wallis, 'the last of the old men who were acquainted with what had passed...forty years [before]', 593. It was the anonymous review (by Leibniz) in the 1705 *Acta* of Newton's *De Quadratura*, countering the prefaced assertion there that the method of fluxions 'was invented by degrees in 1665 & 1666' with the representation that, rather, Newton had but 'substituted' fluxions for Leibnizian differences, which 'gave a beginning to the present controversy: for, after Keill in 1708 'retorted the accusation' in an article in the *Phil. Trans.*, Leibniz was in turn led to complain twice in 1711 to the Royal Society's secretary, Sloane, of 'calumny' by a 'novice', 593. In consequence of which the Society 'appointed a numerous Committee to search old Letters & Papers & report their opinion upon what they found'; by which 'it appeared to them that Mr Newton had the Method in or before 1669 & Mr Leibnitz [not] before 1677', 595.

The main points of Newton's method summarized, and some fallacies regarding it corrected: the (infinitesimal) 'first' increment *o* contrasted with the (finite) fluxional speed $\dot{x}$ (= 1 when *x* 'flows' uniformly), 595. The 'moment' $o\dot{x}$ it is which is the equivalent of the Leibnizian difference *dx*, while ☒ represents the same as ∫*x*, but 'Mr Newton does not place his method in forms of symbols nor confine himself to any particular sort for fluents or fluxions': other instances of his variety of calculus notations ('the oldest in their several kinds by many years'), 596. 'The method of fluxions has all the advantages of the differential & some others. It is more elegant because there is but one infinitely little quantity, *o*...more natural & geometrical because founded upon the first ratios of nascent quantities wch have a being in geometry [while] Nature generates quantities by continual flux', 597. It is also 'of a greater use & certainty, being adapted either [in loose form] to the ready finding out of a Proposition by such approximations as will create no error in the conclusion, or [in rigorous synthetic mode] to the demonstrating it exactly', whereas Leibniz' is 'only for finding it out', 598. 'By the help of this new Analysis Mr Newton found out most of the Propositions in his *Principia*' but 'demonstrated [them] synthetically that the systeme of the heavens might be founded upon good Geometry': which 'makes it difficult for unskillful men to see the [prior] Analysis', 598. The (Bernoullis') criticism of the (badly phrased) terminal scholium of the *De Quadratura* as an indication that Newton 'did not then understand the method of second, third & fourth differences': answered with the rejoinder that if he did not understand it when (in 1704) he added the scholium to the book 'it must have been because he had then forgot', 599. His accurate geometrical interpretation earlier in *Principia*, 2, X of the (vanishingly small) second difference of an ordinate, 600. Other instances of his previous use of second and higher fluxions, notably in the excerpts from the (fuller) *De Quadratura* sent by him to Wallis in 1692 (→ *Opera*, **2**, 1693: 391–6): 'If ye word *ut* wch in that Scholium may have been accidentally omitted... be restored, the Objection will vanish', 601. 'Thus much concerning the nature & history of these methods': summary of Newton's concluding 'Observations' (on his final pages

208–24) where he closes his prosecutor's brief with the affirmation that, because 'second Inventors have no right', Leibniz must 'quit his claim', 602.

APPENDIX 3. TWO VERSIONS OF AN INTENDED STATEMENT
   BY NEWTON ON DISCOVERY OF THE CALCULUS (SUMMER
   1718)   603

[1] (Add. 3968.12: 159ʳ–161ʳ/166ʳ [+165ᵛ]). 'The History of the Method of Moments, called Differences by Mʳ Leibnitz': its origin as a set of 'proofs' by Newton of his absolute priority of discovery, to be introduced by Desmaizeaux into the preface of his coming *Recueil* of Leibniz' letters, 603. The infamous passage in the 1705 *Acta* review (by Leibniz) of the *De Quadratura* where charge is made that Newton 'substituted' fluxional advances of motion for Leibnizian differentials, as once Fabry did to Cavalieri, 604. Newton adduces items from his *epistola posterior* (supported by his 8 November 1676 letter to Collins) to show that already by autumn 1676 he had a developed understanding of his method of fluxions (but, as ever, finds it opportune to backdate to 'summer 1676' the time when he extracted his *De Quadratura* 'out of older papers'), 605. The *De Quadratura*'s prime rule for finding second, third and fourth fluxions, as sent to Wallis in 1692, 'came abroad the next year', having then been 'at least seventeen years in manuscript': its use of 'prickt' letters (a notation 'not only the oldest but the most expedite'), 606. The 1669 *De Analysi*: its supposition of uniformly fluent time as base variable, with an infinitesimal 'moment' for its instantaneous augment, 607. The 'Method of Moments' there applied to find by finite equations and 'converging' series the ordinates, areas and lengths of geometrical and 'mechanical' curves '&c', 607. Collins' letter to Strode in July 1672 cited by Newton as conclusive witness that 'I found the Method gradually in 1665 & 1666', although hearing Barrow 'then read his Lectures about Motion...might put me upon taking these things into consideration', 608. [2] (Add. 3968.31: 451ʳ–458ʳ). An equivalent intended 'Supplement' by Newton to his 1717 'Observations' upon Leibniz' April 1716 letter to Conti. The complexities of the immediate background of this 'open letter' to Desmaizeaux unravelled, 610. Newton's abortive 'designe' in 1671 of publishing his 'Theory of Colours' along with a newly written (never finished) tract 'upon the Method of series & fluxions together': his 'intermission' of mathematical studies during the following five years, 612. Leibniz' first visit to London in 1673 when 'not yet acquainted with the higher Geometry': his subsequent tutoring therein at Paris, personally by Huygens and through correspondence (*via* Oldenburg) with the 'English' (including Gregory), 613. 'Mʳ Leibnitz wrote also [May 1676] for the demonstration of some of my series': a request which drew from Newton his *epistola prior* of 13 June, setting out his (non-fluxional) methods, 614. Leibniz' second visit to London in October 1676, during which (Newton claims) he 'saw my [second] Letter of 24 Octob. & therein had fresh notice of the method..., & in his way home [to Hanover] was meditating how to improve the method of Tangents of Slusius', 614. His 1677 'abbreviation' of Barrow's tangent method, showing how it 'might be improved so as to proceed in æquations involving surds': his publication seven years later in the *Acta* (October 1684) of 'the elements of this method as his own without mentioning the correspondence which he had formerly had with the English about these matters', 615. Jakob Bernoulli's declaration (*Acta* 1691) that Leibniz' calculus was 'founded on that of Dʳ Barrow & differed not from it except in notation & some compendium of operation: L'Hospital's (in preface to his 1696 *Analyse*) that Leibniz 'proceeded where Dʳ Barrow left off', 615. 'I wrote the Book of *Quadratures* in 1676, except the Introduction & Conclusion': 'most' of its propositions are in the earlier 1666 and 1671 tracts 'tho not in the same words & some but in equations', 616. From the 1669 *De Analysi* it is clear that 'I [then] knew this method so [far] as to find the area of any figure accurately if it may be or

at least by approximation *in infinitum*', 616. 'By the inverse method of fluxions I found in 1677 [!] the demonstration that Planets move in Ellipses...& in 1683 [!] at the importunity of Dr Halley I resumed the consideration thereof', 617. 'In writing the Book of *Principles* I made much use of fluxions direct & inverse, but did not set down the calculations in the Book itself because it was written by composition, as all Geometry ought to be'; nonetheless it was *presque tout de ce calcul* (L'Hospital), 618. Leibniz' articles on resisted/celestial motion in the *Acta* for January/February 1689: these are 'nothing else then the [related] Sections of the *Principles* reduced into another order & form of words, enlarged by an erroneous Proposition' and given 'erroneous Demonstration', 619. Wallis' printing of Newton's 1692 letters the next year in Vol. 2 of his *Opera*: 'the first time that the use of letters with pricks & a Rule for finding second third & fourth fluxions were published...& for fluents I [there] put the fluxion with an oblique stroke upon it', 620. In the early 1690's 'the Differential Method grew into reputation', and L'Hospital's 1696 *Analyse* 'being an Introduction to [it] made it spread much more then before', 621. But Wallis (1695) and Fatio (1699) laid claim for Newton being 'oldest inventor by many years', so that during this period Leibniz was 'never in quiet possession', 621. The 1705 *Acta* review of the *De Quadratura*, representing Newton in his fluxions as but a 'substituter' *à la* Fabry of continuous 'progresses' of motion for Leibniz' (primary) Cavalierian discrete 'differences', 622. Answered by Keill in passing in his 1708 *Phil. Trans.* article; and more fully in his ensuing open letter to Leibniz (May 1711), 623. The Leibnizian July 1713 fly-sheet where 'I was singled out & treated very reproachfully...to make me appear': Leibniz' ensuing provocations in his letters to Conti (1715/16), but his refusal to 'enter the lists with my forlorn hope', 623. 'Now I have declared my opinion I leave every body to his own...out of a desire to be quiet', 624.

APPENDIX 4. A PREFACE TO 'DE QUADRATURA' IN ITS PLANNED REISSUE (1719) WITH THE 'PRINCIPIA': THE RESHAPING OF THIS INTO A REVISED FLUXIONS SCHOLIUM   625

The background to this, and to Appendixes 5–8 following: it is Newton's prime intention here 'to shew that that Book [of *Quadratures*] (which has been accused of Plagiary) was in MS before Mr Leibnitz knew anything of the Differential Method', 625. [1] (Add. 3968.21: 221^v). A first Latin *præfatio* stressing that 'it has seemed appropriate to adjoin the tract on quadrature of curves, drawn from a fuller treatise which I had written in 1671, since I made very much use of it in composing the *Principles*', 626. His fleeting further intention to annex also the *Methodus Differentialis* (of which nothing more is heard), 626. [2] (Add. 3968.5: 20^r). A minimal preliminary recasting of the scholium to the 'fluxions' Lemma II of the *Principia*'s second book, 627. [3] (Add. 3965.10: 145^v/146^r). The scholium amplified to embrace a *Scholium secundum* enlarging upon the prior discoveries by Barrow, James Gregory and Sluse of the general method of drawing tangents, 628. [4] (Add. 3968.5: 22^r). The dual scholia compressed into one, with further honing and refinement, 630. [5] (Add. 3968.5: 36^r). The most finished and elaborate recastings (1719) of the fluxions scholium (but one lacking any essential novelties), 631. The trivially altered version of the 1713 scholium which was in fact printed in 1726 in the *Principia*'s *editio tertia*, 632 (footnote).

APPENDIX 5. 'A FULLER RECOUNTING OF THE PRECEDING SCHOLIUM'   633

This 'Enarratio plenior Scholij præcedentis' (Add. 3968.6: 38^r–46^r) is in the main a rambling regurgitation by Newton of his earlier 'proofs' to be 'first inventor', but also now for the first time quotes extensive Latinized extracts from his (yet unpublished) mid-1660's writings on fluxional analysis (on which his claims to priority of discovery ought to have been founded from the beginning). The novelties of the 1669 *De Analysi* in treating problems 'by the conjoint methods of series and moments' when 'vulgar analysis' is inadequate: the 'fluxion' of a fluent quantity viewed as its (instantaneous) 'speed of flowing', 633. Gregory's 1671 series for the inverse tangent: Leibniz' claim to this and other series after his visit to London in 1673 (when he revealed himself to be 'not yet instructed in higher geometry'), 634. The evidence that Newton had already written his *De Analysi* by 1669: his fuller treatise on series and fluxions of two years later, with its quadratures of curves codified in 'certain general theorems', 637. 'Whence it is abundantly enough clear that in 1676—and a minimum of five or seven years before—I had a general method of reducing problems to fluxional equations and converging series, and set the name of Analysis upon it', 638. In 1684, however, when Leibniz 'published certain Elements of this Analysis [in the *Acta*] without mention of what he had received from Collins and Oldenburg, I set down the previous Scholium to establish my priority regarding this method, and to settle that the more difficult problems synthetically demonstrated in this Book of *Principles* had been found out by dint of this Analysis', 639.

'I have papers more ancient than these, written in 1665 and 1666, in which the dates of writing are not a few times noted': Newton passes to quote in Latin version the basic proposition of his paper of 13 November 1665 'To find y^e [relationship of] velocitys of bodies by [that of] ye lines they describe, together with its resolution and demonstration', 639. 'The method of solving problems by motion embraced in seven propositions in another manuscript of 16 May 1666' (here likewise cited in a Latin rendering), 640. The same seven propositions set out by Newton in his October 1666 tract, together with an eighth teaching how to regress from fluxions to fluents, by squaring curves in geometrical equivalent: 'this is the method called by Leibniz summatory, by me the inverse method of fluxions and moments', 641. Barrow's 'appropriate' contemporary witness, supported by Collins in his July 1672 letter to Strode, that 'this method was indeed found by me in 1665 and 1666', 641. Newton's 1671 treatise, teaching first the reduction of quantities to 'converging' series and the like extraction of both 'simple' (pure) and 'affected' roots, then the method of fluxions and its application to solve several problems (the first two, basic ones are enunciated), 642.

'All these things I have brought forth from old manuscripts to reveal the true origin of the present Lemma [II]', 642. Instances of use of second fluxions and 'moments' in the *Principia*: Book 2, Prop. X ('where I called the second fluxion of a curve the variation of its variation') and XIV, Cas. 3 ('where I called the second moment [!] of an area the difference of its moments'), 643. 'By the aid of second moments I found in 1677 [!] the proof of the Keplerian proposition which is had in Book 1, Prop. XI; and much before by their help the curvature of curves and its variation which I discussed in the 1671 treatise and also in the October 1666 manuscript, wherein I also a number of times used pointed letters': 'also, by considering first moments as fluents I found the solid of least resistance which I mention in the scholium to Book 2, Prop. XXXIV', 643. 'Computations by means of moments of fluents are often contracted by resolving the momentaneously incremented fluent into a converging series, as in Book 1, Prop. XCIII, scholium', 645. 'To the same Analysis pertains also the artifice of drawing a curve through any number of given points, and thereby interpolating any series', 646.

(Add. 3968.9: 105ʳ–106ʳ). The twin classical methods of mathematical analysis (resolution) for 'finding inventions' and synthesis (composition) for 'composing them for the public', 647. 'The Propositions in the following Book were invented by Analysis, but demonstrated by Composition to make [them] Geometrically authentic & fit....But if any man who understands Analysis will reduce the demonstrations from their composition back into Analysis (wch is easy to be done) he will see by what method the Propositions were invented', 647. In Book 2, Lemma II 'I set down the Elements of this Analytic Method in order to make use of it in demonstrating some following Propositions', 649. 'I added a Scholium not to give away the Lemma, but to put Mr Leibnitz in mind of his correspondence [during the 1670's] wth Mr Oldenburgh & Mr Collins' (the details are reiterated in familiar manner), 649. By Book 1, Prop. XVII focal conics exhaust all possibilities of motion from a point under an inverse-square force of 'attraction' (to a centre, understand), 652. Drawing the tangent from the wrong end of an arc 'caused an error in the conclusion' in Prop. X of Book 2 in the first edition of the *Principia*, but 'I rectified the mistake' in the second, 653. 'And there may have been some other mistakes occasioned by the shortness of the time... 17 or 18 months...in which the book was written...', 654.

APPENDIX 7. THREE ENGLISH DRAFTS OF A PREFACE TO THE
'DE QUADRATURA' IN ITS PLANNED RE-EDITION   656

[1] (Add. 3968.5: 28^r–29^v [+ .28: 403^r]/24^r–25^r). Newton elaborates upon his claim in the *Introductio* to the 1704 *editio princeps* to have found the method of fluxions 'gradually in the years 1665 & 1666, now drawing upon the *Commercium Epistolicum* as documentary basis for 'setting this matter in a fair light', 656. Gregory's and Barrow's methods of tangents: Newton's own in the 1671 treatise, in which his general method of fluxions was 'interwoven with another Method wherein æquations are reduced to converging series', 657. 'From the Tract wch I wrote in 1671 I extracted in 1676 the following Book of *Quadratures* & in my Letter of Octob. 24 set down the first Proposition thereof with its inverse verbatim [in anagram!] & represented it the foundation of the method....And then I set down a Theorem [which] implies the knowledge of the first six [ensuing] Propositions', 657. 'In my Letter to Mr Collins dated 8 November 1676 I had relation to the 10th Proposition [which] is deduced from the 5th 6th 7th 8th & 9th', 658. 'By the help of this method I found the Demonstration of Keplers Proposition...', 659. The 'Elements' of the 'Method of Fluxions & Moments' were proved 'synthetically' in Lemma II of Book 2 of the *Principia* in order to be used in demonstrating 'some following Propositions'; but a scholium was added to this to put Leibniz 'in mind of making candid acknowledgement' of his debt to Newton, 659. In 1692 'I sent to Dr Wallis the first Proposition of this Book illustrated with examples in finding first & second fluxions...the first time any Rule was published for finding [higher-order] fluxions or differences', 660. *Principia*, Book 2, Prop. XXXIV is 'the first specimen made publick' of 'the method of maxima & minima in infinitesimals or moments', 661. In the *De Analysi* 'the method of series is interwoven with that of fluxions,...composing one general method of Analysis': 'However, the method is capable of improvements [which] are theirs who make them', 662. (Deleted): 'Since by the help of this Method I wrote the Book of *Principles* I have therefore subjoined the Book of *Quadratures*...add[ing] to the end some Propositions taken from my Letters already published for reducing quantities into converging series & thereby squaring figures', 662. [2] (Add. 3968.5: 26^r/26^v). Minor recasting by Newton of his preceding 'proofs' that he had made the essential of his discoveries in calculus by 1671, and then 'in 1676 I extracted the Book of *Quadratures* from the Tract wch I wrote in 1671 & from other older papers', 664. [3] Further honing by Newton of evidence backing his claim in introduction to the *De Quadratura* to have 'invented the method of fluxions gradually in 1665 & 1666', 666. 'The method of converging series I found first by interpolation...in 1665....The same year I got some light into the method of moments & fluctions. And its probable that Dr Barrows Lectures might put me upon [it]', 668. 'The direct method is sufficiently perfect. The inverse...not yet...nor perhaps ever will be', 669 (footnote).

[1] (private). This polished 'Ad Lectorem' takes over the meatiest portions of Newton's English draft prefaces (see Appendix 7) without introducing any novel documentary support for his claim to have discovered fluxional calculus in 1665/6: Wallis and Collins, *d'outre tombe*, along with the still-living Fatio are called upon once more to give their witness, 670. The *De Quadratura* itself was extracted 'around 1676' from the 1671 treatise: 'Newton used it very much in writing the *Principia*, but, having once analytically investigated its propositions, he demonstrated them by synthesis in accord with the law of the ancients', 672. [2] (Add. 3968.32: 464^r). A set of equivalent footnotes to the text of the *De Quadratura*: Newton summarizes his chain of deduction, arguing that the book 'must' have been written by late 1676 in its essence, reminding that its two main propositions were first published in 1693 by 'our Wallis', 673.

The conjectured background to this piece (Add. 3968.34: 480^r–481^r). The central extract of the opinion of the 'eminent mathematician' given by Johann Bernoulli in his letter to Leibniz of 7 June 1713, as printed anonymously by the latter in his July 1713 fly-sheet, 677. Newton 'explained' his method to Leibniz in his *epistolæ prior/posterior* of June/October 1676 '& invented it ten years before or above... without the use of pricked letters. The Book of *Principles* was writ by Composition & therefore there was no occasion of using the calculus of fluxions in it', 678. 'The only Rule wch Mr Newton has given for finding first, second... & other fluxions is conteined in the first Proposition of the Book of *Quadratures* & is a very true one... without [it] the Series of Quadratures... are not to be attained', 678. 'Mr J. Bernoulli in a Letter [of 5 July 1719] to Mr Newton hath positively declared that he was not the author [of this Letter]', 679. 'Mr Nicholas Bernoulli told Mr Newton in his Unkles name in autumn 1712 that there was an error in the conclusion of the solution [of *Principia*, 2, X]' and 'the Eminent Mathematician... suspected that it lay in second differences'; however, in the second edition Newton 'corrected' the error himself, having come to realize that the tangents which are 'first moments' of the curving motion there and so 'should have been drawn the same way' had 'through inadvertency one of them been drawn the contrary way', 679. There is an error of 'much greater consequence' in Section 15 of Leibniz' 1689 *Tentamen* on celestial motions, where 'it doth not appear that Mr L. knew how to work in second differences', 680. The 'three ancient & able Mathematicians' Barrow, Collins and Wallis 'knew what they wrote' when they found that Newton had the method before 1669, whereas 'the author of the Letter of June 1713 wrote only by conjecture', 680.

In this anonymous editor's 'Præfatio' (Add. 3968.13: 209^r–216^r) to the new printing of the 1712 *Commercium Epistolicum*—hitherto not publicly available except to selected individuals and in a few libraries—along with its 1715 *Phil. Trans.* 'Account' rendered into Latin, it is Newton's main purpose to summarize the sequence of pertinent events over the decade since its first appearance. The first six paragraphs as drastically encapsulated in the published (1722) 'Ad Lectorem', 682, note (3). The tally of Leibniz' correspondence with Oldenburg between February 1673 and the latter's death (late summer 1677): 'the autographs of [most of] these are kept in the old Letter-books at the Royal Society.... Of their trustworthiness there is no least doubt', 683. The July 1713 fly-sheet ('how could its author be other than Leibniz?') with its recourse, sight unseen of the *Commercium*, to the 'judgement of a primary mathematician' (un-named): the latter's tying of a method to its 'analytic symbols'—and consequent dispute over the antiquity of Newton's fluxions because 'pointed' letters were not printed till the 1690's—and his attempt (*via Principia*, 2, X) to show that Newton had a 'false' rule for second differences, 684. Was that 'Mathematical judge' Johann Bernoulli? Both Leibniz and Menke later indicated so, 685. Leibniz' endeavour in the 1713 fly-sheet to render Oldenburg's letter to him of 12 April 1675 (communicating a number of series, Gregory's for the inverse-tangent included) 'suspect': brought to nothing when its original 'autograph' was in 1715 shown to the 'Countess' von Kilmansegge, Conti and 'a number of foreign diplomats' at the Royal Society, together with ones of Leibniz himself ('published in the *Commercium*') acknowledging that he had received the series sent, 685. Leibniz' ensuing attempt (through Chamberlayne in 1714) to suggest that letters 'making for him and against Newton' had systematically been omitted from the *Commercium*, and consequent demand that yet unpublished letters be handed over to him by the Society 'so that he might himself publish a fuller epistolary commerce', 686. But Conti 'after several hours' search through the Society's archives' could find nothing favourable to Leibniz which had been left out, 688. The short-lived flurry of letters between Newton and Leibniz (*via* Conti) from late 1715 on: 'to reply to the *Commercium* will need a work of no less bulk' (Leibniz, April 1716), 689. Leibniz' death in late 1716: his *éloge* in the *Acta* (July 1717), 690. 'Setting aside the delusions of the sick, the whole question to be referred to the ancient letters is: who discovered the method insofar as it is had in writings published by Newton?', 691. Newton's basic notion of the momentaneous increase of a fluent and its fluxional (instantaneous) speed: the 'algorithm or arithmetical calculus' to which the concept gives rise (instances in 'literal' fluxions), 691. The converse finding of fluents from fluxions as 'quadrature of curves': known to Newton in the late 1660's when he wrote his *De Analysi*, 692. The 'squaring' of trinomial equations (outlined to Collins in November 1676) and systematic tabulation of algebraically and conically quadrable areas (set out in the 1676 *epistola posterior* for Leibniz) 'could not have come to be without a method of fluxions', 693. Newton's general method of tangents as sent to Collins in December 1672 (here expounded with some anachronistic generalization in notation): 'The solution embraces Prop. I of the *De Quadratura*', 694. 'It has not yet been proved that Leibniz did not see the *De Analysi* in Collins' hand at the time [of his second visit to London] in late October 1676 when he did see several letters of Newton, Gregory and others on series', 695. The variant last paragraphs of the printed 1722 'Ad Lectorem', 696 (footnote).

LIST OF PLATES

PART 1

SOLUTIONS TO CHALLENGE-PROBLEMS, REVISIONS OF EARLIER RESEARCHES AND GENERAL RETROSPECTIONS
(1697–1722)

INTRODUCTION

When in April 1696 Newton quitted academic life in Cambridge to become at once busily involved in the daily affairs of the London Mint, as its Warden, he could have had no anticipation that only nine months afterwards he would momentarily be drawn back once more into the world of abstract mathematics which he had so recently left behind him. Yet on 29 January 1696/7 'in the midst of the hurry of the great recoinage', as his niece Catherine Barton afterwards set the context,[1] when he 'did not come home till 4 [in the afternoon] from the Tower' there to find awaiting him a printed paper bearing a twin mathematical challenge from a young Groningen professor, Johann Bernoulli (brother of the then much more widely renowned Jakob), addressed generally 'to the sharpest mathematicians flourishing throughout the world',[2] Newton did not hesitate straightaway to attack their problem; indeed (to continue Catherine's story) he 'did not sleep till he had solved it wch was by 4 in the morning'. A heroic feat indeed for a man in his middle 50's, one whose erstwhile acuity in such matters was now grown dull through disuse! But let us postpone our uncritical admiration for the undiminished power of intellect which Newton here revealed, exhausted from his day's work at the Mint though he was, in so speedily and thoroughly despatching in but a long winter's evening two questions which had been posed to humble the mathematical Goliaths of the day—at least till we have inspected the hard historical truth underlying this oft-repeated anecdote.

Although Bernoulli (as indeed Newton, almost certainly) was ignorant of the anticipation when he had first posed this primary problem of his circulated 'Programma' privately to Leibniz in the summer of the previous year,[3] the

---

(1) According to her husband of later years, John Conduitt, who entered this anecdote under his wife's initials 'C.C.' in a loose sheaf of 'Miscellanea' (now King's College, Cambridge. Keynes MS 130.5) penned in prelude to his intended grand 'Memoirs' of Newton in the late 1720's. In 1: note (2) below we quote Conduitt's original pencilled draft of this anecdote as it exists in one of the little green pocketbooks (now Keynes MS 130.6¹) which he carried round with him so as to jot down in it all Newtonian hearsay which came his way. That this story came from his own wife does not necessarily lend authenticity to it, for in 1697 she was yet to join her uncle in London as his housekeeper; it was doubtless a tale, how adorned we cannot now know, which Catherine heard from Newton's lips in after years.

(2) 'Acutissimis qui toto Orbe florent Mathematicis S.P.D. Johannes Bernoulli, Math. P.P...Groningæ.' The text of this challenging proclamation, now known universally by the Latin title of 'Programma' under which it is reprinted in Bernoulli's collected *Opera* (1 (Lausanne/Geneva, 1742): 166–9) is reproduced in full in 1, Appendix 1 below.

(3) In his letter of 9 June 1696 (N.S.) (see *Got. Gul. Leibnitii et Johan. Bernoullii Commercium Philosophicum et Mathematicum*, 1 (Lausanne/Geneva, 1745): 167 [= (ed.) C. I. Gerhardt, *Leibnizens Mathematische Schriften*, 3 (Halle, 1855): 283]) where, having spelled out his 'curio-

first of his twin challenges to the world on New Year's Day 1697—namely, 'Given two points in a vertical plane, to assign to a moving body a path along which, beginning its motion from one and descending by its gravity, it reaches the other in shortest time'[4]—had been formulated by Galileo more than half a century before,[5] but he had been able to prove only that a circular arc afforded a nearer approximation to the unknown fall-curve of minimal time of descent under free fall than the 'obvious' choice of the straight line connecting the two points. Nearly sixty years later, making use of the now well-developed techniques of infinitesimal analysis which were only crudely adumbrated in Galileo's day, Leibniz found no difficulty in determining the first-order differential equation of Bernoulli's required 'brachistochrone',[6] though

sum problema' and told Leibniz of his intention, were no others forthcoming, of publishing his own solution at the close of the year, he commented: 'Miror hoc problema hactenus nemini in mentem venisse.'

(4) 'Datis in plano verticali duobus punctis..., assignare viam ...per quam mobile... a puncto [uno] moveri incipiens et propria gravitate descendens, brevissimo tempore perveniat ad punctum [alterum]' as he phrased it in his letter to Leibniz on 9 June 1696 (*Commercium*, 1: 167 [= *Mathematische Schriften*, 3: 283]). This problem of identifying the *linea Brachystochrona* to use the name which Bernoulli gave to it six weeks afterwards in a following letter to Leibniz on 21 July (*Commercium*, 1: 177 [= *Math. Schriften*, 3: 298]) in preference to Leibniz' own suggested *Tachystoptota* offered by him to Bernoulli on the intervening 16 June (*Commercium*, 1: 172 [= *Math. Schriften*, 3: 288]), was a little differently enunciated in the version published in *Acta Eruditorum* (June 1696): 269 (whose phrasing is quoted in 1, Appendix 1: note (9) below).

(5) In the 'Giornata Terza. De Motu Locali' of his *Discorsi e Dimostrazioni Matematiche, intorno à due nuove Scienze attenenti alla Meccanica & i Movimenti Locali* (Leyden, 1638). In the Latin tract 'De Motu Naturaliter Accelerato' (of the early 1630's?) which he there incorporated (pages 156–235) Galileo was, by a complex succession of equalities and inequalities (Propositions I–XXXVI), able at length to demonstrate that the time of descent from rest along a straight line connecting two given points is longer than that taken to fall along any series of consecutive polygonal chords placed in the vertical circle-arc which passes through the higher given point and has the lower one as its base-point, and hence *à fortiori* longer than the time taken to fall smoothly along the circle-arc itself (since this is the limit of that linkage of consecutive chords when each of these becomes vanishingly small); but he went on mistakenly to surmise from this correct result (in his *Scholium* on page 231) that 'Ex his, quæ demonstrata sunt, colligi posse videtur lationem omnium velocissimam ex termino ad terminum non per brevissimam lineam, nempe per rectam, sed per circuli portionem fieri'.

(6) His deduction to this end from the equivalent dynamical criterion for the brachistochrone that 'elementa curvæ [sint] in ratione composita ex elementorum latitudinis directa simplice et ipsarum altitudinum reciproca subduplicata' he passed back to Bernoulli on 16 June 1696 (N.S.) (*Commercium*, 1: 172–3 [= *Math. Schriften*, 3: 288–9]) but without demonstration of the latter, there describing Bernoulli's *problema* as 'profecto pulcherrimum', one which 'me invitum ac reluctantem, pulchritudine sua, ut pomum Evam ad se traxit'. With a rare good humour Bernoulli in his response on 21 July declared himself willing to allow Leibniz' comparison with the fruit of the Garden of Eden 'si modo ego non pro serpente illo maligno habear, qui hoc pomum obtulit' (*Commercium*, 1: 178 [= *Math. Schriften*, 3: 298]). Gerhardt has printed as a 'Beilage' in *Math. Schriften*, 3: 290–4 the paper 'Invenire lineam Tachystoptotam' in which Leibniz privately at this time set down his proof of the brachistochrone condition which he invoked in his letter to Bernoulli.

he did not at once recognize that this defines the 'tachistoptote' curve (his own preferred designation for the path of swiftest descent) to be the common cycloid. Others in France and England to whom Bernoulli had simultaneously transmitted the problem[7] could not come to grips with it at all. At Paris, Varignon (who had been sent the problem on 15 May) admitted to being 'immediately rebuffed by its difficulty' and unaware of 'anyone, of all those to whom I announced your problem, who has resolved it'; while one of these, L'Hospital, independently wrote back to Bernoulli, praising it as being 'among the most curious and prettiest that anyone has yet proposed', but pleading that it would need to be 'reduced to pure mathematics' before he could apply himself to it 'for physics embarrasses me',[8] and even after a lapse of six months he was able to return second-hand only a false solution by Joseph Sauveur.[9] In Oxford a 'certain Swiss' (very likely Bernoulli's youngest brother Hieronymus, who was then in England) brought the challenge to identify the brachistochrone to John Wallis' eye in September following,[10] but he could only helplessly pass it on in December (the three intervening months of silence on Wallis' part speaks mutely for itself) to his fellow Savilian Professor, David Gregory, who in turn spent two further wasted months in trying to demonstrate that the catenary is the required curve of quickest descent till he was put right the next February by Newton,[11] fresh from his own correct resolution—and not as easily as all that,

---

(7) In his letter to Leibniz on 9 June he sniffily remarked: 'Misi [problema] illud in Galliam et Angliam, visurus num heroes isti mathematici duarum harum nationum, qui olim soli sibi mutuo problemata proponentes et solventes neglectis aliis nationibus de palma mathematica certabant, num inquam et hujus legitimam daturi sint solutionem' (*Math. Schriften*, 3: 283–4; the sentence was diplomatically omitted from the earlier *Commercium* by its editor).

(8) Bernoulli quoted both these negative replies to Leibniz in his letter of 21 July 1696 (N.S.): '...omnes quibus illud proponebatur longe absunt [solutione], testibus litteris Varignonij [24 Maij]: *De tous ceux*, inquit, *à qui j'ai annoncé vôtre probléme, je ne scay encore personne qui l'ait resolu; je l'ai tenté, mais la difficulté m'a tout aussy-tot rebuté.* Etiam ipsi Hospitalio minime displicuit problema: *Ce probleme*, mihi scribit [in epistola 15 Junij], *me paroist des plus curieux et des plus jolies que l'on ait encore proposé, et je serois bien aise de m'y appliquer; mais pour cela il seroit necessaire que vous me l'envoyassiez reduit à la mathematique pure, car le phisique m'embarasse*' (*Commercium*, 1: 177 [=*Math. Schriften*, 3: 298–9]; compare (ed.) O. Spiess, *Der Briefwechsel von Johann Bernoulli*, 1 (Basel, 1955): 148, 319).

(9) In his letter of 31 December 1696 (N.S.); see *Der Briefwechsel von Johann Bernoulli*, 1: 334–6.

(10) On page 30 of Codex E (now Christ Church, Oxford. MS 346) of his papers, where David Gregory first made note of the brachistochrone problem, he added in the margin alongside that 'Hoc probl: Bernoullii nomine propositum est Wallisio Oxonii Sept[ri] 1696 ab Helveto quodam'. To be fair, Wallis was then in his 80th year.

(11) The several leaves of his worksheets where he made his vain attempts, afterwards stitched together by him and indexed C219 (now ULE. Dc. 1.61), document Gregory's efforts to attain this false end between 15 December 1696—shortly after he was informed of the problem by Wallis, we presume—and his receipt of a letter from Newton of 11 February

we may add, since even after further talk with Newton about the matter on 7 March 1696/7 Gregory still remained deeply confused as to how the brachistochrone condition is reduced to yield the cycloid's defining fluxional equation.[12] One might well have assumed that the same unnamed Swiss would have taken equal trouble also to draw Newton's attention in London to Bernoulli's challenge problem several months before its light rephrasing in the circulated 'Programma' reached him in late January 1697, but let us not go beyond the extant historical record in framing such a supposition[13] and certainly fight shy,

following, telling him that the cycloid is the brachistochrone. (Newton's letter is itself lost, but Gregory entered an outline of its content on page 72 of his MS Codex E.) On 13 February 1696/7, doubtless fresh from reading Newton's letter of two days before, Gregory began to write out the first draft (A111[ULE]) of a short paper 'De Ratione Temporis quo grave labitur per rectam data duo puncta conjungentem, ad Tempus brevissimum quo, vi gravitatis, transit ab horum uno ad alterum per arcum Cycloidis', drawing proof of his elegant stated ratio of the times from Propositions XXV and XXVI of Huygens' *Horologium Oscillatorium* (Paris, 1673: 57–8). An improved version of this was read out to the Royal Society on 17 March, and published anonymously in *Philosophical Transactions*, **17**, No. 225 (for 'February' [!] 1696/7): 424–5. Two years later Wallis told Leibniz in a letter of 29 August 1699 that 'quod habetur in [*Transactionibus Philosophicis*] pro Martio [1696/7] supplementum [ad Newtonum], est ipsius Gregorii' (*Leibnizens Mathematische Schriften*, **4** (Halle, 1859): 72; and indeed Gregory's autograph original of his published piece (A22 [ULE. Dk. 1.2¹], verbally amended by him to be exactly the text printed) survives to place his authorship of the unsigned printed article beyond rational doubt. Let us hope that we may as easily demolish the entrenched *idée fixe* that this derives from Newton's own pen. (For the record, Castiglione was the first to make this false connection in his edition of *Isaaci Newtoni Opuscula Mathematica, Philosophica et Philologica*, **1** (Lausanne/Geneva, 1744), inserting Gregory's paper—in sequel to Newton's on Bernoulli's two problems—as *Articulum* II (pages 291–2) of its sixth 'Opusculum', and tucking away in his preceding *Editoris Præfatio* the caution that 'Opusculi sexti...de secundo [Articulo] dubitatur; sed quoniam breve est, ad rem facit, & Auctoris ingenium sapere videtur, afferendum censui, melius me facturum arbitratus, siquid alienum ponerem, quàm si quid *Newtoni* prætermitterem' (see page viii). Forty years later Samuel Horsley unquestioningly accepted Castiglione's tentative authoring—did he even read the latter's prefatory words of warning?—in displaying Gregory's paper likewise in the section 'De problematis Bernoullianis' in the fourth volume (London, 1782; see pages 416–17) of his standard edition of *Isaaci Newtoni Opera quæ exstant Omnia.*)

(12) See his paper A78¹ [ULE], setting down his recollected 'Notata Phys: et Math: Lond: cum Newtono præcipue Martio 1696/7'. Enough does, however, come through the muddle of Gregory's imperfect remembrance of his conversation with Newton to confirm our previous conjecture (see VI: 460–1, note (14)) that Newton himself in the first instance merely applied his 'method of maxima & minima in infinitesimals' (as he dubbed it in 1719; see **2**, Appendix 9 below) to the particular case of fall in least time in much the same manner as twelve years before he had determined the defining differential equation of the solid of revolution of least resistance; compare 1: note (8) following. An assertion by Gregory in the same memorandum of March 1697 that 'Newtonus adhibuit series in hac quæstione' is perhaps a reference to the latter's subsequent employment—whether correctly or no (compare 1, Appendix 2: note (5) following)—of Taylorian expansion of the incremented abscissas of the brachistochrone rounded off to the square power of the augment of the ordinate.

(13) There can, however, be no doubt that Bernoulli in the summer of 1696 intended Newton to be the main English recipient of his challenge, for, in naming to Basnage de

in want of any evidence, of suggesting that any such contact was in fact made with Newton in the autumn of 1696.

By the end of that year well did Johann Bernoulli, with his brother Jakob holding off communicating to him his own correct variant solution, publicly pride himself that none but he and Leibniz had successfully mastered the intricacies of the problem of the brachistochrone: a convincing vindication, this, of the technical power of the new *calculus differentialis* and also of the adroitness of its principal proponents in applying it. What better, fully to bring home to his contemporaries the superiority of Germano-Swiss mathematics, than additionally to set a problem of non-differential kind which the world at large should also be challenged to answer? A 'purely geometrical' one whose solution should, so Bernoulli bared his thoughts three months into the new year, likewise 'uniquely depend upon skill, ingenuity and penetration of mind' and 'demand neither a prolix analysis nor the toil of lengthy computation which often makes you lose patience and succumb beneath the burden of a too painful labour', one which a 'man capable of it' would need only 'half an hour's deep concentration' to resolve where 'another less clever would not do it in a century'.[14] And so in the printed 'Program' of 1 January 1697 in which he

---

Beauval in late March 1697 those to whom he had sent his New-Year 'Programma' two months before, he mentioned that he had sent one of the twin copies of his broadsheet to 'Mr Niewton, qui est un des Mathematiciens de notre temps que j'estime le plus' (see his *Briefwechsel*, 1: 430), thereto adding that 'j'en ay envoyé autant à Mr Wallis, mais, on le dit mort' (*ibid.*). Wallis might just as well have been dead for all the effort he made to attack the problem of identifying the brachistochrone: '[Ego] fui', he wrote in his letter to Leibniz on 29 August 1699 (see note (11) above), 'tam hujus rei negligens, ut ne unam horulam impenderim, sive Problematis Solutionem exquirendo, sive aliorum demonstrationes examinando: contentus utique ex aliorum solutionibus resciscere, quæsitam curvam Cycloidem esse..., securus interim tot egregios viros (inter se consentientes) non esse calculo lapsos omnes.'

(14) So he wrote to Basnage de Beauval in late March 1697: 'Croyez-moy, Monsieur, ces sortes de problemes qui dependent uniquement de l'adresse, du genie et de la penetration de l'esprit ne demandent pas une analyse prolixe ni la peine de chifrer longtems, ce qui fait souvent qu'on perd la patience et qu'on succombe sous le travail trop penible; mais il ne faut qu'une demie heure à mediter attentivement pour en venir à bout. Un homme donc capable de cela peut resoudre dans un jour, ce qu'un autre moins habile ne fera dans un siecle. Ainsi j'ose dire hardiment, que quiconque n'a pas resolu mon probleme le premier jour de son application serieuse, c'est en vain qu'il espere de penetrer le lendemain dans la solution' (*Briefwechsel*, 1: 429-30). Leibniz had, we may notice, observed to Bernoulli on the preceding 23 February (N.S.)—this in immediate reference to the Amsterdam broker Mackreel, who had just disdainfully made it known that he would not be replying to Bernoulli's Programma— that 'Si qui in Batavis Problema tuum [*sc.* brachystochronæ] esse indignum dicunt, næ illi vel rem non intelligunt, vel vulpem imitantur, quæ pyra cum attingere non posset, acerba esse dicebat....Certe si Hugenius viveret & valeret, vix quiesceret nisi Problemate tuo soluto'; and then passed percipiently to forecast: 'Nunc nemo est a quo solutionem facile expectem, nisi a D^{no} Marchione Hospitalio, aut a D^{no} Fratre Tuo [Jacobo], aut a D^{no} Newtono, quibus adderem D^{num} Huddenium Consulem Amstelodamensem...nisi dudum has meditationes seposuisset' (*Commercium*, 1: 247 [= *Math. Schriften* 3: 369-70]). To which Bernoulli responded

re-posed his seven-month-old challenge to determine the curve of swiftest fall, there acceding to Leibniz' plea[15] that the deadline for solutions to be returned to him should be extended till the coming Easter, he adjoined his 'New Year present' of what he hoped would prove just such a second problem, requiring (as a test in the inverse Cartesian analysis by which a geometrical condition is reduced to an equivalent locus property) that there should be found the curve which cuts off in a rotating radius vector two segments, the aggregate of given powers of which, each measured from the vector's pole, shall 'everywhere make one and the same sum'. As Bernoulli forecast, this *problema alterum purè Geometricum* caused no greater difficulty to Leibniz than the preceding dynamical one—indeed, the latter wrote back that he had found its 'ready and convenient and, unless I am mistaken, general solution' while jolting along alone in a carriage between Hanover and the nearby market town of Brunswick[16]—but caused real difficulty to most others who tried their mathematical teeth upon it. He could not have expected that the now ageing Newton in England would, immediately upon receiving his copy of the *Programma*, deal so cleanly and comprehensively with its content.

True, by his niece's account (and discarding the possibility that he had earlier been told—or had himself read—of the primary problem of constructing the brachistochrone) it took Newton, for all his hard thought over the evening and early night of 29/30 January, two dozen times longer than Bernoulli's 'bogey' of half an hour to produce the bones of the response which he set out in a letter dated the next day 'For the Right Honourable Charles Montague Esqr/Chancellour of the Exchequer', his close acquaintance since the latter's student days at Cambridge fifteen years before and now, more pertinently, President of the Royal Society.[17] A year away from academic life had doubtless left Newton intellectually rusty, and it would have taken him some short while to attune his mind to mathematical niceties once again. Our detailed appraisal

---

on 13 March: 'Quidni addis etiam Wallisium (qui nil non solvisse putat) iis quos problemati meo pares existimas? Huic & Newtono utrique bina exemplaria programmatis mei sub nudis involucris in Angliam transmisi' (see *Commercium*, 1: 253 [= *Math. Schriften*, 3: 379]); he gave the same information regarding his chosen English recipients of copies of his printed challenge-sheet to Basnage de Beauval a fortnight later (see the previous note).

(15) For the exact terms of this see 1, Appendix 1: note (8) following.

(16) See his letter to Bernoulli of 23 February 1697 (N.S.): 'Problema quod [Programmati Tuo] subjecisti pure Analyticum nuper, otium nactus in itinere ad nundinas Brunsvicenses, consideravi in curru solus, et viam solvendi reperi... certe expeditam et commodam, et ni fallor generalem' (*Commercium*, 1: 247 [= *Math. Schriften*, 3: 370]). The text of the paper on which Leibniz afterwards set down his detailed solution—which suffers from the same deficiency as Bernoulli's own (on which see 1: note (11) below)—is printed by Gerhardt in *Math. Schriften*, 3: 371–2.

(17) The main portion of this open letter to Montague is reproduced (from the autograph original, now Royal Society. Letter Book N.1.61^b) as the opening document below.

(18) See especially 1: notes (8) and (11) following.

of the intricacies of Bernoulli's two problems and of the generality of Newton's undemonstrated constructions of their solution is given in footnote to its reprinted text below.[18] Here we would make the point that it had for a dozen years past been within his proven power readily to solve both, and we can only think that it was his year's total break with things mathematical which in January 1697 prevented him at once jotting down the essence of their analysis and in but a few minutes more roughing out their consequent synthesis. To the outer world, however, which read only the printed version of Newton's letter put out without author's name soon afterwards in the *Philosophical Transactions*[19] the time and mental effort required to achieve their solution was not at issue. Their precision and quality was, and that brief published article, for all that it was unsigned, very clearly bore the trace of Newton's stamp for those with eyes educated to see it—not least for Bernoulli himself, who from the first (or so he told Leibniz) was 'firmly confident' of its author's identity.[20] (Johann's subsequent affirmation to Basnage de Beauval that 'we know indubitably that the author is the celebrated Mr Newton; and, besides, it were enough to understand so by this sample, as *ex ungue Leonem*'[21] has, ever since its scrambled quotation by Wood-

---

(19) *Philosophical Transactions*, **17**, No. 224 (for January 1696/7): 384–9: 'Epistola missa ad prænobilem virum D. Carolum Mountague Armigerum, Scaccarii Regii apud Anglos Cancellarium, & Societatis Regiæ Præsidem, in qua solvuntur duo problemata Mathematica à Johanne Bernoullo Mathematico celeberrimo proposita.'

(20) Immediately upon receiving a copy of the 'solutio Angli Anonymi' (not directly from Newton, but by way of Basnage de Beauval; see the next note) he wrote to Leibniz on 3 April 1697 (N.S.) to confide that '[eum] Newtonum firmiter credo' (*Commercium*, **1**: 262 [= *Math. Schriften*, **3**: 390]), adding his praise of the 'artificium non inelegans'—Newton's system of generalized Cartesian coordinates (that earlier set out by him as Mode 5 of Problem 4 of his 1671 tract; see III: 140, and compare 1: note (11) below)—there introduced as an ancillary to resolve the problem of intercepts 'ducta scilicet recta positione data, & rectas ex polo egredientes transversim secante, quas ut abscissam, illas vero ut applicatas considerat'. Leibniz more cautiously answered on 15 April that 'Solutionem Anglicam a Newtono esse suspicor' (*Commercium*, **1**: 266 [= *Math. Schriften*, **3**: 394]).

(21) In a 'Lettre de Mr Bernoulli à l'Autheur' published by de Beauval in his *Histoire des Ouvrages de Savans* (June 1697): 452–67 [=*Johannis Bernoullii Opera Omnia*, **1** (Lausanne/ Geneva, 1742): 194–204] where he sketches the background to his 'problême de la plus vîte descente, dans les *Actes* de Leipsic' and passes comment upon the several solutions more recently provoked by his republication of this challenge to determine the brachistochrone in his New-Year's *Programma* (on these see note (23) following) Bernoulli observed in particular (*ibid.*: 454–5) that 'mon problême...tout irresolu...passa en Angleterre, là où j'avois grande esperance qu'il trouveroit un dessin plus favorable, puis qu'il y a dans ce païs-là quelques excellens Geometres qui se servent adroitement de nôtre methode, ou d'une autre tout-à-fait semblable[!] à la nôtre. Effectivement le mois de Janvier des *Transactions Philosophiques* imprimées à Londres, que vous avez eu la bonté de m'envoyer me fait voir que je ne me suis point trompé, y ayant trouvé une construction de la Courbe de la plus vite descente parfaitement convenable à la nôtre. Quoi que l'Auteur de cette construction par un excés de modestie ne se nomme pas, nous savons pourtant indubitablement que c'est le celebre Monsr. Newton: & quand même nous ne le saurions point d'ailleurs, ce seroit assez de le connoître par cet

house and its ensuing, far more widely read mis-citation wholly out of context by Brewster, travelled from mere pedestrian *cliché* to be a universally parroted (but no less spurious) myth.[22]) And within weeks the shrewd guess of a few became

échantillon, comme *ex ungue Leonem*'; and he went on to add that 'Ce savant homme est très-digne de la loüange que j'ai promise à ceux qui donneroient une resolution legitime de ma question. J'avoüe neanmoins que toute grande que je penserois la faire, elle seroit petite à l'egard de celles qu'il s'est deja acquises par la publication de son Ouvrage incomparable [*viz.* the *Principia*], dans lequel il fait voir tant de profondeur & tant de penetration d'esprit, que dès le moment que ce problême me vint en pensée, ces deux excellens Maîtres, savoir Mrs. Leibnits & Newton, se presenterent les premiers à mon esprit comme capables de denoüer le nœud, quand personne autre ne le seroit.' The Latin tag here used by Bernoulli in reference to Newton's unmistakable lion's claw—*Leonem ex unguibus æstim*[*es*] in Erasmus' parent version (*Adagia*, I, ix, 34), itself rendering classical Greek equivalents in Plutarch and Lucian alluding to the sculptor Phidias' famed ability finely to assess the size of a lion given only its severed paw—was already a well-worn *cliché* in his day. (To cite but one parallel instance, James Gregory used it of Tschirnhaus 'who (if I may judge *ex ungue leonem*) surely is a great algebraist... albeit probably inferior to Mr Newton' in a letter to John Collins on 20 August 1675; see S. P. Rigaud, *Correspondence of Scientific Men of the Seventeenth Century*, 2 (Oxford, 1841): 269 [= (ed.) H. W. Turnbull, *James Gregory Tercentenary Memorial Volume* (London, 1939): 325].)

(22) In his *Treatise on Isoperimetrical Problems and the Calculus of Variations* (Cambridge, 1810): 150 Robert Woodhouse, in a quick *résumé* of Newton's published solution of the brachistochrone problem, pithily stated that he 'gave, without proof or the authority of his name, a method of describing the cycloid. But John Bernoulli, from the Work recognised its author: "*ex ungue Leonem*"'; and in justification of this too hasty interpretation of Bernoulli's words quoted in an attached footnote, from his 1742 *Opera*, the pertinent sentence of his 1697 French letter to de Beauval (see the previous note). More than two decades later, David Brewster—whose knowledge of the original context probably went no further than a quick reading of Woodhouse—in his influential [*Life of Sir Isaac Newton* (London, ₁1831): 194→] *Memoirs of the Life, Writings and Discoveries of Sir Isaac Newton*, 2 (Edinburgh, 1855): 21 grandiosely delivered his version of what had now become anecdote, stating that 'although that [solution] of Newton was anonymous, yet Bernoulli recognised in it his powerful mind; "*tanquam*", says he, "*ex ungue leonem*", as the lion is known by his claw'; and he subsequently (*ibid.*: 192) further embellished this into the wildly extravagant flourish that 'When the great geometer of Basle saw the anonymous solution, he recognised the intellectual lion by the grandeur of his claw; and in their future contests on the fluxionary controversy, both he and Leibniz had reason to feel that the sovereign of the forest, though assailed by invisible marksmen, had neither lost a tooth nor broken a claw.' In the present century Bernoulli's dead metaphor has acquired a wider range still of bastard progeny, ranging from mere paraphrase of Brewster's rhetoric (such as L. T. More offers in his *Isaac Newton* (New York, 1934): 475: 'It is said that Bernoulli recognised the author from the sheer power and originality of the work; "*tanquam ex ungue leonem*"') to such dubious recent generalizations as F. E. Manuel's that 'the continentals quickly recognized the lion among mathematicians by his claw—*tanquam ex ungue leonem* (*A Portrait of Isaac Newton* (Cambridge, Massachusetts, 1968): 220). We cannot but think that Bernoulli himself, after having endured the ingratitude (so he regarded it) a decade and a half afterwards of having no acknowledgement made in the *Principia*'s second edition of his timely intervention in autumn 1712 which permitted Newton almost at the last moment in its printing to recast the faulty Proposition X of its second book, and thereafter suffered misunderstanding and vilification for his further pains dynamically to 'correct' his English colleague, would rather in later years have been willing to endorse Dryden's line that 'Such mercy from the British Lyon flows...' (*The Hind and the Panther*, **1**, 289).

common report all over Western Europe that the 'anonymous Englishman' who had authored this pithy response to Bernoulli's challenges was the redoubtable onlie begetter of the *Principia Mathematica*.

With the secret of the identity of the brachistochrone thus publicly broken by Newton before Johann's promised deadline of Easter 1697 for disclosing it to be a cycloid had passed, and with yet other correct solutions to them arrived in the meantime from his brother Jakob and from L'Hospital, Bernoulli rapidly lost interest in his two problems. It was left to Leibniz to publish in the Leipzig *Acta* a *catalogue raisonné* of the solutions received by April, his own and Johann's included.[23] In succeeding years proof that an inverted, vertical cycloid is the brachistochrone under 'simple' gravity came quickly to lie within the comprehension and competence of all of average mathematical ability; even such an ordinary London teacher and practitioner as Richard Sault, who produced a minimal variation upon Jakob Bernoulli's demonstration in the 1697 *Acta* which appeared—with a multifold spatter of typographical errors which must utterly have confused all but those already proficient in possible modes of solution to the problem—in the *Philosophical Transactions* in 1698.[24] Two years later in the

(23) 'G[ot]. G[ul.] L[eibnitii] Communicatio suæ pariter, duarumque alienarum ad edendum sibi primum a Dn. *Jo. Bernoullio*, deinde a Dn. Marchione *Hospitalio* communicatarum solutionum problematis curvæ celerrimi descensus a Dn. *Jo. Bernoullio* Geometris publice propositi, una cum solutione sua problematis alterius ab eodem postea propositi' (*Acta Eruditorum* (May 1697): 201–20). This comprised Johann Bernoulli's celebrated reduction of the brachistochrone condition to determining the path of a light ray through a medium of varying density in the Fermatian model in which its speed, say $v = ds/dt$ (over the infinitesimal arc ds in least infinitesimal time dt)—instantaneously proportional in consequence, by the Snellian law of refraction, to the sine (dy/ds) of its inclination to the vertical—at any point (x, y), is identified with the velocity of fall, $\sqrt{(2gx)} \propto \sqrt{x}$ under constant downwards gravity g, attained by the descending light-corpuscle at a depth x below the horizontal from which it starts at rest (pages 206–10), and also his brother Jakob's deduction from first principles (much, we presume, as Newton in his suppressed analysis) of the equivalent condition: $(1/v) \cdot dy/ds =$ constant for passage in least time (pages 211–14); while essentially equivalent, if not entirely general, solutions to the second 'problema purè Geometricum' by Leibniz himself, Jakob Bernoulli and L'Hospital were also there presented (pages 205, 215–16 and 217–18 respectively; on these see 1: note (11) below). Leibniz referred only in passing to his own manner of finding (so he here continued—compare note (6) above—to name the brachistochrone) the 'tachistoptote' curve. L'Hospital was equally elliptical in sketching his own ingenious if circuitous reduction of the defining extremal condition that $\int (1/v) \cdot ds \ (= \int x \cdot (1/vx) ds)$ be a minimum to that of determining the catenary whose arc-element ds has weight $1/vx$; whence (compare v: 521, note (3)) there must be $dx/dy = \int (1/vx) \cdot ds$, that is, $1/vx = d(dx/dy)/ds = d(ds/dy)/dx$ and so $1/v \propto ds/dy$. (L'Hospital had privately communicated this deduction of the brachistochrone condition $dy/ds \propto v \propto \sqrt{x}$ to Bernoulli on the preceding 25 February (N.S.); see the latter's *Briefwechsel*, 1: 342–3.)

(24) *Philosophical Transactions*, **20**, No. 246 (for November 1698): 425–6: 'Curvæ Celerrimi Descensus investigatio analytica excerpta ex literis R. Sault, Math. D[omini].' Sault made his far from lucid proof public with the plea (*ibid.*: 425) that 'quamvis...problema illud [curvæ brachystochronæ] nunc obsoletum videatur, libentius tamen publici juris faciam, quia

same periodical John Craige expounded a yet further simplified approach to determining the path of swiftest descent,[25] which he claimed to have developed without seeing any other than Newton's undemonstrated construction of its cycloidal curve. Even David Gregory in Oxford (if to phrase it so is not too greatly to disparage his painstaking and hitherto unrewarded long labours to that end) at length attained his own correct private proof of the brachisto-chrone's identity.[26] A genuinely novel variant analysis of the problem was published in 1699 by the *émigré* Swiss mathematician (and Newton's sometime *protégé*) Nicholas Fatio de Duillier through an ingenious but extremely involved argument expounding a 'Two-fold geometrical investigation of the line of briefest descent'.[27] This led at the time to some heated exchanges with Leibniz, but more because of Fatio's accompanying characterization of the latter as 'second inventor' of the calculus than for any deep doubt as to the validity and effectiveness of his reduction of the brachistochrone condition to a curvature property uniquely defining the path to be cycloidal, even though Leibniz bitterly scorned this as being an unnecessarily complicated and round-about

celeberrimus Leibnitius omnes Mathematicos, hujus problematis compotes, enumerare suscepit...'.

(25) This, originally set down in a letter to Hans Sloane of 21 December 1700, appears as 'Problema 2. Invenire Lineam Celerrimi Descensus' of the published 'Reverendi D. Johannes Craig, Epistola ad Editorem continens solutionem duorum problematum' (*Philosophical Transactions*, 22, No. 268 (for January 1700/1): 746–51, especially 750–1); in his prefatory remarks Craige asserted somewhat unbelievably to Sloane that 'Qualem [Analysin] alii adhibuerunt, nescio; cum nulla hujus solutio (nec quæ in vestris [*Transactionibus*], nec quæ in Leipsicis *Actis* eduntur) ad manus meas adhuc pervenerit, præter *Newtónianam*, quæ Analysin non exhibet.' Craige's 'Problema 1' (*ibid.*: 748–9) is, we may add, Newton's fifteen-year-old one 'Invenire Lineam curvam cujus rotatione producatur Solidum rotundum, quod (dum in medio fluido secundum axis sui directionem movetur) minimam patiatur Resistentiam', and is straightforwardly resolved by him through an analogous determination of an infinitesimal minimum (compare VI: 467, note (25)).

(26) In his yet unpublished memorandum A64 [ULE], afterwards indexed by him as 'D.G. Inventio Curvæ celerrimi descensus, scᵗ Cycloidis'. While this certainly appears to be but a minimal variant upon Craige's method of determining the brachistochrone, there is some slight circumstantial evidence that it may have been written by February 1700. Let us *pro tempore* give Gregory credit for originality.

(27) 'Nicolai Fatii Duillierii, R[egiæ] S[ocietatis] S[odalis], Lineæ Brevissimi Descensus Investigatio Geometrica Duplex': this was printed by Fatio in appendix to a tract of his on *Fruit-Walls Improved by Inclining them to the Horizon: Or, A Way to Build Walls for Fruit-trees; whereby they may receive more Sun Shine and Heat than ordinary* (London, 1699) where (pages 38–42) he advocated inclining the 'wall' of fruit trees at an angle, facing the sun, equal to the latitude: 'an inclination very good...for the fruits that are ripe in the months of October or the latter end of September'—an application of mathematics to arboriculture which Newton, good Lincolnshire farmer's son that he was, doubtless took not unseriously (though his library copy of Fatio's book, now Trinity College, Cambridge. NQ.10.33, is unmarked). On Fatio's like-wise appended 'Investigatio Geometrica Solidi Rotundi, in quod Minima fiat Resistentia', where Newton's *Principia* problem of the solid of least resistance is analogously reduced to a defining second-order, curvature condition, again see VI: 467, note (25).

way of arriving at the result.[28] Guided by what impulse or motive we know not, Newton on the other hand was sufficiently taken by the basic merit of Fatio's lengthy reasonings privately to strip them of their spiderish complications, down to their bare web of argument.[29] By 1704, when Charles Hayes produced his widely studied textbook on fluxions, the chase after the curve of swiftest descent was long over, and the prey there securely caged as a mere worked example to a more general consectary outlining Johann Bernoulli's elegant model of least-time fall as an optical ray-path.[30]

An ineffectual attempt was, we may add, made by John Machin in 1718 to determine in analogous Bernoullian style the brachistochrone in a general central-force field, that is, the curve of quickest fall from rest at one point to a second given one by a body continuously drawn to a finite centre under a known law of 'gravity'.[31] What Newton thought of Machin's effort to transcend the

(28) Compare 1, Appendix 2: notes (1) and (12) below.

(29) We reproduce Newton's preliminary calculations (now ULC. Add. 3968.41: 2^r) to this end in 1, Appendix 2, there mending a couple of minor slips in his computation (see *ibid.*: notes (5) and (9)) and fleshing out its skeleton from a transcript of Newton's corrected 'investigatio curvæ celerrimi descensus' which David Gregory wrote down on 1 April 1700, shortly after Newton thus amplified it, in his memorandum C122^2 (now Royal Society. Gregory Volume: 22).

(30) *A Treatise of Fluxions: Or, An Introduction to Mathematical Philosophy. Containing a full Explanation of that Method by which the Most Celebrated Geometers of the present Age have made such vast Advances in Mechanical Philosophy....By Charles Hayes, Gent.* (London, 1704): Sect. V, Prop. X, Consectary I, Example I: 130. In his ensuing Prop. XV (*ibid.*: 142–4) Hayes afterwards outlined Bernoulli's 'Dioptrick' reduction of the brachistochrone condition to the defining differential condition $dy/ds \propto v \propto \sqrt{x}$ (see note (23) above) before passing to spell out at great length, following John Craige (see note (25)), the derivation of this from first principles.

(31) In his 'Inventio Curvæ quam Corpus descendens brevissimo tempore describeret; urgente Vi Centripetâ ad datum punctum tendente, quæ crescat vel decrescat juxta quamvis Potentiam distantiæ à Centro; dato nempe imo Curvæ puncto & altitudine in principio Casus', *Philosophical Transactions*, **30**, No. 358 (for October–December 1718): 860–2. If one penetrates the clouding fog of his undemonstrated geometrical constructions— he appends (page 862) the sole justification for their validity that 'Harum Constructionum Demonstrationes è Celeberrimi D. Newtoni *Quadraturis*, ejusdemque *Philos. Nat. Principiis* (Prop. XXXIX. & sequentibus aliquibus [Lib. I]) petitæ, aliâ datâ occasione ostendetur'—it will be found that (on recasting his underlying argument into standard modern polar notation) Machin there mistakenly supposes that, where the origin $(0, 0)$ is the centre of a force-field of inwards 'pull' $f(r)$ at a distance r out from it, the curve of swiftest fall from rest at $r = A$ shall at its any point (r, ϕ) satisfy the modified Bernoullian condition $r.d\phi/ds \propto v$, where

$$v = \sqrt{\left(2 \int_A^r f(r).dr\right)}$$

is (compare VI: 408, note (307)) the speed there instantaneously attained by the falling body; as it is easy to show, however, by a Newtonian argument for such 'maxima & minima in infinitesimals', the correctly generalized brachistochrone condition is $r.d\phi/ds \propto v/r$ (or, equivalently, $p \propto v$ where p is the tangential polar $r^2.d\phi/ds$). Machin was in consequence

terms of Bernoulli's original problem, and whether he noticed the error in its implicit condition for fall along a polar curve in least time (thus failing to identify the hypocycloid as the brachistochrone in Newton's favoured—if physically untrue[32]—hypothesis that the effective 'pull' of terrestrial gravity beneath the Earth's surface decreases directly as the distance from its centre), we do not know. It would go beyond our present purpose to enter into the systematic researches pursued into variational calculus after Newton's death, above all by Euler and Lagrange in the middle and later decades of the eighteenth century,[33] and in consequence of which general criteria for the existence of extremal properties in plane curves were laid down, and refined techniques for their determination fashioned.

We have previously noticed that during his short stay in Cambridge early in May 1694 David Gregory wrote out *inter alia* an incomplete transcript of Newton's treatise 'de Quadratura Curvarum' in its reshaped and truncated version,[34] and that on a similar visit in August the following year Edmond Halley was granted the yet rarer privilege of taking the autograph of Newton's tract away with him back to London, to copy it at his leisure,[35] though in subsequently apologizing for his slowness in returning these theorems on quadratures, he was quick to reassure their author that 'no one has seen them, nor shall, but by your directions'.[36] These two momentary acts of generosity on Newton's part in letting others share the sight of his unpublished essays into the

---

denied the insight that, just as the hypocycloid

$$\phi = \frac{\rho}{R+\rho}\cos^{-1}\left[\frac{R^2+\rho^2-r^2}{2R\rho}\right] + \sin^{-1}\left[\frac{\rho}{r}\sqrt{\left(1 - \frac{(R^2+\rho^2-r^2)^2}{4R^2\rho^2}\right)}\right]$$

is (compare vi: 388, note (275) and 394, note (284)) the tautochrone in a direct-distance force-field ($f(r) \propto r$), so also, because the sine of its slope is $(r \cdot d\phi/ds =) \, k\sqrt{[(R+\rho)^2 - r^2]}/r$, $k = (R-\rho)/2\sqrt{(R\rho)}$, it is the curve of swiftest fall of a body therein from rest at $r = R+\rho$ to any point (r, ϕ), $r \geqslant R-\rho$, where it attains the speed $v \propto \sqrt{[(R+\rho)^2 - r^2]}$.

(32) See vi: 186, note (186).

(33) Euler's magisterial treatise on the topic is his *Methodus inveniendi Lineas Curvas Maximi Minimive Proprietate gaudentes* (Lausanne/Geneva, 1744 [= *Leonhardi Euleri Opera Omnia*, (1) 24, Zurich, 1952]); compare M. Cantor, *Vorlesungen über Geschichte der Mathematik*, 3 (Leipzig, ₁1898): 830–42, especially 830–1. On Lagrange's subsequent creation of a general 'calculus of variations' on this Eulerian foundation see J. A. Serret's edition of his *Œuvres*, 14 (Paris, 1892) and C. R. Wallner's essay on 'Variationsrechnung' in Cantor's *Vorlesungen*, 4 (Leipzig, 1908): 1066–74. Woodhouse's 1810 monograph on *Isoperimetrical Problems and the Calculus of Variations* (see note (22) above) still has some value as an introductory overview, and has to this end recently been reprinted in facsimile (New York, 1976) under the new title of *A History of the Calculus of Variations in the Eighteenth Century*.

(34) See vii: 508, note (2), and compare *ibid*.: 193–5.

(35) See vii: xxvi and note (69) thereto.

(36) As Halley wrote to Newton on 7 September 1695, a fortnight or so after his return to London (see *Correspondence of Isaac Newton*, 4, 1967: 165).

less elementary realms of exact integration were—the first particularly so—to be regretted by him a decade later, but in the meantime all stayed calm. We will not be surprised that, together with Fatio de Duillier (who had been allowed a tantalizing glimpse of parts of the initial treatise 'De quadratura Curvarum' even while it was being written during the early winter of 1691–2),[37] both Gregory and Halley more than once in the next few years pressed Newton to make his tract public. At length, on 'Sunday 15 Nov. 1702' Gregory was able to enter in his private diary[38] that Newton had 'promised Mr Robarts, Mr Fatio, Capt. Halley & me to publish his Quadratures', along with his 'treatise of the Curves of the 2d Genre',[39] in train to his 'treatise of Light'—the lengthy English work on refractive and diffractive 'Opticks' which he had also drafted in the early 1690's[40] but likewise still withheld from public circulation. A young Scots physician, George Cheyne, well versed in contemporary iatro-mathematical medicine and then newly arrived in London from Edinburgh to practise it, was to do more than anyone to bring that promise rapidly to fruition. Encouraged thereto by his mentor Archibald Pitcairne, Cheyne had on the previous 17 April (while still in Scotland) set down a dozen 'General rules of the inverse Method of Fluxions which I about two years ago perform'd as I could'[41] in a 'paper by way of a Letter...there anent' to him which Pitcairne

---

(37) See VII: 12–13.

(38) W. G. Hiscock, *David Gregory, Isaac Newton and their Circle: Extracts from David Gregory's Memoranda, 1677–1708* (Oxford, 1937): 14. Just two days earlier Gregory had noted a somewhat different report—evidently gathered second-hand as gossip—that 'Mr Newton is to republish his [*Principia*]; & therein give us his methode of Quadratures' (*ibid.*: 13; see also VII: xxvii, note (70)).

(39) That is, the 'Enumeratio Linearum Tertij Ordinis' (see VII: 568–644) which Newton had, except for minimal rewriting and verbal adjustment, completed in the summer of 1695; we have previously remarked (VII: 589, note (3)) that this tract on the listing and classification of cubic curves had till a very late stage before it was sent to the printer in 1703 borne, among other such heads, the title 'Enumeratio Curvarum Secundi Generis'. Francis Robarts had other recorded interest as a proponent of Newtonian calculus, for his name reappears some seven years afterwards as the begetter of the article on 'Fluxions' in John Harris' *Lexicon Technicum: Or, An Universal Dictionary of Arts and Sciences*, 2 (London, ₁1710).

(40) As the handwriting of its autograph fair copy, now ULC. Add. 3970.3: 17r–233r (this afterwards was the copy sent to the printer in late 1703, and is so marked up by him), clearly shows to be so.

(41) So Cheyne declared in a letter to Gregory on 20 January 1702/3 (reproduced by Hiscock in appendix to his *Gregory, Newton and their Circle*: 43–5; see especially p. 45), there taking care to emphasize that 'all these are but a few examples of Mr Newtons (excepting yours [!]) Methods, and that all found out within these 20 years by these or not unlike Methods are but either repetitions of, or easie corollaries from these things which he has either imparted to his friends or the publick' (*ibid.*: 45). He makes a similar observation in the published version of his enclosed theorems on 'Fluxionum Methodus Inversa'; see note (47) below. Pitcairne had, we have already noticed (VII: 5–6), been fifteen years previously a close acquaintance of Gregory's up to the time (1691) of the latter's departure south from

'sent up...to Dr Arbuthnot to be shewn to Mr Newton, and to ask his advice if he thought it worth printing'. The latter's response, however, was the scarcely encouraging one of faint praise:

Mr Newton, (whether it were that he could not give himself the trouble of peruseing it but slightly as I have reason to believe, or not to disoblidge Dr Pitcairne who seemed to put some value on it by offering it to the view of such a great man) thought it not intolerable.

Not to be thus put off, Pitcairne would even with this weak backing have straightaway put Cheyne's paper to a London printer. But Cheyne himself, 'haveing a greater design in my head than this paper can signifie' was 'loath to let it pass so' and, 'not haveing the time to accomplish it' for the moment, he 'prevail'd with the Doctor to let this paper alone till I my self came to London'.[42]

The chill which lightly overlay this first distant contact between Newton and Cheyne was to melt into anger and resentment when the two at length met face to face at the end of 1702. In John Conduitt's staccato jotting[43] subsequently recording the encounter:

When Dr Cheyne came from Scotland Dr Arbuthnott carried him to Sr I. & told him [of] the book he had wrote &c but our [*sc.* Scots] countrymen no money to print—let him come to me. Sr I. offered a bag of money. Dr Cheyne refused—both in confusion—but Sr I. [would] not see him afterwards.

As we may in hindsight detect, but as Newton on the spot seemingly did not grasp (or did he?), Cheyne did not so much want financial help from him to publish his paper as his expert criticism of the content and structure of his rules for determining fluents from given fluxions, so as to be able to amend and

---

Edinburgh to take up the Savilian Chair of Astronomy at Oxford, and in particular—this doubtless a black mark against him in Newton's eyes when he came later to learn of it—also the publisher in 1688 (see VII: 6, note (15)) of Gregory's 'independent' invention of Newton's series quadrature of the general algebraic binomial.

(42) All these quotations are from Cheyne's letter to David Gregory on 20 January 1702/3 (Hiscock, *Gregory*: 43). During the preceding summer, we need scarcely add, Cheyne (not to travel to England till the autumn) was still away north in Edinburgh, while Pitcairne in London had access to Newton only through his fellow Scot (and physician) Arbuthnot.

(43) In a pencilled entry made some time after Newton's death in one of the little pocket-books (now King's College. Keynes MS 130.6^2) which he carried round with him to note down such anecdotes on the spot, there giving 'Hentyn' as his source of information. A subsequent rephrasing of this by him (on a separate sheet, now Keynes MS 130.7) reads more stiltedly: 'Dr Arbuthnott[!] told me he told Sr I. [that] Cheyne had writt an ingenious b[ook] upon Mathematicks—but that his countrymen had not money to print. Bring him to me says Sr I. & when he brought him Sr I. offered Cheyne a bag of money, wch he refused. Sr I. would see him no more.'

refashion them before he put these to the printer.[44] And it was, though Newton certainly never found it easy to forgive such slights, real or imagined, surely not merely Cheyne's spurning of his proffered money which occasioned his apparently spontaneous outburst of 'confusion' and antipathy when first they met. Even the hastiest glance at Cheyne's paper must, for all that it readily acknowledged at the pertinent places his own prior published discoveries in the rational and series quadrature of curves, have given him a severe fright. Its minor slips in phrasing and presentation aside,[45] this survey of 'The Inverse Method of Fluxions, or the More General Laws of Fluent Quantities'—as the revised version of it stubbornly put by Cheyne to press in the early winter of 1702/3 came to be titled[46]—gave a competent and comprehensive survey of recent developments in the field of 'inverse fluxions' not merely in Britain, at the hands of Newton, David Gregory and John Craige, [47] but also by Leibniz

(44) 'This paper will be certainly incorrect', wrote Cheyne with genuine diffidence to Gregory in January 1703 (Hiscock, *Gregory*: 45; compare note (41) above), 'because I have none to Inform [me] of my mistakes & none is fit to correct his own errors'. We may add that when he subsequently sent an advance copy of their printed text to Johann Bernoulli at Groningen (by way of 'un certain Ecossois M. Falconer...passant par ici'), the latter—as he wrote to de Moivre on 15 November 1704 (N.S.); see K. Wollenschläger, 'Der mathematische Briefwechsel zwischen Johann I Bernoulli und Abraham de Moivre' (*Verhandlungen der Naturforschenden Gesellschaft in Basel*, **43**, 1933: 151–317): 179–87, especially 180—found upon reading Cheyne's work, after an initial reluctance to credit that one 'qui m'etoit entierement inconnu' could possibly treat such a topic 'selon son merite', that 'contre le prejugé que j'en avois' its author was indeed 'versé profondement dans cette matiere'. It was, of course, typical of Bernoulli to go on to complain that 'le sincere M. Cheyne, ayant reçu...toutes mes observations sur son livre avec quelques regles qui me paroissent plus aisées que quelques unes des siennes, ...ne tarde pas un moment à faire imprimer le cahier des *Addenda et Adnotanda* composés pour la plus grande partie de ce qu'il avoit tiré de mes remarques' (*ibid.*: 187).

(45) Those in the text as printed off are in their largest part near-trivial typographical errors and all but one or two of the rest are small, immediately correctable slips. Let us seek to strangle at its birth a recently broadcast misconception (*vide* T. M. Brown in Scribner's *Dictionary of Scientific Biography*, **3** (New York, 1971): 244, col. b) that David Gregory subsequently found '429 errors' in it; these were (see Hiscock, *Gregory*: 25, note 1) equally minor verbal and syntactical *errata* in Cheyne's vastly different work on *The Philosophical Principles of Natural Religion* (Oxford, 1705).

(46) *Fluxionum Methodus Inversa; sive Quantitatum Fluentium Leges Generaliores. Ad Celeberrimum Virum, Archibaldum Pitcarnium, Medicum Edinburgensem. A Georgio Cheynæo, M.D. &* [since 18 March 1701/2] *R.S.S.* (London, 1703). 'Necessity', Cheyne excused himself to David Gregory in January of the year of its publication (see note (41)), 'which begets so many bad authors (some of my friends thinking it will contribute towards my interest to print it) has forced me to let [my paper] go and now I am about printing it.'

(47) In Newton's case Cheyne refers broadly (page 59) to 'paucula Exempla *Methodorum Newtonianarum* quæ in binis Voluminibus ultimis *Operum* Cl. Wallisii [2/3 (Oxford, 1693/1699)] & *Principiis Philosophiæ Mathematicis* prostant, illustrantur & applicantur', but was careful to add of these 'magni *Newtoni* reperta' that 'Quæ...cum mecum animo perpendo non possum abstinere me quin dicam, omnia in hisce vel per hasce (aut non absimiles Methodos) ab Aliis (intra hosce viginti quatuor annos [up to 1700, understand; since Newton's *epistola prior et*

and Johann Bernoulli on the Continent,[48] and drew the assemblage together and systematized it with proofs and elaborations of Cheyne's own contrivance. When in particular Newton saw among the last his 1676 series-quadrature of an algebraic binomial[49] along with its extension to higher-order 'nomials' as he had adumbrated it to John Wallis in 1692,[50] fully demonstrated and delivered in a manner 'easily reducible to Newtonian or Gregorian form', his mind must bitterly have raced back a dozen years to the occasion when David Gregory had requested like approval of the publication of his similarly 'independent invention' of the general binomial case of such quadrature after Newton had previously (by way of John Craige) fed its essence to him in a couple of particular instances.[51] And the same Archibald Pitcairne who had fifteen years before first published 'Gregory's' theorem to the world[52] was once more the agent behind Cheyne's present essay on the quadrature of curves. Who could blame Newton if he now jumped to the conclusion that Cheyne, too, was deliberately

posterior to Leibniz in 1676] proximè elapsos) edita, esse solum eorundem ab Ipso diu antea cum Amicis, vel publico communicatarum Repetitiones, aut non difficilia Corollaria'. (He wrote much the same in his private letter to Gregory in January; see note (41) above.) Craige is praised (pages 63–4) as 'primus omnium (quod sciam) publico harum [Transcendentium Quadraturarum] Exempla impertivit, & Methodum huc ducentem in peregregio *Tractatu de Quadraturis* [*vide* pages 1–61 of Craige's *Tractatus Mathematicus de Figurarum Curvilinearum Quadraturis et Locis Geometricis* (London, 1693)] Anno 1693 non obscurè insinuavit', and mention is also made (*ibid.*) of his short supplementary article 'De Figurarum Geometricè irrationalium Quadraturis' in *Philosophical Transactions*, 19, No. 232 (for September 1697): 708–11 (with an 'Additio ad Schedulam' in No. 235 (December 1697): 785–7). On David Gregory's less original contributions to the quadrature of curves see IV: 414–15 and VII: 5–6 (+10, note (33)).

(48) In §IX (pages 46–50) of the printed *Fluxionum Methodus Inversa*, where Cheyne outlines the 'Altera *Newtoni* Methodus' (consisting, as he puts it, 'in Assumptione seriei pro qualibet Quantitate incognitâ ex qua cætera commode derivari possint, & in collatione terminorum homologorum Æquationis ad eruendos terminos assumptæ Seriei') which Newton himself had fleetingly referred to in his letter to John Wallis in autumn 1692 (see VII: 176) as understandable 'absque ulteriori explicatione', he cites also the equivalent method 'utpote facillima, & abunde illustrata' delivered 'à Cl. Leibnitio *Actis Lipsiæ* Aprilis 1693 [: 179–80; compare VII: 182, note (26)]…. Hoc est ad minimum 17 Annis postquam erat à Newtono reperta', there taking two out of his three illustrative examples from Leibniz' article; and in his following §X (pages 50–3) he goes on to expound, in Newtonian fluxional recasting, the Taylorian method of series-quadrature (on which see VII: 19, note (76)) presented 'à Clarissimo viro Johanne Bernoullio *Actis Lipsiæ* [Novembris] 1694 [: 437–41]…quâ Fluxionis datæ Fluens per Seriem infinitam, Generalem quidem, terminis tamen plerumque implicitis, obtinetur'.

(49) Presented in §II (pages 2–6) of the *Fluxionum Methodus Inversa* from 'Noviss. Edit. *Oper. Mathem.* Wallisii Vol. II' [Oxford, 1693: 390]; compare VII: 171–2.

(50) And straightaway printed by Wallis in his *Opera*, 2 (1693): 391; on which see VII: 173.

(51) See VII: 3–6.

(52) In his *Solutio Problematis de Historicis seu Inventoribus* (Edinburgh, 1688); compare VII: 6, note (15).

plagiarizing his own hard-won fluxional discoveries, and had maybe seen (without making the fact known in his paper) a transcript—such as he had earlier let Gregory have—of the 1693 tract 'De Quadratura Curvarum' where[53] he had set out his proof of the generalized quadrature series? But whether or not he saw in Cheyne anything more than brash youthful opportunism, the message to Newton suddenly rang clear: in the ten years since he had penned his revised treatise on quadrature, contemporary techniques for squaring curves had progressed to the point where its propositions were in serious danger of being duplicated, and even a mathematician like Cheyne of no great gift or power found it not difficult to recover and apply its basic method for extracting the root of a fluxional equation as an infinite series in powers of the base fluent.[54] And just as in 1691 he had angrily reacted to Gregory's seeming plagiary by starting to draft (if he never finished) an improved account of his variety of tools for effecting the quadrature of curves, so now, 'provoked by Dr Cheyns book'[55]

---

(53) In its Proposition V, namely; see VII: 520–2.

(54) That whose first-order instance he had permitted Wallis to publish in his *Opera*, 2 (1693): 393–6 (reproduced on VII: 177–80). If we ignore the nonsensical last-minute 'improvement' of its statement which in his appended *Errata*—so wrongly advised (compare note (44) above) by Johann Bernoulli?—he instructed his reader to make, Cheyne gives in §VIII (pages 37–46) of his book the correctly extended algorithm for attaining the proper index of the power of the base variable to be treated at each stage of the derivation term by term of the series expansion of the root of any given fluxional equation, exactly as Newton had enunciated it (see VII: 94, l. 3) in Proposition XII of his initial 1691 treatise 'De quadratura Curvarum'.

(55) The words are David Gregory's, set down by him in his diary on 1 March 1703/4 (see Hiscock, *Gregory, Newton and their Circle*: 15). When Cheyne's mentor Pitcairne—thus informed by we do not know whom—wrote to his Scots colleague Colin Campbell in October 1703 to announce that '2 sheets [of Newton's *Opticks*] are cast off [by the printer] already', he adjoined the confused report: 'He adds to the book all his about quadratures. He has done it in ire, being barbarously...and hanoverianlie [*i.e.* 'Leibnizianly'?] abused for his *Principia* by a German latelie in a vile consubstantial book of nonsense and ill-nature' (Edinburgh/ National Library MS 3440: 20^r; compare A. Thackray, *Atoms and Powers. An Essay on Newtonian Matter Theory* (Cambridge, Mass., 1970): 55). Pitcairne would here seem—intendedly?—to confuse Newton's anger at the 'ingratitude' of his own Scots *protégé* with his much better founded wrath at this period against the Dutchman Johannes Grœning, who in a 6-page appendix to his *Historia Cycloeidis* (Hamburg, 1701) had printed as 'mistakes' a list of what was largely Newton's own *errata* to the first edition of the *Principia*, originally communicated by him to Fatio de Duillier in the early 1690's, sent on by Fatio to Huygens who subsequently passed it to Leibniz, and ultimately (after Huygens' death) reaching Grœning; from Fatio, on 4 July 1705 following, David Gregory drew the 'account' (probably untrue) that the list, along with its several 'unauthorized' accretions at each stage of its onwards route, was 'maliciously published by the contrivance of M. Libnitz' (Hiscock, *Gregory*: 27). (For fuller details see I. B. Cohen, *Introduction to Isaac Newton's 'Principia'* (Cambridge, 1971): 186.) We should not, however, overstate the grim, unyielding sternness of Newton's 'ire'. He was, so de Moivre tells it, ready to smile off the suggestion that John Craige would again take the part of a Scots countryman against him. ('Je crois', wrote de Moivre to Johann Bernoulli on 22 April 1704, 'que M. Craige ne manquera pas de s'interesser pour son compatriote; je dis

—but leaving direct criticism of its content to his able 'adjutant' de Moivre[56]—
he launched into a full revision of his shortened 'Tractatus de Quadratura
Curvarum' of 1693,[57] 'prefixing to it an *Introduction* and subjoyning a *Scholium*
concerneing that Method...by which I had found some general Theorems
about squaring Curvilinear Figures, or comparing them with the Conic Sections,

dernierement à M. Newton les raisons que j'avois de le croire; mais il en sourit et m'assura que
je ne devois pas beaucoup le redouter' (see Wollenschläger, 'Briefwechsel von Johann I
Bernoulli' (note (44) above): 179). Not convinced to be sure, de Moivre in his next letter to
Bernoulli on 13 March 1705—where he reported on his own newly published riposte to
Cheyne's book (see the next note), finding in it (with more than a touch, unusual for him, of
personal spite and ill-will) an excess of 'foiblesse de raisonnement', 'impuissance' and
'mauvaise foy'—again wrote that 'On dit que Craige y a grand part; je le croirois bien; il est
digne de Cheyne et de Craige, ces deux grandes lumieres de l'Ecosse' (*ibid.*: 194).)

(56) In his *Animadversiones in D. Georgii Cheynæi Tractatum de Fluxionum Methodo Inversa*
(London, 1704), that is. De Moivre there attempted, none too successfully, to single out
instances of inadequacy on Cheyne's part in grasping the theorems of Newton and Bernoulli
which he there glossed. His adversary countered these testy 'observations' upon his book with
an equally ill-tempered set of *Rudimentorum Methodi Fluxionum Inversæ Specimina, Adversus A. de
Moivre* (London, 1705). Into the details we here have no room to go. It was, there can be no
doubt, de Moivre's overriding purpose—on direct instruction from Newton himself, Varignon
told Bernoulli in a yet unpublished letter at this time, having heard this as hearsay from an
unnamed 'Scot' (see I. Schneider, 'Der Mathematiker Abraham de Moivre (1667–1754)'
(*Archive for History of Exact Sciences*, **5**, 1968: 177–317): 204)—thoroughly to discredit Cheyne's
character in the eyes of contemporary mathematicians. 'Ne trouvez-vous pas', he wrote to
Johann Bernoulli on 22 April 1704, 'qu'un procédé tel que celui de M. Cheyne est bien
injuste? Il prend partout des uns et des autres ce qui l'accommode, quelquefois sans nommer,
et s'il nomme ce n'est que pour faire croire qu'il suppose ceux mêmes qui l'ont fourni ses idées.
La moindre ombre de generalité qu'il ajoute à un theoreme le fait triompher, et il diroit
volontiers [:] le theoreme n'est plus le vôtre, il est devenu mien' (Wollenschläger, 'Briefwechsel': 178). While willing to some extent to go along with this judgement, Bernoulli
responded in November of that year that his principal complaint with Cheyne in his book
was with its concluding assertion (see note (47) preceding) that everything found out and
made public on infinitesimal calculus over the last quarter of a century was a 'repetition or no
hard corollary' of what Newton had previously, in private or publicly, communicated (see
'Briefwechsel': 183)—which was scarcely the reaction de Moivre had intended to provoke.
(He would have been a deal less happy still had he seen Bernoulli's letter to Leibniz in the
middle of the following April, commenting on his own newly printed countering 'Observations'
on Cheyne's book that (see *Leibnizens Mathematische Schriften*, **3**: 761) 'Scriptum Moyvræi contra
Cheynæum nihil peculiare continet, præter Theoremata quædam pro comparatione et
reductione arearum, quæ ipse quidem magni facit, nobis tamen jam diu familiaria, imo latius
extensa, quam a Moyvræo factum'.) De Moivre, and Newton in his shadow, won the battle
inasmuch as Cheyne, rightly soured by the squabble, thereafter withdrew from such 'barren
and airy studies' for the more real and rewarding realms of medicine and religion. But the
victory was an empty one. For all that Cheyne's talent was indeed for mimicry rather than
original expression, British mathematics in the early 1700's was not rich enough in men even
of his secondary calibre that it could afford to have Newton and his aide so harshly and
selfishly crush it.

(57) That is (as he then intended to publish it) the 'Liber secundus' of his treatise on
'Geometria' reproduced on VII: 507–53.

or other the simplest Figures with which they may be compared' as he put it in his 'Advertisement' to the published *Opticks*[58] on whose end-pages he adjoined the amplified treatise in 1704. All too obscurely for anyone not in the know, we may add, he there also observed that he had taken the occasion to make public his treatise because, having 'some Years ago...lent out a Manuscript containing such Theorems', he had 'since met with some Things copied out of it'.[59]

Newton's surviving preparatory drafts[60] of the introductory pages which in late 1703 he set in preliminary to his 1693 manuscript stand in evidence that, far from dashing them off hastily, he took a great deal of care over preparing their content and precisely wording their text. His initial intention, it is clear, was once more to widen the scope of this curtailed 'Second book of Geometry' (as it had been titled), broadening its single aim—the quadrature (in his preferred geometrical model of integration) of a curve whose ordinate, the 'fluxion' of the 'fluent' area sought inversely from it, is some multinomial algebraic function of the related Cartesian abscissa (base variable)—again to encompass, in the manner of Propositions XI–XIII of his parent 1691 treatise 'de quadratura Curvarum',[61] more general problems of the direct and inverse methods of fluxions in which 'the relationship of the fluent quantities is defined by the motion of lines following assigned laws'.[62] *Inter alia* a new Proposition II 'To solve problems by the finding of fluxions' was to be illustrated by twelve worked problems[63] dealing, along lines already well travelled by Newton in

---

(58) *Opticks: Or A Treatise of the Reflexions, Refractions, Inflexions and Colours of Light* (London, $_1$1704[→facsimile reprint Brussels, 1966]): Advertisement: [ii]. (This was already printed off by 16 February 1703/4 when, as the Royal Society Journal Book records, 'The President'—that is, Newton (since the previous 3 November)—'presented his book of Optics to the Society; Mr Halley was desired to peruse it, and to give an abstract of it...'.) The *Tractatus de Quadratura Curvarum* is thereto annexed (on pages $_2$165–211) as the second of the 'Two Treatises of the Species and Magnitude of Curvilinear Figures' promised in the book's subtitle. The first 'small Tract concerning the Curvilinear Figures of the Second Kind, which was written many Years ago' (set on the immediately preceding pages $_2$139–62) is of course the minimally retouched text of his 1695 enumeration of cubic curves (see VII: 588–64).

(59) There is, we may insist, no foundation at all in historical evidence for L. T. More's arbitrary conjecture—stated by him as fact, and widely parroted by even scholarly writers on Newton since—that this 'reference to matter purloined from a manuscript refers to Leibniz' (*Isaac Newton: A Biography* (New York, 1934): 577). At no time before the squabble over calculus priority broke out in 1713 would Newton have entertained so baseless a notion, nor did he ever even afterwards accuse Leibniz of plagiarizing any portion of the 'De Quadratura' not found (in embryo at least) in his earlier papers on fluxions.

(60) Reproduced in 2, §§1/2 below.

(61) See VII: 70–128.

(62) As Newton phrases it in his intended new Case 2 of Proposition I of the 'De Quadratura'; see 2, §1.2 below.

(63) The enunciations of these are set out in the draft reproduced in 2, §1.3; Newton's ensuing elaborations of solutions to the second, fourth and twelfth of his problems are given in

earlier, then (and for long afterwards) unpublished papers,[64] with maxima and minima; the fluxions of curved lines and their tangents, curvature, inflexions and points of 'straightness'; the cubature of their solids of revolution, and the squaring of the curved surfaces of these; and the central forces engendered by and engendering the motions of bodies in curvilinear paths. Subsequently, with this first flush of enthusiasm fast fading in face of the labour and effort which (it must rapidly have become obvious) was needed to achieve it, Newton set himself the more modest goal of briefly outlining the possibilities of a developed calculus of fluxions in a short prefatory account of the 'Methodus Fluxionum'[65] which differs little from the version finally printed. 'The method of fluxions', he there insisted, 'is'—by virtue, Newton implicitly supposed, of the expansibility of all (continuous) functions as 'Taylor' series of their incremented forms, their successive fluxions appearing in sequence as (proportional to) the coefficients of the corresponding first, second, third, ... augments[66]—'a corollary of the method of converging series, and is by its aid made more copious and perfect'.[67] With the remark made in passing that it was

By discovering that quantities increasing. . .in equal times come to be greater or less according as the greater or less is the speed with which they grow. . .I was to seek a

---

§1.4 thereafter (see also the preliminary versions of these in [1] and [2] of the following Appendix).

(64) See especially Propositions III–V of his 1671 treatise (III: 116–82) and Book 2 of his 'Geometria Curvilinea' (IV: 474–84) of a decade later.

(65) Reproduced in 2, §2.1 below.

(66) In the final version of his 'De Quadratura' Newton delayed presenting this basic article of faith till the concluding scholium, where (see 2, §3.2) it is set in prime position. We will here in passing mention only the less than adequate manner in which he there stated it, leaving himself wide open to Johann Bernoulli's later charge (see 2, §3: note (46)) that he was ignorant of how precisely to frame the individual coefficients in such a 'Taylor' expansion. Aware as we may now publicly be of the passage in Proposition XII of the 1691 treatise 'De quadratura Curvarum' (see VII: 96–8) in which Newton accurately specifies their successive formation, we may smile at Bernoulli's earnest and repeated endeavours to inform the world of what (evidently in all sincerity) he took to be a fundamental weakness in his senior English colleague's grasp of higher-order fluxional derivatives. Let us swiftly adjoin that, since Newton here assumes the increment of his base variable to be vanishingly small, Cauchy's later blockbusting counter-instance to Lagrange's hopes furnished by the nil Taylorian expansion ($x \neq 0$) of e^{-1/x^2} in terms of its (universally zero) n-th order derivatives, $n = 1, 2, 3, \ldots$, at $x = 0$ (see his article 'Sur le développement des fonctions en série et sur l'intégration des équations différentielles. . .', *Bulletin scientifique de la Société Philomathique de Paris* (1822): 49–54 [= *Œuvres complètes de Cauchy*, (2) **2**: 276–82]) does not here forcefully explode.

(67) See 2, §2.1 below. In an immediately following passage which, had it not been excised from the final version of the introduction, might perhaps have saved Leibniz in his subsequent anonymous review of the 'De Quadratura' from making this false juxtaposition (see page 26 below) Newton went on to affirm that Leibniz' infinitesimal *differentiæ* corresponded not to his own *fluxiones quantitatum*, but to the *augmenta momentanea* which are generated at these finite 'speeds' of flow in a specified, indefinitely minute 'instant'.

method of determining quantities from the speeds of motion or increment by which they are generated; and...fell in the year 1665 upon the method of fluxions which I have here employed in the quadrature of curves

and with some slight amelioration of his initial discussion of the ordinate of a curve as the fluxion of its area and the coincidence of the increment of its arc-length with that of the tangent there, and of his ensuing instances of the ways in which other related fluxional increments of geometrical configurations may be compared,[68] the 'Introductio' to the published tract on quadrature[69] was complete. After this the eleven propositions of the 1693 parent text[70] were set to follow without essential change.[71] In a newly composed terminal 'Scholium Generale', finally, Newton reaffirmed his faith[72] that 'fluxions of fluent quantities are as the terms of their infinite series', citing in example the expansion of the simple n-th power of the incremented base fluent, and then appended an inadequate, jerky summary of several ways of reducing a given fluxional equation into an equivalent one free of fluxions, when it is posed in immediately quadrable form.[73] And there, except for the brief reminder that 'if, after you have gathered fluents from fluxions, you should be in doubt regarding the truth of your conclusion, you must conversely derive the fluxions of the fluents found and compare them with the fluxions initially propounded' and the terse concluding flourish that 'on these principles the path is laid to greater things', Newton was done.

It is easy for us in hindsight to see that this published 'Tractatus de Quadratura Curvarum' fell, on its appearance in 1704, awkwardly between two stools. Other than for a very few like Craige, David Gregory, de Moivre and Cheyne, British mathematicians of the day were just not well enough up in the new calculus of limit-increments (whether in Newtonian fluxional or Leibnizian

---

(68) See 2, §2.2/3; preparatory drafts of these are given in 2, §1.1 and in [3] of the ensuing Appendix.

(69) Reproduced in 2, §3.1 below.

(70) See vii: 508–52.

(71) When, however, Newton came to repeat the second table of areas of curves whose ordinates are sequential instances of simple algebraic 'forms'—this he had in 1693, in the Scholium to Proposition X, taken over unchanged from his fluxions treatise of twenty years before—he thought to gain effectiveness in his presentation by advancing the fifth and seventh *Formæ* to be in third and fifth place respectively in the new ordering. We may add that after Roger Cotes brought two small slips in the stated evaluations of Forms 7.1 and 8.4 of the printed listing of integrals to his notice in August 1709 (compare note (104) following) Newton's own subsequent further checking yielded there to be a couple of extraneous factors e in Forms 2.2 and 2.3 (compare 2, Appendix [4]): all was put right in William Jones' revised *editio princeps* of the 'De Quadratura' in 1711 (on which see note (101) below).

(72) Compare note (66) preceding.

(73) See 2, §3.2 below; a preliminary draft of the concluding paragraphs on reducing and resolving simple fluxional equations is reproduced in [7] of the attached Appendix.

differential manner) to absorb the niceties of its involved treatment of the complexities of algebraic integration.[74] Within two years of the tract's publication, it is true, the London mathematics teacher Humphrey Ditton endeavoured creditably to make plain to a wider audience the 'First Principles' of the theory of fluxions and its 'Operations, with some of the Uses and Applications of that Admirable Method; according to the Scheme prefix'd to his Tract of Quadratures by (its First Inventor) the incomparable Sir Isaac Newton' (as the subtitle of his 1706 work[75] put it, with a deferent bow to Newton's recently bestowed knighthood), but the technicalities of the complicated schemes of quadrature expounded in the main body of the 'De Quadratura' were there no more than fleetingly touched upon. On the Continent, on the other hand, Leibniz and Johann Bernoulli and such up-and-coming adherents to their flag as Jakob Hermann (the now dying Jakob Bernoulli's capable pupil at Basel) were in no mood to be given instruction in over-complicated patterns of elementary integration, and it was—leaving aside the subtleties of Newton's Proposition XI (which they seem all to have ignored)—precisely the 'greater things' which Newton had mentioned at the end of his tract but had fought shy of delivering, which they were quickly in unison to expect and soon to demand. To be accurate, Johann Bernoulli, who by one of life's ironies was presented with a copy of the *Opticks* by the 'upstart' Cheyne a little after it appeared in the London bookshops, was mildly hopeful of finding 'fine things' in it, or so he

---

(74) Even de Moivre in his 'Specimina quædam illustria Doctrinæ *Fluxionum* sive exempla quibus Methodi istius Usus & præstantia in solvendis Problematis Geometricis elucidatur' (*Philosophical Transactions*, **19**, No. 216 (for March–May 1695): 52–7) had misunderstood 'illa quæ demonstravit Clarissimus *Newtonus* in pag. 251, 252 & 253 *Princ.Phil.* [₁1687, *viz.* in 'Lib. II, Lem. II' (on which see IV: 521–5)] to mean (see *ibid.*: 54–5, where he implicitly equates $GH [= dx]$ and $\dot{x}$) that the instantaneous increment of a quantity is its fluxion. While in 1697 John Craige could bring out an article 'De Figurarum Geometricè irrationalium Quadraturis' (see note (47) above) where, immediately in counterblast to Johann Bernoulli (who had publicly bestowed faint praise on his modicum of mathematical talent), he developed, without demonstration, a complicated extension to Newton's method which yielded the series quadrature of sequences of such 'geometrically irrational' fluxional figures, even seven years later when Newton put forth his own comprehensive 'Tractatus' on the topic the only beginner's *entrée* to the method and calculus of fluxions was the brief and inadequate 'Specimen of the Nature and Algorithm of Fluxions'—one, to be fair, in which accurate distinction was made between Newton's 'Celerity...of Augmentation and Diminution' and Leibniz' 'Infinitely small Increments or Decrements'—appended by the (then) mathematics teacher John Harris to his widely read and much reprinted *New Short Treatise of Algebra: With the Geometrical Construction of Equations, As far as the Fourth Power or Dimension* (London, ₁1702): 115–36. For a fuller (if somewhat lacunary) account of English understanding of the notion and application of fluxional limit-motion we may, for want of any more up-to-date survey, refer to pages 37–43 of F. Cajori's *A History of the Conceptions of Limits and Fluxions in Great Britain from Newton to Woodhouse* (Chicago/London, 1919).

(75) *An Institution of Fluxions* (London, 1706); for an accurate *précis* of the scope and content of Ditton's book see Cajori's *History of...Fluxions*: 44–7.

stated to de Moivre in November 1704;[76] but he had yet done no more than glance through its two Latin appendices (the English of its main optical text he could not read except very haltingly) when he thought to bring the publication to Leibniz' attention nearly a month later.[77] The latter had, we now know, received some while before from Mencke his own copy of the book for review in the *Acta Eruditorum*, and indeed had by then already written his unsigned recension of its two mathematical tracts. But his authorship of this *critique* he chose to conceal even from Bernoulli,[78] and so he now in reply merely privately confided that, while he looked upon the 'determination of the lines of third degree' (in the companion *Enumeratio Linearum Curvarum Tertii Ordinis*) as both 'correct' and 'not a trifling increment to geometry', there was in Newton's tract on quadrature of curves 'nothing markedly new or difficult'[79]—a view to which Bernoulli readily assented.[80] In like manner, Leibniz devoted his anonymous

---

(76) See Wollenschläger, 'Briefwechsel von Johann I Bernoulli' (note (44)): 177–87. In his letter, having noted Cheyne's gift to him (*ibid.*: 178), he added to de Moivre that he was 'fâché de n'entendre pas assez la langue angloise pour entendre bien son traité [d'optique]. Je m'attacherai au premier loisir à lire ses deux autres traités écrits en latin sur le dénombrement des lignes du 3e ordre, et sur la quadrature des courbes: j'espere d'y trouver de belles choses' (*ibid.*: 187).

(77) Johann Bernoulli to Leibniz, 6 December 1704 (N.S.): 'Accepi...Newtoni opus recens editum, continens tres partes....Nondum mihi vacavit ea perlegere' (*Commercium Philosophicum*, 2: 123 [= *Math. Schriften*, 3: 759]).

(78) In the early summer of 1713, months after the private squabbling between Leibniz and (Newton *cum*) Keill had erupted into the public arena with the appearance under the Royal Society's *imprimatur* of (as we now know it to be) Newton's annotated selection of the *Commercium Epistolicum D. Johannis Collins et Aliorum de Analysi promota*, Leibniz still—as indeed he did to the very end of his life three years later—persisted in hiding from his closest supporter in the dispute over calculus priority that he had had any part in writing what, singly more than anything else, had set the furore off. On the same day in late June of that year, just a month before launching into circulation a printed *Charta volans* (equally unsigned) which savagely decried Newton's competence in calculus, he wrote both to Johann and to his nephew Niklaus in essentially identical terms (see *Mathematischen Schriften*, 3: 913 and 986 respectively) reaffirming that attribution of the authorship of the 1705 *Acta* review was 'falsely done' (in his explicit phrase to Niklaus). Though he never, it would appear, formally requested it, he badly wanted Johann's approval for quoting (without prior permission) the considerable portion of the latter's letter to him of the preceding 7 June (N.S.) which was to appear in his July 1713 flysheet as the anonymous 'judicium primarii Mathematici' in support of his own aspersions, and it is perhaps excusable that he should here have lied not to spoil (as he could easily have done) that end.

(79) Leibniz to Johann Bernoulli, 25 January 1705 (N.S.): 'Opus Newtoni de coloribus profundum videtur.... In Tractatu de Quadraturis Curvarum ordinariarum nihil puto esse, quod nobis sit valde novum aut arduum. Determinatio numeri Linearum tertii gradus et, ut credo, recta est et habenda est pro incremento non contemnendo Geometriæ' (*Commercium*, 2: 124 [= *Math. Schriften*, 3: 760–1]).

(80) Bernoulli to Leibniz, 18 April 1705: 'Verum est, nihil me vidisse in ejus tractatu de Quadraturis, quod nobis valde arduum vel novum esset. Curiosa magis est determinatio numeri linearum tertii gradus' (*Math. Schriften*, 3: 762).

public recension[81] in large part to what we have previously qualified as a colourless and none too percipient assessment of the *Enumeratio*,[82] and then went on:

The very ingenious author, before he comes to the 'Quadrature of Curves' (or rather of curvilinear figures), premisses a short 'Introduction'. The better to understand this, it must be known that when some magnitude grows continuously, as (for instance) a line grows by the flowing of a point which describes it, the instantaneous increments are called 'differences', ones, namely, between the magnitude which was before and that produced by the instantaneous change; and hence is born the 'differential' calculus, and, inverse to it, the 'summatory' one,[83] whose elements have been delivered in these *Acts* by its inventor Mr Gottfried Wilhelm Leibniz, and whose various applications have both by [Leibniz] himself and by the Bernoulli brothers and the Marquess de L'Hospital (whose recent untimely death[84] all who love the promotion of deeper learning ought greatly to grieve). In place of Leibnizian differences, accordingly, Mr Newton employs and has ever employed 'fluxions', which are [defined] to be 'very closely near as the augments of the fluents begotten in the very smallest equal particles of time';[85] and has elegantly used them in his [*Philosophiæ*] *Natur*[*alis*] *Principia Mathematica* and also in other subsequent

---

(81) *Acta Eruditorum* (January 1705): 30–6. This review of Newton's 'Tractatus Duo, de Speciebus & Magnitudine Figurarum curvarum' (as it specified) had been completed by Leibniz more than two months before its public appearance; see Mencke's letter to him on 12 November 1704 (N.S.) (printed in C. I. Gerhardt's edition of *Leibnizens Mathematische Werke* (3). *Supplement Band* (Halle, 1860): 15). Mencke subsequently wrote in Leibniz' name alongside the review in his editor's copy of the *Acta* (now in Leipzig University Library), thus removing any lingering doubts there might be as to the latter's full and unique authorship of it. (This was first brought to public notice by G. E. Guhrauer in his *Gottfried Wilhelm von Leibniz: Eine Biographie*, 1 (Breslau, 1846): 311.)

(82) See VII: 568–9. Leibniz made the equally shallow comment in subsequent, likewise anonymous review in *Acta Eruditorum* (February 1712): 74–7 of Jones' 1711 *Analysis*, where the 'Enumeratio' was reprinted, that 'egregie de Geometris meritus fuisset [Newtonus] si demonstrationem numeri linearum tertii ordinis...unà exhibuisset' (*ibid.*: 76; compare II: 261, note (12)). This drew from Newton, as we remarked in VII: 571–2, note (29), the snappish rejoinder—never made public by him—that 'yᵉ Demonstration is the book it self. The Reader is there taught how to find the *plagæ crurum infinitorum* with the Asymptotes of the *crura Hyperbolica*: & all the variety of the *crura infinita* of the lines of the third sort are there reduced to four cases wᶜʰ give the four equations there set down' (from a preliminary draft of Newton's counterblast to Leibniz' review which exists on Mint Papers II: 88ʳ, now printed in *Correspondence of Newton*, 5, 1975: 383 [as §II]).

(83) These *termini technici* were, of course, introduced to the public by Leibniz himself in his first articles on the new *calculus differentialis* and inverse *calculus summatorius* in *Acta Eruditorum* (October 1684): 467–73 and (June 1686): 292–300 respectively (= (ed.) C. I. Gerhardt, *Leibnizens Mathematische Schriften*, 5 (Halle, 1858): 220–6/226–33) to which he here goes on to refer.

(84) L'Hospital had died the previous 2 February 1704 (N.S.) at the—not inconsiderable—age of 62.

(85) Leibniz quotes *verbatim* from the opening of the third paragraph of Newton's 'Introductio' to the 'De Quadratura'; see pages 122–4 below).

publications, in the manner that Honoré Fabry, too, in his *Synopsis Geometrica* substituted advances of motion for the Cavalierian method.[86]

In sequel he passed to make comparison between the 'sum' (y) of 'all' the 'growing' differences $(dy) = v.dx$—'which [sum] is written $\int v\,dx$'—and the integral (to use Bernoulli's name) 'which is the quadrature of the [Cartesian] curve whose abscissa is x and ordinate v'. However,

since the regress here, from the differences to the quantities, or from the quantities to the sums, or finally from the fluxions to the fluents, cannot always be algebraically achieved..., we need in consequence to inquire both in what cases the quadrature may algebraically succeed, and in what manner, where the algebraical approach fails, some auxiliary one might be employable. In each of these roads, too, Mr Newton has most usefully laboured, elsewhere[87] as well as in the present tract on quadratures, where he applies infinite series which in the case where they break off, that is, terminate, exhibit what is required algebraically. We have spoken of this also in our review of the treatise of Mr Cheyne,[88] a Scots physician resident at London. You can look, too, to Mr Craige

(86) We translate Leibniz' Latin in *Acta* (1705): 34–5, *viz.* 'Ingeniosissimus...Autor, antequam ad *Quadraturas Curvarum* (vel potius figurarum curvilinearum) veniat, præmittit brevem *Isagogen*. Quæ ut melius intelligatur, sciendum est, cum magnitudo aliqua continue crescit, veluti linea (exempli gratia) crescit fluxu puncti, quod eam describit, incrementa illa momentanea appellari *differentias*, nempe inter magnitudinem, quæ antea erat, & quæ per mutationem momentaneam est producta, atque hinc natum esse calculum *differentialem*, eique reciprocum *summatorium*; cujus Elementa ab inventore Dn. Godefrido Guilielmo Leibnitio in his *Actis* sunt tradita, variique usus tum ab ipso, tum a Dnn. Fratribus Bernoulliis, tum & Dn. Marchione Hospitalio, (cujus nuper extincti immaturam mortem omnes magnopere dolere debent, qui profundioris doctrinæ profectum amant,) sunt ostensi. Pro differentiis igitur Leibnitianis Dn. Newtonus adhibet semperque adhibuit *fluxiones, quæ sint quam proxime ut fluentium augmenta æqualibus temporis particulis quam minimis genita;* iisque tum in suis *Principiis Naturæ Mathematicis,* tum in aliis postea editis eleganter est usus, quemadmodum & Honoratus Fabrius in sua *Synopsi Geometrica* motuum progressus Cavallerianæ Methodo substituit.'

(87) No doubt Leibniz has principally in mind the summary of his variety of fluxional techniques which John Wallis inserted in 1693 as pages 391–6 of the second volume of his *Opera Mathematica* (on which see VII: 170–80).

(88) His *Fluxionum Methodus Inversa* (London, 1703), of course, whose publication had provoked Newton to send his own work 'De Quadratura Curvarum' to the press (see pages 17–20). Leibniz' short recension of it—a factual summary of its content in the main, but with careful citation of Cheyne's stated debt (see note (48) preceding) to the prior writings of himself and Johann Bernoulli—had earlier appeared, ever anonymously, in *Acta Eruditorum* (October 1703): 450–2. In a letter to Bernoulli shortly afterwards, on 22 November 1703 (N.S.), Leibniz privately wrote much more sourly and indeed not a little peevishly: 'Cheynæus mihi vix problemata aggressus videtur, alioqui sensisset, quam non facile sit problematibus per finita solvendis viam monstrare per Series infinitas..., ut nonnisi multa arte eligi possit, quæ ad rem faciat. Talium autem artium nulla apud eum vestigia deprehendo. Inepte Newtono vindicare vult Methodum Seriei per arbitrarios coefficientes assumtos et comparatione terminorum determinandos investigandæ; nam ego eam publicavi [in *Acta Eruditorum* (April 1693): 179–80; again compare VII: 182, note (26)] cum nec mihi nec cuiquam alteri, saltem publice, constaret tale quid habere et Newtonum' (*Commercium*, 2: 97–8 [=*Math. Schriften*, 3: 727]). Whatever Newton might (had he ever known of it) have said in response to Leibniz'

(the Scot)'s book 'On Quadratures'[89].... In result, we may refrain from reviewing Mr Newton's theorems, being that they cannot be expressed in a few words; and the same goes for certain theorems of his for reduction to easier quadratures.[90]

Hindsight again allows us to appreciate how, in so thinly characterizing Newton's concept of limit-fluxion for his reader, in so hastily dismissing his effort and ingenuity in applying it to resolve wide classes of particular algebraic quadrature, and above all in so off-handedly setting him to be a Fabry to his own Cavalieri, Leibniz here gratuitously planted an unnoticed booby-trap under his throne of 'inventor' of the infinitesimal calculus: one which he was himself improvidently to trigger six years afterwards when he attempted to arraign John Keill before the bar of the assembled Royal Society for bearing the false testimony[91] that he, Leibniz, it was who had published Newton's 'undeniably

present staking of his claim to independent invention of the general Newtonian method of quadrature *per series infinitas*, he would surely not violently have disagreed with the latter's summing up: 'Esto: sit Cheynæus paulo supra Tyronem, certe facile dare potuit, quæ dedit.... Certe nullam novam Seriem pulchram, nullum Theorema elegans affert' (*ibid.*).

(89) That is, *Tractatus Mathematicus de Figurarum Curvilinearum Quadraturis*, to give this 1693 book of Craige's its fuller title (compare note (47) preceding). In his review Leibniz here went on briefly to cite '& ejusdem Theorema ad Quadraturas pertinens' which had been 'nuper exhibitum in his *Actis*' from Craige's original 'Specimen Methodi Generalis determinandi Figurarum Quadraturas initially published—as an open letter to Cheyne, dated 'Gillingham, 22 Apr. 1703', congratulating him upon the 'incredible increment' by which his (then) newly appeared *Fluxionum Methodus Inversa* had advanced its topic, but passing on to communicate a lengthily expounded technique of his own for quadrature by series (one, it seems to us, of more complexity than use) to illustrate 'multa...ad Methodum illam inversam perficiendam necessaria adhuc invenienda'—in *Philosophical Transactions*, **23**, No. 284 (for March/April 1703): 1346–60.

(90) *Acta* (1705): 35–6 'Sed cum *regressus* hic a differentiis ad quantitates, vel a quantitatibus ad summas, vel denique a fluxionibus ad fluentes, non semper Algebraice fieri possit,...ideo quærendum est, tum quibus casibus quadratura Algebraice succedat, tum quomodo Algebraico successu deficiente aliquid subsidiarum adhiberi queat. In utroque etiam a Dn. Newtono est utilissime laboratum, tum alias, tum in hoc tractatu *de Quadraturis*, ubi series adhibet infinitas, quæ eo casu, quo abrumpuntur, seu finiuntur, quæsitum Algebraice exhibent. De quo etiam dictum est nuper in recensione tractatus Dn. Cheynæi, Medici Scotici Londini degentis. Conferri etiam potest tractatus Dn. Craigii Scoti *de Quadraturis*...; quæ faciunt etiam, ut ipsis Theorematis Newtonianis recensendis supersedeamus, quia paucis exponi non possunt: quemadmodum nec ejusdem Theoremata quædam reductionis ad quadraturas faciliores.'

(91) In his 'Epistola' to Edmond Halley 'de Legibus Virium Centripetarum' (*Philosophical Transactions*, **26**, No. 317 (for 'September and October 1708' but not published till the next year): 174–88) where he had *en passant* fleetingly affirmed (*ibid.*: 185) that 'Hæc omnia sequuntur ex celebratissimâ nunc dierum Fluxionum Arithmeticâ, quam sine omni dubio primus invenit Dominus Newtonus, ut cuilibet ejus Epistolas à Wallisio editas [*sc.* in his *Opera Mathematica*, **2**: 391–6, and in the 'Epistolarum Collectio' in *ibid.*, **3**: 617 ff. where Newton's *epistolæ prior et posterior* to Leibniz in 1676 were (pages 622–9/634–45) first published in full, selected excerpts from these having earlier (see IV: 672, note (54)) been included by Wallis in his *Algebra*] legenti, facile constabit; eadem tamen Arithmetica postea mutatis nomine & notationis modo à Domino Leibnitio in *Actis Eruditorum* edita est'.

first' discovery of the 'arithmetic' of fluxions, 'changing its name and mode of notation'. Keill, we may anticipate, neatly sidestepped Leibniz' charge of malevolence by sending Newton the much more explosive 1705 *Acta* review (did he guess that it was Leibniz who had written it?), desiring him to 'read from pag. 3[4]...to the end'.[92] Thereafter the sparks struck in angry thrust and savage counter-parry flew thick and fast. It was, we may now see, not so much that Fabry's systematic employment in his *Synopsis Geometrica*[93] of 'lines begotten through the motion or fluxion of points' to replace the 'indivisibles of *continua*' which Cavalieri had previously employed in carrying through the arguments of his own more celebrated founding work[94] was in intrinsic terms an unfair analogy to draw in contrasting Newton's appeal to the notion of fluxional limit-increase with Leibniz's equivalent invocation of the infinitesimal differential augment as his preferred basis: there is indeed, though Newton was later to make much of the chief discrepancy between the two that the exact analogue of Leibniz' 'indivisible' difference in his own method is not the 'speed' of flow of the fluent quantity in a moment of 'time'—this itself is finite—but the 'moment' (momentaneous increase) of the quantity generated by that speed in that vanishingly small instant,[95] a substantial core of truth in this comparison which Newton could not honestly deny. But the unpleasant broader signification, however unintended, of Leibniz' unqualified assertion that 'In place of Leibnizian differences Mr Newton employs and has ever employed fluxions' was unmistakably to insinuate that, like Fabry of Cavalieri's 'indivisibles', Newton had had prior knowledge of Leibniz' differential calculus in framing his own variant notation of infinitesimal limit-increment of 'flowing' quantities, consciously modelling his own ensuing method of fluxions upon it. And so Newton in 1711 took it, stimulated thereto or no by Keill's provocative action

---

(92) Keill's original letter is now ULC. Add. 3985.1 (printed in *Correspondence of Newton*, 5: 115).

(93) Published at Lyons in 1669. A comprehensive summary of its content is given by E. A. Fellmann in 'Die Mathematischen Werke von Honoratus Fabry' (*Physis*, **1**, 1959: 5–54): 10–25; see also his and J. O. Fleckenstein's 'Honoratus Fabri, ein ''missing link'' zwischen der Indivisibilienmethode und der Fluxionsrechnung' (in *Verhandlungen der Schweizerischen Naturforschenden Gesellschaft* (Basel, 1957): 57–61).

(94) Bonaventura Cavalieri, *Geometria Indivisibilibus Continuorum Nova Quadam Ratione Promota* (Bologna, ₁1635 [→₂1653]); on this see, for instance, C. B. Boyer, *The Concepts of the Calculus* (New York, ₁1939 [=*History of the Calculus and its Conceptual Development* in the 1959 reprint of its second edition in 1949]): 117–23; D. T. Whiteside, 'Patterns of Mathematical Thought in the later Seventeenth Century' (*Archive for History of Exact Sciences*, **1**, 1961: 179–388): 311–18; and M. E. Baron, *The Origins of the Iinfinitesimal Calculus* (Oxford, 1969): 122–33.

(95) This false hare had, we have previously remarked (see note (74) above), been started in England a decade earlier by de Moivre in a widely influential article on the 'Doctrina Fluxionum' which appeared in the *Philosophical Transactions* in 1695.

in pointing to this offensive passage in the review of the 'De Quadratura'. Let us not here tarry to say more—we give in Part 2 a full report—on the rarely humorous, often secretive and vicious, ultimately sterile conflict over coveted priorities and imagined plagiarism which broke out publicly in 1712 and lasted, with much useless scuffling and furtive jockeying for position, for the next decade (and maybe is still not quite at an end).

In the immediate event Leibniz' unsigned remarks in the *Acta* of January 1705 had no impact. When[96] Jakob Bernoulli in Basel read the anonymous review of Newton's two mathematical pieces, he had eyes only for what it said of the 'Enumeratio', astonished at being anticipated and surpassed in an area of analytical geometry on whose investigation he himself had recently spent so much effort.[97] His brother Johann spoke of it not at all in his letters of the period to Leibniz. In Paris, where L'Hospital had died just a few weeks before the *Opticks* with its attendant tracts was published, Varignon was deep in thought upon the theory of central forces, unwilling to be distracted by things more 'purely' mathematical, and Reynaud, Carré and others who might have read Newton's treatise on quadrature with understanding would seem to have remained in ignorance for a considerable time of its content.[98] Back in England the effectively unaltered Latin texts of both the 'Enumeratio' and the 'De Quadratura' were again appended to Samuel Clarke's Latin translation of the

(96) To reiterate, and minimally to supplement, our earlier account (see VII: 520) of Jakob's over-riding interest in seeing how Newton's researches into the properties and listing of cubic curves would match up to his own (never published) recent ones.

(97) Or so Jakob Hermann wrote to Leibniz from Basel on 4 April (N.S.) following: 'Clariss. noster Bernoullius...Nuper cum multum operæ insumeret percurrendis curvis primi generis supra Sectiones Conicas, plurimarumque curvaturas et varios flexus definiisset, incidit in aliquem[!] *Act. Lips.* Newtoniani libri de Specie et Quadratura Curvarum *Opticæ* suæ per modum appendicis adjecti recensionem continentem, seque a Cel. Newtono præoccupatum attonitus invenit' (*Leibnizens Mathematische Schriften*, 4 (Halle, 1859): 271). Jakob almost certainly never lived to see Newton's tract itself, for on 13 July 1705 (a little more than a month before Bernoulli's death) Hermann again wrote to Leibniz to say that he had tried in vain to acquire a copy of the *Opticks* for his old teacher 'a Belgis meis Amicis' and asking if his correspondent could in Hanover oblige a man 'qui summo eum Tractatum Newtonianum videndi tenetur desiderio' (*ibid.*: 277).

(98) So few copies of the *Opticks* circulated in France, in fact, that two years afterwards Rémond de Montmort took it upon himself to make separate reprint of the text of the appended 'De Quadratura' for private circulation there. 'Oserois je Monsieur', he excused himself to Newton on 16 February 1709 (N.S.) in a letter accompanying a presentation of his own newly published *Essai d'Analyse sur les Jeux de Hazard*, 'me faire un mente aupres de vous d'avoir fait imprimer icy il y a 2 ans votre traitté *de quadraturis* pour en distribuer une centaine aux Scavants de ce pays qui n'en pouvoient faire venir d'Angleterre' (*Correspondence*, 4, 1967: 534). The known surviving copies of this 'second edition' of Newton's tract are, in fact, found bound in with Montmort's *Essai* (Paris, ₁1708); see P. and R. Wallis, *Newton and Newtoniana, 1672–1975: A Bibliography* (London, 1977): 181.

Opticks which was published at London in 1706,[(99)] but this was evidently put out to cater for an anticipated demand on the Continent for a rendition of the parent text out of the vernacular and raised no stir in its home country. With (we may assume) Newton's tacit approval the two mathematical appendages were themselves in 1710 put into English, neither very adequately so, by John Harris in the second volume of his widely purchased 'Technical Lexicon'[(100)]— with what gain to its readers not already familiar with the Newtonian concept of the fluxion of a fluent we do not know. A light correction of the Latin original as 'committed to type in 1704' was included the next year in the compendium of Newton's smaller mathematical pieces brought out by William Jones,[(101)] but this was done rather for completeness' sake than to satisfy any obvious demand to have one more publication of the 'De Quadratura'. With it,

---

(99) *Optice: sive de Reflexionibus, Refractionibus, Inflexionibus & Coloribus Lucis Libri Tres. Authore Isaaco Newtono, Equite Aurato. Latine reddidit Samuel Clarke, A.M. ... Accedunt Tractatus duo ejusdem Authoris de Speciebus & Magnitudine Figurarum Curvilinearum, Latine scripti.* The terminal 'Tractatus de Quadratura Curvarum' is there set on separately numbered pages $_3$1–43. Most of the (very few) corrections entered to its text in the margins of his library copy of the *Optice* (now ULC. Adv. b. 39.4)—particularly those to the latter table of quadratures of curves placed in the Scholium to Proposition X (on which compare note (71) above)—were made good in Jones' 1711 *editio princeps* of the tract (see note (101) following), and the two or three remaining are noticed individually at the pertinent places in our footnotes to 2, §3 below. A companion 'Interleaved' copy of the work which was listed among other 'Books that has notes of Sir Isaac Newton' in the catalogue, now BL. Add. MS 25424, drawn up of his library holdings soon after his death in 1727 (see R. de Villamil, *Newton: The Man* (London, 1931): 104–11, especially 110) has never since been recorded and must be presumed to have been lost. It could not, we suppose, have borne on its inserted sheets any startling new corrections or additions to the text of the 'De Quadratura' (or of the 'Enumeratio' printed along with it).

(100) *Lexicon Technicum. Or, An Universal Dictionary of Arts and Sciences,* 2 (London, $_1$1710). The Englished text of the 'De Quadratura' is therein included (on signatures 5M2^r–5O3^r) in full—with some minimal accretions by Harris at the end—as the article on 'QUADRATURE of Curves. By Sir *Is. Newton*', without indication of its provenance, but (we presume) with Newton's tacit blessing; with its columns cut and repasted to fit a smaller page-size (though this is not there indicated) this text is reproduced in facsimile in *The Mathematical Works of Isaac Newton,* 1 (New York, 1964): 141–60. (On Harris' companion Englishing of the 'Enumeratio' as his article on 'CURVES' see VII: 570, note (23).)

(101) *Analysis per Quantitatum Series, Fluxiones, ac Differentias: cum Enumeratione Linearum Tertii Ordinis* (London, 1711): 41–66. In his 'Præfatio' thereto Jones merely noticed (signature C1^v) that '... annexus est ille ipse Tractatus *De Quadratura Curvarum* ... primum typis mandatus ... Anno 1704 ad finem *Optices* ...', but in his ensuing text he incorporated (compare note (99) above) virtually all Newton's own emendations of the 1704 original printing, and Jones' version was rightly to serve as *editio princeps* of the 'De Quadratura' for the next two centuries. Of the handful of *corrigenda et addenda* penned by Newton himself in his library copy of the *Analysis* (now Trinity College, Cambridge. NQ.8.26) we have already noticed all but one— on which see 2, §3.1: note (25) below—in reproducing the parent 1693 'Geometriæ Liber secundus' (VII: 508–48; see especially notes (36) and (48) thereto). Jones would also appear to have made one editorial change in the 1704 printed text off his own bat (see VII: 526–7, note (45)).

certainly, the small market for such a technical treatise was for many years exhausted.[102]

Not for more than five years after Newton's tract made its first public appearance can we trace anyone who read with any real determination and grasp through the modes of algebraic quadrature there variously expounded. The first of whom we know was the young, newly appointed Cambridge Plumian Professor of Astronomy, Roger Cotes, who while waiting in the late summer of 1709 to receive Newton's corrected 'Copy of it which you was pleased to promise me' prepared himself for his task of bringing out the revised edition of the latter's *Principia* by

examining the 2^d Cor: of Prop. 91 Lib. I and found it to be true by ye Quadratures of ye 1st & 2^d Curves of ye 8th Form of ye second Table in Y^r Treatise *De Quadrat*.[103] At the same time I went over ye whole Seventh & Eighth Forms[,] which agreed with my Computation excepting ye First of ye Seventh & Fourth of ye Eighth....[104]

Encouraged by Newton's ready acceptance of these two 'Theorems'[105] Cotes subsequently went on massively to develop and restructure Newton's twin

(102) The publication of a revised edition of Newton's *Principia*, strong rumour of which began to spread in London after the beginning of 1708 (see note (163) below), did open up a further possibility, it is true: that of attaching the 'De Quadratura' to the re-edition in order to supplement the many appeals in the *Principia*'s first book to 'concessis figurarum curvilinearum quadraturis' (compare VI: 451, note (1)). We shall see below that Newton himself had it in mind in April 1712 to append a modified 'Analysis per Quantitates Fluentes et earum Momenta' to the second edition, though he left this new revision incomplete, and soon afterwards abandoned the idea. (The surviving drafts of this augmented tract on quadrature of curves and infinite series are reproduced in 5 below.) Nor was the point lost on Continental publishers: with the same motive of consulting the reader's convenience (joined of course with the less selfless one of making money out of the publication), when the pirated 1714 *editio ultima* of the *Principia*'s 1713 text passed into a second edition at Amsterdam in 1723 there was adjoined to it a reprint *in toto* of Jones' 1711 *Analysis* (including the 'De Quadratura').

(103) The force of this accurate observation by Cotes has already been brought out in VI: 226, note (32). Cotes' checking of Newton's undemonstrated statement in his *Principia* corollary of the total attraction of a 'Sphæroid' upon an external point in its axis of revolution is preserved in Trinity College, Cambridge MS. R.16.38: 24–5, and his proof is indeed there referred to Newton's *Formæ* 8.1 and 8.2 'Tab. Posterioris Tractatûs de Quadr'.

(104) Cotes to Newton, 18 August 1709 (printed in J. Edleston's *Correspondence of Sir Isaac Newton and Professor Cotes* (London, 1850): 3–4 [=*Correspondence of Isaac Newton*, 5, 1975: 3–4]). Of the two errors in Newton's tabulation of 'conic' integrals which Cotes here correctly identifies, the first was (as we previously indicated in III: 257–8, note (564)) also present in the equivalent 'Order' 6.1 of the like catalogue of quadratures presented in Problem 9 of the 1671 fluxions tract, but the latter—quickly checked out by Newton in the stray calculation reproduced in 2, Appendix [6] below—involved merely the omission of two terms in the numerator of the 'Areæ valor' as correctly set out in Newton's earlier *Ordo* 8.4 (see III: 252); both were soon (*vide* note (101) preceding) to be rectified in Jones' 1711 *editio princeps* of the text of the 'De Quadratura'.

(105) See his reply to Cotes on 11 October 1709 (Edleston, *Correspondence*: 4–5 [=*Correspondence of Newton*, 5: 5]).

listings of area-integrals in the scholium to Proposition X of the 'De Quadra-tura', recasting the comparisons in the latter of these with the areas of segments of ellipses and hyperbolas to be in terms of equivalent (radian and logarithmic) 'measures' of angles and ratios, though it was to be more than four years before he gave public illustration of the power and generality of application of his refined techniques for the quadrature of curves,[106] and the key principle under-lying these subtle and ingenious reductions by means of (as we today would say) the elementary real circular and hyperbolic functions—one invoking the 'exceedingly beautiful theorem'[107] by which $a^{2n} \pm x^{2n}$ and $(a^{2n+1} \pm x^{2n+1})/(a \pm x)$ are split into their component quadratic factors—was only to be dug out of Cotes' worksheets after his death. It would take us too far out of our Newtonian way to go into any detail of how he elegantly and with deft precision applied his powerful analytical tool to resolve almost a hundred distinct 'forms' of algebraic integral, categorize and present these though he did in a manner and notation distilled to the point of obscurity.[108] It is not irrelevant, however, to remark that

(106) In an elegant paper on what he called 'Logometria', iceberg-like in its concealment of its true depth, in *Philosophical Transactions*, **29**, No. 338 (for January–March 1714): 5–45; this was reprinted eight years afterwards by Cotes' cousin, Robert Smith, as the opening pages 1–41 of his edition of what he titled *Harmonia Mensurarum, sive Analysis & Synthesis per Rationum & Angulorum Mensuras promotæ...per Rogerum Cotesium* (Cambridge, 1722), and wherein Smith also included Cotes' varied instancing of this 'harmony' of logarithmic and trigonometric 'measures' in a lavish set of eighteen 'forms' of 'Theoremata tum Logometrica tum Trigono-metrica, quæ datarum Fluxionum Fluentes exhibent per Mensuras' (*ibid.*: 43–76), and a richly illustrated 'Problematum Analysis et Constructio per Formas præcedentes' (*ibid.*: 77–109). In an ensuing aside (*ibid.*: 113) Smith affirms that 'Octodecim illa Theoremata... Autor noster conscripsit ante annum 1714, quo *Logometriam* edidit...'. For a detailed and perceptive analysis of the full tripartite 'Logometria' as published by Smith in 1722 see Chapters 1–3 (pages 3–164) of R. C. Gowing's recent Ph.D. thesis (London, 1977) on 'The Mathematical Work of Roger Cotes, 1682–1716'.

(107) So Robert Smith dubbed it in his editorial 'revocation' of it from the 'ruins' of Cotes' papers after his death in June 1716: 'Quæ [adversaria] quanquam primo intuitu Sibyllæ foliis obscuriora videbantur, quod nullo ordine nec verbo erant explicata: multiplices tamen conjectandi occasiones præbendo, spem fortiter conceptam non fefellerunt. Quippe cum ea sæpius evolverem aliamque ex alia rem lucrarer, revocavi tandem ab interitu Theorema pulcherrimum: fundamentum scilicet methodi modo memoratæ' (*Harmonia Mensurarum*: 113). For Newton's anticipations of Cotes' theorem see IV: 205–8 (and compare our accompanying editorial 'Explanation on *ibid.*: 209–11). R. C. Gowing gives a detailed exegesis of Smith's restoration of the theorem in Chapter 4 (pages 165–93) of his 1977 thesis (on which see the previous note).

(108) 'Theoremata tum Logometrica tum Trigonometrica Datarum Fluxionum Fluentes exhibentia, per Methodum Mensurarum Ulterius extensam. Operis Materiam a Rogero Cotesio accepit, Opus ipsum composuit Robertus Smith...' (*Harmonia Mensurarum*: 111–249). Smith's care and assiduity in ensuring that these hundred and fifty pages of complicated expressions are set with nary a typographical error left uncorrected yields a work than which no finer was, in our eyes, printed in the whole of the eighteenth century. Smith himself in his editorial 'Præfatio' made a good beginning in showing how simple substitution reduces each stated integrand to a fraction whose denominator may then, by Cotes' 'Theorema

Cotes shortly before his death also succeeded in removing from the sixth Form of the latter table of integrals in Newton's 'De Quadratura' a limitation[109] which—a revealing witness, this, to the grasp and penetration of Newton's later editors—remains unnoticed in any subsequent printing of its text.

Over the tract's ensuing history and commentaries upon it we may be very brief. As we have already observed,[110] Benjamin Robins in 1727 gave proof (by

pulcherrimum', be split into its component linear and quadratic factors, and so the integrand itself reduced to a sum of partial fractions having each only one of these linear or quadratic expressions as their denominator and therefore straightforwardly separately integrable by 'logarithmic' or 'trigonometric measure'. In the same year that the *Harmonia* was published Henry Pemberton laboured much more elephantinely to give a beginner's entrance into its subtleties in an open *Epistola ad Amicum J[acobum] W[ilson] de Rogeri Cotesii Inventis, Curvarum Ratione, quæ cum Circulo & Hyperbola Comparationem admittunt* (London, 1722); but de Moivre was the first other than Smith to evince a full grasp of the subtleties of Cotes' 'harmony of measures' in his *Miscellanea Analytica de Seriebus et Quadraturis* (London, 1730). (On the latter compare I. Schneider, 'Der Mathematiker Abraham de Moivre' (see note (56) above): 242.)

(109) Namely the implied restriction for the reductions there effected to be real that there shall be $f^2 \geqslant 4eg$. On 5 May 1716, just a month before he died of a sudden fever, Cotes wrote to William Jones: 'You know Sir Isaac has left his sixth form imperfect, and under a limitation. This...appears as an eyesore to me in so beautiful a work....Pray therefore let him know that I can take off that limitation, and make this form as perfect as the others. And use all the address you have to make him set upon the same thing' (S. P. Rigaud, *Correspondence of Scientific Men of the Seventeenth Century*, **1** (Oxford, 1841): 272. It is certain that Jones did indeed inform Newton of Cotes' success in removing the restriction, for there survives a fragment in Newton's hand (reproduced in 2, Appendix [5] below) where he employs Cotes' stratagem of reducing the quartic denominator of the integrand into a different pair of component trinomial quadratic factors to serve as denominator, one each, of two partial fractions to which the integrand may then be reduced—but only generally so (and Newton stumbled in his own variant attempt to reduce his *Forma* 6.1, $f^2 < 4eg$, to preceding integrals listed in his 'De Quadratura' table). Let us add that Robert Smith had been emboldened to reconstruct the edifice of Cotes' complementary angular and logarithmic—that is, circular and hyperbolic—functions, as he built it out of the débris of the latter's rough papers to be the finished *Harmonia Mensurarum*, by an opening passage in the same letter where Cotes informed Jones that 'I must ...tell you that geometers have not yet promoted the inverse method of fluxion by conic areas, or by measures of ratios and angles, so far as it is capable of being promoted by these methods. There is an infinite field still reserved, which it has been my fortune to find an entrance into [by] a general and beautiful method...' (*ibid.*: 270). And that in completing his Herculean task Smith was sustained by Newton's encouragement; compare his letter to Newton of 23 December 1718 (ULC. Add. 3983.40, now printed in *Correspondence*, **7**, 1977: 28–9). The latter made no recorded comment when Brook Taylor issued one generalized form of Cotesian integral as a challenge to Continental mathematicians early in 1719 (compare Taylor's letter to Smith of the preceding 11 December, printed by Edleston in his *Correspondence of Newton and Cotes*: 233–4)—straightaway to receive responses from Johann Bernoulli and Jakob Hermann (in *Acta Eruditorum* (June 1719): 258–70 and (August 1719): 351–61 respectively).

(110) See VII: 166, note (9). To be strict, Robins' demonstration in *Philosophical Transactions*, **34**, No. 397 (for January–March 1727): 232–4 was only by direct successive evaluation of coefficients in particular cases, and his elegant recursive proof of their general pattern did not appear for another six years (see *ibid.*).

recursion) of its hitherto publicly undemonstrated Proposition XI on the reduction of a simple class of multiple integrals. Little further serious research into its content was pursued. A second English translation of the Latin 'De Quadratura' was published in 1745 by the obscure Aberdeen mathematics professor John Stewart, together with a lavish but tediously overstretched and far from perspicacious commentary upon it.[111] The year before, Castiglione had set its original text, taken over without change from Jones' corrected 1711 version, in the first volume of his collection of Newton's minor published pieces.[112] This was further reprinted by Daniel Melander at Uppsala in 1762; and yet again in 1777 by Samuel Horsley in the first volume of his 'complete' edition of Newton's work, there amplified with a number of footnotes mostly elaborating the obvious.[113] In the nineteenth century the 'De Quadratura' was forgotten, its usefulness as a working tool over, and it has been resurrected in our own time solely for its historical value. It is enough here to make passing mention of its modern translations into German and Russian,[114] and also (incompletely so) into Italian.[114]

Of Newton's other mathematical writings in the decade up to 1710 there is little surviving trace, and nothing to indicate that more ever existed. A few sheets recording, around the early autumn of 1706, a short-lived spasm of

(111) *Sir Isaac Newton's Two Treatises of the Quadrature of Curves, and Analysis by Equations of an infinite Number of Terms, explained: Containing the Treatises themselves, translated into English, with a large Commentary; in which the Demonstrations are supplied where wanting, the Doctrine illustrated, and the whole accommodated to the Capacities of Beginners, for whom it is chiefly designed* (London, 1745): 1–32 and 33–320 respectively.

(112) *Isaaci Newtoni...Opuscula Mathematica, Philosophica et Philologica*, 1 (Lausanne/Geneva, 1744): Opusculum III: 203–44. In his 'Præfatio' Castiglione stated that 'Ipse usus sum editione Parisiensi [presumably Montmort's 1706 offprint from the 1704 *Opticks*; on which see note (99) preceding] collatâ cum eâ quam [see note (101)] Celeberrimus W. Jones curavit divulgandam Londini anno 1711'.

(113) *Isaaci Newtoni Opera quæ exstant Omnia*, 1 (London, 1779): 333–86. In his usual banal, inflated way Horsley characterizes the 'De Quadratura' in his introductory 'Præfatio' (*ibid.*: xiv) as 'Libellus, præ cæteris quidem scientiæ ubertate insignis, idem tamen rerum compressione et amplitudine præ cæteris obscurus[!] et gravis'. Whatever would he have said of Cotes' 'Logometria'?

(114) Into German by G. Kowalewski as 'Newton's Abhandlung über die Quadratur der Kurven (1704)' (= *Ostwald's Klassiker der Exakten Wissenschaften*, **164** (Leipzig, 1908): 3–50); into Russian by D. D. Mordukay–Boltovskoy in his rendition (Moscow/Leningrad, 1937) of the first volume of Castiglione's *Newtoni Opuscula Mathematica*. The partial Italian translation by E. Carruccio in *Periodico di Matematiche*, ₄18, 1938: 1–32 (subsequently appended to the second edition of G. Castelnuovo's *Le Origini del Calcolo Infinitesimale nell'Era Moderna* (Bologna, 1962): 127–62) both arbitrarily modernizes Newton's increment o to be h, and omits the tract's more technical portions—specifically Proposition VI, Cases 2–4 of Proposition VII, Proposition VIII, Corollaries 3, 4 and 6–10 of Proposition IX, and Propositions X and XI (though it reproduces 'integralmente' the terminal scholium which follows thereafter).

interest in revising certain of his earlier geometrical papers (for eventual publication, we presume) and a jejune hotchpotch of miscellaneous jottings of about the same period make up—a number of stray optical and astronomical calculations[115] apart—the grand total of what came at this period from his mathematical pen. We may swiftly pass by, as being of slight importance, the novel but impractical constructions of conics by points[116] which he contrived (so the hand in which he recorded them suggests) at about this time, along with the similarly unfruitful 'symptom' which he adduced to define a conic hyperbola (and, in degenerate instance, also parabolas),[116] if we should look with more respect at his determination of the equation of a circle referred to oblique Cartesian coordinates[117] which filled a small lacuna in contemporary knowledge of the analytical geometry of the general conic. Their context (did we but know it) is surely less interesting to us, nearly three centuries on, than the fact that Newton should have been minded at all to compose these fragments— maybe they were to him mere playful geometrical doodlings?—ten years after to outward appearance he had given over such remote, unworldly studies. But of course the mathematician's brain will not be peremptorily stilled to order. His more sustained efforts to reshape portions of his earlier papers for (it would seem) eventual publication merit our closer attention, we believe, even though all we can find to say of their background is no more than surmise.

Newton's several fresh endeavours,[118] shortly after he had seen his 'De Quadratura' and 'Enumeratio' safely into print, to delineate and differentiate the nature and operational procedures of the general mathematical 'sciences' of arithmetic, algebra, geometry and (rational) mechanics largely duplicate, in their quality and much of their content, his essays of a dozen years before[119] to lay the foundation of a comprehensive treatise on 'Geometria', that is, deductive mathematics in its broadest significance. These present reworkings of his former papers were doubtless now yet again intended to be preliminary to

---

(115) Notably those (ULC. Add. 3970.3: 291ᵛ and 3970.9: 622ᵛ) where he made rough estimate of the power of his postulated optical *vis refractiva* in ratio to that of terrestrial gravity (compare Z. Bechler, 'Newton's law of forces which are inversely as the mass...', *Centaurus* **18**, 1974: 184–222, especially 202, 220–1); and those (as for instance ULC. Add. 3965.11: 150ʳ, on which see 3, §3, Appendix: note (52) below) where he yet again computed anew the elements of the orbit of the comet of 1680/1. We ought perhaps also to mention Newton's frequent more humdrum arithmetical tottings-up in his day-to-day London life; for example, his casting up the balance of the 'Accompts' of the King Street 'Tabernacle' from time to time (see *Correspondence*, **4**, 1967: 377–80), and yet more routinely, among his other duties and chores as Master of the Mint, his periodical computing of current exchange values for British and foreign coins, and the like.

(116) See 3, §1 and §4 below respectively.

(117) See 3, §2, Appendix 2.

(118) Reproduced in 3, §2.1–4 below.

(119) See VII, 2: *passim.*

some equally elaborate work on geometrical analysis and the resolution of problems. Let us forgive in a man now grown old their lack of new insight and their confirmation of now outmoded notions: such present affirmations by him as that 'Algebra is nothing more than the rule of indeterminate position by means of which the rule of false position is emended and rendered more perfect'[120] or his declaration that '*Definitions* are ones of words in regard to their meaning, *Axioms* are prime truths, and *Postulates* are ones of the operations which are easiest and most useful'[121] are subdued trumpetings of the comfortable orthodoxies of a Euclidean past[122] rather than clarion calls to revolutionary advance. So be it.

As a young man, we will recall, Newton had in the summer of 1672 communicated to John Collins a set of recently fashioned 'Problems for construing æquations' by means of conchoidal neusis and intersecting conics[123] which such observations on the foundations of arithmetic and geometry might well have introduced. Whether or not this be more than fortuitous coincidence, about (or a little before) the autumn of 1706 he also set himself to rework these unfinished 'Problems' of thirty-five years before into a short tract 'De constructione Problematum Geometricorum' which in its greatest part but minimally revamps its parent, repeating its preparatory definitions all but word for word, and going on to refine the ensuing constructions, but with the addition of a new 'verging' construction of the roots of a general quartic equation.[124] It is revealing to contrast this restyled and amplified paper with the 'Æquationum constructio linearis'[125] which Newton had in the meanwhile likewise elaborated from his original 'Problems for construing æquations' and added somewhat incongruously to the Lucasian lectures on algebra which he in 1684 deposited in Cambridge University Library, and whose text was now (under William Whiston's shaky guiding hand) passing through the press at Cambridge, soon to be published (in 1707) under the title of *Arithmetica Universalis, sive De Compositione et Resolutione Liber*.[126] Not the least of his dissatisfactions with the printed sheets as they came to him in the summer of 1706 was what now

---

(120) 'Algebra nihil aliud est quam Regula Positionis indeterminatæ p[e]r qua[m] Regula Positionis falsæ emendatur et perfectior redditur' (see 3, §2.1 below). We have previously (v: xii) cited Newton's comment to David Gregory in May 1708 that 'Algebra is the Analysis of the Bunglers in Mathematicks' (Hiscock, *Gregory, Newton and their Circle*: 42).

(121) '*Definitiones* sunt verborum quoad significationem, *Axiomata* sunt veritates primæ & *Postulata* sunt operationum facillimarum & maxime utilium' (see 3, §2, Appendix 1.3).

(122) Compare v: xii and vii: 198–9.

(123) See ii: 291–2; the text of the 'Problems' themselves is given on *ibid*.: 450–516.

(124) See 3, §3 below. Two states of a preparatory draft of the 'De constructione' and the preparatory calculation on which Newton based its concluding 'Prob. 4'— the tract's only essentially new feature—are reproduced in [1]/[2] and [3] of the ensuing Appendix.

(125) See v: 420–90, especially 432–64, 476–90.

(126) See v: 9–12.

appeared as the gross overstuffing of this terminal section on the geometrical construction of equations by 'verging' lines; indeed, in his library copy of the printed book he was[127] fiercely to pare away its fat to a bare bone of undemonstrated conchoidal neuses which are effectively those presented in Problems 2 and 3 of his rewritten tract 'De constructione Problematum Geometricorum'. Was there a direct connection here? However false this plausible inference be to the historical truth, it is tempting to guess that the 'De constructione' was planned to replace the more lavish 'Æquationum constructio linearis' in a radical recasting of the *editio princeps* of the *Arithmetica*—one whose other intended changes were also to consist, in David Gregory's phrase, 'chiefly in putting out'[128]—which was never to be. To hazard more until (if ever) the background and context can be better patched together from documentary evidence unknown to us would be both unwise and intolerable.

None of these unfinished papers and fragmentary scraps were shown to anyone else by Newton during his lifetime, and after his death they all disappeared into the jumbled oblivion of Pellet's unsorted packets of 'loose & foul' mathematical sheets, to be half-heartedly dusted off by Adams a hundred years ago, but only here to have their individuality restored and their text made public. The short tract on 'Methodus Differentialis' which next follows,[129] on the other hand, at least went quickly into print, if it had little or no impact upon Newton's contemporaries and was all but equally forgotten in succeeding decades.

The main Propositions I–IV of this 'Method of [finite] Differences' in fact borrow heavily (often word for word over whole paragraphs) from the analogous schemes for interpolation[130]—by constructing an approximating curve 'of parabolic kind' through the end-points of a number of its given (equally or unequally spaced) ordinates, using the divided differences of these to do so—

---

(127) See v: 422, note (615), and compare *ibid.*: 14.

(128) So he entered in his diary on 1 September 1706 (see Hiscock, *Gregory, Newton and their Circle*: 37) along with a summary of Newton's other criticism of the lay-out of the printed *Arithmetica* (on which compare v: 11) and his own expressed hope that the latter's own 'next edition' would be 'published shortly'.

(129) See 4, §2 below. For want of any surviving finished autograph of the piece (which we are not convinced ever existed at all as a complete entity; see 4, §1: note (1)) the text of the 'Methodus' is here reproduced, with some very few standardizations to Newtonian norm, from the text which William Jones 'wrote out' in late 1710 and published the following February as the terminal tract in his compilation of Newton's several modes of *Analysis per Quantitatum Series, Fluxiones, ac Differentias* (London, 1711): 93–101. Preliminary states of its Proposition V and concluding scholium—provably drafted at most only weeks before Newton passed the full text of the 'Methodus' to Jones—are set out in 4, §1 preceding, while what we take to be his intended illustration of Proposition II in worked instances of particular interpolations by means of central differences is given in the ensuing Appendix.

(130) See IV: 52–68.

which Newton had long before entered in his Waste Book in the autumn of 1676. To these he in about late 1710 added some opening verbal explanation of the way in which the basic array of such divided differences is constructed (this, in several of its careless phrasings, tending perhaps more to confuse the un-tutored reader than to illuminate his understanding[131]) and certain further remarks on the application of his ensuing central-difference formulas, along with a final scholium where he adduces simple rules for intercalating an *ordinata intermedia* among a given set of equidistant ordinates and for approximating the area under the curve of their bounding parabola[132] and also, in his concluding paragraph, he outlines how like approaches through an interpolating hyperbolic arc may be reduced by an appropriate inversion of the ordinates to ones *per parabolam*.[133] Of the technical content and purpose of the 'Methodus' we need not here say more, reserving our detailed comments upon its text to our running commentary in footnote thereon below.

Of its background the external documentary record (as it is known to us) tells nothing, and we can once more only conjecture why Newton in 1710 was stimulated not merely to refine and round out his thirty-year-old scheme of finite-difference interpolation, but so unusually to give the resulting 'Methodus Differentialis' rapid publication in a collection of his short mathematical pieces—Jones' 1711 *Analysis*—whose only other novelty was an unaltered first printing of the tract 'De Analysi' which he had long ago, in the late summer of 1669, communicated (by way of Barrow) to John Collins.[134] To our cynical frame of mind, some deeper motive must have cohabited with any altruistic impulse on Newton's part thus to impart to the world the secrets of his powerful techniques of numerical analysis. Never before during the more than two decades since he had, in Lemma V of the third book of his *Principia*, published his complementary theorem for interpolation through advancing (adjusted) differences of given ordinates [135] had he evinced the need publicly to elucidate

---

(131) Compare 4, §2: notes (4), (6) and (8).

(132) These Newton took over both from the 'regulæ in usum constructorum Tabularum' which he had had it in mind to impart to Oldenburg in October 1676 (see IV: 30, and compare 4, §1: note (13) below) and from his more recent investigations into numerical 'Quadrature by Ordinates' in the middle 1690's where (see VII: 690–702, especially 690–4; and compare 4, §1: note (14) below) he made use of a different, slightly less accurate route *per analogiam*.

(133) Newton's intention here is rather more clearly and fully expressed in the preliminary draft reproduced in 4, §1.3.

(134) For the latter to circulate, in excerpt at least, to interested parties both in England and abroad (including James Gregory in Scotland); see II: 167–8.

(135) See *Principia*, ₁1687: 481–3 [=IV: 70–3]. Writing to Cotes on 25 October 1711, just a few months after he had set the 'Methodus' in print (see note (129) above), William Jones assured his correspondent that 'Dr Gregory'—by referring his reader in 'Lib. V. Prop. XXV [Schol.] of his *Astronomy* [*sc. Astronomiæ Physicæ et Geometricæ Elementa* (Oxford, 1702): 434] to the book of Gabriel Mouton, *De Observationibus Diametrorum Solis et Lunæ apparentium*

its undemonstrated enunciation, let alone to make known the yet more universal 'Regula Differentiarum'[136] which he had composed in 1676 (though he had, if we may believe de Moivre,[137] briefly toyed with the thought of adding some notion of approximate quadrature by equidistant ordinates to the 1704 'De Quadratura Curvarum'). Nor, to be fair, did the initially uncomprehending reception by his contemporaries of his *Principia* lemma greatly encourage him to do so. In the autumn of 1707, however, Roger Cotes (then newly appointed to his Plumian professorship at Cambridge, and not yet committed to the demanding task of seeing the second edition of the *Principia* through the press there) wrote an elegant and distinctive paper on 'the Newtonian method of differences' laying bare its essence '& read it to my Auditors in our Schools in 1709'.[138] One way or another—maybe from his successor, Whiston, in the Lucasian mathematics chair (whom he used in autumn 1709 as his Cambridge go-between in sending Cotes 'the greater part of my *Principia* in order to a new

---

[Leyden, 1670: 384–96, on which compare IV: 4–5, note (5)]' for 'Hæc Methodus per differentias procedendi' (see Cotes to Jones of the previous 30 September in S. P. Rigaud, *Correspondence* (note (109) above), 1: 259)—'had but a very Slender notion of the design, extent & use of Lem. 5. Lib. 3 of the *Principia*' (Edleston, *Correspondence*: 210). In his Propositio XXV itself (*Astronomia*: 432–3) Gregory does indeed give a very confused and unseeing account of how taking successive divided *differentiæ* of a given set of ordinates to the curving apparent path of a comet relates to finding the *curva generis Parabolici* passing through their end-points, making no attempt to reproduce either of the complementary cases of Newton's lemma to that end. One knows enough of Gregory, two and a half centuries on, to be sure that he would have reproduced Newton's formulas of interpolation, with fulsome demonstration of their truth, had he understood them.

(136) See IV: 36–50.

(137) See his letter to Johann Bernoulli on 6 July 1708, where (Wollenschläger, 'Brief-wechsel' (note (44) above): 242–50) he told of his own independent discovery of the method of computing $\int_0^p f_x \cdot dx$ in terms of the successive differences of equidistant 'ordinates' $f_0, f_1, f_2, \ldots, f_p$. De Moivre went on to observe to Bernoulli that 'M. Newton avoit fait cela autrefois, et je me souviens que dans le temps [*viz.* the autumn and early winter of 1703/4] qu'il imprimoit son traité de la quadrature des courbes, il me demanda si je croyois qu'il fut à propos d'imprimer ce qu'il avoit fait là-dessus [in his 'Of Quadrature by Ordinates' we presume; see note (132) above]. Je lui fis entendre que le public recevroit avec plaisir tout ce qui viendroit de lui, cependant il ne trouva pas à propos de l'imprimer' (*ibid.*: 248).

(138) So he wrote when he transmitted it to William Jones on 15 February 17[10/]11 as 'a small acknowledgement in kind for the latter's presentation of a copy of his newly published Newtonian compendium of *Analysis* (see Rigaud, *Correspondence*, 1: 258; but we quote the more accurately transcribed text of Cotes' draft letter printed by Edleston in his *Correspondence of Newton and Cotes*: 207). To his 'De Methodo Differentiali Newtoniano'—this was to be published only after his death by Robert Smith in a miscellany of his mathematical papers (*Opera Miscellanea Rogeri Cotes* (Cambridge, 1722 [in appendix to the *Harmonia Mensurarum*]): 22–32) and is analysed in detail by R. Gowing on pages 255–70 of his 1977 thesis (on which see note (106) preceding)—Cotes afterwards attached a *Postscriptum* where he similarly declared (*ibid.*: 32) that 'Propositiones præcedentes Anno 1707 conscriptas Auditoribus meis Academicis exposui publica Lectione Anno 1709. Nesciebam enim a *Newtono* compositum

edition'[139]—Newton must soon have learnt that a course of lectures on one of his favourite themes of earlier research was being delivered in the halls of his old University. And if so, ever apprehensive of having his priority in discovery even so obliquely challenged, he must on hearing the news have been sorely tempted to sit straight down and revamp his own earlier writings on inter-polation by finite differences in much the way in which, at some time during the next year, he did react in compiling what he now titled to be his 'Methodus Differentialis'.

Or so we may guess. But, whatever be the truth of this, Newton had no need to worry. Although Cotes, on reading the printed 'Methodus', was himself at once stimulated to add a postscript to his 1707 paper presenting an elaborate extension of Newton's single published 'three-eighths' rule for approximate quadrature in terms of four equidistant ordinates,[140] he was too gently respect-ful and honourably principled a man to seek to steal his master's thunder, and refused to accede to William Jones' subsequent entreaty that his own paper (thus amplified) should be printed forthwith in the *Philosophical Transactions* even when assured by his correspondent that 'I don't move this to you without Sr Isaac's approbation, who I find is no less willing to have it done'.[141]

Cotes was here to prove not only Newton's aptest pupil—despite later discouragement by him[142]—but for a number of years his only one, in Britain

---

fuisse [! the 'Methodus' was not yet written...] Tractatum de eodem argumento usque dum ejusdem Exemplar typis impressum Anno 1711 dono accepissem a doctissimo Editore Domino *Jones*, cui etiam mearum Apographum ejus humanitate permotus statim transmisi. Perlectis autem Viri celeberrimi Propositionibus intellexi me in plerisque diversa incessisse via' —in systematizing the advancing-differences formulas of the *Principia*, that is, and not the central-differences ones of Newton's newly published tract.

(139) So Newton wrote to Cotes on 11 October (see Edleston, *Correspondence*: 4).

(140) See 4, §2: note (17) below, and compare VII: note (12). In his *Postscriptum* Cotes described Newton's rule as 'Pulcherrima...imprimis & utilissima Regula' (*Opera Miscellanea*: 32). His ensuing listing of 'Hujusmodi Regularum Series aliquantulum producta' sets out (*ibid.*: 33) the corresponding forms for n equidistant ordinates, $n = 3$ ('Simpson's' rule, on which see VII: 692–3, note (8)) and also 5, 6, 7, ..., 11. We may add that in his letter to Johann Bernoulli on 6 July 1708 (see note (138) above) de Moivre had correctly enunciated these Cotesian rules for the cases $n = 3, 4, 5, 6, 7$ and also (*pace* Wollenschläger, 'Brief-wechsel': 311, note (136))—if we add an omitted denominator ($11975.04 \approx$) 11975 and make good a couple of missing decimal points, at least—for $n = 11$.

(141) Jones to Cotes, 17 September 1711 (Edleston, *Correspondence*: 207–8). 'I am of opinion', returned Cotes on 30 September, 'that it is not of so great use as to deserve to be printed after Sr Isaac's *Methodus Differentialis*' (*ibid.*: 209).

(142) When in January 1712 Cotes sent Jones 'y^e first Proposition [of] what I have further concerning ye subject of differences'—namely his 'Canonotechnia, sive Constructio Tabularum per Differentias' (*Opera Miscellanea*: 35–71) giving elaborate schemes for subtabulation by means of adjusted central differences derived (see pages 272–93 of Gowing's 1977 thesis (note (106) above), and compare IV: 43–5, notes (23) and (25)) from the twin 'Newton–Stirling' and 'Newton–Bessel' interpolation formulas of the 'Methodus'—'whereof the Six first [Pro-

or abroad. No review of the *Analysis* in which the 'Methodus' was published ever appeared in the *Philosophical Transactions*, and the anonymous recension which Leibniz gave of Jones' collection of Newtonian mathematical pieces in the *Acta Eruditorum* in February 1712 contained only a brief and subdued notice of the tract on finite differences,[143] one all but lost amid its more expansive comments upon the 'De Analysi' and the pertinent calculus fragments from Newton's letters to Oldenburg, Collins and Wallis also there first published.[144] Leibniz had in fact no taste for the rough arithmetical roundings-out and crude approximations of practical numerical analysis, sharing with that other mighty front-runner of Continental mathematics of the day, Johann Bernoulli, a common preference for the perfection and 'purity' of the mathematically exact.[145] All those of their contemporaries who did manifest interest in discussing and

positions] are particular & fitted for use & are sufficient for all cases that comonly happen, the other four are general' (from the draft of Cotes' letter printed by Edleston in his *Correspondence of Newton and Cotes*: 220; the original has not survived and may perhaps never have been communicated), he added that 'You may shew it to Sr Isaac if You think it proper' (*ibid.*: 221). If Newton ever saw Cotes' 'Canonotechnia' in any part, he certainly gave him no encouragement to proceed with his plan of printing both his papers on finite differences together in a 'small Volume' at Cambridge, for which 'I have already put ye University to the charge of Types for some new characters which I have occasion to make use of' (*ibid.*: 220). No mention of the last occurs in the surviving records of the Cambridge Press for the period—very possibly it was a private arrangement made through Richard Bentley—and no more is heard of Cotes' proposal.

(143) Namely, 'Ultimo... loco comparet exiguus quidam tractatus, cui *Methodi differentialis* hic imponitur nomen speciali quodam sensu, quemque Cl. Editor [Jones] ex autographo Autoris Newtoni descripsit. Complectitur doctrinam describendi Curvas ex datis differentiis differentiarum Ordinatarum. Innititur problemati ducendi Curvam generis Parabolici per data quotcunque puncta. Quare cum omnes Parabolæ quadrari facillime possint; ejus usum [Newtonus] ostendit in quadranda figura quacunque curvilinea quam proxime, cujus ordinatæ aliquot inveniri possunt. Non enim hic alia re opus est, quam ut per terminos ordinatarum datarum ducatur linea curva generis parabolici' (*Acta Eruditorum* (February 1712): 76; compare II: 261, note (12)). In his unpublished private counter to this unsigned recension of Jones' 1711 *Analysis*, Newton was primarily concerned only to make his historical priority clear: 'It appears by Mr Newtons Letter of 24 Octob 1676 [see *Correspondence of Newton*, 2: 119] that this method... of drawing a Curve line of a Parabolic kind through the ends of any number of Ordinates of any Curve, for squaring the Curve *quamproxime*... was then known to him. It depends upon the differences of the Ordinates & the differences of these differences[,] & therefore is [*sc.* rightly] called the Differential method' (ULC. Add. 3968.32: 461^v; compare II: 270, note (30)).

(144) We have reproduced this main part of Leibniz' 1712 *Acta* review on II: 259–62.

(145) It had been Johann Bernoulli's response—in April 1710!—to de Moivre's communication on 6 July 1708 of his version of the general advancing-differences interpolation formula, together with (see note (137) above) its application to derive the simpler instances of the Cotesian rules for approximate quadrature, to dismiss it all with a polite incuriousness: 'Je ne sçais si je comprens bien le probleme de M. Newton, de décrire une courbe (sans doute géo-metrique) qui passe par tant de points qu'on voudra: De la maniere que je l'entens, il me paroit fort facile, mais je le trouve indéterminé; ainsi je ne vois pas ce que vous voulez dire par les aires que vous dites qu'elles sont toujours quarrable' (Wollenschläger, 'Briefwechsel': 265). De Moivre did not pusue the matter in his own letter of reply.

developing what seemed a very 'English' branch of mathematics took their lead from Lemma V of Book 3 of the *Principia*.[146] Even in England itself, just eight years after the 'Methodus Differentialis' had appeared, the young James Stirling in freshly presenting its central-differences formulas and illustrating their application felt the need to remind his fellow mathematicians that Newton had indeed, in his opinion, there 'brought the method of approximating to its highest pinnacle of perfection'.[147] In after years—though Jones' 1711 compendium was reissued in 1723 in appendix to a widely read 'pirated' reprinting of the *Principia*'s second edition[148]—the memory of just how far Newton had gone in the propositions of his 'Methodus Differentialis' dimmed fast. While in Stirling's lavishly amplified treatment of that theme in 1730 Newton's formulas were still credited to him,[149] they were now lost amid a profusion of other theorems on summation and interpolation of Stirling's own discovery. We should not be surprised that in his influential survey of the historical development of the theory of finite differences[150] a century later Frédéric Maurice, in ignorance of their true provenance, attributed Newton's twin central-differences formulas to Stirling (whose own *Methodus Differentialis* he could not have read) and to F. W. Bessel respectively. And so their appellations continue to credit them in our modern textbooks.

The later eighteenth-century reprintings of the text of the 'Methodus' itself

(146) We have already remarked (IV: 8, note (20)) that Jakob Bernoulli's pupil Jakob Hermann was the first to give public elucidation of Newton's underlying method of approach 'in Lemmate V. Lib. III. *Princ. Phil. Natur.* sine omni analysi & demonstratione tradita' in the short tract 'De Curvis Algebraicis per quotlibet data puncta ducendis' appended by him to his *Phoronomia, sive de Viribus et Motibus Corporum Solidorum et Fluidorum Libri Duo* (Amsterdam, 1716): 389–93. But the *Phoronomia* was little read, even on the Continent, and Hermann's a deal less than lucid 'solutio' (which he claimed to have wrested out of Newton's unproven lemma 'annis 1704 et 1705') aroused no ripple of contemporary attention.

(147) 'Existimo... *Newtonum* perduxisse methodum Approximandi ad summum perfectionis vestigium' he wrote in his 'Methodus Differentialis *Newtoniana* Illustrata' (*Philosophical Transactions*, **30**, No. 362 (for September/October 1719): 1050–70): 1051. We have made fuller quotation of Stirling's words in IV: 8, note (20).

(148) The reprint, namely, of the self-styled *editio ultima* first published (likewise at Amsterdam) in 1714; on which see note (102) preceding.

(149) *Methodus Differentialis: sive Tractatus de Summatione et Interpolatione Serierum Infinitarum* (London, 1730). The 'Stirling' Case 1 and 'Bessel' Case 2 of Newton's Proposition III in the 'Methodus' (which is 'Prob. 1' in the 1676 parent text; see IV: 56–60) reappear in Stirling's *Pars Secunda. De Interpolatione Serierum* as the corresponding 'Casus primus' and 'Casus secundus' of its Proposition XX (pages 104–5/105–6 respectively).

(150) 'Mémoire sur les Interpolations, contenant surtout...la démonstration générale et complète de la méthode de quintisection de *Briggs* et de celle de *Mouton*, quand les indices sont équidifférents, et du procédé exposé par *Newton*, dans ses *Principes* [!], quand les indices sont quelconques' (*Connaissance des Temps, ou des Mouvements Célestes...pour l'an 1847* (Paris, 1844): 181–222 [English translation by T. B. Sprague and J. H. Williams in *Journal of the Institute of Actuaries and Assurance Magazine*, **14**, 1867: 1–36]): 184.

in the collected editions of Castiglione (1744)[151] and Horsley (1779)[152] went unheeded by such contemporary working mathematicians as J. L. Lalande[153] and, most notably, by J. L. Lagrange, soon himself to lay the foundations of the modern theory of interpolation by means of finite differences.[154] In the nineteenth century, when it had ceased to have other than historical interest, a brief and somewhat shaky account of its content was given by Ludwig Oppermann,[155] but not till some sixty years ago did the 'Methodus' receive careful, scholarly assessment or translation into any modern language.[156]

(151) *Opuscula Mathematica*...(note (112) above), **1**: Opusculum V: 271–82.

(152) *Newtoni Opera Omnia* (note (113) above), **1**: 519–28.

(153) Who began his widely read 'Memoire sur les Interpolations, ou sur l'usage des différences secondes, troisièmes, &c dans les Calculs astronomiques' (*Mémoires de Mathematique et de Physique tirées des Registres de l'Académie Royale des Sciences de l'Année M.DCCLXI* (Paris, 1763): 125–39) with the conventional nod that 'La méthode générale des interpolations a été traitée dans toute la généralité possible par M. *Newton* [dans ses] *Philos. Nat. Principia Mathematica*...' and then passed to devote the greatest part of his paper to elucidating the rules for interpolation given by Mouton in his 1670 *Observationes* (on which see note (135) above).

(154) In his influential 'Mémoire sur la Méthode d'Interpolation' in the *Nouveaux Mémoires de l'Académie royale des Sciences et Belles Lettres [de Berlin]. Année MDCCXCII* (= (ed.) J. A. Serret, *Œuvres de Lagrange*, **5** (Paris, 1870): 663–84). Lagrange there (page 664) merely passed the vague comment in introduction that 'Tout le monde connaît la formule de Newton pour trouver une ordonnée quelconque d'une courbe parabolique, par les différences successives des ordonnées équidistantes; c'est celle dont on se sert journellement en Astronomie...'.

(155) 'Notes on Newton's formulæ for interpolation', *Journal of the Institute of Actuaries* [of which he was then editor] **15**, 1869: 145–8, 177–9. Having noticed that Newton in his 'celebrated' Lemma V of the *Principia*'s third book 'proposed and solved the fundamental problem of interpolation by finite differences', Oppermann went on to observe (pages 144–5) that 'In the *Methodus Differentialis*, printed in 1715[!], he treats the same subject more at length, ...beginning in the most direct and elementary way and then—when he had overcome the first difficulties—going on so fast and with such large steps that it requires no slight attention to follow him. His solution of the problem is certainly as general, as elegant and as simple as we might expect.... Stirling [in his 1730 *Methodus*] gives it as his opinion that Newton has not chosen the best manner of treating the problem. Nevertheless I am inclined not only to think that Newton's way indeed is the easiest and best, but even to suppose that he regarded his solution as all but self-evident'.

(156) Notably by D. C. Fraser in his study of 'Newton's Interpolation Formulas' (*Journal of the Institute of Actuaries*, **51**, 1918: 77–106 [reprinted in his *Newton's Interpolation Formulas* (London, 1927): 1–25]) where he gives both a facsimile reprint of the 1711 *editio princeps* of the 'Methodus Differentialis' (see *ibid.*: 85–93) and also a first English translation of its text (*ibid.*: 94–101) which has more recently been reproduced (with Newton's table of divided differences in Proposition I inserted, and his illustrative figures in following propositions interposed) in *The Mathematical Works of Isaac Newton*, **2** (New York, 1967): 165–73; see also Fraser's article on 'Newton and Interpolation' in (ed.) W. J. Greenstreet, *Isaac Newton, 1642–1727* (London, 1927): 45–69, especially 47–51; and M. Miller's more derivative account of 'Newtons Differenzmethode' in *Wissenschaftliche Zeitschrift für Verkehrswesen Dresden*, **2**, 1954: 1–13, A first translation of the 'Methodus' into German was given by A. Kowalewski in his *Newton, Cotes, Gauss, Jacobi. Vier grundlegende Abhandlungen über Interpolation und genäherte Quadratur* (Leipzig, 1917): 1–12. More recent ones into Russian and Italian have been respectively

The revisions and ameliorations introduced by Newton into the second edition of his *Principia* which, under Cotes' painstaking managerial eye, went through press in Cambridge during 1709–13[157] have in their greatest portion none but incidental mathematical interest, and so cannot here concern us. In the sixth volume[158] we have already reproduced from their originals penned by him in the early 1690's the texts of the major changes in its first book which Newton passed on into print in the 1713 *editio secunda*, and have also there[159] more sketchily surveyed the content of the more mathematical changes which he then likewise destined for its two following ones and similarly implemented in the revised printing. These we need not take up again. But two other items—the second of which has near-unchallengeable title to be his last considerable piece of mathematical invention—must not here be so lightly brushed aside.

There can be no doubt that when Newton wrote out the original version of the *Principia*'s first book in 1685 he planned, as an adjunct to his many generalized appeals in it to be 'granted the quadrature of curvilinear figures',[160] to add a lemmatical problem displaying the basic procedures for effecting such squaring of curves; and maybe also to attach to this a tabulation of the simpler types of particular area-integral, evaluable outright algebraically or in terms of specified standard trigonometrical and logarithmic conic quadratures, which would usefully underpin the two or three complicated constructions in his preceding text that he had before been content vaguely to justify as following 'through the quadrature of a certain curve' or not at all.[161] Indeed, he went so far as to compose several drafts of an unnumbered 'Prop.   Prob. ' where he outlined

---

published by D. D. Mordukay–Boltovskoy in his 1937 rendering of the first volume of Castiglione's *Opuscula Mathematica* as *Matematicheskie Raboty* (see note (114) above), and by I. Bertoldi in *Periodico di Matematiche*, ₄35, 1957: 14–43.

(157) The detail of these corrections and additions to the 1687 *editio princeps* is now conveniently set out page by page in the *variorum* text of I. B. Cohen and A. Koyré (*Isaac Newton's 'Philosophiæ Naturalis Principia Mathematica': the third edition (1726) with variant readings*, Cambridge, 1972), where they are tagged under the coding E₂. The pains taken by Cotes in further adjusting and amending those *corrigenda* sent to him into yet further improved form shine out clearly in the numerous letters which passed between him (in Cambridge) and Newton (away in London) during the three and a half years from October 1709 when the *Principia* was 'in press'. (On this see Appendix V (=**2**: 817–26) to the Cohen/Koyré *variorum* edition, where a detailed collation of the Newton–Cotes letters—there cited from Edleston's 1850 text (a recent fuller version is in *Correspondence of Newton*, 5, 1975: *passim*)—is given, tabulating Cotes' editorial efforts proposition by proposition; Cohen gives a more generalized comment in his *Introduction to Newton's 'Principia'* (Cambridge, 1971): 227–40.)

(158) See especially vi: 418–20, 546–52, 564–5.

(159) Compare vi: 358, note (217); 450, note (22); 505–7.

(160) Notably in Propositions XLI and LIII–LVI (*Principia*, ₁1687: 127–31, 157–61 [=vi: 344–8, 404–8]).

(161) See especially Proposition XLI, Corollary 3 and XCI, Corollary 2 (*Principia*, ₁1687: 130–1 and 220–1 respectively); on which compare also vi: 354, note (214) and 226, note (32).

the reduction of the area under a curve defined by a trinomial (implicit) Cartesian equation to directly quadrable form,[162] though he set no version of this in the *editio princeps* which was published in 1687. We will not be surprised, therefore, that when Newton began twenty years afterwards to turn his attention to a new edition of his *Principia* he thought equivalently to fill this want therein of any account of the quadrature of curves by appending to it the text of his lately appeared tract 'De Quadratura Curvarum'—or so David Gregory remarked after recording in his diary on 25 March 1708 that 'Sʳ Isaac...has begun to reprint his *Principia Philosophiæ* at Cambridge.'[163] The new edition of the *Principia* began in earnest in the autumn of the next year after Newton, acting upon Richard Bentley's suggestion, successfully prevailed with Roger Cotes to undertake the task of seeing it through the press. But nothing more is heard of any proposal to attach an *addendum* on squaring curves till late April 1712 when, newly returned from a visit to London where he had conversed with Newton about a variety of matters relating to the forthcoming *editio secunda*, Bentley out of the blue gave Cotes to understand that the former still had 'some thoughts of adding to this Book a small Treatise of Infinite Series & yᵉ Method of Fluxions'[164] and reiterated the news after again visiting Newton in the last days of the following June.[165] Cotes was never to be told more about this projected new tract on infinite series and fluxions, and it thereafter disappears from view completely, but for a while in the early spring of 1712 (we may now know on gathering together its long geographically scattered draftings) Newton had indeed worked hard to fashion for publication just such a work on (so he himself titled it in its Latin original) 'Analysis by means of fluent quantities and their [fluxional] moments'.[166]

Founding its introduction and eleven opening propositions upon (as we

---

(162) These rough draftings are reproduced on VI: 450–5; as we there observed (*ibid.*: 451, note (1)), they all lean heavily on his earlier paper treating 'The Quadrature of all curves whose æquations consist of but three terms' (III: 380–2). No ancillary listings of evaluated area-integrals are found with them.

(163) In sequel to which he adjoined (Hiscock, *Gregory, Newton and their Circle*: 41): 'He is to add his Quadratures & Curves of the second genre [*viz.* the 'Enumeratio'].

(164) As Cotes phrased it in his ensuing letter to Newton on 26 April inquiring about the 'design', which he liked 'very well' (*Correspondence of Newton*, 5: 279; compare Edleston, *Correspondence*: 119).

(165) See Cotes' letter to Newton of 20 July 1712, written 'about three Weeks since Dʳ Bentley return'd from London', where he again commended Newton's 'design', now remarking that 'it were better that the publication of Your Book [the *editio secunda* of the *Principia*] should be deferr'd a little, than to have it depriv'd of those additions' (*Correspondence of Newton*, 5: 315; compare Edleston, *Correspondence*: 118).

(166) The rough preliminary drafts and more finished sheets of this 'Analysis per quantitatum fluentes et earum momenta', disunited almost as soon as Newton penned them, are reassembled for the first time in section 5 below.

would expect) the published 'Tractatus de Quadratura Curvarum'—his minor ameliorations and augmentations of its text are in fact keyed to Jones' newly appeared 1711 *editio princeps*—in his amplified 'Analysis' Newton adjoined to this a newly systematized account of his various techniques of expansion into infinite series. In this Proposition XII, in particular, he displayed the elegance and power of his binomial theorem, setting out the algorithm as he had thirty-six years earlier for Leibniz in his *epistola prior* of June 1676, and passing (now as then) to apply it to deduce the elementary circular and logarithmic expansions which quadrature term by term of the series ensuing for the pertinent simple algebraic integrand permits; then went on to expound his more general method, lengthier and exceedingly more cumbrous, for determining in series the root of a given equation, whether (Proposition XIII) merely algebraic or also (Proposition XIV) involving fluxions, by way of intermediary 'fictional' equations, virtually as he had communicated this to John Wallis in the late summer of 1692;[167] and lastly, in his Proposition XV, he instanced how, by considering the ratios of the coefficients of successive terms in a series thus (or otherwise) obtained, one can often readily identify the pattern of its universal 'progression'.[168] A final section of this tract where he planned to give the reader some notion of his ancillary method of finite differences, showing how it might be applied to yield the approximate quadrature *per parabolam* of the area under a given (smoothly continuous) curve, was apparently never written.[169]

(167) Whence they were printed by Wallis the next year in his *Opera Mathematica*, **2**, 1693: 394–6 [=vII: 177–80]; compare **5**: notes (41) and (67). The particular fluxions-free case of his previous 'Prob. 2' which he now separately presents as Proposition XIV is (see **5**: note (45)) essentially identical with the equivalent 'parallelogram' rule for selecting the 'fictitious' equation at any stage first presented by him in the 1671 tract (on which see III: 50–2).

(168) Newton here draws upon the more elaborate discussion of this problem of identifying the pattern of coefficients of terms in a series which he had given in Chapter 5 of his 1684 'Matheseos Universalis Specimina' (see IV: 576–88, and compare **5**: note (72) below).

(169) See **5**: note (5). We may add that in an 'Ad Lectorem' which he tentatively set to the second edition of his *Principia* in the early summer of 1712 (but quickly thereafter withdrew), having first affirmed that 'Interea dum componerem *Philosophiæ Naturalis Principia Mathematica*, plura Proble[m]ata solvi per figurarum quadraturas [compare note (161) preceding] quas Liber de Quadratura figurarum mihi suppeditavit. Alia proposui solvenda concessis figurarum quadraturis [see note (160)].... Et propterea Librum *de Quadratura Curvarum* subjungere visum est quo figuras vel quadravi vel quadrandas proposui, et qui Analysin meam momentorum exhibet qua sæpissime usus sum', Newton concluded: 'Et cum in exponenda Cometarum Theoria usus sim methodo mea differentiali [*sc.* in Book 3 of the *Principia*; on which see IV: 73, note (5) and also vII: 682–6], visum est etiam eandem methodum subjungere' (ULC. Add. 3968.32: 464ᵛ; this is the back of a 2-folio sheet on whose first leaf (f. 463) Newton has penned a revision of the draft letter published in *Correspondence*, **5**: 383 as addressed to ' ? the editors of the *Acta Eruditorum*' under the tentatively conjectured date of 'February 1712/13' which may well be half a year out). The full text of this preface 'To the Reader' may be read in I. B. Cohen's *Introduction to Newton's 'Principia'*: 349 (where a rendering of it into English is also given); see also **2**, Appendix 6 below.

Its intended content could surely have held no surprise in store for anyone familiar with Newton's already published writings on the topic.

And there Newton abandoned this final revamping by him of his joint method of analysis by fluxional calculus and expansion into infinite series. Almost as soon as they were penned, the sheets on which he outlined and drafted the various sections of the 'Analysis per quantitatum fluentes et earum momenta' were to be dispersed, its opening pages passing into the possession of William Jones (how and why we know not) and the remainder into the like oblivion of the mass of Newton's unsorted private papers, where even the fact of its past existence was to go unrecognized till the present day.

In complete contrast to this, the scurried last-minute recasting of the argument of Proposition X in the *Principia*'s second book in late 1712 has rarely since been allowed to step out of the spotlight of continuing scholarly attention. For more than twenty years after it was originally set forth in the *editio princeps*,[170] it is true, the intricacies of this problem of determining the resistance of a medium so impeding the motion of a projectile through it that the missile shall, under constant downwards gravity, travel a given flight-path provoked no recorded reaction from Newton's contemporaries; still less was the subtle error suspected which spoilt by a numerical factor his deduced ratio of the *resistentia Medii* to the *vis gravitatis*. Not even Huygens and Leibniz, who elsewhere went minutely through the text of the *Principia* on the look-out for slips and inconsistencies, appear to have seen need here to test the validity of Newton's complicated reasoning, each seemingly content to note that his construction from first principles (in his preceding Proposition IV) of the logarithmic trajectory which ensues in the case where the resistance is directly proportional to the velocity agreed *mutatis mutandis* with their own equivalent (and in Huygens' case much earlier) derivations of this fall curve by compounding the separate horizontal and vertical components of the thus impeded motion—and doubtless sympathizing with him that no like exact determination of the flight-path should apparently be possible where the resistance is as the square of the speed.[171] As overseer through press of the first edition it had been Edmond

(170) *Principia*, ₁1687: 260–74. For convenience of reference (and also to permit us to adjoin our running editorial commentary upon its detail) the text of this is reproduced, with indication of its division into pages, in 6, Appendix 1 following.

(171) See vi: 70, note (109) and compare *ibid.*: 90, note (17). Huygens made brief public mention of his pre-discovery (in the late 1660's) of the opening Propositions II–IX of the *Principia*'s second book when in his *Discours de la Cause de la Pesanteur* (appended to his *Traité de la Lumière* (Leyden, 1690): 125–80) he declared (p. 168) that 'J'ay vu avec plaisir ce que Mr. Newton écrit touchant les chûtes & les jets des corps pesants dans l'air, ou dans quelqu' aut[r]e milieu qui resiste au mouvement; m'estant appliqué autrefois à la mesme recherche'; and he afterwards there added his opinion (pp. 173, 175) that 'dans la vraye hypothese de la Resistance, qui est en raison double de la Vitesse...il est extremement difficile, si non du tout

Halley's small contribution to enlightenment to make a single, wholly insignificant verbal amelioration in the secretary script of the second book at this point when it was put to him in 1686. Cotes, upon receiving in October 1709 the bulk of Newton's corrected copy for the *Principia*'s second edition, was presented with no change in the text of Book 2, Proposition X and, ignorant of anything untoward, added no alteration of his own before passing it to be printed. The first notice that something was considerably amiss somewhere in its argument came, such is fate, just a few months afterwards from Johann Bernoulli, then hard at work in Basel in developing alternative Leibnizian reformulations of Newton's preferred geometrical modes of dynamical reasoning—as at least in the 1687 *Principia* he saw these to be—by way of the limit-ratios of corresponding increments of simultaneously 'flowing' line-lengths and angles. Early in August 1710 Bernoulli was emboldened cautiously to write to Leibniz that 'In Proposition 10 on page 260 the problem is, it seems to me, not rightly solved by the author, though...where precisely the error occurs is not readily detected'.[172] In support of his assertion he pointed to the contradiction he had found in its ensuing prime example of resisted motion in a semicircular arc, that by Newton's result the resistance to motion along that arc is at its every point equal to the opposing acceleration of gravity acting along the trajectory, with the impossible consequence that a body ought under the urge of gravity to move at a uniform, unaccelerated speed therein; whereas, he went on, 'by my own method of solution' in this particular case 'I find the resistance to be to the motive force [*sc.* of the component of downwards gravity which acts instantaneously in the direction of the circular motion] in the constant ratio of 3 to 2'.[173] Bernoulli

---

impossible de resoudre ce Probleme'. Leibniz, too, in his private annotations upon the *Principia*'s first edition showed great interest in Propositions V–IX of the second book where—without being able to compound the total, oblique motion of the projectile from these—Newton geometrically constructed the horizontal and (gravitationally accelerated) vertical components of trajectory under resistance varying as the square of the speed (see E. A. Fellmann, *G. W. Leibniz: Marginalia in Newtoni Principia Mathematica* (Paris, 1973): 37–40): but he had no observation of any kind to make upon the immediately following Proposition X.

(172) '[*Principia*, ₁1687] Pag. 260, prop. 10. Problema hoc...mihi non recte ab Auctore solutum videtur. Quanquam autem multis in locis evidentia desideretur, ubi tamen præcise erratum sit non facile detegitur' (*Leibnitii et Bernoullii Commercium Philosophicum et Mathematicum*, **2**: 231 [=*Leibnizens Mathematische Schriften*, **3**: 854]).

(173) 'Quidquid sit, ex solutione Auctoris manifeste contradictorium sequitur. Nam pag. 265. ubi applicat solutionem generalem ad circulum, ...esset...resistentia æqualis vi motrici [gravitatis secundum tangentem curvæ agentis], & sic quantum de velocitate per resistentiam quolibet momento amitteretur, tantumdem per vim motricem resarciretur: hinc ergo velocitas foret æquabilis, & tamen dicit decrescere...; adeoque sibi contradicit. Ego per meum solvendi methodum, qui valde naturalis videtur, invenio...resistentiam ad vim motricem esse ut 3 ad 2, hoc est in constanti ratione sesquialtera' (*Commercium*, **2**: 231-2 [=*Math. Schriften*, **3**: 854-5]). On this 'natural' Leibnizian direct mode of approach of Bernoulli's see 6, Appendix 2: note (6) below.

contented himself with outlining his demonstration of this to Leibniz, but he set down his full supporting calculation five months later in a 'Remark' at the end of an open letter to the Paris *Académie des Sciences*, 'touching upon the manner of finding central forces in media resisting in ratios compounded of their densities and any powers whatever of the speeds of the moving body', which was read out to that gathered assembly in January 1711, but not printed in its public *Mémoires* till three years later.[174] Wherein the fault in Newton's general argument in his *Principia* proposition might 'precisely' lie he left his nephew Niklaus tentatively to pinpoint in an 'Addition' to this, well content, it would appear, to go along with the latter's would-be explanation that Newton's initial geometrical measure of the ratio of resistance to gravity was perfectly accurate, the error entering at a subsequent stage when he came mistakenly (so Niklaus supposed) to correlate the coefficients of powers of the base increment in his following 'Taylorian' expansion of the trajectory's augmented ordinate with the corresponding derivatives of the ordinate with respect to the base.[175] All

(174) As an 'Extrait d'une Lettre de M. Bernoulli, écrite de Basle le 10. Janvier 1711 [N.S.]. touchant la maniere de trouver les forces centrales dans des milieux resistans en raisons composées de leurs densités & des puissances quelconques des vîtesses du mobile', *Memoires de l'Academie Royale des Sciences. Année M.DCCXI* (Paris, ₁1714: 47–54, especially 50–1). See 6, §2.3: note (37) below for our summary of the method which Bernoulli here expounds (and which does not differ in its essence, *mutatis mutandis*, from the structure of Newton's recast geometrical argument).

(175) See the 'Addition de M. (Nicolas) Bernoulli, Neveu de l'Auteur de ce Memoire-cy' (*Memoires...pour l'Année M.DCCXI*: 54–6), where Niklaus reported (p. 54) that 'j'ay jugé qu'il y avoit necessairement quelque méprise dans le raisonnement de M. Newton, parce que je n'en trouvois aucun dans son calcul. J'ay donc été curieux de chercher cette méprise; & en examinant avec soin sa solution générale, j'en ay trouvé l'origine: cette méprise est dans le Corol. 3. pag. 263. à l'endroit où cet Auteur dit, *Et hinc si curva linea definiatur....& valor ordinatim applicatæ resolvatur in seriem convergentem; Problema per primos seriei terminos expeditè solvetur.* À cela près j'ay trouvé cette solution de M. Newton fort exacte. C'est cette méthode de changer les quantités indéterminées & variables en suites convergentes, & de prendre les termes de cette suite pour leurs différentielles respectives, ...qui a conduit M. Newton à des solutions fausses...: car cette maniere de prendre les différentielles...n'est bonne que pour les différentielles du premier degré...' While we may in hindsight dismiss this would-be explanation of Newton's mistake as both spurious and inconsistent, let us not under-estimate its seductiveness to one unable to find a better source for the error in his argument. Since this yielded a false value $\frac{1}{2}S\sqrt{(1+Q^2)}/R^2$ for the ratio of resistance to gravity, where Q, R and S are the coefficients of successive powers of the increment o of the base a in the series expansion of the augmented ordinate e_{a+o} of the projectile curve as $e + Qo + Ro^2 + So^3 + \ldots$, while the true measure of this ratio (as at least Johann Bernoulli established it to be in reworking by his alternative Leibnizian route each of Newton's four instanced *Exempla*) is $\frac{1}{2}(d^3e/da^3)\sqrt{(1+(de/da^2))}/(d^2e/da^2)^2$, how alluringly 'evident' it is to suppose that here—as indeed he had seemed generally to prescribe in the concluding scholium of his 1704 'Tractatus de Quadratura Curvarum' (compare 2, §3.2: note (46) below) in the case of the analogous binomial series, a reference to which Niklaus now went on to point—Newton had simply, in his equivalent geometrical way, made substitution of $Q = de/da$, $R = d^2e/da^2$ and $S = d^3e/da^3$! or until we notice (as Niklaus did not) that, were this indeed so, the expansion of the augmented ordinate must corre-

very plausible at first sight, but in fact a phantom conjured out of the mirage of mere numerical coincidence, if its insubstantial fiction was not widely to be appreciated for many a long year afterwards.

Nothing of this penetrated through to Newton himself till late September 1712 when Niklaus Bernoulli arrived unannounced in London, there to be (in his words) 'courteously received everywhere'—'an honour in England where the people are generally not friendly towards foreigners', his uncle later sourly commented—and not least by Newton, who showed himself 'unwontedly urbane', displaying none of his reputed 'somewhat morose humour' (to which Johann subsequently reacted by neatly turning an Ovidian tag to muse that 'study of mathematics, as was formerly said of the arts, "softens the character and suffers it not to be fierce"...'!).[176] It was, de Moivre was later to inform

---

spondingly have been $e + Qo + \frac{1}{2}Ro^2 + \frac{1}{6}So^3 + \dots$, whence we are trapped in the vicious circle that Newton's measure of resistance to gravity must now be 'corrected' to be

$$\frac{1}{2}(\frac{1}{6}S)\sqrt{(1+Q^2)}/(\frac{1}{2}R)^2 = \frac{1}{3}S\sqrt{(1+Q^2)}/R^2.$$

That the substitutions $Q \to Q$, $R \to 2R$ and $S \to 6S$ do, at a first round in the spiral the other way, convert Newton's 1687 measure to be $\frac{1}{2}(6S)\sqrt{(1+Q^2)}/(2R)^2 = \frac{3}{4}S\sqrt{(1+Q^2)}/R^2$, now increased by the right correction-factor of $\frac{4}{3}$, is a meaningless coincidence—'remarquable' was to be Lagrange's double-edged epithet in his *Théorie des Fonctions Analytiques* (on which see note (200) following)—which thoroughly bamboozled all but the acutest and unprejudiced of those who subsequently sought to determine the truth of the matter. When, to anticipate, Newton himself learnt that there was a *faux pas* in the argument of his proposition, he knew that the error must lie somewhere in the reasoning by which he had, in his Corollary 2 (*Principia*, $_1$1687: 263), deduced the parent geometrical ratio $\frac{1}{4}CF(FG-kl)/FG^2$ whence the faulty ancillary measure $\frac{1}{3}S\sqrt{(1+Q^2)}/R^2$ was accurately derived, and it did not take him long to detect that his blunder had been incautiously *en route* to replace the vanishingly small linelet fg by its 'equal' FG in the difference $fg-kl$ (see 6, Appendix 2.1: note (13)); though not till after he had opted to use a differing mode of resolution in the recast Proposition X as it appeared in the *editio secunda* did he fully appreciate—or so a stray computation, itself with a numerical lapse, would suggest (compare 6, Appendix 3: note (22))—that the 'missing' factor of $\frac{3}{2}$ for which the Bernoullis and himself had been hunting is precisely the ratio of the infinitesimal differences $fg-kl$ and $FG-kl$.

It is, finally, only fair to Niklaus Bernoulli to notice that his uncle may have published this 'Addition' to his own critical comments on Newton's discussion of resisted motion without seeking his express permission to do so. Five years after their joint paper appeared he was certainly concerned to assure James Stirling (as the latter informed Newton himself in a letter from Venice in mid-August 1719) that 'what was printed in the *Acta Paris.* relating to [*Principia*, $_1$1687:] 10 Prop. lib. 2, was wrote before he had been in England$_{[,]}$ sent to his friends as his private opinion of the matter, and afterwards published without so much as his knowledge' (King's College, Cambridge. MS 104 [=*Correspondence*, 7, 1977: 54]).

(176) See the draft of Johann Bernoulli's reply to Niklaus on 23 November 1712 (N.S.) (now in Basel University Library. MS L Ia 22.9) where he responds to the lost letter in which Niklaus gave him his impressions of London and those whom he had met there: 'es frewet mich sehr, dass Ihr an aller orten so höflich seyt empfangen worden, welches in Engelland billich für eine ehr zu halten, da die Einwohner sonsten nicht gar leutselig gegen die fremden sich zu erzeigen pflegen; In sonderheit verwundere ich mich über die ungewohnte urbanitet dess H. Newtons alss der den namen hatt eines ziemlichen morosen humors zu seyn, worauss

Johann, in fact he who had upon his arrival introduced Niklaus to Newton, afterwards twice dining with them both at the latter's house,[177] and also he who had first broken to Newton's own ears the news of the Bernoullis' bombshell that there was a serious, undetected error of reasoning in the *editio secunda* of the *Principia*, in the sheets of its second book which had already been printed off in Cambridge (though Niklaus could hardly have known this when he met de Moivre). 'Your nephew', he wrote a fortnight or so after the sequence of events he described,

having told me that he had an objection against an argument in Mr Newton's book regarding the motion of a body which describes a circle in a resisting medium, and having communicated this objection to me, I at once put it to Newton on his behalf. Mr Newton said that he would examine it, and two or three days later, I passing by his house, he told me that the objection was good and that he had corrected the argument; indeed, he showed me his correction, and it proved [*sc.* in its amended result] to conform with your nephew's calculation. He thereupon added that it was his intention to see your nephew to thank him, and begged me to take him to where he was: which I did. For the rest, Mr Newton assures me that this error proceeds simply from having considered a tangent the wrong way, but that the basis of his calculation and the series of which he made use stand rightly as they are.[178]

dan zu sehen, dass von dem *Studiū Mathematicū* in wahrheit kan gesagt werden, was sonsten von den guten künsten ins gemein gesagt wird, dass nämlich dasselbige Studium *emollit mores nec sinit esse feros*'.

(177) De Moivre to Johann Bernoulli, 18 October 1712: 'Ç'a été avec une très agréable surprise que j'ai vû M. votre neveu en ce pays-ci.... J'ai eu l'honneur de l'introduire chez M. Newton et chez M. Halley.... Nous avons vû M. Newton trois fois ensemble, et il nous a fait l'amitié de nous inviter deux fois à diner chez lui' (Wollenschläger, 'Briefwechsel zwischen Johann I Bernoulli und de Moivre' (note (44) above): 270–1).

(178) *Ibid.*: 271: 'je ne dois pas omettre une chose remarquable; M. votre neveu m'ayant dit qu'il avoit une objection contre une conclusion du livre de M. Newton au sujet du mouvement d'un corps qui décrit un cercle dans un milieu resistant, et m'ayant communiqué cette objection, je la fis voir aussitôt à M. Newton de sa parte: M. Newton dit qu'il l'examineroit, et deux ou trois jours après, ayant passé chez lui, il me dit que l'objection étoit bonne, et qu'il avoit corrigé la conclusion; en effet il me montra sa correction, et elle se trouva conforme au calcul de M. votre neveu; là-dessus il ajouta qu'il avoit dessein de voir M. votre neveu pour l'en remercier, et me pria de le mener chez lui, ce que je fis. Au reste M. Newton assure que cette erreur procede simplement d'avoir considéré une tangente à rebours, mais que le fondement de son calcul et les suites dont il s'est servi doivent subsister'—to which de Moivre added that Niklaus 'm'a fait la grace de me donner votre méthode générale pour déterminer la raison de la resistance à la force centripete dans un milieu resistant: ...je trouve votre calcul si simple et si net que j'en suis enchanté'. Doubtless Newton, too, was shown Johann Bernoulli's Leibnizian mode of solution by his nephew—and if he was not, de Moivre would at least have been quick to inform him of its existence and general nature—but whether either saw at this time the text of Niklaus' own 'Addition' to his uncle's paper is not clear. Had Newton glimpsed the nephew's tentative explanation of the error in his *Principia* proposition, however, we might suppose that he would have reacted more angrily and explicitly to its insinuation that he had confused the coefficients in his Taylorian expansions with the corresponding higher orders of 'différentielles'.

A 'simple' mistake. . . . The seasoned historian learns not to believe in such handy, made-to-measure explanations. And to be sure the extensive surviving corpus of Newton's preparatory checkings and more elaborate redraftings of Proposition X of Book 2 of his *Principia*[179] discloses, when ordered in proper sequence, a far more arduous and complicated transit to eventual enlightenment than this meagre admission by Newton suggests. That he in fact succeeded in gaining a correct resolution of the problem of resisted motion through an alternative argument considering the change in speed in equal times over two successive infinitesimal arcs traversed the same way by the impeded projectile—and not by repair of his initial scheme equivalently invoking the backwards and forwards motion of the missile in equal vanishingly small times from some one common point in its trajectory—was, it is now clear, a contingent circumstance born out of mere practical expediency, and in no way the dictate of mathematical necessity. And if Newton did not realize this when he made his first laconic report on his correction of the mistake in his 1687 *editio princeps*, he must have come strongly to suspect as much soon afterwards when he made further tests of the validity of his original mode of approach.[180]

In briefly outlining what these hitherto unprinted private papers of his reveal of Newton's struggle to success in mastering the mathematical and dynamical subtleties of his problem of generally resisted motion, let us steer clear of technical detail. On failing in his first efforts to mend the structure of his initial argument comparing prior and subsequent states of the projectile's motion around a centre point in its impeded onrush,[181] Newton quickly turned for better help to the alternative approach of conceiving the missile analogously to traverse successive infinitesimal arcs of its flight-path in one onward-going motion. And having in this way, after a further false step or two,[182] at length succeeded in his aim, he at once dashed off straightaway below his rough,

(179) Here first published from the autograph originals in section 6 below. For a detailed listing, see the introductory Analytic Table of Contents.

(180) See 6, Appendix 3; and compare note (175) above. Newton's original mistake (we refer to his 1687 figure) had been to suppose that the deviations *fg* and *FG* at either end of the projectile arclet $\widehat{gCG}$ are, because these are generated in equal times in fall under constant gravity away from the prime tangential path *fCF* of rectilinear resisted motion, to be taken (to sufficient approximation at least) as equal in length; whereas, on account of the continuously changing direction of motion over the arc $\widehat{gCG}$, these deviations in fact differ measurably in their magnitude at a third-order infinitesimal level. To compute the ratio of *fg* to *FG* we need in fact only to make the small change in the structure of the initial mode of argument which is outlined in 6, Appendix 3: note (22).

(181) See 6, Appendix 2.1.

(182) In the preliminary tentatives reproduced in 6, (Appendix 2.2/3→) §1.1–5. Newton's stumbling block throughout was his continued failure (compare *ibid.*: notes (9), (16), (34), (46) and (58)) adequately to evaluate the change in motion over two successive arclets of the missile path, rather than any deficiency in his mathematical reasoning therefrom.

victorious computation[183] the hasty draft of a note to Niklaus Bernoulli covering his enclosed 'solution of ye Problem about the density of resisting Mediums set right' and desiring him to 'shew it to your Unkle & return my thanks to him for sending me notice of ye mistake'.[184] That should have been the end of the matter, but it was far from so to be. Newton, it is true, duly shaped and honed his first crude solution into a form[185] fit to put to the critical inspection of the Bernoullis, nephew and uncle alike, but there is every reason [186] to doubt that he did pass this on to Niklaus in London; and, whatever, Johann away in Basel certainly never saw any version of its text. Instead, Newton went swiftly on, first to fashion[187] a variation of its argument where the continuously curving infinitesimal arcs of the projectile curve are approximated by the linked polygonal chain of their yet more minute subtending chords (each of a second order of the infinitesimal), at the single vertices of which the forces of resistance and gravity act in discrete impulses 'once and instantaneously' to change the speed and course of the missile; and then[188] still more radically to reframe his reasoning by now analysing the change in motion of the projectile over successive portions of its curving path the infinitesimal horizontal advances along which are equal. The phrases of this last dextrous attaining of a correct measure of the ratio of the medium's resistance to the force of downwards gravity were to be retained word for word by Newton in the revised text of Proposition X which three months later Cotes received from him early in January 1713 and was forthwith mortised (with some slight paring away of its scholium to fit snugly the space available) into the awaiting printed pages of the *Principia*'s second edition.[189] For an after-

---

(183) That in 6, §1.6 below.

(184) ULC. Add. 3965.12: 219^v; see 6, §1.6: note (76) below, and compare *Correspondence of Newton*, **5**, 1975: 348–50.

(185) See 6, (Appendix 2.4→) §2.1.

(186) Compare 6, §2: note (2) below.

(187) See 6, (Appendix 2.5–7→) §2.2.

(188) See 6, (Appendix 2.8→) §2.3.

(189) On whose pages 232–44 it was published the next summer. In acknowledging his receipt of 'Your alteration of Prop. X, Lib. II' Cotes' reaction on 13 January was to be 'well satisfied with it', but he betrayed only a superficial understanding of what was at issue in his attendant observation that 'You have increased the [ratio of] Resistance [to gravity] in the proportion of 3 to 2, which is the only change in Your Conclusions, arising from hence (as I apprehend it) that in the new Figure LH is to NI as Roo to $Roo + 3So^3$, whereas in y^e former Figure kl was to FG as Roo to $Roo + 2So^3$' (Edleston, *Correspondence of Newton and Cotes*: 147 [=*Correspondence of Newton*, **5**: 369–70]); on this see 6, §2.3: note (32) below. Cotes was in fact (compare 6, §2.3: note (48) following) able to find room for all but a short paragraph in the refashioned text where Newton, in the scholium to the main proposition, sketchily outlined the alternative mode of approach whereby the forces of resistance and gravity are conceived to act in a succession of interspersed, discrete impulses each *simul et semel*. The ensuing retailoring of the sheets of the *editio secunda* already printed off necessitated the printing of new pages 231–40 (that is, leaves Gg4/Hh1–4); and the replacement for folio Gg4 is pasted in on the

thought where he had it in mind to specify how the Taylorian expansion of the incremented ordinate yields the standard formulas for the curvature of a Cartesian curve[190] there was no room at all.

However easy we may find it to excuse the discourtesy through the pressure of his other interests at this time,[191] in terms solely of *Realpolitik* Newton's omission in the autumn of 1712 to convince Johann Bernoulli as to where the error in his original proposition truly lay was a short-sightedness which was to spoil not only their own future personal relationship but also, in its wider blindnesses, such *entente cordiale* as there might have been between the commonalty of British and Continental mathematicians. In that same autumn, as it chanced, Bernoulli was himself deeply occupied in elaborating his variant Leibnizian resolution of the problem of resisted motion in a 'short script' destined for the Leipzig *Acts*, this he rushed into print when his nephew reported that Newton had, immediately he was told of the common numerical factor by which its illustrative examples of circular, parabolic and hyperbolic missile paths were defective, made haste accordingly to correct his preceding general measure of the ratio of resistance to gravity whence those particular instances depend, and to set the whole proposition right in the new edition of the *Principia* about to issue from the press.[192] In the first part of this article on motion under gravity through a

---

stub of the superseded leaf (cut out after binding of the book) in all the several dozen copies of the edition which we have ourselves seen. A minimal alteration in the final paragraph of the scholium (on *Principia*, $_2$1713: 244 (=signature Ii2^v]) was economically left by Cotes to be noticed among other late *corrigenda* to the volume listed on its last page (on which see 6, §2.3: note (52) below).

(190) The three surviving drafts of this intended *addendum* to 'pag. 240' are reproduced in 6, Appendix 4.

(191) In late September 1712 Newton was deeply occupied, above all else, in composing the preface and the numerous accompanying footnotes to his forthcoming selected edition, under the Royal Society's *imprimatur* (and without mention of his rôle as 'onlie begetter' of it), of the *Commercium Epistolicum D. Johannis Collins et Aliorum de Analysi promota* whereon he founded his public claim to be first 'inventor' of the calculus, and in so doing exhibited the deficiences —as he saw them—in Leibniz' counter-claim to priority in its discovery. Many of his surviving autograph draftings of this 'Ad Lectorem' and the ensuing text-notes are indeed, as we have repeated occasion there to mention (see 6, §1: note (76); §2: notes (23), (45) and (51); Appendix 2: notes (16), (28), (43), (47), (77); and Appendix 3: note (20)), found interspersed in his worksheets with his contemporary calculations for, and preparatory revisions of, Proposition X of the *Principia*'s second book. He was also at this time engaged in compiling the added paragraphs on comets which he sent to Cotes on 21 October 1712 (see *Correspondence of Newton*, 5: 350) and which the latter duly passed into print on pages 461–3 of the *editio secunda*.

(192) So we gather from his tart response to his nephew Niklaus on 23 November 1712 (N.S.) when told by the latter that he had informed Newton of the error lurking in the argument of his *Principia* proposition: 'Ich hätte wohl wünschen mögen, darauff Er ein Correction in seiner newen edition gemacht und euch davon eine Copie zugestellet, dan ich forchte es seye just eine von meinen remarques als zum exempel betreffend die resistenz, so Er unrecht in dem *circulo* und anderen *curvis* determiniret, darüber ich etwas zu schreyben under händen habe

resisting medium, in the amended form in which it appeared the next February, he had no compunction in now heavily supporting his nephew's criticism—here made public for the first time—that Newton's 1687 argument was in its basics accurate, if 'somewhat contorted and difficult to understand', and that its lapse was to be sought 'in the manner of its application' whereby the coefficients of successive terms in the series expansion of the projectile curve's incremented ordinate were (so Bernoulli maintained) erroneously equated with its related derivatives.[193] What was to have much more damaging consequences, however, was when four months later he privately characterized this

---

umb in den *actis Lips.* zu publizieren welches aber zimlich schlecht stehen würde wann mich der H. Newton mit seiner correction prævenierte' (from the unpublished original letter, now Basel University Library. MS L Ia 22.9; we have quoted a preceding passage from this in note (176) above). The same day he wrote a deal more circumspectly to de Moivre in London to say that he was 'impatient de voir la nouvelle impression des *Princ. math. phil. natur.* de M. Newton, que mon neveu me marque qu'elle sera achevé dans ce present [!] mois; comme je suis sur le point d'achever un petit ecrit contenant quelquesunes de mes remarques que je fis autrefois sur la vieille edition, et que je publierai peut-estre cet ecrit dans les *Actes* de Leipsic, je serois bien aise de voir si j'aurois eu le bonheur de me rencontrer avec M. Newton dans les nouvelles additions et corrections que l'on trouvera, à ce qu'on me dit, dans cette seconde edition. Mon neveu me mande qu'il a eu l'honneur de faire à M. Newton une petite[!] objection, laquelle lui doit avoir donné lieu d'ajoûter quelque correction à sa nouvelle edition de ses *Principes*; je ne sçais si c'est justement quelqu'une de mes remarques que mon neveu a vû ici; quoiqu'il en soit, j'espere qu'il ne sera pas fasché que le public en profite: c'est toujours avec beaucoup de veneration que je parle de ce grand homme et de son ouvrage incomparable' (Wollenschläger, 'Briefwechsel': 276).

(193) 'De motu Corporum Gravium, Pendulorum, & Projectilium in mediis non resistentibus & resistentibus supposita Gravitate uniformi & non uniformi...Demonstrationes Geometricæ; Autore Joh. Bernoulli' (*Acta Eruditorum* (February/March 1713): 77–95/115–32): 93–5: Theorema V, Scholium: 'Error iste qui tanto Viro excidit, non quidem in ipsa ejus solutione latet, quam justam esse & ab omni paralogismi vitio immunem deprehendi quanquam non parum detortam & intellectu difficilem; sed quærendus ille est in ipso applicandi modo, qui in eo laborat, quod pag. 263 in serie quæ exprimit *DG* terminum quemlibet sumat pro aliqua indeterminatæ *DG* differentiali seu[!] ut ipse vocat fluxione tanti gradus quantæ dimensionis existit littera *o* in ipso termino, quod in primo & secundo verum esse potest, in reliquis vero minime.... Non satis capio qua rationis specie inductus fuerit Vir sagacissimus, ut crederet terminos serierum harum...eosdem esse cum terminis per differentiationis continuationem collectis.... Vacillant certe omnia quæ in sequentibus ex illa regula deducit pro determinanda ratione resistentiæ ad gravitatem [quam] ubique justo minorem facit...in ratione sesqui-altera...'. A few days before the first instalment of his article came out, Bernoulli likewise privately insisted in a letter to de Moivre on 18 February (N.S.) that 'La méprise de M. Newton dont mon neveu vous a communiqué ma remarque faite depuis longues années sur la raison de la resistance à la force centripete, semble tirer son origine du...principe...des tangentes mal conciliées; car je trouve que la regle...qu'il donne en general est bonne, mais que sa maniere de l'appliquer aux exemples n'est pas legitime, si bien que ce n'est pas seulement à la page 265 où il s'agit du cercle [compare note (173) above], mais aussi pag. 268, 269, 274 [*viz.* where Newton works his Examples 2–4 of the conic parabola, conic hyperbola and generalized hyperbola] qu'il determine mal la raison de la resistance à la pesanteur en la faisant par tout $\frac{3}{2}$ fois trop petit' (Wollenschläger, 'Briefwechsel': 282).

'mistake' of Newton's to Leibniz as being one piece of 'evidence by which we may allowably infer that the calculus of fluxions was not born before the differential calculus'.[194] Just seven weeks afterwards, without staying to gain Bernoulli's prior permission, Leibniz incorporated this confidence in his anonymous Latin fly-sheet, the infamous *Charta volans*, as the 'recent observation' of an unnamed 'certain eminent mathematician',[195] and the simmering disquiet with each other between Newton's adherents and the supporters of Leibniz' priority and supremacy broke out into an open squabbling, running emotional fight: one in which, by and large, neither side was prepared to allow the other any concession, whether past documented fact and rational argument therefrom demanded it or no. In vain did Newton urge through his mouth-piece John Keill that the true seat of error in the original argument of Proposition X of his second book was 'in a slip caused by accidentally drawing the tangent to the wrong side of the point of contact' in his resisted flight-path, and excuse himself that 'it was very easy to commit a mistake of this nature, especially in a book written in so short a time as the Book of *Principles* was'.[196] Bernoulli, in

(194) See his letter to Leibniz on 7 June 1713 (N.S.): 'Alterum indicium, quo conjicere licet, Calculum fluxionum non fuisse natum ante Calculum differentialem, hoc est, quod veram rationem fluxiones fluxionum capiendi...per gradus ulteriores Newtonus nondum cognitam habuerit, ...patet ex ipsis *Princ. Phil. Nat.* pag. 263, ubi pro differentiis...primo, secundo, tertio, quarto &c alicujus potestatis ex gr. x^n, judicat ponendos esse secundum, tertium, quartum &c terminos [Bernoulli means 'coefficientes potestatum incrementi o'] ipsius $\overline{x+o}^n$ in seriem expansæ' (*Commercium*, **2**: 309–10 [= *Math. Schriften*, **3**: 911]). In support thereafter of his conjectured inference Johann went on, borrowing from his nephew's 'Addition' two years before (see note (175)) to his own 1711 'Lettre' on resisted motion, to cite Newtons' maladroit phrases in the terminal scholium of his 'De Quadratura Curvarum' where (again see 2, §3.2: note (46) below) he had indeed seemed to equate the coefficients of the powers of the increment o in the series expansion of the n-th power of $x+o$ with the corresponding fluxional derivatives (with respect to x) of the power x^n.

(195) 'quemadmodum ab eminente quodam Mathematico dudum notatum est' in Leibniz' exact words. The full text of the *Charta volans* is now conveniently reproduced in *The Correspondence of Isaac Newton*, **6**, 1976: 15–17, together with brisk, up-to-date editorial commentary upon its detail and information regarding the several rapid reissues of its text in Latin original and in German and French translations, the last (in *Journal Literaire*, **2**.2 (November/December 1713): 448–53) introduced (*ibid.*: 443–8) by a set of 'Remarques sur le différent entre M. de *Leibnitz*, et M. *Newton*'—from Leibniz' own pen, it came quickly to be recognized—which formed something less than the 'raport véritable de ce qui s'est passé' which they claimed to present.

(196) So Keill countered Bernoulli the next year in his 'Réponse' in the *Journal Literaire* (**4**, 1714: 319–58, especially 343–7) 'aux Auteurs des Remarques sur le Différent entre M. de *Leibnitz* & M. *Newton*' (see the previous note), there steadfastly insisting (p. 345) that 'On avoit dit à son Neveu, dans le tems qu'il étoit à Londres, que l'erreur ne consistoit pas dans la methode, mais dans une méprise causée par accident, en tirant la Tangente vers le faux côté du point d'attouchement' and going on (p. 347) to declare yet more firmly still: 'Le Canon que donne M. *Newton*, dans la seconde Edition de ses *Principes*, pour déterminer la proportion de la Resistance à la force de la Gravité...est déduite de la Methode des Suites Convergentes,

particular, was never to his death prepared seriously to consider the possibility that Newton might indeed, when he composed his *Principia* proposition in the middle 1680's, already have acquired—as of course he long before had done!—an exact and sensitive mastery of higher-order derivatives and their geometrical representation. And he resolutely persisted, with a hostility increasing in proportion with his inability to persuade, in stubbornly maintaining his belief in his nephew's invented explanation of Newton's blunder.

Let us not give undue prominence to the arid charges and counter-charges which passed between Bernoulli and Keill during the last years of the latter's life, but despatch even their brief summary to the forgettable greyness of a footnote.[197] Yet let us also regret that Bernoulli turned his blind eye to what it

---

&...les Conclusions qu'on en tire sont exactement conformes à celles de M. *Bernouilli*...'. These points he urged no less vigorously in his further elaborate 'Défense du Chevalier *Newton*' in the *Journal Literaire* (**8**, 1716: 418–33; see especially 428–33) against the joint 'attaques' upon the *Principia* by Johann Bernoulli and his nephew in the Paris *Memoires* for 1710 and 1711—or rather Newton did, since much of the 'Défense' stemmed originally from his own pen. (So, for instance, Newton's English original of the paragraph on Keill's printed pages 427–8 was drafted by him on the verso of Keill's letter to him of 28 October 1716, now ULC. Add. 3985.16.) Newton subsequently had it in mind, indeed, to make a similar riposte to Bernoulli's aspersions in a set of 'Observations' in his own right upon the 1713 *Charta volans*; the hitherto unpublished text of these is reproduced in **2**, Appendix 9 below.

(197) In the closing page of his 1716 'Défense du Chevalier *Newton*' (on which see the previous note) Keill had, rephrasing a point made to him by Newton himself in a letter on 11 May 1714 (Edleston, *Correspondence*: 174–5 [=*Correspondence of Newton*, **6**: 128–9]; compare the unpublished draft original which survives on ULC. Add. 3968.4: 14^r/14^v), developed an argument invoking 'la Différence de la Différence des Ordonnées' to conclude it to be 'certain que [dans la 10. Proposition du 2. Livre] la seconde Différence [de l'Ordonnée *DG*] est égale à la somme des quantitez *FG* & *kl*, ce que Monsieur *Newton* a affirmé, & non égale à *FG* seule, comme le veulent Messrs. *Bernoully*', whence 'M. *Newton* ne s'est point trompé sur les secondes différences'. To which Johann Bernoulli, writing under the cloak of an anonymous *alter ego* which Christian Wolf's careless—or was it intentional?—editorial retention of a 'meam' at one place laid open for everyone with eyes to see, at once riposted in an open 'Epistola pro eminente Mathematico, Dn. *Johanne Bernoullio*, contra quendam ex Anglia antagonistam scripta' published in *Acta Eruditorum* (July 1716): 290–315, there (pp. 301–6) renewing Niklaus' conjectured inference as to the 'true' error in Newton's original *Principia* proposition and now challenging his Scots 'antagonist' to prove '*Newtonum* jam tum temporis, cum *Principia* sua scriberet, scivisse vel animadvertisse, quod *FG*+*kl* sit secunda differentia ipsius *DG*, siquidem ex hactenus dictis satis superque[!] constet, illum sumsisse *FG*+*kl* pro duplo differentiæ secundæ' (*ibid.*: 306–7). And this challenge was reiterated the next year in a separate 'Responsio' to Keill's 'Défense' in *Acta Eruditorum* (October 1718): 454–66 by Bernoulli's well-tutored spokesman Johann Heinrich Kruse, who further required Keill that 'ostendat..., si potest, ... *Newtonum FG*−*kl* pro tertia differentiæ parte sumsisse' (*ibid.*: 465). Keill made a short preliminary response to Bernoulli's 1716 'Epistola' in an answering 'Lettre' in *Journal Literaire*, **10**.2, 1719: 261–87, where *inter alia* he stressed to Johann that, while 'votre Parent [*Nicolas*], pendant son séjour en Angleterre, a fait voir à Monsieur Newton, que son erreur venoit de sa méprise touchant les *Termes de la Series*', Newton himself 'lui assura alors qu'il n'y avoit point de méprise en cela' (see *ibid.*: 272–3). The following year, on pages 16–17

suited neither his purposes nor his preference publicly to recognize. Any unprejudiced reading by him of the text of the revised Proposition X, as published in the *Principia*'s second edition in the summer of 1713, should have led him at once to see that Newton's adjusted general measure of the ratio of resistance to gravity yields, when properly applied (as it there was) in the four ensuing worked examples of projectile curve, results in these particular instances whose truth his own parallel Leibnizian computations confirmed; and hence that his nephew's ill-founded assignation of the error in the proposition's deficient initial version was not merely inconsistent in itself, but also impossible in any absolute sense as well.[198] Certainly, after Pierre Remond de Montmort's communication to him in late January 1718 of (what he chose to regard as) a private challenge from Keill to solve the problem of determining the path of a projectile through a medium resisting its passage according as the square of its instantaneous speed 'in the most simple supposition of gravity and the density of the medium being uniform', when he was inspired himself explicitly to obtain the exact Leibnizian equal of Newton's revised *Principia* measure[199] he ought in all fairness publicly

---

of his more carefully pondered Latin *Epistola ad Virum Clarissimum Joannem Bernoulli* (London, 1720), Keill—having made an apt citation of Psalm XXXIV, 13 on his title-page which he might himself better have obeyed in some of his own more bitter taunts—heavily drove the nail home, once more emphasizing '*Newtonum*, in [Exempl. 1] Prop. X, Lib. II. *Princip.* posuisse $\frac{n^2oo}{2e^3}$ pro quantitate *FG*, quæ est portio ordinatæ inter Curvam & Tangentem:...

Evidenter igitur patet te quantitatem *FG*, non autem *FG + kl* pro differentia secunda agnovisse'

and now likewise insisting '*Newtonum* posuisse terminum Seriei $\frac{an^2o^3}{2e^5}$ pro quantitate *FG − kl*,

Tu tamen expressis verbis affirmas illum eundem terminum pro differentia tertia posuisse, [unde] manifestò...patet te quantitatem *FG − kl* tanquam Differentiam tertiam æstimasse....Hic Error in [Newtoni] *Quadraturis* circa differentias nihil tibi prodierit'; and in overkill—he had gleefully announced his discovery of this passage in Bernoulli's printed writings where 'I find he has all along been mistaken about 2ᵈ Differences' in a letter to Newton on 24 June 1719 (see *Correspondence*, 7, 1977: 48)—he went on to exhibit a parallel instance in Bernoulli's 'Solution du Probleme sur les Isopérimétres' in the 1706 Paris *Memoires* (see his *Opera*, 1 (Lausanne/Geneva, 1742): 427–8) where the latter had indeed, in assigning his preferred Leibnizian equivalents to infinitesimal linelets, analogously erred 'bis in una linea circa differentias secundas'. Only the timely death of his 'antagonist' on 31 August 1721 saved the world at large from enduring the biting counterblast (the manuscript of this 'Keillius Heauton Timoroumenos' survives in Basel University Library) which Bernoulli was goaded to gnash out in further retort. Thereafter he wisely chose to let the altercation cool into a restful oblivion.

(198) Again see note (175) above.

(199) See 6, §2.3: note (50) below, where we cite Bernoulli's equivalent Leibnizian measure (involving first-, second- and third-order derivatives) in the precise terms in which he phrased it the next year in *Acta Eruditorum* (May 1719): 225. As we there show, the resolution into integrable form of the general problem of resisted motion under gravity which he went on to deduce (*ibid.*: 225–6) follows entirely straightforwardly from the basic reduced equation which Newton set out and lightly illustrated in the two newly added paragraphs with which he

to have retracted his previous advocacy of his nephew's thoroughly mistaken conjecture as to where the error in the original argument of Proposition X lay; whereas he continued to talk of the 'hallucination' which he had 'demonstrated' in its reasoning, and relentlessly to mock Keill's 'crassly contradictory' efforts to show him his own mistake. And in that disparagement most non-British mathematicians over the next decades concurred. At the very end of the eighteenth century, however, in an illustration of his new-found general theory of analytical functions wherein derivatives are defined in terms of the corresponding coefficients of such Taylorian expansions into series as Newton had invoked in his *Principia* proposition more than a hundred years before, Lagrange magnificently redressed the balance in what ought at once to have been universally accepted as a near-definitive examination of Newton's mode of resolution of the problem of motion resisted as some power (or combination of powers) of the projectile's instantaneous velocity.[200] (Let us readily admit that in our own exegesis below of the considerably more tangled skein of development revealed in the Newton's private papers we have been continually indebted to its guiding lead, even if we have here and there dared lightly to improve upon the niceties of its detailed explanation.) Those nineteenth-century scholars who further clarified this important chapter in the early history of theoretical ballistics well knew the value of Lagrange's expert testimony on the matter.[201]

opened the scholium to Book 2, Proposition X in the *Principia*'s second edition ($_2$1713: 240). Whether Bernoulli himself made the connection or no, we can find no excuse for his accompanying jeer (*Acta*: 219) that his 'problema [est] nihil aliud quam inversum Prop. X. *Phil. Nat. Princip.* p. 232 Edit. poster. ubi Newtonum hallucinatum demonstraveram', if we may pardon (in context) his ensuing insult that 'Insanientis certe est, crasse adeo sibi contradicere, ut [Keillius] 5 vel 6 lineis unum idemque modo nullius, modo magni æstimet momenti' (*ibid.*)

(200) See his *Théorie des Fonctions Analytiques, contenant les Principes du Calcul différentiel, dégagés de toute Considération d'Infiniment Petits ou d'Évanouissans, de Limites ou de Fluxions, et réduits à l'Analyse Algébrique des Quantités Finies* (Paris, Prairial an V [=May–June 1797]): Seconde Partie, 'Application de la Théorie...à la Mécanique', §§202–5: 244–51; the gist of Lagrange's elegant detection there of 'la véritable source de la méprise où Newton est tombé dans la première édition des *Principes*' is given in 6, Appendix 1: note (6) below. He was soon afterwards, in one of history's curious vagaries, taken to task by the not uncapable mathematician Jean Trembley (in his 'Observations sur une discussion relative à la Théorie de la résistance des milieux', *Mémoires de l'Académie Royale des Sciences et Belles-Lettres [de Berlin], Année MDCCXCVIII* (Berlin, 1801): 60–75), not for discounting the belief of the Bernoullis and—in Lagrange's words—'tous ceux qui en ont parlé depuis' that Newton's mistake lay in wrongly interpreting the coefficients of his series expansions, but on the contrary for supposing that there was an error of any kind at all in the original argument of Proposition X of Book 2 as it was set out in the *Principia*'s first edition! Lagrange, gentleman that he was, courteously but firmly set Trembley right in §17–24 of the augmented Chapitre IV 'De la question où il s'agit de trouver la résistance que le milieu doit opposer pour que le projectile décrive une courbe donnée....' in the *Seconde partie* of his revised *Théorie* (Paris, $_2$1813) [=(ed. J. A. Serret) *Œuvres de J. L. Lagrange*, 9 (Paris, 1881): 360–76].

(201) We mean above all that most splendidly careful and knowledgeable of Newton's

A pity that more recent students of the history of dynamics have joined with biographers of the Bernoullis, uncle and nephew, in returning to accept without gainsay or criticism Johann Bernoulli's partial and distorted estimate of Newton's contribution *vis-à-vis* his own.[202] May the situation not endure.

This *tour de force*, whose niceties no one in his day— not even he himself, one is almost tempted to say!—properly appreciated, was to be the last occasion on

---

editors, Joseph Edleston, who in an aside to Newton's letter to Keill on 20 April 1714 where Bernoulli is accounted to be 'mightily mistaken when he thinks that I there [in y^e 10th Proposition of the second book of the *Principia*] make use of the method of fluxions [when] tis only a branch of y^e method of converging series...' (*Correspondence*: 171 [=*Correspondence of Newton*, 6: 108]) rightly commented in mild criticism of Lagrange's analysis (see the previous note) that 'He has not, however, pointed out in what respect Newton's [first] geometrical expression is erroneous, or at what step of the demonstration the fallacy is introduced. The error consists in substituting FG (which $= Ro^2 + So^3...$) for fg (which $= Ro^2 + 2So^3...$)...' (*ibid.*: note *). (It is but a slip of the pen, let us remark, when A. R. Hall in a gloss upon Edleston's observation in his 'Correcting the *Principia*' (*Osiris*, **13**, 1958: 291–326): 320 states that Newton 'had, from page 262 onwards, taken $FG (= Ro^2 + So^3)$ in place of $fg (= Ro^2 - So^3)$'.) Some thirty years later J. C. Adams made what is essentially Edleston's point in a rough jotting on a slip of paper inserted between pages 260 and 261 in his personal copy (now ULC. Adams.b.68.3) of the first edition of the *Principia*, there affirming that 'This proof [of Proposition X] is defective in consequence of Newton having omitted to take into account the difference between FG and fg due to the difference between the direction, in which the resistance acts at different parts of the arc GCg'. On this point see our own running commentary to Newton's 1687 text in 6, Appendix 1: notes (6) and (8) following.

(202) Thus, for instance, J. O. Fleckenstein in his 'Kurze Biographie' of *Johann und Jakob Bernoulli* (*Elemente der Mathematik*, Beiheft **6**, 1949) writes (p. 22) that 'Das von Newtons getreuem "Kettenhund", dem schottischen Kämpen Keill, gestellte Problem der ballistischen Kurve löste Johann für das allgemeine Widerstandsgesetz $R = av^n$ (*A.E.* 1719), nachdem Newton in den *Principia* es nur für den einfachsten Fall $n = 1$ bewältigen konnte'; and in his more recent article on 'BERNOULLI, Nikolaus I' in Scribner's *Dictionary of Scientific Biography*, **2** (New York, 1970): 56–7 blithely continues to affirm the gospel that 'It was he who pointed out Newton's misunderstanding of the higher-order derivatives, which had caused Newton's errors with the inverse problem of central[!] force in a resisting medium'. In his much more detailed and probing survey of *Ballistics in the Seventeenth Century* (Cambridge, 1952) A. R. Hall states in conclusion (pp. 156–7): 'The subsequent history of mathematical ballistics relates the many attempts to integrate approximately the fundamental differential equation given by Johann Bernoulli.... To rewrite this solution in usable terms...was the work of a period in which the art of calculation had advanced far beyond the rudimentary practices of the age of Newton'. It is undeniable historical fact that Bernoulli did impose his impress upon his student Leonhard Euler and those others who during the following century sought to give explicit solutions, exact (where that might be) or approximate, to his Leibnizian differential equation of resisted projectile motion; but we see no reason why of necessity a parallel stream of fluxionary investigation could not (as we sketch in 6, §2.3: note (50) below) have issued from Newton's equivalent equation in the scholium to Proposition X of Book 2 in the *Principia*'s second edition, nor can we in the tangle of the jerking onward advance—and sometimes regress—in techniques of computation and resolution trace any neat dichotomy such as Hall posits between the particular geometrical methods of Newton's seventeenth century and the generalized analyses slowly developed on their basis during that which followed.

which Newton came creatively to grips with a mathematical topic of any real difficulty. With the background to the publication by him in 1713 of the *Commercium Epistolicum*, under the Royal Society's *imprimatur* but in fact wholly authored by Newton,[203] we refuse here to be concerned, though in the second part of this volume we cannibalize the mass of his still mostly unpublished papers regarding the calculus squabble to show just how widely and zealously he went in his gargantuan efforts to sustain and substantiate his claim to have been 'first inventor of the infinitesimal method'. In the endlessly tortuous and squabbling *mêlée* which ensued, one matter alone has intrinsic right to be included in an edition of Newton's mathematical papers. Early in November 1715, with blurred memory of their exchange of letters on the question twenty years before,[204] Leibniz wrote to Johann Bernoulli asking to be reminded of a problem requiring a descent to 'differentio-differentials' since it might serve, 'as one whose sources will not be immediately apparent', to be put to Keill and the other English mathematicians 'to understand what they are capable of'[205] Bernoulli speedily responded, and a month afterwards Leibniz was able to 'test the pulse of our English analysts' by including with a letter of his to the Venetian nobleman Conti (then briefly resident in London) an open challenge: 'To find a line which cuts at right angles all curves of a determined sequence of one and the same kind—for example, all hyperbolas having the same vertex and same centre—and that by a general way.'[206] As

(203) In Appendix 1.4 to Part 2 following we reproduce, from his autograph, Newton's draft of the report of the Society's committee—of which as P.R.S. he was *de facto* chairman, of course, but from which he as prime claimant in the dispute should have in equity have disassociated himself—and in the following Appendix 1. (5→)6 his English original of the *Commercium*'s Latin 'Ad Lectorem'.

(204) See 7: notes (1) and especially (16) below.

(205) Leibniz to Bernoulli, 4 November 1715 (N.S.): 'Keilius [in his 1714 'Réponse' to the *Charta volans*; see note (196) above] responsione indignus est, sed...e re erit...adjicere problemata quædam, unde intelligamus, quid ipsi [Angli] possint. Cum illam novam differentiandi rationem considerasses..., excogitaveras inde applicationem sic satis generalem, qua, si bene memini, problemata quædam, quæ ad differentio-differentiales descendere solent, intra differentiales primi gradus coërcentur. Ego nunc non bene memini, ...itaque rogo ut si commodum est, iterum communices. Inserviet enim fortasse ad aliquod problema proponendum, cujus non statim apparebunt fontes' (*Commercium*, **2**, 1745: 358–9 [=*Math. Schriften*, **3**, 1856: 948]).

(206) We translate the French of this enclosed 'Billet...pour tâter un peu le pouls à nos Analystes Anglois' as it is printed by Pierre Desmaizeaux in his *Recueil de Diverses Pieces, Sur la Philosophie, la Religion Naturelle, l'Histoire, les Mathematiques, &c. Par Mrs. Leibniz, Clarke, Newton, & autres Autheurs célèbres*, **2** (Amsterdam, 1720): 11, namely 'Trouver une ligne...qui coupe à angles droits toutes les courbes d'une suite déterminée d'un même genre; par exemple, toutes les Hyperboles...qui ont le même Sommet & le même Centre; & cela par une voye generale'. In his response to Leibniz' plea to be reminded of this topic in their earlier correspondence Bernoulli had, 'ut petito Tuo satisfacerem', sent to him on 23 November (see *Commercium*, **2**: 361 [=*Math. Schriften*, **3**: 949]) an 'excerptum' from the lengthy postscript

soon as he received this *défi* (about the beginning of January 1716, we would suppose) Conti lost no time in circulating it, and a number of leading British mathematicians—the 'prime antagonist' Keill and his youthful Oxford colleague James Stirling (then still a student at Balliol), and also de Moivre, John Machin, Henry Pemberton and Brook Taylor among others at London—found no difficulty in determining, and each in their several elegant ways constructing, the family of logarithmic curves intersecting everywhere at right angles Leibniz' instanced series of hyperbolas of varying *latus rectum*, and this in a manner indicating how their prior reduction had been to eliminate the latter, independent parameter between the Apollonian defining 'symptom' of the hyperbolas and their equivalent Cartesian differential equation, to obtain a general derivative condition for the whole class of such hyperbolas which could then readily be recast to yield a like differential definition of the orthogonal family.[207]

When Newton himself learnt of Leibniz' challenge to his English mathematical pulse, his blood raced no less rapidly then anyone else's. Although in a cancelled sentence of the reply which he drafted for Conti in late February (this more directly concerned to rebut Leibniz' newly extended claim to absolute priority in time in discovering the 'infinitesimal algorithm' and angrily to reject

---

of his letter of 14 August 1697 (*Commercium*, **1**: 330–3 [= *Math. Schriften*, **3**: 462–5]) 'continens duas Methodos ex nova illa differentiandi secundum parametrorum variabilitatem ratione erutas', and Leibniz in fact elected now to challenge the 'English'—he ever forgot that Keill was Scottish—with but the simplest instance of Bernoulli's second 1697 problem: 'Construere curvas, sive similes sive non similes, in dato angulo sive invariabili sive data lege variabili secantem'.

(207) See 7: note (1) below for an outline of the solution in this special instance. Keill sent his elegantly constructed solution in a letter to Newton in late February 1716 (see S. P. Rigaud, *Correspondence of Scientific Men of the Seventeenth Century*, **2**: 421–4; its text is more accurately transcribed in *Correspondence of Newton*, **6**: 282–3) with the observation that the curve orthogonal to Leibniz' posited family of hyperbolas 'will be an oval Figure'. Stirling published his equivalent solution the next year on pages 15–19 of the Appendix to his little book on *Lineæ Tertii Ordinis Neutonianæ* (Oxford, 1717). Machin's was, in its initial, not entirely accurate form, put by him to William Jones in late January 1716 (see Rigaud, *Correspondence*, **1**: 268–9) and its corrected version was read out to the Royal Society on 2 February following. Pemberton sent his resolution (argued at tedious length, need we say?) to Keill on 11 February, furnishing a further supplement ten weeks afterwards on 24 April; the unpublished letters of his in which they are contained are now in ULC. Res. 1893.2. Neither de Moivre's nor Taylor's resolutions of the example instanced by Leibniz appear to have survived, but Machin reported upon the latter's construction to Halley on 2 February that 'I find it so exceedingly neat and simple, that I am quite out of love with my own' (Rigaud, *Correspondence*, **1**: 269). In reporting to Leibniz in March 1716 that 'Votre Probleme a été resolu fort aisément en peu de tems' Conti named de Moivre among 'Plusieurs Geometres à Londres & à Oxford' who had given it a 'solution...générale [qui] s'etend à toutes sortes de Courbes soit Geometriques, soit Mécaniques', adding that he had stressed to him that 'Le Probleme est un peu équivoquement proposé: ...il faudroit fixer l'idée d'une suite de Courbes; par Exemple supposer qu'elles ayent la même soûtangente pour la même Ab[s]cisse; ce qui conviendra non seulement aux Sections Coniques, mais à une infinité d'autres...' (Desmaizeaux, *Recueil*, **2**: 14–15).

the latter's accompanying insinuations against his 'somewhat strange [notions in natural] philosophy') he disingenuously excused himself from there responding to it because he had 'left off Mathematicks 20 years ago' and went on more reasonably to add that 'I...look upon solving of Problems as a very unfit argument to decide who was the best Mathematician or invented any thing above 50 years ago',[208] he began in secret to prepare a comprehensive answer to the problem of determining orthogonals to given curves.[209] He did not, unlike so many of his compatriots, mistake the hyperbolic instance adduced by Leibniz for the whole—indeed he all but ignored this or any other particular example of Leibniz' challenge—but he found the greatest difficulty in precisely clarifying its general sense. What, we can follow him in wondering, is the exact nature of the 'sequence' of curves 'of one and the same kind' which is presumed to be given by its terms? De Moivre had thought to fix this ill-defined concept by supposing that what was meant was any family of curves sharing the same subtangent for any particular, common abscissa.[210] This first-level generalization

(208) ULC. Add. 3968.38: 564ʳ, printed in *Correspondence of Newton*, 6: 285.

(209) The several successive English and Latin drafts of this would-be 'Solutio Generalis' of Leibniz' Bernoullian challenge are set out in section 7 below. Let us try to kill off a now deeply entrenched confusion of Newton's response to this problem of constructing orthogonals with the answer which he gave to the twin questions of Bernoulli's *Programma* in late January 1697. This, like many another myth of its kind, has its origin with David Brewster, who in his *Memoirs of the Life, Writings and Discoveries of Sir Isaac Newton*, 2: 21 mangled the phrasing of Leibniz' present challenge and intermixed with it a jumbled remembrance of Conduitt's anecdote of Newton's earlier success in identifying the brachistochrone (see page 3 above) nonchalantly to recount—we withhold the succession of editorial exclamations which we are tempted to interlard in parenthesis—that 'One of the last mathematical efforts of our author was made, with his usual success, in solving...in 1716...[the] problem... to determine the curve which should cut at right angles an infinity of curves of a given nature, but expressible by the same equation. Newton received this problem about five [*sic*] o'clock in the afternoon, as he was returning from the Mint; and though the problem was difficult, and he himself fatigued with business, he reduced it to a fluxional equation before he went to bed'. It is a revealing illustration of how nonsenses may further accrete when such shabby inventions are mindlessly parroted to cite how L. T. More refashions this tale of Brewster's to be an 'instance of Newton's [well known] ability to solve mathematical problems "at sight"' in his *Isaac Newton: a Biography* (New York, 1934), there writing (pp. 474–5) that 'his phenomenal mastery of mathematics persisted to the end of his life, and only needed a spur to call it forth.... In 1716, Leibniz...sent a problem...to the Abbé Conti "for the purpose of feeling the pulse of the English analysts". The problem...was to determine the equation of a curve which will intersect at right angles an infinite number of curves of a given nature, expressible by a given method[!]. It is credibly reported that Newton received this problem about five o'clock, while returning home from the Mint weary with the day's business, and that he solved it the same night before going to bed.'

(210) As one possibility of making precise Leibniz' notion of the 'courbes d'une suite déterminée d'un même genre', at least. See Conti's letter to Leibniz in March 1716 where (the full passage is cited in note (207) preceding) he remarks that the latter's challenge problem was 'un peu équivoquement proposé'. Eight years before, ' nuperrime in *Acta Erudit.* 1698 pag. 471 incidens, ubi [see 7: note (13) below] Cl. Joh. Bernoullius dicit, generalem se invenisse

Newton, too, readily formulated and dealt with.[211] But in other cases where the general property of the given series of curves is not so conveniently characterized he too swiftly jumped to conclude that here, also, a solution would quickly be forthcoming through his 'universal analysis conflated of the interwoven methods of fluxions and series'.[212] Try as he might to master the subtleties of defining the condition of orthogonality between two separate families of curves, the individual members of each of which are severally identified by an independently varying parameter, he was for the first time in his life at a loss to choose an effective mode of approach, and instead he was put feebly and inadequately to uttering loose generalities and vaguely stating a construction of the 'curvity' of the intersecting curve at any point which is in essence circular.[213] Whether or not he was aware of this grave deficiency in his 'General Solution' of Leibniz' Bernoullian problem, when he published this in March 1716[214] he chose not to set his name to it: a precaution which was to save him from the public disgrace of having Bernoulli vent his glee that the lion of British mathematics should have failed properly to come to grips with a problem which he himself had despatched with ease more than twenty years earlier.[215]

---

methodum secandi ordinatim positione datas sive algebraicas sive transcendentes curvas in angulo recto sive obliquo, invariabili seu data lege variabili' he wrote to Leibniz early in January 1708 (see *Leibnizens Mathematische Schriften*, 4: 323), Jakob Hermann had in 'foresight' preferred to generalize the other condition in Leibniz' stated example, illustrating his (correct) general resolution of the primary problem of constructing orthogonals with Bernoulli's 1698 instance of parent *curvæ logarithmicæ* $x = a \log (y/Y)$ all passing 'per commune punctum' $(0, Y)$ and having their subtangents $y\,dx/dy = a$ as their independent parameter, whence the family of orthogonals is $y^2 - 2x^2 = 2y^2 \log y + b$, b free, 'ut habet Dn. Bernoullius in supra citato loco' (see *ibid.*: 323–4).

(211) See 7, Appendix: note (5).

(212) We translate his Latin assertion, rejected at a later stage in his efforts to attain a general resolution of Leibniz' problem that 'solutio...generalior evadet per Newtoni Analysin universalem quæ ex methodis fluxionum & serierum intertextis conflatur' (compare 7, Appendix: notes (22) and (33)).

(213) See 7: notes (3) and (21). Even were it valid, such a reduction of the problem to determining the centre of curvature of any orthogonal curve in the vicinity of a given point of it (as the limit-intersection, namely, of tangents to 'consecutive' members of the parent family of curves at their meet with the orthogonal) would have been inefficient when, without passing to differential conditions of higher order, specification of the intrinsic defining tangential property of the family of orthogonal interesecting curves is readily made. Newton's is in no way 'the best method of attack' (as D. J. Struik puts it in his 'Outline of a History of Differential Geometry' (*Isis*, **19**, 1933: 92–120/161–91): 98), and really does no more than furnish an unwanted proof that (for simply continuous parent curves) the family of orthogonals always exists.

(214) As 'Problematis Mathematicis *Anglis* nuper propositi Solutio Generalis' in *Philosophical Transactions*, **29**, No. 347 (for January–March 1716): Art. III: 399–400. This anonymous Latin text is effectively that which Newton attained in the finished drafts reproduced in [5]/[6] of the Appendix to section 7 below.

(215) In his private correspondence with Leibniz during 1694–6, a summary of which he published in *Acta Eruditorum* (October 1698): 469–72. Compare 7: notes (1) and (16) below.

So far, in fact, was Bernoulli this time[216] to be from detecting evidence of any leonine paw behind this anonymous piece of mathematical reasoning—one which, in his words, had not attained the goal of his problem 'even by a finger-tip'—that he guessed to Leibniz that its 'witless' author was (the by him much despised) Brook Taylor.[217] And when in the early spring of 1716 Conti at Leibniz' direction passed on to the 'English analysts' a yet more exacting test of their capabilities in the form of a double problem, likewise contrived by Bernoulli, which again involved the construction of orthogonals to parent curves for which but a parametric differential condition was now supplied,[218] we can find no evidence that Newton even attempted to solve it, though he afterwards commented anonymously on Leibniz' preceding letter to Conti on 14 April (N.S.) when he published it[219] that it was 'a Problem harder then the former', particularly when it was to be solved 'without converging series'. Let us not hold the frailty and myopia of his ageing intellect against him: just a few years earlier the outcome of his wrestling with Bernoulli's problems on constructing

(216) As, we need scarcely remind, he had claimed nineteen years before it was easy to do in the case of a like unsigned 'échantillon' of the author's claw in the *Philosophical Transactions*; see note (21) preceding.

(217) Bernoulli to Leibniz, 11 November 1716 (N.S.): 'Nihil intelligo ex scripto Anglicano..., neque capio quid generalis illa ab Anonymo jactata solutio contribuat ad casum aliquem specialem solvendum. Vellem tentasset exemplum, quod a me suggestum Anglis proposueras, sed id ipsum quod Anglus anonymus (quem Taylorum esse credo) ne apice quidem digiti attigit, satis arguit quod aqua ipsi hæserit; quis enim sibi imaginabitur, Anglum illum ad provocationem Tuam non statim explorasse vires suas in solvendo isto exemplo particulari?' (*Commercium*, 2: 395–6 [= *Math. Schriften*, 3: 972]). Leibniz, who died at Hanover only three days after Bernoulli dispatched his letter from Basel, almost certainly never saw it. He himself had remarked to Bernoulli on the preceding 23 October that 'imperfecta videtur Anglorum solutio, cum recurrat ad differentias secundi gradus in re præstabili per primas' (*Commercium*, 2: 390 [= *Math. Schriften*, 3: 970–1])—on which point see note (213) above.

(218) See 7: note (25) below, where we both summarize the historical context of this generalization of the problem of constructing the curves orthogonal to cycloids of common base and vertex—and so joining points of 'synchronous' fall in these in his terminology—which Bernoulli had put to the world nineteen years before (in *Acta Eruditorum* (May 1697): 211, in annex to his celebrated optical model for identifying the parent *curvæ brachystochronæ*), and also sketch the intricacies of its Bernoullian solution by points to evince what it was that Newton had failed even to glimpse. The new 'double' problem was given a variety of resolutions on the Continent, notably by Jakob Hermann (see note (220) following) and not least by Johann Bernoulli's son, Niklaus II, who later gave a lavish overview of the whole topic in his 'Exercitatio Geometrica de Trajectoriis orthogonalibus, continens varias earum, tum inveniendarum, tum construendarum methodos, sua vel Demonstratione vel Analysi munitas; cum præmissa discussione quarundam ejusdem Problematis solutionum' (*Acta Eruditorum* (May 1720): 223–61 [= *Johannis Bernoullii Opera*, 2 (Lausanne/Geneva, 1742): 423–72]).

(219) In *Philosophical Transactions*, 30, No. 359 (January/February 1718/19): 927–8. For purity's sake, however, we quote the ensuing phrases from Newton's orthographically slightly variant draft of this unsigned annotation (which survives on ULC. Add. 3967.4: 38ᵛ).

orthogonal curves would have been vastly different and surely crowned with success. By a kind fate, let us add, it was left for Brook Taylor (he of whom Bernoulli thought so little) partially to salvage his country's badly flagging mathematical prestige in late 1717 by putting into print an accurate solution of the new double problem: if this was in its fundaments no more than an elegant variation (employing fluxions and expansions into series, in place of Leibnizian differentials) upon a pithily condensed 'Scheme' of solution given some months previously by Jakob Hermann in the *Acta Eruditorum*, it was yet both lucidly presented and cogently argued.[220] Newton seems to have been well content to stand by it.

The rest is swiftly told. From the mathematical pen of a man who at the close of 1717 began his 76th year of life we would expect no further major new insights and discoveries to have come, nor do we find any. His writings over the final decade till his death in March 1726/7 reveal, indeed, the sad truth of how increasingly fast his once near-absolute technical grasp withered away with his ageing physical frame. Let us not stoop minutely to document from his senescent jottings every fine detail of how increasingly difficult Newton found it to carry through a straightforward arithmetical or algebraic computation,[221] nor tiresomely insist on the dull and jejune quality of the totality of what of broader mathematical significance he elsewhere touched stiffly and repetitively upon. That would, to our mind, pointlessly mock the feebleness of a man who had once achieved so much that is of a rich, enduring brilliance and profundity.

One exception we make. Because of their self-analytical and methodological interest it seems appropriate here to reproduce the opening portions of hitherto

---

(220) 'Solutio Problematis à Dom^{no}. *G. G. Leibnitio* Geometris *Anglis* nuper propositi. Per *Brook Taylor* LL.D. & R.S. Secr.' (*Philosophical Transactions*, **30**, No. 354 (for October–December 1717): 695–701). Hermann had presented the nucleus of his solution, severely compressed into the 28 lines of an 'Exemp. IV', in his wider 'Schediasma de Trajectoriis datæ Seriei Curvis ad angulos rectos occurrentibus...' (*Acta Eruditorum* (August 1717): 348–52 [=*Johannis Bernoulli Opera*, **2**: 279–81]; see especially 351–2), though in a 'Supplementum' to his article the next year (*Acta* (July 1718): 335–6 [=*Opera*, **2**: 279–81]) he adjoined an 'Exemplum' where he outlined a Bernoullian construction by points of the required family of orthogonal curves whose first-order defining differential equation he had previously established. (Again see 7: note (25) for our equally pithy summary of the technical details.)

(221) In about late 1720, to cite just one instance, Newton made very heavy weather of checking—in the midst of inserting some minor amelioration (on ULC. Add. 3965.13: 374^r) to the phrasing of the 'Enarratio plenior' of the *Principia*'s 'fluxions' scholium (that to Book 2, Lemma II) whose text is reproduced in **2**, Appendix 5—that the ordinate raised at the focus of any conic of main axis $2q$ is '$r = \frac{1}{2}$ Lat rect', namely '$4a - \frac{4aa}{q}$ in Ellipsi' where $2a$ is the distance of the focus from the nearer vertex (so that the minor semi-axis, $p = \sqrt{(qr)}$, is $\sqrt{(q^2-(q-2a)^2)} = \sqrt{(4qa-4a^2)}$ and therefore $(r/q) p^2 = r^2$, *recte* by Apollonian definition) but '$r = 4a$ in Parabola' where $q \to \infty$.

unpublished drafts of an intended preface by Newton to a new edition of the *Principia* which he planned to bring out in the late 1710's.[222] Half a dozen years before, when Newton first read the anonymous review (by Leibniz, we now know) which was given of his 'De Analysi' in the *Acta Eruditorum* in 1712,[223] his attention had been caught not least by a concluding sentence which declared that 'Leibniz...has promoted the use of [continuous] transit beyond geometry on into physics itself...'.[224] Stung by its silent implication that he himself had not so applied his own notions of fluxional increase, he had in his private counter to the review asserted, as we have have already noticed, that 'by this infinitesimal Analysis' he had in fact 'found many of the Propositions in his *Principia*..., but he did not propose them as Geometrical Propositions till he had rigorously demonstrated them...by the direct method w^ch [the Ancients] called Synthesis or Composition'; and he added that 'in order [so] to demonstrate them he spent a whole section [in his first book] in demonstrating Lemmas by the Method of the first & last Ratios'.[225] The elaboration and refining of this contention that he had made wide preparatory use of fluxional analysis in originally finding the propositions of his *Principia*, but that in composing them for public view he had suppressed this preliminary mode of investigation in favour of rigorous separate proof *à l'ancienne*, now forms the principal theme of his new preface to the work.

We have elsewhere,[226] on the basis of an examination of the text of the *Principia* itself and of manuscripts ancillary to it, urged our own conclusion that on the contrary the great bulk of its propositions were derived essentially in the same manner as they are set out in print, although a very few—notably the constructions of resisted motion expounded at the beginning of the second

---

(222) See Section 8 below. In our note (2) thereto we conjecture (for what it matters) that he may have been stimulated to do so in part by reading the Varignonian recasting of the main propositions of the *Principia* which Jakob Hermann set out in the pages of his *Phoronomia, sive de Viribus et Motibus Corporum Solidorum et Fluidorum Libri Duo* (Amsterdam, 1716); be it coincidence or not, Newton was certainly provoked enough by Hermann's book to jot down on the back of one draft of his present *Principia* preface a detailed listing of what out of his own earlier work was there treated and what of it was omitted (see page 443, note (2)).

(223) *Acta Eruditorum* (February 1712): 74–7; the portion of this review of Jones' 1711 *Analysis* which relates to the 'De Analysi' (therein first published) is reprinted on II: 259–62.

(224) 'Leibnitius noster...nunc infinite parvam [exprimendi rationem], nunc motum seu continuum transitum, sive fluxum adhibuit, prout visum est commodius; & usum transitus hujus ultra Geometriam ad Physicam ipsam promovit, nova quadam consideratione inventa, quam *Legem continuitatis* vocat...' (*Acta* (1712): 77 [=II: 262]).

(225) See ULC. Add. 3968.32: 462^r; we have quoted the full text of Newton's retort in II: 272, note (34).

(226) In our 1969 Gibson Lecture on 'The Mathematical Principles underlying Newton's *Principia Mathematica*' (Glasgow, 1970; reprinted with revisions and annotations in *Journal for the History of Astronomy*, **1**, 1970: 116–38).

book[227]—are undeniably presented in a synthetic guise which inverts the original mode of their invention. Here in his intended preface to the planned reissue of his *opus maximum* around 1717–18 Newton found it equally difficult to pinpoint instances where he had indeed thirty years before gone to the pains of giving distinct synthetic demonstration of a previously analysed argument; and in default he could but loosely and vaguely claim that 'the synthetic method of fluxions and moments occurs *passim* in the following treatise'.[228] Nor, since the great majority of the *Principia*'s reasonings are pursued by arguments invoking the limit-ratios of vanishingly small infinitesimals of related *continua*, did he find it possible to point to any specimens of the use there of his joint 'analytical method' of fluxions and infinite series other than two applications of the simple binomial expansion[229] and the Taylorian developments of incremented/decremented ordinates in Proposition X of the second book, together with a proud but misguided citation of his use of 'second moments' in the ensuing

---

(227) Namely Propositions II/III, V–IX and XI–XIV, which treat the motions defined (in modern Leibnizian notation) by $dv_x/dt = -\rho(v_x)$ and $dv_y/dt = -\rho(v_y) \pm g$, where the instantaneous speeds $v_x (= dx/dt)$ and $v_y (= dy/dt)$ are the horizontal and vertical components of the missile velocity v attained in time t from some given moment of emission, g is the constant, downwards acting force of gravity, and $\rho(v)$ is the decelerating resistance to the projectile's onward rush, taken to vary as v or v^2 or some combination of these respectively. The proofs of his propositions which Newton sets down in the *Principia*—as in the sixth and seventh Problems of the parent tract 'De motu Corporum' in the simplest case where $\rho(v) = v$ (compare VI: 66–7, note (98) and 69, note (108))—merely establish, in effect by a converse differentiation from first principles, the validity of posited 'synthetic' constructions manifestly derived from a prior 'analytical' integration of the pertinent equation of motion.

(228) See the fourth paragraph of section 8.4 below. We should not, let us warn, confuse these classical notions of mathematical analysis and synthesis with their analogues which Newton claimed to find in scientific inference and confirmation of hypothesis, where, he argued in a celebrated passage in the concluding query of his *Opticks*, 'Analysis consists in making experiments & observations & in arguing by them from compositions to ingredients & from... effects to their causes & from particular causes to more general ones, till the Argument end in the most general: The Synthesis consists in assuming the cases discovered & established, as Principles; & by them explaining the Phænomena proceeding from them, & proving the explanations' (we cite his original English version, now ULC. Add. 3970.3: 286ʳ, of what appeared—in Samuel Clarke's Latin rendering—as the penultimate paragraph of the newly added terminal 'Qu. 23' in *Optice* (London, ₁1706): 347; the text in the further augmented second English edition in 1717, where this is renumbered to be Query 31, reads a little differently). For recent comment upon the subtleties of this wider methodological meaning of the true contrasted approaches, and assessment of its Newtonian verities, we may refer the reader to J. E. McGuire, 'Newton's "Principles of Philosophy"...' (*British Journal for the History of Science*, 5, 1970: 178–86) and to H. Guerlac, 'Newton and the Method of Analysis' (in *Dictionary of the History of Ideas: Studies of Selected Pivotal Ideas*, 3 (New York, 1973): 378–91, especially 385–6: 'Newton's Scientific Method' and 387–90: 'Newton's Mathematical Way').

(229) Namely in Proposition XLV and the scholium to Proposition XCIII of the first book; on which see 8.1: note (13) below.

Proposition XIV.[230] A poor basis on which to found a dogma, one might think. But let the reader judge what degree of credence is to be placed in these far from objective assessments by Newton in his old age of the half-forgotten logical and methodological principles, if such there ever really were, on which he had so long before designed and executed his masterwork. It is enough for us to place these draft prefaces in the public domain, allowing Newton to speak up for himself.

And so to his small final spurt of mathematical activity: the several minor ameliorations which he effected in the *editio prima* of his Lucasian lectures on *Arithmetica Universalis* when he himself, then in his 79th year, saw its revised Latin edition through press at London in 1722, making good his dissatisfactions with the text which Whiston had passed into print at Cambridge fifteen years before.[231] We have already noticed these changes to the primary edition in our commentary on the manuscript text as we reproduced it in the fifth volume:[232] most of what needed to be amended was merely the negative result of Whiston's too trusting and unseeing acceptance of minor slips by Newton in the original deposited copy of his *lectiones*, and we shall not here dwell upon them a second time. To complete the record, however, we adjoin two sets of rough calculations which he made around the close of 1721 to the same general end of improving the text of his only published work on algebra. The first group[233] elaborates, with individual variations, the method of 'exterminating' a common variable between two simultaneous equations which (we may surmise) he had used half a century earlier to derive the four rules encapsulating the end-product of such elimination between pairs of quadratics, cubics and quartics which he enunciated, without proof, in his algebraic lectures.[234] The latter is the checking

---

(230) See 8.3: note (44), where we examine in detail how it was that, thirty years after he 'synthesized' a correct solution therein of the problem of constructing the resisted motion defined (in the modernized terms of note (227) preceding) by

$$dv_y/dt(= v_y \, . \, dv_y/dy) \propto -v_y^2 - 2kv_y \pm g,$$

Newton could so radically mistake the sense of the finite *differentia momentorum* which he had there invoked in demonstrating its truth.

(231) Compare v: 12–14.

(232) See v: 54–508, especially 184–5, note (216) and 421–2, note (615).

(233) That set out in 9, §1 below, with the addition of a few interpolated connectives which (we trust) the better bring out the sequence and import of the underlying argument. The final computation here, [9] in our listing, was subsequently further refined to become the *schema* for eliminating between two given cubic equations which is reproduced on v: 518–19.

(234) See v: 126. The first two of these are already found in Newton's 1670 'Observations' on Kinckhuysen's *Algebra* (see II: 408) and the second two are analogous extensions. Newton never himself, may we make it clear, attained any neat expression of these rules close to the modern determinant form in which their equivalent is stated in II: 408–9, notes (78)/(79) and in v: 126, note (117) respectively.

computation[235] whereby, having almost at the last minute caught in the printed sheets of his 1722 *Arithmetica* a lapse in his exposition of the mechanism for manually tracing the classical cissoid by a pivoted, sliding jointed ruler,[236] he went on correctly to recast his erroneous following construction of the roots of a cubic equation by aid of the more general conchoidal cubic curves which can analogously be described.[237] No one else, so far as we are aware, had previously noticed this not inconsiderable *faux pas* on Newton's part. It was a just and charitable fate—aided, as it might perhaps have been, by the sharp eyes of Machin or de Moivre or some other London colleague who read through the proof-sheets—which thus allowed Newton the opportunity quietly to retrieve his mistake on a reprinted leaf of his 1722 *Arithmetica* which attracted no contemporary comment.

With that last minimal act Newton's mathematical life was ended.[238] All honour to him whose dying spark of fire not even the chill and infirmity of extreme old age could wholly quench!

---

(235) The two states of this, the former vitiated still, are reproduced in 9, §2.

(236) As he had set this construction down on pages 236–7 of his deposited Lucasian *lectiones* (see v: 464–6).

(237) See 9, §2: note (13), and compare v: 468–9, note (686).

(238) We take no account here of his minimal reactions to Henry Pemberton's untiringly abundant proffering of suggestions for ways of improving the mathematical quality of the *Principia* as it passed through press in London during 1723–6 (see *Correspondence of Newton,* 7, 1978: 248–345, *passim*)—the only change of any great importance which Newton was thereby led to make was elegantly to remove the *non sequitur* which (compare vii: 253, note (16)) spoilt Lemma XVI of the first book in previous editions, this in response to a long, involved argument of Pemberton's (see *Correspondence*, 7: 258–61) which itself was in deep error in attempting to locate the oversight—nor of his undated letter to John Lacy (*ibid.*: 361–2) where, in answering a query from the latter regarding the proper way to measure the surface-area of *loca montosa*, he firmly stated his preference for the mathematically exact method over one only approximate 'ubi error omnis quamvis [minimus] est etiamnum in defectu versatur'.

1

THE TWIN PROBLEMS
OF JOHANN BERNOULLI'S
'PROGRAMMA' SOLVED[(1)]

[29/30[(2)] January 1696/7]
From the original draft[(3)] in the Royal Society's Letter Book

... Hactenus Bernoullus.[(4)] Problematum verò solutiones sunt hujusmodi.

Probl. I.[(5)]

Investiganda est curva Linea ADB in qua grave a dato quovis puncto A ad datum quodvis punctum B vi[(6)] gravitatis suæ citissimè descendet.

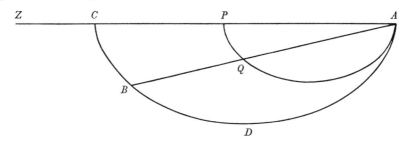

(1) Newton's response as sent in a letter dated 30 January 1696/7, hours only after he had contrived his solutions to Bernoulli's problems, to his friend and patron Charles Montagu, Chancellor of the Exchequer and then P.R.S. This was read out to the assembled Royal Society on 24 February following, and subsequently published (under a transparent cloak of anonymity) in the *Philosophical Transactions*, **17**, No. 224 [for January 1696/7]: §III: 384–9 as the second part (*ibid.*: 388–9) of an 'Epistola missa ad prænobilem virum D. *Carolum Mountague* Armigerum, Scaccarii Regii apud Anglos Cancellarium, & Societatis Regiæ Præsidem, in qua solvuntur duo problemata Mathematica à Johanne Barnoullo [*sic*] Mathematico celeberrimo proposita'. For the benefit of the reader—though Montagu himself perhaps received its printed twin (see Appendix 1, note (2))—this open letter also cited (pages 384–8) the text of Bernoulli's challenge of New Year's Day 1697 'to the sharpest mathematicians in the whole world' from the copy of its folio half-sheet which Newton found waiting for him on 29 January at his house in Jermyn Street on returning home late in the afternoon (or so his niece Catherine Barton thirty years afterwards recalled; see the next note) tired from a hard day's work as Warden of the London Mint. We have in our preceding introduction painted the backcloth to this printed 'Proclamation' by Bernoulli of two problems, one re-posing a seven-month-old challenge of which Newton may have earlier been made aware, the other generalizing one which Bernoulli had likewise discussed in print half a year before. Ignoring the latter, the standing hagiographical tradition fulsomely praises Newton for his surpassing genius in

Translation

...Thus far Bernoulli.[4] Solutions to the problems, however, are of this kind.

Problem I.[5]

There needs to be found out the curved line ADB in which a heavy body shall, under the force[6] of its own weight, most swiftly descend from any given point A to any given point B.

---

speedily solving the prior problem of determining the brachistochrone curve the same evening that he received it. We, contrarily, would comment that while each of Bernoulli's problems was in its own way a probing test of contemporary mathematical prowess, both had been well within Newton's power for a dozen years and more, and it could have required in 1697 no great strain on his part to give their complete solution; in our view, indeed, it was only the mental rustiness ensuing from a year's total break with things mathematical which necessitated his then taking a whole twelve hours to come up with what but a few months before he would have attained in minutes.

(2) At the bottom of his copy of Bernoulli's flysheet (on which see Appendix 1: note (2) following) Newton has written: 'Chartam ex Gallia missam accepi Jan. 29. 169⅞'. We accept the essential truth of John Conduitt's record of his wife's testimony thirty years later that 'Sʳ I.N. when the problem in 1697 was sent by Bernoulli...came tired from the Mint at 4 o'clock, but did not sleep till he had solved it—wᶜʰ he did by 4 in the morning' (King's College, Cambridge. Keynes MS 130.6¹: 5ᵛ/6ʳ). This behaviour was not, of course, in any way out of character. Just a few pages earlier in the same slim green notebook where he made this jotting, Conduitt has pencilled, again appending his wife's initials 'C[atherine] C[onduitt]' as his authority, the recollection that Newton 'When come from the Tower [Mint] would stand still a great while if any paper or book sent him before he would eat' (*ibid.*: 3ᵛ).

(3) N.1.61b. The full letter, filled out with the text of Bernoulli's 'Programma' according to Newton's manuscript indication, is printed in *The Correspondence of Isaac Newton*, 4, 1967: 220–4, but without any mathematical commentary. Hans Sloane (at this time, with Halley away in the north of England busily conducting the affairs of the Chester Mint, acting Clerk to the Royal Society) has added at the head of Newton's script both the date 'Jan. 30. 169⅞'—which, since the letter was read to the Society only more than three weeks later, we may perhaps doubt—and a draft title, '3. Epistola, præhonorabili viro D. Carolo Mountague Armig. Scaccarij Cancellario & S.R. Præsidi inscripta, qua solvuntur duo problemata mathematica a Johanne Bernoullo Mathematico celeberrimo proposita', little differing from that published (see note (1) above).

(4) Understand in his printed 'Programma' (whose text we reproduce in Appendix 1 following). Newton began his open letter to Montagu with the remark 'Accepi, Vir Amplissime, ['ex Gallia' is here deleted] hesterno die duo Problematum a Joanne Bernoullo Mathematicorum acutissimo propositorum exemplaria Groningæ edita in hæc verba' (I received yesterday, most noble Sir, [from France] a pair of the problems propounded by that sharpest of mathematicians Johann Bernoulli, published at Groningen in these words), and without ado passed to quote the text of Bernoulli's folio fly-sheet. We may guess that the French intermediary who passed to Newton his copy of the 'Programma' was either Varignon (who had received several from Groningen in mid-January for distribution in France) or, less likely, L'Hospital, but there can be no doubt that the instruction to do so came directly from Bernoulli himself; see Appendix 1: note (2).

(5) Newton follows the sequence of the challenge problems as they are set out in the 'Programma'. By his own ensuing declaration (see note (9) below) he had not seen Bernoulli's

Solutio.

A dato puncto *A* ducatur recta infinita *APCZ* horizonti parallela et super eadem recta describatur tum Cyclois quæcunq *AQP* rectæ *AB* (ductæ et si opus est productæ) occurrens in puncto *Q*, tum Cyclois alia *ADC* cujus basis[7] et altitudo sit ad prioris basem[7] et altitudinem respectivè ut *AB* ad *AQ*. Et hæc Cyclois novissima transibit per punctum *B* et erit Curva illa linea in qua grave a puncto *A* ad punctum *B* vi gravitatis suæ citissime perveniet. Q.E.I.[8]

Prob. II.

Problema alterum, si recte intellexi, (nam quæ in *Actis Lips.*[9] ab Auctore citantur ad id spectantia, nondum vidi,) sic proponi potest. *Quæritur Curva KIL ea lege ut si recta PKL a dato quodam puncto P, ceu Polo, utcunq ducatur, et eidem Curvæ in punctis duobus K et L occurrat, potestates duorum ejus segmentorum PK et PL a dato illo puncto P ad occursus illos ductorum, si sint æque altæ (id est vel qua-drata, vel cubi vel quadrato-quadrati &c.) datam summam PK^q + PL^q vel PK^cub + PL^cub &c (in omni rectæ illius positione) conficiant.*

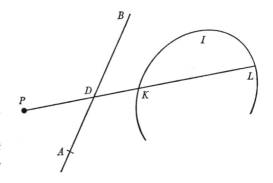

'Supplementum Defectus Geometriæ Cartesianæ...' (*Acta Eruditorum* (June 1696): 264 ff.), and so was unaware that this 'prime' problem of determining the brachistochrone had been there first published, as a 'Problema novum ad cujus solutionem Mathematici invitantur' (*ibid.*: 269), in sequel to Bernoulli's particular solution of the simplest case, where the power is unity, of 'Prob. II' following (on which see Appendix 1: note (13)).

(6) Understand this to be constant and acting vertically downwards.

(7) That is, *AC* and *AP* respectively.

(8) Newton, it will be evident, here assumes without any supporting argument that the curve of fall (from rest) in least time is an inverted cycloid, with a vertex at the initial point *A* and its base set along the horizontal *AZ* through *A*, and thereafter in constructing the arc *A͡DB* of fall from *A* to the general point *B* makes elegant use of the property that all cycloids are similar. The sheets on which Newton worked this problem over the long night of 29/30 January are not (to our knowledge) extant, but his correct inference that the cycloid is indeed the required *linea brevissimi descensus*—the *brachystochrona*, to use the appellation coined by Johann Bernoulli in the privacy of his correspondence with Leibniz the previous summer, rather than Leibniz' own preferred, but still-born, *tachystoptota* (see Leibniz' letter of 16 June 1696 (N.S.) and Bernoulli's response on 21 July in Gabriel Cramer's edition of their *Commercium Philosophicum et Mathematicum*, 1 (Lausanne/Geneva): 172, 177 [= (ed. C. I. Gerhardt) *Leibnizens Mathematische Schriften*, 3 (Halle, 1855): 288, 299])—has two complementary parts: the reduction of the posited kinematical requirement to an equivalent mathematical condition; and the manipulation of this deduced condition into a standard defining 'symptom' of the cycloid. As we have already observed in VI: 460–1, note (14), Newton certainly attained the

Solution.

From the given point A draw the unbounded straight line $APCZ$ parallel to the horizontal and upon this same line describe both any cycloid AQP whatever, meeting the straight line AB (drawn and, if need be, extended) in the point Q, and then another cycloid ADC whose base[7] and height shall be to the previous one's base[7] and height respectively as AB to AQ. This most recent cycloid will then pass through the point B and be the curve in which a heavy body shall, under the force of its own weight, most swiftly reach the point B from the point A. As was to be found.[8]

Problem II.

The second problem, if I have understood it right (for the citations by the author regarding it in the Leipzig *Acta*[9] I have not yet seen), can be proposed in this fashion: *There is sought a curve KIL with the restriction that, if from any given point P as pole a straight line PKL be arbitrarily drawn to meet this same curve in the two points K and L, the powers of its two segments PK and PL drawn from that given point P to those meets shall—taking these to be equally elevated (that is, either squares, or cubes, or fourth powers, and so on)—together (in every position of that straight line) make a given sum, PK^2+PL^2 or PK^3+PL^3 and so on.*

---

first goal by a variant of the 'method of maxima & minima in infinitesimals' (so he came in his old age to name it; compare **2**, Appendix 9 below) which he had employed to determine the solid of revolution of least resistance to motion in the direction of its axis. The muddled record by David Gregory (ULE. A78[1]) of his imperfect remembrance of a conversation which he had with Newton about the brachistochrone a month afterwards 'Londini 7 Martij 169$\frac{6}{7}$' confirms that, where x is the vertical distance fallen from $A(0, 0)$ to $B(x, y)$, so that the orbital speed at B is proportional to $\sqrt{x}$, then Newton's defining condition for the slope of the fall-path at B—and *mutatis mutandis* for that of every infinitesimal portion of the arc $\widehat{AB}$—is that the sine of its inclination to the vertical shall be proportional to the body's speed there; whence $\dot{y}/\sqrt{[\dot{x}^2+\dot{y}^2]} = \sqrt{[x/a]}$, a constant, and accordingly $\dot{y}/\dot{x} = \sqrt{[x/(a-x)]}$ is the fluxional Cartesian equation of the *curva brachystochrona*, which is consequently (as Newton would, a quarter of a century earlier at least, at once have noticed without any need for further computation; compare III: 162 and 163, note (307)) a cycloid described by the rolling along beneath APC ($x = 0$) of a circle of diameter a, the chord through B of which there touches the curve. Three years later, when Newton came to simplify a reduction by Fatio de Duillier of the brachistochrone condition to yield the radius of curvature of the fall-curve of least time at its general point, his preliminary step was to obtain just this first-order fluxional condition before passing, by a 'derivation' of this by first principles, to determine the horizontal chord of the osculating circle at this point. His preliminary computation (ULC. Add. 3968.41:2^r), filled out from a contemporary memorandum of David Gregory's (C122^2 = Royal Society. Gregory Volume: 22) which expands upon its argument, is reproduced in Appendix 2.

(9) 'Supplementum Defectus Geometriæ Cartesianæ circa Inventionem Locorum', *Acta Eruditorum* (June 1696): 264–7, especially 266, where Johann Bernoulli had posed the simplest instance of the generalized problem to which Newton here responds, requiring namely to find (on recasting his statement into present Newtonian terms) the curve *KIL* such that, if *PKL*

Solutio.

Per datum quodvis punctum A ducatur recta quævis infinita positione data ADB rectæ mobili PKL occurrens in D, et nominentur AD x et PK vel PL y,sintcȝ Q et R quantitates ex quantitatibus quibuscuncȝ datis et quantitate x quomodo-cuncȝ constantes[10] et relatio inter x et y definiatur per hanc æquationem $yy + Qy + R = 0$. Et si R sit quantitas data, Rectangulum sub segmentis PK et PL dabitur. Si Q sit quantitas data summa segmentorum illorum (sub signis proprijs conjunctorum) dabitur. Si $QQ - 2R$ detur, summa quadratorum $(PK^q + PL^q)$ dabitur. Si $Q^3 - 3QR$ data sit quantitas, summa cuborum $(PK^{cub} + PL^{cub})$ dabitur. Si $Q^4 - 4QQR + 2RR$ data sit quantitas summa quadrato-quadratorum $(PK^{qq} + PL^{qq})$ dabitur. Et sic deinceps in infinitum. Efficiatur itacȝ ut R, Q, $QQ - 2R$, $Q^3 - 3QR$ &c datæ sint quantitates & Problema solvetur. Q.E.F.[11]

Ad eundem modum Curvæ inveniri possunt quæ tria vel plura abscindent

---

rotates round the fixed pole P to meet it in (but) two points K and L, then the sum of the two segments PK and PL (understanding their sense to be preserved) is the same in all positions. Hindered, however, by an inadequate manner of relating the vector lengths PK, PL to the polar angle through which PKL revolves (see note (11) following), Bernoulli had there been able to arrive only at the equivalent of the particular family of curves defined by

$$[\phi =] \tan^{-1}((x - a)/b) = l(y - y^2)^n$$

(all, needlessly, passing in consequence through the pole P), and not Newton's following fully general defining equation $y^2 - ky + f(x) = 0$, where k is an arbitrary constant, and $f(x)$ is any (admissible) function of the line-length $AD = x$ coordinate to PK, $PL = y$. We may add the comment that Newton similarly grasped the essence of Bernoulli's generalized problem in the ensuing 'Programma' better than its author himself; see note (11) following.

(10) That is, in more modern Bernoullian terminology, Q and R are to be any allowable functions of x. In the present algebraic context, however, Newton doubtless has in mind only simple rational polynomials of form $\sum_i a_i x^i$.

(11) Well primed by his experience with such matters over the past three decades, Newton plunges straight to the heart of Bernoulli's problem, realizing, first, that the way in which the rotation of the *radius vector* PKL generates the pairs of points on the curve KIL to be determined is entirely arbitrary; and hence that it is required merely to specify the condition for the sum of the powers $p = 2, 3, 4, ...$ of the directed lengths $PK = y_1$ and $PL = y_2$ thus generated to be constant—namely, that the quadratic equation $y^2 + Qy + R = 0$ of which these are the roots shall be such as to have $y_1^p + y_2^p$ or $(y_1 + y_2)^p - p(y_1 + y_2)^{p-2} y_1 y_2 + \frac{1}{2}p(p-3)(y_1 + y_2)^{p-4}(y_1 y_2)^2...$ constant, where $y_1 + y_2 = -Q$ and $y_1 y_2 = R$. Such equations connecting the powers of the roots of a general equation—in the paradigm case where it is of eighth degree, at least—with their symmetrical functions which are its coefficients he had attained as one of his earliest mathematical discoveries in May 1665 (see 1: 519), and he had now only to write them out here in suitably simplified form (*viz.* setting $p = Q$, $q = R$ and $r = s = t = ... = z = 0$). As a bonus, and unknowingly putting to shame all Bernoulli's hard work the previous year on page 265 of his 'Supplementum' where he sought most diligently (if awkwardly) to arrive at this generalization of 'prop. 35 & 36 lib. 3 Euclid. [*Elem.*]' (which is the particular instance $y^2 - 2sy \cos \phi + s^2 - r^2 = 0$ of the circle of radius r and with centre distant s from P), Newton

Solution.

Through any given point A let there be drawn any unbounded straight line ADB given in position, meeting the mobile line PKL in D; let AD be named x and PK or PL y, and Q and R be quantities compounded in any manner soever of any given quantities whatever and the quantity x,[10] and let the relationship between x and y be defined by this equation $y^2 + Qy + R = 0$. Then if R be a given quantity, the rectangle beneath the segments PK and PL will be given; if Q be a given quantity the sum of those segments (conjoined under their proper signs) will be given; if $Q^2 - 2R$ be given, the sum of their squares $(PK^2 + PL^2)$ will be given; if $Q^3 - 3QR$ be a given quantity, the sum of their cubes $(PK^3 + PL^3)$ will be given; if $Q^4 - 4Q^2R + 2R^2$ be a given quantity, the sum of their fourth powers$(PK^4 + PL^4)$ will be given; and so forth indefinitely. Make it, consequently, that R, Q, $Q^2 - 2R$, $Q^3 - 3QR, \ldots$ are given quantities and the problem will be solved. As was to be done.[11]

In much the same way curves can be found which shall cut off three or more

adds that if R is constant, then so is (its equal) $y_1 y_2$. To connect the vector-lengths PK, $PL = y$ to the rotation of their common radius round P, he does not (however 'natural' this may appear to modern eyes) employ the polar angle itself—say $\widehat{EPL} = \phi$, where (corresponding to $AE = a$) $PE = b$ is the perpendicular let fall from P to the straight line AK given in position— but rather introduces, following Mode 5 of Problem 4 of his 1671 tract (see III: 140) where he had adumbrated this variant upon his generalization of oblique Cartesian coordinates, the line-length $AD = x (= a + b \tan \phi)$ as correlate variable. All very splendidly and concisely done. In contrast, the three Continental mathematicians whose solutions to Johann Bernoulli's problem (varying little one from the other and so, we may guess, not wholly independent) were published by Leibniz in his joint 'Communicatio' thereon in the *Acta Eruditorum* the next May —namely, Leibniz himself (p. 205), Johann's brother Jakob (pp. 215–16) and L'Hospital (*p.* 219)—severally made heavy weather of it, each following the lead of Bernoulli's 1696 'Supplementum' (compare note (9) above) to use as surrogate for the polar angle ϕ the oblique distance $z = (y/m) \sin \phi$ of K, L from PE (meeting it at the fixed angle $\sin^{-1} m$); thence by a primitive functional analysis determining $z = hy(y^p - y^{2p})$ as the type of their general solution, set simultaneously to satisfy the initial requirement $y_1^p + y_2^p = 1$ and also (by their construction of the oblique distances z parallel to one another) $y_1/z_1 = y_2/z_2$ for each value of the polar angle ϕ. In the minimally generalized Newtonian equivalent this solution would be $y^{2p} + ky^p + A = 0$, where A is an arbitrary (rational algebraic) function of $AD = x$. But this equation defines a family of curves KIL which is met by the *radius vector PKL* rotating round the pole P in (up to) $2p$ real points, whereas, as Newton implicitly provides for in his present solution, Bernoulli's problem demands only the subset of this family which is determined by the general equation $y^2 + Qy + R = 0$, Q and R 'compounded' of x, which is quadratic in y. From this viewpoint Newton's stated equations are the condition for $y^{2p} + ky^p + A = 0$ to have the factor $y^2 + Qy + R = 0$: whence, as an evident corollary, they arise directly by equating to the constant $\frac{1}{2}k$ the non-surd part of $(\frac{1}{2}Q \pm \frac{1}{2}\sqrt{[Q^2 - 4R]})^p$, $p = 2, 3, 4, \ldots$. Since, unlike them, Johann Bernoulli never published his own solution to this second problem of his 'Programma', we cannot know if he here saw further than his brother Jakob or his two other close colleagues. Let us in charity accept the opinion of his most recent editor that 'Bernoulli hat darüber nichts publiziert, vermutlich weil er die Überlegenheit von Newtons Lösung einsah' (Otto Spiess, *Der Briefwechsel von Johann Bernoulli*, **1** (Basel, 1955): 331, note 16).

segmenta similes proprietates habentia. Sit æquatio $y^3 + Qyy + Ry + S = 0$ ubi Q, R et S quantitates significant ex quantitatibus quibuscunqȝ datis et quantitate x utcunqȝ constantes, et Curva abscindet segmenta tria. Et si S data sit quantitas contentum solidum illorum trium dabitur. Si Q sit quantitas data, summa trium illorum dabitur. Si $QQ - 2R$ sit data quantitas, summa quadratorum ex tribus illis dabitur.[12]

---

(12) Understand 'Et sic deinceps in infinitum' (And so on indefinitely). Here the equations of I: 519 are applied little differently by now setting

$$p = Q, \quad q = R, \quad r = S \quad \text{and} \quad s = t = \ldots = z = 0.$$

And in much the same way as we outlined in the previous note they also ensue as the condition for $y^{3p} + ky^{2p} + Ay^p + B = 0$ to have the factor $y^3 + Qy^2 + Ry + S = 0$, where A, B, Q, R and S are each arbitrary functions of the coordinate variable x.

segments having similar properties. Let the equation be $y^3 + Qy^2 + Ry + S = 0$, where Q, R and S denote quantities compounded in any manner soever from any given quantities whatever and the quantity x. Then if S be a given quantity, the 'solid content' (triple product) of those three will be given; if Q be a given quantity, the sum of those three will be given; if $Q^2 - 2R$ be a given quantity, the sum of the squares of those three will be given.[12]

APPENDIX 1. THE TEXT OF BERNOULLI'S PRO-CLAMATION 'TO THE SHARPEST MATHEMATICIANS IN THE WHOLE WORLD' ON NEW YEAR'S DAY, 1697.[1]

From Newton's copy[2] of the broadsheet in the Royal Society, London

Acutissimis qui toto Orbe florent Mathematicis S.P.D.[3]
Johannes Bernoulli, Math. P.P.[3]

Cum compertum habeamus vix quicquam esse quod magis excitet generosa ingenia ad moliendum quod conducit augendis scientiis, quàm difficilium pariter & utilium quæstionum propositionem, quarum enodatione tanquam singulari si qua aliâ via ad nominis claritatem perveniant sibique apud posteritatem æterna extruant monumenta; Sic me nihil gratius Orbi Mathematico facturum speravi quam si imitando exemplum tantorum Virorum *Mersenni*, *Pascalii*, *Fermatii*, præsertim recentis illius Anonymi Ænigmatistæ Florentini[4]

---

(1) New style, that is; the equivalent date in the Julian calendar still used in England at this time is 22 December 1696. The text of this celebrated 'Programma', as Mencke dubbed it in February 1697 when he cited the text of its 'Problema alterum purè Geometricum' for the convenience of the readers of his *Acta Eruditorum* where Bernoulli had originally published its first 'Problema novum' in June the previous year (the title was confirmed by Cramer when he reprinted the full proclamation as item No. XXXIII of this edition of Johann Bernoulli's *Opera*, **1** (Lausanne/Geneva, 1742): 166–9), is here reproduced mainly to show the exact form in which Newton received its two enclosed challenge-problems when he solved them just over a month after it was first issued. Since we have in preceding pages already dwelt at length upon their mathematical niceties in our footnotes commenting on Newton's published solutions, we here confine ourself to sketching in the background and quoting relevant textual passages. A somewhat plodding and not entirely accurate English rendering of Bernoulli's Latin may be found in (ed. J. F. Scott) *The Correspondence of Isaac Newton*, **4** (Cambridge, 1967): 224–6.

(2) Royal Society. Classified Papers I, 30. At the bottom of this printed folio half-sheet (its twin was doubtless passed on to Charles Montagu as an enclosure to his letter of 30 January where he set out his undemonstrated solutions; compare 1: note (1) preceding) Newton has written 'Chartam hanc ex Gallia missam accepi Jan. 29. 169$\frac{6}{7}$'. If the 'ex Gallia' —which he afterwards deleted at the equivalent place in his letter to Montagu (see note (4) on page 73 above)—is not simply his mistake for 'ex Hollandia', then Newton would seem to have received his copy of the 'Programma' from some intermediary in Paris, doubtless Varignon (who had received several from Groningen in mid-January for distribution in France). But there can be no doubt that the communication of it was made directly at its author's instruction. 'Mr Niewton', wrote Bernoulli to Basnage de Beauval three months later on 30 March (N.S.), 'est un des Mathematiciens de nôtre temps que j'estime le plus; il fait mention dans son écrit d'avoir reçû deux exemplaires de mon imprimé, or c'est justement à luy que j'en ay envoyé deux sous une simple couverte' (*Der Briefwechsel von Johann Bernoulli*, **1**, Basel, 1955: 430).

(3) 'S[alutem] p[lurimam] d[icit]' and 'Math[ematicæ] P[rofessor] P[ublicus]' (at Groningen in Holland) respectively.

(4) Vincenzo Viviani, namely, who in April 1692 had issued, much like Bernoulli now, a printed proclamation conveying an 'Ænigma Geometricum De Miro Opificio Testudinis

aliorumque qui idem ante me fecerunt, præstantissimis hujus ævi Analystis proponerem aliquod problema, quo quasi Lapide Lydio[5] suas methodos examinare, vires intendere & si quid invenirent nobiscum communicare possent, ut quisque suas exinde promeritas laudes à nobis publicè id profitentibus consequeretur.

Factum autem illud est ante semestre in *Actis Lips. m. Jun.* pag. 269.[6] ubi tale problema proposui cujus utilitatem cum jucunditate conjunctam videbunt omnes qui cum successu ei se applicabunt. Sex mensium spatium à prima publicationis die Geometris concessum est, intra quod si nulla solutio prodiret

---

Quadrabilis Hemisphæricæ a D. Pio Lisci pusillo Geometra'—that is, 'postremo Discipulo Galilei' on unravelling the anagram—'propositum'. (The full text of Viviani's broadsheet, taken from John Wallis' copy of it (now Royal Society. Classified Papers I, 29), is printed in *Philosophical Transactions*, **17**, No. 196 [for January 1692/3]: 585–6; let us quickly pass over Wallis' following crude and ineffective stab (*ibid.*: 587–92) at a solution.) Viviani's eloquently expressed challenge to find this 'quadrable hemispherical tortoise'—placing it persuasively in an invented classical setting as a purportedly still existing 'Templum augustissimum, ichnographia circulari, ALMÆ GEOMETRIÆ dicatum, quod à Testitudine intus perfectè hemisphærica, operitur: sed in hac, fenestrarum quatuor æquales areæ (circum, & supra basim hemisphæræ ipsius dispositarum) tali configuratione, amplitudine, tantaque industria ac ingenii acumine sunt extructæ, ut, his detractis, superstes curva Testudinis superficies, pretioso opere musivo ornata, Tetragonismi verè Geometrici est capax'—aroused considerable contemporary interest and a widespread response: Archimedean solutions (involving the simple 'geometrical integration' of a sine), much as Viviani himself contrived, were quickly forthcoming from Leibniz (see C. I. Gerhardt's edition of his *Mathematische Schriften*, **5** (Halle, 1858): 270–8, the essence of which privately printed reply was published in *Acta Eruditorum* (June 1692): 275–9); Jakob Bernoulli (in *Acta Eruditorum* (August 1692): 370–1 [= *Opera*, **1** (Geneva, 1744): No. LII: 512–15]); Christiaan Huygens (see his *Œuvres complètes*, **10** (The Hague, 1905): 336–8); and, with typical British dilatoriness, David Gregory (in *Philosophical Transactions*, **18**, No. 207 [for January 1693/4]: 25–9). In essence, when the hemispherical surface is pierced with windows (at its base) whose surrounds are the (halves of) Eudoxian 'horse-fetters' cut out by tangent cylinders of diameter equal to the hemisphere's radius, the surface-area of the 'tortoise shell' remaining is the square of the hemisphere's diameter. We need not be equally long in mentioning Pascal's challenges in 1658 to solve a variety of problems to do with the cycloid (see P. Humbert, *Cet effrayant génie.... L'Œuvre scientifique de Blaise Pascal* (Paris, 1947): 199–232 for a recent broad survey of this aspect of his mathematical genius of which there is yet no adequate technical secondary account); or Fermat's various *défis* to Wallis and other English mathematicians over 1657–8 to resolve his problems in number theory (compare IV: 76–7, note (9); 92–3, note (61); and 96–7, note (70)).

(5) Just as jewellers use this Lydian basanite as a touchstone for testing the purity of gold.

(6) Namely, at the end of the article in the *Acta Eruditorum* 'm[ensis] Jun[ii MDCXCVI]' where (see note (14) below) he discussed the simplest instance of his following 'Problema alterum purè Geometricum'. Having phrased his 'Problema novum' of identifying the brachistochrone (see note (9) below), Bernoulli there warned: 'Ut harum rerum amatores instigentur & propensiori animo ferantur ad tentamen hujus problematis, sciant non consistere in nuda speculatione, ut quidem videtur, ac si nullum haberet usum; habet enim maximum etiam in aliis scientiis quam in mechanicis, quod nemo facile crediderit. Interim (ut forte quorundam præcipiti judicio obviam eam) quanquam recta...sit brevissima inter [duos] terminos..., non tamen illa brevissimo tempore percurritur; sed est curva...Geometris notissima, quam ego nominabo, si elapso hoc anno nemo alius eam nominaverit.'

in lucem, me meam exhibiturum promisi: Sed ecce elapsus est terminus & nihil solutionis comparuit; nisi quod Celeb. *Leibnitius* de profundiore Geometriâ præclarè meritus me per literas[7] certiorem fecerit, se jam feliciter dissolvisse nodum pulcherrimi hujus uti vocabat & inauditi antea problematis, insimulque humaniter rogavit,[8] ut præstitutum limitem ad proximum pascha extendi paterer, quo interea apud Gallos Italosque idem illud publicari posset nullusque adeo superesset locus ulli de angustiâ termini querelæ; Quam honestam petitionem non solum indulsi, sed ipse hanc prorogationem promulgare decrevi, visurus num qui sint qui nobilem hanc & arduam quæstionem aggressuri post longum temporis intervallum tandem Enodationis compotes fierent. Illorum interim in gratiam ad quorum manus *Acta Lipsiensia* non perveniunt, propositionem hîc repeto.[9]

Problema Mechanico-Geometricum de Linea Celerrimi descensûs.

Determinare lineam curvam data duo puncta in diversis ab horizonte distantiis & non in eadem rectâ verticali posita connectentem, super qua mobile propriâ gravitate decurrens & à superiori puncto moveri incipiens citissime descendat ad punctum inferius.

Sensus problematis hic est; ex infinitis lineis quæ duo illa data puncta conjungunt, vel ab uno ad alterum duci possunt eligatur illa, juxta quam si incurvetur lamina tubi canalisve formam habens, ut ipsi impositus globulus & liberè dimissus iter suum ab uno puncto ad alterum emetiatur tempore brevissimo.

Ut vero omnem ambiguitatis ansam præcaveamus, scire B.L.[10] volumus, nos

---

(7) Of 16 June 1696 (N.S.) where, in prelude to setting down the correct defining *æquatio differentialis* '$dy:dx = \sqrt{x}:\sqrt{(b-x)}$' of the 'linea *Tachistoptota*' (as he there proposed to call the curve of swiftest descent) but not identifying this outright as the cycloid which (compare *Acta Eruditorum* (June 1686): 297) he was aware it was, Leibniz spoke of Bernoulli's newly published 'Problema inveniendæ lineæ...celerrimi descensus' as 'profecto pulcherrimum', one which 'me invitum ac reluctantem, pulchritudine sua, ut pomum Evam, ad se traxit' (*Got. Gul. Leibnitii et Johan. Bernoullii Commercium Philosophicum et Mathematicum*, 1 (Lausanne/Geneva, 1745): 172 [= (ed. C. I. Gerhardt) *Leibnizens Mathematische Schriften*, 3 (Halle, 1855): 288]).

(8) In following letters on 23 August, where he made the suggestion to Bernoulli 'Si vis ut Italis Gallisque [!] aliquod spatium relinquatur examinandi Problematis Tui, ne ansam habeant excusandi sese, prorogandus nonnihil terminus erit....Quid ergo, si expectes usque ad finem Anni a prima publicatione *Lipsiensi* computati, id est ad Junium Anni sequentis?' (*Commercium*: 197–8 [= *Math. Schriften*, 3: 322–3]); and six weeks afterwards on 6 October, where he shortened the period of grace by three months, informing the problem's author that 'Quia approbas, scripsi ad Italos & Gallos, ut Pascha proximum [*sc.* anni 1697] pro termino solutionum statuatur' (*Commercium*: 206 [= *Math. Schriften*, 3: 331]).

(9) Bernoulli in fact here rephrases the 'Problema novum ad cujus solutionem Mathematici invitantur' as he had set it down in the *Acta* for June 1696 (p. 269), there requiring only '*Datis in plano verticali duobus punctis A & B assignare Mobili M, viam AMB, per quam gravitate sua descendens & moveri incipiens a puncto A, brevissimo tempore perveniat ad alterum punctum B*'.

10) 'B[enignum] L[ectorem].'

hic admittere *Galilæi*[11] hypothesin de cujus veritate sepositâ resistentiâ jam nemo est saniorum Geometrarum qui ambigat, *Velocitates* scilicet *acquisitas gravium cadentium esse in subduplicata ratione altitudinum emensarum*, quanquam aliàs nostra solvendi methodus universaliter ad quamvis aliam hypothesin sese extendat.

Cum itaque nihil obscuritatis supersit obnixè rogamus omnes & singulos hujus ævi Geometras accingant se promtè, ten[t]ent, discutiant quicquid in extremo suarum methodorum recessu absconditum tenent; Rapiat qui potest præmium quod Solutori paravimus, non quidem auri non argenti summam quo abjecta tantum & mercenaria conducuntur ingenia, à quibus ut nihil laudabile sic nihil quod scientiis fructuosum expectamus, sed cùm virtus sibi ipsi est merces pulcherrima, atque gloria immensum habet calcar, offerimus præmium quale convenit ingenui sanguinis Viro, consertum ex honore, laude & plausu, quibus magni nostri Apollinis perspicacitatem publicè & privatim, scriptis & dictis coronabimus, condecorabimus & celebrabimus.

Quod si verò festum paschatis præterierit nemine deprehenso qui quæsitum nostrum solverit, nos quæ ipsi invenimus publico non invidebimus; Incomparabilis enim *Leibnitius* solutiones tum suam tum nostram ipsi jam pridem commissam protinus ut spero in lucem emittet,[12] quas si Geometræ ex penitiori

---

(11) Read 'Hugenii' more accurately. While it is true that the following hypothesis (that speed of fall of a body from rest under 'simple' constant gravity acting vertically downwards is proportional to the square root of the vertical distance fallen) is implicit in the theorems of the short tract 'De Motu naturaliter Accelerato' which Galileo had incorporated in the 'Giornata Terza' of his *Discorsi e Dimostrazioni Matematiche intorno à due nuoue scienze Attenenti alla Mecanica & i Movimenti Locali* (Leyden, 1638), it was explicitly enunciated—with a lacuna in its proof anticipated by James Gregory in 1672 (see III: 425, note (15))—for the first time by Huygens in Propositio VIII of the 'Pars Secunda' of his *Horologium Oscillatorium sive De Motu Pendulorum ad Horologia aptato Demonstrationes Geometricæ* (Paris, 1673): 33–4. We have earlier observed (III: 396, note (16)) that William Brouncker independently in 1662 drew this fundamental measure of the 'energy' kinematically stored in fall under gravity from the second corollary to Proposition XIX of Book 1 of Simon Stevin's *De Beghinselen der Weeghconst* (Leyden, ₁586).

(12) Leibniz duly obliged four months afterwards in a rather lengthy 'Communicatio suæ pariter, duarumque alienarum ad edendum sibi primum a Dn. *Jo. Bernoullio*, deinde a Dn. *Marchione Hospitalio* communicatarum solutionum problematis curvæ celerrimi descensus a Dn. *Jo. Bernoullio* Geometris publice propositi, una cum solutione sua problematis alterius [*sc.* the 'Problema purè Geometricum' below] ab eodem postea propositi' (*Acta Eruditorum* (May 1697): 201–23). He did not in fact there give his own method of solution to Bernoulli's problem of the path of swiftest fall, but, 'cum cæteris consentiat [mea solutio], mihique quæsiti determinatione contento rem porro illustrare non vacaverit', contented himself merely with asserting 'lineam quæsitam esse figuram segmentorum circularium repræsentatricem [sive] Cycloidem vulgarem' (*ibid.*: 205). The deficiency was more than made good in Johann Bernoulli's ingenious and elegant ensuing appeal to the optical analogy of the 'Curvatura radii in diaphanis non uniformibus' where the path of least time is that of a light-ray traversing a medium whose *raritas* (and so density) is such that the speed v of light at a distance x below the horizontal from which it departs is instantaneously equal to that, $\sqrt{(2gx)}$, which it would

quodam fonte petitas perspexerint, nulli dubitamus quin angustos vulgaris Geometriæ limites agnoscant, nostraque proin inventa tanto pluris faciant, quanto pauciores eximiam nostram quæstionem soluturi extiterint etiam inter illos ipsos qui per singulares quas tantopere commendant methodos, interioris Geometriæ latibula non solum intimè penetrâsse, sed etiam ejus pomœria Theorematis suis aureis, nemini ut putabant cognitis, ab aliis tamen jam longè priùs editis mirum in modum extendisse gloriantur.

Problema alterum purè Geometricum,
quod priori subnectimus & strenæ[13] *loco Eruditis proponimus.*

Ab *Euclidis* tempore vel Tyronibus notum est; Ductam utcunque à puncto dato rectam lineam, à circuli peripheriâ ita secari ut rectangulum duorum segmentorum inter punctum datum & utramque peripheriæ partem interceptorum sit eidem constanti perpetuo æquale. Primus ego ostendi in eod. *Actor. Jun.* 265.[14] hanc proprietatem infinitis aliis curvis convenire, illamque

---

acquire by fall from rest under constant gravity g through the same vertical distance; whence (by the sine law in this Fermatian hypothesis of the transmission of light) $dy/ds \propto v$, and therefore '$dy = dx \sqrt{\dfrac{x}{a-x}}$, ex qua concludo: curvam *Brachystochronam* esse *Cycloidem vulgarem*' (*ibid.*: 209). And for good measure the *Acta*'s editor Mencke adjoined the equivalent but much more long-winded straightforward mathematical solution of Johann's brother Jakob (*ibid.*: 211–14). On the effectively identical—but uniformly incomplete—solutions by Leibniz himself, Jakob Bernoulli and L'Hospital to Johann's 'Problema alterum' which were likewise there presented (*ibid.*: 205, 215–16 and 217–18 respectively) see our more detailed comment in 1: note (11) preceding. As a further bonus to the reader unacquainted with the *editio princeps* of it, at the end of Leibniz's 'Communicatio' the *Acta*'s editor, Mencke, reprints (*ibid.*: 223–4) Newton's anonymously published resolution of the two challenge problems 'ex Transactionibus Philos. Anglic. M[ensis] Jan. 169⁶⁄₇' (on which compare 1: note (1) above); we may remark that the *Index Autorum* added to the volume in December 1697 correctly attributes this (p. 562) as 'Newton (Isaaci) Solutio duorum problematum Matematicorum a Jo. Bernoullio propositorum', but this merely spells out to the world at large what had for many months been an open secret to the *cognoscenti*.

(13) A 'New Year's gift', that is—much as Johannes Kepler had on 1 January 1610 made over the original of his essay *De Nive Sexangula* (Frankfurt, ₁1611) to the Reichshofrat of the Imperial Court at Prague.

(14) In his 'Supplementum Defectus Geometriæ Cartesianæ circa Inventionem Locorum', *Acta Eruditorum* (June 1696): 264–7, where it had been his general thesis (*ibid.*: 264) that 'Notum est, potissimam [*Geometriæ*] Cartesianæ partem conscriptam esse pro inveniendis locis & natura curvarum, tanquam materia in Geometricis summi momenti: Notum vero etiam est, quod methodus Auctoris semper supponat, certam quandam dari relationem punctorum in curva ad puncta in recta positione data quam axem vocat, pro qua relatione invenit æquationem algebraïcam, unde curvæ naturam & constructionem determinat. Verum si ex sola mutua relatione ipsorummet curvæ punctorum (nullis aliis consideratis vel datis) natura vel constructio curvæ eruenda sit, non video qua ratione id obtinere liceat per regulas Cartesii'. In illustration of the deficiency of Descartes' techniques in his *Geometrie* for determining curvilinear loci from given defining conditions, Bernoulli took as his primary instance (*ibid.*: 265)

adeo circulo non esse essentialem: Arrepta hinc occasione, proposui Geometris determinandam curvam vel curvas, in quibus non rectangulum sed solidum sub uno & quadrato alterius segmentorum æquetur semper eidem; sed à nemine hactenus solvendi modus prodiit; exhibebimus eum quandocunque desiderabitur: Quoniam autem non nisi per curvas transcendentes quæsito satisfacimus, en aliud per merè algebr[a]icas in nostra est potestate.

Quæritur Curva, ejus proprietatis, ut duo illa segmenta ad quamcunque potentiam datam elevata & simul sumta faciant ubique unam eandemque summam.

Casum simplicissimum existente sc. numero potentiæ 1, ibidem in *actis* pag. 266. jam solutum dedimus,[15] generalem verò solutionem quam etiamnum premimus, Analystis eruendam relinquimus.[16]

Dabam Groningæ ipsis Cal. Jan. 1697.

---

the 'decantatissima proprietas circuli demonstrata prop. 35 & 36 lib. 3 [*Elem.*] Euclid.' which he here again takes up, adding: 'Quis est qui non statim pronuntiaret hanc proprietatem esse circulo essentialem, & concluderet reciproce: *Ergo curva faciens rectangulum sub segmentis cujusvis rectæ per punctum datum ductæ dato æquale, est peripheria circuli*[?] Ego vero contrarium statuo, & dico sic argumentantem enormiter hallucinari, utpote reperiens, quod infinita dentur genera curvarum, quæ non minus quam circumferentia circuli dicta proprietate gaudeant, cum hoc solo discrimine, quod in hac ob similitudinem partium suarum punctum datum ubivis, in cæteris vero certo tantum loco possit sumi.' Because, however, of the relative cumbrousness and opacity of the system of coordinates in which Bernoulli chose to develop his argument, and through his yet imperfect appreciation of the full possibilities of (algebraically) relating two magnitudes, the defining equation of the requisite family of curves—

$$y = ax(x^{\alpha-1} + x^{1-\alpha}) + bx(x^{\beta-1} + x^{1-\beta}) + \dots,$$

where x and y are his preferred 'disguised polar' coordinates of their arbitrary point such that the product of the root pair $x_1 \times x_2 =$ constant—which he obtained lacks both the clarity and generality of Newton's solution (on which see 1: note (11) preceding).

(15) Bernoulli's second illustration in June 1696 of the defect of Descartes' *Geometrie* in not prescribing how to determine a curve from a given defining condition, there introduced as a 'problema quod olim mihi, si bene memini, aliquis non infimæ notæ Geometra'—perhaps Tschirnhaus?—'proposuerat; erat autem tale. *Quæritur qualis sit curva ABCDE ita ut ducta per verticem A tangente AC, & alia utcunque secante curvam in punctis B & D, summa segmentorum DA & BA sit ubique dupla ipsius tangentis AC*'. His proffered solution (*ibid.*) '$y = x.\overline{x-xx}^n$' is again far from general.

(16) For the benefit of his readers—who had already in June 1696 (see notes (6) and (9) above) been given an advance view of Bernoulli's first problem—Mencke reprinted these last three paragraphs in his *Acta Eruditorum* (February 1697): 95–6 as '*Johannis Bernoulli Mathem. P.P. Problema alterum pure Geometricum, priori* (videantur Acta Eruditorum A. 1696 pag. 269) *subnexum, ac peculiari programmate Groningæ Cal. Jan. 1697. Eruditis propositum*'. On Newton's solution to this generalized problem of intercepts, and also on the less adequate essays at its resolution by Leibniz, Jakob Bernoulli and L'Hospital (compare note (12) above) again see our comments in 1: note (11) preceding.

APPENDIX 2. NEWTON'S SIMPLIFIED PROOF OF FATIO'S REDUCTION OF THE CONDITION OF FALL IN LEAST TIME ALONG AN ARC OF THE BRACHISTOCHRONE TO A CURVATURE PROPERTY OF THE CYCLOID.[1]

[March[2] 1700]

From the original worksheet[3] in the University Library, Cambridge

[*Grave cadente a puncto dato A ad punctum datum M tempore brevissimo: quæritur linea ANM quam describit cadendo.*

---

(1) In the early spring of 1699, in appendix to an elaborate and considerably ingenious discourse by him on improving the yield of fruit-bushes by inclining their 'walls' (banked-up beds) to the horizon the better, on his preferred mathematical theory to catch the sun's rays, Newton's erstwhile close Swiss *émigré* friend Nicholas Fatio de Duillier (on whom see VII: xiv–xv) published at London a 22-page *Lineæ Brevissimi Descensus Investigatio Geometrica Duplex. Cui addita est Investigatio Geometrica Solidi Rotundi, in quod Minima fiat Resistentia.* We have already (see VI: 466–7, note (25); 480, note (45)) epitomized Fatio's latter 'inquiry' where he gave a complicated geometrical demonstration of Newton's differential defining condition for the solid of revolution of least resistance to uniform motion in the direction of its axis, along with first publication of parametric equations for drawing its transverse section by points—and also, let us here anticipate, added (p. 18) his forthright testimony to the priority 'by several years' of Newton's invention of the calculus which scorned—and was (compare note (12) below) vastly to enrage—the 'second inventor' Leibniz, portraying him as sedulously and less than modestly 'everywhere attributing the discovery of this calculus to himself'. (The long-term repercussions of these words are taken up in Part 2 following.) In his preceding twin investigation of the curve of swiftest descent under 'simple' constant gravity acting vertically downwards—the second of these approaches (pp. 11–12) is but the variant upon the first (pp. 5–10) which employs plane chordal trapezia to fill the place of curved infinitesimal cylindrical surfaces, measuring these in his prior analysis by a 'Simpson' approximation (see VII: 692–3, note (8)) of an equivalent order of accuracy—Fatio had represented the body's instantaneous speed v ($\propto \sqrt{x}$ where x is the vertical distance fallen from rest) by a line-length inversely proportionate to it and perpendicular to the fall-path at the related point, and was thereby enabled to express the condition that the time of descent $\int (1/v) \, . \, ds$ over an infinitesimal arc with fixed end-points shall be minimal in a correspondingly complicated geometrical form; whence, effectively by a 'differentiation' from first principles of the underlying equation $(1/v) \, . \, dy/ds = k$, constant, he with much cumbersome computation arrived at the defining condition $u \, . \, dy/ds = 2x$ for the brachistochrone, where $u[= (ds/dx)^3/(d^2y/dx^2)]$ is the radius of curvature at its point (x, y): 'Itaque quantum Corpus grave, per Lineam Brevissimi Descensus cadens, a Linea horizontali...a qua cadendi initium fecit, remotum est; tantum præcise Centrum Curvitatis Lineæ Brevissimi Descensus attollitur supra eandem Horizontalem.... Quæ notissima est Cycloidis Proprietas' (p. 9). We may plausibly surmise that David Gregory, if he did not directly ask Newton to simplify Fatio's complex and long-winded demonstration, at least nudged the latter to interest himself in doing so. Newton's distillation of Fatio's argument to an essence attaining its result in little more than a dozen lines is preserved in the corpus of his autograph papers only on a loose scrap-sheet where he roughly set down the bare, unexplained and not wholly correctly articulated bones of his preliminary calculations. Happily, among the many memoranda of Gregory's stemming from his frequent meetings

Lineæ hujus chordam quamvis infinite parvam *EG* biseca in *F*, et age horizontales *EB*, *FNC*, *GD* hisȝ perpendiculares *AD*, *EL*, *NI*. Quærenda est sagitta *FN* ubi grave illud per chordas *EN*, *NG* tempore brevissimo descendit.][4]

[Cape] $AB = x. \ BC = o = CD.$

$BE = y.$ [Erit]

$\quad H[N] = \dot{y}o = IK.$[5]

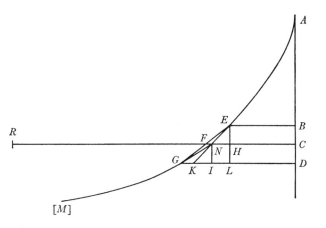

[*M*]

---

with Newton in London at this period there exists one (C122² = Royal Society. Gregory Volume: 22), dated by him '1 Aprile 1700' and afterwards indexed under the head 'Newtoni investigatio curvæ celerrimi descensûs', whose text (though its detail betrays traces of Gregory's own style and preferred notation for indices) clearly exhibits the substance of a lost parent in which Newton both mended the venial slip in his preparatory computations which we point to in note (9) below, and also fleshed out their skeleton to be a verbally connected demonstration of Fatio's result. In here reproducing those autograph calculations we have been free in borrowing from Gregory's memorandum in setting the mathematical scene and in interlarding their jejune equations the better to bring out the nuances of their logical sequence.

(2) We presume that Newton began to draft his *purée* of Fatio's argument little before he passed its corrected, verbally amplified revision to David Gregory on 1 April (see the previous note). A firm *terminus ante quem non* for its composition is of course early spring 1699 when Fatio first made public his 'Investigatio Duplex' into the nature of the curve of swiftest fall. (Gregory had already studied the parent book concerning 'Garden Walls' by 3 March 1698/9, in advance of its publication it may be; see W. G. Hiscock, *David Gregory, Isaac Newton and their Circle. Extracts from David Gregory's Memoranda 1677–1708* (Oxford, 1937): 10.) Newton's library copy (now Trinity College, Cambridge. NQ.10.33) of Fatio's *Fruit-walls Improved by Inclining them to the Horizon* and its attendant appendix is, we may note, wholly unmarked by his pen, though it shows signs of having been assiduously read.

(3) ULC. Add. 3968.41: 2ʳ, the first recto of a folded quarto sheet on whose three remaining blank sides (*ibid.*: 2ᵛ–3ᵛ) Newton long afterwards in spring 1714 jotted down a first draft (revised on *ibid.*: 124ʳ) of emendations to John Keill's submitted manuscript of his response to Leibniz's 1713 Bernoullian *charta volans*. (Thus corrected, and rendered by de Moivre into French, Keill's 'Réponse' appeared in the *Journal Literaire de La Haye*, 4.2 (July/August 1714): 320–58; see especially 335. Newton passed his corrections on to Keill in his letter of 11 May 1714; see (ed.) J. Edleston, *Correspondence of Sir Isaac Newton and Professor Cotes* (London, 1850): 175.)

(4) We take this enunciation and introductory description of the accompanying figure from Gregory's memorandum C122² (on which see note (1) above), making only the minimal adjustments needed to key it to Newton's present (original) accompanying one.

(5) Newton is here more than a little muddled. Since in his figure (compare our blow-up over the page of its central portion, in which we have inserted the vanishingly small arclet $\widehat{ENG}$ as a broken line) there is, corresponding to $BE = y \equiv y_x$,

$KG[=2NF]=\ddot{y}oo.$ [adeoꝗ] $GI=\dot{y}o+\ddot{y}oo.$[5]   [Ponantur]

$$GL=2p=2\dot{y}o+\ddot{y}oo.^{(5)} \quad FN=q.^{(6)}$$

[et tempus illud fit ut]

$$\frac{\sqrt{oo+pp-2pq+qq}}{\sqrt{x}}+\frac{\sqrt{oo+pp+2pq+qq}}{\sqrt{x+o}}=\mathrm{Min[imo]}^{(7)}=R+S.$$

[positis] $R^2=\dfrac{oo+pp-2pq+qq}{x}.$ [et] $SS=\dfrac{oo+pp+2pq+qq}{x+o}.$

[Unde capiendo fluxiones] $2R\dot{R}=\dfrac{-2p\dot{q}+2q\dot{q}}{x}.$ $2S\dot{S}=\dfrac{+2p\dot{q}+2q\dot{q}}{x+o}.$ [evadit]

$$\frac{-p\dot{q}+q\dot{q}}{Rx}+\frac{p\dot{q}+q\dot{q}}{Sx+So}=0.^{(8)}$$

---

to $O(o^3)$

$$CN = y_{x+o} = y+\dot{y}o+\tfrac{1}{2}\ddot{y}o^2 \quad\text{and}\quad DG = y_{x+2o} = y+2\dot{y}o+2\ddot{y}o^2;$$

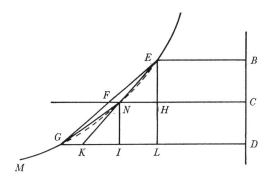

whence $HN = IK = \dot{y}o+\tfrac{1}{2}\ddot{y}o^2$, $IG = \dot{y}o+\tfrac{3}{2}\ddot{y}o^2$ and $LG = 2\dot{y}o+2\ddot{y}o^2$. These errors in assigning second-order increments do not, fortunately, invalidate Newton's ensuing argument since the precise value of $p(=\dot{y}o+\ddot{y}o^2, \textit{recte})$ does not there come into consideration.

(6) Namely, $(\tfrac{1}{2}LG-HN =)$ $\tfrac{1}{2}\ddot{y}o^2+O(o^3)$. In sequel Newton will, keeping the lengths $EL = 2o$ and $LG = 2p$ fixed, allow N to move freely in CF by varying this *sagitta* of the arc $E\widehat{N}G$.

(7) Newton takes the lengths of the arcs $\widehat{EN}$ and $\widehat{NG}$ to differ insignificantly (to $O(o^3)$, it is understood) from those of their chords $EN = \sqrt{[o^2+(p-q)^2]}$ and $NG = \sqrt{[o^2+(p+q)^2]}$; and the speeds of the body falling along them to be those attained (in fall from rest at A) at their initial points E and N, and hence proportional to the square roots of $AB = x$ and $AC = x+o$ respectively. He then, maintaining x, o and p fixed but varying $FN = q$, seeks the position of N in the horizontal CF at which the total time of fall $E \to N \to G$ is least.

(8) This equation to zero of the fluxion, with respect to variable q, of the sum of $R = EN/\sqrt{x}$

[Quare] $-Sp\dot{q}o+Sq\dot{q}o-Sp\dot{q}x+Sq\dot{q}x+Rxp\dot{q}+Rxq\dot{q}=0.$ [id est]

$$\frac{\sqrt{oo+pp+2pq+qq}}{\sqrt{x+o}}\text{ in }\overline{-po+qo-px+qx}=\frac{\sqrt{oo+pp-2pq+qq}}{\sqrt{x}}\times\overline{xp+xq}.$$

[seu] $\sqrt{oo+pp+2pq+qq}\text{ in }\overline{-p+q}\times\overline{x+o}\times\sqrt{x+o}$

$$=\sqrt{oo+pp-2pq+qq}\text{ in }\overline{p+q}\times x\times\sqrt{x}.^{(9)}$$

[hoc est deletis infinitissime parvis$^{(10)}$]

$$\sqrt{oo+pp+2pq}\text{ in }\overline{q-p}\text{ in }\overline{\left.\frac{x+o}{x}\right|}^{\frac{3}{2}}=\sqrt{oo+pp-2pq}\text{ in }\overline{q+p}.$$

[ideoǫ quadrando]

$$\overline{oo+pp+2pq}\text{ in }\overline{pp-2pq}\text{ in }\overline{1+\frac{3\times o}{x}}=\overline{oo+pp-2pq}\text{ in }\overline{pp+2pq}.$$

[id est] $oopp-2,oopq+p^4\overline{\underset{+}{-}}2p^3q\text{ in }\overline{1+3\frac{o}{x}}=oopp+2,oopq+p^4\overset{+}{\underset{-}{}}2p^3q.$

---

and $S = NG/\sqrt{(x+o)}$ is manifestly the condition for the aggregate of the times over $EN+NG$ to attain a local extreme value, evidently the minimum one which Newton seeks. He assumes without further fuss—and as it is correct so to do in this instance (for a rigorous justification see, for example, L. A. Pars, *An Introduction to the Calculus of Variations* (London, 1962): 125–7)—that this local minimality of time of fall over *ENG* is a sufficient requirement for the total time of fall from rest at *A* over the curve $E\widehat{N}M$ to the given terminal point *M* to be the least possible. At once, on dividing through Newton's equation by $\dot{q}(=\frac{1}{2}\ddot{y}o^2\neq 0)$ and replacing R and S by their respective values $\sqrt{[o^2+(p-q)^2]}$ and $\sqrt{[o^2+(p+q)^2]}$, there is

$$(p-q)/\sqrt{[o^2+(p-q)^2]}.\sqrt{x}=(p+q)/\sqrt{[o^2+(p+q)^2]}.\sqrt{(x+o)};$$

whence, since o is indefinitely small in comparison with x, Newton's minimality condition determines (compare VI: 460, note (14)) that $[\partial(\sqrt{[o^2+p^2]}/\sqrt{x})/\partial p=]\,p/\sqrt{[o^2+p^2]}.\sqrt{x}$ is constant, that is, because $o/p=\dot{x}/\dot{y}$, there is $\dot{y}/\sqrt{[\dot{x}^2+\dot{y}^2]}\propto\sqrt{x}$. In sequel he proceeds in effect to differentiate (with respect to the base variable x) this fluxional equation defining the brachistochrone's general point $N(x,y)$, straightaway obtaining Fatio's curvature condition by way of the *sagitta* $NF = q = \frac{1}{2}\ddot{y}o^2$.

(9) The factors '$\times\overline{x+o}$' and '$\times x$' on the opposite sides of this equation—tentatively introduced by carets in the manuscript, manifestly as an afterthought, and erroneously maintained by Newton to blot his succeeding computation—need to be deleted. In the next three equations, correspondingly, the surd factor '$\left.\overline{\frac{x+o}{x}}\right|^{\frac{3}{2}}$' and its square '$1+3\frac{o}{x}[+O(o^2)]$' must be reduced by their quotient to be '$\left.\overline{\frac{x+o}{x}}\right|^{\frac{1}{2}}$' and '$1+\frac{o}{x}$' respectively.

(10) Powers of the vanishingly small variable q, namely.

[et per reductionem] $3ppoo + 3p^4 = 4pqxo + 6pqoo$.[11] [Unde posito quod circulus æqualiter curvus cum linea ad N secat CN in R, erit] $\dfrac{4xo}{3p} = \dfrac{EF^q}{q} = [N]\,R$.[12]

---

(11) Read '$ppoo + p^4 = 4pqxo$' *tout court* on removing from Newton's equation the extraneous factor 3 by which (compare note (9)) its left-hand side is erroneously multiplied, and also deleting (as being indefinitely smaller than the rest) the final term '$6pqoo$'.

(12) Since $EF^2 = o^2 + p^2$ to $O(o^3)$, and the chord NR of the circle having the same curvature as the brachistochrone ANM at N is, in the limit where the chord EG which it bisects comes to be vanishingly small, equal to EN^2/NF. When Newton's earlier slip (see notes (9) and (11) preceding) is set right, there correctly ensues $NR = (o^2 + p^2)/q = 4xo/p$. It follows that the centre, O say, of curvature at N lies at a distance $OS = \frac{1}{2}NR.(p/o) = 2x = 2AC$, and hence that the radius ON is bisected by the horizontal AP through the initial point A at which the body begins to fall—'adeoǥ punctum N et centrum curvaminis O æqualiter distant hinc inde a linea horizontali AP data, qua proprietate cyclois innotescit. Est ergo curva ANM cyclois. Q.E.I.' ends, *mutatis mutandis*, David Gregory's Newtonian memorandum C122[2]. (On this see

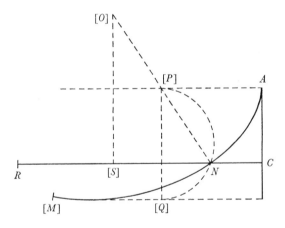

note (1) above where we also quote Fatio's own little-differing parent conclusion.) In an unfairly hostile review of Fatio's 'Lineæ Brevissimi Descensus Investigatio...' in *Acta Eruditorum* (November 1699): 510–13 Leibniz, privately much inflamed at being there relegated to be but a 'secundus Inventor' of the calculus (compare note (1) preceding), took him sarcastically to task for troubling to demonstrate this well-known defining property of the cycloid 'discovered by Huygens'. (However, this 'cycloidis proprietas' is touched upon only fleetingly in Propositio VI of the 'Pars Tertia' of Huygens' *Horologium Oscillatorium* (Paris, 1673): 66–7, and—had Leibniz but known it—was first set out explicitly by Newton in Problem 5 of his 1671 fluxional tract; see Corol. 1 on III: 162.) Six months later in a 'Responsio ad Dn. Nic. Fatii Duillerii Imputationes' (*Acta* (May 1700): 198–214) he did a *volte-face*, now castigating Fatio for arriving at an equivalent second-order curvature condition in the case of the solid of revolution of least resistance, in preference to the first-order defining differential equation which Newton—and after him L'Hospital and Johann Bernoulli and others—had earlier attained: 'quod perinde est ac si quis problema planum ad sectiones Conicas, immo altiores [curvas] referat' (*ibid.*: 201; on the general background see more fully VI: 467–8, note (25)). In fact, the passage from Fatio's second-order curvature property defining the brachistochrone

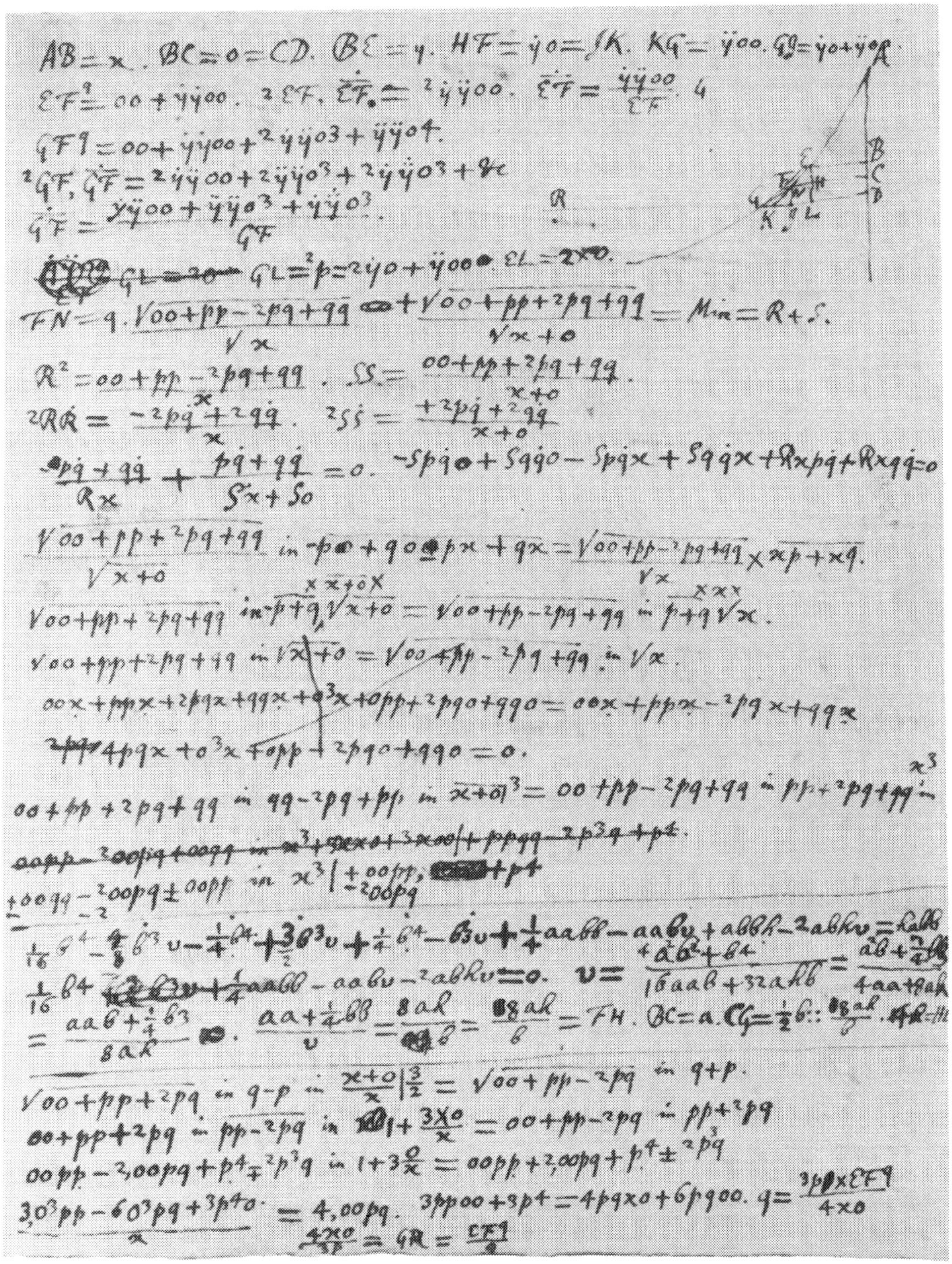

Plate I. The brachistochrone determined from its radius of curvature (**1, 1**, Appendix 2).

to be the cycloid to the tangential one implied in Leibniz's preferred first-order equation is entirely straightforward: for since, in our preceding figure, there is $OS = \dot{y}(1+\dot{y}^2)/\ddot{y} = 2x$, and hence $x^{-1} = 2(\dot{y}^{-1}-\dot{y}/(1+\dot{y}^2))\,\ddot{y}$, there ensues by direct integration $x/a = \dot{y}^2/(1+\dot{y}^2)$ and so $\dot{y} = \sqrt{[x/(a-x)]}$; whence, by a second integration, the curve of swiftest fall is (given that $x = y = 0$ at the origin A) the cycloid defined by the Cartesian equation

$$CN = y = \tfrac{1}{2}a\,\cos^{-1}(1-2x/a)-\sqrt{[x(a-x)]},$$

where $AC = x$ is the distance vertically fallen and $PQ = a$ is the diameter of the circle which generates the cycloid by rolling along under the *linea horizontalis AP*.

2

THE 'DE QUADRATURA CURVARUM' REVISED FOR PUBLICATION[1]

[late 1703][2]

From the autograph originals in the University Library, Cambridge

§1. *Preliminary tentatives at augmenting the 1693 tract.*[3]

[1][4]   *Prop. II. Prob. II.*[5]

Data quacunꝗ relatione fluentium quantitatum invenire relationem fluxionum.
Solvitur Problema per methodum rationum primarum & ultimarum, ut in exemplis sequentibus.

Exempl. 1.[6] Si recta *PG* circa polum *P* revolvitur & interea secat rectas duas

---

(1) Briefly to resume our introductory account of the circumstances which led Newton to revamp his 1693 tract on algebraic quadrature (vii: 508–52), let us remind that when his newly augmented 'Tractatus de Quadratura Curvarum' made its public bow in April 1704 on the coat-tails of his *Opticks*—as the second of the 'Two Treatises of the Species and Magnitude of Curvilinear Figures' appended to it—he was to say on the second page of the 'Advertisement' to his main work: 'In a Letter written to Mr. *Leibnitz* in the year 1676. and published by Dr *Wallis* [*sc.* the *epistola posterior* of 24 October 1676, printed by Wallis in the 'Epistolarum quarundam Collectio' added by him to his *Opera Mathematica*, 3 (Oxford, 1699): 634–45], I mentioned [compare vii: 48–52] a Method by which I had found some general Theorems about squaring Curvilinear Figures, or comparing them with the Conic Sections, or other the simplest Figures with which they may be compared. And some Years ago I lent out a Manuscript containing such Theorems, and having since met with some Things copied out of it, I have on this Occasion made it publick, prefixing to it an *Introduction* and subjoyning a *Scholium* concerning that Method.' We need here no more than fleetingly notice that the un-named person who had without their author's approval circulated theorems 'copied out' of his manuscript tract on quadrature of curves (or so Newton unjustly jumped beyond the evidence to conclude) was the Scots iatromathematical physician George Cheyne, who had in the spring of 1703—directly counter to Newton's expressed wish, it would appear—published a well-stocked if not greatly original *Fluxionum Methodus Inversa; sive Quantitatum Fluentium Leges Generaliores* in which he surveyed the broad splay of recent researches into the 'inverse method of fluxions' (that is, integration techniques), dwelling in particular upon Newton's own favourite realm of quadrature by expansion into infinite series and indicating elaborations of his own which (probably not to his prior knowledge) duplicated parallel extensions in Newton's still unpublished 1691 and 1693 treatises 'De quadratura Curvarum'. So wilful a plagiarism (or so he evidently regarded it) of his own hard-won calculus theorems by a man to whom he had taken an instant dislike at their first meeting did indeed provoke Newton in the later

Translation

[1][4]   *Proposition II, Problem II.*[5]

Given any relationship whatever of fluent quantities, to find the relationship of their fluxions.

The problem is solved by the method of first and last ratios, as in the following examples.

Example 1.[6] If the straight line *PG* revolves round the pole *P* and all the while

months of 1703 to revise his 1693 text on quadrature of curves for publication with but minimal reshaping other than for a newly composed 'Introductio' and a concluding scholium where he heralded the path onwards from its basic propositions 'to greater things'. We here reproduce in §§1/2 the essential portion of Newton's surviving drafts of this added corpus of material where he honed his notion of the concept of 'instantaneous' increase by continuous flow and groped his way with gathering sureness to the outline of geometrical modes of fluxion which he sketched in preface to his earlier severely algebraic discussion of quadrature by finding the 'fluent' defining equations of curves, given their ordinates; and adjoin in §3, from the autograph copy put to the printer in the early winter of 1703/4, the finished text of both this 'Introductio' and the concluding portion of the 'Tractatus de Quadratura Curvarum' as it was published to the world.

(2) When Archibald Pitcairne wrote on 1 October 1703 to a fellow Scot, Colin Campbell, to inform him that '2 sheets [of Newton's *Opticks*] are cast off already', he gave him the news that 'He adds to the book all his about quadratures' (National Library of Scotland. MS 3440: 20ᵣ). Newton could not, of course, have begun to augment his 1693 tract on quadrature of curves for publication till after he had seen the full, printed version of Cheyne's *Fluxionum Methodus Inversa* which appeared earlier in the year, and which (see the preceding introduction) then stimulated him to do so. The finished autograph of the revised 'De Quadratura' (like that of the companion 'Enumeratio') was put to press at the beginning of the winter: 'Hyeme præterita', wrote Newton in sending a copy of the completed *Opticks* (and its attendant mathematical treatises) to an unknown foreign recipient on 26 May 1704, 'librum de rebus Opticis... olim scriptum in lucem edidi cujus exemplar ad te mitto' (see Lot 163 in Sotheby's *Catalogue of the Newton Papers sold by Order of the Viscount Lymington* [13/14 July] 1936: 37).

(3) In these initial ameliorations of the 1693 parent text of the 'De Quadratura' Newton first, in [1], drafts a fresh Proposition II incorporating illustrative examples of applying the method of 'first and last ratios' to resolving geometrical problems; then, in [2], begins to append its equivalent to his original Prop. I as an additional 'Cas. 2'; and also, in [3], lays out a broad scheme for a new Prop. II in which the 'solving of problems by the finding of fluxions' is exemplified, on the pattern of his 1671 tract of more than thirty years before, in twelve subsidiary problems, three of which are elaborated in [4] following.

(4) Add. 3960.7: 87, a page carefully penned by Newton in the first instance, but afterwards more roughly interlineated.

(5) Understand that this newly contrived second proposition is to be inserted in the parent text (see VII: 516) immediately after the opening 'Prop. I. Prob. I' which is its algebraic counterpart.

(6) Newton initially set the more general second example following to be his prime illustration of finding the relationships of geometrical fluxions, but we here advance to its correct place in sequence the special case of it—already twenty years earlier presented by him as Proposition 30 of the first book of his 'Geometria Curvilinea' (see IV: 474)—which he appended in revise at the bottom of the manuscript page.

posit[ione] datas *AB AC* in *G* et *H*: quæruntur fluxiones ipsarum *AG* et *AH*.
Rotetur recta *PG* circa polum suum *P* donec puncta *G* et *H* perveniunt ad *g* et
h et fluxiones ipsarum *AG* et *AH* erunt ad invicem in prima ratione incrementorum nascentium *Gg* et *Hh*.[7]

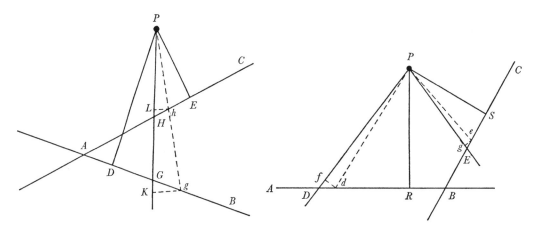

Exempl. [2].[8] Rectæ *PD*, *PE* angulum *DPE* magnitudine datum continentes
rotantur circa polum positione datum *P* et interea secant rectas positione datas
AB, *BC* in *D* et *E*: comparandæ sunt fluxiones rectarum *AD* et *BE*.[9]

Rotetur angulus *DPE* donec puncta *D* et *E* perveniant ad *d* et *e* et fluxio
ipsius *AD* erit ad fluxionem ipsius *BE* in prima ratione partium nascentium
Dd et *Ee*[,] quæ ratio sic invenitur. Demittantur ad *PD* et *PE* perpendicula *df*
et *eg* et a Polo *P* in rectas *AB* et *B*[*C*] demittantur perpendicula *PR*, *PS*. Ob
sim[ilia] tri[angula] *Dfd DRP* est *DP.PR*::*Dd.df*. ob sim. tri. *Pfd*, *P*[*ge*]
est P*d.Pe*::*df.eg*, et ob sim. tri. *Ege*, *ESP* est *PS.PE*::*eg.E*[*e*][,] et componendo
has tres proportiones est *DP*, P*d*, *PS.PR*, P*e*, *PE*::*Dd.E*[*e*]. Coeant jam puncta
d et *e* cum punctis *D* et *E*[,] et ultima ratio linearum evanescentium *Dd* et *Ee*
(seu prima nascentium) erit ratio *DP*q, *PS* ad *PR*, *PE*q et in hac ratione est fluxio
ipsius *AD* ad fluxionem ipsius *BE*.

Simili argumento[10] erit fluxio
ipsius *PD* ad fluxionem ipsius *PE*
in ratione composita ex rationibus
DR ad *PR*, *DP* ad *PE* et *PS* ad *ES*.
hoc est ut

DR × *DP* × *PS* ad *PR* × *PE* × *ES*.[11]

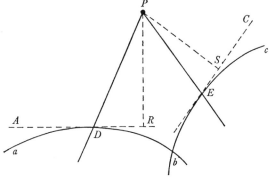

Et si vice rectarum *AB*, *BC* adhi-
bentur Curvæ quæcunꝗ *aDb*, *bEc*
et ad puncta *D* et *E* in quibus rectæ
PD PE datum angulum *DPE* continentes incidunt in Curvas du[c]antur curva-

cuts in G and H two straight lines AB, AC given in position: there are required the fluxions of AG and AH. Rotate the line PG round its pole P till the points G and H reach g and h, and then the fluxions of AG and AH will be to one another in the first ratio of the nascent increments Gg and Hh.[7]

Example 2.[8] The straight lines PD, PE containing the angle $\widehat{DPE}$ given in size rotate round the pole P given in position and all the while cut in D and E the straight lines AB, BC given in position: the fluxions of the straight lines AD and BE[9] are to be compared.

Let the angle $\widehat{DPE}$ rotate till the points D and E reach d and e, and then the fluxion of AD will be to the fluxion of BE in the first ratio of the nascent parts Dd and Ee: which ratio is found in this manner. Let fall the perpendiculars df and eg to PD and PE, and from the pole P let fall the perpendiculars PR and PS to the lines AB and BC. Because of the similar triangles Dfd, DRP there is $DP:PR = Dd:df$, because of the similar triangles Pfd, Pge there is $Pd:Pe = df:eg$, and because of the similar triangles Ege, ESP there is $PS:PE = eg:Ee$; and by compounding these three proportions there is

$$DP \times Pd \times PS : PR \times Pe \times PE = Dd : Ee.$$

Now let the points d and e come to coincide with the points D and E, and the last ratio of the vanishing lines Dd and Ee (that is, the first one of these nascent) will be the ratio $DP^2 \times PS$ to $PR \times PE^2$, and in this ratio is the fluxion of AD to the fluxion of BE.

By a similar argument[10] the fluxion of PD will be to the fluxion of PE in a ratio compounded of the ratios of DR to PR, of PD to PE and of PS to ES, that is, as $DR \times PD \times PS$ to $PR \times PE \times ES$.[11]

And if instead of the straight lines AB, BC there be adduced any curves aDb, bEc and at the points D and E in which the straight lines PD, PE containing the given angle $\widehat{DPE}$ are incident at the curves there be drawn the curves' tangents

---

(7) The lines PD and PE in Newton's accompanying figure are of course the perpendiculars let fall from P to AB and AC respectively. It follows that, since *ob similia triangula* there is $Gg/Kg = PG/PD$, $Kg/Lh = PK/PL$, $Lh/Hh = PE/PH$ and hence *connectendo rationes* $Gg/Hh = PG \times PK \times PE/PD \times PL \times PH$, in the limit as Phg comes to coincide with PHG there ensues $\mathrm{fl}(AG):\mathrm{fl}(AH) = PG^2/PD:PH^2/PE$.

(8) The manuscript—where (see note (6) above) this generalized *exemplum* is set in primary place—reads '1', but we adjust the number to take account of the preceding particular case (where $\widehat{DPE}$ is a straight line) which Newton introduced in its preliminary revision.

(9) 'velocitates punctorum D et E' (the speeds of the points D and E) was first written.

(10) 'ratione' (reasoning) is replaced to avoid confusion with the immediately following use of the Latin word in its technical mathematical sense.

(11) For before the rotating angle $\widehat{dPe}$ came to coincide with $\widehat{DPE}$ there was (by similar triangles) $Df/fd = DR/PR$, $fd/ge = Pd/Pe$ and $ge/Eg = PS/ES$, so that

$$Df/Eg = DR \times Pd \times PS/PR \times Pe \times ES.$$

rum tangentes *AD* et *EC*, et ad tangentes illas demittantur perpendicula *PR*, *PS*: fluxiones curvarum *aD*, *bE* erunt ad invicem ut $DP^q \times PS$ ad $PE^q \times PR$, et fluxiones rectarum *PD*, *P*[*E*] erunt ad invicem ut $DR \times DP \times PS$ and $PR \times PE \times ES$.[12]

---

[2][13] *Prop.* [*I*]. *Prob.* [*I*].[14]

Data relatione[15] *fluentium quantitatum compar*[*ar*]*e fluxiones earum inter se.*

Cas. 1. Si relatio illa per æquationem definitur, multiplicetur omnis...[16] nova.

Solutionis Explicatio.

Sunto *a*, *b*, *c*, *d* &c...[16] dant *m* = [*n* =] 1.

Cas. 2. Si relatio fluentium quantitatum definitur per motum linearum secundum leges assignatas, fluxiones primo intuitu quandoϙ innotescunt. Ut si ad basem[17] communem applicentur plurium Curvarum Ordinatæ quæ ad idem basis punctum semper insistentes & motu uniformi super basi[17] delatæ describunt areas illarum Curvarū: patet primo intuitu quod fluxiones arearum sunt ut Ordinatæ, ideoϙ per Ordinatas exponi possunt, & quod baseos pars ad Ordinatas abscissa fluit uniformiter. Quoties autem in investigatione fluxionum hæretur, recurrendum est ad methodum rationum primarum et ultimarum ut in exemplis sequentibus.[18]

Exempl. 1. Comparandæ sunt fluxiones Lineæ curvæ *AC*, Abscissæ *AB* & Ordinatæ *BC*. Patet primo intuitu quod hæ fluxiones sunt inter se ut Curvæ tangens *CT*, subtangens *TB*[19] et Ordinata *BC*: sed si hoc non pateret problema

---

(12) We have here absorbed into our preceding text a following caption 'Exempl. 2'; after penning this, Newton broke off to insert the special case of his initial 'Exemplum 1' which takes its place in the revised sequence (see note (6) above), introducing it with the fuller observation: 'Fluxiones quantitatum primo intuitu quandoϙ innotescunt; ut quod fluxiones arearum curvarum sunt ut ordinatim applicatæ si modo Abscissæ uniformiter fluunt[,] et quod fluxiones Abscissarū Ordinatarū et Linearū Curvarū sunt inter se ut Subtangentes Ordinatæ et Tangentes earundem Curvarum. Siquando in investigatione fluxionum hæretur, recurrendum est ad methodum rationum primarum et ultimarum ut in sequentibus exemplis' (The fluxions of quantities are on occasion to be known at first glance; so it is that the fluxions of the areas of curves are as the ordinates provided the abscissas flow uniformly, and that the fluxions of the abscissas and ordinates of curves and of their lengths are to one another as the subtangents, the ordinates and the tangents of the same curves. But if ever you stick in the discovery of fluxions, you must have recourse to the method of first and last ratios, as in the following examples).

In [1] in the following Appendix we reproduce a rough skeletal calculation (ULC. Add. 3970.3: 477ʳ) in which Newton adumbrates yet other fluxional inter-relationships between the elements of a similar rotating 'mobile engine'—all very elegant, but without broad significance.

AD and *EC*, and then to those tangents there be let fall the perpendiculars *PR* and *PS*: the fluxions of the curve-arcs $\widehat{aD}$, $\widehat{bE}$ will be to one another as $DP^2 \times PS$ to $PE^2 \times PR$, and the fluxions of the lines *PD*, *PE* will be to one another as $DR \times PD \times PS$ to $PR \times PE \times ES$.[12]

---

[2][13]   *Proposition I, Problem I.*[14]

Given a relationship[15] *of fluent quantities, to compare their fluxions with each other.*

Case 1. If that relationship is defined by an equation, multiply every. . .[16] a fresh (equation).

Explanation of the solution.

Let *a, b, c, d,* . . . be . . . . . .[16] they yield $m = n = 1$.

Case 2. If the relationship of the fluent quantities is defined by the motion of lines following assigned laws, the fluxions are on occasion to be known at first glance. If, for instance, to a common base[17] there be applied the ordinates of several curves which, ever standing at the same point of the base and carried along upon the base[17] with a uniform motion, describe the areas of those curves: it is evident at first glance that the fluxions of the areas are as the ordinates, and can in consequence be expressed by the ordinates; and that the part of the base cut off at the ordinates is uniformly fluent. Each time you stick in the discovery of fluxions, however, you must have recourse to the method of first and last ratios, as in the following examples.[18]

Example 1. There are to be compared the fluxions of the curve-line *AC*, its abscissa *AB* and the ordinate *BC*. It is manifest at first inspection that these fluxions are to one another as the curve's tangent *CT*, its subtangent *TB*[19] and

---

(13) Add. 3960.7: 80, a recasting of the essence of [1] preceding (on the verso of the same folded folio sheet) now set by Newton to be a new Case 2 of the opening proposition of the 'De Quadratura'.

(14) In copy of the heading to [1] the manuscript reads 'Prop. II. Prob. II', but our emendation is beyond doubt what Newton intended here to write.

(15) 'quacunqʒ' (any. . .whatever) is deleted.

(16) Understand the text of Proposition I of the 'De Quadratura' in its 1693 version (vii: 512–16).

(17) Initially 'Abscissa(m)' and 'abscissæ' (abscissa).

(18) This repeats all but word for word the transitional draft on the manuscript's recto (Add. 3960.7: 87) which we have quoted in note (12) above.

(19) In the manuscript Newton inserted, and straightaway deleted as unnecessary, the explicit instruction: 'Ducatur recta *CT* curvam tangens in *C* et Abscissæ occurrens in *T*' (Let there be drawn the straight line *CT* touching the curve at *C* and meeting the abscissa in *T*).

sic solveretur. Progrediatur Ordinata *BC* de
loco suo *BC* ad locum quemvis novum *bc* et
ad *bc* demisso perpendiculo *CD*,[20] augmenta
Curvæ, Abscissæ & Ordinatæ erunt *Cc*, *CD*, &
Dc et fluxiones quæsitæ erunt inter se in prima
ratione horum augmentorum nascentium.[21]
hoc est in ratione Tangentis *CT*, subtangentis
TB et Ordinatæ *BC*.[22]

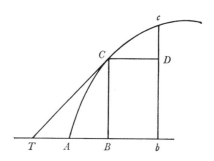

   Exempl. 2.[23]

---

[3][24]   *Prop. II.*[25]

   Per inventionem Fluxionum solvere Problemata.

Res facilius exemplis addiscitur quam præceptis.[26]

   Prob. 1. Invenire maxima et minima. Solut[*io*]. Quantitas ubi maxima [vel
minima][27] est cessat fluere. Quære ergo fluxionem ejus et pone æqualem
nihilo.[28]

   Prob. 2. Invenire fluxiones linearum Curvarum. Solut. Componuntur ex fluxionibus
rectarum[29] quæ ad Curvam terminantur.

   Prob. 3. Ducere Curvarum tangentes. Solut. Quære fluxionem Curvæ et ejus

---

   (20) Parallel to the base *AB*, that is; it is understood that the angle $\widehat{ABC}$ of ordination is
right.

   (21) 'vel quod perinde est in ultima ratione evanescentium' (or, what is exactly the same,
in the last ratio of the vanishing ones) is cancelled. Newton thereafter first concluded: 'Per
puncta *C*, *c* agatur *cCE* Abscissæ occurrens in *E* et ob similitudinem triangulorum *CDc*, *EBC*
augmenta illa *Cc*, *CD*, *Dc* erunt inter se ut sunt *EC*, *EB*, *BC*' (Through the points *C*, *c* draw
cCE meeting the abscissa in *E*, and because of the similarity of the triangles *CDc*, *EBC* those
augments *Cc*, *CD*, *Dc* will then be to one another as are *EC*, *EB*, *BC*). In his manuscript figure
he has deleted the corresponding extension of the chord *cC* as far as its intersection with the
base *AB* in *E*.

   (22) This of course yields (where the angle $\widehat{ABC}$ is no longer restricted to be right) the
general solution to the Cartesian problem of tangents virtually as Newton penned it nearly
forty years earlier, on 20 May 1665, as 'An universall theorem for drawing tangents to
crooked lines' (see 1: 279), and as he rephrased it in both English and Latin versions on
innumerable occasions thereafter. By the 1700's its construction had become no more than
a textbook exercise, but this yet deserved a more precise justification than the loose appeal to
the 'first ratios' of the nascent increments of abscissa, ordinate and arc-length which Newton
here—after a momentary thought of invoking the limit-chord *Cc* (see the previous note)—
alone presents. He subsequently realized as much, and in his final version (see §3.1 below) of
this prime geometrical illustration of the working of his method of 'first/last ratios' vindicated
his present assumption that the vanishingly small arc-increment $\widehat{Cc}$ may be taken to be represented by the contemporaneous fluxional augment of its tangent *TC* by again introducing the
former's chord *Cc* (and its extension to the base) with which both come 'ultimately' to coincide.

   (23) Newton here breaks off, but we may safely assume that—as in the redrafted version of

its ordinate *BC*. But were this not evident, the problem would be solved in this manner: let the ordinate *BC* advance from its place *BC* to any new place *bc* and, with the perpendicular *CD* let fall to *bc*,[20] the augments of the curve, its abscissa and its ordinate will be $\widehat{Cc}$, *CD* and *Dc*; and the required fluxions will then be to one another in the first ratio of these nascent increments,[21] that is, in the ratio(s) of the tangent *CT*, the subtangent *TB* and the ordinate *BC*.[22]

Example 2.[23]

---

[3][24]   *Proposition II.*[25]

To solve problems by the finding of fluxions.

The matter is more readily learnt from examples than precepts.[26]

Problem 1. To find maxima and minima. Solution. When a quantity is greatest (or least)[27] it ceases to 'flow' (vary). Therefore seek out its fluxion and put it equal to nothing.[28]

Problem 2. To find the fluxions of curved lines. Solution. They are compounded of the fluxions of straight lines[29] which terminate at the curve.

Problem 3. To draw the tangents of curves. Solution. Seek out the fluxion of the

---

this 'Cas. 2' which (see §2.2 following) he inserted in his tentative introduction to his treatise 'Methodus Fluxionum'—he meant there to follow 'Exemplum 1' (and may be also 'Exemplum 2'?) of [1] preceding.

(24) Add. 3970.3: 360ʳ. The date of this piece, and also that of [4] ensuing (which follows immediately upon it in the manuscript), is confirmed by the fact that on the same sheet Newton went on—manifestly in the last months of 1703—to draft the final paragraph of the concluding Observation XI of the third book of his *Opticks* (→₁1704: ₂131).

(25) This new 'Prop. II. [Prob.]' was, we may reasonably conjecture, intended to follow on the augmented first proposition of the 'De Quadratura' in (2) preceding. The twelve subsidiary *Problemata* which it enunciates and, in part, sketchily resolves are, it will be evident, taken over in the main from Newton's 1671 tract; Problems 9 and 10 are newly contrived, but represent minimal variations upon traditional Archimedean and half-a-century-old Huygenian techniques of cubature of solids of revolution and the 'planification' of their curved surfaces; while Problems 11 and 12 are drawn from Newton's efforts in the 1680's and early 1690's to measure the central forces under which bodies traverse given (plane) curves round a point force-centre. We specify the individual antecedents in following footnotes.

(26) The ensuing problems are here reordered into the sequence of Newton's recast numbering of them in the manuscript. Initially Problem 2 came after Problem 6, and Problems 9/10 came last.

(27) We make good a trivial omission on Newton's part.

(28) Problem 3 of the 1671 tract (III: 116–18). This criterion does not of course distinguish the true maxima and minima of a quantity from the points (of inflexion in the corresponding Cartesian graph) where it is momentarily stationary in its onward growth or decline.

(29) 'vel curvarum' (or curves) Newton would have added had he here (compare VII: 456–8) other than a rectilinear defining mechanism in mind. As his amplification of it in [4] below makes clear, this is but a rider to Problem 3 following.

exponens Curvam tanget, [adeoq͛] secundum ejus determinationem, age rectam tangentem.[30]

Prob. 4. Invenire centra curvaminis. Solut. Centrum curvaminis est concursus ultimus duorum perpendiculorum coeuntium.[31]

Prob. 5. Invenire Curvaturam maximam vel minimam.[32] *Solut.* Quære fluxionem radij Curvaminis et pone nullam.

Prob. 6. Invenire puncta rectitudinis et flexûs contrarij.[33] *Solut.* Pone radium curvaminis infinitum. Vel quære fluxionem secundam Ordinatim Applicatæ et pone nullam.

Prob. 7. Invenire Curvas[34] *quæ quadrari possunt.*[35] *Solut.* Assume relationem quamvis inter aream et abscissam. Quære fluxionem areæ & huic proportionalem pone Ordinatim applicatam.

Prob. 8. Invenire Lineas Curvas quibus rectas æquare licet.[36] *Solut.* Centra curvaturæ locantur in hujusmodi curvis.

Prob. 9. Invenire solida rotunda quæ cubari possunt.[37] Solida rotunda voco quæ ex curvis circa axem rotatis generantur. *Solut.* Assume relationem inter solidum et abscissam. Quære fluxionem solidi, et hujus radici quadraticæ proportionalem fac Ordinatam Curvæ.[38] Nam hujus curvæ revolutione generabitur solidum quæsitum.

Prob. 10. Invenire solida rotunda quorum superficies quadrari possunt. Solut.[39]

Prob. 11. Corporum quæ in Curvis revolvuntur invenire vires centrifugas. Solut. Sunt directe ut quadrata velocitatis & reciproce ut Radij curvaturæ.[40]

---

(30) In solution of this carry-over from Problem 4 of his 1671 tract (III: 120–48) Newton originally specified: 'Ex fluxionibus rectarum quæ ad Curvam terminantur compone determinationem Curvæ' (From the fluxions of the straight lines which are terminated at the curve compound the instantaneous direction of the curve). For the notion of a finite 'exponent' of the infinitesimal 'determination' of a curve's motion which Newton here invokes, see the 'Exempla prima' of finding the tangential direction of points moving along generalized Cartesian, bipolar and otherwise co-ordinated curves at the end of his 'Geometriæ Liber primus' of ten years before (VII: 458–68).

(31) Problem 5 of the 1671 tract, here—as the amplified version in [4] below confirms—restricted once more to the case (compare III: 150–64) where the curve is defined by Cartesian coordinates. Newton has deleted in sequel—as obvious?—the equivalent specification 'vel est punctum quiescens perpendiculi mobilis' (or is the point [instantaneously] at rest in the mobile perpendicular), and has also suppressed the alternative determination of the curvature centre as that of the osculating circle whose chord he defined (compare VII: 108) by stating: 'Vel sic. Fluxio secunda [*read* Dimidium fluxionis secundæ *correctly*] ordinatim applicatæ est ad fluxionem Curvæ ut hæc fluxio ad chordam circuli [eandem habentis curvaturam cum curva] in Ordinata versus' (Or thus: [half] the second fluxion of the ordinate is to the fluxion of the curve as this fluxion to the chord of the circle [having the same curvature as the curve] in the ordinate's direction).

(32) Cognate Question 4 at the end of Problem 5 of the 1671 tract (III: 180–2).

(33) Cognate Questions 2/3 in the 1671 tract's Problem 5 (III: 178–80).

(34) '*quotcunq͛*' (*any number of*) is deleted.

(35) Problem 7 of the 1671 tract (III: 194–6).

curve, and its exponent will touch the curve; consequently in its instantaneous direction draw the straight line touching the curve.[30]

Problem 4. To find centres of curvature. Solution. The centre of curvature is the ultimate (limit-)meet of two normals as they coincide.[31]

Problem 5. To find maximum or minimum curvature.[32] *Solution.* Discover the fluxion of the radius of curvature and set it nil.

Problem 6. To find points of straightness and of contrary flexion.[33] *Solution.* Set the radius of curvature infinite. Or seek the second fluxion of the ordinate and put it nil.

Problem 7. To find[34] *curves which can be squared.*[35] *Solution.* Assume any relationship you will between the area and the abscissa. Seek the fluxion of the area and proportional to this set the ordinate.

Problem 8. To find curved lines to which it is allowable to equate straight ones. [36] *Solution.* The centres of curvature are located in curves of this sort.

Problem 9. To find round solids which can be cubed.[37] (I call solids 'round' which are generated from curves rotated round an axis.) *Solution.* Assume a relationship between a solid and its abscissa; then seek the fluxion of the solid and proportional to the square root of this make the ordinate of the curve.[38] For the required solid will be generated by the revolution of this curve.

Problem 10. To find round solids whose surfaces can be squared. Solution.[39]

Problem 11. Where bodies revolve in curves, to find their centrifugal forces. Solution. They are directly as the squares of the speed and reciprocally as the radii of curvature.[40]

---

(36) Originally Newton began to write just '*Invenire Longitudi*[*nes Curvarum*]' (*To find the lengths of curves*), and thereafter successively altered this to read '*Invenire Lineas Curvas quibus æquales rectæ exhiberi possunt*' (*To find curved lines, straight ones equal to which can be furnished*), '*...quarum longitudines inveniri possunt*' (*...whose lengths can be found*), '*...quæ rectificari possunt*' (*...which are rectifiable*) and lastly '*...quæ in rectas verti possunt*' (*...which can be turned into straight ones*). This reduction of the general rectification of curves to the construction of their evolutes is Problem 10 of the 1671 tract (III: 292–302).

(37) Initially '*quarum contenta inveniri possunt*' (*whose* [solid] *content can be found*).

(38) The factor of proportionality is $1/\sqrt{\pi}$ since the volume of the solid of revolution formed by rotating the Cartesian curve defined by perpendicular coordinates x and y round the x-axis is $\int \pi y^2 . dx$, and hence the fluxion of that volume (with respect to that of the uniformly fluent abscissa x) is π times the square of the ordinate y.

(39) Here, little differently, since the surface-area of the solid of revolution formed by rotating the curve defined by perpendicular coordinates x and y round the x-axis is $\int 2\pi y . ds$, where s is the curve's related arc-length (and hence $\dot{s} = \sqrt{[1+\dot{y}^2]}$ on taking x to be uniformly fluent), we need but to assume any relationship, $z \equiv z_x$ say, between that surface-area and the abscissa, and then the fluxional equation $\dot{z}^2 = 4\pi y^2(1+\dot{y}^2)$ will determine the ordinate y. We will not be surprised that Newton fights shy of setting this resolution of his problem out in words.

(40) Newton's generalization of the simple Huygenian measure of the *vis centrifuga*, away from the centre, induced by constraining a body to rotate uniformly in a circle (on which see

Prob. 12. Invenire vires centripetas[41] *quibus Corpora*[42] *in datis Curvis circa* [*data*] *centra virium*[42] *revolventur.* *Solut.* Si centrum curvaturæ C [ad punctum L curvæ *LM*] sit centrum virium, vires erunt reciproce ut cubus radij curvaturæ *LC*. Sin centrum virium sit punctum quodvis aliud positione datum *P*, ad *LP*, *LC* demitte perpendicula *CD*, *PQ* et vires centripetæ erunt reciproce ut $LD \times LQ^q$.[43]

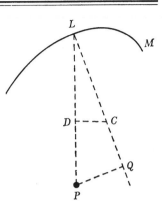

[4][44] *Prob. 2. Invenire fluxiones Linearum curvarum. Solut.*
Componi solent ex fluxionibus rectarum quæ ad curvas terminantur.

Ut si Curvæ alicujus *AC*, sit *AB* vel *GC* Abscissa, *Bb* vel *Cc* exponens fluxionis ejus, *BC* ordinata & *CD* exponens fluxionis ejus. Comple parallelogrammum

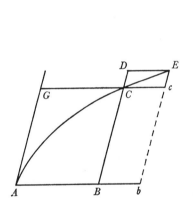 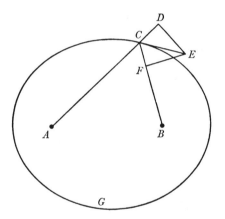

DCcE et ejus diagonalis *CE* erit exponens fluxionis lineæ Curvæ.[45] Si[46] a punctis datis *A*, *B* ad Curvam aliquam *CG* concurrant rectæ *AC*, *BC* et sit *CD* exponens

---

vi: 36–9); in this extension, attained explicitly by him some dozen years before (see vi: **131**, note (86) and 548, note (25)), each infinitesimal portion of the curve traversed is understood to be replaced from instant to instant by the related arc of the osculating circle.

(41) In the first instance Newton here carelessly repeated 'centrifugas' (centrifugal) from the enunciation of the previous Problem 11. Where the force-centre coincides with the centre of curvature, of course, the two forces are equal in magnitude but opposite in the direction of their action. And, to be sure, Newton's following measure of the central force to an arbitrary force-centre is determined as the component of the negative Huygenian *vis centrifuga* which acts simultaneously through the curvature centre; see note (43).

(42) Afterwards specified by Newton in an interlineation in the manuscript to be *L* and *C* respectively, as in his accompanying figure and following text. We omit to reproduce these superfluous designations, considering them to be no more than a private reminder interpolated when he lightly recast the ensuing resolution of the present problem on the next leaf of the manuscript (see [4] below).

Problem 12. To find the centripetal[41] *forces by which bodies*[42] *shall revolve in given curves round (given) centres of force.*[42] *Solution.* If the centre of curvature C at the point L of the curve LM be the centre of force, the forces will be reciprocally as the cube of the radius of curvature CL; but if the centre of force be any other point P given in position, to PL and CL let fall the perpendiculars CD and PQ, and the central forces will then be reciprocally as $DL \times QL^2$.[43]

---

[4][44] *Problem 2. To find the fluxions of curved lines. Solution.* It is usual to compound these of the fluxions of straight lines which terminate at the curve.

Of some curve AC, for instance, let AB (or GC) be the abscissa and Bb (or Cc) the exponent of its fluxion, BC the ordinate and CD the exponent of its fluxion: complete the parallelogram $DCcE$, and its diagonal CE will be the exponent of the fluxion of the curved line.[45] While if from the given points A, B the straight lines AC and BC should meet at some curve CG, and CD be the exponent of the

---

(43) Whence, on letting fall the perpendicular PR ($= QL$) from P onto the tangent LR at L, and extending LP to its meet with the osculating circle in S (so that $LD = \frac{1}{2}LS$), the centripetal force towards P inducing a body to travel in the curve $\widehat{ML}$ will also be inversely as $PR^2 \times LS$. This measure can be deduced as a straightforward consectary of the one set by Newton in prime place in his 1687 *Principia* as Proposition VI of its first book—as indeed he had done a dozen years earlier (see VI: 550) in an added Corollary 5 thereto which was not to appear in print till the *editio secunda* ($_2$1713: 42)—but it is most likely that he would here wish to derive it as a direct sequel to his preceding Problem 11: namely, because the tangential polar PR is (by the Keplerian law of areas; compare note (55) below) inversely pro-

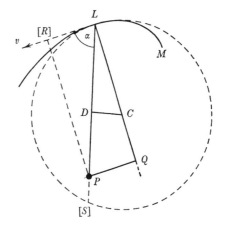

portional to the instantaneous speed, v say, at L of the body traversing $\widehat{ML}$, the Huygenian *vis centrifuga* (apparently) induced at L by the motion is $v^2/CL \propto PR^{-2}/CL$, and this is also the negative of the component of the central force (to P) which is directed through C; hence, where PL is inclined at angle α to the tangent LR, the latter central force is proportional to

$$(v^2/LC . \sin \alpha \propto) \ 1/PR^2 \times LD.$$

(44) Add. 3970.3: 360^r–361^r. This augmented revise of Problems 2, 3 and 12 in [3] preceding follows immediately upon it in the manuscript.

(45) On this standard construction of the tangent problem in oblique Cartesian coordinates by constructing the 'exponent' of the instantaneous direction of 'flow' of a curve see note (30) above. There follows the parallel construction of the problem where the curve is defined by bipolars; compare VII: 464.

(46) Initially 'Rursus si' (Again if). We retain the sense (marking the transition to an example illustrating the solution of the problem in a different system of coordinates) in our English translation.

fluxionis rectæ *AC* et *CF* exponens defluxionis[47] rectæ *BC*, ad has exponentes erige perpendicula *DE*, *FE* concurrentes in *E* et acta *CE* erit exponens fluxionis Lineæ curvæ.

Ut si[48] *AC* Parabola sit, & ejus Abscissa *AB* dicatur *x* et Ordinata *BC y*, & latus rectum *r*; sitcȝ $rx = yy$. erit per Prop 1 $r\dot{x} = 2y\dot{y}$ et inde

$$2y \cdot x^{(49)} :: \dot{x} \cdot \dot{y} :: Cc \cdot cE :: {}^{(50)}$$

Prob. 4. Invenire centra Curvaminis. Solut. Ut exponens fluxionis secundæ Ordinatæ ad quadratum fluxionis lineæ Curvæ ita fluxio curvæ ad line[am rectam radio curvaminis æqualem.][51]

Ut si Figura *PRQ* sit Ellipsis axem habens majorem *PQ* & minorem *RS*. Cape *RH* ad *PC* ut *PC* ad *RC* et erit *H* centrum curvaturæ Ellipseos ad verticem *R*[,] deinde ad Ellipseos punctum quodvis aliud *B* erecto perpendiculo *BVE* axem *PQ* secante in *V* cape *VE* ad *CH* ut $BA^q \times BV$ ad RC^{cub} et erit *E* centrum curvaturæ Ellipseos ad ejus punctum *B*.[52] [Cape] $PX = \frac{1}{2}$ Lat. rect.[53] et *X* [erit] centr[um] curv[aturæ] ad *P*.

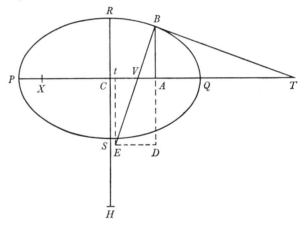

Prob. 12. Invenire vires centripetas quibus Corpora in Curvis datis circa data virium centra revolventur. Solut. In Curva quavis *ATB* circa polum *P* revolvatur corpus *T*.[54] Ad curvam erigatur perpendiculum *TRC* sitcȝ *C* centrum curvaturæ quā Curva habet ad punctū *T* et demissis in *TC* et *TP* perpendiculis *PR* et *CD* erit velocitas Corporis reciproce ut *TR*,[55] et vis centripeta reciprocè ut TR^q in *TD*.[56]

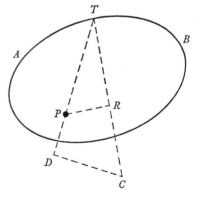

---

(47) Only of course where (as Newton shows it in his figure) the onwards motion in the curve at *C* instantaneously decreases the length of the polar *BC*; in the contrary case read 'fluxionis' (fluxion) simply.

(48) Understand in Newton's first figure, illustrating the case of Cartesian coordinates.

(49) This ratio should be '$2x \cdot y$' ($2x:y$).

(50) Read '$\ldots :: 2y \cdot r$' ($2y:r$) elegantly to complete the construction.

fluxion of the line *AC*, *CF* the exponent of the (backwards)[47] fluxion of the line *BC*: to these exponents erect the perpendiculars *DE*, *FE* meeting in *E* and, when it is drawn, *CE* will be the exponent of the fluxion of the curved line.

For example, if[48] *AC* be a parabola, and its abscissa *AB* be named *x*, its ordinate *BC* *y* and its *latus rectum* *r*, and there be $rx = y^2$, then by Proposition 1 will there be $r\dot{x} = 2y\dot{y}$ and thence $2y : x^{[49]} = \dot{x} : \dot{y} = Cc : cE =$ [50]

Problem 4. To find centres of curvature. Solution. As the exponent of the second fluxion of the ordinate to the square of the fluxion of the curved line, so is the fluxion of the curve to a line [equal to the radius of curvature.][51]

For instance, if the figure *PRQ* be an ellipse having major axis *PQ* and minor one *PS*: take *RH* to *PC* as *PC* to *RC*, and *H* will be the centre of curvature of the ellipse at the vertex *R*; and then at any other point *B* of the ellipse, having erected the normal *BVE* cutting the axis *PQ* in *V*, take *VE* to *CH* as $BA^2 \times BV$ to RC^3, and *E* will be the centre of curvature of the ellipse at its point *B*.[52] Take *PX* equal to half the *latus rectum*[53] and *X* will be the centre of curvature at *P*.

Problem 12. To find the centripetal forces by which bodies shall revolve in given curves round given centres of force. Solution. Let the body *T* revolve in any curve *ATB* round the pole *P*.[54] To the curve raise the normal *TRC* and let *C* be the centre of the curvature which the curve has at the point *T*, and then, when to *CT* and *PT* there are let fall the perpendiculars *PR* and *CD*, the speed of the body will be reciprocally as *RT*,[55] and the centripetal force reciprocally as $RT^2 \times DT$.[56]

---

(51) Newton understands that the ordinate of the curve is applied at right angles to the base. Where—as he sets it to be in his preliminary computations for the ensuing example of the Apollonian ellipse (see the next note)—the ordinate *y* stands thus erect upon the uniformly fluent abscissa *x*, this yields the familiar measure $(1 + \dot{y}^2)^{\frac{3}{2}}/\ddot{y}$ for the radius of curvature at the point (x, y) of the curve.

(52) The calculations deriving this proportion which Newton jotted down in his manuscript (f. 360ᵛ) immediately before the present paragraph are reproduced in [2] of the Appendix.

(53) Namely, $\frac{1}{2}RS^2/PQ = RC^2/PC$. In this special instance the ellipse has the same curvature as the parabola of same focus *X*, vertex *P* and *latus rectum*.

(54) Newton first began to add 'requiritur vis centripeta qua corpus in [curva revolvitur]' (there is required the centripetal force whereby a body [revolves] in a [curve]).

(55) For if the body traversing the curve *BT* moves 'instantaneously' from *T* to the indefinitely close point *t* in its orbit, the vanishingly small time of passage will by the Keplerian area law (see vi: 34–7) be proportional to the infinitesimal triangle $\triangle PTt = \frac{1}{2}\widehat{Tt} \times RT$; whence the body's speed at *T* (that is, $\widehat{Tt}$ divided by the time of the body's passage over its length) will be inversely proportional to *RT*. More directly, *RT* is equal to the tangential polar let fall from *P* perpendicular to the tangent to the curve at *T*, which is in turn (again by the Keplerian area law; compare the second of the corollaries which Newton added in the 1690's (see vi: 548) to Proposition VI of the *Principia*'s first book) inversely proportional to the instantaneous orbital speed at *T*.

(56) See note (43) above for our detailed comment upon this modified Newtonian measure of central force in the earlier version of this 'Prob. 12' in [3] preceding.

§2. *First versions of an enlarged introduction to Newton's
'Method of Fluxions'*.[1]

[1] [2]   METHODUS FLUXIONUM.[3]

Quantitates mathematicas non ut ex indivisibilibus vel partibus quam-minimis vel infinitè parvis constantes[4] sed ut motu continuo descriptas hic considero. Lineæ describuntur ac describendo generantur non per appositionem partium[5] sed per motum continuum punctorum, superficies per motum linearum, solida per motum superficierum, anguli per rotationem laterum, tempora per fluxum continuum[6] & sic in cæteris. Hæ geneses[7] in rerum natura locum habent & in motu corporum quotidie peraguntur & coram oculis exhibentur. Et ad hunc modum Veteres ducendo rectas mobiles in rectas immobiles genesin docuerunt rectangulorum.[8]

Considerando igitur quod quantitates æqualibus temporibus crescentes et crescendo genitæ,[9] pro velocitate majori vel minori qua crescunt ac generantur, evadunt majores vel minores; methodum quærebam determinandi quantitates ex velocitatibus motuum vel incrementorum quibus generantur, & has motuum vel incrementorum velocitates nominando *fluxiones* quantitatum genitarum, & quantitates genitas nominando *fluentes*, incidi anno 1665[10] in methodum fluxionum quam olim cum amicis[11] communicavi & qua hic usus sum in quadratura Curvarum.

---

(1) In these intermediate drafts Newton hones the opening 'Introductio' to his tract on quadrature of curves to be effectively the final form of it (§3.1 below) which he put to the printer in late 1703.

(2) Add. 3960.13: 207[+Add. 3960.7: 88]/209/211; significant minor variants in the preliminary casting on Add. 3960.7: 88–9/Add. 3970.3: [360ᵛ+] 361ᵛ are cited at the pertinent places in following footnotes.

(3) A tentative retitling soon to be abandoned in Newton's ultimate opening to his tract (see §3.1 below) in renewed preference for his earlier head 'De Quadratura Curvarum'.

(4) In draft (Add. 3960.7: 88) Newton first wrote '...ex indivisibilibus constantes vel ex partibus minimis compositas' (consisting of indivisibles or composed of least parts), thereafter changing the 'minimis' (least) to be successively 'infinite parvis' (infinitely small) and then 'quam-minimis' (least possible), after which he recast the whole phrase to read 'ex partibus indivisibilibus vel quibuscunꝗ quamminimis vel infinite parvis constantes' (consisting of indivisible parts or any least possible or infinitely small ones whatever); with the superfluous 'quibuscunꝗ' (any...whatever) suppressed, this last was initially copied by him into the present revised text.

(5) 'infinite parvarum' (infinitely small) is deleted in Newton's draft.

(6) Not necessarily, we would add, uniform for Newton. In the correct form of Proposition X of the second book of his *Principia* which Johann Bernoulli led him to compose in the autumn of 1712 for last-minute insertion in its second edition, he found it convenient (see 6, §2.3 below) to take as uniformly fluent base variable not the time of passage of a projectile over its curved path in a resisting medium, but the horizontal component of its motion.

(7) Newton initially began 'quan[titatum]' (of quantities), as in his draft (Add. 3960.7: 88)

Translation

[1][2]   THE METHOD OF FLUXIONS.[3]

Mathematical quantities I here consider not as consisting of indivisibles, either parts least possible or infinitely small ones,[4] but as described by a continuous motion. Lines are described and by describing generated not through the apposition of parts[5] but through the continuous motion of points; surface-areas are through the motion of lines, solids through the motion of surface-areas, angles through the rotation of sides, times through continuous flux,[6] and the like in other cases. These geneses[7] take place in the physical world and are daily enacted in the motion of bodies visibly before our eyes. And in much this manner the ancients, by 'drawing' mobile straight lines into stationary lines, taught the genesis of rectangles.[8]

By considering, then, that quantities increasing and begotten by increase in equal times[9] come to be greater or lesser in accord with the greater or less speed with which they grow and are generated, I was brought to seek a method of determining quantities out of the speeds of motion or increment by which they are generated, and, naming these speeds of motion or increment 'fluxions' of the quantities generated, and the quantities so born 'fluents', I fell in the year 1665[10] upon the method of fluxions which I have formerly imparted to friends[11] and which I have here employed in the quadrature of curves.

---

where he continued by stating that these geneses 'omnino naturales sunt & coram oculis exhiberi possunt' (are utterly natural and can display themselves before our eyes); the similar emphasis that these 'non sunt fictitiæ sed' (are not fictitious but) is deleted in sequel in his present revise.

(8) In his draft Newton first wrote little differently: 'Sic enim Veteres ducendo rectam mobilem in rectam immobilem generarunt aream rectangulam' (For in this way the ancients, by 'drawing' a mobile straight line into a stationary one, generated a rectangular area). We may add that, while this notion of 'generating' a rectangle by the uniform translation of a straight line (at right angles to itself) has some foundation in ancient Greek 'mechanical' practice (compare VII: 294–6), it was not zealously pursued till Newton's own century by such 'indivisibilist' geometers as Cavalieri, Torricelli and Roberval. Newton himself had made the concept basic in his construction of the fluxion of a geometrical area in his 1666 tract (see 'Example 3ᵈ' on I: 412), and five years later set the generation of rectangular area in this way explicitly out as 'Theor. 5' of the *addendum* to his 1671 tract (see III: 344).

(9) 'in eodem tempore desumptæ' (conceived in the same time) was first inadequately written in the draft (Add. 3960.7: 88).

(10) Newton first began to pen 'ante annos 1665 [? et 1666]' (before the years 1665 [? and 1666]) in his draft; at a late stage in the printing of the 'De Quadratura' he again expanded this phrase to be 'annis 1665 & 1666' (in the years 1665 and 1666) and so—a trivial capitalization of the first word (by the printer, we would guess) apart—this often subsequently quoted affirmation by Newton of his priority in discovering the calculus was published to the world in 1704 (*Opticks*: ₂166).

(11) Does Newton here refer to Barrow and Collins (to whom he communicated his 'De Analysi' in the summer of 1669; see II: 167–8)? or to John Wallis (to whom he imparted

[12]Eodem recidit methodus in qua quantitates tanquam ex differentijs innumeris & quamminimis per additionem continuam[13] genitæ spectantur quamqӡ ideo *differentialem* vocant. Nam differentiæ illæ seu augmenta momentanea sunt ut quantitatum fluxiones ideoqӡ pro fluxionibus usurpari possunt tanquam earum exponentes.[14] Animum quidem advertenti ad methodos quibus Archimedes superficiem Sphæræ et Parabolæ invenit quasqӡ Cavallerius, Wallisius, Fermatius, Barrowus & alij coluerunt,[15] in proclivi erat partes quantitatum quamminimas quas differentias vocant in computis usurpare, sed ad rerum naturam attendendo[16] prior obvenit consideratio fluxionum, propterea quod quantitates mathematicæ per motum quidem continuum describi ac generari possunt, non autem per additionem continuam partium minimarum, quodqӡ differentiæ sunt[17] exponentes fluxionum & semper pro fluxionibus usurpantur easqӡ significant[,] fluxiones autem non spectantur ut exponentes differentiarum. In eo autem præstat methodus fluxionum quod fluxiones non tantum per differentias prædictas seu incrementa momentanea exponi possunt sed etiam per quantitates quascunqӡ quæ fluxionibus illis proportionales sunt, et eæ exponentes eligi et in computis adhiberi quibus operationes et Schemata[18] simpliciora reddūtur: ut cùm fluxiones arearum per ordinatim applicatas exponimus, & sic effugimus descriptionem duarum ordinatarum ad differentiam designandam.

Fluxiones[19] exponi possunt & mensurari vel per incrementa quantitatum fluentium s[i]ngulis temporis momentis[20] genita, vel per alias quascunqӡ quantitates incrementis hisce jam jam nascentibus proportionales. Unde aliqui[21] incrementa illa fluxiones nominant.

Mobilium[22] velocitates sunt ut partes spatiorum descriptorum singulis temporis particulis æqualibus & infinite parvis confectæ,[23] vel ut mathe-

---

portions of his 1691 treatise 'De quadratura Curvarum' in the late summer of the following year; see VII: 170, note (1))? It usually suited his purposes in after years to assume without question that he meant only the former. In his draft text, we may notice, he went momentarily beyond historical fact initially to refer to '...methodum fluxionum de qua olim scripsi ad Cl. Leibnitium' (the method of fluxions regarding which I formerly wrote to the famous Leibniz), but speedily struck out the untruth.

(12) Because of their especial interest we here interpolate, as a 'cancelled passage', the text of the four paragraphs next ensuing in Newton's prior draft on Add. 3960.7: 88.

(13) 'perpetuam' (perpetual) is replaced.

(14) Newton has deleted in sequel 'Eaqӡ de causa idem præstatur per utramqӡ methodum' (And for that reason the same end is achieved by each method).

(15) This same quintet of Archimedes, Cavalieri, Fermat, Wallis and Barrow—often with the addition of James Gregory (see, for instance, Appendix 1.2 to Part 2 following)—is often cited by Newton when he speaks of his precursors in the invention of the fluxional calculus. Note the absence of the name of Descartes (to whom singly Newton was indebted most of all)!

(16) Initially 'rerum naturam spectanti' (in regarding...). Newton had first started to write more clumsily 'sed rei natura postulat prim[o quod]' (but physical nature primarily demands [that]).

[12]Substantially the same is the method wherein quantities are regarded as begotten, as it were, from innumerable minimal-possible differences through continual[13] addition, and which is consequently called 'differential'. For those differences or instantaneous augments are as the quantities' fluxions and consequently can be used in place of the fluxions as their exponents.[14] Anyone paying heed, to be sure, to the methods by which Archimedes discovered the surface-area of the sphere and the (plane) area of the parabola, and which Cavalieri, Wallis, Fermat, Barrow and others have cultivated,[15] was inclined to employ in his computations the minimal parts of quantities which they call differences; but in giving consideration[16] to physical nature there came to be a prior regard for fluxions, seeing that mathematical quantities can be described and generated through continuous motion indeed, but not through the continued addition of minimal parts, and that differences are[17] exponents of fluxions and may always be employed in place of fluxions and denote them, while fluxions are not looked upon as exponents of differences. The superiority of the method of fluxions lies, however, in that fluxions can be expressed not merely through the aforesaid differences, that is, instantaneous increments, but also by means of any quantities whatever which are proportional to those fluxions, and the exponents chosen and applied in calculations whereby operations and diagrams[18] are rendered simpler: when, for instance, we express the fluxions of areas by ordinates and thus evade the description of two ordinates to denote the difference.

Fluxions[19] can be expressed and measured either by the increments of fluent quantities begotten in individual moments[20] of time or by any other quantities whatever proportional to these just barely nascent increments. Whence some[21] name those increments fluxions.

The speeds of mobiles[22] are as the parts of the distances described which are completed[23] in individual, equal and infinitely small particles of time, or (to

---

(17) 'tantum' (merely) is deleted.

(18) Originally 'operationes computandi et figuras describendi' (the operations of computing and describing figures).

(19) Newton first added 'quantitatum genitarum quas fluentes appello' (of the generated quantities which I call fluents).

(20) Tentatively altered to be 'particulis æqualibus' (equal particles) before a return to 'momentis' (moments) was made.

(21) Notably Abraham de Moivre in his influential 'Specimina quædam illustria Doctrinæ Fluentium...' (*Philosophical Transactions*, **19**, No. 216 [for March–May 1695]: 52–7, especially 54–5; compare note (74) of our preceding introduction) and also the 'plagiarist' George Cheyne, who on pages 19–20 *et seq.* of the *Fluxionum Methodus Inversa* whose appearance in 1703 provoked Newton to prepare the present augmented version of his own tract 'De Quadratura Curvarum' for the press employed $F:y\dot{z}$ as an exact analogue of the Leibnizian integral $\int y\,.\,dz$.

(22) 'Projectorum' (Projectiles) was first written.

(23) Initially 'descriptæ' (described) and 'verius' (more truly) respectively.

matice[23] loquar sunt in prima ratione partium nascentium. Et similiter quantitatum quæ crescunt et augentur fluxiones seu crescendi velocitates sunt in prima ratione augmentorum quæ singulis momentis generantur.[24]

    Itacз quantitates indeterminatas.... et abscissa z.[25]

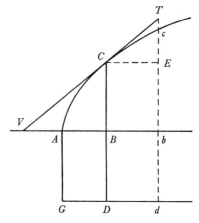

    [26]Fluxiones sunt quam proxime[27] ut fluentium partes æqualibus temporis particulis quàm minimis genitæ, et ut accuratè[28] loquar sunt in primâ ratione partium nascentium; exponi autem possunt per lineas quascuncз quæ sunt ipsis proportionales. Ut si areæ ABC, $ABDG$ ordinatis BC, BD super basi AB uniformi cum motu progredientibus describantur, harum arearum fluxiones erunt inter se ut Ordinatæ describentes BC et BD et per ipsas exponi possunt. Progrediatur Ordinata BC de loco suo BC in locum novum bc.[29] Compleatur parallelogrammum[30] $BCEb$, ac ducatur VCT quæ Curvam tangat in C ipsiscз bc et BA productis occurrat in T et V; et Abscissæ AB, Ordinatæ BC et lineæ curvæ ACc partes modo genitæ erunt Bb, Ec et Cc, et in harum partium nascentium ratione prima sunt latera trianguli CET, ideocз fluxiones ipsarum AB, BC et AC sunt ut trianguli illius CET latera CE, ET & CT, et per eadem latera exponi possunt, vel quod perinde est per latera trianguli consimilis VBC.[31]

    Methodus fluxionum Corollarium est methodi serierum convergentium et ejus ope amplior redditur & perficitur.[32] Sit æquatio $v = x^{\frac{m}{n}}$. Fluat x uniformiter

    (24) Newton has cancelled a concluding sentence: 'Et quemadmodum velocitates sic etiam fluxiones per quantitates quascuncз ipsis proportionales exponi possunt' (And, just as speeds, so also fluxions can be expressed by any quantities whatever proportional to them).

    (25) Understand the second paragraph of the 1693 parent text (VII: 508–10), here trivially augmented by the conjunction 'Itacз' (And so) which smooths the way to its present insertion; the ensuing propositions of the 'Geometriæ Liber secundus' (VII: 512 ff) were doubtless here also intended to follow. This paragraph was also initially copied into the revised text at this point (Add. 3960.13: 207/209), but subsequently delayed—with the now obsolete interpolated conjunction omitted—to be once more (see §3.1 below) the opening paragraph of the main text of the 'De Quadratura Curvarum'.

    (26) In the draft version (Add. 3960.7: 88) Newton here first began: 'Quantitatum quæ motu continuo crescunt et augentur fluxiones seu crescendi velocitates sunt ut quantitatum illarum partes quæ singulis temporis particulis æqualibus et quàm minimis crescendo generantur, vel ut mathematicè loquar sunt in prima ratione partium illarum nascentium. Has partes aliqui differentias vocant et pro fluxionibus usurpant cum fluxiones sint [quantitates finitæ]' (Of quantities which grow and increase by a continuous motion the fluxions or speeds of growing are as the parts of those quantities which are generated by the growth in single, equal and least possible parts of time, or, to speak mathematically, are in the first ratio of those nascent parts. Some call these parts differences and employ them in place of fluxions, though fluxions are (finite quantities]). While Leibniz, who of course introduced the notion, and the

speak in mathematical terms[23]) are in the first ratio of the nascent parts. And, similarly, of quantities which grow and increase the fluxions or speeds of increase are in the first ratio of the augments which are generated in single moments.[24]

And so [I consider] indeterminate quantities... ...[25] and its abscissa z.

[26]Fluxions are very closely near[27] as the parts of their fluents begotten in the very smallest equal particles of time—to speak accurately,[28] indeed, they are in the first ratio of the nascent parts; they can, however, be expressed by any lines whatever which are proportional to them. If, for instance, the areas (ABC), $(ABDG)$ be described by the ordinates BC, BD advancing upon the base AB with uniform motion, the fluxions of these areas will be to one another as the describing ordinates BC and BD and they can be expressed by them. Let the ordinate BC advance from its place BC into the new place bc,[29] complete the parallelogram[30] $BCEb$ and draw VCT to touch the curve at C and meet bc and BA extended in T and V: then the parts of the abscissa AB, ordinate BC and curve-arc AC now begotten will be Bb, Ec and Cc, and in the first ratio of these nascent parts are the sides of the triangle CET; consequently the fluxions of AB, BC and $\widehat{AC}$ are as the sides CE, ET and CT of that triangle CET, and can be expressed by means of these same sides or, what is the same, by the sides of the triangle VBC similar to it.[31]

The method of fluxions is a corollary of the method of converging series and is by its aid made more copious and perfected.[32] Take the equation $v = x^{m/n}$.

Bernoulli brothers Jakob and Johann had made profitable appeal to the 'indivisible' *differentia* of a magnitude to develop their several calculus algorithms and techniques of resolution, it was (compare note.(21) above) rather in Britain that the confusion of this with the finite Newtonian *fluxio* of a quantity had occurred.

(27) This very necessary qualification is omitted in the draft on Add. 3960.7: 89 (as in the initial version cited in the preceding note).

(28) Newton here first specified 'vel ut mathematicè' (or, to [speak] mathematically) as in his two draft versions on Add. 3960.7: 88 (see note (26) above)/89.

(29) The potentially misleading insistence that this is—initially—'finito quovis intervallo a loco priori distantem' (distant from the previous place by any finite interval) is deleted.

(30) For convenience, Newton shows this in his accompanying figure as a rectangle.

(31) In his draft Newton here again (compare note (26) above) went momentarily on to remark that 'Fluentium partes (quales sunt *CE*, *Ec*, *Cc*) aliqui ['differentias vocan[t]' is deleted] pro fluxionibus usurpant. Sunt autem quantitates magnitudinis indetermin[at]æ &' (The parts of fluents, such as *CE*, *Ec*, *Cc*, some [call differences] employ in place of fluxions. They are, however, quantities of indeterminate magnitude, and...) before breaking off. He afterwards set to follow at this point the further illustrations of the geometrical method of first and last ratios whose first versions (elaborated on the style of the parent examples in §1.1/2 preceding) are reproduced in [2]/[3] below.

(32) In prior draft on Add. 3960.7: 89 Newton wrote little differently that 'Methodus Fluxionum et Methodus Serierum convergentium sunt inter se affinitate maxima conjunctæ' (The method of fluxions and the method of converging series are conjoined one to the other in

& fluendo fiat $x+o$ addito augmento quam minimo o & simul convertatur v in $v+p$, id est convertatur æquatio $v=x^{\frac{m}{n}}$ in æquationem $v+p=\overline{x+o}\big|^{\frac{m}{n}}$, & resolvendo binomium $x+o^{(33)}$ in seriem convergentem fiet

$$v+p=x^{\frac{m}{n}}+\frac{m}{n}ox^{\frac{m-n}{n}}+\frac{mm-mn}{2nn}oox^{\frac{m-2n}{n}}+\frac{m^3-3mmn+2mnn}{6nn}o^3x^{\frac{m-3n}{n}}+\&c.^{(34)}$$

Et posito quod [ipsius x aug]mentum $o^{(35)}$ sit exponens fluxionis quantitatis uniformiter fluentis x, termini seriei jam inventæ erunt quantitas altera v et ejus augmenta seu exponentes fluxionū ejus; nempe primus terminus $x^{\frac{m}{n}}$ erit quantitas fluens v, secundus $\frac{m}{n}ox^{\frac{m-n}{n}}$ erit augmentum primum seu exponens fluxionis primæ $\dot{v}$, tertius $\frac{mm-mn}{2nn}oox^{\frac{m-2n}{n}}$ erit augmentum secundum seu exponens fluxionis secundæ $\ddot{v}$, quartus $\frac{m^3-3mmn+2mnn}{6nn}o^3x^{\frac{m-3n}{n}}$ erit augmentum tertium seu exponens fluxionis tertiæ $\dddot{v}$, & sic deinceps in infinitum. Hæc augmenta seu fluxionum exponentes Mathematici[36] jam vocant quantitatum indeterminatarum *differentias*, et ipsarum z, y, x, v differentias primas sic designant $d\bar{z}$, $d\bar{y}$, $d\bar{x}$, $d\bar{v}$. Simpliciora sunt symbola $\dot{z}$, $\dot{y}$, $\dot{x}$, $\dot{v}$ & generaliùs accipiuntur,[37] per quæ utiꝗ non tantum ipsarum z, y, x, v augmenta momentanea seu differentias primas sed et fluxiones et fluxionum exponentes quascunꝗ intelligere licet.[38] Nam fluxiones sunt quantitates finitæ et veræ ideoꝗ symbola sua

the closest affinity). Yet earlier, on Add. 3970.3: 361^v, he had opened with the autobiographical observation that 'Eodem anno [1665] incidi etiam in methodum serierum convergentium' (In the same year [1665] I fell also upon the method of converging series) before similarly adding 'Nam hæ duæ methodi sunt inter se affinitate maxima conjunctæ' (To be sure, these two methods are conjoined one to the other in closest affinity). In this latter, initial version the following example of the binomial series expansion was first tentatively set as a '*Prop. 3. Binomium in seriem convergentem resolvere*' (. . . *To resolve a binomial into a converging series*), evidently to follow the restyled 'Prop. II. Prob. II' in §1.1 above, but was speedily transposed to its permanent place in the 'Introductio'.

(33) Read '$\overline{x+o}\big|^{\frac{m}{n}}$'. We silently mend Newton's slip, here blindly copied from his preliminary draft on Add. 3970.3: 361^v, in our English version *en face*.

(34) The denominator of the coefficient of the last term specified in this expansion, here again unthinkingly copied by Newton from his draft, should be '$6n^3$'; we make appropriate adjustment in our English rendering.

(35) 'sit infinite parvum, quodꝗ hoc augmentum' (be infinitely small, and that this augment) is deleted.

(36) 'Germani' (Germans) is replaced in the draft (on Add. 3970.3: 361^v)—accurately so, since it was the Swiss-born Bernoulli brothers who had done even more than its prime author to develop the full riches of the Leibnizian 'differential' calculus over the dozen years before late 1703. Newton's following designation of the *differentiæ primæ* of z, y, x, v accurately represents a variant of the simpler notation dz, dy, dx, dv employed by Leibniz in his pioneering

Let x flow uniformly and in its flowing come to be $x+o$ by the addition of the minimal augment o, and simultaneously let v be converted to be $v+p$, that is, let the equation $v = x^{m/n}$ be converted into the equation $v+p = (x+o)^{m/n}$; then by resolving the binomial $(x+o)^{m/n}$ into a converging series there will come

$$v+p = x^{m/n} + (m/n)\, ox^{(m-n)/n} + ((m^2-mn)/2n^2)\, o^2 x^{(m-2n)/n}$$
$$+ ((m^3 - 3m^2n + 2mn^2)/6n^{[3]})\, o^3 x^{(m-3n)/n} + \dots.$$

And, supposing that the augment o of $x^{(35)}$ be the exponent of the fluxion of the uniformly flowing quantity x, the terms of the series just now found will be the other quantity v and its augments, that is, the exponents of its fluxions; namely, the first term $x^{m/n}$ will be the fluent quantity v, the second one $(m/n)\, ox^{(m-n)/n}$ will be the first augment or the exponent of its first fluxion $\dot{v}$, the third $((m^2-mn)/2n^2)\, o^2 x^{(m-2n)/n}$ will be the second augment or the exponent of its second fluxion $\ddot{v}$, the fourth $((m^3-3m^2n+2mn^2)/6n^{[3]})\, o^3 x^{(m-3n)/n}$ will be the third augment or the exponent of its third fluxion $\dddot{v}$, and so on indefinitely. These augments or exponents of fluxions mathematicians[36] nowadays call 'differences' of indeterminate quantities, and denote first differences thus: $d\bar{z}, d\bar{y}, d\bar{x}, d\bar{v}$. Simpler and more generally accepted[37] are the symbols $\dot{z}, \dot{y}, \dot{x}, \dot{v}$ by which of course we are free to understand not merely the instantaneous augments of z, y, x, v, that is, their first differences, but also their fluxions and any exponents whatever of the fluxions. [38]For fluxions are finite quantities and

'Nova methodus pro maximis et minimis, itemcg tangentibus, quæ nec fractas, nec irrationales quantitates moratur, & singulare pro illis calculi genus' (*Acta Eruditorum* (October 1684): 467–73) [= (ed. C. I. Gerhardt) *Leibnizens Mathematische Schriften*, **5** (Halle, 1858): 220–6]) of which Leibniz himself briefly made use in the early 1690's (compare *Mathematische Schriften*, **5**: 241, 256).

(37) Not without justice, Leibniz had already expressed a contrary opinion in reporting to Huygens in September 1694 his first impressions of Newton's dotted symbols, then newly made public in Wallis' Latin *Algebra*: 'ie pense que la consideration des differences...est plus propre à éclaircir l'esprit; ayant encor lieu dans les series ordinaries des nombres, et repondant en quelque façon aux puissances et aux racines' (*Œuvres complètes de Christiaan Huygens*, **10**, 1905: 675; compare VII: 181, note (26)).

(38) In original conclusion to his draft on Add. 3970.3: 361ᵛ Newton wrote: 'Igitur differentiæ in qua Methodus fundatur quam *differentialem* vocant nihil aliud sunt quam termini serierum nostrarum convergentium, et Methodus illa Corollariũ est Methodi serierum. Hanc methodum serierum exposui in Epistolis anno 1676 ad Celeberrimũ Leibnitium scriptis & simul significavi me compotem esse methodi alterius quam literis transpositis celabam hanc sententiam involventibus; *Data æquatione fluentes quantitates involvente invenire fluxiones & contra*' (Thus, the differences on which what they call the 'differential' method is founded are none else than the terms of our converging series, and that method is a corollary of the method of series. This method of series I expounded in letters written in the year 1676 to the renowned Leibniz, and at the same time I signified that I had in my power a second method which I concealed in transposed letters enfolding this sentence: *Given an equation involving fluent quantities, to find the fluxions, and vice versa*). He then inserted before this: 'Nam præstat aliquando fluxiones per quantitates finitas exponere quam per infinite parvas, & schemata hac ratione

habere debent, et quoties commodè fieri potest præstat ipsas per lineas[39] finitas coram oculis exponere quàm per infinitè parvas. Methodum convertendi quantitates indeterminatas[40] in series convergentes, atqᵽ adeo inveniendi Differentias, exposui in Epistolis anno 1676 ad Celeberrimum Leibnitium scriptis, & simul significavi me compotem esse methodi[41] tractandi Problemata difficiliora quam literis transpositis celare conabar[42] hanc sententiam involventibus, *Data Æquatione fluentes quotcunqᵽ quantitates involvente invenire fluxiones, & vice versa.* Hujus Problematis pars prior sic solvitur.

Prop. I. Prob. I.

Quantitatum fluentium invenire fluxiones.

Solutio.

Si detur æquatio quantitates fluentes involvens, multiplicetur omnis æquationis terminus per indicem dignitatis quantitatis cujusqᵽ fluentis quam involvit et in singulis multiplicationibus mutetur dignitatis latus in suam fluxionem, et aggregatum factorum omnium sub proprijs signis erit æquatio[43] [nova.]

··· ··· ···

[2][44] Eodem recidit si sumantur fluxiones in ultima ratione partium evanescentium. Agatur recta *Cc* et producatur eadem ad *K*. Redeat ordinata *bc* in

simpliciora evadunt; ut cum fluxiones arearū per Ordinatas exponimus' (For it profits on occasion to express fluxions by finite quantities rather than by infinitely small ones, and schemes come by this means to be simpler; as when we express fluxions of areas by the ordinates). This interpolation he then recast on the preceding f. 360ᵛ to read: 'Nam fluxiones sunt quantitates veræ, differentiæ autem nullam habent certam et veram quantitatem. Harum symbola in computis accuratis non tam partes fluentium significare debent quàm partium nascentium proportiones primas, id est proportiones fluxionum. Et quoties hæ proportiones innotescunt præstat fluxiones per quantitates finitas coram oculis exponere quam per infinitè parvas. Nam et schemata hac ratione simpliciora evadere solent; ut cum fluxiones arearum per ordinatas exponimus neglectis parallelogrammis infinite parvis quæ sunt arearū differentiæ, vel cum fluxiones Lineæ curvæ et ejus abscissæ et ordinatæ per tangentem subtangentem & ordinatam exponimus' (For fluxions are true quantities, while differences have no true and certain quantity. The symbols of these ought in accurate computations to signify not so much the parts of fluents as the first proportions of the nascent parts, that is, the proportions of the fluxions. And each time these proportions are known, it is better to express the fluxions by visibly existing finite quantities than by ones infinitely small. Also, to be sure, schemes usually prove by this means to be simpler; as when we express the fluxions of areas by the ordinates, neglecting the infinitely small parallelograms which are the differences of the areas, or when we represent the fluxions of a curved line and of its abscissa and ordinate by the tangent, subtangent and ordinate). With the first sentence amplified to state that 'fluxiones sunt quantitates veræ ideoqᵽ symbola sua habere debent' (fluxions are true quantities and conse-

real, and consequently ought to have their own symbols; and each time it can conveniently so be done, it is preferable to express them by finite lines[39] visible to the eye rather than by infinitely small ones. The method of converting indeterminate quantities[40] into converging series, and hence of finding differences, I exposed in letters written in the year 1676 to the renowned Leibniz, and at the same time signified that I was master of a[41] method of treating more difficult problems which I took pains to conceal[42] in transposed letters enfolding this sentence: *Given an equation involving any number of fluent quantities, to find the fluxions, and vice versa.* The first part of this problem is solved as follows.

Proposition I, Problem I.

To find the fluxions of fluent quantities.

Solution.

If there be given an equation involving fluent quantities, multiply every term in the equation by the index of each fluent quantity which it involves, and in the separate multiplications change a 'side' (root) of the power into its fluxion, and the aggregate of all the products under their proper signs will be a (fresh) equation.[43]

. . .   .

[2][44]   It comes to the same if the fluxions be taken in the last ratios of the vanishing parts. Draw the straight line Cc and produce it to K. Let the ordinate

quently ought to have their own symbols) and the last one omitted, Newton copied this last version into his present manuscript page (Add. 3960.13: 211) and then straightaway cancelled this, too, in favour of what now follows.

(39) Initially 'quantitates alias' (other...quantities).

(40) In his draft (Add. 3970.3: 361ᵛ) Newton began more succinctly with the equivalent phrase 'Conversionem quantitatum' (The conversion of quantities).

(41) 'novæ' (new) was added in the draft.

(42) Originally 'celabam' (concealed) and then 'designabam' (represented).

(43) With this catchword at the bottom of page 211 marking the transition to a further sheet which has long since disappeared (if indeed it was ever written) the manuscript text abruptly ends. We complete Newton's sentence in line with the corresponding 'Solutio' in the 1693 parent tract (vii: 512). The ensuing 'Explicatio', 'Demonstratio' and 'Explicatio plenior' of Proposition I of the 'De Quadratura' would evidently follow with only trivial change.

(44) Add. 3960.7: 90/89. This interpolation in [1] preceding (see note (31) above) is a rough draft entered on the two later pages of a manuscript sheet (*ibid.*: 87–90) on which Newton earlier penned his first, tentative (and quickly discarded) recastings and amplifications of Propositions II and then I of the 1693 'De Quadratura' which are reproduced in §1.1/2 above.

locum suum priorem *BC* et coeuntibus punctis
c,C, recta *CK* coincidet cum tangente *CT*, &
triangulum evanescens *CEc* jam jam in ultima
sua forma evadet simile triangulo *CET* et ejus
latera evanescentia[45] *CE*, *Ec* & *Cc* erunt ultimò inter se ut sunt trianguli alterius *CET*
latera *CE*, *ET* et *CT*, & in hac ratione sunt
fluxiones linearum *AB*, *BC* et *AC*. Si puncta
C, c parvo quovis intervallo ab invicem distant,
recta *CK* parvo intervallo a tangente distabit;
ut recta *CK* cum tangente *CT* coincidat &
rationes ultimæ linearum *CE*, *Ec* et *Cc* inveniantur[46] debent puncta *C, c* coire & omninò

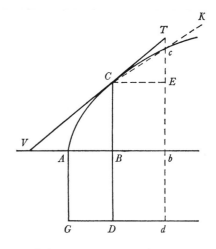

coincidere. Errores quàm minimi in rebus mathematicis non sunt admittendi.[47]

Cæterum ut modus colligendi fluxiones ex partibus fluentium plenius pateat
lubet exempla plura subjungere.

[48]Recta *PCB* circa datum positione polum *P* revolvens secat alias duas
positione datas rectas *AB*, *AC*: quæritur proportio fluxionis lineæ *AB* ad
fluxionem lineæ *AC*. Transeat recta illa
PCB de loco suo *PCB* in locum novum
Pcb et ad rectas *AB*, *AC*, *Pcb* demittantur
perpendicula *PM*, *PN*, *BD*, *CE*: et[49] ob
sim[ilia] tri[angula] *bBD*, *bPM* & *cCE*,
cPN erit *Bb*.*BD*::*bP*.*PM* et

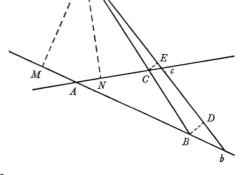

$$BD.CE::BP.CP \quad \text{et} \quad CE.Cc::PN.cP.$$

Et conjungendo rationes fit

$$Bb.Cc::bP \times BP \times PN.PM \times CP \times cP.$$

Coeant jam lineæ *BP*, *bP* et *Bb* erit ad
Cc in ultima sua ratione, hoc est ut

$BP^q \times PN$ ad $CP^q \times PM$, sive ut $\dfrac{BP^q}{PM}$ ad $\dfrac{CP^q}{PN}$, et hac ratione est fluxio lineæ *AB* ad
fl[uxionem] lineæ *AC*.[50]

Et simili argumentatione colligitur quod fluxio rectæ *PB* sit ad fluxionem rectæ

PC ut $\dfrac{MB \times PB}{PM}$ ad $\dfrac{NC \times PC}{PN}$[51] et fluxio ipsius *PB* ad fluxionem ipsius *AC* ut

$\dfrac{MB \times PB}{PM}$ ad $\dfrac{PC^q}{PN}$.[52]

---

(45) 'partes evanescentes' (...parts) is replaced.
(46) 'habeantur' (be had) was first written.

bc go back into its former place *BC* and, as the points *c* and *C* come together, the straight line *CK* will coincide with the tangent *CT*, and so the vanishing triangle *CEc* will as it attains its last form end up akin to the triangle *CET*, and its vanishing sides[45] *CE*, *Ec* and $\widehat{Cc}$ will ultimately be to one another as are the sides *CE*, *ET* and *CT* of the other triangle *CET*, and in this proportion are the fluxions of the lines *AB*, *BC* and *AC*. If the points *C* and *c* are at any small distance apart from each other, the straight line *CK* will be a small distance away from the tangent; in order that the line *CK* shall coincide with the tangent and so the last ratios of the lines *CE*, *Ec* and $\widehat{Cc}$ be discovered,[46] the points *C* and *c* must come together and entirely coincide. The most minute errors are not in mathematical matters to be accepted.[47]

For the rest, so that the manner of gathering fluxions out of the parts of fluents may be more clearly evident, it is apposite to subjoin several examples.

[48]A straight line *PCB* revolving round the pole *P* given in position cuts two other straight lines *AB*, *AC* given in position: there is required the ratio of the fluxion of the line *AB* to the fluxion of the line *AC*. Let the line *PCB* pass from its place *PCB* to the new place *Pcb*, and to the straight lines *AB*, *AC*, *Pcb* let fall the perpendiculars *PM*, *PN*, *BD*, *CE*: then,[49] because the triangles *bBD*, *bPM* and also *cCE*, *cpN* are similar, there will be $Bb::BD = bP:PM$ and

$$BD:CE = BP:CP \quad \text{and} \quad CE:Cc = PN:cP$$

and by combining these ratios there comes

$$Bb:Cc = bP \times BP \times PN : PM \times CP \times cP.$$

Now let the lines *BP*, *bP* coincide, and *Bb* will be to *Cc* in their last ratio, viz. as $BP^2 \times PN$ to $CP^2 \times PM$, that is, as BP^2/PM to CP^2/PN, and in this ratio is the fluxion of the line *AB* to the fluxion of the line *AC*.[50]

And by a like reasoning it is gathered that the fluxion of the straight line *PB* shall be to the fluxion of the straight line *PC* as $MB \times PB/PM$ to $NC \times PC/PN$,[51] and the fluxion of *PB* to the fluxion of *AC* as $MB \times PB/PM$ to PC^2/PN.[52]

---

(47) The first paragraph in [3] below was afterwards set here to follow, the harsh transition smoothed by the connecting phrase 'Simili argumento' (By a similar argument).

(48) The remainder of this preliminary draft insertion, except for the final paragraph, was at once replaced by the improved version of it which here follows in [3].

(49) The rough computations (preserved on Add. 3970.3: 480^r) on which Newton founded his ensuing argument are reproduced in [3] of the following Appendix.

(50) Whence, since $PB/PM:PC/PN = \sin\widehat{ACB}:\sin\widehat{ABC} = AB:AC$, there is more simply fl(AB):fl$(AC) = PB \times AB:PC \times AC$. This simplification was to be introduced by Newton himself into his revised version of the present fluxionary *exemplum* which we reproduce in [3] below.

(51) That is (see the previous note), $MB \times AB:NC \times AC$.

(52) Or, more simply (compare notes (50) and (51)),

$$\text{fl}(PB):\text{fl}(AC) = MB \times AB:PC \times AC.$$

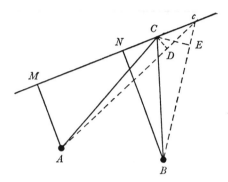

Circa polos positione datos A, B revolvantur rectæ duæ $AC\ BC$ ad rectam tertiam positione datam MN perpetuò concurṛentes ad C: quæritur proportio fluxionis rectæ AC ad fluxionem rectæ BC. Moveantur rectæ illæ AC, BC a locis suis AC, BC in loca nova Ac, Bc et a punctis A, B, C ad rectas MN, Ac, Bc demitte perpendicula AM, BN, CD, CE. Et erit $Dc.Cc::Mc.Ac$ et $Cc.Ec::Bc.Nc$. & connectendo rationes $Dc.Ec::Mc\times Bc.Nc\times Ac$.

Redeant jam rectæ $Ac\ Bc$ in loca sua priora et lineæ evanescentes Dc, Ec erunt ultimò ad invicem ut $MC\times BC$ ad $NC\times AC$. Hæ autem lineæ Dc et Ec evanescent[e]s sunt ultimo ad fluentium partes $Ac-AC$ et $Bc-BC$ in ratione æqualitatis, ac proinde partes evanescentes $Ac-AC$ et $Bc-BC$ sunt ultimò ad invicem ut $MC\times BC$ ad $NC\times AC$ et in eadem ratione sunt fluxiones linearum AC et BC.

Iisdem positis quæritur proportio fluxionis anguli MAC ad fluxionem anguli NBC. Est $CD.Cc::AM.Ac$ et $Cc.CE::Bc.BN$. et connectendo rationes fit $CD.CE::AM\times Bc.Ac\times BN$. Unde $\dfrac{AM\times Bc}{AC}\cdot\dfrac{Ac\times BN}{BC}::\dfrac{CD}{AC}\cdot\dfrac{CE}{BC}::$sinus Anguli CAD ad sinum anguli CBE. Coeant jam puncta C et c et sinus illi evanescentes (quibus proportionales sunt angulorum MAC, NBC partes evanescentes CAD, CBE) erunt ultimo ut $\dfrac{AM\times BC}{AC}$ ad $\dfrac{AC\times BN}{BC}$ sive ut $\dfrac{BC^q}{BN}$ ad $\dfrac{AC^q}{AM}$. Et in hac ratione sunt fluxiones angulorum illorum MAC, NBC.

Quæritur deniꝗ proportio inter fluxionē quantitatis cujusvis x & fluxionem dignitatis ejus x^n. Fluat x donec fiat $x+o$. Resolvatur binomij hujus dignitas $\overline{x+o}\mid^n$ in seriem infinitam $x^n+nox^{n-1}+\dfrac{nn-n}{2}\times oox^{n-2}+\dfrac{n^3-3nn+2n}{6}o^3x^{n-3}$ &c[53]

et augmentum dignitatis x^n erit ut series $nox^{n-1}+\dfrac{nn-n}{2}oox^{n-2}+$ &c.[54]

---

[3][55] Si circulus centro mobili [B] radio fluente [BC] descriptus ducatur ad angulos rectos[56] uniformi cum motu in longitudinem Abscissæ [AB], fluxio

---

(53) We mend a hasty slip by Newton in the manuscript, where the numerator of the coefficient of the last term specified in this binomial expansion is erroneously set to be '$2n^3-3nn+n$'; it seems trivial to record this momentary straying of a 2 in our English version.

(54) To complete Newton's intention, understand in continuation 'adeoꝗ est ad augmentum o quantitatis x ut $nx^{n-1}+\dfrac{nn-n}{2}ox^{n-2}$ &c ad 1. Evanescant jam augmenta illa et earum ratio ultima erit ut nx^{n-1} ad 1: ideoꝗ fluxio ipsius x^n est ad fluxionem ipsius x in eadem ratione' (and hence is to the augment o of the quantity x as $nx^{n-1}+\frac{1}{2}(n^2-n)ox^{n-2}...$ to 1. Now let those augments vanish, and their last ratio will be as nx^{n-1} to 1: in consequence, the fluxion of x^n is to the fluxion of x in the same ratio), as Newton phrased it in his revise (§3.1 below).

Round the poles A, B given in position let there revolve the two straight lines AC, BC perpetually meeting at the third straight line MN given, at C: there is required the ratio of the fluxion of the line AC to the fluxion of the line BC. Let the lines AC, BC move from their places AC, BC into new positions Ac, Bc, and from the points A, B, C to the straight lines MN, Ac, Bc let fall the perpendiculars AM, BN, CD, CE. There will then be $Dc\!:\!Cc = Mc\!:\!Ac$ and $Cc\!:\!Ec = Bc\!:\!Nc$, and by conjoining the ratios $Dc\!:\!Ec = Mc \times Bc\!:\!Nc \times Ac$. Now let the lines Ac, Bc return to their former places and the vanishing lines Dc and Ec will be ultimately one to the other as $MC \times BC$ to $NC \times AC$. These vanishing lines Dc and Ec, however, are ultimately to the parts $Ac-AC$ and $Bc-BC$ in a ratio of equality, and in consequence the vanishing parts $Ac-AC$ and $Bc-BC$ are ultimately one to the other as $MC \times BC$ to $NC \times AC$, and in the same ratio are the fluxions of the lines AC and BC.

With the same suppositions there is required the ratio of the fluxion of the angle $\widehat{MAC}$ to the fluxion of $\widehat{NBC}$. There is

$$CD\!:\!Cc = AM\!:\!Ac \quad \text{and} \quad Cc\!:\!CE = Bc\!:\!BN,$$

and by conjoining the ratios there comes $CD\!:\!CE = AM \times Bc\!:\!Ac \times BN$. Whence $AM \times Bc/AC\!:\!Ac \times BN/BC = CD/AC\!:\!CE/BC$, that is, $\sin \widehat{CAD}$ to $\sin \widehat{CBE}$. Now let the points C and c coincide, and those vanishing sines (to which are proportional the vanishing parts $\widehat{CAD}$, $\widehat{CBE}$ of the angles $\widehat{MAC}$, $\widehat{NBC}$) will then be ultimately as $AM \times BC/AC$ to $AC \times BN/BC$, that is, as BC^2/BN to AC^2/AM. And in this ratio are the fluxions of the angles $\widehat{MAC}$ and $\widehat{NBC}$.

There is required, finally, the ratio between the fluxion of any quantity x you will and the fluxion of its power x^n. Let x flow till it becomes $x+o$ and resolve the power $(x+o)^n$ of this binomial into the infinite series

$$x^n + nox^{n-1} + \tfrac{1}{2}(n^2-n)\, o^2 x^{n-2} + \tfrac{1}{6}(n^3 - 3n^2 + 2n)\, o^3 x^{n-3} \ldots:$$

the augment of the power x^n will be as the series

$$nox^{n-1} + \tfrac{1}{2}(n^2-n)\, o^2 x^{n-2} + \ldots.^{(54)}$$

[3][55]   If a circle described on the mobile centre B and with fluent radius BC be drawn at right angles[56] with uniform motion along the length of the abscissa

---

(55) Add. 3967.1: 8^r. These roughly drafted further additions and improvements to the text of [2] preceding are entered by Newton, on the blank space of a sheet earlier used by him to draft revisions to his 1687 *Principia*, in sequel to checking computations on the *Scholium Generale* following Proposition XL of its second book ($_1$1687: 339–54, especially 344; in the second edition ($_2$1713: 287) these are introduced at the corresponding *Scholium* which is there advanced more logically to follow Proposition XXXI of the second book). We here fill in a number of gaps left by Newton in his manuscript for the insertion of the references to his accompanying figures and the minor corollaries to his textual argument which we now (for the reader's convenience and to avoid needless disruption of the flow of Newton's reasoning) make good.

(56) To the plane of the paper, that is.

solidi geniti [*ABC*] erit ut circulus ille generans & fluxio superficiei hujus solidi erit ut perimeter circuli illius & fluxio lineæ curvæ [*AC*] conjunctim.[57]

Recta PB circa polum datum P revolvens secat aliam positione datam rectam AB ad datum punctum A terminatam: quæritur proportio fluxionis lineæ PB ad fluxionem lineæ AB.

Pergat *PB* de loco suo *PB* in locum novum *Pb*. Centro *P* intervallo *PB* describatur circulus secans rectam *AB* in *E* et rectam *P*[*b*] productam in *D* et *d*[,] et *Bb* (augmentum lineæ *AB*) erit ad *Db* (augmentum lineæ *PD*) ut *bd* ad *bE*. Redeat jam linea *Pb* in locum suum priorem *PB* et augmenta *Db* et *Bb* evanescentia erunt ultimò ut recta *BE* et circuli diameter [*D*]*d*[,] vel si ad *AB* demittatur perpendiculum *PM* ut *MB* ad *PB*, ideoq̃ fluxio ipsius *PB* est ad fluxionem ipsius *AB* ut *MB* ad *PB*.

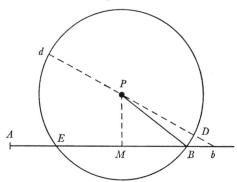

Pergat *PB* de loco suo *PB* in locum novum *Pb*. In recta *Pb* capiatur *PD* æqualis *PB*. Jungatur *BD* et huic parallela agatur recta *PE* ipsi *AB* occurrens in *E* et ad *AB* demittatur perpendiculum *PM*. Et rectarum fluentium *PB*, *AB* augmenta *Db*, *Bb* erunt ad invicem ut sunt rectæ *Pb*, *Eb*. Redeat jam recta *Pb* in locum suum priorem *PB* et augmenta

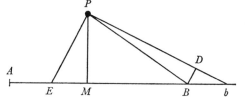

Db, *Bb* evanescentia erunt ultimò ad invicem ut sunt *PB*, *EB* existente angulo *BPE* recto, hoc est ut sunt *MB*, *PB*: ideoq̃ in hac ratione sunt fluxiones rectarum *PB* [*A*]*B* ad invicem.

Unde si recta *PCB* circa Polum datum *P* revolvens secet curvas duas positione datas *GB* et *HC*[58] in *B* et *C*: ad puncta sectionum *B* et *C* ducantur curvarum tangentes *BA* et *CA* concurrentes in *A* et fluxio curvæ *GB* erit ad fluxionem curvæ *HC* ut *PB* × *AB* ad *PC* × *AC*.[59]

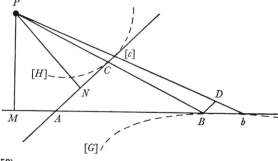

Et cum [fluxio rectæ *PB*] sit ad [fluxionem curvæ *GB*, hoc est ad fluxionem rectæ *AB*] ut [*MB*] ad [*PB*] ut supra: erit convertendo rationes [fluxio rectæ *PB*] ad [fluxionem curvæ *HC*] ut [*MB* × *AB*] ad [*PC* × *AC*].[59]

AB, the fluxion of the solid (*ABC*) begotten will be as that generating circle, and the fluxion of the surface of this solid will be as the perimeter of that circle and the fluxion of the curve-line $\widehat{AC}$ jointly.[57]

A straight line PB revolving round the given pole P intersects another straight line AB given in position and terminated at the given point A: there is sought the ratio of the fluxion of the line PB to the fluxion of the line AB.

Let *PB* proceed from its place *PB* to the new position *Pb*, and with centre *P* and radius *PB* describe a circle cutting the line *AB* in *E* and the line *Pb* (produced) in *D* and *d*: *Bb* (the augment of the line *AB*) will then be to *Db* (the augment of the line *PD*) as *bd* to *bE*. Now let the line *Pb* go back into its former place *PB*, and the vanishing augments *Db* and *Bb* will be ultimately (in ratio) as the straight line *BE* and the circle's diameter *Dd*, that is, if to *AB* there be let fall the perpendicular *PM*, *MB* to *PB*; and consequently the fluxion of *PB* is to the fluxion of *AB* as *MB* to *PB*.

Let *PB* proceed from its place *PB* to the new position *Pb*. In the straight line *Pb* take *PD* equal to *PB*, join *BD* and parallel to this draw the line *PE* meeting *AB* in *E*, and to *AB* let fall the perpendicular *PM*. And then the augments *Db*, *Bb* of the fluent lines *PB*, *AB* will be to one another as the lines *Pb* and *Eb* are. Now let the straight line *Pb* return into its former place *PB*, and the augments *Db*, *Bb* as they vanish will ultimately be one to the other as *PB* and *EB* are with the angle $B\widehat{P}E$ come to be right, that is, as *MB* and *PB* are: in this ratio, consequently, are the fluxions of the lines *PB* and *AB* to one another.

Whence if a straight line *PCB* as it revolves round the given pole *P* shall cut in *B* and *C* the two curves *GB* and *HC*[58] given in position: at the points *B* and *C* of intersection draw the curves' tangents *BA* and *CA* meeting one another in *A*, and the fluxion of the curve $\widehat{GB}$ will then be to the fluxion of the curve $\widehat{HC}$ as $PB \times AB$ to $PC \times AC$.[59]

And since the fluxion of the line *PB* is to the fluxion of the curve $\widehat{GB}$, that is, the fluxion of the line *AB*, as *MB* to *PB* (as above): on converting the ratios, the fluxion of the line *PB* will be to the fluxion of the curve $\widehat{HC}$ as $MB \times AB$ to $PC \times AC$.[59]

---

(57) More directly than in his equivalent Problems 9 and 10 in §1.3 preceding, Newton invokes the standard Archimedean formulas that the volume of the solid generated by the revolution of the curve $\widehat{AC}$ round the axis *AB* (to which its ordinate *BC* stands at right angles) is —in the more familiar Leibnizian equivalent—proportional to $\int BC^2 . d(AB)$, and its surface-area is proportional to $\int BC . d(\widehat{AC})$ (the constants of proportionality being of course π and 2π respectively). This addition to [2] is evidently (as in the final version in §3.1 below) to be inserted immediately after its first paragraph. What now follows is (compare note (48) above) meant to replace the corresponding following instances of comparison of geometrical fluxions there set out.

(58) We insert suitable depictions of these curves $\widehat{GB}$ and $\widehat{HC}$ in broken line in Newton's accompanying figure (in whose manuscript original they are not found).

(59) Compare notes (50) and (52) above respectively. Observe that these simplified arguments no longer require the perpendiculars *PM* and *PN*, and they are not found in the final version of the text in §3.1 below.

§3.   *The final text of the 'De Quadratura Curvarum':*
its new introduction and refurbished conclusion.[1]

[1][2]   DE QUADRATURA CURVARUM.[3]

INTRODUCTIO.

Quantitates Mathematicas non ut ex[4] partibus quàm minimis constantes
sed ut motu continuo descriptas hic considero. Lineæ describuntur ac describendo
generantur non per appositionem partium sed per motum continuum punctorum, superficies per motum linearum, solida per motum superficierum, anguli
per rotationem laterum, tempora per fluxum continuum,[5] & sic in cæteris.
Hæ Geneses in rerum natura locum verè habent & in motu corporum quotidie
cernuntur. Et ad hunc modum Veteres ducendo rectas mobiles in longitudinem
rectarum immobilium genesin docuerunt rectangulorum.

Considerando igitur quod quantitates æqualibus temporibus[6] crescentes &
crescendo genitæ, pro velocitate majori vel minori qua crescunt ac generantur,
evadunt majores vel minores; methodum quærebam determinandi quantitates
ex velocitatibus motuum vel incrementorum quibus generantur; et has motuum
vel incrementorum velocitates nominando *Fluxiones* et quantitates genitas
nominando *Fluentes* incidi anno 1665[7] in Methodum Fluxionum qua hic usus
sum in Quadratura Curvarum.

Fluxiones sunt quam proximè ut Fluentium augmenta[8] æqualibus temporis

---

(1) This (Add. 3962.1 : 1ʳ–4ʳ/19ʳ–26ʳ, with on f. 27ʳ an accompanying half-page of figures—
originally, in Newton's usual manner, inset by him in the verbal text, but afterwards there
crossed out and separately redrawn by him to suit the printer's convenience) made up,
together with the henceforth included pages (*ibid.*: 5ʳ–18ʳ[+18 *bis*/18 *ter* now lost]/19ʳ
[=VII: 512–48]) of the parent 1693 tract, the finished copy of the 'Tractatus de Quadratura
Curvarum' which was put by Newton to press in late 1703 (or perhaps the first weeks of the
new year). The printed version duly made its public appearance the next spring as the second
of the 'Two Treatises of the Species and Magnitude of Curvilinear Figures' appended by him
to his *Opticks: Or, A Treatise of the Reflexions, Refractions and Inflexions of Light* (London, ₁1704:
₂165–211, with the accompanying Figures 1–9 drawn on a separate sheet, 'Quadr. Tab. I.',
thrown in at the end). As we have already observed (VII: 508, note (1)) in preamble to reproducing the main 1693 text, the copy-reader has indicated in the margins of the manuscript the
principal (provisional) signature/page-divisions of the printed sheets, keying these to a left-
hand square bracket inserted in the body of Newton's words to mark the precise break: in
the second paragraph below, for instance, a '[' introduced between 'quærebam' and
'determinandi' joins with a marginal interpolation 'Yy 6/166' to pinpoint that the latter
word begins page 166 of the printed 'De Quadratura'. (Again compare D. F. McKenzie, *The
Cambridge University Press: 1696–1712*, 1 (Cambridge, 1966): 118.) In his library copies of the
1706 Latin *Optice* and of William Jones' 1711 compendium of Newton's minor mathematical
works (*Analysis per Quantitatum Series, Fluxiones, ac Differentias...*)—these are now ULC. Adv.
b.39.4 and Trinity College. NQ.8.26 respectively—Newton has made a few marginal
additions and corrections to the 1704 *princeps* text of the 'De Quadratura' there republished,

Translation

[1][2]   ON THE QUADRATURE OF CURVES.[3]

INTRODUCTION.

Mathematical quantities I here consider not as consisting of[4] least possible parts, but as described by a continuous motion. Lines are described and by describing generated not through the apposition of parts but through the continuous motion of points; surface-areas are through the motions of lines, solids through the motion of surface-areas, angles through the rotation of sides, times through continuous flux,[5] and the like in other cases. These geneses take place in the reality of physical nature and are daily witnessed in the motion of bodies. And in much this manner the ancients, by 'drawing' mobile straight lines into the length of stationary ones, taught the genesis of rectangles.

By considering, then, that quantities increasing and begotten by increase in equal times[6] come to be greater or lesser in accord with the greater or less speed with which they grow and are generated, I was led to seek a method of determining quantities out of the speeds of motion or increment by which they are generated; and, naming these speeds of motion or increment 'fluxions' and the quantities so born 'fluents', I fell in the year 1665[7] upon the method of fluxions which I have here employed in the quadrature of curves.

Fluxions are very closely near as the augments[8] of their fluents begotten in the

---

not all of which have found their way into print; these we take notice of at the pertinent places in our following footnotes, sketching the background to their genesis where this is known.

(2) Add. 3962.1: 1ʳ–4ʳ. Newton here knits together the component drafts in §2 preceding, introducing a few further small ameliorations. Significant variants in his discarded transitional drafts on Add. 3960.13: 215→210 and Add. 3963.4: 60ʳ [3960.13: 208] are cited below.

(3) The 1704 *princeps* printed version has the amplified title 'TRACTATUS DE QUADRATURA CURVARUM' (A TREATISE ON THE QUADRATURE OF CURVES); if it was not authored by him, the emendation was presumably made with Newton's full approval.

(4) 'indivisibilibus vel' (indivisibles or) was initially copied here by Newton from his prior draft, and then deleted.

(5) Let us again insist (compare §2.1 note (6)) that for Newton this continuous flux of real time was not necessarily to be taken as the uniform base fluent in his mathematical discussion of the 'motion of bodies'.

(6) Newton tentatively altered this to be 'temporis particulis' (particles of time) and then changed it back.

(7) Expanded to be 'Annis 1665 & 1666' (in the years 1665 and 1666) in the printed version (*Opticks*, ₁1704: ₂166); compare §2.1: note (10) preceding.

(8) This replaces 'partes' (parts), initially here transcribed from the predraft (§2.1); several other like substitutions are made in sequel by Newton in the manuscript. In his library copy of Jones' 1711 *Analysis* (see note (1)) he afterwards inserted in front an additional 'momenta seu' (moments, that is).

particulis quàm minimis genita, et ut accuratè
loquar sunt in prima ratione augmentorum
nascentium; exponi autem possunt per lineas
quascunæ quæ sunt ipsis proportionales. Ut
si areæ *ABC*, *ABDG* Ordinatis *BC*, *BD* super
basi *AB* uniformi cum motu progredientibus
describantur, harum arearum fluxiones erunt
inter se ut[9] Ordinatæ describentes *BC* et *BD*,
et per Ordinatas illas exponi possunt, propterea quod Ordinatæ illæ sunt ut arearum
augmenta nascentia. Progrediatur Ordinata
BC de loco suo *BC* in locum quemvis novum
bc. Compleatur parallelogrammum *BCEb*,

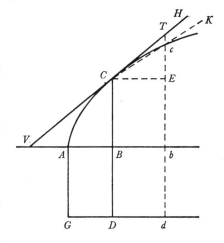

ac ducatur recta *VCTH* quæ Curvam tangat in *C* ipsisꝗ *bc* et *BA* productis
occurrat in *T* et *V*: et Abscissæ *AB*, Ordinatæ *BC*, & Lineæ Curvæ *ACc* augmenta modò genita erunt *Bb*, *Ec* et *Cc*; et in horum augmentorum nascentium
ratione prima sunt latera trianguli *CET* ideoꝗ fluxiones ipsarum *AB*, *BC* & *AC*
sunt ut trianguli illius *CET* latera *CE*, *ET* & *CT* et per eadem latera exponi
possunt, vel quod perinde est per latera trianguli consimilis *VBC*.

Eodem recidit si sumantur fluxiones in ultima ratione partium evanescentium. Agatur recta *Cc* et producatur eadem ad *K*. Redeat Ordinata *bc* in locum
suum priorem *BC*, et coeuntibus punctis *C* et *c*, recta *CK* coincidet cum tangente
CH, et triangulum evanescens *CEc* in ultima sua forma evadet simile triangulo
CET, et ejus latera evanescentia *CE*, *Ec* et *Cc* erunt ultimò inter se ut sunt
trianguli alterius *CET* latera *CE*, *ET* et *CT*, & propterea in hac ratione sunt
fluxiones linearum *AB*, *BC* et *AC*. Si puncta *C* & *c* parvo quovis intervallo ab
invicem distant recta *CK* parvo intervallo a tangente *CH* distabit. Ut recta *CK*
cum tangente *CH* coincidat et rationes ultimæ linearum *CE*, *Ec* et *Cc* inveniantur, debent puncta *C* & *c* coire et omninò coincidere. Errores quàm minimi
in rebus mathematicis non sunt contemnendi.[10]

Simili argumento si circulus centro *B* radio *BC* descriptus in longitudinem
Abscissæ *AB* ad angulos rectos uniformi cum motu ducatur, fluxio solidi geniti
ABC erit ut circulus ille generans, et fluxio superficiei ejus erit ut perimeter
Circuli illius et fluxio lineæ curvæ *AC* conjunctim. Nam quo tempore solidum
ABC generatur ducendo circulum illum in longitudinem Abscissæ *AB*, eodem
superficies ejus generatur ducendo perimetrum circuli illius in longitudinem
Curvæ *AC*.[11]

---

(9) A tentatively interlineated phrase 'augmenta nascentia id est ut' (the nascent augments,
that is, as) is here deleted.

(10) Much as in §2.2 preceding, in his transitional draft on Add. 3960.13: 215 Newton
here went on to write in an intervening paragraph: 'Cæterum ut modus inveniendi

very smallest equal particles of time: to speak accurately, indeed, they are in the first ratio of the nascent augments, but they can, however, be expressed by any lines whatever which are proportional to them. If, for instance, the areas (ABC), $(ABDG)$ be described by the ordinates BC, BD advancing upon the base AB with uniform motion, the fluxions of these areas will be to one another as[9] the describing ordinates BC and BD, and can be expressed by those ordinates, for the reason that those ordinates are as the nascent augments of the areas. Let the ordinate BC advance from its place BC into any new place bc, complete the parallelogram $BCEb$, and draw the straight line $VCTH$ to touch the curve at C and meet bc and BA extended in T and V: then the augments of the abscissa AB, ordinate BC and curve-arc $\overset{\frown}{AC}$ now begotten will be Bb, Ec and $\overset{\frown}{Cc}$, and in the first ratio of these nascent augments are the sides of the triangle CET; consequently the fluxions of AB, BC and $\overset{\frown}{AC}$ are as the sides CE, ET and CT of that triangle CET, and can be expressed by means of these same sides or, what is the same, by the sides of the triangle VBC similar to it.

It comes to the same if the fluxions be taken in the last ratio of the vanishing parts. Draw the straight line Cc and produce it to K. Let the ordinate bc go back into its former place BC and, as the points c and C come together, the straight line CK will coincide with the tangent CH, and so the vanishing triangle CEc will as it attains its last form end up akin to the triangle CET and its vanishing sides CE, Ec and $\overset{\frown}{Cc}$ will ultimately be to one another as are the sides CE, ET and CT of the other triangle CET: in this proportion in consequence are the fluxions of the lines AB, BC and $\overset{\frown}{AC}$. If the points C and c are at any small distance apart from each other, the straight line CK will be a small distance away from the tangent CH; in order that the line CK shall coincide with the tangent CH and so the last ratios of the lines CE, Ec and $\overset{\frown}{Cc}$ be discovered, the points C and c must come together and entirely coincide. The most minute errors are not in mathematical matters to be scorned.[10]

By a similar argument, if a circle described on centre B and with the radius BC be drawn along the abscissa's length AB at right angles and with a uniform motion, the fluxion of the solid (ABC) begotten will be as that generating circle, and the fluxion of its surface will be as the perimeter of that circle and the fluxion of the curve-line $\overset{\frown}{AC}$ jointly. For in the time that the solid (ABC) is generated by drawing the circle along the length of the abscissa AB its surface is generated by drawing the perimeter of that circle along the length of the curve $\overset{\frown}{AC}$.[11]

---

fluxiones clarius pateat, lubet alia exempla subjungere' (For the rest, so that the method of finding fluxions may be more clearly evident, it is appropriate to subjoin other examples). This is not present in the preliminary revise on Add. 3960.13: 210.

(11) See §2.3: note (57) above. This paragraph is not found in the transitional draft on Add. 3960.13: 215.

Recta PB circa polum datum P revolvens secet aliam positione datam rectam AB: quæritur proportio fluxionum rectarum illarum AB et PB. Progrediatur recta *PB* de loco suo *PB* in locum novum *Pb*. In *Pb* capiatur *PC* ipsi *PB* æqualis, et ad *AB* ducatur *PD* sic, ut angulus *bPD* æqualis sit angulo *bBC*; et ob similitudinem triangulorum *bBC, bPD* erit augmentum *Bb* ad augmentum *Cb* ut *Pb* ad *Db*. Redeat jam *Pb* in locum suum priorem *PB* ut augmenta illa evanescant, et evanescentium ratio ultima, id est ratio ultima *Pb* ad *Db* ea erit quæ est *PB* ad *DB* existente angulo *PDB*[12] recto, et propterea in hac ratione est fluxio ipsius *AB* ad fluxionem ipsius *PB*.

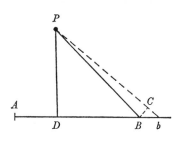

Recta PB circa datum Polum P revolvens secet alias duas positione[13] *datas rectas AB et AE in B et E: quæritur proportio fluxionum rectarum illarum AB et AE.* Progrediatur recta revolvens *PB* de loco suo *PB* in locum novū *Pb* rectas *AB, AE* in punctis *b* et *e* secantem, et rectæ *AE* parallela[14] *BC* ducatur ipsi *Pb* occurrens in *C*, et erit *Bb* ad *BC* ut *Ab* ad *Ae* et *BC* ad *Ee* ut *PB* ad *PE*, et conjunctis rationibus *Bb* ad *Ee* ut *Ab* × *PB* ad *Ae* × *PE*. Redeat jam linea *Pb* in locum suum priorem *PB*, et augmentum evanescens *Bb* erit ad augmentum evanescens *Ee* ut *AB* × *PB* ad *AE* × *PE*, ideoq̄ in hac ratione est fluxio rectæ *AB* ad fluxionem rectæ *AE*.

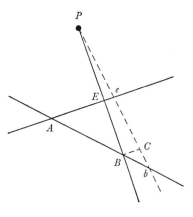

Hinc si recta revolvens *PB* lineas quasvis curvas positione datas secet in punctis *B* et *E*, et rectæ mobiles *AB, AE* Curvas illas tangant in sectionum punctis *B* et *E*: erit fluxio Curvæ quam recta *AB* tangit ad fluxionem Curvæ quam recta *AE* tangit ut *AB* × *PB* ad *AE* × *PE*. Id quod etiam eveniet si recta *PB* Curvam aliquam positione datam perpetuò tangat in puncto mobili *P*.

Fluat quantitas x uniformiter et invenienda sit fluxio $\underline{quantitatis\ x^n}$. Quo tempore quantitas *x* fluendo evadit $x+o$, quantitas x^n evadet $\overline{x+o}|^n$, id est per methodum serierum infinitarum $x^n + nox^{n-1} + \dfrac{nn-n}{2}oox^{n-2} + \&\text{c}$. Et augmenta *o* et

$$nox^{n-1} + \frac{nn-n}{2}oox^{n-2} + \&\text{c}$$

sunt ad invicem ut 1 et $nx^{n-1} + \dfrac{nn-n}{2}ox^{n-2} + \&\text{c}$. Evanescant jam augmenta illa,

---

(12) Ever equal to $\widehat{BCb}$, that is.

(13) The second half of this word, 'tione', begins a new manuscript page (f. **3**ʳ) whose discarded first version has strayed a long way in the present ordering of his mathematical papers to be Add. 3963.4: 60ʳ. We briefly notice the latter's variants in ensuing footnotes.

The straight line PB revolving round the given pole P shall intersect another straight line AB given in position: there is required the ratio of the fluxions of those lines AB and PB. Let the line *PB* advance from its place *PB* into the new position *Pb*; in *Pb* take *PC* equal to *PB*, and to *AB* draw *PD* such that the angle $b\widehat{P}D$ is equal to $b\widehat{B}C$: then, because the triangles *bBC* and *bPD* are similar, the augment *Bb* will be to the augment *Cb* as *Pb* to *Db*. Now let *Pb* return to its former place *PB* so that those augments shall come to vanish, and their last ratio as they do so—the last ratio of *Pb* to *Db*, that is—will be that had by *PB* to *DB* where the angle $P\widehat{D}B^{(12)}$ is right; and in this ratio, accordingly, is the fluxion of *AB* to the fluxion of *PB*.

The straight line PB revolving round the given pole P shall intersect in B and E two other straight lines AB and AE given in position: there is required the ratio of the fluxions of those lines AB and AE. Let the revolving line *PB* advance from its place *PB* into the new position *Pb* where it intersects the lines *AB*, *AE* in the points *b* and *e*, and parallel to the line *AE*[14] draw *BC* meeting *Pb* in *C*, and there will then be *Bb* to *BC* as *Ab* to *Ae*, and *BC* to *Ee* as *PB* to *PE*, and on combining these ratios *Bb* to *Ee* as $Ab \times PB$ to $Ae \times PE$. Now let the line *Pb* go back into its former place *PB*, and the vanishing augment *Bb* will be to the vanishing augment *Ee* as $AB \times PB$ to $AE \times PE$; in this ratio, consequently, is the fluxion of the straight line *AB* to the fluxion of the straight line *AE*.

Hence if the revolving straight line *PB* shall cut any curved lines given in position in the points *B* and *E*, and the mobile straight lines *AB* and *AE* touch those curves at the points *B* and *E* of intersection: the fluxion of the curve which the line *AB* touches will be to the fluxion of the curve which the line *AE* touches as $AB \times PB$ to $AE \times PE$. The same will also occur if the straight line *PB* shall perpetually touch at its mobile point *P* some curve given in position.

Let the quantity x flow uniformly and the fluxion of the quantity x^n need to be found. In the time that the quantity *x* comes in its flux to be $x+o$, the quantity x^n will come to be $(x+o)^n$, that is (when expanded) by the method of infinite series

$$x^n + nox^{n-1} + \tfrac{1}{2}(n^2 - n)\, o^2 x^{n-2} + \ldots;$$

and so the augments o and $nox^{n-1} + \tfrac{1}{2}(n^2 - n)\, o^2 x^{n-2} + \ldots$ are one to the other as

---

(14) In his initial revise of §2.2 on Add. 3960.13: 215 Newton first (absent-mindedly?) copied the figure illustrating his more cumbrous original resolution of the present problem of fluxional comparison where he appealed to the perpendicular *BD* let fall from *B* to *Pb*, that analogously dropped from *PB*'s intersection (there designated *C*) with the second given line, and the two normals *PM* and *PN* from *P* to *AB* and *AC*; but he quickly inserted the parallel *BD*, deleted all four perpendiculars, and, lastly, in the preliminary redraft on Add. 3960.13: 210 relettered the points *C*, *c* and *D* to be *E*, *e* and *C*, exactly as they are now designated.

et eorum ratio ultima erit 1 ad nx^{n-1}: ideoqȝ fluxio quantitatis x est ad fluxionem quantitatis x^n ut 1 ad nx^{n-1}.[15]

Similibus argumentis per methodum rationum primarum et ultimarum colligi possunt fluxiones linearum seu rectarum seu curvarum in casibus quibuscunqȝ, ut et fluxiones superficierum, angulorum et aliarum quantitatum.[16] In finitis autem quantitatibus Analysin sic instituere, et finitarum nascentium vel evanescentium rationes primas vel ultimas investigare, consonum est Geometriæ Veterum:[17] et volui ostendere quod in Methodo Fluxionum non opus sit[18] figuras[19] infinitè parvas in Geometriam introducere.

Ex Fluxionibus invenire Fluentes Problema difficilius est, et solutionis primus gradus est Quadratura Curvarum;[20] de qua sequentia olim[21] scripsi.

DE QUADRATURA CURVARUM.[22]

Quantitates indeterminatas ut motu perpetuo crescentes vel decrescentes, id est ut fluentes vel defluentes in sequentibus considero designoqȝ literis z, y, x, v,

---

(15) In the discarded preliminary sheet Add. 3963.4: 60^r this paragraph was in the first instance omitted, the preceding and following paragraphs passing one into the other without intervention; in afterthought the deficiency was made good by Newton's subsequent squeezing in of a '† *Fluat quantitas x . . .*' (*Let the quantity x flow . . .*) indicating that the related paragraph on Add. 3960.13: 208 (which completes the unfinished first version in §2.2 preceding as we specified in note (54) thereto) is to be intercalated.

(16) In a first, cancelled opening to this paragraph on Add. 3960.13: 210 Newton wrote: 'Similibus argumentis colligitur quod fluxio arcus est ad fluxionem sinus ut radius ad cosinum, et ad fluxionem tangentis in duplicata ratione sinus[!] ad radium. Et hinc fluxiones arcuum et angulorum calculum analyticum ingredi possunt' (By similar arguments it is gathered that the fluxion of an arc is to the fluxion of its sine as the radius to the cosine, and to the fluxion of the tangent in the doubled ratio of the sine to the radius. And hence fluxions of arcs and angles can enter analytical calculus). He then instead more accurately affirmed: 'Similibus argumentis per methodum rationum primarum et ultimarum colligi possunt fluxiones linearum seu rectarū seu curvarum in casibus quibuscunqȝ et fluxiones angulorum. Nam fluxiones angulorum sunt ut fluxiones arcuum angulos ad æquales distantias subtendentium, et fluxio arcus est ad fluxionem sinus ut radius ad cosinum et ad fluxionem tangentis in duplicata ratione cosinus ad radium' (By similar methods there can through the method of first and last ratios be gathered the fluxions of lines—whether straight or curved—in any cases whatever, and the fluxions of angles. For fluxions of angles are as the fluxions of arcs subtending the angles at equal distances; and the fluxion of an arc is to the fluxion of its sine as the radius to the cosine, and to the fluxion of its tangent in the doubled ratio of the cosine to the radius). In the further refashioning on the discarded sheet Add. 3963.4: 60^r the latter sentence was suppressed and the first one augmented to read as in its final version above. Here, of course, he harks briefly back to Proposition 14 of his 'Geometria Curvilinea' (IV: 440) and the related passage on 'Fluxiones linearum et angulorum' (VII: 450–2) in his 'Geometriæ Liber primus' ten years afterwards.

(17) In his discarded sheet (Add. 3963.4: 60^r) Newton initially concluded much more bitingly: 'a qua utiqȝ non temere est recedendum, præsertim cum figuræ infinite parvæ in Geometria[m] nondū recipiantur. Fluxiones sunt quantitates finitæ et certæ & finitas certasqȝ per infinite parvas & incertas exponere ac designare minus est artificiosum & erroribus magis

1 and $nx^{n-1} + \frac{1}{2}(n^2 - n)\, ox^{n-2} + \ldots$. Now let those augments come to vanish and their last ratio will be 1 to nx^{n-1}; consequently the fluxion of the quantity x is to the fluxion of the quantity x^n as 1 to nx^{n-1}.[15]

By similar arguments there can by means of the method of first and last ratios be gathered the fluxions of lines, whether straight or curved, in any cases whatever, as also the fluxions of surface-areas, angles and other quantities.[16] In finite quantities, however, to institute analysis in this way and to investigate the first or last ratios of nascent or vanishing finites is in harmony with the geometry of the ancients,[17] and I wanted to show that in the method of fluxions there should be [18] no need to introduce infinitely small figures[19] into geometry.

From the fluxions to find the fluents is a more difficult problem. The first step of its solution is the quadrature of curves,[20] and on this I formerly[21] wrote the following:

On the Quadrature of Curves.[22]

In the sequel I consider indeterminate quantities as increasing or decreasing by a perpetual motion, that is, as onwards or backwards flowing. I denote them

---

obnoxium' (from which, of course, you must not rashly depart, particularly since infinitely small figures are not yet received into geometry. Fluxions are finite and certain quantities, and to express and denote certain and finite ones by infinitely small and uncertain ones is less adroit and more susceptible to errors). On Newton's general reluctance in his old age to go against what he looked to as the elegance, simplicity and synthetic power of classical Greek geometry see vii: 198–9.

(18) This replaces a more dogmatic 'est' (is).

(19) Initially 'quantitates' (quantities).

(20) In his discard sheet (Add. 3963.4: 60^r) Newton originally added 'seu inventio Arearum ex fluxionibus per Ordinatas describentes designatis' (that is, the finding of areas from fluxions denoted by their describing ordinates), continuing thereafter with the linking phrase 'De hac re' (On this matter...).

(21) No earlier than the autumn of 1692, when Newton passed to John Wallis an early version (see vii: 173) of the theorem on series quadrature which he subsequently subsumed into his 'Geometriæ Liber secundus' as its Proposition V (see vii: 520); for this latter parent tract on quadrature of curves we have assumed a date of composition in early 1693 (which, if not quite accurate, can only be months out; compare vii: 508, note (2)). We have earlier (vii: 15–16) traced the stages by which, to underline his priority of 'invention' of the calculus, Newton came in the decade after 1712 to insist that he wrote his 'Book of Quadratures' in '1676 in summer'—a backdating by sixteen years and more which is now become so deeply entrenched in the received historical account as to be well-nigh inexpungeable. (The myth is most recently perpetuated by J. E. Hofmann in the *Akademie* edition of Leibniz' *Mathematischer ...Briefwechsel: 1672–1676* (*Sämtliche Schriften und Briefe* (3) 1, Berlin, 1976), where (p. xxiii, note 5) the excerpts from Newton's 1691 treatise 'De quadratura Curvarum' (vii: 48–128) which were printed by Wallis in his *Opera Mathematica*, 2, 1693: 391–6 [=vii: 171–80] are described as 'Auszüge aus Newtons *Quadratura Curvarum* von 1676, die erst 1704 als Anhang zur *Optics* [*sic*] gedruckt zugänglich wurden'.)

(22) Originally (on both the discarded sheet Add. 3963.4: 60^r and afterwards on f. 3^r of the present revision) just 'QUADRATURA CURVARUM' (THE QUADRATURE OF CURVES), then subse-

et earū fluxiones seu celeritates crescendi noto ijsdem literis punctatis $\dot{z}, \dot{y}, \dot{x}, \dot{v}$. Sunt et harum fluxionum fluxiones seu mutationes magis aut minus celeres quas ipsarum z, y, x, v fluxiones secundas nominare licet . . . .[23] Et notandum est quod quantitas quælibet prior in his seriebus est[24] area figuræ curvilineæ cujus ordinatim applicata rectangula est quantitas posterior et abscissa est z: uti $\sqrt{az-zz}$[25] area curvæ cujus ordinata est $\sqrt{az-zz}$ et abscissa z. Quo autem spectant hæc omnia patebit in Propositionibus quæ sequuntur.[26]

. . .   .

---

[2][27]   *Scholium.*[28]

Ubi quadrandæ sunt figuræ; ad Regulas hasce generales semper recurrere nimis molestum esset: præstat Figuras quæ simpliciores sunt et magis usui esse possunt semel quadrare et quadraturas in Tabulam referre, deinde Tabulā consulere quoties ejusmodi Curvam aliquam quadrare oportet. Hujus autem generis sunt Tabulæ duæ sequentes, in quib[u]s z denotat Abscissam, y Ordinatam rectangulam & t Aream Curvæ quadrandæ, & d, e, f, g, h, η sunt quantitates datæ cum signis suis $+$ et $-$.[29]

TABULA CURVARŪ SIMPLICIORUM QUÆ QUADRARI POSSUNT.[30]

Curvarum formæ.[31]	*Curvarum areæ*
Forma prima.	
$dz^{\eta-1}=y.$	$\dfrac{d}{\eta}z^{\eta}=t.$

---

quently (on f. 4^r)—this was to be the final title of the full tract in the published version (see note (3) above)—'TRACTATUS DE QUADRATURA CURVARUM' (A TREATISE ON THE QUADRATURE OF CURVES) before again being in part curtailed.

(23) We omit to reproduce the central portion of Newton's opening paragraph (which is exactly as in the 1693 parent on VII: 510 other than for the simplified denotations of fluents and fluxions specified in notes (8) and (9) thereto), but take it up again at its end.

(24) Newton afterwards inserted 'ut' (as) in minimal correction on Add. 3960.13: 209, and so it is printed in the *editio princeps* (*Opticks*, ₁1704: ₂171).

(25) To correspond with his preceding late insertion (see the previous note) Newton here added a parallel 'ut' (as) in his library copy of Jones' 1711 *Analysis* (note (1) above), but the interpolation appears in no subsequent printed edition.

(26) There now ensues in the manuscript (Add. 3962.1: 5^r *et seq.*) the lightly touched-up text of Propositions I–X of the parent 1693 tract on quadrature of curves, essentially as it is printed on VII: 512–48 except that the final paragraph of Proposition I is (see VII: 517, note (22)) now suppressed in favour of the more comprehensive—if still shallow and superficial—account of the 'integration' of elementary algebraic fluxional equations which he will present in his freshly composed terminal scholium to the 'De Quadratura' (in §3.2 which now follows).

by the letters z, y, x, v and mark their fluxions, that is, speeds of increase, by the same letters with points on: $\dot{z}, \dot{y}, \dot{x}, \dot{v}$. There are also fluxions of these fluxions, or changes in their degree of speed, which it is allowable to name second fluxions of $z, y, x, v \ldots \ldots$ [23] And note that any prior quantity whatever in these series is[24] the area of a curvilinear figure whose rectangular ordinate is the latter quantity and its abscissa is z: for instance, $\sqrt{az-z^2}$ is [25] the area of the curve whose ordinate is $\sqrt{az-z^2}$ and abscissa z. What, however, is the regard of all these things will become evident in the propositions which follow.[26]

$\cdots$ $\cdots$ $\cdots$

---

[2][27]   *Scholium.*[28]

Where figures are to be squared, it would be excessively troublesome always to recur to these general rules. It is better once for all to square the figures which are simpler and can be of more use, and to list their quadratures in a table: we can thereafter consult the table each time we have to square some curve of this kind. Of this type are the two following tables, wherein z denotes the abscissa, y the rectangular ordinate and t the area of the curve to be squared, while d, e, f, g, h, η are given quantities with their signs $+$ and $-$.[29]

A TABLE OF THE SIMPLER CURVES WHICH CAN BE SQUARED.[30]

Forms[31] *of curves.*	*Areas of the curves.*
First form.	
$dz^{\eta-1}=y.$	$(d/\eta)\,z^\eta=t.$

---

(27) Add. 3962.1: 19^r–26^r. In this concluding portion of the text of the augmented 'De Quadratura Curvarum' as published in 1704 Newton minimally refashions (and trivially reorders) the scholium to Proposition X of the 1693 original, where (see VII: 548–50) he re-presented his twin tables of exact and conically evaluable integrals from Problem 9 of his 1671 tract, and also in a brand new scholium to the ensuing Proposition XI of the parent tract (VII: 550–2) attaches a sketch of the ways of resolving the simplest fluxional equations (of first order) which are either in a directly quadrable form or are immediately castable into one.

(28) To Proposition X, that is; see the previous note.

(29) A but slightly rephrased and amplified revise of the corresponding introductory paragraph on VII: 548.

(30) Under this minimally augmented title—and now (see the next note) renaming his earlier 'orders' of curves to be 'forms'—Newton again (compare VII: 548–9, notes (88) and (89)) repeats the parallel columns of the 'Catalogus Curvarum aliquot ad rectilineas figuras relatarum' which he had originally set more than thirty years earlier to be the prior tabular illustration of Problem 9 of his 1671 fluxional tract (see III: 236–8).

(31) The 'ordines' (orders) of the parent 1693 tabulation (VII: 548) is here replaced, and the same change is effected in the rest of the table, *mutatis mutandis.*

Forma secunda.

$$\frac{dz^{\eta-1}}{ee+2efz^{\eta}+ffz^{2\eta}}=y.$$

$$\frac{dz^{\eta}}{\eta ee+\eta efz^{\eta}}=t,\ \text{vel}\ \frac{-d}{\eta ef+\eta ffz^{\eta}}=t.$$

Forma tertia.

1. $dz^{\eta-1}\sqrt{e+fz^{\eta}}=y.$
   $$\frac{2d}{\eta f}R^3=t,\quad\text{existente}\quad R=\sqrt{e+fz^{\eta}}.$$

2. $dz^{2\eta-1}\sqrt{e+fz^{\eta}}=y.$
   $$\frac{-4e+6fz^{\eta}}{15\eta ff}dR^3=t.$$

3. $dz^{3\eta-1}\sqrt{e+fz^{\eta}}=y.$
   $$\frac{16ee-24efz^{\eta}+30ffz^{2\eta}}{105\eta f^3}[dR^3]=t.$$

4. $dz^{4\eta-1}\sqrt{e+fz^{\eta}}=y.$
   $$\frac{-96e^3+144eefz^{\eta}-180effz^{2\eta}+210f^3z^{3\eta}}{945\eta f^4}dR^3=t.$$

Forma quarta.

1. $\dfrac{dz^{\eta-1}}{\sqrt{e+fz^{\eta}}}=y.$
   $$\frac{2d}{\eta f}.R=t.$$

2. $\dfrac{dz^{2\eta-1}}{\sqrt{e+fz^{\eta}}}=y.$
   $$\frac{-4e+2fz^{\eta}}{3\eta ff}dR=t.$$

3. $\dfrac{dz^{3\eta-1}}{\sqrt{e+fz^{\eta}}}=y.$
   $$\frac{16ee-8efz^{\eta}+6ffz^{2\eta}}{15\eta f^3}dR=t.$$

4. $\dfrac{dz^{4\eta-1}}{\sqrt{e+fz^{\eta}}}=y.$
   $$\frac{-96e^3+48eefz^{\eta}-36effz^{2\eta}+30f^3z^{3\eta}}{105\eta f^4}dR=t.$$

Tabula Curvarum simpliciorum quæ cum Ellipsi et Hyperbola comparari possunt.

Sit jam *aGD* vel *PGD* vel *GDS* Sectio Conica cujus area ad Quadraturam Curvæ propositæ requiritur, sitꝗ ejus centrum *A*, Axis *Ka*, Vertex *a*, semiaxis conjugatus *AP*, datum Abscissæ principium *A* vel *a* vel *α*, Abscissa *AB* vel *aB* vel

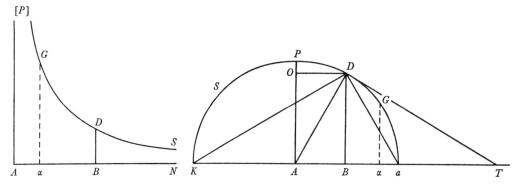

$\alpha B = x$, Ordinata *BD* = *v*, et Area *ABDP* vel *aBDG* vel *αBDG* = *s*, existente *αG* Ordinata ad punctum *α*. Jungantur *KD*, *AD*, *aD*. Ducatur Tangens *DT*

Second form.

$dz^{\eta-1}/(e^2 + 2efz^\eta + f^2z^{2\eta}) = y.$ $dz^\eta/\eta e(e + fz^\eta) = t$, or $-d/\eta f(e + fz^\eta) = t.$

Third form.

1. $dz^{\eta-1}\surd(e + fz^\eta) = y.$ $(2d/\eta f) R^3 = t,$ where $R = \surd(e + fz^\eta).$

2. $dz^{2\eta-1}\surd(e + fz^\eta) = y.$ $(d(-4e + 6fz^\eta)/15\eta f^2) R^3 = t.$

3. $dz^{3\eta-1}\surd(e + fz^\eta) = y.$ $(d(16e^2 - 24efz^\eta + 30f^2z^{2\eta})/105\eta f^3) R^3 = t.$

4. $dz^{4\eta-1}\surd(e + fz^\eta) = y.$ $(d(-96e^3 + 144e^2fz^\eta - 180ef^2z^{2\eta} + 210f^3z^{3\eta})/945\eta f^4) R^3 = t.$

Fourth form.

1. $dz^{\eta-1}/\surd(e + fz^\eta) = y.$ $(2d/\eta f) R = t.$

2. $dz^{2\eta-1}/\surd(e + fz^\eta) = y.$ $(d(-4e + 2fz^\eta)/3\eta f^2) R = t.$

3. $dz^{3\eta-1}/\surd(e + fz^\eta) = y.$ $(d(16e^2 - 8efz^\eta + 6f^2z^{2\eta})/15\eta f^3) R = t.$

4. $dz^{4\eta-1}/\surd(e + fz^\eta) = y.$ $(d(-96e^3 + 48e^2fz^\eta - 36ef^2z^{2\eta} + 30f^3z^{3\eta})/105\eta f^4) R = t.$

A TABLE OF THE SIMPLER CURVES
WHICH CAN BE COMPARED WITH THE ELLIPSE AND HYPERBOLA.

Now let *aGD* or *PGD* or *GDS* be the conic section whose area is required for the quadrature of the curve proposed, and let its centre be *A*, its axis *Ka*, vertex *a*, conjugate semi-axis *AP*, the given origin of its abscissa *A* or *a* or *α*, the abscissa

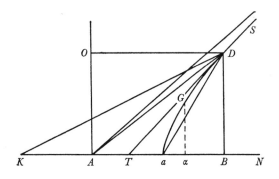

AB or *aB* or *αB* = *x*, ordinate *BD* = *v*, and its area (*ABDP*) or (*aBDG*) or (*αBDG*) = *s*, where *αG* is the ordinate at the point *α*. Join *KD*, *AD*, *aD*, draw the tangent *DT* meeting the abscissa *AB* in *T*, and complete the rectangle

occurrens Abscissæ *AB* in *T*, et compleatur parallelogrammum *ABDO*. Et siquando ad quadraturam Curvæ propositæ requiruntur areæ duarum Sectionum Conicarum, dicatur posterioris Abscissa ξ, Ordinata Υ, et Area σ.[32]

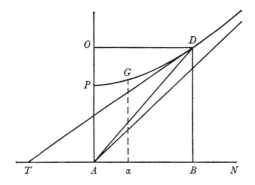

(32) To his reader's considerable profit, Newton here amplifies the terse heading (see VII: 550) which ten years before he had thought adequate to introduce his previous repeat of the 1671 tract's 'Catalogus Curvarum aliquot ad Sectiones Conicas relatarum' (III: 244–54), once more (compare III: 242) giving a prior explanation of the accompanying diagrams of conics to which its content is keyed. Because we have already (in footnotes on III: 256–8) presented a detailed analysis of the equivalent individual entries in the earlier 'Catalogus', we need here make only brief further comment thereon. In a slightly superior rearrangement the fifth and seventh of the previous 'orders of curves' are now (compare note (36) below) advanced to be respectively third and fourth in the new sequence of 'forms', the third and fourth 'orders' now

ABDO. And should ever there be required for the quadrature of the curve proposed the areas of two conics, let the latter one's abscissa be called ξ, its ordinate Υ and its area σ.[32]

---

following on in fifth and sixth place. Otherwise the table was printed in 1704 as it is here reproduced in sequel, though several small slips, and one notable inadequacy remained to mar the perfection of its detail. We have already remarked (III: 257–8, note (564)) that when Roger Cotes came afterwards, in the summer of 1709, to test the accuracy of the published 'Tabula' he rightly queried its Forms 7.1 and 8.4, the rest agreeing with his computation. Upon checking for himself, Newton easily confirmed (see note (39) below, and [6] of the Appendix) that he had carelessly omitted two terms in transcribing the parallel Order 8.4 from his 1671 table; and he readily accepted (see note (38)) Cotes' suggested insertion making good a like lacuna in the preceding (Order 6.1→) Form 7.1. At this time also—so we infer from our examination of the handwriting of the fragment (reproduced in [4] of the Appendix) on which he jotted down his recomputation of Form 2.1—he went over his other entries and in the second and third instances of his 'Forma secunda' caught three extraneous factors e (see note (35) below). All these *errata et emendanda* he noted in the margins of his library copy of the 1706 *Optice* (note (1) above), at the pertinent pages of the reissue of the 'De Quadratura' appended thereto, and they were put right in the corrected text of it which Jones published in his 1711 *Analysis*. The limitation $f^2 \geqslant 4eg$ to Form 6, finally, was at length removed by Cotes, a little before his death, in the spring of 1716 (see note (37) below); but, from his own subsequent, not wholly effective attempt (see [5] of the Appendix) to take the restriction away, it would appear that Newton himself—for a time at least—was only aware in broad outline of the adroit and elegant way in which Cotes filled in this yawning gap in the coverage of his 'Forma sexta'.

Curvarum Formæ.	Sectionis Conicæ		Curvarum Areæ.
	Abscissa.	Ordinata.	
Forma prima.			
1. $\dfrac{dz^{\eta-1}}{e+fz^\eta}=y.$	$z^\eta=x.$	$\dfrac{d}{e+fx}=v.$	$\dfrac{1}{\eta}s=t=\dfrac{\alpha GDB}{\eta}$ [33].
2. $\dfrac{dz^{2\eta-1}}{e+fz^\eta}=y.$	$z^\eta=x.$	$\dfrac{d}{e+fx}=v.$	$\dfrac{d}{\eta f}z^\eta-\dfrac{e}{\eta f}s=t.$
3. $\dfrac{dz^{3\eta-1}}{e+fz^\eta}=y.$	$z^\eta=x.$	$\dfrac{d}{e+fx}=v.$	$\dfrac{d}{2\eta f}z^{2\eta}-\dfrac{de}{\eta ff}z^\eta+\dfrac{ee}{\eta ff}s=t.$
Forma secunda.			
1. $\dfrac{dz^{\frac12\eta-1}}{e+fz^\eta}=y.$	$\sqrt{\dfrac{d}{e+fz^\eta}}=x.$	$\sqrt{\dfrac{d}{f}-\dfrac{e}{f}xx}=v.$	$\dfrac{2xv\div4s}{\eta}=t=\dfrac{4}{\eta}ADGa.$ [34]
2. $\dfrac{dz^{\frac32\eta-1}}{e+fz^\eta}=y.$	$\sqrt{\dfrac{d}{e+fz^\eta}}=x.$	$\sqrt{\dfrac{d}{f}-\dfrac{e}{f}xx}=v.$	$\dfrac{2de}{\eta f}z^{\frac\eta2}$ [35] $\dfrac{4es-2exv}{\eta f}=t.$
3. $\dfrac{dz^{\frac52\eta-1}}{e+fz^\eta}=y.$	$\sqrt{\dfrac{d}{e+fz^\eta}}=x.$	$\sqrt{\dfrac{d}{f}-\dfrac{e}{f}xx}=v.$	$\dfrac{2de}{3\eta f}z^{\frac32\eta}-\dfrac{2dee}{\eta ff}z^{\frac\eta2}$ [35] $+\dfrac{2eexv-4ees}{\eta ff}=t.$

The notes (33)–(39) for pp. 136–42 are on pp. 148–51.

Forms of curves.	The conic's		Areas of the curves.
	abscissa.	ordinate.	
First form.			
1. $dz^{\eta-1}/(e+fz^\eta) = y.$	$z^\eta = x.$	$d/(e+fx) = v.$	$(1/\eta)\,s = t = (\alpha GDB)/\eta.$ [33]
2. $dz^{2\eta-1}/(e+fz^\eta) = y.$	$z^\eta = x.$	$d/e(e+fx) = v.$	$(d/\eta f)\,z^\eta - (e/\eta f)\,s = t.$
3. $dz^{3\eta-1}/(e+fz^\eta) = y.$	$z^\eta = x.$	$d/(e+fx) = v.$	$(d/2\eta f)\,z^{2\eta} - (de/\eta f^2)\,z^\eta + (e^2/\eta f^2)\,s = t.$
Second form.			
1. $dz^{\frac{1}{2}\eta-1}/(e+fz^\eta) = y.$	$d^{\frac{1}{2}}/\sqrt(e+fz^\eta) = x.$	$\sqrt(d/f - (e/f)\,x^2) = v.$	$|2xv - 4s|/\eta = t = 4(ADGa)/\eta.$ [34]
2. $dz^{\frac{3}{2}\eta-1}/(e+fz^\eta) = y.$	$d^{\frac{1}{2}}/\sqrt(e+fz^\eta) = x.$	$\sqrt(d/f - (e/f)\,x^2) = v.$	$(2de/\eta f)\,z^{\frac{1}{2}\eta [35]} + (4es - 2exv)/\eta f = t.$
3. $dz^{\frac{5}{2}\eta-1}/(e+fz^\eta) = y.$	$d^{\frac{1}{2}}/\sqrt(e+fz^\eta) = x.$	$\sqrt(d/f - (e/f)\,x^2) = v.$	$(2de/3\eta f)\,z^{\frac{3}{2}\eta} - (2de^2/\eta f^2)\,z^{\frac{1}{2}\eta [35]} + (2e^2xv - 4e^2s)/\eta f^2 = t.$

The notes (33)–(38) for pp. 137–43 are on pp. 148–9.

Forma tertia.

1. $\frac{d}{z}\sqrt{e+fz^\eta}=y.$ $\frac{1}{z^\eta}=xx.$ $\sqrt{f+exx}=v$ $\frac{4de}{\eta f}$ in $\frac{v^3}{2ex}-s=t=\frac{4de}{\eta f}$ in $aGDT$, vel in $APDB\div TDB.$

     Vel sic $\frac{1}{z^\eta}=x.$ $\sqrt{fx+exx}=v.$ $\frac{8dee}{\eta f}$ in $s-\frac{1}{2}xv-\frac{fv}{4e}+\frac{ffv}{4eex}=t=\frac{8dee}{\eta ff}$ in $aGDA+\frac{ffv}{4eex}.$

2. $\frac{d}{z^{\eta+1}}\sqrt{e+fz^\eta}=y.$ $\frac{1}{z^\eta}=xx.$ $\sqrt{f+exx}=v.$ $-\frac{2d}{\eta}s=t=\frac{2d}{\eta}APDB$, seu $\frac{2d}{\eta}aGDB.$

     Vel sic $\frac{1}{z^\eta}=x.$ $\sqrt{fx+exx}=v.$ $\frac{4de}{\eta f}$ in $s-\frac{1}{2}xv-\frac{fv}{2e}=t=\frac{4de}{\eta f}\times aGDK.$

3. $\frac{d}{z^{2\eta+1}}\sqrt{e+fz^\eta}=y.$ $\frac{1}{z^\eta}=x.$ $\sqrt{fx+exx}=v.$ $-\frac{d}{\eta}s=t=\frac{d}{\eta}\times -aGDB$ vel $BDPK.$

4. $\frac{d}{z^{3\eta+1}}\sqrt{e+fz^\eta}=y.$ $\frac{1}{z^\eta}=x.$ $\sqrt{fx+exx}=v.$ $\frac{3dfs-2dv^3}{6\eta e}=t.$

Forma quarta.

1. $\frac{d}{z\sqrt{e+fz^\eta}}=y.$ $\frac{1}{z^\eta}=xx.$ $\sqrt{f+exx}=v.$ $\frac{4d}{\eta f}$ in $\frac{1}{2}xv\div s=t=\frac{4d}{\eta f}$ in PAD vel in $aGDA.$

     Vel sic $\frac{1}{z^\eta}=x.$ $\sqrt{fx+exx}=v.$ $\frac{8de}{\eta ff}$ in $s-\frac{1}{2}xv-\frac{fv}{4e}=t=\frac{8de}{\eta ff}$ in $aGDA.$

2. $\frac{d}{z^{\eta+1}\sqrt{e+fz^\eta}}=y.$ $\frac{1}{z^\eta}=xx.$ $\sqrt{f+exx}=v.$ $\frac{2d}{\eta e}$ in $s-xv=t=\frac{2d}{\eta e}$ in POD, vel in $AODGa.$

     Vel sic $\frac{1}{z^\eta}=x.$ $\sqrt{fx+exx}=v.$ $\frac{4d}{\eta f}$ in $\frac{1}{2}xv\div s=t=\frac{4d}{\eta f}$ in $aDGa.$

3. $\frac{d}{z^{2\eta+1}\sqrt{e+fz^\eta}}=y.$ $\frac{1}{z^\eta}=x.$ $\sqrt{fx+exx}=v.$ $\frac{d}{\eta e}$ in $3s\div 2xv=t=\frac{d}{\eta e}$ in $3aDGa\div \triangle aDB.$

4. $\frac{d}{z^{3\eta+1}\sqrt{e+fz^\eta}}=y.$ $\frac{1}{z^\eta}=x.$ $\sqrt{fx+exx}=v.$ $\frac{10dfxv-15dfs-2dexxv}{6\eta ee}=t.$

Third form.

1. $dz^{-1}\sqrt{(e+fz^\eta)} = y.$ $z^{-\eta}=x^2.$ $\sqrt{(f+ex^2)}=v.$ $(4de|\eta f)(v^3|2ex-s)=t=(4de|\eta f)\times(aGDT'$
   or $\times|(APDB)-(TDB)|.$

Or thus: $z^{-\eta}=x.$ $\sqrt{(fx+ex^2)}=v.$ $(8de^2|\eta f^2)(s-\frac{1}{2}xv-\frac{1}{4}fv|e+\frac{1}{4}f^2v|e^2x)=t$
   $=(8de^2|\eta f^2)((aGDA)+\frac{1}{4}f^2v|e^2x).$

2. $dz^{-\eta-1}\sqrt{(e+fz^\eta)}=y.$ $z^{-\eta}=x^2.$ $\sqrt{(f+ex^2)}=v.$ $-(2d|\eta)s=t=(2d|\eta)(APDB)$ or $(2d|\eta)(aGDB).$

Or thus: $z^{-\eta}=x.$ $\sqrt{(fx+ex^2)}=v.$ $(4de|\eta f)(s-\frac{1}{2}xv-\frac{1}{2}fv|e)=t=(4de|\eta f)(aGDK).$

3. $dz^{-2\eta-1}\sqrt{(e+fz^\eta)}=y.$ $z^{-\eta}=x.$ $\sqrt{(fx+ex^2)}=v.$ $-(d|\eta)s=t=(d|\eta)\times-(aGDB)$ or $(BDPK).$

4. $dz^{-3\eta-1}\sqrt{(e+fz^\eta)}=y.$ $z^{-\eta}=x.$ $\sqrt{(fx+ex^2)}=v.$ $d(3fs-2v^3)|6\eta e=t.$

Fourth form.

1. $dz^{-1}|\sqrt{(e+fz^\eta)}=y.$ $z^{-\eta}=x^2.$ $\sqrt{(f+ex^2)}=v.$ $(4d|\eta f)|\frac{1}{2}xv-s|=t=(4d|\eta f)\times(PAD)$ or $\times(aGDA).$

Or thus: $z^{-\eta}=x.$ $\sqrt{(fx+ex^2)}=v.$ $(8de|\eta f^2)(s-\frac{1}{2}xv-\frac{1}{4}fv|e)=t=(8de|\eta f^2)\times(aGDA).$

2. $dz^{-\eta-1}|\sqrt{(e+fz^\eta)}=y.$ $z^{-\eta}=x^2.$ $\sqrt{(f+ex^2)}=v.$ $(2d|\eta e)(s-xv)=t=(2d|\eta e)\times(POD)$ or $\times(AODGa).$

Or thus: $z^{-\eta}=x.$ $\sqrt{(fx+ex^2)}=v.$ $(4d|\eta f)|\frac{1}{2}xv-s|=t=(4d|\eta f)\times(aDGa).$

3. $dz^{-2\eta-1}|\sqrt{(e+fz^\eta)}=y.$ $z^{-\eta}=x.$ $\sqrt{(fx+ex^2)}=v.$ $(d|\eta e)|3s-2xv|=t=(d|\eta e)\times|3(aDGa)-\triangle aDB|.$

4. $dz^{-3\eta-1}|\sqrt{(e+fz^\eta)}=y.$ $z^{-\eta}=x.$ $\sqrt{(fx+ex^2)}=v.$ $d(10fxv-15fs-2ex^2v)|6\eta e^2=t.$

Forma quinta.

1. $\dfrac{dz^{\eta-1}}{e+fz^\eta+gz^{2\eta}}=y.$
$\qquad \sqrt{\dfrac{d}{e+fz^\eta+gz^{2\eta}}}=x.$
$\qquad \sqrt{\dfrac{d}{g}+\dfrac{ff-4eg}{4gg}xx}=v.$
$\qquad \dfrac{xv-2s}{\eta}=t.$

Vel sic, $\sqrt{\dfrac{dz^{2\eta}}{e+fz^\eta+gz^{2\eta}}}=x.$
$\qquad \sqrt{\dfrac{d}{[e]}+\dfrac{ff-4eg}{4ee}xx}=v.$
$\qquad \dfrac{2s-xv}{\eta}=t.$

2. $\dfrac{dz^{2\eta-1}}{e+fz^\eta+gz^{2\eta}}=y.$
$\left\{\begin{array}{l}\sqrt{\dfrac{d}{e+fz^\eta+gz^{2\eta}}}=x.\\[1em] fz^\eta+gz^{2\eta}=\xi.\end{array}\right.$
$\left\{\begin{array}{l}\sqrt{\dfrac{d}{g}+\dfrac{ff-4eg}{4gg}xx}=v.\\[1em] \dfrac{1}{e+\xi}=\Upsilon.\end{array}\right.$
$\qquad \dfrac{d\sigma+2fs-fxv}{2\eta g}=t.$

Forma sexta, ubi scribitur p pro $\sqrt{ff-4eg}$. [37]

1. $\dfrac{dz^{\frac{1}{2}\eta-1}}{e+fz^\eta+gz^{2\eta}}=y.$
$\left\{\begin{array}{l}\sqrt{\dfrac{2dg}{f-p+2gz^\eta}}=x.\\[1em] \sqrt{\dfrac{2dg}{f+p+2gz^\eta}}=\xi.\end{array}\right.$
$\left\{\begin{array}{l}\sqrt{d+\dfrac{-f+p}{2g}xx}=v.\\[1em] \sqrt{d+\dfrac{-f-p}{2g}\xi\xi}=\Upsilon.\end{array}\right.$
$\qquad \dfrac{2xv-4s-2\xi\Upsilon+4\sigma}{\eta p}=t.$

2. $\dfrac{dz^{\frac{3}{2}\eta-1}}{e+fz^\eta+gz^{2\eta}}=y.$
$\left\{\begin{array}{l}\sqrt{\dfrac{2dez^\eta}{fz^\eta-pz^\eta+2e}}=x.\\[1em] \sqrt{\dfrac{2dez^\eta}{fz^\eta+pz^\eta+2e}}=\xi.\end{array}\right.$
$\left\{\begin{array}{l}\sqrt{d+\dfrac{-f+p}{2g}xx}=v.\\[1em] \sqrt{d+\dfrac{-f-p}{2g}\xi\xi}=\Upsilon.\end{array}\right.$
$\qquad \dfrac{4s-2xv-4\sigma+2\xi[\Upsilon]}{\eta p}=t.$

Fifth form.

1. $dz^{\eta-1}/(e+fz^\eta+gz^{2\eta}) = y.$

$\qquad d^{\frac{1}{4}}/\sqrt{(e+fz^\eta+gz^{2\eta})} = x.$ $\sqrt{(d/g + (\tfrac{1}{4}(f^2-4eg)/g^2)\,x^2)} = v.$ $(xv-2s)/\eta = t.$

Or thus: $d^{\frac{1}{4}}z^\eta/\sqrt{(e+fz^\eta+gz^{2\eta})} = x.$ $\sqrt{(d/e + (\tfrac{1}{4}(f^2-4eg)/e^2)\,x^2)} = v.$ $(2s-xv)/\eta = t.$

2. $dz^{2\eta-1}/(e+fz^\eta+gz^{2\eta}) = y.$

$\qquad\left\{\begin{array}{l} d^{\frac{1}{4}}/\sqrt{(e+fz^\eta+gz^{2\eta})} = x. \\ fz^\eta + gz^{2\eta} = \xi. \end{array}\right.$ $\left\{\begin{array}{l}\sqrt{(d/g + (\tfrac{1}{4}(f^2-4eg)/g^2)\,x^2)} = v. \\ 1/(e+\xi) = \Upsilon.\end{array}\right.$ $(d\sigma + 2fs - fxv)/2\eta g = t.$

Sixth form, where p is written in place of $\sqrt{(f^2-4eg)}$. [37]

1. $dz^{\frac{1}{2}\eta-1}/(e+fz^\eta+gz^{2\eta}) = y.$

$\qquad\left\{\begin{array}{l}\sqrt{(2dg/(f-p+2gz^\eta))} = x. \\ \sqrt{(2dg/(f+p+2gz^\eta))} = \xi.\end{array}\right.$ $\left\{\begin{array}{l}\sqrt{(d-(\tfrac{1}{2}(f-p)/g)\,x^2)} = v. \\ \sqrt{(d-(\tfrac{1}{2}(f+p)/g)\,\xi^2)} = \Upsilon.\end{array}\right.$ $(2xv-4s-2\xi\Upsilon+4\sigma)/\eta p = t.$

2. $dz^{\frac{3}{2}\eta-1}/(e+fz^\eta+gz^{2\eta}) = y.$

$\qquad\left\{\begin{array}{l}\sqrt{(2dez^\eta/((f-p)\,z^\eta+2e))} = x. \\ \sqrt{(2dez^\eta/((f+p)\,z^\eta+2e))} = \xi.\end{array}\right.$ $\left\{\begin{array}{l}\sqrt{(d-(\tfrac{1}{2}(f-p)/g)\,x^2)} = v. \\ \sqrt{(d-(\tfrac{1}{2}(f+p)/g)\,\xi^2)} = \Upsilon.\end{array}\right.$ $(4s-2xv-4\sigma+2\xi\Upsilon)/\eta p = t.$

Forma septima.

1. $\dfrac{d}{z}\sqrt{e+fz^\eta+gz^{2\eta}}=y.$
$\begin{cases} z^\eta=x. \\ \dfrac{1}{z^\eta}=\xi. \end{cases}$
$\left.\begin{array}{l}\sqrt{e+fx+gxx}=v. \\ \sqrt{g+f\xi+e\xi\xi}=\Upsilon.\end{array}\right\}$
$\dfrac{4dee\xi\Upsilon+2def\Upsilon^{(38)}-2dffv-8deev+4dfgs}{\eta p}=t.$

2. $dz^{\eta-1}\sqrt{e+fz^\eta+gz^{2\eta}}=y.$ $z^\eta=x.$ $\sqrt{e+fx+gxx}=v.$ $\dfrac{d}{\eta}s=t=\dfrac{d}{\eta}$ in $\alpha GDB.$

3. $dz^{2\eta-1}\sqrt{e+fz^\eta+gz^{2\eta}}=y.$ $z^\eta=x.$ $\sqrt{e+fx+gxx}=v.$ $\dfrac{d}{3\eta g}v^3-\dfrac{df}{2\eta g}s=t.$

4. $dz^{3\eta-1}\sqrt{e+fz^\eta+gz^{2\eta}}=y.$ $z^\eta=x.$ $\sqrt{e+fx+gxx}=v.$ $\dfrac{6dgx-5df}{24\eta gg}v^3+\dfrac{5dff-4deg}{16\eta gg}s=t.$

Forma octava.

1. $\dfrac{dz^{\eta-1}}{\sqrt{e+fz^\eta+gz^{2\eta}}}=y.$ $z^\eta=x.$ $\sqrt{e+fx+gxx}=v.$ $\dfrac{8dgs-4dgxv-2dfv}{4\eta eg-\eta ff}=t=\dfrac{8dg}{4\eta eg-\eta ff}$ in $\alpha GDB\pm\triangle DBA.$

2. $\dfrac{dz^{2\eta-1}}{\sqrt{e+fz^\eta+gz^{2\eta}}}=y.$ $z^\eta=x.$ $\sqrt{e+fx+gxx}=v.$ $\dfrac{-4dfs+2dfxv+4dev}{4\eta eg-\eta ff}=t.$

3. $\dfrac{dz^{3\eta-1}}{\sqrt{e+fz^\eta+gz^{2\eta}}}=y.$ $z^\eta=x.$ $\sqrt{e+fx+gxx}=v.$ $\dfrac{3dff-2dff\;s\;\;xv-2defv}{-4deg+4deg\quad 4\eta egg-\eta ffg}=t$

4. $\dfrac{dz^{4\eta-1}}{\sqrt{e+fz^\eta+gz^{2\eta}}}=y.$ $z^\eta=x.$ $\sqrt{e+fx+gxx}=v.$
$\dfrac{36defg+8degg-28defg+10deef\;\;v}{\;}$
$\dfrac{-15deff\;s\;-2dffg\;xv\;+10deff\;xv\;-16deeg\;v}{[24\eta eg^3-6\eta ffgg]}=t.^{(39)}$

Seventh form.

1. $dz^{-1}\sqrt{(e+fz^\eta+gz^{2\eta})} = y.$ $\quad \begin{cases} z^\eta = x. \\ z^{-\eta} = \xi. \end{cases}$ $\quad \begin{rcases} \sqrt{(e+fx+gx^2)} = v. \\ \sqrt{(g+f\xi+e\xi^2)} = \Upsilon. \end{rcases}$

$$d(4e^2\xi\Upsilon + 2ef\Upsilon^{(38)} - 2f^2v - 8e^2\sigma + 4fgs)/\eta b = t.$$

2. $dz^\eta\sqrt{(e+fz^\eta+gz^{2\eta})} = y.$ $\quad z^\eta = x.$ $\quad \sqrt{(e+fx+gx^2)} = v.$

$$(d/\eta)\,s = t = (d/\eta) \times (\alpha GDB).$$

3. $dz^{2\eta}\sqrt{(e+fz^\eta+gz^{2\eta})} = y.$ $\quad z^\eta = x.$ $\quad \sqrt{(e+fx+gx^2)} = v.$

$$(d/3\eta g)\,v^3 - (df/2\eta g)\,s = t.$$

4. $dz^{3\eta}\sqrt{(e+fz^\eta+gz^{2\eta})} = y.$ $\quad z^\eta = x.$ $\quad \sqrt{(e+fx+gx^2)} = v.$

$$d(6gx-5f)\,v^3/24\eta g^2 + d(5f^2-4eg)\,s/16\eta g^2 = t.$$

Eighth form.

1. $dz^{\eta-1}/\sqrt{(e+fz^\eta+gz^{2\eta})} = y.$ $\quad z^\eta = x.$ $\quad \sqrt{(e+fx+gx^2)} = v.$

$$d(8gs - 4gxv - 2fv)/\eta(4eg-f^2) = t$$
$$= d(8g/\eta(4eg-f^2) \times ((\alpha GDB) \pm \triangle DBA).$$

2. $dz^{2\eta-1}/\sqrt{(e+fz^\eta+gz^{2\eta})} = y.$ $\quad z^\eta = x.$ $\quad \sqrt{(e+fx+gx^2)} = v.$

$$d(-4fs + 2fxv + 4ev)/\eta(4eg-f^2) = t$$

3. $dz^{3\eta-1}/\sqrt{(e+fz^\eta+gz^{2\eta})} = y.$ $\quad z^\eta = x.$ $\quad \sqrt{(e+fx+gx^2)} = v.$

$$d((3f^2-4eg)\,s - (2f^2-4eg)\,xv - 2efv)/\eta(4eg-f^2)\,g = t.$$

4. $dz^{4\eta-1}/\sqrt{(e+fz^\eta+gz^{2\eta})} = y.$ $\quad z^\eta = x.$ $\quad \sqrt{(e+fx+gx^2)} = v.$

$$d\frac{\left((36eg-15f^2)fs + (8eg-2f^2)gx^2v + (-28eg+10f^2)fxv + (10f^2-16eg)ev\right)}{6\eta(4eg-f^2)g^2} = t.$$

Forma nona.

1. $\dfrac{dz^{\eta-1}\sqrt{e+fz^\eta}}{g+hz^\eta}=y.$ $\qquad\sqrt{\dfrac{df}{h}+\dfrac{eh-fg}{h}xx}=v.$ $\qquad\sqrt{\dfrac{d}{g+hz^\eta}}=x.$

$$\frac{\begin{smallmatrix}4fg\\-4eh\end{smallmatrix}\,s\,\begin{smallmatrix}-2fg\\+2eh\end{smallmatrix}\,xv+\dfrac{2dfv}{x}}{\eta fh}=t.$$

2. $\dfrac{dz^{2\eta-1}\sqrt{e+fz^\eta}}{g+hz^\eta}=y.$ $\qquad\sqrt{\dfrac{df}{h}+\dfrac{eh-fg}{h}xx}=v.$ $\qquad\sqrt{\dfrac{d}{g+hz^\eta}}=x.$

$$\frac{\begin{smallmatrix}4egh\\-4fgg\end{smallmatrix}\,s\,\begin{smallmatrix}-2egh\\+2fgg\end{smallmatrix}\,xv+\tfrac23 dh\dfrac{v^3}{x^3}-2dfg\dfrac{v}{x}}{\eta fhh}=t.$$

Forma decima.

1. $\dfrac{dz^{\eta-1}}{g+hz^\eta\sqrt{e+fz^\eta}}=y.$ $\qquad\sqrt{\dfrac{df}{h}+\dfrac{eh-fg}{h}xx}=v.$ $\qquad\sqrt{\dfrac{d}{g+hz^\eta}}=x.$

$$\frac{2xv-4s}{\eta f}=t=\frac{4}{\eta f}ADGa.$$

2. $\dfrac{dz^{2\eta-1}}{g+hz^\eta\sqrt{e+fz^\eta}}=y.$ $\qquad\sqrt{\dfrac{df}{h}+\dfrac{eh-fg}{h}xx}=v.$ $\qquad\sqrt{\dfrac{d}{g+hz^\eta}}=x.$

$$\frac{4gs-2gxv+2d\dfrac{v}{x}}{\eta fh}=t.$$

Ninth form.

1. $\dfrac{dz^{\eta-1}\sqrt{(e+fz^\eta)}}{g+hz^\eta} = y.$ $\quad \sqrt{\dfrac{d}{g+hz^\eta}} = x.$ $\quad \sqrt{\dfrac{df+(eh-fg)\,x^2}{h}} = v.$ $\quad \dfrac{(eh-fg)\,(-4s+2xv)+2dfv|x}{\eta fh} = t.$

2. $\dfrac{dz^{2\eta-1}\sqrt{(e+fz^\eta)}}{g+hz^\eta} = y.$ $\quad \sqrt{\dfrac{d}{g+hz^\eta}} = x.$ $\quad \sqrt{\dfrac{df+(eh-fg)\,x^2}{h}} = v.$ $\quad \dfrac{g(eh-fg)\,(4s-2xv)+\frac{2}{3}dhv^3|x^3-2dfgv|x}{\eta fh^2} = t.$

Tenth form.

1. $\dfrac{dz^{\eta-1}}{(g+hz^\eta)\sqrt{(e+fz^\eta)}} = y.$ $\quad \sqrt{\dfrac{d}{g+hz^\eta}} = x.$ $\quad \sqrt{\dfrac{df+(eh-fg)\,x^2}{h}} = v.$ $\quad (2xv-4s)|\eta f = t = (4|\eta f)\,(ADGa).$

2. $\dfrac{dz^{2\eta-1}}{(g+hz^\eta)\sqrt{(e+fz^\eta)}} = y.$ $\quad \sqrt{\dfrac{d}{g+hz^\eta}} = x.$ $\quad \sqrt{\dfrac{df+(eh-fg)\,x^2}{h}} = v.$ $\quad (4gs-2gxv+2dv|x)|\eta fh = t.$

Forma undecima.

1. $dz^{-1}\sqrt{\dfrac{e+fz^\eta}{g+hz^\eta}}=y.$ $\qquad \left\{\begin{array}{l}\sqrt{g+hz^\eta}=x.\\[1ex]\sqrt{h+gz^{-\eta}}=\xi.\end{array}\right.$ $\qquad \left\{\begin{array}{l}\sqrt{\dfrac{eh-fg}{h}+\dfrac{f}{h}xx}=v.\\[2ex]\sqrt{\dfrac{fg-eh}{g}+\dfrac{e}{g}\xi\xi}=\Upsilon.\end{array}\right.$ $\qquad \dfrac{dxv^3z^{-\eta}-4dfs-4de\sigma}{\eta fg-\eta eh}=t.$

2. $dz^{\eta-1}\sqrt{\dfrac{e+fz^\eta}{g+hz^\eta}}=y.$ $\qquad \sqrt{g+hz^\eta}=x.$ $\qquad \sqrt{\dfrac{eh-fg}{h}+\dfrac{f}{h}xx}=v.$ $\qquad \dfrac{2d}{\eta h}s=t.$

3. $dz^{2\eta-1}\sqrt{\dfrac{e+fz^\eta}{g+hz^\eta}}=y.$ $\qquad \sqrt{g+hz^\eta}=x.$ $\qquad \sqrt{\dfrac{eh-fg}{h}+\dfrac{f}{h}xx}=v.$ $\qquad \dfrac{dhxv^3{-3dfg \atop -deh}s}{2\eta fhh}=t.$

Eleventh form.

1. $dz^{-1}\sqrt{\dfrac{e+fz^\eta}{g+hz^\eta}} = y.$

$\left.\begin{array}{l}\sqrt{(g+hz^\eta)} = x. \\[4pt] \sqrt{(h+gz^{-\eta})} = \xi.\end{array}\right.$
$\left.\begin{array}{l}\sqrt{\dfrac{(eh-fg)+fx^2}{h}} = v. \\[8pt] \sqrt{\dfrac{(fg-eh)+e\xi^2}{g}} = \Upsilon.\end{array}\right\}$

$\dfrac{d(xv^3z^{-\eta} - 4fs - 4e\sigma)}{\eta(fg-eh)} = t.$

2. $dz^{\eta-1}\sqrt{\dfrac{e+fz^\eta}{g+hz^\eta}} = y.$

$\sqrt{(g+hz^\eta)} = x.$

$\sqrt{\dfrac{(eh-fg)+fx^2}{h}} = v.$

$(2d/\eta h)\,s = t.$

3. $dz^{2\eta-1}\sqrt{\dfrac{e+fz^\eta}{g+hz^\eta}} = y.$

$\sqrt{(g+hz^\eta)} = x.$

$\sqrt{\dfrac{(eh-fg)+fx^2}{h}} = v.$

$d(hxv^3 - (3fg+eh)\,s)/2\eta fh^2 = t.$

NOTES TO PAGES 136–147

(33) Understand in the first of the figures above. Here and at corresponding places below Newton afterwards added alongside a reference to the pertinent 'Fig' in the pull-out sheet to which (see note (1) above) all the diagrams illustrating the text of the 'De Quadratura' were subsequently transferred. Since in the present reproduction of the manuscript we retain these figures in their original place, there seems to us no need to repeat the awkward keyings of textual reference to corresponding figure which the later transposition of these to a more distant, collective site came to necessitate.

(34) Let us remind (see III: 256, note (553)) that $\div$ is (a slightly modified version of) Barrow's sign denoting the absolute difference of the magnitudes between which it is placed; and so we again 'translate' it in our English rendering. A subsequent computation whereby Newton confirmed the accuracy of this evaluated 'area' is reproduced in [4] of the Appendix: its handwriting would, we have already suggested, make it strictly contemporary with the parallel check ([6] following) which Roger Cotes stimulated him to make of Form 8.4 in August 1709.

(35) In his library copy of the 1706 *Optice* (note (1) above) in which the 'De Quadratura' made its second public appearance, Newton afterwards—doubtless at the same time that (see the previous note) he checked the accuracy of the preceding Form 2.1, whether in August 1709 or no—deleted the extraneous factors e in these terms (here faithfully copied from the equally faulty second and third cases of the 'Ordo secundus' in his parent 1671 'Catalogus'; see III: 244), to make them read '$\dfrac{2de}{\eta f}z^{\frac{\eta}{2}}$' and '$\dfrac{2d}{3\eta f}z^{\frac{3}{2}\eta} - \dfrac{2de}{\eta ff}z^{\frac{\eta}{2}}$' respectively. These emendations were subsequently incorporated by William Jones into the lightly corrected text of the 'De Quadratura' which he published in his 1711 *Analysis per Quantitatum Series, Fluxiones, ac Differentias*.

(36) In the manuscript (f. 20^r) Newton initially went on here to copy out the two cases of the analogously positioned 'Ordo tertius' of his 1671 table (III: 246) and then, evidently appreciating the slight improvement in sequence of presentation which results upon promoting the earlier 'Ordo quintus' (III: 248) and 'Ordo septimus' (*ibid.*: 250) to be now in third and fourth place respectively, delayed this first 'Forma tertia' to be successively 'quarta' (fourth) and thereafter 'quinta' (fifth): in which final position he recopied it below (on the next recto side, f. 21^r, in the manuscript) without textual variation.

(37) Presupposing that $f^2 \geqslant 4eg$; when this is not so Newton's following reduction fails. The lacuna was not filled for another dozen years and more till Roger Cotes succeeded in taking off the limitation by factorizing the denominator $e + fz^\eta + gz^{2\eta}$ as

$$(e^{\frac{1}{2}} + \lambda z^{\frac{1}{2}\eta} + g^{\frac{1}{2}}z^\eta)\,(e^{\frac{1}{2}} - \lambda z^{\frac{1}{2}\eta} + g^{\frac{1}{2}}z^\eta), \quad \lambda = \sqrt{(2\sqrt{(eg)} - f)},$$

and thereby splitting the general 'Curvarum Forma sexta' $z^{\frac{1}{2}\theta\eta - 1}/(e + fz^\eta + gz^{2\eta})$, $\theta = 1, 2, \dots$, into partial fractions, each individually integrable by cases of the preceding Form 5. On setting $\eta \to 2\eta$ for convenience—'You see I happened to alter the appearance of the form a little, ... but this makes no real difference as to the matter in hand' he rightly informed William Jones on 5 May 1716 in the letter (*Correspondence of Scientific Men of the Seventeenth Century*, 1 (Oxford, 1841): 270–4, especially 272) in which he sent a 'transcript' (*ibid.*: 273) of his mode of reduction 'set down for my memory about it'—the integrand becomes

$$dz^{\theta\eta - 1}/(e + fz^{2\eta} + gz^{4\eta}) = dgz^{\theta\eta - 1}/(r + sz^\eta + gz^{2\eta})\,(r - sz^\eta + gz^{2\eta}),$$

where $r = e^{\frac{1}{2}}g^{\frac{1}{2}}$ and (now) $s = g^{\frac{1}{2}}\sqrt{(2r - f)}$. This Cotes adroitly splits into the equal aggregate

$$\frac{dg}{2rs}\left(\frac{sz^{\theta\eta - 1} + gz^{(\theta + 1)\eta - 1}}{r + sz^\eta + gz^{2\eta}} + \frac{sz^{\theta\eta - 1} - gz^{(\theta + 1)\eta - 1}}{r - sz^\eta + gz^{2\eta}}\right)$$

which is straightforwardly evaluable by Cases θ and $\theta+1$ of Form 5: in particular Form 6.1 reduces in this way to Forms 5.1 and 5.2. (There are niceties here about the variations possible in the signs of e, g—both which must be the same of course—and of f; these Cotes goes on punctiliously to distinguish, but they need not here detain us.) Cotes had, however, fallen coldly out of favour with Newton during the last months, from the late summer of 1712 on, in which he prepared the *editio secunda* of the *Principia* for publication (see *Correspondence of Isaac Newton*, **5**, 1975: xxvi–xxix) and he had never returned into the latter's good graces: how now tactfully to inform the author of the 'De Quadratura' that he had not only detected what 'appears as an eyesore to me in so beautiful a work' but also neatly excised it? To Jones, therefore, he wrote to 'beg your assistance and management in an affair, which I cannot so properly undertake myself, expecially by letters. You know Sir Isaac has left his sixth form imperfect, and under a limitation.... The very great respect and honour, which is due to him upon all accounts, makes me wish it were removed by himself. Pray therefore let him know that I can take off that limitation, and make this form as perfect as the others. And use all the address you have to make him set upon the same thing, and let me know that he has done it. My design is...that he hearing from you what I had done, did himself, at your request, reconsider his sixth form, and very easily ma[k]e it perfect'. Cotes died just a month later, probably without ever knowing Newton's response to this delicate feeler. The surviving scrap (reproduced in [5] of the Appendix) on which Newton did indeed come to reconsider his *Forma* 6.1 makes it clear that Jones, in transmitting the news of Cotes' success in removing its lacuna, kept faith with the latter by not passing on the detail of his reduction, but merely dropped the hint that it depended on doubling the index η of the base variable and then factorizing the ensuing denominator $e+fz^{2\eta}+gz^{4\eta}$. Even thus aided, Newton did not find it easy to 'perfect' his sixth form, but stuck at the inadequate splitting of the integrand $dz^{\theta\eta-1}/(e+fz^{2\eta}+gz^{4\eta})$ into the components (so we again Cotesianly write them)

$$\frac{dg}{2s}\left(\frac{z^{(\theta-1)\eta-1}}{r-sz^\eta+gz^{2\eta}}-\frac{z^{(\theta-1)\eta-1}}{r+sz^\eta+gz^{2\eta}}\right):$$

this of course labours under the deficiency that (when $\theta=0$) Form 6.1 is reduced to a non-existent Form '5.0'. The evidence does not show whether he ever knew of Cotes' variant splitting into partial fractions which transcends this new limitation. Certainly, Cotes' letter to Jones in May 1716 remained unknown to Newton's subsequent eighteenth-century editors Castiglione and Horsley, and the present deficiency has never been made good in any re-issue of the 1704 *editio princeps* of the 'De Quadratura'.

(38) In this faithful copy of the 'areæ valor' in the corresponding *Ordo* 6.1 of his parent 1671 table Newton again (see III: 257–8, note (564)) omits the two additional terms '$-2dfgxv+4degv$' which his numerator here lacks. As we there remarked, Roger Cotes brought this deficiency—together with the miscopyings in Form 8.4 below—to Newton's notice in his first known letter to him on 18 August 1709, having 'some days ago' been 'examining the 2^d Cor: of Pro 91 Lib I [*Princip.*] and found it to be true [see VI: 226, note (32)] by ye Quadratures of ye 1st & 2^d Curves of ye 8th Form of y^e second Table in Y^r Treatise *De Quadrat*. At the same time I went over y^e whole Seventh & Eighth forms which agreed with my Computation excepting y^e First of y^e Seventh & Fourth of y^e Eighth' (Joseph Edleston, *Correspondence of Sir Isaac Newton and Professor Cotes* (London, 1850): 3–4 [=*Correspondence of Isaac Newton*, **5**, 1975: 3–4]). The two missing terms were subsequently entered by Newton's pen in the margin alongside at this point in his library copy of the 1706 *Optice* (on which see note (1)), and due correction of the printed text was afterwards made in the light revamping of the 'De Quadratura' in William Jones' 1711 *Analysis*. (In his 1709 letter Cotes himself elegantly replaced the augmented quintet of terms $4dee\xi\Upsilon+2def\Upsilon-2dffv+4degv-2dfgxv$ by the equivalent pair

$$4de\Upsilon^3/\xi-2dfv^3/x;$$

on this again see III: 258, note (564).)

In Tabulis hisce, series Curvarum cujusqʒ formæ utrinqʒ in infinitum continuari possunt. Scilicet in Tabula prima, in numeratoribus arearum formæ tertiæ et quartæ, numeri coefficientes initialium terminorum (2, −4, 16, −96, 868[40] &c) generantur multiplicando numeros −2, −4, −6, −8, −10 &c in se continuò, & subsequentium terminorum coefficientes ex initialibus derivantur multiplicando ipsos gradatim, in Forma quidem tertia, per −$\frac{3}{2}$, −$\frac{5}{4}$, −$\frac{7}{6}$, −$\frac{9}{8}$, −$\frac{11}{10}$ &c, in quarta verò per −$\frac{1}{2}$, −$\frac{3}{4}$, −$\frac{5}{6}$, −$\frac{7}{8}$, −$\frac{9}{10}$, &c. Et Denominatorum coefficientes 1, 3, 15, 105 &c prodeunt multiplicando numeros 1, 3, 5, 7, 9 &c in se continuò.

In secunda verò Tabula, series Curvarum formæ primæ, secundæ, quintæ, sextæ, nonæ et decimæ ope solius divisionis, et formæ reliquæ ope Propositionis tertiæ & quartæ,[41] utrinqʒ producuntur in infinitum.

Quinetiam hæ series mutando signum numeri η variari solent. Sic enim, e.g. Curva $\frac{d}{z}\sqrt{e+fz^\eta} = y$,

evadit $\frac{d}{z^{\frac{1}{2}\eta+1}}\sqrt{f+ez^\eta}\,[=y]$.[42]

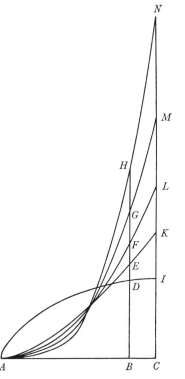

Prop. XI. Theor. VIII.[43]

Sit *ADIC* Curva quævis Abscissam habens *AB*=z et Ordinatam *BD*=y, et sit *AEKC* Curva alia cujus Ordinata *BE* æqualis est prioris areæ *ADB* ad unitatem applicatæ, et *AFLC* Curva tertia cujus Ordinata *BF* æqualis est secundæ areæ *AEB* ad unitatem applicatæ, et *AGMC* Curva quarta cujus Ordinata *BG* æqualis est tertiæ areæ *AFB* ad unitatem applicatæ, et *AHNC* Curva quinta cujus Ordinata *BH* æqualis est quartæ areæ *AGB* ad unitatem applicatæ[,] et sic deinceps in infinitum.

Sunto *A*, *B*, *C*, *D*, *E* &c Areæ Curvarum Ordinatas habentium y, zy, z^2y, z^3y, z^4y et Abscissam communem z.

(39) We make good Newton's careless omission of the denominator in the manuscript. It is more important to notice that he here mistranscribes the (correct) numerator of the corresponding *Ordo* 8.4 in the parent 1671 table (see III: 252): the coefficient of *s* should be '36*defg* −15*df*³', that of *xv* should be '−28*defg*+10*df*³', and that of *v* should be '+10*deff*−16*deeg*'. Further to complicate matters, the last was (mis?)printed in the 1704 *princeps* edition (*Opticks*, ₁1704: ₂203) as '+10*deeg*−16*deef*', a garbling which passed unchecked into the reissue of the 'De Quadratura' in appendix to the 1706 *Optice*. When (see the previous note) Cotes brought the slip to Newton's attention in August 1709, the latter made the rough checking computation which is reproduced in [6] of the Appendix; and then, freshly convinced of the correctness of his original—and Cotes' 'revised'—numerator, he appropriately altered the text of his

In these tables the series of curves of each form can be continued indefinitely in either direction. In the first table, specifically, in the numerators of the areas of the third and fourth forms the numerical coefficients of the initial terms $(2, -4, 16, -96, [7]68, ...)$ are generated by multiplying the numbers -2, -4, -6, -8, -10, ... continually into one another, and the coefficients of the subsequent terms are derived from the initial ones by multiplying these step by step in the case of the third form by $-\frac{3}{2}$, $-\frac{5}{4}$, $-\frac{7}{6}$, $-\frac{9}{8}$, $-\frac{11}{10}$, ..., but in that of the fourth by $-\frac{1}{2}$, $-\frac{3}{4}$, $-\frac{5}{6}$, $-\frac{7}{8}$, $-\frac{9}{10}$, ... And the denominators' coefficients $1, 3, 15, 105, ...$ ensue by multiplying the numbers $1, 3, 5, 7, 9, ...$ into one another continually. In the second table, however, the series of curves of the first, second, fifth, sixth, ninth and tenth forms are extended by the help of division alone, while the remaining forms are produced indefinitely in either direction by the aid of Propositions III and IV.[41]

These series may, furthermore, usually be varied by changing the sign of the number η. Thus, for instance, the curve $dz^{-1}\sqrt{(e+fz^\eta)} = y$ comes to be $dz^{-\frac{1}{2}\eta-1}\sqrt{(f+ez^\eta)} = y$.[42]

Proposition XI, Theorem VIII.[43]

Let ADI be any curve having abscissa $AB = z$ and ordinate $BD = y$; and let AEK be another curve whose ordinate BE is equal to the former one's area (ADB) divided by unity, and AFL a third curve whose ordinate BF is equal to the second one's area (AEB) divided by unity, and AGM a fourth curve whose ordinate BG is equal to the third one's area (AEB) divided by unity, and AHN a fifth curve whose ordinate BH is equal to the fourth one's area (AGB) divided by unity, and so on without end. Take $A, B, C, D, E, ...$ to be the areas of the curves having ordinates $y, zy, z^2y, z^3y, z^4y, ...$ and common abscissa z.

---

library copy of the 1706 *Optice*, whence the accurately adjusted version passed into print in Jones' 1711 *Analysis*. Correct replacement of ef here by f^2 is made in the English version.

(40) Read '768' correctly; here, too, Newton repeats a venial slip of his pen originally made at the corresponding point in his 1671 tract's Problem 9 (see III: 260, note (575)). Printed correction of the error—which neither Newton nor Jones seem to have noticed—was first publicly made by John Colson in his English rendering of the latter treatise (*The Method of Fluxions and Infinite Series; with its Application to the Geometry of Curve-lines* (London, 1736): 107).

(41) See VII: 518/520–6 respectively.

(42) We trivially make good a minimal omission in the manuscript which—following a now familiar pattern—passed likewise to mar the 1704 and 1706 printed editions of the 'De Quadratura', but was afterwards repaired by Newton's pen in his library copy of the latter (on which see note (1)) and publicly set right in the re-edition of the text in William Jones' 1711 *Analysis*.

(43) A word-for-word repeat of the equivalent proposition in the parent 1693 text (VII: 550–2), now augmented by a short 'Corol.' and a final following scholium whose clumsily phrased first paragraph was (see note (46) below) soon to rouse Johann Bernoulli's miscomprehending scorn. We may here forgo a lengthy technical commentary in following footnotes since this would but repeat what we have said before (see VII: 166, note (9)/169, note (21); 551–3, notes (95)–(97)).

Detur Abscissa quævis $AC=t$, sitcp $BC=t-z=x$, et sunto P, Q, R, S, T areæ Curvarum Ordinatas habentium y, xy, xxy, x^3y, x^4y et Abscissam communem x.

Terminentur autem hæ areæ omnes ad Abscissam totam datam AC, nec non ad Ordinatam positione datam & infinite productam CI: et erit arearum sub initio positarum prima $ADIC=A=P$, secunda $AEKC=tA-B=Q$. Tertia $AFLC=\dfrac{ttA-2tB+C}{2}=\frac{1}{2}R$. Quarta $AGMC=\dfrac{t^3A-3ttB+3tC-D}{6}=\frac{1}{6}S$. Quinta $AHNC=\dfrac{t^4A-4t^3B+6ttC-4tD+E}{24}=\frac{1}{24}T$. [&c].

Corol. Unde si Curvæ quarum Ordinatæ sunt y, zy, z^2y, z^3y &c vel y, xy, x^2y, x^3y &c quadrari possunt, quadrabuntur etiam Curvæ $ADIC$, $AEKC$, $AFLC$, $AGMC$ &c et habebuntur Ordinatæ BE, BF, BG, BH areis Curvarum illarum proportionales.

Scholium.[44]

Quantitatum fluentium fluxiones esse primas, secundas, tertias, quartas, aliascp diximus supra.[45] Hæ fluxiones sunt ut termini serierum infinitarum convergentium. Ut si z^n sit quantitas fluens et fluendo evadat $\overline{z+o}|^n$, deinde resolvatur in seriem convergentem

$$z^n+noz^{n-1}+\frac{nn-n}{2}\,ooz^{n-2}+\frac{n^3-3nn+2n}{6}\,o^3z^{n-3}+\&c:$$

terminus primus hujus seriei erit quantitas illa fluens, secundus nox^{n-1} erit ejus incrementum primum seu differentia prima[46] cui nascenti proportionalis est

---

(44) In his preliminary draft of this concluding scholium which is reproduced (from Add. 3962.3: 61^v) in [7] of the Appendix, Newton has not yet inserted the following first paragraph over which there was subsequently to be so much Bernoullian fuss and scorn; but in compensation he there presents a worked instance of the preceding proposition in which the curve AD is set to be the Apollonian hyperbola defined by $y = a^2/z$, though he fails to completely specify the curve ensuing from its triple integration—namely,

$$\int_1^t \int_1^z \int_1^z a^2/z.dz^3 = a^2(\tfrac{1}{2}t^2\log t-\tfrac{3}{4}t^2+t-\tfrac{1}{4}).$$

(45) In the preamble on page 130 above. (For a full reproduction of Newton's text see vii: 510.)

(46) Taken at their face-value, these three carelessly adjoined phrases could be taken to mean that Newton assumed the i-th derivative of z^n to be $\binom{n}{i} z^{n-i}$, $i = 1, 2, 3, \ldots$; and more generally, where $y \equiv y_x$ is augmented to become $y_{x+o}=y+Qo+Ro^2+So^3+\ldots$, that he took the coefficients Q, R, S, ... in this series-expansion to be respectively the first, second, third, ... derivatives of y with respect to the base x. In the event, wishing to explain how the expression $\frac{1}{2}S\sqrt{(1+Q^2)}/R^2$ which Newton had obtained in the 1687 *editio princeps* of his *Principia* (Book 2, Proposition X) for the ratio of resistance to gravity in a projectile path is convertible into the (correct) equivalent Leibnizian differential form $\frac{1}{2}(d^3y/dx^3)\sqrt{(1+(dy/dx)^2)}/(d^2y/dx^2)^2$ whose validity he himself had been able to confirm only in a particular instance, Johann

Let there be given any abscissa $AC = t$, and set $BC(= t-z) = x$, and take $P, Q, R, S, T, \ldots$ to be the areas of the curves having ordinates y, xy, x^2y, x^3y, x^4y, ... and common abscissa x.

Let all these areas terminate, however, at the whole given abscissa AB and also at the ordinate CI given in position and indefinitely extended. The first of the areas initially posited will then be $(ADIC) = A = P$, the second one $(AEKC) = tA - B = Q$, the third one $(AFLC) = \frac{1}{2}(t^2A - 2tB + C) = \frac{1}{2}R$, the fourth one $(AGMC) = \frac{1}{6}(t^3A - 3t^2B + 3tC - 3D) = \frac{1}{6}S$, the fifth one will be $(AHNC) = \frac{1}{24}(t^4A - 4t^3B + 6t^2C - 4tD + E) = \frac{1}{24}T$, and so on.

Corollary. Whence if the curves whose ordinates are y, zy, z^2y, z^3y, ... or y, xy, x^2y, x^3y, ... can be squared, the curve-areas $(ADIC)$, $(AEKC)$, $(AFLC)$, $(AGMC)$, ... will also be squared, and there will be had the ordinates BE, BF, BG, BH, ... proportional to those curve-areas.

Scholium.[44]

That fluent quantities possess first, second, third, fourth and other fluxions we have stated above.[45] These fluxions are as the terms of infinite converging series. Thus, if z^n be a fluent quantity and shall by its flowing come to be $(z+o)^n$, and thereafter be resolved into the converging series

$$z^n + noz^{n-1} + \tfrac{1}{2}(n^2 - n)\, o^2 z^{n-2} + \tfrac{1}{6}(n^3 - 3n^2 + 2n)\, o^3 z^{n-3} + \ldots:$$

the first term of this series will be the fluent quantity; the second one no^{n-1} will be its first increment or first difference,[46] and proportional to this as it is

---

Bernoulli in 1713 (as we specify in 6, Appendix 1, note (14) below) indeed seized on the present shaky parallel to maintain that Newton had there too committed this fundamental blunder of equating $Q = dy/dx$, $R = d^2y/dx^2$, $S = d^3y/dx^3$, ... In his 1687 proposition, however, Newton had fallen into an error of a very different kind—there failing to make allowance for a third-order component of tangential resistance (see 6, Appendix 1 below)—and when, prodded to do so by Bernoulli, he made due correction (see 6, §§1/2) he found that his earlier expression had 'merely' to be increased by a factor of $\frac{3}{2}$ to be now $\frac{3}{4}S\sqrt{(1+Q^2)}/R^2$: whence ensues an accurate translation into differential equivalent on setting (in his preferred fluxional form) $Q = \dot{y}/\dot{x}$, $R = \frac{1}{2}\ddot{y}/\dot{x}^2$ and $S = \frac{1}{6}\dddot{y}/\dot{x}^3$. Now knowing that Newton had set down the general 'Taylor' expansion $y_{x+o} - y = (\dot{y}/\dot{x})\,o + \frac{1}{2}(\ddot{y}/\dot{x}^2)\,o^2 + \frac{1}{6}(\dddot{y}/\dot{x}^3)\,o^3 + \ldots$ a dozen years before in Corollary 4 to 'Cas. 3' of Proposition XII of his 1691 treatise 'De quadratura Curvarum' (compare VII: 98–9, note (109)), we may appreciate Newton's intense exasperation at here being—as it seemed to him—so wilfully misunderstood by Bernoulli. When Johann's nephew returned home to Basel in October 1712 from his stay in London during which he had put to Newton his uncle's counter-instance to the truth of the 1687 version of *Principia*, **2**, X he bore with him a copy of Jones' 1711 *Analysis* in which Newton's pen had none too satisfactorily mended the sense of the present paragraph of the 'De Quadratura' by inserting an 'ut' (as) before each of the two latter phrases 'ejus incrementum secundum seu differentia secunda' and 'ejus incrementum tertium seu differentia tertia'. (See Johann Bernoulli's ensuing letter to Leibniz on 7 June 1713 (N.S.), where—in a section not printed in the scurrilous *Charta volans* by which its recipient swiftly made public its central content—he grumpily noticed that 'in exemplari quod mihi dono [Newtonus] misit per Agnatum meum, ibi calamo adscripsit altera vice voculam *ut...*,

ejus fluxio prima, tertius $\frac{nn-n}{2}\,o[o]\,z^{n-2}$ erit ejus incrementum secundum seu differentia secunda[46] cui nascenti proportionalis est ejus fluxio secunda, quartus $\frac{n^3-3nn+2n}{6}\,o^3z^{n-3}$ erit ejus incrementum tertium seu differentia tertia[46] cui nascenti fluxio tertia proportionalis est[,] & sic deinceps in infinitum.

Exponi autem possunt hæ fluxiones[47] per Curvarum Ordinatas BD, BE, BF, BG, BH &c. Ut si Ordinata $BE\left(=\frac{ADB}{1}\right)$ sit quantitas fluens, erit ejus fluxio prima ut Ordinata BD. Si $BF\left(=\frac{AEB}{1}\right)$ sit quantitas fluens, erit ejus fluxio prima ut Ordinata BE et fluxio secunda ut Ordinata BD. Si $BH\left(=\frac{AGB}{1}\right)$ sit quantitas fluens, erunt ejus fluxiones, prima secunda tertia et quarta, ut Ordinatæ BG, BF, BE, BD.

Et hinc in æquationibus quæ quantitates tantum duas incognitas[48] involvunt, quarum una est quantitas uniformiter fluens et altera est fluens quælibet alterius fluentis, inveniri potest fluens illa altera per quadraturam Curvarum. Exponatur enim fluxio ejus per Ordinatam BD, et si hæc sit fluxio prima, quæratur area $ADB=BE\times1$, si fluxio secunda quæratur area $AEB=BF\times1$, si fluxio tertia, quæratur area $AFB=BG\times1$ &c: et area inventa erit exponens fluentis quæsitæ.[49]

Sed et in æquationibus quæ fluentem et ejus fluxionem primam sine altera fluente, vel duas ejusdem fluentis fluxiones, primam et secundam, vel secundā et tertiam, vel tertiam et quartam &c, sine alterutra fluente involvunt: inveniri

scribendo nunc *erit ut ejus* &c. Adeo ut errorem suum non animadverterit, nisi brevi ante, & forte non-nisi post adventum Agnati mei in Angliam' (*Got. Gul. Leibnitii et Johan. Bernoullii Commercium Philosophicum et Mathematicum*, **2** (Lausanne/Geneva, 1745): 310–11 [= (ed. C. I. Gerhardt) *Leibnizens Mathematische Schriften*, **3**.2 (Halle, 1856): 911–12]). Newton has made analogous penned insertion of 'ut' in his library copy of the 1706 *Optice* (see note (1)) at the pertinent place; and the text thus augmented is written out in full on a scrap which he tucked into his interleaved copy (ULC. Adv. b. 39.1) of the *Principia*'s first edition—evidently to be with pages 264–5, but now incongruously bound in between pages 304 and 305; whence it is reproduced by I. B. Cohen in *Isaac Newton's 'Philosophiæ Naturalis Principia Mathematica': the third edition (1726) with variant readings*, **2** (Cambridge, 1972): 802.) In anonymous 'Account of the [his!] Book entituled *Commercium Epistolicum Collinii & aliorum, De Analysi promota;* published by order of the Royal-Society...' (*Philosophical Transactions*, **29**, No. 342 [for January/February 1714/15]: 173–224; compare **2**, Appendix 2 below), having in the interim endured further charges from Leibniz and the Bernoullis that he had a mistaken notion of higher-order derivatives, Newton subsequently issued a well-pondered statement which ought to have ended the squabble: 'It has been represented that Mr. *Newton*, in the Scholium at the End of his Book of Quadratures, has put the third, fourth, and fifth Terms of a converging Series respectively equal to the second, third, and fourth Differences of the first Term, and therefore did not then understand the Method of second, third, and fourth Differences. But in the first Proposition of that Book he shewed how to find the first, second, third and following Fluxions *in infinitum*; and therefore when he wrote that Book, which was before the Year 1676 [!], he

nascent is its first fluxion; the third $\frac{1}{2}(n^2-n)\,o^2z^{n-2}$ will be its second increment or second difference,[46] and proportional to this as it is nascent is its second fluxion; the fourth $\frac{1}{6}(n^3-3n^2+2n)\,o^3z^{n-3}$ will be its third increment or third difference,[46] and to this as it is nascent its third fluxion is proportional; and so on indefinitely.

These fluxions[47] can, however, be expressed by the curves' ordinates *BD*, *BE*, *BF*, *BG*, *BH*. Should, for instance, the ordinate *BE* $(=(ADB)/1)$ be the fluent quantity, its first fluxion will be as the ordinate *BD*. If *BF* $(=(AEB)/1)$ be the fluent quantity, its first fluxion will be as the ordinate *BE* and its second fluxion as the ordinate *BD*. If *BH* $(=(AGB)/1)$ be the fluent quantity, then its first, second, third and fourth fluxions will be (respectively) as the ordinates *BG*, *BF*, *BE*, *BD*.

And hence in equations involving but two unknowns,[48] one of which is a uniformly flowing quantity and the second is any fluxion you please of another fluent quantity, the second fluent can be found through the quadrature of curves. For express its fluxion by means of the ordinate *BD*, and if this be the first fluxion ascertain the area $(ADB) = BE\times1$, if it is the second fluxion ascertain the area $(AEB) = BF\times1$, if the third fluxion, ascertain the area $(AFB) = BG\times1$, and so on: then the area when found will be the exponent of the required fluent.[49]

But also in equations involving a fluent and its first fluxion without the second fluent, or two fluxions of the same fluent—the first and second, or second and third, or third and fourth, and so on—without either of the fluents, it is possible

did understand the Method of all the Fluxions, and by consequence of all the Differences. And if he did not understand it when he added that Scholium to the End of the Book, which was in the Year 1704, it must have been because he had then forgot it.... Nor is it likely, that in the Year 1704 when he added the aforesaid Scholium to the End of the Book of Quadratures, he had forgotten not only the first Proposition of that Book, but also the last Proposition upon which that Scholium was written. If the word (*ut*), which in that Scholium may have been accidentally omitted between the words (*erit*) and (*ejus*), be restor'd, that Scholium will agree with the two Propositions and with the rest of his Writings, and the Objection will vanish' (*ibid.*: 206–7, 208). Though Bernoulli continued to taunt Newton on the matter for several years more—notably in his equally anonymously authored 'Epistola pro eminente Mathematico, Dn. Johanne Bernoullio, contra quendam ex Anglia antagonistam [*viz.* Newton's 'champion' John Keill in *Journal Literaire*, **4**, 1714: 319–58] scripta' (*Acta Eruditorum* (July 1716): 296–315, especially 302–6)—he rested his case on a shifting sand of unfounded assumption, and even he had ultimately (with ill grace, we need not stress) to acquiesce in the truth that Newton's grasp of the general higher-order derivative was without essential blemish.

(47) Newton first began to write 'quanti[tates]' (quantities).

(48) Initially 'indeterminatas' (indeterminates).

(49) As we have already stated (note (44) above), in his preliminary draft on Add. 3962.3: 61ᵛ (reproduced in [7] of the following Appendix) Newton here adjoined the incompletely worked instance where *ADI* is an Apollonian hyperbola having *ABC* and the parallel to *BD* through *I* as its asymptotes.

possunt fluentes per quadraturam Curvarum. Sit æquatio $aa\dot{v}=av+vv$, existente $v=BE$, $\dot{v}=BD$, $z=AB$ & $\dot{z}=1$, et æquatio illa complendo dimensiones fluxionum evadet $aa\dot{v}=av\dot{z}+vv\dot{z}$, seu $\dfrac{aa\dot{v}}{av+vv}=\dot{z}$. Jam fluat v uniformiter et sit ejus fluxio $\dot{v}=1$ et erit $\dfrac{aa}{av+vv}=\dot{z}$, et quadrando curvam cujus Ordinata est $\dfrac{aa}{av+vv}$ & Abscissa v,[50] habebitur Fluens z. Adhæc sit æquatio $aa\ddot{v}=a\dot{v}+\dot{v}\dot{v}$ existente $v=BF$, $\dot{v}=BE$, $\ddot{v}=BD$ et $z=AB$ et per relationem inter $\ddot{v}$ et $\dot{v}$ seu BD et BE invenietur relatio inter AB et BE ut in exemplo superiore. Deinde per hanc relationem invenietur relatio inter AB et BF quadrando Curvam AEB.

Æquationes quæ tres incognitas quantitates involvunt aliquando reduci possunt ad æquationes quæ duas tantum involvunt, et in his casibus fluentes invenientur ex fluxionibus ut supra. Sit æquatio $a-bx^m=cxy^n\dot{y}+dy^{2n}\dot{y}\dot{y}$. Ponatur $y^n\dot{y}=\dot{v}$ et erit $a-bx^m=cx\dot{v}+d\dot{v}\dot{v}$. Hæc æquatio quadrando Curvam cujus Abscissa est x et Ordinata $\dot{v}$[51] dat aream v, et æquatio altera $y^n\dot{y}=\dot{v}$ regrediendo ad fluentes dat $\dfrac{1}{n+1}y^{n+1}=v$. Unde habetur Fluens y.

Quinetiam in æquationibus quæ tres incognitas involvunt et ad æquationes quæ duas tantum involvunt reduci non possunt, Fluentes quandoꝗ prodeunt per quadraturam Curvarum. Sit æquatio

$$\overline{ax^m+bx^n}|^p=rex^{r-1}y^s+sex^r\dot{y}y^{s-1}-f\dot{y}y^t,$$

existente $\dot{x}=1$. Et pars posterior $rex^{r-1}y^s+sex^r\dot{y}y^{s-1}-f\dot{y}y^t$, regrediendo ad fluentes, fit $ex^ry^s-\dfrac{f}{t+1}y^{t+1}$, quæ proinde est ut area Curvæ cujus Abscissa est x et Ordinata $\overline{ax^m+bx^n}|^p$, et inde datur fluens y.

[52]Sit æquatio $\dot{x}\times\overline{ax^m+bx^n}|^p=\dfrac{d\dot{y}y^{n-1}}{\sqrt{e+fy^n}}$. Et fluens cujus fluxio est $\dot{x}\times\overline{ax^m+bx^n}|^p$ erit ut area Curvæ cujus Abscissa est x et Ordinata est $\overline{ax^m+bx^n}|^p$. Item fluens cujus fluxio est $\dfrac{d\dot{y}y^{n-1}}{\sqrt{e+fy^n}}$ erit ut area Curvæ cujus Abscissa est y et Ordinata $\dfrac{dy^{n-1}}{\sqrt{e+fy^n}}$, id est (per Casum 1 Formæ quartæ Tab. I) ut area $\dfrac{2d}{\eta f}\sqrt{e+fy^\eta}$. Pone ergo $\dfrac{2d}{\eta f}\sqrt{e+fy^\eta}$ æqualem areæ Curvæ cujus Abscissa est x et Ordinata $\overline{ax^m+bx^n}|^p$ et habebitur fluens y.[53]

---

(50) By Form 6.1 of the latter table of integrals evaluable *via* conics in the Scholium to Proposition X preceding, on setting $z^\eta \to v$ and also making $d = a^2$, $e = 0$, $f = a$ and $g = 1$. But it is here simpler to split the 'ordinate' $a^2/v(a+v)$ into the component partial fractions $a/v - a/(a+v)$, and then integrate each of these 'hyperbola-areas' separately with respect to the base v, to attain $a \log (v/(a+v))$ as the required fluent.

(51) Namely, by resolving the preceding equation into the directly quadrable equivalent

to find the fluents by the quadrature of curves. Let the equation be $a^2\dot{v} = av + v^2$, where $v = BE$, $\dot{v} = BD$, $z = AB$ and $\dot{z} = 1$, and on filling out the dimensions of its fluxions that equation will come to be $a^2\dot{v} = av\dot{z} + v^2\dot{z}$, that is, $a^2\dot{v}/(av+v^2) = \dot{z}$. Now let v flow uniformly and let its fluxion $\dot{v} = 1$, and there will be

$$a^2/(av+v^2) = \dot{z},$$

and by squaring the curve whose ordinate is $a^2/(av+v^2)$ and abscissa v[50] the fluent z will be had. Further, let the equation by $a^2\ddot{v} = a\dot{v} + \dot{v}^2$, where now $v = BF$, $\dot{v} = BE$, $\ddot{v} = BD$ and $z = AB$, and through the relationship between $\ddot{v}$ and $\dot{v}$, that is, BD and BE, there will be ascertained the relationship between AB and BE as in the previous example; and then by means of this relationship that between AB and BF will be found by squaring the curve-area (AEB).

Equations involving three unknown quantities can on occasion be reduced to equations involving merely two, and in these cases the fluents will be ascertained from their fluxions as above. Take the equation $a - bx^m = cxy^n\dot{y} + dy^{2n}\dot{y}^2$. Put $y^n\dot{y} = \dot{v}$ and there will be $a - bx^m = cx\dot{v} + d\dot{v}^2$: on squaring the curve whose abscissa is x and ordinate $\dot{v}$[51] this equation yields the area v, while the other equation $y^n\dot{y} = \dot{v}$ on regressing to fluents gives $(1/(n+1))\,y^{n+1} = v$; whence there is had the fluent y.

Even indeed in equations involving three unknowns and irreducible to ones involving but two the fluents may sometimes be forthcoming through the quadrature of curves. Take the equation

$$(ax^m + bx^n)^p = rex^{r-1}y^s + sex^r\dot{y}y^{s-1} - f\dot{y}y^t,$$

where $\dot{x} = 1$: the latter part $rex^{r-1}y^s + sex^r\dot{y}y^{s-1} - f\dot{y}y^t$ will on regressing to fluents become $ex^ry^s - (f/(t+1))\,y^{t+1}$, and this in consequence is as the area of the curve whose abscissa is x and ordinate $(ax^m + bx^n)^p$; and thereby the fluent y is yielded.

[52]Let the equation be $\dot{x}(ax^m + bx^n)^p = d\dot{y}y^{\eta-1}/\sqrt{(e+fy^\eta)}$ and the fluent whose fluxion is $\dot{x}(ax^m + bx^n)^p$ will be as the area of the curve whose abscissa is x and ordinate $(ax^m + bx^n)^p$; the fluent whose fluxion is $d\dot{y}y^{\eta-1}/\sqrt{(e+fy^\eta)}$ will be as the area of the curve whose abscissa is y and ordinate $dy^{\eta-1}/\sqrt{(e+fy^\eta)}$, that is (by Case 1 of the fourth form in Table I) as the area $(2d/\eta f)\sqrt{(e+fy\eta)}$. Therefore put $(2d/\eta f)\sqrt{(e+fy^\eta)}$ equal to the area of the curve whose abscissa is x and ordinate $(ax^m + bx^n)^p$, and there will be yielded the fluent y.[53]

---

form $\dot{v}(= y^n\dot{y}) = (1/2d)\,(-cx \pm \sqrt{[c^2x^2 - 4d(a - bx^m)]})$. Newton here generalizes his corresponding example on Add. 3962.3: 61$^\mathrm{v}$ (see [7] of the Appendix), in which $m = 2$ and $n = 1$

(52) Newton's rough prior casting of the four final paragraphs following survives on Add. 3962.5: 66$^\mathrm{r}$. As ever, we cite in pertinent ensuing footnotes the significant variations in its text from the revised version (on Add. 3962.1: 26$^\mathrm{r}$) here reproduced.

(53) In his draft on Add. 3962.5: 66$^\mathrm{r}$ Newton began to add yet a further example: 'Sit æquatio

$$axx\dot{x} + 2bx\dot{x}y = cxx\dot{y} + \dot{y}yy \quad \text{seu} \quad y = \frac{bx\dot{x}}{\dot{y}} \pm \sqrt{bbxx\frac{\dot{x}\dot{x}}{\dot{y}\dot{y}} + axx\frac{\dot{x}}{\dot{y}} - cxx}.$$

Et nota quod Fluens omnis quæ ex fluxione prima colligitur augeri potest vel minui quantitate quavis non fluente.[54] Quæ ex fluxione secunda colligitur augeri potest vel minui quantitate quavis cujus fluxio secunda nulla est. Quæ ex fluxione tertia colligitur augeri potest vel minui quantitate quavis cujus fluxio tertia nulla est. Et sic deinceps in infinitum.

Postquam verò fluentes ex fluxionibus collectæ sunt,[55] si de veritate Conclusionis dubitatur, fluxiones fluentium inventarū vicissim colligendæ sunt & cum fluxionibus sub initio propositis comparandæ. Nam si prodeunt æquales Conclusio rectè se habet: sin minus, corrigendæ sunt fluentes[56] sic, ut earum fluxiones fluxionibus sub initio propositis æquentur. Nam et Fluens pro lubitu assumi potest et assumptio corrigi ponendo fluxionem Fluentis assumptæ æqualē fluxioni propositæ et terminos homologos inter se comparando.

Et his principijs[57] via ad majora sternitur.[58]

[hoc est] $\dot{y}y = bx\dot{x} \pm x\sqrt{bb\dot{x}\dot{x} + a\dot{x}\dot{y} - c\dot{y}\dot{y}}$. [Pone] $\frac{\dot{y}}{\dot{x}} = \dot{z}$. [ut et] $\frac{y}{x} = z[!]$. [Fit] $a + 2bz = c\dot{z} + \dot{z}zz$

[adeoꝗ $\dot{z} = \dfrac{a + 2bz}{c + zz}$]'. We need not insist on the slip which Newton here rashly made in seeking to reduce his given equation to straightforwardly quadrable form, since he himself broke off to cancel this wayward instance before completing it as we have surmised in our square-bracketed interpolation.

(54) 'quavis data' (by any given) was first written. Newton removes in his replacement any ambiguity that the 'given' quantity may be fluent.

(55) Initially, in the draft on Add. 3962.5: 66ʳ, Newton began: 'Postquam a fluxionibus ad fluentes factus est regressus' (After regress has been made from fluxions to fluents).

And note that every fluent which is gathered from a first fluxion can be increased or diminished by any non-fluent[54] quantity; one derived from a second fluxion can be increased or diminished by any quantity whose second fluxion is nil; one gathered from a third fluxion can be increased or diminished by any quantity whose third fluxion is nil; and so on indefinitely.

But if, after you have gathered fluents from fluxions,[55] you should be in doubt regarding the truth of your conclusion, you must conversely derive the fluxions of the fluents found and compare them with the fluxions initially propounded. For if they turn out equal, the conclusion is had right; while if they do not, you need to correct the fluents[56] in such a way that their fluxions shall be equal to the fluxions initially proposed. Of course, a fluent can also be assumed at will and the assumption corrected by setting the fluxion of the assumed fluent equal to the fluxion propounded and then comparing corresponding terms with one another.

And on these principles[57] the path is laid to greater things.[58]

---

(56) 'per additionem vel subductionem datorum' (through the addition or subtraction of givens) is deleted at this point in the draft, after which Newton went on to write 'sic, ut fluxiones cum fluxionibus...congruant' (in such a way that their fluxions shall accord with the fluxions...).

(57) 'initijs' (beginnings) was first written in the draft on Add. 3962.5: 66ʳ.

(58) This transition to the unspecified subtleties of advanced fluxional calculus—as already displayed in Propositions XI–XIII of his 1691 treatise 'De quadratura Curvarum' (VII: 70–128)?—is reminiscent of the parallel sentence (see III: 116) in which, 'Jactis hisce sequentium fundamentis', Newton more than thirty years earlier had passed from the corresponding introductory portion of his 1671 fluxions tract 'ad Problemata magis particularia'.

APPENDIX. PRELIMINARY AND CHECKING COMPUTATIONS FOR THE 1704 'DE QUADRATURA' AND AN INITIAL VERSION OF ITS FINAL SCHOLIUM.[1]

From the originals in the University Library, Cambridge

[1][2]   *Exempl. 2.* In machinis mobilibus.[3]

[Angulus *ACD* circa datum polum *C* volvitur et interea secant crura *CA*, *CD* rectam *Aa* et curvam quamcunꝗ *Dd*. Rotetur angulus donec puncta *A* et *D* perveniant ad *a* et *d*: comparandæ sunt fluxiones rectarum *CA* et *CD*.]

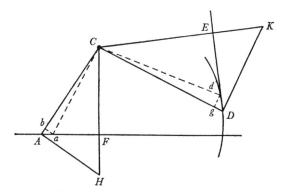

[Demittantur ad *CA* et *CD* perpendicula *ab* et *dg*, ut et ad *Aa* et *Dd* perpendicula *CF* et *CE*. Tum ob similia triangula est] $Ca.ab::Cd.dg = \dfrac{ab,Cd}{Ca}$. [et] $AF.CF::Ab.ba = \dfrac{CF,Ab}{AF}$.

[unde] $Ca.ab::Cd.dg = \dfrac{CF,Ab,Cd}{Ca,AF}$. [adeoꝗ] $CE.ED::dg.gD = \dfrac{ED,CF,Ab,Cd}{CE,Ca,AF}$.

[vel reducendo] $Ab.Dg::CE, Ca, AF.DE, CF, Cd$. [Ergo punctis *a* et *d* cum punctis *A* et *D* ultimò coeuntibus fit] $\dot{C}A.\dot{C}D^{[4]}::CE, CA, AF.DE, CF, CD$. [vel concinnius, ad *CA* et *CD* erectis perpendiculis *AH* et *DK* respectivè, evadit]

$$\dot{C}A.\dot{C}D::\dfrac{CA,AF}{CF} = AH.\dfrac{DE,CD}{CE} = DK.$$

---

(1) These miscellaneous fragments and minor preparatory drafts are here gathered together solely for convenience (though of course they have their own internal ties). Following footnotes key them individually to pertinent places in §§1–3 preceding, and also specify their context (where this is known) and points of interest.

(2) Add. 3970.3: 477ʳ, rough computations (whose bare skeleton is here considerably fleshed out by editorial interpolations to clarify their sense and inter-connection) penned by Newton at the head of a folded sheet (*ibid.*: 477ʳ–478ᵛ) which he subsequently used to draft a preliminary version and initial revise of the 'Observations' and attendant 'Queries' of the third book of his 1704 *Opticks*.

(3) 'In mobile engines' we might translate, recalling the name which Newton gave to an analogous (if a deal more sophisticated) *machina mobilis* forty years earlier; see 1: 264. The present rotating-angle mechanism is effectively that which he shortly afterwards set as his main 'Exemplum [1→] 2' to illustrate the new 'Prop. II. Prob. II' on modes of geometrical fluxion which he tentatively inserted in the parent 1693 text of the 'De Quadratura' (compare §1.1: note (12) preceding), though he here computes an additional fluxional relationship (that of *FA* to *CA* and to *CD*) of which he subsequently made no use.

(4) That is, in Newton's earlier notation for such fluxions of line-lengths, 'fl *CA*.fl *CD*'. And *mutatis mutandis* for his other dotted fluxions in the sequel.

[Adhæc, quia $CA^q = AF^q +$ CF^q ubi perpendiculum CF non fluit, erit capiendo fluxiones] $\dot{FA}.\dot{CA}::CA.AF.$ [adeoq̃] $\dot{FA}.\dot{CD}::CE, CA^q.DE, CF, CD.$

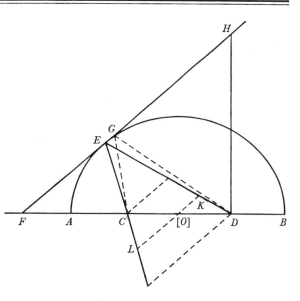

[Sit AEB semiellipsis cujus diameter AB et foci C, D. Tangat FH ellipsin ad ejus punctum quodvis E et ducantur CE, CD. Jam ubi punctum G pervenit ad E erit] $DF.CF::\square EDG.\square ECG.$[5]

[hoc est] $\square ECG$ ut $\dfrac{CF}{DF}\square EDG.$

[necnon] $\angle ECG$ ut $\dfrac{\square ECG}{EC^q}.$[6]

[unde] $\angle ECG$ ut $\dfrac{CF\times\square EDG}{DF, EC^q}=\dfrac{\square EDG}{ED, EC}=\dfrac{\square EDG}{EK^q-KD^q}.$ [vel] $\angle ECG\times EK^q-KD^q$ ut $\square EDG.$ [ductâ KL per centrum O tangenti FE parallelâ].[7]

---

[2][8] [Ellipseos $PRQS$ cujus axis major PQ, minor RS sit Abscissa $CA=x$ et Ordinata $AB=y$ et pone æquationem esse] $axx+byy=c.$ [Erit per Prop. 1]

---

(5) Read 'area EDG. area ECG'. Each of these two ratios (because the infinitesimal triangles share as common base the vanishingly small increment EG of the ellipse's arc $\widehat{AE}$, and the focal *radii vectores* ED and EC are equally inclined to the tangent FEH to the ellipse at E) is the same as that of the perpendicular distances of D and of C from FEH.

(6) For (in the limit as G coalesces with E) the area of the triangle ECG becomes indistinguishable from that of the circle-sector $(ECG') = \frac{1}{2}EC^2.\widehat{ECG}$, where $\widehat{EG'}$ is the arc of the circle of centre C and through E which meets CG in G'.

(7) Whence (see VI: 556, and compare *ibid.*: 46, 140) $EK = EL = \frac{1}{2}(EC+ED) = AO$, the ellipse's semi-axis. But this adds little to the result already attained above that the infinitesimal focal sector $(EDG) \propto EC\times ED.\widehat{ECG}$ (where the factor of proportionality is of course $\frac{1}{2}$): one which Newton applied in the scholium to Proposition XXXI of the first book of his published *Principia*—or so we may with confidence restore his suppressed supporting argument (compare VI: 172, note (164) and 321, note (159))—to develop his refinement of Boulliau's 'upper-focus' equant approximation to planetary motion in a Keplerian ellipse. Unwilling perhaps to pursue the technical ramifications in an introductory treatise on quadrature, he made no attempt to refine this second illustration of the power of his notion of geometrical 'flow' in any other draft of the 1704 'De Quadratura' known to us. The intended rôle of the line DH in Newton's figure (meant to be drawn perpendicular to ADB?) is mysterious: we reproduce it, at its exact inclination, merely for completeness' sake.

(8) Add. 3970.3: 360^v, preparatory calculations for the instance of the ellipse which is set to illustrate Problem 4, on constructing the radius of curvature at a general point of a curve referred to Cartesian coordinates, in §1.4 preceding.

$a\dot{x}\dot{x}+b\dot{y}\dot{y}=0.$ [et inde] $\dot{x}.\dot{y}::$

$[-]by.ax::AT.AB[::AB.AV].$

[Iterum per idem Prop. 1 fit]
$a\dot{x}\dot{x}+b\dot{y}\dot{y}+b\dot{y}\ddot{y}=0.$ [Sit jam E
centrum curvaturæ ad B et erit]

$$\frac{\dot{x}\dot{x}+\dot{y}\dot{y}}{\ddot{y}}=DB.^{(9)}$$

[hoc est quia] $\dfrac{-a\dot{x}\dot{x}}{by}=\dot{y}.$

[adeoqʒ $a\dot{x}\dot{x}+\dfrac{aax\dot{x}\dot{x}}{byy}=-by\ddot{y}$ fit]

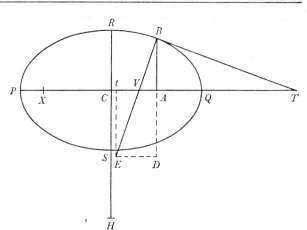

$$\frac{\dot{x}\dot{x}+\dfrac{aaxx}{bbyy}\dot{x}\dot{x}}{-\dfrac{a\dot{x}\dot{x}}{by}-\dfrac{aaxx\dot{x}\dot{x}}{bbyyy}}=BD=\frac{bbyy+aaxx}{-aby-\dfrac{aaxx}{y}}=\frac{bbyy+aaxx}{abyy+aaxx}y=\frac{bbyy+aaxx}{ac}y.$$

[unde] $AD^{(10)}=\dfrac{bb-ab}{ac}y^3.$ $\left[\text{itaqʒ } VE=\dfrac{bb-ab}{ac}yy \text{ in } BV.\right]^{(11)}$

[Pone] $CQ=\sqrt{\dfrac{c}{a}}=d.$ $CR=\sqrt{\dfrac{c}{b}}=e.$ [Erit] $add=bee=c.$

[itaqʒ] $\dfrac{-c}{dd}+\dfrac{c}{ee}=b-a=\dfrac{dd-ee}{ddee}c.$ [necnon] $\dfrac{cee}{dd}=\dfrac{ac}{b}.$ [quare] $\dfrac{dd-ee}{e^4}y^3=AD.$

[Unde] $\dfrac{dd-ee}{[e^4]}\times yy\times BV=VE.$ [vel] $\dfrac{CQ^q-CR^q}{CR^{qq}}\times BA^q\times BV=VE.$ [sive] VE

est ut $BA^q\times BV.^{(12)}$ [Prodit ergo] $\dfrac{CQ^qBA^q-CR^qBA^q+CR^{qq}}{CR^{qq}}BV=BE.$

[Dicantur Ellipseos latus transversum q et latus rectum r. Erit $CQ=\frac{1}{2}q.$

$CR=\frac{1}{2}\sqrt{qr}.$ hoc est] $\dfrac{4c}{qq}=a.$ $\dfrac{4c}{qr}=b.$ [adeoqʒ] $\dfrac{4c}{qq}xx+\dfrac{4c}{qr}yy=c.$ [sive] $rxx+qyy=\dfrac{qqr}{4}.$

[vel] $\dfrac{r}{q}xx+yy=\frac{1}{4}qr.^{(13)}$ [Fit etiam] $\dfrac{\frac{1}{4}qqyy-\frac{1}{4}qryy+\frac{1}{16}qqrr}{\frac{1}{16}qqrr}BV=BE.$ [seu quia]

---

(9) That is, the projection upon the abscissa AB of the curvature radius
$$EB=(1+\dot{y}^2/\dot{x}^2)^{\frac{3}{2}}/(\ddot{y}/\dot{x}^2)=(EB/DB).(\dot{x}^2+\dot{y}^2)/\ddot{y},$$
where $ED/BD=BA/AT=\dot{y}/\dot{x}.$

(10) Namely, $DB-AB$ (or y) where, on eliminating x, $DB=(b^2y^2+a(c-by^2))\,y/ac.$

(11) Since $EV/VB=DA/AB$ (or y).

(12) In particular, where B coincides with R (and so E comes to lie at H) there is $CH\propto RC^3$, and therefore '$VE.CH::BA^q\propto BV.RC^{\text{cub}}$' as Newton specifies (there without any justification) in his construction of the curvature radius at the ellipse's general point B in 'Prob. 4' in §1.4 preceding.

$\frac{1}{4}rrxx = \frac{1}{16}qqrr - \frac{1}{4}qry.$ [est] $\frac{\frac{1}{4}qqyy + \frac{1}{4}rrxx}{\frac{1}{16}qqrr} BV = BE = \frac{4qqyy + 4rrxx}{qqrr} BV.$ [sive quia]

$q^3 r - 4qqyy = 4qrxx.$ [est] $\frac{q^3 r - 4qrxx + 4rrxx}{qqrr} BV = BE = \frac{q^3 - 4qxx + 4rxx}{qqr} BV.$

[Cape $PA = z.$ adeoç $AQ = q - z$ et erit] $x = z - \frac{1}{2}q.$ [unde]

$$\frac{qqr + 4qqz - 4qrz + \overline{4r - 4q} \times zz}{qqr} BV = BE.\ [\text{hoc est}]$$

$$BE = BV + \frac{4q - 4r}{qqr} \times PAQ \times BV = BV + \frac{4q - 4r}{qrr} yy \times BV.^{(14)}$$

$[3]^{(15)}$ $Db . BD :: Mb . MP.$

$BD . CE :: PB . PC.$ $CE . Ec :: PN . Nc|$

$Db . Ec :: MB, BP, PN . MP, PC, NC$

$:: \dfrac{MB, PB}{PM} . \dfrac{NC, PC}{PN} . \|$

$CE . Cc :: PN . Pc . |$

$Db . Cc :: MB, PB, PN . MP, PC^q$

$[::] \dfrac{MB, PB}{MP} . \dfrac{PC^q}{PN}.$

[Hoc est] $Db . Ec . Cc . Bb :: \dfrac{MB, PB}{PM} . \dfrac{NC, PC}{PN} . \dfrac{PC^q}{PN} . \dfrac{PB^q}{PM}.^{(16)}$

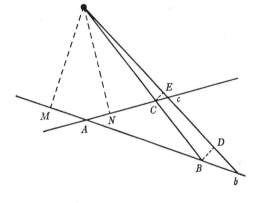

---

(13) The standard Cartesian defining equation of the ellipse of *latus rectum r* and main axis *q*, referred to its centre as origin of coordinates.

(14) Because $(AB^2 =) y^2 = (r/q) . PA \times AQ.$

(15) Add. 3970.3: 480^r, in the original sequence a verso page of a worksheet on which Newton has also jotted (f. 479^v) a calculation relating to Proposition VI of Book 1, Part 1 of his *Opticks* ($_1$1704: $_1$58) and a preliminary redraft (f. 480^v) of part of Observation VIII of its Book 2 (*ibid.*: $_2$100), and, most notably, where he has penned out two versions (ff. 480^v, 479$^{r/v}$) of a provisional preface to the *Opticks* which he never used (but which, with some few errors and omissions in transcription, have recently been printed by J. E. McGuire in *British Journal for the History of Science*, 5, 1970: 184–6). The text has no accompanying figure, but we make good the lack by borrowing that which Newton afterwards set to illustrate his revised argument on Add. 3960.7: 90 (see §2.2 preceding).

(16) More simply, as Newton himself was afterwards equivalently to remark (compare §2: notes (50) and (52) above), since $PB/PM : PC/PN = \text{cosec } \widehat{ABC} : \text{cosec } \widehat{ACB} = AB : AC$, there is

$$Db : Ec = MB \times AB : NC \times AC \text{ and } Db : Cc = MB \times AB : PC \times AC,$$

so that $Db : Ec : Cc : Bb = \bar{} MB \times AB : NC \times AC : PC \times AC : PB \times AB.$

[4][17] $\dfrac{dz^{\frac{1}{2}\eta-1}}{e+fz^\eta}\dot{z}=$ moment[o] Areæ. [Per Formam 2.1 ubi] $\dfrac{dz^{\frac{1}{2}\eta}}{e+fz^\eta}=y.$

[ponendum est] $\sqrt{\dfrac{d}{e+fz^\eta}}=x.$ [et] $\sqrt{\dfrac{d}{f}-\dfrac{e}{f}xx}=v=\sqrt{\dfrac{dz^\eta}{e+fz^\eta}}.$ [itacp] $\dfrac{dz^{\frac{1}{2}\eta}}{e+fz^\eta}=xv=\square.$

[Hoc est] $dz^{\frac{1}{2}\eta}=e\square+f\square z^\eta.$ [unde] $\frac{1}{2}\eta d\dot{z}z^{\frac{1}{2}\eta-1}=e\overset{\cdot}{\square}+f\square z^\eta+\eta f\dot{z}\square z^{\eta-1}.$

[sive] $\dfrac{\frac{1}{2}\eta de\dot{z}z^{\frac{1}{2}\eta-1}-\frac{1}{2}\eta df\dot{z}z^{\frac{3}{2}\eta-1}}{\overline{e+fz^\eta}|^2}=\overset{\cdot}{\square}.$ [Porro] $d=exx+fxxz^\eta.$

[unde] $0=2e\dot{x}x+2f\dot{x}xz^\eta+\eta fx^2\dot{z}z^{\eta-1}.$ [vel] $\dot{x}=\dfrac{-\eta fx\dot{z}z^{\eta-1}}{2e+2fz^\eta}.$

[adeocp] $\dot{x}v=\dot{s}=\dfrac{-\frac{1}{2}\eta f\square\dot{z}z^{\eta-1}}{e+fz^\eta}=\dfrac{-\frac{1}{2}\eta df\dot{z}z^{\frac{3}{2}\eta-1}}{\overline{e+fz^\eta}|^2}.$

[Quare] $\dfrac{\overline{2xv-4s}}{\eta}\left[\text{seu}\ \dfrac{2}{\eta}\overset{\cdot}{\square}-\dfrac{4}{\eta}\dot{s}\right]=\dfrac{de\dot{z}z^{\frac{1}{2}\eta-1}-df\dot{z}z^{\frac{3}{2}\eta-1}}{\overline{e+fz^\eta}|^2}+\dfrac{2df\dot{z}z^{\frac{3}{2}\eta-1}}{\overline{e+fz^\eta}|^2}$

$=\dfrac{de\dot{z}z^{\frac{1}{2}\eta-1}+df\dot{z}z^{\frac{3}{2}\eta-1}}{\overline{e+fz^\eta}|^2}=\dfrac{d\dot{z}z^{\frac{1}{2}\eta-1}}{e+fz^\eta}.$ [18]

[5][19] $\left[\dfrac{dz^{\eta-1}}{e+fz^{2\eta}+gz^{4\eta}}\right].$ [Quia] $\dfrac{t}{e+fz^\eta+gz^{2\eta}}-\dfrac{t}{e-fz^\eta+gz^{2\eta}}=\dfrac{-2tfz^\eta}{ee+\dfrac{+2eg}{-ff}z^{2\eta}+ggz^{4\eta}}.$

---

(17) Add. 3960.13: 221, a computation confirming the accuracy of Form 2.1 in the latter table of conically evaluable 'areas of curves' which, in copy of the corresponding 'Catalogus' in his 1671 tract (III: 244–54, especially 244), he appended in scholium to Proposition X of the parent 'De Quadratura Curvarum' more than twenty years afterwards (compare VII: 550–1, note (92)). By its handwriting it was written at the same date as the parallel check on Form 8.4 (see [6] below) which Roger Cotes' letter to him on 18 August 1709 spurred Newton to make, and we may surmise that he was led in sequel to test the correctness of the other entries in his table, thereby detecting the extraneous factors in his ensuing Forms 2.2 and 2.3 whose corrected versions were first made public in William Jones' 1711 re-edition of the *princeps* printing of the 'De Quadratura' (see §3.2: note (35) above).

(18) Whence, by reversing the argument, Form 2.1 in Newton's table (see page 136 above) is established to be correct.

(19) Add. 3964.8: 15[r], a stray scrap (by its handwriting written in the middle 1710's) in which Newton endeavours to complete his reduction of Form 6.1 of the table of conically evaluable 'areas of curves' appended in scholium to Proposition X of the published 'De Quadratura', now taking account of the instance (before omitted in his implied requirement that $\sqrt{(f^2-4eg)}$ be real) where $|f|<2\sqrt{(eg)}$. From what we have already stated (in §3.2: note (37) preceding) it will be clear that this unfinished computation was occasioned by Newton's hearing word through William Jones in May 1716 of Roger Cotes' success in reducing the integrand $dz^{\frac{1}{2}\eta-1}/(e+fz^\eta+gz^{2\eta})$ in this relatively intractable case, together with the hint that the latter had proceeded by doubling the index η of the base variable z and then factorizing the ensuing trinomial quartic in z^η into its pair of component trinomial quadratics.

[adeoꝗ capiendo $ee = a$. $2eg - ff = b$. $gg = c$. seu $e = a^{\frac{1}{2}}$. $g = c^{\frac{1}{2}}$ itaꝗ $f = \sqrt{2a^{\frac{1}{2}}c^{\frac{1}{2}} - b}$]

$$\frac{t}{a^{\frac{1}{2}} + \sqrt{2\sqrt{ac}) - b}\Big| z^{\eta} + c^{\frac{1}{2}}z^{2\eta}} - \frac{t}{a^{\frac{1}{2}} - \sqrt{2\sqrt{ac}) - b}\Big| z^{\eta} + c^{\frac{1}{2}}z^{2\eta}}[=]\frac{-2tfz^{\eta}}{a + bz^{2\eta} + cz^{4\eta}}.$$

[erit substituendo e, f et g pro a, b et c atꝗ reducendo]

$$\frac{tz^{-1}}{e^{\frac{1}{2}} + \sqrt{2\sqrt{eg}) - f}\Big| z^{\eta} + g^{\frac{1}{2}}z^{2\eta}} - \frac{tz^{-1}}{e^{\frac{1}{2}} - \sqrt{2\sqrt{eg}) - f}\Big| z^{\eta} + g^{\frac{1}{2}}z^{2\eta}}[=]\frac{-2tfz^{\eta-1}}{e + fz^{2\eta} + gz^{4\eta}}. \qquad (20)$$

[Pone igitur] $-2tf = d$.[(21)]

---

[6][(22)] [Ubi Ordinata est] $\dfrac{4\eta e z^{2\eta-1} + 5\eta f z^{3\eta-1} + 6\eta g z^{4\eta-1}}{2\sqrt{e + fz^{\eta} + gz^{2\eta}}} = y$. [Erit Area]

$z^{2\eta}\sqrt{e + fz^{\eta} + gz^{2\eta}} = t$. [Quia autem per Formam 8.2 et 8.3 ponendo] $z^{\eta} = x$.

$\sqrt{e + fx + gxx} = v$.[(23)] [sive $xxv = t$.]

[ubi] $\dfrac{4\eta e z^{2\eta-1}}{\sqrt{\ \ })} \ [=]y$.   [area est] $\dfrac{-16\eta efs + 8\eta efxv + 16\eta eev}{4\eta eg - \eta ff}$.

$$\left[\frac{5\eta f z^{3\eta-1}}{\sqrt{\ \ })}\right] \qquad \frac{{+15\eta f^3 \atop -20\eta efg}s \ {-10\eta f^3 \atop +20\eta efg}xv - 10\eta effv}{4\eta egg - \eta ffg}.$$

[erit] $\dfrac{{36\eta efg \atop -15\eta f^3}s \ {-28\eta efg \atop +10\eta f^3}xv \ {-16\eta eeg \atop +10\eta eff}v}{4\eta egg - \eta ff \text{ in } 2} + xxv = $ areæ ipsius $\dfrac{3\eta g z^{4\eta-1}}{\sqrt{e + fz^{\eta} + gz^{2\eta}}}$.

---

(20) The numerator here should be '$-2t\sqrt{2\sqrt{eg}) - f}\big| z^{\eta-1}$' of course.

(21) In sequel to our correction in the previous note, read '$-2t\sqrt{2\sqrt{eg}) - f}\big| = d$' *recte*. In consequence Newton's preferred 'Cotesian' reduction of the integrand $dz^{\eta-1}/(e + fz^{2\eta} + gz^{4\eta})$ splits it into the partial fractions $(1/2\lambda)\,(dz^{-1}/(e^{\frac{1}{2}} - \lambda z^{\eta} + g^{\frac{1}{2}}z^{2\eta}) - dz^{-1}/(e^{\frac{1}{2}} + \lambda z^{\eta} + g^{\frac{1}{2}}z^{2\eta}))$ where $\lambda = \sqrt{(2\sqrt{(eg)} - f)}$. The auxiliary integrands $dz^{-1}/(e^{\frac{1}{2}} \pm \lambda z^{\eta} + g^{\frac{1}{2}}z^{2\eta})$ are, however, instances of a Form '5.0' which does not appear in Newton's 'De Quadratura' table, and he doubtless saw as much when he here broke off his computation. There is nothing to show that he afterwards attained the analogous—and acceptable—reduction of Form 6.1, where $|f| < 2\sqrt{(eg)}$, which Cotes himself had (see §3.2: note (37) preceding) communicated to William Jones on 5 May 1716.

(22) Add. 4005.15: 83^r, the roughly jotted computation made by Newton in August 1709 when, after Cotes had (see §3.2: notes (38) and (39) preceding) drawn his attention to the several errors in the numerator of the 'areæ valor' in Form 8.4 of his 'De Quadratura' table of conically evaluable integrals, he set himself independently to confirm the truth of Cotes' emended numerator (which—see III: 252—was also that of the corresponding *Ordo* 8.4 in his own parent 1671 table).

(23) And likewise understanding in the sequel that s is the 'area' $\int v \, . \, dx$.

[Quare in Forma 8.4 ubi $\dfrac{dz^{4\eta-1}}{\sqrt{e+fz^{\eta}+gz^{2\eta}}}=$ Ordinatæ y, erit]

$$\dfrac{\begin{matrix}36efgd \\ -15f^3d\end{matrix}^{\,s}\begin{matrix}+8degg \\ -2dffg\end{matrix}\,xxv\begin{matrix}-28defg \\ +10df^3\end{matrix}\,xv\begin{matrix}-16deeg \\ +10deff\end{matrix}\,v}{24\eta eg^3-6\eta ffgg}$$ [$=$Areæ t rectè].[24]

---

[7][25]   *Scholium.*

[26]Ordinatæ secundæ *BE* fluxio prima est ut Ordinata prima[27] *BD*. Et Ordinatæ tertiæ *BF* fluxio prima est ut Ordinata secunda *BE* et fluxio secunda ut Ordinata prima *BD*. Et Ordinatæ quartæ *BG* fluxio prima est ut Ordinata tertia *BF*, secunda ut Ordinata secunda *BE*, ac tertia ut Ordinata prima *BD*. Et sic deinceps in infinitum. Unde si proponatur Æquatio quævis duas tantum incognitas quantitates involvens quarum una (puta *z*) est quantitas uniformiter fluens et altera est fluxio aliqua Fluentis alterius *y*: inveniri potest illa altera Fluens per quadraturam Curvarum.

Exempl. 1. Proponatur Æquatio $z\ddot{y}=aa$,[28] & quæratur *y*. Sit uniformiter fluens $z=$ Abscissæ *AB* et fluxio tertia $\dddot{y}=BD_{[,]}$ secunda $\ddot{y}=BE$, prima $\dot{y}=BF$ et fluens $y=BC$, et sunto *A, B, C* areæ curvarum Ordinatas habentium $\dddot{y}$, $z\ddot{y}$, $z^2\ddot{y}$.

id est $\dfrac{aa}{z}$, *aa, aaz*. Et erit *A* area Hyperbolæ Logarithmo Asymp[to]ti re-

---

(24) In agreement with Cotes and *Ordo* 8.4 in the 1671 table; see note (22) above and §3.2: note (39) preceding.

(25) Add. 3962.3: 61ᵛ, a first rough drafting of the scholium to Proposition XI of the 'De Quadratura' in which Newton, as a *finale* to his text, passes briefly to consider the simplest reductions of fluxional equations to quadrable form. This lacks the opening paragraph of the revise (pages 152–4 above) where Newton—to his later frustration and despair at being so calculatedly misunderstood by the Leibnizians—clumsily presented the 'Taylor' expansion of the binomial $(x+o)^n$, but in some compensation he here adduces a worked example of the preceding Proposition XI, and then goes on to give two instances of the way in which a given first-order fluxional equation can be resolved into integrable form by a mere change of variable.

(26) Newton initially began: 'Ordinatarum *BD, BE, BF, BG, BH* & [c] quælibet præcedens est ut fluxio sequentis. Ideoɋ Ordinatæ...'.

(27) Originally 'exponitur per Ordinatam primam'.

(28) That is, in equivalent multiple-integral form (see VII: 552–3, note (96)),

$$y=\int_k^t\int_k^z\int_k^z a^2/z\,.\,dz^3$$

where *k* is some given lower bound to the fluent areas.

spondens,[29] $B = maaz$,[30] & $C = \dfrac{naazz}{2}$[30] adeoɋ

$$\frac{ttA - 2tB + C}{2} = \frac{ttA}{2} - tmaaz + \frac{naazz}{4} = \text{Areæ } AFLC = BG = y.\,[31]$$

Si æquatio duas tantum involvit quantitates incognitas quarum una sit fluxio alterius: Pone reciprocum fluxionis pro fluxione alterius quantitatis fluentis et quære illam alteram fluentem ut supra.[32] Ut si æquatio sit $\ddot{v} - a\dot{v}\ddot{v} = bb$,[33] pro $\ddot{v}$ scribe $\dfrac{1}{\dot{y}}$ et æquatio evadet $\dfrac{1}{\dot{y}} - a\dot{v}\ddot{v} = bb$. seu $\dot{y} = \dfrac{1}{bb + a\dot{v}\ddot{v}}$ [,] ubi si $\dot{v}$ fluat uniformiter invenietur y.[34]

Æquatio quæ tres involvat fluentes quandoɋ reduci potest ad æquationem quæ duas tantum involvit. Ut si fuerit $ax^2 - 2by\dot{y} + ccy^2\dot{y} = 0$, pro $y\dot{y}$ scribe $\dot{v}$ et fiet $ax^2 - 2b\dot{v} + ccv\dot{v} = 0$. Inde vero prodit v. Et æquatio $y\dot{y} = \dot{v}$ dat $\frac{1}{2}yy = v$.[35]

---

(29) In Leibnizian equivalent, $A = \int_{k}^{z} a^2/z \,.\, dz = a^2 \log (z/k)$, where k is the lower bound of the 'asymptote' z.

(30) The arbitrary 'weighting' factors m and n, a late insertion by Newton in the manuscript, are evidently his clumsy attempt to take account of the additional terms accruing from the lower bounds in the triple integration of $\dddot{y}$ to yield y. Correctly, '$B = aaz - aak$' and '$C = \frac{1}{2}aazz - \frac{1}{2}aakk$' where k is the common lower bound.

(31) Rather, on setting the common lower bound of the 'areæ' to be unity, there ensues

$$y = \int_{1}^{t}\int_{1}^{z}\int_{1}^{z} a^2/z \,.\, dz^3 \left[= \int_{1}^{t}\int_{1}^{z} a^2 \,.\, \log z \,.\, dz^2 = \int_{1}^{t} a^2(z \log z - z + 1) \,.\, dz \right]$$

$$= \tfrac{1}{2}(t^2A - 2tB + C) = a^2(\tfrac{1}{2}\log t^2 \log t - \tfrac{3}{4}t^2 + t - \tfrac{1}{4})$$

since now $(z \to t)$ $A = a^2 \log t$, $B = a^2(t-1)$ and $C = \frac{1}{2}a^2(t^2-1)$.

(32) If, that is, there is given $(\dot{z}/\dot{y} =)\ \dot{z} = f(z)$, then change the base variable $y \to z$ and find $\dot{y}(= \dot{y}/\dot{z}) = 1/f(z)$, which yields the general solution of the fluxional equation as the 'quadrature' $y = \int 1/f(z) \,.\, dz$. Newton's present awkward phrasing of this simple rule of inversion is much improved in his revised statement of it (page 154 above) 'complendo dimensiones fluxionum' before successively setting $\dot{y}$ and then $\dot{z}$ equal to unity.

(33) That is, to reduce this second-order equation to the standard first-order form to which Newton's preceding rule can (see the previous note) be applied, $\dot{z} - az^2 = b^2$ where $z = \dot{v}$; whence comes the ensuing replacement of $\ddot{v} = \dot{z}$ (or $\dot{z}/\dot{y}$) $= b + az^2$ by $(\dot{z}/\dot{y}$ or) $1/\dot{y}$.

(34) Specifically, $y = \int 1/(b + az^2) \,.\, dz$, a quadrature at once attained—in analytical equivalent, in terms of the related circular and hyperbolic functions—through Form 2.1 of the latter table of conically evaluable area-integrals appended in scholium to the preceding Proposition X (see page 136 above) on there setting $\eta = 2$ and also $d = 1, e = b, f = a$.

(35) More precisely, $\frac{1}{2}y^2 = v - k$, k an arbitrary constant of integration. The quadrature of $\dot{v} = bc^{-2} \pm \sqrt{(b^2c^{-4} - ac^{-2}x^2)}$ is, we need scarcely add, directly obtainable (compare the first reduction of Form 3.2 on page 138 above when $\eta = -2$ and so $z = x$) as the area, $\int y \,.\, dx$, under the conic defined by the Cartesian equation $ax^2 - 2by + c^2y^2 = 0$.

3

MISCELLANEOUS WRITINGS ON MATHEMATICS[1]

[1705–6?]

From the original drafts in the University Library, Cambridge

§1. A NOVEL CONSTRUCTION OF THE GENERAL CONIC BY POINTS: THE ELLIPSE TREATED AS A SIMPLE AFFINE TRANSFORMATION OF THE CIRCLE.[2]

[1] Detur triangulum *AVv* rectangulum ad *V*.[3] Ab ipsius *Vv* puncto quovis *B* ducantur *BE* et *BD* ipsis *Av* et *AV* parallelæ respective quarum *BE* occurrat ipsi *AV* in *E* & *BD* sit medium proportionale inter *VB* et *AE*. Et perinde ut

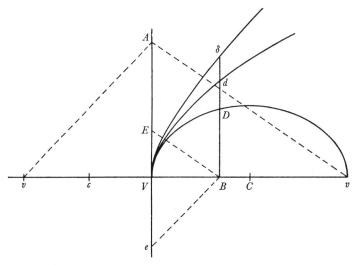

ad partes *V* versus *A*
punctum *E* cadit ad partes contrarias ⎬ figura *VDv* quam punctum *D*[4] perpetuo
in ipso puncto *V*

Ellipsis.
tangit erit Hyperbola.[5]
Parabola.

---

(1) Other than that—as Newton's handwriting style in the manuscripts and the proximity of these to other, datable ones of the period confirm—these autographs were drafted over the

Translation

[1]   Let there be given the triangle AVv right-angled at V.[3] From any point B of Vv draw BE and BD parallel to Av and AV respectively, of which let BE meet AV in E and let BD be the mean proportional between VB and AE. And, correspondingly as the point E falls

$$\left\{\begin{array}{l} \text{on the side of } V \text{ towards } A \\ \text{on the opposite side} \\ \text{at the point } V \text{ itself} \end{array}\right\},$$

the figure VDv which the point D[4] perpetually touches will be $\left\{\begin{array}{l} \text{an ellipse} \\ \text{a hyperbola} \\ \text{a parabola} \end{array}\right\}$.[5]

---

same narrow period around 1705–6, no common theme runs through this assorted miscellany of mathematical jottings and more connected revisions of earlier papers on geometry. It might be (as we suggest in §2: note (1) below) that their main portion was to form a revision of the lavish work on 'Geometria' on whose first books he had spent so much effort in the early 1690's (see VII, 2: *passim*), now augmented by the reworked abridgement of the 'Problems for construing æquations' (II: 450–516) of thirty years earlier which is reproduced in §3 under the here inverted Latin title of 'De constructione Problematum Geometricorum'; or maybe (as we alternatively conjecture in §3: note (1) below) the latter was composed as an intended replacement to the equivalent geometrical constructions which, both long-windedly and less than appositely, terminate (see V: 432–90) the deposited text of Newton's Lucasian lectures on algebra during 1673–83 which was at this time, under William Whiston's shaky editorial hand and with his own grudging, less than happy acquiescence (see V: 9–14), at the press in Cambridge. Till—if ever—the background to these present writings can more surely be patched together on the basis of documentary evidence, no more may be said without departing into unfounded hypothesis and speculation.

(2) The text of these two constructions is penned by Newton on the back page (Add. 3962.5: 67ᵛ) of the same folded manuscript sheet on which he had previously (at the end of 1703 or in the first days of 1704) set down the preliminary draft (*ibid.*: 66ʳ) of the four final paragraphs of the published 'De Quadratura Curvarum' whose main variants we have cited in 2, §3.2 preceding (see its notes (52) *et seq.*). While neither lacks its individual elegances, they both remain but schoolmasterish exercises in the elementary geometry of conics, in the manner of what, as a Cambridge undergraduate forty years earlier, he had (see 1: 29–45) copied from Schooten's *Exercitationes Mathematicæ* and the related *Commentarii* to his (second) Latin edition of Descartes' *Geometrie*. What ultimate purpose they were meant to serve is not clear.

(3) Newton previously began to the same effect: 'Dentur VA, Vv angulum rectum AVv continentes' (Let there be given VA, Vv containing the right angle $\widehat{AVv}$).

(4) In the case of the ellipse, understand. Where the constructed conic is a parabola or a hyperbola, Newton denotes this locus-point by d and δ respectively.

(5) For, where $Vv = q$ and $AV = r$, there is by Newton's construction $AE = (r/q).BV$; so that the locus-point D is defined by the Apollonian 'symptom' $BD^2 = (r/q).VB \times Bv$, and the curve VD is a conic of *latus rectum r*. This, as Newton equivalently specifies, will be an ellipse, parabola or hyperbola according as $AE/AV(=Bv/Vv) \lesseqgtr 1$, that is, as v lies (as shown in Newton's figure) at a finite distance to the right of the vertex V, or at infinity, or at a finite distance to the left of V.

[2] [6]Si centro C diametro AB describatur Circulus $ADBE$ et omnes ejus ordinatim applicatæ FG secentur in g in ratione data[5] punctum g locabitur in Ellipsi AgB cujus Axis principalis est AB et Vertices principales sunt A et B et Centrum C & Axis conjugatus est Ordinatarum maxima duplicata de et Latus rectum est[7] ad hunc axem ut hic Axis ad Axem principalem.[8]

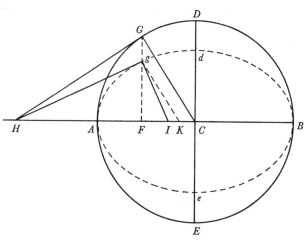

In Axe AB producto capiatur CH ad CA ut CA ad CF et acta Hg Ellipsin tanget in g propterea quod acta HG circulum tanget in G.[9]

Capiatur FK ad FC et FI ad FK ut Fg ad FG, et gI erit perpendicularis ad Ellipsin in g. Nam gK erit parallela GC ideoq triangulum FgK erit simile triangulis FGC & FHG et triangulum FgI simile triangulo FHg.[10]

§2. RENEWED ATTEMPTS TO DISTINGUISH THE INDIVIDUAL AREAS AND METHODS OF MATHEMATICAL INVESTIGATION: THE 'SCIENCE' OF AXIOMATIC DEDUCTION AND THE 'ART' OF RESOLVING PROBLEMS.[1]

(6) In a cancelled first version of the ensuing paragraph Newton initially gave a separate statement of its Archimedean construction of the ellipse affinely from the circumscribing circle —this he had long ago taken from the fourth book of Schooten's *Exercitationes* (see 1: 45, note (74))—in the following words: 'Si recta data AB bisecetur in C et centro C intervallo AC describatur Circulus $AefB$. et ab ejus puncto quovis e ad diametrum AB demittatur normaliter ordinata de, eaq secetur in E in ratione data, punctum E locabitur in Ellipsi cujus axis est linea data AB, et vertices principales sunt A et B, centrumq est C et' (If the given straight line AB be bisected at C and with centre C, radius AC the circle $AefB$ be described, and from any point e of it the ordinate de be let fall at right angles to the diameter 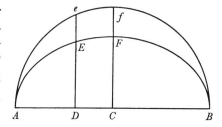 AB, let that be cut at E in a given ratio: the point E will be placed in an ellipse whose axis is the given line AB and principal vertices are A and B, and C is its centre, and...). On the model of what comes after, we may appropriately complete Newton's unfinished sentence by continuing '...et si ad centrum illud C erigatur perpendiculum Cf Ellipsem secans in F, erit latus rectum ad $2CF$ ut $2CF$ ad AB' (and, if at that centre C there be raised the perpendicular Cf cutting the ellipse in F, the *latus rectum* will be to $2CF$ as $2CF$ to AB).

[2] [6]If with centre C and diameter AB there be described the circle $ADBE$ and every one of its ordinates FG be cut at g in a given ratio, the point g will be located on the ellipse AgB whose main axis is AB and principal vertices are A and B, having centre C and with its conjugate axis the doubled greatest ordinate de, while its *latus rectum* is[7] to this axis as this axis to the principal one.[8]

In the axis AB extended take CH to CA as CH to CF, and when Hg is drawn it will touch the ellipse at g, seeing that when HG is drawn it will touch the circle at G.[9]

Take FK to FC and FI to FK as Fg to FG, and gI will be perpendicular to the ellipse at g: for gK will be parallel to GC, and consequently the triangle FgK will be similar to the triangles FGC and FHG, and the triangle FgI similar to the triangle FHg.[10]

---

(7) By definition, of course.

(8) Namely, AB.

(9) Newton had, we have seen (vi: 574), earlier made good use of this conservation of the subtangent FH in the affine transition of the locus-point $G \rightarrow g$ in his unpublished ameliorations of the *Principia*'s opening propositions some dozen years before. The property—at once evident if the figure is viewed as the oblique projection of a cylinder erect on the base circle $ADBE$, with HGg the tangential plane to the cylinder through the point H in the latter's plane —was familiar to those, Descartes, Fermat, Beaugrand and others, who gave general solution of the problem of drawing tangents to (algebraic) curves in the late 1630's.

(10) In the Cartesian equivalent where $CF = x$ and $Fg = y$ are the coordinates of $g(x, y)$, on setting $AB = q$ and its *latus rectum* $= r$, so that $de = \sqrt{(qr)}$, the defining equation of the ellipse will be $y^2 = (r/q)\,(\frac{1}{4}q^2 - x^2)$; in consequence, the subnormal at g is

$$FI = y \cdot dy/dx = -(r/q)\,x,$$

where $r/q = Cd^2/CD^2 = (Fg/FG)^2 = (FK/FC)^2 = FI/FC$.

(1) It will be evident that these renewed efforts by Newton to differentiate and delineate the nature, principles and basic operations and techniques of arithmetic, algebra, geometry and (theoretical) mechanics—the only mathematical 'sciences' of numerical quantity and spatial magnitude of which he (like his contemporaries) could conceive—follow closely in the steps of his equally abortive essays a dozen years earlier to draft a treatise on 'Geometria', where he had dealt with these matters at some length (see vii, **2**: *passim*, but especially 200–8 [+ 220–4], 286–94 [+ 338–44] and 382–400). The tone and content of these pieces suggests that they were again to be a preamble to some more sustained work on geometrical analysis and the resolution of problems. If so—and if he went on with that purpose—the following tract 'De Constructione Problematum Geometricorum' (§3 below) in which, at about this time (autumn 1706; compare Appendix 1: note (2)), he revamped his earlier 'Problems for construing æquations' (II: 450–516) suggests itself, *faute de mieux*, as the piece which they were to introduce. Except for the internal evidence of the texts themselves we know nothing to support or gainsay the conjecture. The reader must decide for himself whether the juxtaposition is more than fortuitous. (It might equally well be, as we indicate in §3: note (1) below, that the latter revise was drafted by Newton to take the place of the equivalent, unfinished termination of his deposited Lucasian lectures on algebra whose text was then at the press in Cambridge.)

[1][2]   Arithmetica circa numeros et res numerabiles, Geometria circa mensuras et res omnes longas latas aut profundas quæ mensurandæ veniunt[3] & Mechanica circa vires et motus locales versantur.[4] Et Quæstionū solutiones sunt arithmeticæ quæ fiunt per solas operationes Arithmeticas nempe addendi subducendi multiplicandi dividendi et extrahendi radices, Geometricæ quæ fiunt per operationes Mechanicas[5] ducendi lineas & construendi figuras accurate vi postulatorum, & Mechanicæ quæ fiunt per operationes alias mechanicas applicandi vires[6] ac dirigendi corpora in motibus secundum leges assignatas. Est uticჳ Geometria species aliqua Mechanicæ, et vi postulatorum accurata redditur.[7] Quantitates autem Geometricæ et Mechanicæ aliæcჳ omnes quatenus numerari possunt vel per numeros exponi tractantur in Arithmetica, et similiter quæ Geometricæ non sunt, quatenus tamen per quantitates Geometricas exponi possunt, tractantur etiam in Geometria.

Methodus autem solvendi Quæstiones tam in Arithmetica quam in alijs scientijs duplex est, per compositionem scilicet[8] et per resolutionem. Componimus pergendo directe a datis numeris ad quæsitos: resolvimus assumendo quæsitum numerum tanquam datum et inde derivando datū aliquem numerum tanquam quæsitum ut ad Theorema vel corollarium aliquod perveniamus ex cujus veritate numerum quæsitum vere obtineamus. Prioris methodi expla̅[9] habemus in Regulis[10] proportionalium quæ in Arithmetica vulgari tradi solent[,] nempe in Regula aurea directa et inversa, simplici et composita, et in Regula sortium.[11] Alterius methodi exempla habemus in Regula Positionis falsæ, in Quæstionibus quas Diophantus in *Arithmetica* sua tractavit[12] et in Algebra, quam uticჳ Arithmeticam speciosam vocant. Nam Algebra nihil aliud est quam Regula Positionis indeterminatæ pr quā Regula Positionis falsæ emendatur et perfectior redditur.[13] In Regula Positionis falsæ pro numero quæsito ponimus

---

(2) Add. 3963.15: 172r. The preparatory casting of this (*ibid.*: 171v), penned on the back of a copy of the news-sheet 'Postscript to the Post=Boy' for 26 September 1706, is reproduced in Appendix 1.1 below.

(3) Newton first wrote 'quæ mensurari vel etiam per mensuras exponi possunt' (which can be measured or at least expressed by measures), afterwards changing the 'possunt' (can) to be 'deb[e]ant' (ought to).

(4) Compare vii: 338–9.

(5) The tautological 'Geometricas' (geometrical' is here replaced in the manuscript.

(6) 'et generandi motus' (and generating motions) is deleted in sequel.

(7) Compare vii: 288–90.

(8) Newton first wrote 'uticჳ' (to be sure). In his earlier exegeses (see vii: 240–2, 306–8) on Pappus' preamble to the seventh book of his *Mathematical Collection* which is the *locus classicus* for this Greek notion of the necessity for prior analysis and then, *per conversionem*, subsequent synthesis of a problem, he had treated its subtleties a deal more circumspectly.

(9) 'exempla.'

(10) 'aureis' (golden) is cancelled. On this traditional 'Rule of Three' by which a fourth proportional to three given magnitudes is found, see D. E. Smith's *History of Mathematics*, 2

Translation

[1][2]   Arithmetic has to do with numbers and numerable things; geometry with measures and all things long, wide and deep which come needing to be measured;[3] mechanics with forces and motions from place to place.[4] And arithmetical solutions of questions are ones which are achieved through the operations of arithmetic alone, viz: adding, subtracting, multiplying, dividing and extracting roots; geometrical ones are those accomplished by the mechanical[5] operations of drawing lines and constructing figures accurately by dint of postulates; and mechanical are those effected through other mechanical operations of applying forces[6] and directing bodies in motions following assigned lines. Of course, geometry is a particular species of mechanics, and is rendered accurate by force of postulates.[7] Geometrical and mechanical quantities, however, and all others are, insofar as they can be numbered or expressed by numbers, treated in arithmetic; and likewise those not geometrical are yet, insofar as they can be expressed by geometrical quantities, treated also in geometry.

The method, however, of solving questions both in arithmetic and in other sciences is dual: through composition, namely,[8] and through resolution. We compose by proceeding directly from given numbers to ones sought: we resolve by assuming a sought number as given and therefrom deriving some given number as though it were sought, so that we may attain a theorem or some corollary from whose truth we may truly obtain the number sought. Of the first method we have examples in the[10] rules of proportionals which are customarily delivered in ordinary arithmetic, namely, in the direct and inverse golden rules, simple and compound, and in the rule of lots.[11] Of the second method we have examples in the rule of false position, in the questions which Diophantus treated in his *Arithmetic*,[12] and in algebra—which to be sure they call specious arithmetic; for algebra is nothing else than the rule of indeterminate position by means of which the rule of false position is emended and rendered more perfect.[13] In the rule of false position in place of the number

---

(Boston, 1925): 486. Newton himself had appealed to it in the 'Prænoscenda' to his 'Compendium Trigonometriæ' (see IV: 134, note (11)).

(11) The related *Regula consortii* (Rule of partnership) by which a given quantity is shared into component parts which are in a given proportion one to another; see Smith's *History*, 2: 555–6, where it is remarked (*ibid.*: 556) that 'English writers, following the Italian practice [of employing the term *compagnie*], often used the word "company", although in general... "fellowship" was preferred'.

(12) Compare T. L. Heath's analysis of 'Diophantus' Methods of Solution' in Chapter IV of his *Diophantus of Alexandria. A Study in the History of Greek Algebra* (Cambridge, ₂1910): 54–98, especially 95–8 where he examines the applications of παρισότητος ἀγωγή in the *Arithmetic*. Newton had himself employed like variants upon the general *Regula falsi* to some limited success in his own explorations into Diophantine 'Problemata Numeralia' (IV: 74–108).

(13) This last clause replaces 'et a Regula falsæ Positionis in eo solo differt quod in hac

symbolum[14] determinatæ et plerumcӡ falsæ significationis, et errorem positionis per operationes Arithmeticas corrigimus:[14] in Algebra pro eodem numero quæsito ponimus symbolū non falsæ sed indeterminatæ significationis et significationem determinamus per operationes Arithmeticas.[15] Quinetiam pro numeris cognitis ponere licet symbola signif[ic]ationis indeterminatæ et sic investigare Theorema generale quo Quæstio[16] quibuscuncӡ numeris proposita solvatur. Pergit Arithmetica vulgaris ad operationes tantum quæ et faciliores sunt et in rebus humanis frequentius occurrunt. Longius pergit Algebra[,] quæ uticӡ Analysis dici solet et est Analyseos pars illa quæ in Arithmetica vulgari desideratur. Hæc Arithmeticam complet eamcӡ universaliorem reddit, et propterea Arithmeticam et Algebram tanquam unam scientiam conjunctim explicabo.[17]

---

[2][18] Numerantur res omnes,[19] mensurantur magnitudines et moventur corpora; et artes numerandi mensurandi et movendi dicuntur Arithmetica Geometria et Mechanica.

Harum scientiarum Principia sunt Definitiones verborum quoad significationem, Axiomata seu veritates primæ et Postulata operationum facillimarum et maxime utilium.[20] Et hæc Principia debent esse quam paucissima.[21] Ideocӡ definimus tantum nomina rerum quæ in arte tractantur, ponimus tantum axiomata quæ notiora sunt et ab alijs veritatibus derivari nequeunt, et postulamus tantum quod Artifex accurate didicerit operationes facillimas earum quæ rebus humanis utilitatem aliquam afferre possunt.

In Arithmetica (siquis hanc scientiam per Propositiones demonstrative tractaret) definire liceret nomina numerorum et operationum et terminorum proportionalium; postulare additionem & subductionem unitatis; & axiomata ponere sequentia.

1. Quæ alicui tertio æqualia sunt, æqualia sunt inter se.

2. Si æqualia æqualibus adduntur summæ sunt æqualia.

3. Quæ ad aliquod tertium eandem habent proportionem æqualia sunt inter se.

Regula ponimus numerum certum pro ignoto in A[lgebra]' (and differs from the rule of false position in this alone, that in the present rule we put a fixed number in place of the unknown one in Algebra).

(14) These replace 'numerum' (number) and 'investigamus' (investigate) here first written in the manuscript.

(15) Newton initially added 'donec numerus ille innotescat' (till that number comes to be ascertained).

(16) Originally 'omnis istiusmodi Quæstio' (every question of the sort).

(17) Newton breaks off to begin again in [2] following.

(18) Add. 3963.15: 173ʳ–174ʳ. A quickly abandoned transitional draft (*ibid.*: 174ᵛ) in

sought we put a symbol[14] of determinate and for the most part false significa-tion, and correct[14] the error in positing by means of arithmetical operations: in algebra for the same number sought we put a symbol not of false but of indeterminate signification, and we determine its meaning by means of arith-metical operations.[15] We are, to be sure, free to put symbols of indeterminate signification in place of known numbers and in this way track down a general theorem by which a question[16] proposed in any numbers whatever may be solved. Ordinary arithmetic goes on merely to operations which are both rather easy and occur the more frequently in human affairs. Greater is the range of algebra, which of course is usually called analysis, and it is the part of analysis which is wanting in ordinary arithmetic: this completes arithmetic, rendering it more universal, and on that account I shall expound arithmetic and algebra together jointly as a single science.[17]

---

[2][18]   All things[19] are numbered, magnitudes measured and bodies moved; and the arts of numbering, measuring and moving are called arithmetic, geometry and mechanics.

The principles of these sciences are *definitions* of words as regards their meaning, *axioms* or prime truths, and *postulates* of the easiest and most useful operations.[20] And these principles ought to be as few as possible.[21] In conse-quence we define merely the names of things which are treated in the art, we posit merely the axioms which are better known and cannot be derived from other truths, and we postulate merely that the practitioner shall carefully have learnt the easiest of the operations which can bring some usefulness in human matters.

In arithmetic (were anyone to treat this science in demonstrative fashion by means of propositions) one would be free to define the names of numbers and operations and terms in proportion; to postulate the addition and subtraction of unity; and to posit the following axioms:

1. Things which are equal to some third are equal to one another.

2. If equals are added to equals the sums are equals.

3. Things which have the same proportion to some third are equal to one another.

---

which Newton first began to rephrase [1] preceding is reproduced in Appendix 1.2 below, while a first state of the opening of the present revise (on f. 173ʳ) follows in Appendix 1.3.

(19) Initially just 'quantitates' (quantities).

(20) 'in rebus practicis' (in practical matters) is deleted in sequel.

(21) The necessary qualification 'si modo sufficiant' (provided they suffice) was first added and then cancelled (as being self-evident?). In his 'Geometriæ Liber primus' a dozen years before (see VII: 290) Newton had spoken of the 'ignominy' of postulating a superfluity of such basic principles.

4. Numerorum æquimultiplices sunt ad invicem ut numeri.

Hæc pauca sufficiunt.

In Geometria[22] si numerus Postulatorum Euclidis augendus esset, quarto loco postularem: *Inter datas duas lineas rectam ducere longitudine datam quæ ad datum punctum converget.*[23] Hæc operatio post eas quas Euclides postulavit est omnium facillima et in rebus practicis utilissima & late patens,[24] cum Problemata omnia solida expediat et ad naturam planorum reducat, ampliatis finibus Geometriæ Veterum.[25] *Datum Conum dato plano secare* operatio est longe difficilior et in rebus practicis (quibus utiꝗ in omni problematum genere apprime consulendum est) prorsus inutilis; et præterea Postulatum prius requirit, nempe, *Dato axe et angulo conum describere.* Nam conus in Geometria primitiva definitur quidem ut subjectum Theorematum, et mensuratur ubi forte dari contingat: sed non describitur per postulatum aliquod et absꝗ descriptione non semper datur ad constructionem problematum.[26] Similiter lineæ omnes curvæ per defini[ti]ones suas in Geometriam recipiuntur: sed earum descriptio difficilis et ad solutionem Problematum inutilis ideoꝗ frustra postulatur.[27]

In Mechanica postulare licet; *Datum corpus data acie in directum lata secare: Datum corpus data acie circa datum axem rotante secare: & Datum corpus data vi in datam plagam movere.*

In solutione Problematum, Arithmetica latissime patet. Nam quantitates Geometricæ et Mechanicæ aliæꝗ omnes per numeros exponi possunt, & Quæstiones circa quantitates illas subinde per Arithmeticam tractari, idꝗ resolvendo et componendo. Sed compositio & resolutio Geometrica simplicior et elegantior esse solet.[28] Adinventa est enim Geometria ut in mensurandis terris per ejus simplicitatem effugiamus computandi tædium, neꝗ aliquid in

---

(22) 'definiendæ sunt magnitudines et figuræ' (are to be defined magnitudes and figures), Newton first went on to write.

(23) In his deposited Lucasian lectures on algebra Newton had been very ready to admit this conchoidal lemma, 'cum Mathematicorum Principe Archimede alijsꝗ Veteribus', in the geometrical construction of 'solid' problems (see v: 428–30). But when he came to draft his treatise on 'Geometria' in the early 1690's, with his gaze more narrowly centred on Euclidean purity, he had (see vii: 388) classed this among other 'mechanical' postulates which it would (in his phrase) 'contaminate' geometry to introduce. It is interesting that he is again more liberal—and practical—in his choice of how most appropriately to broaden the confines of classical 'plane' geometry.

(24) Originally '...utilissima sive praxin expeditam et accuratam spectes sive usum late patentem' (most useful...whether you have regard to the expediency and accuracy of its practice or the wide scope of its application).

(25) On the variety of such reduction of the construction of 'solid' problems (those effectable by interesting conics, and hence solved by determining the roots of the general cubic/quartic equations in the Cartesian equivalent) by conchoidal neusis to that of 'plane' ones (accom-

4. Equimultiples of numbers are to each other as the numbers.
These few suffice.

In geometry,[22] if the number of Euclid's postulates were to need to be augmented, I would postulate in fourth place: *Between two given lines to draw a straight line given in length which shall be directed through a given point.*[23] This operation is, after those which Euclid postulated, the easiest of all and the most useful in practical matters and also broad in its scope[24] inasmuch as it expedites all 'solid' problems, reducing them to the nature of 'plane' ones, so enlarging the bounds of the geometry of the ancients.[25] *To cut a cone by a given plane* is an operation far more difficult and one in practical terms (to which in every kind of problem we must principally have regard, of course) utterly useless; and, moreover, it requires a prior postulate, viz: *With given axis and angle to describe a cone.* To be sure, a cone is indeed defined in elementary geometry as the subject of theorems, and is measured when it happens by chance to be given; but it is not described by means of some postulate, and without such description it is not always given for the construction of problems.[26] Similarly all curved lines are through their definitions received into geometry; but their description is difficult and useless to the solution of problems, and its postulation is consequently to no purpose.[27]

In mechanics it is lawful to postulate: *To cut a given body by a given knife-edge carried straight through*; *To cut a given body by a given knife-edge rotating round a given axis*; and *To move a given body by a given force in a given direction.*

In the solution of problems the range of arithmetic is very wide. For geometrical and mechanical quantities and all others can be expressed by means of numbers, and questions about those quantities thereafter treated by arithmetic, and that by resolving and by composing. But geometrical resolution and composition is usually simpler and more elegant.[28] Geometry was, to be sure, contrived in order by its simplicity to evade the tediousness of computation in

---

plishable by means of the Euclidean straight-edge and compasses) which Newton himself had earlier contrived, see the 'Problemata generalia' which terminate his 'Problems for construing æquations' (II: 468–516)—these are encapsulated in the final Problems 2–4 of the 'De constructione Problematum' (§3 below) in which, around this time, he recast his thirty-year-old tract—and also the analogous conchoidal constructions of 'solid' equations which he set in the early 1680's to conclude the deposited copy of his Lucasian lectures on algebra (see v: 430 ff.).

(26) Newton first wrote more simply and effectively 'sed non postulatur et absc̡ postulato non describitur' (but it is not postulated, and without postulate it is not described).

(27) Compare VII: 292–8, 382–8.

(28) Newton initially began to add '& constructiones sim[pliciores dat?]' (and [yields?] simpler constructions) before breaking off. In sequel he has likewise cancelled the sentence 'Artifex est qui compositionem et resolutionem utramc̡ calleat' (The master-mathematician is one who is skilled in composition and resolution both). Compare VII: 288–90 and 306–8.

Geometria utile est vel laude dignum quod expeditius & melius per Arithmeticam confit.[29]

Arithmetica vulgaris post operationes primas addendi subducendi multiplicandi dividendi et extrahendi radices, componit problemata faciliora per regulas quasdem proportionaliū[30] et ad resolutionem problematum vix[31] progreditur nisi forte per Regulam Positionis falsæ. Resolutio in Algebra doceri solet et inde Algebra vulgo nominatur Analysis. Hæc uticӈ pro positionibus falsis utitur positionibus indeterminatis, ponendo nempe pro numero quæsito symbolum aliquod significationis indeterminatæ et inde computum ineundo quasi numerus quæsitus daretur et alius aliquis numerus quæreretur. Quinetiam pro numeris datis ponuntur symbola significationis indeterminatæ, et ineundo computum prodeunt Theoremata generalia in quibus deinceps pro symbolis determinatis scribere licet numeros quoscuncӈ datos et sic Problema in numeris quibuscuncӈ propositum gēraliter[32] solvere. Igitur Arithmetica vulgaris & Algebra se mutuo perficiunt et simul sumptæ unam constituunt scientiam quæ Arithmetica generalis dici meretur, et propterea utramcӈ conjunctim explicabo.

---

[3][33] Numerantur res omnes, mensurantur magnitudines, moventur corpora, et ex his operationibus oriuntur[34] Arithmetica Geometria et Mechanica. Hæ Scientiæ sunt quatenus per Definitiones Axiomata et Theoremata veritatem docent, Artes vero quatenus per Postulata & Problematum constructiones praxin tradunt & ostendunt.

Mechanica inter Scientias mathematicas per Axiomata & demonstrationes veritatum plurimarum locum obtinuit, inter Artes mathematicas non item. Praxis ejus in Postulatis[35] non fundatur sed pure manualis est. Geometria et Arithmetica tam Artes Mathemati[c]æ sunt quam Scientiæ.

Geometria quatenus scientia est, magnitudines et figuras omnes considerat quæ definiri possunt, et per Axiomata et Theoremata proprietates earum et proportiones indicat: quatenus ars est, postulat quod Artifex operationes omnium primas et facillimas et in praxi mensurandi maxime necessarias expeditas et utiles accurate didicerit, et per Postulata et Constructiones

---

(29) Compare vii: 286–8, 472.

(30) Much as in [1] preceding (see page 172 above), Newton originally went on to specify 'quas directam ['auream' was first written] et inversam simplicem et compositam vocant' (which they call direct [golden] and inverse, simple and compound).

(31) 'vix ultra compositionem' (scarcely ever...beyond composition) in a cancelled prior equivalent phrase.

(32) 'generaliter.'

(33) Add 3963.15: 175ʳ–176ʳ, a final recasting of [2] preceding by way of the preliminary revise (*ibid.*: 173ᵛ/174ʳ) reproduced in Appendix 1.4 below.

measuring territory, nor is there anything useful or praise-worthy in geometry which is accomplished better or more promptly by means of arithmetic.[(29)]

Ordinary arithmetic, after the primary operations of adding, subtracting, multiplying, dividing and extracting roots, composes the easier problems through certain rules of proportionals,[(30)] and hardly ever passes to the resolution of problems[(31)] unless perchance by way of the rule of false position. Resolution is usually taught in algebra, and in consequence algebra is ordinarily given the name of 'analysis'. In place of false positions, of course, it uses indeterminate ones, putting, namely, for the number sought some symbol of indeterminate signification, and thence proceeding with the computation as though the number sought were given and some other number were sought. Indeed, symbols of indeterminate signification may be put for given numbers, and on carrying through the computation there ensue general theorems, wherein we are in turn free to write in place of the determinate symbols any given numbers whatever and in this way give general solution of a problem proposed in any numbers whatever. Ordinary arithmetic and algebra, therefore, mutually complete one another, and taken together constitute a single science which merits being called 'general arithmetic', and on that account I will develop both jointly.

[3][(33)]   All things are numbered, magnitudes measured, and bodies moved, and out of these operations there arise[(34)] arithmetic, geometry and mechanics. These are sciences inasmuch as they teach the truth by means of definitions, axioms and theorems, but arts (skills) insofar as they deliver and exhibit its practice by means of postulates and constructions of problems.

Mechanics holds a place among the mathematical sciences through its axioms and demonstrations of a very great many truths, but not so among the mathematical arts. Its practice is not founded on postulates[(35)] but is purely manual. Geometry and arithmetic are both mathematical arts and sciences.

Geometry, inasmuch as it is a science, considers all magnitudes and figures which can be defined, and discloses their properties and proportions by means of axioms and theorems; insofar as it is an art, it postulates that its practitioner shall carefully have learnt the prime and easiest operations of all, and those most necessary, expedient and useful in the technique of measurement, and by means of postulates and constructions of problems thereby demonstrated it

(34) Newton first wrote 'et scientiæ'—initially 'Artes'—'numerandi mensurandi et movendi dicuntur' (and the sciences (arts) of numbering, measuring and moving are called).

(35) 'et operationibus inde demon[stratis]' (and on operations thereby demonstrated) is deleted in sequel.

Problematum inde demonstratas determinat et exponit mensuras Figurarum et Magnitudinum omnium quæ per Definitiones proponuntur.

Arithmetica quatenus scientia est, leges et proprietates numerorum et proportionum per axiomata et Theoremata ostendit:[36] quatenus Ars est, docet operationes addendi subducendi Multiplicandi Dividendi & extrahendi radices (subintellectis scilicet postulatis necessarijs,) et per has operationes numeros omnes quæsitos et omnes eorum proportiones investigat ac determinat, magnitudines, motus, vires, tempora[37] aliasꝗ omnes quantitates per numeros exponendo, problemata Geometrica Mechanica cæteraꝗ omnia quæ circa rerum quantitates et proportiones proponi possunt considerat ac determinat.

In Arithmetica vulgari sola praxis in usus humanos edoceri solet: specimina Theoriæ in libris septimo octavo & nono *Elementorum* Euclidis ut et in quinto ac decimo, habentur. Omnis enim Proportio est numerus Consequentium in Antecedente et propterea Theoria[38] Proportionalium ad Arithmeticam pertinet. Nam et Euclides Theoriam[38] illā per multiplicationes explicuit ac demonstravit.

Geometria cum Arithmetica affinitatem habet maximam et ejus operationibus libere utitur. Sed operationes Geometricæ simpliciores & breviores esse solent et magis elegantes et eo nomine præferuntur. Ubi tales operationes non occurrunt præstat adhibere Arithmeticam.

Geometræ[39] ex quo sectiones Conorum inventæ sunt, Problemata[40] in tria distinxerunt genera; planum, solidum et lineare: et lineare quidem ad Mechanicam retulerunt, solidum vero ad Geometriā vi hujus Postulati, *Datum conum dato plano secare.* Conus utiꝗ in Theoriam primitivam vi Definitionis suæ receptus fuerat et inde ab authoribus hujus Postulati pro dato habitus est, cum tamen absꝗ postulato describi non potest Geometrice & absꝗ descriptione non datur practice[41] ad Constructionem problematū. Et prætere[a] Datum conum dato plano secare est operatio longe difficillima et in rebus practicis (quibus utiꝗ semper consulendum est) prorsus inutilis. Postulas Tyronem hoc accurate didicisse, cum tamen Geometræ peritissimi hoc facere nunquam soleant. Rectas et Circulos describunt sæpissime, Conos vero secare nunquam discunt.[42]

---

(36) This replaces 'tradit' (delivers) in the manuscript. In a cancelled prior opening sentence Newton had stated more simply that 'Arithmetica quantitates omnes per numeros exponit et determinat' (Arithmetic exhibits and determines all quantities by means of numbers).

(37) 'pondera' (weights) was also initially added here.

(38) Originally 'doctrina(m)' (doctrine).

(39) 'Veteres' (The ancients) was first written.

(40) Initially specified as ones 'mensurandi' (of measuring).

(41) Newton first wrote with unnecessary emphasis 'in praxi Geometrica' (in geometrical practice).

(42) A lone draft sentence on the manuscript's back (f. 176ᵛ) originally affirmed, before being converted substantially into its present form, that 'Postulare sectionem Coni eodem

determines and sets forth measures of all figures and magnitudes which are proposed by the definitions.

Arithmetic, inasmuch as it is a science, discloses[36] the laws and properties of numbers and proportions by means of axioms and theorems: insofar as it is an art, it teaches the operations of adding, subtracting, multiplying, dividing and extracting roots (with, obviously, the necessary postulates understood), and by means of these operations it tracks down and determines all numbers sought and all their proportions, and by expressing magnitudes, motions, forces, times[37] and all other quantities through numbers it considers and determines geometrical, mechanical and all the other problems which can be propounded about the quantities and proportions of things.

In ordinary arithmetic the practice alone is, for its human uses, usually taught with thoroughness; specimens of the theory are had in the seventh, eighth, and ninth books of Euclid's *Elements*, as also in the fifth and tenth. Every ratio is, of course, the number of latter members in the antecedent one, and on that ground the theory[38] of proportionals pertains to arithmetic. And, to be sure, even Euclid explained and demonstrated that theory[38] by means of multiplications.

Geometry has the greatest affinity with arithmetic and makes liberal use of its operations. But geometrical operations are usually simpler and shorter and more elegant, and on that head to be preferred. Where such operations are not met with, then it is of advantage to employ an arithmetical one.

Geometers,[39] ever since conic sections were discovered, have distinguished problems[40] into three kinds: plane, solid and linear. The linear one, indeed, they referred to mechanics; but the solid kind they assigned to geometry by dint of this postulate: *To cut a given cone by a given plane.* The cone had, of course, been received into elementary theory by force of its definition, and thence was had as a given by the authors of this postulate, even though without the postulate it cannot be described geometrically and without a description it is not given as a practical technique[41] for the construction of problems. Moreover, the operation of cutting a given cone by a given plane is by far the most difficult and in practical terms (to which, of course, we need ever to pay regard) an absolutely unusable one. You postulate that the novice has learnt how accurately to do this, even though the most expert geometers are accustomed never to do so: they do describe straight lines and circles exceedingly often, but they never learn how to cut cones.[42]

---

recidit ac si postulares Artificem didicisse: cum tamen Geometræ peritissimi hoc facere nunquam soleant, Rectas et circulos describunt sæpissime, Conos vero nunquam secant' (To postulate the sectioning of a cone comes to the same as if you were to postulate that the practised man has learnt how (to do so); though, however, the most expert geometers are accustomed never to do this, they do describe straight lines and circles, but they never cut cones).

In Theoria Geometriæ fingere licet Conos quo[s]vis vel sectiones quasvis Conicas dari et a circulis quibuscunɋ datis secari et inde ideas quasdam formare quantitatum quæsitarum[,] et ad hujusmodi figmenta vel ideas vel Constructiones ideales sufficiunt Conorum & Sectionum Conicarum Definitiones, absɋ Postulatis.[43] Sunt enim hæ constructiones Theoremata potius quam Problematum solutiones[,] vel ut verius dicam Porismata sunt mediam habentia naturam inter Theoremata et Problemata.[44] Proponi enim possunt in forma Theorematum dicendo, Si talis Conus daretur ac tali plano secaretur et Sectio tali secaretur circulo et ab intersectionibus ducerentur tales rectæ, hæ rectæ forent longitudines quæsitæ. Hujusmodi igitur Constructiones ideales e Geometria minime rejecerim sed inter Porismata potius numerarim quæ ad veras Problematum solutiones investigandas conducere possint. Et sic Constructiones[45] per lineas magis compositas sunt etiam Porismata.

Problemata Geometrica in genera duo distinguere licet, rectilinea quæ solvi possunt per lineas rectas, curvilinea quæ solvi debent per curvas. Rectilinea vero solvuntur per Postulata tria sequentia. *A dato puncto ad datum punctum rectam lineam ducere*, et *producere ad distantiam datæ longitudini æqualem*, et *datum angulum data longitudine subtendere quæ ad datum punctum converget*. Facillima sunt hæc Postulata et usibus practicis aptissima, et præterea ad solutionem Problematum omnium planorum et solidorum sufficiunt. At major fuerit solvendi copia si maneant Euclidis Postulata tria & addatur hoc quartum. *Inter lineas duas quasvis positione datas longitudinem datam ponere quæ ad datum punctum converget*. Problemata quæ per hæc postulata solvi non possunt solventur in idea per intersectiones curvarū, practice verò per Arithmeticam & Geometriam conjunctim.[46]

Scientiarum Mathematicarum methodus est duplex, Synthesis et Analysis. vel Compositio et Resolutio. Arithmetica vulgaris fere tota est in compositione. Tractat enim facillima tantum problemata pergendo a datis ad quæsita per Regulas Proportionalium præterquam in Regula Positionis Falsæ. Arithmetica speciosa quæ et Algebra dici solet loco positionū falsarū utitur positionibus indeterminatis & ab his positionibus computum init per Analysin, unde et Analysis dicitur. Geometria propriam habet Compositionem et Resolutionem,

---

(43) 'describendi figuras' (of describing figures) is cancelled in sequel. Let us leave to others more capable the analysis of the Newtonian intermix of the (Cambridge) Platonist 'ideal' and Cartesian mental 'cogitation'—with an added dash of Lockean empiricism, it may well be—which here briefly manifests itself.

(44) As, of course, Pappus defines a porism in the prolegomenon to the seventh book of his *Mathematical Collection*; see VII: 231, note (4). There seems no classical authority for Newton's present interpretation that a porism is any more 'ideal' a construction than that of any particular problem to which it may, as a theoretical rider, be applied. A dozen years earlier he had held a far different and more practical notion: 'Porisma est Propositio qua ex conditionibus Problematis datum aliquod ad ejus resolutionem utile colligimus' he wrote in his 'Inventio Porismatum' (VII: 230).

In geometrical theory it is allowable to imagine that any cones or any conic sections you wish are given and that these are cut by any given circles whatever, and we are thence permitted to form certain ideas of the quantities sought; and in the case of imaginative notions or ideas or ideal constructions of this sort the definitions of cones and conic sections suffice, without postulates.[43] These constructions are, to be sure, theorems rather than the solutions of problems, or, to speak more truthfully, they are porisms having a middle nature between theorems and problems.[44] They can, to be specific, be proposed in the form of theorems by saying: If such and such a cone were to be given and be cut by such and such a plane, and the section were to be cut by such and such a circle, and from the intersections such and such straight lines were to be drawn, then these lines would be the lengths sought. Ideal constructions of this sort, therefore, I would not in the least have rejected from geometry, but would rather have numbered them among porisms in that they can contribute to the discovery of the true solutions of problems. And in this fashion[45] constructions by means of lines more elaborate are also porisms.

It is permissible to distinguish geometrical problems into two kinds: rectilinear ones which can be solved by straight lines, and curvilinear ones which have to be solved by curves. Rectilinear ones, however, are solved by the three following postulates: *To draw a straight line from a given point to a given point*; *To extend it to a distance equal to a given length*; and *To subtend a given angle by a given length which shall be directed through a given point*. These postulates are the easiest ones and the most fit to practical uses, and in addition they suffice for the solution of all plane and solid problems. But there will be a greater resource for solving if Euclid's three postulates remain and there be added this fourth one: *Between any two lines given in position to set a given length which shall pass through a given point*. Problems which cannot be solved by way of these postulates will be solved in notion through the intersections of curves, but in practice through arithmetic and geometry jointly.[46]

The method of the mathematical sciences is dual: synthesis and analysis, or composition and resolution. Ordinary arithmetic is almost wholly a matter of composition: for it treats merely the easiest problems by proceeding from givens to things sought by means of rules of proportionals, except in the case of the rule of false position. Specious arithmetic, which is usually also called algebra, in place of false positions uses indeterminate ones, and pursues the computation from these positings by means of analysis—whence it also is called analysis. Geometry has its own (mode of) composition and resolution, one which is

---

(45) 'omnes' (all) is deleted here.

(46) Initially, in a cancelled prior equivalent sentence, 'per Arithmeticam' (through arithmetic) *tout court*. On Newton's conchoidal addition to Euclid's three postulates, again compare VII: 382–8.

quæ computis[47] Arithmeticis simplicior esse solet & elegantior sed difficilius addiscitur. Verbis enim enunciatur sine computo, & quicquid verbis solis enunciari non potest, id [in] Geometria utile quidem esse potest, sed elegans non est. Tota Geometriæ vis et perfectio in operationum simplicitate consistit.

Arithmetica vulgaris utitur figuris significatione determinatis[,] Algebra pro numeris ponit etiam species vel symbola significationis indeterminatæ ut dictum est, et inde Arithmetica symbolica vel speciosa dicitur. Conclusiones autem sic prodeunt generales in quibus utiꝗ pro speciebus substituendo numeros quosvis datos[48] Problema in quibuscunꝗ numeris propositum generaliter solvatur.

Igitur Arithmetica vulgaris et Algebra[49] ut Synthesis & Analysis conjungi debent et ex ambabus Arithmeti[c]a una generalior componi et propterea utramꝗ conjunctim explicabo. Euclides in septimo *Elementorum* Additionem, Subtractionem, Multiplicationem, Divisionem et extractiones radicum numerorū possibiles esse postulavit, id est Artificem hæc ex Arithmetica vulgari didicisse & præstare posse. Nos praxin totam docebimus.[50]

---

[4][51]   Veteres Problemata omnia plana solverunt per Prop. 58, 59, 60, 84, 85, 86, 87 *Datorum* Euclidis, ut Gregorio olim significavi,[52] & omnia solida solverunt quærendo puncta quædam per quæ locus solidus transiret & locum per puncta illa describendo. Norunt enim Locum solidum per puncta quinꝗ describere ut ex Prop. [13] Lib. [VIII] Pappi[53] manifestum est: Apollonius in Libris suis *de*

---

(47) This replaces the non-classical (post-Augustan) equivalent 'computationibus' in the manuscript.

(48) 'figuras quascunꝗ datas' (and given figures whatever) was first written: why (if this was not a mere slip of Newton's pen) is not clear.

(49) Initially 'Arithmetica vulgaris et speciosa' (common and 'specious' arithmetic).

(50) These last two sentences are a late insertion in the manuscript (f. 175ᵛ), where they replace a cancelled last clause of the paragraph's opening one which reads 'et missis demonstrationibus quæ satis faciles esse solent computandi praxin docebo' (and, setting aside the demonstrations which are usually easy enough, I shall now teach the practice of computing). The 'single more general arithmetic', conjoining ordinary arithmetic and algebraic analysis, which Newton here heralds is of course the *arithmetica universalis* whose methods and variety of applications, geometrical and mathematically scientific, he had a quarter of a century earlier sought to expound from his Lucasian professorial rostrum at Cambridge (see v: 54–490). Here he proceeded no further in the elaboration of its practice than roughly to pen (on the ensuing ff. 175ᵛ/176ʳ) a confused set of introductory axioms, postulates and definitions regarding numbers integral, fractional and commensurable—these derive in the main from related *axiomata, postulata* and *definitiones* in Isaac Barrow's Latin edition of Euclid's *Elements*—and the elementary arithmetical-*cum*-algebraical operations of addition, subtraction, multiplication and division performable upon them. These listed *principia arithmetica* are reproduced in full in Appendix 1.5.

(51) Add. 3963.14: 156ʳ/156ᵛ; a first, cancelled draft of this piece on the same manuscript sheet (*ibid.*: 156ʳ) is reproduced in Appendix 1.6 following. Whether or not it was in fact written in conjunction with [1]–[3] preceding, this recasting by Newton of his survey a dozen

customarily simpler and more elegant than arithmetical calculations, but is mastered with more difficulty: for it is enunciated in words and without computation, and whatever cannot be enunciated in words alone, while it can indeed be useful in geometry, is not elegant. The whole strength and perfection of geometry consists in the simplicity of its operations.

Ordinary arithmetic employs figures determined in their signification. Algebra puts in place of numbers also 'species' (variables) or symbols of indeterminate signification, as they are called, and therefrom is named symbolic or specious arithmetic. The conclusions resulting in this way are general ones since, of course, on substituting in place of the variables in them any given numbers you wish[48] a problem proposed in any numbers whatever shall be generally solved.

In consequence, ordinary arithmetic and algebra[49] ought to be conjoined as synthesis and analysis, and a single, more general arithmetic compounded from both; and on that ground I shall develop one and the other together jointly. Euclid in the seventh (book) of the *Elements* postulated that addition, subtraction, multiplication, division and the extractions of roots of numbers are possible; that is, that the practitioner has learnt these from ordinary arithmetic and can deploy them. We shall now explain the entirety of their practice.[50]

---

[4][51]   The ancients solved all 'plane' problems by means of Propositions 58, 59, 60, 84, 85, 86, 87 of Euclid's *Data*, as I some time ago pointed out to Gregory,[52] and solved all 'solid' ones by ascertaining certain points through which a 'solid' locus (conic) should pass and then describing the locus through those points. For of course they knew how to describe a solid locus through five points, as is manifest from Proposition 13 of Book VIII of Pappus;[53] while Apollonius in his books *On determinate section, On cutting off a ratio* and *On cutting*

---

years earlier of the 'Analysis Geometrica' of the classical Greek mathematicians (VII: 200–12) serves to complement his observations there of the efficacy of implementing the traditional Euclidean 'plane' postulates so as to permit the construction of 'solid' problems by conchoidal neusis (compare [2]: note (25) above), and so we may not unfittingly set it here as an *addendum*. A date of composition in or shortly after 1704 is strongly suggested by Newton's handwriting style in the manuscript, and this is virtually guaranteed by the fact that his first use of the sheet was to draft (at the head of f. 156ʳ) the last paragraph of his 'Enumeratio Linearum Tertii Ordinis' shortly before he put it to press in the early winter of 1703/4 (see VII: 644, note (134)). A stray computation on the sheet's verso (at the foot of f. 156ᵛ) of the defining equation of a circle in oblique Cartesian coordinates is, for want of a better place, reproduced in Appendix 2 below.

(52) During David Gregory's visit to Cambridge in early May 1694, when Newton showed him *inter alia* the original draft (see VII: 220–6) of his 'Analysis Geometrica'; compare VII: 222, note (10).

(53) Compare IV: 276–8. A parallel remark to this effect is made in his parent 'Analysis Geometrica' (see VII: 206, and compare *ibid*.: note (26)).

sectione determinata, de section[e] rationis & *de sectione spatij* puncta invenire docuit per quæ locus describendus esset.

Nempe si rectæ quatuor a puncto communi ad alias quatuor positione datas rectas in datis ∠ is (54) ducendæ sunt et inveniendum esset tale punctum commune ut rectangulum sub duabus ductis sit ad rectangulum sub alijs duabus ductis in ratione data: invenire liceret ejusmodi puncta quotcunꝗ per Librum *de sectione determinata*.(55)

Porro si rectæ duæ a puncto communi ducendæ essent una ad datum punctum & altera in dato angulo ad rectam tertiam positione datam, ita ut posterior ductarum datam habeat rationem ad rectā quartam positione datam & ad datum punctum et priorem ductarum hinc inde termina[tam], punctum illud locabitur in sectione conica(56) & ejusmodi puncta quotcunꝗ inveniri possunt per Librum *de sectione Rationis*.(57)

Item si posterior ductarum cum recta illa quarta(58) datum spatium comprehendat, punctum locabitur in Sect[ione] con[ica](59) et ejusmodi puncta

---

(54) 'angulis'.

(55) Newton has deleted the ensuing explanatory sentence: 'Talia autem puncta quatuor habentur in concursibus rectarum positione datarum, & inveniendum est quintum per Librum *de sectione determinata*, deinde per puncta illa quinꝗ describendus est Locus solidus' (Four such points, however, are had at the meets of the four lines given in position, and the fifth needs to be found by the book *On determinate section*, and then the 'solid' (conic) locus described through those five points). The Apollonian *sectio determinata* cut out by the 4-line conic locus will, of course, be pairs of points which are (in Desargues' terminology, though Newton never read the latter's rare *Brouillon Project* of 1639 where this *terminus technicus* was introduced) in involution. (See VII: 190, note (26); and compare *ibid.*: 323, note (114).) This connection of the Euclidean *locus ad quatuor lineas* with Apollonius' (now perished) book is yet more fully spelled out by Newton in the like passage in his earlier 'Analysis Geometrica' (see VII: 206–8); but, attractive and plausible though the tie may seem to modern eyes to be, we must again insist (as in VII: 208, note (28)) that no textual evidence survives from antiquity which begins to substantiate it, and the juxtaposition must remain untestable historical speculation.

(56) An involved and cumbrous enunciation which ill-advisedly forgoes the clarity attained in the parent draft (see Appendix 1.6) by keying its statement to an accompanying figure. In terms of the latter, Newton supposes that from the point D two straight lines DB and DC are let fall, the first passing through the fixed point P and meeting the given straight line AB in B, the second inclined to the second given straight line EC at the fixed angle ECD: if (says Newton) the length CD bears a given ratio to the segment AB (terminated at the given endpoint A), then the locus of D so defined will be a conic—a hyperbola, in fact, whose asymptotes are parallel to the given lines AB and EC. This is readily established by setting the meet E of these given lines to be the origin of the oblique system of Cartesian coordinates in which

$EC = x$ and $CD = y$ determine the locus-point $D(x, y)$; for, on taking the two given points

off a space taught how to find points through which the locus needed to be described.

Specifically, if four straight lines be drawn from a common point at given angles to four other straight lines given in position, and it were needed to find the (locus of the) common point such that the rectangle contained by two of the lines drawn shall be to the rectangle contained by the two others drawn in a given ratio: it would be permissible to find any number of points of this type by means of the book *On determinate section*.[55]

Moreover, if two straight lines had needed to be drawn from a common point, one to a given point and the other in a given angle to a third straight line given in position, such that the latter of the lines drawn shall have a given ratio to a fourth straight line given in position and terminated on each end at a given point and the first of the lines drawn, then that point will be located in a conic[56] and any number of points of the sort can be found by means of the book *On cutting off a ratio*.[57]

Likewise, if the latter of the drawn lines shall comprise with the fourth straight line[58] a given 'space' (product), then the point will be placed in a conic[59] and any number of points of the sort will be found by means of the book

---

to be $P(a, b)$ and $A(c, d)$, the line PD as it rotates round P will meet AB ($y = dx/c$) in $B(X, dX/c)$ such that $(X-a)/(x-a) = (dX/c-b)/(y-b)$ or $X = c(bx-ay)/(dx-cy+bc-ad)$; whence, on putting $EA = \sqrt{(c^2-2cd.\cos \widehat{ECD}+d^2)} = e$ for simplicity, there ensues

$$AB = e((b-d) x - (a-c) y - bc+ad)/(dx-cy+bc-ad).$$

The condition $AB/CD = \lambda$, constant, accordingly decrees the locus (D) to be a hyperbola with its asymptotes parallel to $y = 0$ and to $dx-cy = 0$.

(57) For where, in the figure of the previous note, CD is an arbitrary transversal, this will meet the hyperbolic locus (D) in points D determined (for fixed EC) by the *sectio rationis* $AB/CD = \lambda$, constant. Since the ensuing equation which thereby fixes $CD = y$ is quadratic in y (again see the previous note), two pairs of points D will be defined in each transversal—and so Apollonius geometrically constructs them in their several cases in his work. (Compare T. L. Heath, *A History of Greek Mathematics*, **2** (Oxford, 1921): 176–9. In September 1706, we would remind, Edmond Halley's *princeps* Latin rendering, *Apollonii Pergæi de Sectione Rationis Libri Duo ex Arabico MS!º Latine versi….Opus Analyseos Geometricæ studiosis apprime Utile*, had appeared only a few weeks earlier 'E Theatro Sheldoniano' at Oxford.) Newton's own proof that the locus (D) is a conic would probably proceed by reversing this argument, appealing to his favourite criterion that an algebraic curve of n-th degree can be met by a straight line in n points; because D can pass to infinity both when $CD = \infty$ (and so $AB = \infty$) and also when C is at infinity in EC, the conic (D) is again proved to be a hyperbola with its asymptotes parallel to AB and EC respectively. In the manuscript he has deleted a final clause 'deinde Sectio Conica per eorum quinqꝫ describi' (and then the conic described through five of them), subsuming its sense into the parallel second sentence of the following paragraph.

(58) Originally 'recta altera abscissa' (the other straight line cut off).

(59) In the terms of note (56) preceding, the variant restriction $AB \times CD = \mu$, constant, evidently again determines the locus (D) to be a hyperbola, now with asymptotes parallel to AP (of Cartesian equation $(b-d) x - (a-c) y = bc-ad$) and to EC ($y = 0$).

quotcunꝗ invenientur per Librum *de sectione spatij*.[60] Habitis autem punctis quinꝗ Conica Sectio per eadem describi debet.

Hac methodo Veteres loca solida ad Constructionem Problematum solidorum composuerunt, et nos eorum vestigijs insistendo construximus Problemata omnia quærendo puncta quædam curvarum & describendo Curvas per puncta inventa.[61] Quæ methodus ut antiquissima sic omnium simplicissima[62] & optima videtur.

––––––––––––––––––––––––

(60) Here, much as in note (57), where *CD* is an arbitrary transversal, this will meet the locus (*D*) in points *D* now determined by the *sectio spatii* $AB \times CD = \mu$, constant. Since the Cartesian equation of (*D*) is again quadratic in *y*, two pairs of points *D* will (for fixed $EC = x$) once more be defined in each such transversal—and so Apollonius, too, doubtless gave full geometrical construction of them in their component cases. (Compare Heath, *Greek mathematics*, **2**: 179–80; and see Halley's reconstruction of Apollonius' lost work *De Sectione Spatii* at the close of his *Apollonii Pergæi de Sectione Rationis Libri Duo*... (see note (57) above): 139–68.) Again we may assume that Newton's own preferred demonstration would proceed in the reverse sequence, departing from the twin points cut out by the conic locus (*D*) in each arbitrary transversal by the *sectio spatii* $AB \times CD = \mu$ to show that, since *D* may pass to infinity both when $CD = \infty$ (and hence *B* comes to coincide with *A*) and also when *C* is at infinity in *EC*, this is a hyperbola having asymptotes parallel to *AP* and *EC* respectively.

(61) In a cancelled preceding sentence Newton initially wrote in more explicit manner: 'Nos Veterum vestigijs insistendo docuimus descriptionem Conicarum sectionum per data quinꝗ & descriptionem curvarum secundi generis per puncta septem data quorum unum sit punctum duplex, & quomodo per curvas sic descriptas Problemata omnia non plusquam

On cutting off a space.[60] Once five points are obtained, however, the conic ought to be described through these.

By this method the ancients composed 'solid' loci in order to attain the construction of 'solid' problems; and we, following in their track, have constructed all problems by ascertaining particular points of curves and describing the curves through the points so found.[61] This method would seem both the most ancient and also the simplest[62] and best of all.

---

quadrato-quadrato cubica solvenda sunt' (We, by following in the footsteps of the ancients, have taught the description of conics through five given points and that of curves of second kind (cubics) through seven given points, one of which shall be a double point, and how by means of curves thus described all problems of not more than square-square-cube (seventh) grade are to be solved). We need only quickly remind that he here refers to the last three sections, §§ XXXII–XXXIV of his newly printed 'Enumeratio Linearum Tertij Ordinis' (see VII: 638–44) where he gave public exposition of his favourite organic description of curves by rotating fixed-angles, and went on (as he here says) to apply the general conic and cubic with double point thus generated—and indeed such other cubics as the cubic parabola and simple hyperbolism of a parabola which are not restricted to possessing a (real) *punctum duplex*—to construct equations of pertinent degree. Since a cubic with double point can be made to pass only through seven distinct points (the *punctum duplex* counting for two coincident ones), it will of course construct only problems which have no more than seven separate solutions.

(62) 'commodissima' (most convenient) is rightly here deleted by Newton. The adjective 'simple' is not, as Newton well knows (compare v: 424), uniquely defined in mathematical context, and he begs the question of its present validity.

APPENDIX 1.
PRIOR VERSIONS OF NEWTON'S REFLECTIONS
UPON THE NATURE OF ARITHMETIC AND GEOMETRY,
AND ON THE RESOLUTION OF PROBLEMS.[1]

[1][2] Arithmetica circa numeros' et res numerabiles, Geometria circa mensuras et res mensurabiles, Mechanica circa vires et motus locales versātur.[3] In Arithmeti[c]a per numeros exponimus,[4] dein quæstiones circa res illas per operationes Arithmeticas solvimus, id est per numerorum additionem sub-ductionem multiplicationem divisionem et extractiones radicum. Et similiter in Geometria res omnes per lineas vel areas vel contenta solida exponere licet & quæstiones circa res illas per operationes Geometricas ducendi lineas et con-struendi figuras solvere. Numeri per quos species ejusdem generis vel individua ejusdem speciei numeramus semper sunt integri. Numeri vero per quos quanti-tatem rei vel qualitatis gradum designamus (id est antecedentes proportionū quantitatis vel gradus ad unum) possunt esse vel partes integrorum vel integris incommensurabiles.[5] Et hi omnes numeri per characteres determinatæ signifi-cationis notari solent et in Arithmetica vulgari tractantur. Numeri vero qui vel indeterminati sunt vel ignorantur vel ignorari finguntur per species indeter-minatæ significationis notari solent et in Algebra tractari. Arithmetica vulgaris

---

(1) As ever, these preliminary and transitional drafts of his more finished texts in §2 preceding are here reproduced, with notice in the main but of the further verbal variants in their manuscript, because of their several individual points of interest.

(2) Add. 3963.15: 171ᵛ. This first version of [1] in §2 above is, in a rare intrusion of con-temporary event into the 'timeless' pages of Newton's private papers, penned on the blank back of a single-sheet 'Postscript to the/Post-Boy/London, Sept. 26 [1706]'— 'Printed for Abel Roper at the Black-Boy in Fleetstreet'—giving 'Advice' of the arrival at Spithead from Spain of 'her Majesty's Ship the Fowey Pink, a Fifth Rate, of thirty two Guns, ...sent Express from Sir John Leake, whose Squadron he left in Altea-Bay, homeward-bound, [with the report] That the Castle of Alicant[e] surrendred to her Majesty's Forces [at the beginning of August], fourteen days after the taking of the Town'. (On Leake see *The Dictionary of National Biography* (2) **13** (London, ₁1908): 758–61, especially 760. The related 'Numb. 1775' of *The Post-Boy* to which this supplement was appended had appeared the previous day.)

(3) In a tentative revise of this opening sentence entered at the other end of the news-sheet's verso Newton affirmed in amplification of the two latter phrases: 'In Mechanica, corpora movemus & figuras describimus: In Geometria figuras definimus & ['lineas mensu-ramus' and then 'proportiones enarramus' are deleted] per simplic[iss]imas operationes ex Mechanica postulatas mensuramus et nonnunquam construimus, et figurarum constructiones proprietates & proportiones demonstramus'.

(4) Thus shortened from 'In Arithmetica numeramus et res numeratas per numeros exponimus', after which Newton first went on to add 'sive res illæ sint species ejusdem generis sive individua ejusdem speciei sive partes æquales ejusdem quantitatis'. Most of the latter is re-woven into the text a few lines below.

(5) Initially just 'vel fracti vel surdi'.

simplicior est quam Algebra et circa quæ[s]tiones faciliores versari solet et per
methodum compositionis progredi: Algebra est Arithmeticæ gradus perfectior
& ab ignotis quantitatibus ad notas plerumcʒ regreditur ut ad Theorema vel
Corollarium quocuncʒ demum modo perveniatur quod ad solutionem quæst-
ionis conducit.[6]

---

[2][7]   Arithmetica circa rerum omnium Numeros, Geometria circa Magni-
tudi[n]um mensuras, Mechanica circa corporum vires et motus versantur.
Arithmetica[8] circa numeros vel determinatos vel indeterminate spectatos[8]
versatur. Prioris generis est Arithmetica vulgaris quæ uticʒ quæstiones faciliores
tantum tractat idcʒ per synthesin.[9] Posterioris generis est Arithmetica in
speciebus quæ et Algebra et Analysis dici solet. In Arithmetica vulgari desideratur methodus analytica[10] & vices ejus per Regulam Falsæ Assumptionis
aliquatenus suppletur.[11] Ut Arithmetica perficiatur et universalis reddatur,
debet methodus utracʒ adhiberi,[12] et propterea omissis assumptionibus falsis
et earum loco assumptionibus indeterminatis adhibitis Arithmeticam utramcʒ
conjunctim explicabo. Sic enim conclusiones quæ in Arithmetica vulgari
particulares esse solent, generales reddentur & in Theoremata vertentur.[13]

---

[3][14]   [15]Numerantur omnes quantitates, mensurantur magnitudines &
moventur corpora, et artes numerandi mensurandi et movendi dicuntur
Arithmetica Geometria et Mechanica. Harum scientiarum principia sunt
Definitiones Axiomata et Postulata. *Definitiones* sunt verborum quoad significationem, *Axiomata* sunt veritates primæ[16] & *Postulata* sunt operationum

---

(6) Newton has deleted a final sentence 'Unde Algebra dici solet Analysis in speciebus'.

(7) Add. 3963.15: 174ᵛ, a tentative first revision, quickly abandoned, of the text in §2.1
preceding (compare §2.2: note (18)).

(8) 'vulgò' and 'tantum' respectively were here added in a cancelled preceding version of
the present sentence.

(9) The equivalent 'per compositionem' is replaced.

(10) 'seu Regula Assumptionis indeterminatæ' is deleted in sequel.

(11) Newton has cancelled a following sentence where he went on to assert 'Melius est
assumptiones indeterminatas adhibere quam falsas'.

(12) Initially 'conjungi', after which Newton passed to affirm '& pro falsis assumptionibus
indeterminatæ assumptiones'—he first began to specify 'Algebra quæ nihil aliud est quam
Regula a[ssumptionis indeterminatæ]'—'in[troduci]'.

(13) Newton here breaks off to start yet again in [3] immediately following.

(14) Add. 3963.15: 173ʳ, a first state of the opening of §2.2 above (which follows after it on
the same manuscript page).

(15) Much as in §2.1 preceding and [1]/[2] above, Newton initially began by stating:
'Arithmetica circa numeros, Geometria circa magnitudines & figuras, Mechanica circa vires
et motus versantur.'

(16) 'id est eæ quæ a prioribus derivari nequeunt' is deleted in sequel in the manuscript.

facillimarum & maxime utilium.[17] Et hæc Principia debent esse quam paucissima: ideoqȝ definimus tantum nomina rerum quæ in arte tractantur, assumimus tantum axiomata quæ ex alijs veritatibus derivari nequeunt et postulamus tantum quod Artifex accurate didicerit operationes facillimas & ex his colligimus Theoremata et constructiones Problematum.[18] Sic in Mechanica postulare licet 1 *Datum corpus dato cono*[19] *secare.* 2 *Datum Corpus circa datum axem data acie formare.* 3 *Datum Corpus data vi secundum datam rectam movere.* In Geometria, ad Euclidis Postulata addere licet[20] *Datam longitudinem inter datas duas lineas ponere quæ ad datum punctum converget.* Hoc ab Euclidis Postulatis derivari nequit, æque facilis est, utilis in rebus practicis[,] ad construction[e]m omnium solidorum problematum sufficit,[21] et per operationem faciliorem construi non potest, ideoqȝ postulari debet. Eo autem postulato et concesso problemata omnia solida ad plana reducentur. Datum Conum dato plano secare problema difficilius est & in rebus practicis nulli usui inservit ideoqȝ postulari non debet.[22]

---

[4][23]   Arithmetica Geometria & Mechanica scientiæ sunt quatenus per Definitiones Axiomata et Theoremata veritatem docent, artes vero quatenus per Postulata Problemata et operationes manuales praxin tradunt. Et Mechanica quidem inter scientias Mathematicas per demonstrationes veritatum plurimarum locum obtinuit, inter artes Mathematicas non item.[24] Ejus praxis non traditur accurate per Postulata et Demonstrationes[25] sed pure manualis est.[26] Geometria per Definitiones Axiomata et Theoremata contemplatur proprietates et mensuras figurarum et magnitudinum omĩum, planarum et solidarum, rectilinearum et curvilinearum, et sic in scientiam amplissimam et nobilissimam evasit; et præterea praxin mensurandi et

---

(17) Newton first added: '*Theoremata* sunt veritatum consequentium quæ a primis demonstrantur, & *Problemata* sunt operationum quæ a facillimis demonstrando consequuntur.'

(18) Compare vii: 290.

(19) 'data superficie' is replaced in the manuscript.

(20) Originally 'post Euclidis postulata, facillimum est'.

(21) See §2.2: note (25) above.

(22) Compare vii: 292–8.

(23) Add. 3963.15: 173ᵛ/174ʳ, two transitional drafts bridging the texts of §2.2 and §2.3 preceding; compare §2.3: note (33). We reproduce the text of the latter draft, citing significant variations in the former in related footnotes.

(24) Tentatively altered in the manuscript (f. 174ʳ) to be 'nondum numeratur', and then changed back.

(25) 'inde petitas' is deleted.

(26) Newton first added 'et inaccurata'. His prior version of the two preceding sentences reads (f. 173ᵛ): 'Et Mechanica quidem per Definitiones Axiomata et Theoremata in scientiam Mathematicam aliqua ex parte evasit, sed praxin per Postulata nondum cœpit ostendere, ideoqȝ ars est tantum manualis.'

mensuras delineandi[27] per Axiomata tria Euclidea tradidit & eatenus inter
artes mathematicas numerari meretur. Arithmetica vulgaris magis Ars est
quàm Scientia. Tradi enim solet per operationes practicas addendi subducendi
multiplicandi dividendi et extrahendi radices,[28] omissis veritatum demonstrationibus in quibus praxis fundatur.

In Elementis Euclidis multa tractantur quæ ad Arithmeticam pertinent.
Nam liber septimus octavus et nonus proprietates numerorum continent[,]
& quæ in quinto ac decimo de proportionibus linearum docentur rectius in
Arithmetica de proportionibus omnium quantitatum doceri debuerit. Quo-
niam vero Arithmetici elementa scientiæ suæ minimè demonstraverant, posuit
Euclides hæc in Elementis Geometriæ.

Numerus rerum similium & non similium (quales esse solent individua
ejusdem speciei vel species ejusdem generis[29]) semper est integer: numerus
rerum æqualium continere potest partem aliquam unius ex æqualibus et inde
oriuntur numeri fracti et surdi et alij omnes qui sunt unitatibus incommensurabiles.

Numerus æqualium designat proportionem omnium ad unum, et omnis
proportio antece[de]ntis ad consequens est numerus consequentium in antecedente. Est enim numerus æqualium ad unum ut numerus ad unitatem; et in
omni proportione consequens spectandum est ut unum aliquod cognitum,[30]
et antecedens ut quantitas quæ innotescit ex proportione ejus ad consequens.[31]
Quantitas autem quælibet pro uno haberi potest et inde oritur proportionalitas
ad consequentia diversa. Antecedentia utiꝗ sunt ad consequentia sua in eadem
proportione ubi continent eundem numerum suorum consequentium. In omni
proportione numerus consequentium in antecedente est antecedens respectu
consequentis hoc est proportio antecedentis ad consequens, ideoꝗ numeri sunt
proportionum indices.

---

(27) Originally 'et sub mensuris ponendi'.

(28) Newton initially went on to adjoin 'et ad determinationem problematum circa
quantitatum proportiones aptissima est', and then more fulsomely concluded by writing 'et
cum ad omnes ejus operationes peragendas sufficiat postulare additionem et subductionem
unitatis, Postulata ob simplicitatem negliguntur. Patet quidem per Euclid[i]s libros [inf]ra
memoratos quod Elementa Arithmeticæ per Definitiones Axiomata et Theoremata tradi
possent, sed in Arithmetica vulgari quæ utiꝗ in usus humanos componitur omitti solet
Theoria numerorum et proportionum, & sola praxis computandi in usus humanos edoceri'.

(29) Did Newton have his classification of the general cubic curve into its four canonical
forms, and their sub-classification into component *genera*, each divided into individual subsidiary *species*, here especially in mind?

(30) '...locum habet unius' was first more opaquely written.

(31) Newton afterwards (at the head of f. 174ʳ) thought to amend the two preceding
sentences to read more shortly 'Numerus rerum æqualium est ad rem unam ex æqualibus ut
numerus ille ad unitatem. Et in omni proportione numeri sunt proportiones rerum ad unum';
but he swiftly changed his mind.

In Arithmetica exponendo magnitudines per numeros Problemata omnia Geometrica tractantur[,] sed praxis Geometrica solet esse simplicior & elegantior, ideoqg præfertur. Ubi hoc non obvenerit præstat adhibere Arithmeticam.[32]

Geometræ ex quo Sectiones Conorum adinventæ sunt Problemata mensurandi in tria genera distribuerunt, plana vocantes quæ in plano per rectas et circulos[,] solida quæ per sectionem Coni et linearia quæ per descriptiones linearum nondum cognitas solvuntur. [33]Et linearia quidem retulerunt ad Mechanicam[,] solida vero ad Geometriam per hoc postulatum: *Datum Conum dato plano secare.* Sed Conus non datur in Geometria per postulatum aliquod, et si daretur sectio ejus adeo difficilis est ut longe facilius sit problemata solida per Arithmeticam solvere. Inutile est igitur hoc Postulatū, et ejus loco præstat operationem aliquam facillimam & expeditissimam [et] ad res practicas maxime accommodatam postulare quo tædium computationum Arithmeticarum in solutione Problematum solidorum vitemus.

Problemata igitur in duo distinxerim genera, Rectilinea dixerim quæ per hæcce quatuor Postulata solvuntur: *A dato puncto ad datum punctum rectam lineam ducere,* et *continuo producere,* et *in ea ad datum punctum longitudinem datam ponere,* et *angulum datum data longitudine subtendere quæ ad datum punctum converget.* His Postulatis Problemata omnia plana et solida solvuntur.[34] Sed major erit solvendi copia si ad Postulata tria Euclidis addatur hoc quartum: *Inter datas duas lineas datam ponere longitudinem quæ ad datum punctum converget.*

Ut veteres Sphæram Cilyndrum et Conum vi Definitionum sine Postulatis in

---

(32) In prior draft (on f. 173ᵛ) Newton adjoined a second antithesis: 'Problemata in Arithmetica ad æquationes deduci solent & pro numero dimensionum æquationis in gradus vel genera distingui. In Geometria vero rectius deducuntur per Analysin Geometricā ad operationes describendi lineas per Postulata.'

(33) Newton's prior version of the sequel reads (f. 173ᵛ): 'Rectius duo tantum statuenda sunt genera, Geometricum quod per Postulata Geometrica et lineare vel Mechanicum quod per operationes Mechanicas ob praxin difficilem et inutilem nondum postulatas construitur. Arithmetica genus utrumqg comprehendit reducendo Problemata ad æquationes. Siquando Constructiones Mechanicæ obvenerint quarum simplicitate et elegantia tædium computationum Arithmeticarum vitemus, licebit has in Geometriam per Postulatum novum introducere. Si tales non obvenerint præstabit Arithmeticam adhibere, quam operationibus perplexis & ineligantibus Geometriam corrumpere.' This he then attenuated and augmented to state, much as in the following paragraph of his subsequent revise: 'Rectius forte duo tantum statuissent genera, rectilineum et curvilineum si modo ad Euclidis postulata tria hoc quartum addidissent *Datum angulum data longitudine subtendere ad datum punctum convergente.* Vel *Inter datas duas lineas datam longitudinem ponere quæ ad datum punctum converget.*'

(34) See §2.2: note (24) above. Newton began to add in complement at this point that 'Curvilinea [*sc.* problemata] quarum solutio manualis descriptionem curvarum linearum requirit [? solventur per inventionem punctorum reducendo eorum conditiones ad æquationes]', but there broke off. In redraft in §2.3 preceding he was subsequently to state that 'Problemata quæ per hæc postulata solvi non possunt solventur in idea per intersectiones curvarum, practice verò per Arithmeticam & Geometriam conjunctim'.

Geometriam receperunt, sic lineæ omnes curvæ & figuræ curvis lineis et curvis superficiebus terminatæ vi Definiti[on]um suarum sine descriptione postulata in Geometri[am] recipi possunt.

[5][35] Arithmetica non solum numeros abstractos sed etiam res omnes numeratos et rerum proportiones considerat, et ejus Elementa per hujusmodi Principia tradi possunt.

Def [*initiones*].[36]

1 Unum est res omnis absolute et per se spectata.

2 Numeri sunt rerum quantitates comparativæ respectu unius seu proportiones rerum ad unum, & unitas est proportio unius ad unum.

3[37] Numerus integer est qui ex unitatibus integris constat.

4 Numeri partes aliquotæ sunt quæ inter se æqualia sunt et simul sumptæ numerum componunt.

5. Numerus fractus est qui ex unitatis partibus aliquotis constat numerum integrum non componentibus.

6 Mensura numeri est ejus pars quælibet aliquota.

7 Commensurab[i]les sunt numeri quos eadem pars metitur.[38]

8 Incommensurabiles qui nullam habent communem mensuram.

9[39] Ineffabiles qui sunt numeris integris incommensurab[i]les.

[10][40] Antecedens rationis est res quælibet[41] quæ cum re alia quoad quantitatem comparatur.

---

(35) Add. 3963.15: 175ᵛ/176ʳ. In this cancelled final portion of the text whose introductory part (*ibid.*: ff. 175ʳ/175ᵛ) is set out in §2.3 preceding, Newton—firmly modelling himself on the equivalent *definitiones*, *axiomata* and *postulata* in Isaac Barrow's edition of Euclid's *Elements* (*Euclidis Elementorum Libri XV breviter demonstrati, Operâ Is. Barrow, Cantabrigiensis*..., Cambridge, ₁1655)—Newton roughly lists the definitions, axioms and postulates comprising the principles on which he proposes to found the practice of his *arithmetica una generalior* conjoining arithmetic and algebraic analysis (see §2.3: note (50) above).

(36) In the manuscript Definitions 1 and 2 are entered below 3, 4 and 5 following, but we here reorder them in accord with Newton's numbered sequence.

(37) Newton initially went on to enunciate: 'Numerus unorum est proportio totius ad un[um].'

(38) Originally, in parallel with the next definition of its converse, 'qui communem habent mensuram'. Compare Euclid, *Elements* X, Def. 1: 'Commensurabiles magnitudines dicuntur, quas eadem mensura metitur' (Barrow, *Euclidis Elementorum*: 190); the ensuing Def. II announces that 'Incommensurabiles...sunt, quorum nullam communem mensuram contingit reperi'.

(39) Originally 'Numerus surdus est qui nec integer est nec fractus sed unitati incommensurabilis', and then 'Numerus ineffabilis est qui cum unitate nullam habet communem mensuram'. *Numeri ineffabiles* are of course the irrational ἄλογοι of *Elements* X, Def. IV.

(40) We insert this number in line with the preceding. Newton would evidently have numbered all his succeeding *definitiones*, *axiomata* and *postulata* in similar ordinal sequence had he come to polish his text rather than abruptly cancel it.

(41) Initially 'est quantitas'.

Consequens rationis est res altera quacum Antecedens comparatur.

Ratio Antecedentis ad Consequens est numerus Co[n]sequentium in Antecedente.[42] Hæc est ratio directa.

Ratio inversa est numerus [Anteced]entium in [Consequ]ente.

Ratio ex duabus rationibus composita est numerus numeri Consequentium in Antecedente.

Ratio duplicata est numerus numeri Consequentium in Antecedente ubi numeri sunt æqu[ales].

Continue proportionalia[43] dicuntur quæ sunt in eadem ratione, primū ad secundum, secundum ad tertium, tertium ad quartum et sic deinceps.

Media proportionalia dicuntur quæ sunt continue proportionalia inter duo extrema.

In continue proportionalibus ratio primi ad tertium quartum quintum dicitur ratio duplicata triplicata quadruplicata primi ad secundum: et ratio primi ad secundum dicitur ratio subdulplicata subtriplicata subquadruplicata rationis primi ad tertium quartum quintum.

Multiplicare numerum per numerum est invenire numerum numeri.

Dividere numerum est invenire numerum qui per divisorem multiplicatus facit numerum dividendum.

Extrahere radicem[44] numeri est invenire numerum qui per seipsum multiplicatus facit numerum primum.

---

(42) Euclid—who delays his own introduction of the notion of a 'numerus' to the opening *Definitiones* of his seventh book (Barrow, *Euclidis Elementorum*...: 140–1)—has in *Elements* V, Def. III (*ibid.*: 90) to be content with imprecisely stating that 'Ratio est duarum magnitudinum ejusdem generis mutua quædam secundùm quantitatem habitudo', clarifying the obscurity of this by his following Def. V: 'Rationem habere inter se magnitudines dicuntur; quæ possunt se mutuò superare'. In compensation, where Newton's present enunciations define only integers and fractions, and omit consideration of all real 'numeri' not rationally commensurable with unity, Euclid can pass in his Def. VI (*ibid.*: 91) to state the celebrated Eudoxian provision, by like excess, equality, or deficiency of equimultiples over a given ratio, for magnitudes (rational or no) to be 'in eadem ratione'. Newton's ensuing definitions of inverse and compound ratio closely imitate those in *Elements* V, Def. XIII and X respectively (*ibid.*: 92, 91 respectively).

(43) Newton here needs a previous definition equivalent to *Elements* V, Def. VII: 'Eandem habentes rationem proportionales vocentur' (Barrow, *Euclidis Elementorum*...: 91). Euclid's text lacks any explicit notion of 'continuè proportionales', but the deficiency is made good by Barrow in an editorial insertion after *Elements* V, Def. X in his edition (*ibid.*: 92).

(44) A needlessly restrictive 'quadr[at]icam' is deleted in sequel in the manuscript. Unlike the preceding multiplication and division of numbers (whose definitions correspond *mutatis mutandis* to *Elements* VII, Def. XV and XXIII) root-extraction is not explicitly framed by Euclid, though the notions of 'latus quadrati' and 'latus cubi' are introduced by Barrow in his glosses on VII, Def. XVIII and XIX (*Euclidis Elementorum*...: 141) and the following *Postulatum* III makes the blanket demand that 'Additio, subtractio, multiplicatio, divisio, extractionesꝗ radicum...conceduntur...tanquam possibilia'.

Axiomata.[45]

Quæ eidem æqualia [sunt,] et inter se sunt æqualia.

Si æqualibus æqualia adjecta sunt, tota sunt æqualia.

Si ab æqualibus æqualia ablata sunt[,] tota sunt æqualia.

Quæ ad idem vel æqualia eandem habent rationem æqualia sunt inter se.

Omne totum æquale est om[n]ibus suis partibus simul sumptis.

Si totum est ad totum [46]ut ablatum ad ablatum, erit et reliquum ad reliquum in eadem ratione.

Postulata.

Dato numero addere unitates.

Dato numero subducere unitates.[47]

---

[6][48]   Problemata omnia plana Veteres[49] per Prop [27, 28] & [29] Lib VI *Elem* & Prop [58, 59, 60, 84, 85, 86 & 87] *Datorum* Euclidis solverunt (ut Gregorio olim[50] significavi) & omnia solida quærendo puncta quædam per quæ Locus solidus transiret & Locum per puncta illa describendo. Norunt enim Locum solidū per data quincȝ puncta describere (ut ex Prop. [13] Lib. [VIII] Pappi manifestū est), ut & puncta invenire per Librum Apollonij *de sectione determinata*. Nam Liber ille continet casus omnes[51] inveniendi punctū a quo si quatuor rectæ ad alias totidem positione datas rectas in datis angulis ducantur rectangulum sub duabus ductis sit ad rectangulū sub aljis duabus ductis in ratione data. Ejusmodi quatuor puncta semper inveniebantur in concursibus

---

(45) The first three of these are the corresponding *Axiomata* 1–3 of Euclid's first book (Barrow, *Euclidis Elementorum...*: 6–7), while the fourth subsumes the ensuing Axiom 8: 'quæ sibi mutuò congruunt, ea inter se sunt æqualia' (*ibid.*: 8). Newton's fifth 'self-evident truth' is rather but a definition of 'omne totum', and his last a derivable consequence of this.

(46) A needlessly restrictive 'in ratione data' is here deleted.

(47) Newton broke off at this point to cancel the entire preceding list of definitions, axioms and postulates; they do indeed form, to our modern eyes, an inaptly Euclidean preparatory to his promised delivery of the practice of his 'arithmetica una generalior' (on which again see §2.3: note (50) above). One is here fleetingly reminded of the universal computer proposed by Alan Turing, but we may be sure that, had Newton here continued, he would have gone on to frame other postulates in a narrowly Euclidean style, blazing the way to the application of his other defined arithmetical operations of multiplication, division and root-extraction much as in *Elements* VII, Postulate 3 (compare note (44) preceding).

(48) Add. 3963.14: 156r, a first version of §2.4 above (whose manuscript text follows immediately after on *ibid.*: 156r/156v).

(49) Initially 'Euclides', the onlie true begetter of the *Elements*.

(50) In May 1694, when David Gregory first met Newton in Cambridge; see VII: 222, note (10), and compare §2.4: note (52) preceding.

(51) Newton first went on to specify 'per quæ Locus solidus transeat in Problemate illo celeberrimo de quatuor [lineis]'.

rectarum positione datarum. Quintum punctum inveniebant per Librum sectionis determinatæ, deinde per puncta illa quincɜ locum describeba[n]t.[52]

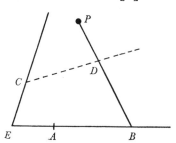

Ad solutionem Problematum solidorum spectat etiam Liber Apollonij *de Sectione spatij* ut et liber alter *de sectione rationis*. Dentur positione rectæ duæ *EB*, *EC* et in recta *EB* detur punctum *A*. [53]Et ad rectam *EC* in dato quovis angulo *ECD* ordinatim applicetur recta $CD_{[,]}$ deinde a dato puncto *P* agatur recta *DPB* abscindens[54] rectas *CD*, *AB* vel in data ratione ad invicem vel datum spatium comprehendentes[,] & punctum *D* locabitur in sectione conica.[55] Docet autem Apollonius in libris *de sectione rationis* et [*de*] *sectione spatij* quomodo punctum *D* inveniri potest.[56] Et inventis ejusmodi punctis quincɜ conica sectio per eadem describenda est.

APPENDIX 2.
DERIVATION OF THE DEFINING EQUATION OF A CIRCLE IN OBLIQUE CARTESIAN COORDINATES.[1]

From a stray jotting[2] in the University Library, Cambridge

---

(52) See §2.4: note (55). Lengthways in the margin alongside Newton began to adjoin 'Constructiones Problematum hic traditæ componuntur ad mod[um præcedentium?]', but there broke off.

(53) Newton first went on equivalently to state: 'A dato polo *P* agatur recta quævis et in ea quæratur punctum *D* a quo si recta *DC* ad rectam *EC* in dato angulo *ECD* ducatur, recta illa ducta *DC* vel sit ad rectam *AB* in data ratione vel cum recta illa *AB* datum spatium comprehendat[,] et punctum *D* locabitur in Sectione Conica'.

(54) This was tentatively replaced in the manuscript by an equivalent 'secans', but straightaway restored.

(55) Specifically, a hyperbola in either case; see §2.4: notes (56) and (59) preceding for a succinct Cartesian proof.

(56) On taking, that is, the line *CD* to be an arbitrary transversal, and thence by an Apollonian *sectio rationis* and *sectio spatii* respectively locating the pair of points *D* in which this meets the conic locus (*D*); compare §2.4: notes (57) and (60) above.

(1) Lest this should seem an empty textbook exercise in the autumn of 1706 when (so we suppose) Newton roughed out the following quick computation, we would remind that Descartes had not considered this most general Cartesian equation of the circle in the analysis of the general 4-line conic locus which he set out in the second book of his *Geometrie* (see his *Discours de la Methode*..., Leyden, ₁1637: 328–9 = *Geometria*, ₁1649: 32–5[→₂1659: 28–32]); nor was the omission made good by Florimond de Beaune, Frans van Schooten or (as far as we know) any of the other seventeenth-century mathematicians who afterwards broadened and systematized the basis of plane Cartesian 'analytical' geometry. It would appear that Newton himself here attains this most general defining equation for the first time.

(2) At the bottom of the verso of the manuscript leaf (Add. 3963.14: 156) which he other-

[Cape] $AG = x$. $GH = y$. [ut et]

$HK = dy = v$. $KG = ey$. [adeoꝗ]

   $AK = w = x - ey$.[(3)] [Finge]

   $w^2 + v^2 = aw + bv + c$.[(4)]

[prodit]

   $x^2 - 2exy + eeyy = ax - aey + c$.
   $\qquad\quad + ddyy \qquad\quad + bdy$

[Hoc est, quia $dd + ee = 1$,[(5)] fit]

$$x^2 - 2exy + yy = ax \begin{matrix} + bd \\ - ae \end{matrix} y + c.$$

[æquatio invenienda].[(6)]

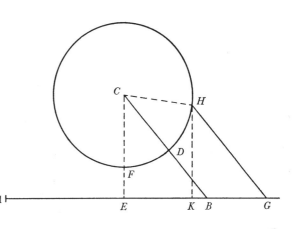

wise used to set out the two versions of his presentation of the 'Analysis Geometrica' of the ancients which are reproduced in §2, Appendix 1.6 and §2.4 preceding, and also to draft (at the head of f. 156ʳ) the last paragraph of his published 'Enumeratio Linearum Tertii Ordinis' (on which see VII: 644, note (134)). We take this scrap to be contemporary with the former rather than with the latter; but the difference is merely one of thirty months or so, and it is of only marginal significance to dwell upon it.

(3) Understand, of course, that HK is the perpendicular from H onto AG, so that $d = \sin \widehat{AGH}$ and $e = \cos \widehat{AGH}$, and accordingly $d^2 + e^2 = 1$. The coupled equations $\begin{cases} w = x - ey \\ v = dy \end{cases}$ define, we need not say, the transformation from the perpendicular coordinates $AK = w$, $KH = v$ to the oblique ones $AG = x$, $GH = y$ which Newton will in sequel employ to broaden the general Cartesian equation of the circle (where the coordinates are restricted to be perpendicular to one another). He had a quarter of a century earlier been much more familiar with the reverse transformation $\begin{cases} x = w + (e/d)v \\ y = (1/d)v \end{cases}$, making elegant use of it (on the pattern of Descartes' parallel reduction of the conic's general equation; compare IV: 219, note (9)) to derive the canonical forms of the cubic's defining Cartesian equation (see IV: 362–8). The ordinates CB and CE in Newton's accompanying figure—they are very clearly drawn respectively parallel to HG and perpendicular to AG in the manuscript—play no rôle in the ensuing computation, but evidently define the circle's centre C in the two coordinate systems by way of the related abscissas AB and AE.

(4) The most general equation of the circle, that is, when this is referred to the perpendicular coordinates $AK = w$, $KH = v$. At once $AE = \frac{1}{2}a$ and $EC = \frac{1}{2}b$ fix its centre C, while its radius is $CF = CH = \sqrt{(\frac{1}{4}(a^2 + b^2) + c)}$.

(5) See note (3) above.

(6) Whence the most general defining equation of the circle, where this is referred to the oblique coordinates $AG = x$, $GH = y$, proves to be of form $x^2 - 2xy \cdot \cos \widehat{AGH} + y^2 = ax + b'y + c$.

§3. THE CONSTRUCTION OF GEOMETRICAL PROBLEMS BY NEUSES.[1]

From the original[2] in the University Library, Cambridge

De constructione Problematum Geometricorum.

Definitiones.[3]

1. Extensum vel magnitudo data lege moveri dicitur, quando motus ejus ab alijs extensis mutuo contactu determinatur.

2. Positio in qua mobile sistendum est dari dicitur ubi aliud occurrit positione datum extensum vel extensi punctum quod mutuo contactu potest eidem obsistere.

Postulata.[3]

1. Extensum quodlibet quacunꝗ data lege movere.

2. Mobile in quacunꝗ data positione sistere.

3. Extensa describere quæ per cursus moventium vel positiones quiescentium designantur.

Problemata.

Prob. 1.

Lineam rectam [4] ducere.

Circum data duo quævis puncta *A, B* revolvatur solidum quodcunꝗ *EFGH* (per Post. 1) et per ejus puncta omnia quiescentia ducatur linea *CABD*, per Post. 3. Dico factum.

---

(1) An amended restyling of the set of ‘Problems for construing æquations’ (II: 450–516) which Newton sent to John Collins some thirty-five years earlier, in or shortly after 1670. The preliminary drafts of this piece which are reproduced (from the originals now Add. 3963.11: 115^r–116^v) in [1] and [2] of the Appendix reveal yet more sharply how narrowly he modelled it upon its parent. All that is here essentially new is an inserted scholium justifying the Archimedean neuses employed in the three terminal problems geometrically to construct the (real) roots of the reduced/general cubic equations, and those of a quartic having an odd number of roots of same sign. Newton initially also had it in mind to consider at least the complementary case of the quartic which has an even number of roots of the same sign—he made tentative provision for this contrary circumstance in the unfinished (and there inapplicable) ‘Cas. 3’ of ʋhe draft ‘Prob. 4’ which is reproduced (from Add. 3965.11: 151^r) in [3] of the Appendix—but what else he may once have intended to go on to include we shall probably never know. Once written, the tract vanished straight into the disordered limbo of Newton's private papers, leaving no external recorded trace, and it is here publicly resurrected for the first time. The internal evidence of the text, other than broadly to confirm a date for its composition of around 1705 (see especially note (52) of the Appendix), gives us no firmer information regarding its *raison d'être*. It may be simply (as we suggested in §2: note (1) preceding) that this formed but one aspect of a broader intention by Newton to revive his unpublished geometrical researches and have them printed in some appropriate collective compendium (as William Jones was to gather his shorter writings on calculus and infinite series in his

Translation

ON THE CONSTRUCTION OF GEOMETRICAL PROBLEMS.

Definitions.[3]

1. An extended entity or magnitude is said to move with a given law when its motion is determined from other extended ones by mutual contact.

2. The position in which a moving body shall be halted is said to be given when there occurs another extended entity, or point of one, given in position which can stop it by mutual contact.

Postulates.[3]

1. Any extended entity you please may move with any given law whatever.

2. A moving body may halt in any given position whatever.

3. Extended entities may describe the paths marked out through the courses of moving things or by the positions of stationary ones.

Problems.

Problem 1.

To draw a straight line.[4]

Around any two given points *A, B* let there revolve (by Postulate 1) any solid *EFGH* whatever, and through all its stationary points draw (by Postulate 3) the line *CABD*. I say it is done.

─────────────────────────────

Analysis in 1711). If immediate stimulus there was from without, we may conjecture that Newton designed this revamping of his earlier set of 'Problems' to replace the equivalent, too prolixly presented geometrical constructions of the roots of cubic and quartic equations by Archimedean neuses and intersecting conics which he appended twenty years before to the deposited text of his Lucasian lectures on algebra (see v: 432–70) whose printed version was, under William Whiston's feeble editorial hand, coming off the press in Cambridge during 1705–6; we have already indicated (see v: 9–14) how dissatisfied Newton was with the printed *Arithmetica Universalis*, and have specified (v: 422, note (615)) how he pared away to the bare bone the fat—and in some ways the muscle—of the appended 'Æquationum Constructio linearis', rejecting this very marginal head which Whiston gratuitously inflated to be a half-title, in his private library copy of the book. This, like all other circumstantial surmise, must remain mere speculation until such time as more precise documentary evidence regarding the background to the present tract should turn up.

(2) Add. 3963.11: 113ʳ–114ᵛ/117ʳ/118ʳ. The manuscript text is a carefully penned piece with very few interlineations or, except for the subsequent cancellation of the scholium to Problem 4 and of the non-planar Problems 10–13, substantial deletions.

(3) Except for the minimal retouchings made in draft (see [1] of the Appendix) these repeat the corresponding introductory *Definitiones* and *Postulata* of the parent 'Problems for con-struing æquations' (II: 450–2).

(4) Forgetting that he has here split 'Prob. 1' of the parent text into two, Newton momentarily anticipated his present 'Prob. 2' by adding the adverb '*continuo*' (*continually*) but speedily deleted it.

Si negas; sit *AKB* linea recta circum præfatam *CABD* revolvens, & sit *AkB* eadem recta in alia positione. Et rectæ *AKB*, *AkB* comprehendent spatium, contra Ax 14 primi *Elementorum*.[5]

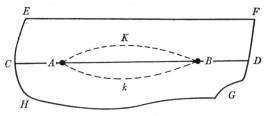

Prob. 2.

Lineam rectam continuò producere.

Ducatur alia recta (per Prob. 1) quæ pro regula ducendi rectas adhiberi possit, et moveatur eadem ea lege ut ejus puncta duo quævis incidant semper in lineam ducendam per Post. 1, et interea per totum moventis cursum ducatur linea per Post. 3. Dico factum.

Prob. 3.

Per data duo puncta rectam ducere.

Recta quævis ducta et quantum satis est producta moveatur lege quacunꝗ (per Post. 1) donec eadem punctis illis duobus datis occurrat (per Post. 2). Et per ejus puncta quiescentia ducatur linea quæ hinc inde ad puncta illa duo data terminetur, ℘ Post. 2 & 3. Dico factum.

Prob. 4.[6]

Planum per data tria puncta A, B, C describere.

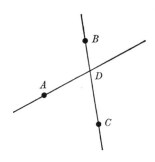

Per duo quælibet punctorum *B* et *C* duc rectam indefinitam *BC* (per Prob. 2 & 3.) Dein circa tertium punctum *A* ceu polum revolvatur alia similiter ducta *AD* quæ lineam priorem *BC* perpetuo tangat (per Post. 1:) Et (per Post. 3) describatur superficies quam hæc linea cursu suo designat. Dico factum.[7]

Nam cum rectæ *BC* puncta duo *B* et *C* sint in plano describendo (per hypothesin) ea tota cum eodem plano coincidet (per Def. 4 & 7 et Ax. 14 lib. 1 *Elem*) ideoꝗ concursus rectarum *AD* et *BC* semper erit in eodem plano. Recta autem *AD* convenit etiam cum eodem plano in puncto *A* (per hypothesin) & propterea tota co[incid]it cum eodem plano (per Def 4 et 7 & Ax. 14 lib. 1 *Elem*) & revolvendo planum describit.[8]

Quæ sequuntur, in dato plano construenda esse intellige.

(5) In Isaac Barrow's Latinization, 'Duæ rectæ lineæ spatium non comprehendunt' (*Euclidis Elementorum* Libri XV. breviter demonstrati (Cambridge, ₁1655): 8; compare II: 453, note (8)).

(6) Effectively 'Prob. 2' on II: 452/454.

(7) Initially 'Et hæc [alia recta *AD*] describet planum quod per puncta *A*, *B* & *C* transibit' (And this [other straight line *AD*] will describe the plane which shall pass through the points *A*, *B* and *C*).

If you deny it, let *AKB* be a straight line revolving round the above-mentioned one *CABD*, and let *AkB* be the same straight line in another position. Then the straight lines *AKB*, *AkB* will comprehend a space, contrary to Axiom 14 of the first (book) of the *Elements*.[5]

Problem 2.

Continually to extend a straight line.

Draw (by Problem 1) another straight line, so as to be able to apply it as a ruler for drawing straight lines, and (by Postulate 1) move it with the law that any two points of it shall ever fall onto the line to be drawn, and all the while (by Postulate 3) let the line be drawn through the total course of the moving one. I say it is done.

Problem 3.

Through two given points to draw a straight line.

Let any straight line, drawn and far enough extended, move (by Postulate 1) with any law whatever till it shall (by Postulate 2) meet the two given points. Then through its stationary points draw (by Postulates 2 and 3) a line which shall terminate on either end at those two given points. I say it is done.

Problem 4.[6]

To describe a plane through three given points A, B, C.

Through any two of the points you please, *B* and *C*, draw (by means of Problems 2 and 3) the unlimited straight line *BC*. Then, round the third point *A* as pole let there revolve another one similarly drawn, *AD*, so as (by Postulate 1) perpetually to touch the previous line *BC*, and (by Postulate 3) describe the surface which this line shall trace out in its course. I say it is done.[7]

For, since the two points *B* and *C* of the straight line *BC* are (by hypothesis) in the plane to be described, it will (by Definitions 4 and 7 and Axiom 14 of Book 1 of the *Elements*) wholly coincide with this same plane, and consequently the meeting-point of the straight lines *AD* and *BC* will ever be in this plane. The line *AD*, however, meets this same plane (by hypothesis) in the point *A*, and accordingly (by Definitions 4 and 7 and Axiom 14 of Book 1 of the *Elements*) wholly coincides with this plane, and so by revolving describes the plane.[8]

Understand that the following ones are to be constructed in a given plane.

---

(8) In sequel Newton has cancelled a 'Schol.' which reads: 'Regularum trium acies, si ad invicem successive applicentur et perfecte congruant, sunt rectæ. Et corporum trium superficies, si ad invicem successive applicentur & perfecte congruant, sunt planæ. Si non congruunt atteri debent inter se, donec perfecte congruant' (The edges of three rulers, if they be successively applied one to another and agree perfectly, are straight lines. And the surfaces of three bodies, if they be successively applied to one another and agree perfectly, are plane. Should they not agree, they ought to be rubbed away one against another till they do agree perfectly).

Prob. 5.

Dato centro et intervallo circulum describere.

A centro ad datum intervallum duc rectam per Prob. 3. et in dato plano circa centrum illum ceu polum convolve (per Post. 1) & aream describe (per Post. 3) quam recta revolvens cursu suo designat. Dico factum.

Nam figura descripta circulus erit per Def. 15 lib. 1 *Elem.*[9]

Prob. 6.

Ad datum punctum in recta quavis per punctum illud transeunte ponere longitudinem datæ cuivis rectæ æqualem.[10]

Moveatur data recta lege quacunᗡ (per Post. 1) donec ejus terminus alteruter incidat in punctum datum & terminus alter in rectam positione datam (per Post. 2.) Et problema solvetur.

Prob. 7.

Inter datum punctum et datam positione lineam ponere longitudinem datam.

Moveatur longitudo illa lege quacunᗡ (per Post. 1) donec ejus terminus alteruter incidat in punctum datum et terminus alter in lineam datam per Post. 2.

Schol.

Simili angulorum et figurarum translatione describi possunt anguli angulis et figuræ figuris similes et æquales. Angulus quoᗡ rectus semel constructus pro norma[11] construendi rectos angulos adhiberi potest. Et his Principijs hactenus expositis Problemata omnia Euclidea expedite construi possunt; ut ijs immorari non sit opus. Sequentia ad constructionem Problematum solidorum[12] spectant.

Prob. 8.

Inter datas positione duas lineas ponere datam longitudinem quæ ad datum punctum converget.

Terminis suis tangat longitudo illa lineas datas & tangendo moveatur (per

---

It would indeed have been not a little incongruous to refer in the present highly theoretical context of geometrical construction to a practical technique for grinding a 'fiducial' straight-edge or plane surface, one long known to makers of optical glass and, we may remind, to Newton in his youth (see 1: 460, where forty years before he went on to give hints on how to maintain uniform pressure between the several parts of the bodies in contact).

(9) In Barrow's Latin rendering, 'Circulus est figura plana, sub una linea comprehensa, quæ peripheria appellatur, ad quam ab uno puncto eorum, quæ intra figuram sunt posita, cadentes omnes rectæ lineæ inter se sunt æquales' (*Euclidis Elementorum Libri XV...*, ₁1655: 2).

(10) Newton initially copied his preliminary draft on f. 115ᵛ (see [1] of the Appendix) to enunciate more shortly, much as in the parent 'Prob. 4' on II: 454, the demand 'Ad datum

Problem 5.

With given centre and radial distance to describe a circle.

From the centre at the given radial distance draw a straight line by means of Problem 3, and in the given plane (by Postulate 1) rotate it round that centre as pole and so (by Postulate 3) describe the area which the revolving straight line shall trace out in its course. I say it is done.

For the figure described will be a circle by Definition 15 of Book 1 of the *Elements*.[9]

Problem 6.

At a given point in any straight line passing through that point to set a length equal to any given straight line.[10]

Let the given straight line move (by Postulate 1) with any given law whatever till (by Postulate 2) either one of its end-points shall fall upon the given point and its other one onto the straight line given in position. The problem will then be solved.

Problem 7.

Between a given point and a line given in position to set a given length.

Let the length move (by Postulate 1) with any law whatever till (by Postulate 2) either one of its end-points shall fall on the given point and the other one upon the given line.

Scholium.

By a similar transporting of angles and figures there can be described angles congruent to angles and figures congruent to figures. A right angle, too, once constructed can be employed as a pattern[11] for constructing right angles. And on these principles so far expounded all Euclidean problems can speedily be constructed; there is in consequence no need to dwell on them. The sequel has regard to the construction of 'solid' problems.[12]

Problem 8.

*Between two lines given in position to set a given length
which shall be inclined through a given point.*

Let the length at its end-points touch (by Postulate 1) the given lines and, so

---

punctum in data recta ponere datam longitudinem' (At a given point in a given straight line to set a given [line-]length).

(11) A 'set-square' we would say.

(12) This is true only of the conchoidal neusis whose construction by tentatives is set out in Problem 8. Newton has forgotten that in Lemma XXVI of his 'De motu Corporum Liber primus' he had outlined in corollary (see VI: 299 [= *Philosophiæ Naturalis Principia Mathematica*

Post. 1) donec eadem, si opus est producta, incidat in punctum datum (per Post 2.)

Prob. 9.

Datas positione tres lineas linea recta secare cujus partes interceptæ datarum erunt longitudinum.

Inter duas datarum linearum moveatur pars una terminis suis tangens utramꝗ, (per Post 1) donec terminus partis alterius incidat in lineam tertiam per Post. 2.

[13]Prob. 10.

Dato centro et intervallo sphæram describere.

Hoc centro et intervallo describatur circulus per Prob. 5 et hic circulus circum quamvis ejus diametrum ceu axem revolvatur per Post. 1, et describatur figura solida quæ per cursum circuli designatur per Post. 3.

Prob. 11.

Dato axe et intervallo cylindrum describere.

Circum axem ad intervallum illud revolvatur linea recta axi parallela per Post. 1, et describatur figura solida cujus superficiem recta illa revolvendo designat per Post. 3.

Prob. 12.

Dato axe et angulo conum describere.

Circum axem revolvatur angulus per Post. 1, et describatur figura solida quam angulus revolvendo designat per Post. 3.

Prob. 13.

Dato plano datum conum secare.

Datum planum producatur per conum[14] (per Prob. 2 et 4) et partes coni sectione distinctas describantur per Post. 3.

Schol.

A mensuratione agrorum Geometria originem duxit, & huic negotio Postulata Euclidea sufficiunt: ideoꝗ Geometræ antiquissimi hæc sola adhibuerunt.[15] Sphæræ cylindri et coni definitiones posuerunt ut horum solidorum investigarent proprietates, descriptiones vero non postularunt quia figuris hisce ad

---

(London, ₁1687): 97]) a reduction of Problem 9 to a Euclidean ruler-and-compasses construction, whence the problem cannot be employed in turn to resolve any other but a 'plane' one.

(13) Newton's purpose in cancelling the four Problems 10–13 next following was evidently a late decision to restrict the scope of his 'De Constructione' to embrace only planar entities.

touching, move (by Postulate 2) till it, produced if need be, shall fall on the given point.

Problem 9.

To cut three lines given in position by a straight line, the intercepted parts of which shall be of given lengths.

Let (by Postulate 1) one part move between two of the given lines, touching both at its end-points, till (by Postulate 2) the end-point of the other part shall fall on the third line.

[13]Problem 10.

To describe a sphere, given the centre and radial distance.

With this centre and radius describe a circle by means of Problem 5, and (by Postulate 1) let this circle revolve round any diameter of it as axis, then (by Postulate 3) describe the solid figure which is traced out through the passage of the circle.

Problem 11.

To describe a cylinder, given the axis and radial distance.

Let there (by Postulate 1) around the axis at that distance revolve a straight line parallel to the axis, then describe (by Postulate 3) the solid figure whose surface that line traces out by revolving.

Problem 12.

To describe a cone, given the axis and (vertex) angle.

Let (by Postulate 1) the angle revolve round the axis, and describe (by Postulate 3) the solid figure which the angle shall trace out by revolving.

Problem 13.

To cut a given cone by a given plane.

Extend the given plane (by means of Problems 2 and 4) through the cone[14] and the parts of the cone separated by the section shall be described by Postulate 3.

Scholium.

Geometry drew its origin from the mensuration of fields, and for this business the Euclidean postulates suffice; and these alone were in consequence employed by the very ancient geometers.[15] They posited the definitions of the sphere, cylinder and cone that they might investigate the properties of these solids, but did not postulate their descriptions because they then as yet did not use these

---

(14) 'donec conum secat' (till it cuts the cone) is here replaced.
(15) Compare Newton's equivalent assertions of a dozen years earlier on vii: 286–8.

constructiones Problematum nondum essent usi.[16] Tandem circa tempora Platonis et Euclidis, cum Problema duplicandi cubum proponeretur et per sectionem Coni solveretur,[17] et elementa sectionum conicarum scribi inciperent, postulatum est in his Elementis *Datum conum dato plano secare*, quasi conus per solam suam Definitionem daretur in Geometria.[18] Sufficiunt Definitiones figurarum ad Theo[re]mata de ijs investiganda, non autem ad figuras ipsas describendas. Habentur quidem in intellectu per Definitiones, non autem in rerum natura. Linea recta et circulus non habentur in Geometria per Definitiones suas. Ut habeantur postularunt Geometræ easdem describere. Et similiter ut habeatur conus qui secetur, debet ejus descriptio vel postulari vel per Postulata doceri: aliter constructio Problematum per sectiones conorum non erit geometrica.[19] Præterea, Conum describere & descriptum dato plano secare sunt operationes difficiles & erroribus obnoxiæ, ideoꝗ Geometræ hodierni[20] Sectionem in plano sine constructione solidorum, describere docent. Archimedes vero majori cum judicio pro principio habuit operationem ponendi inter datas duas lineas longitudinem datam & ad datum punctum convergentem. Nam per hoc Postulatum Problemata omnia solida expeditius & accuratius solventur quam per constructiones & sectiones solidorum vel sectionum descriptiones in plano.[21] Quinetiam constructiones Archimedeæ sunt Euclideis magis affines, ut conferenti constructiones Problematum quarundam præcedentium nempe sexti septimi octavi et noni inter se, atꝗ etiam constructiones duorum sequentium inter se, manifestum erit.

1. Angulum datum bifariam secare, idꝗ continue.[22]

Detur angulus *BAC* et hujus lateri alterut[ri] *AC* parallelam age *BF* lateri alteri occurrentē in *B*. In *BF* cape *BD* æqualem *AB* et acta *AD* bisecabit angu-

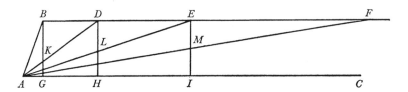

lum *BAC*. In eadem *BF* cape *DE* æqualem *AD* et acta *AE* bisecabit angulum *DAC*. In eadem *BF* cape *EF* æqualem *AE* et acta *AF* bisecabit angulum *EAC*, et sic deinceps in infinitum.

---

(16) Compare VII: 296.

(17) Namely by Menæchmus, whose two variant conic constructions of the problem (preserved by Eutocius) are pithily set out by T. L. Heath in his *History of Greek Mathematics*, **1** (Oxford, 1921): 253–5.

(18) Compare VII: 286, 292–4.

(19) Compare VII: 290–2, 298.

figures for the construction of problems.[16] At length around the time of Plato
and Euclid, since the problem of duplicating the cube came to be proposed and
solved by the section of a cone,[17] and elements of conic sections began to be
written, the postulate was made in these elements *To cut a given cone by a given
plane*, as though the cone were given in geometry by its definition alone.[18]
Definitions of figures suffice for discovering theorems about them; not, however,
for describing the figures themselves. They are indeed had in thought through
their definitions, but not so in physical reality. The straight line and circle are
not had in geometry through their definitions. That they may so be had,
geometers postulated to describe them. And similarly, that there may be had
a cone to be cut, its description ought either to be postulated or taught through
postulates: otherwise, the construction of problems by means of conic sections
will not be geometrical.[19] Moreover, to describe a cone and to cut it, once
described, by a given plane are difficult operations and ones liable to error, and
for that reason today’s geometers[20] teach how to describe a conic in the plane
without a construction of solids. Archimedes, indeed, with greater discernment
had as his principle the operation of setting between two given lines a given
length inclined through a given point; for through this postulate all ‘solid’
problems will be more speedily and accurately solved than by means of constructions and sections of solids or through the descriptions of conics *in plano*.[21]
To be sure, the Archimedean constructions are more akin to the Euclidean
ones, as will be apparent to anybody who compares the constructions of certain
of the preceding problems—namely, the sixth, seventh, eight and ninth—with
each other, and also those of the following two:

1. To cut a given angle into two, and that continually.[22]

Let there be given the angle $B\widehat{A}C$ and parallel to one or other side AC of this
draw BF meeting the second side in B. In BF take BD equal to AB, and AD
when drawn will bisect the angle $B\widehat{A}C$. In the same BF take DE equal to AD,
and AE when drawn will bisect the angle $D\widehat{A}C$. In the same BF take EF equal
to AE, and AF when drawn will bisect the angle $E\widehat{A}C$; and so on indefinitely.

---

(20) Understand Frans van Schooten (compare I: 29–45 *passim*, and see also II: 8, note (21)),
Philippe de La Hire (see VI: 271, note (70)) and, above all, Newton himself.

(21) Compare VII: 302, 382–4. Analogous remarks on the expediency and accuracy of the
Archimedean construction of ‘solid’ problems by conchoidal neusis are found in the terminal
section of Newton’s Lucasian lectures on algebra, dealing analogously with the ‘Æquationum
constructio linearis’; see V: 426–8, 470.

(22) ‘Prob. 17’ of the parent ‘Problems for construing æquations’ (see II: 464). Newton’s
minimally amplified restyling of its (here unjustified) construction closely follows that which
he states for the analogous problem of repeatedly trisecting a given angle which follows.

2. Angulum datum trifariam secare idqᵹ continue.[23]

Detur angulus *BAC* et hujus lateri alterutri *AC* parallelam age *BF* lateri alteri occurrentem in *B*. A puncto *B* ad rectam *AC* demitte perpendiculum *BG*[24] et inter rectas *BG* et *BF* ponatur longitudo *KD* quæ sit duplæ *AB* æqualis et convergat ad punctum *A*, & angulus *DAC* erit anguli *BAC* pars tertia.[25] A puncto *D* ad rectam *AC* demitte perpendiculum *DH*[24] et inter rectas *DH* et *DF* ponatur longitudo *LE* quæ sit duplæ *AD* æqualis et convergat ad punctum *A* et angulus *EAC* erit anguli *DAC* pars tertia.[25] A puncto *E* ad rectam *AC* demitte perpendiculum *EI*[24] et inter rectas *EI* et *EF* ponatur longitudo *MF* quæ sit duplæ *AE* æqualis et convergat ad punctum *A*, et angulus *FAC* erit anguli *EAC* pars tertia. Et sic deinceps in infinitum.

Si descriptiones curvarum per motum localem in Geometriam introducendæ sunt, hoc optime præstabitur per Postulata tria a nobis posita. Et hæc sola ad operationes omnes Geometricas abunde sufficiunt.

Si motus omnis localis ad Mechanicam referatur et Postulata tria Euclidea ob antiquitatem retinenda censeantur, adjici potest quarto loco[26] Postulatum Archimedeum, per quod utiqᵹ absqᵹ difficili curvarum descriptione construentur Problemata omnia solida. Et hac ratione Geometria veterum manebit pura & incontaminata.

Quod si Problemata quæ solida dicuntur per sectiones solidorum solvenda sint, ad Postulata addenda erunt Postulata duo nova, nempe Conum dato angulo describendi & conum descriptum dato plano secandi. Et postquam per hæc Postulata construxeris Problema octavum & Problema nonum potes[27] alia omnia solida per hæc duo construere.

Sive igitur cum Archimede postulemus confectionem Problematis octavi et per hoc Postulatum construamus Problema nonum; sive per alia Postulata construamus utrumqᵹ; possumus hæc duo ad alia construenda commode adhibere.

Quare cum Mathematici hodierni Problemata omnia per operationes Arithmeticas ad æquationes deducere ament, & Æquationes (quæ quidem sunt conclusiones plane Arithmeticæ) per operationes Geometricas construere: non pigebit constructiones cubicorum & biquadraticorum ad hæc duo Problemata[28]

---

(23) 'Prob. 17' of the parent text (II: 464). The ensuing construction repeats virtually word for word the initial recasting of its statement which Newton drafted on f. 116ʳ, where (see [2] of the Appendix) the neusis is set as 'Prob. 24' of a somewhat more ample (and optimistic) preliminary listing. For Newton's own justification of the exactness of this angle-trisection again see II: 464; and for the Archimedean background compare V: 460–2.

(24) These perpendiculars *BG* (meeting *AD* in *K*), *DH* (meeting *AE* in *L*) and *EI* (meeting *AF* in *M*) are carelessly omitted from the manuscript figure (at the bottom of f. 114ʳ), but we see no need to insist on the deficiency in our reproduction of it.

(25) In each case a less specific 'trifariam secabitur' (will be cut into three) is replaced.

(26) Again compare VII: 388–90, 394.

2. *To cut a given angle into three, and that continually.*[23]

Let there be given the angle $B\widehat{A}C$ and parallel to one or other side AC of this draw BF meeting the second side in B. From the point B to the straight line AC let fall the perpendicular BG,[24] and between the straight lines BG and BF place the length KD which shall be equal to twice AB and be inclined through the point A, and the angle $D\widehat{A}C$ will be a third part of $B\widehat{A}C$.[25] From the point D to the straight line AC let fall the perpendicular DH,[24] and between the straight lines DH and DF, place the length LE which shall be equal to twice AD and be inclined through the point A, and the angle $E\widehat{A}C$ will be a third part of $D\widehat{A}C$.[25] From the point E to the straight line AC let fall the perpendicular EI,[24] and between the straight lines EI and EF place the length MF which shall be equal to twice AE and be inclined through the point A, and the angle $F\widehat{A}C$ will be a third part of $E\widehat{A}C$. And so on indefinitely.

If descriptions of curves through local motion are to be introduced into geometry, this will best be accomplished by means of the three postulates posited by us. And these alone amply suffice for all geometrical operations.

If all local motion be referred to mechanics and the three Euclidean postulates reckoned, because of their venerableness, as needing to be kept, there can be added in fourth place[26] the Archimedean postulate, by means of which, of course, without a difficult description of curves all 'solid' problems will be constructed. And in this manner the geometry of the ancients will remain pure and uncontaminated.

But if the problems which are called 'solid' should be required to be solved through sections of solids, to the postulates there will need to be added two new ones: namely, that of describing a cone with a given angle, and that of cutting a cone, once described, by a given plane. And, after you have by means of these postulates constructed Problems 8 and 9, you will be able[27] to construct all other solid ones through these two.

Whether, therefore, with Archimedes we postulate the accomplishing of the eighth problem and by means of this postulate construct the ninth one, or whether we construct both through other postulates, we can employ these two with advantage to construct others.

On which ground, seeing that today's mathematicians love to bring everything down by means of arithmetical operations to equations, and construct equations (which, of course, are conclusions thoroughly arithmetical) by geometrical operations, it will not be an infliction to reduce the construction of cubic and quartic ones to these two problems.[28] In equations, however, we

(27) A correctly hesitant 'possis' (you might be able) is wrongly here replaced. On the inadequacy of Problem 9 to construct 'solid' problems see note (12) above.

(28) '*Elementis* Euclidis addita' (when added to the *Elements* of Euclid) is deleted. Newton's

reducere. In æquationibus vero supponimus quantitates datas per numeros *p*, *q*, *r*, *s* signis suis + et − affectos significari et quantitatem quæsitam exponi per radicem *x*.

Prob. 1.[29]

Æquationem quadraticam xx = px+q modis infinitis construere.

Duc quamvis rectam *KCB*, et in ea cape *KA* cujusvis longitudinis, quam dic *n*. Dein cape *KB* æqualem $\frac{q}{n}$, idcȝ versus *A* si habeatur +*q*, aliter ad contrarias partes puncti *K*. Biseca *AB* in *C*, et centro *C* radio *AC* fac circulum *AX*, cui inscribe rectam *AX* æqualem *p*, et eandem utrincȝ produc donec alium circulum quem centro *C* radio *CK* describes, secet in punctis *Y* et *Z*. Et erunt *AY* et *AZ* radices æquationis construendæ.[30]

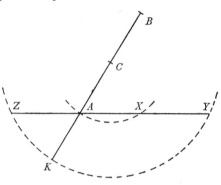

Et nota quod radices affirmativæ cadunt ad partes versus *X* si habeatur +*p*, et negativæ ad partes contrarias. Et contra, si habeatur −*p*.

Nota etiam quod aliquis e divisoribus termini *q* commode assumetur[31] pro *n*, et quod ubi habetur +*q*, differentia inter *n* et $\frac{q}{n}$ non debet esse minor quam *p*.

Prob. 2.[32]

Æquationem cubicam x³ = qx+r modis infinitis construere.

Duc quamlibet rectam *KC*, et in ea cape *KA* cujusvis longitudinis, quam dic *n*. Dein cape *KB* æqualem $\frac{q}{n}$, idcȝ versus *A* si habeatur −*q*; aliter ad contrarias

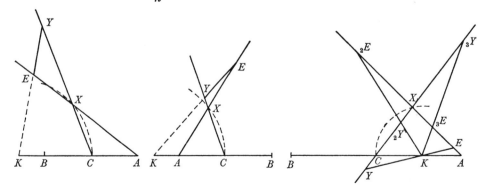

partes puncti *K*. Biseca *AB* in *C*, et centro *K* radio *KC* fac circulum *CX*, cui inscribe rectam *CX* æqualem $\frac{r}{nn}$, & produc utrincȝ. Item junge *AX* et produc utrincȝ. Denicȝ inter has rectas *CX* et *AX* inscribe rectam *EY* ejusdem longi-

suppose that given quantities are denoted by the numbers p, q, r, s affected with their signs $+$ and $-$, and that the quantity sought is expressed by the root x.

Problem 1.[29]

To construct the quadratic equation $x^2 = px + q$ in an infinity of ways.

Draw any straight line KCB, and in it take KA of any length—call it n. Next take KB equal to q/n; this towards A if q is had positive, otherwise on the opposite side of the point K. Bisect AB in C, and with centre C, radius AC construct the circle AX; in this inscribe the straight line AX equal to p, and extend it either way till it cuts in the points Y and Z another circle which you shall describe with centre C and radius CK. Then AY and AZ will be the roots of the equation to be constructed.[30]

And note that positive roots fall on the side towards X if p be had positive, negative ones on the opposite side; and conversely so if p be had negative.

Note also that some one of the divisors of the term q may be conveniently assumed[31] for n, and that when q is had positive the difference between n and q/n ought not to be less than p.

Problem 2.[32]

To construct the cubic equation $x^3 = qx + r$ in an infinity of ways.

Draw any straight line KC, and in it take KA of any length—call it n. Next take KB equal to q/n; this towards A if q be had negative, otherwise on the opposite side of the point K. Bisect AB in C, and with centre K and radius KC trace the circle CX; in this inscribe the straight line CX equal to r/n^2, and extend it either way. Likewise, join AX and extend it either way. Finally, between these lines CX and AX inscribe a straight line EY of the same length as the line CA such that

---

first phrasing of this clause affirmed slightly differently 'non pigebit constructiones per hæc duo Problemata in sequentibus ostendere' (it will not be irksome to exhibit constructions by means of these two problems in the sequel). In fact—as Newton could do no other (compare the previous note)—only the conchoidal neusis adduced in Problem 8 is applied, in Problems 2 and 4 following, geometrically to construct the roots of the general reduced cubic and (a particular class of) quartic equation.

(29) The first of the 'problemata infinitis modis soluta' which conclude the parent text (see II: 468), here repeated virtually word for word.

(30) As before (compare II: 469, note (58)), it will be evident that Newton's geometrical construction yields the required equation in the form $x(x-p) = n(q/n)$, n free.

(31) Newton initially copied 'adhibeatur' (employed) from the equivalent 'Nota 2' of the parent 'Prob. 1' (II: 468).

(32) The second of the 'problemata infinitis modis soluta' which terminate the parent text (see II: 470–4), repeated virtually word for word except that the unimplemented '*Modus 3. Idem adhuc aliter*' (on which see II: 475, note (69)) is here ignored.

tudinis cum recta CA, ita ut hæc EY transeat per punctum K; et longitudo XY erit radix æquationis. Radices autem affirmativæ sunt quæ cadunt versus punctum C si habeatur $-r$, & contra si habeatur $+r$.[33]

Corol. [34]Hinc duæ mediæ proportionales inter quantitates quaslibet, n et $\dfrac{r}{nn}$ inveniri possunt. Nam si desit terminus q, id est si puncta K et B coincidant, erunt AX, YX, KE et CX continue proportionales.

Idem aliter.

Duc quamlibet rectam KC, et in ea cape KA cujusvis longitudinis[,] quam dic n. Dein cape KB æqualem $\dfrac{q}{n}$, idcĝ versus A si habeatur $+q$, aliter ad con-

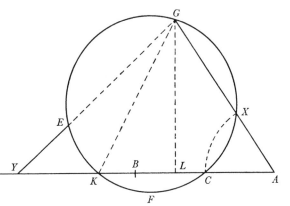

trarias partes puncti K. Biseca AB in C, et centro A radio AC fac circulum CX, cui inscribe rectam CX æqualem $\dfrac{r}{nn}$, et per puncta K, C et X describe circulum. Junge AX et produc donec iterum secet circulum KCX, puta in G. Deniĝ inter hunc circulum et rectam KC utrinĝ productam inscribe rectam EY ejusdem longitudinis cum recta AC & convergentem ad punctum G. Et acta EC erit radix æquationis. Radices autem affirmativæ cadunt in majori circuli segmento KGC si habeatur $+r$, et in minori KFC si habeatur $-r$.[35]

---

(33) Exactly as with 'Modus 1' of the parent 'Prob. 2' we may (compare II: 471, note (62)) define the position of the fixed point K in regard to the given lines AX and CX by putting $KX = a$, $KC = b$, $CX = c$, $CA = d$, and then specify the possible locations of the interposed line $EY = CA$ passing through K by determining the related unknowns $XY = x$ and $KE = y$. At once, by Menelaus' theorem (since the transversal AXE intersects the triangle KCY) there is $XY \times EK \times CA = CX \times EY \times KA$ or $xy = c(b+d)$, and also

$$2KC \cdot \cos \widehat{KCX} = (b^2+c^2-a^2)/c = (b^2+(c+x)^2-(d+y)^2)/(c+x)$$

or

$$x^2 + (c+(a^2-b^2)/c)\,x + a^2 - d^2 - 2dy - y^2 = 0;$$

whence on eliminating y and taking out the ensuing factor $x+c = 0$ (corresponding to the particular case in which EY coincides with AC) there results

$$x^3 + ((a^2-b^2)/c)\,x^2 + (b^2-d^2)\,x - c(b+d)^2 = 0.$$

Accordingly, on identifying this with the canonical reduced cubic equation $x^3 - qx - r = 0$, there must be $a-b = 0$, $b^2-d^2 = -q$ and $c(b+d)^2 = r$; so that, as Newton posits in his construction, $KX = KC$ and also, on taking $KA = b+d = n$ to be arbitrary, $KB = b-d = -q/n$ and $CX = c = r/n^2$. For Newton's geometrically synthesized 'Demonstratio' of this preliminary algebraic analysis see II: 470–2.

this EY shall pass through the point K. The length XY will then be a root of the equation. Positive roots, however, are those which fall towards the point C if r be had negative; the converse if r be had positive.[33]

Corollary. [34]Hence two mean proportionals between any quantities n and r/n^2 you please can be found. For if the term $q(x)$ should be lacking, that is, if the points K and B coincide, then AX, YX, KE and CX will be in continued proportion.

The same another way.

Draw any straight line KC and in it take KA of any length—call it n. Next take KB equal to q/n; this towards A if q be had positive, otherwise on the opposite side of the point K. Bisect AB in C, and with centre A and radius AC effect the circle CX; in this inscribe the straight line CX equal to r/n^2, and through the points K, C and X describe a circle. Join AX and extend it till it again meets the circle, say in G. Finally, between this circle and the line KC extended either way inscribe a straight line EY of the same length as the line AC and inclined through the point G. Then EC when drawn will be a root of the equation. Positive roots, however, fall within the greater segment KGC of the circle if r be had positive, and in the lesser one KFC if r be had negative.[35]

---

(34) In a first phrasing of the following observation Newton more closely imitated the analogous 'Coroll:' to the parent 'Prob. 2' (see II: 472), writing: 'Si terminus q desit, id est si puncta K et B coincidant; erunt AK, YX, KE et CX continue proportionales. Et inde duæ mediæ proportionales inveniri possunt' (If the term in q be lacking, that is, if the points K and B coincide, then AK, YX, KE and CX will be continued proportionals. And thereby two mean proportionals can be found). In his original 'Problems for construing æquations' Newton had given separate enunciation of this special case of the preceding neusis construction in the 'Idem aliter' to his 'Prob. 15. Invenire duo media proportionalia...' (see II: 460, and compare *ibid.*: 461, note (34)).

(35) As with the complementary 'Modus 2' of the parent 'Prob. 2' we may again (compare II: 473, note (66)) fix the geometrical framework in its relative position by setting $KE = a$, $CA = b$, $AX = c$ (so that $AG = b(a+b)/c$) and $KG = d$, and then specify the possible locations of the line $EY = AX$ through G interposed between the given line KCA and the circle drawn through K, C, X, G by determining the related unknown line-lengths $EC = x$ (not shown by Newton in his figure) and $KY = y$. Because the triangles CEY and GKY are similar, there is at once $CE:EY = GK:KY$ and so $xy = cd$; also, since $GY = y(y+a)/c$, there is

$$2GK.\cos \widehat{GKY} = (d^2+y^2-(y(y+a)/c)^2)/y = -(n^2+d^2-(bn/c)^2)/n,$$

where $KA = a+b = n$, and consequently

$$y^4+2ay^3+(a^2-c^2)\,y^2+(n(b^2-c^2)-c^2d^2/n)\,y-c^2d^2 = 0;$$

whence, on taking out the factor $y+n = 0$ (corresponding to EY coincident with XA), there comes $y^3+(a-b)\,y^2+(b^2-c^2)\,y-c^2d^2/n = 0$ and thence, on substituting $y = cd/x$, finally $x^3-(n(b^2-c^2)/cd)\,x^2+(a^2-b^2)\,x-ncd = 0$. Accordingly, when comparison with the canonical equation $x^3-qx-r = 0$ is made, there ensues $b^2-c^2 = 0$, $a^2-b^2 = -q$ and $ncd = r$, so that, as Newton ordains, $CA = AX$, $KB = a-b = -q/n$ and $CX = cd/n = r/n^2$. The original 'Modus 2' is completed by a geometrical demonstration *à l'ancienne* (see II: 472–4) which Newton here forgoes.

Coroll. Hinc etiam duæ medi[æ] proportionales inter quantitates quascunæ datas n et $\dfrac{r}{nn}$ inveniri possunt. Nam si desit terminus q, id est si puncta K et B coincidant, rectæ AK, CE, KY et CX erunt continue proportionales.[36]

Prob. 3.

Æquationem quamcunæ cubicam $x^3 = px^2 + qx + r$ modis infinitis construere.[37]

Multiplicetur æquatio $x^3 - px^2 - qx - r = 0$ per $x + m = 0$ assumendo quantitatem m pro lubitu. Sit autem signum ipsius diversum a signo ipsius r.[38] Et æquationem prodeuntem $x^4 \begin{smallmatrix} -p \\ +m \end{smallmatrix} x^3 \begin{smallmatrix} -q \\ -mp \end{smallmatrix} xx \begin{smallmatrix} -r \\ -mq \end{smallmatrix} x - mr = 0$ construe per Propositionem sequentem.

Prob. 4.

Æquationem quamcunæ biquadraticam $x^4 = px^3 + qxx + rx + s$ construere, cujus terminus primus[39] & ultimus ejusdem sunt signi.[40]

Cas. 1. Si habeatur $-q$, ducatur $CX = \frac{1}{2}p$ et $GX = \dfrac{2s}{r}$ idæ ad contrarias partes si sint ejusdem signi, aliter ad [easdem]. Erige perpendiculum CK et angulum rectum XCK subtende recta $KX = \sqrt{\dfrac{rr}{4s} - q}$. Comple parallelogrammum $KGXF$ et inter rectas GX, XF utrinæ productas inscribe $EY = \dfrac{r}{2\sqrt{s}}$ et erit XY radix æquationis.[41]

Cas. 2. Si habeatur $+q$, ducatur $CX = \dfrac{r}{2\sqrt{s}}$ et $GX = \sqrt{\dfrac{4s}{pp}}$. Erige perpendiculum CK et angulum

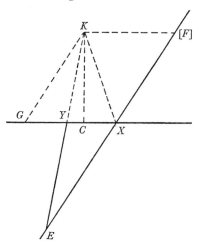

---

(36) A verbal amplification of the more pithy original 'Coroll.' on II: 474. This also appeared in the parent 'Problems for construing æquations' as an 'Idem aliter' in its own right to the initial 'Prob. 15' (see II: 462, and compare *ibid.*: 462–3, note (35)).

(37) Newton initially here repeated the enunciation of 'Prob. 3' of the parent text (see II: 474), namely '*Æquationem tertio termino carentem $x^3 = pxx + r$ [infinitis modis] construere*' (*To construct the equation $x^3 = px^2 + r$ lacking its third term [in an infinity of ways]*), but straightaway cancelled this heading. He doubtless there had it in mind, as in 'Prob. 2' preceding, to add (without demonstration) the two complementary constructions by rectilinear and circular neusis which follow on II: 474–6.

(38) Newton intends the opposite, *viz.* 'Sit autem signum ipsius [m] idem cum signo ipsius r' (but let the sign of m be the same as that of r). Problem 4 can be applied to construct the resulting quartic equation only when the product mr is positive.

Corollary. Hence also can there be found two mean proportionals between any given quantities n and r/n^2 whatever. For if the term (in) q be lacking, that is, if the points K and B coincide, the straight lines AK, CE, KY and CX will be in continued proportion.[36]

Problem 3.

To construct any cubic equation $x^3 = px^2 + qx + r$ *whatever in an infinity of ways.*[37]

Multiply the equation $x^3 - px^2 - qx - r = 0$ by $x + m = 0$, assuming the quantity m at pleasure (but let its sign be different from that of r[38]). And construct the resulting equation

$$x^4 - (p-m)\,x^3 - (q+mp)\,x^2 - (r+mq)\,x - mr = 0$$

by the following proposition.

Problem 4.

To construct any quartic equation $x^4 = px^3 + qx^2 + rx + s$ *whatever,
the first*[39] *and last terms in which are of same sign.*[40]

Case 1. If q be had negative, draw $CX = \tfrac{1}{2}p$ and $GX = 2s/r$; this on opposite sides should they be of same sign, otherwise on the same side. Erect the perpendicular CK, and subtend the right angle $\widehat{XCK}$ by the straight line $KX = \sqrt{(r^2/4s - q)}$. Complete the parallelogram $KGXF$, and between the lines GX, XF extended either way inscribe $EY = r/2\sqrt{s}$: then XY will be a root of the equation.[41]

Case 2. If q be had positive, draw $CX = r/2\sqrt{s}$ and $GX = 2\sqrt{s}/p$. Erect the perpendicular CK and subtend the right angle $\widehat{XCK}$ by the straight line

---

(39) Namely $+x^4$; in other words, the constant s in the quartic's equation (as it is ordered) must be positive, whence it must not have an even number of roots of the same sign. Where a given quartic does have a pair or pairs of roots of same sign, its geometrical construction will have to be effected either by a less than elegant and more complicated refashioning of the following rectilinear neuses (compare note (62) of the Appendix) or by returning to the fussy and none-too-accurate determining of the meets of intersecting circle and ellipse/hyperbola which he presented in 'Prob. 6' and 'Prob. 7' of the parent text (see II: 490–8) and which elsewhere he now rejects.

(40) Newton's preliminary computations for the following neusis (as they survive on Add. 3965.11: 150^r) are reproduced in [3] of the Appendix. We adduce their gist in footnote to each of the two cases of construction which he proceeds to distinguish.

(41) On setting $KF = a$, $KX = b$, $CX = c\,(\leqslant b)$, $EY = d$ and $XY = x$, the equality $\sqrt{(KX^2 + XY^2 - 2CX \times XY)} = (KY \text{ or}) \, EY \times GY/XY$ at once yields $\sqrt{(x^2 - 2cx + b^2)} = d(a-x)/x$; that is, on squaring and reordering $x^4 = 2cx^3 - (b^2 - d^2)\,x^2 - 2ad^2x + a^2d^2$. And when this is identified with the given equation $x^4 = px^3 + qx^2 + rx + s$ there comes $2c = p$, $b^2 - d^2 = -q$, $2ad^2 = -r$ and $a^2d^2 = s$, whence conversely $CX = c = \tfrac{1}{2}p$, $KF = GX = a = -2s/r$ and $KX = b = \sqrt{(d^2 - q)}$ where $EY = d = \sqrt{(r^2/4s)}$. Compare notes (54) and (59) of the Appendix.

rectum XCK subtende recta $XK = \sqrt{\frac{1}{4}pp + q}$. Comple parallelogrammum $KGXF$ et inter rectas GX, XF utrincg productas inscribe $EY = \frac{1}{2}p$. et erit EK radix æquationis.[42]

---

(42) Since there is $EK = GX \times EY/XY$, the substitution $x \to ad/x$ at once produces the quartic equation $x^4 = 2dx^3 + (b^2 - d^2)\, x^2 - 2acdx + a^2d^2$; and when this is identified with $x^4 = px^3 + qx^2 + rx + s$ there ensues $2d = p$, $b^2 - d^2 = q$, $2acd = -r$ and $a^2d^2 = s$, so that now $CX = c = \sqrt{(r^2/4s)}$, $KF = GX = a = \sqrt{(4s/p^2)}$ and $KX = b = \sqrt{(d^2 + q)}$ where $EY = d = \frac{1}{2}p$. Compare notes (57) and (60) of the Appendix.

The manuscript text here terminates (at the bottom of f. 118^r). Newton would no doubt have wished to add a further 'Prob. 5' exhibiting some equally elegant construction by conchoidal neusis of the complementary case of the quartic equation $x^4 = px^3 \pm qx^2 \pm rx - s$, but no evidence exists that he attempted to contrive one.

per puncta K, C et X describe circulum. Junge AX et produc ~
doare iterum secet circulum KCX, puta in G. Denique inter hunc
circulum et rectam KC utrinq producta inscribe rectam
EY ejusdem longitudinis cum recta AC & convergentem ad punctum
G. Et acta EC erit radix æquationis. Radices autem affirmativæ
cadunt in majori circuli segmento KGC si habeatur $+r$. et in
minori KFC si habeatur $-r$.

Coroll. Hinc etiam duæ mediæ proportionales inter quantitates
quascunq datas, n et $\frac{r}{nn}$ inveniri possunt. Nam si dictæ terminis
q, id est si puncta K et B coincidant, rectæ Ak, CE, KY et CX erunt
continue proportionales.

Prob. 3

~~Æquationem tertio termino carentem~~
$$x^3 = pxx + r \quad \text{construere}$$
Æquationem quamcunq cubicam $x^3 = px^2 + qx + r$ modis
infinitis construere.

Multiplicetur æquatio $x^3 - px^2 - qx - r = 0$ per $x \pm m = 0$
assumendo quantitatem m pro libito. Sit autem signum ipsius diversum
a signo ipsius r. Et æquationem productam $x^4 - px^3 - qxx - rx - m \atop +m \ -mp \ -mq \ -mq$
$= 0$ construe per Propositionem sequentem.

Prob. 4

Æquationem quamcunq biquadraticam $x^4 = px^3 + qxx + rx + s$
construere, cujus terminus primus & ultimus ejusdem
sunt signi.

Cas. 1. Si habeatur $-q$. ducatur
$CX = \frac{1}{2}p$ et $GX = \frac{s}{r}$ ad contrarias
partes si sint ejusdem signi, aliter ad
contrarias. Erige perpendiculum CK et
angulum rectum XCK subtende recta
$kX = \sqrt{\frac{rr}{4s} - q}$. Comple parallelogrammum
$KGXF$ et inter rectas GX XF utrinq
productas, inscribe $EY = \frac{r}{2\sqrt{s}}$ et erit XY radix æquationis.

Cas. 2. Si habeatur $+q$, ducetur $CX = \frac{r}{2\sqrt{s}}$ et $GX = \frac{\sqrt{4s}}{pp}$
Erige perpendiculum CK et angulum rectum XCK subtende
recta $XK = \sqrt{\frac{1}{4}pp + q}$. Comple parallelogrammum $KGXF$
et inter rectas GX, XF utrinq productas inscribe $EY = \frac{1}{2}p$.
et erit EK radix æquationis.

Plate II. Geometrical construction of the real roots of a quartic equation (**1, 3, §3**).

$XK=\sqrt{(\frac{1}{4}p^2+q)}$. Complete the parallelogram $KGXF$, and between the lines GX, XF extended either way inscribe $EY=\frac{1}{2}p$: then EK will be a root of the equation.[42]

APPENDIX. PRELIMINARY DRAFTS AND COMPUTATIONS FOR THE 'CONSTRUCTION OF GEOMETRICAL PROBLEMS'.[1]

From the originals in the University Library, Cambridge

[1][2]   DE CONSTRUCTIONE PROBLEMATUM.

Definitiones

1. Extensum vel magnitudo lege data moveri dicitur[3] quando motus ejus ab alijs extensis mutuo contactu determinatur.

2. Positio in qua movens sistendum est dari dicitur ubi occurrit aliud positione datum extensum quod mutuo contactu potest eidem obsistere.

Postulata

1. Extensum quodlibet quacunꝗ data lege[4] movere.

2. Movens in quacunꝗ data positione sistere.

3. Et extensa[5] describere quæ per cursus moventium vel positiones quiescentium designantur.

Prob. 1.

Lineam rectam ducere.[6]

Circum data duo quævis[7] puncta ceu polos revolvatur solidum longitudinis

---

(1) These comprehend, firstly, Newton's initial scheme for an expansion of the opening portion of the earlier 'Problems for construing æquations' (II: 450–66) into an extended treatise on such geometrical construction, the enunciation of whose 28 problems are set out in [1], followed by, in [2], his provisional revision of this into a tighter tract having but 16 problems; and also, in [3], his preliminary calculations for the terminal Problem 4 in the final section of the finished tract reproduced in §3 preceding, together with his first verbal drafting of this.

(2) Add. 3963.11: 115ʳ–116ʳ. As in [2] following, we here largely confine our editorial commentary to citing verbal variants in the manuscript text and to making pertinent back-reference to the parent 'Problems for construing æquations'.

(3) Initially 'in data ratione movetur' as in the equivalent *Definitio* 1 on II: 450.

(4) Newton first here copied 'ratione' from the parent text (compare II: 452).

(5) Initially altered to be 'lineas' and then changed back.

(6) Newton originally embraced the following problem also in the joint enunciation '*Lineam rectam longitudinis cujuscunꝗ ducere & continuo producere*'; compare *Problema 1* on II: 452, whose construction is repeated in the sequel with but trivial verbal variant.

(7) 'utcunꝗ distantia' is deleted. Understanding the figure of the corresponding 'Prob. 1' of his parent text (see II: 452), Newton in a subsequent interlineation specified these two points—and the line drawn through them in sequel—as '*A, B*'.

cujuscunq & per ejus puncta omnia quiescentia ducatur linea per Postulat[um] 3. Dico factum.

Si negas sit[8]

Prob. 2.

Lineam rectam continuo producere.

Ducatur alia recta per Prob. 1, et moveatur eadem ea lege ut ejus puncta duo quævis incidant semper in lineam producendam per Post. 1, et interea per totum moventis cursum ducatur linea per Post. 3. Dico factum.

Prob. 3.

Per data duo puncta rectam ducere.

Recta quævis ducta & quantum satis est producta moveatur lege quacunq per Post. 1 donec eadem punctis duobus datis occurrat per Post. 2 et per ejus puncta quiescentia ducatur linea, quæ hinc inde ad puncta illa duo data terminetur per Post. 2 et 3. Dico factum.

Prob. 4.

Planum per data tria puncta describere.[9]

Per duo punctorum ducatur linea recta per Prob. 3 & producatur eadem pro lubitu per Prob. 2. Applicetur alia recta tam ad punctum tertium quam ad lineam jam ductam per Post. 2[10] & revolvatur eadem circum hoc punctum tertium ceu polum ea lege ut lineam illam per puncta duo ductam perpetuo tangat per Post. 1. Dico factum.

[11]*Prob. 5.*

Dato centro et intervallo in dato plano circulum describere.[12]

---

(8) Three lines are left blank in the manuscript for the later insertion of this *reductio* proof from the parent text, *viz.* (see II: 452) 'Si negas sit alia quævis recta circa præfatam volvens...: et rectæ...comprehendent spatium contra Ax. 14 primi *Elem*'. The two problems which follow are newly contrived, but merely round out the basic 'Prob. 1' on II: 452.

(9) Essentially 'Prob. 2' on II: 452–4.

(10) A reference to 'Prob. 3' is replaced here.

(11) In the manuscript (f. 115ᵛ) Problems 5, 6, 7, 8, 9 were initially numbered 5, 13, 14, 6 and 7(+8); Problem 10 is a late insertion not originally present; and Problems 13, 14, 20 and 21 were first 13B, 10, 9B and 9C respectively. We omit from our text (in which Newton's problems are reordered in the sequence of their final numberings) an original '*Prob: 9.* Inter datum punctum et datam lineam ponere datam longitudinem' which was briefly reborn as a first 'Prob. VIII' in the revised listing in [2] (see note (38) below) before being cancelled.

(12) 'Prob. 3' of the parent text (see II: 454).

[*Prob.*] *6.*

Super data recta terminata in dato plano triangulum æquilaterum constituere.[13]

Prob. 7.

Angulum rectum construere.[14]

Schol. Simili angulorum et figurarum translatione describi possunt anguli angulis et figuræ figuris similes et æquales. Angulus quoq3 rectus semel constructus[15] pro norma constituendi[15] rectos adhiberi possit. Et his principijs Problemata omnia Euclidea expedite construuntur.

Prob. 8.

A dato puncto in dato plano rectam ducere datæ rectæ æqualem.[16]

Prob. 9.

Ad datū punctum in data recta ponere datam longitudinem.[17]

Prob. [10].[18]

Datum rectam bisecare.

Prob. 11.

A dato puncto rectam ducere datæ rectæ perpendicularem.[19]

Idem aliter.

---

(13) Initially enunciated, in copy of 'Prob. 10' of the parent text (see II: 458), as 'Datis lateribus triangulum constituere'.

(14) There seems little point in doing so until, as in the two simplest Euclidean instances of it (*Elements* I, 11 and 12, which Newton introduces in sequel as his 'Prob. 12' and 'Prob. 11' respectively), particular conditions under which the right angle is set to be constructed are specified.

(15) Newton first wrote 'descriptus' and 'ducendi' respectively.

(16) 'Prob. 4' of the parent 'Problems' (see II: 454). As Newton there stresses in his subsequent scholium, this comprehends the next problem as its primary case.

(17) This combines (compare note (12) above) an initial '*Prob. 7. Ad datam rectam longitudinem addere datæ cuivis rectæ æqualem*' and '*Prob. 8. A data recta longitudinem auferre datæ cuivis rectæ æqualem*'.

(18) Numbered '9' by Newton in the manuscript, but we take advantage of his advancing of what was initially '*Prob. 10. Angulum rectum construere*' to seventh place to make his listing here consistent. He would presumably resolve the problem by constructing the perpendicular bisector of the given line as in *Elements* I, 10.

(19) 'Prob. 9' of the parent text (see II: 458), there solved by constructing the semicircle on a diameter one end-point of which is the given point and the other any point in the given line. The alternative construction of the following '*Idem aliter*' would doubtless be that of *Elements* I, 12, where the foot of the perpendicular required is found, by way of 'Prob. 10' preceding, as the mid-point of the chord intercepted in the given line by an arbitrary circle drawn with the given point as its centre.

Prob. 12.

A data recta ad quodvis ejus punctum perpendiculum erigere.[20]

Prob. 13.

Ad rectam positione datam in dato angulo rectam aliam per
datum punctum ducere.[21]

Prob. 14.

A dato puncto rectam ducere datæ rectæ parallelam.[22]

Prob. 16.[23]

Rectam lineam in data ratione secare.

Prob. 17.

Invenire quartam proportionalem.[24]

Prob. 18.[25]

Invenire continue proportionales in ratione data.

Prob. 19.

Invenire mediam proportionalem.[26]

Prob. 20.

Inter datas duas lineas ponere datam longitudinem quæ ad
datum punctum converget.[27]

---

(20) 'Prob. 8' of the parent 'Problems' (see II: 456).

(21) Case 2 of 'Prob. 4' on II: 454.

(22) 'Prob. 7' of the parent text (II: 456).

(23) With the delaying of 'Prob. 15' to be 'Prob. 18' (see note (25) below) the manuscript lacks a 'Prob. 15'. The construction of the present problem would evidently (as in *Elements* VI, 10) proceed by drawing an arbitrary straight line divided in the given ratio through one of the end-points, and then a parallel to the line joining the other end-points from the section point in the auxiliary line to the given one.

(24) 'Prob. 13' on II: 458, one again readily constructed by drawing parallels (as in *Elements* VI, 2).

(25) Initially set to precede the two previous problems in the manuscript (f. 116ʳ). The construction is manifestly by a repeated application of 'Prob. 17' in the particular case where the two middle terms of the proportion are equal.

(26) 'Prob. 14' on II: 460 (whose construction is effectively that of *Elements* VI, 13).

(27) The general Archimedean neusis whose construction by continuously moving the given line till it shall lie through the given point is postulated in 'Prob. 5' on II: 456. Equivalent constructions by a conchoid and through the meets of a circle and hyperbola were afterwards adduced by Newton in his Lucasian lectures on algebra (see V: 428–30).

Prob. 21.

Datas tres lineas linea recta secare cujus partes interceptæ
datarum erunt longitudinum.[28]

[Prob.] 22.

Invenire duo media proportionalia.[29]

[Prob.] 23.

Datum angulum bifariam secare[30] *idq̃ continuo.*

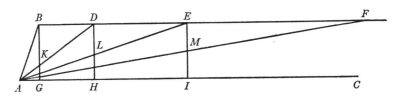

[Prob.] 24.

Datum angulum trifariam secare idq̃ continuo.[31]

Detur angulus *BAC* et hujus lateri alterutro *AC* age parallelam *BF* lateri
alteri *AB* occurrentem in *B*. A puncto *B* ad rectam *AC* demitte perpendiculum
BG et inter *BG* et *BF* ponatur longitudo *KD* quæ sit duplæ *AB* æqualis et convergat ad punctum *A*: et angulus *BAC* trifariam secabitur. Demitte perpendiculum *DH* et inter *DH* et *DF* ponatur longitudo *LE* quæ sit duplæ *AD* æqualis
et convergat ad punctum *A*, et angulus *DAC* trifariam secabitur. Demitte[32]

---

(28) An interesting 'plane' problem, not found in the parent 'Problems for construing
æquations', but which Newton had elegantly constructed in 1685 as the degenerate case of
Lemma XXVI of his 'De motu Corporum Liber primus'—that sketchily outlined in its
'Corol.' (see VI: 288, and compare *Philosophiæ Naturalis Principia Mathematica*, ₁1687: 97)—
where the interposed *triangulum specie et magnitudine datum* collapses into its greatest side.

(29) A variety of neusis constructions of this are set out in 'Prob. 15' of the parent text
(II: 462–4).

(30) Initially just '*bisecare*', as in the parent 'Prob. 16' on II: 464.

(31) The following construction is but a verbal reshaping of the equivalent 'Prob. 17' on
II: 464.

(32) Newton breaks off, finding no more room to complete the statement of his construction
on his manuscript page (f. 116ʳ, where it is squashed in after the enunciation of Problem 16,
above and around two other deleted ones—provisionally numbered '20' and '21'—requiring
respectively 'Invenire summam quadratorum' and 'Invenire differentiam quadratorum').
Understand in conclusion, much as before: 'Demitte perpendiculum *EI* et inter *EI* et *EF*
ponatur longitudo *MF* quæ sit duplæ *AE* æqualis et convergat ad punctum *A*, et angulus
EAC trifariam secabitur. Et sic deinceps in infinitum.'

[*Prob.*] 25.

Anguli recti partem quintam invenire.[33]

[*Prob.*] 26.

Anguli recti partem septimam invenire.[34]

[*Prob.*] 27.

A dato puncto rectam ducere quæ circulum describendum tanget.[35]

[*Prob.*] 28.

A dato puncto rectam ducere quæ curvam descriptam tanget.[36]

[2][37]   *Prob.* [*V*]

Dato centro et intervallo in dato plano circulum describere.

Prob. VI.

Ad datū punctū in data recta ponere datam longitudinem.

Prob. [*VII.*]

Ad datam rectam longitudinem addere datæ cuivis rectæ æqualem.

---

(33) 'Prob. 19' on II: 466, where it is given twin constructions which are essentially *Elements* IV, 10.

(34) 'Prob. 20' on II: 466, whose elegant neusis construction is here doubtless to be repeated in some verbally rephrased form.

(35) Newton would, we suppose, here employ the circle having the given external point and the given circle's centre as its diametral end-points directly to construct the points of tangency (see VI: 228), rather than make use of the slightly circuitous method expounded in *Elements* III, 17.

(36) This general problem of drawing tangents to a curve was doubtless here to be constructed by a Robervallian composition of the component motions determining the curve's instantaneous direction. We need not dwell upon the variety of solutions which Newton had previously (see particularly I: 377–82; IV: 476–82; VII: 456–68) given to this fundamental problem of geometrical fluxions.

(37) Add. 3963.11: 115ᵛ/116ᵛ, an immediate revise of [1] preceding in which, to avoid confusion, the problems are now ordered by Roman numerals. We omit the introductory *Definitiones* and *Postulata* which, together with the opening Problems I–IV, are to be understood to carry over without change in their statement.

Prob. IIX.[38]

Inter datas duas lineas ponere datam longitudinem quae ad datum punctum converget.[39]

Prob. IX.

*Datas tres lineas linea recta secare cujus partes interceptæ
datarum erunt longitudinum.*[39]

[40]*Prob. XI.*

Dato centro et intervallo sphæram describere.

Hoc centro et intervallo describatur circulus & hic circulus circum quamvis ejus diametrum revolvatur per Post. 1 & describatur figura solida quæ per cursum circuli designatur per Post. 3.

Prob. XII.

Dato axe et intervallo cylindrum describere.

Circum axem ad intervallum illud revolvatur linea recta axi parallela per Post. 1, & describatur figura solida cujus superficiem recta illa revolvendo d[e]signat per post. 3.

---

(38) That is, 'VIII' in more conventional equivalent. We have already mentioned (note (12) above) that Newton here momentarily took over from his first draft, where it equally finds no place, an initial 'Prob. VIII' requiring *Inter datum punctum et datam lineam ponere datam longitudinem* before cancelling it—doubtless having come to realize that its construction is a corollary to Problem V which serves no distinct purpose in its own right.

(39) Problems 20 and 21 respectively in the listing in [1] preceding. In a further revise sketchily adumbrated at the bottom of f. 116ʳ these are renumbered as 'Prob. 1' and 'Prob. 2' respectively, and introduced by an inserted scholium to the opening Problems I–V that 'His Principijs problemata omnia plana, sequentibus omnia solida construuntur'. Newton here, we may add, overlooks that, because (as he had shown more than twenty years before; see note (28) above) the latter problem is accomplishable by a 'plane' ruler-and-compasses construction, it cannot in turn be applied to construct any problem which is of a grade higher than plane. The slip is, as we have seen, carried over into the finished text in §3 preceding.

(40) The manuscript lacks any 'Prob. X'. The next four problems in solid geometry begin a new page (f. 116ᵛ) in the manuscript, and were there originally numbered 1–4. It may be that Newton meant these to follow a second inserted scholium at the bottom of the previous page (f. 116ʳ) where—in sequel to the tentative recasting of the two preceding problems which we record in note (39)—he began to write 'Ut harmonia inter constructiones solidorum Problematum per hoc Principium et constructiones planorum Problematum per principia planorum innotescat [? etiam sequentia adhibere præstat]' before breaking off in mid-sentence. Two small interlineations in the stated construction of Problem XII below, upon which it would be tediously pointless otherwise to insist, serve to confirm one's general impression that all four are here newly reborn out of his earlier, somewhat vaguer efforts to found elementary solid geometry upon a firm Euclidean basis (compare VII: 382 ff.).

Prob. XIII.

Dato axe et angulo conum describere.

Circum axem revolvatur angulus per Post. 1 et describatur figura solida quam angulus revolvendo designat per post. 3.

Prob. XIV.

Dato plano datum conum secare.

Datum planum producatur donec conum secet per Pro[b.] 2 & 4.

Schol.[41]

A mensurandis terris Geometria originem duxit[42] & huic negotio Postulata Euclidea suffecerint: ideoq Geometræ antiquissimi[43] hæc sola adhibuerunt. Sphæræ cylindri et Coni definitiones[44] posuerunt ut horum solidorum investigarent proprietates. Descriptiones vero non postularunt quia figuris hisce ad constructiones Problematum nondum essent usi.[45] Tandem circa tempora Platonis cum Problema duplicandi cubū proponeretur & per Sectionem Coni solveretur: postularunt aliqui Conum datum dato plano secare, quasi Conus per solam suam Definitionem daretur in Geometria. Sufficiunt Definitiones figurarum ad Theoremata de ijs investiganda, non autem ad figuras ipsas describendas. Linea recta et circulus non habentur in Geometria per Definitiones suas. Ut habeantur, postularunt Geometræ easdem describere. Et similit[e]r ut habeatur Conus ad constructionem problematum debet ejus descriptio vel postulari vel per Postulata doceri: aliter constructio non erit Geometrica. Ad has igitur constructiones duo requiruntur Postulata, *Conum describere*, [et] *descriptum dato plano secare.*[46] Et hæ operationes sunt difficiles[47] & erroribus obnoxiæ, et aptiores quæ ex principijs simplicioribus doceantur quam quæ pro Principijs habeantur. Ideoq recentiores Geometræ sectionem in

---

(41) A first version of the scholium on pages 206–8 above where Newton supports his ensuing preference (compare v: 422–8, 468–70) for using Archimedean rectilinear neuses in constructing 'solid' problems, rather than equivalently determining the meets of auxiliary circles and conics, both by appealing to classical authority in urging the logical unsatisfactoriness of employing curves whose definition requires prior postulate of the sectioning of a cone (compare vii: 290–2, 298, 382–4) and also by hinting at the practical difficulties of accurately drawing such constructing conics *in plano.*

(42) Newton first began little differently: 'Geometria in mensurandis terris initium habuit'.

(43) 'primi' is replaced.

(44) 'etiam' was here tentatively afterwards inserted and then deleted.

(45) Initially 'figuras hasce adhibuerunt'.

(46) Newton originally added a third postulate '& *sectionem ad problemata solvenda adhibere*', later cancelling it no doubt because of its triviality.

(47) An incomplete preceding 'comp[lexæ]' is here deleted.

plano[48] sine constructione solidorum describere docent. Archimedes vero majori cum judicio pro Principio habuit longitudinem datam & ad datum punctum convergentem inter datas duas lineas ponere. Et per hoc Principium Problemata omnia solida et nonnulla plusquam solida facilius et accuratius, & non minus ge[o]metrice solvi putamus quam per constructiones & sectiones solidorum vel sectionum descriptiones in plano. Quinetiam constructiones Archimedeæ[49] sunt Euclideis magis affines[,] ut conferenti constructiones Problematis VI VII VIII & IX inter se, atcs etiam constructiones sequentium duorum problematum inter se, manifestum erit.

Prob. XV.

Angulum datum bifariam secare idcs continuo.[50]

Prob. XVI.

Angulum datum trifariam secare idcs continuo.[50]

Si descriptiones Curvarum in Geometriam introducendæ sunt, hoc optime præstabitur per postu[l]ata tria a nobis posita et hæc sola ad operationes omnes Geometricas sufficiunt. Si Postulata tria Euclidea ob antiquitatem retinenda sunt, adjici potest quarto loco Postulatum Archimedeum ponendi inter duas positione datas longitudinem datam quæ ad datum punctum converget & per hoc Postulatum abscs descriptione curvarum construētur Problemata omnia solida. Si Postulatum datos conos datis planis secandi ob antiquitatem etiam retinendū sit, addendum erit Postulatum quintum conos describendi et per hæc duo Postulata construendum est Prob. VIII ubi lineæ datæ rectæ sunt et tum demum per hoc Problema construentur solida omnia. Cum autem Prob. IX ubi lineæ datæ rectæ sunt per Prob. VIII construi possit[,] nos per hæc duo Problema[ta] constructionem omnium solidorum in sequentibus docebimus.[51]

---

(48) A not necessarily true specification 'per operationes mechanicas' is struck out in sequel.

(49) So changed from 'Archimedis', which in turn replaces 'per hæc duo Problemata'. Newton here intends Problems 'IIX' and IX above—and hence would seem to have already determined to omit the intervening Problems XI–XIV (which he took over into his final text in §3 preceding, there to cancel them)—but only the first of these can be applied to construct 'solid' problems (see note (39)).

(50) In the manuscript (f. 116ᵛ) a short space is left after each of these enunciations for the intended later insertion of a précis of their construction; for which compare 'Prob. 16' and 'Prob. 17' of the parent 'Problems for construing æquations' (II: 464).

(51) Understand that there are now to follow the four final problems of the 'De constructione' where Newton sets out his preferred modes of geometrically constructing the (real) roots of quadratic, cubic and quartic equations.

[3]$^{(52)}$ [In figura$^{(53)}$ pone] $KF = a$.

$KX = b$. $CX = c$.$^{(54)}$ $EY = d$. $XY = x$.

[necnon] $KG = e$. $GC = f$. $GY = v$.

[Est] $KG^q + GY^q - 2CGY = KY^q$ [seu]

$$ee + vv - 2fv = KY^2.$$

[Ut et $YX.EY :: GY.KY$ id est]

$$v - a . d :: v . KY \left[= \frac{dv}{v-a} \right].$$

[Quare] $\dfrac{ddvv}{vv - 2av + aa} = ee + vv - 2fv.$

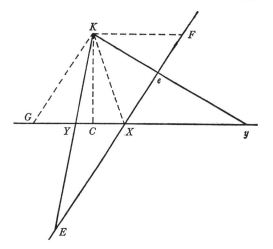

[et reducendo prodit]

$$v^4 {-2f \atop -2a} v^3 + [4] af \, {+ee \atop +aa \atop -dd} vv \, {-2aee \atop -2aaf} v + aaee [= 0].^{(55)}$$

(52) Add. 3965.11: 150^r/151^v, preparatory computations for the terminal 'Prob. 4' of the 'De constructione Problematum Geometricorum', together with a first draft of its verbal text. Below the calculations on f. 150^r Newton has set down an observation, relating (though this is not said) to the comet of 1680/1 which was his prime example of such solar parabolic orbit in his *Principia* (compare VI: 504–6, note (26)), that 'Nod[us descendens erat] in ♑ 2$^{[gr]}$.2'. Distantia Aphelij a Nodo in plano orbis 9$^{[gr]}$. 22$^{[']}$. 48$^{["]}$. rectius 9. 22. 24'. These values for the longitude of the comet's first intersection with the ecliptic and for the uncorrected angular distance between this descending node and the comet's point of (true) perihelion are those which Halley had communicated to him on 21 October 1695 in a postscript stating that, by his computation, 'The [ascending] node of the Comet being ♋ 2°. 2'. the Angle in the plain of the Orb between the Node and [Peri]helion is 9°. 22'. 48″...' (E. F. MacPike, *The Correspondence and Papers of Edmond Halley* (Oxford, 1932): 96 = *The Correspondence of Isaac Newton*, **4**, 1967: 183). Why Newton here thought it 'more correct' to reduce Halley's angle by 24' is not clear to us. His renewed interest in the 1680/1 comet at this time was doubtless again inspired by Halley, who had recently, in June 1705, published both at Oxford and in London his celebrated *Astronomiæ Cometicæ Synopsis* (where his earlier calculated longitude of the *nodus ascendens* is maintained, but that of the *perihelion in orbe* is a little altered to be ♐ 22° 29' 30″). We may add that in the second edition of his *Principia* ($_2$1713: 458–9) Newton somewhat disingenuously referred only to a first recalculation of Halley's whose result had been passed by the latter to him on 7 October 1695 (MacPike, *Halley*: 93 = *Correspondence of Newton*, 4: 173), according to which, upon maintaining the nodes 'in ♋ & ♑ 1^{gr}· 53″ as in the *editio princeps* ($_1$1687: 494), '*Halleius* noster...distantiam Perihelii a Nodo ascendente, in Orbita Cometæ mensuratam, invenit esse 9^{gr}· 20″.

(53) Understand *ex conditionibus Problematis* that the fixed line-length $EY(= ey)$ is to be inclined between the given straight lines FX and GX so as to pass through the given point K, which Newton fixes in position by its oblique distances $KF = GX$ and $KG = FX$ from those lines (so completing the parallelogram $KFXG$). It will be evident from the sequel that KC is the perpendicular from K to GX.

[Pone etiam $KE = z$. $KY =$]$z - d = t$. [Est insuper $KX^q + XY^q - 2CXY = KY^q$ seu $bb + xx - 2cx = tt$. Ut et $KE . KF :: EY . XY$ sive] $z = t + d \left[. a :: d . x = \dfrac{ad}{z} \right]$. [Evadit]

$$\dfrac{-tt}{+bb} z^2 - [2]cadz + aadd = 0. ^{(56)} \quad \text{[et reducendo prodit]}$$

$$t^4 = -2dt^3 \begin{matrix} --dd \\ +bb \end{matrix} tt \begin{matrix} +bbd \\ -[2]cad \end{matrix} [t] \begin{matrix} +bbdd \\ +aadd \\ -[2]acdd \end{matrix} . ^{(57)}$$

Prob. [4].

Æquationem biquadraticam $x^4 = px^3 + qx^2 + rx + s$ construere.

Cas. 1. Si habeatur $+s$ & $-q^{(58)}$ ducatur $CX = \frac{1}{2}p$ & $GX = -\dfrac{2s}{r}$. Erige per-

---

(54) Where of course, since KX is the hypotenuse of the triangle KCX, there is $b \geqslant c$. Newton's argument departing from this first choice of givens KF, KX, CX and interposed EY, to attain the quartic equation in XY whose roots determine the required end-points Y in GX, is not preserved, though it was (see note (59) below) to furnish the underlying *raison d'être* of Case 1 of 'Prob. 4' following. It is readily restored by deducing that, since

$$\text{`} KY^q = KX^q + CX^q - 2CXY \quad \text{seu} \quad bb + xx - 2cx \text{'}$$

and also '$XY . EY :: GY . KY = \dfrac{d, \overline{a - x}}{x}$', there results from eliminating KY between these two equations the quartic condition $x^4 = 2cx^3 - (b^2 - d^2) x^2 - 2ad^2x + a^2d^2$, whose real roots define the (2 or 4) possible sites of Y. Here, however, Newton passes to outline the analogous argument which from the givens KF, EY, KG and GC obtains the equivalent equation connecting the related unknown line-segment GY: one evidently derivable directly from the above quartic by making the substitutions $b = \sqrt{(e^2 + a^2 - 2af)}$, $c = \sqrt{(a^2 - f^2)}$ and $x = a - v$.

(55) We here trivially mend an erroneous term '$+2af$' in Newton's coefficient of v^2 in the manuscript.

(56) So we correct the manuscript text, where Newton both conjured up a phantom extra term '$-dd$' in the coefficient of z^2 and also omitted the necessary numerical factor in that of z. We see no point in following through the consequences of these slips in computation as they distort the resulting equation in t which follows, and have accordingly made appropriate adjustment in our reproduction of it. The errors are not further perpetuated and were manifestly (see the next note) soon corrected by Newton himself.

(57) Alternatively, on setting $t = z - d$ in the previous equation (or, equivalently, by substituting $x = ad/z$ in the quartic derived in note (54)) there likewise results the condition $z^4 = 2dz^3 + (b^2 - d^2) z^2 - 2acdz + a^2d^2$ determining the (2 or 4 possible) real values of KE which correspondingly fix the position of the other required end-points E in FX. This less complicated analogous quartic equation will provide the basis for Case 2 of the draft 'Prob. 4' which now follows (on f. 151^v in the manuscript; see note (60) below).

(58) These two complementary cases were not initially distinguished by Newton, who combined their attendant neusis constructions as a single '*Cas. 1.* Si habeatur $+s$ ducatur...' followed by an unqualified 'Vel sic. Ducatur...'.

pendiculum *CK* et angulum rectum *XCK* subtende recta $KX=\sqrt{\dfrac{rr}{4s}-q}$. Comple

parallelogrammum *KGXF* et inter rectas *GX*, *XF* utrinꝗ productas inscribe

$EY=\dfrac{r}{2\sqrt{s}}$ et erit *XY* radix æquationis.[59]

Cas. 2. Si habeatur $+s$ et $+q$[58] ducatur $CX=\dfrac{r}{2\sqrt{s}}$ et $GX=\sqrt{\dfrac{4s}{pp}}$. Erige per-

pendiculum *CK* et angulum rectum *XCK* subtende recta $KX=\sqrt{q+\tfrac{1}{4}pp}$. Comple
parallelogrammum *KGXF* et inter rectas *GX*, *XF* utrinꝗ productas inscribe
$EY=\tfrac{1}{2}p$ et erit *EK* radix æquationis.[60]
Cas. [*3*][61] Si habeatur $-s$ et[62]

---

(59) On identifying the coefficients of the equation now to be constructed with those of the
quartic derived in note (54) above, there results $2c = p$, $b^2-d^2 = -q$, $-2ad^2 = r$ and $a^2d^2 = s$,
whence $c = \tfrac{1}{2}p$ and also $a = -2s/r$, so that $d^2 = r^2/4s$ and consequently $b^2 = r^2/4s-q$.

(60) Here, on identifying the coefficients p, q, r, s with those of the quartic in z derived in
note (57), there ensues much as before $2d = p$, $b^2-d^2 = +q$, $-2acd = r$ and $a^2d^2 = s$, so that
$c^2 = r^2/4s$ and $d = \tfrac{1}{2}p$, and therefore $b^2 = \tfrac{1}{4}p^2+q$, $a^2 = 4s/p^2$.

(61) The superseded initial numbering '*2*' (on which see note (58) above) is left unamended
in the manuscript.

(62) Newton here breaks off, doubtless coming to realize that, since (where all four are real)
three of the possible positions of *Y* (in *GX*) and correspondingly of *E* (in *FX*) must, from
evident geometrical considerations, lie to one side of *X* with the fourth by itself on the other
side, the product of the related line-segments *XY* or *KE* must always be negative in sign;
whence neither of these can be roots of a quartic of form $x^4 \pm px^3 \pm qx^2 \pm rx+s = 0$. This
difficulty can—when not all the roots of the quartic are complex!—be got round by suitably
augmenting or diminishing the roots till one or three are of one sign and the fourth root is of
the other (or, in geometrical equivalent, by no longer choosing *X* to be the meet of the given
lines *FE* and *GY* in 'Cas. 1', and so on). But this destroys the basic simplicity of the neusis
construction, and Newton does not persevere with this 'Cas. 3' in his revised 'De constructione
Problematum Geometricorum' (§3 preceding).

§ 4. THE APOLLONIAN 'SECTIO RATIONIS' RESOLVED INTO A QUADRATIC EQUATION, AND THE GENERAL REDUCED QUARTIC CONSTRUCTED BY A PAIR OF INTERSECTING PARABOLIC LOCI.[1]

From a stray jotting[2] in the University Library, Cambridge

[1]

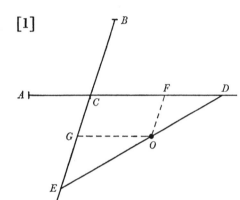

[Pone] $AF=a$. $CF=b$. $FO=c$.

$FD=x$.   [adeoqß] $\dfrac{bc}{x}=GE$. $BG=p$.

[Erit] $d.e::AD=a+x.BE=p+\dfrac{bc}{x}$.

[seu facta reductione]

$$xx=\frac{dp-ea}{e}x+\frac{dbc}{e}.^{(3)}$$

[2] [Pone] $AB=x$. $BC=y$. $Ag=a$.
$gO=b$.   [adeoqß] $Ch=b+y$. $Bg=a-x$.
$Ae=c$.   [sive] $eB=c+x$. $eg=c+a=d$.
$gk=e$.   [Fit] $fg,hC=b,\overline{a-x}.^{(4)}$ [seu]

$fg=\dfrac{ba-bx}{b+y}$.   [itaqß] $ef=d-\dfrac{ba-bx}{b+y}$.

$dB=\dfrac{ec+ex}{d}.^{(5)}$ [vel] $Cd=\dfrac{ec+ex+dy}{d}$.

[atqß] $Af^{(6)}=\dfrac{ay+bx}{b+y}$.

[Cape jam $Cd.Af::p.d$ et evadet]

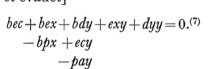

$$bec+bex+bdy+exy+dyy=0.^{(7)}$$
$$-bpx \ +ecy$$
$$-pay$$

(1) We know nothing of the background to these unattached calculations, and can make conjecture as to their purpose only from the sparse internal evidence of their text. A reduction of the ancient problem of the *sectio rationis* to algebraic equation leads on, in true roving mathematical spirit, to the consideration of generalizations of the problem which are given similar algebraic translation, and the two simplest parabolic instances are combined to yield a

Translation

[1]   Put $AF = a$, $CF = b$, $FO = c$, $FD = x$ (and hence $GE = bc/x$), $BG = p$. There will be $d:e = (AD \text{ or}) \ a+x : (BE \text{ or}) \ p+bc/x$, that is, when reduction is made, $x^2 = ((dp - ea)/e) \ x + dbc/e$.[3]

---

[2]   Put $AB = x$, $BC = y$, $Ag = a$, $gO = b$ and hence $Ch = b+y$, $Bg = a-x$; also $Ae = c$ or $eB = c+x$, together with $eg = c+a = d$, $gk = e$. There comes $fg \times hC = b(a-x)$[4] or $fg = (ba - bx)/(b+y)$ and so $ef = d - (ba-bx)/(b+y)$, $dB = (ec + ex)/d$[5] or $Cd = (ec + ex + dy)/d$, while Af[6] $= (ay + bx)/(b+y)$. Now take $Cd : Af = p : d$ and there will prove to be

$$bec + (be - bp) \ x + (bd + ec - pa) \ y + exy + dy^2 = 0.[7]$$

---

quartic lacking its second term. Given (see the next note) that their date of composition is narrowable to the latter part of 1705, it is tempting to connect these computations with the preceding tract 'De Constructione Problematum Geometricorum' and suggest that Newton had it in mind to reverse this sequence, employing the two parabolic loci he had come to define geometrically to construct the roots of the general reduced quartic equation. But this is mere guesswork on our part.

(2) This is entered by Newton at the head of a folio sheet (Add. 3970.3: 291ᵛ) where he subsequently penned rough numerical calculations and then a preliminary English draft of what, rendered into Latin by Samuel Clarke, afterwards appeared as Query 22 in his *Optice: sive de Reflexionibus, Refractionibus, Inflexionibus & Coloribus Lucis* (London, ₁1706): 319–21. (See Z. Bechler, 'Newton's law of forces which are inversely as the mass...', *Centaurus* 18, 1974: 184–222, especially p. 186 where the last two lines of the present manuscript text are caught in a photo-reproduction of the following optical computations.) Their ink and handwriting style gives every reason to suppose that these mathematical calculations were set down little in advance of their optical brethren, namely, towards the end of 1705.

(3) Apollonius' problem of the *sectio rationis* (on which see §2: note (57) preceding)—here that of drawing a straight line through the fixed point O to intersect the given straight lines AC and BC in D and E respectively such that AD is in a given ratio to BE—reduced to an equivalent quadratic equation in $DF = x$, where F is the oblique projection (parallel to BG) of O upon AC. Newton does not bother to specify the two solutions which are evidently possible.

(4) That is, $gO \times Bg$.

(5) Namely $gk \times eB/eg$.

(6) Or $Ag - fg$.

(7) The Cartesian equation (referred to origin A and coordinates $AB = x$, $BC = y$) of the locus (C)—evidently a hyperbola having asymptotes

$$y = b(p/e - 1) \quad \text{and} \quad dy + ex + ec - pa + (bd/e) \ p = 0$$

in the non-degenerate case—whose defining property is that the oblique distance Cd (parallel to DA) of C from the given line ek is ever in given ratio to the intercept Af cut off in the given line Ag by the rotating vector joining C to the fixed pole $O(a, b)$.

[Si recta] *ge* [fit] infinita $= Ae = d = c.$[8] [erit $Bd = gk = e$, adeoqᴣ] $Cd = e + y.$
[manente] $Af = \dfrac{ay + bx}{b + y}$. [Quare posito jam quod $Cd . Af :: p . q,$[9] fit]

$$yy + ey + be - \frac{pb}{q} x = 0.^{(10)}$$
$$+ by$$
$$- \frac{pa}{q} y$$

If $gk = 0$[11] & $\dfrac{p}{q} = \dfrac{b}{a},$[12] then $yy = \dfrac{bb}{a} x = rx.$

If $AD = x$. $DC = y.$[13] $gk = e$ & $\dfrac{p}{q} = \dfrac{b}{a},$[14] then $xx + ex + be - \dfrac{bb}{a} y = 0.$

Sit $\dfrac{bb}{a} = t$ & erit[15] $\dfrac{y^4}{rr} + \dfrac{eyy}{r} + be - \dfrac{bb}{a} y = 0.$ Or $y^4 * + eryy - r^2 ty + rrbe = 0.$[16]

---

(8) We would nowadays prefer equivalently to state that, in the limit as the point e passes to infinity in the line Af, the ratio of $eg = d$ to $eA = c$ tends to unity. Newton gives no special figure to illustrate this particular case, but we adjoin a suitably amended version of the one he sets to illustrate the general locus.

(9) Since he now takes $eg = d$ to be infinite, Newton can of course no longer employ d as one of the members by which he denotes the (finite) ratio of Cd to Af.

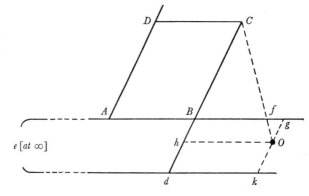

(10) With its term *exy* now become vanishingly small with respect to dy^2, the preceding Cartesian equation degenerates to be that of a parabola having its axis parallel to Ag ($y = 0$).

(11) That is, $e = 0$. In this further degenerate case dk comes to coincide with Bg (see the accompanying figure), so that the parabolic locus (C) is here defined by the constancy of the ratio of BC to Af.

(12) Whence (compare the previous note) $Af : BC = Ag : gO$. The ensuing equation may be directly derived from this proportion without recourse to algebra; for, since

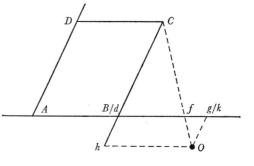

$$Af = AB + (Bf \text{ or}) BC \times Bg/Ch$$
$$= (gO \times AB + Ag \times BC)/(BC + gO),$$

there is $Ag/gO = Af/BC = (gO \times AB/BC + Ag)/(BC + gO) = gO \times AB/BC^2.$

If the straight line *ge* comes to be infinite, and so equal to *Ae*, or $d = c$,[8] there will be $Bd = gk = e$ and hence $Cd = e + y$, there remaining $Af = (ay + bx)/(b + y)$. As a consequence, now supposing that $Cd : Af = p : q$,[9] there comes to be

$$y^2 + (e + b - pa/q)\, y + be - (pb/q)\, x = 0.^{(10)}$$

If $gk = 0$[11] and $p/q = b/a$,[12] then $y^2 = (b^2/a)\, x = rx$.

If $AD = x$, $DC = y$,[13] $gk = e$ and $p/q = b/a$,[14] then $x^2 + ex + be - (b^2/a)\, y = 0$.

Let $b^2/a = t$ and there will[15] be $(1/r^2)\, y^4 + (e/r)\, y^2 + be - (b^2/a)\, y = 0$, or $y^4[+0y^3] + ery^2 - r^2ty + r^2be = 0.$[16]

---

(13) On interchanging the previous coordinates *x* and *y*, that is.

(14) So that (in the figure of note (8) above) there is $Af : Cd = Ag : gO$. This particular instance of the preceding locus is, we need scarcely insist, again a parabola with its axis parallel to *Ag* (now $x = 0$).

(15) On substituting $x = y^2/r$ in the most recent equation.

(16) This reduced quartic equation will of course, when solved, determine the meets of the parabolas $y^2 = rx$ and $x^2 + ex - ty + be = 0$ derived above. Conversely—a point which could not have been missed by Newton, and may well have been the prime motive for his making the present calculation (compare note (1) preceding)—the real roots of this quartic may evidently be discovered by constructing these parabolas from their defining geometrical 'symptoms' as above, and then plotting their points of intersection. Had this thought occurred to him, however, he would at once have seen how much easier is the equivalent construction through the meets of the parabola $y^2 = rx$ and the circle $x^2 + y^2 + (e - r)\, x - ty + be = 0$ along the lines pioneered seventy years earlier by Descartes in the third book of his *Geometrie* (on which see I: 494, note (11) and II: 491, note (101)).

4

THE 'METHOD OF [FINITE] DIFFERENCES'[1]

[late 1710][2]

§1. NEWTON'S PRELIMINARY AUGMENTATIONS OF THE 1676 PARENT TEXT.[3]

From the original drafts in the University Library, Cambridge

[1][4]   *Prop. V.*

Data serie Ordinatarum æqualibus ab invicem intervallis distantium, invenire Ordinatas intermedias.

[5]Erigantur ordinatæ intermediæ et produ[c]antur eædem donec occurrant curvæ generis Parabolici per terminos Ordina[ta]rum omnium datarum descriptæ.

---

(1) Alone out of the numerous products of Newton's mathematical pen which are known to us, the autograph (such as it might have been) of this familiar piece outlining general formulas for interpolation—and thence quadrature—by means of finite differences has seemingly perished. The work was first set in print by William Jones as the closing text (pages 93–101) in the collection of Newton's tracts and letter-extracts which he published under the omnium-gatherum title of *Analysis per Quantitates, Series, Fluxiones, ac Differentias: cum Enumeratione Linearum Tertii Ordinis* (London, 1711), there pithily remarking in editorial prelude that 'Coronidis loco subjicitur Tractatulus, cui titulo est, *Methodus Differentialis*, quem Cl. Auctoris permissu ex ejus autographo descripsi' (*Præfatio*: [c1ᵛ]) and adding for the sake of the un-tutored reader the information that 'Complectitur autem Doctrinam describendi Curvas ex datis Differentiis differentiarum Ordinatarum. Hæc *Methodus Differentialis* innititur Problemati ducendi Curvam Parabolici generis per data quotcunq puncta; de quo Cl. Auctor olim mentionem fecerat in Epistola sua ad *D. Oldenburgum* [24 Octob.] 1676 missa [where, see *The Correspondence of Isaac Newton*, 2, 1960: 119 and compare IV: 37, note (3), Newton had qualified it as 'fere ex pulcherrimis quæ solvere desiderem']; & cujus solutionem dedit in Lem. 5. Lib. 3. *Princip. Philos.* [₁1687: 481–3 (=IV: 70–3)][₃] non tamen prorsus eandem quoad Constructionem cum ea quam impræsentiarum tradimus'—as indeed is patently true, since Newton's formulas in his *Principia* are founded on the tabulation of advancing differences of the given arguments, not on the central differences which here provide an equivalent basis for interpolation. The transcript of the 'Methodus Differentiarum' which Jones 'wrote down' and subsequently used as printer's copy for the printed tract is, we assume, that which still survives in private possession, and whose text (differing only trivially from the printed version in minimal recastings to suit preferred editorial and typographical conventions) is here set out in §2. Three surviving preliminary draftings by Newton of what became its two final propositions (initially these were conjoined as a single 'Prop. V') and the concluding scholium precede

Translation

[1][4]   *Proposition V.*

Given a series of ordinates distant at equal intervals from one another, to find inter-mediate ordinates.

[5]Erect the intermediate ordinates [at their places] and extend them till they meet the curve of parabolic kind drawn through the end-points of all the given ordinates.

---

this in §1. Their handwriting style agrees with the bounds which may externally be fixed to the date of the third of these fragments (see §1.3: note (17) below) in pinpointing their com-position to a time in, or at least shortly before, 1710; and we may confidently infer—contrary to the long-standing tradition which advances the writing of the tract to 1676, building too rigidly on Newton's reference in his *epistola posterior* of that year to Leibniz (again see *Corre-spondence*, **2**: 119) to his having contrived a 'commoda expedita et generalis solutio...
Problematis *Curvam geometricam describere quæ per data quotcunq̃ puncta transibit*'—that the 'Methodus', in the form in which the world has known it since 1711, was put together out of earlier writings specially for publication by Jones in his *Analysis*. We know of no similar transitional drafts connecting its main Propositions I–IV with the parent scheme for inter-polation by central divided differences (see IV: 52–68) which Newton did set down, in his Waste Book, about October 1676, and which they so faithfully copy. Perhaps, indeed, there never was any full one. It would have been entirely in character for him to have given Jones only the verbal enunciations of Proposition I and II, and merely to have outlined how Propositions III and IV were to be modelled upon his original 'Prob. 1' and 'Prob. 2', leaving to his editor the labour of writing out a complete text of the augmented tract. This title, we may add, is presumably Newton's, although Jones would have been only too ready to provide one had he been asked (compare III: 33, note (3)).

(2) See the previous note. A period of several weeks must, we assume, have elapsed between Jones putting the newly composed 'Methodus' to the printer for publication in his *Analysis* and the book's appearance in mid-February 1711 (when—see S. P. Rigaud, *Correspondence of Scientific Men of the Seventeenth Century*, **1** (Oxford, 1841): 257 [=*Correspondence of Newton*, **5**, 1975: 94]—Roger Cotes wrote to Jones thanking him for his 'most valuable & acceptable Gift' of a volume 'collecting...those curious & usefull Treatises [of Sir Isaac's] which were before too much dispersed').

(3) These successive preliminary versions of what became the terminal Propositions V/VI and concluding scholium of the printed 'Methodus'—there, as here, ancillary to the general formulas for interpolation by central differences expounded in the opening Propositions I–IV in the image of their 1676 parent (see IV: 54–66)—are of no surpassing intrinsic importance, but they do provide witness to the painstaking trouble to which Newton, even in his late 60's, was prepared to go in refurbishing and sharpening the detail of his mathematical papers for publication. As ever in such cases, we here take note of textual variants in the main, delaying our detailed exegesis of technical points to our presentation of the finished content of the 'Methodus' in §2 following.

(4) Add. 4005.15: 71ʳ/71ᵛ, a slim slip of paper (evidently cut—or maybe torn—off the bottom of a larger sheet) which carries only the roughly drafted following text.

(5) Newton initially went on equivalently to specify: 'Per terminos ordinatarum omnium datarum describatur Curva generis Parabolici. Hæc enim abscindet Ordinatas quæsitas in locis suis erectas' (Through the end-points of all the given ordinates describe a curve of parabolic kind. For this will cut off the required ordinates when these are erected at their places).

Schol.

Ex hoc Theoremate generali derivari[6] possunt Theoremata particularia quibus Problema expeditius solvetur. Et utilis est hæc methodus ad Tabulas construendas per interpolationem serierum ... ... reducetur.[7]

Deinde curva satis accurate quadrabitur per generalia Theoremata superius exposita, vel etiam per compendia ad casus particulares applicata, qualia sunt quæ sequuntur.

    Cas. 1.[8]

[2][9]   *Prop. V.*

Datis aliquot terminis seriei cujuscunqȝ[10] ad data intervalla dispositis invenire terminum quemvis intermedium quamproxime.

Ad datam rectam et data intervalla erigantur termini dati in dato angulo & per eorum puncta extima per Propositiones præcedentes ducatur linea curva generis Parabolici. [11]Hæc enim abscindet terminos omnes intermedios per seriem totam.

                Schol.

Utilis est hæc methodus....[12] Ut si quatuor sint Ordinatæ A, B, C, D ad æqualia intervalla dispositæ Ordinata intermedia nova in medio omnium erit

$$\frac{9B + 9C - A - D}{16}^{[13]}$$

et Area inter Ordinatam primam et quartam erit[14]

[Et nota quod ubi]...reducetur.[15] Et tum demum[16]

---

    (6) 'formari' (be formed) was first written.

    (7) Understand some preliminary version—evidently entered by Newton on a separate sheet which no longer survives—of the two opening paragraphs of the terminal scholium of the 'Methodus' (see pages 252–3 below), but one which did not (compare the next note) adduce any particular 'compendious' rules of numerical quadrature applicable in individual cases.

    (8) The draft here breaks off, but we may reasonably suppose that Newton intended to adjoin two or more of the easier cases 'Of Quadrature by Ordinates' (see VII: 690–4) which he had successively derived *per analogiam* from the simple trapezoidal approximation fifteen years before, and whose 'Cas. 3'—that of the 'three-eighths rule'—is set as unique instance in the revised scholium in [2] next following, as in that which concludes the published 'Methodus' (see page 252 below).

    (9) Add. 3965.12: 236^v, a stray jotting which is manifestly the immediate revise of [1] preceding.

    (10) Initially, in a less radical reshaping of the equivalent enunciation in [1], Newton specified: '*Data serie Ordinatarum*' (*Given a series of ordinates*); and then here in first draft of the ensuing phrase wrote '*una cum earum intervallis*' (*together with their intervals*).

    (11) In a cancelled first sequel Newton went on to instruct: '*Erigatur etiam ad eandem rectam & in eodem angulo terminus alius ad intervallum quodcunqȝ*' (Erect also to the same straight line and at the same angle another term at any interval whatever)—doubtless

Scholium.

From this general theorem can be derived[6] particular theorems whereby a problem shall be more speedily solved. And this method is useful for constructing tables by the interpolation of sequences, [as also...]...[its quadrature] will be reduced [to that of a second curve through fewer ordinates].[7]

Thereafter, a curve will be squared accurately enough by way of the general theorems exhibited above, or through short cuts adapted to particular cases, such as the following ones.
Case 1.[8]

[2][9]   *Proposition V.*

Given some number of terms of any series whatever[10] arrayed at given intervals, to find any intermediate term you will with close approximation.

To a given straight line and at the given intervals erect the given terms at a given angle, and through their outermost points by means of the preceding propositions draw a curved line of parabolic kind. [11]For this, indeed, will cut off all the intermediate terms throughout the series.

Scholium.

This method is useful....[12] If, for instance, there be four ordinates A, B, C, D arrayed at equal intervals, the new intermediate ordinate in the middle of all will be $\frac{1}{16}(9B+9C-A-D)$[13] and the area between the first and fourth ordinate will be[14]

And note that, when...will be reduced....[15] And then at last[16]

intending to add, much as in [1] above, 'et producatur isdem donec curvæ occurrat' (and extend this till its meet with the curve).

(12) Understand, much as in the revised 'Methodus' (§2 below): 'Utilis est hæc methodus ad Tabulas construendas per interpolationem Serierum, ut et ad solutiones Problematum quæ a quadraturis Curvarum dependent, præsertim si Ordinatarum intervalla et parva sint & æqualia inter se, et Regulæ computentur et in usum reserventur pro dato quocunqჳ numero Ordinatarum' (This method is useful for constructing tables by the interpolation of sequences, as also for the solution of problems which depend on the quadratures of curves, particularly if the intervals between the ordinates be both small and equal to one another, and rules be computed and stored for use for any given number of ordinates whatever). The ancillary sheet from which Newton drew this sentence is (compare note (7) preceding) seemingly lost.

(13) A replacement in afterthought for $\frac{'2A+B+C+2D'}{6}$. Newton's computations at the head of the manuscript scrap reveal how he obtained this erroneous formula, assuming a cubic interpolating parabola of form $y \equiv y_x = e+fx+gx^2+hx^3$, and then (as earlier on VII: 674) forgetfully taking a, b, c, d to be the non-equidistant ordinates y_2, y_1, y_{-1} and y_{-2} respectively; whence

$$\frac{'a+d}{2}[=e+g].\frac{c+b}{2}[=e+4g]',$$

[3][17] Dentur puncta quinqȝ *A, B, C, D,*
E et ab ijs ad rectam quamcunqȝ datam[18]
demittantur perpendicula *AF, BG, CH, DI,*
EK secantia rectam aliam *fk* in punctis *f,*
g, h, i et *k.* Sit autem *fk* parallela rectæ *FK,*
& in perpendiculis demissis capiantur
puncta *a b c d* et *e* ea lege ut *aF, fF* et *AF;*
bG, gG et *BG; cH, hH* et *CH; dI, iI* et *DI; eK,*
kK et *EK* sint continue proportionales. Per
puncta *a b c d* et *e* ducatur curva generis
Parabolici, et sit hujus Ordinata *Mn.* Secet

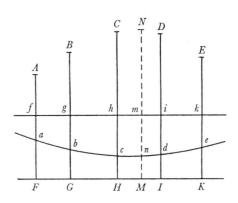

hæc rectam *fk* in *m* & producatur ad *N* ut sit *NM* ad *mM* ut *mM* ad *nM.* Et
punctum [*N*] tanget curvam generis Hyperbolici[19] quæ transibit per data
quinqȝ puncta *A, B, C, D, E.*

---

so that '$\dfrac{c+b-a-d}{2}$ $[=-3g]$' and consequently the *ordinata intermedia* $y_0 = e$ is

$$'\frac{c+b}{2}+\frac{c+b-a-d}{6}\,'.$$

In subsequent adjustment of these initial computations (which are terminated in the manuscript by a characteristic vertical stroke of his pen) Newton altered *a* and *d* to denote the
properly equidistant outer ordinates y_3 and y_{-3}, so that now '$\dfrac{a+d}{2}$ $[=e+9g]$' and consequently
'$\dfrac{b+c}{2}+\dfrac{b+c-a-d}{16}$ $[=e]$' (again as on VII: 674, *mutatis mutandis*). He had, we may remind, first
obtained this corrected rule for approximately intercalating a term intermediate between four
given equidistant ones more than thirty years earlier, at one time intending to include it as
'Reg. 4' (see IV: 30) of those he planned to impart in October 1676, on a sheet afterwards
rejected by him from his *epistola posterior* to Leibniz, for Oldenburg to place generally 'in usum
Calculatorum' (*ibid.*: 28); and more recently (see VII: 680–1) specified its extension to the
case where six—and, in general, any higher even number—of ordinates are given, developing
this as a trivial corollary to the simplest instance $x = 0$ of the 'Newton–Bessel' central-
differences interpolation formula which he set down in his original 1676 scheme as 'Cas. 2'
of his 'Prob. 1' (see IV: 58–60) and now restyled to be the corresponding Case 2 of Problem III
of his 'Methodus Differentialis'.

(14) Understand '...erit $\dfrac{A+3B+3C+D}{8}R$, existente R intervallo inter ordinatas' (will
be $\frac{1}{8}(A+3B+3C+D)R$, where R is the interval between the ordinates [*A* and *D*]), or
some similar equivalent to Newton's rephrasing of this in his published 'Methodus' (see page 252
below). This 'three-eighths' rule for approximate quadrature by four equidistant ordinates
had been set by him some fifteen years before as 'Cas. 3' of his short tract 'Of Quadrature by
Ordinates' (see VII: 692), and was very possibly intended to be the prime instance of the
'compendia per casus particulares applicata' of such quadrature in the predraft in [1] preceding
(compare our note (8) to it). More directly, in the terms of Newton's present approximating
cubic parabola $y \equiv y_x = e+fx+gx^2+hx^3$ (see the previous note) and the given ordinates

Dentur puncta quinqₔ A B C D E et ab ijsdem ad rectam quamvis positione datam demittantur perpendicula AF BG CH DJ EK, et in his perpendiculis si opus est productis capiantur ~~ordinatæ~~ AF, BG, CH, DJ et EK ~~proportionales~~
~~dignitatibus~~ dignitati cujuscunqₔ ipsorum AF, BG, CH, DJ EK ~~proportionales~~
æquales et per puncta a b c d et e ducatur curva generis Parabolici

Sit AB abscissa et BGD ordinata, _communi_ curvarum
duarum CDE, FGH, ijsdem occurrens in D et G
et relatio inter longitudines

Sunt CDE, FGH curvæ duæ abscissam habentes
communem AB et ordinatas in eadem recta jacentes A
BD, BG. Et relatio inter has ordinatas definiatur per æquationem
quamcunqₔ. Dentur puncta quotcunqₔ per quæ ~~æquatio~~ Curva CDE
transire debet. et per æquationem illam dabuntur puncta totidem
nova per quæ curva FGH transibit. Pro Propositiones superiores describet
Curva FGH generis Parabolici quæ puncta illa omnia nova transeat, et per æquationem illam dabitur curva CDE quæ per
puncta totidem primo data transibit.

Plate III. A preliminary drafting of the terminal scholium to the 'Methodus Differentialis'
(**1**, 4, §1.3).

[3]$^{(17)}$ Let there be given five points A, B, C, D, E and from these to any given straight line $^{(18)}$ whatever let fall the perpendiculars AF, BG, CH, DI, EK cutting another straight line fk in the points f, g, h, i and k. Let, however, fk be parallel to the line FK, and in the perpendiculars so dropped take points a, b, c, d and e with the restriction that aF, fF and AF; bG, gG and BG; cH, hH and CH; dI, iI and DI; and eK, kK and EK shall be in continued proportion. Through the points a, b, c, d and e draw a curve of parabolic kind, and let its ordinate be Mn. Let this cut the line fk in m, and extend it to N so that NM shall be to mM as mM to nM. The point N will then touch a curve of hyperbolic kind$^{(19)}$ which shall pass through the five given points A, B, C, D, E.

$a = y_3$, $b = y_1$, $c = y_{-1}$ and $d = y_{-3}$—the distance between the first and last of which is $R = 6$ units—there is at once $\int_{-3}^{3} y_x . dx = 6e + 18g$, that is, $R(e + 3g) = \frac{1}{4}R(\frac{1}{2}(a+d) + \frac{3}{2}(b+c))$.

(15) Here understand that the second paragraph of the scholium of the published 'Methodus' (taken over from its draft on an accompanying sheet which has failed to survive; compare notes (7) and (12) above) is here to follow, essentially as on page 252 below.

(16) The manuscript here breaks off, and Newton passes to draft the preliminary version of the final paragraph of the 'Methodus' which is reproduced in [3] following. He probably intended to add only some neutral remark like 'Et tum demum problemata solventur' (And then at last the problems will be solved).

(17) Add. 4005.15: 72^r/72^v, a drafting of the final paragraph of the 'Methodus'—one all but identical with the revised text cited in note (22) below—entered on the opened-out cover of a letter directed (f. 72^v) 'For / Sr Isaak Newton / at / Chelsea', evidently by a correspondent who also resided at London. (The hand is perhaps that of John Woodward, who wrote a stiffly formal letter to Newton from Gresham College on 30 May 1710 (ULC. Add. 3965.12: 288, printed in *Correspondence*, 5, 1975: 44) which its recipient equally thriftily—and despisingly?—re-used for a *Principia* draft; but in any event the envelope was addressed before late September 1710 when Newton moved residence from the village of Chelsea, as it then was, to the house in St Martin's Street 'near Leicester fields' where he was to reside, except for a brief sojourn in Kensington, till his death.) Newton's first use of the blank spaces of the letter-cover, we may add, was to jot down his rough calculations deducing, from an annual mean advance of lunar apogee of '40°. 39'. 54″ in 365 [days]'—a daily progress, that is, of '6'. 41″. 4‴. 46^{IV} = 401″$\llcorner$079452'—and an eccentricity of the earth's solar orbit of $16\frac{15}{16}$ parts in 1000, an *æquatio maxima medii motus Apogæi* (over the period of 3 months = '91$\frac{1}{4}$ [days]' in which it attains its maximum value from zero) of '619$^{(''')}$$\llcorner$8858875' $\approx$ 10' 20″: greatly different from the directly computed 'observed' value of 19'. 44″ for this semi-annual equation of the moon's mean apogee motion which Newton sets down alongside. One may here trace the origin of the fudge by which he afterwards in the new scholium to Prop. XXXV of Book 3 in the *Principia*'s second edition ($_2$1713: 421; compare the preliminary version of this scholium (Trinity College, Cambridge. R. 16.38: 169^r–171^r [=*Correspondence of Newton*, 5: 287–91, especially 288]) sent by him to Cotes in April 1712) increased this deficient theoretical value to one close to truth by multiplying it in the ratio of 2° 56' 9″ to '1°. 56'. 25″$\llcorner$8' (as here it is set): the values, namely, of the moon's *æquatio maxima centri* in the hypotheses that the sun's perturbance is inversely in the tripled and doubled powers respectively of its distance from the earth. But let us pass quickly on.

(18) '*FGHIK*' was first superfluously specified.

(19) Of necessity, since the points of this new curve corresponding to those in which the parabola (n) through a, b, c, d, e intersects the base line FK will lie at infinity.

[Vel generalius.] Dentur puncta quinqꝫ *A B C D E* et ab ijsdem ad rectam quamvis positione datam demittantur perpendicula *AF BG CH DI EK* et in his perpendiculis si opus est productis capiantur[20] *aF, bG, cH, dI* et *eK* dignitati cuicunqꝫ ipsorum *AF, BG, CH, DI* et *EK* æquales[21] et per puncta *a, b, c, d* et *e* ducatur curva generis Parabolici.[22]

---

(20) 'ordinatæ' (ordinates) is deleted.

(21) 'proportionales' (proportional) is replaced. Newton intends that the hyperbola (*N*) and parabola (*n*) shall be related by the general condition $nM = k \cdot NM^{-i}$, where *i* is some positive integer, thereafter taking $k = 1$ for simplicity.

(22) There follows in the manuscript (f. 72^v) a revision of the preceding text which—other than for a cancelled first opening 'Sit *AB* abscissa et *BGD* ordinata communis curvarum duarum *CDE, FGH*, ijsdem occurrens in *D* et *G*[,] et relatio inter longitudines [*BG* et *BD* definiatur per æquationem quamcunqꝫ?]' (Let *AB* be the abscissa and *BGD* the common ordinate of the two curves *CDE, FGH*, meeting these in *D* and *G* [respectively], and the relationship between the lengths [*BG* and *BD* be defined by any equation whatever...]) and also the variant accompany-

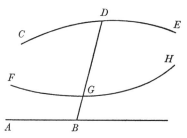

ing figure which we reproduce alongside—is all but word for word identical with the conclusion of the last paragraph of the 'Methodus'. Compare §2: notes (19) and (20) below.

Or more generally. Let the five points A, B, C, D, E be given and from them to any straight line given in position let fall the perpendiculars AF, BG, CH, DI, EK, and in these perpendiculars, extended if need be, take[20] aF, bG, cH, dI and eK equal[21] to any power whatever of AF, BG, CH, DI, and EK; then through the points a, b, c, d and e draw a curve of parabolic kind.[22]

§2. THE REMODELLED TRACT AS PRINTED.[1]

From a fair copy by William Jones[2] in private possession

METHODUS DIFFERENTIALIS.

Prop. I.[3]

Si figuræ curvilineæ Abscissa componatur ex quantitate quavis data A et quantitate indeterminata x, et Ordinata constet ex datis quotcunɋ quantitatibus b, c, d, e &c in totidem terminos hujus progressionis Geometricæ x, x², x³, x⁴ &c respective ductis,[4] et ad Abscissæ puncta totidem data erigantur Ordinatim applicatæ: dico quod Ordinatarum differentiæ primæ dividi possint per earum intervalla, et differentiarum sic divisarum differentiæ dividi possint per Ordinatarum binarum intervalla, et sic deinceps in infinitum.

Etenim si pro Abscissæ parte indeterminata x ponantur quantitates quævis datæ p, q, r, s, t &c successivè, et ad Abscissarum sic datarum terminos erigantur Ordinatæ α, β, γ, δ, ϵ &c. Hæ Abscissæ et Ordinatæ et Ordinatarum differentiæ divisæ per Abscissarum differentias (quæ utiɋ sunt Ordinatarum intervalla) & quotorum differentiæ divisæ per Ordinatarum alternarum differentias, et sic deinceps, exhibentur per Tabulam sequentem.[5]

---

(1) Namely, in Jones' gathering of Newton's opuscula on *Analysis per Quantitatum Series, Fluxiones, ac Differentias* (London, 1711): 93–101; on this publication see our detailed comment in § 1: note (1). Here we more narrowly observe in preliminary to the text of the 'Methodus' that its four principal propositions are minimally restyled and verbally augmented repetitions of corresponding elements of the scheme for series interpolation by central differences of both equally and unequally distant given arguments which (see IV: 52–66) Newton originally set down in his Waste Book about October 1676 (compare *ibid.*: 54, note (2)). Only the two concluding Propositions V/VI and terminal scholium are newly composed—their preparatory draftings and revisions (as these have survived) are reproduced in §1. 1–3 preceding—but they, too, present *compendia* for numerical interpolation and approximate quadrature which have similar precedents in the middle 1670's (see IV: 30–4) and 1690's (see VII: 690–4).

(2) We have above (again see §1: note (1)) suggested that this is the very transcript—or at least a copy retained by its editor in case of its loss?—which Jones 'wrote down' from Newton's autograph (in whatever state of completion that might have been) in the latter part of 1710, and speedily put to the printer. Only a poor photostat of this manuscript has been available to us, but there appear to be no printer's paging marks (of the type entered half a dozen years earlier in the margins of the press copies of Newton's 'De Quadratura' and 'Enumeratio'; compare VII: 508, note (2) and 588, note (2)) imposed upon it. In making the following reproduction of its text we have taken a very few liberties in restoring the punctuation and capitalization to authentic Newtonian form where (notably in the final paragraph; see note (19) below) this is known; and have also, in its Propositions III and IV, elevated the subscript numerals (as in 'A_3' and 'A_5B_5') which, both in Jones' manuscript and the 1711 *editio princeps*, distinguish individual points and ordinates in a sequence—and for which we know no precedent in any extant Newton autograph—to the superscript equivalents ('A^3', 'A^5B^5') of their 1676 parent: an 'index' notation which, however confusing to modern eyes, Newton continued to employ in the 1690's (see VII: 682–6) in preference to the left-hand numerals—

Translation

THE METHOD OF DIFFERENCES

Proposition I.[3]

If the abscissa of a curvilinear figure be composed of any given quantity A and the indeterminate quantity x, and its ordinate consist of any number of given quantities b, c, d, e, ... multiplied respectively into an equal number of terms of the geometric progression x, x², x³, x⁴, ...[4] *and if ordinates be erected at the corresponding number of given points in the abscissa: I assert that the first differences of the ordinates are divisible by the intervals between them, and that the differences of the differences so divided are divisible by the intervals between every second ordinate, and so on indefinitely.*

For indeed, if in place of the indeterminate part x of the abscissa there be successively put any given quantities $p, q, r, s, t, ...$, and at the end-points of the abscissas thus yielded there be erected ordinates $\alpha, \beta, \gamma, \delta, \epsilon, ...$, then these abscissas and ordinates and differences of ordinates divided by the differences of abscissas (which of course are the intervals between the ordinates) and the differences of the quotients divided by the differences of alternate ordinates, and so on, may be displayed through the following table:[5]

---

were these an editorial emendation of Halley's?—introduced analogously in Lemma V of Book 3 of his *Principia* (₁1687: 481–2; compare ɪv: 71–2). (We of course continue in our English version to translate Newton's superscripts into modern Jonesian subscript equivalents.)

(3) A mere verbal elaboration of the sense of the table of divided differences which (see ɪv: 62) introduces 'Prob. 2' in the parent 1676 scheme (essentially identical to Proposition IV of the present 'Methodus') where it is applied.

(4) That is, corresponding to the abscissa $A+x$, Newton assumes that the general ordinate of his 'curvilinear figure' is expressible in the parabolic polynomial form

$$[a+]bx+cx^2+dx^3+ex^4+...,$$

exactly as in his parent scheme (ɪv: 62, top) except that he here carelessly neglects to specify the ordinate's initial term a, its value when the abscissa is A. He will repair the omission in his ensuing table, but only at the cost of confusing a with A! (See note (6) below.)

(5) As in his original table, where (compare ɪv: 61, note (24))

$$f_{A+x} = a+bx+cx^2+dx^3+ex^4[+\&c],$$

Newton successively computes the first, second, third and fourth divided differences of $\alpha = f_{A+p}, \beta = f_{A+q}, \gamma = f_{A+r}, \delta = f_{A+s}$ and $\epsilon = f_{A+t}$, viz.

$$\zeta = (f_{A+p}-f_{A+q})/(p-q) = (p,q), \quad \eta = (q,r), \quad \theta = (r,s), \quad \kappa = (s,t);$$

$$\lambda = ((p,q)-(q,r))/(p-r) = (p,q,r), \quad \mu = (q,r,s), \quad \nu = (r,s,t);$$

$$\xi = ((p,q,r)-(q,r,s))/(p-s) = (p,q,r,s), \quad \pi = (q,r,s,t);$$

and
$$\sigma = ((p,q,r,s)-(q,r,s,t)) = (p,q,r,s,t).$$

Abscissæ.	*Ordinatæ.*
$A+p$	$A^{(6)}+bp+cp^2+dp^3+ep^4=\alpha$
$A+q$	$A^{(6)}+bq+cq^2+dq^3+eq^4=\beta$
$A+r$	$A^{(6)}+br+cr^2+dr^3+er^4=\gamma$
$A+s$	$A^{(6)}+bs+cs^2+ds^3+es^4=\delta$
$A+t$	$A^{(6)}+bt+ct^2+dt^3+et^4=\epsilon$

Divisor.	*Diff. Ord.*	*Quoti per divisionem prodeuntes.*
$p-q$)	$\alpha-\beta$	$b+c\times\overline{p+q}+d\times\overline{pp+pq+qq}+e\times\overline{p^3+p^2q+pq^2+q^3}=\zeta$
$q-r$)	$\beta-\gamma$	$b+c\times\overline{q+r}+d\times\overline{qq+qr+rr}+e\times\overline{q^3+q^2r+qr^2+r^3}=\eta$
$r-s$)	$\gamma-\delta$	$b+c\times\overline{r+s}+d\times\overline{rr+rs+ss}+e\times\overline{r^3+r^2s+rs^2+s^3}=\theta$
$s-t$)	$\delta-\epsilon$	$b+c\times\overline{s+t}+d\times\overline{ss+st+tt}+e\times\overline{s^3+s^2t+st^2+t^3}=\kappa$
$p-r$)	$\zeta-\eta$	$c+d\times\overline{p+q+r}+e\times\overline{pp+pq+qq+pr+qr+rr}=\lambda$
$q-s$)	$\eta-\theta$	$c+d\times\overline{q+r+s}+e\times\overline{qq+qr+rr+qs+rs+ss}=\mu$
$r-t$)	$\theta-\kappa$	$c+d\times\overline{r+s+t}+e\times\overline{rr+rs+ss+rt+st+tt}=\nu$
$p-s$)	$\lambda-\mu$	$d+e\times\overline{p+q+r+s}=\xi$
$q-t$)	$\mu-\nu$	$d+e\times\overline{q+r+s+t}=\pi$
$p-t$)	$\xi-\pi$	$e=\sigma.$

Prop. II.[7]

Ijsdem positis et quod numerus terminorum b, c, d, e &c sit finitus: dico quod Quotorum ultimus æqualis erit ultimo terminorum b, c, d, e &c et quod per Quotos reliquos dabuntur termini reliqui b, c, d, e &c, et his datis dabitur Linea Curva generis Parabolici quæ per Ordinatarum omnium terminos transibit.

Etenim in Tabula superiore Quotus ultimus σ æqualis erat termino ultimo e. Et hic terminus ductus in summam datam $\overline{p+q+r+s}$ et ablatus de Quoto ξ relinquit terminum penultimum d. Et quantitates jam datæ

$$d\times\overline{p+q+r}+e\times\overline{pp+pq+qq+pr+gr+rr}$$

si auferantur de Quoto λ, relinquunt terminorum antepenultimum c. Et quantitates jam datæ $c\times\overline{p+q}+d\times\overline{pp+pq+qq}+e\times\overline{p^3+ppq+pqq+q^3}$ si auferantur de Quoto ζ, relinquunt terminum b. Et simili computo si plures essent

---

(6) Read 'a' in each case, as in the parent table on IV: 62; we mend our facing English version accordingly. In his following Proposition II Newton will (see note (8) below) compound this slip in transcription by a further confusion of ordinate a with abscissa A.

(7) Filling a small lacuna in his 1676 scheme, where (see IV: 60) the application of the

Abscissas.	Ordinates.
$A+p$	$[a]+bp+cp^2+dp^3+ep^4=\alpha.$
$A+q$	$[a]+bq+cq^2+dq^3+eq^4=\beta.$
$A+r$	$[a]+br+cr^2+dr^3+er^4=\gamma.$
$A+s$	$[a]+bs+cs^2+ds^3+es^4=\delta.$
$A+t$	$[a]+bt+ct^2+dt^3+et^4=\epsilon.$

Divisor	Ord. diffs.	Quotients resulting by the division.
$p-q$)	$\alpha-\beta$	$b+c(p+q)+d(p^2+pq+q^2)+e(p^3+p^2q+pq^2+q^3)=\zeta.$
$q-r$)	$\beta-\gamma$	$b+c(q+r)+d(q^2+qr+r^2)+e(q^3+q^2r+qr^2+r^3)=\eta.$
$r-s$)	$\gamma-\delta$	$b+c(r+s)+d(r^2+rs+s^2)+e(r^3+r^2s+rs^2+s^3)=\theta.$
$s-t$)	$\delta-\epsilon$	$b+c(s+t)+d(s^2+st+t^2)+e(s^3+s^2t+st^2+t^3)=\kappa.$
$p-r$)	$\zeta-\eta$	$c+d(p+q+r)+e(p^2+pq+q^2+pr+qr+r^2)=\lambda.$
$q-s$)	$\eta-\theta$	$c+d(q+r+s)+e(q^2+qr+r^2+qs+rs+s^2)=\mu.$
$r-t$)	$\theta-\kappa$	$c+d(r+s+t)+e(r^2+rs+s^2+rt+st+t^2)=\nu.$
$p-s$)	$\lambda-\mu$	$d+e(p+q+r+s)=\xi.$
$q-t$)	$\mu-\nu$	$d+e(q+r+s+t)=\pi.$
$p-t$)	$\xi-\pi$	$e=\sigma.$

Proposition II.[7]

With the same suppositions, and taking the number of terms b, c, d, e, ... to be finite, I assert that the last of the quotients will be equal to the last of the terms b, c, d, e, ..., and that the remaining terms b, c, d, e, ... will be yielded by means of the remaining quotients; and that once these are determined there will be given the curve of parabolic kind which shall pass through the end-points of all the ordinates.

For, to be sure, in the above table the last quotient σ was equal to the last term e; and this term multiplied into the given sum $p+q+r+s$ and taken away from the quotient ξ leaves the last term but one d; and if the quantities $d(p+q+r)+e(p^2+pq+q^2+pr+qr+r^2)$ be taken away from the quotient λ, they leave the last term but two c; and if the quantities

$$c(p+q)+d(p^2+pq+q^2)+e(p^3+p^2q+pq^2+q^3)$$

now given be taken away from the quotient ζ, they leave the term b. And by a

---

preceding table of divided differences to the general problem *Datis quotcunꝗ punctis Curvam describere quæ per omnia transibit* is regarded as being self-evident, Newton in this newly composed proposition specifies how thereby a curve of 'parabolic kind' can be determined to pass through the end-points of any given number of ordinates.

termini, colligerentur omnes per Quotorum Ordines totidem. Deinde quanti-tates datæ $bp+cpp+dp^3+ep^4$ [&c] si subducantur de Ordinata prima α, relinquunt Abscissæ terminum primum A.[8] Et quantitas

$$A^{(8)}+bx+cx^2+dx^3+ex^4+\&c$$

est Ordinata Curvæ generis Parabolici quæ per Ordinatarum omnium datarum terminos transibit, existente Abscissa $A+x$.[9]

Ex his Propositionibus quæ sequuntur facile colligi possunt.

Prop. III.[10]

Si Recta aliqua AA^9 in æquales quotcunq̃ partes AA^2, A^2A^3, A^3A^4, A^4A^5 &c dividatur et ad puncta divisionum erigantur parallelæ AB, A^2B^2, A^3B^3 &c: invenire curvam Geometricam generis Parabolici quæ per omnium erectarum terminos B, B^2, B^3 &c transibit.

Erectarum AB, A^2B^2, A^3B^3 &c quære differentias primas b, b^2, b^3 &c. secundas c, c^2, c^3 &c. tertias d, d^2, d^3 &c. et sic deinceps usq̃ dum veneris ad ultimam differentiam, quæ hic sit i. Tunc incipiendo ab ultima differentia excerpe medias differentias in alternis columnis vel Ordinibus differentiarum...... est impar, posterius ubi par.

---

(8) Newton—this cannot be Jones' error—should here have written 'Ordinatæ terminum primum a' and 'a' respectively, and so we adjust our English version on the facing page. A like correction is silently made by D. C. Fraser in his English translation of the 'Methodus' (*Journal of the Institute of Actuaries*, **51**, 1918: 94–101, especially 94 [= *Newton's Interpolation Formulas* (London, 1927): 18]), but without mending the corresponding slip in the preceding table of divided differences from which (see note (6) above) Newton's present confusions derive. None of these oversights, we might add, are rectified by Newton himself in his library copy (now Trinity College, Cambridge. NQ.8.26) of Jones' 1711 *Analysis*; we must presume that he himself remained ignorant of having perpetrated them till his death.

(9) Since (in the contracted notation of note (5) above) there is

$$(f_{A+x}-\alpha)/(x-p) = b+c(x+p)+d(x^2+px+p^2)+e(x^3+px^2+p^2x+p^3)... \equiv (p,x),$$

$$((p,x)-\zeta)/(x-q) = c+d(x+p+q)+e(x^2+(p+q)\,x+p^2+pq+q^2)... \equiv (p,q,x),$$

$$((p,q,x)-\lambda)/(x-r) = d+e(x+p+q+r)... \equiv (p,q,r,x)$$

and $((p,q,r,x)-\xi)/(x-s) = e... \equiv (p,q,r,s,x),$

when the 'final' fourth differences are constant, and so the ordinate of the approximating parabola is $f_{A+x} = a+bx+cx^2+dx^3+ex^4$, there results (as Newton states) $(p,q,r,s,x)=e=\sigma$, constant for all x. At once, on 'unwrapping' this scheme stage by stage there ensues the equivalent expression

$$f_{A+x} = \alpha+(x-p)\,(\zeta+(x-q)\,[\lambda+(x-r)\,\{\xi+(x-s)\,\sigma\}])$$

in terms of the linked chain, departing from $\alpha = f_{A+p}$, of successive advancing divided differences $\zeta = (p,q)$, $\lambda = (p,q,r)$, $\xi = (p,q,r,s)$ and $\sigma = (p,q,r,s,t)$. This is the 'master' formula for interpolation which Newton had already made public in Case 2 of Lemma V of the third book of his *Principia* ($_1$1687: 482; see IV: 72–3, and compare *ibid.*: note (5)). Any other

similar computation, were there more terms, they would all be gathered by way of the corresponding number of ordinates. Then, if the given quantities $bp+cp^2+dp^3+ep^4...$ be subtracted from the first ordinate α, they leave the first term $[a]$ of the [ordinate]. And the quantity $[a]+bx+cx^2+dx^3+ex^4+...$ is the ordinate of a curve of parabolic kind which shall pass through the end-points of all the given ordinates, the abscissa being $A+x$.[9]

From these propositions the ones which follow can easily be gathered.

Proposition III.[10]

If some straight line A_1A_9 be divided into any number of equal parts A_1A_2, A_2A_3, A_3A_4, A_4A_5, ... and at the points of division there be erected the parallels A_1B_1, A_2B_2, A_3B_3, ...: to find a geometrical curve of parabolic kind which shall pass through the end-points B_1, B_2, B_3, ... of all the erected lines.

Of the erected lines A_1B_1, A_2B_2, A_3B_3, ... seek the first differences b_1, b_2, b_3, ...; the second ones c_1, c_2, c_3, ...; the third ones d_1, d_2, d_3, ...; and so on, till you come to the last difference, which here let be i_1. Then, beginning from the last difference, take out the middle differences in alternate columns or ranks of differences ... ... is odd, the latter when it is even.

---

linked path through the array of divided differences—indeed any linear combination of such paths—will evidently yield variant expressions for the 'ordinate' f_{A+x} to be intercalated between the given terms α, β, γ, δ, ϵ. And when in particular a mean horizontal path is chosen, we attain the generalized formulas for interpolation by central differences now known as 'Newton–Stirling' and 'Newton–Bessel' which their primary author originally set down in the twin cases of 'Prob. 2' in his 1676 scheme (see IV: 62–8) and will here re-present virtually word for word in the corresponding 'Cas. 1' and 'Cas. 2' of Proposition IV following. It was perhaps not Newton's intention in the first instance so narrowly to repeat his earlier problem in the 'Methodus' in all its full niceties of detail. Among his papers now in the University Library at Cambridge there is preserved a manuscript scrap (Add. 3964.8: 29^r), in William Jones' hand but very possibly transcribed to Newton's dictation (or maybe from an autograph of his which has perished?), in which the derivation of these general central-difference formulas 'per Constructionem superiorem Problematis describendi Curvam per data quotcunꝗ puncta' is much more sketchily adumbrated, without explicit citation of the formulas themselves. For its several minor interests, and because of its likely direct Newtonian association, we reproduce the fragment in full in the Appendix.

(10) A minimal restyling of 'Prob. 1' in the parent 1676 scheme (IV: 54–60) in which Newton first set down this general theorem for interpolating by central differences where the given 'ordinates' between which intercalation is to be made lie at equal distances of argument along the base 'abscissa'. We see no need to repeat any portion of the original text which passes over into the corresponding Proposition III entirely unchanged, and in the same way refer the reader to our earlier footnotes for a running commentary upon the technicalities of the twin formulas for such central-difference interpolation which, in the full text of the 'Methodus', Newton now passes to state.

Cas. 1.[11]

In Casu priori sit A^5B^5 iste medius terminus... ...longitudo ordinatim applicatæ PQ.

Cas. 2.[11]

In Casu posteriori sint A^4B^4 et A^5B^5 duo medii termini ... ... signa eorum mutanda sunt.

Prop. IV.[12]

Si recta aliqua in partes quotcunꝗ inæquales AA^2, A^2A^3, A^3A^4, A^4A^5 &c dividatur et ad puncta divisionum erigantur parallelæ AB, A^2B^2, A^3B^3 &c: invenire Curvam Geometricam generis Parabolici quæ per omnium erectarum terminos B, B^2, B^3 &c transibit.

Sunto puncta data B, B^2, B^3, B^4, B^5, B^6, B^7 &c et ad Abscissam quamvis AA^7 demitte Ordinatas perpendiculares BA, B^2A^2 &c. Et fac ... ... vel medium Arithmeticum inter duas medias si numerus eorum est par.

Cas. 1.[13]

In Casu priori sit A^4B^4 ista media ordinatim applicata ... ... longitudo ordinatim applicatæ PQ.

Cas. 2.[13]

In Casu posteriori sint A^4B^4 et A^5B^5 duæ mediæ ordinatim applicatæ ... ... $=PQ$.

Prop. V.[14]

Datis aliquot terminis seriei cujuscunꝗ ad data intervalla dispositis, invenire terminum quemvis intermedium quamproxime.

Ad rectam positione datam erigantur termini dati in dato angulo interpositis datis intervallis, et per eorum puncta extima per Propositiones præcedentes ducatur linea Curva generis Parabolici. Hæc enim continget terminos omnes intermedios per seriem totam.

---

(11) These complementary cases repeat wholly without change the texts of the corresponding 'Cas. 1' and 'Cas. 2' of the parent 'Prob. 1' in which (see IV: 56–8/58–60 respectively) Newton originally set down his statement of these 'Newton–Stirling' and 'Newton–Bessel' formulas, and we omit here needlessly to do more than indicate their presence in the 'Methodus'.

(12) The following enunciation is, in replacement of Newton's original vague direction 'Curvam Geometricam describere quæ per data quotcunꝗ puncta transibit' (see IV: 62–8), newly reframed on the model of that which introduces the preceding Proposition III, but the ensuing text follows virtually word for word that of the parent 'Prob. 2' (IV: 62–8).

(13) As the corresponding 'Cas. 1' and 'Cas. 2' of the original 'Prob. 2' (see IV: 64–6/66–8 respectively) whose texts they copy without change other than in capitalizing a few individual

Case 1.[11]

In the former case let $A_5 B_5$ be that middle term .. .. the length of the ordinate PQ.

Case 2.[11]

In the latter case let $A_4 B_4$ and $A_5 B_5$ be the two middle terms. .. .. their signs are to be changed.

Proposition IV.[12]

If some straight line be divided into any number of unequal parts $A_1 A_2$, $A_2 A_3$, $A_3 A_4$, $A_4 A_5$, ... and at the points of division there be erected the parallels $A_1 B_1$, $A_2 B_2$, $A_3 B_3$, ...: to find a geometrical curve of parabolic kind which shall pass through the end-points B_1, B_2, B_3, ... of all the erected lines.

Let those given points be B_1, B_2, B_3, B_4, B_5, B_6, B_7, ... and to any abscissa $A_1 A_7$ let fall the perpendicular ordinates $B_1 A_1$, $B_2 A_2$, ... Then make .. .. or alternatively the arithmetic mean between the two middle-most if their number is even.

Case 1.[13]

In the former case let $A_4 B_4$ be that middle ordinate .. .. the length of the ordinate PQ.

Case 2.[13]

In the latter case let $A_4 B_4$ and $A_5 B_5$ be the two central ordinates ... ... $PQ = ...$

Proposition V.[14]

Given some number of terms of any series whatever arrayed at given intervals, to find any intermediate term you will with close approximation.

To a straight line given in position erect the given terms at a given angle, interposing the given intervals, and through their outermost points by means of the preceding propositions draw a curved line of parabolic kind. For this, indeed, will touch all the intermediate terms throughout the series.

---

words, these state the generalization of the preceding 'Newton–Stirling' and 'Newton–Bessel' formulas for interpolation by central divided differences where the given 'ordinates' to be intercalated no longer lie at equal distances one from another along the base line; on which see also notes (5) and (7) of the Appendix.

(14) A straightforward specification, only minimally altered in its phrasing from the revised preliminary draft in §1.2 preceding, of the way in which the previous construction of a parabola through the end-points of any number of given ordinates set out in given position along a base line is to be applied to the general problem of interpolating any given terms whose difference in argument is likewise given.

Prop. VI.[15]

Figuram quamcunq Curvilineam quadrare quamproxime cujus Ordinatæ aliquot inveniri possunt.

Per terminos Ordinatarum ducatur linea Curva generis Parabolici ope Propositionum præcedentium. Hæc enim figuram terminabit quæ semper quadrari potest, et cujus Area æquabitur Areæ figuræ propositæ quamproxime.

Scholium.

Utiles sunt hæ Propositiones ad Tabulas construendas per interpolationem Serierum, ut et ad solutiones Problematum quæ a quadraturis Curvarum dependent, præsertim si Ordinatarum intervalla et parva sint & æqualia inter se, et Regulæ computentur et in usum reserventur pro dato quocunq numero Ordinatarum.[16] Ut si quatuor sint Ordinatæ ad æqualia intervalla sitæ, sit A summa primæ et quartæ, B summa secundæ & tertiæ, et R intervallum inter primam et quartam, et Ordinata nova in medio omnium erit $\dfrac{9B-A}{16}$[16] et Area tota inter primam et quartam erit $\dfrac{A+3B}{8}R$.[17]

Et nota quod ubi Ordinatæ stant ad æquales ab invicem distantias, sumendo summas Ordinatarum quæ ab Ordinata media hinc inde æqualiter distant et duplum Ordinatæ mediæ componitur Curva nova cujus Area per pauciores Ordinatas determinatur et æqualis est Areæ Curvæ prioris quam invenire oportuit. Quinetiam si pro Ordinatis novis sumantur summa Ordinatæ primæ et secundæ, et summa tertiæ & quartæ, et summa quintæ & sextæ, et sic deinceps; vel si sumantur summa trium primarum Ordinatarum, et summa trium proximarum, et summa trium quæ sunt deinceps [&c]; vel si sumantur summæ quaternarum Ordinatarum, vel summæ quinarum: Area Curvæ novæ æqualis erit Areæ Curvæ primo propositæ. Et sic habitis Curvæ quadrandæ Ordinatis quotcunq quadratura ejus ad quadraturam curvæ alterius per pauciores ordinatas reducetur.[18]

---

(15) The equally evident application of the preceding construction of an approximating parabola to solution *quam proximè* of the problem 'Of Quadrature by Ordinates' which Newton had broached—and then resolved by a not entirely adequate variant method of argument *per analogiam*, building up more complex rules for such numerical quadrature successively out of simpler ones and ultimately from the primitive trapezoidal approximation—some fifteen years earlier (see VII: 690–4, and compare *ibid.*: 696, note (14), where we contradicted his affirmation that 'Hæ sunt quadraturæ Parabolæ quæ per terminos Ordinatarum omnium transit').

(16) Although this elegant rule for approximating the *ordinata intermedia*, given four equidistant ordinates, had been set down by him as but the first of several such 'regulæ in usum Calculatorum' which he intended to communicate to Oldenburg—if not to be passed on to

Proposition VI.[15]

To square to a close approximation any curvilinear figure whatever, some number of whose ordinates can be ascertained.

Through the end-points of the ordinates draw a curve of parabolic kind with the aid of the preceding problems. For this will bound a figure which can always be squared, and whose area will be equal to the area of the figure proposed with close approximation.

Scholium.

These propositions are useful for constructing tables by the interpolation of sequences, as also for the solution of problems which depend on the quadratures of curves, particularly if the intervals between the ordinates be both small and equal to one another, and rules be computed and stored for use for any given number of ordinates whatever.[16] If, for instance, there be four ordinates positioned at equal intervals, let A be the sum of the first and the fourth, B the sum of the second and third, and R the interval between the first and fourth, and then the new ordinate in the midst of all will be $\frac{1}{16}(9B-A)$ [16] and the total area between the first and fourth will be $\frac{1}{8}(A+3B)\,R$.[17]

And note that, when the ordinates stand at equal distances from one another, on taking the sums of the ordinates which are equally distant from the middle ordinate on its either side along with twice the middle ordinate there is formed a new curve whose area is determined by fewer ordinates and equal to the area of the previous curve which it was required to find. Furthermore, if in place of the new ordinates there be taken the sum of the first and second ordinates, and the sum of the third and fourth ones, and the sum of the fifth and sixth, and so on; or if there be taken the sum of the first three ordinates, and the sum of the next three, and the sum of the three which are then after, and so on; or if there be taken the sums of the ordinates four at a time, or their sums five at a time: the area of the new curve will be equal to the area of the curve first proposed. And in this way, where there are had any number of ordinates of the curve to be squared, its quadrature will be reduced to that of a second curve through fewer ordinates.[18]

Leibniz—in his *epistola posterior* in October 1676 (see IV: 30), we have previously remarked (in §1.2: note (13)) that Newton found some difficulty in correctly deriving it anew in prelude to his prior draft of this present citation of it.

(17) The familiar 'three-eighths' rule of numerical quadrature which he had first derived fifteen years before in 'Cas. 3' of his tract 'Of Quadrature by Ordinates' (see VII: 692, and compare §1.2: note (14) here preceding).

(18) This replacing of groups of individual ordinates by their aggregate may—except, it is evident, in the first case—entail some loss in accuracy of approximation even after applying an appropriate correction factor (compare VII: 695–6, notes (12) and (13)).

Per data vero puncta quotcunq non solum Curvæ
lineæ generis Parabolici sed etiam curvæ aliæ in-
numeræ diversorum generum duci possunt.[19]
Sunto *CDE, FGH* curvæ duæ abscissam habentes
communem *AB* et ordinatas in eadem recta jacentes
BD, BG. Et relatio inter has ordinatas definiatur
per æquationem quamcunq. Dentur puncta quotcunq per quæ Curva *CDE* transire debet, et per
æquationem illam dabuntur puncta totidem nova
per quæ curva *FGH* transibit. Per Propositiones

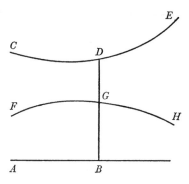

superiores describatur Curva *FGH* generis Parabolici quæ per puncta illa
omnia nova transeat, et per æquationem eandem[20] dabitur curva *CDE* quæ
per puncta omnia[20] primo data transibit.

---

(19) We lightly amend Jones' punctuation and capitals in the remainder of this final paragraph to accord with those of Newton's autograph draft (now ULC. Add. 4005.15: 72ᵛ); except
for the two minor alterations of single words cited in the next note and a trivial redrawing of
Newton's original figure (for which see §1.3: note (22) above) there is no other difference
between the two versions.

Through any number of given points, however, there can be drawn not only curved lines of parabolic kind, but also innumerable other curves of different kinds.[19] Let *CDE*, *FGH* be two curves having the common abscissa *AB* and ordinates *BD*, *BG* lying in the same straight line, and let the relationship between these ordinates be defined by means of any equation whatever. Let there be given any number of points through which the curve *CDE* must pass, and through that equation there will be given an equal number of new points through which the curve *FGH* shall pass. By means of the previous propositions describe a curve *FGH* of parabolic kind to pass through all the new points, and through that[20] equation there will be yielded the curve *CDE* which shall pass through all the[20] points first given.

---

(20) In his preliminary draft of this final sentence (see the previous note) Newton here wrote 'illam' (the) and 'totidem' (the equal number of) respectively. The gain in clarity is minimal.

APPENDIX. A (DISCARDED?) NEWTONIAN ADDENDUM TO PROPOSITION II OF THE 'METHODUS DIFFERENTIALIS'.[1]

From the original[2] in the University Library, Cambridge

Jam[3] si tria tantum dantur puncta quæ in directum non jacent,[4] Parabola quæ per eadem transibit erit Primi Generis, nempe Conica, et quantitates λ, μ, ν, &c prodibunt æquales & earum differentiæ evanescent, & ex datis quantitatibus β, $\dfrac{\zeta+\eta}{2}$ et λ dabitur Parabola.[5]

Si quatuor tantum dantur puncta quæ nec in Linea recta, nec in Parabola Conica jacent;[6] Parabola quæ per eadem transibit erit Secundi Generis, et quantitates ξ, π, &c prodibunt inter se æquales, et earum differentiæ evanescent, & ex datis quantitatibus $\dfrac{\beta+\gamma}{2}$, η, $\dfrac{\lambda+\mu}{2}$, ξ dabitur Parabola.[7] Et sic deinceps in infinitum.

Dabitur autem Parabola in his Casibus per Constructionem superiorem[8] Problematis describendi Curvam per data quotcunq puncta.

---

(1) This stray fragment on interpolation by means of central (mean) divided differences—employed, in the Cartesian geometrical model, to construct an approximating parabola of 'first' (quadratic) and 'second' (cubic) kinds in the simplest cases cited—is penned in the hand of William Jones, editor of the published 'Methodus', but we may reasonably suppose that it was likewise written out directly at Newton's instruction (from some much rougher autograph jotting, perhaps, which was then thrown away) with the intention that it should flesh out the spare frame of Proposition II. If this be so, we will appreciate how its duplication of what he afterwards, in narrow repetition of his 1676 parent 'Prob. 2' set out more generally and explicitly in his ensuing Proposition IV (see page 250 above, and compare IV: 62–8) would, in the total context of the work, have seemed both clumsy and pointless, and so led him to discard it from his finished text.

(2) ULC. Add. 3964.8: 29^r, a loose scrap placed (at least since the well-intentioned but confused efforts of his cataloguers to order the Portsmouth papers a century ago) without any indication of its origin among a loose gathering of mathematical fragments in Newton's hand.

(3) Understand the parent context where (in the notation of §2: note (5) preceding) the given 'ordinates' $\alpha = f_{A+p}$, $\beta = f_{A+q}$, $\gamma = f_{A+r}$, $\delta = f_{A+s}$, ... are to be interpolated by the approximating *curva generis parabolici* $f_{A+x} = a + bx + cx^2 + dx^3 + ...$ whose coefficients $a, b, c, d, ...$ are to be evaluated in terms of the successive ranks of divided differences

$$\zeta = (\alpha-\beta)/(p-q) \equiv (p,q), \quad \eta = (q,r), \quad \theta = (r,s), ...;$$

$$\lambda = (\zeta-\eta)/(p-r) \equiv (p,q,r), \quad \mu = (q,r,c), ...;$$

$$\xi = (\lambda-\mu)/(p-s) \equiv (p,q,r,s), ...$$

$$\begin{array}{ccccc}
\alpha & & & & \\
& \zeta & & & \\
\beta & & \lambda & & \\
& \eta & & \xi & \\
\gamma & & \mu & & \cdot\cdot \\
& \theta & & \cdot\cdot & \\
\delta & & \cdot\cdot & & \\
& \cdot\cdot & & & \\
\end{array}$$

As before (see §2: note (9)), any chain—or linear combination of chains—of progression rightwards through the accompanying array, beginning with any of the given terms $\alpha, \beta, \gamma, \delta, ...$

and taking in one each of the following columns of differences $\zeta, \eta, \theta, \ldots; \lambda, \mu, \ldots; \xi, \ldots$, will determine a related expression for the 'ordinate' f_{A+x} to be intercalated. Compare the instances adduced in notes (5) and (7) below.

(4) In which case the interpolating curve degenerates to some straight line whose ordinate $f_{A+x} = a + bx$ is to be specified by any one of the given terms $\alpha, \beta, \gamma, \ldots$ and their constant divided differences $\zeta = \eta = \ldots$.

(5) Whose ordinate $f_{A+x} = a + bx + cx^2$ can (see note (3) above) at once be expressed in either of the forms

$$f_{A+x} = \beta + (x-q) \, (\zeta + (x-p) \, \lambda)$$

or

$$f_{A+x} = \beta + (x-q) \, (\eta + (x-r) \, \lambda)$$

and hence as the arithmetic mean of these, namely:

$$f_{A+x} = \beta + (x-q) \cdot \tfrac{1}{2}(\zeta+\eta) + (x-q) \, (x - \tfrac{1}{2}(p+r)) \, \lambda,$$

the form here required. This is, we need scarcely remind after what has gone before, the generalized 'Newton–Stirling' formula (here instanced in its simplest case) for interpolation by central differences which in the published 'Methodus' Newton will re-present, in narrow repetition of its 1676 parent (see IV: 64–6), in Case 1 of the ensuing Proposition IV (see page 250 above).

(6) When the second-order differences $\lambda, \mu, \ldots$ in the corresponding interpolation array (see note (3) above) will be equal, and hence the third-order ones $\xi, \ldots$ zero, so that here only the (mean) central differences $\tfrac{1}{2}(\beta+\gamma)$, η and $\tfrac{1}{2}(\lambda+\mu) = \lambda = \mu$ need to be employed.

(7) The ordinate $f_{A+x} = a + bx + cx^2 + dx^3$ of this cubic approximating parabola can (again see note (3) above) be expressed in either of the forms

$$f_{A+x} = \beta + (x-q) \, (\eta + (x-r) \, [\lambda + (x-p) \, \xi])$$

or

$$f_{A+x} = \gamma + (x-r) \, (\eta + (x-q) \, [\mu + (x-c) \, \xi])$$

and consequently as their arithmetic mean

$$f_{A+x} = \tfrac{1}{2}(\beta+\gamma) + (x - \tfrac{1}{2}(q+r)) \, \eta + (x-q) \, (x-r) \cdot \tfrac{1}{2}(\lambda+\mu) + (x-q) \, (x-r) \, (x - \tfrac{1}{2}(p+s)) \, \xi :$$

which is the cubic instance of the generalized 'Newton–Bessel' formula for interpolation by central differences whose full version Newton states, repeating its 1676 parent (see IV: 66–8) all but word for word, in Case 2 of Proposition IV of the published 'Methodus' (see page 250 above).

(8) Understand that set out in Proposition II of the preceding 'Methodus' (pages 246–8 above).

5

THE
'DE QUADRATURA CURVARUM'
YET FURTHER RECAST[1]

[mid-April? 1712]

From the originals in private possession and in the University Library, Cambridge

[1][2]   Analysin[3] qua Propositiones in Libris *Principiorum* investigavi visum est jam subjungere ut Lectores eadem instructi Propositiones in his libris traditas

---

(1) A little after Easter 1712 Richard Bentley, freshly returned to Cambridge after conversing with its author about the coming second edition of his *Principia*, gave Roger Cotes to understand that Newton then had 'some thoughts' of appending to it 'a small Treatise of Infinite Series & ye Method of Fluxions'—or so Cotes spoke of it to Newton on 26 April 1712 in a letter (*Correspondence of Isaac Newton*, **5**, 1975: 279) seeking further information about it. Though Newton again mentioned his proposal to Bentley about the beginning of July following (see Cotes' letter to Newton three weeks afterwards on 20 July, where (*Correspondence*, **5**: 315) he acknowledged Bentley's new report of it to him) he never referred to this intention in his own correspondence with his Cambridge editor, and nothing more outwardly was heard of it. But this new treatise on infinite series and fluxional analysis, if as so many times before it was never fully completed, did achieve an existence more substantial than that of a fleeting, superseded thought in Newton's mind, and the surviving portion of the 'Analysis per quantitates fluentes et earum momenta' which he penned in or shortly after April 1712 is here reproduced from the original sheets, now (as since long before Newton's death) geographically dispersed and here for the first time re-united in print. It will be evident that he takes as his basis (in [1]) a lightly recast version of the text of the 'De Quadratura Curvarum' as it had been most recently reissued by William Jones in his compendium of Newtonian *Analysis per Quantitatum Series, Fluxiones, ac Differentias* the year before (see the next note), adjusting its rôle to be that of an ancillary source lending weight and precision to his several general and specific appeals in the *Principia*—notably in Propositions XLI and XCI of its first book—to be 'granted the quadrature of curvilinear figures'. To this now familiar mainstay of his modes of exact algebraic integration there are (in [2]) attached, on the shaky, evasively half-truthful ground (see note (3)) that this was the analysis by which he had 'sought out propositions in the books of the *Principia*', five additional propositions broadening its scope to be once more close to that of the parent 'De quadratura Curvarum' of 1691–2 (VII: 48–128). These adduce (in Proposition XII) the formula for expanding a binomial into an equivalent infinite series as he had set it down and exemplified it for Leibniz in his *epistola prior* in 1676, together with (Proposition XIV) its application to the term-by-term quadrature of binomial integrands, thus obtaining (Proposition XVI) the series expansions of the basic circular and logarithmic functions; and also (in Proposition XV) repeat his procedure for extracting the fluent 'root' of a given fluxional equation term by term as a series of advancing powers of the base variable, virtually word for word as Wallis had earlier printed it in his Latin *Algebra* in 1693, now setting

Translation

[1]⁽²⁾ The analysis⁽³⁾ by which I sought out propositions in the books of the *Principia* I have decided now to subjoin so that readers instructed in it may be

---

in rider to it (in Proposition XIV) the analogous technique for the series extraction of the root of an ordinary algebraic equation in two variables. An intended *finale* where (see note (5)) he planned to apply his method of finite differences 'closely to approximate the area of a figure from a few of its ordinates given'—in the fashion, doubtless, of Proposition VI of his published 'Methodus Differentialis'—would appear never to have materialized on paper. On the date of composition see notes (2) and (49) following.

(2) This scheme for amending and minimally amplifying the text of the published 'De Quadratura Curvarum' to be the main portion of a new tract of 'Analysis per quantitates fluentes et earum momenta' is, except for one small addition to Proposition III which we here introduce (see note (18) following) from a stray scrap in Cambridge University Library, taken from a folded quarto sheet on which Newton has also penned drafts of his intended author's preface to the *Principia*'s second edition whereto he planned to append the 'Analysis', and also of the self-styled 'Committee' report on Leibniz' claim to priority in the invention of the calculus which we reproduce in Appendix 1.4 to Part 2 of this volume (and which, with some inessential alterations and a few additions, was read out in the Committee's name to the assembled Royal Society on 24 April 1712). From this subsequently discarded 'Ad Lectorem' (whose full text is now printed—under a date of '?Autumn 1712' which seems to us too late by half a year—in *The Correspondence of Isaac Newton*, 5: 112–13) we here set in proem, exactly as it is placed in the manuscript, the second paragraph where Newton introduces the sub-joined 'Analysis' to the reader of the *editio secunda, auctior et emendatior* of his *Principia*. The page references in the sequel are to the minimally corrected text of the 'De Quadratura' newly published by Jones in his collection of Newton's tracts and letter fragments on *Analysis per Quantitatum Series, Fluxiones, ac Differentias* (London, 1711): 41–66.

(3) 'fluentium' (of fluents) is deleted. By his ensuing phrase Newton invites his reader to infer that all the propositions in the *Principia* were discovered by the fluxional analysis which he will go on to append—a false generalization, as we have earlier underlined (see VI: 24) in our prefatory observations on the original drafts, as they now survive, of the *Principia*'s main dynamical theorems. Though he could on occasion (see Section 8 below) be more cautiously concrete in specifying the particular propositions to which he here refers, at other times when it suited his purposes he could make his present inference clear beyond ambiguity: 'The Propositions in the following Book', he wrote in preface to an intended re-edition of the *Principia* in the middle 1710's which came to nothing, 'were invented by Analysis: but... I composed what I invented...to make it Geometrically authentic & fit for the publick.... But if any man who understands Analysis, will reduce the Demonstrations of the Propositions from their [geometrical] composition back into Analysis (wch is easy to be done) he will see by what method of Analysis the Propositions were invented' (ULC. Add. 3968.9: 101^r, reproduced in full in **2**, Appendix 6 below). That 'analytical' equivalents to many of the theorems derived by infinitesimal geometry in the *Principia* may readily be found—in some instances, notably Proposition VI of the first book (on which compare VI: 42, note (30)), this is not so 'easy to be done'—is of course no proof that Newton had such algebraic analogues in mind when he contrived his geometrical counterparts, and there are very few propositions in the book where he appealed explicitly to a prior scheme of quadrature of the areas under given curves. We here witness the birth of a deeply entrenched later myth; compare D. T. Whiteside, 'The Mathematical Principles underlying Newton's *Principia Mathematica*' (*Journal for the History of Astronomy*, **1**, 1970: 116–38): 118–20.

facilius examinare possint et earum numerum inventis novis augere. Hujus An[a]lyseos partes varias chartis sparsis olim cum amicis[4] communicavi, & partes illas hic conjunxi ut methodus tota simul legatur, facilius & melius intelligatur, et magis prosit. Partium ultima est Methodus differentialis,[5] qua utiꝗ area figuræ ex paucis ordinatis per earum differentias colligitur quamproxime. Nam inventio arearum ex ordinatis describentibus, ad hanc Analysin omnino pertinet. . . .[6]

ANALYSIS PER QUANTITATES FLUENTES ET EARUM MOMENTA.

Introductio.

p. 41.[7] [Quantitates mathematicas &c *usꝗ ad*] *l. 20.*—et has motuum vel incrementorum velocitates nominando *Fluxiones*, quantitates genitas nominando *Fluentes*, & incrementa momentanea nominando *Momenta* incidi paulatim Annis 1665 & 1666 in Methodū[8] quam hic describo.

Fluxiones et momenta sunt quamproxime ut Fluentium augmenta &c.

p. 44. l. 16.[9] *Dele* Ex Fluxionibus invenire Fluentes . . . . . . olim scripsi.

p. 45. l. 1.[10] *dele* DE QUADRATURA CURVARUM. [*ut et*] *Q*[u]antitates indeterminatas ut motu . . . . . . considero designoꝗ [literis *z, y, x, v*, & earum fluxiones] *et scribe* Quantitates indeterminatas & fluentes seu motu perpetuo crescentes vel decrescentes designo literis *z, y, x, v* & earum momenta vel fluxiones seu celeritates crescendi [&c].

---

(4) Understand Henry Oldenburg (to whom he addressed his *epistolæ prior et posterior* for Leibniz in 1676) and John Wallis (to whom in the late summer of 1692 he passed on two principal propositions extracted from his 1691 treatise 'De quadratura Curvarum'); compare notes (24) and (67) following.

(5) The method of interpolation by finite differences, that is, with (as Newton goes on to state) special emphasis on its application to approximate the area under a given portion of a curve in terms of the base abscissa and a number of its ordinates spaced out at equal distances along it. Newton's prime illustration, we may be sure, would have been one or more of the simpler of the Cotesian formulas for such 'Quadrature by Ordinates' which he had deduced in 1695 somewhat cumbrously—and not entirely exactly—*per analogiam* (see VII: 690–6), but now anchoring these to Proposition VI of his published 'Methodus Differentialis' (4, §2 preceding) whereby they are accurately derived (compare 4, §1.2: note (14)) as the quadratures of the parabolas of pertinent 'parabolic kind' drawn to pass through the outer endpoints of the given equidistant ordinates.

(6) There intervenes in the manuscript a short paragraph in which Newton gives a gracious acknowledgement of his debt to Cotes for seeing the *Principia*'s second edition through press: 'In his omnibus edendis Vir doctissimus D. Rogerius Cotes Astronomiæ apud Cantabrigienses Professor operam navavit, prioris *Principiorum* editionis errata ['sphalmata' was first written] correxit & me submonuit ut nonnulla ad incudem revocarem. Unde factum est ut hæc editio priore sit emendatior'. Subsequently, as 1712 drew to its end, Newton grew increasingly cold to Cotes—not least for failing to spot the deep-seated error in Proposition X, Book 2 whose

the more easily able to assess the propositions delivered in these books and add fresh findings to their number. I some while ago imparted to friends[4] various portions of this analysis in scattered papers, and those parts I have here combined so that the method shall be read together as a whole, the more easily and better to be understood and to greater effect. The last of the parts is the method of differences,[5] whereby of course the area of a figure is gathered to a very near approximation from a few of its ordinates through their differences; for the finding of areas from their describing ordinates entirely pertains to this analysis.... [6]

ANALYSIS BY FLUENT QUANTITIES AND THEIR MOMENTS.

Introduction.

p. 41.[7] [Mathematical quantities...*up to*] *l. 20*...and, naming these speeds of motion or increment 'fluxions', the quantities so born 'fluents', and the momentary increments 'moments', I fell gradually in the years 1665 and 1666 on the method[8] which I here describe.

Fluxions and moments are very closely near as the augments of their fluents....

p. 44. l. 16.[9] *Delete*: From the fluxions to find the fluents .. .. I formerly wrote (the following).

p. 45. l. 1.[10] *Delete*: ON THE QUADRATURE OF CURVES. *and also*: (In the sequel) I consider indeterminate quantities as...by a (perpetual) motion......
I denote them by the letters z, y, x, v and mark their fluxions *and write*: Quantities which are indeterminate and fluent, that is, increasing or decreasing by a perpetual motion, I denote by the letters z, y, x, v and mark their moments or fluxions, that is, speeds of increase....

---

late correction cost Newton himself both a great deal of effort (as we shall see in the next section) and a loss of face with Johann Bernoulli which was to spoil all their future relations— and the brief 'Auctoris Præfatio' to the published *editio secunda* (a slightly amplified revise, dated 'Mar. 28. 1713', of the first paragraph, here omitted, of the present draft preface of a year earlier) neglects to make any mention even of Cotes' name.

(7) Of Jones' 1711 *Analysis*, that is; compare note (2) above. What follows is a minimal augmentation of the second of the opening paragraphs in the 'Introductio' to the published *Tractatus de Quadratura Curvarum* (see page 122 above), intercalating the implicit definition of fluxional 'moments' as 'momentaneous increases' of fluents (in correspondingly 'instantaneous' moments of the base variable of 'time').

(8) The superfluous specification 'Fluxionum' (of fluxions) is deleted here in the manuscript.

(9) Understand '*et seq.*'; Newton's cancellation is that of the last paragraph of the introduction to his published 'De Quadratura' (Jones, *Analysis*: 44, ll. 16–18; compare page 128 above).

(10) The start of the main text of the published 'De Quadratura' (compare page 128 above). Newton's instruction commands that the opening half-title be expunged, and that the beginning of the first sentence be slightly recast.

p. 47.[11] [*Adde*]

Corol. 1.[12] Hinc quantitates maximæ et minimæ inveniri possunt. Nam quantitas fit maxima vel minima ubi cessat augeri vel diminui. Inveni[a]tur igitur fluxio ejus et ponatur eadem æqualis nihilo.

Corol. 2.[13] Hinc etiam tangentes curvarum duci possunt. Capiatur utiꝗ subtangens in ea ratione ad ordinatam quam fluxio abscissæ habet ad fluxionem Ordinatæ.

Corol. 3.[14] Hinc etiam curvatura lineæ ad quodvis ejus punctum inveniri potest. A puncto illo duc rectas duas, alteram tangentem & alteram perpendicularem Curvæ, et erit fluxio partis Abscissæ inter Tangentem illam et punctum in Abscissa datum ad fluxionem partis Abscissæ inter Perpendiculum illud et punctum in abscissa datum ut pars Abscissæ inter Ordinatam et Tangentem ad partem Abscissæ inter Perpendiculum illud et rectam quæ a centro curvaturæ ad abscissam perpendiculariter demittitur.[15]

Corol. 4.[16] Hinc deniꝗ vis centripeta ad datum quodvis punctum tendens qua corpus in linea quacunꝗ Curva movebitur, inveniri potest. Nam vis illa ex invento curvaturæ centro dabitur per Corol. 3.[17] Prop. VI Lib. 1 *Princip.*

---

(11) That is, at the end of the 'Explicatio plenior' of Proposition I of the 'De Quadratura' (compare VII: 514–16, but remember that the last paragraph there was—see *ibid.*: 517, note (22)—omitted from the 1704 printed version). Newton's four added corollaries are manifestly a late attempt to rescue the main portion of the problems illustrating the application of fluxions which he had initially intended (see 2, §1.3 preceding) to incorporate as Proposition II of the *editio princeps*. The non-trivial variants in a superseded preliminary drafting of these which follows immediately after this head in the manuscript are cited in ensuing footnotes.

(12) 'Prob. 1' on page 98 above. In his preliminary draft (see the previous note) Newton here went on to specify that 'Si quantitas ubi fit maxima vel minima desideretur, ponatur ejus fluxio nulla' (If a quantity be desired where it comes to be a maximum or a minimum, set its fluxion nil).

(13) 'Prob. 3' on pages 98–100 above, with the 'determination' (instantaneous tangential direction) of a curve now constructed by way of the subtangent. Newton's preliminary draft (see note (11)) begins with the more precise enunciation 'Si recta ducenda est quæ Curvam in dato puncto tangat, capiatur subtangens...' (If a straight line needs to be drawn which shall touch a curve at a given point, take the subtangent...).

(14) Compare 'Prob. 4' on page 100 above. In his new-found ensuing construction of a problem which he had first resolved nearly fifty years earlier Newton proceeds clumsily to state, without hinting at its *raison d'être*, a by no means evident proportion which defines the perpendicular projection upon a curve's base of the centre of curvature at any point of it.

(15) Where, of necessity, understand that the ordinate of the curve is at right angles to its base abscissa. Newton's preliminary draft (see note (11)) affirms no less awkwardly that 'Si curvatura lineæ ad quodvis ejus punctum desideretur: a puncto illo duc Tangentem et Perpendiculum Curvæ & cum sit fluxio partis abscissæ...ut est pars abscissæ...ad abscissam perpendiculariter demittitur, ex fluxionibus illis habebitur centrum curvaturæ' (If the curvature of a line at any point of it be desired: from that point draw the tangent and the normal to the curve, and then, since the fluxion of the part of the abscissa...is to...as the part of the abscissa...is to...let fall...perpendicularly to the abscissa, from those fluxions the centre of curvature will be had). If in the figure set by him to accompany his following

p. 47.[11] *Add*:

Corollary 1.[12] Hence greatest and least quantities can be found. For a quantity comes to be greatest or least when it ceases to increase or diminish. Therefore find its fluxion and set it equal to nil.

Corollary 2.[13] Hence, also, tangents of curves can be found. Take of course the subtangent to the ordinate in the ratio which the fluxion of the abscissa has to the fluxion of the ordinate.

Corollary 3.[14] Hence, too, can the curvature of a line at any point of it be found. From that point draw two straight lines, one tangent to the curve and the other perpendicular to it, and then the fluxion of the part of the abscissa between that tangent and the point given in the abscissa will be to the fluxion of the part of the abscissa between that perpendicular and the point given in the abscissa as the part of the abscissa between the ordinate and the tangent to the part of the abscissa between that perpendicular and the straight line which is let fall from the centre of curvature perpendicularly to the abscissa.[15]

Corollary 4.[16] Hence, finally, there can be found the centripetal force tending to any given point you will, whereby a body shall move in any curved line whatever. For, once the centre of curvature is ascertained, that force will be yielded therefrom by means of Corollary 3[17] to Proposition VI of Book 1 of the *Principia*.

---

addition to page 66 of Jones' *Analysis* the angle $\widehat{ABC}$ of ordination be right, and the normal CP is extended to the centre O of curvature at C, with V now the tip of the perpendicular let fall from O onto AB, then Newton's defining proportion is $\mathrm{fl}(TA):\mathrm{fl}(AP) = TB:PV$, so that in more familiar Leibnizian equivalent there results $PV = TB \cdot d(AP)/d(TA)$. On setting $AB = x$ and $BC = y$ to be the Cartesian coordinates of the general point $C(x, y)$ of the curve, so that $TA = y \cdot dx/dy - x$ and $AP = y \cdot dy/dx + x$, and hence

$$d(TA)/dx = -y \cdot d((dy/dx)^{-1})/dx = -y \cdot (dx/dy)^2 \cdot d^2y/dx^2,$$

while $d(AP)/dx = (dy/dx)^2 + 1 + y \cdot d^2y/dx^2$, it readily follows that $PV = \rho \cdot dy/ds - BP$, where $OC = \rho = -(ds/dx)^3/(d^2y/dx^2)$. Newton's derivation of the proportion by considering the related infinitesimal augments Bb of the abscissa AB, Cc of the arc $\widehat{VC}$, Tt of the subtangent AT, and Pp of the subnormal AP (and defining the curvature centre O as the 'last' meet of the indefinitely close normals at C and c as these come to coincide), is preserved on the verso of the manuscript sheet; we reproduce it in [1] of the Appendix, fattening out its sparse frame with our usual editorial interlardings to bring out the logical connection of its parts.

(16) Compare 'Prob. 12' on page 102 above, where (see our note (43) there accompanying) Newton's present appeal to the corollaries to be added to its Book 1, Proposition VI in the *Principia*'s second edition (see the next note) is spelled out. In his preliminary draft (see note (11) above) Newton asserted little differently that 'Si vis centripeta desideretur qua corpus in linea quacunꝗ curva moveri possit: ex invento curvaturæ centro dabitur vis illa per Corol. 3...' (If the centripetal force be desired whereby a body can move in any curved line whatever: once the centre of curvature is found, that force will be yielded therefrom by means of Corollary 3...).

(17) Strictly, this should read '5'. In the expansion of the sole corollary to Proposition VI of the *Principia*'s Book 1 in the first edition ($_1$1687: 45; compare vi: 132–4) to be the five which illustrate its application in the *editio secunda* ($_2$1713: 42; compare vi: 546–50) 'Corol. 3' is the

[*p. 48.*][18]   *Ad Prop III* [*adde*] *Corol.* Si Ordinata sit pz^η, Area erit $\dfrac{p}{\eta+1}\, z^{\eta+1}$.

p. 64. l. 8.[19] *Dele*: Hæ fluxiones sunt ut termini .. .. & sic deinceps in infinitum.

p. 66. lin 7.[20] *Adde* Curvæ cujusvis
VC sit abscissa *AB*, ordinata *BC*, tangens *CT* et perpendiculum *CP* et detur
angulus *ABC*, et si relatio duarum[21]
quarumvis ex lineis quinq *BC*, *BT*,
BP, *CP* et *CT* detur per æquationem
quamvis, relatio inter eas omnes dabitur, et inde dabitur relatio inter abscis-

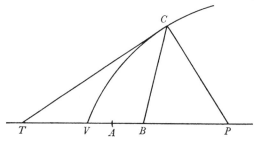

sam *AB* et ordinatam *BC* per quadraturam Curvæ cujus abscissa est *BC* et
ordinata est ut $\dfrac{BT}{BC}$. Nam fluxio ipsius *AB* est ad fluxionem ipsius *BC* ut *BT* ad
BC; ideoq si *BC* sit abscissa Curvæ et fluat uniformiter & fluxio ejus exponatur
per datam quamvis *N*, fluxio ipsius *AB* exponetur per $\dfrac{BT}{BC}N$. Et si hæc fluxio
fiat Ordinata ejusdem Curvæ, fluens exponetur per ejus aream.[22]

---

particular case of 'Corol. 5' in which 'Orbis vel circulus est, vel angulum contactus cum circulo
quam minimum continet, eandem habens curvaturam eundemq radium curvaturæ ad
punctum contactus'.

(18) We here interpolate this minimal corollary to Proposition III, keying it to the appropriate page in Jones' 1711 edition of the 'De Quadratura', from ULC. Add. 3960.6: 94,
where it is entered as an afterthought in an *addendum* to the 'Demonstratio' of Proposition XIII
of the 'Analysis' in [2] below (see our note (39) thereto). A preliminary draft of the corollary
on Add. 3960.6: 86 preceding affirms a little more elaborately that 'Si abscissa sit *z* &
Ordinata sit az^θ, area erit $\dfrac{a}{\theta+1}\, z^{\theta+1}$'.

(19) The second line of the terminal scholium of the 'De Quadratura'; compare page 152
above. In ignorance of the storm which was already beginning to rumble between Leibniz,
Johann Bernoulli and their adherents in their prejudiced assessment of his capabilities in
calculus, and which little more than a year later was (see 2, §3.2: note (46)) to flash out at the
clumsy verbal correlation which he had here set out between the successive terms in the series
expansion of the binomial increment of a given power of a fluent and its corresponding
fluxional 'moments', Newton instructs in sequel that the second sentence of the first paragraph
of the scholium and the whole of its second paragraph shall be cancelled. In their place he
will append a more precisely stated equivalent enunciation of the 'Taylor' expansion of an
incremented power of a variable in the scholium to Proposition XII which follows immediately
after the termination of the published 'De Quadratura' in the expanded text of the 'Analysis';
compare [2]: note (30) below.

(20) That is, the first line (beginning 'Et nota quod fluens omnis. . .'; compare page 158
above) of the last paragraph but two of the published 'De Quadratura'. Understand, however,
that the following addition is to be interposed immediately before this, as a separate paragraph.
A first version of this illustration of the fluxional inter-relationship of ordinate, tangent and
normal at a point on a curve, making the restriction that the angle $\widehat{ABC}$ between abscissa and

p. 48.[18] *To Proposition III add*:
Corollary. If the ordinate be pz^η, the area will be $pz^{\eta+1}/(\eta+1)$.

p. 64. l. 8.[19] *Delete*: These fluxions are as the terms... ...and so on indefinitely.

p. 66. l. 7.[20] *Add*: In any curve VC let the abscissa be AB, the ordinate BC, the tangent CT and the perpendicular CP, and let there be given the angle $\widehat{ABC}$: then, if the relationship of any two[21] out of the five lines BC, BT, BP, CP and CT be given by means of any equation, the relationship between them all will be given, and thence will be given the relationship between the abscissa AB and the ordinate BC through the quadrature of a curve whose abscissa is BC and ordinate is as BT/BC. For the fluxion of AB is to the fluxion of BC as BT to BC; and in consequence, if BC be the abscissa of the curve and should flow uniformly, its fluxion being expressed by any given (quantity) N, the fluxion of AB will be expressed by $(BT/BC)\,N$; and so, if this fluxion should become the ordinate of the same curve, its fluent will be expressed by its area.[22]

---

ordinate shall be right (on which particular instance see also our next note) is reproduced in [2] of the Appendix.

(21) Given the angle $\widehat{ABC}$, any two of the following linear elements manifestly suffice to fix the configuration $CTBP$ in absolute dimension, and hence their ratio—derivable, as Newton says, from any more general (homogeneous) *relatio* specified to connect them—will determine it in species. In particular, since $TC = \sqrt{(BC^2 + TB^2 - 2BC \times TB \cdot \cos\widehat{ABC})}$, while

$$CP = TC.\tan\widehat{BTC} \quad \text{and} \quad BP = BC.\sin\widehat{ABC}/\cos\widehat{BTC}$$

where $\sin\widehat{BTC} = (BC/TC).\sin\widehat{ABC}$, the ratio between any two others is straightforwardly reducible to that between BC and TB. Yet more simply, in the special case when $\widehat{ABC}$ is a right angle there is, as Newton himself stated in his prior version (see [2] of the Appendix) where this restriction is made, at once $BT:BC = BC:BP = TC:CP$.

(22) 'atcp adeo per quadraturam hujus Curvæ invenietur' (and therefore will be found through the quadrature of this curve) Newton initially concluded an otherwise effectively identical immediately preceding draft of this passage. Similarly, since $TC = BC.\mathrm{fl}(\widehat{VC})/\mathrm{fl}(BC)$, the curve's general arc-length is the fluent whose fluxion, with respect to BC as base variable, is TC/BC.

Already in his *epistola posterior* to Leibniz in October 1676 Newton had cited the special instance of this 'Inversum Problema de tangentibus' where the angle $\widehat{ABC}$ is right as one whose resolution did not require him to apply his general method of reduction to an infinite series: 'quando in triangulo rectangulo [TBC] quod ab...axis parte [TB] & tangente [TC] ac ordinatim applicata [BC] constituitur, relatio duorum quorumlibet laterum per æquationem quamlibet definitur, Problema solvi potest abscp mea methodo generali'; but he was quick to adjoin that 'ubi pars axis [AB] ad punctum aliquod [A] positione datum terminata ingrediatur vinculum [*sc.* æquationis] tunc res aliter se habere solet'. (See *Correspondence*, **2**, 1960: 129.) We have earlier remarked (see IV: 576–7, note (147)) that Leibniz did not himself at first fully comprehend the distinction which his correspondent drew between the simple defining *relatio* of form $TB[= BC.d(AB)/d(BC)] = f(BC)$—which, as Newton here (within the limitations of his understanding of what such a functional relationship of TB and BC might be) effectively asserts, is directly quadrable as $AB = \int f(BC)/BC.d(BC)$—and the (usually) far less tractable generalized *relatio* $TB = f(AB, BC)$.

[2][23]   *Prop. XII.*[24]

Dignitatem Binomij in seriem numero terminorum infinitam resolvere.

Solutio.

Sit binomium $P+PQ$, & index dignitatis ejus $\frac{m}{n}$ & erit

$$\overline{P+PQ}\big|^{\frac{m}{n}} = P^{\frac{m}{n}} + \frac{m}{n}AQ + \frac{m-n}{2n}BQ + \frac{m-2n}{3n}CQ + \frac{m-3n}{4n}DQ + \&c.$$

Ubi P et Q possunt esse quantitates quælibet vel simplices vel compositæ, et A, B, C, D &c designant terminos seriei,[25] nempe A terminum primum $P^{\frac{m}{n}}$, B terminum secundum $\frac{m}{n}AQ$, C terminum tertium $\frac{m-n}{2n}BQ_{[,]}$ & sic deinceps in infinitum.

Explicatio per exempla.[26]

Exempl. 1. Est

$$\sqrt{c^2+x^2}\,(\text{seu } \overline{c^2+x^2}\big|^{\frac{1}{2}}) = c + \frac{x^2}{2c} - \frac{x^4}{8c^3} + \frac{x^6}{16c^5} - \frac{5x^8}{128c^7} + \frac{7x^{10}}{256c^9} - \&c.$$

Nam, in hoc casu, est $P=c^2$, $Q=\frac{x^2}{c^2}$, $m=1$, $n=2$, $A(=P^{\frac{m}{n}}=\overline{cc}\big|^{\frac{1}{2}})=c$,

$$B\left(=\frac{m}{n}AQ\right)=\frac{x^2}{2c}, \quad C\left(=\frac{m-n}{2n}BQ\right)=-\frac{x^4}{8c^3},$$

& sic deinceps.

Exempl. 2. Est

$$\sqrt{5}: \overline{c^5+c^4x-x^5}\ (\text{i.e. } \overline{c^5+c^4x-x^5}\big|^{\frac{1}{5}}) = c + \frac{c^4x-x^5}{5c^4}[+]\frac{-2c^8x^2+4c^4x^6-2x^{10}}{25c^9} + \&c,$$

ut patebit substituendo in allatam Regulā, 1 pro m, 5 pro n, c^5 pro P, & $\dfrac{c^4x-x^5}{c^5}$

---

(23) ULC. Add. 3960.6: 71–81/3960.7: 113/115. The preparatory schemes for these additional Propositions XII–XVI which are reproduced below (from the originals in Add. 3960.6: 83–5 and 91–2) in [3]–[6] of the following Appendix will reveal the extent of the trouble to which Newton went in thus augmenting the text of his published 'De Quadratura'.

(24) Newton here presents his series expansion of the general binomial in the form in which he had thirty-six years before communicated it to Leibniz in his *epistola prior* of 13 June 1676 (see *Correspondence*, 2: 21–3) and in which John Wallis had nine years afterwards first published it (see IV: 672, note (54). In initial draft (see [3] of the Appendix) he had led gently up to this main run through the nursery slopes of reductions to infinite series from first principles *per divisionem* and *per extractionem radicis quadraticæ* much as in his earlier 1671 tract (see III: 36, 40), but abandoned these preparatories in his ensuing reworkings of the proposition (see [4] and [5] of the Appendix).

[2][23]   *Proposition XII.*[24]

To resolve a power of a binomial into a series unlimited in the number of its terms.

Solution

Let the binomial be $P+PQ$ and the index of its power m/n, and there will be

$$(P+PQ)^{m/n} = P^{m/n} + (m/n)\,AQ + (\tfrac{1}{2}(m-n)/n)\,BQ$$
$$+ (\tfrac{1}{3}(m-2n)/n)\,CQ + (\tfrac{1}{4}(m-3n)/n)\,DQ + \dots,$$

where P and Q can be any quantities you please, simple or composite, and $A, B, C, D, \dots$ denote the terms of the series,[25] namely, A the first term $P^{m/n}$, B the second term $(m/n)\,AQ$, C the third term $(\tfrac{1}{2}(m-n)/n)\,BQ$, and so on indefinitely.

Explanation through examples.[26]

Example 1. There is $\sqrt{(c^2+x^2)}$, that is $(c^2+x^2)^{\frac{1}{2}}$,

$$= c + \tfrac{1}{2}x^2/c - \tfrac{1}{8}x^4/c^3 + \tfrac{1}{16}x^6/c^5 - \tfrac{5}{128}x^8/c^7 + \tfrac{7}{256}x^{10}/c^9 - \dots.$$

For in this case there is $P = c^2$, $Q = x^2/c^2$, $m = 1, n = 2$, and so

$$A(= P^{m/n} = (c^2)^{\frac{1}{2}} =)\,c, \quad B(= (m/n)\,AQ) = \tfrac{1}{2}x^2/c,$$
$$C(= (\tfrac{1}{2}(m-n)/n)\,BQ) = -\tfrac{1}{8}x^4/c^3,$$

and so on.

Example 2. There is $\sqrt[5]{(c^5+c^4x-x^5)}$, that is, $(c^5+c^4x-x^5)^{\frac{1}{5}}$,

$$= c + \tfrac{1}{5}(c^4x-x^5)/c^4 - \tfrac{2}{25}(c^8x^2-2c^4x^6+x^{10})/c^9 + \dots,$$

as will be evident on substituting in the rule adduced 1 in place of m, 5 for n, c^5

---

(25) 'proxime præcedentes' (immediately preceding), manifestly rendered superfluous by the more elaborate specification which follows, is here deleted. In crucial cases, no doubt, Newton would also have his reader respect the further limitation $|Q| < 1$ necessary for the binomial series to converge when the index m/n is not a positive integer (and so the series itself does not terminate after a finite number of terms). Whether he ever appreciated that this same condition is also sufficient for convergence is not clear. We have remarked (see IV: 609, note (48)) how ill-formed his notion of a *series divergens* yet was when he came to write his tract 'De computo serierum' in 1684, and he nowhere went beyond his cloudy statements there to achieve a clearer notion of series convergence. Here he is content to repeat from his 1676 *epistola prior* to Leibniz, in Example 2 following, the loose criterion that '*x* valde parvum sit...ut series convergat' in the instance where $Q = (c^4x-x^5)/c^5$.

(26) These worked applications of the preceding general binomial expansion repeat, without change of word or symbol, the nine *Exempla* which Newton had set to illustrate its use in his *epistola prior* to Leibniz in June 1676 (see *Correspondence*, 2: 21–3; and compare IV: 667, note (34)). We need not here dwell on their detail, clearly expounded as it is by Newton himself.

pro Q. Potest etiam $-x^5$ substitui pro P, & $\dfrac{c^4x+c^5}{-x^5}$ pro Q, et tunc evadet

$$\sqrt{5}:\overline{c^5+c^4x-x^5}=-x+\frac{c^4x+c^5}{5x^4}+\frac{2c^8x^2+4c^9x+[2]c^{10}}{25x^9}+\&\text{c}.$$

Prior modus eligendus est si x valde parvum sit, posterior si valde mag[n]um, ut series convergat.[27]

Exempl. 3. Est $\dfrac{N}{\sqrt{3}:\overline{y^3-aay}}$ (hoc est $N\times\overline{y^3-aay}|^{-\frac13}$) æquale

$$N\times\overline{\frac1y+\frac{a^2}{3y^3}+\frac{2a^4}{9y^5}+\frac{14a^6}{81y^7}}+\&\text{c}.$$

Nam $P=y^3$, $Q=\dfrac{-aa}{yy}$, $m=-1$, $n=3$. $A(=P^{\frac{m}{n}}=y^{3\times-\frac13})=y^{-1}$, hoc est $\dfrac1y$.

$$B\left(=\frac{m}{n}\,AQ=\frac{-1}{3}\times\frac1y\times\frac{-aa}{yy}\right)=\frac{aa}{3y^3},\quad\&\text{c}.$$

Et N quantitatem quamcunꝗ seu simplicem seu compositam significat.

Exempl. 4. Radix cubica ex quadrato-quadrato ipsius $d+e$, (hoc est $\overline{d+e}|^{\frac43}$) est
$d^{\frac43}+\dfrac{4ed^{\frac13}}{3}+\dfrac{2e^2}{9d^{\frac23}}-\dfrac{4e^3}{81d^{\frac53}}+\&\text{c}.$ Nam $P=d$, $Q=\dfrac{e}{d}$, $m=4$, $n=3$, $A(=P^{\frac{m}{n}})=d^{\frac43}$, &c.

Exempl. 5. Eodem modo simplices etiam potestates eliciuntur. Ut si quadrato-cubus ipsius $[d+e$ (hoc est$]$ $\overline{d+e}|^5$, seu $\overline{d+e}|^{\frac51}$) desideretur: erit juxta Regulam,
$P=d$, $Q=\dfrac{e}{d}$, $m=5$ & $n=1$; adeoꝗ $A(=P^{\frac{m}{n}})=d^5$, $B\left(=\dfrac{m}{n}\,AQ\right)=5d^4e$, et sic
$C=10d^3e^2$, $D=10d^2e^3$, $E=5de^4$, $F=e^5$ et $G=\left(\dfrac{m-5n}{6n}\,FQ\right)=0$. Hoc est

$$\overline{d+e}|^5=d^5+5d^4e+10d^3e^2+10d^2e^3+5de^4+e^5.$$

Exempl. 6. Quinetiam Divisio, sive simplex sit, sive repetita, per eandem Regulam perficitur. Ut si $\dfrac{1}{d+e}$ (hoc est $\overline{d+e}|^{-1}$ sive $\overline{d+e}|^{\frac{-1}{1}}$) in seriem simplicium terminorum resolvendum sit: erit juxta Regulam, $P=d$, $Q=\dfrac{e}{d}$, $m=-1$, $n=1$, & $A(=P^{\frac{m}{n}}=d^{\frac{-1}{1}})=d^{-1}$ seu $\dfrac1d$,

$$B\left(=\frac{m}{n}\,AQ=-1\times\frac1d\times\frac{e}{d}\right)=-\frac{e}{d^2},\quad\&\text{ sic }\quad C=\frac{ee}{d^3},\quad D=-\frac{e^3}{d^4}\quad\&\text{c}.$$

Hoc est $\dfrac{1}{d+e}=\dfrac1d-\dfrac{e}{d^2}+\dfrac{e^2}{d^3}-\dfrac{e^3}{d^4}+\&\text{c}.$

for P and $\dfrac{(c^4x-x^5)}{c^5}$ for Q. You can also substitute $-x^5$ in place of P and

$-\dfrac{(c^4x+c^5)}{x^5}$ for Q, and then there will turn out

$$\sqrt[5]{(c^5+c^4x-x^5)} = -x+\frac{1}{5}\frac{(c^4x+c^5)}{x^4}+\tfrac{2}{25}(c^8x^2+2c^9x+c^{10})+\dots.$$

The first mode needs to be chosen if x be extremely small, the latter one if it be extremely large, in order that the series shall converge.[27]

Example 3. There is $\dfrac{N}{\sqrt[3]{(y^3-a^2y)}}$, that is, $N(y^3-a^2y)^{-\frac{1}{3}}$, equal to

$$N\left(\frac{1}{y}+\frac{1}{3}\frac{a^2}{y^3}+\frac{2}{9}\frac{a^4}{y^5}+\frac{14}{81}\frac{a^6}{y^7}+\dots\right).$$

For $P=y^3$, $Q=-\dfrac{a^2}{y^2}$, $m=-1$, $n=3$, and so $A(=P^{m/n}=y^{3\times-\frac{1}{3}})=y^{-1}$, that

is $\dfrac{1}{y}$, $B\left(=\left(\dfrac{m}{n}\right)AQ=\dfrac{-1}{3}\times\dfrac{1}{y}\times-\dfrac{a^2}{y^2}\right)=\dfrac{1}{3}\dfrac{a^2}{y^3}$, and so on. But N denotes any quantity whatsoever, simple or composite.

Example 4. The cube root of the 'square-square' (fourth power) of $d+e$, that is, $(d+e)^{\frac{4}{3}}$, is $d^{\frac{4}{3}}+\frac{4}{3}ed^{\frac{1}{3}}+\dfrac{2}{9}\dfrac{e^2}{d^{\frac{2}{3}}}-\dfrac{4}{81}\dfrac{e^3}{d^{\frac{5}{3}}}+\dots$. For $P=d$, $Q=\dfrac{e}{d}$, $m=4$, $n=3$, $A(=P^{m/n})=d^{\frac{4}{3}}$, and so on.

Example 5. In the same way simple powers, too, are elicited. If, for instance, the 'square-cube' (fifth power) of $d+e$, that is, $(d+e)^5$ or $(d+e)^{5/1}$, were

desired: there will, in accordance with the rule, be $P=d$, $Q=\dfrac{e}{d}$, $m=5$ and

$n=1$; and hence $A(=P^{m/n})=d^5$, $B\left(=\left(\dfrac{m}{n}\right)AQ\right)=5d^4e$, and in this way

$C=10d^3e^2$, $D=10d^2e^3$, $E=5de^4$, $F=e^5$ and $G\left(=\left(\dfrac{1}{6}\dfrac{m-5n}{n}\right)FQ\right)=0$; that is,

$$(d+e)^5 = d^5+5d^4e+10d^3e^2+10d^2e^3+5de^4+e^5.$$

Example 6. And to be sure, division, be it simple or repeated, is achieved by

the same rule. If, for instance, $\dfrac{1}{(d+e)}$, that is, $(d+e)^{-1}$ or $(d+e)^{-1/1}$, has to be

resolved into a series of simple terms: there will, according to the rule, be $P=d$,

$Q=\dfrac{e}{d}$, $m=-1$, $n=1$; and so $A(=P^{m/n}=d^{-1/1})=d^{-1}$, that is, $\dfrac{1}{d}$,

$B\left(=\left(\dfrac{m}{n}\right)AQ=-1\times\left(\dfrac{1}{d}\right)\times\dfrac{e}{d}\right)=-\dfrac{e}{d^2}$, and likewise $C=\dfrac{e^2}{d^3}$, $D=-\dfrac{e^3}{d^4}$, and so on;

that is, $\dfrac{1}{(d+e)}=\dfrac{1}{d}-\dfrac{e}{d^2}+\dfrac{e^2}{d^3}-\dfrac{e^3}{d^4}+\dots.$

(27) How great 'valde' is Newton again declines to specify in this unique comment upon the convergence of the series expansion of a binomial. Compare note (25).

Exempl. 7. Sic et $\overline{d+e}|^{-3}$ (hoc est unitas ter divisa per $d+e$, vel semel per cubum ejus,) evadit $\dfrac{1}{d^3} - \dfrac{3e}{d^4} + \dfrac{6e^2}{d^5} - \dfrac{10e^3}{d^6} + \&c.$

Exempl. 8. Et $N \times \overline{d+e}|^{\frac{-1}{3}}$ (hoc est, quantitas quælibet N divisa per radicem cubicam ipsius $d+e$,) evadit $N \times \overline{\dfrac{1}{d^{\frac{1}{3}}} - \dfrac{e}{3d^{\frac{4}{3}}} + \dfrac{2e^2}{9d^{\frac{7}{3}}} - \dfrac{14e^3}{81d^{\frac{10}{3}}}} + \&c.$

Exempl. 9. Et $N \times \overline{d+e}|^{\frac{-3}{5}}$ (hoc est, N divisum per radicem quadrato-cubicam ex cubo ipsius $d+e$, sive $\dfrac{N}{\sqrt{5} : \overline{d^3 + 3d^2e + 3de^2 + e^3}}$) evadit

$$N \times \overline{\dfrac{1}{d^{\frac{3}{5}}} - \dfrac{3e}{5d^{\frac{8}{5}}} + \dfrac{12e^2}{25d^{\frac{13}{5}}} - \dfrac{52e^3}{125d^{\frac{18}{5}}}} + \&c.$$

Demonstratio.

Regula proposita recte exhibet dignitates vel potestates Binomij ubi dignitatum indices $\dfrac{m}{n}$ sunt numeri integri, ut computanti statim patebit. Et Regula quæ ad innumera æqualia intervalla recte se habet, recte se habebit in locis intermedijs.[28]

Scholium.

Si Binomij Nomen secundum sit momentum Nominis primi & dignitas Binomij in seriem, juxta Regulam allatam, resolvatur; termini seriei erunt ut momenta dignitatis; nempe terminus secundus B ut momentum primum, tertius C ut momentum secundum, quartus D ut momentum tertium, & sic deinceps. Et[29] $1 \times B$ erit momentum primum dignitatis, $1 \times 2C$ momentum secundum, $1 \times 2 \times 3D$ momentum tertium, $1 \times 2 \times 3 \times 4E$ momentum quartum, & sic porrò in infinitum.[30]

Prop. XIII.[31]

Quantitatem fluentem, cujus fluxio aliqua est dignitas binomij quantitatem aliam uniformiter fluentem involventis, in serie numero terminorum infinita invenire.

---

(28) This appeal to a loose principle of continuity—one on which he had, in imitation of his mentor John Wallis, founded his own youthful researches into interpolating integral sequences, from which his earliest formulation of the binomial theorem had emerged as an unsought corollary (see 1: 96–134)—would, unfortunately, no longer be regarded as an acceptable basis for a demonstration by those of Newton's younger contemporaries, de Moivre, Brook Taylor and notably Roger Cotes, who were now coming to demand more precise criteria for rigorous proof. (Compare note (46) of the Appendix.) In sequel Newton began initially to copy out '*Corol. 1.* Hinc areæ Curvi[linearum...]' from the draft which is reproduced in [5] of the Appendix, but quickly cancelled these few opening words. It was at this point, we may therefore presume, that he decided to expand this prime corollary to be—by way of the intermediate draft printed in [6] of the Appendix—the separate Proposition XIII which follows.

Example 7. So too $(d+e)^{-3}$, that is, unity thrice divided by $d+e$, or once by its cube, turns out to be $1/d^3 - 3e/d^4 + 6e^2/d^5 - 10e^3/d^6 + \dots$.

Example 8. And $N(d+e)^{-\frac{1}{3}}$, that is, an arbitrary quantity N divided by the cube root of $d+e$, proves to be $N(1/d^{\frac{1}{3}} - \frac{1}{3}e/d^{\frac{4}{3}} + \frac{2}{9}e^2/d^{\frac{7}{3}} - \frac{14}{81}e^3/d^{\frac{10}{3}} + \dots)$.

Example 9. And $N(d+e)^{-\frac{3}{5}}$, that is, N divided by the 'square-cube' (fifth) root of the cube of $d+e$, or $N/\sqrt[5]{(d^3 + 3d^2e + 3de^2 + e^3)}$, comes to be

$$N(1/d^{\frac{3}{5}} - \frac{3}{5}e/d^{\frac{8}{5}} + \frac{12}{25}e^2/d^{\frac{13}{5}} - \frac{52}{125}e^3/d^{\frac{18}{5}} + \dots).$$

Demonstration.

The propounded rule correctly displays the 'dignities' or powers of the binomial when the indices m/n of the powers are integers, as will at once be evident to one who makes the computation. And a rule which holds correctly at innumerable equal intervals will hold correctly in the intermediate places.[28]

Scholium.

If the second member in a binomial be the moment of the first member, and a power of the binomial be, in accordance with the rule adduced, resolved into a series, the terms of the series will be as the moments of the power; specifically, the second term B as the first moment, the third one C as the second moment, the fourth one D as the third moment, and so on. In fact[29] $1 \times B$ will be the first moment of the power, $1 \times 2C$ the second moment, $1 \times 2 \times 3D$ the third moment, $1 \times 2 \times 3 \times 4E$ the fourth moment, and so forth infinitely.[30]

Proposition XIII.[31]

To find a fluent quantity, some fluxion of which is a power of a binomial involving another uniformly fluent quantity, in a series unlimited in the number of its terms.

---

(29) Newton began in sequel to affirm, much as in his draft scholium in [5] of the Appendix, that 'si termini seriei continuo multiplicentur per terminos hujus progressionis

$$1 \times 2 \times 3 \times 4 \times 5 \times 6 \ \&c,$$

secundus B per 1, [tertius C per 1×2, quartus D per $1 \times 2 \times 3$, & sic deinceps, tum habebuntur momenta dignitatis binomij]' (if the terms of the series be continually multiplied by the terms of this progression: $1 \times 2 \times 3 \times 4 \times 5 \times 6 \times \dots$, the second B by 1, [the third C by 1×2, the fourth D by $1 \times 2 \times 3$, and so on, then will there be had the moments of the binomial's power]), but broke off to make the more precise equivalent enunciation which now follows.

(30) In this case $Q = o\dot{P}$, the 'moment' of the *nomen primum* P with respect to a base variable of 'time' whose 'instantaneous' increment is unity; whence it will be clear that Newton here specifies—unconsciously anticipating Johann Bernoulli's objection the next year (see 2, § 3.2: note (46) preceding) that he did not know how to—the 'Taylor' expansion of the incremented power $(P + o\dot{P})^{m/n}$ as $P^{m/n} + oB + o^2C + o^3D + o^4E + \dots$, correctly stating that B, $2C$, $6D$, $24E, \dots$ are the first, second, third, fourth, $\dots$ fluxions of $P^{m/n}$. (Compare Newton's more cryptic prior enunciation of these correspondences in his equivalent scholium to 'Prop. XI' in [4] of the Appendix; and also our note (29) thereto.)

(31) Newton's draft of this proposition, here but lightly refashioned, is printed (from the original now Add. 3960.6: 92/93) in [6] of the Appendix.

Solutio.

Resolvatur dignitas binomij in seriem per Propositionem præcedentem, & multiplicetur terminus unusquisq̃ seriei per fluentem alteram, ac dividatur idem terminus per indicem dignitatis quam fluens illa altera jam habet in hoc termino,[32] et dignitas binomij sit ejus fluxio prima. Et operatione semel, bis vel ter repetita habebitur Fluens ubi dignitas binomij est ejus fluxio secunda, tertia vel quarta. Et sic deinceps in infinitum.

Explicatio per exempla.[33]

Proposito circulo cujus radius est unitas, sit x abscissa, $\sqrt{1-xx}$ Ordinata,[34] & 1 momentum Abscissæ, & momentum areæ quam Ordinata describit erit Ordinata illa $\sqrt{1-xx} = \overline{1-xx}\,|^{\frac{1}{2}} = 1 - \dfrac{1}{2}xx - \dfrac{1}{8}x^4 - \dfrac{1}{16}x^6 - \dfrac{5}{128}x^8 - \dfrac{7x^{10}}{256} - $ &c. Et inde Area[35] (per hanc Propositionem) prodit

$$x - \tfrac{1}{6}x^3 - \tfrac{1}{40}x^5 - \tfrac{1}{112}x^7 - \tfrac{5}{1152}x^9 - \tfrac{7}{2816}x^{11} - \&\text{c.}$$

Momentum verò arcus cujus sinus est x, erit

$$\frac{1}{\sqrt{1-xx}} = \overline{1-xx}\,|^{-\frac{1}{2}} = 1 + \frac{1}{2}x^2 + \frac{3}{8}x^4 + \frac{5}{16}x^6 + \frac{35}{128}x^8 + \frac{63}{256}x^{10} + \&\text{c.}$$

Et inde prodit arcus ille $= x + \tfrac{1}{6}x^3 + \tfrac{3}{40}x^5 + \tfrac{5}{112}x^7 + \tfrac{35}{1152}x^9 + \tfrac{63}{2816}x^{11} + \&$c.

Proposito circulo cujus diameter est unitas, sit x Abscissa, $\sqrt{x-xx}$ Ordinata, & 1 momentum Abscissæ; et momentum Areæ quam Ordinata describit erit Ordinata illa $\sqrt{x-xx}^{(36)} = x^{\frac{1}{2}} - \tfrac{1}{2}x^{\frac{3}{2}} - \tfrac{1}{8}x^{\frac{5}{2}} - \tfrac{1}{16}x^{\frac{7}{2}} - \tfrac{5}{128}x^{\frac{9}{2}} - \&$c. Et inde Area (per hanc Propositionem) prodit $\tfrac{2}{3}x^{\frac{3}{2}} - \tfrac{1}{5}x^{\frac{5}{2}} - \tfrac{1}{28}x^{\frac{7}{2}} - \tfrac{1}{72}x^{\frac{9}{2}} - \tfrac{5}{704}x^{\frac{11}{2}} - \&$c. Momentum verò arcus[37] erit $\tfrac{1}{2}x^{-\frac{1}{2}} + \tfrac{1}{4}x^{\frac{1}{2}} + \tfrac{3}{16}x^{\frac{3}{2}} + \tfrac{5}{32}x^{\frac{5}{2}} + \tfrac{35}{256}x^{\frac{7}{2}} + \tfrac{63}{512}x^{\frac{9}{2}} + \&$c. Et inde arcus prodit $x^{\frac{1}{2}} + \tfrac{1}{6}x^{\frac{3}{2}} + \tfrac{3}{40}x^{\frac{5}{2}} + \tfrac{5}{112}x^{\frac{7}{2}} + \tfrac{35}{1152}x^{\frac{9}{2}} + \tfrac{63}{2816}x^{\frac{11}{2}} + \&$c.

Si quæratur Area quam Ordinata normaliter erecta $\dfrac{1}{1+xx}$ describit, existente

---

(32) We are instructed, in other words, to integrate each of the component terms of the series of algebraic powers separately.

(33) Newton initially here proceeded to illustrate his general proposition by repeating the slightly generalized versions of the following examples which he had compiled for his pre-paratory draft (note (31) above)—but omitting his faulty, concluding attempt there to evaluate the area of the 'quadratrix' where (see note (60)) of the Appendix) he erroneously departed from the reciprocal of the *momentum arcus* defining the *figura arcuum* which (see *ibid.*: note (63)) he goes on to square—before passing in present sequel to cite their equivalents in which the radius or diameter (as appropriate) is for simplicity now put to be unity.

(34) Understand 'normaliter erecta' (raised at right angles), *sc.* to the abscissa.

(35) Namely, $\frac{1}{2}(\sin^{-1}x + x\sqrt{[1-x^2]})$, as recourse to a geometrical figure will readily show; compare I: 108, 124.

Solution.

Resolve the power of the binomial into a series by means of the preceding proposition, and multiply each single term of the series by the other fluent while also dividing the term by the index of the power which the other fluent already has in this,[32] and the power of the binomial shall be its first fluxion. And with the operation once, twice or thrice repeated there will be had a fluent where the power of the binomial is its second, third or fourth fluxion. And so on indefinitely.

Explanation through examples.[33]

Where there is propounded a circle whose radius is unity, let x be the abscissa, $\sqrt{(1-x^2)}$ the ordinate[34] and 1 the moment of the abscissa, and the moment of the area which that ordinate describes will then be that ordinate

$$\sqrt{(1-x^2)} = (1-x^2)^{\frac{1}{2}} = 1 - \tfrac{1}{2}x^2 - \tfrac{1}{8}x^4 - \tfrac{1}{16}x^6 - \tfrac{5}{128}x^8 - \tfrac{7}{256}x^{10} - \dots.$$

And thence the area[35] turns out (by this proposition) to be

$$x - \tfrac{1}{6}x^3 - \tfrac{1}{40}x^5 - \tfrac{1}{112}x^7 - \tfrac{5}{1152}x^9 - \tfrac{7}{2816}x^{11} - \dots.$$

The moment, however, of the arc whose sine is x will be

$$1/\sqrt{(1-x^2)} = (1-x^2)^{-\frac{1}{2}} = 1 + \tfrac{1}{2}x^2 + \tfrac{3}{8}x^4 + \tfrac{5}{16}x^6 + \tfrac{35}{128}x^8 + \tfrac{63}{256}x^{10} + \dots.$$

And thence there results that arc equal to

$$x + \tfrac{1}{6}x^3 + \tfrac{3}{40}x^5 + \tfrac{5}{112}x^7 + \tfrac{35}{1152}x^9 + \tfrac{63}{2816}x^{11} + \dots.$$

Where there is propounded a circle whose diameter is unity, let x be the abscissa, $\sqrt{(x-x^2)}$ the ordinate, and 1 the moment of the abscissa, and the moment of the area which the ordinate describes will then be that ordinate

$$\sqrt{(x-x^2)}^{(36)} = x^{\frac{1}{2}} - \tfrac{1}{2}x^{\frac{3}{2}} - \tfrac{1}{8}x^{\frac{5}{2}} - \tfrac{1}{16}x^{\frac{7}{2}} - \tfrac{5}{128}x^{\frac{9}{2}} - \dots.$$

And thence the area ensues (by this proposition) to be

$$\tfrac{2}{3}x^{\frac{3}{2}} - \tfrac{1}{5}x^{\frac{5}{2}} - \tfrac{1}{28}x^{\frac{7}{2}} - \tfrac{1}{72}x^{\frac{9}{2}} - \tfrac{5}{704}x^{\frac{11}{2}} - \dots.$$

The moment, however, of the arc[37] will be

$$\tfrac{1}{2}x^{-\frac{1}{2}} + \tfrac{1}{4}x^{\frac{1}{2}} + \tfrac{3}{16}x^{\frac{3}{2}} + \tfrac{5}{32}x^{\frac{5}{2}} + \tfrac{35}{256}x^{\frac{7}{2}} + \tfrac{63}{512}x^{\frac{9}{2}} + \dots,$$

and thence there results the arc

$$x^{\frac{1}{2}} + \tfrac{1}{6}x^{\frac{3}{2}} + \tfrac{3}{40}x^{\frac{5}{2}} + \tfrac{5}{112}x^{\frac{7}{2}} + \tfrac{35}{1152}x^{\frac{9}{2}} + \tfrac{63}{2816}x^{\frac{11}{2}} + \dots.$$

If there be sought the area which the ordinate $1/(1+x^2)$ erected at right

---

(36) '$= \overline{x-xx}|^{\frac{1}{2}}$', of course. The corresponding area is in exact form

$$\tfrac{1}{8}(\sin^{-1}2\sqrt{[x-x^2]} - 2(1-2x)\sqrt{[x-x^2]}).$$

(37) Namely $1/2\sqrt{[x-x^2]} = \tfrac{1}{2}(x-x^2)^{-\frac{1}{2}}$. As reference to a geometrical figure will at once again visually make clear, the arc itself has the exact expression $\sin^{-1}\sqrt{x}$.

x abscissa & 1 momento ejus; hæc Ordinata in seriem resoluta evadet

$$1 - x^2 + x^4 - x^6 + x^8 - \&c.$$

Et inde (per hanc Propositionem) elicitur area $x - \frac{1}{3}x^3 + \frac{1}{5}x^5 - \frac{1}{7}x^7 + \frac{1}{9}x^9 - \&c.$[38]

Demonstratio.

Si Figuræ cujusvis curvilineæ Abscissa sit x et Ordinata normalis x^n, area erit $\frac{1}{n+1}x^{n+1}$ per Prop. [III].[39] Et si Ordinata ex pluribus ejusmodi partibus componatur, pars unaquæcɓ per eandem Propositionem dabit aream ab ipsa descriptam,[40] et aggregatum omnium Arearum sub signis proprijs $+$ et $-$ erit area tota quam Ordinata tota describit.

Prop. XIV.[41]

Ex Æquatione quantitates duas fluentes involvente,[42] extrahere alter[utr]am in serie numero terminorum infinita.

Resolutio.

In æquatione quacuncɓ affecta sit z quantitas [una][43] fluens & y quantitas altera fluens quam extrahere oportet. Termini omnes ex eodem æquationis latere consistentes æquentur nihilo, & dignitas quantitatis y (si opus sit) exaltetur vel deprimatur sic ut ejus index nec alicubi negativus sit nec tamen altior quam ad hunc effectum requiritur. Et sit kz^λ terminus infimæ dignitatis eorum qui necɓ per y necɓ per ejus dignitatem quamcuncɓ[44] multiplicantur. Sit $lz^\mu y^\alpha$ terminus alius quilibet in quo y habetur, et omnes ordine terminos percurrendo collige ex singulis seorsim numerum $\dfrac{\lambda - \mu}{\alpha}$ sic, ut tot habeas ejusmodi

---

(38) The familiar 'Gregory' expansion of the inverse tangent; compare III: 34–5, note (5). (In correction of our assertion there regarding the independent Hindu invention of this series, it is now known that the Leibnizian *reductio per divisionem* by which Nilakaṇṭha—as Newton here—obtained it in about 1502 itself derives, according to an anonymous contemporary Sanskrit commentary on Bhāskara II's *Lilāvati*, from the equally obscure earlier Keralese mathematician Mādhava (*c.* 1340–1425); see R. C. Gupta, 'The Mādhava–Gregory series', *Mathematical Education*, 7, 1973: 67–70. To Mādhava also we must now (*pace* our earlier note (112) on II: 237) attribute the Leibnizian iteration which yields by repeated integration, term by term, the series expansions of the sine and cosine; see A. K. Bag, 'Mādhava's sine and cosine series', *Indian Journal of the History of Science*, 11, 1976: 54–7, and also C. T. Rajagopal and M. S. Rangachari, 'On an untapped source of medieval Keralese mathematics', *Archive for History of Exact Sciences*, 18, 1978: 89–102, especially 91–9.)

(39) Understand the corollary newly added in [1] preceding to Proposition III of the printed 'De Quadratura Curvarum', *ad pag. 48* in Jones' 1711 edition; compare note (18) above.

(40) 'ipsi competentem' (pertaining to it) is replaced.

(41) As with the corresponding Proposition XII in his initial scheme for extending the propositions of the printed 'De Quadratura' in [4] of the Appendix (compare our notes (31)

angles describes, x being the abscissa and 1 its moment, this ordinate when resolved into a series will prove to be $1 - x^2 + x^4 - x^6 + x^8 - \ldots$, and thence there will (by this proposition) be elicited the area $x - \frac{1}{3}x^3 + \frac{1}{5}x^5 - \frac{1}{7}x^7 + \frac{1}{9}x^9 - \ldots$[38]

Demonstration.

If in any curvilinear figure the abscissa be x and the ordinate normal to it x^n, the area will (by Proposition III[39]) be $x^{n+1}/(n+1)$. And if the ordinate be composed of several parts of this kind, each individual part will (by the same Proposition) yield the area described by it,[40] and the aggregate of all the areas under their proper signs $+$ and $-$ will be the total area which the total ordinate describes.

Proposition XIV.[41]

Out of an equation involving two fluent quantities[42] *to extract one or the other in a series unlimited in the number of its terms.*

Resolution.

In any affected equation whatever let z be one [43] fluent quantity and y the other fluent which it is required to extract. Let all the terms be positioned on the same side of the equation and equal to nothing, and the power of the quantity y (if there be need) raised or lowered so that its index shall neither anywhere be negative, nor yet be higher than is required to this effect. Then let kz^λ be the term of lowest power of those which are multiplied neither by y nor by any power whatever[44] of it, and $lz^\mu y^\alpha$ any other term you please in which y is had; and by running through all the terms in order gather from each one separately the number $(\lambda - \mu)/\alpha$, so that you have as many numbers of this sort as there are

---

and (33) thereto), this is the special instance of Proposition (XIII→) XV following—a word-for-word repetition of the 'Prob. II' which Newton had (see note (67) below) sent to John Wallis twenty years previously—where no fluxional quantities enter the given equation in y and z whose root $y \equiv y_z$ is to be expanded as an infinite series. The several alterations and omissions necessary accurately to specify the resolution in this particular case are mapped out in the 'Resolutio' given by Newton of that prior Proposition XII, which is here taken over by him without change; the ensuing 'Operatio secunda/tertia et sequentes' are here straightforwardly adapted from their more general equivalents in the parent 1692 'Prob. II' in the same fashion. At two places (see notes (42) and (43) following) where Newton has not himself fully effected the change-over, we have ourselves in our edited text made requisite adjustment of the manuscript.

(42) We omit a following phrase '*quarum alterutra æquabiliter fluit*' (*one or other of which is equably fluent*) which in this non-fluxional particular case of Proposition XV is not a little incongruous, expanding Newton's '*alteram*' in sequel to fill its place.

(43) For the same reason as in the preceding note we here excise Newton's ill-chosen abverb '*æquabiliter*' (equably). There is no present gain in restricting the base variable z to be uniformly 'flowing', and it is surely mere unthinking oversight on Newton's part that he here continues to tie it to be so.

(44) Initially just '*dignitatem aliquam*' (some power).

numeros quot sunt termini. Horum numerorum maximus[45] vocetur v, et z^v erit dignitas primi termini seriei. Pro ejus coefficiente ponatur a, et in æquatione resolvenda[46] scribe az^v pro y, ac termini omnes resultantes in quibus z ejusdem est dignitatis ac in termino kz^λ sub signis proprijs collecti ponantur æquales nihilo. Nam hæc æquatio debite reducta dabit coefficientem a. Sic habes az^v terminum primū seriei.

Operatio secunda.

Pro reliquis omnibus hujus seriei terminis nondum inventis pone p, et habebis æquationem $y = az^v + p$. In Resolvenda, pro y scribe hunc ejus valorem & habebis Resolvendam novam, ubi p officium præstet ipsius y. Et ex hac Resolvenda primum extrahes terminum seriei p eodem modo atcg terminum primum seriei totius $y = az^v + p$ ex Resolvenda prima extraxisti.

Operatio tertia et sequentes.

Dein tertiam Resolvendam eadem ratione invenies atcg secundam invenisti, et ex ea terminum tertium Seriei totius extrahes. Et similiter Resolvendam quartam invenies et ex ea quartum seriei terminum, & sic [in] infinitum. Series autem sic inventa erit radix Æquationis quam extrahere oportuit.

Explicatio per Exemplum.[47]

Ex æquatione $y^3 + c^2 y + czy - 2c^3 - z^3 = 0$[48] extrahenda sit quantitas fluens y.

---

(45) On substituting $y = az^v + O(z^{v+1})$ the general term $lz^\mu y^\alpha$ comes to be of order $z^{\mu + \alpha v}$, so that $\mu + \alpha v \geqslant \lambda$, where λ is the lowest index of the terms kz^λ in the given equation which are free of y; accordingly $v \geqslant (\lambda - \mu)/\alpha$, and hence at this first stage we need, as Newton prescribes, concern ourselves only with those terms $lz^\mu y^\alpha$ in y in the equation for which $(\lambda - \mu)/\alpha$ is greatest, setting this to be the index v of the first term az^v in the series expansion of the root $y \equiv y_z$. This algebraic algorithm corresponds exactly, of course, to Newton's earlier geometrical rule for isolating the terms of the reduced 'fictitious' equation whose root is the term az^v to $O(z^{v+1})$—that originally expounded by him in his 1671 treatise on methods of series and fluxions (see III: 50–2) and communicated five years afterwards to Leibniz in his *epistola posterior* in October 1676 (see *Correspondence*, 2: 126–7)—whereby the general term $lz^\mu y^\alpha$ is displayed in a rectangular Cartesian array of unit cells, occupying the one whose bottom-left corner is the point (α, μ), and a straight line is swung round the corresponding corner-point $(0, \lambda)$ of the bottom-most one, occupied by kz^λ, of the column of terms in z alone till it first touches one or more of the other cells (α, μ) occupied; in which position its slope $(\lambda - \mu)/\alpha$ is manifestly greatest. (Though Newton here fails to notice the fact—and perhaps never fully appreciated it?—any other lower vertex of the convex polygon formed by connecting the corner-points of occupied cells will equally serve as a pivot-point, and the slope of the polygon's side thus similarly determined—as the greatest value of $(M - \mu)/(\alpha - A)$ where the line is drawn to connect the corner-points (A, M) and (α, μ) of the cells occupied by the 'vertex' term $Lz^M y^A$ and the general term $lz^\mu y^\alpha$ respectively—will likewise yield the first term Az^N, $N = \text{Max} \left[(M - \mu)/(\alpha - A) \right]$, of a viable series-approximation to a root of the given equation. Compare our note (29) on III: 50–1.) As Newton goes on to state, equation to zero of the coefficients $\sum_\alpha (la^\alpha) + k$ of z^v in the ensuing

terms. Call the greatest[45] of these numbers ν, and z^ν will be the power of the first term of the series. For its coefficient put a, and in the equation to be resolved[46] write az^ν in place of y, and put equal to nothing all the resulting terms, collected under their proper signs, in which z is of the same power as in the term kz^λ; this equation will of course, when appropriately reduced, yield the coefficient a. In this way you have the first term az^ν of the series.

Second operation.

For all the rest of the terms of the series not yet found put p, and you will have the equation $y = az^\nu + p$. In the resolvend equation in place of y write this its value, and you will have a new resolvend in which p does duty for y. And out of this resolvend you will extract the first term of the series p in the same manner as you extracted the first term of the total series $y = az^\nu + p$ out of the first resolvend.

Third and following operations.

Then you will find the third resolvend by the same method as you found the second, and out of it extract the third term of the total series. And in like manner you will find the fourth resolvend, and from it the fourth term of the series, and so on indefinitely. Once thus found, however, the series will be the root of the equation which it was required to find.

Explanation through an example.[47]

Out of the equation $y^3 + c^2y + czy - 2c^3 - z^3 = 0$[48] let there need to be extracted

---

æquatio fictitia yields as its (real) roots the coefficient a of the first term az^ν in the series expansion $y = y_z$ sought.

(46) Newton here momentarily preferred the participle 'resultante' (resulting), but—foreseeing no doubt the imminent verbal clash with 'resultantes' in his next clause—at once returned to the equivalent gerundive which he had originally opted for in his draft 'Prop. XII' (see [4] of the Appendix).

(47) We will not be surprised that in this newly contrived elaborate illustration of his preceding general rule Newton chooses, with his usual parsimony, to take as his *exemplum* what is effectively the unaltered cubic equation—initially (see the next note) it was identically the same—which forty years before he had adduced in his 1671 tract (see III: 54–6) to instance the equivalent geometrical rule there expounded (*ibid.*: 50–2; compare note (45) above) analogously to isolate the 'fictitious' equations arising at each new stage; the unexplained skeleton of its worked series-solution he had passed on to Leibniz five years afterwards in his *epistola prior* of 13 June 1676 (see *Correspondence*, **2**: 24), adding thereto the following October a brief notice of his ancillary 'parallelogram' rule (see *ibid.*: 126–7). He now seeks to obtain a further term in the series evaluation of the root, that in the 5th power of the base variable ($x\rightarrow$) z, but a small numerical slip in his computation (the original rough calculation, preserved on Add. 3965.12: 218^r, is reproduced in [7] of the Appendix) vitiates all his hard work, yielding an erroneous coefficient for it (compare note (60) below).

(48) So changed by Newton from his worked example of forty years before (with $x \rightarrow z$) $y^3 + a^2y + azy - 2a^3 - z^3 = 0$ (on which see the previous note). The constant a was evidently

[49]Terminus infimæ dignitatis eorum qui non multiplicantur per y est $-2c^3$. Hic est igitur kz^λ, existente $\lambda = 0$. Et numeri $\dfrac{\lambda - \mu}{\alpha}$ ex terminis[50] per y multiplicatis collecti sunt 0, 0[51] & -1, quorum maximus 0 est ν, et az^ν est a. Et in æquatione resolvenda scribendo az^ν seu a pro y, & ponendo terminos resultantes

$$a^3 + c^2 a - 2c^3$$

in quibus dignitas ipsius z est ν seu 0, æquales nihilo, prodibit $a = c$.[52] Sic habes primum seriei terminum $az^\nu = c$.

Operatio secunda.

Pro reliquis seriei terminis pone p, et[53] substituendo $c + p$ pro y prodibit Resolvenda nova $p^3 + 3cpp + czp + 4ccp + ccz - z^3 = 0$, ubi p vicem subit ipsius y, et terminus ccz est kz^λ existente $\lambda = 1$. Et numeri $\dfrac{\lambda - \mu}{\alpha}$ ex terminis[54] per p multiplicatis collecti, sunt $\frac{1}{3}$, $\frac{1}{2}$, 0, & 1, quorum maximus 1 est ν. Et scribendo az^ν seu az in Resolvenda pro p, & colligendo terminos resultantes in quibus index dignitatis z est λ seu 1, prodit $4ccaz + ccz = 0$, et $a = -\frac{1}{4}$. Sic habes secundum seriei terminum $az^\nu = -\frac{1}{4}z$.

---

replaced, here initially by g before this in turn was superseded by c, to avoid confusion with the coefficient of the term az^ν which he sets himself to specify in the 'Operatio prima' of his preceding general account of his method of extracting the root of an equation as an infinite series.

(49) Newton initially went on thoughtlessly to copy out the here extraneous supposition 'Pone $\dot{z} = 1$' (Put $\dot{z} = 1$) from his 1692 'Prob. II' (Wallis, *Opera*, 2: 394 [=vii: 178]) before hastily suppressing it. At this point, obviously to avoid carrying over further superfluities of this kind and to allow him the better to rephrase his present transposed 'Explicatio', Newton left off attempting the short cut of directly recasting the earlier printed text to fit the instance of a non-fluxional equation, and roughed out the sequel (including the ensuing 'Demonstratio' and both terminal corollaries) on the blank inner pages, now Add. 3965.8: 87v/88r, of a business letter (*ibid.*: 88v) written to him by John Anstis a few weeks earlier on 24 March 1711/12, and of no value to him other than as scrap paper once its request asking him to draw up a report 'to be signed at the Mint on Wednesday next [*sc.* 26 March]' had been conveyed. Such minor variants in this preliminary drafting as are not wholly insignificant are, as usual, cited by us in following footnotes. The text of Anstis' letter (reproduced in *Correspondence*, 5: 254 with appropriate editorial commentary) has no mathematical importance in itself, but the other jottings entered by Newton on all four pages of its folded sheet (f. 87r–88v) —in part at least before he drafted the companion 'Explicatio per Exemplum' which here concerns us, as the sequence of writing in the manuscript clearly reveals—serve yet more narrowly than its preamble (see note (2) above) to pinpoint the date of composition of the 'Analysis' as mid-April 1712: these are in fact successive drafts and related checking computations for Newton's revision of Proposition XXXVII in Book 3 of his *Principia* (as it was to appear in print the next year in the second edition), replacing the value of 35° which he assumed for the angular departure of greatest winter/summer tides from solstice in the initial version he sent to Cotes on 8 April 1712 (see *Correspondence*, 5: 264–7, especially 265) by that of

the fluent quantity y. [49]The term of lowest power of those which are not multiplied by y is $-2c^3$. This is therefore kz^λ, with $\lambda = 0$. And the numbers $(\lambda - \mu)/\alpha$ collected from the terms[50] multiplied by y are 0, 0,[51] and -1, the greatest of which, 0, is ν, and so az^ν is a. And by entering in the equation to be resolved az^ν, that is, a, in place of y, and setting the resulting terms $a^3 + c^2a - 2c^3$ in which the power of z is ν, that is, 0, equal to nothing, there will result $a = c$.[52] In this way you have the first term of the series $az^\nu = c$.

Second operation.

For the rest of the terms of the series put p, and[53] on substituting $c + p$ in y's place there will ensue the new equation to be resolved

$$p^3 + 3cp^2 + czp + 4c^2p + c^2z - z^3 = 0,$$

where p takes on the rôle of y, and the term c^2z is kz^λ, with now $\lambda = 1$. And the numbers $(\lambda - \mu)/\alpha$ collected from the terms[54] multiplied by p are $\frac{1}{3}, \frac{1}{2}, 0$, and 1, the greatest of which, 1, is ν. Then by entering in the resolvend az^ν, that is, az, in p's place and gathering the resulting terms in which the index of the power of z is λ, or 1, there ensues $4c^2az + c^2z = 0$, and so $a = -\frac{1}{4}$. In this way you have the second term of the series $az^\nu = -\frac{1}{4}z$.

$37°$ (as it is in the *editio secunda*), but at first (on f. 87^r) carelessly retaining 0.819152 for its cosine before (on f. 88^v) making due correction of the oversight substantially as he wrote it up in the amended paper 'inclosed' by him with his next letter to Cotes a fortnight afterwards on 22 April (see *Correspondence*, **5**: 274–5).

(50) Namely of form $lz^\mu y^\alpha$.

(51) In his draft (see note (49)) Newton originally set these to be '$\frac{0}{3}, \frac{0}{1}$' and also added the further number '$-\frac{3}{0}$' corresponding to the term $-z^3$. He would have done better to omit this infinite value since, corresponding to the case where the ruler in the geometrical equivalent (see note (45)) pivots to lie vertically along the cells containing $-2c^3$ and $-z^3$, this isolates the 'fictitious' equation $-2c^3 - z^3 = 0$, which determines the particular solution $z = -c\sqrt[3]{2}$ and so $(y \neq 0)$ $y = \pm c\sqrt/(\sqrt[3]{2} - 1)$. But of course Newton is here intent less on completeness than on making adequate illustration of his method of extracting a solution in series.

(52) Neglecting, as always (compare IV: 542–4), the complex roots $a = \frac{1}{2}c(-1 \pm \sqrt{-7})$. Newton here initially began to copy from his preliminary draft 'Et scribendo az^ν seu a pro y, æquatio resolvenda evadit $a^3 + c^2a + cza - 2c^3 - z^3 = 0$. Hujus autem [termini $a^3 + c^2a - 2c^3$ in quibus dignitas ipsius z est ν seu 0, positi æquales nihilo dant $a = c$]' (And by writing az^ν, that is, a, in place of y, the equation to be resolved comes to be $a^3 + c^2a + cza - 2c^3 - z^3 = 0$. When, however, its terms $a^3 + c^2a - 2c^3$ in which the power of z is ν, that is, 0, are set equal to nil they yield $a = c$).

(53) In his rough draft (see note (49)) Newton has cancelled the unnecessary intervening phrase 'habebis æquationem $y = [c] + p$ &' (you will have the equation $y = c + p$, and so) at this junction.

(54) Here, *mutatis mutandis* as above (see note (50)), understand that these are of the general forms $lz^\mu p^\alpha$, $lz^\mu q^\alpha$ and $lz^\mu r^\alpha$ respectively.

Operatio tertia.

Pro reliquis terminis pone q,[55] et substituendo $-\frac{1}{4}z+q$ pro p prodit Resolvenda nova $q^3 - \frac{3}{4}zqq + \frac{3}{16}zzq - \frac{65}{64}z^3 + 3cqq - \frac{1}{2}czq - \frac{1}{16}czz + 4ccq = 0$. Ubi q vicem subit ipsius y et terminus $-\frac{1}{16}cz^2$ est kz^λ existente $\lambda = 2$. Et numeri $\dfrac{\lambda - \mu}{\alpha}$ ex terminis[54] per q multiplicatis collecti, sunt $\frac{2}{3}.\frac{1}{2}.0.1.1.2$, quorum maximus 2 est ν. Et scribendo az^ν seu az^2 in Resolvenda pro q & colligendo terminos resultantes in quibus index dignitatis z est λ seu 2, prodit $4ccaz^2 - \frac{1}{16}cz^2 = 0$, seu $a = \dfrac{1}{64c}$. Sic habes tertium seriei terminum $\dfrac{z^2}{64c}$.

Operatio quarta.

Pro reliquis terminis pone r[55] et substituendo $\dfrac{z^2}{64c} + r$ pro q prodit Resolvenda

nova $\quad r^3 + \dfrac{3zz}{64c}rr + \dfrac{3z^4}{4096cc}r + \dfrac{z^6}{262144c^3} - \dfrac{3z}{4}rr - \dfrac{3z^3}{128c}r - \dfrac{3z^5}{4096cc}$

$$+ 3crr + \dfrac{9zz}{32}r + \dfrac{15z^4}{4096c} - \dfrac{cz}{2}r - \dfrac{131z^3}{128} + 4ccr = 0,^{(56)}$$

ubi r vicem subit ipsius y et terminus $-\dfrac{131}{128}z^3$ est kz^λ existente $\lambda = 3$. Et numeri $\dfrac{\lambda - \mu}{\alpha}$ ex terminis[54] per r multiplicatis collecti, sunt $1, \frac{1}{2}, -1, 1, 0, \frac{3}{2}, 1, 2, 3$, quorum maximus 3 est ν. Et sc[r]ibendo az^ν seu az^3 in resolvenda pro r et colligendo terminos resultantes in quibus index dignitatis z est λ seu 3 prodit $-\dfrac{131z^3}{128}[+]4accz^3 = 0$ seu $a = [+]\dfrac{131}{512cc}$. Sic habes quartum seriei terminum $[+]\dfrac{131z^3}{512cc}$.

Et sic pergitur in infinitum prodeunte

$$y = c - \tfrac{1}{4}z + \frac{z^2}{64c} + \frac{131z^3}{512cc} + \frac{509z^4}{16384c^3} - \frac{4159z^5}{262112c^4} - \&c.^{(57)}$$

Et notandum est quod [58]post inventionem termini unius duorumve termini

---

(55) Newton here initially took over the respective intervening phrases 'et erit $p = -\frac{1}{4}z+q$' and 'erit $q = \dfrac{z^2}{64c} + r$' from his preliminary draft before again (compare note (53)) cancelling so minimal an amplification of what goes before.

(56) In his draft Newton incorrectly computed the coefficients of the terms in z^3r/c, z^5/c^2, z^2r and z^4/c, and it clearly cost him not a little trouble to set these right.

Third operation.

For the rest of the terms put q,[55] and on substituting $-\frac{1}{4}z+q$ in p's place there ensues the fresh resolvend

$$q^3 - \tfrac{3}{4}zq^2 + \tfrac{3}{16}z^2q - \tfrac{65}{64}z^3 + 3cq^2 - \tfrac{1}{2}czq - \tfrac{1}{16}cz^2 + 4c^2q = 0,$$

in which q takes on the rôle of y, and the term $-\frac{1}{16}cz^2$ is kz^λ, with now $\lambda = 2$. And the numbers $(\lambda - \mu)/\alpha$ collected from the terms[54] multiplied by q are $\frac{2}{3}, \frac{1}{2}, 0, 1, 1, 2$, the greatest of which, 2, is ν. Then by entering az^ν, that is, az^2, in the resolvend in q's place and gathering the resulting terms in which the index of the power of z is λ, or 2, there ensues $4c^2az^2 - \frac{1}{16}cz^2 = 0$, or $a = \frac{1}{64}/c$. So you have the third term of the series $\frac{1}{64}z^2/c$.

Fourth operation.

For the rest of the terms put r,[55] and on substituting $\frac{1}{64}z^2/c + r$ in q's place there ensues the new resolvend

$$r^3 + \tfrac{3}{64}(z^2/c)\,r^2 + \tfrac{3}{4096}(z^4/c^2)\,r + \tfrac{1}{262144}z^6/c^3 - \tfrac{3}{4}zr^2 - \tfrac{3}{128}(z^3/c)\,r$$

$$-\tfrac{3}{4096}z^5/c^2 + 3cr^2 + \tfrac{9}{32}z^2r + \tfrac{15}{4096}z^4/c - \tfrac{1}{2}czr - \tfrac{131}{128}z^3 + 4c^2r = 0,[56]$$

where r takes on the rôle of y, and the term $-\frac{131}{128}z^3$ is kz^λ, with now $\lambda = 3$. And the numbers $(\lambda - \mu)/\alpha$ collected from the terms[54] multiplied by r are $1, \frac{1}{2}, -1$, $1, 0, \frac{2}{3}, 1, 2, 3$, the greatest of which, 3, is ν. And by writing az^ν, that is, az^3, in the resolvend in r's place and gathering the resulting terms in which the index of the power of z is λ, or 3, there ensues $-\frac{131}{128}z^3 + 4ac^2z^3 = 0$, or $a = \frac{131}{512}/c^2$. So you have the fourth term of the series $\frac{131}{512}z^3/c^2$.

And in this way you may proceed indefinitely, there ensuing

$$y = c - \tfrac{1}{4}z + \tfrac{1}{64}z^2/c + \tfrac{131}{512}z^3/c^2 + \tfrac{509}{16384}z^4/c^3 - \tfrac{4159}{262112}z^5/c^4 - \dots.[57]$$

But note that [58]after the discovery of a term or two the remaining ones

---

(57) The coefficient of the term in z^5/c^4 should be $-\dfrac{1825}{131072}$; see note (60) following. Newton concluded this sequence of worked 'operations' in his draft with a succinct 'Et sic deinceps in infinitum' (And so on indefinitely).

(58) In his draft version (see note (49)) Newton initially went on to write 'postquā tres vel quatuor termini extracti [sunt, termini reliqui successive prodeunt] ordinando Resolvendam secundum dimensiones ipsius p vel q vel r et dividendo terminum ultimum per coefficientem penultimi, & mutando signū' (after three or four terms [are] extracted, [the remaining terms successively result] by ordering the resolvend according to the dimensions of p or q or r, and dividing the last term by the coefficient of the penultimate one and then changing sign); and subsequently altered its latter half to read 'applicando terminum infimæ dignitatis z non multiplicatum per p, vel q, vel r ad coefficientem infimæ dignitatis p vel q vel r, et signū mutan[d]o' (by dividing the term in the lowest power of z not multiplied by p or q or r [&c] by the coefficient of the lowest power of p or q or r, and changing sign). In taking over this

reliqui successive prodeunt applicando terminos infimos Resolvendæ non multiplicatos per p vel q vel r ad terminos multiplicatos per p vel q vel r unius dimensionis & mutando signa. Qua ratione plures etiam termini simul prodire solent.[59] Sic in operatione quarta dividendo

$$-\frac{131}{128}z^3 + \frac{15}{4096c}z^4 - \frac{3}{4096cc}z^5 \quad \text{per} \quad 4cc - \tfrac{1}{2}cz + \frac{9}{32}zz$$

prodit

$$r = \frac{131z^3}{512cc} + \frac{509z^4}{16384c^3} - \frac{4159z^5}{262112c^4} - \&\text{c.}^{[60]}$$

Et in operatione prima, si quantitas a fuerit radix æquationis affectæ, potest quælibet æquationis ejus radix in numeris extrahi, et pro a substitui. Et si radix illa sit impossibilis, debet quantitas z quantitate aliqua data augeri vel minui qua radix illa evadat realis.[61] Et si quantitas z est valde magna[,] pro ea scribi debet $\frac{1}{z}$ in Resolvenda, et post extractionem radicis, quantitas z pro $\frac{1}{z}$ restitui.[62]

---

amended phrasing into his revised text he again originally specified 'ordinando Resolvendam secundum dimensiones p vel q vel r, et dividendo terminum ultimum per coefficientem penultimi' (by ordering the resolvend according to the dimensions of p or q or r, and dividing the last term by the coefficient of the penultimate one), further recasting the latter clause to be 'et fingendo terminos duos ultimos æquales esse nihilo' (and conceiving the two last terms to be equal to nothing) before finally settling on the expression of his meaning which here follows. In his draft, we may add, Newton passed to instance his initial, more limited application of this mode of rounding off by noting: 'Sic in operatione secunda dividendo terminum ccz per $4cc$ coefficientem termini $[+]4ccp$ & signum mutando prodit $-\tfrac{1}{4}z$ qui secundus est terminus seriei. Et in operatione tertia dividendo terminum $-\tfrac{1}{16}czz$ per $4cc$ coefficientem termini $4ccq$ et signum mutando, prodit $\frac{zz}{64c}$ qui tertius est terminus seriei. Et in operatione quarta dividendo $\frac{131}{128}z^3$ per $4cc$ coefficientem [ipsius] $4ccr$ & signum mutando, prodit $-\frac{131z^3}{512cc}$ qui quartus est terminus seriei' (Thus in the second operation, on dividing the term c^2z by the coefficient, $4c^2$, of the term $+4c^2p$ and changing the sign, there results $-\tfrac{1}{4}z$ which is the second term of the series. And in the third operation, on dividing the term $-\tfrac{1}{16}cz^2$ by the coefficient, $4c^2$, of the term $4c^2q$ and changing the sign, there results $\tfrac{1}{64}z^2/c$ which is the third term of the series. And in the fourth operation, on dividing $\frac{131}{128}z^3$ by the coefficient, $4c^2$, of $4c^2r$ and changing the sign, there results $-\frac{131}{512}z^3/c^2$ which is the fourth term of the series).

(59) More precisely, as Newton initially stated at this point in his draft, 'postquam aliquot termini seriei inventi sunt potest numerus eorum duplicari' (after some few terms of the series have been discovered, their number can be doubled). At the third stage, for instance, where $y = c - \tfrac{1}{4}z + q$ with

$$q^3 + (3c - \tfrac{3}{4}z)\,q^2 + (4c^2 - \tfrac{1}{2}cz + \tfrac{3}{16}z^2)\,q + (-\tfrac{1}{16}cz^2 - \tfrac{65}{64}z^3) = 0,$$

since q is of $O(z^2)$, and so q^2 of $O(z^4)$, there is to this latter order

$$q = -(-\tfrac{1}{16}cz^2 - \tfrac{65}{64}z^3)/(4c^2 - \tfrac{1}{2}cz...) = \tfrac{1}{64}z^2/c + \tfrac{131}{512}z^3/c^2....$$

successively ensue on dividing the lowest terms of the resolvend not multiplied by p or q or r, by the terms multiplied by p or q or r of one dimension, and then changing signs. By this method further terms also usually result at the same time.[59] Thus in the fourth operation, on dividing $-\frac{131}{128}z^3 + \frac{15}{4096}z^4/c - \frac{3}{4096}z^5/c^2$ by $4c^2 - \frac{1}{2}cz + \frac{9}{32}z^2$ there ensues

$$r = \tfrac{131}{512}z^3/c^2 + \tfrac{509}{16384}z^4/c^3 - \tfrac{4159}{262112}z^5/c^4 - \ldots \text{[60]}$$

And in the first operation, had the quantity a been the root of an affected equation, any root you please of the equation can be extracted in numbers, and substituted for a. And should that root be 'impossible', the quantity z must be increased or diminished by some given quantity whereby the root shall come to be real.[61] While if the quantity z is extremely large, you ought in its place to write $1/z$ in the resolvend and, after extraction of the root, restore the quantity z in place of $1/z$.[62]

Newton himself proceeds to double the three terms in the series expansion of y obtained at the fourth stage.

(60) Here (compare the previous note) since r is of $O(z^3)$, and so r^2 of $O(z^6)$, to the latter order there is

$$(4c^2 - \tfrac{1}{2}cz + \tfrac{9}{32}z^2 \ldots)\, r + (-\tfrac{131}{128}z^3 + \tfrac{15}{4096}z^4/c - \tfrac{3}{4096}z^5/c^2 \ldots) = 0;$$

whence r is evaluated to $O(z^6)$ by dividing its coefficient in this equation into the last term 'et mutando signum'. Newton's computational skill was, unfortunately, never a match to his mathematical insight, and here too he falters: the coefficient of the last term (in z^5/c^4) should be $-\frac{1825}{131072}$. His rough accompanying calculation, preserved on Add. 3965.12: 218^r (and reproduced in its essentials in [7] of the Appendix) reveals that he failed to correct a carelessly doubled denominator halfway through it.

In his draft (see note (49)) he initially went on here to state: 'Deinde alius etiam terminus adjungi potest si in termino antepenultimo pro *qq* vel *rr* vel *ss* substituatur quadratum termini primi Quotientis. Ut in hoc exemplo quadratum ipsius $-\frac{131}{128}z^3$ pro *rr*' (Thereafter yet another term can be adjoined if in the last term but two there be substituted in place of q^2 or r^2 or s^2 the square of the first term in the quotient. As in the present example the square of $-\frac{131}{128}z^3$ in place of r^2). If we remember also now to include in the reckoning the other terms $+\frac{1}{262144}z^6/c^3$ and $-\frac{3}{128}(z^3/c)\,r$ also of $O(z^6)$ in the equation in r at the fourth stage of operation, then the observation is exact; but whether the ensuing complication makes this further step of computational value is dubious. In deleting the remark Newton evidently soon concluded that it was not worth the extra trouble.

(61) Newton originally wrote 'possibilis' (possible) in feebler counterpoint in his draft. As we have insisted before in commenting upon an analogous statement in his 'Matheseos Universalis Specimina' in 1684 (see IV: 543, note (58)), such an augmentation or diminution of the base variable z by some constant k so as to let the primary *aequatio fictitia* have a corresponding real root is, however, of no ultimate avail since the ensuing series expansion $y = y_{z \pm k}$ will, though its terms indeed have real (non-complex) coefficients, now ineluctably diverge to infinity.

(62) Thus minimally altered by Newton from his draft (see note (49)) where he equivalently wrote 'pro ea scribe $\frac{1}{z}$ in Resolvenda, et ubi radix extracta est, iterum scribe $\frac{1}{z}$ pro z' (in its place write $1/z$ in the resolvend, and, when the root is extracted, again write $1/z$ in z's place).

Demonstratio.

Si quantitas z sit valde parva (id enim supponitur ut series convergat) errores p, q, r, &c diminuentur in infinitum, & quavis data quantitate minores tandem evadent.

Corol. 1.[63] Hinc si radix y sit Ordinata figuræ cujusvis curvilineæ,[64] existente z Abscissa; inveniri potest area per Ordinatam descripta, ut in Propositione præcedente. Inveniri etiam potest alia quæcunqß quantitas fluens cujus momentum quodlibet est y.

Corol. 2 Et si habeatur fluens alterutra[65] in serie quacunqß convergente, extrahi potest Fluens altera[65] in alia serie. Ut si Hyperbolæ abscissa sit z & Ordinata $\dfrac{1}{1+z}$, & area per Ordinatam descripta

$$y = z - \tfrac{1}{2}z^2 + \tfrac{1}{3}z^3 - \tfrac{1}{4}z^4 + \tfrac{1}{5}z^5 - \tfrac{1}{6}z^6 + \&c,$$

extrahendo radicem ex primis sex vel septem terminis prodibit abscissa

$$z = y + \tfrac{1}{2}yy + \tfrac{1}{6}y^3 + \tfrac{1}{24}y^4 + \tfrac{1}{120}y^5 + \tfrac{1}{720}y^6 + \&c.^{[66]}$$

Prop. XV.[67]

Ex Æquatione fluxionem radicis involvente radicem extrahere.

Solutio.

Termini omnes, ex eodem æquationis latere [68]consistentes, æquentur nihilo, et ipsarum y & $\dot{y}$ dignitates (si opus sit) exaltentur vel deprimantur, sic ut earum indices nec alicubi negativi sint, nec tamen altiores quam ad hunc effectum

---

(63) In his draft of this 'Corol.' Newton initially appealed yet more closely to the opening instance in his 'Explicatio per exempla' of the preceding Proposition XIII, there first passing to affirm more broadly that 'Hinc ex dato Curvæ cujusvis arcu vel area inveniri potest abscissa & ordinata' (Hence, given the arc or area of any curve, from it there can be found the abscissa and ordinate), and adding in would-be illustration the instance 'Ut si in circulo abscissa sit x, ordinata $\sqrt{1-xx}$, [adeoqß] area $1 \times y = x - \tfrac{1}{6}xx - \tfrac{1}{40}x^4 - \tfrac{1}{112}x^6 - \tfrac{5}{1152}x^8 -$ &c [!], & arcus $z = x + \tfrac{1}{6}x^3 + \tfrac{3}{40}x^5 + \tfrac{5}{112}x^7 + \tfrac{35}{1152}x^9 +$&c, per extractiones fluentis x prodibit $x = y$', before abruptly breaking off in mid-equation. The reversions of the area $y = \int_0^x \sqrt{(1-x^2)} \, . \, dx$ and arc-length $z = \int_0^x 1/\sqrt{(1-x^2)} \, . \, dx [= \sin^{-1}x]$ which he here intended to display are manifestly $x = y + \tfrac{1}{6}y^3 + \tfrac{13}{120}y^5 + \dots$ and $x[= \sin z] = z - \tfrac{1}{6}z^3 + \tfrac{1}{120}z^5 - \dots$ respectively.

(64) In first draft just 'Curvæ' (curve) *tout court.*

(65) Originally specified as 'quantitas y' (the quantity y) and 'quantitas z' (the quantity z) respectively in the draft.

(66) The full working by which the first five terms in this given series expansion of $y = \int_0^z 1/(1+z) \, . \, dz [= \log(1+z)]$ are made to yield the corresponding first five terms of the inverse series $z = z_y [= e^y - 1]$ had of course—with the trivial interchange of variables y and z —been set down by Newton in his 1671 tract forty years before; see III: 58.

Demonstration.

If the quantity z be extremely small (that is supposed, of course, so that the series shall converge), the errors p, q, r, ... will diminish indefinitely, and at length come to be less than any given quantity.

Corollary 1.[63] Hence if the root y be the ordinate of any curvilinear figure,[64] z being the abscissa, the area described by the ordinate can be found, as in the preceding proposition. There can also be found any other fluent quantity whatever, whose moment is any y you please.

Corollary 2. And if there be had either one of the fluents[65] in any converging series whatever, the other fluent[65] can be extracted in another series. If, for instance, in a hyperbola the abscissa be z and ordinate $1/(1+z)$, and so the area described by the ordinate $y = z - \frac{1}{2}z^2 + \frac{1}{3}z^3 - \frac{1}{4}z^4 + \frac{1}{5}z^5 - \frac{1}{6}z^6 + ...$, by extracting the root out of the first six or seven terms there will ensue the abscissa

$$z = y + \tfrac{1}{2}y^2 + \tfrac{1}{6}y^3 + \tfrac{1}{24}y^4 + \tfrac{1}{120}y^5 + \tfrac{1}{720}y^6 + ... \text{[66]}$$

Proposition XV.[67]

Out of an equation involving the fluxion of the root to extract the root.

Solution.

Let all the terms, positioned on the same side of the equation, be equal to nothing, and the powers of y and $\dot{y}$ (if need be) raised or lowered such that their indices shall neither anywhere be negative, nor yet be higher than is required to

---

(67) This is a word-by-word repeat of the second of the two problems which Newton originally communicated privately to John Wallis in the late summer of 1692, taking as its source (compare note (70) below) the text which the latter made haste to introduce into the second volume of his own collected *Opera Mathematica* (then in press in Oxford) and duly published the next year, and which we ourselves have reproduced therefrom in vii: 177–80. We have previously observed (see note (41) above) that this generation term by term of the series expansion of the root $y = y_z$ of a given equation involving two 'fluents' y and z together with their first-order fluxions (of which $\dot{z}$ is, for convenience, taken to be unity) served as model for the preceding Proposition XIV which is its particular case where the given equation involves no fluent quantities. The extension of it—not quite correctly here formulated (see note (71) below)—to the general higher-order case, *mutatis mutandis*, is now freshly adumbrated by Newton in his final paragraph along the lines of 'Cas. 1' of Proposition XII of his 1691 treatise 'De quadratura Curvarum' where (see vii: 92–4) he first developed it. Notice that he returns in the following enunciation to his 1692 title, discarding the slightly more explicit one which he had set on the corresponding 'Prop. XIII' in the draft which we reproduce in [4] of the Appendix.

(68) In the manuscript (Add. 3960.7: 113) Newton embraces all but the last paragraph of the sequel in a '.... terminos plures', understanding the printed text of his 1692 'Prob. II' published by Wallis in his *Opera Mathematica*, **2**, 1693: 394, l. 2–395, *lin. ult.* To evince its full sweep we think it fitting here again (as on vii: 177–80) to reproduce the full text, now of course complementing it by our English version *en face.*

requiritur; & sit kz^λ terminus infimæ dignitatis eorum qui necg per y necg per ejus fluxionem $\dot{y}$ necg per earum dignitatem quamvis multiplicantur. Sit $lz^\mu y^\alpha \dot{y}^\beta$ terminus alius quilibet, & omnes ordine terminos percurrendo collige ex singulis seorsim numerum $\dfrac{\lambda - \mu + \beta}{\alpha + \beta}$ sic, ut tot habeas ejusmodi numeros quot sunt termini. Horum numerorum maximus vocetur v, et z^v erit dignitas primi termini seriei.[69] Pro ejus coefficiente ponatur a, et in æquatione resolvenda scribe az^v pro y, & vaz^{v-1} pro $\dot{y}$; ac termini omnes resultantes in quibus z ejusdem est dignitatis ac in termino kz^λ sub proprijs signis collecti, ponantur æquales nihilo. Nam hæc æquatio debite reducta dabit coefficientem a. Sic habes az^v terminum primum seriei.

Operatio secunda.

Pro reliquis omnibus hujus seriei terminis nondum inventis pone p, et habebis æquationem $y = az^v + p$, & inde etiam per Prob. I æquationem $\dot{y} = vaz^{v-1} + p$. In resolvenda pro y & $\dot{y}$ scribe hos eorum valores et habebis Resolvendam novam, ubi p officium præstat ipsius y; et ex hac Resolvenda primum extrahes terminum seriei p eodem modo atcg terminum primum seriei totius $y = az^v + p$ ex Resolvenda prima extraxisti.

Operatio tertia et sequentes.

Dein tertiam Resolvendam eadem ratione invenies atcg secundam invenisti, et ex ea terminum tertium seriei totius extrahes. Et similiter Resolvendam quartam invenies, et ex ea quartum seriei terminum, & sic in infinitum. Series autem sic inventa erit radix æquationis quam extrahere oportuit.

Exemplum.

Ex æquatione $y^2\dot{z}^2 - z^2\dot{z}\dot{y} - dd\dot{z}\dot{z} + dz\dot{z}^2 = 0$ extrahenda sit radix y. Pone $\dot{z} = 1$, et æquatio evadet $y^2 - z^2\dot{y} - dd + dz = 0$, quæ est Resolvenda. Jam vero terminus infimus in quo nec y necg $\dot{y}$ reperitur, est dd, qui ipsi kz^λ æquatus dat $\lambda = 0$. Terminis reliquis y^2, $-z^2\dot{y}$ pone $lz^\mu y^\alpha \dot{y}^\beta$ æqualem successive, et inde in primo casu habebis $\mu = 0$, $\alpha = 2$, $\beta = 0$ et in secundo $\mu = 2$, $\alpha = 0$ & $\beta = 1$: et hinc $\dfrac{\lambda - \mu + \beta}{\alpha + \beta}$ fit in primo casu 0, in secundo -1. Unde v est 0, et az^v & vaz^{v-1} sunt a et 0; quarum ultimæ duæ a et 0 in Resolvenda pro y & $\dot{y}$ scriptæ producunt $aa + 0z^2 - dd + dz = 0$, et termini aa & $-dd$ in quibus index dignitatis z est λ seu 0, positi æquales nihilo dant $a = d$. Unde primus seriei terminus az^λ evadit d.

---

(69) For, much as before in the case where $\beta = 0$ (see note (45) above), on substituting $y = az^v + O(z^{v+1})$ the general term $lz^\mu y^\alpha \dot{y}^\beta$ comes to be of order $z^{\mu + \alpha v + \beta(v-1)}$; whence if λ is the lowest power of z in the y-free terms there is $\mu + \alpha v + \beta(v-1) \geqslant \lambda$, and so

$$v \geqslant (\lambda - \mu + \beta)/(\alpha + \beta).$$

this effect; and let kz^λ be the term of lowest power of those which are multiplied neither by y nor by its fluxion $\dot{y}$ nor by any power of them. Let $lz^\mu y^\alpha \dot{y}^\beta$ be any other term you please, and by running through all the terms in order gather from each one separately the number $(\lambda - \mu + \beta)/(\alpha + \beta)$, so that you have as many numbers of this sort as there are terms. Call the greatest of these numbers ν, and z^ν will be the power of the first term of the series.[69] For its coefficient put a, and in the equation to be resolved write az^ν in place of y and $\nu az^{\nu-1}$ in place of $\dot{y}$, and put equal to nothing all the resulting terms, collected under their proper signs, in which z is of the same power as in the term kz^λ; this equation, of course, when properly reduced will yield the coefficient a. So you have the first term az^ν of the series.

Second operation.

For all the rest of the terms of this series not yet found put p, and you will have the equation $y = az^\nu + p$, and thence also by means of Problem I the equation $\dot{y} = \nu az^{\nu-1} + p$. In the resolvend equation in place of y and $\dot{y}$ write these their values, and you will have a new resolvend in which p does duty for y; and out of this resolvend you will extract the first term of the series p in the same manner as you extracted the first term of the total series $y = az^\nu + p$ out of the first resolvend.

Third and following operations.

Then you will find the third resolvend by the same method as you found the second, and from it extract the third term of the total series. And in like manner you will find the fourth equation to be resolved, and from it the fourth term of the series, and so on indefinitely. Once it is thus found, however, the series will be the root of the equation which it was required to extract.

Example.

Out of the equation $y^2 \dot{z}^2 - z^2 \dot{z} \dot{y} - d^2 \dot{z}^2 + dz \dot{z}^2 = 0$ let the root y need to be extracted. Put $\dot{z} = 1$ and the equation will come to be $y^2 - z^2 \dot{y} - d^2 + dz = 0$: this is the resolvend. Now, however, the lowest term in which neither y nor $\dot{y}$ is found is d^2, and this when equated to kz^λ yields $\lambda = 0$. To the remaining terms y^2, $-z^2 \dot{y}$ set $lz^\mu y^\alpha \dot{y}^\beta$ equal successively, and therefrom in the first case you will have $\mu = 0$, $\alpha = 2$, $\beta = 0$ and in the second one $\mu = 2$, $\alpha = 0$ and $\beta = 1$; and hence $(\lambda - \mu + \beta)/(\alpha + \beta)$ becomes in the first case 0, in the second -1. In consequence ν is 0, and so az^ν and $\nu az^{\nu-1}$ are a and 0. The two last of these, a and 0, when entered in the equation in place of y and $\dot{y}$ produce $a^2 + 0z^2 - d^2 + dz = 0$, and the terms a^2 and $-d^2$ in which the index of the power of z is λ, or 0, when set equal to nothing yield $a = d$. Whence the first term az^λ of the series proves to be d.

Operatio secunda.

Pro terminis reliquis pone p, et habebis æquationem $y=d+p$, et inde (per Prob. I) $\dot{y}=\dot{p}$; qui valores in Resolvenda pro y et $\dot{y}$ substituti dant Resolvendam novam $2dp+pp-zzp+dz=0$, ubi p et $\dot{p}$ vicem subeunt ipsarum y et $\dot{y}$. Terminus unicus in quo nec p neqȝ $\dot{p}$ reperitur est dz, qui cum termino kz^{λ} collatus dat $\lambda=1$. Terminis reliquis $2d\dot{p}$, pp & $-zz\dot{p}$ pone $lz^{\mu}p^{\alpha}\dot{p}^{\beta}$ æqualem successive, et inde in primo casu habebis $\mu=0$, $\alpha=1$ & $\beta=0$; in secundo $\mu=0$, $\alpha=2$ & $\beta=0$; et in tertio $\mu=2$, $\alpha=0$ & $\beta=1$. Et hinc $\frac{\lambda-\mu+\beta}{\alpha+\beta}$ evadit in primo casu 1, in secundo $\frac{1}{2}$, in tertio 0. Unde ν est 1, et az^{ν} & $\nu az^{\nu-1}$ sunt az et a. Termini duo ultimi az et a in Resolvenda pro p et $\dot{p}$ respective scripti producunt $2daz+a^{2}z^{2}-az^{2}+dz=0$. Et termini $2daz+dz$ in quibus index dignitatis z est λ seu 1, positi æquales nihilo dant $a=-\frac{1}{2}$. Unde az^{λ} terminus primus seriei p fit $-\frac{1}{2}z$.

Operatio tertia.

Pro terminis reliquis nondum inventis pone q et habebis æquationem $p=-\frac{1}{2}z+q$, et inde (per Prob. I) $\dot{p}=-\frac{1}{2}+\dot{q}$; qui valores pro p et $\dot{p}$ in Resolvenda novissima substituti producunt Resolvendam novam

$$2dq-zq+qq+\tfrac{3}{4}zz-zz\dot{q}=0,$$

ubi q & $\dot{q}$ vices supplent ipsorum y et $\dot{y}$. Terminus unicus in quo neqȝ q nec $\dot{q}$ reperitur est $\frac{3}{4}zz$, qui cum az^{λ} collatus dat $\lambda=2$. Terminis reliquis $2dq$, $-zq$, $+qq$, $-zz\dot{q}$ pone $lz^{\mu}q^{\alpha}\dot{q}^{\beta}$ æqualem successive, et inde in primo casu habebis $\mu=0$, $\alpha=1$, & $\beta=0$; in secundo, $\mu=1$, $\alpha=1$, $\beta=0$; in tertio, $\mu=0$, $\alpha=2$, $\beta=0$; in quarto $\mu=2$, $\alpha=0$, $\beta=1$: et inde $\frac{\lambda-\mu+\beta}{\alpha+\beta}$ evadit in primo casu 2, in secundo tertio et quarto 1. Et hinc ν est 2, vel az^{ν} et $\nu az^{\nu-1}$ sunt az^{2} et $2az$: qui valores in Resolvenda pro q & $\dot{q}$ substituti dant $2daz^{2}-az^{3}+aaz^{4}+\frac{3}{4}zz-2az^{3}=0$, et termini $2dazz+\frac{3}{4}zz$ in quibus index dignitatis z est λ seu 2, positi æquales nihilo dant $a=-\frac{3}{8d}$. Unde az^{λ} terminus primus seriei q evadit $-\frac{3zz}{8d}$.

Operatio quarta.

Pro reliquis seriei terminis nondum inventis pone r, et habebis æquationes $q=-\frac{3zz}{8d}+r$ et $\dot{q}=-\frac{3z}{4d}+\dot{r}$, et inde Resolvendam novam

$$2dr+\frac{9z^{3}}{8d}[-]zr+\frac{9z^{4}}{64dd}-\frac{3zzr}{[4]d}+rr-zz\dot{r}=0;$$

et ex ea per Methodum superiorem habebis $-\frac{9z^{3}}{16dd}$ terminum primum seriei r. Et sic pergitur in infinitum.

Second operation.

For the rest of the terms put p, and you will have the equation $y = d+p$ and therefrom (by Problem I) $\dot{y} = \dot{p}$; and these values when substituted in the resolvend in place of y and $\dot{y}$ yield the new resolvend $2d\dot{p}+p^2-z^2p+dz=0$, in which p and $\dot{p}$ take on the rôle of y and $\dot{y}$. The unique term in which neither p nor $\dot{p}$ is found is dz, and when this is collated with the term kz^λ it yields $\lambda = 1$. To the remaining terms $2d\dot{p}$, p^2 and $-z^2p$ set $lz^\mu p^\alpha \dot{p}^\beta$ successively equal, and therefrom in the first case you will have $\mu=0$, $\alpha=1$ and $\beta=0$; in the second $\mu=0$, $\alpha=2$ and $\beta=0$; and in the third $\mu=2$, $\alpha=0$ and $\beta=1$. Hence $(\lambda-\mu+\beta)/(\alpha+\beta)$ comes in the first case to be 1, in the second to be $\frac{1}{2}$, and in the third 0; and in consequence ν is 1, and so az^ν and $\nu az^{\nu-1}$ are az and a. The two last terms az and a when written in the resolvend in place of p and $\dot{p}$ respectively produce

$$2daz+a^2z^2-az^2+dz=0.$$

And when the terms $2daz+dz$ in which the index of the power of z is λ, or 1, are set equal to nothing they yield $a = -\frac{1}{2}$. Whence the first term az^λ of the series p becomes $-\frac{1}{2}z$.

Third operation.

For the rest of the terms not yet found put q and you will have the equation $p = -\frac{1}{2}z+q$, and therefrom (by Problem I) $\dot{p} = -\frac{1}{2}+\dot{q}$; and when these values are substituted for p and $\dot{p}$ in the most recent resolvend they produce the new resolvend $2d\dot{q}-zq+q^2+\frac{3}{4}z^2-z^2\dot{q}=0$, where q and $\dot{q}$ take over the rôles of y and $\dot{y}$. The unique term in which neither q nor $\dot{q}$ is found is $\frac{3}{4}z^2$, and this when collated with az^λ yields $\lambda=2$. To the remaining terms $2d\dot{q}$, $-zq$, $+q^2$, $-z^2\dot{q}$ set $lz^\mu q^\alpha \dot{q}^\beta$ successively equal, and therefrom in the first case you will have $\mu=0$, $\alpha=1$, and $\beta=0$; in the second one $\mu=1$, $\alpha=1$, $\beta=0$; in the third $\mu=0$, $\alpha=2$, $\beta=0$; in the fourth $\mu=2$, $\alpha=0$, $\beta=1$: and thence $(\lambda-\mu+\beta)/(\alpha+\beta)$ comes in the first case to be 2, and in the second, third and fourth ones 1. In consequence ν is 2, or az^ν and $\nu az^{\nu-1}$ are az^2 and $2az$; and when these values are substituted in the resolvend in place of q and $\dot{q}$ they give $2daz^2-az^3+a^2z^4+\frac{3}{4}z^2-2az^3=0$, and when the terms $2daz^2+\frac{3}{4}z^2$ in which the index of the power of z is λ, or 2, are set equal to nothing they yield $a = -\frac{3}{8}/d$. Whence the first term az^λ of the series q proves to be $-\frac{3}{8}z^2/d$.

Fourth operation.

For the rest of the terms of the series not yet found put r, and you will have the equations $q = -\frac{3}{8}z^2/d+r$ and $\dot{q} = -\frac{3}{4}z/d+\dot{r}$, and therefrom the new resolvend equation $2d\dot{r}+\frac{9}{8}z^3/d-zr+\frac{9}{64}z^4/d^2-\frac{3}{4}(z^2/d)\,r+r^2-z^2\dot{r}=0$; and from it by means of the above method you will obtain $-\frac{9}{16}z^3/d^2$ as the first term of the series r. And in this way you may proceed indefinitely.

Est igitur radix extrahenda

$$y = d + p = d - \tfrac{1}{2}z + q = d - \tfrac{1}{2}z - \frac{3zz}{8d} + r = d - \tfrac{1}{2}z - \frac{3zz}{8d} - \frac{9z^3}{16dd} - \&\mathrm{c}.$$

Et operationem continuando producere licet radicem ad terminos plures.

Et eadem methodo[70] radices æquationum fluxiones secundas, tertias, quartas $(\ddot{y}, \dddot{y}, \ddddot{y},)$ aliasq̃ involventium extrahi possunt. Ut si æquatio fluxionem secundam involvat, index v est maximus numerorum $\dfrac{\lambda - \mu + \beta + \gamma}{\alpha + \beta + \gamma}$[71] et termini z^v coefficiens a invenietur scribendo in Resolvenda az^v pro y, vaz^{v-1} pro $\dot{y}$, et $\overline{vv - v} \times az^{v-2}$ pro $\ddot{y}$.

Prop. XVI.[72]

Problemata per inventionem serierum numero terminorum infinitarum resolvere.

Solutio.

Pro terminis seriei quantitates indeterminatæ usurpandæ sunt et ex conditionibus Problematis gradatim determinandæ. Et inter computandum negligendi sunt termini omnes in quibus quantitas uniformiter fluens non est infimæ dignitatis.

Explicatio per Exempla.

Exempl. 1.[73] Inveniendus sit sinus[74] ex arcu dato. Sit radius 1, arcus uniformiter fluens z ejus[q̃] fluxio 1, sinus vel abscissa x, & ordinata $\sqrt{1 - xx}$, et ex natura

---

(70) Newton here went unthinkingly on to copy out a 'dicit Newtonus' (says Newton) interposed by Wallis at this point in the parent printed text (*Opera Mathematica*, **2**: 396, l. 1 [=vii: 180, l. 4]), but of course hastily deleted the interjection.

(71) In the English version we make good Newton's careless omission of the coefficient 2 of the last quantity in the numerator of this fraction. Here substitution of $y = az^v + \dots$ brings the general term—now understood to be $lz^\mu y^\alpha \dot{y}^\beta \ddot{y}^\gamma$ of course—to be of dimension

$$\mu + \alpha v + \beta(v - 1) + \gamma(v - 2) \geqslant \lambda,$$

where λ is the lowest power of z among the y-free terms; whence $v \geqslant (\lambda - \mu + \beta + 2\gamma)/(\alpha + \beta + \gamma)$. And the like holds true in cases where the given equation involves any higher order of fluxions of its fluents, as Newton had affirmed generally in 'Cas. 1' of Proposition XII of his 1691 treatise 'De quadratura Curvarum' (see vii: 92–4; and compare our note (97) thereto).

(72) This amplified presentation of the second 'methodus generalior', consisting 'in assumptione seriei pro quantitate qualibet incognita ex qua cætera commodè derivari possunt, et in collatione terminorum homologorum æquationis resultantis ad eruendos terminos assumptæ seriei', whose enunciation—even that jumbled in an anagram—Newton had passed to Leibniz in his *epistola posterior* in October 1676 (see *Correspondence*, **2**: 129; and compare ii: 191, note (25)) was here initially tacked on by him to the previous proposition as a 'Schol.' briefly affirming that 'Proposita quæstione in qua quantitas aliqua ignota desideratur in serie infinita convergente, pro terminis seriei quantitates indeterminatæ usurpari ['sigillatim assumi' was first written] possunt et ex conditionibus Problematis gradatim determinari' (Where a question is proposed in which some unknown quantity is desired [expanded] in a converging infinite series, in place of the terms of the series indeterminate quantities can be employed [assumed one by one] and then determined step by step from the

The root to be extracted is therefore

$$y=d+p=d-\tfrac{1}{2}z+q=d-\tfrac{1}{2}z-\tfrac{3}{8}z^2/d+r=d-\tfrac{1}{2}z-\tfrac{3}{8}z^2/d-\tfrac{9}{16}z^3/d^2-\ldots.$$

And by continuing the procedure you are free to extend the root to further terms.

And by the same method[70] the roots of equations involving second, third, fourth fluxions $(\ddot{y},\dddot{y},\ddddot{y})$ and other ones can be extracted. If, for instance, it involves the second fluxion, the index v is the greatest of the numbers

$$(\lambda-\mu+\beta+[2]\,\gamma)/(\alpha+\beta+\gamma)^{(71)}$$

and the coefficient a of the term in z^v will be found by writing in the resolvend az^v in place of y, vaz^{v-1} in place of $\dot{y}$, and $(v^2-v)\,az^{v-2}$ in place of $\ddot{y}$.

Proposition XVI.[72]

To resolve problems through the finding of series unlimited in the number of their terms.

Solution

For the terms of the series indeterminate quantities are to be employed and determined step by step from the conditions of the problem. And in the course of the computation all terms are to be neglected in which the uniformly fluent quantity is not of the lowst power.

Explanation through examples.

Example 1.[73] *There needs to be found the sine*[74] *in terms of its arc given.* Let the radius be 1, the uniformly fluent arc z and its fluxion 1, the sine or abscissa x, and the ordinate $\sqrt{(1-x^2)}$, and from the nature of the problem there will be

---

conditions of the problem). His ensuing illustrations of this general method of deriving series are notably less adventurous than those which he set to exemplify his earlier detailed account of its operation in 'Cap. 5' of his 'Matheseos Universalis Specimina' in 1684 (see IV: 576–88); here he contents himself with obtaining from first principles the simple sine and exponential series whose expansions he had previously, as he explained in his *epistola prior* to Leibniz in June 1676 (see *Correspondence*, 2: 25, 27), deduced by reversing those for the inverse-sine and natural logarithm determined by him in the winter of 1664–5 by direct reduction of an integrand to a series of terms thereafter individually integrated (see I: 108–10, 112–13). In his following scholium (which terminates the text of the 'Analysis' as we have it, lacking the unwritten further proposition(s) where he originally—see note (1) above—planned to deliver some précis of his method of finite differences) Newton will, as a sophisticated would-be illustration of his guide-rule for identifying the pattern of a series development by examining the individual ratios of its successive terms, adduce yet a further, trigonometrical expansion— erroneous coefficients and all!—from his 1676 *epistola prior* to Leibniz (see note (85) below; and compare *Correspondence*, 2: 25).

(73) We omit to cite *verbatim* a first casting of this example which Newton has cancelled immediately before in the manuscript (Add. 3960.7: 113), thinking it more than enough to point to its main discrepancies in following footnotes.

(74) Understand '*in circulo*' (*in a circle*) of course, as Newton indeed himself specified in a deleted preliminary phrase which he subsequently here neglected to incorporate.

problematis erit 1 ad $\sqrt{1-xx}::\dot{z}.\dot{x}::1.\dot{x}$ seu $\sqrt{1-xx}=\dot{x}$, et $1-xx=\dot{x}\dot{x}$.
Assumatur[75] $az^v[+\&\mathrm{c}]=x$ et erit $1-aaz^{2v}[-\&\mathrm{c}]=vva^2z^{2v-2}[+\&\mathrm{c}]$. Et collatis
infimæ dignitatis[76] terminis 1 et vva^2z^{2v-2} fit $v=1=a$, et $az^v=z$. Est igitur
primus seriei terminus z.

Pro termino secundo pone jam az^v et erit $x=z+az^v+\&\mathrm{c}$[77] et
$$1-zz-2az^{v+1}+\&\mathrm{c}=\dot{x}\dot{x}=1+2vaz^{v-1}+\&\mathrm{c}.$$
Et collatis terminis[78] fit $v=3$ et $a=-\frac{1}{6}$ et series $=z-\frac{1}{6}z^3+\&\mathrm{c}$.

Pro termino tertio pone jam az^v et erit $x=z-\frac{1}{6}z^3+az^v+\&\mathrm{c}$[79] et
$$1-z^2+\tfrac{1}{3}z^4+\&\mathrm{c}=\dot{x}\dot{x}=1-z^2+\tfrac{1}{4}z^4+2vaz^{v-1}+\&\mathrm{c}.$$
Et collatis terminis fit $v=5$ et $a=\frac{1}{120}$ et series $=z-\frac{1}{6}z^3+\frac{1}{120}z^5-\&\mathrm{c}$. Et sic
pergitur in infinitum.

Exempl. 2.[80] *Invenienda sit linea curva*[81] *cujus subtangens datur longitudine.* Sit z
abscissa æquabiliter fluens, 1 fluxio ejus, y Ordinata & s subtangens seu pars
abscissæ inter Ordinatam et Tangentem. Et cum subtangens (ex natura
curvarum[)] sit ad Ordinatam ut momentum abscissæ ad momentum ordi-
natæ, erit $s\dot{y}=y\dot{z}=y$, et $\dot{y}=\frac{1}{s}y$. Quo cognito solvetur Problema[82] assumendo
seriem[83] pro y et terminos ejus gradatim determinando ut sequitur. Sit
$az^n+bz^{n+1}+cz^{n+2}+dz^{n+3}+ez^{n+4}+\&\mathrm{c}=y$[84] & erit eadem series $=s\dot{y}=$
$$nsaz^{n-1}+\overline{n+1}[\times]sbz^n+\overline{n+2}\times scz^{n+1}+\overline{n+3}\times sdz^{n+2}+\overline{n+4}\times sez^{n+3}+\&\mathrm{c}$$
et collatis terminis fit $n=0$, $a=sb$, $b=2sc$, $c=3sd$, $d=4se$, &c adeoӡ
$$y=a+\frac{a}{s}z+\frac{a}{2ss}z^2+\frac{a}{6s^3}z^3+\frac{a}{24s^4}z^4+\&\mathrm{c},$$
ubi quantitas a pro lubitu assumi potest.[85]

---

(75) 'primus seriei terminus' (the first term of the series) was inserted here in a cancelled
preliminary version of the present sentence where Newton needlessly first went on to state
'et erit $1.\sqrt{1-aaz^{2v}}::\dot{z}.va\dot{z}z^{v-1}::1.[vaz^{v-1}]$'.

(76) Initially just 'infimis' (lowest).

(77) In his draft (see note (73)) Newton here interposed 'et hujus seriei $z+az^v$ momentum
$[\dot{x}\text{vel}]\dot{z}+va\dot{z}z^{v-1}$ erit ad arcus z momentum $\dot{z}[=1]$ ut $\sqrt{1-z^2-2az^{v+1}-aaz^{2v}}$ ad 1, seu
$1+vaz^{v-1}=\sqrt{[1-]z^2-2az^{v+1}-aaz^{2v}}$ (and of this series $z+az^v$ the moment $(\dot{x}\text{ for})\dot{z}+vaz^{v-1}\dot{z}$
will be to the moment $\dot{z}(=1)$ of the arc z as $\sqrt{(1-z^2-2az^{v+1}-a^2z^{2v})}$ to 1, or $1+vaz^{v-1}$, will
be equal to $\sqrt{(1-z^2-2az^{v+1}-a^2z^{2v})}$).

(78) 'inferiorum dignitatum' (of lower powers), that is, as Newton properly specified in his
draft; whence the term $-2az^{v+1}$ is neglected because of its dimension $(v+1=)$ 4. In his
preceding equation he has already similarly omitted the terms $+vvaaz^{2v-2}$ and $-aaz^{2v}$ 'ob
parvitatem' (because of their smallness) as he initially phrased it in his draft.

(79) So that $\dot{x}=1-\frac{1}{2}z^2+vaz^{v-1}$ '$+\&\mathrm{c}$'. In his draft Newton again (compare note (77)
above) first wrote more elaborately 'et seriei momentum $[\dot{x}=]1-\frac{1}{2}z^2+vaz^{v-1}$ in $\dot{z}$ erit ad
arcus momentum $\dot{z}$ ut $\sqrt{1-z^2+\frac{1}{3}z^4-2az^{v+1}[\&\mathrm{c}]}$ ad 1, seu' (and the moment
$$\dot{x}=(1-\tfrac{1}{2}z^2+vaz^{v-1})\,\dot{z}$$
of the series will be to the moment $\dot{z}$ of the arc as $\sqrt{(1-z^2+\frac{1}{3}z^4-2az^{v+1}...)}$ to 1, that is). In

$1 : \sqrt{(1-x^2)} = (\dot{z}$ or$)$ $1 : \dot{x}$, that is, $\sqrt{(1-x^2)} = \dot{x}$, and so $1-x^2 = \dot{x}^2$. Assume[75] $az^v + \ldots = x$ and there will be $1-a^2z^{2v} - \ldots = v^2a^2z^{2v-2} + \ldots$. And when the terms 1 and $v^2a^2z^{2v-2}$ of lowest power[76] are collated there comes $v = 1 = a$, and so $az^v = z$. The first term of the series is therefore z.

Now put az^v for the second term and there will be $x = z + az^v + \ldots$[77] and so $1 - z^2 - 2az^{v+1} + \ldots = \dot{x}^2 = 1 + 2vaz^{v-1} + \ldots$. And when terms[78] are collated there comes $v = 3$ and $a = -\frac{1}{6}$, and so the series $= z - \frac{1}{6}z^3 + \ldots$.

Now put az^v for the third term and there will be $x = z - \frac{1}{6}z^3 + az^v + \ldots$[79] and so $1 - z^2 + \frac{1}{3}z^4 + \ldots = \dot{x}^2 = 1 - z^2 + \frac{1}{4}z^4 + 2vaz^{v-1} + \ldots$. And when terms are collated there comes $v = 5$ and $a = \frac{1}{120}$, and so the series $= z - \frac{1}{6}z^3 + \frac{1}{120}z^5 - \ldots$. And in this way you may proceed indefinitely.

Example 2.[80] *There needs to be found a curved line*[81] *whose subtangent is given in length.* Let z be the equably flowing abscissa, 1 its fluxion, and s the subtangent or portion of the abscissa between the ordinate and tangent. Then, since the subtangent (from the nature of curves) is to the ordinate as the moment of the abscissa to the moment of the ordinate, there will be $s\dot{y} = y\dot{z} = y$ and so $\dot{y} = y/s$. Once this is recognized, the problem will be solved[82] by assuming a series[83] for y and determining its terms step by step as follows. Let there be

$$az^n + bz^{n+1} + cz^{n+2} + dz^{n+3} + ez^{n+4} + \ldots = y^{[84]}$$

and then this same series will be equal to $s\dot{y} =$

$$nsaz^{n-1} + (n+1)\,sbz^n + (n+2)\,scz^{n+1} + (n+3)\,sdz^{n+2} + (n+4)\,sez^{n+3} + \ldots.$$

And when terms are collated there comes $n = 0$, $a = sb$, $b = 2sc$, $c = 3sd$, $d = 4se$, ..., so that $y = a(1 + z/s + \frac{1}{2}z^2/s^2 + \frac{1}{6}z^3/s^3 + \frac{1}{24}z^4/s^4 + \ldots)$, where the quantity a can be assumed at will.[85]

---

sequel the term $-2az^{v+1}$ of dimension $(v+1 =)$ 6 is, like others of yet higher order in (equivalent) power of z, discarded 'ob parvitatem'.

(80) Newton's rough prior casting of this second illustrative example survives on a separate worksheet (now in private possession) where he has also set down preliminary calculations for the scholium which follows, and a first version of its enunciation of the series expansion for the chord of a multiple angle. We cite their significant points in ensuing footnotes.

(81) The (inverse of the) *curva logarithmica* (logarithmic curve), namely, as Newton himself afterwards specifies; see note (87) below.

(82) Newton here initially began to copy out 'inve[nietur $\dot{y}$]' ($\dot{y}$ will be found) from his draft of this *exemplum* (see note (80)), but broke off to amend this unsatisfactory phrase.

(83) 'pro lubitu' (at pleasure) is added in the draft.

(84) This replaces a cancelled 'Sit igitur $y = az^v + \&c$' (Let, therefore, $y = az^v + \ldots$) immediately preceding in the manuscript (Add. 3960.7: 115). As Newton rightly here saw, there is in this case no need laboriously to go separately through the single steps of determining the individual terms in the series *gradatim*, one by one.

(85) The now familiar series expansion of the exponential function $y = ae^{z/s}$ which Newton had derived in his *epistola prior* to Leibniz in 1676 by reversing that of the inverse function $z = ab \log(1 + y'/a)$, $ab = s$, $y' = y - a$ (with the implicit restriction $-a < y \leqslant a$); (see *Correspondence*, 2: 27–8, and also *ibid.*: 60–1, where Leibniz on 17/27 August returned his own claim to independent discovery of the expansion 'ex seriebus regressuum pro Hyperbola'.

Potuit autem Problema solvi sine inventione seriei, per ea quæ dicta sunt in Scholio Propositionis XI.[86] Nam cum sit $s\dot{y} = y\dot{z}$ fluat y uniformiter & sit ejus fluxio $\dot{y} = 1$ et erit $\dfrac{s}{y} = \dot{z}$. Sit y abscissa & $\dot{z}$ ordinata figuræ & erit z area ejus, id est area hyperbolæ cujus ordinata est $\dfrac{s}{y}$.

Quinetiam absqȝ his methodis Problema solvitur. Nam si augmenta momentanea abscissæ sint æqualia & subtangens detur longitudine, augmenta vel decrementa momentanea Ordinatæ erunt in proportione geometrica, hoc est, si abscissa augeatur in ratione logarithmi, Ordinata augebitur vel diminuetur in ratione numeri.[87]

Scholium.[88]

Postquam termini aliquot seriei[89] per hujusmodi operationes inventi sunt, regula progressionis terminorum investigari debet ut series pro lubitu continuari possit, ut fit in exemplis sequentibus. Pro terminis seriei ponantur A, B, C, D, E &c, nempe A pro termino primo, B pro secundo, C pro tertio & sic deinceps. Et $\sqrt{1-xx}$ erit

$$1 - \tfrac{1}{2}Ax^2 - \tfrac{1}{4}Bx^2 - \tfrac{3}{6}Cx^2 - \tfrac{5}{8}Dx^2 - \tfrac{7}{10}Ex^2 - \&\mathrm{c}.$$

Et in circulo radio 1 descripta ex sinu recto x, arcus[90] erit

$$x + \frac{1 \times 1}{2 \times 3}Axx + \frac{3 \times 3}{4 \times 5}Bxx + \frac{5 \times 5}{6 \times 7}Cxx + \frac{7 \times 7}{8 \times 9}Dxx + \&\mathrm{c}$$

& [ex cosinu x][91] area arcu & sinibus recto et verso comprehensa erit

$$x + \frac{-1 \times 1}{2 \times 3}Ax^2 + \frac{1 \times 3}{4 \times 5}Bx^2 + \frac{3 \times 5}{6 \times 7}Cx^2 + \frac{5 \times 7}{8 \times 9}Dx^2 + \&\mathrm{c}.^{[92]}$$

---

(86) Now understand, of course, the amended form of the corresponding terminal scholium to the published tract *De Quadratura Curvarum* as Newton has specified in [1] (on which compare note (19) above).

(87) Originally, in his draft casting (see note (80) preceding), Newton concluded to the same effect by writing 'adeoqȝ si abscissa sit ut Logarithmus, ordinata erit ut ejus numerus' (and hence if the abscissa be as the logarithm, the ordinate will be as its number). In this use of the auxiliary hyperbola $y\dot{z} = s$ to construct the logarithmic curve

$$z = \int_a^y s/y \,.\, dy = s \log (y/a)$$

which exactly models the previous series expansion of $y = ae^{z/s}$ there is nothing which Newton had not already grasped nearly fifty years before when, in the autumn of 1664 (see 1: 458), he first constructed the *curva logarithmica* geometrically to depict the exponential growth of capital when its compounded interest accrues 'instantaneously' from moment to moment. On the earlier, independent inventors of this curve in the seventeenth century—notably Descartes, Torricelli and James Gregory—see 1: 376–7, note (47).

(88) In the style of Chapter 2 of his 1684 'Matheseos Universalis Specimina' (see IV: 544; compare also the drafts of this on *ibid.*: 634–41) Newton proceeds to illustrate how the 'progression' of a given series may often more easily be determined by examining that of the ratio of its successive terms, though in his examples he here contents himself with setting down the pattern of the coefficients merely from inspection.

The problem could have been solved without the finding of a series through what was stated in the scholium of Proposition XI.[86] For since there is $s\dot{y}=y\dot{z}$, let y flow uniformly and its fluxion $\dot{y}$ be equal to 1, and there will be $s/y=\dot{z}$. Let y be the abscissa and $\dot{z}$ the ordinate of a figure, and z will be its area—the area, that is, of the hyperbola whose ordinate is s/y.

The problem may, furthermore, be solved independently of these methods. For if the momentary augments of the abscissa be equal and the subtangent given in length, the momentary augments or decrements of the ordinate will be in geometrical proportion; in other words, if the abscissa increase in the ratio of the logarithm, the ordinate will increase or diminish in the ratio of the number.[87]

Scholium.[88]

After a number of terms of a series[89] have been found through operations of the present sort, the rule of the progression of the series ought to be investigated in order that the series can be continued at will, as happens in the following examples.

For the terms of the series let there be put $A, B, C, D, E, \ldots$; namely, A for the first term, B for the second one, C for the third, and so on. Then $\sqrt{(1-x^2)}$ will be $1-\frac{1}{2}Ax^2-\frac{1}{4}Bx^2-\frac{3}{6}Cx^2-\frac{5}{8}Dx^2-\frac{7}{10}Ex^2-\ldots$. And in a circle described with unit radius the arc[90] will, in terms of right sine x, be

$$x+\frac{1\times 1}{2\times 3}\,Ax^2+\frac{3\times 3}{4\times 5}\,Bx^2+\frac{5\times 5}{6\times 7}\,Cx^2+\frac{7\times 7}{8\times 9}\,Dx^2+\ldots,$$

while (in terms of cosine x)[91] the area comprised by the arc and the right and versed sines will be

$$x+\frac{-1\times 1}{2\times 3}\,Ax^2+\frac{1\times 3}{4\times 5}\,Bx^2+\frac{3\times 5}{6\times 7}\,Cx^2+\frac{5\times 7}{8\times 9}\,Dx^2+\ldots.\text{[92]}$$

---

(89) Initially just 'series aliqua' (some series).

(90) That is, $\sin^{-1}x$.

(91) We insert a necessary proviso in Newton's text; even so the 'area' $\int_0^x \sqrt{(1-x^2)}\,.dx$ whose series expansion he goes on to state is more properly that 'arcu, sinu & cosinu comprehensa'. It is readily determined that 'ex sinu recto x' the 'area arcu & sinibus recto et verso comprehensa' is $\frac{1}{2}(\sin^{-1}x-x\sqrt{[1-x^2]}) = \frac{1}{3}x^3+\frac{1}{10}x^5+\frac{3}{56}x^7+\frac{5}{144}x^9+\ldots$, that is,

$$'\frac{1}{3}x^3+\frac{1\times 3}{2\times 5}\,Ax^2+\frac{3\times 5}{4\times 7}\,Bx^2+\frac{5\times 7}{6\times 9}\,Cx^2+\&c'.$$

(92) Newton's prior reductions of the ratios '$[-]\frac{1}{6}.\frac{3}{20}.\frac{5}{14}$ [&c]' of the successive terms in this series to be '$\frac{[-1,]1}{2,3}.\frac{1,3}{4,5}.\frac{3,5}{6,7}$ [&c]', as here stated with their pattern of formation made clear, still survive on the worksheet where he also roughed out his preceding 'Exempl. 2' of the main proposition (compare note (80) above).

Et in circulo cujus diameter est 1 ex sinu verso x arcus[93] erit

$$x^{\frac{1}{2}} + \frac{1 \times 1}{2 \times 3} Ax + \frac{3 \times 3}{4 \times 5} Bx + \frac{5 \times 5}{6 \times 7} Cx + \frac{7 \times 7}{8 \times 9} Dx + \&\text{c.}$$

et area arcu et sinibus recto et verso comprehensa[94]

$$\frac{2}{3} x^{\frac{3}{2}} + \frac{-1 \times 3}{2 \times 5} Ax + \frac{1,5}{4,7} Bx + \frac{3 \times 7}{6 \times 9} Cx + \frac{5 \times 9}{8 \times 11} Dx + \&\text{c.}$$

Et vicissim ex arcu z sinus rectus est

$$z \;[-]\frac{1}{2 \times 3} Az^2 \;[-]\frac{1}{4 \times 5} Bz^2 \;[-]\frac{1}{6 \times 7} Cz^2 \;[-]\frac{1}{8 \times 9} Dz^2 [-] \&\text{c.}^{[95]}$$

et sinus complementi $1 - \dfrac{1}{1 \times 2} Az^2 - \dfrac{1}{3 \times 4} Bz^2 - \dfrac{1}{5 \times 6} Cz^2 - \dfrac{1}{7 \times 8} Dz^2 - \&\text{c.}^{[96]}$

Et si arcus capiendus sit in ratione data ad datum arcum, & ratio illa sit n ad 1, sitꝗ d diameter circuli & x chorda arcus dati, chorda arcus quæsiti erit

$$nx + \frac{1-nn}{2 \times 3dd} xxA + \frac{9-nn}{4 \times 5dd} xxB + \frac{25-nn}{6 \times 7dd} xxC + \frac{36-nn}{8 \times 9dd} xxD$$

$$+ \frac{49-nn}{10 \times 11dd} xxE + \&\text{c.}^{[97]}$$

Et si logarithmus dicatur l, numerus ejus erit ut[98]

---

(93) Namely $\frac{1}{2} \int_0^x 1/\sqrt{(x-x^2)} \, . \, dx = \sin^{-1}\sqrt{x}$. Since Newton wrote the following expansion straight out on his draft sheet without further ado, it will be clear that he here merely transposed $x \to x^{\frac{1}{2}}$ in his preceding series for the inverse sine.

(94) That is, $\int_0^x \sqrt{(x-x^2)} \, . \, dx$. On his draft sheet (see note (80) above) Newton took some trouble in establishing that the successive ratios '$[-]\frac{3}{10} . \frac{5}{28} . \frac{7}{18} . \frac{360}{704} = \frac{9,5}{88}$' are indeed

'$\dfrac{-1,3}{2,5} . \dfrac{1,5}{4,7} . \dfrac{3,7}{6,9} . \dfrac{5,9}{8,11}$' respectively.

(95) In the manuscript Newton has carelessly set all the signs to be '+'. The slip has its origin in his draft sheet, where the successive coefficients in the series expansion of $\sin z$ (after the first) are put to be '$+\dfrac{1}{2,3}$', '$+\dfrac{1}{2,3,4,5}$', '$+\dfrac{1}{2,3,4,5,6,7}$' and so on.

(96) Here, too, Newton set the successive coefficients in the explicit series expansion of $\cos z$ to be '$+\dfrac{1}{1,2}$', '$+\dfrac{1}{2,3,4}$', '$+\dfrac{1}{2,3,4,5,6}$' and so on in his worksheet, but has here made correction for the wrong signs in alternate terms. In sequel he began to adjoin 'Et tangens z [est $z+\frac{1}{3}Az^3+...$]' (And tangent z [is $z+\frac{1}{3}Az^3+...$]), but broke off—wisely so, since (unless

And in a circle whose diameter is unity the arc[93] will, in terms of versine x, be

$$x^{\frac{1}{2}}+\frac{1\times1}{2\times3}\,Ax+\frac{3\times3}{4\times5}\,Bx+\frac{5\times5}{6\times7}\,Cx+\frac{7\times7}{8\times9}\,Dx+\ldots,$$

while the area comprised by the arc and the right and versed sines[94] will be

$$\frac{2}{3}\,x^{\frac{3}{2}}+\frac{-1\times3}{2\times5}\,Ax+\frac{1\times5}{4\times7}\,Bx+\frac{3\times7}{6\times9}\,Cx+\frac{5\times9}{8\times11}\,Dx+\ldots.$$

Conversely, in terms of arc z the right sine is

$$z-\frac{1}{2\times3}\,Az^2-\frac{1}{4\times5}\,Bz^2-\frac{1}{6\times7}\,Cz^2-\frac{1}{8\times9}\,Dz^2-\ldots,^{(95)}$$

and the 'sine of the complement' (cosine)

$$1-\frac{1}{1\times2}\,Az^2-\frac{1}{3\times4}\,Bz^2-\frac{1}{5\times6}\,Cz^2-\frac{1}{7\times8}\,Dz^2-\ldots.^{(96)}$$

And if an arc needs to be taken in given ratio to a given arc, then let the ratio be as n to 1, and d the diameter of the circle and x the chord of the given arc, and the chord of the required arc will be

$$nx+\frac{1-n^2}{2\times3d^2}\,x^2A+\frac{9-n^2}{4\times5d^2}\,x^2B+\frac{25-n^2}{6\times7d^2}\,x^2C+\frac{36-n^2}{8\times9d^2}\,x^2D+\frac{49-n^2}{10\times11d^2}\,x^2E+\ldots.^{(97)}$$

And if a logarithm be called l, its number will be as[98]

---

and until one introduces the notion of a 'Bernoulli' number) the successive coefficients (and *a fortiori* their ratios) in the series $\tan z = z+\frac{1}{3}z^3+\frac{2}{15}z^5+\frac{17}{315}z^7+\ldots$ obey no obvious algebraic rule of formation.

(97) Newton has thoughtlessly copied from his *epistola prior* of 1676 to Leibniz (see *Correspondence*, **2**: 25), where he first publicly communicated this series expansion of $\text{chd}\,n\theta = d\sin\frac{1}{2}n\theta$ in powers of $x = \text{chd}\,\theta = d\sin\frac{1}{2}\theta$, the erroneous numerators '$36-nn$' and '$49-nn$' of the coefficients of the two last terms—which should of course (compare IV: 668, note (37)) be '$49-nn$' and '$81-nn$' respectively, in accord with the general 'progression'

$$((2i-1)^2-n^2)/2i(2i+1)d^2, \quad i = 1, 2, 3, 4, 5, \ldots.$$

(98) The manuscript terminates abruptly at this point (not quite at the foot of Add. 3960.7: 115). In immediate sequel it is clear that we should understand the series expansion of the exponential function e^l cast into the recursive form '$1+lA+\frac{1}{2}lB+\frac{1}{3}lC+\frac{1}{4}lD+\&c$'. We have seen that in his preamble to the 'Analysis per quantitates fluentes et earum momenta' Newton planned to devote an 'ultima pars'—one or more separate propositions, we assume—to expounding his method of finite differences, particularly its application approximately to square the area under a given curve in terms of a given number of its equidistant ordinates. (See our detailed comment theoreon in note (5) preceding.) There is, we may again remark, nothing to show that he ever set a word of this intended *addendum* down on paper.

APPENDIX. PRELIMINARY COMPUTATIONS AND DRAFTS FOR THE 'ANALYSIS'.[1]

From the originals in private possession and the University Library, Cambridge

[1][2]

[Est] $ST.TC::Pp.PO.$[3]

[adeoꝗ]

$\qquad ST.TB::Pp.VO.$[4]

[Unde] $ST.Pp::TC.PO$
$\qquad\qquad ::TB.VO::BC.PV.$[4]

[Hoc est]

$\qquad [T]t.Pp::TB.VP.$[5]

[quia][6] $Tt.TS::TB.BC$
$\qquad\qquad ::BC.BP::OV.VP.$

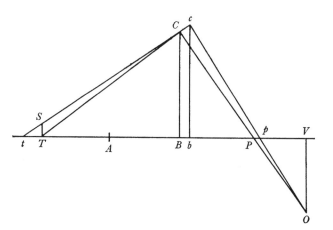

[2][7]   Sit VC linea quævis curva, AB abscissa, BC Ordinata,[8] CT tangens ejus & CP perpendiculum: et si relatio duarum quarumvis ex lineis quinꝗ $BC\ BT$

---

(1) These jotted prior calculations and superseded preparatory drafts are once more here adjoined to the preceding main text because of the several minor illuminations which they individually cast upon its composition. As ever, we do not insist upon the many points where they duplicate its content and verbal phrasing.

(2) Newton's computation in preliminary to the new Corollary 3 to Proposition I, where (see note (15) of 5 preceding) he states without proof a proportion defining the location in the base AB of the perpendicular projection V onto it of the centre O of curvature at the arbitrary point C of a given curve (not here drawn). In prior construction he determines the centre O as the limit-meet of the normals to the curve at C and at the indefinitely close point c, which meet the base in P and p respectively, and also draws the corresponding (perpendicular) ordinates BC, bc and tangents TC, tc at the points C and c, raising the perpendicular TS at T till its occurrence with tc.

(3) Because the infinitesimal triangles TSC and PpO are equiangular and therefore similarly proportioned.

(4) Since the triangles CTB and POV are similar, so that $TC:TB:CB = OP:OV:PV$.

(5) Where, that is, $Tt = d(AT)$ and $Pp = d(AP)$, with (in the limit as c comes to coincide with C, and so t, p with T and P respectively) $d(AT):d(AP) = \mathrm{fl}(AT):\mathrm{fl}(AP)$. We have set out in note (15) of the preceding main text the analogous Leibnizian derivation of this defining proportion.

(6) Since, in the limit as tc comes to coalesce with TC, the angle $\widehat{btc}$ is indistinguishably different from $B\widehat{T}C$, and hence the infinitesimal triangle TtS is, in the 'last' moment before it vanishes, similar to the triangle BTC.

(7) A superseded first version of Newton's addition *ad pag. 66, lin. 2* of the 1711 'De Quadratura' (see page 264 above) in which the ordinate BC is restricted—as in the sentence of his 1676 *epistola posterior* to Leibniz where (see our note (22) to the preceding main text of the 'Analysis') he originally broached this special case—to stand at right angles to the abscissa AB.

(8) Understand 'ad abscissam perpendicularis' since Newton's following proportion $BP:BC = BC:TB = CP:TC$ manifestly holds true only when $\widehat{ABC}$ is a right angle.

BP CT CP detur per æquationem quamlibet, dabitur relatio inter omnes. Et cum
sit *BP.CB*:: *CB.BT*:: *CP.CT*:: $\overline{\dot{CB}}.\overline{\dot{AB}}$
fluat *CB* uniformiter & sit ejus fluxio 1
& habebitur $\overline{\dot{AB}}$. Sit jam curvæ alicujus
abscissa *AD=CB* & Ordinata $DE=\overline{\dot{AB}}$,
et fluens *AB* erit ut area hujus Curvæ,[9]
adeoœ per quadraturam hujus Curvæ dabitur.

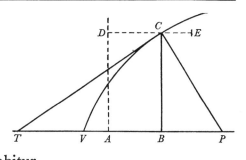

[3][10]   *Prop. 12.*

Quantitates fractas & radicales in series approximantes convertere &
inde Figurarum areas elicere.

Fractiones in series approximantes reducuntur per Divisionem & quantitates
radicales per extractionem Radicum; perinde instituendo has operationes in
speciebus ac institui solent in decimalibus numeris.[11] Ut si $\dfrac{aa}{b+x}$ in seriem
reducenda sit: forma operationis erit ut sequitur.[12]

$$b+x \Big)\ aa+0\ \Big(\frac{aa}{b}-\frac{aax}{bb}+\frac{aaxx}{b^3}-\frac{aax^3}{b^4}+\&c$$

$$\underline{aa+\frac{aax}{b}}$$

$$0-\frac{aax}{b}+\quad 0$$

$$\underline{-\frac{aax}{b}-\frac{aaxx}{bb}}$$

$$0\ +\frac{aaxx}{bb}\ \dots\ \&c.^{(13)}$$

---

(9) The locus of the end-point *E*, that is. In modern Leibnizian equivalent, since
$$DE = d(AB)/d(BC)$$
by construction, on setting *BC* = *AD* there results *AB* = ∫*DE*.*d(AD)*.

(10) ULC. Add. 3960.6:83/85, a first scheme for enlarging the eleven propositions of the
published 'De Quadratura' by a single 'Prop. 12' which embraces the content of Propositions
XII–XIV in the augmented 'Analysis' (compare pages 266–84 above).

(11) Save for the trivial replacement of an 'istas' by 'has', this opening sentence is taken
over word for word by Newton from the second paragraph of his 1676 *epistola prior* to Leibniz (see
Correspondence, **2**:21, ll. 4–6), whence also (compare note (24) to the preceding main text of the
'Analysis') he borrows his following statement of the general binomial expansion. In his *epistola*
he gave no instances of these reductions to 'converging' series *per divisionem* and *per extractionem
radicum*; in sequel he goes to his parent 1671 tract on series and fluxions to fill in the gap.

(12) Newton's prime example of reduction *per divisionem* in his 1671 tract (see III: 36). Seeing
no need here to repeat its full display, we have slightly contracted the manuscript text.

(13) Newton has cancelled an unfinished following phrase in which he began to assert that

Si $\sqrt{1-xx}$ in seriem vertenda est, hæc erit operationis forma.[14]

$$1-xx)\ 1-\tfrac{1}{2}xx-\tfrac{1}{8}x^4-\tfrac{1}{16}x^6-\&c$$

$$\underline{1}$$

$$0-xx$$

$$\underline{-xx+\tfrac{1}{4}x^4}$$

$$0-\tfrac{1}{4}x^4$$

$$\underline{-\tfrac{1}{4}x^4+\tfrac{1}{8}x^6+\tfrac{1}{64}x^8}$$

$$0-\tfrac{1}{8}x^6-\tfrac{1}{64}x^8 \quad [\&c].$$

[15]Sed hæ reductiones[16] multum abbreviantur per Theorema sequens.

$$\overline{P+PQ}|^{\frac{m}{n}}=P^{\frac{m}{n}}+\frac{m}{n}\,AQ+\frac{m-n}{2n}\,BQ+\frac{m-2n}{3n}\,CQ+\frac{m-3n}{4n}\,DQ+\&c.$$

...   .   et mutando signum Quoti.[17]

Et idem terminus eodem fere modo invenitur in secundo Diagrammate. Sed hic præcipua difficultas est in inventione primi termini Radicis. Id quod

---

'Hujusmodi divisiones Wallisius noster exercere cœpit anno [?]'. Not quite three years later, in his anonymous 'Account of the Book entituled *Commercium Epistolicum Collinii & aliorum...*' (*Philosophical Transactions*, **29**, No. 342 [for January/February 1714/15]: 173–224; the text is all but fully reproduced from Newton's surviving autograph drafts in **2**, Appendix 2 below), he was to specify that '*Dr. Wallis* in his [*Mathesis Universalis: sive*] *Opus Arithmeticum* published A.C. 1657 [as the first treatise of his *Operum Mathematicorum Pars Prima*]: Cap. 33...reduced the Fraction $\dfrac{A}{1-R}$ by perpetual Division into the Series $A+AR+AR^2+AR^3+AR^4+\&c$'. We need here but lightly insist that this reduction *per divisionem* to an infinite series was common knowledge to the *calculatores* of fourteenth-century Oxford and Paris, and also to the Keralese mathematicians who just a little later anticipated Mercator, James Gregory and Leibniz in applying such a reduction to an integrand before squaring the ensuing series term by term to derive the expansions of the elementary real functions. In Wallis' own century Gregory St Vincent had given an elaborate discussion of the infinite geometrical series in Book 2 of his *Opus Geometricum* (Antwerp, 1647). Further to qualify the justice of Newton's remark would take a large monograph; we may be grateful that he here chose to delete it.

(14) The following is styled on the instance $\sqrt{(a^2+x^2)}$ of such reduction *per extractionem radicis (quadraticæ)* in the parent 1671 tract (see III: 40). Newton here seemingly takes it as evident that x^2 must be less than unity for the resulting series to converge.

(15) Newton again takes up the text of his *epistola prior* (at *Correspondence*, **2**: 21, 1. 7).

(16) 'extractiones radicum' are specified in the 1676 *epistola*: in truth, while it embraces the preceding reduction *per divisionem* within its generalized grasp, Newton's following binomial *Theorema* can scarcely be said 'much ' to 'shorten' its application.

(17) Here understand the several opening pages of the 1676 *epistola prior* where (*Correspondence*, **2**: 21,1.10–23,1.6) Newton defines the quantities P, Q, A, B, C, ... in his general binomial expansion, and applies it to nine worked examples; and then passes (*ibid.*: 23, 1. 7–24, 1. 22, *plus* the late-inserted phrase 'ut -1 per 10 aut 0,061 per 11,23 et mutando signum quoti' cited (p. 42) in note (10) thereto) to set out his technique for the term-by-term extraction of the roots of 'affected' equations, much as in his 1671 tract five years before (see III: 42–4, 54–6).

methodo generali sic perficitur.[18] Descripto angulo recto BAC, latera ejus BA, CA divido in partes æquales; et inde normales erigo ... ... *vide Epist. 24 Oct. 1676*[19] ...... per coefficientem radicis p, q, r aut s.

[20]Postquam Ordinata figuræ cujuscunqʒ Curvilineæ in serie approximante habetur, si area figuræ desideratur, inveniendæ sunt areæ quæ a singulis Ordinatæ partibus describuntur; et harum omnium aggregatum sub signis suis $+$ et $-$ erit area tota. Inveniuntur autem areæ singulæ per Corol. Prop. V.[21]

Ut si Abscissa sit x et Ordinata sit $\dfrac{1}{1+xx}=y$, dividendo prodibit

$$y = 1 - x^2 + x^4 - x^6 + x^8 - \&c$$

et area $= x - \frac{1}{3}x^3 + \frac{1}{5}x^5 - \frac{1}{7}x^7 + \frac{1}{9}x^9 - \&c$. Vel si terminus xx ponatur in divisore primus, prodibit $y = x^{-2} - x^{-4} + x^{-6} - x^{-8} + \&c$ et area

$$= -x^{-1} + \tfrac{1}{3}x^{-3} - \tfrac{1}{5}x^{-5} + \tfrac{1}{7}x^{-7} - \&c.$$

Priori modo computes cum x sit [22]minor quam 1, posteriore cum x sit [22]major.

Si Ordinata sit $\sqrt{aa-xx}=y$, extrahendo radicem prodibit

$$y = a - \frac{x^2}{2a} - \frac{x^4}{8a^3} - \frac{x^6}{16a^5} - \frac{5x^8}{128a^7} - \&c$$

et inde (per Corol. Prop. V) area erit

$$ax - \frac{x^3}{6a} - \frac{x^5}{40a^3} - \frac{x^7}{112a^5} - \frac{5x^9}{1152a^7} - \&c.$$

---

(18) In his 1676 *epistola* at this point (*Correspondence*, **2**: 24, l. 24) Newton had gone on to write 'sed hoc brevitatis gratia jam prætereo'. Here he jumps straight to the passage in his ensuing *epistola posterior* to Leibniz in October 1676 (see *Correspondence*, **2**: 126, ll. 22 ff.) where he subsequently made good this omission in his earlier account.

(19) In this excerpt from his *epistola posterior* to Leibniz (*Correspondence*, **2**: 126, l. 22–127, l. 22) Newton once more expounds his parallelogram rule for isolating the 'fictitious' equations whose solution yields, at each separate stage, the pertinent term in the step-by-step extraction of the root(s) of the fuller given equation as an infinite series. We may guess that he would here also have adduced, from the more detailed account of the method set out by him in his 1671 treatise (see III: 51–4), the worked example of the cubic $y^3 + a^2y + axy - 2a^3 - x^3 = 0$ (*ibid.*: 54–6) which he fought shy five years afterwards of elaborating for his German correspondent.

(20) What follows is (as the several cancellations and interlineations in the manuscript underscore) here freshly contrived. In his revised 'Analysis' this application of the binomial theorem to expand an integrand into an infinite series, subsequently to be 'squared' term by term, was afterwards to be set by Newton (see pages 270–4 above) as a separate following 'Prop. 13'.

(21) Newton has deleted in sequel the unnecessary specification 'ponendo sc. $e = 1 = R$ et $f = 0 = g = h = b = c = d$'.

(22) The exaggeration 'longe' is rightly here suppressed.

Sit area illa $=z$ et extrahendo radicem affectam z, prodibit abscissa

$$x = \frac{z}{a} + \frac{z^3}{6a^5} + \frac{13z^5}{120a^9} + \frac{493z^7}{5040a^{13}} + \&c.$$

Scholium

Eadem ratione qua Ordinatæ deducuntur ex areis per Prop. II, III, et IV, et areæ ex ordinatis per Prop. V, VI, et XII possunt[23]

---

[4][24]   *Prop. XI.*[25] *Prob.*

Dignitatem Binomij[26] *in seriem numero terminorum infinitam resolvere.*

Sit P nomen prius[,] Q nomen posterius divisum per P, $\frac{m}{n}$ index dignitatis &

$\overline{P+PQ}|^{\frac{m}{n}}$ dignitas et erit

$$\overline{P+PQ}|^{\frac{m}{n}} = P^{\frac{m}{n}} + \frac{m}{n}AQ + \frac{m-n}{2n}BQ + \frac{m-2n}{3n}CQ + \frac{m-3n}{4n}DQ + \&c.$$

Ubi P et Q possunt esse quantitates vel simplices vel compositæ, et A, B, C, D &c designant terminos seriei proxime præcedentes, A terminum primū $P^{\frac{m}{n}}$, B terminum secundum $\frac{m}{n}AQ$, C terminum tertium $\frac{m-n}{2n}BQ$ & sic deinceps in infinitum.

Exempl. 1. Est $\sqrt{cc+xx}$ .... commode instituuntur.[27]

---

(23) Newton here breaks off, but we may evidently complete his sense by adding 'etiam ordinatæ ex areis ut et areæ ex ordinatis deduci'.

(24) ULC. Add. 3960.6: 94/92, a revised statement of his 1676 binomial *Theorema*—once more illustrated (see note (27)) by the nine worked examples of his *epistola prior*, and with its application to 'areas and fluents' and to 'universal analysis' outlined (see notes (28) and (29)) —to which Newton now appends, in Propositions XII*, XIII and XIV, an amended version of the generalized technique for extracting the root of a given equation in two 'fluents' whose first-order fluxional instance he had transmitted to John Wallis in the later summer of 1692 (see note (33) below).

(25) A careless slip for 'XII', we suppose. (Newton gives no other indication of intending to suppress the corresponding Proposition XI of his published 'De Quadratura', at least.) He initially qualified his following statement of the general binomial expansion as a 'Lemma' before changing his mind to make it a 'Prob[lema]'. The subtlety of Newton's distinction evades us, especially since in the parent *epistola prior* he had (see *Correspondence*, **2**: 21) dubbed it a 'Theorema'—and to those with a more exact classical grounding it would almost certainly have been a 'Porisma'!

(26) A superfluous 'cujuscunɋ' is deleted.

(27) Understand again the nine worked illustrative examples of the 1676 *epistola* (*Correspondence*, **2**: 21, l. 6–23, l. 3) together with Newton's ensuing affirmation to Leibniz there (*ibid.*: 23, ll. 4–6) that 'Per eandem Regulam Geneses Potestatum, Divisiones per Potestates aut per quantitates radicales, et extractiones radicum altiorum in numeris etiam commodè instituuntur'.

Corol. 1. De areis & fluentibus.[28]

Corol. 2. De Analysi universali.[29]

Schol. De momentis primis secundis tertijs &c. Si $1 \times B$, $1 \times 2C$, $1 \times 2 \times 3D$, $1 \times 2 \times 3 \times 4E$ [&c] $= \dot{y}, \ddot{y}, \dddot{y}, \ddddot{y}$ [&c]. Si $PQ = \dot{P}$.[30]

Prop. XII. Prob.[31]

Ex æquatione duas fluentes involvente quarum alterutra æqualiter fluit, extrahere alteram in serie numero terminorum infinita.[32]

---

(28) This corollary (like that which follows) is a late interlineation in the manuscript. We assume that it was here destined to embrace the suppressed 'Prop. XII. Prob.' which Newton went on initially to pen (see note (31)).

(29) This vague subhead could mean everything or almost nothing. It is a not unreasonable guess that Newton here planned to give some brief illustration of the 'universal analysis' by series which his binomial expansion opened up, applying it (it might be) to resolve some geometrical problem intractable by the 'vulgar analysis' such as that which he had adduced as prime *exemplum* in Chapter 3 of his 'Matheseos Universalis Specimina' nearly thirty years before (see IV: 554–8).

(30) As in his parallel scholium to the reworked equivalent 'Prop. XII' in [5] following, Newton here intends of course the 'moment' $o\dot{P}$ of the *nomen primum P*; whence, on taking the base fluent to be x and setting $P^{m/n} = y \equiv y_x$, there is

$$y_{x+o} = (P + o\dot{P})^{m/n} = A + oB + o^2C + o^3D + o^4E + \ldots,$$

where

$$A = P^{m/n} = y, \quad B = (m/n)\,P^{m/n-1}\,\dot{P} = \dot{y}, \quad C = \binom{m/n}{2} P^{m/n-2}\,\dot{P}^2 = \tfrac{1}{2}\ddot{y},$$

$$D = \binom{m/n}{3} P^{m/n-3}\,\dot{P}^3 = \tfrac{1}{6}\dddot{y}, \quad E = \binom{m/n}{4} P^{m/n-4}\,\dot{P}^4 = \tfrac{1}{24}\ddddot{y}, \quad \ldots.$$

As we now know, but Johann Bernoulli and his Continental contemporaries did not (and refused—see 2, §3.2: note (46) preceding—to credit), Newton had stated the general 'Taylor' expansion of which this is but a particular instance twenty years earlier in Proposition XII of his 1691 'De quadratura Curvarum' (compare VII: 98, note (107)).

(31) In initial amplification of what he afterwards seemingly relegated to be but a 'Corol. 1' to his preceding Proposition XI (compare note (27) above) Newton here first required '*Aream Curvilineæ cujuscunꝗ cujus ordinata non est latus Æquationis affectæ in serie numero terminorū infinita exhibere*', thereafter adjoining as its resolution: '*Inveniatur Ordinata in serie numero terminorum infinita per Propositionem præcedentem. Et per Corollarium Prop III quære areas singulis seriei Terminis respondentes & aggregatum arearum sub signis suis + et − erit Area quæsita*'. By his restriction that the *ordinata*, y say, shall not be given—implicitly, that is —as the 'side' (root) of an 'affected' (general algebraic) equation he understands, of course, that it shall be (straightforwardly reducible to be) a simple binomial power of the *abscissa z* multiplied, it may be, by a simple power of the *nomen*; namely $y = az^k(z^l + b)^{m/n}$, where the *quantitates a, b* and the indices k, l are arbitrarily given, but the numerator m and denominator n of the binomial's index are, in generalization of the prime *Theorem* for series quadrature which he passed to Leibniz in his 1676 *epistola posterior* (see *Correspondence*, 2: 115; and compare VII: 26, 50) and afterwards subsumed into Proposition V of the published 'De Quadratura' (see VII: 520), co-prime integers.

(32) This superseded enunciation is not cancelled by Newton in the manuscript, but he indicates well enough that it is *de trop* by attaching a superscript asterisk (see the next note) to the first of the ensuing Propositions XII–XIV among which he hastens to divide up its content.

Prop. XII.[33]

Ex æquatione affecta radicem fluentem in serie numero terminorum simplicium infinita extrahere.

Resolutio.

In æquatione quacunqȝ affecta sit z quantitas æquabiliter fluens et y quantitas altera fluens quam extrahere oportet.[34] Termini omnes ex eodem æquationis latere consistentes æquentur nihilo & dignitas quantitatis y (si opus sit) exaltetur vel deprimatur sic ut ejus index nec alicubi negativus sit nec tamen altior quam ad hunc effectum requiritur[,] et sit kz^λ terminus infimæ dignitatis eorum qui neqȝ per y neqȝ per ejus dignitatem quamcunqȝ[35] multiplicantur. Sit $lz^\mu y^\alpha$ terminus alius quilibet, et omnes ordine terminos percurrendo collige ex singulis seorsim numerum $\dfrac{\lambda-\mu}{\alpha}$ sic, ut tot habeas ejusmodi numeros quot sunt termini. Horum numerorum maximus vocetur ν, et z^ν erit dignitas primi termini seriei. Pro ejus coefficiente ponatur a et in æquatione resolvenda scribe az^ν pro y[36] ac

---

Here, perhaps, he initially intended to cite only the first-order instance—that of Proposition XIII below—as he had passed it on to Wallis in the late summer of 1692 (see note (34) following) before suddenly noticing that it could be applied *mutatis mutandis* to derive the series expansion, as he proceeds to sketch in the *Resolutio* of Proposition XII, of an ordinary 'affected' equation, thus rendering unnecessary his previous special appeal here (see notes (16) and (18) above) to the equivalent *ad hoc* procedure for such series extraction of an equation's root which he had communicated to Leibniz in his 1676 *epistolæ prior/posterior*.

(33) The manuscript reads 'Prop. XII*'; on which see the previous note. Newton initially passed to enunciate the demand '*Latus Æquationis affectæ in serie numero terminorum infinita exhibere*' and then to specify:

$$\text{'Sit æquatio }\quad a+bx\quad +cxx\quad +dx^3+\&c\text{'}$$
$$+fyx\quad +gyxx\quad +hyx^3$$
$$+kyyx^2\quad +lyyx^3$$
$$+my^3x^3$$

before breaking off to write in their place what here follows.

(34) This replaces an initial concluding phrase 'et extrahenda sit altera fluens quantitas y'. In a cancelled first opening sentence Newton originally began: 'Sunto a, b, c, d &c quantitates datæ, sitqȝ z abscissa & y ordinata...' and then altered this to read '...et y et z quantitates indeterminatæ in æquatione[,] et exhibenda sit altera fluens quantitas y in serie infinita'. In sequel he proceeds to adapt to the present particular instance where the given equation involves no fluxions of its fluent quantities the mode of resolution of the 'Prob. Ex æquatione fluxionem radicis involvente radicem extrahere' as, in minimal refashioning of Proposition XIII of his 1691 treatise 'De quadratura Curvarum' (see VII: 92–6), he had communicated it to John Wallis in September 1692, to be published by the latter the next year in his *Opera Mathematica*, **2**: 394 [= VII: 177–8]. A revealing initial slip a few lines below (see note (36)) makes it clear that the transpositions from the 1693 printed text here involved were carried through by Newton purely mentally without intermediate assistance of any amended transcript of it.

(35) 'aliquam' is replaced; the 1693 parent text here has 'quamvis'.

(36) Newton here began unthinkingly to copy out also 'et $\nu az^{\nu-1}$ [pro $\dot y$]' from the 1693 text (see note (33)) before checking himself.

termini omnes resultantes in quibus *z* ejusdem est dignitatis ac in termino kz^λ sub proprijs signis collecti ponentur æquales nihilo. Nam hæc æquatio debite reducta dabit coefficientem *a*. Sic habes az^ν terminum primum seriei.

Operatio secunda.[37]

Corol. 1. De invenienda area.[38]
Corol. 2. De transmutatione serierum.[39]

Prop. XIII.

Ex æquatione primum radicis momentū involvente radicem extrahere.[40]

Prop. XIV.

Ex æquatione primum et secundum radicis momentū involvente radicem extrahere.[41]

---

(37) Understand, *mutatis mutandis*, the 'Operatio secunda' and 'Operatio tertia et sequentes' of Newton's 1692 'Prob. II' (Wallis, *Opera Mathematica*, 2: 394 [= VII: 178]) in the special case when no fluxions of its related 'fluent' variables are present in the given equation. The fully written out texts of these second, third and ensuing operations appear in the corresponding Proposition XIV of the finished 'Analysis', where—by way of a not essentially variant preliminary draft on Add. 3965.8: 87^v/88^r (see note (49) on page 278 preceding)—Newton rounds out his general exposition of this algebraic rule for attaining the series expansion $y \equiv y_z$ of a root of any given equation by appending the elaborately explained working of an illustrative *exemplum*: that (trivial alterations in denoting letters apart) which he had set forty years before to instance his equivalent geometrical rule for identifying, on a rectangular array of cells containing the terms of the given equation, the analogous 'fictitious' equations which arise at each stage (see *ibid.*: note (45)).

(38) Where, that is, the abscissa *z* and ordinate *y* of the curve whose area has to be determined are inter-connected by some implicit relationship, and a real root $y \equiv y_z$ of the algebraic equation mirroring that relationship is, by the preceding sequence of operations, derived as a 'converging' series of powers of *z* which thereafter needs but to be integrated term by term to obtain the quadrature sought. Compare Newton's more pithy phrasing of this in the corresponding prime corollary to Proposition XIV of the revised 'Analysis' (see page 284 above).

(39) To improve the convergence of the series $y \equiv y_z$ which ensures from the preceding, of course. In this second corollary—which was to find no place in the finished 'Analysis'—Newton had it in mind, so we suppose, briefly to set out (and doubtless instance in some particular case) the *Regula transmutationum* which he had expounded at length in the like-titled second chapter of his 1684 treatise 'De computo serierum' (see IV: 604–12).

(40) Initially '*Ex æquatione fluxionem radicis involvente radicem extrahere*' in exact repeat of the heading (see VII: 177) set by Newton on his 1692 'Prob. II', whose text he here understands to follow word for word, together with its worked 'Exemplum' of extracting the root $y \equiv y_z$ of the equation $y^2 - zy^2\dot{y} - d^2 + dz = 0$ (see *ibid.*: 178–80); compare Proposition XV of the finished 'Analysis' (pages 284–90 above).

(41) The extension of the preceding proposition in which the general term of the given equation is $lz^\mu y^\alpha \dot{y}^\beta \ddot{y}^\gamma$ and the index ν is to be chosen as the greatest of the numbers

$$(\lambda - \mu + \beta + 2\gamma)/(\alpha + \beta + \gamma)$$

[5][42]   *Prop. XII.*[43]

Dignitatem Binomij in seriem numero terminorum infinitam extrahere.

Solutio.

Designet m indicem dignitatis cujusvis quæ per multiplicationem producitur, $-m$ indicem dignitatis cujusvis quæ per divisionem producitur, n indicem dignitatis cujusvis quæ per extractionem radicis alicujus producitur, $\dfrac{m}{n}$ cum signo suo indicem dignitatis binomij, et $P+PQ$ binomium: Et erit

$$\overline{P+PQ}\Big|^{\frac{m}{n}} = P^{\frac{m}{n}} + \frac{m}{n}AQ + \frac{m-n}{2n}BQ + \frac{m-2n}{3n}CQ + \frac{m-3n}{4n}DQ + \&c.$$

Explicatio.[44]

Demonstratio.

Regula proposita recte exhibet dignitates binomij ubi dignitatū indices $\dfrac{m}{n}$ sunt numeri integri ut computanti statim patebit.[45] Dignitas $\overline{a+ab}\big|^2$ (per Regulam) fit $a^2+2a^2b+a^2b^2$, quadratum utiꝗ ex $a+ab$. Et $\overline{a+ab}\big|^3$ fit $a^3+3a^3b+3a^3b^2+a^3b^3$, cubus utiꝗ ex $a+ab$, & $\overline{a+ab}\big|^4$ fit

$$a^4+4a^4b+6a^4b^2+4a^4b^3+a^4b^4,$$

quadrato-quadratum ex $a+ab$, & sic deinceps in infinitum.[46] Regula autem

---

ensuing in each case; compare 'Cas. 1' of Proposition XII of Newton's 1691 treatise 'De quadratura Curvarum' (see VII: 92–4). In the final 'Analysis' this—and its yet further higher-order elaborations—is subsumed by Newton in a single short paragraph which he there inconspicuously adjoins to the preceding proposition (see page 290 above).

(42) Add. 3960.6: 91/92. We omit Newton's opening reminder to himself—or is this where he first framed the title of his augmented tract 'De Quadratura Curvarum'?—that the following revised 'Prop. XII' is, together with the ensuing Propositions XIII/XIV whose revised enunciations he here sets in sequel, intended to form part of a new-styled 'Analysis per quantitates fluentes et earum momenta'.

(43) A straightforward revision of 'Prop. XI'—its number now rightly increased by a unit (see note (24) above)—in (4) preceding.

(44) '*per exempla*', that is, as in the corresponding Proposition XII of the finished 'Analysis'. Newton once more (compare notes (17) and (27) above) understands the nine worked examples which he had set to illustrate his binomial expansion in his *epistola prior* to Leibniz in June 1676.

(45) Initially 'ut examinatio statim patebit' in a rough preparatory version of the present *Demonstratio* jotted down by Newton at the top of the manuscript page in a convenient blank space.

(46) These are, of course, the familiar polynomial expansions of $a^m(1+b)^m$, $m = 2, 3, 4, \ldots$. At the head of the manuscript page Newton has set out a private reminder of the term-by-term development of $(1+b)^m$ as $A+B+C+D+E\ldots$, namely that '$A=1$. $B=mbA$. $C=\dfrac{m-1}{2}bB$.

quæ ad innumera æqualia intervalla recte se habet, recte se habebit in locis intermedijs.[47]

Corol. 1.[48]   Hinc areæ Curvilinearum & longitudines arcuum per series numero terminorum infinitas computari possunt.

Sit radius r, abscissa x & Ordinata

$$\sqrt{r^2-xx}=(\text{per Reg})\ r-\frac{1}{2}\frac{xx}{r}-\frac{1}{8}\frac{x^4}{r^3}-\frac{1}{16}\frac{x^6}{r^5}-\frac{5}{128}\frac{x^8}{r^7}-\frac{7}{256}\frac{x^{10}}{r^9}-\&c$$

et Area[49] per Cor. Prop. 1 erit

$$rx-\frac{1}{6}\frac{x^3}{r}-\frac{1}{40}\frac{x^5}{r^3}-\frac{1}{112}\frac{x^7}{r^5}-\frac{5}{1152}\frac{x^9}{r^7}-\frac{7}{2816}\frac{x^{11}}{r^9}-\&c.$$

Sit Abscissa x et Ordinata

$$\sqrt{x-xx}=x^{\frac{1}{2}}-\tfrac{1}{2}x^{\frac{3}{2}}-\tfrac{1}{8}x^{\frac{5}{2}}-\tfrac{1}{16}x^{\frac{7}{2}}-\tfrac{5}{128}x^{\frac{9}{2}}-\tfrac{7}{256}x^{\frac{11}{2}}-\&c$$

et Area[49] erit $\tfrac{2}{3}x^{\frac{3}{2}}-\tfrac{1}{5}x^{\frac{5}{2}}-\tfrac{1}{28}x^{\frac{7}{2}}-\tfrac{1}{72}x^{\frac{9}{2}}-\tfrac{5}{704}x^{\frac{11}{2}}-\tfrac{7}{1664}x^{\frac{13}{2}}-\&c.$

Sit Abscissa x et Ordinata

$$\frac{1}{1+x^2}=\overline{1+x^2}\,|^{\frac{-1}{1}}=1-xx+x^4-x^6+x^8-x^{10}\ \&c$$

et Area[49] erit $x-\tfrac{1}{3}x^3+\tfrac{1}{5}x^5-\tfrac{1}{7}x^7+\tfrac{1}{9}x^9-\tfrac{1}{11}x^{11}\ \&c.$

Corol. 2. Hinc etiam Analysis longe universalior redditur. Nam si area figuræ

$D=\dfrac{m-2}{3}Cb.\ E=\dfrac{m-3}{4}Db.$ [&c]', and also penned in alongside the frustum $m=1,2,3,\dots,7$ of the corresponding triangle of binomial coefficients in the form:

1	1b	bb					
1	2b	1	b^3				
1	3b	3	1	b^4			
1	4b	6	4	1	b^5		
1	5b	10	10	5	1	b^6	
1	6b	15	20	15	6	1	b^7
1	7b	21	35	35	21	7	1.

(47) Newton evokes the Wallisian principle of continuity which he had appealed to half a century earlier in first extending the validity of the binomial expansion to non-integral values of (positive) index; see 1: 104 ff. His younger contemporaries, however, were no longer willing to place their faith in so vague and loosely controllable a mathematical tenet, and to them the present 'demonstration' of the truth of the expansion for all real values of the index (and, it is understood, for $|b|<1$) would no longer have seemed rigorous. Already, even as Newton wrote the present paragraph, Roger Cotes was developing in his still unpublished private papers (Trinity College, Cambridge. R.16.38: 365–7) a more satisfactory 'Demonstratio Theorematis Newt: pro extractione Radicum &c' which deduced it effectively as a Taylor expansion. Time stands still for no man....

(48) In further revise in [6] following Newton will wisely hive off this unwieldy prime corollary to form the nucleus of a separate Proposition XIII.

(49) Namely, $\tfrac{1}{2}(r^2\sin^{-1}[x/r]+x\sqrt{[r^2-x^2]})$, $\tfrac{1}{8}(\sin^{-1}2\sqrt{[x-x^2]}-2(1-2x)\sqrt{[x-x^2]})$ and $\tan^{-1}x$ respectively.

cujusvis quæstionem ingrediatur potest area illa in serie[50] infinita computari & series pro area substitui. Vel si quantitas quælibet fluens quæstionem ingrediatur &[51] relatio inter momentum fluentis hujus et quan[ti]tatem aliam quamvis fluentem per æquationem habeatur: potest area figuræ computari cujus ordinata est momentum illud et abscissa est fluens altera; & area sic inventa pro fluente quæ quæstionem ingreditur substitui.

Schol.[52] Si binomij nomen secundum sit momentum nominis primi, & dignitas binomij in seriem juxta Reg[u]lam allatam resolvatur, termini seriei erunt ut momenta dignitatis, nempe terminus secundus *B* ut momentum primum, tertius *C* ut momentum secundum, et sic deinceps. Et si termini seriei continuo multiplicentur per terminos hujus progressionis $1 \times 2 \times 3 \times 4 \times 5$ &c. secundus *B* per 1, tertius *C* per 1×2, quartus *D* per $1 \times 2 \times 3$ & sic deinceps, tum habebuntur momenta dignitatis binomij: scilicet *B* momentum primum, $2C$ momentum secundum, $6D$ momentum tertium, $24E$ momentum quartum, & sic porro in infinitum.

Prop. XIII.[53]

Æquationis affectæ radicem fluentem in seriem numero terminorum infinitā convertere.

Prop. XIV.[53]

Æquationis affectæ primum radicis fluentis momentum involventis, radicem in seriem numero terminorum infinitam convertere.

---

[6][54]   *Prop. XIII.*

Quantitatem fluentem in serie numero terminorum infinita invenire, cujus fluxio aliqua[55] est dignitas binomij quantitatem alteram uniformiter fluentem involventis.

Solutio.

Resolvatur binomium in seriem perpetuam per Propositionem præcedentem

---

(50) 'per seriem' is replaced.

(51) Newton first went on to qualify '& ex alterâ quantitate ejus momentum uniformiter fluente determinetur'.

(52) Newton spells out more fully the fluxional correspondences of the successive terms in the binomial expansion $(P+o\dot{P})^{m/n} = P^{m/n} + oB + o^2C + o^3D + o^4E + \ldots$ which he had more tersely enumerated in the prior version of this scholium in [4] preceding; compare note (30) above.

(53) These are but more elaborately phrased enunciations of the corresponding Propositions XII/XIII in [4] preceding, their numbers now (see note (24) above) rightly increased by a unit. Newton's omission of the earlier Proposition XIV from his present revised scheme presumably indicates that it was no longer to have separate existence in the 'Analysis', being henceforth subsumed into the previous one (the present XIV); compare note (41).

(54) Add. 3960. 6: 92/93. Newton amplifies Corollary 1 to Proposition XII in [5] preceding to be a separate Proposition XIII (understanding that its Propositions XIII/XIV shall again follow, their numbers again being increased by a unit, as in the final 'Analysis').

(55) Newton originally went on to specify '*cujus fluxio quævis*'—initially '*momentum quodvis*'— '*est binomium quodcunҩ propositam fluentem involvens*'.

& multiplicetur terminus unusquisq$^{(56)}$ seriei per fluentem alteram ac dividatur idem terminus per indicem dignitatis quam fluens illa altera jam habet in hoc termino, et aggregatum terminorum prodeuntium erit fluens quæsita si modo binomium sit ejus fluxio prima.$^{(57)}$ Et operatione semel vel bis vel ter repetita habebitur fluens ubi binomium est ejus fluxio secunda vel tertia vel quarta. Et sic deinceps in infinitum.

Explicatio per exempla.

In circulo cujus radius est *r* sit *x* Abscissa & $\sqrt{rr-xx}$ Ordinata, et quæratur area quam Ordinata describit. Quoniam momentum hujus areæ est ut Ordinata, resolvo Ordinatam in seriem infinitam per Propositionem præcedentem. Et fiet

$$\sqrt{rr-xx} = \overline{rr-xx}\,|^{\frac{1}{2}} = r - \frac{xx}{2r} - \frac{x^4}{8r^3} - \frac{x^6}{16r^5} - \frac{5x^8}{128r^7} - \frac{7x^{10}}{256r^9} - \&c.$$

Et inde (per hanc Propositionem) prodit area

$$rx - \frac{x^3}{6r} - \frac{x^5}{40r^3} - \frac{x^7}{112r^5} - \frac{5x^9}{1152r^7} - \frac{7x^{11}}{2816r^9} - \&c.$$

Aufer aream trianguli $\frac{1}{2}x\sqrt{r^2-xx}$, seu

$$\tfrac{1}{2}rx - \frac{1}{4}\frac{x^3}{r} - \frac{1}{16}\frac{x^5}{r^3} - \frac{1}{32}\frac{x^7}{r^5} - \frac{5}{256}\frac{x^9}{r^7} \&c$$

et manebit sector $\frac{1}{2}rx + \dfrac{1}{12}\dfrac{x^3}{r} + \dfrac{3}{80}\dfrac{x^5}{r^3} + \dfrac{5}{224}\dfrac{x^7}{r^5} + \dfrac{35}{2304}\dfrac{x^9}{r^7} + \&c.$

Divide per $\frac{1}{2}r$ et prodibit arcus

$$x + \frac{x^3}{6rr} + \frac{3x^5}{40r^4} + \frac{5x^7}{112r^6} + \frac{35x^9}{1152r^8} + \&c.$$

Si diametro *d* describatur circulus, & quæratur area quam Ordinata $\sqrt{dx-xx}$ describit, erit

$$\sqrt{dx-xx} = \overline{dx-xx}\,|^{\frac{1}{2}} = d^{\frac{1}{2}}x^{\frac{1}{2}} - \frac{x^{\frac{3}{2}}}{2d^{\frac{1}{2}}} - \frac{x^{\frac{5}{2}}}{8d^{\frac{3}{2}}} - \frac{x^{\frac{7}{2}}}{16d^{\frac{5}{2}}} - \frac{5x^{\frac{9}{2}}}{128d^{\frac{7}{2}}} - \&c,$$

et area quæsita $\dfrac{2}{3}d^{\frac{1}{2}}x^{\frac{3}{2}} - \dfrac{x^{\frac{5}{2}}}{5d^{\frac{1}{2}}} - \dfrac{x^{\frac{7}{2}}}{28d^{\frac{3}{2}}} - \dfrac{x^{\frac{9}{2}}}{72d^{\frac{5}{2}}} - \dfrac{5x^{\frac{11}{2}}}{704d^{\frac{7}{2}}} - \&c.$

Si quæratur area quam Ordinata normaliter erecta $\dfrac{1}{1+xx}$ describit [existente *x* abscissa],$^{(58)}$ hæc Ordinata in seriem resoluta evadet

$$1 - xx + x^4 - x^6 + x^8\ [-]\&c,$$

& inde (per hanc Propositionem) elicitur area

$$x - \tfrac{1}{3}x^3 + \tfrac{1}{5}x^5 - \frac{x^7}{7} + \frac{x^9}{9} - \&c.$$

---

(56) An equivalent 'singulus' is replaced.

(57) Originally (compare note (55)) altered to be 'momentum primum' and then changed back again.

(58) We make trivial insertion of a clarifying phrase which Newton himself introduces at this point in his final version of this supplementary Proposition XIII (see pages 272–4 above).

Si radio r describatur circulus & quæratur longitudo arcus cujus sinus est x, exponatur momentum ipsius x per numerum 1, & momentum arcus erit

$$\frac{r}{\sqrt{rr-xx}} \; \left[\text{seu } r \times \overline{rr-xx}\right]^{\frac{-1}{2}} = 1 + \frac{x^2}{2r^2} + \frac{3x^4}{8r^4} + \frac{5x^6}{16r^6} + \frac{35x^8}{128r^8} + \frac{63x^{10}}{256r^{10}} + \&\text{c. Et inde}$$

arcus fit $x + \dfrac{x^3}{6r^2} + \dfrac{3x^5}{40r^4} + \dfrac{5x^7}{112r^6} + \dfrac{35x^9}{1152r^8} + \dfrac{63x^{11}}{2816r^{10}} + \&$c ut supra.$]^{(59)}$

Si circulus diametro d describatur & quæratur arcus cujus sinus versus est x,$^{(60)}$ exponatur momentum ipsius x per numerum 1, et momentum arcus erit

$$\frac{2\sqrt{dx-xx}}{d}\,^{(61)} \; [\text{id est}] \; \frac{2x^{\frac{1}{2}}}{d^{\frac{1}{2}}} - \frac{x^{\frac{3}{2}}}{d^{\frac{3}{2}}} - \frac{x^{\frac{5}{2}}}{4d^{\frac{5}{2}}} - \frac{x^{\frac{7}{2}}}{8d^{\frac{7}{2}}} - \frac{5x^{\frac{9}{2}}}{64d^{\frac{9}{2}}} - \&\text{c}$$

et arcus
$$\frac{4x^{\frac{3}{2}}}{3d^{\frac{1}{2}}} - \frac{2x^{\frac{5}{2}}}{5d^{\frac{3}{2}}} - \frac{x^{\frac{7}{2}}}{14d^{\frac{5}{2}}} - \frac{x^{\frac{9}{2}}}{36d^{\frac{7}{2}}} - \frac{5x^{\frac{11}{2}}}{352d^{\frac{9}{2}}} - [\&\text{c}] \; \text{seu}^{(62)}$$

Si quæratur area Figuræ cujus abscissa est sinus versus & ordinata arcus in circulo quovis: Sit d diameter circuli & x abscissa figuræ, & Ordinata erit

$$\frac{4x^{\frac{3}{2}}}{3d^{\frac{1}{2}}} - \frac{2x^{\frac{5}{2}}}{5d^{\frac{3}{2}}} - \frac{x^{\frac{7}{2}}}{14d^{\frac{5}{2}}} - \frac{x^{\frac{9}{2}}}{36d^{\frac{7}{2}}} - \frac{5x^{\frac{11}{2}}}{352d^{\frac{9}{2}}} - \&\text{c}^{(63)}$$

ut supra, & area
$$\frac{8x^{\frac{5}{2}}}{15d^{\frac{1}{2}}} - \frac{4x^{\frac{7}{2}}}{35d^{\frac{3}{2}}} - \frac{x^{\frac{9}{2}}}{63d^{\frac{5}{2}}} - \frac{x^{\frac{11}{2}}}{198d^{\frac{7}{2}}} - \frac{5x^{\frac{13}{2}}}{2288d^{\frac{9}{2}}} - \&\text{c.}$$

Est autem hæc figura Quadratrix,$^{(64)}$ et area ejus est fluens quantitas cujus fluxio prima est ut arcus ille & fluxio secunda ut binomij dignitas $\left.\overline{\dfrac{4dx-4xx}{dd}}\right|^{\frac{1}{2}}.^{(65)}$

---

(59) In the manuscript Newton himself leaves a gap for the later insertion of the series expansions which we here intercalate.

(60) That is, $\frac{1}{2}d(1-\cos\theta)$ where the arc subtends an angle θ at the centre of the circle; whence the arc itself is $(\frac{1}{2}d.\theta =) \frac{1}{2}d.\cos^{-1}(1-2x/d)$, and consequently its derivative with respect to the *sinus versus* x is $\frac{1}{2}d/\sqrt{(dx-x^2)}$.

(61) By a curious carelessnesss Newton here interchanges the numerator and denominator of the correct value for the *momentum arcus* (on which see the previous note). The ensuing series which accurately expand this erroneous reciprocal and then integrate it term by term are not, accordingly, to the present purpose.

(62) Newton breaks off without adding any alternative expression for this expansion of $2d^{-1}\int_0^x \sqrt{[dx-x^2]}.dx$. Though he carries this latter into his next paragraph (see the following note) we may be sure that his ingrained good sense soon let him see that the required arc is in fact $d.\sin^{-1}(x^{\frac{1}{2}}/d^{\frac{1}{2}}) = d^{\frac{1}{2}}x^{\frac{1}{2}} + \frac{1}{6}x^{\frac{3}{2}}/d^{\frac{1}{2}} + \frac{3}{40}x^{\frac{5}{2}}/d^{\frac{3}{2}} + \ldots$.

(63) Newton carries over his mistaken series expansion for the arc from the preceding paragraph. We need not insist that the area which he proceeds to derive for his sinusoidal *figura* is equally erroneous.

(64) This is untrue. Where y is the ordinate of the figure, the Cartesian equation of the quadratrix with respect to a base circle of diameter d is $\frac{1}{2}d - x = y.\cot(y/\frac{1}{2}d)$, whereas the present curve—which Newton had in Example 5 of Problem 8 of his 1671 tract dubbed *figura arcuum* (see III: 240)—has the defining equation $y = \frac{1}{2}d.\sin^{-1}\sqrt{(x/d)}$. Why Newton could confuse the two curves will readily be explained in regard to the figure which (see II: 240) accompanies Newton's accurate determination of the area of the quadratrix's general segment in his 'De Analysi' in 1669: there, on setting the circle's radius $VA = \frac{1}{2}d$ and the abscissa

Schol.

Per hujusmodi series Analysis multum ampliatur.[66] Nam si quando area curvilinea vel longitudo curvæ alicujus vel Fluens quæcunꝗ quæstionem ingrediatur, et habeatur ejus momentum aliquod: potest momentum illud in seriem infinitam resolvi & Fluens inde[67] inveniri in alia serie & hæc series pro Fluente in Quæstione usurpari & in æquationes substitui.

[7][68]   [Dividendo per $4cc - \frac{1}{2}cz + \frac{9}{32}zz$ prodit]

$$-\frac{131}{128}z^3 - \frac{15z^4}{4096c} - \frac{3z^5}{4096cc} \quad \left[+\frac{131z^3}{512cc} + \frac{509}{16384c^3}z^4 - \frac{4159}{262112c^4}z^{5\,(70)}\right.$$

$$+\left[\frac{131}{128}\right] - \frac{131}{1024c} + \frac{1179}{16384cc}$$

$$-\frac{509}{4096c}z^4 \qquad \frac{1167}{16384cc}z^5$$

$$+\left[\frac{509}{4096c}\right] - \frac{509}{65528}^{(69)} \qquad \begin{array}{c} 4668 \\ [509] \end{array}$$

$$+\frac{4159}{65528}$$

$VG = x$, the defining property of the cycloid which Newton employs sets

$$AB = CD = y = \widehat{VK} = \tfrac{1}{2}d.\sin^{-1}\!\sqrt{(x/d)},$$

but DK passes through the circle's centre A (and not the end-point G of the abscissa).

(65) That is, $\sqrt{(dx-x^2)}/\tfrac{1}{2}d$. But this should again (compare notes (60) and (61) above) be the reciprocal $\tfrac{1}{2}d/\sqrt{(dx-x^2)}$.

(66) 'augetur' is replaced.

(67) Newton initially specified 'ex serie illa'.

(68) Add. 3965.12: 218^r, a scrapsheet from which we extract Newton's rough computation rounding off, up to terms in z^5, the series expansion $c - \tfrac{1}{4}z + \tfrac{1}{64}z^2/c + \dots$ of the root y of the equation $y^3 + c^2y + czy - 2c^3 - z^3 = 0$ which, much as in his *epistola prior* to Leibniz in 1676 (see note (37) above), he sets to illustrate the working of Proposition XIV of his 'Analysis per quantitates fluentes'. (The manuscript sheet was subsequently used by him to draft the paragraph relating the calculated elements of the comet of 1683 which he adduced to 'confirm the truth of our theory'—as laid down in the first edition of the *Principia* at the end of Book 3— along with other additions in the summer of 1712; transmitted to Cotes on 14 October following (see *Correspondence*, 5: 347), this duly appeared in print the next year with only trivial verbal emendation on p. 478 of the *editio secunda*). We ignore two numerical slips in the computation which Newton himself here caught and corrected, but reproduce a third which (see the two following notes) vitiate the final term in the series expansion as he carried it over into his main text (see page 282 above).

(69) The denominator of this fraction should be '32768', thus leaving '$\frac{1825}{32768}$' in the next line. Newton presumably here computed 16384×2 as 4×2 and then 1638×4.

(70) On changing the signs in the quotient, as Newton's rounding-off requires (see note (60) on page 283 preceding). The numerical coefficient of the last term in z^5/c^4 is, however, wrong; on dividing the corrected term $+1825z^5/32768c^2$ in the bottom line of the computation (see the previous note) by $4c^2$ there properly results '$-\dfrac{1825}{131072c^4}z^5$ [&c]'.

6

PROPOSITION X OF THE PRINCIPIA'S SECOND BOOK REWORKED[1]

[autumn 1712]

From originals in the University Library, Cambridge

§1. TOTTERING STEPS TOWARDS ACHIEVING A VALID ARGUMENT.[2]

[1][3] [Sit *AK* planum illud[4] plano Schematis perpendiculare, *ALK* linea curva, *C* corpus in ipsa motum, & *CF* recta ipsam tan]gens in *C*. Temporibus

---

(1) When in late September 1712 Johann Bernoulli's nephew, Niklaus I, arrived in London ready to disclose to Abraham de Moivre (who lost no time in imparting to Newton himself) his uncle's discovery two years before that an independent check from first principles revealed that 'Exempl. 1' of the tenth proposition of Book 2 of the *Principia*—and hence, therefore, the argument of the parent proposition from which this particular case of resisted motion in a semi-circle was correctly deduced (though neither Johann or Niklaus were themselves properly ever to detect wherein its error lies)—was wrong by a numerical factor, Newton applied himself urgently first to check the validity of the objection, and then to recast his original general argument into a corrected form which would yield Bernoulli's result as its prime example. The erroneousness of his 1687 'Exempl. 1' he quickly confirmed by reducing it, again under Bernoulli's guiding hand, to a *non sequitur* (see Appendix 2.1: note (6) below). The magnitude of the lengthy mental effort which it cost him to mend the argument of his 1687 text can only properly be gauged from the physical bulk of the great many surviving worksheets and attendant scraps of calculation where he endeavoured in a variety of ways to do so, and whose main portion is reproduced in the following pages. We have de Moivre's statement in a letter to Johann Bernoulli on 18 October, a fortnight or so after this initial breakthrough, that it was only 'deux ou trois jours' after Newton first said he would examine the matter that 'il me dit que l'objection [de M. Bernoulli] étoit bonne, et qu'il avoit corrigé la conclusion [de sa proposition]; en effet il me montra sa correction, et elle se trouva conforme au calcul de M. votre neveu' (K. Wollenschläger, 'Der mathematische Briefwechsel zwischen Johann I Bernoulli und Abraham de Moivre' (*Verhandlungen der Naturforschenden Gesellschaft in Basel*, **43**, 1933: 151–317): 270–4, especially 271); plus his report of Newton's assurance that 'cette erreur procede simplement[!] d'avoir considéré une tangente à rebours, mais que le fondement de son calcul et les suites dont il s'est servi doivent subsister' (*ibid.*). And to be sure, while the latter assertion is not strictly true—his original expression of the resistance over successive arcs of the projectile path by the difference of the segments of the single tangent drawn opposite ways from their common point is not itself mistaken in its validity, and is very simply adapted (had Newton but seen this) to yield the correct numerical increase of the 1687 value for the ratio of resistance to gravity to be half as much again (see Appendix 2.1: note (13) below)—Newton's private papers make it abundantly clear that he saw his way through to correcting the argument of his *editio princeps* only after he adopted the alternative approach of

Translation

[1][3]   Let AK be that plane[4] perpendicular to the plane of the figure, ALK the curved line, C the body moved in it and CF the straight line touching it at C.

considering the pair of tangents drawn the same way from corresponding end-points of the infinitesimal arcs successively traversed. Discovery, however, of a variant way of attaining the true goal was here not enough: short of reprinting the two hundred and forty or so pages following (or, yet less satisfactorily still, of messily running over into duplicated pages terminating at their end in an ugly bridging blank), the recast argument had of necessity to be cut down and tailored exactly to fit the gap now left on pages 232–44 of the already printed-off main text of the *editio secunda* after discarding the unchanged repeat of the 1687 text of the present proposition which initially filled that space. Not till three months later, in fact, did Newton communicate a finished replacement to his editor, Roger Cotes, in Cambridge. Here, in §1, we reproduce the preliminary sequence of draftings by way of which he attained the initial version (§1.6) of the correct reasoning which he straightaway transmitted in person to Niklaus Bernoulli about the beginning of October 1712; and in sequel, in §2, the further restylings and remouldings whereby over the ensuing weeks—and for safety's sake contriving an *idem aliter* where the continuous curve of the projectile path is approximated by the linked chain of its infinitesimal chords—he shaped and honed that corrected mode of argument to be the *tour de force* which he made public in the summer of 1713 in the second edition of his *Principia*.

(2) At the end of September, namely, or just possibly during the first few days of October; see the previous note. In Appendix 2.1 below we reproduce Newton's preparatory check on the accuracy with which in the initial version of his proposition—the text (*Principia*, ₁1687: 260–74) is, for convenience of reference and of commentary, reprinted in the preceding Appendix 1—he had applied his faulty measure there of the ratio of resistance to gravity: having confirmed that his correctly computed result in 'Exempl. 1' does indeed yield Bernoulli's nonsensical corollary, in this particular case of a resisted semi-circular path, that the total motion of the projectile is unaccelerated, he here went on to make a number of fruitless probings of his general reasoning to try to locate its flaw (of which we instance one). In Appendix 2.2/3 we print our edited text of two roughly jotted preliminary attempts, effectively equivalent one to the other in their broad structure, where Newton first thought to attain his goal by an alternative argument taking into account the tangential motions the same way from corresponding end-points of the arcs successively traversed, but in each case failed accurately to determine the increment of motion due to gravity in its proper ratio to the decrement due to resistance. We here take up the unfolding story of his behind-scenes effort to correct his 1687 text (as his surviving manuscripts would appear fully to tell it) at the point where, making a renewed attempt to master the niceties involved in making comparison of the forces of resistance and gravity by way of the linear increments which these generate in the same infinitesimal time, he for the first time attempts directly to amend his original statement of it in the *editio princeps*.

(3) Add. 3965.12: 196ʳ. Newton begins this recasting of the parent proposition *in medio* at the head of page 261 in the *Principia*'s first edition (see page 374 below): to round out its text we have interpolated a lightly amended version of his opening sentence at the bottom of page 260. The manuscript figure is of the right-hand quadrant only, and its points O, d, g, B, C, D, F and H were originally there denoted by A, B, C, D, E, F, k and n respectively, being referred—in the absence of the ordinate mp here shown—to a desultory opening computation

$$'AB = x, \ BD = o, \ BC = e, \ DE = e + ao + boo + co^3 + \&c = g.$$
$$AD = z, \ DF = p. \ DE = g. \ FG = g + ap + bpp + cp^3 + \&c \ [!].$$
$$En = Gk = boo + co^3 = n'.$$

æqualibus describat corpus arcus gC, CG et ad $A[K]$ demittantur perpendicula gd, GD. Et producta DG occurrat tangenti CF in F. Sit $\overline{Bm=Bd}$ et erecta perpendicularis [mn] occurrat tangenti CF in p, et compleatur $\overline{\text{pgrm}}$[5] $GFp[q]$.[6]

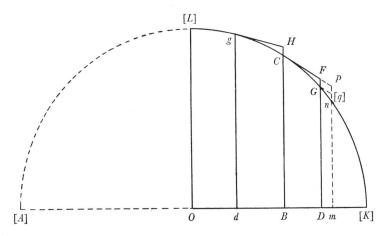

Jam ob tempora æqualia æquales sunt lineolæ CH, GF a gravitate[7] genitæ. Si nulla esset resistentia, corpus in fine temporis reperiretur in n. Per resistentiam fit ut corpus reperiatur in $G_{[,]}$ ideoq lineola [q]G[8] vel pF per resistentiam generatur estq resistentia ad gravitatem ut Fp ad FG[9] id est ut $\dfrac{Dm \times gH}{dB}$[10] ad

$FG=CH$, seu $Dm \times gH$ ad $Bd \times CH$ vel $\dfrac{pn-FG}{2} \times gH$ ad CH^q.[11]

Exempl. 1.[12] [In semicirculo ALK sit diametrum] $AK=2n$. [ideoq ponendo] $OB=a$. $BC=e$. Bd vel $B[m]=o$. [fit] $aa+ee=nn$.

For simplicity's sake we have omitted to reproduce a further ordinate *il* drawn (as in the *editio princeps*) parallel to BC at the distance $iB = BD$ from it, since no use of it is made in the present remoulded argument; and have also for consistency's sake, because the figure has again to do double duty both for the general case and the particular instance of the semi-circle in 'Exempl. 1', added the left-hand quadrant shown in broken line. The preceding statement of the problem survives, it is understood, unchanged, the body being supposed to move (in the vertical plane of the paper) under the joint action of a constant force of gravity straight downwards 'directe ad planum Horizontis' and of the medium's resistance directly opposed to its onrush in the instantaneous direction of its motion.

(4) 'Horizontis' (of the horizontal), namely, as Newton specifies it in his preceding enunciation; compare the previous note.

(5) Read 'parallelogrammum'.

(6) In his manuscript figure Newton by an oversight takes n to mark both the intersection of mp with the arc $\widehat{GK}$, and also the meet with mp of the parallel through G to CFp; for distinction's sake the latter intersection is here denoted by q.

(7) Strictly, both the deviations CH and GF will also share equal third-order components of the decelerating resistance to motion along the arclets $\widehat{gC}$ and $\widehat{CG}$ traversed in successive equal times; see Appendix 2.1: note (13) below.

In equal times let the body describe the arcs $\widehat{gC}$, $\widehat{CG}$, and to AK let fall the perpendiculars gd, GD; and let DG produced meet the tangent CF in F. Let $Bm = dB$, and the erected perpendicular mn meet the tangent CF in p, then complete the parallelogram $GFpq$.[6] Now because of the equal times the linelets CH and GF begotten by gravity[7] are equal. Were the resistance nil, the body at the end of the time would be found at n. Through the resistance it comes that the body is found at G, and consequently the linelet qG,[8] or pF, is generated by the resistance; and so the resistance is to gravity as Fp to FG,[9] that is, as $Dm \times gH/dB$[10] to $FG = CH$, or as $Dm \times gH$ to $dB \times CH$ or $\frac{1}{2}(pn - FG) \times gH$ to CH^2.[11]

Example 1.[12] In semicircle ALK let the diameter $AK = 2n$, and consequently on setting $OB = a$, $BC = e$ and dB or $Bm = o$ there comes to be $a^2 + e^2 = n^2$. There

---

(8) Newton here wrote 'nG'; compare note (6) preceding.

(9) This should be '$2FG$', the exponent of the increment in speed, $g\theta$, due to gravity g acting over the time θ in which the arc $\widehat{CG}$ is traversed by the missile, during which its deceleration in speed due to resistance is accurately represented by the decrement in the tangential direction; compare Appendix 2.2: note (25) below.

(10) More precisely, $Fp = (Dm/dB) \times fC$ where (as in the 1687 text; see the figure on page 374 below) f is the meet of dg with the extension beyond C of the tangent FC; but the substitution of fC by gH is allowable within the limits of accuracy which Newton here assumes.

(11) Read '$2CH^q$' ($2CH^2$) *recte*, on mending the slip which we indicated in note (9) preceding. Newton here assumes—accurately so to a near enough approximation—that the infinitesimal arc $\widehat{CG}$ traversed in resisted motion from C coincides with the corresponding portion of the parabolic arc $\widehat{Cn}$ travelled in unresisted motion; whence there is $pn:FG = Cp^2:CF^2$ and so $(pn - FG):FG = (Cp^2 - CF^2):CF^2$, that is, (since Fp is supposed indefinitely small in comparison with CF)

$$\tfrac{1}{2}(pn - FG):(FG \text{ or}) HC = Fp:(CF \approx) Cp = Dm:(Bm \text{ or}) dB.$$

In the analytical equivalent introduced in the following 'Exempl. 1' (here as in the *editio princeps*) it follows that, corresponding to the series expansion of the augmented ordinate $DG = e_{a+o}$ as $e + Qo + Ro^2 + So^3 + \ldots$ (where $Q \equiv Q_a = de/da$, $R \equiv R_a = \frac{1}{2}d^2e/da^2$ and

$$S \equiv S_a = \tfrac{1}{6}d^3e/da^3 = \tfrac{1}{3}dR/da),$$

there is likewise, on setting

$Bm = dB = p$, $mn = e_{a+p} = e + Qp + Rp^2 + Sp^3 + \ldots$ and $dg = e_{a-p} = e - Qp + Rp^2 - Sp^3 \ldots$;

so that $np = Rp^2 + Sp^3 + \ldots$ and $CH = R_{a-p}p^2 + S_{a-p}p^3 + \ldots$, that is, $Rp^2 - 2Sp^3 + \ldots$, and therefore Newton's present (unamended) measure for the ratio of resistance to gravity is $\frac{1}{2}(3Sp^3 \ldots) \cdot p\sqrt{(1 + Q_{a-p}^2)}/(Rp^2 + \ldots)^2 = \frac{3}{2}S\sqrt{(1 + Q^2)}/R^2$ in the limit as p comes to vanish. Through making a trivial numerical slip in applying this to the particular case of the semicircular path $e = +\sqrt{(n^2 - a^2)}$ in his ensuing reworked 'Exempl. 1', Newton was not at once to notice that the result is double the true value which Bernoulli had independently found. When, in the fuller version ([2] following) which he elaborated upon this same basis, he did discover the error in his calculation of 'Exempl. 1', Newton omitted to notice that the mistake was readily rectified by doubling (as it should be) the exponent FG of the downwards thrust of gravity, but put himself vainly (in [3] ensuing) to distinguish the resisted and unresisted arcs $\widehat{CG}$ and $\widehat{Cn}$.

(12) We reproduce only the essentials of the following computation, ignoring the several rough intermediate multiplications and divisions of terms and the incomplete evaluations of line-lengths which pack out the manuscript here (at the bottom of f. 196^r).

[Quare erit]

$$mn = e - \frac{ao}{e} - \frac{nnoo}{2e^3} - \frac{anno^3}{2e^5} - \&\text{c.}$$

[necnon]

$$dg = e + \frac{ao}{e} - \frac{nnoo}{2e^3} + \frac{anno^3}{2e^5} \ [-\&\text{c}].$$

[adeoqȝ] $dg^3 = e^3 + 3eao.$ [ut et] $pn = \dfrac{nnoo}{2e^3} + \dfrac{anno^3}{2e^5}.$

[et similiter] $CH = \dfrac{nnoo}{2dg^3} + \dfrac{anno^3}{2dg^5} = \dfrac{nnoo}{2e^3 + 6eao} + \dfrac{anno^3}{2e^5} = \dfrac{nnoo}{2e^3} - \dfrac{anno^3}{e^5}.$

[itaqȝ] $pn - CH = \dfrac{3anno^3}{2e^5}.$ [Fit igitur]

$$\frac{pn - CH}{2} \times gH \text{ ad } CH^q :: \frac{3anno^3}{4e^5} \times \sqrt{00 + \frac{aa}{ee}00} \cdot \frac{n^4o^{4\,(13)}}{2e^6}$$

$$:: \frac{3an^3o^4}{4e^6} \cdot \frac{n^4o^4}{2e^6} :: 3a \cdot 2n :: \text{resi[s]t. grav.}^{(14)}$$

---

[2][15] [Sit *AK* planum illud plano Schematis perpendiculare, *ACK* linea curva, *C* corpus in ipsa motum, & *CF* recta ipsam tan]gens in *C*. Ad planum

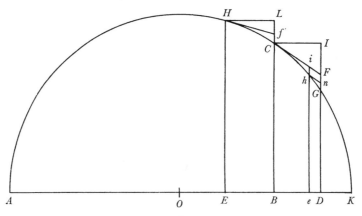

horizontale *AK* demittantur perpendicula *HE*, *CB*, *he*, *GD* ea lege ut intervalla *EB*, *BD* sint æqualia, & arcus *HC*, *Ch* a corpore moto *C* describantur temporibus æqualibus. Agantur *Hf*, *CiF* curvam descriptam tangentes in punctis *H* et *C*, et perpendiculis *BC*, *eh* et *DG* productis occurrentes in *f*, *i* et *F*. Et compleantur parallelogramma *BCID*, *Fihn*. Si corpus resistentiam nullam pateretur tempora quibus transit inter perpendicula *EH* et *BC*, *BC* et *DG* æqualibus intervallis ab invicem distantia, forent æqualia, et corpus his temporibus æqualibus vi gravitatis descend[end]o a tangentibus æquales describeret altitudines *fC* et *Fn*, ideoqȝ in fine temporum reperiretur in loco *n*. Per resistentiam fit ut corpus in fine temporum reperiatur in loco *h*. Ideoqȝ lineola *nh* vi resistentiæ et lineola

will therefore be $mn = e - (a/e) o - \frac{1}{2}(n^2/e^3) o^2 - \frac{1}{2}(an^2/e^5) o^3 - \ldots$, and also

$$dg = e + (a/e) o - \frac{1}{2}(n^2/e^3) o^2 + \frac{1}{2}(an^2/e^5) o^3 \ldots,$$

so that $dg^3 = e^3 + 3eao \ldots$, and again $pn = \frac{1}{2}(n^2/e^3) o^2 + \frac{1}{2}(an^2/e^5) o^3$, and similarly

$$CH = \frac{1}{2}(n^2/dg^3) o^2 + \frac{1}{2}(an^2/dg^5) o^3 = \frac{1}{2}n^2 o^2/(e^3 + 3eao) + \frac{1}{2}(an^2/e^5) o^3$$
$$= \frac{1}{2}(n^2/e^3) o^2 - (an^2/e^5) o^3,$$

and accordingly $pn - CH = \frac{3}{2}(an^2/e^5) o^3$. There comes therefore to be

$$\frac{1}{2}(pn - CH) \times gH : CH^2 = \frac{3}{4}(an^2/e^5) o^3 \sqrt{(o^2 + (a^2/e^2) o^2)} : \frac{1}{2}(n^4/e^6) o^{4(13)}$$
$$= \frac{3}{4}(an^3/e^6) o^4 : \frac{1}{2}(n^4/e^6) o^4 = 3a : 2n$$

equal to the resistance to gravity.[14]

----

[2][15]   Let AK be that plane perpendicular to the plane of the figure, ACK the curved line, C the body moved in it and CF the straight line touching it at C. To the horizontal plane AK let fall the perpendiculars HE, CB, he, GD with the stipulation that the intervals EB, BD be equal and the arcs $\overset{\frown}{HC}$, $\overset{\frown}{Ch}$ be described by the moving body C in equal times. Draw Hf, CiF touching the curve described at the points H and C, and meeting the perpendiculars BC, eh, DG produced in f, i, F. And complete the parallelograms $BCID$, $Fihn$. Were the body to suffer no resistance, the times in which it passes between the perpendiculars EH and BC, and BC and DG, standing at equal distances from one another, would be equal and the body, descending in these equal times by the force of gravity from tangents, would describe equal heights fC and Fn, and would consequently at the end of the times be found at n. Through the resistance it comes that the body is at the end of the times found at the place h. Consequently the linelets nh and Fn or ih are, by the force of the resistance and that of gravity respec-

----

(13) This square of $CH = \frac{1}{2}(n^2/e^3) o^2 - \ldots$ should (with terms in higher powers of o ignored) read $\frac{'n^4 o^4'}{4e^6}$! This numerical slip neatly—to Newton's immediate joy in the next line, and to his subsequent confusion in [2] where he catches it—balances out his earlier error in putting the exponent of the downwards gravity to be only half its just value (see note (9) preceding).

(14) We may imagine Newton's delight in attaining this correct result, and picture him dashing straight on to compose (in [2]) the more rounded out and elaborate version of his present argument, only too quickly to detect the numerical mistake (see the previous note) which turned sour the sweetness of his illusory success.

(15) Add. 3965.12: 197^v/197^r. Newton proceeds to mould his seemingly triumphant recast approach in [1] preceding into a full-scale revision of the 1687 text—to emerge crestfallen at the end when a corrected calculation of the application of his new-found ratio of resistance to gravity in the semi-circular path of his prime *Exemplum* yields a value twice as large as the true one. The draft again takes up the *editio princeps* at the top of page 261, starting in mid-word almost at the close of its opening sentence, which for clarity's sake we here make good by an editorial interpolation founded on Newton's revised beginning to the redraft in [6] below.

Fn vel *ih* vi gravitatis simul generantur,[16] estɋ resistentia ad gravitatem ut

hn[17] ad *hi* vel *Cf* adeoɋ resistentia est ut $\dfrac{hn}{Cf}$.[17]

Est autem tempus ut $\sqrt{Cf}$ et velocitas ut descripta longitudo *Ci* directe et

tempus inverse seu [ut] $\dfrac{Ci}{\sqrt{Cf}}$, et resistentia eiɋ proportionalis $\dfrac{hn}{Cf}$, ut Medij densitas

et quadratum velocitatis, adeoɋ Medij densitas ut resistentia directe &

quadratum velocitatis inverse, id est ut $\dfrac{hn}{Ci \times Ci}$.   Q.E.I.

Corol. 1. Est resistentia ad gravitatem ut $\frac{1}{2}nG \times Ci$ ad Cf^{quad}. Nam resistentia erat ad gravitatem ut *hn* ad $Cf_{[,]}$ hoc est ut $hn \times Cf$ ad Cf^{quad}. Et *Ci* est ad *hn* ut *Cf* vel *nF* ad $\frac{1}{2}nG$.[18]

Corol. 2. Et Medij densitas est ut $\dfrac{nG}{Ci \times Ci}$.

Corol. 3. Cùm *Ci* et *CF* ob infinite minorem *iF* sint ad invicem in ratione

æqualitatis, erit densitas Medij ut $\dfrac{nG}{2CF, Cf}$, et resist[entia] ad grav[itatem] ut

$\frac{1}{2}nG \times CF$ ad Cf^{q}, et velocitas ut $\dfrac{CF}{\sqrt{Cf}}$.

Et hinc si Curva [linea definiatur per relationem inter basem seu abscissam *AB* & ordinatim applicatam *BC* (ut moris est) & valor ordinatim applicatæ resolvatur in seriem convergentem: Problema per primos seriei terminos expedite solvetur; ut in Exemplis sequentibus].[19]

Exempl. 1. [Sit linea *ACK* . . . . . . & curvatura Curvarum.][20]

[*At nota*] *p. 264 l. 12. pro* exhibet *scribe* determinat.

Præterea cum *Ci* et *CF* ob infinite minorem *iF* s[i]nt ad invicem in ratione æqualitatis, erit *Ci* latus quadratum ex CI^{q} et $IF^{q}{}_{[,]}$ hoc est ex BD^{q} et quadrato termini secundi.[21] Et scribendo *BE* pro *BD* seu $-o$ pro $+o$ valor *DG* convertitur in valorem *EH*. Et in valore ipsius *IF* scribendo $a-o$ pro *a* et *EH* pro e[22] habebitur *Lf*. Et inde simul prodeunt *Cf* vel *Fn* et *nG*. Terminos autem in quibus *o* est plusquam trium dimensionum semper negligo ut infinite minores quam qui in hoc Problemate considerandi veniant. Itaɋ si designatur *DG* universaliter hac serie[23] $BC + Qo + Roo + So^{3}$, erit *IF* æqualis Qo, *CF* æqualis $\sqrt{oo + QQoo}$, *FG* æqualis $Roo + So^{3}$,

$$EH = e - Qo + Roo - So^{3} \quad \& \quad CL = -Qo + Ro^{2} - So^{3}.$$

---

(16) Newton falls into the same subtle error as before (see [1]: note (9) above). In fact, corresponding to the linelet *Fn* generated by the force of gravity, the force of resistance will engender a decrement of only $\frac{1}{2}Fi = \frac{1}{2}nh$.

(17) In consequence, for '*hn*' here read '$\frac{1}{2}hn$' *recte*.

(18) Newton again makes the assumption (compare [1]: note (11) preceding) that, to

tively,[16] simultaneously generated, and so the resistance is to gravity as hn[17] to hi, or Cf, and hence the resistance is as hn[17]$/Cf$.

The time, however, is as $\sqrt{Cf}$, and the speed as the described length Ci directly and the time inversely, or as $Ci/\sqrt{Cf}$, while the resistance—and hn/Cf proportional to it—is as the density of the medium and the square of the speed, and hence the density of the medium is as the resistance directly and the square of the speed inversely, that is, as hn/Ci^2. As was to be found.

Corollary 1. The resistance is to gravity as $\frac{1}{2}nG \times Ci$ to Cf^2. For the resistance was to gravity as hn to Cf, that is, as $hn \times Cf$ to Cf^2, while Ci is to hn as Cf, or nF, to $\frac{1}{2}nG$.[18]

Corollary 2. And the density of the medium is as nG/Ci^2.

Corollary 3. Since Ci and CF are, because of iF being indefinitely small, to each other in a ratio of equality, the density of the medium will be as $\frac{1}{2}nG/CF \times Cf$, and the resistance to gravity as $\frac{1}{2}nG \times CF$ to Cf^2, and the speed as $CF/\sqrt{Cf}$.

And hence if the curved line be defined by the relationship between its base or abscissa AB and ordinate BC (as is customary) and the value of the ordinate be resolved into a converging series, then the problem will promptly be solved by means of the first terms of the series; as in the following examples.

Example 1. Let the line ACK ...... and the curvature of curves.[20]

(But note: on p. 264, l. 12 in place of 'exhibits' read 'determines'.)

Moreover, since Ci and CF are, because of iF being indefinitely small, in a ratio of equality to each other, Ci will be the square root of $CI^2 + IF^2$, that is, of BD^2 and the square of the second term.[21] And by writing BE in place of BD or $-o$ for $+o$ the value of DG is converted into the value of EH. And by in the value of IF writing $a-o$ for a and EH for e[22] there will be had Lf. And thereby at one go there ensue Cf, or Fn, and nG. The terms, however, in which o is of more than three dimensions I ever neglect as infinitely less than the ones to come to be considered in the present problem. Accordingly, if DG be denoted universally by this series[23] $BC + Qo + Ro^2 + So^3$, there will be IF equal to Qo, CF equal to $\sqrt{(o^2 + Q^2o^2)}$, FG equal to $Ro^2 + So^3$, $EH = e - Qo + Ro^2 - So^3$ and so

---

sufficient accuracy, the infinitesimal arc $\overset{\frown}{Ch}$ of the curve $\overset{\frown}{ACK}$ of resisted motion coincides (along its length) with the corresponding parabolic arc $\overset{\frown}{CG}$ of unresisted free fall under gravity; whence $Ci^2:CF^2 = ih(\text{or } Fn):FG(\text{or } Fn+nG)$, so that $Ci:CF = Fn:(Fn+\frac{1}{2}nG)$, and therefore $Ci:(CF-Ci \text{ or}) hn = Fn:\frac{1}{2}nG$.

(19) We fill out the manuscript's jejune 'Et hinc si Curva &c' from the text of the *editio princeps*, as Newton intends, to furnish a proper lead-in to what follows.

(20) Understand the three opening paragraphs of this 'Exempl. 1' as they are printed in the *editio princeps* (*Principia*, $_1$1687: 263, l. 10 – 264, l. 15 [= pages 378–9 below]).

(21) Namely, in the series expansion of the incremented ordinate $DG = e_{a+o}$ in terms of powers of the base increment $BD = o$; whence $IF = Qo$.

(22) That is, BC.

(23) 'his terminis' (by these terms) was first, less adequately, written.

Et in valore ipsius *IF* scribendo *OE* pro *OB*[24] et *EH* pro *BC* habebitur *Lf* quæ subducta de *CL* relinquit *Cf* vel *Fn*. Et hæc subducta de *FG* relinquit *nG*.

Sic in problemate jam solvendo[25] erit *IF* seu $Qo = -\dfrac{ao}{e}$.

Ci [vel *CF*] seu $\sqrt{oo + QQoo} = \sqrt{oo + \dfrac{aaoo}{ee}} = \dfrac{no}{e}$. *FG* seu $Roo + So^3 = -\dfrac{nnoo}{2e^3} - \dfrac{anno^3}{2e^5}$.

[itaꝗ]
$$Lf = \frac{\overline{a - o} \times o}{e + \dfrac{ao}{e} - \dfrac{nnoo}{2e^3}} = \frac{ao}{e} - \frac{nnoo}{e^3} + \frac{3anno^3}{2e^5}. \quad [26]$$

[ut et] $Cf = CL - Lf = \dfrac{nnoo}{2e^3} - \dfrac{anno^3}{e^5} = [ih \text{ vel}] \; Fn.$ [adeoꝗ] $nG^{[27]} = \dfrac{3anno^3}{2e^5}$. [Unde

evadit] Resist. ad Grav. ut [$\frac{1}{2}nG \times Ci$ ad Cf^{quad} vel] $\dfrac{3anno^3}{4e^5} \times \dfrac{no}{e}$ ad $\dfrac{n^4o^4}{4e^6}$ seu $3a$ ad n.[28]

---

[3][29] [...Fingatur autem corpus in progressu impediri a Medio.][30] Temporibus æqualibus describat Corpus sine resistentia et gravitate spatia æqualia *AF*, *FG*,[31] per gravitatem solam sine resistentia arcus Parabolicos *AH*, *HI*, per gravitatem et resistentiam arcus *AD*, *DE*, et *FH* vel *BD*[32] erit spatium quod corpus vi gravitatis cadendo describit prima temporis parte & *GI* vel *CE*[32] erit spatium quod corpus vi gravitatis cadendo describit toto tempore[,] et ob duplum tempus erit hoc spatium quadruplum spatij

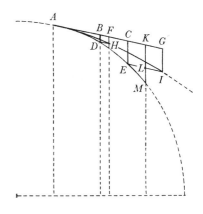

---

(24) That is, $a - o$ in place of a.

(25) In a cancelled initial presentation of the following calculation, doubtless abandoned by him as fulsome (and rightly so), Newton first continued at this point: 'si scribatur $-\dfrac{a}{e}$ pro Q,

$o\sqrt{1 + \dfrac{aa}{ee}}$ seu $\dfrac{no}{e}$ pro $o\sqrt{1 + QQ}$ [seu] *Ci* vel *CF*, $-\dfrac{nn}{2e^3}$ pro R, $-\dfrac{ann}{2e^5}$ pro S' (if there be written $-a/e$ for Q, $o\sqrt{(1 + a^2/e^2)}$ or $(n/e)\,o$ for $o\sqrt{(1 + Q^2)}$, that is, *Ci* or *CF*, $-\frac{1}{2}n^2/e^3$ for R, and $-\frac{1}{2}an^2/e^5$ for S).

(26) That is, $oQ_{a-o} = Qo - 2Ro^2 + 3So^3 - ...$ in the general case. Newton's division which derives this in the present instance, where $Q \equiv Q_a = -(a/e)\,o$, is separately preserved on Add. 3968.41: 145^v.

(27) Namely, $FG - Fn = 3So^3...$ in the general case.

(28) Which—with the square of $Cf = \frac{1}{2}(n^2/e^3)\,o^2...$ now correctly calculated (compare [1]: note (13))—thoroughly deflates the 'victory' which by this variant argument he had thought to have won over a problem which yet remained stubborn in its own in-built resistance to his mathematical attack! Thoroughly dismayed as he here must have been, but undeterred, Newton now passes, in [3] ensuing, to see if there is fruit in making distinction between the parabolic path $\widehat{CG}$ of free fall and the corresponding arc $\widehat{Ch}$ of the curve traversed in resisted

$CL = -Qo + Ro^2 - So^3$. And in the value of IF by writing OE in place of OB[24] and EH in place of BC there will be had Lf, which when taken from CL leaves Cf, or nF. And this when taken from GF leaves nG.

Thus in the problem now to be solved[25] there will be

$$IF \text{ (that is, } Qo) = -(a/e)\, o,$$

$$Ci \approx CF \text{ (or } \sqrt{(o^2 + Q^2 o^2)}) = \sqrt{(o^2 + (a^2/e^2)\, o^2)} = (n/e)\, o,$$

and

$$FG \text{ (or } Ro^2 + So^3) = -\tfrac{1}{2}(n^2/e^3)\, o^2 - \tfrac{1}{2}(an^2/e^5)\, o^3;$$

accordingly

$$fL = (a-o)\, o/(e + (a/e)\, o - \tfrac{1}{2}(n^2/e^3)\, o^2) = (a/e)\, o - (n^2/e^3)\, o^2 + \tfrac{3}{2}(an^2/e^5)\, o^3,\text{[26]}$$

while also $Cf = CL - fL = \tfrac{1}{2}(n^2/e^3)\, o^2 - (an^2/e^5)\, o^3 = hi$, or nF, so that

$$nG\text{[27]} = \tfrac{3}{2}(an^2/e^5)\, o^3.$$

Whence there results the resistance to gravity as $\tfrac{1}{2}nG \times Ci$ to Cf^2 or

$$\tfrac{3}{4}(an^2/e^5)\, o^3 \times (n/e)\, o \text{ to } \tfrac{1}{4}(n^4/e^6)\, o^4,$$

that is, $3a$ to n.[28]

---

[3][29] ...Imagine, however, that the body is in its progress impeded by the medium.[30] In equal times let it without the resistance and gravity describe the equal distances AF, FG;[31] through gravity alone without resistance the parabolic arcs $\overset{\frown}{AH}$, $\overset{\frown}{HI}$; through gravity and the resistance the arcs $\overset{\frown}{AD}$, $\overset{\frown}{DE}$; and FH, or BD,[32] will be the distance which the body in falling by the force of gravity describes in the first part of the time, and GI, or CE,[32] will be the distance which the body in falling by the force of gravity describes in the total

---

motion. He is yet blind to the fact that this untoward doubling of the true ratio of resistance to gravity in the semi-circular instance arises from his failure properly to relate the increments generated by the two in equal infinitesimal times (on which see note (16) above).

(29) Add. 3965.12: 196^v. Newton attempts a more radical recasting of [2] on the facing manuscript page, but now commits two independent errors (see notes (32) and (34) following) which only partially cancel each other out in Newton's ensuing measure of the ratio of resistance to gravity (on which see note (36) below).

(30) The manuscript text begins at the third sentence of the opening paragraph of the first edition (*Principia*, $_1$1687: 261, l. 7). We insert in prelude to it a light reshaping of the preceding sentence (*ibid.*: ll. 1–7) which better bridges the gap to what goes before (and is understood by Newton here to stand unaltered).

(31) Not a little confusingly, Newton now drastically changes his previous denotations of points in his accompanying figure (which we have here extended in broken line to show more closely its correspondence with his previous ones): F continues to mark its former point, but A is what was before C (and the points D and H, correspondingly, are what Newton denoted by h and n respectively in his figure in [2] preceding), while the configuration $CEGIKLM$ is here, of course, wholly new.

(32) The equalities $BD = FH$ and $CE = GI$ here silently assumed by Newton hold true, in fact, only to the order of the square of AB. He will come to grief below (see note (36) following) in thinking that these magnitudes may be substituted indiscriminately one for the other in evaluating the difference $LM = KM - KL$, of $O(AB^3)$.

prioris *FH* (per Lem. [XI] Libr. 1).[33] Compleantur parallelogramma *BFDH* & *CGIE* et erit *HD* lineola per vim resistentiæ genita in prima parte temporis et *IE* lineola per resistentiam genita in toto tempore. Et hæc lineola (per Lem. [XI] Libr. 1)[33] est quadrupla prioris. Et resistentia erit ad gravitatem ut *DH* ad *BD*[34] vel *EI* ad *EC*.[34] Cape *BK* æqualem *AB* et *CK* erit dupla ipsius *BF* vel *DH* ideoφ resistentia erit ad gravitatem ut $\frac{1}{2}EL$ ad *BD*, id est ut $\frac{AB-BC}{2}$ ad *BD*.[35]

Est autem tempus ut $\sqrt{BD}$ et velocitas ut descripta longitudo *AD* vel *AB* directe et tempus $\sqrt{BD}$ inverse seu [ut] $\frac{AB}{\sqrt{BD}}$, et resistentia eiφ proportionalis $\frac{EL}{2BD}$ ut Medij densitas et quadratum velocitatis, adeoφ Medij densitas ut resistentia directe et quadratum velocitatis inverse[,] id est ut $\frac{EL}{2AB^q}$.

Corol. 1. Resistentia est ad gravitatem ut $LM \times AB$ ad $8BD^q$. Nam resistentia erat ad Gravitatem ut $\frac{1}{2}EL$ ad *BD* hoc est ut $EL \times BD$ ad $2BD^q$. et $\frac{1}{4}LM$ est ad [$\frac{1}{4}EC$ seu] $\frac{1}{4}KL=BD$ ut *EL* ad *BC*, [itaφ] *LM* est ad *EL* ut *EC* seu *4BD* ad *BC* vel *AB*.[36]

Corol. 2. Medij densitas est ut $\frac{LM}{8AB \times BD}$.

Corol. 3. Et hinc si curva &c.[37]

Exempl. 1. Sit Linea ... ... *a* ad *e*.[38] Termini sequentes $\frac{nnoo}{2e^3} + \frac{anno^3}{2e^5}$ &c

---

(33) This Lemma (see VI: 116) lays down generally that the vanishingly small linear increment generated by any force is proportional to the square of the infinitesimal time in which it acts; whence in twice the time a length four times as long is generated.

(34) Read '2*BD*' and '2*EC*' on making proper comparison of the contemporaneous increments due respectively to the resistance and to gravity; compare [1]: note (9).

(35) Whence this ratio should be $\frac{1}{2}(AB-BC):2BD$. The slip is carried through into the sequel (see the next note). In the analytical equivalent (compare [1]: note (11) preceding) there is here, *mutatis mutandis*, $AB = o\sqrt{(1+Q^2)}$ and $BD = Ro^2 + So^3 + \ldots$, while also $AB:AF:AG = o:p:2p$, so that $AB:BF:CG = o:(p-o):4(p-o)$ and consequently

$$AC = AB.(2p-4(p-o))/o = AB.(2-2(p-o)/o);$$

accordingly, since also $BD = Rp^2 - 2Sp^3 \ldots$ (compare Appendix 2.1: note (13) below) and hence $p^2 = o^2 + 3(S/R) o^3 \ldots$ or $p = o + \frac{3}{2}(S/R) o^2 \ldots$, the corrected measure

$$\frac{1}{4}(AB-BC)/BD = \frac{1}{4}(2AB-AC)/BD$$

yields $\frac{1}{4}(2(p-o)\ldots)\sqrt{(1+Q^2)}/(Ro^2+\ldots) = \frac{3}{4}S\sqrt{(1+Q^2)}/R^2$ for the true ratio of resistance to gravity. Newton himself, unfortunately, now proceeds to introduce a second error in his ensuing computation.

(36) For, because the resisted path $\overset{\frown}{ADM}$ is effectively a parabola (to sufficient accuracy here) and also $AK = 2AB$ (by construction), the tangent at *M*—that is, since $CK = 2BF$ is infinitesimal in comparison with *AK*, the extension of the chord *ME*—will pass through *B*, and therefore $BC:CE = EL:LM$. The assumption, however, that *CE* is equal to *4BD*, that is,

$$4(Ro^2 + So^3 \ldots),$$

time, and, because the time is double, the distance will (by Lemma XI of Book 1)[33] be four times the previous one. Complete the parallelograms $BFHD$ and $CGIE$, and then HD will be the linelet born through the force of resistance in the first part of the time, and IE the linelet begotten through the resistance in the total time; and this latter linelet is (by Lemma XI of Book 1)[33] four times the previous one. And the resistance will be to gravity as DH to BD,[34] or EI to EC.[34] Take BK equal to AB, and CK will be twice BF or DH; and consequently the resistance will be to gravity as $\frac{1}{2}EL$ to BD, that is, as $\frac{1}{2}(AB-BC)$ to BD.[35]

The time, however, is as $\sqrt{BD}$, and the speed as the described length AD or AB directly and the time $\sqrt{BD}$ inversely, or as $AB/\sqrt{BD}$, while the resistance— and $\frac{1}{2}EL/BD$ proportional to it—is as the density of the medium and the square of the speed, and hence the density of the medium as the resistance directly and the square of the speed inversely, that is, as $\frac{1}{2}EL/AB^2$.

Corollary 1. The resistance is to gravity as $LM \times AB$ to $8BD^2$. For the resistance was to gravity as $\frac{1}{2}EL$ to BD, that is, as $EL \times BD$ to $2BD^2$, and $\frac{1}{4}LM$ is to ($\frac{1}{4}EC$ or) $\frac{1}{4}KL = BD$ as EL to BC, so that LM is to EL as EC, that is, $4BD$, to BC or AB.[36]

Corollary 2. The density of the medium is as $\frac{1}{8}LM/AB^2$.

Corollary 3. And hence if the curve....[37]

Example 1. Let the line (ACK) be ... ... a to e.[38] The following terms

whereby he computes LM—the excess of $KM = R(2o)^2+S(2o)^3... = 4Ro^2+8So^3...$ over CE—to be $4So^3...$, is mistaken; accurately, CE is

$$R(2o-2(p-o))^2+S(2o-2(p-o))^3... = 4Ro^2+8So^3-8Ro(p-o)...,$$

whence Newton's equation of this to $4Ro^2+4So^3...$ implies $p = o+\frac{1}{2}(S/R) o^2...$ whereas *recte* (see the previous note) $p = o+\frac{3}{2}(S/R) o^2...$, thus producing $CE = 4Ro^2-4So^3...$ and in consequence $LM(= KM-CE) = 12So^3...$, three times the length ensuing by Newton's computation. Allowing, further, for his improvident earlier doubling of the fraction $\frac{1}{4}(AB-BC)/BD$ which correctly expresses the ratio of resistance to gravity (see the two previous notes), we will not be surprised that his resulting evaluation of $\frac{1}{8}LM \times AB/BD^2$ as $\frac{1}{8}(4So^3...).o\sqrt{(1+Q^2)}/(Ro^2...)$, that is, $\frac{1}{2}S\sqrt{(1+Q^2)}/R^2$ in the limit as o vanishes, leads him back circuitously to the measure of his *editio princeps*, only two-thirds of the true one. And so he himself finds when he makes the calculation in the particular instance of the semicircular path in 'Exempl. 1' following.

(37) Understand that Corollary 3 of the original proposition (*Principia*, $_1$1687: 263) is to follow unchanged. At this point Newton recurs to the scheme of his *editio princeps*, henceforth referring to his present denotation of the points of his figure only in brackets until his last couple of lines, when he again confusingly understands only his above figure. (But we should not forget that the present piece is only a rough, unfinished draft never intended for publication.)

(38) Here are to go in the two opening paragraphs and first three sentences of the third of 'Exempl. 1' in the *editio princeps* of the *Principia* ($_1$1687: 263–4, l. 4; see pages 378–9 below).

In sequel Newton first went on to copy, as there, 'Terminus tertius qui hic est $\left[\dfrac{nnoo}{2e^3}\right]$' before

breaking off to make the following minimal rephrasing of the remainder of the paragraph.

designant lineolam *FG* (*BD*)[(39)] quæ jacet inter tangentem et Curvam adeoᴄꝫ determinat angulum contactus *FCG* seu curvaturam quam Curva linea habet in *C*. Si lineola illa *FG* minuatur in infinitum, termini subsequentes evadent infinite minores tertio ideoᴄꝫ negligi possunt. Terminus quartus qui hic est $\dfrac{anno^3}{2e^5}$ determinat variationem Curvaturæ quintus variationem variationis, & sic deinceps. Unde &c [...curvatura] Curvarum.

Præterea *CF* (*AB*)[(39)] est latus quadratum ex *CI^q* et *IF^q* hoc est ex *BD^q* et quadrato termini secundi[,] id est in hoc exemplo $\sqrt{oo + \dfrac{aaoo}{ee}}$ seu $\dfrac{no}{e}$. Et scribendo

$2o$ pro $2BD (= \quad {}^{(40)})$ prodit $KM = \dfrac{2nnoo}{e^3} + \dfrac{4anno^3}{e^5}$ [,] et ablato *KL* seu $4BD$ hoc est

$\dfrac{2nnoo}{e^3} + \dfrac{2anno^3}{e^5}$ [(41)] manet $\dfrac{2anno^3}{e^5} = LM$. Unde Medij densitas est ut $\dfrac{a}{2ne}$, et Resistentia ad Gravitatem ut *a* ad *n*.[(42)]

[4][(43)] [Sit *AK* planum illud plano Schematis perpendiculare, *LCK* linea curva, & *C* corpus in ipsa motum. Fingatur autem corpus in progressu impediri a Medio.][(44)] Æqualibus temporis momentis[(45)] describat corpus spatia *CG*, *Gg* sitᴄꝫ arcus *Gh* æqualis *C*[*G*], et differentia *gh* erit decrementum spatij quod viribus gravitatis et resistentiæ conjunctis momento temporis generatur. Ducantur rectæ *CF Gf* curvam descriptam tangentes in punctis *C* et *G*.

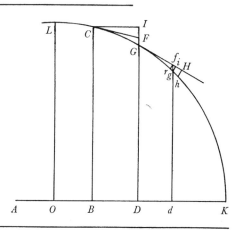

(39) The '*FG*' and '*CF*' relate to the figure in the 1687 text (for which see Appendix 1 below), while the '*BD*' and '*AB*' are their respective equivalents in the present redrawn and relettered diagram; compare note (31) preceding. For convenience we adjoin in our English version a scheme indicating the elements of the former which are here pertinent, augmented (in broken line) by the third ordinate which lies through the points *K*, *L* and *M* of the latter one.

(40) There is no correspondent to $2BD$—that is, *BN* in our 'English' figure (as we have interpolated the accompanying text there to specify)—in Newton's present scheme, and so he has had perforce here to leave a blank. Understand this to be, of course, the horizontal distance of *KLM* from the prime ordinate through *A* (that is, *BC* in the 1687 text).

(41) Read $[4Ro^2 - 4So^3 =] \; '\dfrac{2nnoo}{a^3} - \dfrac{2anno^3}{e^5}'$ *recte*, whence on taking this away from *KM* there ought in sequel to remain $[12So^3 =] \; '\dfrac{6anno^3}{e^5} = LM'$; compare note (36) preceding.

(42) And so Newton is once more led back to the 'impossible' result of his 1687 text in this particular instance of resisted motion in a semi-circle—obtained, it is true, this time *via* two contrary errors (see notes (35) and (36) above) which do not together wholly compensate for each other's fault. Undeterred by this fresh failure, he presses on to try yet a further avenue

$\frac{1}{2}(n^2/e^3)\,o^2 + \frac{1}{2}(an^2/e^5)\,o^3$ designate the linelet $FG\,(BD)$[39] which lies between the tangent and the curve, and hence determines the angle of contact $\widehat{FCG}$, and thus the curvature which the curved line has at C. If that linelet FG now be diminished indefinitely, the subsequent terms will prove to be infinitely less than the third, and can accordingly be neglected. The fourth term—here $\frac{1}{2}(an^2/e^5)\,o^3$—determines the variation of the curvature, the fifth the variation of the variation, and so on. Whence... the curvature of curves.

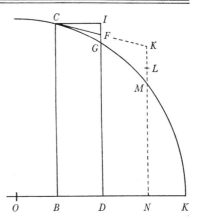

Furthermore, $CF\,(AB)$[39] is the square root of $CI^2 + IF^2$, that is, of BD^2 and the square of the second term—in the present example, namely, $\sqrt{(o^2 + (a^2/e^2)\,o^2)}$ or $(n/e)\,o$. And upon writing $2o$ in place of $2BD([BN])$[40] there results

$$KM = 2(n^2/e^3)\,o^2 + 4(an^2/e^5)\,o^3,$$

and when KL (or $4BD$), that is, $2(n^2/e^3)\,o^2 + 2(an^2/e^5)\,o^{3}$[41] is taken away there remains $LM = 2(an^2/e^5)\,o^3$. Whence the density of the medium is as $\frac{1}{2}a/ne$, and the resistance to gravity as a to n.[42]

---

[4][43]   Let AK be that plane perpendicular to the plane of the figure, LCK the curved line and C the body moved in it. Imagine, however, that the body is in its progress impeded by the medium.[44] Let the body in equal moments of time[45] describe the distances $\widehat{CG}$, $\widehat{Gg}$ and let the arc $\widehat{Gh}$ be equal to $\widehat{CG}$, and the difference $\widehat{gh}$ will be the decrement of the (arc-)distance which is generated by the forces of gravity and resistance jointly in a moment of time. Draw the straight lines CF, Gf touching the described curve at the points C and G. And

---

of approach in [4] and [5] following, one which will finally lead him to his goal in [6] after further adjustment.

(43) Add. 3965.12: 200$^\text{r}$. Newton now for the first time attempts to construct the equation of motion in the instantaneous tangential direction, conceiving the projectile's deceleration at any point of its path—as measured by the difference in length of its vanishingly small arcs described in equal infinitesimal times—to be compounded of the accelerative component in that direction of the force of downwards gravity—denoted by the equal distances fallen away from the tangent in those times—and of the combating action the opposite way of the medium's resistance to the missile's motion through it; whence the latter resistance is equal (but opposite in sense) to the sum of the body's acceleration and the component (acting the same way) of gravity.

(44) Newton again, as in [3] preceding (compare note (30) thereto), commences his recasting of the 1687 text *in medio* at the third sentence of its first paragraph (*Principia*: 261, l. 7); and yet once more we interpolate a lightly restyled version of the two opening ones to afford a less drastic entrance into what follows.

(45) Originally just 'Temporibus æqualibus' (in equal times), as in [3].

Et ad planum Horizonti parallelum AOK demittantur perpendicula CB, GD, gd et producantur DG, dg donec tangentibus illis occurrant in F et f. Et æqualia erunt spatia FG, fg quæ corpus vi gravitatis cadendo æqualibus illis temporis momentis describit. In arcu Cg capiatur Gr æqualis tangenti Gf, et erit rg[46] incrementum arcus vi gravitatis in momento temporis, ideoɋ rh[47] erit decrementum arcus ex resistentia Medij in eodem momento, et resistentia erit ad gravitatem ut rh ad rg.[48]

Pro arcuum differentia gh scribatur vel differentia chordām[49] CG, Gg vel differentia tangentium CF $[G]f$ et sit differentia illa D et pro rg scribatur $\dfrac{Dd \times gf}{GF}$ vel $\dfrac{BD \times GF}{CF} = d$, et erit resistentia ad gravitatem ut $D+d$ ad d.[50]

Est autem tempus ut $\sqrt{GF}$ et velocitas ut descripta longitudo CG vel CF directe et tempus $\sqrt{GF}$ inverse, et resistentia ut Medij densitas et quadratum velocitatis, adeoɋ Medij densitas ut resistentia directe et quadratum velocitatis inverse, id est ut $\dfrac{D+d}{d}$ directe & GF[51] inverse, sive ut $\dfrac{D+d}{d \times GF}$.   Q.E.[I].

    ...   .[52]

Exempl. 1. Sit Linea ACK semicirculus ... ... et evadet

$$DG = e - \frac{ao}{e} - \frac{nnoo}{2e^3} - \frac{anno^3}{2e^5} \ \&c \ldots \ldots \text{curvatura Curvarum.}[53]$$

Sit jam $OD = b$. $DG = f$ et $Dd = p$ et erit $dg = f - \dfrac{bp}{f} - \dfrac{nnpp}{2f^3} - \dfrac{bnnp^3}{2f^5}$. et

$fg = \dfrac{nnpp}{2f^3} + \dfrac{bnnp^3}{2f^5}$.   Est autem $CF = \sqrt{oo + \dfrac{aaoo}{ee}} = \dfrac{no}{e}$ et $Gf = \sqrt{pp + \dfrac{bbpp}{ff}} = \dfrac{np}{f}$.

[Itaɋ] $FG = \dfrac{nnpp}{2f^3} + \dfrac{bnnp^3}{2f^5} = fg = \dfrac{nnoo}{2e^3} + \dfrac{anno^3}{2e^5}$.   et $\dfrac{ffpp + bp^3}{f^5} = \dfrac{eeoo + ao^3}{e^5}$.

---

(46) This should be '$2rg$' *recte*, when proper comparison of the decrement

$$\overset{\frown}{CG}(\text{or } FH) - \overset{\frown}{Gg} = gh$$

is made with $(\rho - g.rg/fg)\,\theta^2$, where ρ is the medium's resistance and $g\theta^2 = 2fg$ is twice the vertical distance fallen from the tangent GH under downwards gravity g in the infinitesimal time θ in which the missile traverses the arc $\overset{\frown}{Gg}$ of its resisted path; compare [1]: note (9) above, and Appendix 2.2: note (25) following.

(47) In the terms of the previous note, this should be $(\rho\theta^2 =)\ 2rg + gh$, that is, '$rg + rh$'.

(48) Newton's slip for 'fg' of course. Over and above this *lapsus calami*—which is, unfortunately, perpetuated in the sequel—the correct ratio should (see the two previous notes) be $(\tfrac{1}{2}\rho\theta^2 : \tfrac{1}{2}g\theta^2 =)\ \tfrac{1}{2}(rg + rh) : fg$.

(49) 'chordarum'.

(50) Again read 'fg'; see note (48).

(51) Further confusion! Newton means $\dfrac{{}^{\prime}CF^q{}^{\prime}}{GF}$ (CF^2/GF).

to the plane *AOK* parallel to the horizon let fall the perpendiculars *CB, GD, gd* and extend *DG, dg* till they meet those tangents in *F* and *f*. Then will there be equal the distances *FG, fg* which the body will on falling by the force of gravity in those equal moments of time describe. In the arc $\widehat{Cg}$ take $\widehat{Gr}$ equal to the tangent *Gf*, and then will *rg*[46] be the increment of the arc by the force of gravity in a moment of time, and consequently *rh*[47] will be the decrement of the arc ensuing from the resistance of the medium in the same moment, and so the resistance will be to gravity as *rh* to *rg*.[48]

For the difference *gh* of the arcs write either the difference of the chords *CG* and *Gg* or the difference of the tangents *CF* and *Gf*, and let that difference be *D*; while in place of *rg* write $Dd \times gf/GF$ or $BD \times GF/CF$, which let equal *d*: the resistance will then be to gravity as $D+d$ to *d*.[50]

The time, however, is as $\sqrt{GF}$, and the speed as the described length *CG* or *CF* directly and the time $\sqrt{GF}$ inversely, and the resistance as the density of the medium and the square of the speed, and hence the density of the medium as the resistance directly and the square of the speed inversely, that is, as $(D+d)/d$ directly and GF[51] inversely, or as $(D+d)/d \times GF$. As was to be found.

. . .   .[52]

Example 1. Let the line *ACK* be a semicircle . . . . . . and there will prove to be $DG = e - (a/e)\,o - \frac{1}{2}(n^2/e^3)\,o^2 - \frac{1}{2}(an^2/e^5)\,o^3\ldots,$ . . . . the curvature of curves.[53]
Now let $OD = b$, $DG = f$ and $Dd = p$, and there will be

$$dg = f - (b/f)\,p - \tfrac{1}{2}(n^2/f^3)\,p^2 - \tfrac{1}{2}(bn^2/f^5)\,p^3$$

and so $fg = \tfrac{1}{2}(n^2/f^3)\,p^2 + \tfrac{1}{2}(bn^2/f^5)\,p^3$. However,

$$CF = \sqrt{(o^2 + (a^2/e^2)\,o^2)} = (n/e)\,o \quad \text{and} \quad Gf = \sqrt{(p^2 + (b^2/f^2)\,p^2)} = (n/f)\,p.$$

Accordingly

$$FG = \tfrac{1}{2}(n^2/f^3)\,p^2 + \tfrac{1}{2}(bn^2/f^5)\,p^3 = fg = \tfrac{1}{2}(n^2/e^3)\,o^2 + \tfrac{1}{2}(an^2/e^5)\,o^3,$$

---

(52) We here omit a following paragraph in the manuscript where Newton began, out of sequence, to redraft the latter portion of the third paragraph of the ensuing 'Exempl. 1' (*viz. Principia*, $_1$1687: 264, ll. 4–11), beginning 'Terminus tertius & sequentes $\dfrac{nnoo}{2e^3} + \dfrac{anno^3}{2e^5} + \&c$ designabunt lineolam *FG* quæ jacet inter tangentem et curvam, adeoq determ[i]nant angulum contactus *FCG* seu curvaturam quam curva linea habet in *C*. Si lineola illa *FG* minuatur in infinitum, termini subsequentes evadent infinite minores tertio ideoq negligi possunt, & solus terminus tertius curvaturam determinabit. Terminus quartus [&c]'. This minimal rephrasing was taken over word for word by Newton into his finished revise (§2.1 below), and we will not here dwell upon it.

(53) Understand the first three paragraphs of the original text of this prime instance (*Principia*, $_1$1687: 263–4, l. 15 [= pages 378–9 following]), with the trivial replacement of 'fiet' (*ibid.*: 263, l. 16) by an equivalent 'evadet'.

[At] $b = a + o. f = e - \dfrac{ao}{e}$. [Fit igitur] $\dfrac{eepp - 2aopp + ap^3}{e^5 - 5ae^3o} = \dfrac{eeoo + ao^3}{e^5}$. [sive]

$$eepp - 2aopp + ap^3 = eeoo - 4ao^3.$$

[adeoœ]

$$oo . pp :: ee - 2ao + ap . ee - 4ao :: ee - ao . ee - 4ao :: ee . ee - 3ao$$

[vel] $BC . BC - 3IG$.[54]

[5][55]   Curvam *LNCGgK* tangant rectæ *NE*, *CF* in punctis *N* et *C* et quæratur Medij densitas qua corpus in hac curva moveri potest. Sit *NE* tangens quam corpus in momento temporis describit et *EC* altitudo quam describeret cadendo in eodem momento, et in fine momenti corpus reperietur in *C*. Sit *CF* tangens quam corpus describeret in momento proximo et *FG* altitudo quam describeret in eodem momento cadendo et in fine momenti corpus reperietur in *G*.[56] Et differentia arcuum *NCCG* vel quod perinde est, differentia tangentium *NE*, *CF* erit decrementum spatij quod corpus singulis momentis describit.[57]

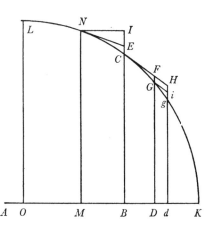

(54) Whence at once '$o . p :: ee . ee - \frac{3}{2}ao$' and consequently '$p = o - \dfrac{3a}{2ee}oo$' (*recte*, on ignoring terms of the order of o^3), so that $|gh| = (Gf - CF$ or) $np/f - no/e$, that is,

$$no(1 - \tfrac{3}{2}(a/e^2)o)/e(1 - (a/e^2)o) - no/e = \tfrac{1}{2}(an/e^3)o^2$$

is (to the same order) equal to $rg = (a/n)fg$. Whether or not he foresaw that setting the resistance to be to gravity as rh/rg here, in this semi-circular instance, yields the impossible result that the resistance must everywhere be twice the gravity, and so itself constant (and hence countering a uniform speed of the missile's motion), Newton at this point broke off to rephrase his argument anew on the next manuscript leaf (f. 201), which we now pass to reproduce in [5]. Even had he mended his careless slip of writing 'rg' for 'fg' in his basic ratio (see note (48) above) he would, we might add, have attained the unacceptable value $gh/fg = 2a/n$ in the present example.

In the general case, if (with Newton) we expand the incremented ordinate $DG = e_{a+o}$ into the series $e + Qo + Ro^2 + So^3 + \dots$ (where, as ever, $Q \equiv Q_a = de/da$, $R \equiv R_a = \frac{1}{2}dQ/da$ and $S \equiv S_a = \frac{1}{3}dR/da$), so that $CF = o\sqrt{(1+Q^2)}$ and $FG = Ro^2 + So^3 \dots$, then correspondingly when the augmented base $OD = a + o = b$ receives the further increment $Dd = p$ there is $Gf = p\sqrt{(1 + Q_b^2)}$, that is, $p\sqrt{(1 + (Q + 2Ro + \dots)^2)} = p\sqrt{(1+Q^2)} . (1 + 2QRo/(1+Q^2)\dots)$ and also $fg = R_b p^2 + S_b p^3 + \dots = (R + 3So + \dots)p^2 + (S + \dots)p^3$; whence, on equating the distances FG and fg fallen away from the initial tangential direction in equal times, there comes to be $p^2 + (S/R)p^3 \dots = (Ro^2 + So^3 \dots)/(R + 3So \dots) = o^2 - 2(S/R)o^3 \dots$ or $p^2 = o^2 - 3(S/R)o^3 \dots$, and consequently $p = o - \frac{3}{2}(S/R)o^2 \dots$. (Compare [3]: note (36), where the present successive base increments p and o are interchanged.) It follows that

$$Gf = o\sqrt{(1+Q^2)} + (-\tfrac{3}{2}S\sqrt{(1+Q^2)}/R + 2QR/\sqrt{(1+Q^2)})o^2 \dots,$$

and so $(f^2p^2 + bp^3)/f^5 = (e^2o^2 + ao^3)/e^5$. But $b = a - o$ and $f = e - (a/e)o$. There comes therefore $(e^2p^2 - 2aop^2 + ap^3)/(e^5 - 5ae^3o) = (e^2o^2 + ao^3)/e^5$ or

$$e^2p^2 - 2aop^2 + ap^3 = e^2o^2 - 4ao^3,$$

and hence

$$o^2 : p^2 = e^2 - 2ao + ap : e^2 - 4ao = e^2 - ao : e^2 - 4ao = e^2 : e^2 - 3ao,$$

or $BC : (BC - 3IG)$.[54]

---

[5][55]   Let the straight lines NE, CF touch the curve $LNCGgK$ at the points N and C, and let there be sought the density of the medium whereby the body can move in this curve. Let NE be the tangent which the body describes in a moment of time, and EC the height which it would describe by falling in the same moment, and at the end of the moment the body will be found at C. Let CF be the tangent which the body would describe in the next moment, and FG the height which it would describe in the same moment by falling, and at the end of the moment the body will be found at G.[56] And the difference of the arcs $\overarc{NC}$ and $\overarc{CG}$, or, what is effectively the same, the difference of the tangents NE and CF, will be the decrement of the distance which the body describes in the separate moments.[57] Call this difference D and let E be the excess of arc $\overarc{NC}$

---

and hence $gh = CF - Gf = (\frac{3}{2}S\sqrt{(1+Q^2)}/R - 2QR/\sqrt{(1+Q^2)})o^2...$, while also

$$rg = (Dd/Gf) \times fg = (Q_b/\sqrt{(1+Q_b^2)}).FG = (QR/\sqrt{(1+Q^2)})\,o^2...;$$

accordingly, Newton's intended ratio $rh/fg = (rg+gh)/fg$ of resistance to gravity (again see note (48)) yields, in the limit as the incremented ordinates DG and dg come to coalesce with BC (that is, as o and p each vanish), the erroneous corresponding measure $\frac{3}{2}S\sqrt{(1+Q^2)}/R^2 - Q/\sqrt{(1+Q^2)}$. (The true expression $\frac{3}{4}S\sqrt{(1+Q^2)}/R^2$ results straightforwardly, we need not say, from the corrected ratio $\frac{1}{2}(rg+rh)/fg$.)

(55) Add. 3965.12:201^r, a minor reshaping of the argument in [4] preceding (on the second leaf of the same folded manuscript sheet). Newton impercipiently repeats both his previous slips—of carelessly expressing the action of gravity, in proportion to that of resistance, by its component in the direction of motion, and of failing to assign the proper numerical coefficients in the equation of forces acting in that direction (see notes (58) and (59) below)—before uselessly attempting to reduce to computationally more amenable form the mistaken ratio of resistance to gravity which he thereby derives.

(56) In a cancelled first continuation Newton initially went on to repeat his basic argument in [1] and [2] above before again taking up the modified approach of [4] preceding, there writing: 'Producatur tangens CF ad H ut sit CH æqualis tangenti NE, et tangentium differentia FH erit decrementum spatij singulis momentis descripti ex resistentia oriundum adeoq resistentia erit ad gravitatem ut FH ad FG' (Produce the tangent CF to H so that CH be equal to the tangent NE, and the difference FH of the tangents will be the decrement of the [arc-] distance described in the single instants which must arise from the resistance, and hence the resistance will be to gravity as FH to FG). *Recte*, of course, this ratio should be FH to $2FG$; compare [1]: note (9) above.

(57) In sequel Newton went on to specify 'Et hoc decrementum oritur a viribus gravitatis et resistentiæ conjunctis' (And this decrement arises from the forces of gravity and resistance conjoined) and began to adjoin 'Gravitas auget [et resistentia minuit...]' (Gravity increases [and resistance diminishes...]) before breaking off to cancel the whole.

Dicatur hæc differentia D et sit E excessus arcus NC supra tangentem NE, vel quod perinde est sit $E = \dfrac{IE \times CE}{[NE]}$. et erit $E^{(58)}$ incrementum spatij singulis momentis descripti, ex gravitate oriundum, D decrementum ejus ex resistentia et gravitate oriundum, et $D + E^{(58)}$ decrementum ejus ex resistentia sola oriundum, adeoq̃ resistentia erit ad gravitatem ut $D + E$ ad E.$^{(59)}$

Est autem tempus ut $\sqrt{EC}$ et velocitas ut descripta longitudo NC vel NE directe & tempus $\sqrt{EC}$ inverse seu [ut] $\dfrac{NE}{\sqrt{EC}}$, et resistentia ut Medij densitas et quadratum velocitatis adeoq̃ Medij densitas ut resistentia directe & quadratum velocitatis inverse$_{[,]}$ id est ut $\dfrac{D+E}{E} \times \dfrac{EC}{NE^q}$.$^{(60)}$ Q.E.I.

Corol. 1. Capiatur $Bd = BM$ et erigatur $\perp^{(61)}$ dgH Curvæ occurrens in g et tangenti ejus CF productæ in $[H]$ et compleatur pgrm$^{(61)}$ $GF[H]i$. et erit$^{(62)}$ $[H]i + \tfrac{1}{2}ig \cdot \tfrac{1}{2}ig :: [H]C \cdot [H]F$. Et $[H]i + \tfrac{1}{2}ig \cdot Hi :: C[H] \cdot CF$. seu$^{(63)}$

$$\frac{CE + [Hg]}{2} \cdot CE :: C[H] \cdot CF.^{(64)}$$

[6]$^{(65)}$ Sit AK planum illud plano schematis perpendiculare; ACK linea curva;$^{(66)}$ C corpus in ipsa motum & FC recta ips[a]m tangens in C. Fingatur autem Corpus C progredi ab A ad K per lineam illam ACK, et interea impediri a Medio resistente. Et a puncto C ad rectam AK demittatur perpendiculum CB et in recta illa

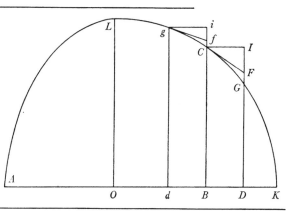

---

(58) This should be '$2E$' in each case; compare [1]: note (9) and [4]: note (46) preceding.

(59) Newton intends 'CE' here, and not its projection in the tangential direction at C, but repeats his careless slip in [4] preceding at the corresponding place (see our note (48) thereto). On further introducing the numerical coefficients here also omitted (see the previous note) the correct ratio of resistance to gravity is in fact as $D + 2E$ to $2CE$. And the sequel needs correspondingly to be adjusted.

(60) While Newton now attaches—as he did not in [4] preceding (see note (51) thereto)— the proper factor proportional to the reciprocal of the square of the speed

$$NE/\sqrt{(2CE/g)} \;\propto\; NE/\sqrt{(CE)}$$

where g is the downwards gravity, this measure of the density needs (see the previous note) yet further correction to be accurately '$\,\mathrm{ut}\; \dfrac{D+2E}{2CE} \times \dfrac{CE}{NE^q}\,$' that is, as $(D+2E)/NE^2$.

(61) Read 'perpendiculum' and 'parallelogrammum' respectively.

(62) Again presupposing (compare [1]: note (11); [2]: note (18); [3]: note (36)) that the resisted path $\widehat{CGg}$ is, to sufficient accuracy, a parabola; whence $Hg : (FG$ or$)$ $Hi = HC^2 : FC^2$, and therefore $(\sqrt{(Hg \times Hi)} \approx)$ $Hi + \tfrac{1}{2}ig : Hi = HC : FC$.

over the tangent NE—or, what is effectively the same, let $E = IE \times CE/NE$—and then $E^{(58)}$ will be the increment of the distance described in separate moments arising from gravity, D its decrement arising from the resistance and gravity, and $D+E^{(58)}$ its decrement arising from the resistance alone, and hence the resistance will be to gravity as $D+E$ to E.$^{(59)}$

The time, however, is as $\sqrt{EC}$, and the speed as the described length $\widehat{NC}$, or NE, directly and the time $\sqrt{EC}$ inversely, that is, as $NE/\sqrt{EC}$; while the resistance is as the density of the medium directly and the square of the speed inversely, and hence the density of the medium as the resistance directly and the square of the speed inversely, that is, as $((D+E)/E) \times EC/NE^2$.$^{(60)}$ As was to be found.

Corollary 1. Take $Bd = MB$ and erect the perpendicular dgH meeting the curve in g and its tangent CF produced in H, and complete the parallelogram $GFHi$. There will then $^{(62)}$ be $(Hi + \frac{1}{2}ig):\frac{1}{2}ig = CH:FH$, and so $(Hi + \frac{1}{2}ig):Hi = CH:CF$, that is,$^{(63)}$ $\frac{1}{2}(CE+Hg):CE = CH:CF$.$^{(64)}$

---

[6]$^{(65)}$ Let AK be that plane perpendicular to the plane of the figure, ACK the curved line,$^{(66)}$ C the body moved in it and FC the straight line touching it at C. Imagine, however, that the body C advances from A to K by way of that line ACK and is all the while impeded by the resisting medium. And from the point C to the straight line AK let fall the perpendicular CB, take in that straight line

---

(63) Because $Hi = (FG$ or) RC and $ig = Hg - Hi$, that is, $Hi + \frac{1}{2}ig = (Hi + Hg)$.

(64) At this point, evidently conscious that this adaptation of his previous elaboration in [3] above here leads him nowhere, Newton breaks off to begin the more radically reconstructed mode of approach by which he will, in [6] ensuing, at long last achieve his goal. Subsequently, ever thrifty in his economical re-use of what was then become mere scrap, he employed the final few inches of paper left blank at the foot of the page to draft a neat variant proof (cited in Appendix 2.8 note (70)) that the surrounding medium offers no resistance to motion in the simple 'Galileian' parabola which is the second example set by him to illustrate his problem.

(65) Add. 3965.12: 219^r(+220^r)/219^v. Deterred by his repeated failure to mend his 1687 argument by way of its basic assumption of arcs successively traversed in equal instants of time, Newton here in lieu supposes that it is the projection of the resisted motion upon the horizontal which uniformly increases in 'æqualibus temporis momentis'. After a slight initial fumble (see note (68) following) he was thereby able for the first time correctly to express the change in speed in the missile's motion due to the joint action of the resistance of the medium and of the component of gravity in the instantaneous tangential direction, and thence to deduce the true ratio of resistance to gravity which he here states, in a version straightaway simplified, in his Corollary 1. After confirming in the primary instance of a semi-circular resisted path—the growing roughness of the manuscript's handwriting and (see note (75) below) the increasing ellipticity in its verbal exposition jointly afford a revealing glimpse of his desperate effort here to be in agreement—that his general ratio does indeed in this special case yield the result independently obtained by Johann Bernoulli and passed to him by Niklaus as indication that the argument of his 1687 text was at fault, Newton is at last free to pen (with a near-audible sigh of relief) the quick letter to Johann's nephew whose draft we cite in our concluding note (76), enclosing with it a copy of his present recast scheme of argument.

(66) As in the *editio princeps* the accompanying figure is meant to depict the semi-circle of 'Exempl. 1', for which it also does duty, but we faithfully reproduce the proportions of the

sumantur hinc inde lineolæ æquales *BD, Bd* & erigantur perpendicula *DG, dg* Curvæ occurrentia in *G* ac *g*. Producatur *DG* donec tangenti *CF* occurrat in *F* & compleatur parallelogrammum *CBDI* & agatur recta *gf* Curvam tangens in *g* & perpendiculo *BC* producto occurrens in *f*. Et tempora quibus corpus describit arcus *gC, CG* erunt in subduplicata ratione altitudinum *fC, FG* quas corpus temporibus illis cadendo a tangentibus describere posset, & velocitates ut longitudines descriptæ *gC, CG* directe et tempora inverse.

Quare exponantur tempora per $\sqrt{Cf}$ & $\sqrt{FG}$[67] & velocitates per $\dfrac{gC}{\sqrt{Cf}}$ & $\dfrac{CG}{\sqrt{FG}}$, vel quod perinde est per $\dfrac{gf}{\sqrt{Cf}}$ & $\dfrac{CF}{\sqrt{FG}}$; & decrementum velocitatis tempore $\sqrt{FG}$ exponetur per $\dfrac{gf}{\sqrt{Cf}} - \dfrac{CF}{\sqrt{FG}}$;[68] hoc decrementum oritur ex resistentia et gravitate conjunctis. Resistentia decrementum auget[,] gravitas diminuit. Est autem $\dfrac{2FG}{\sqrt{FG}}$ seu $2\sqrt{FG}$[69] velocitas quam corpus tempore $\sqrt{FG}$ cadendo & altitudinem *FG* describendo acquireret. Et hæc velocitas est ad velocitatem quam gravitas eodem tempore addit velocitati corporis in arcu *CG* ut *FG* ad *CG−CF* sive ad $\dfrac{IF \times FG}{CF}$, ideoq velocitas quam gravitas addit velocitati corporis est $\dfrac{2IF \times FG}{CF \times \sqrt{FG}}$ seu $\dfrac{2IF \times \sqrt{FG}}{CF}$. Et hæc velocitas addita prædicto velocitatis decremento ex resistentia et gravitate oriundo componit decrementum velocitatis ex resistentia

---

free-hand drawn manuscript diagram, in which the left-hand quadrant *ALO* has been squashed affinely into the elliptical shape depicted. Initially, it would appear, Newton wrote '*LK* linea curva' (*LK* the curved line) and, correspondingly, drew just the right-hand quadrant; then amended this to be as it is here set, but without finding space enough to the right of what he had already written to afford the augmented figure its full horizontal extension.

    (67) Since the times of free fall from rest under gravity *g* through the distances *Cf* and *FG* are in fact $\sqrt{(2Cf/g)}$ and $\sqrt{(2FG/g)}$ respectively, Newton here supposes a scaling factor of $\sqrt{(\frac{1}{2}g)}$.

    (68) Much as in his preliminary listing on f. 197ʳ of the main lines of the present mode of argument (there keyed to a differently lettered figure whose variant markings of points we see no use in here specifying), Newton initially here (on f. 219ʳ) went on to conclude: 'et incrementum velocitatis ex gravitate sola'—acting in the instantaneous direction of the body's motion, that is—'exponetur per $\dfrac{CG-CF}{\sqrt{FG}}$, vel quod perinde est, per $\dfrac{FI\sqrt{FG}}{CF}$, et decrementum velocitatis ex resistentia sola per $\dfrac{gf}{\sqrt{Cf}} - \dfrac{CF}{\sqrt{FG}} + \dfrac{FI\sqrt{FG}}{CF}$, et velocitas cadendo genita per $\sqrt{GF}$' (and the increment in speed from gravity alone will be represented by $(\widehat{CG-CF})/FG$, or, what is exactly the same, by $(FI/CF)\sqrt{FG}$, and so the decrement in speed from resistance alone by $gf/\sqrt{Cf}-CF/\sqrt{FG}+(FI/CF)\sqrt{FG}$, and the speed engendered in falling by $\sqrt{FG}$). The resulting expression $gf/\sqrt{(Cf \times FG)}-CF/FG+FI/CF$ for the ratio of resistance to gravity is of course

on either hand equal linelets DB, Bd, and erect perpendiculars DG, dg meeting the curve in G and g. Extend DG till it meets the tangent CF in F, complete the rectangle $CBDI$, and draw the straight line gf touching the curve at g and meeting the perpendicular BC produced in f. Then the times in which the body describes the arcs $\widehat{gC}$, $\widehat{CG}$ will be in the halved ratio of the heights fC, FG which the body were by falling from the tangents able to describe, and the speeds as the lengths $\widehat{gC}$, $\widehat{CG}$ described directly and the times inversely. In consequence, represent the times by $\sqrt{Cf}$ and $\sqrt{FG}$,[67] and so the speeds by $\widehat{gC}/\sqrt{Cf}$ and $\widehat{CG}/\sqrt{FG}$ or, what is effectively the same, by $gf/\sqrt{Cf}$ and $CF/\sqrt{FG}$, and the decrement of the speed in time $\sqrt{FG}$ will be expressed by $gf/\sqrt{Cf} - CF/\sqrt{FG}$:[68] this decrement arises from the resistance and gravity jointly, the resistance increasing it and gravity diminishing it. Further, $2FG/\sqrt{FG}$ or $2\sqrt{FG}$[69] is the speed which the body would acquire by falling in time $\sqrt{FG}$ and covering the height FG; and this speed is to the speed which gravity adds in the same time to the speed of the body in the arc $\widehat{CG}$ as FG to $CG - CF$, that is, $IF \times FG/CF$, and consequently the speed which gravity adds to the speed of the body is

$$2IF \times FG/(CF \times \sqrt{FG}) \quad \text{or} \quad 2IF \times \sqrt{FG}/CF.$$

And so this speed, when added to the previously mentioned decrement of speed arising from the resistance and gravity, makes up the decrement of the speed

---

erroneous; we leave the reader to confirm that it again produces the faulty measure

$$\tfrac{3}{2}S\sqrt{(1+Q^2)}/R^2 - QR/\sqrt{(1+Q^2)}$$

which was deduced analogously in [4]: note (54) above, and which in the prime instance of the semi-circle $e = \sqrt{(n^2 - a^2)}$ correspondingly dictates the resistance there to be to gravity as '$2a$ ad n'—and so Newton himself observed, after a rapid checking computation, on his preliminary worksheet f. 197^r. At this point, on going back through the steps of his prior general reasoning he at last caught the mistake which had, in one form or another, bedevilled all his previous attempts in [1]–[5] to frame a correct argument, and straightaway inserted coefficients '2' at the pertinent places of his preceding sentence, changing it to read 'et incrementum velocitatis ex gravitate sola exponetur per $\dfrac{2CG - 2CF}{\sqrt{FG}}$, vel quod perinde est, per $\dfrac{2FI\sqrt{FG}}{CF}$, et decrementum velocitatis ex resistentia sola per $\dfrac{gf}{\sqrt{Cf}} - \dfrac{CF}{\sqrt{FG}} + \dfrac{2FI\sqrt{FG}}{CF}$, et velocitas cadendo genita per $2\sqrt{GF}$'; whence the ratio of resistance to gravity ensues *recte* to be as $\tfrac{1}{2}gf/\sqrt{(Cf \times FG)} - \tfrac{1}{2}CF/FG + FI/CF$ to unity. Then on his next manuscript leaf (f. 220^r) he straightaway refashioned this crucial sentence, augmenting it to make clear why the 'incrementum velocitatis ex gravitate' needs thus to be doubled (compare the next note), and this replacement we here print in sequel, following Newton's instruction in the original (by way of a referent '†') to do so.

(69) Since $2FG$ is the distance which the body would cover in the time $\sqrt{FG}$ were it to move uniformly with the speed which, in falling from rest at F under the downwards pull of gravity, it attains at the point G. In the terms of note (67) preceding, the same results on scaling down the terminal velocity $\sqrt{(2g.FG)}$ in the 'exponent' ratio $\sqrt{(\tfrac{1}{2}g)}$.

sola oriundum $\dfrac{gf}{\sqrt{Cf}}-\dfrac{CF}{\sqrt{FG}}+\dfrac{2IF\times\sqrt{FG}}{CF}$. Proindeqβ resistentia est ad vim gravi-

tatis ut $\dfrac{gf}{\sqrt{Cf}}-\dfrac{CF}{\sqrt{FG}}+\dfrac{2IF\sqrt{FG}}{CF}$ ad $2\sqrt{FG}$ sive ut $\dfrac{gf}{2\sqrt{Cf}\times\sqrt{FG}}-\dfrac{CF}{2FG}+\dfrac{IF}{CF}$ ad 1.[70]

Est autem resistentia ut Medij densitas et quadratum velocitatis conjunctim[,] & propterea densitas Medij ut resistentia directe & quadratum velocitatis

inverse, id est ut $\overline{\dfrac{gf}{2\sqrt{Cf}\times\sqrt{FG}}-\dfrac{CF}{2FG}+\dfrac{IF}{CF}}\times\dfrac{FG}{CG^q}$.   Q.E.I.

Corol. 1. Resistentia est ad gravitatem ut $\dfrac{gf}{Cf+FG}-\dfrac{CF}{2FG}+\dfrac{IF}{CF}$ ad 1. Nam pro

$2\sqrt{Cf\times FG}$[71] scribere licet $Cf+FG$.

Corol. 2. Si curva linea [definiatur per relationem inter basem seu abscissam] ... ut in Exemplis sequentibus.[72]

Exempl. 1. Sit linea *ACK* semicirculus super diametro *AK* descriptus, & requiratur Medij densitas quæ faciat ut projectile in hac linea moveatur.

Bisecetur semicirculi diameter *AK* in *O*, et dic *OK n*, *OB a*, *BC e* et *BD* vel *Bd o*; et erit DG^q seu

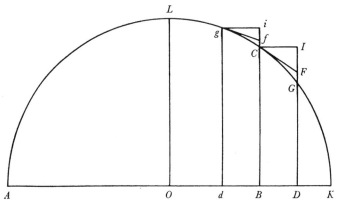

OG^q-OD^q æquale $nn-aa-2ao-oo$ seu $ee-2ao-oo$; et radice per methodum

nostram extracta, fiet $DG=e-\dfrac{ao}{e}-\dfrac{oo}{2e}-\dfrac{aaoo}{2e^3}-\dfrac{ao^3}{2e^3}-\dfrac{a^3o^3}{2e^5}-$ &c. Hic scribatur *nn*

pro $ee+aa$ et evadet $DG=e-\dfrac{ao}{e}-\dfrac{nnoo}{2e^3}-\dfrac{anno^3}{2e^5}-$ &c.

---

(70) *Et voilà!* In the analytical equivalent in which the general point $C(a, e)$ of the path $\widehat{LCK}$ is defined by some given relationship $e = e_a$ between the abscissa $OB = a$ and ordinate $BC = e$, on expanding the incremented ordinate $DG = e_{a+o}$ as the series $e + Qo + Ro^2 + So^3 + ...$ (where, as ever, $Q \equiv Q_a = de/da$, $R \equiv R_a = \frac{1}{2}dQ/da$, $S \equiv S_a = \frac{1}{3}dR/da$, ...) there is once more $IF = Qo$, $GF = Ro^2 + So^3 + ...$ and so $CF = o\sqrt{(1+Q^2)}$; while corresponding to the decremented ordinate $dg = e_{a-o}$ there is likewise

$$Cf = R_{a-o}o^2 + S_{a-o}o^3 + ... \text{ or } (R-3So...)\, o^2 + (S-...)\, o^3... = Ro^2 - 2So^3...$$

and $gf = o\sqrt{(1+Q_{a-o}^2)}$ or $o\sqrt{(1+(Q-2Ro...)^2)} = o\sqrt{(1+Q^2)} - 2(QR/\sqrt{(1+Q^2)})\, o^2...$;

so that $\sqrt{(Cf\times GF)} = \sqrt{(R^2o^4 - RSo^5...)} = Ro^2 - \frac{1}{2}So^3...$,

and therefore the ratio of resistance to gravity comes at last correctly to be

$$\frac{o\sqrt{(1+Q^2)} - 2(QR/\sqrt{(1+Q^2)})\, o^2...}{2(Ro^2 - \frac{1}{2}So^3...)} - \frac{o\sqrt{(1+Q^2)}}{2(Ro^2 + So^3...)} + \frac{Q}{\sqrt{(1+Q^2)}} = \frac{3S\sqrt{(1+Q^2)}o^4...}{4R^2o^4 + 2RSo^5},$$

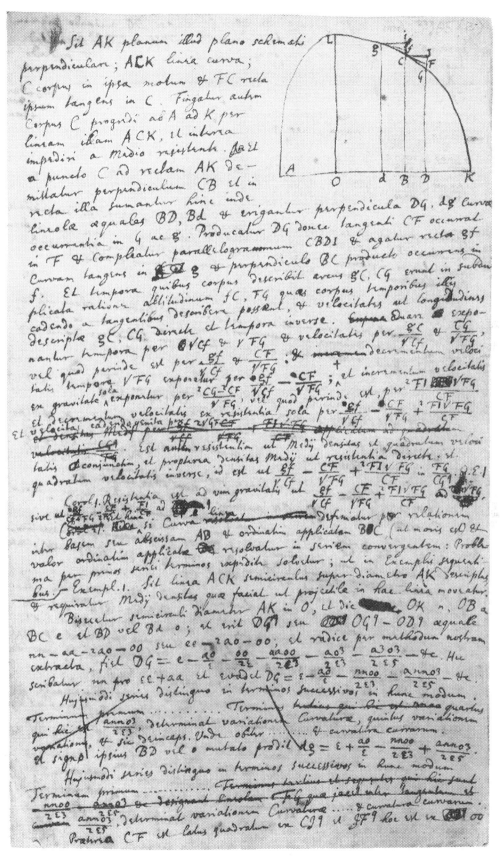

Plate IV. The problem of resisted motion under gravity at last given accurate resolution (**1**, 6, §1.6).

arising from the resistance alone $gf/\sqrt{Cf} - CF/\sqrt{FG} + 2IF \times \sqrt{FG}/CF$. As a result, the resistance is to the force of gravity as $gf/\sqrt{Cf} - CF/\sqrt{FG} + 2IF \times \sqrt{FG}/CF$ to $2\sqrt{FG}$, or as $gf/2\sqrt{(Cf \times FG)} - CF/2FG + IF/CF$ to 1.[70]

The resistance, however, is as the density of the medium and the square of the speed jointly, and accordingly the density of the medium is as the resistance directly and the square of the speed inversely, that is, as

$$(gf/2\sqrt{(Cf \times FG)} - CF/2FG + IF/CF) \times FG/CG^2.$$

As was to be found.

Corollary 1. The resistance is to gravity as $gf/(Cf + FG) - CF/2FG + IF/CF$ to 1. For in place of $2\sqrt{(Cf \times FG)}$[71] it is allowable to write $Cf + FG$.

Corollary 2. If the curved line be defined by a relationship between the base or abscissa ... as in the following examples.[72]

Example 1. Let the line ACK be a semicircle described upon diameter AK, and let there be required the density of the medium which shall make a projectile move in this line.

Bisect the semicircle's diameter AK at O, and call $OK = n$, $OB = a$, $BC = e$ and BD or $dB = o$: then will DG^2, that is, $OG^2 - OD^2$, be equal to

$$n^2 - a^2 - 2ao - o^2 \quad \text{or} \quad e^2 - 2ao - o^2;$$

and, when the root is extracted by our method, there will come

$$DG = e - (a/e)\,o - \tfrac{1}{2}(1/e + a^2/e^3)\,o^2 - \tfrac{1}{2}(a/e^3 + a^3/e^5)\,o^3 - \dots.$$

Here write n^2 in place of $e^2 + a^2$ and there will prove to be

$$DG = e - (a/e)\,o - \tfrac{1}{2}(n^2/e^3)\,o^2 - \tfrac{1}{2}(an^2/e^5)\,o^3 - \dots.$$

---

that is, $\tfrac{3}{4}S\sqrt{(1 + Q^2)}/R^2$ in the limit as the infinitesimal increment (decrement) $BD = dB = o$ becomes zero. For any particular resisted path, let us observe, the values of Cf and gf are readily computable from first principles by direct calculation of Q_{a-o} and R_{a-o} without pausing separately to evaluate their equivalents $Q - 2Ro + 3So^2\dots$ and $R - 3So + \dots$ respectively; and so Newton determines these in sequel in his prime instance of the semicircle in 'Exempl. 1' following (see note (75) below).

In a cancelled final sentence of the manuscript paragraph Newton initially went on to notice that 'Hic pro $2\sqrt{Cf} \times \sqrt{FG}$ scribi potest $Cf + FG$. Et sic gravitas erit ad resistentiam [!] ut $\dfrac{gf}{Cf + FG} - \dfrac{CF}{2FG} + \dfrac{IF}{CF}$ ad 1' (Here in place of $2\sqrt{Cf} \times \sqrt{FG}$ there can be written $Cf + FG$. And thus [resistance] will be to [gravity] as $gf/(Cf + FG) - CF/2FG + IF/CF$ to 1)—and make two further reductions of the ratio to would-be simpler terms which in fact increase its complexity (and in the latter of which he errs in confusing Cf with CF)—before delaying this remark to be a separate 'Corol. 1' to his main argument. Correspondingly, '$Cf + FG$' was first written in place of '$2\sqrt{Cf} + \sqrt{FG}$' in the ensuing paragraph.

(71) That is, $\sqrt{((Cf + GF)^2 - (-Cf + GF)^2)}$, where (see the previous note) $GF = Ro^2 + So^3\dots$ and $Cf = Ro^2 - 2So^3\dots$, and so their difference $3So^3$ will, in the limit as o vanishes, come to be infinitely less than their sum $2Ro^2 - So^3\dots$.

(72) Understand the text of 'Corol. 2' as it is printed in the *editio princeps* of the *Principia* ($_1$1687: 263 [= pages 376–8 below]).

Hujusmodi series distinguo in terminos successivos in hunc modum. Terminum primum... .... [Terminus quartus qui hic est] $\frac{anno^3}{2e^5}$ determinat variationem Curvaturæ... & curvatura Curvarum.[73]

Præterea CF est latus quadratum ex CI^q et IF^q hoc est ex oo et $\frac{aaoo}{ee}$ sive $\frac{nnoo}{ee}$, ideoq̃ est $\frac{no}{e}$. Et linea DG mutando signum ipsius o, vertitur in lineam

$$dg = e + \frac{ao}{e} - \frac{nnoo}{2e^3} + \frac{anno^3}{2e^5} \ \&c. \text{ Unde habetur } Ci = \frac{ao}{e} - \frac{nnoo}{2e^3} + \frac{anno^3}{2e^5} \ \&c. \text{ Et lineæ}$$

GF et CF scribendo Od pro OB[74] et dg pro BC vertuntur in Cf et gf. Et inde prodeunt $Cf = \frac{nnoo}{2e^3} - \frac{anno^3}{e^5}$ et $gf = \frac{no}{e} - \frac{anoo}{e^3}$.[75]

$$\left[\text{Unde } \frac{gf}{Cf+GF} - \frac{CF}{2GF} = \right] \quad \frac{noee - anoo}{\dfrac{nnoo}{2} - \dfrac{anno^3}{ee}} - \frac{no}{\dfrac{nno^2}{ee} + \dfrac{anno^3}{e^4}}$$
$$+ \dfrac{nnoo}{2} + \dfrac{anno^3}{2ee}$$

$$[\text{sive}] \qquad \frac{ee - ao}{no - \dfrac{anoo}{2ee}} - \frac{e^4}{noee + anoo} = \frac{[+]\frac{1}{2}aeenoo}{eennoo} = +\frac{a}{2n}$$

$$\left[\text{ut et } \frac{FI}{CF} = \frac{a}{n}. \text{ Fit resistentia ad gravitatem ut } \frac{a}{2n} \right] + \frac{a}{n} \text{ ad } 1 \text{ seu } 3a \text{ ad } 2n.[76]$$

---

(73) Here understand the third paragraph of 'Exempl. 1' in the 1687 text [= pages 378–9 below], with the minor replacement of 'exhibet' (p. 264, l. 12) by 'determinat'.

(74) That is, '$a-o$ pro a' ($a-o$ in place of a).

(75) As his preliminary computation on f. 201^v places beyond doubt—except that he there made the slip of calculating Cf to be $R_{a+o}o^2 + S_{a+o}o^3 + O(o^4)$, that is,

$$\text{`} \frac{nnoo, \overline{ee - 2ao} + ano^3}{2e^5 - 10e^3 ao} = \frac{nnoo}{2e^3} + \frac{4anno^3}{2e^5} \text{'}$$

—Newton derives these values of the lengths of Cf and gf as

$$(R_{a-o}o^2 + S_{a-o}o^3 =) \ \tfrac{1}{2}(n^2/e_{a-o}^3) \, o^2 + \tfrac{1}{2}(an^2/e_{a-o}^5) \, o^3 \quad \text{and} \quad (o\sqrt{(1 + Q_{a-o}^2)} =) \ no/e_{a-o}$$

respectively, in each case rounding off the ensuing series in ascending powers of o to their first two terms which are alone pertinent to the remaining stages of the computation. At this crucial stage in his computation he abandons verbal explanation in his anxiety to gain the end

Series of this sort I distinguish into their successive terms in this fashion. The first term ... . The fourth term, here $\frac{1}{2}(an^2/e^5)\,o^3$, determines the variation of the curvature...and the curvature of curves.[73]

Moreover, CF is the square root of $CI^2 + IF^2$, that is, of $o^2 + (a^2/e^2)\,o^2$ or $(n^2/e^2)\,o^2$, and is consequently $(n/e)\,o$. And the line DG, by changing the sign of o, turns into the line $dg = e + (a/e)\,o - \frac{1}{2}(n^2/e^3)\,o^2 + \frac{1}{2}(an^2/e^5)\,o^3\dots$. Whence there is had $Ci = (a/e)\,o - \frac{1}{2}(n^2/e^3)\,o^2 + \frac{1}{2}(an^2/e^5)\,o^3\dots$. And the lines GF and CF, on writing Od in place of OB[74] and dg in place of BC, turn into Cf and gf. And thence there result $Cf = \frac{1}{2}(n^2/e^3)\,o^2 - (an^2/e^5)\,o^3$ and $gf = (n/e)\,o - (an/e^3)\,o^2$.[75] Whence

$$gf/(Cf + GF) - CF/2GF$$

$$= (e^2no - ano^2)/(n^2o^2 - \tfrac{1}{2}(an^2/e^2)\,o^3) - no/((n^2/e^2)\,o^2 + (an^2/e^4)\,o^3),$$

that is,

$$(e^2 - ao)/(no - \tfrac{1}{2}(an/e^2)\,o^2) - e^4/(e^2no + ano^2) = \tfrac{1}{2}ae^2no^2/e^2n^2o^2 = \tfrac{1}{2}a/n,$$

and also $FI/CF = a/n$. The resistance comes, therefore, to be to the gravity as $\frac{1}{2}a/n + a/n$ to 1, that is, $3a$ to $2n$.[76]

---

result, and the mathematical calculations themselves are set down *staccato*. We have made suitable editorial interpolations to bring out their sense.

(76) Correct at last! With tangible relief Newton passed straightaway to jot down a minimal augmentation of his preceding paragraph—in which, between '...negligi possunt.' and 'Terminus quartus...' (*Principia*, ₁1687: 264, l. 11), there was now to be inserted the additional sentence 'In hoc Problemate pro lineola *FG* adhibemus terminum tertium et quartum' (In this problem in place of the linelet *FG* we employ the third and fourth terms)— and then, immediately beneath, dashed off the draft of a short letter to Niklaus Bernoulli: 'I send you inclosed the solution of yᵉ Probleme about the density of resisting Mediums set right. I desire you to shew it to your Unkle & return my thanks to him for sending me notice of yᵉ mistake' (f. 219ᵛ; see also *Correspondence of Isaac Newton*, 5, 1975: 348). This 'solution' was, we may presume, the carefully tidied and slightly augmented revision of the present paper which is reproduced (from the original at Add. 3965.12: 190ʳ–191ʳ) in §2.1 next following. But whether Johann's nephew in fact received his intended copy of it there is (see §2: note (2) below) every reason to doubt.

Subsequently, we may add, Newton used the considerable blank space remaining at the bottom of f. 219ᵛ (continuing on f. 220ᵛ) to draft the preliminary reworking of the latter half of §2.1 which is set out in Appendix 2.4, and at a yet later time—one more witness to what was never far from the forefront of his mind in this autumn of 1712—he penned on a still vacant area at the foot of f. 220ʳ his rough initial essay at what were to be notes * and ** on page 104 of his anonymous edition (soon to go to press under the Royal Society's *imprimatur*) of the *Commercium Epistolicum D. Johannis Collins et Aliorum de Analysi promota* by which he sought 'impartially' to document his historical claim to be first inventor of the calculus, against Leibniz' counter-assertion of his own prior and essentially independent discovery.

§2. REFINING THE CORRECTED ARGUMENT, CONTRIVING A VARIANT MODE OF PROOF, AND MOULDING THE WHOLE TO FILL THE SAME PRINTED SPACE.[1]

[1][2]   *Prop. X.   Prob. III.*

Tendat uniformis vis gravitatis directe ad planum Horizontis, sitқ resistentia ut Medij densitas et quadratum velocitatis conjunctim: requiritur tum Medij densitas in locis singulis, quæ faciat ut corpus in data quavis linea curva moveatur, tum corporis velocitas et Medij resistentia in locis singulis.

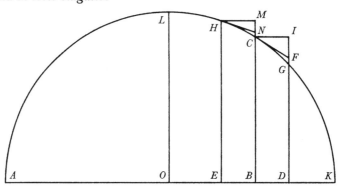

Sit *AK* planum illud plano schematis perpendiculare, *ALCK* linea curva, *C* corpus in ipsa motum, et *CF* recta ips[a]m tangens in *C*. Fingatur corpus *C* progredi ab *L* ad *K* per lineam illam *ACK*, et interea impediri a Medio resistente.

---

(1) Having by the beginning of October 1712 (in the rough first casting reproduced in §1.6 preceding) finally attained a correct measure of the resistance to gravity in a projectile's given path of fall through an impeding medium, Newton now proceeds at a deal more leisurely pace, first (in [1]) to write out its derivation in a neater and verbally amplified form; then (in [2]) to fashion an alternative mode of argument whereby the opposing forces of resistance and of gravity—hitherto presumed continuous in their action—are assumed equivalently to be impressed 'instantaneously' as discrete impulses at successive moments of time; and lastly (in [3]) further to refashion the primary approach through continuously acting resistance and gravity into a form which differs only in inessential fine details from that which three months afterwards, early in January 1713, he sent to Cotes in Cambridge for inclusion in the second edition of the *Principia*.

(2) Add. 3965.12: 190ʳ–191ʳ, an augmented elaboration (by way of a much rougher intermediate outline on f. 194 which lacks the present last paragraph and concluding sentence of the penultimate one) of §1.6 preceding. From the unusually careful way in which Newton has penned the original, we must presume that this is the 'solution' of the problem of resisted motion 'set right' which in his first flush of generosity—until at least his dour native unforthcomingness took hold—he intended (see §1: note (76)) to communicate to Niklaus Bernoulli (then still in London) for onward transmission to his uncle Johann in Basel. Abraham de Moivre, our sole reporter on the spot of what in fact took place, informed Johann a few days

Translation

[1]⁽²⁾   *Proposition X, Problem III.*

Let the uniform force of gravity tend directly to the plane of the horizon, and let the resistance be as the density of the medium and the square of the speed jointly: there is required both the density of the medium at individual places which shall make a body move in any given curved line, and the speed of the body and the resistance of the medium at individual places.

Let AK be that plane perpendicular to the plane of the figure, $ALCK$ the curved line, C the body moved in it and CF a straight line touching it at C. Imagine the body C to advance from L to K by way of that line ACK and all the

---

after the event that 'Monsieur Newton, …ayant…corrigé la conclusion, …me montra sa correction, et elle se trouva conforme au calcul de M. votre neveu; là-dessus il ajouta qu'il avoit dessein de voir M. votre neveu pour l'en remercier, et me pria de le mener chez lui, ce que je fis' (Wollenschläger, 'Briefwechsel' (§1: note (1)): 271). When he came face to face with Niklaus, Newton reassured him in a general way that the error in his 1687 argument proceeded—in de Moivre's words (*ibid.*)—'d'avoir considéré une tangente à rebours' (*viz.* taking the motion in the arc $\overset{\frown}{gC}$ to be approximated by that along the tangent at its end-point C and not that along the one at g, correspondingly as the tangent at C approximates the motion along the successive arc $\overset{\frown}{CG}$, in the terms of his original figure), but that 'le fondement de son calcul et les suites dont il s'est servi doivent subsister'. No script, however, where Newton penned out a corrected working of his problem would seem then to have changed hands; and a few weeks later, when Niklaus described his London visit to Johann, he wrote only briefly and vaguely that his[!] 'objection against Mr Newton' had led the latter to make 'a correction to his second edition', which Johann had to infer concerned 'the [problem of] resistance' which, he had shown, was 'incorrectly determined' in the case of the circle and the 'other curves' in which Newton had instanced his solution. (Niklaus' letter of ?early November 1712 to his uncle reporting upon his stay in London is lost, but we paraphrase the pertinent portion of Johann's yet unpublished reply on 23 November (N.S.), now in Basel University Library (MS. L Ia 22.9), where he wrote: 'Ich hätte wohl wünschen mögen zu wüssen worin Ewer Objection wider H[errn] Newton bestanden, darauff Er eine Correction in seiner newen edition gemacht und euch davon eine Copie zugestellet, dan ich forchte es seye just eine von meinen remarques als zum exempel betreffend die resistenz, so Er unrecht in dem circul und anderen *curvis* determinieret', adding the revealing aside that 'darüber ich etwas zu schreyben under händen habe umb in den *actis Lips.* zu publizieren [→ 'De motu Corporum gravium, Pendulorum, & Projectilium in mediis non resistentibus & resistentibus supposita Gravitate uniformi & non uniformi atque ad quodvis datum punctum tendente, et de variis aliis huc spectantibus, Demonstrationes Geometricæ', *Acta Eruditorum* (February/March 1713): 77–95/115–32, especially 91–3; on which see Appendix 1: note (14) below] welches aber zimlich schlecht stehen würde wann mich der H. Newton mit seiner correction prævenierte'. Johann Bernoulli wrote in similar terms to de Moivre the same day (see Wollenschläger, 'Briefwechsel': 276) that he was 'impatient de voir la nouvelle impression des *Princ. Math. phil. natur.* de M. Newton, que mon neveu me marque qu'elle sera achevée dans ce présent mois [November 1712]; comme je suis sur le point d'achever un petit écrit contenant quelquesunes de mes remarques que je fis autrefois sur la vieille édition, et que je publierai peut-être cet écrit dans les Actes de Leipsic, je serois bien aise de voir si j'aurois eu le bonheur de me rencontrer avec M. Newton dans les nouvelles additions et corrections que l'on trouvera, à ce qu'on me dit, dans cette seconde

A puncto C ad rectam AK horizonti parallelam demittatur perpendiculum CB, et in recta illa sumantur hinc inde lineolæ æquales EB, BD, et erigantur perpendicula EH, DG curvæ occurrentia in H et G. Producatur DG donec tangenti CF occurrat in F, compleatur parallelogrammum $CBDI$, et agatur recta HN curvam tangens in H & perpendiculo $[BC]$ producto occurrens in N. Et tempora quibus corpus describit arcus HC, CG erunt in subduplicata ratione altitudinum NC, FG quas corpus temporibus illis describere posset a tangentibus cadendo, et velocitates erunt ut longitudines descriptæ HC, CG directe et tempora inverse. Exponantur tempora per $\sqrt{CN}$ et $\sqrt{FG}$ et velocitates per $\frac{HC}{\sqrt{CN}}$ et $\frac{CG}{\sqrt{FG}}$, vel quod perinde est, per $\frac{HN}{\sqrt{CN}}$ et $\frac{CF}{\sqrt{FG}}$: et decrementum velocitatis tempore $\sqrt{FG}$ factum exponetur per $\frac{HN}{\sqrt{CN}} - \frac{CF}{\sqrt{FG}}$. Hoc decrementum oritur a resistentia corpus retardante & gravitate corpus accelerante, et augeri debet ea velocitatis parte quam gravitas generat, ut habeatur decrementum velocitatis a resistentia sola oriundum. Gravitas in corpore cadente & spatium FG cadendo describente generat velocitatem $\frac{2FG}{\sqrt{FG}}$ seu $2\sqrt{FG}$, at in corpore arcum CG describente generat velocitatem $\frac{2CG - 2CF}{\sqrt{GF}}$, id est $\frac{2FI\sqrt{FG}}{CF}$. Est enim CF ad FI ut FG ad $CG - CF$.[3] Addatur hæc velocitas ad decrementum prædictum et habebitur decrementum velocitatis ex resistentia sola oriundum, nempe

$$\frac{HN}{\sqrt{CN}} - \frac{CF}{\sqrt{FG}} + \frac{2FI\sqrt{FG}}{CF}.$$

Proindeꝗ resistentia est ad gravitatem ut $\frac{HN}{\sqrt{CN}} - \frac{CF}{\sqrt{FG}} + \frac{2FI\sqrt{FG}}{CF}$ ad $2\sqrt{FG}$, sive ut $\frac{HN}{2\sqrt{CN \times FG}} - \frac{CF}{2FG} + \frac{FI}{CF}$ ad 1. Et cum resistentia sit ut densitas Medij et quadratum velocitatis conjunctim, densitas Medij erit ut resistentia directe et

édition. Mon neveu me mande, qu'il a eu l'honneur de faire à M. Newton une petite[!] objection, laquelle lui doit avoir donné lieu d'ajoûter quelque correction à sa nouvelle édition de ses *Principes*; je ne sçais si c'est justement quelqu'une de mes remarques que mon neveu a vû ici...'.) In later years neither Niklaus nor his uncle ever spoke of having been shown a preliminary version of the corrected argument of the *editio secunda*. Let us add the clincher that the present text was, almost before its ink was dry, further amended in Newton's usual perfectionist manner both by minor replacements and interlineations in the manuscript itself, and then by the more radical recasting of its latter half which is reproduced in Appendix 2.4 below. Newton would never have handed over to the Bernoullis' microscopic inspection an already superseded version of his revised reasoning!

(3) Clearly dissatisfied with his wording of this sentence, Newton subsequently roughed

while to be impeded by the resisting medium. From the point C to the straight line AK, parallel to the horizon, let fall the perpendicular CB, and in that line on either hand take equal linelets EB and BD, and erect the perpendiculars EH and DG meeting the curve in H and G. Extend DG till it meets the tangent CF in F, complete the rectangle $CBDI$, and draw the straight line HN touching the curve at H and meeting the perpendicular BC produced in N. And the times in which the body describes the arcs $\widehat{HC}$ and $\widehat{CG}$ will be in the halved ratio of the heights NC and FG which the body would in those times be able to describe by falling from the tangents, and the speeds will be as the lengths $\widehat{HC}$ and $\widehat{CG}$ described directly and the times inversely. Represent the times by $\sqrt{CN}$ and $\sqrt{FG}$, and the speeds by $\widehat{HC}/\sqrt{CN}$ and $\widehat{CG}/\sqrt{FG}$, or, what is effectively the same, by $HN/\sqrt{CN}$ and $CF/\sqrt{FG}$, and then the decrement of the speed attained in time $\sqrt{FG}$ will be represented by $HN/\sqrt{CN}-CF/\sqrt{FG}$. This decrement arises from the resistance retarding the body and gravity accelerating it, and must be increased by the part of the speed which gravity generates in order to have the decrement of the speed arising from the resistance alone. Gravity in a falling body which, as it falls, describes the distance FG generates the speed $2FG/\sqrt{FG}$, that is, $2\sqrt{FG}$, but in the body describing the arc $\widehat{CG}$ it generates the speed

$$2(\widehat{CG}-CF)/\sqrt{FG},$$

that is, $2FI\times\sqrt{FG}/CF$; for there is CF to FI as FG to $\widehat{CG}-CF$.[3] Add this speed to the above-mentioned decrement and there will be had the decrement of the speed arising from the resistance alone, namely

$$HN/\sqrt{CN}-CF/\sqrt{FG}+2FI\times\sqrt{FG}/CF.$$

Accordingly, the resistance is to gravity as $HN/\sqrt{CN}-CF/\sqrt{FG}+2FI\times\sqrt{FG}/CF$ to $2\sqrt{FG}$, or as $HN/2\sqrt{(CN\times FG)}-CF/2FG+FI/CF$ to 1. And since the resistance is as the density of the medium and the square of the speed jointly, the

---

out on the verso of a loose folio worksheet (now Add. 3968.41: 119) an elaboration where he would have it read '[Gravitas in corpore cadente &] spatium *FG* cadendo describente generat velocit[at]em qua duplum illud spatium 2*FG* eodem tempore describi posset[,] ut ex demonstratis Galilæi notum est, id est velocitatem quæ exponitur per spatium 2*FG* applicatum ad tempus $\sqrt{FG}$[,] hoc est velocitatem $2\sqrt{FG}$: at in corpore arcum *CG* [describente] generat tantum velocitatem quæ sit ad hanc velocitatem ut *CG*−*CF* ad *FG* vel *FI* ad *CF*, id est velocitatem $\dfrac{2FI}{CF}\sqrt{FG}$' ([Gravity in a falling body which] in falling describes the distance *FG* generates a speed whereby twice that distance, 2*FG*, were able to be described in the same time, as is known from Galileo's demonstrations; that is, a speed which is expressed through the distance 2*FG* divided by the time $\sqrt{FG}$, or a speed $2\sqrt{FG}$: but in the body describing the arc $\widehat{CG}$ it generates merely a speed which is to this one as $\widehat{CG}-CF$ to *FG* or *FI* to *CF*, that is, a speed $2(FI/CF)\sqrt{FG}$). On Newton's precise knowledge of the 'demonstrata Galilæi' which he here cites, see note (14) below.

quadratum velocitatis inverse, id est ut $\dfrac{HN}{2\sqrt{CN \times FG}} - \dfrac{CF}{2FG} + \dfrac{FI}{CF}$ in $\dfrac{FG}{CG^q}$.
Q.E.I.[4]

Corol. 1. Resistentia est ad gravitatem ut $\dfrac{HN}{CN+FG} - \dfrac{CF}{2FG} + \dfrac{FI}{CF}$ ad 1. Nam pro $2\sqrt{CN \times FG}$ scribere licet $CN+FG$.[5]

Corol. 2. Densitas Medij est ut $\dfrac{HN}{CN+FG} - \dfrac{CF}{2FG} + \dfrac{FI}{CF}$ in $\dfrac{FG}{CF^q}$.

Corol. 3. Et hinc si curva linea definiatur per relationem inter basem seu abscissam *AB* et ordinatim applicatam *BC* (ut moris est) et valor ordinatim applicatæ resolvatur in seriem convergentem: Problema per primos seriei terminos expedite solvetur ut in Exemplis sequentibus.

Exempl. 1. Sit linea *ACK* semicirculus super diametro *AK* descriptus, et requiratur Medij densitas quæ faciat ut projectile moveatur in hac linea.

Bisecetur semicirculi diameter *AK* in [*O*] et dic *OK n*, *OB a*, *BC e*, et *BD* vel *BE o*, et erit DG^q seu $OG^q - OD^q$ æquale $nn-aa-2ao-oo$, seu $ee-2ao-oo$, et radice per methodum nostram extracta fiet

$$DG = e - \frac{ao}{e} - \frac{oo}{2e} - \frac{aaoo}{2e^3} - \frac{ao^3}{2e^3} - \left[\frac{a^3o^3}{2e^5}\right] - \&\text{c.}$$

Hic scribatur *nn* pro $aa+ee$ et evadet $DG = e - \dfrac{ao}{e} - \dfrac{nnoo}{2e^3} - \dfrac{anno^3}{2e^5} - \&\text{c.}$

Hujusmodi series distinguo in terminos successivos in hunc modum. Terminum primum appello in quo quantitas infinite parva *o* non extat; secundum in quo quantitas illa extat unius dimensionis; tertium in quo extat duarum; quartum in quo trium est[,] et sic in infinitum. Et primus terminus qui hic est *e*, denotabit semper longitudinem ordinatæ *BC* insistentis ad indefinitæ quantitatis initium *B*; secundus terminus qui hic est $\dfrac{ao}{e}$, denotabit differentiam inter *BC* & *DF*, id est lineolam *IF* quæ absciditur complendo parallelogrammum *BCID*, atq̃ adeo positionem tangentis *CF* semper determinat: ut in hoc casu capiendo *IF* ad *IC* ut est $\dfrac{ao}{e}$ ad *o* seu *a* ad *e*. [6]Terminus tertius & sequentes $\dfrac{nnoo}{2e^3} + \dfrac{anno^3}{2e^5} + \&$c designabunt lineolam *FG* quæ jacet inter tangentem et curvam, adeoq̃ determinant angulum contactus *FCG*, seu curvaturam quam Curva linea habet in *C*. Si lineola illa *FG* minuatur in infinitum, termini subsequentes evadent infinite minores tertio ideoq̃ negligi possunt, & solus terminus tertius curvaturam

---

(4) Except for renaming the decremented ordinate *dg* now to be *EH* (and correspondingly altering the lower-case denotations of the old points *f* and *i* here to be *N* and *M* respectively), this is all virtually word for word as in §1.6 preceding. What hereafter follows in the ensuing paragraphs is more considerably amplified and improved in its statement (and will be yet

density of the medium will be as the resistance directly and the square of the speed inversely, that is, as $(HN/2\sqrt{(CN \times FG)} - CF/2FG + FI/CF) \times FG/CG^2$. As was to be found.[4]

Corollary 1. The resistance is to gravity as $NH/(CN+FG) - CF/2FG + FI/CF$ to 1. For in place of $2\sqrt{(CN \times FG)}$ it is allowable to write $CN+FG$.[5]

Corollary 2. The density of the medium is as

$$(HN/(CN+FG) - CF/2FG + FI/CF) \times FG/CF^2.$$

Corollary 3. And hence if the curved line be defined by a relationship between the base or abscissa AB and ordinate BC (as is customary) and the value of the ordinate be resolved into a converging series, the problem will promptly be solved through the first terms of the series, as in the following examples.

Example 1. Let the line ACK be a semicircle described upon diameter AK, and let there be required the density of the medium which shall make a projectile move in this line.

Bisect the semicicle's diameter AK in O and call $OK = n$, $OB = a$, $BC = e$ and BD or $EB = o$, and then DG^2, that is, $OG^2 - OD^2$, will be equal to

$$n^2 - a^2 - 2ao - o^2,$$

that is, $e^2 - 2ao - o^2$, and when the root is extracted by our method there will come $DG = e - (a/e)\,o - \frac{1}{2}(1/e + a^2/e^3)\,o^2 - \frac{1}{2}(a/e^3 + a^3/e^5)\,o^3 - \dots$. Here write n^2 in place of $a^2 + e^2$ and there will prove to be

$$DG = e - (a/e)\,o - \tfrac{1}{2}(n^2/e^3)\,o^2 - \tfrac{1}{2}(an^2/e^5)\,o^3 - \dots.$$

Series of this sort I distinguish into their successive terms in this manner. The first term I call that in which the indefinitely small quantity o does not occur; the second that in which it occurs of one dimension; the third that in which it occurs of two; the fourth that in which it occurs of three, and so on indefinitely. And the first term, here e, will ever denote the length of the ordinate BC standing at the beginning, B, of the indefinite quantity; the second term, here $(a/e)\,o$, will denote the difference between BC and DF, i.e. the linelet IF which is cut off by completing the rectangle $BCID$, and hence ever determines the position of the tangent CF—in this case, specifically, by taking IF to IC as $(a/e)\,o$ to o, that is, a to e. [6]The third and following terms, $\frac{1}{2}(n^2/e^3)\,o^2 + \frac{1}{2}(an^2/e^5)\,o^3 + \dots$, will designate the linelet FG which lies between the tangent and the curve, and hence determines the angle of contact $F\widehat{C}G$, or the curvature which the curve has at C; should that linelet FG be diminished infinitely, the subsequent terms will prove to be infinitely less than the third, and can consequently be neglected, the third term alone determining the curvature. The fourth term, here

more radically changed in the draft revision of this latter portion which is reproduced in Appendix 2.4 below).

(5) See §1: note (71) above, *mutatis mutandis*.

(6) The following sentence is copied without deviation from the preliminary draft on f. 200^r.

determinabit. Terminus quartus qui hic est $\dfrac{anno^3}{2e^5}$ exhibet variationem Curvaturæ; quintus variationem variationis, & sic deinceps.[7] Unde obiter patet usus non contemnendus harum serierum in solutione Problematum quæ pendent a tangentibus & curvatura Curvarum.

Præterea CF est latus quadratum ex CI^q & IF^q hoc est ex $oo+\dfrac{aaoo}{ee}$ sive ex $\dfrac{nnoo}{ee}$,

ideoq est $\dfrac{no}{e}$. Et linea DG mutando signum ipsius o vertitur in lineam EH, et lineolæ GF ac CF scribendo OE pro OB et EH pro BC vertuntur in lineolas CN et HN. Indeq prodeunt

$$EH=a+\frac{ao}{e}-\frac{nnoo}{2e^3}+\frac{anno^3}{2e^5}-\&c,\quad CN=\frac{nnoo}{2e^3}-\frac{anno^3}{e^5},$$

$$HN=\frac{no}{e}-\frac{anoo}{e^3},\quad\text{et}\quad\frac{HN}{CN+FG}-\frac{CF}{2FG}+\frac{FI}{CF}=\frac{3a}{2n}.\ ^{(8)}$$

Est igitur resistentia ad gravitatem ut $\dfrac{3a}{2n}$ ad 1 sive $3a$ ad $2n$, et densitas Medij ut $\dfrac{3a}{2n}\times\dfrac{FG}{CF^q}$ id est ut $\dfrac{a}{e}$. Et velocitas $\dfrac{CF}{\sqrt{FG}}$ fit $\sqrt{2e}$. Ideoq si corpus C certa cum velocitate, secundum lineam ipsi OK parallelam, exeat de loco L, et Medij densitas in locis singulis C sit ut $\dfrac{OB}{OC}$, et resistentia etiam in loco aliquo C sit ad vim gravitatis ut $3OB$ ad $2OK$; corpus illud describet circuli quadrantem LCK. Q.E.I.

At si corpus idem de loco A secundum lineam ipsi AK perpendicularem egrederetur, sumenda esset OB seu a ad contrarias partes centri O, et propterea signum ejus mutandum esset & scribendum $-a$ pro $+a$. Quo pacto prodiret Medij densitas ut $\dfrac{-a}{e}$. Negativam autem densitatem (hoc est quæ motus corporum accelerat) Natura non admittit, & propterea naturaliter fieri non potest ut corpus ascendendo ab A describat circuli quadrantem AL. Ad hunc effectum deberet corpus a Medio impellente accelerari[,] non a resistente impediri.[9]

---

(7) See Appendix 1: note (20) below, and compare VII: 113, note (146).

(8) Newton's checking computations to this end are preserved on Add. 3968.41: 11ʳ/ 3968.12: 176ᵛ and on the reverse of the folio (now in private possession) on which he set down the draft of the 'Royal Society Committee' report on Leibniz' claim to calculus priority which is reproduced in **2**, Appendix 1.4 below. In essence these successively calculate:

$$\text{`}\ \frac{NH}{CN+FG}=\frac{\overline{eeno-anoo}\times e^5}{e^3,eennoo-\dfrac{anno^3}{2ee}}=\frac{e^4-ae^2o}{eeno-\frac{1}{2}anoo}=\frac{ee}{no}-\frac{a}{2n}.$$

$\frac{1}{2}(an^2/e^5)\,o^3$, exhibits the variation of the curvature; the fifth the variation of the variation, and so on.[7] Whence, incidentally, there is opened up a use of these series not to be disdained in the solution of problems which depend on tangents and the curvature of curves.

Moreover, CF is the square root of CI^2+IF^2, that is, of $o^2+(a^2/e^2)\,o^2$, that is, of $(n^2/e^2)\,o^2$, and consequently is $(n/e)\,o$. And the line DG, by changing the sign of o, turns into the line EH, while the linelets GF and CF, by writing OE in place of OB and EH in place of BC, turn into the linelets CN and HN; and thence there ensue

$$EH = a+(a/e)\,o-\tfrac{1}{2}(n^2/e^3)\,o^2+\tfrac{1}{2}(an^2/e^5)\,o^3-\ldots,\quad CN = \tfrac{1}{2}(n^2/e^3)\,o^2-(an^2/e^5)\,o^3,$$

$$HN = (n/e)\,o-(an/e^3)\,o^2 \ \text{ and so }\ HN/(CN+FG)-CF/2FG+FI/CF = \tfrac{3}{2}a/n.[8]$$

Therefore the resistance is to gravity as $\tfrac{3}{2}a/n$ to 1, or $3a$ to $2n$, and the density of the medium as $\tfrac{3}{2}(a/n)\times FG/CF^2$, that is, as a/e; and the speed $CF/\sqrt{FG}$ comes to be $\sqrt{(2e)}$. Consequently, if the body C should go off from the point L with a certain speed, following a line parallel to OK, and the density of the medium in individual places be as OB/OC, and the resistance also in any arbitrary point C be to the force of gravity as $3OB$ to $2OK$: that body will describe the circle-quadrant $\overarc{LCK}$. As was to be found.

But if the same body should set off from the place A following a line perpendicular to AK, you would have needed to take OB, or a, on the opposite side of the centre O, and on that account to change its sign and to have written $-a$ in place of $+a$; and on this basis there would have resulted a density of the medium (varying) as $-a/e$. But a negative density—one, that is, which accelerates the motions of bodies—is not admitted in nature, and for that reason it cannot naturally happen that a body, in ascending from A, shall describe the circle-quadrant $\overarc{AL}$. To this effect a body would have needed to be accelerated by an impelling medium, not impeded by a resisting one.[9]

[necnon] $\qquad \dfrac{CF}{2FG} = \dfrac{no}{e}\times\dfrac{e^5}{eennoo+anno^3} = \dfrac{e^4}{eeno+anoo} = \dfrac{ee}{no}-\dfrac{a}{n}.$

[adeoϛ] $\qquad \dfrac{HN}{CN+FG}-\dfrac{CF}{2FG} = \dfrac{aee}{2nee+an^3o} = \dfrac{a}{2n}.$ [ut et] $\dfrac{FI}{CF} = \dfrac{a}{n}$,

(9) The ensuing Examples 2, 3 and 4 are understood to follow from similar application of the corrected measure $HN/(CN+FG)-\tfrac{1}{2}CF/FG+FI/CF$ of resistance to gravity in the case of the 'Galileian' parabola—which remains, of course, the unresisted fall-path—and of the conic/higher-order hyperbolas, where the expressions derived in the *editio princeps* will now likewise need to be increased in the ratio of 3 to 2. In sequel, Newton passed these minimal ameliorations by, returning at once to his preceding corollaries to rework them, on ff. (200^v→) 219^v/220^v, into the form which is printed in Appendix 2.4 below, before going on yet more radically to recast his general argument (by way of the preliminary draft in Appendix 2.8) into its final version in [3] following.

[2][10]   *Idem aliter.*

Invenienda sit Medij densitas qua Projectile moveatur in linea curva *PRQ*
progrediendo ab *R* versus *Q*. Et sit [P]*ABQ* basis vel abscissa Curvæ sintꝗ *BF*,
CG, *DH*, *EI* ordinatæ quatuor ad æquales ab invicem distantias infinite parvas

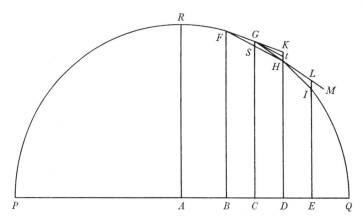

BC, *CD*, *DE* basi insistentes in punctis *B*, *C*, *D* et *E* & Curvæ occurrentes in
punctis *F*, *G*, *H* et *I*. Agantur chordæ *FG*, *GH*, *HI*. Et producantur chordæ
FG & *GH* donec prior occurrat ordinatæ *DH* in *K* et posterior occurrat ordinatæ
EI in *L*. Concipe autem projectile moveri in chordis illis infinite parvis *FG*, *GH*,
HI, alijsꝗ similibus innumeris per totam Curvæ longitudinem *PQ*, et resistentiam ac gravitatem agere tantum in punctis *F*, *G*, *H*, *I*, & actionibus suis
mutare motum corporis in solis illis punctis & efficere ut co[r]pus postquam
descripsit chordam[11] *FG* non longius pergat in hac chorda producta versus *K*
sed in puncto *G* cursum mutet et recta pergat ad *H*[,] et ubi chordam *GH*
descripsit non longius pergat in hac chorda versus *L* sed cursum mutet et recta
pergat in chorda *HI* & sic deinceps in infinitum. Ac tempora quibus chordæ
GH, *HI* describuntur erunt in subduplicata ratione altitudinum *KH*, *LI* per
quas corpus vi impulsuum gravitatis temporibus illis descendit recedendo a rectis
FGK, *GHL*. Ducatur chorda *FH* ordinatam *GC* secans in *S*: et recta[12] huic
parallela per punctum *G* ducta bisecabit rectam *KH* puta in *t* et curvam satis
accurate tanget in *G*. Et si corpus moveretur[13] in ipsa curva describendo non
chordas arcuum sed ipsos arcus *FG*, *GH*, *HI*, & simul per uniformem et continuam vim gravitatis impelleretur deorsum, hoc corpus interea dum describit
arcū *GH*, vi gravitatis descenderet a tangente *Gt* per spatium *tH*, id est per
spatium $\frac{1}{2}KH$.[14] Sunt igitur altitudines *KH LI* duplo majores altitudinibus

(10) Add. 3965.12: 198ʳ–199ʳ. Having temporarily become stuck in his endeavours (see
Appendix 2.4 below, and especially our concluding note (46) thereto) to refine the measure of
the ratio of resistance to gravity attained in (§1.6 →) §2.1 preceding, Newton here successfully
tries an alternative approach—its kernel appears in stark clarity in the two preliminary

[2]⁽¹⁰⁾   *The same another way.*

Let there need to be found the density of a medium whereby a projectile shall move in the curved line *PRQ*, advancing from *R* towards *Q*. Let *PABQ* be the base or abscissa of the curve, and *BF*, *CG*, *DH*, *EI* four ordinates at equal, infinitely small distances *BC*, *CD*, *DE* from each other, standing on the base at the points *B*, *C*, *D* and *E* and meeting the curve in the points *F*, *G*, *H* and *I*. Draw the chords *FG*, *GH*, *HI*; and extend the chords *FG* and *GH* till the former meets the ordinate *DH* in *K* and the latter meets the ordinate *EI* in *L*. Now conceive that the projectile moves in the infinitely small chords *FG*, *GH*, *HI* and innumerable other similar ones throughout the length of the curve $\widehat{PQ}$, and that the resistance and gravity act merely at the points *F*, *G*, *H*, *I*, by their actions changing the motion of the body at those points alone, and achieving that the body, after it has described the chord⁽¹¹⁾ *FG*, shall no longer proceed in the extension of this chord towards *K*, but shall change its course at the point *G* and proceed straight towards *H*; and, when it has described the chord *GH*, shall no longer pass on in this chord towards *L*, but change its course and proceed in the chord *HI*, and so on indefinitely. And the times in which the chords *GH*, *HI* are described will be in the halved ratio of the heights *KH*, *LI* through which the body by dint of impulses of gravity in those times descends, falling away from the straight lines *FGK*, *GHL*. Draw the chord *FH* cutting the ordinate *GC* in *S*; and the straight line⁽¹²⁾ drawn parallel to this through the point *G* will bisect the straight line *KH*, in *t* say, and accurately enough touch the curve at *G*. And were the body to move⁽¹³⁾ in the curve itself, describing not the chords of the arcs but the arcs $\widehat{FG}$, $\widehat{GH}$, $\widehat{HI}$ and at the same time be impelled downwards through a uniform and continuous force of gravity, this body would, all the while it describes the arc $\widehat{GH}$, descend by the force of gravity from the tangent *Gt* through the distance *tH*, that is, through the distance $\frac{1}{2}KH$.⁽¹⁴⁾ The heights *KH* and *LI*,

---

schemes of computation reproduced in Appendix 2.5/6—whereby the forces of resistance and gravity are now conceived to act not continuously over infinitesimal moments of time but 'simul et semel' at the beginning of each such successive instant (compare VI: 540, note (8)).

(11) In momentary forgetfulness of his present approximation of the curve $\widehat{FGHI}$ by the chain of infinitesimal chords *FG/GH/HI*, Newton here first slipped to write 'arcum' (arc).

(12) Specified in the manuscript as '*Gt*', but the identification is rendered both superfluous and clumsy in its anticipation of Newton's subsequent explicit introduction of the point *t* of this tangent's intersection with *KH* (in the phrase 'puta ut *t*' interleaved by him as an afterthought). As ever, of course, he here assumes that the vanishingly small arc $\widehat{FGH}$ is (to sufficient approximation) parabolic, so that the tangent at its mid-point *G* can accurately be taken as parallel to its chord *FH*, and therefore $Ht = SG(=\frac{1}{2}HK)$.

(13) Initially, in a cancelled preceding phrase whose differences are otherwise insignificant, 'pergeret' (...proceed) was written.

(14) Newton first went on to add 'et interea velocitatem acquireret qua uniformiter continuata corpus eodem tempore describere[t] duplum illu[d] spatium₍ₛ₎ id est spatium

quas corpus vi gravitatis cadendo describeret interea dum describit arcus GH, HI; ideoq altitudines illæ KH LI sunt ut quadrata temporum quibus chordæ GH, HI describuntur si modo corpus moveatur in chordis. Velocitates autem sunt ut chordæ descriptæ et [inverse ut] tempora describendi, id est ut $\frac{GK}{\sqrt{KH}}$ et $\frac{HI}{\sqrt{LI}}$. Et velocitas corporis in loco quovis G æqualis est velocitati corporis in eodem loco Parabolam in vacuo describentis cujus vertex locus ille G, diameter est GC, et latus rectum est $\frac{2GH^q}{HK}$. [15]

Concipe jam quod corpus ubi chordam GH descripsit recta progrediatur versus M eadem cum velocitate. Et si capiatur HM ad GH ut tempus describendi arcum HI ad tempus describendi arcum $GH_{[,]}$ id est ut $\sqrt{LI}$ ad $\sqrt{KH}$: corpus perveniet ad punctum M quo tempore per impulsus resistentiæ et gravitatis in loco H perveniret ad I. Et propterea corpus per impulsum resistentiæ in puncto H ita retardatur ut vice longitudinis HM describat tantum longitudinem HL & per impulsum gravitatis ita acceleratur deorsum ut recedendo a recto tramite HL simul describat altitudinem LI. Est igitur resistentia ad gravitatem ut LM ad LI. Et densitas Medij ut $\frac{LM}{LI} \times \frac{LI}{GH^q}$ id est reciproce ut $\frac{GH^q}{LM}$.

Quare si inveniantur longitudines Ordinatarum BF, CG, DH, EI, et pro KH ponatur dupla sagitta GS id est $2CG-BF-DH$, & similiter pro LI ponatur $2DH-CG-EI$. et capiatur LM ad GH ut $\sqrt{LI}-\sqrt{KH}$ ad $\sqrt{KH}$ id est ut $\sqrt{LI \times KH}-KH$ ad KH sive ut $\frac{LI-KH}{2}$ ad KH (nam pro $\sqrt{LI \times KH}$ [16] scribi potest $\frac{LI+KH}{2}$) solvetur Problema. [17]

---

totum KH, ut Galilæus demonstravit' (and in the meanwhile would acquire a speed whereby, if it were uniformly continued, the body would in the same time describe twice that distance, that is, the total distance KH, as Galileo has demonstrated). While the vellum worksheet, ULC. Add. 3958.2:45, on which he penned his comparison of the earth's diurnal centrifugal force with that of its gravity shows that he was already then, in the middle 1660's, familiar (by way of Salusbury's 1661 translation) with the passage in the Second Day of Galileo's *Dialogo... sopra i due Massimi Sistemi del Mondo* where this speed rule was lightly adumbrated in its mean-speed form (compare J. W. Herivel, *The Background to Newton's 'Principia'* (Oxford, 1966):183), we may again lightly insist that he had no direct knowledge—even *via* Salusbury's rare 1665 English version—of the *Discorsi e Dimostrazioni Matematiche...Attenenti alla Mecanica & i Movimenti Locali* (Leyden, ₁1638) where, in 'Theor. I' of the propositions on uniformly accelerated motion included in its 'Giornata Terza', Galileo afterwards came to give this 'Merton rule' more rigorous demonstration; indeed, in his present phrasing Newton much more closely follows Christiaan Huygens' formulation of the theorem in his *Horologium Oscillatorium sive de Motu Pendulorum ad Horologia aptato Demonstrationes Geometricæ* (Paris, 1673), where in Propositio V of the 'Pars Secunda. De Descensu Gravium...' it is asserted: 'Spatium

therefore, are twice as big as the heights which a body would in falling by the force of gravity describe all the while it describes the arcs $\widehat{GH}$, $\widehat{HI}$; and consequently those heights *KH*, *LI* are as the squares of the times in which the chords *GH*, *HI* are described should it but be that the body moves in the chords. The speeds, however, are as the chords described and (inversely as) the times of their description, that is, as $GK/\sqrt{KH}$ and $HI/\sqrt{LI}$. And the speed of the body at any place *G* is equal to the speed of a body in the same place describing in a vacuum a parabola whose vertex is that place *G*, diameter is *GC* and *latus rectum* is $2GH^2/HK$.[15]

Now conceive that when the body has described the chord *GH* it shall go straight on towards *M* with the same speed. Then, if *HM* be taken to *GH* as the time of describing the arc $\widehat{HI}$ to the time of describing the arc $\widehat{GH}$, that is, as $\sqrt{LI}$ to $\sqrt{KH}$, the body will arrive at the point *M* in the time in which it would through impulses of resistance and gravity at the place *H* have reached *I*. Accordingly, let the body be so slowed at the point *H* through an impulse of resistance that instead of the length *HM* it describes merely the length *HL*, and so accelerated downwards through an impulse of gravity that in falling away from the straight path it simultaneously describes the height *LI*: in consequence, the resistance is to gravity as *LM* to *LI*; and the density of the medium as $(LM/LI) \times LI/GH^2$, that is, reciprocally as GH^2/LM.

Wherefore, if the lengths of the ordinates *BF*, *CG*, *DH* and *EI* be found out, and in place of *KH* there be put twice the *sagitta GS*, that is $2CG - BF - DH$, and similarly in place of *LI* there be put $2DH - CG - EI$, and if *LM* be taken to *GH* as $\sqrt{LI} - \sqrt{KH}$ to $\sqrt{KH}$, that is, as $\sqrt{(LI \times KH)} - KH$ to *KH*, or as $\frac{1}{2}(LI - KH)$ to *KH* (for in place of $\sqrt{(LI \times KH)}$[16] there can be written $\frac{1}{2}(LI + KH)$), the problem will be solved.[17]

---

peracto certo tempore, à gravi è quiete casum inchoante, dimidium esse ejus spatii quod pari tempore transiret motu æquabili, cum celeritate quam acquisivit ultimo casus momento' (*ibid.*: 29).

(15) That is, $(GH^2/Ht \approx)\ Gt^2/Ht$.

(16) Since $4LI \times KH = (LI + KH)^2 - (LI - KH)^2$, where the difference $LI - KH$ is infinitesimal in proportion to *LI* and *KH*, and so to their sum; compare §1.6: note (71) above.

(17) In a subsequent computation on Add. 3965.10: 136ᵛ Newton afterwards confirmed that where $CG = e$ and, corresponding to the increment $CD = o$ of the base $AC(= a)$, the augmented ordinate *DH* has the series expansion $e + fo + go^2 + ho^3 + \ldots$—the coefficients $f(= de/da)$, $g(= \frac{1}{2}df/da)$ and $h(= \frac{1}{3}dg/da)$ are, of course, what he elsewhere calls *Q*, *R* and *S* respectively—this mode of solution yields generally that 'Resistentia erit ad Gravitatem ut $3h\sqrt{1+ff}$ ad $4gg$. Velocitas ut $\sqrt{\dfrac{1+ff}{g}}$. et Medij densitas ut $\dfrac{h}{g\sqrt{1+ff}}$.' (See Appendix 2.7 following, where we reproduce the gist of this checking calculation.) Here, however, he passes to apply the present geometrical construction of the ratio of resistance to gravity, and of the density of the medium, severally to each of the four *Exempla* which he had set to illustrate his

Exempl. 1. Sit PRQ semicirculus centro A diametro PQ descriptus, et nominentur AP vel $AQ=n$, $AC=a$, $BC=CD=DE=o$ & $CG(=\sqrt{nn-aa})=e$. Et prodibit $DH=\sqrt{ee-2ao+oo}=e-\dfrac{ao}{e}-\dfrac{nnoo}{2e^3}-\dfrac{anno^3}{2e^5}-$ &c.

[adeoqʒ] $EI=e-\dfrac{2ao}{e}-\dfrac{2nnoo}{e^3}-\dfrac{4anno^3}{e^5}$. $BF=e+\dfrac{ao}{e}-\dfrac{nnoo}{2e^3}+\dfrac{anno^3}{2e^5}$.

$HK=\dfrac{nnoo}{e^3}$. $IL=\dfrac{nnoo}{e^3}+\dfrac{3anno^3}{e^5}$. $GH=\sqrt{oo+\overline{GC-DH}^q}=\dfrac{no}{e}+\dfrac{anoo}{2e^3}$.

[ut et] $LM=\dfrac{3anoo}{2e^3}$. [Fit igitur] Resistentia ad Gravitatem ut $\dfrac{3anoo}{2e^3}$ ad $\dfrac{nnoo}{e^3}$ seu $3a$ ad $2n$. Velocitas ut $\sqrt{e}$. Densitas Medij ut $\dfrac{3a}{2ne}$ id est ut $\dfrac{a}{e}$.

Exempl. 2. Sit linea $PRCK$ Parabola axem habens AR horizonti AK perpendicularem, et requiratur Medij densitas quæ faciat ut projectile in ipsa moveatur.

In hac figura lineæ HK et IL invenientur æquales, ideoqʒ lineæ GH et HM æquales erunt & puncto M in puncto L incidente, linea LM eiqʒ proportionalis resistentia erit nulla. Et nulla Medij densitate projectile movebitur in Parabola uti olim Galilæus demonstravit.[18]

Exempl. 3. Sit linea $PRGQ$[19] Hyperbola, Asymptoton habens NX plano horizontali AK perpendicularem, et quæratur Medij densitas quæ faciat ut Projectile noveatur in hac linea.

Sit AX Asymptotos altera, ordinatim applicatæ DH productæ occurrens in V, et ex natura Hyperbolæ, rectangulum XV in VH dabitur. Datur autem ratio DN ad VX, et propterea datur etiam rectangulum DN in $V[H]$.[20] Sit illud bb; & completo parallelogrammo $DNXZ$, dicatur CN a, CD o, NX c, et ratio data VZ ad ZX vel DN ponatur esse $\dfrac{m}{n}$. Et

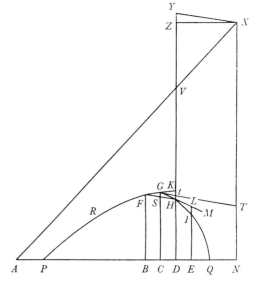

erit DN æqualis $a-o$, VH æqualis $\dfrac{bb}{a-o}$, VZ æqualis $\dfrac{m}{n}\overline{a-o}$ et DH seu $NX-VZ-VH$ æqualis $c-\dfrac{m}{n}a+\dfrac{m}{n}o-\dfrac{bb}{a-o}$. Resolvatur $\dfrac{bb}{a-o}$ in seriem conver-

proposition in the *editio princeps* of the *Principia* (see Appendix 1 below), confirming *ab initio* in each instance—or, more strictly, in Examples 1, 3 and 4, since the surrounding medium

Example 1. Let *PRQ* be a semicircle described with centre *A* and diameter *PQ*, and name *PA* or $AQ = n$, $AC = a$, $BC = CD = DE = o$

and
$$CG(= \sqrt{(n^2 - a^2)}) = e.$$

There will then ensue

$$DH = \sqrt{(e^2 - 2ao + o^2)} = e - (a/e)\, o - \tfrac{1}{2}(n^2/e^3)\, o^2 - \tfrac{1}{2}(an^2/e^5)\, o^3 - \ldots,$$

and hence
$$EI = e - 2(a/e)\, o - 2(n^2/e^3)\, o^2 - 4(an^2/e^5)\, o^3 \ldots,$$

$$BF = e + (a/e)\, o - \tfrac{1}{2}(n^2/e^3)\, o^2 + \tfrac{1}{2}(an^2/e^5)\, o^3 \ldots, \quad HK = (n^2/e^3)\, o^2 \ldots,$$

$$IL = (n^2/e^3)\, o^2 + 3(an^2/e^5)\, o^3 \ldots,$$

$$GH = \sqrt{(o^2 + (GC - DH)^2)} = (n/e)\, o + \tfrac{1}{2}(an/e^3)\, o^2 \ldots$$

and also $LM = \tfrac{3}{2}(an/e^3)\, o^2 \ldots$. The resistance comes, therefore, to be to gravity as $\tfrac{3}{2}(an/e^3)\, o^2$ to $(n^2/e^3)\, o^2$, or $3a$ to $2n$; the speed as $\sqrt{e}$; the density of the medium as $\tfrac{3}{2}a/ne$, that is, as a/e.

Example 2. Let the line *PRCK* be a parabola having its axis *AR* perpendicular to the horizontal *AK*, and let there be required the density of the medium which shall make a projectile move in it.

In this instance the lines *HK* and *IL* in the figure will be found to be equal, and consequently the lines *GH* and *HM* will be equal, and so, with the point *M* coincident with the point *L*, the line *LM* will be nothing, and the resistance proportional to it nil. And so with nil density of the medium the projectile will move in the parabola, as Galileo once demonstrated.[18]

Example 3. Let the line $PRGQ$[19] be a hyperbola, having asymptote *NX* perpendicular to the horizontal plane *AK*, and let there be required the density of the medium which shall make a projectile move in this line.

Let *AX* be the other asymptote, meeting the ordinate *DH* produced in *V*, and from the nature of the hyperbola the product $XV \times HV$ will be given. Now there is given the ratio of *ND* to *VX*, and in consequence there is also given the product $ND \times HV$. Let that be b^2, and, having completed the rectangle *DNXZ*, call $NC = a$, $DC = o$, $NX = c$, and set the given ratio *VZ* to *XZ* or *ND* to be m/n. Then will then be $ND = a - o$, $HV = b^2/(a - o)$, $VZ = (m/n)(a - o)$ and *DH*, or $NX - VZ - HV$, $= c - (m/n)\, a + (m/n)\, o - b^2/(a - o)$. Resolve $b^2/(a - o)$ into the

---

continues to offer no resistance to motion in the parabolic path of Example 2—that the resistance as he had evaluated it in his 1687 text needs to be increased by half as much again.

(18) See Appendix 1: note (26), and compare VI: 106, note (35).

(19) The manuscript lacks an accompanying figure; that which we here supply to fill the want is founded on the diagram which illustrates the corresponding 'Exempl. 3' in the *editio princeps* (*Principia*, $_1$1687: 267; see page 382 below).

(20) Newton wrote '*VG*' in a momentary slip of his pen.

gentem $\dfrac{bb}{a}+\dfrac{bb}{aa}o+\dfrac{bb}{a^3}oo+\dfrac{bb}{a^4}o^3+$ &c et fiet DH æqualis

$$c-\frac{m}{n}a-\frac{bb}{a}+\frac{maa-nbb}{naa}o-\frac{bb}{a^3}oo-\frac{bb}{a^4}o^3-\text{&c.}$$

Pro $c-\dfrac{m}{n}a-\dfrac{bb}{a}$[21] scribatur e, et habebuntur Ordinatæ quatuor

$$BF=e+\frac{nbb-maa}{naa}o-\frac{bb}{a^3}oo+\frac{bb}{a^4}o^3-\text{&c.}\quad CG=e.$$

$$DH=e+\frac{maa-nbb}{naa}\,o-\frac{bb}{a^3}oo-\frac{bb}{a^4}o^3.\quad EI=e+\frac{2maa-2nbb}{naa}\,o-\frac{4bb}{a^3}oo-\frac{8bb}{a^4}o^3.$$

Et inde $\quad HK=\dfrac{2bboo}{a^3}.\; IL=\dfrac{2bboo}{a^3}+\dfrac{6bbo^3}{a^4}.\;$ [&] $IL-HK=\dfrac{6bbo^3}{a^4}.$

Estꝗ $\quad DN\,.\,GT$[22]$::CD\,.\,GH=\dfrac{GT}{a}\,o.\;$ et $\;KH\,.\,\dfrac{LI-KH}{2}::GH\,.\,LM=\dfrac{3GT}{2aa}\,oo.$

Adeoꝗ Resistentia ad Gravitatem ut $\dfrac{3GT}{2aa}$ ad $\dfrac{2bb}{a^3}$ seu $3GT$ ad $4VH$.[23] Velocitas autem est ut $\dfrac{GH}{\sqrt{HK}}$ id est ut GT in $\sqrt{a}$, et Medij densitas ut Resistentia $\dfrac{3GT}{4VH}$ directe et quadratum velocitatis $a\times GT^q$ inverse, id est[24] ut $\dfrac{1}{GT}$. Et velocitas in loco quovis G ea est quacum corpus in Parabola pergeret verticem G diametrum GC & latus rectum $\dfrac{2GH^q}{HK}$ seu $\dfrac{GT^q}{VH}$ habente.

Exempl. 4. Ponatur indefinite quod linea $PRGQ$ Hyperbola sit centro X, asymptotis AX, NX, ea lege descripta ut constructo rectangulo $XZDN$ cujus latus ZD secet Hyperbolam in H & Asymptoton ejus in V, fuerit VH reciproce ut ipsius ZX vel DN dignitas aliqua ND^n cujus index est numerus n: et quæratur Medij densitas qua projectile progrediatur in hac curva.

---

(21) That is, CG.

(22) Understand this to be (as we show it in the accompanying figure) the extension of Gt, drawn parallel to the chord FH (and so bisecting HK in t), till its meet T with NX. This will of course be tangent to the infinitesimal arc $\overgroup{FH}$ of the resisted path at its mid-point G.

(23) Newton's preliminary calculations to this end survive on Add. 3968.41: 119^v (there entered, starting from the page bottom, in a blank space below a preliminary draft of a passage in the unpublished lengthy original English preface to the *Commercium Epistolicum* which is reproduced in **2**, Appendix 1.5 below; see our note (47) thereto). Since these are more detailed than the present bare statement of their result, we may here give their gist, namely:

$$\text{'}CG-DH=\frac{-maa+nbb}{naa}o+\text{&c.}\quad GH=\frac{\sqrt{mma^4-2mnaabb+nnb^4+nna^4}}{naa}o+\text{&c.}$$

converging series $b^2/a + (b^2/a^2)\, o + (b^2/a^3)\, o^2 + (b^2/a^4)\, o^3 + \ldots$ and there will come to be

$$DH = c - (m/n)\, a - b^2/a + ((ma^2 - nb^2)/na^2)\, o - (b^2/a^3)\, o^2 - (b^2/a^4)\, o^3 - \ldots,$$

For $c - (m/n)\, a - b^2/a^{(21)}$ write e, and there will be had the four ordinates

$$BF = e + ((nb^2 - ma^2)/na^2)\, o - (b^2/a^3)\, o^2 + (b^2/a^4)\, o^3 - \ldots,$$

$$CG = e, \quad DH = e + ((ma^2 - nb^2)/na^2)\, o - (b^2/a^3)\, o^2 - (b^2/a^4)\, o^3 \ldots$$

and $\qquad EI = e + 2((ma^2 - nb^2)/na^2)\, o - 4(b^2/a^3)\, o^2 - 8(b^2/a^4)\, o^3 \ldots.$

And thence $\quad HK = 2(b^2/a^3)\, o^2 \ldots, \quad IL = 2(b^2/a^3)\, o^2 + 6(b^2/a^4)\, o^3 \ldots$

and so $IL - HK = 6(b^2/a^4)\, o^3 \ldots.$ There is, further, $DN : GT^{(22)} = CD : \mathrm{GH}$ or $GH = (GT/a)\, o$, and $HK : \frac{1}{2}(IL - HK) = GH : LM$ or $LM = \frac{3}{2}(GT/a^2)\, o^2.^{(23)}$ And hence the resistance is to gravity as $\frac{3}{2}GT/a^2$ to $2b^2/a^3$, that is, $3GT$ to $4HV$. While the speed is as $GH/\sqrt{HK}$, that is, as $GT \times \sqrt{a}$; and the density of the medium as the resistance $3GT/4HV$ directly and the square of the speed $a \times GT^2$ inversely, that is,$^{(24)}$ as $1/GT$. And the speed at any place G is that with which the body would proceed in a parabola having vertex G, diameter GC and *latus rectum* $2GH^2/HK$, that is, GT^2/HV.

Example 4. Suppose indefinitely that the line $PRGQ$ be a hyperbola described with centre X and asymptotes AX, NX with the stipulation that, on constructing the rectangle $XNDZ$ whose side DZ shall cut the hyperbola in H and its asymptote in V, there be HV reciprocally as some power of (XZ or) ND, ND^n, whose index is the number n: then let there be sought the density of the medium whereby a projectile shall move onward in this curve.

$$[KH =]\ \frac{2bboo}{a^3} \cdot [\tfrac{1}{2}, \overline{LI - KH} =]\ \frac{6bbo^3}{a^4}\ (::a.3 \times o)$$

$$:: GH . LM = \frac{3\sqrt{mma^4 - 2mnaabb + nnb^4 + nna^4}}{na^3}\ oo.$$

[unde] $\qquad$ Resist . grav :: $3\sqrt{mma^4 - 2mnaabb + nnb^4 + nna^4} . 2nbb.$

[Etiam] $DN = a[-o] . XY :: o . \dfrac{o, XY}{a} = GH.$ [ut et] $\dfrac{2boo}{a^3} \cdot \dfrac{6bbo^3}{a^4} :: \dfrac{XY, o}{a} \cdot \dfrac{3XY, oo}{[2]aa} = LM.$

[adeoчз] $\qquad \dfrac{3XY, oo}{2aa} \cdot \dfrac{2bb, oo}{a^3} :: 3XY . \dfrac{4bb}{a} :: \mathrm{Resist . Grav} :: 3XY . 4VH = 2YG\ [sic]$'

(on minimally adjusting the last line). There follows on the worksheet a drafting of the two next sentences which in its final version does not differ from that which here ensues other than for insignificant verbal honings; we need not dwell on the near-trivial point that the *latus rectum* $2GH^2/HK$ of the parabola approximating the arc $\overset{\frown}{FGH}$ was initially there specified to be

$$`\frac{2GT^q, oo, a^3}{aa, 2bboo} = \frac{GT^q, a}{bb}`.$$

(24) Since $VH = b^2/a$ is proportional to $1/a$.

Pro CN, CD, NX scribantur n, o, c respective sitᴓ VZ ad ZX vel DN ut d ad f,

& VH æqualis $\dfrac{bb}{DN^n}$: et erit DN æqualis $a-o$, VH æqualis $\dfrac{bb}{\overline{a-o}^n}$, $VZ = \dfrac{d}{f}$ in $a-o$,

& DH seu $NX - VZ - VH$ æqualis $c - \dfrac{da}{f} + \dfrac{do}{f} - \dfrac{bb}{\overline{a-o}^n}$. Resolvatur terminus ille

$\dfrac{bb}{\overline{a-o}^n}$ in seriem infinitam $\dfrac{bb}{a^n} + \dfrac{nbbo}{a^{n+1}} + \dfrac{nn+n}{2a^{n+2}} bbo[o] + \dfrac{n^3+3nn+2n}{6a^{n+3}} bbo^3$ &c & pro

terminis $c - \dfrac{da}{f} - \dfrac{bb}{a^n}$ [21] scribatur e: ac fiet DH æqualis

$$e + \frac{do}{f} - \frac{nbbo}{a^{n+1}} - \frac{nn+n}{2a^{n+2}} bboo - \frac{n^3+3nn+2n}{6a^{n+3}} bbo^3 \ \&c.$$

Et inde fiet

$$EI = e + \frac{2do}{f} - \frac{2nbbo}{a^{n+1}} - \frac{2nn+2n}{a^{n+2}} bboo - \frac{4n^3+12nn+8n}{3a^{n+3}} bbo^3.$$

et

$$BF = e - \frac{do}{f} + \frac{nbbo}{a^{n+1}} - \frac{nn+n}{2a^{n+2}} bboo + \frac{n^3+3nn+2n}{6a^{n+3}} bbo^3.$$

Et inde $\quad HK = \dfrac{nn+n}{a^{n+2}} bboo. \quad IL = \dfrac{nn+n}{a^{n+2}} bboo - \dfrac{n^3+3nn+2n}{a^{n+3}} bbo^3.$

& $\qquad\qquad IL - HK = \dfrac{n^3+3nn+2n}{a^{n+3}} bbo^3.$

Estᴓ $\qquad\qquad DN . GT^{(22)} :: CD . GH = \dfrac{GT}{a} o.$

Et $\qquad\qquad KH . \dfrac{IL-HK}{2} :: GL . LM = \dfrac{n+2}{2aa} oo, \ GT.$

Adeoᴓ resistentia ad Gravitatem ut $\dfrac{n+2}{2aa} GT$ ad $\dfrac{nn+n}{a^{n+2}} bb$ seu $\dfrac{n+2}{2nn+2n} GT$ ad VH.[25]

---

[3][26]   *Prop. X. Prob. III.*

Tendat uniformis vis gravitatis directe ad planum Horizontis, sitᴓ resistentia ut Medij densitas et quadratum velocitatis conjunctim: requiritur tum Medij densitas in locis

---

(25) Newton here leaves off, leaving the remainder of his manuscript sheet (ff. 199ʳ [bottom half]/199ᵛ) blank—evidently with the intention of adjoining yet other things. As we have already mentioned in note (17) above, Newton jotted down on Add. 3965.10: 136ᵛ (virtually as we reproduce its edited text in Appendix 2.7 following) a rough confirmation that his present alternative mode of solution to his problem does indeed yield the corrected general measure of the ratio of resistance to gravity (in terms of the coefficients of the series expansion of the augmented ordinate in powers of the related base increment), but this he never wrote up in a finished form. Yet more fragmentary is the abandoned preliminary computation immediately above (on the same f. 136ᵛ) for what was manifestly destined to be the further worked example of a yet more complicated logarithmico-hyperbolic path of resisted fall.

For NC, DC, NX write n, o, c respectively, and let VZ be to XZ or ND as d to f, and $HV = b^2/ND^n$: then will there be

$$ND = a-o, \quad HV = b^2/(a-o)^n, \quad VZ = (d/f)\,(a-o)$$

and DH, that is, $NX - VZ - HV$, $= c - (d/f)\,a + (d/f)\,o - b^2/(a-o)^n$. Resolve the term $b^2/(a-o)^n$ into the infinite series

$$b^2/a^n + n(b^2/a^{n+1})\,o + \tfrac{1}{2}(n^2+n)\,(b^2/a^{n+2})\,o^2 + \tfrac{1}{6}(n^3 + 3n^2 + 2n)\,(b^2/a^{n+3})\,o^3 \ldots$$

and in place of the terms $c - (d/f)\,a - b^2/a^{n}$[21] write e, and there will come to be $DH = e + (d/f - nb^2/a^{n+1})\,o - \tfrac{1}{2}(n^2+n)\,(b^2/a^{n+2})\,o^2 - \tfrac{1}{6}(n^3 + 3n^2 + 2n)\,(b^2/a^{n+3})\,o^3 \ldots$. And thence will come

$$EI = e + 2(d/f - nb^2/a^{n+1})\,o - 2(n^2+n)\,(b^2/a^{n+2})\,o^2 - \tfrac{4}{3}(n^3 + 3n^2 + 2n)\,(b^2/a^{n+3})\,o^3 \ldots$$

and

$$BF = e - (d/f - nb^2/a^{n+1})\,o - \tfrac{1}{2}(n^2+n)\,(b^2/a^{n+2})\,o^2 + \tfrac{1}{6}(n^3 + 3n^2 + 2n)\,(b^2/a^{n+3})\,o^3 \ldots;$$

and thence $HK = (n^2+n)\,(b^2/a^{n+2})\,o^2 \ldots$,

$$IL = (n^2+n)\,(b^2/a^{n+2})\,o^2 - (n^3 + 3n^2 + 2n)\,(b^2/a^{n+3})\,o^3 \ldots,$$

and so $IL - HK = (n^3 + 3n^2 + 2n)\,(b^2/a^{n+3})\,o^3 \ldots$. Now there is

$$ND:GT^{(22)} = DC:GH \quad \text{or} \quad GH = (GT/a)\,o,$$

and $\quad HK:\tfrac{1}{2}(IL - HK) = GL:LM \quad \text{or} \quad LM = \tfrac{1}{2}(n+2)\,(GT/a^2)\,o^2;$

and hence the resistance is to gravity as $\tfrac{1}{2}(n+2)\,.\,GT/a^2$ to $(n^2+n)\,b^2/a^{n+2}$, that is, $(\tfrac{1}{2}(n+2)/(n^2+n))\,GT$ to HV.[25]

---

[3][26]   *Proposition X, Problem III.*

Let the uniform force of gravity tend directly to the plane of the horizon, and let the resistance be as the density of the medium and the square of the speed jointly: there is required both the density of the medium in individual places which shall make a body move

---

Perhaps this intended illustration also was meant to be adjoined to the present piece. Let the reader look at Appendix 2.9 (where, for lack of any more appropriate place, we reproduce the essence of its text) and decide for himself.

(26) Add. 3965.12: 192^r–193^v, the definitive version of Newton's mended solution of the general problem of resisted motion under constant downwards gravity, now finally recast from [1] preceding (by way of the intermediate draft reproduced in Appendix 2.8 below) with the variant approach of [2] briefly encapsulated in the opening paragraph of the augmented ensuing scholium—the latter in fact, for want of room in the space available, not to appear at all in the published *editio secunda* of the *Principia*. With minimal verbal polishing and a few terminal additions (which, for completeness' sake, we include in our present text) this is 'the tenth Proposition of the second Book corrected' (now Trinity College. R.16.38: 262–5) which

singulis quæ faciat ut corpus in data quavis linea curva moveatur, tum corporis velocitas et Medij resistentia in locis singulis.

Sit PQ planum illud plano schematis perpendiculare; $PFHQ$ linea plano huic occurrens in punctis P et Q; [27]G, H, I, K loca quatuor corporis in hac curva ab F ad Q pergentis; & GB, HC, ID, KE ordinatæ quatuor parallelæ ab his punctis

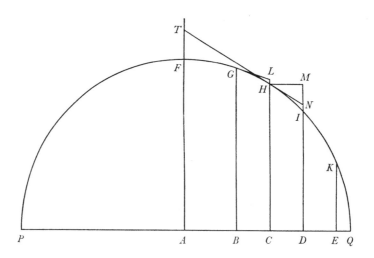

ad horizontem demissæ & lineæ horizontali PQ ad puncta B, C, D, E insistentes; & sint BC, CD, DE distantiæ Ordinatarum inter se æquales. A punctis G et H ducantur rectæ GL, HN curvam tangentes in G et H et ordinatis CH, DI sursum productis occurrentes in L et N et compleatur parallelogrammum $HCDM$. Et tempora quibus corpus describet arcus GH, HI erunt in subduplicata ratione altitudinum $LH NI$ quas corpus temporibus illis describere posset, a tangentibus cadendo: et velocitates erunt ut longitudines descriptæ GH, HI directe et tempora inverse. Exponantur tempora per T et t & velocitates per $\dfrac{GH}{T}$ et $\dfrac{HI}{t}$:

et decrementum velocitatis tempore t factum exponetur per $\dfrac{GH}{T} - \dfrac{HI}{t}$. Hoc decrementum oritur a resistentia corpus retardante et gravitate corpus accelerante. Gravitas in corpore cadente et spatium NI cadendo describente, generat velocitatem qua duplum illud spatium eodem tempore describere potuisset (ut Galilæus demonstravit[28]), id est velocitatem $\dfrac{2NI}{t}$: at in corpore arcum HI describente auget arcum illum sola longitudine $HI - HN$ seu $\dfrac{MI \times NI}{HI}$ ideoq́ generat tantum velocitatem $\dfrac{2MI \times NI}{t \times HI}$. Addatur hæc velocitas ad decrementum prædictum & habebitur decrementum velocitatis ex resistentia sola

in any given curved line, and also the speed of the body and resistance of the medium at individual places.

Let PQ be that plane perpendicular to the plane of the figure; $PFHQ$ a line meeting this plane in the points P and Q; [27]G, H, I, K four places of the body proceeding in this curve from F to Q; and GB, HC, ID, KE four parallel ordinates let fall from these points to the horizon and standing on the horizontal line PQ at the points B, C, D, E, and let the distances BC, CD, DE from one another be equal. From the points G and H draw straight lines GL, HN touching the curve at G and H and meeting the ordinates CH and DI extended upwards in L and N, and complete the rectangle $HCDM$. Then the times in which the body shall describe the arcs $\overset{\frown}{GH}$, $\overset{\frown}{HI}$ will be in the halved ratio of the heights LH, NI which the body could in those times describe in falling from the tangents; and the speeds will be as the lengths $\overset{\frown}{GH}$, $\overset{\frown}{HI}$ described directly and the times inversely. Express the times by T and t, and the speeds by $\overset{\frown}{GH}/T$ and $\overset{\frown}{HI}/t$; and then the decrement of the speed achieved in time t will be expressed by $\overset{\frown}{GH}/T - \overset{\frown}{HI}/t$. This decrement arises from the resistance slowing the body and gravity accelerating it. Gravity in a falling body which, as it falls, describes the distance NI generates a speed with which it could have covered twice that distance in the same time (as Galileo demonstrated[28]), *i.e.* the speed $2IN/t$; but in a body describing the arc $\overset{\frown}{HI}$ it increases that arc by the length $\overset{\frown}{HI} - HN$ alone, that is, by $IM \times IN/\overset{\frown}{HI}$, and in consequence generates merely the speed $2IM \times IN/t \times \overset{\frown}{HI}$. Add this speed to the previously noted decrement and there will be had the

---

Newton eventually sent to his editor, Roger Cotes, on 6 January 1712/13 with the scanty comment that 'it will require the reprinting of a sheet and a quarter from pag. 230 to pag. 240. There is a wooden cut [of the figure] belonging to it wch I intend to send you by the next Carrier. I think this Proposition as it is now done will take up much the same space as before' (*ibid.*: 261, first published by Edleston in his edition of *The Correspondence of Sir Isaac Newton and Professor Cotes* (London, 1850): 145 [= *Correspondence of Isaac Newton*, 5, 1975: 361]). So lightly he dismissed his hard sweat over many days to mend and reconstruct the detail of his solution to a problem in a variety of ways, as only he knew how!

(27) Newton initially went on to specify, much as in [1] preceding (compare the intervening draft in Appendix 2.8 below): 'H corpus in ipsa motum, et [H]N recta ipsam tangens in [H]. Fingatur autem corpus H progredi ab F ad Q per lineam illam curvam $PFHQ$, et interea impediri a Medio resistente' (H the body moved in it, and HN the straight line touching it at H. Imagine, however, the body H to advance from F to Q by way of that curved line $PFHQ$, and all the while to be impeded by the resisting medium). In the manuscript original he carelessly omitted wholly to adapt the text of this cancelled passage to refer accurately to his re-lettered accompanying figure, where the denotations A, B, C, D, E of points in the base PQ and those, F, G, H, I, K, of corresponding points in the arc $\overset{\frown}{FHQ}$ itself now logically mirror in their advancing sequence the successive positions of the projectile H in its resisted flight from F to Q. A few similar trivial lapses on Newton's part are silently corrected by us into their intended referents in sequel.

(28) See [2]: note (14) above, where we suggest that Newton might here more appropriately have written 'ut Hugenius demonstravit'.

oriundum, nempe $\dfrac{GH}{T} - \dfrac{HI}{t} + \dfrac{2MI \times NI}{t \times HI}$. Proindeq̃ cum gravitas eodem tempore

in corpore cadente generet velocitatem $\dfrac{2NI}{t}$, Resistentia erit ad gravitatem ut

$\dfrac{GH}{T} - \dfrac{HI}{t} + \dfrac{2MI \times NI}{t \times HI}$ ad $\dfrac{2NI}{t}$, [29] sive ut $\dfrac{t, GH}{T} - HI + \dfrac{2MI \times NI}{HI}$ ad $2NI$.

[30]Jam pro Abscissis BC,[31] CD, $[C]E$ scriba[n]tur $-o$, o, $2 \times o$, pro ordinata CH scribatur P, et pro MI scribatur series quælibet $Qo + Roo + So^3 + $ &c. Et seriei termini omnes post primum, nempe $Roo + So^3 + $ &c erunt NI, & Ordinatæ DI, EK et BG erunt $P - Qo - Roo - So^3$, $P - 2Qo - 4Ro[o] - 8So^3$ & $P + Qo - Roo + So^3$ respective. Et quadrando Differentias Ordinatarum $BG - CH$ et $CH - DI$ & ad quadrata prodeuntia addendo quadrata ipsarum BC, CD, habebuntur arcuum GH, HI quadrata $oo + QQoo - 2QRo^3$ &[c] et $oo + QQoo + 2QRo^3 + $ &c. Quorum radices $o\sqrt{1+QQ} - \dfrac{QRoo}{\sqrt{1+QQ}}$ & $o\sqrt{1+QQ} + \dfrac{QRoo}{\sqrt{1+QQ}}$ sunt arcus GH et HI.

Præterea si ab Ordinata CH subducatur semisumma Ordinatarum BG et DI, et ab Ordinata DG subducatur semisumma Ordinatarum CH et EK, manebunt arcuum GI et HK sagittæ Roo et $Roo + 3So^3$. Et hæ sunt lineolis LH et NI proportionales[32] adeoq̃ in subduplicata[33] ratione temporum infinite parvorum T et t, ideoq̃[34] ratio temporum $\dfrac{t}{T}$ est $\sqrt{\dfrac{R + 3So}{R}}$ seu $\dfrac{R + \frac{3}{2}So}{R}$. Et $\dfrac{t}{T} GH - HI + \dfrac{2MI \times NI}{HI}$

substituendo ipsorum $\dfrac{t}{T}$, GH, HI, MI et NI valores jam inventos evadit

---

(29) This is the fundamental equation of motion in the tangential direction at H, as will perhaps more readily appear to modern eyes on naming the forces of resistance and gravity acting instantaneously upon the missile at H to be (in our usual notation) ρ and g respectively, and also setting $V = \widehat{GH}/T$ and $v = \widehat{HI}/t$ to be the (mean) speeds over the successive arcs $\widehat{GH}$ and $\widehat{HI}$: for then, since $MI/HI = MN/HN$ to adequate accuracy, there comes, on compounding (with Newton) the various increments and decrements of speed here generated, to be $\rho t = V - v + (MN/HN) \cdot 2NI/t$ where $(NI = \frac{1}{2}gt^2$ and so) $2NI/t = gt$, and consequently $-(v-V)/t = \rho - (MN/NH) \cdot g$; that is, the deceleration of the body at H is instantaneously hereby compounded of the medium's direct resistance to its motion and the contrary acceleration of its speed due to the component of downwards gravity acting along the tangent HN.

(30) We may notice (for what it is worth) that Newton's preparatory draft of the next sentence and a half—in exactly the same words (and with the same slip '..., DE scribatur' which we mend in immediate sequel)—is found at the bottom of Add. 3965.10: 136^v.

(31) More precisely, in line with Newton's following assignation of its length as $-o$, this should read 'CB', and so Cotes—whom we follow in our English version—changed it to be in the fair copy of the present text which (see note (26) preceding) he received in January 1713.

(32) In fact $LH = Ro^2 - 2So^3$ and $NI = Ro^2 + So^3$, and these are indeed (on ignoring higher powers of o) in the given ratio $(Ro^2 - 2So^3)/Ro^2 = (Ro^2 + So^3)/(Ro^2 + 3So^3) = 1 - 2(S/R)o$ to the corresponding *sagittæ*; and this holds true generally since, corresponding to the tangential

decrement of the speed arising from the resistance alone, namely

$$\widehat{GH}/T - \widehat{HI}/t + 2IM \times IN/t \times \widehat{HI}.$$

Accordingly, since gravity in the same time in a falling body generates the speed $2IN/t$, the resistance will be to gravity as $\widehat{GH}/T - \widehat{HI}/t + 2IM \times IN/t \times \widehat{HI}$ to $2IN/t$,[29] that is, as $\widehat{GH} \times t/T - \widehat{HI} + 2IM \times IN/\widehat{HI}$ to $2IN$.

[30] Now for the abscissas CB, CD, CE write $-o$, o and $2o$, for the ordinate CH write P, and in place of IM write any series $Qo + Ro^2 + So^3 + \dots$ you please. Then all the terms of the series after the first, namely, $Ro^2 + So^3 + \dots$, will be IN; and the ordinates DI, EK and BG will be

$$P - Qo - Ro^2 - So^3 \dots, \quad P - 2Qo - 4Ro^2 - 8So^3 \dots \quad \text{and} \quad P + Qo - Ro^2 + So^3 \dots$$

respectively. And by squaring the differences of ordinates $BG - CH$ and $CH - DI$, and to the resulting squares adding the squares of BC and CD, there will be had the squares of the arcs $\widehat{GH}$ and $\widehat{HI}$,

$$o^2 + Q^2o^2 - 2QRo^3 \dots \quad \text{and} \quad o^2 + Q^2o^2 + 2QRo^3 + \dots;$$

the roots of which,

$$o\sqrt{(1+Q^2)} - QRo^2/\sqrt{(1+Q^2)} \quad \text{and} \quad o\sqrt{(1+Q^2)} + QRo^2/\sqrt{(1+Q^2)},$$

are the arcs $\widehat{GH}$ and $\widehat{HI}$. Moreover, if from the ordinate CH there be taken the half-sum of the ordinates BG and DI, and from the ordinate DG be taken the half-sum of the ordinates CH and EK, there will remain the *sagittæ* of the arcs $\widehat{GI}$ and $\widehat{HK}$, viz. Ro^2 and $Ro^2 + 3So^3$: these are proportional to the linelets HL and IN,[32] and hence in the doubled[33] ratio of the infinitely small times T and t, so that[34] the ratio of the times t/T is $\sqrt{(R + 3So)}/\sqrt{R}$, or $(R + \frac{3}{2}So)/R$. And so $\widehat{GH} \times t/T - \widehat{HI} + 2IM \times IN/\widehat{HI}$ comes, by substituting for t/T, $\widehat{GH}$, $\widehat{HI}$, IM and

---

deviation $R_{a+(n-1)o}o^2 + S_{a+(n-1)o}o^3 = Ro^2 + (3n-2) So^3$ where the base increment is no, the related *sagitta* is $\frac{1}{2}(e_{a+(n-1)o} + e_{a+(n+1)o}) - e_{a+no} = Ro^2 + 3nSo^3$, with

$$(Ro^2 + (3n-2) So^3)/(Ro^2 + 3nSo^3) = 1 - 2(S/R) o, \text{ constant},$$

on again neglecting higher powers of o. But to demonstrate this proportionality of tangential deviation to corresponding *sagitta* we need first to evaluate both. It is curious that Newton should here assume the constancy of this ratio as 'evident' in seeking to forgo computation of the deviations LH and NI in favour of their more directly calculable 'arrows'.

(33) We render in our English version the sense of the 'duplicata' which Newton meant—here as in his preparatory draft (see Appendix 2.8: note (64))—to write. The obtrusive prefix 'sub' was afterwards struck out by Cotes in the fair copy sent to him in January 1713 (see note (26) above), and the mote did not endure to impair the clarity of vision of the reader of the *Principia*'s ensuing *editio secunda*.

(34) Afterwards trivially altered to be 'et inde' (and thence) in the copy sent to Cotes for publication.

$\dfrac{3Soo}{2R}\sqrt{1+QQ}$.[35] Et cum $2NI$ sit $2Roo + \&c_{[,]}$ Resistentia jam erit ad gravitatem

ut $\dfrac{3Soo}{2R}\sqrt{1+QQ}$ ad $2Roo$, id est ut $3S\sqrt{1+QQ}$ ad $4RR$.[36]

Velocitas autem ea est quacum corpus de loco quovis H secundū tangentem HN egrediens, in Parabola diametrum HC et latus rectum $\dfrac{HN^q}{NI}$ seu $\dfrac{1+QQ}{R}$ habente, deinceps in vacuo moveri potest.[37]

---

(35) On ignoring terms of the order of o^3, that is. Newton's checking computation survives at the bottom of Add. 3965.10: 136^v, evaluating $(t/T)\,\widehat{GH} - \widehat{HI} + 2MI \times NI/HI$, that is,

$$(1 + \tfrac{3}{2}(S/R)\,o)\,(o\sqrt{(1+Q^2)} - QRo^2/\sqrt{(1+Q^2)}) - (o\sqrt{(1+Q^2)} + QRo^2/\sqrt{(1+Q^2)})$$
$$+ 2QRo^2/\sqrt{(1+Q^2)}$$

to be   '$o\sqrt{}) - \dfrac{QRoo}{\sqrt{})} + \dfrac{3Soo\sqrt{}}{2R} - \dfrac{3So^3Q}{2\sqrt{}} - o\sqrt{}) - \dfrac{QRoo}{\sqrt{})} + \dfrac{2QRoo}{\sqrt{})}$',

therein rejecting mutually cancelling terms and suppressing the relatively negligible $-\tfrac{3}{2}QSo^3/\sqrt{(1+Q^2)}$ to achieve his final '$= \dfrac{3Soo\sqrt{1+QQ}}{2R}$'.

(36) Newton initially went on to conclude his paragraph with a first version of Corollary 1 below, stating: 'Ordinata quævis AF et tangens HN productæ concurrant in T, et erit CD ad HN ut AC ad HT, adeoᵍ $\dfrac{HT}{AC} = \dfrac{HN}{CD} = \sqrt{1+QQ}$' (Let any ordinate AF and the tangent NH meet, when produced, in T, and there will then be CD to HN as AC to HT, and hence $HT/AC = HN/CD = \sqrt{(1+Q^2)}$).

(37) For in the terms of note (29) preceding, since the projectile in its curving path from H deviates from its tangential motion along HN under the downwards pull of gravity g through the distance $NI = \tfrac{1}{2}gt^2$ in the infinitesimal time t in which the vanishingly small arc $\widehat{HI}$ of its trajectory is covered, the square of its instantaneous speed $v = \widehat{HI}/t \approx HN/t$ at H is therefore $HN^2/(2NI/g) \propto HN^2/NI = \lim\limits_{o\,\to\,\mathbf{zero}} [o^2(1+Q^2)/(Ro^2 + \ldots)] = (1+Q^2)/R$, the *latus rectum* of the parabola which osculates the projectile path over its infinitesimal arc $\widehat{GHI}$, having for its diameter the line (*viz.* HC) drawn through its vertex H parallel to NI.

From this value, $v = \sqrt{(\tfrac{1}{2}g \cdot (1+Q^2)/R)}$ in explicit form, for the body's instantaneous speed, Newton's preceding ratio for the ratio of resistance ρ to gravity g can be directly obtained—without any need to make prior calculation of the lengths of the arclets $\widehat{GH}$, $\widehat{HI}$ and the ratio T/t of the times in which these are successively traversed—from the basic equation of motion at H, namely (again see note (29) above) $\rho - (MN/HN) \cdot g = -(v-V)/t$, that is, $-v(v-V)/HN$; for at once, since $MN = Qo$ and so $HN = o\sqrt{(1+Q^2)}$ where $HM = CD = o$ is the increment $d(AC)$ of the abscissa $AC = a$, and $v-V$ is the related increment of the speed v, there is in Leibnizian equivalent $\rho - (Q/\sqrt{(1+Q^2)}) \cdot g = -v \cdot dv/da \cdot \sqrt{(1+Q^2)}$ and consequently $\rho/g = (Q - \tfrac{1}{2}d(v^2/g)/da)/\sqrt{(1+Q^2)}$; whence, because Q, R, S are the coefficients of the Taylorian expansion $e + Qo + Ro^2 + So^3 + \ldots$ of the incremented ordinate $DI = e_{a+o}$ (and so $Q = de/da$, $R = \tfrac{1}{2}d^2e/da^2 = \tfrac{1}{2}dQ/da$, $S = \tfrac{1}{6}d^3e/da^3 = \tfrac{1}{3}dR/da$), there is

$$\tfrac{1}{2}d(v^2/g)/da = \tfrac{1}{4}d((1+Q^2)/R)/da = \tfrac{1}{4}(4Q - (1+Q^2) \cdot 3S/R^2)$$

and consequently $\rho/g = \tfrac{3}{4}S\sqrt{(1+Q^2)}/R^2$, as before. Except for the last reduction to Newton's corrected expression of this ratio in the *Principia*'s second edition—but with the elegance that

IN the values now found, to be $\frac{3}{2}(S/R)\sqrt{(1+Q^2)}\,o^2$.[35] And since $2IN$ is $2Ro^2 + ...$, the resistance will now be to gravity as $\frac{3}{2}(S/R)\sqrt{(1+Q^2)}\,o^2$ to $2Ro^2$, that is, as $3S\sqrt{(1+Q^2)}$ to $4R^2$.[36]

The speed, however, is that with which a body, setting out from any place *H* in the direction of the tangent *HN*, can thereafter move *in vacuo* in a parabola having diameter *HC* and *latus rectum* HN^2/NI, that is, $(1+Q^2)/R$.[37]

---

the speed v was there determined by setting the component $g/\sqrt{(1+Q^2)}$ of gravity normal to the curve at *H* equal to the *vis centripeta* v^2/r, where $r(=(1+Q^2)^{\frac{3}{2}}/2R)$ is the radius of curvature at the point (compare VI: 131, note (86), and also 548–9, note (25))—this mode of solution had been attained two years earlier by Johann Bernoulli, and had indeed then led him on to detect Newton's error in 'Exempl. 1' in its original evaluation of the ratio of resistance to gravity in the instance of the semicircle $e = \sqrt{(n^2 - a^2)}$ to be (on accurately applying the deficient general measure $\frac{1}{2}S\sqrt{(1+Q^2)}/R^2$) but a/n. Here of course there is $r = n$ by geometrical definition, while straightforwardly $Q = de/da = -a/e$, so that $\sqrt{(1+Q^2)} = -n/e$ and consequently $d(v^2/g)/da = d(n/\sqrt{(1+Q^2)})/da = -de/da$, whence there is $\rho/g = (Q - \frac{1}{2}(-Q))/\sqrt{(1+Q^2)} = \frac{3}{2}a/n$ *recte*; and so Bernoulli found it, privately communicating the particular result—and an outline of its derivation—to Leibniz on 12 August 1710 (N.S.) (see (ed. C. I. Gerhardt) *Leibnizens Mathematische Schriften*, **3** (Halle, 1856): 855). Five months afterwards Bernoulli sent the gist of the general formula $\rho = (gQ - v.dv/da)/\sqrt{((1+Q^2))}$ to Paris in a letter (addressed to Varignon?) of 10 January 1711 (N.S.) which was read out eighteen days later to the assembled members of the *Académie des Sciences* (though not to be printed till 1714 in its *Mémoires de Mathématique & de Physique pour l'Année M.DCCXI* [→ Paris, ₂1730: 47–54]), and where he again pointed in a concluding 'Remarque' to Newton's 'méprise...dans l'application qu'il a faite, *Pag. 265*. de sa solution du *Prob. 3. pag. 260*. au cercle...dans ses *Princ. Math.*' (see *ibid.*: 50–1). In this form the solution to the direct problem of resisted motion under a general 'force centripete de pesanteur' g was in early October 1712 shown by Johann's nephew Niklaus (during the latter's stay in London) personally to both Newton and Abraham de Moivre; in consequence of which de Moivre wrote to Johann on the following 17 December to report that 'M. Newton ...souhaite que je sçache, si vous voudriez qu'on imprimât un théoreme de vous, que Monsieur votre neveu nous a communiqué pour déterminer la raison de la résistance à la force centripete, ou si vous aimez mieux attendre que son traité [the *Principia* in its *editio secunda auctior et emendatior*] paroisse, qui sera vers la fin du mois de février prochain...', adding on his own account that 'je l'ai examiné plusieurs fois, et il me paroit très beau et digne de vous' (Wollenschläger, 'Briefwechsel' (§1: note (1)): 277). But—even as Newton was honing his mended Proposition X into its present final version—Bernoulli himself, doubtless spurred on yet more fiercely by de Moivre's news that his nephew had imparted his theorem on resisted motion under gravity (to a finite force-centre in a straightforward generalization) to Newton, was hard at work in Basel elaborating the geometrical construction of its equivalent form $\rho/g = (Q - \frac{1}{2}d(r/\sqrt{(1+Q^2)})/da)/\sqrt{(1+Q^2)}$ to be the basic 'Problema I' of what he described in his reply to de Moivre on 18 February 1713 (N.S.) as 'un écrit de 40 pages...qui porte pour titre: "De motu corporum gravium, pendulorum, et projectilium in mediis non resistentibus et resistentibus, supposita gravitate uniformi et non uniformi, atque ad quodvis datum unum punctum tendente, et de variis [aliis] huc spectantibus, demonstrationes geometricæ" [et que] j'ai envoyé au commencement de cette année à Leipsic...pour être publié dans les *Actes* [→ *Acta Eruditorum* (February/March 1713): 77–95/115–32].... Vous y [= §§35–38 (+*Additio*): 115–18 (+132)] trouverez la démonstration de mon théoreme que mon neveu vous a communiqué pour déterminer la raison de la résistance à la force centripete, dont j'ai donné l'analyse il y a 3[!] ans à l'Académie des Sciences de Paris dans une piece [qui] se trouvera dans les *Mémoires* de l'Académie de l'an 1710[!] nouvellement imprimé, à ce qu'on m'écrit,

[Et] Resistentia [est] ut Medij densitas & quadratum velocitatis conjunctim & propterea Medij densitas ut Resistentia directe & quadratum velocitatis inverse, id est ut $\dfrac{3S\sqrt{1+QQ}}{4RR}$ directe & $\dfrac{HI^q}{t}$ seu $\dfrac{1+QQ}{R}$ inverse, hoc est ut

$\dfrac{S}{R\sqrt{1+QQ}}$.   Q.E.I.[38]

Corol. 1. Si tangens *HN* producatur utrinꝗ donec occurrat Ordinatæ cuilibet

AF in *T*: erit $\dfrac{HT}{AC}$ æqualis $\sqrt{1+QQ}$, adeoꝗ in superioribus pro $\sqrt{1+QQ}$ scribi

potest. Qua ratione Resistentia erit ad gravitatem ut $3S \times HT$ ad $4RR \times AC$,

Velocitas erit ut $\dfrac{HT}{AC\sqrt{R}}$ et Medij densitas erit ut $\dfrac{S \times AC}{R \times HT}$.

Corol. 2. Et hinc si Curva linea *PFHQ* definiatur per relationem inter basem seu abscissam *AC* et Ordinatim applicatam *CH* (ut moris est) et valor ordinatim applicatæ resolvatur in seriem convergentem: Problema per primos seriei terminos expedite solvetur[,] ut in exemplis sequentibus.

Exempl. 1. Sit linea *PFHQ* semicirculus super diametro *PQ* descriptus, & requiratur Medij densitas quæ faciat ut Projectile in hac linea moveatur.

Bisecetur semicirculi diameter *PQ* in *A* et dic *AQ n*, *AC a*, *CH e*, et *CD o*: et erit

DI^q seu $AQ^q - AD^q = nn - aa - 2ao - oo$ seu $ee - 2ao - oo$, et radice per methodum

nostram extracta fiet $DI = e - \dfrac{ao}{e} - \dfrac{oo}{2e} - \dfrac{aaoo}{2e^3} - \dfrac{ao^3}{2e^3} - \dfrac{a^3o^3}{2e^5}$ &c. Hic scribatur *nn*

pro $ee + aa$ et evadet $DI = e - \dfrac{ao}{e} - \dfrac{nnoo}{2e^3} - \dfrac{anno^3}{2e^5}[-]$ &c.

Hujusmodi series distinguo in terminos successivos in hunc modum. Terminum primum appello in quo quantitas infinite parva *o* non extat; secundum in [quo] quantitas illa est unius dimensionis; tertium in quo extat duarum,

---

mais si vous croyez digne qu'on l'imprime séparément dans vos *Transactions*, je vous en laisse le maître, pour le faire paroître en telle maniere que vous jugerez à propos' (Wollenschläger, 'Briefwechsel': 282–3). There could of course, as Bernoulli very well knew, be no question of giving a simultaneous printing in the *Philosophical Transactions* of a lengthy article which bid fair to render the *editio secunda* of the *Principia* obsolete even before it was published. When shown Bernoulli's letter by de Moivre, Newton's reaction was to seek to dampen the growl of his thunder. 'I heare', he wrote on 31 March to his editor, Roger Cotes, in Cambridge, 'that Mr Bernoulli has sent a Paper of 40 pages to be published in the *Acta Leipsica* relating to what I have written upon the curve Lines described by Projectiles in resisting Mediums. And therein he partly makes Observations upon what I have written & partly improves it. To prevent being blamed by him or others for any disingenuity in not acknowledging my oversights or slips in the first edition I beleive it will not be amiss to print next after the old *Præfatio ad Lectorem*, the following Account of this new Edition. *In hac secunda Principiorum Editione, multa sparsim emendantur & nonnulla adjiciuntur....*' (Trinity College, Cambridge. R.16.38: 284 [=*Correspondence of Isaac Newton*, 5, 1975: 400]). That Newton did not in this newly inserted *Auctoris Præfatio* mention either Bernoulli's name or the 'Libri secundi Prop. X'

While the resistance is as the density of the medium and square of the speed jointly, and accordingly the density of the medium is as the resistance directly and the square of the speed inversely, that is, as $3S\sqrt{(1+Q^2)}/4R^2$ directly and $\widehat{HI}^2/t$ or $(1+Q^2)/R$ inversely—in other words, as $S/R\sqrt{(1+Q^2)}$. As was to be found.[38]

Corollary 1. If the tangent HN be produced either way till it shall meet any ordinate AF you please in T, there will then be TH/AC equal to $\sqrt{(1+Q^2)}$, and hence can be written in the above in its place. For which reason the resistance will be to gravity as $3S \times TH$ to $4R^2 \times AC$, the speed will be as $TH/AC\sqrt{R}$, and the density of the medium as $S \times AC/R \times TH$.

Corollary 2. And hence, if the curved line $PFHQ$ be defined by means of a relationship between the base or abscissa AC and the ordinate CH (as is customary) and the value of the ordinate be resolved into a converging series, then the problem will be promptly solved through the first terms of the series, in the following examples.

Example 1. Let the line $PFHQ$ be a semicircle described upon the diameter PQ, and let there be required the density of the medium which shall make a projectile move in this line.

Bisect the semicircle's diameter PQ in A and call $AQ = n$, $AC = a$, $CH = e$ and $CD = o$: then will there be DI^2, that is, $AQ^2 - AD^2 = n^2 - a^2 - 2ao - o^2$, or $e^2 - 2ao - o^2$, and with the root extracted by our method there will come to be $DI = e - (a/e)\,o - \frac{1}{2}(1/e + a^2/e^3)\,o^2 - \frac{1}{2}(a/e^3 + a^3/e^5)\,o^3\ldots$. Here write n^2 in place of $e^2 + a^2$ and there will prove to be $DI = e - (a/e)\,o - \frac{1}{2}(n^2/e^3)\,o^2 - \frac{1}{2}(an^2/e^5)\,o^3 - \ldots$.

Series of this sort I distinguish into their successive terms in this fashion. The first term I call that in which the infinitely small quantity o does not occur; the second one that in which the quantity occurs of one dimension; the third that in which it occurs of two dimensions, the fourth that in which it is of three, and

which the latter had (by way of his nephew) led him to reconstruct with so much trouble was at once a meanness—if not spite—and a cowardice which stored up in Bernoulli's mind a hoard of bitter recrimination which, once publicly displayed, both lived on to regret. (The tale is briefly told in our preceding introduction.) It was not, we may here anticipate, seemingly till six years afterwards in 1719 that Bernoulli, taunted by what he took to be a challenge from John Keill to solve the inverse problem (see note (50) below) of determining the path $e = e_a$ of a projectile through a medium which resists its onrush instantaneously as some given power of its speed, at length made equivalent reduction of his resistance equation to derive from it Newton's measure $\rho/g = \frac{3}{4}S\sqrt{(1+Q^2)}/R^2$—or more precisely, in his preferred Leibnizian terms, $\rho/g = \frac{1}{2}(d^3e/da^3)\,(ds/da)/(d^2e/da^2)$, where $\widehat{FH} = s$ is the arc-length of the missile path corresponding to the base distance $AC = a$ travelled.

(38) As in the preliminary version reproduced in Appendix 2.8 (see our note (62) thereto), these last two paragraphs of the main text are found in reverse order in the manuscript, but we obey Newton's subsequent instruction to invert their sequence, minimally recasting the original opening 'Est autem Resistentia ut...' (The resistance, however, is as...) to be as it is in the lightly corrected copy of the present proposition sent to Cotes in January 1713 for publication.

quartum in quo trium est, & sic in infinitum. Et primus terminus qui hic est e denotabit semper longitudinem ordinatæ CH insistentis ad initium[39] indefinitæ quantitatis o; secundus terminus qui hic est $\frac{ao}{e}$, denotabit differentiam inter CH et DN, id est lineolam MN quæ abscinditur complendo parallelogrammum $HCDM$, atcp adeo positionem Tangentis HN semper determinat: ut in hoc casu capiendo MN ad HM ut est $\frac{ao}{e}$ ad o, seu a ad e. Terminus tertius qui hic est $\frac{nnoo}{2e^3}$ designabit lineolam IN quæ jacet inter tangentem et Curvam adeocp determinat angulum contactus IHN seu curvaturam quam curva linea habet in H. Si lineola illa IN finitæ est magnitudinis, designabitur per terminum tertium una cum sequentibus in infinitum. At si lineola illa minuatur in infinitum, termini subsequentes evadent infinite minores tertio, ideocp negligi possunt. Terminus quartus qui hic est $\frac{anno^3}{2e^5}$, determinat variationem curvaturæ, quintus variationem variationis, & sic deinceps.[40] Unde obiter patet usus non contemnendus harum serierum in solutione Problematum quæ pendent a Tangentibus et Curvatura curvarum.

Conferatur jam series $e - \frac{ao}{e} - \frac{nnoo}{2e^3} - \frac{anno^3}{2e^5}$ &c qua ordinata DI in Problemate jam solvendo designatur cum serie $P - Qo - Roo - So^3$ [&c] qua designabatur in inventione solutionis, et perinde pro $P\,Q\,R$ et S scribantur $e, \frac{a}{e}, \frac{nn}{2e^3}$ & $\frac{ann}{2e^5}$, et pro $\sqrt{1+QQ}$ scribatur $\sqrt{1+\frac{aa}{ee}}$ seu $\frac{n}{e}$, et prodibit Medij densitas ut $\frac{a}{ne}$, hoc est (ob datam n) ut $\frac{a}{e}$ seu $\frac{AC}{CH}$, id est, ut Tangentis longitudo illa HT quæ ad semidiametrum AF ipsi PQ normaliter insistentem terminatur: & resistentia erit ad gravitatem ut $3a$ ad $2n$ id est ut $3AC$ ad circuli diametrum PQ; velocitas autem erit ut $\sqrt{2CH}$.[41] Quare si corpus[42] justa cum velocitate secundum lineam ipsi PQ parallelam exeat de loco F, et Medij densitas in singulis locis H sit ut longitudo tangentis HT, & resistentia etiam in loco aliquo H sit ad vim gravitatis ut $3AC$ ad PQ, corpus illud describet circuli quadrantem FHQ.   Q.E.I.

At si corpus idem de loco P secundum lineam ipsi PQ perpendicularem

---

(39) The point C at which the linelet $CD = o$ begins.

(40) See Appendix 1: note (20) below, where we also remark that when he came subsequently to review the text of the *Principia*'s *editio secunda* he tentatively framed an addition *Ad pag.* 240—after the fifth paragraph of the ensuing scholium as it is here augmented—where he showed how the linelet $NI = Ro^2 + \ldots$ yields, jointly with the tangent $HN = o\sqrt{(1+Q^2)}$, the chord $HN^2/NI = (1+Q^2)/R$ of the circle which osculates the curve $\overarc{FHQ}$ over its infinitesimal arc $\overarc{HI}$; whence at once the radius of curvature of the curve at the point H is $(1+Q^2)^{\frac{3}{2}}/2R$,

so on indefinitely. And the first term, here e, will ever denote the length of the ordinate CH standing (on the base) at the beginning[39] of the indefinite quantity o; the second term, here $(a/e)\,o$, will denote the difference between CH and DN, that is, the linelet NM which is cut off by completing the rectangle $HCDM$, and hence ever determines the position of the tangent HN: in this case, specifically, by taking NM to HM as is $(a/e)\,o$ to o, or a to e. The third term, here $\frac{1}{2}(n^2/e^3)\,o^2$, will designate the linelet IN which lies between the tangent and the curve, and hence determines the angle of contact $\widehat{IHN}$, or the curvature which the curved line has at H: if that linelet IN is of finite size, it will be denoted by the third term together with the ones following on to infinity; but if that linelet be infinitely diminished, the subsequent terms will prove to be infinitely less than the third, and can consequently be neglected. The fourth term, here $\frac{1}{2}(an^2/e^5)\,o^3$, determines the variation of the curvature, the fifth the variation of the variation, and so on.[40] Whence, incidentally there is opened up a use not to be despised of these series in the solution of problems which depend on tangents and the curvature of curves.

Now compare the series $e-(a/e)\,o-\frac{1}{2}(n^2/e^3)\,o^2-\frac{1}{2}(an^2/e^5)\,o^3\ldots$ whereby the ordinate DI in the problem now to be solved is denoted with the series $P-Qo-Ro^2-So^3\ldots$ by which it was designated in finding out the solution, and likewise in place of P, Q, R and S write e, a/e, $\frac{1}{2}n^2/e^3$ and $\frac{1}{2}an^2/e^5$, and so for $\sqrt{(1+Q^2)}$ write $\sqrt{(1+a^2/e^2)}$, that is, n/e; and there will result the density of the medium as a/ne, that is (because n is given) as a/e or AC/CH, in other words as the length TH of the tangent which terminates at the radius AF standing at right angles to PQ; while the resistance will be to gravity as $3a$ to $2n$, that is, as $3AC$ to the circle's diameter PQ; and the speed will be as $\sqrt{(2CH)}$.[41] In consequence, should a body[42] go forth from the place F with a proper speed along a line parallel to PQ, and the density of the medium in individual places H be as the length of the tangent TH, and the resistance also at an arbitrary place H be to the force of gravity as $3AC$ to PQ, the body will describe the circle-quadrant $\widehat{FHQ}$. As was to be found.

But should the same body set off from the place P along a line perpendicular

---

that is, $(ds/da)^3/(d^2e/da^2)$ in the more familiar Leibnizian equivalent (compare note (37) preceding). No such insertion was made, there or at any other place, in the *Principia*'s third edition in 1726, but its three surviving drafts are reproduced in Appendix 4.

(41) This is rounded off to be just 'ut $\sqrt{CH}$' (as $\sqrt{(CH)}$) in the corrected copy of the present text which Newton later (see note (26) preceding) forwarded to Cotes to be printed.

(42) As Newton himself was to do in the copy which he subsequently sent to Cambridge for publication, we here omit an outmoded ensuing 'C'—that is, 'H' in the present figure (see note (27) preceding)—which he at this point carelessly copied into the manuscript from his preliminary draft (Appendix 2.8 following).

egrederetur et in arcu circuli moveri inciperet, sumenda esset *AC* seu *a* ad contrarias partes centri *A* & propterea signum ejus mutandum esset & scribendum −*a* pro +*a*. Negativam autem densitatem, hoc est quæ motus corporum accelerat, Natura non admittit; et propterea naturaliter fieri non potest ut corpus ascendendo a *P* describat circuli quadrantem *PF*. Ad hunc effectum deberet corpus a Medio impellente accelerari,[5] non a resistente impediri.

    Exempl. 2. Sit linea *PFHQ*[43] Parabola &c[44]

[*Pag. 268. lin. 10, lege* ut 3*XY* ad 2*YG*.][45]

[*Pag. 269. lin. 8, lege* ut 3*S* in $\dfrac{XY}{a}$ ad 4*RR*, id est ut *XY* ad $\dfrac{2nn+2n}{n+2}$ *VG*.][45]

Pag. 269. lin. 11. After the words habente. **Q.E.I.** *insert these two Paragraphs.*[46]

Schol.

    Fingere liceret quod projectile pergeret[47] in arcuum *GH*, *HI*, *IK* chordis, et in solis punctis *G*, *H*, *I*, *K* per vim gravitatis & vim resistentiæ agitaretur,[47] perinde ut in Propositione prima Libri primi corpus per vim centripetam intermittentem agitabatur, deinde chordas in infinitum diminui ut vires redderentur[47] continuæ. Et solutio Problematis hac ratione facillima evaderet.[48]

---

    (43) So we 'translate' the manuscript reading '*ALCK*' transcribed by Newton from his draft, which is keyed to a differently lettered figure. The oversight was carried over by him into the copy which he subsequently transmitted to Cambridge to be printed, but was there caught by Cotes before it could pass into the *editio secunda*.

    (44) Understand that the remainder of the text of this second worked example of the 'Galileian' unresisted parabolic path is here to follow exactly as in the first edition (*Principia*, ₁1687: 266) except for the few transliterations necessitated by its now being referred to the figure on page 356 above, whose points are (see note (27) thereto) differently denoted.

    (45) These corrections to the 1687 text of the ensuing hyperbolic Examples 3 and 4— immediate consequences of course (compare Appendix 1: note (35)) of Newton's present multiplication by a like factor of $\frac{3}{2}$ of his original, deficient general value $\frac{1}{2}S\sqrt{(1+Q^2)}/R^2$ for the ratio of resistance to gravity—are here inserted in the manuscript text (which lacks any version of them) from the equivalent place on the last page (Trinity College. R.16.38: 265) of the augmented copy of the mended proposition which he sent to Cotes early in January 1713. Newton's preliminary framings of these minor adjustments, set out—immediately below a first roughing-out of the first of the two paragraphs which form in sequel a new opening to the ensuing scholium (compare note (47) below)—on the back (Add. 3968.37: 548ᵛ) of a draft of the English preface to his edition of the *Commercium Epistolicum* then also in the press, are identically worded.

    (46) This instruction was afterwards altered in the copy sent to Cotes to read equivalently: '*pag. 269. lin. 12. lege sequentia*' (at page 269, line 12 read the following).

    (47) In the version passed on to Cotes for publication these phrases were minimally changed to read 'Fingere...projectile pergere' (...imagine the projectile to proceed), '&...agitari' (and to be disturbed) and 'reddantur' (shall be rendered) respectively. In his initial framing of this paragraph on Add. 3968.37: 548ᵛ (see note (45) above) Newton first began little differently: 'Fingere licet quod corpora pergerent...' (It is permissible to imagine that bodies proceed...).

to PQ and begin to move in the arc of the circle, you would need to take AC, that is, a, on the opposite side of the centre A, and accordingly to change its sign, writing $-a$ in place of $+a$. A negative density however—one, that is, which accelerates the motions of bodies—is not admitted in nature; and accordingly it cannot naturally happen that a body in ascending from P shall describe the circle-quadrant $\overset{\frown}{PF}$: to this effect a body would have needed to be accelerated by an impelling medium, not impeded by a resisting one.

Example 2. Let the line $PFHQ$[43] be a parabola....[44]

Page 268, line 10. Read as $3XY$ to $2YG$.[45]

Page 269, line 8. Read as $3S \times XY/a$ to $4R^2$, that is, as XY to

$$(2(n^2+n)/(n+2))\ VG.[45]$$

Page 269, line 11. After the words having.... As was to be found. *insert these two paragraphs.*

Scholium.

It would have been permissible to imagine that the projectile were to proceed[47] in the chords GH, HI, IK of the arcs, and be disturbed[47] through the force of gravity and the force of resistance at the sole points G, H, I and K, exactly as in Proposition I of the first book a body was disturbed through an intermittent centripetal force; and then the chords to be infinitely diminished so that the forces were rendered[47] continuous. And the solution of the problem by this means would have turned out very easy.[48]

---

(48) Newton alludes to the variant mode of solution set out in his 'Idem aliter' (§2.2 preceding) whereby the forces of gravity and resistance are conceived to act on the projectile instantaneously in discrete unit-impulses, 'simul et semel', at successive infinitesimal moments of time. Unfortunately, on the lone page 240 of the *editio secunda* yet remaining unassigned in the re-set proposition there proved to be not quite room sufficient to include the whole of the scholium's augmented beginning (which, let us remind, included not only the two additional paragraphs here newly inserted but also the first two of the old opening). Space enough could have been contrived in one or more of several obvious ways—in his covering letter to Cotes on 6 January 1712/13 (Trinity College. R.16.38: 261 [=*Correspondence*, 5: 361]) Newton himself made the practical suggestions that an extra line might, as needed, be added on each of the preceding pages 231–9 still in the printer's frame, and also that the text might be packed more closely in around the 'cuts' (as could easily have been done, at the cost of resetting the next page and a half, with the sprawling block on page 238 illustrating the hyperbolic Examples 3/4)—but Cotes elected instead, when the matter came to the crunch, to suppress the present paragraph *in toto* to gain room for what comes after, ultimately presenting the excision as a *fait accompli* to its author (who, if he noticed the omission, is not known to have objected) and in the meanwhile mildly responding to him that 'Some things in Your Paper I have altered, [but] they are not worth Your Notice...' (*ibid.*: 266 [=*Correspondence*, 5: 370]). The paragraph thus denied its rightful place in the *Principia*'s second edition disappeared straight into an oblivion from which it was retrieved only in the middle of the following century through the care of Joseph Edleston, who published its text as communicated to Cotes, in a footnote to his

Eadem ratione qua prodijt densitas Medij ut $\dfrac{S \times AC}{R \times HT}$ in Corollario primo: si Resistentia ponatur ut Velocitatis V dignitas quælibet V^n cujus index est numerus n; prodibit densitas Medij ut$\dfrac{S}{R^{\frac{4-n}{2}}} \times \left.\overline{\dfrac{AC}{HT}}\right|^{n-1}$ [(49)]. Et propterea si curva inveniri potest ea lege ut $\dfrac{S}{R^{\frac{4-n}{2}}}$ sit ad $\left.\overline{\dfrac{HT}{AC}}\right|^{n-1}$, vel $\dfrac{S^2}{R^{4-n}}$ ad $\overline{1+QQ}|^{n-1}$, in data ratione: corpus movebitur in hac curva in uniformi Medio cum resistentia quæ sit ut velocitatis dignitas V^n.[(50)] Sed redeamus ad curvas simpliciores.

Quoniam [motus non fit in Parabola nisi in Medio non resistente &c].

$\left[\textit{Pag. 270, lin. 9 \& 14. lege } \dfrac{2nn+2n}{n+2} \textit{ pro } \dfrac{3nn+3n}{n+2} .\right]$[(51)]

$\left[\text{In the errata put } \textit{Pag. 274. lin. 4 lege } \dfrac{2nn-2n}{n-2} \textit{ VG.}\right]$[(52)]

---

first printing of the latter's letter of reply to Newton from which we have just now quoted, in his edition of the *Correspondence of Sir Isaac Newton and Professor Cotes* (London, 1850): 147, note *. He did not of course know of the full argument of the parent 'Idem aliter', whose manuscript remained closeted for another twenty years and more in the private possession of Lord Portsmouth at Hurstbourne Park along with Newton's other scientific papers (see 1: xxviii) before at length being brought to Cambridge, in whose University Library it has this past century found a permanent resting-place (even if its significance there too went unappreciated by the outer world).

(49) For there is (see note (37) above) $V = \sqrt{(\frac{1}{2}g \cdot (1+Q^2)/R)}$, where g is the constant force of downwards gravity; whence, on putting δ to be the medium's density at H, the resistance $\frac{3}{4}gS\sqrt{(1+Q^2)}/R^2$ is proportional to $\delta V^n \propto \delta((1+Q^2)/R)^{\frac{1}{2}n}$, and consequently there ensues $\delta \propto R^{\frac{1}{2}n-2}S/(1+Q^2)^{\frac{1}{2}(n-1)}$. In here presenting this result Newton introduces the elegance of replacing $(1+Q^2)^{\frac{1}{2}}$ by the equal ratio $HN/(HM$ or) $CD = TH/AC$, but this is mere geometrical decoration. In his preliminary derivation, on Add. 3968.37: 548^v, of this general measure of the variation of density along the projectile path (see Appendix 2.10) he stated it without any such cloaking embellishment, and then went on to display the pattern of the particular forms which it assumes in the instances where successively $n = 0, 1, 2, 3, ..., 7$. On Add. 3965.12: 201^v, where Newton subsequently mapped out the summary of this scheme which he now gives, he initially was content solely to adduce the prime case $n = 2$ of the sequel, affirming, namely, that 'Cum densitas Medij per Corol. 1 fuerit ut $\dfrac{S \times AC}{R \times HT}$, si Curva inveniri possit ea lege ut $S \times AC$ sit ad $R \times HT$ in data ratione, movebitur Corpus in hac curva in uniformi medio' (Since the density of the medium was, by Corollary 1, as $S \times AC/R \times TH$, if a curve might be found obeying the law that $S \times AC$ be in a given ratio to $R \times TH$, the body will move in this curve in a uniform medium); and only afterwards there added more generally: 'Et si Velocitas dicatur V et n sit index dignitatis ejus et Resistentia sit ut Medij de[n]sitas et V^n, erit medij densitas ut Resistentia directe & V^n inverse, id est ut $\dfrac{3S\sqrt{1+QQ}}{4RR}$ directe & $\left.\overline{\dfrac{1+QQ}{R}}\right|^{\frac{n}{2}}$ inverse,

In the same manner as the density of the medium resulted to be as

$$(S/R) . (AC/TH)$$

in Corollary 1: if the resistance be put as any power of the speed V you please, V^n, whose index is the number n, the density of the medium will result to be as $(S/R^{2-\frac{1}{2}n}) . (AC/TH)^{n-1}$.[49] Accordingly, if a curve can be found obeying the law that $S/R^{2-\frac{1}{2}n}$ be to $(TH/AC)^{n-1}$, or S^2/R^{4-n} be to $(1+Q^2)^{n-1}$, in a given ratio, then the body will move in this curve in a uniform medium subject to a resistance which shall be as the power V^n of the speed.[50] But let us return to simpler curves.

Because motion does not take place in a parabola except in a non-resisting medium....

Page 270, lines 9 and 14. Read $2(n^2+n)/(n+2)$ *in place of* $3(n^2+n)/(n+2)$.[51]
In the *errata* put: '*Page 274, line 4. Read* $2(n^2-n) . VG/(n-2)$.'[52]

---

hoc est ut $\dfrac{3S \times \overline{1+QQ}^{\frac{-n+1}{2}}}{4R^{\frac{4-n}{2}}}$, (And if the speed be called V, and n the index of its power, and so the resistance as the medium's density and V^n, the medium's density will then be as the resistance directly and V^n inversely, that is, as $\frac{3}{4}S\sqrt{(1+Q^2)}/R^2$ directly and $((1+Q^2)/R)^{\frac{1}{2}n}$ inversely, or as $\frac{3}{4}S(1+Q^2)^{\frac{1}{2}(-n+1)}/R^{\frac{1}{2}(4-n)})$.

(50) The draft of this sentence on f. 201^v (in sequel to that cited in the previous note) begins slightly differently by positing 'Et propterea si hujusmodi Curva inveniri potest ut SS in R^{n-4} sit ad $\overline{1+QQ}|^{n-1}$ in ratione data, ...' (And accordingly, if a curve can be found of this sort that $S^2 \times R^{n-4}$ shall be to $(1+Q^2)^{n-1}$ in a given ratio, ...). Newton's condition for motion through a uniformly dense medium resisting as some given nth power of the transient body's speed—in a path defined, namely, by the relationship $e = e_a$ connecting the abscissa $AC = a$ and ordinate $CH = e$ of its general point $H(a, e)$, with Q, R and S the coefficients of the powers o, o^2 and o^3 in the series expansion of the incremented ordinate $DI = e_{a+o}$ corresponding to the augmented abscissa $AD = a+o$—can be looked directly upon as a criterion for testing whether any particular curve $\overgroup{FHI}$ can be traversed through a uniform medium under any specified law of resistance. And so, certainly, he himself went straightaway on to apply it, in the brief lines of calculation which are reproduced (with a deal of editorial fleshing out) in Appendix 2.11 following, to fail the logarithmic curve

$$e \propto - \int_b^a (x+cx^2)^{-1} . dx = \log(c+1/a) - \log(c+1/b),$$

b the maximum horizontal range attained (at the origin) from the firing point $(b, 0)$, as a path traversable in the primary case where the medium resists as the square of the projectile's speed; and thereafter abortively to make trial of the more general curve $e \propto \int_b^a (x^m+cx^{m+1})^n . dx$ as a possible trajectory in the less physically realistic case where the medium resists the flight of the missile uniformly at all points (and so $n = 0$). But from the inverse viewpoint, since (as Newton himself already well appreciated—see VII: 98—if he could never make it clear to the Bernoullis that he had long before attained this key insight) the coefficient of the power o^i in the Taylorian series expansion of e_{a+o} is the adjusted corresponding derivative $(1/i!) d^ie/da^i$, $i = 1, 2, 3, ...$

successively, so that here (compare note (37) above) there is

$$Q = de/da, \quad R = \tfrac{1}{2}d^2e/da^2 = \tfrac{1}{2}dR/da \quad \text{and} \quad S = \tfrac{1}{6}d^3e/da^3 = \tfrac{1}{3}dR/da,$$

this same criterion encapsulates a third-order derivative condition from which one may work back to determine, for any given index n, the totality of possible projectile curves traversible through a medium resisting as the nth power of the speed. In the simplest instance $(n = 1)$ where the resistance is proportional to the speed the condition requires that

$$R^{-\frac{3}{2}}S(=\tfrac{1}{3}R^{-\frac{3}{2}}.dR/da)$$

be constant, whence a first integration yields $R^{-\frac{1}{2}} \propto m - na$ or $2R(= dQ/da) \propto (m-na)^{-2}$, and a second integration produces $Q(= de/da) \propto k - n/(m-na)$, so that finally there ensues

$$e \propto ka + l + \log(m - na):$$

the defining Cartesian equation of the general point (a, e) of the *curva logarithmica* which, *mutatis mutandis*, Newton had already geometrically attained in autumn 1684 in the concluding scholium of his treatise 'De motu Corporum'—there, unbeknownst, following in the steps of Christiaan Huygens in then still unpublished researches of sixteen years earlier—by compounding from first principles the horizontal and vertical components of the resisted motion under downwards gravity which he had separately obtained in his preceding Problems 6 and 7 (see VI: 68–72, and our note (113) thereto), and whose minimally reshaped construction he afterwards printed as Proposition IV of the second book of his *Principia* ($_1$1687: 241–3 [= VI: 85–7]). Determination of the general trajectory where the resistance is proportional to the square of the speed, and for which no parallel separation of the motion into like resisted (and downwards gravitationally accelerated) components is possible, is far less easy; indeed, no comparable explicit solution of the instance $n = 2$ of Newton's condition—which he himself put to be '$eS = R\sqrt{1+QQ}$', where e is some constant parameter (see Appendix 2.11: note (84)) —is possible, other than by successive approximation by series and iterations not themselves easy to contrive, in terms of the elementary algebraic and circular/hyperbolic functions which (whether in analytical form or framed in the model of conic-areas) he alone admitted into his workaday mathematical scheme of things. And yet when David Gregory came to visit him in Cambridge early in May 1694 Newton stated his belief that the solution of finding the projectile trajectory in this 'true' hypothesis of the medium's resistance to its motion was 'in his power' not merely where the gravity is constant and straight downwards, but also acts more generally as the inverse square of the distance from a finite force-centre. (See Gregory's long revised memorandum C42, now in Edinburgh University Library, in which he summarized the content of his talks 'Maio 1694' with Newton, there reporting *inter alia* that the latter 'Propositioni X Lib: II [*Princip.*] subnexus est aliud problema quo semita projecti investigatur in vero rerum systemate, hoc est posita gravitate reciproce ut quadratum distantiæ a centro et resistentia directe ut quadratum velocitatis, quod nunc in potestate esse credit'; compare H. W. Turnbull's English translation in *The Correspondence of Isaac Newton*, **3**, 1961: 384. Similarly worded accounts of what Newton claimed 'se nunc nosse' are found in Gregory's further memoranda C33 and C44, now in the Royal Society, of 'adnotata Phys: et Math: cum Newtono Cantabrigiæ 4. 5. 6. 7 Maij 1694'; see *Correspondence*, **3**: 313, 335.) What then did Newton have it in mind to do in supplement to his published Proposition X? Perhaps a shade over-generous to our hero, we find it easy to believe that he had then, in the instance $n = 2$ at least, taken the short further step which obtains the solution of the presently stated differential condition in parametric form, *viz.* on transposing this to be $R^{\frac{1}{2}n-1}S \propto (1+Q^2)^{\frac{1}{2}(n-1)}R$; whence

a first integration produces $R^{\frac{1}{2}n} \propto \int (1+Q^2)^{\frac{1}{2}(n-1)}.dQ \equiv I_Q$, say, and consequently

$$R(=\tfrac{1}{2}dQ/da) \propto (I_Q)^{2/n};$$

and therefrom, by a second integration, $a \propto \int (I_Q)^{-2/n}.dQ$ where

$$e\left(= \int Q.da\right) \propto \int Q(I_Q)^{-2/n}.dQ.$$

Whether this approaches the historical truth or no, when in January 1697 Newton received—and (see 1 above) solved—Johann Bernoulli's challenge to identify the *curva brevissimi descensus* we again have David Gregory's testimony in a memorandum dated '20 feb^{rii} 169$\frac{7}{8}$.' (A90, now in Edinburgh University Library; reproduced in *Correspondence*, **4**, 1967: 266) that 'Newtonus vicissim propositurus erat Bernoullio et Leibnitio Problema de via projecti cum resistentia est in duplicata ratione velocitatis, quod perperam solverat Leibnitius in *Actis Lipsiæ*'. (On the last point see VI: 70–1, note (109); where let us withdraw our false conjecture that 'Newton seemingly remained ignorant of [Leibniz'] 'Schediasma' for a quarter of a century till John Keill brought it to his notice'. Gregory, noting the gist of a conversation with Newton 'Londini 7 Martij 169$\frac{6}{7}$' in a yet unpublished memorandum (ULE/A78^1), reports the latter as already complaining that 'Libnitius fallitur dum iter projectilis determinat. . .in medio ubi resistitur in ratione duplicata velocitatis. Error male provenit, quod hoc de Compositione Motus conficiat'.) Had Newton in 1697 posed to Bernoulli—or indeed to Leibniz or (in the words of Bernoulli's *Programma*) generally 'Acutissimis qui toto Orbe florent Mathematicis'—this counter-challenge to determine the path of a projectile through a medium resisting as the square of its speed, we may well doubt that the latter was then well practised enough in the subtleties of reducing the equation of such resisted motion to its equivalent third-order differential form, and in the techniques whereby this might be integrated to yield its general solution. Things stood wholly differently twenty years later when in late January 1718 Pierre Rémond de Montmort quoted to Bernoulli a like provocative sentence from a letter written by John Keill the previous summer to Brook Taylor (and passed forthwith by its recipient to Montmort) where Keill, tired of Bernoulli's blanket claims for the superiority of Leibnizian mathematical analysis, snappishly countered that 'if he would apply his skill to something of use I desire he would solve the problem Mr Leibnitz attempted but erroneously mistook and could not solve$_{[,]}$ to find the curve a Projectile describes in the air in the most simple supposition of gravity and the density of the medium being both uniform: but the resistance in duplicate proportion of the velocity'. (So Montmort quoted it back to Keill himself in late October 1718 in a letter, loose in Packet 2 of the 'Lucasian papers' (ULC. Res. 1893a), which is now printed in *Correspondence of Isaac Newton*, **7**, 1977: 11; Keill himself was displeased that this sentence of his letter should have been transmitted to Bernoulli, but whether this was wholly 'contrary to his intention'—as Edleston surmises in his comment upon Montmort's letter in his edition of the *Correspondence of Sir Isaac Newton and Professor Cotes, including Letters of other Eminent Men*. . .(London, 1850): 187, note ‡—is arguable.) Quick to seize this opportunity of scoring off the man who had over the preceding five years grown to be the arch 'antagonist' and defender of the Newtonian supremacy in all things mathematical, Bernoulli now hastened to reduce his equivalent 1713 expression (see note (37) above) for the resistance to projectile motion under constant downwards gravity, deducing in his preferred Leibnizian form—without any mention of Newton's prior attainment of its equal in the *Principia*!—the prime result of the present, corrected proposition that 'positis [gravitate] g & $dy[= da]$ constantibus

[erit] $ddp = \dfrac{\mp 2[\rho]\, dp^2}{g\sqrt{1+pp}}$' where p is the trajectory's slope (*viz.* Newton's Q, whence $dp = 2R.dy$

and $d^2p = 6S.dy^2$); and was able, from his supposition that the resistance ρ is equal to $\frac{1}{2}v^{2m} = \frac{1}{2}(g(1+p^2).dy/dp)^m$, at once to conclude that '$dp^{m-1}\,ddp = \mp g^{m-1}dy^m\,dp \times \overline{1+pp}^{\,m-\frac{1}{2}}$

& integrando $\dfrac{dp^m}{m} = \mp g^{m-1}\,dy^m \int dp \times \overline{1+pp}^{\,m-\frac{1}{2}}$'. And so he was six months afterwards to publish his general solution to the 'Problema. Construere Curvam (concessis quadraturis), quam corpus uniformiter grave tendens perpendiculariter ad horizontem describit in medio uniformiter denso; supposita resistentia in quacunque multiplicata ratione velocitatis' at the end of his anti-Keillian 'responsio ad non-neminis provocationem' in *Acta Eruditorum* (May 1719): 216–26 [= *Opera*, **2** (Lausanne/Geneva, 1742): 393–402]; see especially 224–5. The full, parametral solution 'du probleme de Mr. Keill, pris dans un sens general' was subsequently set out by Bernoulli, effectively as we have (in Newtonian equivalent) outlined it above, in a letter to Montmort on the following 13 July (N.S.)—'Vous verrez', he wrote, 'que toute ceste analyse n'est qu'une chaine d'égalités deduites de la formule generale pour la determination des resistances, que j'ai donnée dans les journaux de Leipsic de 1713 p. 118 & 119'—and this, cast into Latin, he made public two years later in the *Acta* (May 1721): 228–30 [= *Opera*, **2**: 513–16] as the 'Operatio analytica per quam deducta est...solutio'. That Bernoulli should in his 'Responsio' (page 219) think to castigate as a 'crassly contradictory hallucination' the very proposition of the *Principia* in which Newton first stated—if deficiently so in its *editio princeps*—his geometrical equivalent of the fundamental differential condition on which Bernoulli now founded his solution of Keill's 'challenge' is a circumstance which we can attribute only to unseeing ill-will on his part.

(51) We again borrow a needed correction from the augmented text which Newton passed to Cotes on 6 January 1712/13. The draft on Add. 3968.37: 548^v (on which see note (45) above) reads just '*p. 270. lin. 9, 14.* $\dfrac{2nn+2n}{n+2}$'. There follows immediately after this a first version 'injuriam Newtono illatam repellendo' of a phrase which appears in line 3 of the printed 'Ad Lectorem' to the *Commercium Epistolicum D. Johannis Collins* passing at this time through press as 'injuriam D. *Newtono* oblatam propulsans'. Since advance copies of the *Commercium* were (see its Journal Book) brought in to the Royal Society by Newton's presidential order on 8 January 'to be delivered to each person of the Committee appointed for that purpose, to examine it before publication', the amended preface had been put to the printer some little while before; and we infer that an equal time intervened between Newton drafting the present final version of his *Principia* proposition and subsequently transmitting its augmented fair copy to Cotes on 6 January, two days before this.

(52) This, of course, to save resetting the whole leaf Ii2 in the *Principia*'s *editio secunda* in order merely thus (twice) to adjust an erroneous coefficient '3' to be '2'. And so it was done (compare Appendix 1: note (50)).

APPENDIX 1. THE ORIGINAL TEXT
AS PRINTED IN 1687.[1]

From the *Principia*'s *editio princeps*[2]

|| *Prop. X. Prob. III.* ||[260]

Tendat uniformis vis gravitatis directe ad planum Horizontis, sitꝗ resistentia ut medii densitas & quadratum[3] *velocitatis conjunctim: requiritur tum Medii densitas in locis singulis, quæ faciat ut corpus in data quavis linea curva moveatur, tum corporis velocitas in iisdem locis.*

---

(1) Though the first edition of the *Principia* has twice been reproduced in photo-facsimile in recent years (London, 1953; Brussels, 1965), and the text of the present Proposition X of its second book was more than two centuries ago reprinted in parallel column alongside that of the corrected 1713 *editio secunda* by the editor, Gabriel Cramer, of Johann Bernoulli's *Opera Omnia*, 1 (Lausanne/Geneva, 1742 [→ (facsimile offset) Hildesheim, 1968]): 481–93 in an 'Excerptum' making plain their differences, in visual testimony of the profit here derived by Newton from Bernoulli's intervention in autumn 1712, we make no apology for once more printing in full Newton's original 'solution' of the problem of determining the motion of a projectile under constant downwards gravity through a medium directly opposing its onrush from point to point according to its own varying density and the moving body's instantaneous speed. Not only does this allow us conveniently to adduce the *ipsissima verba* in which Newton presented the flawed 1687 version upon whose main foundation he twenty-five years later built the succession of recastings reproduced in Appendix 2 below and §§1/2 preceding; but it permits us, following in Lagrange's footsteps (see note (6) below), to indicate in our commentary precisely wherein the defects of this initial attempted solution lie.

(2) *Philosophiæ Naturalis Principia Mathematica* (London, ₁1687): 260–74. Other than for Humphrey Newton's secretary copy of the finished text (Royal Society, MS LXIX, coded as 'M' in A. Koyré and I. B. Cohen, ... *The Third Edition (1726) with Variant Readings* (Cambridge, 1972), where the *variorum* text of the present Proposition X of Book 2 is set out on pp. 376–94)—this has only the single non-trivial variant which we cite in note (18) following—there survive no preliminary worksheets or drafts relating to this portion of the *editio princeps* which might allow us any independent insight into Newton's mind as he broached the solution of the general problem of resisted projectile motion which he here essays. In reproducing these printed pages we have indicated their division in our outer margins, and have also for the most part retained the minor standardizations of Newton's punctuation which the printer introduced into Humphrey Newton's press copy in setting this up in type. We have, however, everywhere reduced to lower-case *o* the flurry of capitalized increments *O* which, by way of Humphrey Newton's careless secretarial pen, confusingly and inconsistently bespatter its 'Exempl. 4'. (In an all too familiar fashion the slip eluded Cotes' notice when he went through the *Principia*'s text in preparation for the second edition, and was equally unseeingly passed by Pemberton into its *editio ultima* in 1726, whence it is perpetuated in all subsequent editions.)

(3) An unnecessary restriction upon the power of the speed entering the law of resistance: one too rigidly according with the purview of the 'Sect. II. De motu corporum quibus resistitur in duplicata ratione velocitatum' of the second book of the *Principia* wherein—in sequel to the preceding Propositions V/VI/VII and VIII/IX which treat the two special cases (of nil gravity and of straight upwards/downwards fall respectively) where the resisted motion

||26[1][5] Sit *AK* planum illud plano Schematis perpendiculare; *ACK* linea curva;[4] *C* corpus in ipsa motum; & *FCf* recta ipsam tan||gens in *C*. Fingatur autem corpus *C* nunc progredi ab *A* ad *K* per lineam illam *ACK*, nunc vero regredi per eandem lineam; & in progressu impediri a Medio, in regressu æque promoveri, sic ut in iisdem locis eadem semper sit corporis progredientis & regredientis velocitas. Æqualibus autem temporibus describat corpus progrediens arcum quam minimum *CG*, & corpus regrediens arcum *Cg*; & sint *CH*, *Ch* longitudines æquales rectilineæ quas corpora de loco *C* exeuntia his temporibus absqɜ Medii & Gravitatis actionibus describerent: & a punctis *C*, *G*, *g* ad planum horizontale *AK* demittantur perpendicula *CB*, *GD*, *gd*, quorum *GD* ac *gd* tangenti occurrant in *F* & *f*. Per Medii resistentiam fit ut corpus progrediens vice longitudinis *CH* describat solummodo longitudinem *CF*; & per vim gravitatis transfertur corpus de *F* in *G*: adeoɜ lineola *HF* vi resistentiæ, & lineola *FG* vi gravitatis simul generantur.[6] Proinde (per Lem. X. Lib. I.[7]) lineola *FG* est ut vis gravitatis &

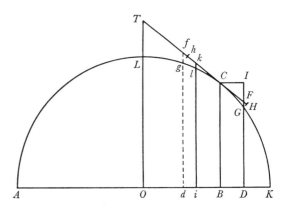

is rectilinear, and in consequence readily determinable in simpler *ad hoc* manner—Newton places his general problem. In the latter of the two paragraphs (see page 368) which he afterwards in October 1712 inserted at the opening of the terminal scholium he rightly relaxed this needless limitation, there positing the resistance to be 'ut velocitatis dignitas quælibet'.

(4) In the accompanying figure (which Newton, ever frugal, sets to do double duty by illustrating both the general case and that of 'Exempl. 1' following) this path of the projectile *C* through the resisting medium is depicted as a semicircle of centre *O* and diameter *AK*. We should understand it to be any 'smoothly' continuous curve drawn in the (vertical) plane of the paper.

(5) This page is by error numbered '262' in the printed original.

(6) Here is born the confusion which blights Newton's succeeding argument: the 'linelet' *FG* is generated not merely by the force of vertically downwards gravity, but also through the component (here negative) of the force of resistance to the motion along *CFH* which acts in the same downwards direction. To summarize the lengthy and percipient analysis of this mode of approach given by J. L. Lagrange in his *Théorie des Fonctions Analytiques* (Paris, Prairial An V [= May–June 1797]): Seconde Partie, §§202–5: 244–51 (much augmented in the corresponding Chapitre IV of the 2e Partie of the revised edition (Paris, ₂1813) [= (ed. J. A.

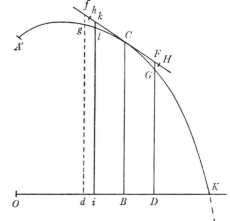

quadratum temporis conjunctim, adeoꝗ (ob datam gravitatem) ut quadratum temporis; & lineola *HF* ut resistentia & quadratum temporis, hoc est ut resistentia & lineola *FG*. Et inde resistentia fit ut *HF* directe & *FG* inverse, sive ut $\frac{HF}{FG}$. Hæc ita se habent in lineolis nascentibus. Nam in lineolis finitæ magnitudinis hæ rationes non sunt accuratæ.

Et simili argumento est *fg* ut quadratum temporis, adeoꝗ ob æqualia tempora æquatur ipsi *FG*;[(8)] & impulsus quo corpus regrediens urgetur est ut $\frac{hf}{fg}$. Sed impulsus corporis regredientis ‖ & resistentia progredientis ipso motus initio ‖[262] æquantur, adeoꝗ & ipsis proportionales $\frac{hf}{fg}$ & $\frac{HF}{FG}$ æquantur; & propterea ob æquales *fg* & *FG* æquantur etiam *hf* & *HF*, suntꝗ adeo *CF*, *CH* (vel *Ch*[(9)]) & *Cf* in progressione Arithmetica, & inde *HF* semidifferentia est ipsarum *Cf* & *CF*; & resistentia quæ supra fuit ut $\frac{HF}{FG}$, est ut $\frac{Cf-CF}{[2]FG}$.[(10)]

Serret) *Œuvres*, **9**, Paris, 1881: 360–76])—see especially §204: 257–8 (= ₂1813: §§20/21)—if we suppose that the moving body C traverses the tangent CF under the directly retarding resistance ρ in the vanishingly small time θ, and all that time is subject also to the constant downwards 'pull' of gravity g, then, where the related increments of the base $OB = x$ and ordinate $BC = y$ are $(BD =)$ o and p respectively, there is at once $\dot{x} = o/\theta$ and (on setting $\dot{y}/\dot{x} = dy/dx = Q$) $\dot{y} = Qo/\theta$, so that $\dot{y} - Q\dot{x} = 0$ and hence $\ddot{y} - Q\ddot{x} - \dot{Q}\dot{x} = 0$; whence the instantaneous speed v at C is $\sqrt{(\dot{x}^2 + \dot{y}^2)} = \dot{x}\sqrt{(1 + Q^2)}$, and the Eulerian equations of motion are $\ddot{x} = -\rho/\sqrt{(1 + Q^2)} = -\rho\dot{x}/v$ and $\ddot{y} = -\rho Q/\sqrt{(1 + Q^2)} - g = Q\ddot{x} - g$, so that

$$\ddot{y} - Q\ddot{x} = \dot{Q}\dot{x} = -g$$

and therefore $\dot{y} - Q\dot{x} = \dot{Q}\dot{x} = (-g\ddot{x}/\dot{x}$ or$)$ $g\rho/v$. Further $o = \dot{x}\theta + \frac{1}{2}\ddot{x}\theta^2 + \frac{1}{6}\dddot{x}\theta^3 + ...$, and so $CF = o\sqrt{(1 + Q^2)} = v\theta - \frac{1}{2}\rho\theta^2...$, whence by inversion

$$\theta = (\sqrt{(1 + Q^2)}/v)\, o + \frac{1}{2}(\rho/v^3)\,(1 + Q^2)\, o^2...;$$

while similarly $p = \dot{y}\theta + \frac{1}{2}\ddot{y}\theta^2 + \frac{1}{6}\dddot{y}\theta^3 + ...$ Accordingly, there is

$$FG = -(p - Qo) = \frac{1}{2}g\theta^2 - \frac{1}{6}(g\rho/v)\,\theta^3 + O(\theta^4).$$

(7) *Principia*, ₁1687: 32 (= VI: 116).

(8) Since in the terms of note (6) preceding (because the arcs $\overset{\frown}{gC}$ and $\overset{\frown}{CG}$ are supposed traversed in equal times) there is manifestly $fg = \frac{1}{2}g(-\theta)^2 - \frac{1}{6}(g\rho/v)(-\theta)^3 + O(\theta^4)$, the linelets fg and FG differ in fact by the third-order term $\frac{1}{3}(g\rho/v)\,\theta^3$ which cannot, as Newton supposes, in the sequel be neglected. To the order of θ^2, of course, the two are indistinguishable.

(9) The tangent-lengths hC and CH are the distances traversed each way from C in equal times when the medium offers no resistance to the body's motion; in the terms of note (6) above there is $hC = CH = v\theta$, manifestly (as Newton goes on equivalently to affirm) the arithmetic mean of $fC = v\theta + \frac{1}{2}\rho\theta^2$ and $CF = v\theta - \frac{1}{2}\rho\theta^2$.

(10) We insert a numerical coefficient lacking—as it here may—in the denominator of the fraction as printed, so as accurately to pave the way into Corollary 2 below. Thus amended, Newton's fraction exactly represents the ratio of the resistance to gravity, for since (once more in the terms of note (6) above) $HF = hf = \frac{1}{2}(fC - CF) = \frac{1}{2}\rho\theta^2$ and $FG = \frac{1}{2}g\theta^2...$, at once $\rho/g = \lim_{\theta \to 0} [(fC - CF)/2FG]$.

Est autem resistentia ut Medii densitas & quadratum velocitatis. Velocitas autem ut descripta longitudo CF directe & tempus $\sqrt{FG}$ inverse, hoc est ut $\frac{CF}{\sqrt{FG}}$, adeoɋ quadratum velocitatis ut $\frac{CF^q}{FG}$. Quare resistentia, ipsiɋ proportionalis $\frac{Cf-CF}{FG}$ est ut Medii densitas & $\frac{CF^q}{FG}$ conjunctim; & inde Medii densitas ut $\frac{Cf-CF}{FG}$ directe & $\frac{CF^q}{FG}$ inverse, id est ut $\frac{Cf-CF}{CF^q}$. Q.E.I.[11]

Corol. 1. Et hinc colligitur quod si in *Cf* capiatur *Ck* æqualis *CF*, & ad planum horizontale *AK* demittatur perpendiculum *ki* secans curvam *ACK* in *l*, fiet Medii densitas ut $\frac{FG-kl}{CF\times\overline{FG+kl}}$. Erit enim *fC* ad *kC* ut $\sqrt{fg}$ seu $\sqrt{FG}$[12] ad $\sqrt{kl}$, & divisim *fk* ad *kC*, id est *Cf—CF* ad *CF* ut $\sqrt{FG}-\sqrt{kl}$ ad $\sqrt{kl}$, hoc est (si ducatur terminus uterɋ in $\sqrt{FG}+\sqrt{kl}$) ut $FG-kl$ ad $kl+\sqrt{FG\times kl}$, sive ad $FG+kl$. Nam ratio prima nascentium $kl+\sqrt{FG\times kl}$ & $FG+kl$ est æqualitatis. Scribatur itaɋ

$$\frac{FG-kl}{FG+kl} \text{ pro } \frac{Cf-CF}{CF},$$

& Medii densitas quæ fuit ut $\frac{Cf-CF}{CF^{\text{quad}}}$ evadet ut $\frac{FG-kl}{CF\times\overline{FG+kl}}$.[13]

‖[263]   ‖*Corol. 2.* Unde cum $2HF$ & $Cf-CF$ æquentur, et FG & kl (ob rationem æqualitatis) componunt $2FG$, erit $2HF$ ad CF ut $FG-kl$ ad $2FG$; & inde HF ad FG, hoc est resistentia ad gravitatem, ut rectangulum CF in $FG-kl$ ad $4FG^{\text{quad}}$.[13]

---

(11) We make good a trivial slip in the *editio princeps* at this point from the 'Errata Sensum turbantia' on its concluding page (signature Ooo4^r).

(12) Newton's crucial *faux pas*. If *fg* is left unreplaced by *FG* the measures of the density of the medium and of the ratio of its resistance to the force of gravity which Newton proceeds to derive in this corollary and the next one are exact; see the following note.

(13) Since *fg*, *FG* and *kl* differ from each other only by terms of order θ^3 (see notes (6) and (8) preceding), *fg* cannot, as Newton supposes, here be validly replaced by *FG* in the difference $fg-kl$. When no such substitution is made, his present argument accurately derives the true measures, $(fg-kl)/CF(FG+kl)$ and $CF(fg-kl)/4FG^2$ respectively, of the density of the resisting medium and of the ratio of its resistance to gravity. With foreknowledge of the Lagrangian deductions from the pertinent equations of motion which we outlined in note (6) above, and in anticipation of the Taylorian expansions of the incremented ordinates y_{x+o} which Newton adduces in his primary *Exemplum* in sequel, it is a simple matter to amend the equivalent analytical expressions for these measures which he himself there deduces from his present erroneous geometrical ones. For, on positing (with Newton) the expansion of $DG = y_{x+o}$ into the series $y+Qo+Ro^2+So^3...$, there follows $FG = Ro^2+So^3... = \frac{1}{2}g\theta^2-\frac{1}{6}(g\rho/v)\,\theta^3...$ where $\theta = (\sqrt{(1+Q^2)}/v)\,o+\frac{1}{2}(\rho/v^3)\,(1+Q^2)\,o^2...$, so that

$$R = \tfrac{1}{2}(g/v^2)\,(1+Q^2) \quad \text{and} \quad S = \tfrac{1}{3}(g\rho/v^4)\,(1+Q^2)^{\frac{3}{2}},$$

and hence $fg = \frac{1}{2}g\theta^2+\frac{1}{3}(g\rho/v)\,\theta^3... = Ro^2+2So^3...$; while correspondingly, since likewise $il = y_{x-o} \equiv y-Qo+Ro^2-So^3...$, there is $kl = Ro^2-So^3....$ In the limit as $\theta \to 0$, therefore, there results $\dfrac{fg-kl}{CF(FG+kl)} = \dfrac{3S}{2R\sqrt{(1+Q^2)}}$ and $\dfrac{CF(fg-kl)}{4FG^2} = \dfrac{3S\sqrt{(1+Q^2)}}{4R^2}$.

Corol. 3. Et hinc si curva linea [*ACK*] definiatur per relationem inter basem seu abscissam *AB* & ordinatim applicatam *BC* (ut moris est[14]) et valor ordi-

To Newton in the autumn of 1712, when Niklaus Bernoulli verbally put to him his uncle's objection that the value of the ratio of resistance to gravity obtained from his 1687 expression ought, in the particular case of the resisted semicircular path treated in 'Exempl. 1', to be increased by half as much again, it did not seem in the least way clear how he was to adjust the argument of his *editio princeps* so as to squeeze out of its amended form the factor $\frac{3}{2}$ by which both his original measure of density and that of the ratio of resistance to gravity were deficient. (Johann Bernoulli himself, as Niklaus no doubt told Newton, had been able to detect the error in the semi-circle case only by independently computing the correct result from first principles; he could find no fault with the preceding general argument of Newton's 1687 Proposition X, and subsequently gave his support to his nephew's misguided—if ingenious and persuasive—claim that Newton had 'failed' to notice that the 'true' expansion of the incremented ordinates 'should' have been $y_{x\pm o} = y \pm Qo + \frac{1}{2}Ro^2 \pm \frac{1}{6}So^3...$, whence the 1687 coefficients R and S 'ought' to have been entered as $2R$ and $6S$ respectively. See our preceding introduction.) In his initial efforts to correct his reasoning in the *editio princeps* he first confirmed Bernoulli's related objection (see Appendix 2.1: note (6)) that, in the case of motion in a semicircular arc which is Newton's 'Exempl. 1', the projectile's speed must impossibly be both uniform and gravitationally accelerated; but he could not immediately rid himself of his false supposition that, because *fg* is indistinguishable from *FG* to the order of θ^2, the latter may without appreciable error be everywhere substituted in the former's place. In lieu, he recast his original argument to treat the change in motion over the arcs $\widehat{gC}$ and $\widehat{CG}$ when these are considered to be successively traversed by the projectile moving the *same* way from $g \to C$ and from $C \to G$; and, after a considerable battle with the attendant complication of correctly introducing the component of gravity (see Appendix 2.2/3 and §1.1–5 above), at length successfully attained his objective (see §1.6 preceding). Thereafter, he further shaped and rounded out his corrected argument into the polished revise (§2.3) which Cotes lightly tailored to fit the space available for it on pages 232–44 of the *Principia*'s second edition.

(14) Namely, in the standard system of perpendicular Cartesian coordinates in which the abscissa $AB = a$ and ordinate $BC = e$ (as Newton proceeds to denote these lines using Fermatian variables) define the general point $C(a, e)$ of a given curve on positing the appropriate analytical relationship $e = e_a$ between them. Though he was not (in any manuscript known to us) to make explicit verbal statement of the fact till he penned Proposition XII of his 1691 treatise *De quadratura Curvarum* some half dozen years later (see VII: 98 and our notes (107) and (109) thereto), it seems harsh to deny that Newton was already fully aware that the successive coefficients in the expansion of the incremented ordinates $DG = e_{a+o}$ and $il = e_{a-o}$ into an equivalent 'converging' series are proportional to the corresponding fluxional derivatives of e with respect to a: specifically, where $e_{a\pm o} = e \pm Qo + Ro^2 \pm So^3...$, then $Q = \dot{e}(= de/da)$, $R = \frac{1}{2}\ddot{e}(= d^2e/da^2)$, $S = \frac{1}{6}\dddot{e}(= \frac{1}{6}d^3e/da^3)$, and so on. Insight into the general form of these coefficients is not, of course, needed in determining the series-expansions of the incremented ordinates $e_{a\pm o}$ in the case of particular curves of defining Cartesian equation $e = e_a$, nor does Newton linger on the point in obtaining from first principles the pertinent 'converging series' for the four instances of the semicircle $e = +\sqrt{(n^2-a^2)}$, the parabola $e = (ca-a^2)/b$, the common Apollonian hyperbola $e = c - (m/n)\,a - b^2/a$ and its higher-order generalization $e = c - ka - b^2/a^n$ which he proceeds to cite in exemplification of his general measures for the density of the medium through which the projectile passes in these curves, and for the resistance which opposes its passage. Knowledge of the precise form of the relationships $Q = de/da$, $R = \frac{1}{2}dQ/da$ and $S = \frac{1}{3}dR/da$ (in some equivalent) is, however, of crucial importance when—as Newton himself went on for the first time publicly to do (see §2.3: note (50) above) in the two opening paragraphs by which he augmented the ensuing scholium to his present Proposition X in the *Principia*'s second edition—we seek, departing from the (cor-

natim applicatæ resolvatur in seriem convergentem: Problema per primos seriei terminos expedite solvetur, ut in Exemplis sequentibus.

Exempl. 1. Sit Linea *ACK* semicirculus super diametro *AK* descriptus, & requiratur Medii densitas quæ faciat ut Projectile in hac linea moveatur.

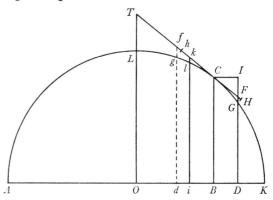

Bisecetur semicirculi diameter *AK* in *O*, et dic *OK n*, *OB a*, *BC e*, & *BD* vel *Bi*[(15)] *o*: & erit *DG^q* seu *OG^{q(16)}* − *OD^q* æquale *nn* − *aa* − 2*ao* − *oo* seu *ee* − 2*ao* − *oo*; & radice per methodum nostram extracta fiet

$$DG = e - \frac{ao}{e} - \frac{oo}{2e} - \frac{aaoo}{2e^3} - \frac{ao^3}{2e^3} - \frac{a^3o^3}{2e^5} \ \&\text{c.}$$

Hic scribatur *nn* pro *ee* + *aa* & evadet

$$DG = e - \frac{ao}{e} - \frac{nnoo}{2e^3} - \frac{anno^3}{2e^5} - \&\text{c.}^{(17)}$$

Hujusmodi Series distinguo in terminos successivos in hunc modum. Terminum primum appello in quo quantitas infinite[(18)] parva *o* non extat,

---

rected) measure of resistance $\rho = \frac{3}{4s}gS\sqrt{(1+Q^2)}/R^2$, to resolve the inverse problem of determining the path of a projectile through a medium resisting as some *n*th power of the body's instantaneous speed (that is, $\rho = kv^n$ where $v^2 = \frac{1}{2}g(1+Q^2)/R$). This may well have been the underlying reason why Johann Bernoulli, who in his long and perceptive article 'De motu Corporum gravium, Pendulorum, & Projectilium in mediis non resistentibus & resistentibus supposita Gravitate uniformi...' (*Acta Eruditorum* (February/March 1713): 77–95/115–32) independently attained the general correction of Newton's measure in a relatively intractable equivalent geometrical form (*ibid.*: Theorema IV: 91–3), was to be so adamant in his assertion that Newton's sloppy verbal account of the 'Taylor' expansion of the binomial $(z+o)^n$ in the terminal scholium of his 1704 *Tractatus de Quadratura Curvarum* cloaked a deeply mistaken notion of higher-order derivatives. (On this last, and its tiresome reiteration in the squabble from 1713 onwards over calculus priority, see 2, §3: note (46) preceding.)

(15) There is $iB = BD$ since by hypothesis $kC = CF$.

(16) Doubtless to avoid further congestion of lines around the point *G* in his figure, Newton has not there shown this radius *OG*.

(17) Corresponding to the defining Cartesian equation $BC = \sqrt{(OC^2 - OB^2)}$ or

$$e_a = +\sqrt{(n^2 - a^2)}$$

of the semicircular path of the projectile $C(a, e)$ this is the series expansion of the incremented ordinate $e_{a+o} = e + Qo + Ro^2 + So^3 + ...$ where in fact (compare note (14) preceding, and Newton's following paragraph) $Q = de/da$, $R = \frac{1}{2}d^2e/da^2$ and $S = \frac{1}{6}d^3e/da^3$.

(18) In strict truth this should be 'indefinite', and so indeed had Humphrey Newton penned it in the fair secretary copy of the text of the present proposition which (see note (2) above) went to the printer in spring 1686. It is not known whether it was Newton himself or his editor, Edmond Halley, who made the present unsatisfactory replacement (in the ensuing printer's proof, we assume), if indeed it was more than a simple printer's slip.

secundum in quo quantitas illa extat unius dimensionis, tertium in quo extat duarum, quartum in quo trium est, & sic in infinitum. Et primus terminus, qui hic est *e*, denotabit semper longitudinem ordinatæ *BC* insistentis ad indefinitæ quantitatis [*o*] initium *B*; secundus termi‖nus, qui hic est $\frac{ao}{e}$, denotabit ‖[264] differentiam inter *BC* & *DF*, id est lineolam *IF*, quæ absconditur complendo parallelogrammum *BCID*, atcȝ adeo positionem Tangentis *CF* semper determinat:[19] ut in hoc casu capiendo *IF* ad *IC* ut est $\frac{ao}{e}$ ad *o* seu *a* ad *e*. Terminus tertius, qui hic est $\frac{nnoo}{2e^3}$, designabit lineolam *FG* quæ jacet inter Tangentem & Curvam, adeocȝ determinat angulum contactus *FCG*, seu curvaturam quam curva linea habet in *C*.[20] Si lineola illa *FG* finitæ est magnitudinis, designabitur per terminum tertium una cum subsequentibus in infinitum. At si lineola illa minuatur in infinitum, termini subsequentes evadent infinite minores tertio, ideocȝ negligi possunt. Terminus quartus, qui hic est $\frac{anno^3}{2e^3}$, exhibet variationem Curvaturæ; quintus variationem variationis, & sic deinceps. Unde obiter patet usus non contemnendus harum Serierum in solutione Problematum quæ pendent a Tangentibus & curvatura Curvarum.

---

(19) This *secundus terminus* [−](*a*/*e*) *o* in the series expansion of the incremented ordinate $e_{a+o} = +\sqrt{(n^2 - (a+o)^2)}$ is $Qo = (de/da) o$, that is,

$$BD.d(BC)/d(OB) \text{ or } BD.(IF/CI) = IF;$$

whence the linelet *FG* is (*DG*−*BC*−*IF* or) $e_{a+o} - e - Qo = Ro^2 + So^3 + \ldots$, where (see note (17) preceding) $R = \frac{1}{2}d^2e/da^2$, $S = \frac{1}{6}d^3e/da^3$, and so on.

(20) When, at least, the linelet *FG*—that is, $\frac{1}{2}(d^2e/da^2) o^2$ on ignoring terms in o^3 and higher powers of *o* as 'indefinitely' small—stands, like the neighbouring parent ordinate *BC*, at right angles to the vanishingly small arclet $\overset{\frown}{CG}$, and therefore *de*/*da* = 0: in all other positions the *exponens curvaturæ* will more generally be $FG/(CF/CI)^3 \propto \frac{1}{2}(d^2e/da^2)/(1 + (de/da)^2)^{\frac{3}{2}}$, the reciprocal of the diameter of the circle osculating the curve at *C*. Newton was to make this very necessary restriction explicit when, half a dozen years later, he introduced his present measure of the horn-angle between tangent and curve into Proposition XIII of his 1691 treatise 'De quadratura Curvarum' (see VII: 112–14) along with the following notion of likewise expressing variations in such *curvatura* through the successive coefficients, *S*, *T*, ... of the powers of *o* in the series $(FG - Ro^2 =) So^3 + To^4 + \ldots$, there specifying that these are respectively the higher-order derivatives $\frac{1}{6}d^3e/da^3$, $\frac{1}{24}d^4e/da^4$, and so on. (Compare VII: 112–13, note (146).) Yet another quarter of a century on, when he came in the middle/late 1710's to look over the *editio secunda* of his *Principia* for a new edition of its text, he thought for a time—not least to substantiate his claim elsewhere to have independently attained such a general measure for the curvature of a curve in his October 1666 tract on fluxions—to specify how the *lineola FG* = Ro^2, taken in conjunction with the tangential deviation *IF* = *Qo*, does in fact straightforwardly yield the formula $(1 + Q^2)^{\frac{3}{2}}/[2]R$, that is, $(1 + (de/da)^2)^{\frac{3}{2}}/(d^2e/da^2)$, for the radius of curvature at *C*(*a*, *e*). Such an addition was not made in the *editio tertia* in 1726, but in Appendix 4 following we reproduce three surviving drafts of an *addendum* to this effect intended to be inserted, at (see note (37) below) a not altogether happy place, in the ensuing scholium.

Præterea CF est latus quadratum ex CI^q & IF^q, hoc est ex BD^q & quadrato termini secundi. Estqz $FG+kl$ æqualis duplo termini tertii, & $FG-kl$ æqualis duplo quarti. Nam valor ipsius DG convertitur in valorem ipsius il & valor ipsius FG in valorem ipsius kl scribendo Bi pro BD seu $-o$ pro $+o$. Proinde cum FG sit $-\frac{nnoo}{2e^3}-\frac{anno^3}{2e^5}$ &c erit $kl=-\frac{nnoo}{2e^3}+\frac{anno^3}{2e^5}$ &c. Et horum summa est $-\frac{nnoo}{e^3}$, differentia $-\frac{anno^3}{e^5}$. Terminum quintum & sequentes[21] hic negligo ut infinite minores quam qui in hoc Problemate considerandi veniant. Itaqz si designetur Series[22] universaliter his terminis $\mp Qo - Roo [\mp] So^3$ &c, erit CF æqualis $\sqrt{oo+QQoo}$, $FG+kl$ æqualis $2Roo$, & $FG-kl$ æqualis $2So^3$. Pro CF, $FG+kl$ &

‖[265] $FG-kl$ scribantur ‖ hi earum valores, & Medii densitas quæ erat ut $\dfrac{FG-kl}{CF \text{ in } \overline{FG+kl}}$ jam fiet ut $\dfrac{S}{R\sqrt{1+QQ}}$.[23] Deducendo igitur Problema unumquodqz ad seriem convergentem, & hic pro Q, R & S scribendo terminos seriei ipsis respondentes; deinde etiam ponendo Resistentiam Medii in loco quovis G esse ad Gravitatem ut $S\sqrt{1+QQ}$ ad $2RR$,[23] & velocitatem esse illam ipsam quacum corpus de loco C secundum rectam CF egrediens, in Parabola diametrum CB & latus rectum $\dfrac{1+QQ}{R}$ habente deinceps moveri posset,[24] solvetur Problema.

Sic in Problemate jam solvendo si scribantur $\sqrt{1+\dfrac{aa}{ee}}$ seu $\dfrac{n}{e}$ pro $\sqrt{1+QQ}$, $\dfrac{nn}{2e^3}$ pro R, & $\dfrac{ann}{2e^5}$ pro S, prodibit Medii densitas ut $\dfrac{a}{ne}$, hoc est (ob datam n) ut $\dfrac{a}{e}$ seu $\dfrac{OB}{BC}$, id est ut Tangentis longitudo illa CT quæ ad semidiametrum OL ipsi AK normaliter insistentem terminatur; et resistentia erit ad gravitatem ut a ad n, id est ut OB ad circuli semidiametrum OK, velocitas autem erit ut $\sqrt{2BC}$. Igitur si corpus C certa cum velocitate secundum lineam ipsi OK parallelam exeat de

---

(21) Namely '$+To^4+$&c', to continue Newton's assignation of coefficients in the series expansion of e_{a+o} as $e+Qo+Ro^2+So^3....$

(22) Understand those for $IG = (BC-DG$ or$)$ $e-e_{a+o}$ and for $BC-il = e-e_{a-o}$.

(23) Duly corrected (see note (13) above) by a factor of $\frac{3}{2}$ these ratios should be '$\dfrac{3S}{2R\sqrt{1+QQ}}$' and '$ut$ $3S\sqrt{1+QQ}$ ad $4RR$' respectively.

(24) Whence explicitly, on setting the force of gravity as before to be g, the body's speed v at C is assigned to be $\sqrt{(\frac{1}{2}g.(1+Q^2)/R)}$. For at once, since $FG = Ro^2+...$ is $\frac{1}{2}g\theta^2-...$ where (as in note (6) above) θ is the time taken by the body to traverse the arc $\widehat{CG} = o\sqrt{(1+Q^2)}+...$, there is (in the limit as o and θ each become vanishingly small) $v = o\sqrt{(1+Q^2)}/\theta$ where $o/\theta = \sqrt{(\frac{1}{2}g/R)}$.

loco *L*, & Medii densitas in singulis
locis *C* sit ut longitudo tangentis
CT, & resistentia etiam in loco
aliquo *C* sit ad vim gravitatis ut *OB*
ad *OK*: corpus illud describet circuli
quadrantem *LCK*. Q.E.I.

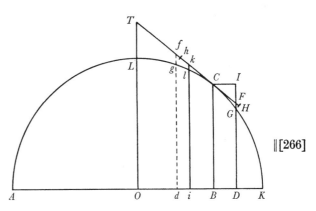

‖[266]

At si corpus idem de loco *A* secundum lineam ipsi *AK* per‖pen-
dicularem egrederetur, sumenda
esset *OB* seu *a* ad contrarias partes
centri *O*, & propterea signum ejus
mutandum esset, & scribendum −*a* pro +*a*. Quo pacto prodiret Medii den-

sitas ut −$\frac{a}{e}$. Negativam autem densitatem (hoc est quæ motus corporum

accelerat) Natura non admittit, & propterea naturaliter fieri non potest ut
corpus ascendendo ab *A* describat circuli quadrantem *AL*. Ad hunc effectum
deberet corpus a Medio impellente accelerari, non a resistente impediri.

Exempl. 2. Sit linea *ALCK* Parabola axem habens *OL* horizonti *AK* perpendicularem, & requiratur Medii densitas quæ faciat ut projectile in ipsa moveatur.

Ex natura Parabolæ rectangulum *ADK* æquale est rectangulo sub ordinata
DG & recta aliqua data: hoc est, si dicatur recta illa *b*, *AB a*, *AK c*, *BC e*[25] &
BD o, rectangulum *a*+*o* in *c*−*a*−*o* seu *ac*−*aa*−2*ao*+*co*−*oo* æquale est rectang-

ulo *b* in *DG*, adeoq̄ *DG* æquale $\frac{ac-aa}{b}+\frac{c-2a}{b}o-\frac{oo}{b}$. Jam scribendus esset hujus

seriei secundus terminus $\frac{c-2a}{b}o$ pro *Qo*, & ejus coefficiens $\frac{c-2a}{b}$ pro *Q*; tertius

item terminus [−]$\frac{oo}{b}$ pro *Roo*, & ejus coefficiens [−]$\frac{1}{b}$ pro *R*. Cum vero plures

non sint termini, debebit quarti termini *So*³ coefficiens *S* evanescere, & propter-

ea quantitas $\dfrac{S}{R\sqrt{1+QQ}}$ cui Medii densitas proportionalis est, nihil erit.

Nulla igitur Medii densitate movebitur Projectile in Parabola, uti olim demonstravit *Galilæus*.[26]   Q.E.I.

Exempl. 3. Sit linea *AGK* Hyperbola Asymptoton habens *NX* plano horizontali

---

(25) So that the defining Cartesian equation of the parabola is $e = e_a \equiv a(c-a)/b$.

(26) And Thomas Harriot, who twenty years before Galileo considered also resisted motion
in tilted parabolic paths (see VI: 7, note (17)), even more so; Newton here repeats his earlier
affirmation in scholium to the 'Leges Motus' at the beginning of Book 1 of the *Principia* (₁1687:
20) that 'adinvenit *Galilæus*... motum projectilium fieri in Parabola, conspirante experientia,
nisi quatenus motus...per aeris resistentiam aliquantulum retardantur' (compare VI: 106,
note (35)).

AK perpendicularem, & quæratur Medii densitas quæ faciat ut Projectile moveatur in hac linea.

‖[267] Sit *MX* Asymptotos altera ordinatim applicatæ *DG* pro‖ductæ occurrens in *V*, & ex natura Hyperbolæ rectangulum *XV* in *VG* dabitur. Datur autem ratio *DN* ad *VX*, & propterea datur etiam rectangulum *DN* in *VG*. Sit illud *bb*, &

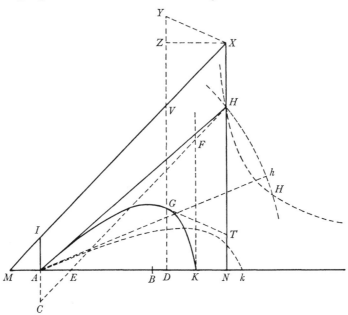

completo parallelogrammo *DNXZ*, dicatur *BN a*, *BD o*, *NX c*, & ratio data *VZ* ad *ZX* vel *DN* ponatur esse $\frac{m}{n}$.[27] Et erit *DN* æqualis $a-o$, *VG* æqualis $\frac{bb}{a-o}$,

VZ æqualis $\frac{m}{n}\overline{a-o}$ & *GD* seu $NX-VZ-VG$ æqualis $c-\frac{m}{n}a+\frac{m}{n}o-\frac{bb}{a-o}$.

Resolvatur terminus $\frac{bb}{a-o}$ in seriem convergentem $\frac{bb}{a}+\frac{bb}{aa}o+\frac{bb}{a^3}oo+\frac{bb}{a^4}o^3$ &c,

& fiet *GD* æqualis $c-\frac{m}{n}a-\frac{bb}{a}+\frac{m}{n}o-\frac{bb}{aa}o-\frac{bb}{a^3}o^2-\frac{bb}{a^4}o^3$ &c. Hujus seriei terminus

secundus $\frac{m}{n}o-\frac{bb}{aa}o$[28] usurpandus est pro *Qo*, tertius cum signo mutato $\frac{bb}{a^3}o^2$[28]

---

(27) Whence on setting the ordinate *BC* (not shown in Newton's accompanying figure) to be *e*, as before, the defining Cartesian equation of the hyperbolic path $\overset{\frown}{AGK}$ of general point $C(a, e)$ is $e = e_a \equiv c-(m/n)\,a-b^2/a$.

(28) As N. R. Hanson first observed—but failed adequately to amend—in a short note in *Scripta Mathematica*, **26**, 1961: 83–5, Newton here makes a muddle in assigning the values of the coefficients $Q(= de/da)$ and $R(= \frac{1}{2}d^2e/da^2)$ in the series expansion $e-Qo+Ro^2-So^3+...$ of the (here decremented) ordinate $DG = e_{a-o}$. Read correctly: '...terminus secundus cum signo mutato $-\frac{m}{n}o+\frac{bb}{aa}o$ usurpandus est pro *Qo*, tertius $-\frac{bb}{a^3}o^2$ pro *Ro²*, ...'.

pro Ro^2, & quartus cum signo etiam mutato $\dfrac{bb}{a^4}o^3$ pro So^3, eorumcɜ coefficientes

$\dfrac{m}{n}-\dfrac{bb}{aa}$, $\dfrac{bb}{a^3}$[(29)] & $\dfrac{bb}{a^4}$ scribendæ sunt ‖ in Regula superiore pro Q, R & S. Quo facto ‖[268]
prodit medii densitas ut

$$\dfrac{\dfrac{bb}{a^4}}{\dfrac{bb}{a^3}\sqrt{1[+]\dfrac{mm}{nn}-\dfrac{2mbb}{naa}+\dfrac{b^4}{a^4}}} \quad \text{seu} \quad \dfrac{1}{\sqrt{aa+\dfrac{mm}{nn}aa-\dfrac{2mbb}{n}+\dfrac{b^4}{aa}}},$$

id est, si in VZ sumatur VY æqualis VG, ut $\dfrac{1}{XY}$. Namcɜ aa & $\dfrac{mm}{nn}aa-\dfrac{2mbb}{n}+\dfrac{b^4}{aa}$
sunt ipsarum XZ & ZY[(30)] quadrata. Resistentia autem invenitur in ratione ad Gravitatem quam habet XY ad YG, & velocitas ea est quacum corpus in Parabola pergeret verticem G diametrum DG & latus rectum $\dfrac{YX^{\text{quad.}}}{VG}$ habente.[(31)]

---

(29) Here, correspondingly, read '$-\dfrac{m}{n}+\dfrac{bb}{aa}$, $-\dfrac{bb}{a^3}$'.

(30) That is, since the increment DB is assumed to be vanishingly small, and YX is drawn parallel to the tangent GT to the infinitesimal arc $\overset{\frown}{CG}$, 'NB & NB in Q'. More precisely, as D comes to coincide with B there is $ZX=a$ and $ZY=-(m/n)a+b^2/a=-VZ+GV$.

(31) Newton understands his previous 'semicircular' figure, with the infinitesimal arc $\overset{\frown}{CG}$ of the trajectory $\overset{\frown}{ACK}$ (where DG lies infinitely close to the parallel ordinate BC) taken to coincide with that of the parabola (G) in which $FG \propto CF^2$ 'hugging' it; this has *latus rectum* $r = CF^2/FG = (1+Q^2)/R$ in the limit as $CF = o\sqrt{(1+Q^2)}$ and $FG = Ro^2+So^3+...$ each pass to be vanishingly small, while the body's speed over $\overset{\frown}{CG} \approx CF$ is

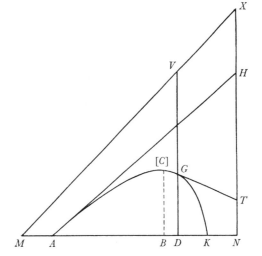

$$CF/\sqrt{(FG/\tfrac{1}{2}g)} = \sqrt{(\tfrac{1}{2}gr)}$$

in the same limit. In the present instance of the hyperbola $BC = c-(m/n)NB-b^2/NB$, where YX is constructed equal and parallel to the trajectory's tangent $GT = ND\sqrt{(1+Q^2)}$, Newton's values for the density of the medium, the ratio of its resistance to gravity, and the projectile's speed at $C(\approx G)$ readily result on substituting (in his respective measures

$$S/R\sqrt{(1+Q^2)}, \quad \tfrac{1}{2}S\sqrt{(1+Q^2)}/R^2$$

and $\qquad \sqrt{(\tfrac{1}{2}g.(1+Q^2)/R)}$

of these) the quantities $Q = -m/n+b^2/a^2$, $R = -b^2/a^3$ and $S = b^2/a^4$ which are the coefficients in the series expansion of the decremented ordinate $DG = e_{a-o} = BC-Qo+Ro^2-So^3+...$; that is, on going over (with him) from $NB = a$ to the decremented abscissa $ND = a-o$ with which it coincides in the limit as $BD = o$ vanishes, by entering

$$Q = -m/n+b^2/ND^2 = -(NT-DG)/ND, \quad R = -b^2/ND^3 = -GV/ND^2$$
$$\text{and} \quad S = b^2/ND^4 = -R/ND.$$

Ponatur itaqʒ quod Medii densitates in locis singulis G sint reciproce ut distantiæ XY quodqʒ resistentia in loco aliquo G sit ad gravitatem ut XY ad YG, & corpus de loco A justa cum velocitate emissum describet Hyperbolam illam AGK. Q.E.I.

Exempl. 4. Ponatur indefinite quod linea AGK Hyperbola sit, centro X Asymptotis MX, NX ea lege descripta ut constructo rectangulo $XZDN$ cujus latus ZD secet Hyperbolam in G & Asymptoton ejus in V, fuerit VG reciproce ut ipsius ZX vel DN dignitas aliqua ND^n, cujus index est numerus n; & quæratur Medii densitas qua Projectile progrediatur in hac curva.[32]

Pro $[B]N$, BD, NX scribantur a, o, c[33] respective, sitqʒ VZ ad ZX vel DN ut d ad e & VG æqualis $\dfrac{bb}{DN^n}$, & erit DN æqualis $a-o$, $VG = \dfrac{bb}{\overline{a-o}^n}$, $VZ = \dfrac{d}{e}$ in $a-o$,

& GD seu $NX - VZ - VG$ æqualis $c - \dfrac{d}{e}a + \dfrac{d}{e}o - \dfrac{bb}{\overline{a-o}^n}$. Resolvatur terminus ille

$\dfrac{bb}{\overline{a-o}^n}$ in seriem infinitam $\dfrac{bb}{a^n} + \dfrac{nbb}{a^{n+1}}o + \dfrac{nn+n}{2a^{n+2}}bbo^2 + \dfrac{n^3+3nn+2n}{6a^{n+3}}bbo^3$ &c ac fiet

‖[269] GD æqualis $c - \dfrac{d}{e}a - \dfrac{bb}{a^n} \| + \dfrac{d}{e}o - \dfrac{nbb}{a^{n+1}}o - \dfrac{nn+n}{2a^{n+2}}bbo^2 - \dfrac{n^3+3nn+2n}{6a^{n+3}}bbo^3$ &c. Hujus

seriei terminus secundus[34] $[-]\dfrac{d}{e}o[+]\dfrac{nbb}{a^{n+1}}o$ usurpandus est pro Qo, tertius

$[-]\dfrac{nn+n}{2a^{n+2}}bbo^2$ pro Ro^2, $[\&]$ quartus[34] $\dfrac{n^3+3nn+2n}{6a^{n+3}}bbo^3$ pro So^3. Et inde Medii

densitas $\dfrac{S}{R \times \sqrt{1+QQ}}$ in loco quovis G fit $\dfrac{n+2}{3\sqrt{a^2 + \dfrac{dd}{ee}a^2 - \dfrac{2dnbb}{ea^n}a + \dfrac{nnb^4}{a^{2n}}}}$, adeoqʒ

si in VZ capiatur VY æqualis $n \times VG$, est reciproce ut XY. Sunt enim a^2 &

$\dfrac{dd}{ee}a^2 - \dfrac{2dnbb}{ea^n}a + \dfrac{nnb^4}{e^{2n}}$ ipsarum XZ & ZY quadrata. Resistentia autem in

eodem loco G sit ad Gravitatem ut S in $\dfrac{XY}{a}$ ad $2RR$, id est XY ad $\dfrac{3nn+3n}{n+2}VG$. Et

---

(32) Whence, more generally, the defining Cartesian equation of the point $C(a, e)$ of the hyperbolic path $\widehat{ACGK}$ is now put to be $BC = e = e_a \equiv c - (d/e)a - b^2/a^n$. (The trivial replacement here of the coefficient of the term in $-a$ in the preceding example which is its prime case merely, of course, permits Newton to reserve the constant n for its more familiar rôle as index of a general power.)

(33) Here and in sequel the printer has set these three letters as capitals, but for consistency's sake—and, we believe, as Newton himself intended (compare note (2) above)—we everywhere reduce these to lower-case.

(34) Read 'cum signo mutato' in each case. In the printed original—whose text is adjusted in no subsequent edition—Newton here falls again into his earlier confusion in ascribing the signs of the coefficients Q, R and S of the successive powers of o in the series expansion of the decremented ordinate $DG = e_{a-o}$.

velocitas ibidem ea ipsa est quacum corpus projectum in Parabola pergeret verticem G, diametrum GD & Latus rectum $\dfrac{1+QQ}{R}$ seu $\dfrac{2XY^{\text{quad}}}{nn+n \text{ in } VG}$ habente.[35] Q.E.I.

Scholium.

Quoniam motus non fit in Parabola nisi in Medio non resistente, in Hyperbolis vero hic descriptis fit per resistentiam perpetuam, perspicuum est quod linea quam Projectile in Medio uniformiter resistente describit, propius accedit ad Hyperbolas hasce quam ad Parabolam. Est utiꝗ linea illa Hyperbolici generis, sed quæ circa verticem magis distat ab Asymptotis; in partibus a vertice remotioribus propius ad ipsas accedit quam pro ratione Hyperbolarum quas hic descripsi. Tanta vero non ‖ est inter has & illam differentia quin illius loco ‖[270] possint hæ in rebus practicis non incommode adhiberi. Et utiliores forsan futuræ sunt hæ quam Hyperbola magis accurata & simul magis composita. Ipsæ vero in usum sic deducentur.

Compleatur parallelogrammum $XYGT$, & ex natura harum Hyperbolarum facile colligitur quod recta GT tangit Hyperbolam in G[36] ideoꝗ densitas Medii in G est reciproce ut tangens GT & velocitas ibidem ut $\sqrt{\dfrac{GT^{q}}{GV}}$, resistentia autem ad vim gravitatis ut GT ad $\dfrac{3nn+3n}{n+2}\,GV$.[37]

---

(35) That is, explicitly, the body's speed over the infinitesimal arc $\overset{\frown}{CG}$ is

$$\sqrt{(\tfrac{1}{2}g \cdot 2YX^2/(n^2+n) \cdot GV)}$$

where g is the constant downwards force of gravity. These generalized values for the density of the medium, the ratio of its resistance to gravity, and the projectile's velocity result in the same way as in the preceding particular case $n = 1$ (see note (31) above) on making substitution of the coefficients

$$Q = -d/e + nb^2/a^{n+1}, \quad R = -\tfrac{1}{2}n(n+1)\,b^2/a^{n+2} \quad \text{and} \quad S = \tfrac{1}{6}n(n+1)\,(n+2)\,b^2/a^{n+3}$$

in the series expansion $DG = e_{a-o} = BC - Qo + Ro^2 - So^3 + \dots$; whence, on replacing $NB = a$ by the decremented abscissa $ND = a - o$ (as Newton again confusingly specifies his construction), there is $YZ = ND\sqrt{(1+Q^2)}$, $Q = -d/e + nb^2/ND^{n+1}$, $R = -\tfrac{1}{2}n(n+1) \cdot GV/ND^2$ and $S = -\tfrac{1}{3}(n+2)\,R/ND$. On increasing the preceding defective measure $\tfrac{1}{2}S\sqrt{(1+Q^2)}/R^2$ correctly in the ratio of 3 to 2 (see note (13) above), Newton in his *editio secunda* here (*Principia*, ₂1713: 239, l. −3)—and *mutatis mutandis* in his previous 'Exempl. 3' (where $n = 1$) correspondingly— had but numerically to adjust the ratio of the resistance to gravity to be as XY to '$\dfrac{2nn+2n}{n+2}\,VG$'.

(36) Since it is constructed parallel to XY, of slope $YZ/ND = Q = de/da$, where $e = c - (d/e)\,a - b^2/a^n$; see notes (31) and (35) preceding.

(37) Read '$\dfrac{2nn+2n}{n+2}\,GV$' on correcting the deficient general measure from which (see note (35)) this value for the ratio of resistance to gravity in the hyperbolic trajectory of 'Exempl. 4' preceding is derived; and so Newton adjusted it in the *editio secunda* of the *Principia* (₂1713:

Proinde si corpus de loco A secundum rectam AH projectum describat Hyperbolam AGK, & AH producta occurrat Asymptoto NX in H actaq AI occurrat alteri Asymptoto MX in I: erit Medii densitas in A reciproce ut AH, & corporis velocitas ut $\sqrt{\dfrac{AH^q}{AI}}$, ac resistentia ibidem ad Gravitatem ut AH ad $\dfrac{3nn+3n}{n+2}$ in AI.[38] Unde prodeunt sequentes Regulæ.

Reg. 1. Si servetur Medii densitas[39] in A & mutetur angulus NAH, manebunt longitudines AH, AI, HX. Ideoq si longitudines illæ in aliquo casu inveniantur, Hyperbola deinceps ex dato quovis angulo NAH expedite determinari potest.

Reg. 2. Si servetur tum angulus NAH tum Medii densitas in A & mutetur velocitas quacum corpus projicitur, servabitur longitudo AH & mutabitur AI in duplicata ratione velocitatis reciproce.

Reg. 3. Si tam angulus NAH quam corporis velocitas in A gravitasq acceleratrix servetur, & proportio resistentiæ in A ad gravitatem motricem augeatur in ratione quacunq: augebitur proportio AH ad AI in eadem ratione, manente Parabolæ latere recto eiq proportionali longitudine $\dfrac{AH^q}{AI}$, & propterea ∥[271] minuetur AH in eadem ratione, & AI minuetur in ratione illa du∥plicata. Augetur vero proportio resistentiæ ad pondus ubi vel gravitas specifica sub æquali magnitudine fit minor, vel Medii densitas major, vel resistentia ex magnitudine diminuta diminuitur in minore ratione quam pondus.

Reg. 4. Quoniam densitas Medii prope verticem Hyperbolæ m[aj]or est quam in loco A, ut servetur densitas mediocris debet ratio minimæ tangentium GT ad Tangentem AH inveniri & densitas in A, per Regulam tertiam, diminui in ratione paulo minore quam semisummæ Tangentium ad Tangentem AH.

Reg. 5. Si dantur longitudines AH, AI & describenda sit figura AGK: produc HN ad X ut sit HX æqualis facto sub $n+1$ & AI, centroq X & Asymptotis MX, NX per punctum A describatur Hyperbola ea lege ut sit AI ad quamvis VG ut XV^n ad XI^n.

Reg. 6. Quo major est numerus n, eo magis accuratæ sunt hæ Hyperbolæ in ascensu corporis ab A, & minus accuratæ in ejus descensu ad [K], & contra.[40]

---

240). It was at this none too pertinent place that Newton afterwards thought to append his derivation of the length of the radius of curvature at C by the formula $(1+Q^2)^{\frac{3}{2}}/2R$ (see Appendix 4 below) in amplification of his previous oblique statement that $R(=\frac{1}{2}d^2e/da^2)$ is a measure of the curvature of the curve $\overparen{ACK}$ there; on which see note (20) above.

(38) This particular value of the preceding ratio of resistance to gravity when the body C is at A needs (see the previous note) likewise to be adjusted to be 'ut AH ad $\dfrac{2nn+2n}{n+2}$ in AI'; and so it was repaired in the *editio secunda* of the *Principia* ($_2$1713: 240, *lin. ult.*).

(39) Understand 'ut et corporis velocitas'.

Hyperbola Conica mediocrem rationem tenet, estqɜ cæteris simplicior. Igitur si Hyperbola sit hujus generis, & punctum *K* ubi corpus projectum incidet in rectam quamvis *AN* per punctum *A* transeuntem quæratur: occurrat producta *AN* Asymptotis *MX, NX* in *M* & *N*, & sumatur *NK* ipsi *AM* æqualis.

Reg 7. Et hinc liquet methodus expedita determinandi hanc Hyperbolam[41] ex Phænom[e]nis. Projiciantur corpora duo similia & æqualia eadem velocitate in angulis diversis *HAK, ha[k]* incidentqɜ in planum Horizontis in *K* et *k*, & notetur proportio *AK* ad *ak*. Sit ea *d* ad *e*. Tum erecto cujusvis longitudinis perpendiculo *AI*, assume utcunqɜ longitudinem *AH* vel *Ah* & inde collige graphice longitudines *AK, Ak* per Reg. 6. Si ratio *AK* ad *Ak* sit eadem cum ratione *d* ad *e*, longitudo *AH* recte assumpta fuit. Sin minus cape in recta infinita

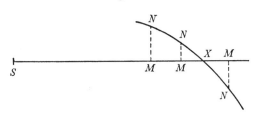

SM longitudinem *SM* æqualem assumptæ *AH* & erige perpendiculum *MN* æ‖quale rationum differentiæ $\dfrac{AK}{Ak} - \dfrac{d}{e}$ ‖[272] ductæ in rectam quamvis datam.[42] Simili methodo ex assumptis pluribus longitudinibus *AH* invenienda sunt plura puncta *N*: & tum demum si per omnia agatur Curva linea regularis *NNXN*, hæc abscindet *SX* quæsitæ longitudini *AH* æqualem.[43] Ad usus Mechanicos sufficit longitudines *AH, AI* easdem in angulis omnibus *HAK* retinere. Sin figura ad inveniendam resistentiam Medij accuratius determinanda sit, corrigendæ sunt semper hæ longitudines per Regulam quartam.

---

(40) Since for given initial firing speed and direction at *A*, and hence fixed inclination of *IVX* to *ADN*, there is $GV(=b^2/ND^n) \propto XV^{-n}$, increasing the index *n* (> 0, of course, for a hyperbolic trajectory $\widehat{AGK}$) will make the ascending portion of the projectile path steeper and shallower, and its ensuing part less rounded and more nearly vertical, and the whole more closely approaching the sides of the angle $A\widehat{H}N$. Whether, particularly for the low missile speeds of musket and cannon shot obtaining in the practice of ordnance in the later seventeenth century, the first better approximates the empirical truth, and whether, for the relatively high resistance afforded by the (usually) damp maritime air of England, the latter deviates more from it, we may well doubt. Certainly, the simple conic hyperbola (*n* = 1), that of his 'Exempl. 3' preceding, for which Newton goes on to plump as maintaining a reasonable mean between these two see-sawing extremes has, for all its mathematical simplicity, not very much to commend it as an accurate representation of the resisted projectile trajectory, even as an approximation to be refined 'per Reg. 4' (as he states at the end of the next paragraph).

(41) 'Conicam', that is.

(42) Since the ratios *AK/Ak* and *d/e* are themselves pure numbers, without the requisite linear geometrical dimension.

(43) It will evidently be accurate enough to draw the *curva regularis* $\widehat{NNXN}$ free-hand, 'æquo manus motu' as Newton had phrased it a year or so before in Chapter 3 of his 'Matheseos Universalis Specimina' (see IV: 560 and 561, note (112)) in specifying an analogous geometrical rule of false position, without appealing to more precise methods of passing a curve 'smoothly' through the error-points *N*.

Reg. 8. Inventis longitudinibus *AH, HX,* si jam desideretur positio rectæ *AH* secundum quam Projectile data illa cum velocitate emissum incidit in punctum quodvis *K*: ad puncta *A* & *K* erigantur rectæ *AC, KF* horizonti perpendiculares, quarum *AC* deorsum t[e]ndat & æquetur ipsi *AI* seu ½*HX.* Asymptotis *AK, KF*

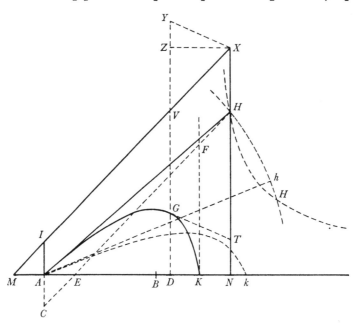

descibatur Hyperbola cujus Conjugata[44] transeat per punctum *C,* centroq̃
‖[273]   *A* & intervallo *AH* describatur Circulus secans Hyperbolam illam in ‖ puncto *H,* et projectile secundum rectam *AH* emissum incidet in punctum *K.* Q.E.I. Nam punctum *H* ob datam longitudinem *AH* locatur alicubi in circulo descripto. Agatur *CH* occurrens ipsis *AK* & *KF,* illi in [*E*] huic in *F,* et ob parallelas *CH, MX* & æquales *AC, AI* erit *AE* æqualis *AM,* et propterea etiam æqualis *KN.*[45] Sed *CE* est ad *AE* ut *FH* ad *KN,* & propterea *CE* & *FH* æquantur. Incidit ergo punctum *H* in Hyperbolam Asymptotis *AK, KF* descriptam cujus conjugata transit per punctum *C,* atq̃ adeo reperitur in communi intersectione[46] Hyper-

_____

(44) The hyperbola's branch (not drawn by Newton in his figure) which lies diametrically opposite that, of centre *K,* passing through the two points *H*; the point *C* lies in this branch since, as Newton goes on to state, *CA = AI* and so *AE = (MA =) KN,* whence the intercepts *CE* and *FH* cut off between the hyperbola and its asymptotes *AK* and *FH* are equal. (See the next note.)

(45) By Apollonius, *Conics* II, 8/16: a familiar elementary property of the hyperbola which Newton had learnt while still an undergraduate from his reading of Schooten's commentary on the second book of Descartes' *Geometrie* (see 1: 42) and afterwards extended *mutatis mutandis* to curves of higher algebraic kind (see II: 93; IV: 356).

(46) Or rather 'in communibus intersectionibus' since, as Newton shows in his figure (and verbally affirms in sequel), there are two points *H* of intersection, in general distinct, each of which defines a trajectory $\widehat{AGK}$ of the given horizontal range *AK.*

bolæ hujus & circuli descripti. Q.E.D. Notandum est autem quod hæc operatio perinde se habet sive recta *AKN* horizonti parallela sit sive ad horizontem in angulo quovis inclinata; quodɕ ex duabus intersectionibus *H*, *H* duo prodeunt anguli *NAH*, *NAH* quorum minor[47] eligendus est; & quod in Praxi mechanica sufficit circulum semel describere, deinde regulam interminatam *CH* ita applicare ad punctum *C* ut ejus pars *FH* circulo & rectæ *FK* interjecta æqualis sit ejus parti *CE* inter punctum *C* & rectam [*A*]*K* sitæ.

Quæ de Hyperbolis dicta sunt facile applicantur ad Parabolas. Nam si *XAGK* Parabolam designet quam recta *XV* tangat in vertice *X*,[48] sintɕ ordinatim applicatæ *IA*, *VG* ut quælibet ab- scissarum *XI*, *XV* dignitates XI^n, XV^n,[49] agantur *XT*, *TG*, *HA*, quarum *XT* parallela sit *VG* et *TG*, *HA* parabolam tangant in *G* & *A*: et corpus de loco quovis *A* secundum rectam *AH* productam justa cum velocitate projectum describet hanc Parabolam, si modo densitas Medij in locis singulis *G* sit reciproce ut tangens *GT*. Velocitas autem in *G* ea erit quacum Projectile pergeret, ‖ in spatio non resistente, in Parabola Conica verticem *G*, diametrum *VG* deorsum productam & latus rectum $\sqrt{\dfrac{2TG^q}{nn-n \times VG}}$ habente. Et resistentia

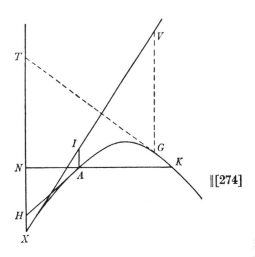

‖[274]

in *G* erit ad vim Gravitatis ut *TG* ad $\dfrac{3nn-3n}{n-2}VG$.[50] Unde si *NAK* lineam

---

(47) This is not shown by Newton in his figure, but starts off from *A* in the direction of the lower of the two points *H*. But why he should here decree that the shallower of the two trajectories $\overset{\frown}{AK}$ possible must invariably be selected is not at all clear. (When an obstacle of some kind intervenes between the firing-point *A* and target *K* there is evidently practical advantage in lofting the projectile—a mortar-shell, say— along the higher trajectory which is depicted in the accompanying figure.) Newton himself subsequently saw no need thus arbitrarily to restrict his choice, and the present phrase does not appear in his second and third (and hence all subsequent) editions of the *Principia*.

(48) Understand that this (general higher-order) parabola has for its accompanying diameter the extension of *TNX* below *X*.

(49) Whence the defining 'symptom' of the parabolic trajectory $\overset{\frown}{XGK}$ is $GV \propto XV^n$, in contrast with that, $GV \propto XV^{-n}$, for the preceding hyperbolic path. It follows without further ado that the results obtained previously for the latter, hyperbolic trajectory adapt themselves to the present parabolic one merely by substituting $n \rightarrow -n$.

(50) Read '$\dfrac{2nn-2n}{n-2}VG$' *recte* on bumping up the parent measure for the ratio of resistance to gravity by the factor $\frac{3}{2}$ (see note (13) above); the adjustment was made in the *editio secunda*, but only as a last-minute addition to the concluding 'Corrigenda' (*Principia*, ₂1713: signature Rrr2ᵛ). This and the preceding results for the density of the medium and the projectile's speed

horizontalem designet, et manente tum densitate Medij in A tum velocitate quacum corpus projicitur mutetur utcunqɜ angulus NAH, manebunt longitudines AH, AI, HX et inde datur Parabolæ vertex X & positio rectæ XI, et sumendo VG ad IA ut XV^n ad XI^n dantur omnia Parabolæ puncta G per quæ Projectile transibit.[51]

readily ensue from Newton's corresponding evaluations in the case of the earlier hyperbolic trajectory on substituting $-n$ in place of n (see the previous note). From first principles, alternatively, because—on extending VG downwards, much as before, till its meet D with the horizontal through X, and understanding DG to be the incrementation of the ordinate BC with respect to the increment BD of the base— the lines XV and DV are in given proportion to the base XD, it follows, on setting $XB = a$ and $BC = e$ to be the (mutually perpendicular) coordinates of its general point $C(a, e)$, that the defining Cartesian equation of the trajectory $\overset{\frown}{XACK}$ is $e = e_a \equiv ka - la^n$, where k is the slope of the tangent XV at the vertex X and

$$l = GV/XD^n = (1+k^2)^{\frac{1}{2}n}.(GV/XV^n).$$

(Since the firing-point $A(NA, XN)$ is by definition on the path, there will also be $l = (XN - k.NA)/NA^n$.) The 'Taylor' series expansion of $DG = e_{a+o}$ in powers of the increment $BD = o$ will then straightforwardly yield the coefficients

$$Q(= de/da) = k - nla^{n-1}, \quad R(= \tfrac{1}{2}d^2e/da^2) = -\tfrac{1}{2}n(n-1) \, la^{n-2}$$

and

$$S(= \tfrac{1}{6}d^3e/da^3) = -\tfrac{1}{6}n(n-1)(n-2) \, la^{n-3};$$

that is, by transporting the construction (with Newton) to the incremented ordinate DG which is 'ultimately' coincident with BC,

$$Q = k - (nl.XD^{n-1} \text{ or}) \, n.GV/XD, \quad R = -\tfrac{1}{2}n(n-1).GV/XD^2 \quad \text{and} \quad S = \tfrac{1}{3}(n-2) \, R/XD.$$

Substitution of these values in Newton's previously derived general measures

$$S/R\sqrt{(1+Q^2)} = (S/R).XD/GT, \quad \tfrac{1}{2}S\sqrt{(1+Q^2)}/R^2 = \tfrac{1}{2}(S/R^2).GT/XD$$

and

$$\sqrt{(\tfrac{1}{2}g.(1+Q^2)/R)}$$

for the density of the medium, the ratio of its resistance to gravity, and the body's speed respectively yields at once the expressions

$$\tfrac{1}{3}(n-2)/GT \propto 1/GT, \quad |-\tfrac{1}{3}(n-2)/n(n-1)|.GT/GV \quad \text{and} \quad \sqrt{(\tfrac{1}{2}gr)}$$

(where *latus rectum* $r = GT^2/R.XD^2 = |-1/\tfrac{1}{2}n(n-1)|.GT^2/GV$) here assigned by him for these.

(51) Newton could well have gone on at this point to develop an analogous higher-order generalization of the tilted Apollonian parabola which James Gregory had in 1672—retracing, we now know, the steps of Thomas Harriot three-quarters of a century before him—adduced, as the path which a falling body traverses under a constant uni-directional resistance, in his little tract *De Motu Penduli & Projectorum* which Newton himself knew well. (See VI: 7, notes (16) and (17).) But here he breaks off, content to have noticed the generalization of the simple, unresisted Galileian parabola which maintains the verticality of its axis.

APPENDIX 2. INITIAL ATTEMPTS
(LATE SEPTEMBER? 1712) TO ADJUST THE
DEFECTIVE 1687 ARGUMENT.[1]

From originals in the University Library, Cambridge and in private possession

[1][2]   [Positis] $OB = a.\ BC = e.$

[ut et] $BD = o = Bi.\ Bd = p.$[3]

[erit] $\dfrac{ao}{e} = IF.\ GF = \dfrac{nno^2}{2e^3} + \dfrac{anno^3}{2e^5}.$[4]

[adeoq̃] $CF = \sqrt{oo + \dfrac{aaoo}{e[e]}} = \dfrac{no}{e}.$

[Jam existente] $gf = [GF,\ \text{fit}]$

$$\dfrac{nnpp}{2e^3} - \dfrac{annp^3}{2e^5}\ [+\&\mathrm{c}] = \dfrac{nnoo}{2e^3} + \dfrac{anno^3}{2e^5} + \&\mathrm{c}.$$

[sive] $pp - \dfrac{ap^3}{ee} = oo + \dfrac{ao^3}{ee}.$ [hoc est] $oo.pp :: ee - ap.ee + ao.$

---

(1) When about the end of September 1712 Abraham de Moivre communicated to Newton the shattering counterblast to the truth of Proposition X of the *Principia*'s second book (as it was even then being reprinted without change) which Johann Bernoulli transmitted through his nephew Niklaus by his independent demonstration from first principles that the result of its 'Exempl. 1' (dealing with resisted motion in a semicircular arc) was a third too small, we have de Moivre's subsequent report to Bernoulli (see §1: note (1) above) that Newton took 'two or three days' to satisfy himself that the objection was valid, and to produce an alternative general argument—where tangents to the two successive infinitesimal arcs of the projectile path whose comparison was basic to both original and amended versions were now drawn the same way from corresponding end-points (and not, as earlier, opposite ways from their common point)—which merely(!) bumped up the general measure derived, in Corollary 2, for the ratio of resistance to gravity by the same numerical factor $\frac{3}{2}$ which rectified the deficiency in its prime *Exemplum*. During those few days, in fact, Newton's pen filled sheet after sheet of paper in abortive attempts first to make minor repair to the edifice of the argument of his *editio princeps* (on which see our running commentary to the reprinting of its text in Appendix 1 preceding) and then in essaying a variety of increasingly more radical reconstructions, each in their several ways equally defective, before gradually coming piece by piece to lay the foundations of the soundly built replacement of it which he put to the world at large in his *editio secunda*, adroitly cementing it into the hole left there by the excision of its parent. The later stages in this rebuilding are set out in §§1/2 preceding: here we reproduce the more fragmentary record of his initial efforts to comprehend wherein lay the error in the argument of his 1687 version—he at length did so, it would appear, but only afterwards when he returned to probe it (see Appendix 3 following)—and then to draft alternatives to it.

(2) ULC. Add. 3965.10: 109^r/103^r; the former page is reproduced in photocopy as Plate III (facing page 64) in A. R. and M. B. Hall, *Unpublished Scientific Papers of Isaac Newton. A selection from the Portsmouth Collection in the University Library, Cambridge* (Cambridge, 1962). Newton here checks the accuracy of the computation by which the result in 'Exempl. 1' of resisted motion in

[itaᵹ] $$o.p::\sqrt{ee-ap}=e-\frac{ap}{2e}.\sqrt{ee+ao}=e+\frac{ao}{2e}::e.e+\frac{ao}{e}.^{(5)}$$

[seu] $$o.p-o::e.\frac{ao}{e}::BC.IF::CF.kf=Cf-CF=2FH=\frac{naoo}{e^3}.$$

[Unde] $$\frac{naoo}{2e^3}=FH.\frac{nnoo}{2e^3}=FG::a.n.$$

Ergo $GH\perp FH$ et corpus non acceleratur.$^{(6)}$ [Rursus quia] $\frac{naoo}{2e^3}=Cf-CF.$

[fit] $\frac{a}{n}=\frac{Cf-CF}{FG}^{(7)}$ ut resistentia. [et] $\frac{a}{2ne}\left[=\frac{Cf-CF}{CF^q}\right]$ ut densitas.

---

the circle quadrant $\overset{\frown}{LCK}$, of radius $OK = OC = n$, is derived from the preceding general measure $FH/FG = [\frac{1}{2}](fC-CF)/FG$ as deduced in the *editio princeps* (see pages **374–80** above). In our edited version of the roughly written manuscript text we make one or two small transpositions the better to convey its sense; compare the next note.

(3) Newton sets this 'opening' explanation of terms after '$Ct = FG$' below, in fact. As in the *editio princeps*, understand that the circle-arcs $\overset{\frown}{gC}$ and $\overset{\frown}{CG}$ are traversed in equal, indefinitely small times.

(4) Ignoring terms of the order of o^4, that is. Similar suppositions are made *mutatis mutandis* in sequel.

(5) Since $p = o + O(o^2)$.

(6) For FH will be at once the increment of the component of gravity acting in the direction of motion along the tangent CH (inclined at $\overset{\frown}{BCH} = \overset{\frown}{BOC} = \cos^{-1}(a/n)$ to the vertical), and also the decrement in the body's motion due to the contrary resistance of the medium. Which is impossible, because the constant downwards acceleration of the body must have a component in the direction CH which is proportional to $\cos \overset{\frown}{BCH} = a/n$, while if the body's total motion in that direction is not accelerated there can be no variation in the resistance of the medium to that uniform progress. This *reductio ad absurdum* of the ratio obtained in 'Exempl. 1' of the *editio princeps* for resistance to semicircular motion in a plane perpendicular to the horizontal had, as his nephew Niklaus doubtless informed Newton at their meeting in London in early October 1712, already been made by Johann Bernoulli, and privately communicated by him to Leibniz in a letter of 12 August 1710 (N.S.) (first printed in *Got. Gul. Leibnitii et Johan. Bernoullii Commercium Philosophicum et Mathematicum*, **2** (Lausanne/Geneva, 1745): 231–2 [= (ed. C. I. Gerhardt) *Leibnizens Mathematische Schriften*, **3** (Halle, 1856): 854–5]). Bernoulli made his objection publicly known a few months afterwards in his long article 'De motu Corporum gravium, Pendulorum, & Projectilium in mediis non resistentibus & resistentibus, supposita Gravitate uniformi...' (*Acta Eruditorum* (February/March 1713): 77–95/115–32): 93: §32, where he laid heavy stress on what he called a 'manifesta contradictio'.

(7) The denominator of this 1687 measure of the ratio of resistance to gravity ought (see Appendix 1: note (10)) to read '$2FG$'. When, however, as Johann Bernoulli first remarked in 1710 (again see his letter to Leibniz of 12 August that year) and as his nephew informed Newton—in what detail we do not know—at London in the last days of September 1712, we argue directly from first principles in the present case of the semicircle $e = +\sqrt{(n^2-a^2)}$ for which (in Leibnizian terms) there is $da:de:ds = e:-a:n$ where $\overset{\frown}{LC} = s$, since the instantaneous speed $v = ds/dt$ at C (where t is the time of passage over $\overset{\frown}{LC}$) is $\sqrt{(\frac{1}{2}g.(1+Q^2)/R)} = \sqrt{(-ge)}$, while the medium's resistance ρ counteracts both the curvilinear acceleration $dv/dt = v.dv/ds$ and the

[Cape] $GF = \dfrac{CF^q}{2GD} = $ dato. [Erit] CF^q ut GD [id est] CF ut $GD^{\frac{1}{2}}$. Velocitas ut $GD^{\frac{1}{2}}$. [Quia hic est o^2 ut e^3 erit] BD ut $GD^{\frac{3}{2}}$. [et] GI ut $OD, DG^{\frac{1}{2}}$, utcg decrementum $\square^{\text{ti (8)}}$ velocitatis.[9]

[Est] $\dfrac{Cf - CF}{2} = HF$. $[FG=] \dfrac{CF^q}{2CB}$ [hoc est] resist . Grav. ::

$$\dfrac{\sqrt{fg} \times \overline{fd+dg} - \sqrt{FG} \times \overline{FD+GD}}{2} . FG.$$

[sive] $\sqrt{fd+dg} - \sqrt{FD+DG} . 2\sqrt{FG}^{(10)} :: \sqrt{dg} - \sqrt{DG} . \sqrt{2FG}$.

[Pone $DG = e + Qo + Roo + So^3$ &c.[11] Evadit $Q = -\dfrac{a}{e}$. $R = -\dfrac{nn}{2e^3}$. $S = -\dfrac{ann}{2e^3}$.

adeocg resistentia ad gravitatem ut] $2So^3\sqrt{oo+QQoo}$ ad $4RRo^4$.[12] [id est]

---

component $-g . de/ds$ of gravity g acting in the instaneous direction of motion, there ensues $\rho/g = -d(\frac{1}{2}v^2/g - e)/ds = \frac{3}{2}de/ds$, that is, $\frac{3}{2}a/n$ *recte*. It is Newton's purpose in the sequel to recover this adjusted result from a correspondingly corrected general measure for the ratio of resistance to gravity for an arbitrary given curve.

(8) Read 'quadrati'.

(9) Since $d(v^2) = -d(ge) \propto de$. How to go on from here is not evident, and Newton in sequel goes back to his 1687 measure $\frac{1}{2}(fC-CF)/FG$ of the ratio of resistance to gravity.

(10) Newton here makes the error of dividing top and bottom in the fraction in the previous line by $\sqrt{fg} \approx \sqrt{FG}$. In his ensuing terms, however, there is (see Appendix 1: note (13)) $fg = Ro^2 + 2So^3 + \ldots$ and $FG = Ro^2 + So^3 + \ldots$, and the ratio $1 + (s/R)$ $o \ldots$ of these cannot be neglected, so that this would-be simplication is invalid. For what it is worth, the further reduction which follows is accurate since it is valid here to replace $\sqrt{(fd+gd)} - \sqrt{(FD+GD)}$, that is,

$$\sqrt{(2gd+FG)} - \sqrt{(2GD+FG)} = (\sqrt{(2gd)} - \sqrt{(2GD)}) \, (1 - \tfrac{1}{4}FG/\sqrt{(dg \times DG)} \ldots),$$

by $\sqrt{(2gd)} - \sqrt{(2GD)}$ *tout court*. But, not least (we suppose) because it would evidently prove harder than ever to resolve the resulting limit-ratio $(\sqrt{dg} - \sqrt{DG})/\sqrt{(2FG)}$ to a standard equivalent algorithm in terms of Q, R and S, Newton there wisely leaves over, returning to the parent fraction $\frac{1}{2}(fC-CF)/FG$ to essay (see note (12) below) yet another reduction of it to a computationally more amenable form.

(11) The series expansion of the incremented ordinate e_{a+o} into terms in advancing powers of o, namely, where $e = e_a$ is the Cartesian equation of the curve LCK of general point $C(a, e)$, so that $Q = de/da$, $R = \frac{1}{2}d^2e/da^2$ and $S = \frac{1}{6}d^3e/da^3$ (see Appendix 1: note (14)); whence $CF = o\sqrt{(1+Q^2)}$ and $FG = Ro^2 + So^3 \ldots$. Analogously, corresponding to the decrement $Bd = -p$, there will be $dg = e_{a-p} = e - Qp + Rp^2 - Sp^3 \ldots$; whence $fC = p\sqrt{(1+Q^2)}$ and $fg = Rp^2 - Sp^3$.

(12) That is '$S\sqrt{1+QQ}$ ad $2RR$' on dividing through the antecedent and consequent members of this ratio by the common factor $2o^4$. This evaluation would seem to arise by reducing the parent fraction $(fC-CF)/2FG$ to be $(fC^2-CF^2)/2FG(fC+CF)$ or

$$(fC^2/CF^2-1) . CF/4FG = (p^2-o^2) . o\sqrt{(1+Q^2)}/4(Ro^2 + \ldots)o^2,$$

where, on equating $fg = Rp^2 - Sp^3 \ldots$ to $FG = Ro^2 + So^3 \ldots$, there is in consequence

$$p^2 = o^2 + 2(S/R) o^3 \ldots.$$

Having again attained his defective 1687 measure, when in sequel he applies it to the semicircular case of 'Exempl. 1' he of course duly repeats the 'impossible' result which he there before found (on which see note (6) above).

$$\frac{ann}{2e^5} \times \frac{n}{e} \text{ ad } \frac{n^4}{2e^6} \text{ [seu] ut } \frac{a}{2} \text{ ad } \frac{n}{2}. \text{ [Unde] Grav.resist::} [n.a.]$$

[Tangat gt curvam LCK ad g. Erit] $Ct = FG$.[13]

Grav.Resist::$FG.FH = \frac{1}{2}f[k]$.[14] [Jam quia]

$$fg.[kl] :: [fC^q . kC^q = \overline{fC - fk}^q \text{ sive] } \tfrac{1}{2}fC . \tfrac{1}{2}fC - f[k].$$

---

(13) This equation—a late insertion penned in blacker ink to the left of the specification of quantities which we have here advanced to be the opening line of the piece (see note (3) above, and also the printed photo-facsimile of the manuscript cited in note (2) preceding)—marks the point at which Newton first considered the deviation from straight over the arc $\widehat{gC}$ by considering, in parallel with that, CF, to the arc $\widehat{CG}$, the tangent gt from its second end-point C. He was here perhaps wiser than he knew, since the downwards components of resistance (which he had hitherto neglected to take into account in asserting the equality of fg and FG) now act the same way: specifically, continuing to understand that the infinitesimal arcs $\widehat{gC}$ and $\widehat{CG}$ are traversed in equal times, θ say, we may show (compare note (6) to Appendix 1 preceding) that, on ignoring terms of the order of θ^4, the projectile in its successive motion from g and C respectively is, under the joint action of the constant downwards pull of gravity g and the component in that direction of the medium's resistance ρ opposing the missile's onrush at C, diverted from its initial tangential paths gt and CF through the equal distances $tC = FG = \frac{1}{2}g\theta^2 - \frac{1}{6}(g\rho/v)\,\theta^3$, where v is the projectile's instantaneous speed. Had he thought to make the connection, Newton could here straightforwardly have gone on to make correct deduction of the properly adjusted expression for the ratio of resistance to gravity from the basic geometrical measure $\frac{1}{2}(fC - CF)/FG$ stated by him in his *editio princeps*. For, corresponding to the base increment $BD = o$, the series expansion of the incremented ordinate $DG = e_{a+o}$ as $e + Qo + Ro^2 + So^3 + \ldots$ (where there is $Q \equiv Q_a = de/da$, $R \equiv R_a = \frac{1}{2}d^2e/da^2 = \frac{1}{2}dQ/da$ and $S \equiv S_a = \frac{1}{6}d^3e/da^3 = \frac{1}{3}dR/da$) again yields $CF = o\sqrt{(1+Q^2)}$ and $FG = Ro^2 + So^3 + \ldots$; while likewise, corresponding to the decrement $Bd = -p$, there is $dg = e_{a-p} = e - Qp + Rp^2 - Sp^3 + \ldots$ and consequently

$$tC = R_{a-p}p^2 + S_{a-p}p^3 + \ldots = (R - 3Sp + \ldots)\,p^2 + (S - \ldots)\,p^3 = Rp^2 - 2Sp^3 \ldots.$$

Whence, on equating the deviations FG and tC, there ensues

$$p^2 = o^2 + 3(S/R)o^3\ldots \quad \text{or} \quad p = o + \tfrac{3}{2}(S/R)\,o^2\ldots,$$

so that $\frac{1}{2}(fC - CF)/FG = \frac{1}{2}(p - o)\sqrt{(1+Q^2)}/(Ro^2 + \ldots)$ comes, in the limit as o vanishes, to be $\frac{3}{4}S\sqrt{(1+Q^2)}/R^2$ *recte*. Newton himself would not appear to have seen this relatively simple way of mending the argument of his *editio princeps* till after he had come (in the variant fashions of §2.1/2) preceding) more radically to recast his original reasoning, when he returned privately (see Appendix 3 following) to consider anew wherein its faultiness lay, there (in Appendix 3.3) obtaining the ratio of o to p equivalently *ex sagittis*.

In immediate sequel (on f. 109^r), unable yet to shake himself free from the *idée fixe* that fg and FG have equal length, he set himself vainly to derive a correct argument from the thoroughly confused premiss '$GF = C[t] = g[f]$', properly maintaining that '$Cf - CF =$ decr[emento] mot[us]' measures the resistance over $\widehat{gCG}$. We see no need to reproduce these ineffective computations in their full detail, thinking it enough instance of their quality to print one typical passage (almost at the page bottom) where a late replacement of fg by FG (see note (15) following) leads Newton inescapably on to duplicate once more the result attained in the 1687 text.

(14) That is, $\frac{1}{2}(fC - CF)$, since $iB = BD$ and so $kC = CF$.

[fit] $fg \cdot fg - [kl] :: fC \cdot 2f[k]$. [unde] $f[k] = \dfrac{fg - [kl]}{2fg} fC$.

[Itacɟ] Grav. R[es]ist :: $FG \cdot \dfrac{fg - [kl]}{4fg} fC :: 4FG^q \cdot \overline{FG - [kl]}$ in $fC^{(15)} ::$

$$\dfrac{n^4 o^4}{e^6} + \dfrac{2an^4 o^5}{e^8} \cdot \dfrac{anno^3}{e^5} \times \dfrac{no}{e} :: n \cdot [a].$$

---

[2][16]   Sit A altitudo ad quam corpus C veloci-   [L]
tate sua sursum versa ascendat sine resistentia[17]
et erit velocitas corp[or]is in C ad vim quam
corpus sine resistentia movendo interea acquiret
dum pergit a C ad G in subduplicata ratione
$A + IG$ ad A sive ut $A + \frac{1}{2}IG$ ad A. Quare si
velocitas in C fuerit V, incrementum velocitatis
in G sine resistentia foret $\dfrac{\frac{1}{2}IG}{A} V^{(18)}$ & decre-

mentum ex resistentia et grav[itate][19] $\dfrac{cf - CF}{CF} V$,

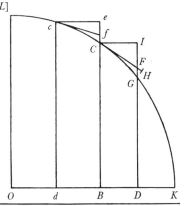

---

(15) The quiet replacement by FG, here, of fg in the preceding numerator $fg - kl$ equivalently repeats the basic error of the 1687 argument (see Appendix 1: note (8)); whence, *de rigeur*, the impossible result obtained in 'Exempl. 1' of the *editio princeps* again ensues when Newton in the next line applies his 'new' measure to compute the ratio of resistance to gravity in the instance of the semi-circular projectile path.

(16) From the recto of a folio sheet in private possession; Newton afterwards used the space left blank at the page bottom to draft the two long footnotes ∗ on pages 70 and 90 of his anonymously edited compendium of the *Commercium Epistolicum D. Johannis Collins et Aliorum de Analysi promota* (London, 1713) which he put to press in November 1712. We omit a few opening lines where Newton again duplicated his calculation in 'Exempl. 1' of the *editio princeps* to confirm that in the case of the semicircle $e = \sqrt{(n^2 - a^2)}$ the measure '$S\sqrt{1 + QQ}$ ad $2RR$' which he had obtained in 1687 does indeed here lead to the erroneous result 'Resist ad Grav. ... :: $a.n$', and take up the manuscript text at the point where—employing for the first time his new-found insight (see note (13) to [1] preceding) that the tangents to the arcs $\overgroup{cC}$ and $\overgroup{CG}$ ought better to be drawn 'the same way' from the corresponding end-points c and C—he broaches a modification of his original mode of reasoning.

(17) Using Newton's present denotations of V for the missile's speed at C, and t for the equal infinitesimal times in which it successively describes the arcs $\overgroup{cC}$ and $\overgroup{CG}$, it is readily shown that, where (as ever) g is the constant downwards force of gravity, then
$$\sqrt{(2gA)} = V = (\overgroup{CG}/t \approx) CF/t$$
and therefore, since $GF = \frac{1}{2}gt^2$, $A = \frac{1}{4}CF^2/GF = \frac{1}{4}(DG + DF)$, that is, $\frac{1}{2}BC$ in the limit as the incremented ordinate DG (and so DF also) comes to coincide with BC. But the explicit value of A is not here pertinent, because it will in the sequel be more conveniently replaced by its equal $CF^2/4FG$.

(18) That is, $IG \cdot g/V = (IG/GF) \cdot gt$ where $gt = 2GF/t = 2(GF/CF)$. V is the downwards increment in the projectile's speed due to the pull of gravity g in the time t in which $\overgroup{CG}$ is traversed.

(19) Namely $-(\overgroup{CG} - \overgroup{cC})/t = -(CF - cf)/t$.

[itaqʒ] ex resistentia sola $\dfrac{IG}{2A}+\dfrac{cf-CF}{CF}\times V$ et ex gravitate sola $GF,V.$[20] Est

ergo resistentia ad gravitatem ut $\dfrac{IG}{2A}+\dfrac{cf-CF}{CF}$ ad $GF.$[21] [Jam posito t pro tem-

poribus æqualibus quo corpus pergit a c ad C ut et a C ad G erit] $2Cf.Cc::t.\dfrac{Cc,t}{2Cf}$.

[adeoqʒ] $tt.\dfrac{Cc^q,tt}{4Cf^q}::Cf.A=\dfrac{Cc,Cc}{4Cf}=\tfrac{1}{2}CB.$ [Quare est] Resistentia ad Gravitatem

ut $\dfrac{2IG,Cf}{Cc,Cc}+\dfrac{cf-CF}{CF}$ ad $GF=Cf.$

[Unde fit]

    Resistentia ut $\dfrac{2IG,Cf}{Cc,Cc}+\dfrac{cf-CF}{CF}$.

    □[22] Velocitatis ut $\dfrac{Cc,Cc}{Cf}$.

    Densitas ut $\dfrac{2IG,Cf^q}{Cc^4}+\dfrac{Cf,cf-Cf,CF}{CF,Cc,Cc}$ sive ut $\dfrac{R,Cf}{Cc,Cc}$. [23]

Vel sic. [Est] $CF.IF::GF.\dfrac{IF,GF}{CF}=\text{increment[o] tangentis}$[24] ex gravitate.

Decrementum ex $\overline{\text{resisten[tia]}-\text{gravitate}}$ [est] $cf-CF.$[25] [Ergo] Decrementum

ex resistentia [est] $cf-CF+\dfrac{IF,GF}{CF}$. [26] [adeoqʒ] Resistentia ad Gravitatem ut

$cf-CF+\dfrac{IF,GF}{CF}$ ad $FG.$[27]

---

(20) This should (see note (18)) read '$\dfrac{2GF}{CF},V$'.

(21) Whence this should be $\dfrac{`2GF'}{CF}$, thus yielding the ratio of resistance to gravity here to be correctly as $(\tfrac{1}{2}IG\times CF/A$ or$)$ $2(IG/CF)\times GF+cf-CF$ to $2GF$, where

$$IG/CF=IF/CF=\cos\widehat{BCF}$$

in the limit as DGF passes to coincide with BC.

(22) Read 'Quadratum'.

(23) Where, of course, R denotes the 'R[esistentia]'.

(24) Namely CF.

(25) That is, $-(\widehat{CG}-\widehat{cC})=-d(\widehat{LC})$. This, however, represents the decrement $t.dV$ in *speed* of motion ensuing from the countering pulls of resistance and gravity along the tangent CF. The corresponding downwards 'incrementum velocitatis ex gravitate' is $t.gt=2GF$, and the component of this along CF will be $2(IF/GF)\times GF$. This subtlety will be the remaining obstacle to Newton's attaining the correct measure of resistance to gravity, one not easily overcome by him.

(26) This should (see the previous note) be '$+\dfrac{2IF,GF}{CF}$'.

(27) And consequently this ratio must be *recte* 'ut $cf-CF+\dfrac{2IF,GF}{CF}$ ad $2GF$' (as in note (21) above). We omit to reproduce the rough jottings following in the manuscript where Newton

[3]^(28) [Sit] $EB=BD$. OE, OB, OD abscissæ. EH, BC, DG Ordinatæ. HN, CF tangentes. Per terminos seriei invenientur HN, CN, CF, FG et habebitur $HC-HN=\dfrac{\overline{EH-BN}\ \text{in}\ CN}{HN}$ et addendo HN habebitur HC. Sit

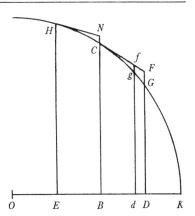

$$Cf.CF::\sqrt{}CN.\sqrt{}FG$$

& acta *fgd* ordinatis parallela, erit $fg=CN$,^(29) & arcus HC, Cg erunt synchroni, & eorum decrementum momentaneum erit $HC-Cg$ vel (quod perinde est) $HN-Cf$. Ab hoc decremento gravitas detrahit $Cg-Cf$ vel $HC-HN$,^(30) ideoqȝ resistentia generat

$$HN-Cf+HC-HN=HC-Cf,^{(31)}$$

et erit resistentia ad gravitatem ut $HC-Cf^{(31)}$ ad *fg* vel CN.^(32)

---

began tentatively to recast his 1687 text by instructing that 'a puncto G in tangentem demittatur perpendiculum Gn', thus obtaining the 'incrementum tangentis ex gravitate' directly as Fn. Overleaf Newton has checked yet once more that the series expansion of the incremented ordinate $DG = \sqrt{}(e^2-2ao-o^2)$ is indeed as he evaluated it to be in his 1687 text; and then went on briefly to convert the coordinates of C to be the truly Cartesian '$OB = x$. $BC = y$', now supposing that $DG = y_{x+o}$ has the Taylorian expansion '$y+bo+coo+do^3$ [&c]', whence '$\dot{x} = o$. $-bo = IF$. $-coo = FG$. $CF = q = o\sqrt{1+bb}$', but the calculations tail off uselessly in an attempt to evaluate the erroneous measure 'Grav. Resist::$FG.FH = \dot{CF}+Fn$' with '$\dot{CF}$'($= cf-CF$) computed, by a slip, as the true fluxion $\dot{q}$. Even had he calculated aright from his false measure, Newton would have found that

$$cf = o\sqrt{(1+b^2)}-2(bc/\sqrt{(1+b^2)})\,o^2..., \quad CF = (1-\tfrac{3}{2}(d/c)\,o...)\,o\sqrt{(1+b^2)}$$

and therefore $\dot{CF} = (\tfrac{3}{2}d\sqrt{(1+b^2)}/c-2bc/\sqrt{(1+b^2)})\,o^2+O(o^3)$, while $Fn = (b/\sqrt{(1+b^2)})\,co^2$, so that the ensuing ratio of resistance to gravity would have been $\tfrac{3}{2}d\sqrt{(1+b^2)}/c^2-b/\sqrt{(1+b^2)}$.

(28) ULC. Add. 3968.18: 261^r. This minor recasting of [2] preceding is jotted down on the back of a transcript in Newton's hand of Leibniz' letter to Hans Sloane of 29 December 1711 (N.S.)—the original is now British Library. MS Sloane 4043: 19/20 (and reproduced in its context in *The Correspondence of Isaac Newton*, 5, 1975: 207)—whence it was first printed a few weeks later in Newton's *Commercium Epistolicum D. Johannis Collins et Aliorum...*: 118/119.

(29) Understanding that the arcs $\overset{\frown}{HC}$ and $\overset{\frown}{Cg}$ are traversed in successive equal 'instants' of time (compare note (13) above).

(30) This should (compare note (25) above) be '$2\overline{Cg-Cf}$ vel $2\overline{HC-HN}$'.

(31) Whence read '$HN-Cf+2HC-2HN = 2HC-Cf-HN$ vel $HC+Cg-2Cf$' here.

(32) This should (again see note (25)) read '$2fg$ vel $2CN$'. Accurately adjusted, therefore, the ratio of resistance to gravity will be as $\tfrac{1}{2}(\overset{\frown}{HC}+\overset{\frown}{Cg})-Cf$ to *fg*. In the analytical equivalent in which (as in the 1687 text) $OB = a$ and $BC = e = e_a$ are the coordinates of the general point $C(a, e)$ of the trajectory, and the increment $BD = o$ yields the series expansion

$$e+Qo+Ro^2+So^3+...$$

of the ordinate $DG = e_{a+o}$, there is correspondingly $EH = e_{a-o} = e-Qo+Ro^2-So^3+...$ and so $CN = Ro^2-2So^3+...$; whence, if we name $Bd = p$, so that $fg = Rp^2+Sp^3+...$, the equality

In circulo cujus centrum O, radius OK, est $HN = \sqrt{CN \times 2CB}$ & $Cf = \sqrt{fg \times 2gd}$[33]

& $\quad HN - Cf = \sqrt{CN} \text{ in } \overline{\sqrt{2CB} - \sqrt{2gd}} \quad$ et $\quad HC = HN + \dfrac{EH - BN}{HN} \times CN.$

[itaꝗ] $\qquad HC - Cf = \sqrt{CN} \text{ in } \overline{\sqrt{2CB} - \sqrt{2gd}} + \dfrac{EH - BN}{HN} CN.$

[atꝗ accuratius][34] $HC - Cf = CN \text{ in } \dfrac{\sqrt{2CB + CN} - \sqrt{2gd + CN}}{\sqrt{CN}} + \dfrac{OB}{OC}.$

[hoc est] $\qquad\qquad = CN \text{ in } \dfrac{2CB + CN - 4\sqrt{CB, gd} + 2CN, 2CB}{HN}$[35]

---

[4][36]   *Corol. 1.* Resistentia est ad gravitatem ut $\dfrac{HN}{CN + FG} - \dfrac{CF}{2FG} + \dfrac{FI}{CF}$ ad 1.
Nam pro $2\sqrt{CN \times FG}$ scribere licet $CN + FG$.

---

of the deviations CN and fg from their tangents HN and Cf (see note (25) preceding) gives $p^2 = o^2 - 3(S/R) o^3 \dots$ and so $p = o - \frac{3}{2}(S/R) o^2 \dots$. Accordingly,

$$HN = o\sqrt{(1 + Q^2_{a-o})} = o\sqrt{(1 + (Q - 2Ro + 3So^2 - \dots)^2)},$$

that is, $o\sqrt{(1 + Q^2 - 4QRo + \dots)} = o\sqrt{(1 + Q^2)} - 2(QR/\sqrt{(1 + Q^2)}) o^2 \dots$, and correspondingly

$$Cf = p\sqrt{(1 + Q^2)} = o\sqrt{(1 + Q^2)} - \tfrac{3}{2}(S\sqrt{(1 + Q^2)}/R) o^2 \dots,$$

while also $\widehat{HC} - HN = \widehat{Cg} - Cf = CN \cdot Q/\sqrt{(1 + Q^2)}$. In consequence, therefore, in the limit as $BD = o$ comes to vanish, Newton's present measure $(\widehat{HC} - Cf)/fg$ yields the same erroneous value $\frac{3}{2}S\sqrt{(1 + Q^2)}/R^2 - Q/\sqrt{(1 + Q^2)}$ for the ratio of resistance to gravity as would, *mutatis mutandis*, result (see note (27) above) from that of [2] preceding, while of course the adjusted measure $(\frac{1}{2}(\widehat{HC} + \widehat{Cg}) - Cf)/fg$ correctly produces $\frac{3}{4}S\sqrt{(1 + Q^2)}/R^2$.

(33) More accurately, since $HN^2 = CN \times (BC + BN)$ for the circle osculating the trajectory over its arc $\widehat{HC}$, there is $HN = \sqrt{(CN \times (2BC + CN))}$; and correspondingly

$$Cf = \sqrt{(gf \times (2dg + gf))},$$

that is (because the deviations CN and gf are equal), $\sqrt{(CN \times (2dg + CN))}$. And so Newton will amend them in the sequel.

(34) See the previous note.

(35) Newton breaks off before completing the radical with its needed terms ' $+2CN$, $2fg + CN^q$ ' (and with the last ratio ' $+OB/OC$ ' still to adjoin). There now follow—if we have rightly restored the chronological sequence of these manuscript drafts—the fuller and more verbally finished attempts by him to correct the argument of his 1687 text which we have set out in §1 preceding, and which culminate at long last (at the end of §1.6) in the magic moment when for the first time he attains his goal.

(36) Add. 3965.12: 219^v/220^v, filled out (as the manuscript implicitly directs) from the preliminary draft on *ibid.*: 200^v. Newton here endeavours to reduce the general measure of the ratio of resistance to gravity derived in §2.1 above—and slightly simplified in its Corollary 1 (here repeated without change)—into a form which is easier to evaluate when applied in particular instances; but he fails to remark that two of the three terms in the resulting equivalent measure in 'Corol. 2' cancel each other out, leaving (see note (39) following) a yet simpler reduced ratio still.

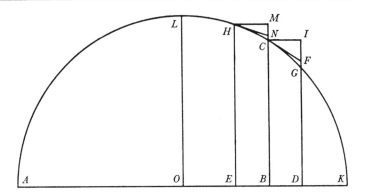

Corol. 2.[37] Si pro $FG-CN$ scribatur A et pro $CF-HN$ scribatur B, erit Resistentia ad Gravitatem ut $\dfrac{HN}{2FG-A}-\dfrac{CF}{2FG}+\dfrac{FI}{CF}$ ad 1, id est ut[38]

$$\frac{CF,A}{4FG^q}-\frac{B}{2FG}+\frac{FI}{CF}$$

ad 1.[39] Ubi,[40] si $IF-MN$ dicatur D, scribi potest $\dfrac{IF,D}{CF}$ pro B.

Corol. 3. Et hinc si Curva Linea definiatur per relationem inter [basem seu abscissam AB & ordinatim applicatam BC (ut moris est) et valor ordinatim applicatæ resolvatur in seriem convergentem: Problema per primos seriei terminos] expedite solvetur. Sit $BD=o$ et $DG=P-Qo-Ro^2-So^3-$&c[41] & distinguantur hujusmodi series [in terminos successivos in hunc modum. Terminus primus appelletur in quo quantitas infinite parva o non extat;

---

(37) Newton initially phrased this to assert erroneously: 'Si pro $FI-NM$ scribatur A, pro $FG-CN$ scribatur B, et pro CF scribatur C: Resistentia erit ad gravitatem ut

$$\frac{B\times C}{2FG^q}-\frac{A\times FI}{2C\times FG}+\frac{FI}{4C}\ [!]\ \text{ad 1'}.$$

(38) In a cancellation at this point in the draft on f. 200^v Newton, replacing HN by its equal $CF-B$, initially went on to insert the equivalent of the intervening step

$$`\frac{-2FG\times B+CF\times A}{4FG^q-2FG\times A}+\frac{FI}{CF}\ \text{sive'}$$

here (understanding of course that the difference $A=GF-CN$ is infinitely less than GF itself).

(39) Because (on ignoring terms of the order of BD^3 and smaller) the difference $B=CF-HN$ is readily shown—compare Newton's own reduction of the present linelets to their Cartesian equivalents which we cite in note (43) following—to be equal to $2FG\times FI/CF$, this ratio reduces at once (as Newton himself fails here to observe) to be 'ut $CF\times A$ ad $4FG^q$', *tout court*.

(40) Understand, as Newton explicitly stated in his draft at this place, 'cum sit IF ad CF ut $CF-HN$ ad $IF-MN$'.

(41) This, as ever, to allow P, Q, R and S to have positive signs in the particular case of the semicircle in 'Exempl. 1' below. For simplicity in our own exposition we assume the equivalent expansion $e+Qo+Ro^2+So^3+...$ of the incremented ordinate $DG=e_{a+o}$.

secundus in quo quantitas illa extat unius dimensionis; tertium in quo extat duarum; quartum in quo trium est, et sic in infinitum. Et primus terminus qui hic est e, denotabit semper longitudinem ordinatæ BC insistentis ad indefinitæ quantitatis o initium B; secundus terminus Qo denotabit differentiam inter BC et DF[42] id est lineolam IF quæ abscinditur complendo parallelogrammum $BCID$, atgɔ adeo positionem Tangentis CF semper determinat. Tertius terminus & subsequentes $Roo + So^3 + \&c$ designabunt lineolam FG quæ jacet inter Tangentem & Curvam adeogɔ determinat angulum contactus FCG seu curvaturam quam Curva linea habet in C. Si lineola illa minuatur in infinitum, termini subsequentes evadent infinite minores tertio ideogɔ negligi possunt, & solus terminus tertius determinabit curvaturam curvæ. Terminus quartus So^3 determinat variationem curvaturæ, quintus variationem variationis, & sic deinceps. Unde obiter patet usus non contemnendus harum serierum in solutione Problematum quæ pendent a Tangentibus & curvatura curvarum.

Præterea CF^q est latus quadratum ex CI^q et IF^q sive ex $oo + QQoo$, et GD mutando signum ipsius BD vel o mutatur in $EH = P + Qo - Roo + So^3 - \&c$. Et inde habetur $CM = Qo - Roo + So^3 - \&c$. Et] scribendo AE pro AB vel KE pro KB, quantitas IF mutatur in MN. Et simul prodeunt $IF - MN = D$, $MC - MN = NC$, $FG - NC = A$. $\dfrac{IF \times D}{CF} = B$. Et Resistentia ad Gravitatem ut $\dfrac{CF,A}{4FG^q} - \dfrac{B}{2FG} + \dfrac{FI}{CF}$ ad 1.[43] Res exemplis patebit.

Exempl. 1. Sit linea ACK semicirculus super diametro AK descriptus, et requiratur medij densitas quæ faciat ut projectile moveatur in hac linea.

---

(42) Where of course CF is tangent at C.

(43) Which on deleting the equal terms $B/2FG$ and FI/CF again (see note (39) above) reduces to be 'ut $CF \times A$ ad $4FG^q$'. We should not insist too hard on Newton's oversight here, however, for it would at once have been corrected had he thought to convert his present measure to its equivalent expression in terms of the coefficients Q, R and S in the series expansion of the incremented ordinate DG. In a parallel calculation which survives on Add. 3968.41: 119^r he found no difficulty in thus reducing (much as we restored this computation in §1.6 preceding) the primary measure $HN/(CN + FG) - \frac{1}{2}CF/FG + FI/CF$, there successively evaluating: '$IF = Qo$. $FG = Roo + So^3$. $CN = Roo - 2So^3$. $CF = o\sqrt{1 + QQ}$. $HN = o\sqrt{1 + QQ} - \dfrac{2QRoo}{\sqrt{}}$.

[itagɔ]

$$\frac{CF}{2FG} = \frac{\sqrt{1 + QQ}}{2Ro + 2Soo} \cdot \frac{HN}{CN + FG} = \frac{o\sqrt{} - \dfrac{2QRoo}{\sqrt{1 + QQ}}}{2Roo - So^3} = \frac{\sqrt{1 + QQ} - \dfrac{2QRo}{[\sqrt{}]}}{2Ro - So^2}.$$

[adeogɔ earum differentia est]

$$\frac{2Ro\sqrt{} - So^2\sqrt{} - 2Ro\sqrt{} - 2Soo\sqrt{} + 4QR^2o^2[/\sqrt{}]}{4RRoo + 2RSo^3} = \frac{-3S\sqrt{}}{4RR}\left[+ \frac{Q}{\sqrt{1 + QQ}} \text{ seu } \frac{FI}{CF} \right]';$$

whence 'resistentia erit ad gravitatem ut $3S\sqrt{1 + QQ}$ ad $4RR$'. (On the same manuscript page, we may add, Newton has roughed out what were to be notes ∗ on pages 41 and 42 [top] of his *Commercium Epistolicum D. Johannis Collins et Aliorum. . . .*)

Bisecetur semicirculi diameter in O, et dic $OK\,n$, $OB\,a$, $BC\,e$, & BD vel $BE\,o$, et erit DG^q seu $OG^q - OD^q = nn - aa - 2ao - oo$ seu $ee - 2ao - oo$, et radice per methodum nostram extracta fiet $DG = e - \dfrac{ao}{e} - \dfrac{oo}{2e} - \dfrac{aaoo}{2e^3} - \dfrac{ao^3}{2e^3} - \dfrac{a^3o^3}{2e^5}\,[- \&c].$ Hic scribatur nn pro $aa + ee$ et evadet $DG = e - \dfrac{ao}{e} - \dfrac{nnoo}{2e^3} - \dfrac{anno^3}{2e^5} - \&c.$ Unde fit $BC = e.$

$IF = \dfrac{ao}{e}$, $FG = \dfrac{nnoo}{2e^3} + \dfrac{anno^3}{2e^5}$. et $CF = o\sqrt{1 + QQ} = o\sqrt{1 + \dfrac{aa}{ee}} = \dfrac{no}{e}$. Et mutando signum ipsius $[o]$ prodit $EH = e + \dfrac{ao}{e} - \dfrac{nnoo}{2e^3} + \dfrac{anno^3}{2e^5} - \&c$ et $CM = \dfrac{ao}{e} - \dfrac{nnoo}{2e^3} + \dfrac{anno^3}{2e^5}$.

Et scribend[o] $a - o$ pro OE vertitur IF seu $\dfrac{ao}{e}$ in

$$\frac{ao - oo}{EH} = \frac{ao}{e} - \frac{aaoo}{e^3} + \frac{ao^3}{2e^3} - \frac{3a^3o^3}{2e^5}\,^{(44)} = MN.$$

Unde fit

$$CM - MN = \frac{nnoo}{2e^3} - \frac{anno^3}{2e^5} = NC.$$

$$FG - NC = \frac{anno^3}{e^5} = A.^{(45)} \quad IF - MN = \frac{aaoo}{e^3} = D.^{(45)} \quad \frac{anoo}{e^3} = B.^{(46)}$$

---

(44) This should be '$\dfrac{ao}{e} - \dfrac{nnoo}{e^3} + \dfrac{3anno^3}{2e^5}$', whence in sequel '$CM - MN = \dfrac{nnoo}{2e^3} - \dfrac{anno^3}{e^5} = NC$'.

(45) Correctly '$\dfrac{3anno^3}{2e^5} = A$' and '$\dfrac{nnoo}{e^3} = D$'.

(46) *Recte*! even though Newton could not have derived this from his erroneous preceding value for D, as $(IF/CF) \times D$. On substituting Newton's computed values in the reduced expression $A \times CF/4FG^2$ (see note (39)) for this ratio—as he himself would appear to have gone on to compute on Add. 3968.41: 14^v—it follows that 'resistentia erit ad gravitatem ut $\dfrac{anno^3}{e^5} \times \dfrac{no}{e}$ ad $\dfrac{4n^4o^4}{4e^6} :: a . n$', that is, the deficient result of the *editio princeps* once more. On rectifying Newton's value for A (see the previous note) we of course accurately increase this ratio by a factor of $\frac{3}{2}$ to be '$3a$ ad n'.

In yet further abortive endeavours to remodel the present measure of the ratio of resistance to gravity obtained in (§1.6 →) §2.1 preceding, Newton first began to argue (Add. 3968.41: 119^r) that

'Decrementum spatij $-HC + CG = -HN + CF$, ex resist + Grav.

[this should be '$-\frac{1}{2}, \overline{-HC + CG} = \frac{1}{2}, \overline{-HN + CF}$']

Incre[mentum] ex Grav. $= CG - CF\left[= \dfrac{FI}{CF} \times GF\right]$. [adeoφ]

Decr. ex resi[s]t. $= CF - HN + \dfrac{FI, FG}{CF}$ ad descens[um] ex Grav. $= GF$

[hoc est] Resist. Grav$:: \dfrac{CF}{GF} - \dfrac{HN}{GF} + \dfrac{FI}{CF} . 1$';

[5][47] [Posito quod ordinatæ AG, BH, CI, DK ab invicem æqualiter distant, et ductis HN, IPR curvam ad puncta H et I tangentibus, evadent sagittæ]

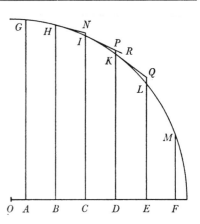

$$IN = 2BH - AG - CI.\quad KP = 2CI - BH - DK.$$

[eritcg

$$IN.KP]::\text{temp[us] in } HI.\text{temp in } IK^{(48)}::HI.IR.$$

[Unde] $\dfrac{HI}{IN}.\dfrac{IK^{(49)}}{KP}::\text{veloc[itas] in } HI.\text{veloc in } IK.$

---

and then (*ibid.*: 11ʳ), after realizing that the arcs $\widehat{HC}$, $\widehat{CG}$ are not traversed in equal times, he directed that there be taken P in $\widehat{CG}$ such that '$CP.CG::\sqrt{CN}.\sqrt{FG}$ & arcus HC, CP erunt synchroni,ʹet $HC - CP$ [read '$\frac{1}{2},\overline{HC-CP}$' *recte*] erit decrementum momentaneum ex resist & grav. Est $HC - HN$ increm. momentaneum ex grav. Ergo $2HC - CP - HN$ decrem. moment. ex resist.' On correcting the last decrement to be '$\frac{3}{2}HC - \frac{1}{2}CP - HN$', this will accurately be in ratio to the 'descensus ex Grav.' FG as the resistance to gravity. (For, if the ordinate VP be drawn through P to meet the tangent CF in W, on calling $BV = p$ there will, with Newton's own notations in 'Corol. 3' assumed, be $\widehat{HC} - \widehat{CP} = HN - CW =$

$$\left(o\sqrt{(1+Q^2)} - 2(QR/\sqrt{(1+Q^2)})\,o^2...\right) - p\sqrt{1+Q^2},$$

where $p/o = \widehat{CP}/\widehat{CG} = \sqrt{CN}/\sqrt{GF} = \sqrt{(Ro^2 - 2So^3...)}/\sqrt{(Ro^2 + So^3...)}$, that is, $1 - \frac{3}{2}(S/R)\,o...$, while $\widehat{HC} - HN = (FI/CF) \times GF$ is $(Q/\sqrt{(1+Q^2)}).(Ro^2 + ...)$; and $\frac{1}{2}(\widehat{HC} - \widehat{CP}) + \widehat{HC} - HN$ will be in ratio to GF as $\frac{3}{4}S\sqrt{(1+Q^2)}/R^2$ to unity in the limit as $BD = o$ becomes vanishingly small.) This appeal to 'synchronous arcs'—an indirect return to the basis of his 1687 argument—was to afford the foundation on which, by way of the preparatory draft in [8] below, Newton built the final version (§2.3 preceding) of its soundly reconstructed reasoning.

(47) Add. 3968.18: 261ʳ, a first, faulty computation of the ratio of resistance to gravity along a given fall-path according to the variant notion, elaborated (and corrected in its detail) in [6] following and fully developed in the *Idem aliter* in §2.2 preceding, that these forces shall act but 'once and instantaneously' at the beginning of each successive moment of time. As ever, we round out the angular outline of its bare calculations with a dress of verbal interpolations which will shape and point the niceties of their sequence. (The original, we may briefly notice, is jotted on the verso of a transcription by him of Leibniz' letter to Hans Sloane on 29 December 1711 (N.S.) claiming priority of invention in calculus, and published by Newton a few weeks afterwards in his anonymously edited printing of related extracts of the *Commercium Epistolicum D. Johannis Collins et Aliorum de Analysi promota* (London, ₁1712): 118–19; see also *The Correspondence of Isaac Newton*, 5, 1975: 207.)

(48) This should read '$\sqrt{IN}.\sqrt{KP}::\text{temp in } HI.\text{temp in } IK$' *recte*.

(49) Whence this ratio should be '$\dfrac{HI}{\sqrt{IN}}.\dfrac{IK}{\sqrt{KP}}$'. The slip is perpetuated in the sequel (see the next note) and carried thereafter into the ensuing computation in the semicircular instance, producing a mistaken doubling of the true result.

(50) Read '$\dfrac{HI\ \text{in}\sqrt{KP} - \sqrt{IN}}{\sqrt{IN}} = \dfrac{HI\ \text{in } KP - IN}{\sqrt{IN},\ \sqrt{KP} + IN}$ hoc est $\dfrac{HI\ \text{in } KP - IN}{2IN}$'.

[adeoꝗ] $PR.PK::\text{Resist}.\text{Grav}::\dfrac{HI,KP-HI,IN}{IN}=\dfrac{HI\,\text{in}\,KP-IN}{IN}^{(50)}.KP$

$::HI\,\text{in}\,\dfrac{AG-3BH+3CI-DK}{IN}.KP::AG-3BH+3CI-DK.\dfrac{IN\times KP}{HI}\,^{(51)}.$

[Sit jam curva semicirculus cujus radius æqualis est n. Cape $OB=a$. $BH=e$. & $AB=BC=CD=o$. Fit $aa+ee=nn$. ut et]

$$AG=e+\frac{ao}{e}-\frac{nnoo}{2e^3}+\frac{anno^3}{2e^5}.$$

$$BH=e.$$

$$CI=e-\frac{ao}{e}-\frac{nnoo}{2e^3}-\frac{anno^3}{2e^5}.$$

$$DK=e-\frac{2ao}{e}-\frac{2nnoo}{e^3}-\frac{4anno^3}{e^5}.$$

[Itaꝗ] $AG-3BH+3CI-DK=\dfrac{3anno^3}{e^5}$. $IN=\dfrac{nnoo}{e^3}$. $KP=\dfrac{nnoo}{e^3}+\dfrac{3anno^3}{e^5}$. $HI=\dfrac{no}{e}$.

[Quare est] $\text{Resist}.\text{Grav}::\dfrac{3anno^3}{e^5}.\dfrac{n^3o^{3(52)}}{e^5}::3a.n.$

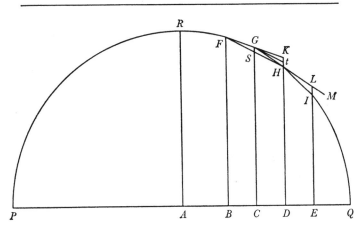

[6][53] [Sit] PRQ Curva. $BC=CD=DE$ [æquales] partes Abscissæ. BF, CG, DH, EI ordinatæ. FG, GH chordæ ordinatis DH, EI productis occurrentes in K et L. HI chorda tertia. [Erit] $HK=2CG-BF-DH$. $IL=2DH-CG-EI$.

---

(51) Whence '$\text{Resist}.\text{Grav}::AG-3BH+3CI-DK.\dfrac{2IN\times KP}{HI}$' correctly.

(52) This needs (see the previous note) to be doubled, yielding

'$\text{Resist}.\text{Grav}::\dfrac{3anno^3}{e^5}.\dfrac{2n^3o^3}{e^5}::3a.2n$'.

(53) Add. 3965.12: 198^r, a lightly remodelled (and now correct!) version of [5] preceding.

Cape *HM* ad *GH* ut tempus describendi chordam *HI* ad tempus describendi chordam $GH_{[,]}$ id est ut $\sqrt{IL}$ ad $\sqrt{HK}$ seu $\sqrt{HK \times IL}$ ad $H[K]$ id est ut $\dfrac{HK+IL}{2}$ ad *HK*. Et Resistentia erit ad Gravitatem ut spatium *ML* ex resistentia amissum ad spatium *LI* gravitate genitum.

    Exempl. Sit *RPQ* semicirculus. [Pone] $AC=a$. $AQ=n$. $CG=e=\sqrt{nn-aa}$.

[necnon] $\qquad\qquad\qquad\qquad BC=CD=DE=o$.

[erit] $\qquad\quad DH=\sqrt{nn-aa-2ao-oo}=e-\dfrac{ao}{e}-\dfrac{nnoo}{2e^3}-\dfrac{anno^3}{2e^5}$.

[itaɋ] $\qquad BF=e+\dfrac{ao}{e}-\dfrac{nnoo}{2e^3}+\dfrac{anno^3}{2e^5}$. $\quad EI=e-\dfrac{2ao}{e}-\dfrac{2nnoo}{e^3}-\dfrac{4anno^3}{e^5}$.

[adeoɋ] $\qquad HK=\dfrac{nnoo}{e^3}$. $\quad IL=\dfrac{nnoo}{e^3}+\dfrac{3anno^3}{e^5}$. $\quad GH=\dfrac{no}{e}$ [&c].

[Unde] $\quad LM.GH::\dfrac{IL-HK}{2}.HK::\dfrac{3anno^3}{2e^5}\cdot\dfrac{nnoo}{e^3}$. [sive] $LM=\dfrac{3anoo}{2e^3}$.

[Quare] Resistentia ad gravitatem ut $\dfrac{3anoo}{2e^3}$ ad $\dfrac{nnoo}{e^3}$ seu $3a$ ad $2n$.[54]

---

[7][55]   [Vel generalius] Sunto $BC=CD=DE=o$. $CG=e$, ac [pone]

$$DH=e-fo-goo-ho^3 \text{ [\&c]}.$$

Et Ordinatæ reliquæ duæ *BF* et *EI* erunt

$$e+fo-goo+ho^3 \quad \& \quad e-2fo-4goo-[8]ho^3.$$

Et inde fit $HK^{(56)}=2goo$. $IL^{(56)}=2goo+6ho^3$. [itaɋ] $\dfrac{IL-HK}{2}=3ho^3$. $GH=o\sqrt{1+ff}$.

[Unde] $LM^{(57)}=\dfrac{3ho^2\sqrt{1+ff}}{2g}$. [Quare] Resistentia ad Gravitatem [sive *LM* ad *IL*]

---

(54) Rightly, of course. There follows immediately after (on ff. 198ʳ–199ʳ) the fully elaborated 'Idem aliter' which is reproduced in §2.2 preceding.

(55) Add. 3965.10: 136ᵛ, a stray computation deriving from the geometrical constructions of the ratio of resistance to gravity and of the medium's density in §2.2 preceding (compare our note (17) thereto) the correct general measures of these in terms of the coefficients in the 'Taylor' expansion of the incremented ordinate. We see no need here to repeat the figure (that on page 346 above) which Newton here understands, or to reproduce the faulty opening calculation which he straightaway discards to pen what now—increased by a few editorial interpolations the better to bring out the sense—here follows.

(56) That is, $2CG-BF-DH$ and $2DH-CG-EI$ respectively.

(57) Namely, $GH(LI-KH)/2KH$ as Newton constructs it on page **348** above.

ut $3h\sqrt{1+ff}$ ad $4gg$. Velocitas $\left[\text{sive } \dfrac{GH}{\sqrt{HK}}\right]$ ut $\sqrt{\dfrac{1+ff}{g}}$. Et Medij densitas $\left[\text{seu } \dfrac{LM}{GH^q}\right]$ ut $\dfrac{h}{g\sqrt{1+ff}}$. Velocitas illa ipsa[58]

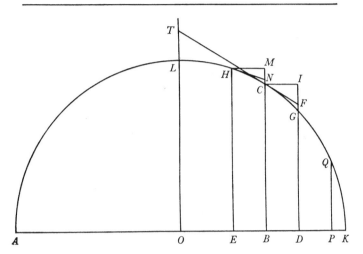

[8][59] [Sit *AK* planum illud plano schematis perpendiculare, *ALCK* linea curva, *C* corpus in ipsa motum, et *CF* recta ipsam tangens in *C*. Fingatur corpus *C* progredi ab *L* ad *K* per lineam illam *ACK*, et interea impediri a Medio resistente. A puncto *C* ad rectam *AK* horizonti parallelam demittatur perpendiculum *CB*, et in recta illa sumantur hinc inde lineolæ æquales *EB*, *BD*, *DP*[60] et

---

(58) Newton evidently intended to adjoin, much as at the corresponding place in 'Exempl. 1' of the *editio princeps* (*Principia*, $_1$1687: 265; see page 380 above), 'Velocitas illa ipsa est quacum corpus de loco *G* secundum rectam *GL* egrediens, in Parabola diametrum *GC* et latus rectum $\dfrac{1+ff}{g}$ habente deinceps moveri posset'.

(59) Add. 3965.10: 135^v–136^v, a preliminary recasting of §2.1 into the finished form of §2.3 preceding, wherein Newton for the first time introduces the clarification of explicitly naming the (unequal) infinitesimal times in which the successive arcs $\overset{\frown}{HC}$ and $\overset{\frown}{CG}$ are traversed in uniform horizontal motion $E \to B \to D$. This intermediate draft takes up the text of §2.1 in the middle of its fifth opening sentence (at 'quas corpus temporibus illis...'), but we here adjoin the beginning which Newton understands—and add the right-most ordinate *PQ* and the tangent *CT* to the original figure (on Add. 3965.12: 190^r) in accord with the demands of the present argument—so that the piece may be a self-contained unit. We see no need to repeat the bulk of the running commentary which we have already made, *mutatis mutandis*, at corresponding places in the final version in §2.3 above, but for the most part restrict our editorial remarks in sequel to certain special (mostly textual) points.

(60) We insert these specifications of the additional base-point *P* and corresponding ordinate *PQ* (in the opening sentences of §2.1 preceding which we otherwise here copy) to agree with Newton's assumptions in what follows. This new ordinate *PQ* is here introduced of course— in a remnant of the alternative approach in §2.2 where the resistance and gravity are conceived to act in instantaneous impulses 'simul et semel' only at the successive points *H, C, G,*...

erigantur perpendicula *EH*, *DG*, *PQ*[60] curvæ occurrentia in *H*, *G* et *Q*. Producatur *DG* donec tangenti *CF* occurrat in *F*, compleatur parallelogrammum *CBDI*, et agatur recta *HN* curvam tangens in *H* & perpendiculo producto occurrens in *N*. Et tempora quibus corpus describit arcus *HC*, *CG* erunt in subduplicata ratione altitudinum *NC*, *FG*] quas corpus temporibus illis describere posset a tangentibus illis cadendo. Et velocitates erunt ut longitudines descriptæ *HC*, *CG* directe & tempora inverse. Exponantur tempora per *T* et *t* et velocitates per $\frac{HC}{T}$ et $\frac{CG}{t}$; & decrementum velocitatis tempore *t* factum exponetur per $\frac{HC}{T} - \frac{CG}{t}$. Hoc decrementum oritur a resistentia corpus retardante & gravitate corpus accelerante & augeri debet ea velocitatis parte quam gravitas generat ut habeatur decrementum velocitatis a resistentia sola oriundum. Gravitas in corpore cadente & spatium *FG* cadendo generante generat velocitatem qua duplum illud spatium eodem tempore describere posset,[61] id est velocitatem $\frac{2FG}{t}$. At in corpore arcum *CG* describente generat tantum velocitatem quæ sit ad velocitatem $\frac{2FG}{t}$ ut *CG* − *CF* ad *FG* seu *GI* ad *CG*[,] id est velocitatem $\frac{2FG \times GI}{t \times CG}$. Addatur hæc velocitas ad decrementum prædictum et habebitur decrementum velocitatis ex resistentia sola oriundum, nempe $\frac{HC}{T} - \frac{CG}{t} + \frac{2FG,GI}{t,CG}$. Proindeqȝ cum Gravitas eodem tempore generet velocitatem $\frac{2FG}{t}$, Resistentia erit ad Gravitatem ut $\frac{HC}{T} - \frac{CG}{t} + \frac{2FG,GI}{t,CG}$ ad $\frac{2FG}{t}$, sive ut $\frac{t}{T}HC - CG + \frac{2FG,GI}{CG}$ ad $2FG$.[62]

Jam pro *BD*, *BP*, *BE* scribe *o*, 2 × *o*, −*o* & pro Ordinata scribe seriem quamvis interminatam $d - eo - foo - go^3[-]$ &c. & O[r]dinatæ reliquæ *EH*, *BC*, *PQ* erunt $d + eo - foo + go^3$, d, $d - 2eo - 4foo - 8go^3$ respective[,] et secundus seriei

---

—to represent the deviation *FG* from the tangential motion along *CF* by the related *sagitta* $DG - \frac{1}{2}(BC + PQ)$; on which see note (63) below.

(61) 'per Lem X lib. 1' is deleted in sequel—rightly so, since this is nothing to the point. Newton intends the reference 'ex demonstratis Galilæi' as he phrased it in his preparatory elaboration of §2.1 at this point, or the 'ut Galilæus demonstravit' which he rephrased this to be in the final version in §2.3; on the accuracy of this reference to Galileo see §2.2: note (14). We have mended a trivial slip of 'gravitate' for 'resistentia' in the previous sentence.

(62) Newton subsequently squeezed in between this and the following paragraphs a separate line 'Est autem ⊙——⊙' reminding himself to advance the next paragraph but one to follow immediately after in the redraft: this he did not immediately do, however, and he had again to insert a similar instruction in the revised version to invert their sequence (see §2.3: note (38))—which in our edited text we have there carried out, in line with what was finally printed.

assumptæ terminus *eo* erit *IF*, ac tertius & sequentes *foo* + *go*³ + &c erunt *FG*. Et quadrando differentias Ordinatarum *HE* − *CB* & *CB* − *GD* et ad quadrata prodeuntia addendo quadrata ipsarum *EB*, *BD*, habebuntur arcuum *HC*, *CG* quadrata *oo* + *eeoo* − 2*efo*³ &c et *oo* + *eeoo* + 2*efo*³ &c. Quorum radices

$$o\sqrt{1+ee} - \frac{efoo}{\sqrt{1+ee}} \,\&c \quad \text{et} \quad o\sqrt{1+ee} + \frac{efoo}{\sqrt{1+ee}} \,\&c$$

sunt arcus ipsi *HC* et *CG*. Præterea si ab ordinata *BC* subducatur semisumma Ordinatarum *EH* et *DG* et ab ordinata *DG* subducatur semisumma ordina[ta]rum *BC* et *PQ*[,] manebunt arcuum *HC* et *CG* sagittæ *foo* et *foo* + 3*go*³. Hæ sunt lineolis *CN* et *FG*[63] proportionales adeoq in subduplicata[64] ratione temporum *T* et *t*. Et pro ratione temporum $\frac{t}{T}$ scribi potest $\sqrt{\frac{f+3go}{f}}$ seu $\frac{2f+3go}{2f}$.

Pro $\frac{t}{T}$, *HC*, *CG*, *FG* et *GI* scribantur eorum valores et prodibit Resistentia ad Gravitatem ut 3*g* $\sqrt{1+ee}$ ad 4*ff*.[65]

Est autem Resistentia ut Medij densitas et quadratum velocitatis et propterea Medij densitas ut resistentia directe et quadratum velocitatis inverse[,] id est ut $\frac{3g\sqrt{1+ee}}{4ff}$ directe et $\frac{CG^q}{t}$ seu $\frac{1+ee}{f}$ inverse, hoc est ut $\frac{g}{f\sqrt{1+ee}}$.

Velocitas autem ea est quacum corpus de loco *C* secundum rectam *CF* egrediens, in Parabola diametrum *CB* et latus rectum $\frac{CF^q}{FG}$ seu $\frac{1+ee}{f}$ habente deinceps in vacuo moveri potest. Q.E.I.

Corol. 1. Si tangens *CF* producatur utrinq donec occurrat Ordinatæ *OL* in *T*, erit $\frac{CT}{OB} = \sqrt{1+ee}$, adeoq Resistentia ad gravitatem ut 3*g* × *CT* ad 4*ff* × *OB* & Medij densitas ut $\frac{g,OB}{f,CT}$.

---

(63) To be precise, these are *fo*² − 2*go*³ and *fo*² + *go*³ respectively (on ignoring terms in *o*⁴ and higher powers of *o*), and are consequently in the constant ratio 1 − 2(*g*/*f*) *o* to the corresponding *sagittæ BC* − ½(*EH* + *DG*) and *DG* − ½(*BC* + *PQ*); compare, *mutatis mutandis*, our fuller comment on this point in §2.3: note (32) above.

(64) Read 'duplicata' *tout court*. The slip is explained if we remark that Newton initially phrased his present sentence to read 'Hæ sunt lineolis *CN* = … et *FG* =… proportionales quorum radices pro temporibus *T* et *t* scribi possunt'. The obtrusive prefix 'sub' was carried over by Newton into his finished version and on into the fair copy of it which he transmitted to Cotes, but was caught by the latter before it gained standing in print in the *Principia*'s second edition. The lapse does not of course affect the sequel.

(65) Newton first concluded by straightaway adjoining without break of paragraph: 'Et Medij densitas ut $\frac{g}{f\sqrt{1+ee}}$. Velocitas autem ea erit quacum corpus de loco *C* secundum rectam *CF* egrediens, in Parabola diametrum *CB* et latus rectum $\frac{CF^q}{FG}$ seu $\frac{1+ee}{f}$ habente deinceps in vacuo moveri posset. Q.E.I.'

Corol. 2.[66] Cum Densitas Medij sit ut $\dfrac{g}{f\sqrt{1+ee}}$, si quantitas *g* datam habeat rationem ad quantitatem $f\sqrt{1+ee}$ (puta rationem *p* ad 1), densitas Medij erit uniformis. Proinde si inveniri potest[67] linea curva cujus hæc sit proprietas ut si a quovis ejus puncto ducatur abscissa horizontalis *o* & Ordinatæ perpendicularis in seriem convergentem resolutæ tres primi termini semper prodeant $eo+foo+pf\sqrt{1+ee}$ in o^3, Corpus in Medio uniformi justa cum velocitate e quovis hujus Curvæ puncto secundum tangentem ejus egrediens perget semper moveri in hac Curva.

Exem[pl]. 1. Sit linea *ACK* semicirculus ...... et curvatura curvarum.[68]

Conferantur jam termini hujus seriei cum terminis seriei superius assumptæ et pro illius terminis[69] *d, e, f, g* scribantur hujus termini $e, \dfrac{a}{e}, \dfrac{nn}{2e^3}, \dfrac{ann}{2e^5}$ et prodibit resistentia ad gravitatem ut $\dfrac{3ann}{2e^5}\sqrt{1+\dfrac{aa}{ee}}$ ad $\dfrac{n^4}{4e^6}$ seu 3*a* ad 2*n*, id est ut 3*OB* ad *AK*, et Medij densitas ut $\dfrac{a}{e}$ seu $\dfrac{OB}{BC}$, id est ut Tangentis longitudo illa *CT* quæ ad semidiametrum *OL* ipsi *AK* normaliter insistentem terminatur. Proinde si corpus *C* certa cum velocitate...non a resistente impediri.[70]

---

(66) In its original statement this 'Corol. 2' affirmed merely that 'Cum velocitas sit ut spatium descriptum *CG* directe & tempus invers[e] & tempus sit ut $\sqrt{GF}$, quadratum velocitatis erit ut $\dfrac{CG^q}{GF}$ [sive] $\dfrac{1+ee}{f}$'. In cancelling it Newton doubtless reckoned that he had already made this point well enough (if only implicitly) in the last paragraph of his preceding main text.

(67) 'Curva cujus abscissa horizontalis *o* a quovis curvæ puncto ducta sit, et ordinata sit $d+eo+foo+\dfrac{f\sqrt{1+ee}}{p}$ [!] in o^3...', Newton first went on before checking himself. In our edited version of his revised continuation we silently adjust his trivial slip of dividing (and not, as he must, multiplying) by the constant *p*.

(68) Understand the first three paragraphs of this prime example of the semicircle as they are printed in *Principia*, $_1$1687: 263–4 [= pages 378–9 above].

(69) Newton means 'coefficientibus' of course. Strictly, however, where the first coefficient *d* is equated with the first term *e* in the expansion of $e_{a+o} \equiv \sqrt{(n^2-(a^2-o^2))}$, there is correspondingly $e=-a/e, f=-\frac{1}{2}n^2/e^3$ and $g=-\frac{1}{2}an^2/e^5$. The double use of *e* here is manifestly confusing, and Newton in his final version (§2.3 preceding) wisely recurs to the letters *P, Q, R, S* which he had employed in his 1687 text to express the Taylorian expansion of the general incremented ordinate. The series $e-fo-go^2-ho^3...$ for this, by which (on f. 136^v following, in the manuscript sequence at least) he derived the equivalent measure $\frac{3}{4}h\sqrt{(1+f^2)}/g^2$ of the ratio of resistance to gravity in the afterthought to §2.2 reproduced in [7], evidently marks a stage on the road back at which Newton saw no need other than briefly to tarry.

(70) Understand that, as in the parent text (§2.1), the concluding portion of 'Exempl. 1' in the *editio princeps* (*Principia*, $_1$1687: 265) is to follow with the minimal change that now 'resistentia...in loco aliquo *C* sit ad vim gravitatis ut 3*OB* ad 2*OK*'. And the same likewise for the remaining Examples 2–4.

[9][71] [Asymptotis *AZ*, *YZ* describatur hyperbola *LPM* habens semidiametrum unitati æqualem. Cape]

$ZB = x$.[72] [unde] $BP = \dfrac{1}{x}$. [Tum po-

nendo] $ALPB$[73] $= z$. [et] $AB = y$. [sit]

$\dfrac{z^{n+1}}{y^n} = BF$ [ordinatæ lineæ curvæ *AFK*

in qua projectile moveatur].[74] [Exi-
stente jam decremento] $BC = o$. [erit]

$$CQ = \frac{1}{x-o} = \frac{1}{x} + \frac{o}{x^2} + \frac{oo}{x^3} + \frac{o^3}{x^4}[+\&c].$$

[ut et] Area *AL*[*Q*] *C*[75]

$$= z + \frac{o}{x} + \frac{oo}{2x^2} + \frac{o^3}{3x^3} + \frac{o^4}{4x^4}[+\&c].$$

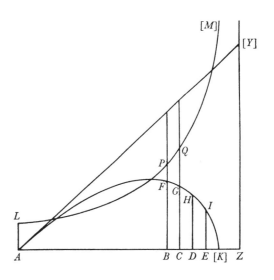

In a stray draft on Add. 3965.12: 200[v], we may add, Newton set down an elegant variant conclusion to his discussion of unresisted parabolic motion in 'Exempl. 2': there, presupposing that the incremented ordinate $DG = e_{a+o}$ of the parabola $e = e_a \equiv (ac - a^2)/b$ has been reduced to its equal series $(ac - a^2)/b + ((c-2a)/b)\, o - (1/b)\, o^2$ as in the 1687 text, Newton went on to argue from first principles that 'Tertius terminus $\frac{oo}{b}$ designat lineolam *FG*, et scribendo *OE* pro *OB* et *EH* pro *BC* non mutatur, ideoҩ designat etiam lineolā $CN_{[,]}$ & propter æquali- tatem harum lineolarum *CN* et *FG* tempora describendi arcus *HC* et *CG* sunt æqualia, et propterea motus corporis secundum horizontem non retardatur. Nulla igitur Medij densitate projectile movebitur in Parabola, uti olim demonstravit Galilæus. Q.E.I.' (He then passed a deal less fruitfully to apply the geometrical measure $HN/(CN+GF) - CF/2GF + FI/CF$ attained in §2.1 directly to compute the ratio of resistance to gravity in the hyperbolic instance $e_a \equiv c - (m/n)\, a - b^2/a$ of the ensuing 'Exempl. 3', correctly there evaluating *GF* and *CF*, but erroneously calculating their 'incremented' forms *CN* and *HN* 'substituendo *NE*[!] pro *NB* seu $a+o$ pro a'—the true substitution is $(NB =)\, a \to (ND =)\, a-o$ of course—and further mangling the reduction of the resulting value of *HN* to its equivalent series expansion in powers of the decrement o, before at length breaking off in some muddle and confusion.) Within the tight confines of the 10-page scissure in the already printed-off sheets of the *editio secunda* which needed to be filled, there was, we may or may not regret, no room spare to indulge the luxury of adducing alternative modes of solution—and still less so to append such further illustrative instances of 'realistic' projectile path as the logarithmico-hyperbolic curve which Newton toyed briefly and ineffectively with in the fragment whose gist we reproduce in [9] now following.

(71) Add. 3965.10: 136[v]. This unfinished sketch of an intended further worked *Exemplum* of a resisted projectile path would—had its definition been mended (see note (74))—doubtless have been furnished with some neat geometrical construction of the ratio of resistance along its arc to the downwards force of gravity, and of the consequent variation of the surrounding medium's density which will permit such a curving motion of a projectile through it, much in the manner of the simpler hyperbolic trajectories examined in Examples 3/4 of the published proposition (and further elaborated in the ensuing scholium). The original jotted calculations (here verbally fleshed out in our usual fashion) are penned at the head of a page on which they

[Ubi $n = 1$, hoc est $BF = \dfrac{zz}{y}$, evadit] Area $AL[Q]C$ quad[rata]

$$= zz + \frac{2,oz}{x} + \frac{zoo}{xx} + \frac{2z,o^3}{3x^3} \ [\&c]$$

$$+ \frac{oo}{xx} \qquad + \frac{o^3}{x^3}$$

[et] Hæc area quad. applicata ad $[AC =] y + o$ [est CG] =

$$\frac{zz}{y} - \frac{z^2 o}{yy} + \frac{z^2 oo}{y^3} - \frac{z^2 o^3}{y^4} \ [\&c].^{(76)}$$

$$\left[+ \frac{2zo}{xy} \quad \cdots \quad \cdots \right]$$

---

are immediately followed by the *addendum* to §2.2 above which is set out in [7] of this Appendix; it could therefore well be (as we have tentatively conjectured in §2.2: note (25)) that this logarithmic-*cum*-hyperbolic curve was meant there also to be adjoined in its finished form.

(72) Previously in his Proposition X, of course, Newton has invariably employed the Fermatian variable a to denote this general abscissa, but he here for once shows his proper colours as a true disciple of Descartes.

(73) This is, $\log (b/x)$ on setting the base length $AZ = b$.

(74) In explicit form, accordingly, the defining Cartesian equation of Newton's posited curved $\widehat{AFG}$ is $e = [\log (b/x)]^{n+1}/(b-x)^n = [\log (b/(b-y))]^{n+1}/y^n$, where $ZB = x$ ($AB = y$) and $BF = e$ are the mutually perpendicular coordinates of its general point $F(x, e) \equiv (y, e)$, and $AZ = x + y = b$ is a given constant base-length. In his accompanying figure (here accurately reproduced in its proportions, with a few of its points additionally named for convenience of reference) Newton depicts this in the familiar shape of a resisted projectile path, falling away from the tangent AY at its initial 'firing' point A to meet the horizontal in a point—we mark it as K, as in the published hyperbolic Examples 3/4—between E and Z. But this cannot be. On expanding $\log (b/x)$ as an infinite series in $y = b - x$ the defining equation reduces to be $e = y^{-n}.(y/b + \frac{1}{2}y^2/b^2 + \frac{1}{3}y^3/b^3 + ...)^{n+1} = y/b^{n+1} + \frac{1}{2}(n+1) y^2/b^{n+2} + ...,$ and hence the slope of the curve at F is $(de/dy =) 1/b^{n+1} + (n+1) y/b^{n+2} + ...$, which increases with $AB = y$ to be infinitely great when $y = b$. (No real points exist on the curve when $y > b$, of course.) In other words, Newton's trajectory climbs ever higher away from its initial tangent AY, asymptotically to close with ZY at an infinite distance above the point Z—as indeed will be geometrically obvious, since the hyperbola-area $(ALPB) = \log (b/(b-y))$ comes to be infinitely great as $y \to b$. Let us comment in the fullness of hindsight that Newton would have done a deal better—as may be it was his intention?—to choose $BF = y^{n+1}/z^n$ for the defining 'symptom' of this variant species of resisted path: below the initial tangent AY (fixed by $ZY = AZ^{n+1}$) this falls smoothly away towards the 'horizon' AZ to be near-vertical in its final portion, tangent in the last to ZY (the point K in the figure here coinciding with Z). In his manuscript text, however, he himself presses on to make further computation regarding the particular instance $n = 1$ of his 'fall' path before abandoning his calculation as the complexities increase.

(75) The decremented hyperbola-area $\log (b/(x-o))$, that is, $\log (b/x)$ [or z] $- \log (1-o/x)$.

(76) And there Newton leaves off his evaluation of the incremented ordinate in this simplest instance of the curve defined by $e = e_y \equiv z^2/y$. The coefficients $Q = -z^2/y^2 + 2z/xy, R = z^2/y^3...$ and $S = -z^2/y^4...$ in this expansion of e_{y+o} as $z^2/y + Qo + Ro^2 + So^3...$ serve as before, of course,

[10][77]   Cum Resistentia sit ad gravitatem ut $3S\sqrt{1+QQ}$ ad $4RR$; sit hæc resistentia ut Medij densitas et velocitatis V potestas V^n: & Medij densitas erit ut resistentia directe & velocitatis potestas V^n inverse, id est ut $\dfrac{3S\sqrt{1+QQ}}{4RRV^n}$. Pro

V scribatur $\sqrt{\dfrac{1+QQ}{R}}$ [78], et Densitas erit ut $\dfrac{3S\sqrt{1+QQ}}{4RR\times\overline{\dfrac{1+QQ}{R}}\Big|^{\frac{n}{2}}}=\dfrac{SR^{\frac{n-4}{2}}}{\overline{1+QQ}\Big|^{\frac{n-1}{2}}}$.

Sit $n=1$, et Densitas erit ut $\dfrac{S}{R^{\frac{3}{2}}}$. Sit $n=3$ et Densitas erit ut $\dfrac{S}{1+QQ \text{ in } R^{\frac{1}{2}}}$. Sit

$n=5$ et Densitas erit ut $\dfrac{SR^{\frac{1}{2}}}{\overline{1+QQ}\Big|^2}$.   Sit $n=4$ et densitas Medij erit ut $\dfrac{S}{\overline{1+QQ}\Big|^{\frac{3}{2}}}$.

Sit $n=2$ et Dens. erit ut $\dfrac{S}{R\sqrt{1+QQ}}$. [Hoc est] Sit $n=0,1,2,3,4,5,6,7$ et

Densitas Medij erit ut $\dfrac{S\sqrt{1+QQ}}{RR}$, $\dfrac{S}{R^{\frac{3}{2}}}$, $\dfrac{S}{R\sqrt{1+QQ}}$, $\dfrac{S}{R^{\frac{1}{2}}\times\overline{1+QQ}}$, $\dfrac{S}{\overline{1+QQ}\Big|^{\frac{3}{2}}}$,

$\dfrac{SR^{\frac{1}{2}}}{\overline{1+QQ}\Big|^2}$, $\dfrac{SR}{\overline{1+QQ}\Big|^{\frac{5}{2}}}$, $\dfrac{SR^{\frac{3}{2}}}{\overline{1+QQ}\Big|^3}$, &c. Or $\dfrac{S}{R^2\times\overline{1+QQ}\Big|^{-\frac{1}{2}}}$, $\dfrac{S}{R^{\frac{3}{2}}}$, $\dfrac{S}{R\times\overline{1+QQ}\Big|^{\frac{1}{2}}}$

$\dfrac{S}{R^{\frac{1}{2}}, \overline{1+QQ}\Big|^1}$, $\dfrac{S}{R^0\times\overline{1+QQ}\Big|^{\frac{3}{2}}}$, $\dfrac{S}{R^{-\frac{1}{2}}\times\overline{1+QQ}\Big|^2}$, $\dfrac{S}{R^{-1}\times\overline{1+QQ}\Big|^{\frac{5}{2}}}$, $\dfrac{S}{R^{-\frac{3}{2}}\times\overline{1+QQ}\Big|^3}$,

&c.

Et universaliter,

Velocitas ut $\overline{1+QQ}\Big|^{\frac{1}{2}}\times R^{-\frac{1}{2}}$,   Resistentia ut $\overline{1+QQ}\Big|^{\frac{1}{2}}\times S\times R^{-2}$,

Densitas ut $S\times R^{\frac{n-4}{2}}\times\overline{1+QQ}\Big|^{\frac{1-n}{2}}$.[79]

to determine the ratio $\frac{3}{4}S\sqrt{(1+Q^2)}/R^2$ of the medium's resistance to gravity, and the related exponents, $\sqrt{((1+Q^2)/R)}$ and $S/R\sqrt{(1+Q^2)}$ respectively, of the body's instantaneous speed and of the medium's density (where the resistance is supposed to be as the square of the speed).

(77) Add. 3968.37: 540ᵛ, a preliminary jotting on a draft of the English preface to the *Commercium Epistolicum* printed in **2**, Appendix 1.5. Newton computes the variation of density in a medium resisting the passage of a body through it in a given curve according as some general *n*-th power of the speed; and thereafter enumerates—haphazardly at first, and then more systematically in two alternative displays of the pattern of its succeeding particular varieties of form—the simpler modes in which this manifests itself when $n = 0, 1, 2, 3, ..., 7$ before again encapsulating its general expression in an equivalent formulation.

(78) Strictly, $V = \sqrt{(\frac{1}{2}g.(1+Q^2)/R)}$ where g is the force of downwards gravity; see §2.3: note (37) above. But here, of course, the neglected factor $\sqrt{(\frac{1}{2}g)}$ can be appropriately absorbed into the constant of proportionality.

(79) Whence, by way of an intermediate draft at Add. 3965.12: 201ᵛ (on which see §2.3: note (49) above), this general formula for the density was remoulded to be as it afterwards appeared in the *Principia's editio secunda* in the opening paragraph of the augmented scholium to Proposition X of the second book (₂1713: 240; compare page 368 above).

[11][80]   [Dic $ZC=x$ et asymptotis AZ, ZX describatur hyperbola LQM cujus ordinata est $CQ=$] $\dfrac{1}{x+cxx}=y$.

[Tum ponendo hyperbolæ aream $ALQC$[81] $=z$, sit $\dfrac{z}{1}=CG$ ordinatæ lineæ curvæ AFG in qua projectile moveatur. Existente jam incremento $CB=o$, fit

$$BP=\frac{1}{\overline{x+o+c,x+o}|^2} \quad \text{seu}]$$

$$\frac{1}{x+cxx+o+2cox+coo}=\frac{1}{x+cxx}-\frac{o+2cox+coo}{x+cxx\times\overline{x+cxx}}$$

$$+\frac{oo+4coox+4ccooxx+2co^3x+4cco^3x}{\overline{x+cxx}|^3}$$

$$-\frac{o^3+6co^3x+12cco^3xx+8c^3o^3x^3}{\overline{x+cxx}|^4}+\&\text{c.}$$

[hoc est] $y+\dot{y}$[o &c][82] $=\dfrac{1}{x+cxx}-\dfrac{o+2cox}{\overline{x+cxx}|^2}+\dfrac{oo+3coox+3ccooxx}{\overline{x+cx}|^3}$

$$-\frac{o^3+4cxo^3+6ccxxo^3+4c^3x^3o^3}{\overline{x+cx}|^4}[+\&\text{c}].$$

(80) These rough computations, testing whether two other types of logarithmic path may possibly be traversed in a medium which resists the passage of a projectile according to the square of the latter's instantaneous speed or in a uniform fashion independent of it, are to be found at the top of Add. 3968.37: 540ᵛ [reversed] immediately before the text, deriving the formula for the variation of density in a medium resisting as a given nth power of the speed, which is reproduced in [10] preceding, and whose particular instances $n = 0$ and $n = 2$ are in anticipation—in the chronological sequence in which Newton penned his manuscript—here understood. As ever where the original is skeletal and bare, we have connected Newton's lines of undescribed calculation with appropriate interposed verbal sinews, and fleshed out the fragmentary shreds of the geometrical configuration to which these are (largely in his mind alone) keyed by adjoining—and relating the steps in the argument to—an adaptation of the diagram which Newton himself employs in the parallel context of [9] above.

(81) In modern analytical equivalent this is

$$\left(\int_x^b y.dx \text{ or}\right)\int_x^b (1/x-c/(1+cx)).dx = \Big[\log x - \log(1+cx)\Big]_x^b$$

on setting $ZA = b$. The projectile curve AG will in consequence be defined by the Cartesian equation $z = z_x \equiv \log(c+1/x) - \log(c+1/b)$. Since at its general point $G(x, z)$ this has the slope $(dz/dx =) y$, it will be touched at the 'firing' point $A(b, 0)$ by the straight line AY ($z = (b-x)/(b+cb^2)$) meeting XZ ($x = 0$) in Y such that $ZY = 1/(1+c.ZA)$.

(82) Understand the 'Taylor' expansion $y+\dot{y}o+\frac{1}{2}\ddot{y}o^2+\frac{1}{6}\dddot{y}o^3+\ldots$ of the ordinate $CQ = y_o$, now regarded as a function of the increment $BC = o$, so that $y_0 = y = 1/x-c/(1+cx)$ and

[adeoɋ] $\text{Area}^{(83)} = \dfrac{1}{x+cxx}\,o - \dfrac{1+2cx}{2 \times \overline{x+cxx}\,|^2}\,o^2 + \dfrac{[1]+3cx+3ccxx}{3 \times \overline{x+cxx}\,|^3}\,o^3 - \&\text{c}.$

[Itaɋ si resistentia ponatur esse ut quadratum velocitatis, prodeunte scilicet]

$$eS = R\sqrt{1+QQ}.^{(84)}$$

[evadit] $\sqrt{\dfrac{1+xx+2cx^3+ccx^4}{\overline{x+cxx}\,|^2}} \;\text{in}\; \dfrac{1+2cx}{2 \times \overline{x+cxx}\,|^2} = \dfrac{e+3ecx+3eccxx}{3 \times \overline{x+cxx}\,|^3}.^{(85)}$

[Vel generalius finge $CQ = y =$] $\overline{x^m+cx^{m+1}}\,|^n =^{(86)}$

$$x^{mn} + ncx^{mn+1} + \frac{[n]n-n}{2} \times c^2 x^{mn+2} + \frac{n^3-3nn+2n}{6}\,c^3 x^{mn+3}\;[+\&\text{c}].$$

[Erit] Ord[inata $CG]^{(87)} =$

$$\frac{1}{mn+1}\,x^{mn+1} + \frac{nc}{mmnn+3mn+2}\,x^{mn+2} + \frac{[n]nnc^3-nc^3}{2m^3n^3+12mmnn+22mn+12}\,x^{mn+3}\;[\&\text{c}].^{(88)}$$

---

consequently

$$\dot y = (dy/dx \text{ or}) -1/x^2 + c^2/(1+cx)^2, \quad \ddot y = (d\dot y/dx \text{ or}) \; 2/x^3 - 2c^3/(1+cx)^3,$$
$$\dddot y = -6/x^4 + 6c^4/(1+cx)^4, \quad \dots.$$

(83) Namely $(BPQC) = \left(\displaystyle\int_x^{x+o} y\,.\,dx \text{ or}\right) \displaystyle\int_0^o y_o\,.\,do.$

The ensuing evaluation of this as $Qo + Ro^2 + So^3 + \dots$ (where $Q = y$, $R = \tfrac12\dot y$, $S = \tfrac16\ddot y$, ...), by expanding y_o into its Taylorian series (see the previous note) and then integrating this term by term, yields of course (since $y = \dot z$, $\dot y = \ddot z$, $\ddot y = \dddot z$, ...) the parallel Taylorian expansion of the equal quantity $BF - CG = z_{x+o} - z$ which Newton in sequel requires.

(84) Where, that is, Q, R and S are the coefficients of the successive powers of o in the preceding series expansion of $z_{x+o} - z$ (see the previous note) and e is some constant. The quantity $S/R\sqrt{(1+Q^2)}$ is (see [10] preceding) the measure of the density of the medium which permits a projectile to traverse the curve $z = z_x$ when the resistance is—as Newton here puts it to be—proportional to the square of the body's speed, and so its constancy is the criterion for such resisted motion through a uniform medium.

(85) 'Q[uod] F[ieri] N[equit]', as Newton would no doubt have tagged this impossibly asymmetric identity had he written out his present argument more formally. He gives up pursuing this present chimera, and passes forthwith from this instance (where $m = 1$, $n = -1$) to consider the more general projectile curve defined by the Cartesian equation

$$z = z_x \equiv \int_0^x y\,.\,dx,$$

where now $y = (x^m + cx^{m+1})^n$.

(86) On expanding this binomial $x^{mn}(1+cx)^n$ into its equivalent series, that is.

(87) Namely $z = \displaystyle\int_0^x y\,.\,dx.$

(88) This would-be integration term by term of the preceding series merits our rare editorial accolade of a double exclamation!! The denominators of the coefficients here are successively $mn+1$, $(mn+1)(mn+2)$ and —as Newton laboriously calculated in a blank space a little way above after an erroneous computation of this triple product in his head $-2(mn+1)(mn+2)(mn+3)$, that is, '$\overline{mn+3} \times 2m^2n^2 + 6mn+4$'.

[Existente igitur resistentia uniformi, vel $S\sqrt{1+QQ}=\frac{4}{3}RR$, fiet]

$$\frac{\sqrt{mmnn+2mn+1} \text{ in } x^{2mn+2}+ee}{mn+1}^{(89)} \times \frac{[n]nnc^3-nc^3}{2,\overline{mn+1}\times\overline{mn+2}\times\overline{mn+3}}$$

$$=\frac{\frac{4}{3}nncc}{\overline{mn+1}\times\overline{mn+1}}.^{(90)}$$

(89) The expression under the radical is gobbledygook for '$mmnn + 2mn + 1 + x^{2mn+2}$'.

(90) Newton blithely substitutes for Q, R and S, in his preceding criterion for uniform resistance to motion in the curve $z_x = \int_0^x y\,.\,dx$, the first three terms in the preceding Wallisian evaluation of this area $(CQMXZ) = CG$ (by expanding the binomial $y = x^{mn}(1+cx)^n$ into its equivalent series in powers of x^{mn+p}, $p = 0, 1, 2, \ldots$ and integrating these term by term), thereby ignoring that these are in fact the successive coefficients of powers of o in the like expansion of the area $(BPQC) = \int_0^o y_o\,.\,do = z_{x+o} - z$ (see note (83) above) upon replacing y_o by its equivalent Taylorian series $y + \dot{y}o + \frac{1}{2}\ddot{y}o^2 + \ldots$ and integrating this term by term (so that $Q = y = \dot{z}$, $R = \frac{1}{2}\dot{y} = \frac{1}{2}\ddot{z}$ and $S = \frac{1}{6}\ddot{y} = \frac{1}{6}\dddot{z}$ *recte*). The oversight still gets him nowhere, of course, since the lone variable x^{2mn+2} obtrudes within the radical on the left-hand side of this otherwise numerical equation. We may imagine the somewhat ratty mood in which he straightaway drew a line across the manuscript page and began to set down his derivation of the formula for the variation of density along a given flight path, in a medium resisting as a general power of the speed, which we have reproduced in [10] above.

APPENDIX 3. THREE DRAFTS OF A
RETROSPECTIVE ATTEMPT (LATE AUTUMN? 1712)
ONCE MORE TO SALVAGE THE 1687 ARGUMENT.[1]

From the original worksheet in the University Library, Cambridge

[1][2]   [Sit *AK* planum illud plano Schematis perpendiculare, *LCK* linea curva, *C* corpus in ipsa motum, *fCF* recta ipsam tangens in *C*. Fingatur autem corpus *C* nunc progredi per lineam illam, nunc vero regredi per eandem lineam; & in progressu impediri a Medio, in regressu æque promoveri, sic ut in ijsdem locis eadem sit corporis progredientis & regredientis velocitas. Æqualibus autem temporibus describat corpus progrediens arcum quam minimum *CG*] et corpus regrediens arcum *Cg*, et a punctis *C*, *G*, *g* ad planum horizontale *AK* demittantur perpendicul[a] *CB*, *GD*, *gd*, quorum *GD* ac *gd* sursum productæ tangenti occurrant in *F* et *f*. Et *FG*, *fg* erunt altitudines quas corpus æqualibus tempori-

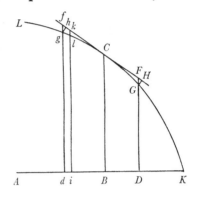

(1) These last (known) endeavours by Newton to preserve essentially intact, with only minor adjustment of the structure of its reasoning, the original form of Proposition X of the *Principia*'s second book, are all found on the single folio sheet (now Add. 3965.10: 135/136) from which we printed Appendix 2.7, 8 and 9 preceding. Because the preliminary revise of [1] which is reproduced in [2] occurs (on f. 136ᵛ) immediately below his prior derivation (see Appendix 2.7) of the general formula for the ratio of resistance to gravity from the equivalent geometrical measure attained in the 'Idem aliter' (in §2.2) to the reconstructed proposition, while his present final attempt in [3] to salvage the 1687 argument precedes (on f. 135ᵛ) the preparatory recasting of the radical reworking in (§1.6 →) §2.1 to be §2.3 which is set out in Appendix 2.8, we infer that this last try to repair the fault in his original proposition without drastically rebuilding it from its foundations upwards—one which may, if (see the concluding footnote) we interpret a careless calculation on Add. 3968.41: 132ᵛ aright, well have been ultimately successful—took place only a short while before (at the very end of 1712?) Newton drafted the finished version of the recast proposition which he sent to Cotes on 6 January 1712/13 to be published in the *editio secunda* (see §2: note (1) preceding). It would, we need not say, have been a face-saving triumph for him to have mended the lapse in his 1687 argument with the little adjustment which version [3] here requires, and we can think it was only through pressure of time—he was at this period both composing his 'Scholium generale' to the *Principia* in its *editio secunda* and seeing his edition of the *Commercium Epistolicum* through the press, in addition to carrying out his normal duties at the Mint and as P.R.S.—that he omitted to work it up into a form fit to replace its defective original in print.

(2) Add. 3965.10: 135ʳ. The roughly penned (and subsequently much corrected and interlineated) original begins abruptly *in medio*, taking up the parent printed text some way into the third sentence of its scene-setting first paragraph. We here make the deficiency good, introducing within our usual editorial square brackets a light tailoring of the omitted opening lines (*Principia*, ₁1687: 260, l. 30 – 261, l. 10), and, for convenience, placing alongside an abstract of

bus vi gravitatis descendendo a tangente, describit, adeoꝗ æquantur inter se.[3]
Et tangenti *Ff* parallela et æqualis est chorda arcus *Gg* & utriꝗ æqualis est arcus
ille quam minimus *gG*. A punctis *g, G* in tangentem demittātur perpendicula
gh, GH: et arcus *gC, CG* æquales erunt tangentis partibus *Ch, CH* seu *Cf−fh* &
CF+FH. Et horum arcuum differentia erit *Cf−CF−2fh*. Hæc differentia[4]
oritur ab actionibus resistentiæ et gravitatis conjunctim interea dum corpus
describit arcū alterutrum. Resistentia motum corporis retardat, gravitas
accelerat & accelerando auget arcum *CG* longitudine *FH*.[5] Addatur hæc
longitudo ad differentiam prædictam & habebitur longitudo quam resistentia
sola generat, nempe *Cf−CF−fh*.[6] Et Gravitas eodem tempore generat
descensum *FG*. Proindeꝗ resistentia est ad gravitatem ut *Cf−CF−fh*[6] ad *FG*.
Hæc ita se habent in lineolis nascentibus; nam in lineolis finitæ magnitudinis
hæc ratio non est accurata.

Est autem resistentia ut Medij densitas [& quadratum velocitatis. Velocitas
autem ut descripta longitudo *CF* directe & tempus $\sqrt{FG}$ inverse, hoc est ut
$\dfrac{CF}{\sqrt{FG}}$, adeoꝗ quadratum velocitatis ut $\dfrac{CF^q}{FG}$. Quare resistentia ipsiꝗ][7] proportionalis $\dfrac{Cf-CF-fh}{FG}$ [est ut Medij densitas & $\dfrac{CF^q}{FG}$ conjunctim; & inde Medij
densitas][7] ut $\dfrac{Cf-CF-fh}{FG}$ directe et $\dfrac{CF^q}{FG}$ inverse, id est ut $\dfrac{Cf-CF-fh}{CF^q}$. Q.E.I.

Newton's accompanying figure, redrawn (since it does not here do double duty for the prime
exemplum of the semicircle) to show a general projectile path *LCK*, and with the perpendiculars
gh, GH from *g, G* to the tangent *fCF* at *C* inserted in accord with his present direction. The last
two paragraphs as they stand in the manuscript (with Newton's suspension dashes not replaced,
as here, by the corresponding portions of the 1687 text) are, we may add, reproduced by I. B.
Cohen in his *Introduction to Newton's 'Principia'* (Cambridge, 1971): 108 but without comment
upon their dynamical context or mathematical truth.

(3) This is (see Appendix 1: note (8) above) true only when rounding-off is made to the order
of the square of the base increment *BD* (*Bd*). Apart from his further lapses, dynamical and
mathematical, Newton will in his Corollary 1 below repeat his error in the *editio princeps* text
(compare *ibid.*: note (13)) of assuming it valid to replace the third-order difference *fg−kl* by its
'equal' *FG−kl*.

(4) Since Newton goes on to consider the actions of the retarding resistance and the component of downwards accelerating gravity only over $\widehat{CG} = \frac{1}{2}\widehat{gCG}$, this should read 'Hujus
differentiæ dimidium', *viz.* $\frac{1}{2}(fC-CF)-fh$.

(5) Understand 'seu *fh*' since *FH* and *fh* are the projections along the tangent *fCF* of the
downwards gravitational deviations *FG* and *fg* from it, which take place in equal times and
may therefore (at the second-order level of accuracy here supposed) be considered equal.

(6) This should accordingly be $\frac{1}{2}(fC-CF)$ *tout court*. The ensuing (correct) measure
$\frac{1}{2}(fC-CF)/FG$ of the ratio of resistance to gravity is of course that attained in the *editio princeps*
(*Principia*, ₁1687: 262; compare Appendix 1: note (10) above).

(7) In each case we replace dashes in the manuscript by the pertinent sections of the parent
printed text (*Principia*, ₁1687: 262), as Newton intends.

Corol. 1. Et hinc colligitur [quod si in *Cf* capiatur *Ck* æqualis *CF*, & ad planum horizontale *AK* demittatur perpendiculum *ki* secans curvam *LCK* in *l*][7] fiet

Medij densitas ut $\dfrac{FG-kl-fh}{CF \text{ in } \overline{FG+kl}}$. [8]

---

[2][9]   Sit *AK* planum illud plano schematis perpendiculare; *ACK* linea curva; *C* corpus in ipsa motum; *fCF* recta ipsam tangens in *C*; & *gC*, *CG* arcus quam minimi æqualibus temporis momentis[10] descripti. A punctis *g, C, G* ad planum horizontale *AK* demittantur perpendicula *gd, CB, GD*, quorum *gd* ac *GD* sursum product[æ] tangenti occurrant in *f* et *F*. Et in tangente capiantur hinc inde longitudines *Ch CH* arcubus *Cg CG* æquales respective. Jam quo tempore corpus describit arcum *CG*, idem absçg gravitate describeret rectam *CF*. Gravitas deducit corpus a tangente deorsum & descendere facit per altitudinem *FG*, et longitudini *CF* quam corpus absçg gravitate describeret addit longitudinem *FH*. Arcuum *gC, CG* differentia *gC−CG*[11] est decrementum spatij quod corpus singulis temporis momentis describit & hoc decrementum oritur a resi[s]tentia et gravitate conjunctim. Gravitas auget descriptum spatium longitudine *FH*, ideoçg decrementum ejus ex resistentia sola oriundum est *gC−CF*, seu *Ch−CF* id est *Cf−CF−FH*.[12] Gravitas autem eodem temporis momento generat altitudinem *FG*. Et propterea resistentia est ad gravitatem ut *Ch−CF* ad *FG*.[13]

---

[3][14]   [Sit *AK* planum illud plano Schematis perpendiculare, *LCK* linea

---

(8) Two exclamation marks are warranted here!! On replacing *CF* by its equal *kC*, the ratio $(fC-CF)/CF$ becomes $(fC-kC)/kC = (\sqrt{fg}-\sqrt{kl})/\sqrt{kl}$ (since, to sufficient accuracy, the vanishingly small arc $\widehat{glC}$ may be taken to be a parabola of vertex *C* and diameter *CB*, whence $fC:kC = \sqrt{fg}:\sqrt{kl}$), that is $(fg-kl)/(\sqrt{fg}+\sqrt{kl})\sqrt{kl} = (fg-kl)/(fg+kl)$ (because *fg* comes to coincide with *kl* in the limit, and only their difference is significant); and in this Newton puts the 'equal' deviation *FG* in place of *fg* (erroneously so when the difference *fg−kl* is thereby replaced; see note (3) above). Even so, he should here have ended up with the fraction '$\dfrac{FG-kl}{CF \text{ in } \overline{FG+kl}} - \dfrac{fh}{CF^q}$'. Perhaps realizing as much, Newton here breaks off to redo his main argument before repairing its present corollary.

(9) Add. 3965.10: 136ᵛ. This light recasting of [1] preceding is squashed in, below the calculations reproduced in Appendix 2.7, at the foot of the left-hand page when the manuscript sheet (ff. 135/136) is opened out to have f. 135ʳ (bearing the text of [1]) on the right.

(10) 'particulis' is replaced.

(11) This should read 'semidifferentia ½, $\overline{gC-CG}$'; compare [1]: note (4).

(12) Whence this should be '½ $\overline{gC-CG}$+FH seu *fh*, id est ½ $\overline{fC-CF}$'.

(13) Correctly (see the preceding notes) 'ut *fC−CF* ad 2*FG*', the measure obtained in the 1687 parent text, which Newton again here seeks uselessly to alter by his faulty present argument.

(14) Add. 3965.10: 135ʳ/135ᵛ (where it follows immediately after the text given in [1] preceding). Newton makes a last effort to salvage the structure of his 1687 argument, now correctly relating the times of passage successively over the equal arcs $\widehat{gC}$ and $\widehat{CG}$ by the square

curva, *C* corpus in ipsa motum, *fCF* recta ipsam tangens] in *C*. Describat autem corpus æqualibus temporis particulis arcus quam minimos *gC*, *CG* ijsq̃ æquales sint *Ch*, *CH* in tangente sumptæ, et a punctis *g*, *C*, *G* ad planum horizontale *AK* demit[t]antur perpendicula *gd*, *CB*, *GD* quorum *gd* ac *GD* sursum productæ tangenti occurrant in *f* et *F*. & Vis gravitatis quo tempore Corpus describit arcum *CG*, descendere facit idem a tangente & altitudinem *FG* descendendo describere.[15] Arcuum *Cg*, *CG* differentia *Cg−CG*[16] oritur ab actionibus resistentiæ et gravitatis conjunctim.[17] Resistentia differentiam auget, gravitas diminuit ac diminuendo [in] singulis tem[poris] particulis aufert lineolam *FH*. Addatur hæc lineola, et hab[eb]itur decrementum arcus singulis temporis particulis descripti a resistentia sola oriundum, nempe *Cg−CG+FH*.[18] Gravitas autem eodem tempore generat descensum *FG* ut supra. Et propterea resistentia est ad gravitatem ut *Cg−CG+FH* ad *FG*.[19]

Sit jam Ordinata quævis *BC=P*, Abscissæ particula *BD=o*, et Ordinata proxima

$$DG = P - Qo - Roo - So^3 - \&c.$$

Et hujus seriei terminus secundus designabit lineolam *FI* quam tangens aufert a parallelogrammi *DBCI* latere *DI*. Termini reliqui *Roo+So³*[*+* &c] designabunt lineolam *FG* quæ tangenti & Curvæ interjacet. Et tertius quidem terminus *Roo* determinabit curva-

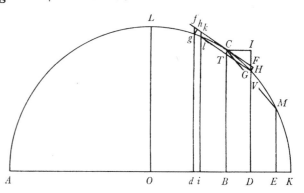

<hr>

roots of the *sagittæ CT* and *GV* at their corresponding end-points *C* and *G* (see note (20) below), but again tripping into a numerical lapse (see note (16)) which once more leads him to depart from the correct basic geometrical measure of resistance to gravity as he had accurately obtained it in the *editio princeps*. A stray line and a half of calculation on Add. 3968.41:132ᵛ, itself spoilt by a like careless numerical slip, would, however, indicate that he did ultimately (to his private, otherwise undocumented satisfaction) complete the repair of his original proposition by reverting to its measure $\frac{1}{2}(fC-CF)/FG$ of the ratio of resistance to gravity, and then calculating the correct analytical equivalent, $\frac{3}{4}S\sqrt{(1+Q^2)}/R^2$, to this by employing his present reasoning *ex sagittis*; see our concluding note (22).

(15) Newton initially went on to specify more fully: 'Et si corpus ijsdem velocitatis gradibus regrederetur in arcu *Cg* et quo tempore pervenit a *G* ad *C* eodem perveniret a *C* ad *g*, vis gravitatis descendere faceret corpus a tangente et altitudinem *fg* describere. Et propter æqualitatem temporum altitudines *FG*, *fg* a gravitate genitæ sunt æquales. Et triangulorum rectangulorum & similium *FGH*, *fgh* latera *FH* et *fh* sunt etiam æqualia. Et arcuum *Cg*, *CG* differentia *Cg−CG* = *Ch−CH* est *Cf−CF−2FH*. Oritur hæc differentia ab actionibus...'.

(16) Again (see [2]: note (11) above) this should read 'semidifferentia $\frac{1}{2}, \overline{gC-CG}$'.

(17) 'in singulis temporis particulis' is deleted in sequel.

(18) Correctly, this should (see [2]: note (12)) be '$\frac{1}{2}, \overline{gC-CG}+FH$, id est $\frac{1}{2}, \overline{fC-CF}$'.

(19) Whence determining, *recte*, that 'resistentia est ad gravitatem ut *gC−CG+2FH*, seu *fC−CF*, ad 2*FG*', as in the *editio princeps*.

turam hujus lineæ curvæ ad punctum C, quartus So^3 determinabit variationem curvaturæ, Quintus variationem variationis & sic deinceps in infinitum. In Abscissa ad utramǧ partem punctorum B, D cape DE, $B[i]$ ipsi BD æquales, et erectis ordinatis $E[M]$ et $[il]$ et in serie pro BE et $B[i]$ scribendo $2 \times o$ et $-o$ prodibit $E[M] = P - 2Q[o] - 4Roo - 8So^3$ et $[il] = P + Qo - Roo + So^3$. Jungantur chordæ lG et $C[M]$ secantes ordinatas BC et DG in $[T]$ et $[V]$, et sagitta $C[T]$ æqualis erit excessui Ordinatæ BC supra $B[T]$ id est supra semisummam Ordinatarum $[il]$ et $DG_{[,]}$ id est quantitati Roo, et simili argumento sagitta $G[V]$ erit $Roo + 3So^3$. Tempora autem quibus corpus describit arcus lC, CG sunt in subduplicata ratione spatiorum quæ corpus his temporibus cadendo describere posset$_{[,]}$ & hæc spatia sunt ut hæ sagittæ.[20] Ideoǧ tempus describendi arcum lC est ad tempus describendi arcum CG vel [describendi] gC ut $\sqrt{Roo}$ ad $\sqrt{Roo + 3So^3}$ id est ut R ad $\sqrt{RR + 3RSo}$ seu R ad $R + \frac{3}{2}So$, et in hac ratione est arcus lC ad arcum gC. Et propterea arcus gC est lC in $\dfrac{R + \frac{3}{2}So}{R}$.

Et Resistentia est ad Gravitatem ut lC in $\dfrac{2R + 3So}{2R} - CG + FH$ ad FG.[21] Est autem $lC^q = [i]B^q + \overline{[il] - BC}|^q$ id est $oo + QQoo - 2QRo^3$ &c et extracta radice fit $lC = o\sqrt{1 + QQ} - \dfrac{QRoo}{\sqrt{1 + QQ}}$. Et simili argumento fit $CG = o\sqrt{1 + QQ} + \dfrac{QRoo}{\sqrt{1 + QQ}}$. Et inde lC in $\dfrac{2R + 3So}{2R} - CG$ fit $-\dfrac{2QRoo}{\sqrt{1 + QQ}} + \dfrac{3Soo}{2R} \times \sqrt{1 + QQ}$. Adde

$$FH = \frac{GI \times CF}{CG} = \frac{Qo \times Roo}{o\sqrt{1 + QQ}} \left[\text{seu } \frac{QRoo}{\sqrt{1 + QQ}} \right]$$

et fiet resistentia ad Gravitatem ut $\dfrac{3Soo\sqrt{1 + QQ}}{2R} - \dfrac{QRoo}{\sqrt{1 + QQ}}$ ad Roo $\Big[$ id est ut

$\dfrac{3S\sqrt{1 + QQ}}{2RR} - \dfrac{Q}{\sqrt{1 + QQ}}$ ad 1$\Big]$.[22]

---

(20) See §2.3: note (32) preceding.

(21) This should be '$\ldots - CG + 2FH$ ad $2FG$', on correcting Newton's parent measure (see note (19) above). His preparatory calculations for what follows survive on Add. 3968.41:132^v (immediately below an unused comment of his on the third paragraph of Leibniz' letter to Wallis of 28 May 1697 (N.S.) as printed in the *Commercium Epistolicum D. Johannis Collins et Aliorum* which he was soon to publish 'for' the Royal Society), where he successively computes: 'temp[us in lC]. temp [in gC] $:: R . \sqrt{RR + 3RSo} = R + \frac{3}{2}So$. [adeoǧ]

$$gC = lC \text{ in } 1 + \frac{3So}{2R} = o\sqrt{1 + QQ} - \frac{QRoo}{\sqrt{1 + QQ}} + \frac{3Soo\sqrt{\ }\ )}{2R}.$$

[ut et] $CG = o\sqrt{\ }\ ) + \dfrac{QRoo}{\sqrt{\ }\ )}$. [fit] $gC - CG = \dfrac{3Soo}{2R\sqrt{\ }\ )} - \dfrac{2QRoo}{\sqrt{\ }\ )}$. Add [e $FH = \dfrac{QRoo}{\sqrt{\ }\ )}$ et $\ldots$].'

(22) On adjusting the parent ratio to be 'ut $gC - CG + 2FH$ ad $2FG$' (see note (19) above), it

APPENDIX 4. AN ADDENDUM (POST-1713) ON THE CURVATURE OF CURVES.[1]

From the original drafts[2] in private possession and in the University Library, Cambridge

[1]   [3]Dixi supra quod si Lineæ Curvæ Abscissa augeatur momento o &

follows, *recte*, that 'fiet resistentia ad gravitatem ut $\dfrac{3Soo\sqrt{1+QQ}}{2R}$ ad $2Roo$, id est ut $3S\sqrt{1+QQ}$ ad $4RR$'. We must allow that Newton himself afterwards attained this final repair of his faulty argument in the *editio princeps*; for in a line and a half subsequently appended by him immediately after the preparatory calculation on Add. 3968.41: 132^v cited in the previous note he recomputed there to be (where now e, f and g fill the place of the previous coefficients Q, R and S):

$$'1 + \frac{3go}{f} \text{ in } o\sqrt{\ \ }) - \frac{efoo}{\sqrt{\ \ }} : -o\sqrt{1+ee} - \frac{efoo}{\sqrt{\ \ }} : +\frac{2foo, eo}{o\sqrt{\ \ }} = \frac{-2efoo}{+\sqrt{\ \ }} + \frac{3goo\sqrt{\ \ }}{f}.$$

[adeoɋ] resistentia ad gravitatem ut $\dfrac{3g\sqrt{1+ee}}{f}$ ad $2f$, seu $3g\sqrt{1+ee}$ ad $2ff$.' It was thereafter but the work of a moment to adjust the careless transliteration here of the previous ratio '$\dfrac{R+\frac{3}{2}So}{R}$' (of the times of passage over the arcs $\overset{\frown}{gC}$ and $\overset{\frown}{IC}$) now to be '$1+\dfrac{3go}{2f}$', and so mend this computation of $\overset{\frown}{gC}-\overset{\frown}{CG}+(2FH$ or$)$ $2CF \times GI/CG$ to yield '$\dfrac{3goo\sqrt{\ \ }}{2f}$'; whence, on dividing by $(2FG=)$ '$2foo$', the ratio of resistance to gravity results accurately to be 'ut $\dfrac{3g\sqrt{1+ee}}{2f}$ ad $2f$, seu $3g\sqrt{1+ee}$ ad $4ff$'. But in its uncorrected state Newton left off the calculation. Why indeed should he bother, in this stray scrap not intended to be communicated to posterity, to ink in a '2' in a denominator when he now knew it ought to be there!

(1) This intended addition to Proposition X, of Book 2 in the *Principia*'s second edition— 'ad pag. 240', Newton specifies in his two revised versions of it ([2] and [3] below), at a less than wholly suitable place in its scholium (see Appendix 1: note (37) preceding)—elaborates his somewhat elliptical observation in its 'Exempl. 1' (*Principia*, $_1$1687: 263–4 [= $_2$1713: 235–6]; compare Appendix 1: note (20)) that the successive coefficients in the Taylorian expansion of the incremented ordinate of a curve, in powers of the related base increment o, determine the curvature of a curve at a point and its several orders of variation. There are no surprises here for those familiar—as his contemporaries were not!—with Newton's previous explorations of the problem of constructing the osculating circle (see especially I: 245–64, 387–8, 419–27; III: 150–6; VI: 606–8) and that of assessing the variation in the measure of curvature which the reciprocal of its radius furnishes (see VII: 108–18), but it is interesting that he was at last, in old age, willing to let the world see the elegant general formula for the radius of curvature which he had written down half a century before in his (yet wholly unpublished) 1666 tract on fluxions, and here newly presents in [3], as ever without adducing any demonstration. Had Newton lived to put this *addendum* into print, his successors might have known something.

(2) Newton's handwriting in these strongly suggests a date of composition of around 1715– 16. It is tempting to connect this planned disclosure of his method of curvature with the brief glimpse of the content of his earliest papers on series and fluxions which he permitted the outside world—in witness that he 'invented' these methods 'in the Year 1665, and improved them in the Year 1666'—in the 'Observations' upon Leibniz' letter to Conti of 9 April 1716

Ordinata deinceps resolvatur in æquationem convergentem[4] qualis est hæc
$P + Qo + Roo + So^3 + To^{4\,(5)} +$ &c, terminus primus P designabit Ordinatam ejus
insistentem Abscissæ nondum auctæ,[6] secundus seriei terminus Qo determinabit
tangentem Curvæ, tertius Roo determinabit ejus Curvaturam, Quartus So^3
variationem Curvaturæ, Quintus To^4 variationem variationis. Et hinc patet me
fluxiones fluxionum olim considerasse cum variatio variationis nihil aliud sit
quam fluxio fluxionis Curvaturæ. Terminus secundus Qo determinat variationem seu fluxionem Ordinatæ, & hæc variatio dat positionem Tangentis.
Tertius determinat motum angularem Tangentis seu variationem aut fluxionem
hujus anguli ubi angulus fluit uniformiter, & hæc fluxio proportionalis est
angulo contactus ut et Curvaturæ hujus Lineæ Curvæ; Quartus determinat variationem seu fluxionem Curvaturæ ubi arcus
fluit uniformiter; Quintus determinat variationem variationis seu fluxionem fluxionis,
id est fluxionem secundam Curvaturæ, et
hæc fluxio proportionalis est fluxioni tertiæ
Anguli quem Tangens continet cum Ordinata, & sic deinceps.

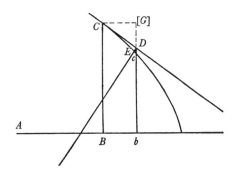

Sit AB Abscissa, BC Ordinata[7] & Bb
momentum Abscissæ. Compleatur [paral-
lelogrammum $CBbG$ ...][8]

---

[2][9]   *Pag 240*[10] *post* —vim gravitatis ut GT ad $\dfrac{2nn+2n}{n+2}\,GV.$ *adde*

Dixi supra quod si Curvæ cujusvis Abscissa momento o augeatur et prodiens
Ordinata resolvatur in Seriem,[5] puta $e \pm Qo - Roo[\pm]So^3$ &c, hujus seriei
terminus secundus Qo dabit Tangentem Curvæ, tertius Roo dabit curvaturam

---

(N.S.) which he inserted in his augmented edition of Joseph Raphson's *History of Fluxions*
(London, ₂1717 [both Latin and English versions]): 11–19 [=*Correspondence of Isaac Newton*, **6**,
1976: 341–8], especially 115–16. But this must remain our undocumented conjecture.

(3) Newton first began: 'In Scholio [read 'Exemplo 1'!] ad hujus Propositionem decimam,
dico quod si Curvæ cujusvis Abscissa AC momento uno augeatur & prodiens Ordinata resolvatur
in seriem convergentem, seriei hujus terminus secundus dabit Tangentem Curvæ, tertius
dabit ejus Curvaturam, quartus dabit variationem Curvaturæ, quintus variationem variationis, & sic deinceps'.

(4) This replaces 'Differentialem'!

(5) In the printed Proposition X Newton makes no explicit mention of this fifth term in the
series expansion of the incremented *ordinata*.

(6) This clause was subsequently deleted.

(7) Understand 'ad rectos angulos'.

(8) Newton breaks off to begin the redraft in [2], without completing his accompanying
figure.

(9) Add. 3965.13: 374ʳ, a first revision of [1].

ejus, quartus So^3 dabit variationem Curvaturæ, quintus variationem variationis & sic deinceps.

Sit *ACcE* Curva, *AB* Abscissa ejus, *Bb* momentum primum Abscissæ $= o = CG$. *BC* Ordinata $= bG$. *GD* momentum primum Ordinatæ[11] $= Qo$. *Dc* momentum secundum Ordinatæ $= Roo$ [&c]. Jungatur *CD* et hæc Curvam tanget in *C*. Ducatur *DE* perpendicularis ad Curvam et erit *DE* ad *Dc* seu *Roo* ut *CG* seu o ad *CD*. Producatur *DE* ad *Q* ut sit *DQ* ad *CD* ut *CD* ad *DE*[12] et erit *DQ* vel *CQ* radius Curvaturæ $=$

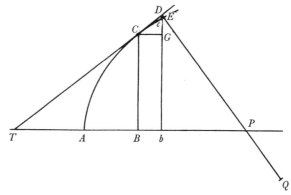

$$\frac{oo + QQoo}{Ro^3}\sqrt{oo + QQoo} = \frac{\overline{1 + QQ}|^{\frac{3}{2}}}{R}.\;^{(13)}$$

[3][14]   *pag. 240.*[10] *Post verba* —vim gravitatis ut *GT* ad $\dfrac{2nn + 2n}{n \times 2}$ *GV adde*

Dixi supra quod tertius seriei terminus *Roo* seu *NI* determinat angulum contactus *IHN* seu curvaturam quam Curva linea habet in *H*. Sit *HIK* Curva quævis cujus curvatura in puncto *H* inveniri debet,[15] et circulus ducatur qui Curvam tangat in puncto *H* et secet in puncto *I* ipsi infinite propinquam. Et hic

---

(10) That is, after the fourth paragraph in the augmented opening to the scholium in the *editio secunda* [= second paragraph in *Principia*,₁1687: 270]; again see Appendix 1: note (37).

(11) Here and in the next line we mend a careless 'Abscissa' in the manuscript. Again understand this ordinate *BC*(*bc*) to be at right angles to the abscissa *AB*(*b*).

(12) Since $CD^2 = DE \times 2DQ$, this should be '2*DE*'.

(13) Here, correspondingly, the denominators need to be doubled to be '2*Ro*³' and '2*R*' respectively. In the Fermatian notation used by Newton in his main proposition, where the (mutually perpendicular) abscissa $AB = a$ and ordinate $BC = e$ determine the general point $C(a, e)$ of the curve *AD*, the coefficients in the series expansion of the incremented ordinate $bc = e_{a \pm o} = e + Qo + Ro^2 \pm \dots$ are successively $Q = de/da$ and $R = \frac{1}{2}d^2e/da^2$, so that Newton's (corrected) formula $(1 + Q^2)^{\frac{3}{2}}/2R$ for the radius of curvature at *C* is merely the familiar Leibnizian one $(1 + (de/da)^2)^{\frac{3}{2}}/(d^2e/da^2)$ writ differently.

(14) Add. 3965.13: 377ʳ. Newton's unfinished final version of his intended *addendum* to Book 2, Proposition X of the *Principia* explaining how the third term Ro^2 in the Taylorian expansion of the incremented ordinate $e_{a \pm o}$ is a measure of the curvature at the point (a, e) of a curve defined (now in general oblique coordinates) by some given Cartesian equation $e = e_a$.

(15) This rightly replaces a too dogmatic initial 'invenienda est'.

circulus eandem habebit curvaturam quam Curva proposita habet in puncto *H*, et ex natura Circuli si *NI* producatur ad *R* ut sit *NR* ad *NH* ut *NH* ad *NI*, dabitur punctum *R* per quod circulus transibit, & inde dabitur circulus.[16]

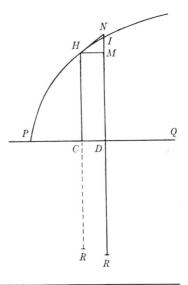

Hoc fundamento[17] sequentem Problematis constructionem olim condidi quod ex Tractatu sub finem anni 1666 conscripto[18] hic describam.

Sit Abscissa *PC* = *x* & Ordinata[19] *CH* = *y*, et æquationis cujuscunɋ qua habitudo ipsarum *x* et *y* ad invicem definitur termini omnes ex eodem æquationis latere dispositi & nihilo æquales facti significentur per literam ꭗ. Et symbolo ꭗ significentur ijdem termini in ordinem redacti secundum dimensiones ips[i]us *x* et per progressionem[20] quancunɋ multiplicati. Et symbolo ꭗ significentur ijdem termini in ordinem redacti secundum dimensiones ipsius *y* et per progressionem[20] quancunɋ multiplicati. Symbolo autem ꭗ significentur idem termini in ordinem redacti secundum dimensiones ips[i]us *x* et multiplicati per duas arithmeticas progressiones arithmeticas intervallo termini unius ab invicem differentes. Et symbolo ꭗ significentur ijdem termini in ordine dispositi secundum dimensiones ipsius *y*

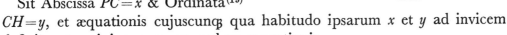

---

(16) Newton here summarizes the construction of the chord of curvature along an ordinate of a curve which he had expounded, in fluxional equivalent, in Proposition XIII of his 1691 treatise 'De quadratura Curvarum' (see vii: 108–10). He could here have passed likewise to specify that, if the coordinates $PC = x$ and $CH = y$ of the point $H(x, y)$ of the given curve are inclined at the angle $\widehat{PCH} = \cos^{-1}k$, then, where $CD = o$ and the decremented $DI = y_{x-o}$ has the series expansion $y - Qo + Ro^2 - \ldots(Q = dy/dx = \dot{y}, \ R = \frac{1}{2}d^2y/dx^2 = \frac{1}{2}\ddot{y})$, so that $CH - DN = Qo$ and $NI = Ro^2\ldots$, there is accordingly $NH = o\sqrt{(1 - 2kQ + Q^2)}$ and hence $NR = o^2(1 - 2kQ + Q^2)/(Ro^2\ldots)$; wherefore in the limit as ($o \to$ zero and so) NR comes to coincide with the chord of the osculating circle this is yielded to be $(1 - 2kQ + Q^2)/R$, that is, $(1 - 2k\dot{y} + \dot{y}^2)/\frac{1}{2}\ddot{y}$ in the fluxions equivalent. The particular case where the ordination angle $\widehat{PCH}$ is right (and so $k = 0$) is, *mutatis mutandis*, that of [2] preceding.

(17) This is only broadly true. Newton in historical fact attained his following theorem (initially in May 1665; see 1: 280–90) from first principles by constructing the centre of curvature at a given point of a curve—like Apollonius (*Conics*, v) and Huygens (compare 1: 271,

et multiplicati per duas progressiones arithmeticas intervallo termini unius ab invicem differentes. Symbolo autem ꝯꞓ significentur ijdem termini in ordinem redacti secundum dimensiones ipsius x et multiplicati per majorem progressionem duarum per quas ꝯꞓ multiplicata fuit, et subinde in ordinem redacti secundum dimensiones ipsius y et multiplicati per majorem duarum progressionum per [quas] ꝯꞓ multiplicata fuit. Et interea Progressiones omnes æquales habeant terminorum differentias & eodem terminorum ordine procedant respectu dimensionum ipsarum x et y. Et linea *HR* erit

$$\frac{ꝯꞓ\ ꝯꞓ\ ꝯꞓ\ yy + ꝯꞓ\ ꝯꞓ\ ꝯꞓ xx}{- ꝯꞓ\ ꝯꞓ\ ꝯꞓ\ y + 2\ ꝯꞓ\ ꝯꞓ\ ꝯꞓ\ y - ꝯꞓ\ ꝯꞓ\ ꝯꞓ\ y}. \qquad (21)$$

Ut si æquatio[22]

---

note (75)) before him did he then but know it—as the limit-meet of the normals to the curve in the point's immediate vicinity. The deeper insight and sophistication of his later researches into curvature were then still mostly in the egg.

(18) The October 1666 fluxional tract whose text is reproduced on I: 400–48. The following formula is that expounded 'in such cases where y is ordinately applyed to x at right angles' as the general construction (there undemonstrated) of the tract's 'Prob. 2^d. To find ye quantity of crookednesse of lines' (*ibid.*: 419–24, especially 421).

(19) Once more this needs to be restricted to be 'ad rectos angulos', as Newton specifies in his original account (see the previous note).

(20) In each case understand 'arithmeticam' (as on I: 421, whose English text Newton here otherwise accurately renders into Latin.

(21) This is 'Theoreme 1st' on I: 421, trivially re-keyed to the present figure. In modern partial-derivative notation, let us remind, where the curve (H) is defined with respect to abscissa $AB = x$ and ordinate (at right angles to it) $BC = y$ by the general algebraic equation $ꝯꞓ \equiv f(x,y) = 0$, Newton's ꝯꞓ, ꝯꞓ and ꝯꞓ, ꝯꞓ, ꝯꞓ are the homogenised first- and second-order derivatives xf_x, yf_y and x^2f_{xx}, xyf_{xy}, y^2f_{yy}; whence his formula constructs HR, the vertical projection of the radius of curvature, to be equal to $-(f_x^2+f_y^2)f_y/(f_x^2f_{yy}-2f_xf_yf_{xy}+f_y^2f_{xx})$.

(22) Newton breaks off without adducing any illustrative example(s) applying his preceding algorithm to a particular case. We might reasonably suppose that he would have gone on to work, *inter alia*, one or other species of the general conic determined by the Cartesian equation $y^2 = rx \pm (r/q)\,x^2$, where r is its *latus rectum* and q its main axis. (Compare I: 422–3; VII: 110–12.) It is perhaps not wholly irrelevant here to add that there survive on Add. 3965.19: 751^v contemporary calculations by him where he begins to employ the main construction of 'Prob. 2^d' of the 1666 tract (see I: 420) in the case of the ellipse '$rx-exx = yy$', passing to compute '$r\dot{x}-2ex\dot{x} = 2y\dot{y}$ [adeoꝗ] $\dot{x}.\dot{y}::y.\frac{1}{2}r-ex$' and thence the subnormal $y\dot{y}/\dot{x} = \frac{1}{2}r-ex$ and the related normal $y\sqrt{(1+\dot{y}^2/\dot{x}^2)}$, and their incremented forms as $x \to x+o$, before making a geometrical slip and leaving off.

7

NEWTON'S RESPONSE TO BERNOULLI'S SECOND CHALLENGE PROBLEM[1]

[c. February 1716][2]

From the original English drafts in the University Library, Cambridge

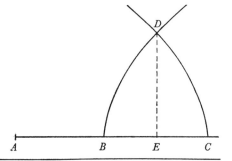

[1][3] If a given series of Curves of the same kind succeeding one another in an uniform manner, according to any general rule[,] one of w[ch] is *BD* are to be cut by another curve *CD* in any given angle right or oblique invariable or variable by any given rule: let *D* be any point of intersection & the two Rules will give the angle of inter-

---

(1) That of constructing a general orthogonal to a given parent sequence of curves: 'Construere curvam datas ordinatim positione curvas sive similes sive non similes in dato angulo sive invariabili sive data lege variabili secantem' as Johann Bernoulli had long before posed it privately to Leibniz on 14 August 1697 (N.S.) along with an outline of its resolution by points. (See C. I. Gerhardt, *Leibnizens Mathematische Schriften*, 3 (Halle, 1856): 464–5.) To Leibniz' request eighteen years later asking (not a little disingenuously) to be reminded of this 'applicatio generalis [novæ differentiandi rationis] qua, si bene memini, problemata quædam quæ ad differentio-differentiales descendere solent, intra differentiales primi gradus coërcentur' —'Inserviet enim fortasse ad aliquod problema proponendum [ad Anglos] cujus non statim apparebunt fontes' he added in this letter of 4 November 1715 (N.S.) (*Mathematische Schriften*, 3: 948)—Bernoulli promptly responded a fortnight later by copying out the pertinent excerpt from his old communication and sending it with the remark 'Si quid invenias, quod in usum Tuum faciet contra Keilium [whom Bernoulli chiefly blamed for fomenting the contemporary furore about calculus priority] aliosvis adversarios, gratissimum mihi erit' (*ibid.*: 949). And so it came that with his letter a few days afterwards on 6 December to the Venetian nobleman Conti, then on a visit to London, Leibniz included a mathematical note 'ut pulsum Anglorum Analystarum nonnihil tentemus' and to see 'quousque suis fluxionibus profecerint' as he reported back to Bernoulli (*ibid.*: 952). In the attenuated form, however, in which Leibniz in his 'Billet' rephrased Bernoulli's problem 'pour tâter un peu le pouls à nos Analystes Anglois' the would-be solver was challenged merely 'Trouver une ligne...*CD* qui coupe à angles droits toutes les courbes d'une suite déterminée d'un même genre; par exemple, toutes les Hyperboles...qui ont le même sommet & le même centre, & cela par une voye generale'. (See Pierre Desmaizeaux, *Recueil de Diverses Pieces sur la Philosophie, la Religion Naturelle, l'Histoire, les Mathématiques, &c. Par Msrs. Leibniz, Clarke, Newton, & autres Autheurs célèbres*, 2 (Amsterdam, [1]1720): 11; Leibniz' essentially identical draft of the 'Billet', still preserved among his manuscripts at Hanover, is printed by C. I. Gerhardt in his *Briefwechsel von G. W. Leibniz mit*

section, the perpendicular to the Curve desired & the center of its Curvity.[4]
And the perpendicular will give the first fluxion of the Ordinate & the curvity

---

Mathematikern, **1** (Berlin, 1899): 267.) As soon as Conti received Leibniz' letter (in the first days of January 1716 it would appear) he made haste to circulate its enclosed problem among the mathematicians at London. The example, however, which Leibniz appended to his rephrasing of Bernoulli's challenge the better (so he thought) to illustrate its terms—to draw, namely, a curve intersecting at right angles each of the family of hyperbolas of (as we may put it) Cartesian equation $y^2 = rx + (r/q) x^2$, having common main axis q and varying parameter r—misled the several 'English analysts' (John Machin, Henry Pemberton, Brook Taylor, but also the Scots Keill and James Stirling, and the French Huguenot expatriate Abraham de Moivre) who attempted its solution into believing it enough merely to take into consideration like classes of simple curves whose parameter is straightforwardly absorbable into their common defining differential equation (*viz.* $dy/dx = (x + \frac{1}{2}q) y/x(x+q)$ in Leibniz' hyperbolic instance) as an independent constant of integration; whence the family of orthogonal curves is immediately determinable in an analogous manner as having a perpendicular slope (*sc.* in Leibniz' example $-dx/dy = (x + \frac{1}{2}q) y/x(x+q)$, yielding in explicit solution the curves whose Cartesian equation is $y^2 = -x^2 - qx + \frac{1}{2}q^2 \log (x + \frac{1}{2}q) + s$, s free). Newton himself, privately forewarned by Conti as he may have been, was alone in detecting the true author behind Leibniz' pen, and in appreciating that Bernoulli had intended a sterner trial of the pulse of 'English' mathematics. Taking his lead from an article in *Acta Eruditorum* (October 1698): 466–73 where (*ibid.*: 470–1) Bernoulli had long before printed an extract from the end of Leibniz' letter to him of 16 December 1694 (N.S.) in which the latter pithily resolved the simple problem of orthogonals, applying it (see *Mathematische Schriften,* **3**: 157) to deduce that the ellipses $\frac{1}{2}y^2 = a^2 - x^2$ and parabolas $y^2 = bx$ mutually intersect at right angles, and had then gone on (see [3] below) to pose the general problem, Newton now proceeds to attempt to outline its mode of solution in a succession of English drafts whose polished Latin version (honed and recast by way of the preliminary castings reproduced in the Appendix) he published anonymously in *Philosophical Transactions,* **29,** No. 347 [for January–March 1716): §III: 399–400: 'Problematis Mathematicis Anglis nuper propositi Solutio Generalis'. Regrettably, despite all his best efforts he here succeeds only in establishing the existence of a solution (and misleadingly suggesting that one needs to consider the curvature of curves—and so the second fluxions of their ordinates—in attaining it), and he is forced in lieu of any more precise answer to conclude with a curt 'Et sic Problema semper deducetur ad æquationes'. Bernoulli rightly saw in those words a basic failure to come to grips with his problem, and so far failed to glimpse any lion's paw behind them that he guessed in his last letter to Leibniz—one to which the latter did not live to respond—that the anonymous author of this article which '[solutionem] ne apice quidem digiti attigit' was Brook Taylor! (See *Leibnizens Mathematische Schriften,* **3**: 972.)

(2) Since Newton answered Leibniz' long 'P.S.' to his letter to Conti on 6 December 1715 (N.S.) (which tartly criticized his general 'Philosophie de la Nature') by an equally lengthy riposte to Conti in late February following (see *Correspondence,* **6,** 1976: 285–8), we assume that his companion reply to its enclosed mathematical 'Billet' was composed around the same time. In any event all of the ensuing drafts must have been penned by him between the beginning of January 1716, when Conti first showed him Leibniz' letter, and his sending to the printer of the finished Latin 'Solutio' which appeared in public in the *Philosophical Transactions* some couple of months later.

(3) A cancelled preliminary passage on Add. 3968.25: 371ʳ in which Newton lays out his approach to a solution.

(4) The tangent to the curve $\overparen{BD}$ of the given family at any point D of it is manifestly the perpendicular there to the member, $\overparen{CD}$, of the orthogonal family which passes through that

will give the second fluxion.[5] And so the Probleme is reduced to equations involving first & second fluxions. Let the Ordinate be represented by the area of another Curve upon y^e same abscissa, & the first fluxion will be represented by the Ordinate of this other curve & the second fluxion by the proportion of the Ordinate to the subtangent. And so the Probleme is reduced to the property of the tangent of a curve.

---

[2][6]   [Suppose] $rr = xx + yy$. [Then] $0 = 2x\dot{x} + 2y\dot{y}$. [and therefore]
$$\dot{x}\dot{x} + x\ddot{x} + \dot{y}\dot{y} + y\ddot{y} = 0.$$

[Putting $\dot{x} = 1$. $\ddot{x} = 0$] $1 + \dot{y}\dot{y} + y\ddot{y} = 0$. [that is] $\dfrac{1 + \dot{y}\dot{y}}{y} = -\ddot{y}$.[7]

---

point, and so defines the latter when the law by which the 'series of Curves of the same kind' $\overset{\frown}{BD}$ 'succeed one another in an uniform manner' is brought into the argument (as Newton here does not attempt to do). But how he can claim that the centre of curvature of $\overset{\frown}{BD}$ at D—that is, the limit-meet of the perpendiculars to $\overset{\frown}{BD}$ at points on it in the immediate vicinity of D—is of itself sufficient to define the location of points on $\overset{\frown}{CD}$ in the vicinity of D, as he needs to do to obtain the second fluxion (derivative) of its ordinate, we cannot see. It seems to us, as to Johann Bernoulli (compare note (1) above), that Newton's present essay at resolving the general problem of orthogonals is not merely muddled and confused but, to be blunt, no beginning of any path to a proper solution at all.

(5) This point will be elaborated in the fragment reproduced in [2] following.

(6) Extracted from Add. 3968.41: 117^r, the second recto of a folded folio sheet which is otherwise crowded with a miscellany of preliminary drafts and jottings for a lengthy set of 'Observations' upon Leibniz' postscript to his letter to Conti on 6 December 1715 (N.S.) to which (see note (1)) Bernoulli's problem was appended, and also upon Leibniz' yet more forthrightly tart second letter to Conti on 9 April 1716 (N.S.) (see *Correspondence of Isaac Newton*, 6: 304–12). (The finished text of these 'Observations', first published by Newton himself in the Appendix (pages 97–123, especially 111–19) added to the 1717 reissue of Raphson's *History of Fluxions* in both English and Latin versions, has often since been reprinted, most recently in *Correspondence*, 6: 341–9.) We have already cited (see III: xv, note (15)) the substance of a stray sentence at the end of the present piece which reads in its full version: 'Upon account of my prowess in these [*sc.* mathematical] matters he'—Isaac Barrow it must be—'procured for me a fellowship in Trinity College in the year 1667 & the Mathematick Profesorship two years after'. As we read Newton's words, he would seem to imply (as we suggested in I: 10–11, note (26)) that Barrow had had no sight of his mathematical researches during his twin *annus mirabilis* of 1664–6 before he returned from his enforced sojourn in Lincolnshire in March 1667. Not till the summer of 1669, may we repeat, does there exist independent contemporary evidence of any degree of intimacy between the two.

(7) Whence '$\dfrac{1 + \dot{y}\dot{y}}{\ddot{y}} = -y = \dfrac{x}{\dot{y}} = \dfrac{\sqrt{yy + xx}}{\sqrt{1 + \dot{y}\dot{y}}}$ or $\dfrac{r}{\sqrt{1 + \dot{y}\dot{y}}}$. That is $r = \dfrac{\overline{1 + \dot{y}\dot{y}}|^{\frac{3}{2}}}{\ddot{y}}$'. It will be evident that this intrinsic fluxional measure of the radius r of a circle, of centre $(0, 0)$, which is defined with respect to perpendicular Cartesian coordinates x, y also evaluates that of the circle $x^2 + y^2 + ax + by + c = 0$, of arbitrary centre $(-\frac{1}{2}a, -\frac{1}{2}b)$, which is set in general position in the plane; and is in consequence the magnitude of the radius of 'curvity' at the point (x, y) on any given Cartesian curve $f(x, y) = 0$ which the latter circle there osculates. And on this understood basis Newton formulates the rule for measuring curvature which he proceeds to expound in the ensuing 'Schol[ium]'.

Schol.

As the Ordinate is to the subperpendicular[8] (or subtangent to the Ordinate) so is the fluxion of the Absciss to the fluxion of the Ordinate. [9]Let the summ of the squares of the fluxions of the Absciss & ordinate be applied to a quantity wch is to the Ordinate as ye Radius of Curvity to the Perpendicular, & the Latus will be the second fluxion of ye Ordinate.[10] When these fluxions make the Ordinate decrease they must be taken with negative signes. How to reduce the equations wch result from hence[11] & to separate the unknown quantities is the business not of this but another method.

---

[3][12]   In the *Acta Eruditorum* for October 1698 pag. 471 Mr John Bernoulli wrote in this manner.[13] *Methodum quam optaveram generalem secandi (Curvas) ordinatim positione datas sive algebraicas sive transcendentes, in angulo recto sive obliquo, invariabili sive data lege variabili, tandem ex voto erui, cui Leibnitio approbatore, ne γρύ addi posset ad ulteriorem perfectionem, et vel ideo tantum quod perpetuo ad æquationem deducat, in qua si interdum indeterminatæ sunt inseparabiles, methodus non ideo imperfectior est, non enim hujus sed alius est methodi indeterminatas separare. Rogamus itaqʒ fratrem[14] ut velit suas quoqʒ vires exercere in re tanti momenti. Suscepti laboris non penitebit si felix successus fructu jucundo compensaverit. Scio relicturum suum quem nunc fovet modum qui in paucissimis tantum exemplis adhiberi potest.[15]* These Gentlemen

---

(8) That is, the subnormal ($y\dot{y}/\dot{x}$ in Newton's preceding terms, where x is the ordinate and y the abscissa at the point on the curve in question).

(9) Newton first went mistakenly on to write here: 'And if you say, as the Perpendicular [$y\sqrt{1+\dot{y}^2}$] to the Ordinate so the Radius of Curvity to a fourth proportional [*viz.* $r/\sqrt{1+\dot{y}^2}$] & to this 4th Proportional apply a quarter [!] of the summ of the squares of the fluxions of the Abscissa & Ordinate, you will have the second fluxion of the Ordinate'.

(10) In terms of his preceding analytical working (compare note (7) above) Newton's cumbrously expressed rule is essentially that we must divide $1+\dot{y}^2$ by $r/\sqrt{(1+\dot{y}^2)}$ to have $\ddot{y}$.

(11) In the case of Newton's would-be 'solution' of Bernoulli's problem in [1], namely, where he rashly—and mistakenly—assumes that an easy reduction to finding the 'center of Curvity' of the curve to be constructed is possible (compare note (4) above).

(12) Add. 3964.6: 5^r/6^r, the most finished English draft of Newton's response to Bernoulli's problem of determining orthogonals. A preliminary version, now Add. 3968.25: 371^r/371^v, is reproduced in *The Correspondence of Isaac Newton*, 6, 290–2; we cite its significant variants from the present revise at pertinent places in following footnotes.

(13) In draft (see the previous note) Newton first wrote to the same effect that 'Mr John Bernoulli in the *Acta Eruditorum* for October 1698 pag. 471 wrote thus' before passing to make like quotation of the following passage from Bernoulli's 'Annotata in Solutiones Fraternas Problematum quorumdam suorum, editas proximo *Actorum* Maio' (*Acta* (1698): 466–73 [= *Johannis Bernoulli...Opera Omnia*, 2 (Lausanne/Geneva, 1742): 262–71, especially 268–9]).

(14) Johann's brother Jakob, that is.

(15) 'The general method I had desired for cutting curves given in sequence and position, be they algebraic or transcendental, at an angle right or oblique, fixed or variable according to a given law, I have at length hunted out in redemption of my pledge; to it, Leibniz assents, not a jot were able to be added to its further perfection, and that simply because it leads ever

had been four or five years about Problemes of this kind,[16] & to give the very
same solution with that of Bernoulli here mentioned might require a spirit of

---

on to an equation. If on occasion the indeterminates in this are inseparable, the method is not
on that ground less than perfect; for it is the burden not of this method but of another one to
separate the indeterminates. We accordingly ask our brother [Jakob] that he be minded to
exert his energies also in a matter of such moment. The toil he undertakes will not be painful
if there be the compensation of a happy outcome with its pleasant fruit. I know that he will
abandon his presently cherished mode (of solution), which can be applied in but a very few
instances.'

(16) Newton's basis for this statement is manifestly the immediately preceding passage in
Bernoulli's article where (*Acta* (October 1698): 469–70 [= *Opera*, 1: 266–8]) he for the first
time publicly introduces the general problem of determining 'lineas, quas vocabo Trajectorias,
quæ alias ordinatim datas normaliter, vel in angulo quovis dato, seu etiam data lege variante,
trajiciunt'. Noticing that 'Diu adeo est, quod hanc materiam seposuerim; quam, ni fallor,
primus in lucem protraxi' Bernoulli there went on to quote an extract from his private letter
'jam quatuor abhinc annis' to Leibniz on 2 September 1694 (N.S.) where he had originally
posed the 'Problema non minus elegans quam utile' of 'Datis...infinitis curvis positione,
invenire curvam quæ omnes ad angulos rectos secat' and also made the fruitful
suggestion that 'hoc Problema insignem usum præstat in determinanda curvatura radiorum
lucis per medium inæqualiter densum transeuntium, juxta hypothesin Hugenii; si quidem
radius nihil aliud sit quam linea undulationes ad angulos rectos secans' (see also *Leibnizens
Mathematische Schriften*, 3: 151–2). In his reply on 16 December following—the pertinent
extract of which Bernoulli also cited in his *Acta* article—Leibniz readily dealt (as we have
remarked in note (1) above) with the simple problem of orthogonals where the parameter of
the given family of curves is absorbable, as an independent constant of integration, into their
common defining differential equation. But as Newton does not here seem to know—how else
could he have written his concluding dismissal of the problem of orthogonals as 'being of little
or no use' and 'remained in the *Acta Eruditorum* many years without a resolution'?—Bernoulli
came a little more than two years afterwards to promote the Fermatian optical analogy yet
further at the end of his article on the 'Curvatura radii in diaphanis non uniformibus' (*Acta*
(May 1697): 206–11 [= *Opera*, 1: 187–93]) where he elegantly derived the *curva brachystochrona*
as the cycloidal least-time path of a light-ray whose speed at each instant is that of the corresponding gravitational fall from rest (see 1, Appendix 1: note (12) preceding). In his last two
paragraphs he there (*ibid.*: 210) put forward the related problem of determining the simple
curva synchrona, that to which bodies descending in least time along their several cycloidal
brachistochrones from rest at a fixed point would arrive after a given interval, adding that
'si attente considerentur ea, quæ supra diximus de radio luminis' it will be clear that this
curve is a Huygenian 'wave'—'eam ipsam...quam *Hugenius* in suo *tract. de Lumine* [*Traité de
la Lumière* (Leyden, 1690)] pag. 44 in schemate suo per lineam *BC* repræsentat, vocatque
undam, quæ...omnes radios ex puncto luminoso emanantes normaliter secat', the problem
of whose construction 'in pure Geometricum reductum' is '*invenire*...*curvam, quæ omnes
Cycloides communis initii normaliter secat*'. The 'very easy' one by points which Johann proceeded
to set down, along with the broader challenge 'Si quis methodum suam in aliis exercere velit,
quærat lineam, quæ ordinatim positione datas curvas (non quidem Algebraicas, quod haud
arduum foret, sed) transcendentes, ex. gr. Logarithmicas super communi axe, & per idem
punctum ductas, ad angulos rectos secat', remained publicly undemonstrated, it is true, for
the next twenty years till he himself made the problem of determining the cycloidal synchrone
a particular case of a yet more difficult *défi* which, at Leibniz' further instigation, he was to
contrive in March 1716 as a further test of the prowess of 'English' mathematicians (see note
(25) below). In the privacy of their personal correspondence over the ensuing months, how-

divination. It may suffice that the following method of solving it is a general one.[17]

The Probleme.

Any[18] *Series of Curves being given of one & the same kind, succeeding one another in form & position in an uniform manner according to any general Rule: to find another Curve or series of Curves w*[ch] *shall cut all the Curves in the first*[19] *Series, in any angle right or blique, invariable or variable according to any Rule assigned.*

The Method of Solution.[20]

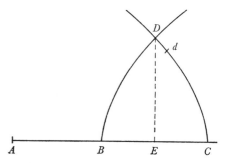

Let *BD* be any one of the Curves in the first Series, *CD* the Curve in the other Series w[ch] is to cut it, *D* the point of intersection, *AE* the common Abscissa of the two Curves, & *ED* their common Ordinate: and the first Rule will give the tangent of the Curve *BD* at the point *D*, & the second Rule will give the angle of intersection & tangent of the other Curve *CD* at the same point *D*. And by the same method the tangent at another point *d*[21] may be found. Let the points

---

ever, Jakob Bernoulli joined his brother Johann and Leibniz in still more deeply exploring the topic of orthogonal curves and the related one of general *curvæ synchronæ*; in his letter to Leibniz on 7 June 1697 (N.S.), in particular, Johann was able at once to answer his brother's query as to what is the path of descent in least time from a given point to a given vertical line by straightaway replying that in this case the vertical must touch the brachistochrone's synchrone, so that the line of swiftest descent is here a semi-cycloid terminating at the vertical (see *Leibnizens Mathematische Schriften,* **3**: 415–16). Nearly twenty years after such a scurry of behind-scenes activity, it is clear that Newton yet knew nothing of its achievements beyond the brief glimpses of it then published in the Leipzig *Acta* and elsewhere—and of all these he had probably only ever read the article by Johann Bernoulli from which he here quotes.

(17) In his draft (f. 371ʳ) Newton wrote slightly differently: 'But the Probleme may be generally solved after the following manner'.

(18) Originally just '*A*' as in the draft.

(19) 'said' is replaced.

(20) Newton slightly amplifies the outline roughed out in [1], but without in any real sense making good its basic *non-sequitur* (on which see the next note).

(21) On the *same* 'other Curve *CD*' understand. But Newton does not specify how this crucial requirement is to be met, and in default his argument remains circular: for (compare [1]: note (4) above) until the curvature of the intersecting curve *CD* at the point *D* is known, its meet in *d* with the 'next' curve of the parent series cannot directly be determined, and thus *a fortiori* neither can there be ascertained (from the tangent to this parent curve through *d*) the direction of the perpendicular at *d* to $\widehat{Dd}$ whose limit-intersection with the perpendicular thereto at *D* yields the centre of its 'curvity'. What the solution of Bernoulli's problem requires, as Newton here fails wholly to state, is that some intrinsic defining differential property of the given parent series of curves shall be determined—be it a tangential one (as in the simple problem of orthogonals, when no recourse to the curvature of the curves is necessary) or one also involving higher-order differences (as in Bernoulli's subsequent problem of March 1716, on

D, d be infinitely near to one another; & the perpendiculars erected upon those tangents at the points *D, d* will convene & intersect at the centre of the curvity wch the curve *CD* hath at the point *D*.[22] Let the Abscissa *AE* flow uniformly, & its fluxion be called 1, & the position of the Tangent of the Curve *CD* will give the first fluxion of its Ordinate *ED*, & the curvity of the same Curve *CD* at the point *D* will give the second fluxion of the same Ordinate. And so the Probleme will be reduced to Equations involving the first & second fluxions of the Ordinate of the Curve desired.[23]

There may be some art in chusing the positions of the Abscissa & Ordinate or other fluents to wch the invention of the Curve *CD* may be best referred & in reducing the Equation & separating the fluent[s] after ye best manner. But nothing more is here desired than a general method of reducing the Probleme to equations.[24] This Probleme being of little or no use hath remained in the *Acta Eruditorum* many years without a resolution, & for the same reason I forbeare to prosecute it further.[25]

---

which see note (25) below)—from which an analogous intrinsic common property of the intersecting curves may, by some specified general mode of solution, be constructed in its turn.

(22) Newton originally (as in his draft) terminated the previous sentence but one with the added phrase '& both the Rules together will give the Radius of Curvity of the other Curve *CD* at the same point *D*', adding his clarification of this abrupt remark in a separate concluding paragraph at the end stating: 'The curvity of the intersecting Curve *CD* at the point *D* is found by taking in the tangent of that Curve *CD* another point *d* infinitely neare to the first point *D*, & finding at this other point *d* the tangent of the Curve in the series which passes through this other point *d* according to the first Rule, & also the tangent of the Curve intersecting it at the same point *d* according to the second Rule; & upon the two intersecting Curves at ye points of intersection *D* & *d* erecting perpendiculars. For these perpendiculars shall intersect one another at the center of the Curvity of the Curve *CD*'. (Compare *Correspondence*, **6**: 292.)

(23) In sequel Newton has deleted '& by reducing the Equations & extracting or separating the fluents (which is not the business of this Method ['Probleme' was first written in the draft]) will be resolved'; its gist is subsumed, of course, in the final clause (in the manuscript a late interlineation manifestly salvaging its content) of the opening sentence of the next paragraph. We may add that his first drafting of this 'Method of Solution' on Add. 3968.25: 371^r affirmed a deal more pithily that 'the two Rules will give the perpendicular to the Curve *CD* & the radius of its curvity at the point *D*, & the position of the perpendicular will give the first fluxion & the curvity the second fluxion of the ordinate of the same curve *CD*. And so the Probleme will be reduced to equations involving fluxions[,] & by separating or extracting the fluents will be resolved'.

(24) Newton initially wrote more vaguely 'of resolving the Probleme', adding thereafter in his draft 'without entring into particular cases'.

(25) This last sentence is not found in the draft on Add. 3968.25: 371^r/371^v. There, however, he initially proceeded to adjoin:

'Mr Leibnits in the *Acta Eruditorum* for May 1700 pag 204, challenged Mr Fatio to solve this Probleme, *Invenire Curvam aut saltem proprietatem tangentium Curvæ quæ Curvas etiam transcendentes ordinatim datas secet ad angulos rectos.* Let *ED* the Ordinate of the Curve *CD* be represented by the area of another Curve upon the same Abscissa AE & the first fluxion of this Ordinate will be represented by the Ordinate of the other Curve & the second fluxion by the proportion of this last Ordinate to ye subtangent of the other curve. And so the property of the Tangent of the other Curve is given.

'Let the Ordinate of the Curve desired be represented by the area of another curve upon the same Abscissa & the first fluxion will be represented by the Ordinate of this other Curve, & the second fluxion by the proportion of the Ordinate to the subtangent. And so the Probleme is reduced to the property of a Tangent' [initially '...to the property of the Tangent of the other Curve & to the Quadrature thereof'].

As Leibniz specified in his 'Responsio ad Dn. Nic. Fatii Duillerii imputationes' (*Acta Eruditorum* (May 1700): 198–214 [= *Mathematische Schriften*, 5 (Halle, 1858): 340–9])—this in its main portion a reply to Fatio de Duillier's complaint that he had not, in January 1697, been judged a worthy recipient of Johann Bernoulli's *Programma* (1, Appendix 1 above) challenging the sharpest minds in the world at large to determine the *curva brachystochrona*, even though he had reduced the condition of fall in least time to a property of the radius of curvature at its general point (much as Newton himself in his private working; see 1, Appendix 2: note (12) preceding) which defined this to be a cycloid—his challenge there to Fatio had been to show his mathematical mettle by attacking the 'problema a Domino Joh. Bernoullio propositum in *Actis Erud. Lips.* Maji 1697, p. 211' where (see note (16) above) Bernoulli had, in sequel to his 'optical' derivation of the defining differential equation of the brachistochrone, posed the question of ascertaining the *curva synchrona* which intersects at right angles all such cycloidal paths of fall from rest at a fixed point. As before, this variant outline by Newton of a would-be general solution to Bernoulli's problem is impotent to approach the cases, such as the latter's instance of the family of cycloids of common vertex and base, where the parameter regulating the sequence of the parent curves cannot be absorbed into their common defining differential equation as an independent constant ensuing through its integration, whence this may (on converting its slope to the perpendicular one) straightforwardly yield the related tangential condition for the family of orthogonal curves. Had he been encouraged by Leibniz' back-reference to go to the page in the *Acta Eruditorum* for May 1697 where Bernoulli set down his undemonstrated construction of the cycloidal synchrone by points, we cannot believe that anything but his inability to fathom its *raison d'être* would have prevented Newton (even in his 74th year) from further 'prosecuting' its subtleties.

In the event, even as Newton's lightly recast Latin version of the preceding 'Method of Solution' of Bernoulli's problem of orthogonals was (by way of the preliminary draftings reproduced in the following Appendix) in course of being set anonymously in print in the *Philosophical Transactions* in March 1716, the latter was himself preparing at Leibniz' suggestion a new challenge which incorporated its author's old question of constructing the cycloidal synchrone as but one particular instance, and so was set unknowingly to sabotage Newton's best effort at a general solution of the problem even before this was delivered to the world at large. Already made aware by Conti that the hyperbolic illustration in his prior 'défi aux Analystes Anglois' to construct orthogonal curves (see note (1)) was widely taken in London to embrace the full scope of what was required to be solved, Leibniz on the previous 31 January had applied to Bernoulli to supply a new example 'quod non particulari aliqua facilitate adjuvari putes, sed ad generalem adigere..., ne dicant ne a nobis quidem sufficientem solutionem dari posse' (*Leibnizens Mathematische Schriften*, 3: 956). Johann responded six weeks later on 11 March with a blockbuster in two parts (*ibid.*: 958):

'*Problema 1°*. Super recta AG tanquam axe ex puncto A construere infinitas curvas, qualis est ABD, ejus naturæ, ut radii osculi ex singulis singularum curvarum punctis B educti, secentur ab axe AG in C in data ratione, ut nempe sit $BO:BC = 1:n$. *2°*. Construendæ sunt trajectoriæ, qualis est $E[B]F$, priores curvas ABD ad angulos rectos secantes'.

To Leibniz he provided the full 'solutio', but only in the form of an undemonstrated construction. The more forcefully to underline the truth of his judgement on Newton's published

attempt at solution of the general problem—
that 'Anglus anonymus...ne apice quidem
digiti attigit (*ibid.*: 972)—we may sketch the
subtleties of the preparatory analysis which
Bernoulli himself suppressed. Take the point
$B(x, y)$ to be determined by the perpendicular
coordinates $AI = x$, $IB = y$, and put the radius
of curvature BO at this point on the parent
curve ABD (of arc-length $\widehat{AB} = s$) to be ρ. From
the condition $BC = n.BO$ which defines this
family of curves there straightforwardly ensues

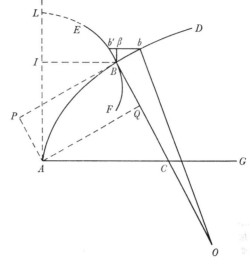

$$-x\,ds/dy = (n\rho \text{ or }) \; n(ds/dy)^3/(d^2x/dy^2)$$
$$= n(ds/dy)^3/(ds/dx)\,(d^2s/dy^2)$$

and so $(d^2s/dy^2)/(ds/dy = -(n/x)\,dx/dy$, whence
$ds/dy = a^n/x^n$; and consequently

$$dy/dx = x^n/\sqrt{(a^{2n}-x^{2n})}, \; a \text{ free,}$$

is the defining differential equation of the sequence of parent curves ABD (all of which, as will
be evident, intersect the base AG at their common vertex A). Let the member, LEF, of the
orthogonal family which passes through B have the related arc-length $\widehat{LB} = s'$, and to the
common increment $B\beta = dx$ of the abscissa AI of parent and orthogonal curves let there
respectively correspond in these the increments of ordinate IB and arc-length $\beta b = dy$,
$\beta b' = dy'$; and $Bb = ds$, $Bb' = ds'$. The proportion $B\beta:\beta b = b'\beta:B\beta$ yields at once the prime
condition $dx/dy = -dy'/dx$ for the curves AB and LB to cut one another at right angles at their
common point $B(x, y = y')$. With respect to the former draw the instantaneous tangential
and normal polars $AP = p = y\,dx/ds - x\,dy/ds$ and $AQ = q = \sqrt{(x^2+y^2-p^2)} = y\,dy/ds+x\,dx/ds$:
since by considering infinitesimals it is readily proved that $dp/ds = q/\rho$, the common second-
order differential defining condition for the parent curves ABD is therefore $-dp/dy = nq/x$.
And because in the curves LBF the tangential and normal polars are $AQ = P = q$ and
$PA = Q = -p$, while the contemporaneous increments of these are Bb and Bb', so that
$dp/dP = (ds/ds' \text{ or}) - dy/dx$, there is in consequence $dP/dx = nP/x$ and thence $P = x^n/b^{n-1}$,
where b varies with individual orthogonal curves. For $n = 1$, when $q = P = x$ and so
$p = -Q = -y$, there ensue twin families of orthogonal circles $x^2 = y(2a-y)$ and $y^2 = x(2b-x)$
of respective radii $(BC =) \; a$ and b. In other cases the Cartesian equation

$$y = \int x^n/\sqrt{(a^{2n}-x^{2n})}\,.\,dx$$

of the parent family yields there to be

$$y + (\sqrt{(a^{2n}-x^{2n})}/x^{n-1} \text{ or}) \; x\,dx/dy = -(n-1)\int a^{2n}/x^n\sqrt{(a^{2n}-x^{2n})}\,.\,dx,$$

so determining that in the orthogonal family the area

$$(n-1)\int a^{2n}/x^n\sqrt{(a^{2n}-x^{2n})}\,.\,dx = (-y+x\,dy'/dx \text{ or}) - Q\,ds'/dx = p\,ds/dy$$

shall be a^n/b^{n-1}, constant for the pair of parameters a, b which define the curves AD and EF
mutually perpendicular at B: whence (as was Bernoulli's solution in March 1716) the latter
curve may in principle always be constructed, given the former. In the instance $n = \frac{1}{2}$, where
the parent curve is the cycloid $y = \frac{1}{2}a \cos^{-1}(1-2x/a) - \sqrt{(x(a-x))}$ generated (as will be
geometrically obvious) by rolling a circle of diameter a along the base AG, because its normal
polar $AQ = \frac{1}{2}\sqrt{(ax)} \cos^{-1}(1-2x/a)$ is the tangential polar $\sqrt{(bx)}$ of the orthogonal synchrone,

it follows, on keeping b fixed, that if in each cycloid (as a varies) we set out along the generating circle the arc-length $(\frac{1}{2}a \cos^{-1}(1-2x/a) =)\sqrt(ab)$ and through its end-point draw a parallel to the base, this will meet the cycloid in the point where it meets the synchrone defined by the particular parameter b. And this is the elegant construction of it by points which Bernoulli had stated without demonstration nineteen years earlier in *Acta Eruditorum* (May 1697): 211.

To conclude what is already an overlong 'fuss' note, Leibniz communicated Bernoulli's new challenge virtually word for word to Conti on 14 April 1716 (N.S.) with the provocative taunt that his 'nouveaux Amis' in England might like to try their hand at solving this more difficult case of the problem of orthogonals where the curves are transcendental 'en attendant qu'ils trouvent le moyen de parvenir à la solution génerale [du] Probleme dont quelques-uns parmi eux ont voulu resoudre des cas particuliers pour en fixer, disent-ils, les idées' and noting that 'Ce Probleme n'est point nouveau. M. *Jean Bernoulli* l'a deja proposé dans le mois de *May* des Actes de *Leipsic* 1697, *p.* 211. Et comme M. *Fatio* méprisoit ce que nous avions fait; on en repeta la proposition pour lui & pour ses semblables, dans les Actes de *May* 1700. *p.* 204. Il peut servir encore aujourd'hui à fair[e] connoitre à quelques-uns, s'ils sont allez aussi avant que nous en Methodes...'. (We cite the text of the letter as Newton himself, anonymously as always, set it in print in *Philosophical Transactions*, **30**, No. 359 [for January/February 1718/19]: 925–7, especially 926; but its content was not widely circulated till Desmaizeaux reprinted it in his *Recueil de Diverses Pieces...* **2** (Amsterdam, ₁1720): 26–8.) Newton probably knew of the new challenge-problem within a couple of weeks after; he certainly was familiar with the minimally altered phrasing of it which Bernoulli put to Montmort on 8 April 1717 (N.S.) since he has added its enunciation in his own hand to the secretary copy of this letter which Montmort straightaway passed on to him (see *Correspondence*, **6**: 384). But his only public reaction was to add the feeble, unsigned comment to Leibniz' letter as it appeared in the *Philosophical Transactions* early in 1719 (see above) that 'M[r] Leibnitz being told that his Probleme [put to Conti in December 1715] was solved, he changed it into a new one of finding both the series to be cut & the other series w[ch] is to cut it.... And the first part of this double Probleme (viz[t] by any given property of a series of Curves to find the Curves) is...harder[!] then the former & of w[ch] a general solution is not yet given' (*ibid.*: 927–8; we quote, however, Newton's autograph on ULC. Add. 3967.4: 38[v], now printed in full—if not quite accurately—in *Correspondence*, **6**: 323, note (1)). Such a 'general solution', however, had already (had Newton but known it) been published a year and a half earlier by Jakob Hermann in his 'Schediasma de Trajectoriis datæ Seriei Curvis ad angulos rectos occurrentibus: continens solutionem generalem Problematis in *Actis Erud.* 1698 p. 471 primum propositi...' (*Acta Eruditorum* (August 1717): 348–52; see also his 'Supplementum' in *Acta* (July 1718): 335–6, and further 'Additamentum' in *Acta* (February 1719): 69–77); and Johann Bernoulli had thence been stimulated to publish his own *gemina constructio* of the 'double Probleme', exactly as he had privately passed it to Leibniz more than two years before, in the interim report 'De Trajectoriis curvas positione datas ad Angulos rectos vel alia data lege secantibus' which his son Niklaus II published in June 1718 (*Acta*: 248–62; see especially §9: 252–3). Niklaus in turn was to print his own working-up of his father's and Hermann's essays soon after as a 'Tentamen solutionis generalis Problematis de construenda Curva, quæ alias ordinatim positione datas ad angulos rectos secat' (*Acta* (June 1719): 295–304). It was left to Brook Taylor to salvage England's pride in an elegant (if somewhat jejunely argued) equivalent fluxional solution *per series infinitas* of the 'Problema à Dom[no] *G. G. Leibnitio* Geometris *Anglis* nuper propositum' which appeared in *Philosophical Transactions*, **30**, No. 354 [for October–December 1717]: 695–701. (All these papers may be found conveniently reprinted together in Johann Bernoulli's *Opera Omnia*, **2** (Lausanne/Geneva, 1742): 275–314.)

APPENDIX. THE 'GENERAL SOLUTION' OF BERNOULLI'S PROBLEM TURNED BY NEWTON INTO LATIN.[1]

From a miscellany of autograph drafts in the University Library, Cambridge, and in private possession

[1][2] *Curvam Lineam ducere quæ Curvas omnes unius cujuscunꝗ generis continua serie secundum legem quamcunꝗ assignatam dispositas ad angulos rectos secabit.*

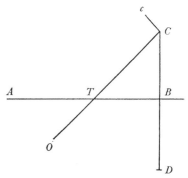

Sit *C* punctum aliquod in Curva ducenda, *CB* Ordinata hujus Curvæ et *AB* Abscissa ejus. Et ex natura Curvarum secandarum habebitur Tangens Curvæ illius secandæ quæ transit per punctum il!ud *C*. Sit ea *CO* secans Abscissam *AB* in *T*. Sit *c* punctum aliud in Curva ducenda, et ex natura Curvarum secandarum habebitur Tangens Curvæ illius quæ transit per punctum illud *c*. Sit ea *cO* occurrens Tangenti priori in *O*. Et ubi linea *Cc* diminuitur in infinitum angulus *OCc* evadet rectus & punctum *O* evadet centrum curvaminis arcus infinite parvi *Cc*.[3]

---

(1) Since these Latin reshapings and honings by Newton of his English originals, reproduced in the preceding pages, add little of significance to their ineffective attempt at a solution of the problem of determining orthogonal curves—and that serving further to cloud his myopic insight into the core of the problem—we restrict our ensuing editorial commentary in the main to noticing minor points of verbal divergence between them. With trivial further changes the final text attained by him in [5]/[6] below was, by way of a last, lightly amended secretary copy of its phrasing (now Add. 3968.25:369ʳ), published anonymously by him as 'Problematis Mathematicis *Anglis* nuper propositi Solutio Generalis' in *Philosophical Transactions*, **29**, No. 347 [for January–March 1716]: §III: 399–400.

(2) ULC. Add. 3960.11:197ʳ, a Latin elaboration of the preliminary solution in [1] above, where Newton tries—and fails—to put firm flesh on the loose bones of its would-be reduction to 'equations involving first & second fluxions'.

(3) True in itself. But, until the curvature of the *Curva ducenda Cc* at *C* is known, we cannot determine the precise location of the indefinitely close point *c*, and so *a fortiori* cannot establish which member of the parent family of *Curvæ secandæ* it is which passes through this latter point, so fixing the slope of its tangent there which coincides with the normal *cO* to the *Curva ducenda* whose limit-intersection *O* with *CO* defines the centre *O* of 'curvity' there. Whence again (compare [3]: note (21) preceding) Newton's argument is irretrievably circular. In sequel he passes to construct the normal at the point *C(x, y)* on the *Curva secanda* determined by some given equation relating (by way of some independent parameter, it is understood) its perpendicular Cartesian coordinates *AB* and *BC*.

Dicatur $AB=x$. $BC=y$. $BT=t=BD$[4] et erit $y.t::\dot{x}=1.\dot{y}$. Ergo $y\dot{y}=t$ & $\frac{1}{2}yy=\boxed{t}$.[5] *Et* $y=\sqrt{2\boxed{t}}$.

Ad Tangentem CO erigatur perpendiculum Cc et ab aliquo ejus puncto c[6]

[2][7]   [Sit] $AB=x$. $BC=y$. $BP=s$.

       $CP=p$. $BT=q$. $CT=t$,

rad[ius] curv[itatis] $=r$.

   [Erit] $q.y::y.s::\dot{x}.\dot{y}$.

[ut et] $q.t::y.p::\dot{x}\dot{x}+\dot{y}\dot{y}.r\ddot{y}$.[8]

[Hoc est] $y.p::\dfrac{pp\dot{x}\dot{x}}{yy}.r\ddot{y}$. [sive] $\dfrac{p^3\dot{x}\dot{x}}{y^3}=r\ddot{y}$.

[itacʒ] $y^3.p^3::\dot{x}\dot{x}.r\ddot{y}$.[9]

Scholium.

Ut Ordinata ad sub-perpendicularem (vel sub-tangens ad Ordinatam) ita est fluxio Abscissæ ad fluxionem Ordinatæ. Applicetur summa quadratorum fluxionum Ordinatæ et Abscissæ ad lineam quæ sit ad Ordinatam ut Radius curvitatis ad Perpendicularem, & Latus erit fluxio secunda Ordinatæ.[10] Hæ

---

(4) Why Newton should want to extend the ordinate CB below the base AB to a distance BD equal to the subtangent BT of the *Curva secanda* is mysterious. In sequel he assumes that— as in Leibniz' instance of the family of hyperbolas of fixed axis and vertex which he set to illustrate the general problem of determining orthogonal curves in posing it to Conti on 6 December 1715 (see [1]: note (1) preceding)—the *Curvæ secandæ* are such as to possess a common subtangent BT which does not alter with their varying parameter, and he makes no further appeal to BD or its end-point D.

(5) As ever with Newton, understand this 'quadratum ipsius t' to be the 'square of t' ($\int t.dx$ in the Leibnizian equivalent). In the case of Leibniz' illustrative instance in posing Bernoulli's challenge to Conti (see the previous note) where (so we may most simply put them to be) the parent hyperbolas—of common axis q and vertex $(0, 0)$ but with variable parameter (*latus rectum*) r—are defined by the equation $y^2 = rx+(r/q)\,x^2$, and so have subtangents $t(= y\,dx/dy) = y^2/r(\frac{1}{2}+x/q) = x(q+x)/(\frac{1}{2}q+x)$, free of r, the orthogonal *curvæ ducendæ* are consequently defined by the condition $y^2 = [-]\,2\int t.dx = -x^2-qx+\frac{1}{2}q^2\log(x+\frac{1}{2}q)+s$, s free.

(6) More generally, where the subtangent BT of the parent *Curva secanda* is not independent of the varying parameter, Newton begins to recast his preceding 'Solutio generalis' by now supposing the indefinitely near point c of the *Curva ducenda* at right angles to this at C to lie upon the perpendicular to the parent curve there. In breaking off in mid-sentence he has evidently come to appreciate the falseness of his presupposition (which dictates that the curvature of the *Curva ducenda* at C shall be zero, unless he was now prepared to go on to consider this orthogonal curve as the limit of a polygon possessing an infinity of rectilinear arc-elements such as Cc).

(7) To a stray calculation on a sheet in private possession we adjoin the bottom part of the page (ULC. Add. 3968.41: 117^r) in which Newton proceeds to cast his original English scholium to his solution of Bernoulli's problem (see 7.2 above) into Latin equivalent.

(8) Read '$y.t::s.p::\dot{x}\dot{x}+\dot{y}\dot{y}.r\ddot{y}$', as Newton will below correct himself.

(9) We take this last proportion from Add. 3968.41: 117^r, whose Latin portion now follows.

fluxiones ubi Ordinatam diminuunt, negativæ ponendæ sunt. (11) Non hujus sed alius est methodi æquationes prodeuntes reducere & indeterminatas separare. Problema hocce cum nullius fere sit usus[,] in *Actis eruditorum* annos plures neglectum et insolutum mansit. Et eadem de causa solutionem ejus non ulterius prosequar.

[*s*] . *p* :: $\dot{x}\dot{x} + \dot{y}\dot{y}$. *rÿ* [recte. Quare scribendum est]

Scholium.

Fluat Abscissa uniformiter & erit Ordinata ad subperpendicularem (vel subtangens ad Ordinatam) ut fluxio Abscissæ ad fluxionem Ordinatæ, et ut subtangens ad Tangentem ita summa quadratorum fluxionum Abscissæ et Ordinatæ ad rectangulum sub radio curvitatis et fluxione secunda Abscissæ.(12) Hæ fluxiones [ubi Ordinatam diminuunt] &c.

---

[3](13)   In *Actis Eruditorum* pro mense Octobri Anni 1698 pag. 471 D. Johannes Bernoulli hæc scripsit. *Methodum quam optaveram generalem secandi [curvas] ordinatim positione datas sive algebraicas sive transcendentes, in angulo recto sive obliquo, invariabili sive data lege variabili, tandem ex voto erui, cui, Leibnitio approbatore, ne γρύ addi posset ad ulteriorem perfectionem, et vel ideo tantum quod perpetuò ad æquationem deducat, in qua si interdum indeterminatæ sunt inseparabiles, methodus non imperfectior est, non enim hujus sed alius est methodi indeterminatas separare. Rogamus itaꝗ fratrem ut velit suas quoꝗ vires exercere in re tanti momenti. Suscepti laboris non penitebit si felix successus fructu jucundo compensaverit. Scio relicturum suum quem nunc fovet modum qui in paucissimis tantum exemplis adhiberi potest.* Hi tres viri celeberrimi jam ab annis quatuor vel quinꝗ circiter sese in solvendis hujusmodi Problematîs exercuerant. Absꝗ spiritu divinandi eādem solutionē cum Bernoulliana tradere difficile fuerit.(14) Sufficit quod solutio sequens sit generalis.

---

(10) Newton here follows his erroneous preceding proportion 'y . p :: $\dot{x}\dot{x} + \dot{y}\dot{y}$. *rÿ*' of course.

(11) A superseded first version of what follows reads little differently in immediate sequel: 'Quomodo æquationes sic prodeuntes reducendæ sunt & quantitates indeterminatæ separandæ non est hujus sed alius methodi'.

(12) A further carelessness on Newton's part. This should of course read 'Ordinatæ'.

(13) This first Latin rendering of Newton's generalized statement of Bernoulli's general problem of orthogonals, and of his 'Method of Solution' of it in the English original (ULC. Add. 3964.6: 5r/6r) reproduced in 7.3 preceding, was penned by him on a loose sheet now in private possession. Some preparatory tentatives at a Latin casting which he scribbled on the back of the English original (*ibid.*: 6v) are cited in following footnotes.

(14) Newton's initial phrasing here asserted that 'Absꝗ spiritu divinandi haud eadem solutio cum Bernoulliana traderetur'.

Problema.

Data quacunꝗ linearum Curvarum[15] *ejusdem generis serie se mutuo quoad formam et positionem secundum Legem*[16] *quamcunꝗ generalem succedentium: invenire aliam Curvarum seriem quæ Curvas priores in angulo*[16] *recto vel obliquo, invariabili vel data Lege variabili secabit.*

Methodus Solutionis.

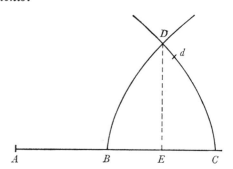

Sit *BD* curva quævis in prima serie, *CD* curva in altera serie quæ Curvam *BD* secare debet, *D* punctum intersectionis, *AE* communis Abscissa Curvarum & *ED* communis Ordinata: Jam [17]Lex prima dabit tangentem Curvæ *BD* ad punctum *D*, & Lex secunda dabit angulum intersectionis & tangentem Curvæ alterius ad idem punctum *D*. Et eadem methodo tangens Curvæ cujusvis in secunda serie ad aliud quodvis punctum *d* inveniri potest. Sit intervallum punctorum *D*, *d* infinite parvum, et perpendicula ad tangentes e punctis *D*, *d* erecta concurrent et se mutuo intersecabunt in centro curvaturæ quam Curva *CD* habet ad punctum *D*. Fluat Abscissa *AE* uniformiter & fluxio ejus vocetur 1; et positio tangentis Curvæ *CD* dabit fluxionem primam Ordinatæ *ED*, et curvatura ejusdem Curvæ *CD* ad punctum *D* dabit secundam fluxionem ejusdem Ordinatæ. Et sic Problema deducitur ad æquationes involventes fluxionem primam et secundam Ordinatæ ad Curvam quæsitam *CD* pertinentes. Quod erat faciendum.

Ars aliqua requiritur ad eligendam positionem aptissimam Abscissæ ad reducendas æquationes, et ad superandas quantitates indeterminatas sed ea non

---

(15) An '*unius et*'—rendering Newton's '*one &*' at this point in his English original—was initially here inserted in the preparatory Latin casting on Add. 3964. 6: 6ᵛ.

(16) In the preliminary Latin enunciation of Bernoulli's *Problema* (see the previous note) Newton here first wrote '*Regulam*' and '*in angulo quocunꝗ*' respectively.

(17) In a freer preliminary Latin version of the sequel Newton initially stated on the verso of his English original (see note (13) above) that 'Lex prima dat tangentes Curvarum in serie prima ad intersectionum puncta quæcunꝗ & Lex secunda ex angulis intersectionum dat perpendicula ad Curvas in secunda serie e punctis intersectionum erecta, & concursus ultimus seu intersectio ulti[ima] duorum perpendiculorum coeuntium est centrum curvaturæ quam Curvæ coeuntes habent in puncto intersectionis. Ducatur Abscissa in situ quocunꝗ commodo fluatꝗ eadem uniformiter et pro ejus fluxione ponatur unitas: Et positio perpendiculi ex coitu prodeuntis dabit fluxionem primam Ordinatæ ad Curvam quæsitam ex coitu prodeuntem₍,₎ et Curvatura Curvæ ejusdem dabit fluxionem secundam Ordinatæ ejusdē. Et sic Problema deducetur ad æquationes involventes fluxionem primam et secundam Ordinatæ, quod erat faciendum'.

sunt hujus methodi.[18] Hoc Problema, cum nullius fere sit usus, in *Actis Erudi-torum* annos plures neglectum et insolutum mansit. Et eadem de causa solutionem ejus non ulterius prosequor.

---

[4][19]   *Scholium.*

Sub-tangens Curvæ quæsitæ est ad Ordinatam ejus (vel Ordinata ad sub-perpendicularem) ut fluxio Abscissæ ad fluxionem Ordinatæ. Et Subtangens est ad Tangentem (vel Ordinata ad Perpendicularem) ut summa quadratorum fluxionum Abscissæ uniformiter fluentis & Ordinatæ ad rectangulum sub radio curvitatis & fluxione secunda Abscissæ.[20] Vel (quod perinde [est]) cubus Ordinatæ est ad cubū perpendicularis, ut quadratum fluxionis Ordinatæ ad rectangulum sub radio curvitatis & fluxione secunda Abscissæ.[20] Hæ fluxiones ubi Ordinatam diminuunt negativæ ponendæ sunt.

Non hujus sed alius est methodi æquationes prodeuntes reducere & in series convergentes ubi opus est convertere.[21] Nam solutio Problematis generalior evadet per Newtoni Analysin[22] universalem quæ ex methodis fluxionum & serierum intertextis conflatur.[22] Problema hocce cum nullius fere sit usus[,] in *Actis Eruditorum* annos plures neglectum et insolutum mansit, et eadem de causa resolutionem ejus non ulterius prosequor.

---

[5][23]   In *Actis Eruditorum* pro mense Octobri Anni 1698, pag. 471, D.

---

(18) In his preparatory Latin draft (in immediate sequel to what we quoted in the previous note, without break of paragraph) Newton phrased this sentence to affirm a deal more curtly that 'Reductio utiꝗ æquationum et separatio indeterminatarum non est hujus sed alterius methodi'.

(19) From a loose sheet in private possession. Newton amplifies the scholium ([2] above) where he adduces the standard measure of the radius of curvature at a point of a curve defined by mutually perpendicular abscissa and ordinate—once again (see the next note) mis-stating the proportions which relate its length to the first and second fluxions of the ordinate with respect to the base—and now incorporates with it a light restyling of the last paragraph of [3] preceding.

(20) Read 'Ordinatæ' in each case. Newton thoughtlessly perpetuates his earlier careless-ness in [2] (see note (12) thereto).

(21) Initially '& ad series infinitas ubi opus est deducere'.

(22) These replace the alternative phrasings 'methodum' and 'componitur' respectively. Newton first of all here wrote just 'Nam resolutio Problematis per series generalior evadet, et plerumꝗ facilior'.

(23) From a second loose sheet in private possession. With the lightly recast terminal scholium in [6] following, this final Latin version was copied by a secretary onto ULC. Add. 3968.25: 369r/369v (the last two paragraphs of this are reproduced in *The Correspondence of Isaac Newton*, **6**, 1976: 292); and thence, with the further addition of a short clarifying phrase to the scholium, passed unchanged into print in the *Philosophical Transactions* in March 1716 (see note (1) above).

Johannes Bernoulli hæc scripsit. *Methodum quam optaveram generalem secandi* [*Curvas*] *ordinatim positione datas . . . . in paucissimis tantum exemplis adhiberi potest.*[24]

Hi tres viri celeberrimi sese jam ab annis quatuor vel quincp circiter in solvendi hujusmodi Problematis exercuerant. Abscp spiritu divinandi eandem solutionem cum Bernoulliana tradere difficile foret. Sufficit quod solutio sequens sit generalis et ad æquationē semper deducat.

[25]*Problema.*

Quæritur methodus generalis inveniendi seriem Curvarum quæ Curvas in serie alia quacuncp data constitutas, ad angulum vel datum vel data lege variabilem secabunt.[26]

Solutio.

Natura Curvarum secandarum[27] dat tangentes earundem ad intersectionum puncta quæcuncp,[28] & anguli intersectionum dant perpendicula Curvarum secantium, et perpendicula duo coeuntia[29] per concursum suum ultimum, dant centrum curvaminis Curvæ secantis ad punctum intersectionis cujuscuncp. Ducatur Abscissa in situ quocuncp commodo, & sit ejus fluxio unitas, & positio perpendiculi dabit fluxionem primam Ordinatæ ad Curvam quæsitam pertinentis, et curvamen[30] hujus Curvæ dabit fluxionem secundam ejusdem Ordinatæ.[31] Et sic Problema semper deducetur ad æquationes. Quod erat faciendum.

---

(24) We omit the bulk of Newton's quotation of Bernoulli, whose full text is exactly as printed in [3] preceding.

(25) The first rough outline and more finished draft of the following recast enunciation of Bernoulli's 'Problema' and Newton's would-be 'Solutio' of it exist on the back (ULC. Add. 3964.6: 6ᵛ) of the English original reproduced in 7.3 preceding, below the preliminary attempts at its Latin transliteration to which we drew attention in [3]: note (13) above. As ever, we cite significant phrases from these preparatory draftings in ensuing footnotes.

(26) In his preliminary draft Newton concluded his enunciation—to better heuristic effect —merely by writing '. . . *secabunt ad angulos rectos*', and answering Bernoulli's full challenge in a separate paragraph at the end of the following 'Solutio' in which he declared not a little bravely that 'Si angulus intersectionis sit obliquus & vel constans vel quacuncp lege variabili, resolutio Problematis erit nihilo difficilior'.

(27) This replaces an equivalent 'in prima serie' in the first rough draft.

(28) Understand 'assumpta', as Newton was careful to specify *mutatis mutandis* in his preparatory draft on Add. 3964.6: 6ᵛ.

(29) Originally 'concurrentia' in both preliminary drafts (see note (25)), but so altered in the latter.

(30) Initially 'Curvatura' in the preparatory draft.

(31) A superfluous 'cujus terminus locatur in Curva' was added and then quickly deleted in the first draft.

Scholium.

Reductio æquationum et separatio indeterminatarum non est hujus sed alterius methodi. Problema hocce cum nullius fere sit usus, in *Actis Eruditorum* annos plures neglectum et insolutum mansit. Et eadem de causa solutionem ejus non ulterius prosequor.

---

[6](32)   *Scholium.*

Non hujus sed alius est methodi æquationes reducere & indeterminatas separare.(33) Problema hocce cum nullius fere sit usus, in *Actis Eruditorum* annos plures neglectum et insolutum mansit. Et eadem de causa solutionem ejus non ulterius prosequor.

---

(32) This recasting of the preceding scholium is found at the foot of the draft version (now in private possession) whose text is reproduced in [4] above.

(33) In the secretary press copy (ULC. Add. 3968.25: 369^v; see note (23) above) Newton has cancelled these last three words, setting in their place (much as in [4]) 'et in series convergentes ubi opus est convertere. Nam solutio Problematis per Newtoni Analysin universalem evadet generalior'. In the final version printed, however, Newton chose to reinstate this phrase '& indeterminatas separare', there following it with the qualification 'absolutè si fieri possit, sin minus per Series infinitas' (*Philosophical Transactions*, **29**: 400).

8

ANALYSIS AND SYNTHESIS: NEWTON'S DECLARATION ON THE MANNER OF THEIR APPLICATION IN THE 'PRINCIPIA'[1]

[late 1710's]

From original drafts,[2] now in private possession and in the University Library, Cambridge, of his preface to an intended re-edition

[1][3] Analysis Veterum Geometrarum in deductione consequentiarum ex

---

(1) In the decade from 1712 during which the squabble over priority of invention in calculus fumed and spluttered, Newton a number of times excused himself for not making much apparent use of his joint methods of fluxions and infinite series in his master-work, the *Principia*, by countering that he had indeed made wide use of their techniques of analysis in finding out many—'most' he grandiosely claimed in late 1714 (in the review of his *Commercium Epistolicum D. Johannis Collins et Aliorum de Analysi promota* which he published anonymously in *Philosophical Transactions*, **29**, No. 342 [for January/February 1714/15]: 173–224 [≈ **2**, Appendix 2 below]; see its page 206)—of the propositions which he set out therein; 'but', he added (*ibid.*), 'because the Ancients for making things certain admitted nothing into Geometry before it was demonstrated synthetically, he demonstrated the Propositions synthetically, that the Systeme of the Heavens might be founded upon good Geometry. And this makes it now difficult for unskilful Men to see the Analysis by which these Propositions were found out'. Elsewhere, he went on: 'But if any man who understands Analysis, will reduce the Demonstrations of the Propositions from their composition back into Analysis (wch is easy[!] to be done) he will see by what method of Analysis the Proposition was invented' (ULC. Add. 3968.9: 105^r, the first page of a contemporary draft preface to an intended re-issue of the *editio secunda* of the *Principia* which is printed in full in **2**, Appendix 6 below). When he came, however, to substantiate these assertions, he must have been dejected to see in just how relatively few of the *Principia*'s 'composed' propositions—if indeed any at all!—he could himself trace any variant preliminary analysis in the Grecian style. And when in the successive versions of a new Latin preface to a planned re-issue of the *Principia*'s *editio secunda* in or shortly after 1716—stimulated thereto, it may be, by the appearance in that year of Jakob Hermann's *Phoronomia* (see the next note)—he tried to bolster his case why he had not made greater use there of his fluxional and series methods, he was put willy-nilly to laying emphasis on the handful of lemmas, theorems and problems where (so he claimed) he had left his 'preceding' analysis without recasting its argument into equivalent synthetically demonstrated form. For their intrinsic interest and possible autobiographical truth we here reproduce pertinent

Translation

[1][3]   The analysis of the ancient geometers seems to have consisted in the

---

excerpts from these hitherto unpublished draftings of his intended *Principia* preface, amending slips in their detail in our ensuing editorial commentary upon their text. We would at the same time point the reader to our own evaluation of 'The Mathematical Principles underlying Newton's *Principia Mathematica*' (*Journal for the History of Astronomy*, 1, 1970: 116–38) where we press our renegade judgement, based on examination of the surviving corpus of his working papers and calculation sheets, that 'The published state of the *Principia*...is *exactly* that in which it was written'. The reader will make up his (or her) own mind.

(2) Newton's handwriting in the manuscripts is all but identical with that in his dated correspondence of the period 1716–18 (see for instance the drafts of his letter to Conti on 26 February 1716 now in ULC. Add. 3968.29/38). A firm *ante quem non* for their composition is fixed by the circumstance that the preliminary draft of their most finished version, that reproduced in [4] below, is penned on a sheet which he initially used to jot down an annotation on Jakob Hermann's *Phoronomia, sive de Viribus et Motibus Corporum Solidorum et Fluidorum Libri Duo*, which was published at Amsterdam in the first days of 1716: this dismissively records, in reaction to a compendium of more than 400 pages in which Hermann conscientiously sought to present in Leibnizian form the essence of recent investigations by Huygens, the Bernoulli brothers Jakob and Johann, Varignon and others (not least himself) into the theory of free, constrained and resisted dynamical motion and its application to the motion of solid and fluid bodies, the—not entirely exact—numerical count that 'Hermannus omisit [in *Princip. Math.*] Sectionem totam primam [seu paginas] 10[,] quartam [et] quintam 46[,] decimam undecimam duodecimam, decimam tertiam [et] decimam quartam 88 Libri primi [hoc est paginas] 144. ac tertiam & quartam 16, et Scholium generale sext[æ] et septimam 45 et nonam 11 secundi [id est pag.] 72. & Librum totum tertium 129. Totum omissum [pag.] 345 [ubi] impressum [4]84'. (In correction the page-totals are, for the second (1713) edition of the *Principia* here considered, too high in the case of Book 1, Sections 4/5 and 10–14 by 8 and 10 pages respectively, while Newton more excusably overstates the size of Book 2, Sections 3/4 by but a single one.) It is interesting that Leibniz, who speedily gave the *Phoronomia* a lengthy and commendatory review in the *Acta Eruditorum* (January 1716): 1–10 in which he drew equal attention to the prior contributions there catalogued of Johann Bernoulli 'qui mirifice hæc studia provexit' and the 'multa etiam, quæ a famigeratissimo Newtono reperta sunt' (for the firm attribution to him of this anonymous recension see Émile Ravier, *Bibliographie des Œuvres de Leibniz* (Paris, 1935): 107), privately commented to Bernoulli two months earlier on 4 November 1715 (N.S.) that 'Multa sunt bona, sed...Videtur nimium Anglis deferre' (see C. I. Gerhardt, *Leibnizens Mathematische Schriften*, 3 (Halle, 1856): 948). How soon after publication Newton acquired the copy of Hermann's book which was in his library at his death we do not know, but we may assume that only a couple of months or so intervened. When he came to read it, one passage in its preface where Hermann spoke of his guiding aims in writing the book must have caught Newton's attention: 'Perspicuitati, quantum potui, litavi; propterea multæ demonstrationes provectioribus Geometris nimis prolixæ fortasse videbuntur.... Interim diffiteri nolo me etiam, quantum potui, elegantiam sectatum esse, atque propterea demonstrationes lineares algebraicis prætulisse, experientia multiplici edoctum, meditationem figurarum sæpissime simpliciores & elegantiores suppeditare solutiones ac constructiones, quam Analysin speciosam. Speciosam dico, nam subinde utor analysi geometrica seu lineari absque symbolis algebraicis procedente, cujus beneficio multa elegantius obtinentur quam calculis analyticis, etsi non semper. Ejusmodi analysi geometrica veteres usos existimo, quemadmodum ex *Euclidis datis* & *Apollonii* libello *de Sectione Rationis* a Celeberrimo

datis donec quæsitum prodiret[4] constitisse videtur. In hunc finem Euclides libros *Datorum* et *Porismatum* et *Locorum Planorum* et Apollonius libros *de Sectione Rationis* & *Sectione Spatij* scripserunt.[5] Ubi vero Quæsitum ex datis non facile consequeretur, vel quærebant Lemmata aut P[o]rismata[6] per quæ Datum aliquod novum colligere possent vel assumebant ignota tanquam data ut inde datum aliquod tanquam ignotum colligerent,[7] ac tandem ex relatione quacunꝗ inter data et quæsitum invertendo ordinem argumentationis quæsitum deducerent.

Analysis recentiorum ex Arithmetica originem habuit, et nihil aliud est quam Arithmetica Universalis ad quantitates seu Geometricas seu alias quascunꝗ applicata. Qui utuntur hac Analysi perraro Propositiones suas componunt.

Ubi Propositio aliqua per Analysin quamcunꝗ prodijt,[8] hæc non statim in Geometriam admittenda est sed demonstratio ejus prius componi debet. Nam Geometriæ vis & laus omnis ab ejus certitudine, certitudo autem a Demonstrationibus syntheticis dependet. Ea de causa Veteres non solebant Analyticas Propositionum suarum inventiones in lucem dare sed inventas semper demo[n]strabant synthetice & hujusmodi demonstrationibus munitas edebant. Eadem de causa et nos Propositiones per Analysin nostram inventas demonstravimus synthetice[9] ut in Geometriam admitterentur & Philosophia[10] Cœlorum quæ in Libro tertio habetur fundaretur in Propositionibus Geometrice demonstratis. [11]Solutionem generalem Problematis XX[X]I[12] Libri primi &

Edmundo Hallæo edito [Oxford, 1706; see VII: 187, note (13)] non obscure colligitur; simili etiam analysi summus *Newtonus* ad stuporem usque usus est in suis *Principiis'* (*Ad Benevolum Lectorem*: [viii]). If it did so, there in itself was stimulus enough for Newton to expand at length —in preface to a planned re-edition of his own work provoked (as we might suppose) by the appearance in print of Hermann's bulky 'Leibnizian *Principia*'—on this theme so close to his own heart, stressing his own preferred viewpoint upon it. But 'tis all pure speculation.

(3) From the second verso (revised) of a loose folded sheet, now in private possession, on which Newton afterwards also penned (on its two middle pages) a preliminary draft of [4] below. On its first verso a superseded initial version of the two opening paragraphs of what here follows reads more shortly: 'Veteres in resolutione Problematum primo colligebant ex datis quicquid occurreret, & in hunc finem Euclides Libros *Datorum* et *Porismatum* & *Contactuum*[!] et *Locorum Planorum* et Apollonius libros *de sectione rationis* & *sectione spatij* &c scripserunt. Si hac methodo quæsitum colligere potuerunt resolvebatur Problema, sin minus assumebant quæsitum tanquam datum ut inde datum aliquod tanquam quæsitum colligerent et ex [originally 'nexu'] relatione inter datum et quæsitum deducerent quæsitum regrediendo. Recentiores relationem illam per computationem Arithmeticam ad quantitates geometricas applicatam colligere didicerunt' (The ancients in the resolution of problems used first to gather from the givens whatever might come to ensue, and to this end Euclid wrote his books of *Data, Porisms, Contacts*[!] and *Plane Loci*, and Apollonius his books *On cutting off a ratio* and *On cutting off a space*, and so on. If by this method they were able to collect what was sought, the problem was resolved; but if not, they used to assume what was sought as though it were a given in order that they might thence gather some given as though it were sought, and so from the (connection) relationship between given and sought deduce the sought by going

deduction of consequences from givens until the thing sought should result.[4]. To this end Euclid wrote his books of *Data*, *Porisms* and *Plane loci*, and Apollonius his books *On cutting off a ratio* and *On cutting off a space*.[5] Where, however, the thing sought would not easily ensue from the givens, they either looked for lemmas or porisms[6] through which some new given might be gatherable, or assumed unknowns as givens so that thereby they might gather some given as though it were unknown,[7] and at length by inverting the sequence of argument deduce the thing sought from whatever relationship between the givens and the sought.

The analysis of those more recent has taken its origin from arithmetic, and is nothing else than universal arithmetic applied to quantities, be they geometrical ones or others whatever. Those who employ this analysis very rarely compose their propositions.

When some proposition has resulted[8] through any analysis whatever, this is not to be admitted into geometry at once, but rather its demonstration ought first to be composed. For all the force and commendation of geometry hangs upon its absolute conviction, and that conviction upon its synthetical demonstrations. On that ground the ancients were not wont to put on public view the analytical contrivances of their propositions, but, once they had found them out, used ever to demonstrate them synthetically, and to set them forth fortified with demonstrations of this kind. For the same reason we, too, synthetically[9] demonstrated the propositions found out through our analysis so that they might be admitted into geometry, and that the 'philosophy'[10] of the heavens which is delivered in the third book be founded on propositions geometrically demonstrated. [11] The general solution of Problem XXXI[12] of the first book and that

---

back. Those more recent have learnt how to collect that relationship by means of an arithmetical computation applied to geometrical quantities).

(4) Merely, it is clear, to evade an ugly verbal repetition, this replaces 'consequeretur' (should be a consequence).

(5) On these 'ancient' Greek works of higher geometrical analysis—only the first and fourth of which have survived into modern times—see our editorial remarks on VII: 186–91, in introduction to the papers (*ibid.*: 200–351) in which Newton in the early 1690's summarized (and, where necessary, reconstructed) their contents and interpreted their individual purposes.

(6) Initially just 'Propositiones' (propositions).

(7) 'ut quod inde consequeretur viderent' (so that they see what might ensue) was first written.

(8) So changed from 'inventa fuerit' (shall have been found). Newton originally began his paragraph by presupposing somewhat less sharply 'Ubi Problema aliquod resolutum fuerit' (When some problem shall have been resolved).

(9) This replaces an equivalent 'per compositionem' (through composition).

(10) The 'Philosophia naturalis' (physical science) promised in the *Principia*'s title, that is. Newton first wrote a little more restrictively—but more accurately?—'Theoria' (theory).

(11) In a preceding cancelled passage Newton first went on to affirm 'Et hinc fit ut Analysis qua Propositiones illæ inventæ fuerunt perraro appareat. Ejus tamen extant vestigia quædam

Problematis III[12] Lib. II exhibuimus analytice ut et Problematis in scholio Prop. XCIII Lib. I propositi.[13] Propositiones reliquas synthetice demonstratas reliquimus[14] sed harum Demonstrationes syntheticæ æque ac inventiones analyticæ ex methodo fluxionum & momentorum derivatæ sunt.[15] Methodus rationum primarum et ultimarum quæ sub initio libri primi in Lemmatibus XI exhibetur, nihil aliud est quam pars methodi fluxionum et momentorum synthetice demonstrata, & præmittuntur hæc Lemmata ut eorum beneficio Propositiones sequentes per methodum analyticam fluxionum et momentorum inventæ synthetice demonstrari possent.[16] Elementa methodi fluxionum & momentorum demonstrantur synthetice in Lem. II Lib. II, et Propositiones

passim in compositione Demonstrationum' (And hence it comes to be that the analysis whereby those propositions were found out is very rarely apparent. Certain traces of it, however, survive *passim* in the composition of the demonstrations) before passing to add, much as in the revised sequel, that 'Sed et in Propositionibus nonnullis Analysis integra describitur: ut Prop. XLV & Schol. Prop. XCIII [Lib. I] & Lem. II & Prop. X Lib. II' (But also in several propositions the entire analysis is written down: as in Prop. XLV and the scholium to Prop. XCIII of Book 1, and in Lemma II and Prop. X of Book 2).

(12) That is, 'Prop. XLV' and 'Prop. X' respectively, as Newton equivalently specified these *Problemata* in his cancelled draft (see the previous note).

(13) In Prop. XLV, Prob. XXXI of the *Principia*'s first book Newton had (see VI: 368–76) employed the opening terms of the expansion of the binomial $(T-X)^n$ 'in seriem indeterminatam per methodum nostram serierum convergentium' (*ibid.*: 372) in solving 'Arithmeticè' the problem of determining the motion of the apsides in disturbed, inverse-square elliptical orbits 'qui sunt circulis maxime finitimi' (see *ibid.*: 368). In the scholium to his following Prop. XCIII (*Principia*, ₁1687: 225–6 = ₂1713: 202) he had outlined a contracted solution of the problem of defining the fall of a body in a straight line under an arbitrary 'lex attractionis' by making a similar reduction of the incremented power $(A+o)^{m/n}$ into a like 'converging' series of terms in rising powers of o: in a loose catalogue 'De methodo fluxionum' which Newton inserted around this time in his interleaved personal copy of the second edition (now ULC. Adv. b. 39.2), there gathering three other instances also of such fluxional analyses in the *Principia* (see notes (16), (18) and (44) below), Newton here correspondingly pronounced of this 'Schol. pag. 202' that 'Methodus solvendi Problemata per series & momenta conjunctim exponitur' (see A. Koyré and I. B. Cohen, *Isaac Newton's 'Philosophiæ Naturalis Principia Mathematica': the third edition (1726) with variant readings* (Cambridge, 1972): 793–4, where the full list is printed). On the manifold intricacies, finally, of Prop. X, Prob. III of the second book—in its faulty original version and its corrected 1713 revision—see pages 312–419 above; we may add that, after Leibniz in his anonymous *charta volans* in the summer of 1713 made public Bernoulli's conclusion that the error (as he thought to detect it) in the initial working of this proposition mirrored its author's defective understanding of second and higher-order fluxions (see 6, §1: note (1) preceding), it most often best suited Newton's purpose disingenuously to claim of its variety of 'Taylor' expansions of 'momentaneously' incremented ordinates that 'Tis only a branch of yᵉ method of converging series that I there make uses of' (as he stated to Keill on 20 April 1714; see *Correspondence*, 6: 108) and that 'les seconds et troisiémes moments ne sont pas considerez dans cet endroit, aussi n'ont ils rien [à] faire avec la methode [des suites convergentes] qu'on y propose de resoudre les Problemes' (as he put it in the draft of a letter of the same period to the editor, Thomas Johnson, of the *Journal Literaire de La Haye*; see *ibid.*: 84).

(14) A superseded first opening here reads slightly differently: 'Et propositiones particulares

of Problem III[12] of Book 2 we have displayed analytically, as also that of the problem propounded in the scholium of Proposition XCIII of Book 1.[13] The rest of the propositions we have left synthetically demonstrated,[14] but their synthetical proof equally well as their analytical discovery was derived by way of the method of fluxions and moments.[15] The method of first and last ratios which is set out near to the beginning of the first book in eleven lemmas is nothing other than a part of the method of fluxions and moments synthetically demonstrated; and these lemmas are premised in order that with their benefit the following propositions found by means of the analytical method of fluxions and moments could be synthetically demonstrated.[16] The elements of the method of fluxions and moments are given synthetic proof in Lemma II of Book 2, and a number of

---

per solutiones istas inventas synthetice demonstrandas reliquimus' (And particular propositions found through those (modes of) solutions have been left by us to be synthetically demonstrated).

(15) Newton initially went on to make a favourite affirmation of his about the *Principia*: 'At methodus Analytica per quam Propositiones hasce invenimus, ex resolutione Demonstrationum syntheticarum passim elucet' (But the analytical method through which we found these propositions shines out everywhere from the resolution of their synthetical demonstrations). In its support he many times—in the draft (ULC. Add. 3968.41: 22ᵛ), for instance, of an 'Account' a year earlier 'of what has been done since the publishing of the *Commercium*' (soon afterwards absorbed by him into the unsigned review of the book which appeared in the *Philosophical Transactions* early in 1715) where in his own English words he declared that although his 'Principles of Philosophy...is written by composition after the manner of the Ancients:...yet the Analysis by wᶜʰ it was invented shines through the composition'—cited an acknowledgement by L'Hospital (or rather his 'ghost' writer Fontenelle at this place) that 'C'est encore une justice dûe au sçavant M. *Newton*, & que M. *Leibniz* luy a renduë luy-même [*Journal des sçavans* du 30. Aoust 1694]: Qu'il avoit aussi trouvé quelque chose de semblable au Calcul différentiel, comme il paroît par l'excellent Livre intitulé *Philosophiæ naturalis principia Mathematica*, qu'il nous donna en 1687. lequel est presque tout de ce calcul' (*Analyse des Infiniment Petits, pour l'Intelligence des Lignes Courbes* (Paris, ₁1696): Preface: [éiiᵛ/éiiiʳ]). In his concluding 'Annotatio' on the 'Judicium Mathematici [*sc.* Bernoulli] 7 *Julii* 1713 datum, & charta volante sine nomine Autoris per orbem sparsum' which he added to the second edition of his *Commercium* Newton was yet again to employ his present metaphor in insisting that in the *Principia* '*Newtonus* Analytico suo fluxionum calculo utendi non habuit frequentem occasionem. Nam liber ille inventus est quidem per Analysin, at scriptus est per Synthesin more veterum ut oportuit.... At Analysis tamen ita elucet per Synthesin illam, ut *Leibnitius* ipse olim agnoverit...' (*Commercium Epistolicum...iterum impressum* (London, 1722): 247).

(16) In his companion list of instances 'De methodo fluxionum' in the *Principia* (see note (13) above) Newton similarly affirmed of this 'Lib. I. Sect. I' that the book 'est de methodo rationum primarum et ultimarum cujus ope sequentia demonstrantur, id est de methodo rationum quas momenta quantitatum fluentium vel nascentia vel evanescentia habent ad invicem. Et hæc est methodus momentorum synthetica. Eadem si Analytice tractetur evadit methodus momentorum Analytica, quam etiam methodum fluxionum voco'. We have already observed (see VI: 108, note (39)) how quickly this initial vision of providing a rigorously demonstrated synthetic basis for the limit-arguments to be used in the ensuing main text of the *Principia* faded into mirage, and how soon and widely he lapsed into the informality of employing 'analytical' infinitesimal reasonings in his proofs.

aliquot[17] sequentes demonstrantur synthetice per hoc Lemma.[18] Composuimus igitur hunc Tractatum[19] maxima ex parte per Methodum fluxionum et momentorū[,] partim investigando per Analysin partim demonstrando per Synthesin. Et qui Propositiones per hanc methodum analytice inventas componere et synthetice demonstrare noverit, is synthetice demonstratas reso[l]vere & analytice investigare haud ignorabit.

---

[2][20] Veteres duplici methodo tractabant res Geometricas, Analysi scilicet et Synthesi, seu Resolutione et Compositione ut ex Pappo liquet.[21] Per Analysin investi[ga]bant Propositiones suas et per Synthesin demonstrabant inventas ut in Geometriam admitterentur. Laus enim Geometriæ in ejus certitudine consistit, ideoq nihil in ipsam prius admitti debet quam reddatur certissimum. Hæc certitudo oritur ex demonstrationibus et Veterum Demonstrationes omnes erant syntheticæ. Algebra nihil aliud est quam Arithmetica[22] ad res Geometricas applicata, et ejus operationes complexæ sunt & erroribus nimis obnoxiæ & a solis Algebræ peritis legi possunt. Propositiones autem in Geometria sic proponi debent ut a plurimis legantur, et mentem claritate sic maxime afficiant, ideoq synthetice demonstrandæ sunt. Utilis est Analysis ad veritates inveniendas, sed certitudo inventi examinari debet[23] per compositionem Demonstrationis & quam fieri potest perspicua clara & omnibus manifesta reddi: præsertim cum Propos[i]tio quæ non demonstratur synthetice, ex mente Veterum non demonstratur, ideoq in Geometriam admitti non debet. Problemata quoq quorum constructiones innotescunt tantum per Analysin,

---

(17) Initially, in a preceding drafting of this present sentence, Newton specified 'septem sequentes' (the seven following)—*viz.* Book 2, Propositions VIII–XIV (*Principia*, ₁1687: 254–82 = ₂1713: 227–52)—and began here, too, to write 'septem' (seven) before changing his mind and leaving the exact number un-named. In fact, only five of the following propositions (VIII, IX, XI, XIV and XXIX) appeal to the preceding 'Lem. II Lib. II'. As we have seen (VI: 336; compare IV: 522, note (1)), in Prop. XXXIX of the first book he had analogously computed the limit-ratio $d(V^2)/d(AD)$ (of the 'nascent' augment of the square of the speed V with respect to base AD in his figure) to be $2V.dV/d(AD)$ by an 'analytical' argument from first principles in which V and AD were put to have contemporaneous increments, I and DE respectively, which 'ultimately' vanish to be nothing.

(18) 'hæc Elementa' (these elements) was first written in Newton's initial drafting. In a preliminary English version of his 'Annotatio' on Bernoulli's 'Judicium' in the second edition of the *Commercium Epistolicum* (see note (15)) Newton afterwards similarly affirmed that while 'in my Principles I had [not] frequent occasion to use my calcu[lu]s of fluxions[,] for after I had invented the Propositions by Analysis I demonstrated them by composition...[yet] yᵉ 2ᵈ lemma of the second book contains the elements of this calculus' (ULC. Add. 3968.41: 20ʳ, where in a casting into Latin which follows he passed uncompromisingly to declare that 'In *Principijs Naturæ Mathematicis* calculo fluentium utendi nulla[!] erat occasio...'). In his contemporary catalogue of instances 'De methodo fluxionum' in the *Principia* (note (13)

the[17] propositions following are synthetically demonstrated by means of this lemma.[18] We in consequence composed this treatise[19] for the greater part by means of the method of fluxions and moments, partly by investigating through analysis, and partly by demonstrating through synthesis. And he who knows how to compose propositions found out analytically by means of this method, and demonstrate these synthetically, will not be ignorant of how to resolve them when synthetically demonstrated and analytically to investigate them.

---

[2][20]   The ancients treated geometrical matters by a dual method, namely analysis and synthesis, or resolution and composition, as is clear from Pappus.[21] Through analysis they discovered their propositions and through synthesis, once they were found, they demonstrated them so that they might be admitted into geometry. For the glory of geometry consists in its absolute certainty, and consequently nothing ought to be admitted into it before it is rendered wholly certain. This certainty arises from the demonstrations, and the ancients' demonstrations were all synthetical. Algebra is nothing other than [22]arithmetic applied to geometrical matters, and its operations are complicated and excessively susceptible to errors, and can be understood by the learned in algebra alone. Propositions in geometry, however, ought to be propounded in such a way that they may be appreciated by the great majority and thus most impress the mind with their clarity, and they need consequently to be synthetically demonstrated. Analysis is useful for finding out truths, but the certainty of a finding ought to be attested[23] through the composition of a demonstration, and so made as transparent, clear and manifest to all as it is possible; particularly so when a proposition which is not demonstrated synthetically is, in the ancients' understanding, not proven and consequently ought not to be admitted into geometry. Problems, also, whose constructions are ascertained merely

---

above) he noted more cautiously of this 'Lib. II. Lem. II [₂1713] pag. 224' that he had shown in it 'quomodo fluentium ex lateribus per multiplicationem divisionem vel extractionem radicum genitarum momenta et fluxiones inveniri possunt'. The lemma has been reproduced in full on IV: 521–5 with our detailed commentary on its content and (see *ibid.*: 523, note (6)) imperfection.

(19) The *Principia*, that is.

(20) ULC. Add. 3968.9: 107ʳ, a first revision of [1] preceding.

(21) In the opening preamble to the seventh book of his *Mathematical Collection*, namely. Newton's accurate transcription of the pertinent passage from Manolessi's second edition of Commandino's Latin translation of Pappus' work (*Pappi Alexandrini Mathematicæ Collectiones a Federico Commandino Vrbinate in Latinum conuersæ, & Commentarijs illustratæ* (Bologna, ₂1660): 240) is reproduced on VII: 248, along with his detailed comment upon the ancients' twin method of prior analytical 'resolution' and converse synthesized 'solution' of geometrical problems there set out in its fullest classical statement; see also VII: 306–8.

(22) Understand 'Universalis' (universal), as in [1] preceding.

(23) Originally just 'duplicatur' (is duplicated).

nondum soluta sunt sed tantum resoluta, nec prius soluta dici debent quam eorum Constructiones demonstrentur synthetice.

His de causis Propositiones in Libris *Principiorū* quas inveni per Analysin demonstravi per Synthesin, si forte Prop. XLV Libri primi & Prop. X Lib. II excipiantur.[24] Mathematicis autem hujus sæculi,[25] qui fere toti versantur in Algebra, genus hocce syntheticum scribendi minus placet, seu quod nimis prolixum videatur & methodo veterum nimis affine, seu quod rationē inveniendi minus patefaciat.[26] Et certe minori cum labore potuissem scribere Analyticè quàm ea componere quæ Analytice inveneram: sed propositum non erat Analysin docere. Scribebam ad Philosophos[27] Elementis Geometriæ imbutos & Philosophiæ naturalis fundamenta Geometrice demonstrata ponebam. Et inventa Geometrica quæ ad Astronomiam et Philosophiam non spectabant, vel penitus præteribam, vel leviter tantum attingebam. Cum autem de Analysi disputetur qua usus sum, visum est hanc paucis exponere.[28]

---

[3][29] Veteres in rebus Mathematicis[30] methodum duplicem colebant, Analysin et Synthesin seu Resolutionem et Compositionem: per Analysin Propositiones investigabant, per Synthesin demon[s]trabant inventas, & nondum demonstratas non admittebant in Geometriam. Laus enim Geometriæ in rerum certitudine constabat. Ideoქ in Libris sequentibus Proposi[tio]nes per Analysin inventas composui ut certissimæ red[d]erentur, & ob certitudinem dignæ essent quæ admitterentur in Geometriam. [31]Sensu enim latiore Geometriam hic voco quæ magnitudines tam motu locali quam alia quacunქ ratione descriptas ac definitas mensurare docet. Analysis speciosa[32] nihil aliud est quam Arith-

---

(24) In both these propositions Newton had (see [1]: note (13)) used 'analytical' reductions into series 'Arithmeticè...per methodum nostram serierum convergentium'.

(25) Not just the 'new' eighteenth century into which Newton has lived on, of course, but 'hujus sæculi præteriti' (this century past) as well. Though his initial cast of phrase in the manuscript—where he more vigorously declared that 'Mathematici...hujus sæculi, qui fere toti in Algebra versantur, conquesti sunt quod has [*sc.* propositiones] non scripseram Algebraice, quasi hoc genus syntheticum scribendi minus placet...' (the mathematicians of the present century, versed wholly in algebra as they are, have complained at my not having written these propositions algebraically, as if this synthetic style of writing is less pleasing)—underlines that he here has Leibniz' and Bernoulli's recent criticisms of his *Principia* chiefly in mind, he is in his general mood once more back in the middle 1680's, again occupied in 'inventing' the propositions of his *Principia* and 'composing' them for the printed page.

(26) Originally 'seu quod ratio inveniendi minus appareat' (or because the manner of discovery be less apparent).

(27) Students of 'natural philosophy' in Newton's understanding. In our English version we have, anachronistic though this translation strictly is, rendered these 'philosophers' of (non-chemical and non-biological) nature by their modern equivalents.

(28) Newton breaks off, leaving this intended brief exposition of his use of analysis in the *Principia* unwritten. It would doubtless have gone much as in the last paragraph of [1] preceding.

through analysis are not yet solved but merely resolved, nor ought to be said to be solved before their constructions shall be demonstrated synthetically.

For these reasons propositions in the books of the *Principles* which I found through analysis I demonstrated through synthesis, Proposition XLV of the first book and Proposition X of Book 2 perchance excepted.[24] To the mathematicians of the present century,[25] however, versed almost wholly in algebra as they are, this synthetic style of writing is less pleasing, whether because it may seem too prolix and too akin to the method of the ancients, or because it is less revealing of the manner of discovery.[26] And certainly I could have written analytically what I had found out analytically with less effort than it took me to compose it. I was writing for scientists[27] steeped in the elements of geometry, and putting down geometrically demonstrated bases for physical science. And the geometrical findings which did not regard astronomy and physics I either completely passed by or merely touched lightly upon. Since, however, there is dispute about the analysis which I used, I have thought fit to set this out in a few words.[28]

---

[3][29]   The ancients used in mathematical[30] matters to practise a dual method, analysis and synthesis, or composition and resolution. Through analysis they discovered propositions, and through synthesis they demonstrated them once found—and when these were not yet demonstrated they did not admit them into geometry; for geometry's title to praise lay in the utter certainty of its matters. And on that account I have in the books which follow composed the propositions found out by analysis in order to render them absolutely certain and so, because of their certainty, worthy to be admitted into geometry. [31]In a rather broad sense, to be sure, I here call geometry that which instructs how to measure magnitudes described and defined not only by local motion but in any other manner whatever. 'Specious' analysis[32] is nothing other than

---

(29) From the recto of a loose leaf (now in private possession) found with the folio sheet from which we drew the text of [1] above. In a cancelled preliminary opening to the piece Newton initially began: 'Veteribus mos erat ['res' and then 'veritates' was first written in sequel] Propositiones Geometricas investigare per Analysin & inventas demonstrare per synthesin [originally 'methodo synthetica'] ut in lucem emitterentur. Ea ratione effectum [est] ut Geometria scientiarum certissima redderetur.' (For the ancients it was customary to investigate (things, truths) geometrical propositions by means of analysis and, when found, to demonstrate them by synthesis (by the synthetic method) so as to put them publicly forth. In this manner it was achieved that geometry was rendered the most certain of sciences).

(30) 'Geometricis' (geometrical) was first, slightly more restrictively specified.

(31) This following sentence was originally tacked on to the previous one with the more hesitant connective 'si modo Geometriam vocare liceat sensu latiore quæ...' (provided it be allowable to call geometry in a rather broad sense that which...).

(32) That which employs general 'specious' (algebraic) variables: in sequel Newton has deleted 'quæ jam in usu est' (which is now in use), repellent as this was to his present predilection for geometrical analysis to admit.

metica ad res Geometricas applicata. In calculis hujus generis errores facile admittuntur. At si Propositiones sic inventæ demonstrentur synthetice hæ duplicem acquirunt certitudinem, et illā quæ ex calculo Ari[th]metico haberi potest et illam quæ ex demonstratione Geometri[c]a producitur, et per talem demonstrationem certitudinem Geometricam acquirunt, et absçp certitudine Geometrica in Geometriam admitti non debent. Hic non tam brevitati verborum et computationum, quam rerum evidentiæ et certitudini consulendum est. Et hinc fit ut in Libris sequentibus computationes Analyticæ vix aut ne vix quidem occurrant nisi forte in Prob. XXXI[(33)] Lib. I et Prob. III[(33)] Lib. II et Scholi[o] ad Prop. XCIII Lib. I & Lemmate II Lib. II ubi agitur de methodis investigandi proposita.

Mobilium vires & velocitates ad Geometriam proprie non pertinent, at per lineas, superficies, solida et angulos exponi possunt et eatenus ad Geometriam reduci. Causas virium non expendo. ... ...[(34)]

---

[4][(35)] Geometræ Veteres[(36)] quæsita investigabant per Analysin, inventa demonstrabant per Synthesin, demonstrata edebant ut in Geometriam reciperentur. Resoluta[(37)] non statim recipiebantur in Geometriam: opus erat solutione per compositionem demonstrationum. Nam Geometriæ vis et laus omnis in certitudine rerum, certitudo in demonstrationibus luculenter compositis constabat. In hac scientia non tam brevitati scribendi quam certitudini

---

(33) That is, 'Prop. XLV' and 'Prop. X' respectively; compare [1]: note (12).

(34) In the two remaining paragraphs of this draft *Principia* preface Newton passed from things mathematical to affirm (we see no need here to render his Latin words into English):

'Sufficit si vires sint qualitates manifestæ licet causæ earum nos lateant, & occultæ dicantur. In libris duobus primis ubi de viribus ad centrum tendentibus agitur, & species virium nondum innotescunt, easdem generali nomine centripetas vocavi. In libro tertio quamprimum didici Lunam gravitare in Terram et gravitate sua retineri in Orbe, vim centripetam Lunæ et reliquorum Planetarum cœpi nominare gravitatem. Eodem modo si vires electricæ vel magneticæ deprehendantur esse causæ aliquorum phænomenωn: hæc phænomena per vires illas recte explicari possunt etiamsi causæ virium nos lateant. Aliàs nullum omnino phænomenωn per causam suam recte explicari posset nisi causa hujus causæ, & causæ prioris redderetur & sic deinceps usçp donec ad causam primam deventum sit.

'Gravitatis causam mechanicam D. Fatio olim excogitavit, sed veram esse non probavit. Hypothesis erat, & in Philosophia experimentali hypotheses non considerantur. Argumenta hic desumuntur ab experimentis per Inductionē. Et argumentum ab inductione, licet Demonstratio perfecta non sit tamen fortius est quam argumentum ab Hypothesi sola[,] et quo plura sint experimenta vel Phænomena a quibus deducitur eo fortius evadit. Hypotheses igitur in hoc Tractatu non fingimus neçp argumenta inde desumimus, cum cedant argumentis ab inductione.'

To delve even superficially into the status of hypothesis and induction in Newton's general methodology of scientific invention would draw us into a tangled web on which there is a massive secondary literature: let us be content to draw to scholarly attention this present

arithmetic applied to geometrical matters. In calculations of this kind errors are easily introduced. But if the propositions so found out be demonstrated synthetically, then these acquire a dual certainty, both that which can be had from arithmetical calculation and that which is produced from geometrical demonstration; through such demonstration they acquire geometrical certainty, and without geometrical certainty they ought not to be admitted into geometry. Here regard must be paid not only to the brevity of words and computations, but also to the evidence and certainty of things. And hence it comes that in the following books analytical computations occur hardly at all—or not even that unless maybe in Problem XXXI[33] of Book 1 and Problem III[33] of Book 2, in the scholium to Proposition XCIII of Book 1 and in Lemma II of Book 2, where the theme is one of methods of investigating things proposed.

The forces and speeds of movable bodies do not properly pertain to geometry, but they can be expressed by means of lines, surface-areas, solids and angles, and to that extent reduced to geometry. The causes of forces I do not ponder. ... ...[34]

---

[4][35]   The ancient geometers[36] investigated things sought through analysis, demonstrated them when found out through synthesis, and published them when demonstrated so that they might be received into geometry. Once analysed[37] they were not straightaway received into geometry: there was need of their solution through composition of their demonstrations. For the force of geometry and its every merit lay in the utter certainty of its matters, and that certainty in its splendidly composed demonstrations. In this science regard must be paid not only to the conciseness of writing but also to the certainty of

---

clear statement of the reason why he does not in his *Principia* 'feign' hypotheses which are not founded on reasoned induction from experimental phenomena. On Nicolas Fatio de Duillier's mechanistic hypothesis 'De la Cause de la Pesanteur'—which he contrived in the late 1680's and outlined to the Royal Society in June 1688 before passing its amplified and corrected version to Newton the next spring ('now I think clear of objections, and...the true one' he wrote to him a little beforehand on 24 February 1689/90; see *Correspondence of Isaac Newton*, **3**, 1961: 390–1)—H. W. Turnbull has written an excellent summary account (*ibid.*: 69–70, note (1)). Though Fatio wrote to de Beyrie in March 1694 claiming that 'Mr Newton convient de l'exactitude de mes demonstrations' (see *ibid.*: 309) he never succeeded in persuading him that gravity had in the physical 'truth' any such mechanical cause: in late December 1691, indeed, David Gregory set down in a private memorandum (C 85, printed *ibid.*: 191) that 'Mr Newton and Mr Hally laugh at Mr Fatios manner of explaining gravity'.

(35) Extracted from Newton's unfinished final version of his intended *Principia* preface on ULC. Add. 3968.9: 109r/109v. Variant phrases and a few lengthier preparatory draftings in his rough preliminary casting of the piece (on the two centre-pages of the folded sheet, now in private possession, from whose verso we drew the text of [1] preceding; see note (3) thereto) are intercalated in following footnotes.

(36) 'primo' (first) is here deleted in the draft.

(37) Initially 'Problematum resolutiones' (The resolutions of the problems).

rerum consulendum est. Ideoqɜ in sequenti Tractatu Propositiones per Analysin inventas demonstravi synthetice.

Geometria Veterum versabatur quidem circa magnitudines; sed Propositiones de magnitudinibus non[n]unquam demonstrabantur mediante motu locali:[38] ut cum triangulorum æqualitas in Propositione quarta libri primi *Elementorum* Euclidis demonstraretur transferendo tr[i]angulum alterutrum in locum alterius. Sed et genesis magnitudinum per motum continuum recepta fuit in Geometria:[39] ut cum linea recta duceretur in lineam rectam ad generandam aream, & area duceretur in lineam rectam ad generandum solidum. Si recta quæ in aliam ducitur datæ sit longitudinis generabitur area parallelogramma. Si longitudo ejus lege aliqua certa continuo mutetur, generabitur area curvilinea. Si magnitudo areæ in rectam ductæ continuo mutetur generabitur solidum superficie curva terminatum. Si tempora, vires, motus et velocitates motuum exponantur per lineas areas solida vel angulos, tractari etiam possunt hæ quantitates in Geometria.[40]

Quantitates continuo fluxu crescentes[41] vocamus fluentes & velocitates crescendi vocamus fluxiones, & incrementa[42] momentanea vocamus momenta, et methodum qua tractamus ejusmodi quantitates vocamus methodum fluxionum et momentorum: estqɜ hæc methodus vel synthetica vel analytica.[43]

Methodus synthetica fluxionum et momentorum in Tractatu sequente passim occurrit, et ejus elementa posui in Lemmatibus undecim primis Libri primi & Lemmate secundo Libri secundi.

Method[i] analyticæ specimina occurrunt in Prop. XLV & Schol. Prop. XCIII Lib. I & Prop. X & XIV[44] Lib. II. Et præterea describitur in Scholio ad

(38) The draft has 'per motum localem' (through local motion) here, but the change was probably made only to add verbal emphasis.

(39) As we earlier remarked regarding Newton's assertion of this view in the 'Introductio' to his printed *Tractatus de Quadratura Curvarum* a dozen years before (see 2, §2.1: note (8) above), while this notion of generating magnitudes through continuous motion has some foundation in ancient Greek 'mechanical' practice, its active pursuit was only undertaken by such geometers of Newton's own century as Cavalieri and Roberval. Such 'flowing geneses' were a natural way of defining the lengths, areas, surfaces and volumes of the figures to which he himself applied his own fluxional analyses in his early papers (see for instance I: 430, III: 344).

(40) This dynamical application has, of course, no classical precedent, but was largely pioneered by Newton himself in the *Principia*.

(41) In the draft Newton initially wrote 'continuo fluxu vel defluxu crescentes vel decrescentes' (increasing or decreasing in a continuous flow onwards or backwards).

(42) And here, correspondingly, he first added 'vel decrementa' (or decrements) in the draft.

(43) In his draft Newton originally here went on to declare that 'Propositiones quæ in hoc Trac[ta]tu sequuntur invenimus per methodum Analyticam, demonstravimus autem synthe[ti]ce ut Philosophia cœlorum in libro tertio exposita fundaretur in Propositionibus Geometrice demonstratis. Et qui methodum compositionis probe intellexerit, is methodum resolutionis ignorare non potest' (The propositions which follow in this treatise we found by means of the analytical method, but we demonstrated them synthetically so that the (natural)

things. And on that account I in the following treatise synthetically demonstrated the propositions found out through analysis.

The geometry of the ancients had, of course, primarily to do with magnitudes, but propositions on magnitudes were from time to time demonstrated by means of local motion:[38] as, for instance, when the equality of the triangles in Proposition 4 of Book I of Euclid's *Elements* were demonstrated by transporting either one of the triangles into the other's place. Also, the genesis of magnitudes through continuous motion was received in geometry:[39] when, for instance, a straight line were drawn into a straight line so as to generate an area, and an area were drawn into a straight line to generate a solid. If the straight line which is drawn into another be of given length, there will be generated a parallelogram area. If its length be continuously changed according to some fixed law, a curvilinear area will be generated. If the size of the area drawn into the straight line be continuously changed, there will be generated a solid terminated by a curved surface. If times, forces, motions and speeds of motion be expressed by means of lines, areas, solids or angles, then these quantities too can be treated in geometry.[40]

Quantities increasing in a continuous flow[41] we call fluents, the speeds of flowing we call fluxions and the momentary increments[42] we call moments, and the method whereby we treat quantities of this sort we call the method of fluxions and moments: this method is either synthetic or analytical.[43]

The synthetic method of fluxions occurs widespread in the following treatise, and I have set its elements in the first eleven lemmas of the first book and in Lemma II of the second.

Specimens of the analytical method occur in Proposition XLV and the scholium to Proposition XCIII of Book 1, and in Propositions X and XIV[44] of

philosophy of the heavens should be founded on geometrically demonstrated propositions. And he who has properly understood the method of composition cannot be ignorant of the method of resolution). Were it not that he passed to repeat this shaky statement in a number of subsequent contexts private and public—for instance (see note (15) above) in the terminal 'Annotatio in Judicium Mathematici' in the second (1722) edition of his *Commercium Epistolicum D. Johannis Collins et Aliorum...*—we might have presumed that, in omitting it from his revised text at this point, he was no longer willing to stand by its ubiquitous accuracy. It is of course (compare his explicit distinction of the two in [1] preceding) his geometrical method of 'first and last ratios' which he assumes to be the synthetic mode of fluxions, while the algebraic 'calculus' of fluxions is the analytical one. The respective occurrences of these in the *Principia* he now proceeds to specify in the two following paragraphs.

(44) The '& XIV' is a late interlineation in the preliminary draft, and is there amplified by a lone phrase on the manuscript's verso (overlooked by Newton in his revise?) where he adjoined 'cas. 3 ubi momenta secunda vocavi differentiam momentorum et momentum differentiæ' (Case 3, where I called second moments the difference of moments and the moment of a difference), and also noted of 'Prob. III Lib. II' that he meant its 'Exempl. I ubi'—see page 364 above—' fluxionem fluxionis curvaturæ vocavi variationem variationis' (where I called the fluxion of the fluxion of the curvature the variation of its variation).

Lem. II Lib. II. Sed et ex Demonstrationibus compositis Analysis qua Propositiones inventæ fuerunt, addisci potest regrediendo.[45]

Scopus Libri *Principiorum* non fuit ut methodos[46] mathematicas edocerem, non ut difficilia omnia ad magnitudines[47] motus & vires spectantia eruerem, sed ut ea tantum tractarem quæ ad Philosophiam naturalem et apprime ad motus cœlorum spectarent: ideoᵦ quæ ad hunc finem parum conducerent, vel penitus omisi, vel leviter tantum attigi, omissis demonstrationibus.[48]

· · ·   · · ·   · · ·[49]

In his contemporary listing of instances 'De methodo fluxionum' in the *Principia* (see note (13) above) he correspondingly docketed this 'Lib. II Prop. 1[4]. [₂1713:] pag. 251' as one where 'argumentum procedit per differentiam momentorum, ideoᵦ ideam tunc habui momentorum secundorum, et primus hanc ideam in lucem edidi'. Newton had already pointed to this would-be proof that 'when he wrote his *Principles of Philosophy*' he 'had extended his method to the consideration of the second fluxions of quant[it]ies' some two years previously in drafting a response to Bernoulli's *judicium* to the contrary (as it had been anonymously published by Leibniz in the 1713 *Charta volans*), one he intended to put to the *Journal Literaire* for printing (see *Correspondence*, **6**: 80–90): 'And particularly in demonstrating the 14ᵗʰ Proposition of the second Book', he there wrote (*ibid.*: 84), '[Mʳ Newton] has these words. *Est igitur differentia momentorum, id est momentum differentiæ arearum* &c. Where *differentia arearum* is the first difference & *momentum differentiæ* is the second difference of the areas'. Unfortunately, in his anxiety to find an example in his *Principia* where he had employed a 'second difference' he here seriously mistook the import of his words of thirty years before. In Case 3 of Book 2, Proposition XIV (and in that of the preceding Proposition XIII which is its rider) he had sought to evaluate—so we may restate his problem in modern Leibnizian equivalent—the distance, s say, fallen straight downwards from rest, in time t, by a body accelerated by uniform gravity, g, acting in the direction of fall and simultaneously slowed by a contrary force of resistance, $\rho_v = v^2 + 2av$ ($< g$ since the body is supposed to continue to accelerate) where $v(= ds/dt)$ is its instantaneous speed of motion; at once, since the equation of motion is $dv/dt = v \cdot dv/ds = g - \rho_v$, that is, $b^2 - (v+a)^2$ on putting $\sqrt{(a^2+g)} = b$, it follows that

$$t = \int_0^v 1/(b^2 - (v+a)^2) \cdot dv = (1/b) \left[\tanh^{-1} ((v+a)/b) \right]_0^v,$$

whence

$$s = \int_0^v v/(b^2 - (v+a)^2) \cdot dv = \left[-\tfrac{1}{2} \log (b^2 - (v+a)^2) - (a/b) \tanh^{-1} ((v+a)/b) \right]_0^v$$
$$= \tfrac{1}{2} \log (g/(g - \rho_v)) - at.$$

Making use of some equivalent prior analysis of the problem, it is clear, in his resolution as set down in his Case 3 Newton constructs the hyperbolic area $I_v \equiv \tfrac{1}{4}(b^3/a) \log (g/(g - \rho_v))$ and the hyperbola-sector $J_v \equiv \tfrac{1}{2}b^2(\tanh^{-1} ((v+a)/b) - \tanh^{-1} (a/b))$ in an elegantly conjoined but straightforward geometrical manner—in his accompanying figure there is (as we need not elaborate) $BA = a, BD = b, AP = v$ and, where Z is some convenient 'data quævis constans', also $CA = \tfrac{1}{4}(b^3/a)/Ab = g/Z$ and $AK = \rho_v/Z$, whence $I_v = (AKNb)$ and $J_v = (EDT)$—and thereafter, on supposing a 'momentum velocitatis' ($PQ =$) dv corresponding to a 'momentum temporis...determinatum' dt, demonstrates from first principles (in terms of the related infinitesimal increments of area) that the 'differentia momentorum' $dI_v - dJ_v$, 'id est momentum differentiæ arearum' $d(I_v - J_v)$, is equal to

$$\tfrac{1}{2}(b^3/a) ((v+a)/(g - \rho_v)) \cdot dv - \tfrac{1}{2}b^2 \cdot dt = \tfrac{1}{2}(b^3/a) v \cdot dt \propto v \cdot dt = ds.$$

Book 2. It is, furthermore, described in the scholium to Lemma II of Book 2. And from their composed demonstrations, also, the analysis by which the propositions were found can be learnt by going backwards.[45]

The purview of the book of *Principles* was not such that I was elaborately to explain its mathematical methods,[46] my aim therein not to root out all the difficulties regarding magnitudes,[47] motions and forces, but merely to treat what should regard physics and, principally, the motions of the heavens; consequently, what would too little contribute to this end I either completely omitted or merely touched lightly upon, omitting the demonstrations.[48]

. . .   .[49]

(Cases 1 and 2 analogously construct the solution of the similar problem of resisted vertical ascent, where the equation of motion is $dv/dt = g + \rho_v = (v+a)^2 + (g-a^2)$, according as $a < \sqrt{g}$ and $a > \sqrt{g}$ respectively.) Since the *differentia* $I_v - J_v = \frac{1}{2}(b^3/a)s$ is not in general vanishingly small, its *incrementum momentaneum* $d(I_v - J_v)$ is not a *momentum secundum arearum* as Newton would, thirty years on, have it be. Whether he ever came to recognize his error in so identifying it we do not know. As late as 1719, in a first English draft of what afterwards became the *Annotatio* on the 'Judicium Mathematici' in the *Charta volans* appended by him to the second edition of his *Commercium Epistolicum* (London, ₂1722)—but does not itself contain the mistaken observation—he could still forthrightly pronounce that 'it plainly appears by Prop. XIV. Lib. II *Princip*. that Mʳ Newton wⁿ he wrote that Book knew how to work in second differences wᶜʰ he there calls *Differentia Momentorum*...' (ULC. Add. 3968.34: 481ʳ; the full text of these hitherto unpublished 'Observations' upon Bernoulli's 1713 judgement is reproduced in **2**, Appendix 9 below).

(45) In sequel Newton has cancelled a further sentence stating that 'Tractatum de hac Analysi ex chartis [he added 'sparsis' in his draft] antea editis desumptum Libro *Principiorum* subjunxi' (A tract on this analysis, taken from (scattered) papers previously published, I have subjoined to the book of *Principles*). This gathering of snippets from his earlier published writings on analytical fluxional calculus which he thought briefly of adjoining to the new re-issue of his *Principia* was doubtless the 'Analysis per quantitates fluentes et earum momenta' (see 5.1/2 above) which he had compiled around April 1712, intending it to play a similar ancillary rôle for the second edition (compare *ibid.*: notes (1) and (2)).

(46) Newton's draft (see note (35) above) adds 'inveniendi' (of finding out).

(47) 'figuras' (figures) is deleted in sequel.

(48) In a cancelled following paragraph in the draft Newton initially went on to state: 'Objectio fuit'—by Johann Bernoulli (see **VI**: 138, note (124))—'quod Corollarium Propositionis XIII Libri primi in editione prîma non demonstraverim. Dicunt enim quod corpus *P* secundum rectam positione datam data cum velocitate a dato loco exiens, vi centripeta cujus lex datur [*sc.* reciproce proportionalis quadrato distantiæ a centro virium] Curvas plures describere posse. Sed hallucinantur. Si mutetur vel positio rectæ vel velocitas corporis₍,₎ mutari potest Curva describenda et ex circulo Ellipsis, ex Ellipsi Parabola vel Hyperbola fieri. Sed positione rectæ et velocitate corporis et lege vis centripetæ manentibus curva alia atçз alia describi non potest. Ideoçз si ex data curva determinatur vis centripeta, vicissim ex data vi centripeta determinabitur curva. In secunda [editione] paucis tantum verbis'—in an expanded 'Corol. 1' (*Principia*, ₂1713: 53, compare Newton's essentially identical preliminary text of this on **VI**: 554–6)—'attigi. In utraçз constructionem hujus Corollarij in Prop. XVII exhibui [*vide* **VI**: 158–60] qua veritas ejus satis elucesceret, & Problema generaliter solutum

[see VI: 344–8] in Prop. XLI Lib. I' (It has been objected that the Corollary of Proposition XIII of Book 1 in the first edition was not demonstrated by me. For they assert that a body *P*, going off along a straight line given in position, at a given speed and from a given place under a centripetal force whose law is given (*viz.* reciprocally proportional to the square of the distance from the force-centre), can describe a great many curves. But they are deluded. If either the position of the straight line or the speed of the body be changed, then the curve to be described can also be changed and from a circle become an ellipse, from an ellipse a parabola or hyperbola. But where the position of the straight line, the body's speed and the law of central force stay the same, differing curves cannot be described. And in consequence, if from a given curve there be determined the (generating) centripetal force, there will conversely from the central force given be determined the curve. In the second edition I touched on this in a few words merely; but in each I displayed a construction of this Corollary in Proposition XVII whereby its truth would adequately come to be apparent, and exhibited the problem generally solved in Proposition XLI of Book 1). And he thereafter passed to state more shortly on the same theme: 'Demonstrationem Corollarij primi Propositionis XIII Lib. I ut abunde satis obviam in editione prima prætermisi, in secunda amico Cantabrigiense'—Roger Cotes in autumn 1709 (again see VI: 148, note (124))—'postulante exp[r]essi paucis, in Editionis utriusq̄ Prop. X[V]II Libri primi casus omnes Corollarij hujus construxi. Et præterea in Prop. XLI Lib. I Regulam generalem demonstratam dedi quæ Corollarium hocce ut casum particularem complectitur' (The demonstration of the first Corollary of Book 1, Proposition XIII I omitted in the first edition as being abundantly obvious enough; in the second, at the demand of a Cambridge friend, I expressed it in a few words; while in Prop. XVII of the first Book I constructed all the cases of this Corollary. And, furthermore, in Proposition XLI of Book 1 I set forth with demonstration a general rule which embraces this Corollary as a particular case). We have earlier (VI: 147, note (124)) cited the essence of Newton's near-equivalent English words from a later draft *Principia* preface on ULC. Add. 3968.9: 101ᵛ whose full text (*ibid.*: 101ʳ/101ᵛ) is reproduced in I. B. Cohen's *Introduction to Newton's 'Principia'* (Cambridge, 1971): 293–4 (see especially 294, ll. 16–20; and compare 2, Appendix 6 below); in the lightly reworked phrasing on the ensuing f. 102ᵛ (where it has subsequently been cancelled) his statement affirms that 'The Demonstration of the first Corollary of the 13ᵗʰ Proposition of the first Book being very obvious, I omitted it in the first Edition...& contented myself with determining in the 17ᵗʰ Proposition what will be the Conic Section described in all cases by a body going from any place with any velocity in any determination [*sc.* of its motion] if attracted by a force reciprocally as the square of the distance'.

(49) The manuscript continues, before breaking off in mid-sentence, for three further paragraphs which relate to more general scientific aspects of the *Principia*, but which we may (without translation and with but a rare parenthetical exclamation) here cite for their wider dynamical interest:

'In Libris duobus primis vires generaliter tractavi [in draft Newton added 'non definiendo cujus sint generis nisi forte ubi de corporibus pendulis & oscillantibus agitur'] easq̄ si in centrum aliquod seu immotum seu mobile tendunt, centripetas (nomine generali) vocavi, non inquirendo in causas vel species virium, sed earum quantitates determinationes & effectus tantum considerando. In Libro tertio quamprimum didici vires quibus Planetæ in orbibus suis retinentur, recedendo a Planetis in quorum centra vires illæ tendunt, decrescere in duplicata ratione distantiarum, & vim qua Luna retinetur in Orbe suo circum Terram, descendendo ad superficiem Terræ æqualem evadere vi gravitatis nostræ, adeoq̄ vel gravitatem esse vel vim gravitatis duplicare: cœpi gravitatem tractare ut vim qua corpora cœlestia in orbibus suis retineantur. [Newton here continued in the draft by specifying 'Et postquam didici insuper quod causa gravitatis₍,₎ quæcunq̄ tandem sit, agat in omnia corpora omnesq̄ omnium partes pro quantitate materiæ in singulis & æquali cum vi et efficacia penetret ad usq̄ centrum corporum omnium, tract[abam] gravitatem ut vim universalem qua Planetæ omnes in se mutuo agant & Sol pro magnitudine sua summa cum vi agat in Planetas omnes alias, efficiatɋ ut in ipsum tendentes describant—'perpetuo' is deleted!—Orbes in ipsum incurvatos perinde ut lapis projectus tendendo in Terram interea dum per aera movetur lineam in Terram incurvatam describit' before there cancelling the passage.] Et in eo versatur Liber iste tertius, ut Gravitatis proprietates, vires, directiones & effectus edoceat.

'Planetas in orbibus fere concentricis & Cometas in orbibus valde excentricis circum Solem revolvi, Chaldæi[!]'—in his draft Newton more loosely and slightly less inaccurately here spoke of 'veteres'—'olim crediderunt. Et hanc philosophiam Pythagoræi in Græciam invexerunt. Sed et Lunam gravem esse in Terram & stellas graves esse in se mutuo, et corpora omnia in vacuo æquali cum velocitate in Terram cadere, adeoɋ gravia esse pro quantitate materiæ in singulis notum fuit [?!] Veteribus [*sc.* Græcis et Romanis]. Defectu demonstrationum hæc philosophia intermissa fuit ['ad nos propagata non fuit & opinioni vulgari de orbibus solidis cessit' follows in the draft]. Eandem non inveni sed vi demonstrationum in lucem tantum revocare conatus sum. Sed et Præcessionem Æquinoxiorum, & fluxum & refluxum maris et motus inæquales Lunæ et orbes Cometarum & perturbationem orbis Saturni per gravitatem ['gravitationem' in the draft] ejus in Jovem'—in his draft Newton initially wrote more precisely 'Saturnum item et Jovem per gravitatem mutuam se invicem prope conjunctionem trahere'—ab ijsdem Principijs consequi, et quæ ab his Principijs consequuntur cum Phænomenis probe [initially '...per eandem Theoriam determina[ri] & cum Observationibus accurate factis optime'] congruere, hic ostensum est. Causam gravitatis ex phænomenis nondum didici.

'Qui leges et effectus Virium electricarum pari successu et certitudine eruerit, philosophiam multum promovebit, etsi forte causam harum Virium ignoraverit. Phænomena primo observanda sunt, dein horum causæ proximæ, & postea causæ causarum eruendæ: ac tandem a causis causarum per phænomena stabilitis, ad earum effectus, argumentando a priori, descendere licebit. Philosophia naturalis non in opinionibus Metaphysicis, sed in Principijs proprijs fundanda est; & hæc [?Principia ex Phænomenis deducanda sunt...]'. A cancelled final Lockean observation affirming 'Et inter Phænomena numerandæ sunt actiones mentis quarum conscij sumus' is amplified in the initial version of the two last sentences which Newton penned in his draft, there stating:

'Quod in Metaphysica docetur, si a revelatione divina deducitur, religio [*i.e.* tenet of faith] est; si a Phænomenis per sensus quinɋ externos, ad Physicã pertinet; si a cognitione actionum internarum mentis nostræ per sensum reflexionis, philosophia est de sola mente humana & ejus ideis tanquam Phænomenis internis & ad Physicam item pertinet. De Idearum objectis disputare nisi quatenus sunt phænomena somnium est. In omni Philosophia incipere debemus a Phænomenis, & nulla admittere rerum principia nullas causas nullas explicationes nisi quæ per phænomena stabiliuntur. Et quamvis tota philosophia non statim pateat, tamen satius est aliquid indies addiscere quam hypotheseωn præjudicijs mentes hominum præoccupare.'

9

MINOR COMPLEMENTS TO THE 'ARITHMETICA UNIVERSALIS'[1]

[1720–1][2]

From rough worksheets in the University Library, Cambridge

§1.[3] REDUCTION OF SIMULTANEOUS ALGEBRAIC EQUATIONS TO ELIMINANTS IN BUT A SINGLE VARIABLE.[4]

(1) We have previously (see v: 12–14) given evidence of the extent of Newton's private dissatisfaction with the printed version of his deposited Lucasian 'lectures' of the 1670's and early 1680's which William Whiston brought out under this title at Cambridge in 1707, pointing in particular to the many improvements in its layout and content which he penned into this his *Arithmetica Universalis, sive De Compositione et Resolutione Arithmetica Liber*. And we there also made clear (see *ibid.*: 14–15, note (60)) that Newton took personal charge of the recast and lightly augmented reissue of the *editio princeps* where most of his projected changes were incorporated, and which at length appeared fifteen years afterwards as its *Editio Secunda. In qua multa immutantur & emendantur, nonnulla adduntur* (London, 1722). The alterations which Newton effected in his second edition of his *Arithmetica*—in their largest part these are textual ameliorations of no grand import—have already been cited at the pertinent places in our running commentary upon our reproduction of the original manuscript of these Lucasian *lectiones* in v: 54–508. Here, in a not wholly unfitting finale to his six preceding decades of mathematical activity, we set out the essence of two groups of ancillary computations by him in which he elaborates his rules in the *Arithmetica* for eliminating an unknown between two given simultaneous equations, and then passes to correct his printed construction of the roots of a cubic equation by conchoidal neusis.

(2) Newton's handwriting in these calculations is one characteristic of his later old age, and we may be safe on that ground alone in broadly assigning a date for their composition of 'around 1720'. Since, however, Pierre Desmaizeaux remarks in the draft of a letter to Conti on 11 September 1720 that already 'La seconde edⁿ. de l'*Alg*. de Mʳ Newton est fort avancée (see *The Correspondence of Isaac Newton*, **7**, 1977: 100), those reproduced in §1 can scarcely be later than this period; while the checking computation in §2 led Newton to redo a page of the sheets of the *Arithmetica*'s second edition after these had been printed off (see v: 469, note (686)), and so we can only be months out in assigning to it a date of '*c*. late 1721'.

(3) Slightly condensed from Newton's more straggling array of computations on Add. 4005.2: [in sequence] 2ᵛ–1ᵛ, and here rounded out (in our usual fashion) with interpolated verbal connectives indicating, as appropriate, the flow of Newton's argument. Subsequently, we may notice, he used the rectos of these folios to cast two drafts of 'A Scheme for establishing the Royal Society' whose more finished state (f. 1ʳ) is printed by David Brewster in his *Memoirs of the Life, Writings and Discoveries of Sir Isaac Newton*, **1** (Edinburgh, 1855): 102–4; a final version (*ibid.*: [4ʳ →] 6ʳ) in which Newton has curiously (perhaps through a slip of his pen) replaced a key word 'Mineralogy' by 'Meteorology' is cited by F. E. Manuel in his

[1]   [Given]

$$\begin{aligned} &a &+e\,y &+h\,yy = y^3 = &l &+q\,y &+t\,yy. \\ &+bx &+fx &+kx & &+mx &+rx &+vx \\ &+cxx &+gxx & & &+nxx &+sxx \\ &+dx^3 & & & &+px^3 \end{aligned}$$

[there results by subduction]

$$\begin{aligned} &a-l &+\overline{e-q}\,y &+\overline{h-t}\,yy = 0. \\ &+\overline{b-m}\,x &+\overline{f-r}\,x &+\overline{k-v}\,x \\ &+\overline{c-n}\,xx &+\overline{g-s}\,xx \\ &+\overline{d-p}\,x^3 \end{aligned}$$

Or$^{(5)}$

$$\left.\begin{array}{ll} l & +q\,y \\ +mx & +rx \\ +nxx & +sxx \\ +px^3 \\ \hline t+vx \end{array}\right\} = yy.$$

[And thence]

$$\begin{array}{l} a &+ey \\ +bx &+fxy \\ +cxx &+gxxy \\ +dx^3 \end{array} \; [+] \; \frac{\begin{array}{l} +hl &+klx \\ +hmx &+kmxx \\ +hnxx &+knx^3 \\ +hpx^3 &+kpx^4 \end{array}}{t+vx} = \frac{\begin{array}{l} ly &+qyy \\ +mxy &+rxyy \\ +nxxy &+sxxyy \\ +px^3y &+tx^3yy \end{array}}{t+vx}$$

Portrait of Isaac Newton (Cambridge, Mass., 1968): 444–5, note 26. There is also at the foot of f. 2^v a stray snippet which declares that 'Cyaxeres (according to Herodotus) had a son called Astyages who in the year of the great eclyps predicted by Thales (Anno Nabonass. 147 [= 601 B.C.]) married Ariene the daughter of Alyates & afterwards succeeded his father Cyaxeres in the throne [of the Medes]'. (Newton afterwards saw fit to go against Herodotus' authority, determining in Chapter IV, §xx of his elaborately spun 'Chronology of Ancient Kingdoms Amended' (ULC. Add. 3988, later published posthumously at London in 1728; see I: xix, note (11)) that in fact Astyages was Cyaxeres' father!)

(4) Thus yielding *mutatis mutandis* in corresponding cases the rules—and in [4] and [8] below an extension of these to the quartic case—which Newton had in the 1670's set down in his Lucasian lectures on algebra (see V: 126) for 'exterminating' a common unknown between pairs of quadratic or cubic equations. In each case his mode of reduction follows, with small individual variations, the same general pattern of successively eliminating at each stage the highest power of y present till at length there remains only y itself conjoined in an equation with the pertinent powers of x and their known coefficients, whence by direct substitution in one or other of the original or derived equations it may totally be eradicated. Newton here adopts, we may notice, the somewhat wasteful procedure of separately specifying the expressions in x which are the coefficients of the simultaneous equations in y to be reduced; this leaves him badly short of letters with which to denote other coefficients as these ensue in his reductions, and in lieu of other alternatives he adopts on occasion the confusing practice of re-using symbols which he has previously employed. The wary reader will not need to be given special warning every time Newton so transposes the denotations of his letters. In [8] and [9] he switches to the clearer convention of introducing capitals to represent compound coefficients, but again is put to re-using these also; his solution in the improved version of [9] which we have earlier set in appendix to our text of the 'Arithmetica' (see V: 518–19) is to bring in yet further lower-case Greek letters to eke out the Roman alphabet.

(5) On transposing $a-l \to l$, $b-m \to m$, $c-n \to n$, ..., $g-s \to s$ and also $h-t \to -t$, $k-v \to -v$, we must understand. (Compare the previous note.)

[or]
$$\left[\frac{\overline{l+mx+nxx+px^3}\,|y}{t+vx}+\right]\frac{q+rx+sxx}{t+vx}\times\frac{\begin{array}{l}l+mx+nxx+px^3\\[2pt]+\overline{q+rx+sxx}\,|y\end{array}}{t+vx}\,.$$

---

[2]   [Given] $\begin{array}{l}a\quad+e\\+bx\ +fx\\+cxx+gxx\\+dx^3\end{array}\!\!\begin{array}{l}+h\\\,^{y}+kx\end{array}yy=y^3.$ [and] $\begin{array}{l}l\ +py\\+mx+qxy\\+nxx\end{array}=yy.$

[there is by reduction] $\begin{array}{l}a\quad+\overline{e-l}\,|y\ +\overline{h-p}\,|yy\\+bx\ +\overline{f-m}\,|xy+\overline{k-q}\,|xyy\\+cxx+\overline{g-n}\,|xxy\end{array}=0.$

[that is] $\left[\begin{array}{l}a\quad+\overline{e-l}\,|y\\+bx\ +\overline{f-m}\,|xy\\+cxx+\overline{g-n}\,|xxy\end{array}\right]\begin{array}{l}+\overline{lh-lp}\\+\overline{hm-pm}\,|x\\+\overline{hn-pn}\,|xx\\+\overline{hp-np}\,|y\\+\overline{hq-nq}\,|xy\end{array}\begin{array}{l}+\overline{kl-ql}\,|x\\+\overline{mk-mq}\,|xx\\+\overline{nk-nq}\,|x^3\\+\overline{kp-pq}\,|xy\\+\overline{kq-qq}\,|xxy\end{array}[=0.]$

---

[3]   [Given]

$\begin{array}{l}a\quad+fy\quad+kyy\ +ny^3\\+bx\ +gxy\ +lxyy+pxy^3\\+cxx+hxxy+mxxyy\\+dx^3+i^3xy\\+ex^4\end{array}=y^4.$ [and] $\begin{array}{l}r\ +vy\\+sx+wxy\\+txx\end{array}=yy.^{(6)}$

---

[4]   [Given] $\begin{array}{l}a\\+bx^4\end{array}\!+cx^3y+dxxyy+exy^3=y^4.$ [and] $\begin{array}{l}r\\+sxx\end{array}\!+txy=yy.$

[By reduction there is][7]

$$a+bx^4+cx^3y+drxx+dsx^4+dtx^3y$$
$$+erxy+esx^3y(+etxxyy\ [\text{or}])+etrxx+etsx^4+ettx^3y$$
$$=rr+2rsxx+ssx^4+2rtxy+2stx^3y(+ttyy\ [\text{or}])+ttr+ttsxx+t^3xy.$$

---

(6) Having posed this general case of eliminating between a simultaneous quartic and quadratic, Newton opts to delay his discussion of it till [8] below, for the moment passing over its complexities to treat but the simple instance of it which he specifies in [4] following.

(7) On substituting in place of the factor y^2 of the last two terms on the left-hand side of the preceding quartic its equivalent value $r+sx^2+txy$.

Put $er-2rt-t^3=f^{(8)}$ & $c+dt+es-2st[+]ett=g^{(8)}$ And it will be

$$\begin{matrix} a & +b & +dr \\ -rr & +ds & -2rs \\ [-ttr] & -ss & -etr \\ & [+]ets & [-tts] \end{matrix} \begin{matrix} x^4 \\ \\ \end{matrix} xx+fxy+gx^3y=0.$$

---

[5]   [Given] $\overset{a}{+bx^3}+cxxy+dxyy=y^3.$ [and] $\overset{e}{+fx^3}+gxxy+hxyy=y^3.$

[By subduction]

$$\overline{a-e+\overline{b-f}}|x^3+\overline{c-g}|xxy+\overline{d-h}|xyy=0. \quad \text{Or} \quad k+lx^3+mxxy=xyy.$$

[Whence]

$$\left.\begin{matrix} a+bx^3+cxxy \\ +dk+dlx^3+dmxxy \end{matrix}\right\}=\frac{ky}{x}+lxxy(+mxy\ [\text{or}])+mk+mlx^3+mmxxy.$$

[That is] $\begin{matrix} a & +b & +c \\ +dk & +dl & +dm \\ -mk-ml & -l \\ & -mm \end{matrix} \begin{matrix} x^3 \\ \\ \end{matrix} xxy=\frac{ky}{x}.$ [or] $\dfrac{\overline{a+dk-mk}|x+\overline{b+dl-ml}|x^4}{\overline{l+mm-c-dm}|x^3+k}=y.$

[Of form] $p+qx^3=rxxy+\dfrac{ky}{x}.$ [or] $\dfrac{px+qx^4}{rx^3+k}=y.$

---

[6]
[Let] $\overset{a}{+bx^3}+cxxy\begin{matrix}+dyy \\ +exyy\end{matrix}=y^3.$ $\overset{f}{+gx^3}+hxxy\begin{matrix}+k \\ +lx\end{matrix}yy=y^3.$

And by subduction $\overset{m}{+nx^3}+pxxy=\overset{q}{+rx}yy.$

[Thence] $\begin{matrix} qa+qcxxy & +qd \\ +qbx^3+rcx^3y & +qex \\ +rax & +rdx \\ +rbx^4 & +rexx \end{matrix} \begin{matrix} \\ yy= \end{matrix} \begin{matrix} my \\ +nx^3y \end{matrix}+pxxyy.$

---

(8) These should strictly read '$-f$' and '$-g$' respectively; but Newton is only concerned here to indicate the form of these simplifying substitutions, leaving the signs of the ensuing compound coefficients to be entered as context requires.

That is[9]

$$\begin{bmatrix} qa+qcxx \\ +qbx^3+rcx^3 \\ +rxx \\ +rbx^4 \end{bmatrix} y \begin{array}{l} +dm+dppxxy \\ +dnx^3 \quad +epx^3y \\ +mex \\ +enx^4 \end{array} = \begin{bmatrix} m \\ +nx^3 \end{bmatrix} y + \end{bmatrix} \frac{mpxx+npx^5+ppx^4y}{q+rx}.$$

[Which is of the form] $\dfrac{p+rx^4y^{(10)}}{qx^5+\ s}$

---

[7]   [Given] $\begin{array}{l} a \ +ey +hyy \\ bx \ +fxy+ixyy \\ cxx+gxxy \\ dx^3 \end{array} =y^3.$ [and] $\begin{array}{l} k \ +ny \\ +lx+pxy \\ +mxx \end{array} =yy.$

[By multiplication and subduction there is] $\begin{array}{l} ky \ +nyy \ -a \\ +lxy \ +pxyy \ -bx \\ +mxxy \ -hyy-cxx \\ \quad -ey \ -ixyy-dx^3 \\ \quad -fxy \\ \quad -gxxy \end{array} =0.$

[That is] $\begin{array}{l} a \ +ny \\ +bx \ +pxy \\ +cxx+qxxy \\ +dx^3 \end{array} = \dfrac{ryy}{+sxyy}.$ [Or] $\begin{array}{l} a \ +ny \\ +bx \ +pxy \\ +cxx \ +qxxy \\ +dx^3 \ -rxy \\ \quad -rk \ -rpxy \\ \quad -rlx \ -snxy \\ \quad -rmxx-spxxy \\ \quad -skx \\ \quad -slxx \\ \quad -smx^3 \end{array} =0.$

---

[8]

[Given] $\begin{array}{l} a+bx+cxx+dx^3+ex^4 \\ \overline{+f+gx+hx^2+ix^3} \text{ in } y \\ \quad \overline{+k+lx+mx^2} \text{ in } yy \\ \qquad \overline{+n+px} \text{ in } y^3 \end{array} =y^4.$ [and] $\begin{array}{l} yy=q+rx+sxx \\ \qquad\qquad +ty \\ \qquad\qquad +vxy \end{array}$

---

(9) On replacing the factor y^2 in the preceding equation by its equivalent value

$$(m+nx^3+px^2y)/(q+rx).$$

(10) Newton breaks off before making his meaning here fully clear. On multiplying through by $q+rx$, the preceding equation evidently has the form $A(x)+B(x).y = C(x)+D(x).y$, which is at once reducible to be $y = (A(x)-C(x))/(D(x)-B(x))$.

[then by reduction]
$$\left.\begin{array}{l} a \quad +b \\ -qq-2qr \end{array}x \begin{array}{l} +c \\ -2qs \end{array}xx \begin{array}{l} +d \\ -2rs \end{array}x^3 \begin{array}{l} +e \\ -ss \end{array}x^4 \right. \quad [=0.]$$
$$-rr$$

$$\left.\begin{array}{l} +f \quad +g \\ -2qt-2qr \end{array}x \begin{array}{l} +h \\ -2rv \end{array}xx \begin{array}{l} +i \\ -2sv \end{array}x^3\right\} \text{ in } y$$
$$-2rt \quad -2st$$

$$\left.\begin{array}{l} +k \quad +lx +mxx \\ -tt-2tvx-vvxx \end{array}\right\} \text{ in } (yy=) \begin{array}{l} q+rx+sxx \\ +ty+vxy \end{array}$$

$$\left.\begin{array}{l} +nq+nrx+nsxx+psx^3 \\ +pqx+prxx \end{array}\right\} \text{ in } y$$

$$\left.\begin{array}{l} +nt+nvx+pvxx \\ +ptx \end{array}\right\} \text{ in } (yy=) \begin{array}{l} q+rx+sxx \\ +ty+vxy \end{array}$$

[That is of form]
$$\begin{array}{l} A+Bx+Cxx+Dx^3+Ex^4 \\ +\overline{F+Gx+Hxx+Ix^3} \text{ in } y \end{array} \quad [=0.]$$
$$+\overline{K+Lx+Mxx} \text{ in } [yy=] q+rx+sxx$$
$$+ty+vxy$$

[Whence]
$$y = \frac{A+Bx+Cxx+Dx^3+Ex^4}{F+Gx+Hxx+Ix^3} \cdot \quad ^{(11)}$$

---

[9]

[Given]
$$\begin{array}{l} a+bx+cxx+dx^3 \\ +\overline{e+fx+gxx} \text{ in } y \\ +\overline{h+ix} \text{ in } yy \end{array} = y^3 = \begin{array}{l} k+lx+mxx+nx^3 \\ +\overline{q+rx+sxx} \text{ in } y \\ +\overline{t+vx} \text{ in } yy. \end{array}$$

[and by subduction]
$$\frac{\begin{array}{l} k+lx+mxx+nx^3 \\ +\overline{p+qx+rxx} \text{ in } y \end{array}}{s+tx} = yy.$$

[There is]
$$\begin{array}{l} a+bx+cxx+dx^3 \\ +\overline{e+fx+gxx} \text{ in } y \end{array} = \overline{y-h-ix} \text{ in } [yy \text{ or}] \frac{\begin{array}{l} k+lx+mxx+nx^3 \\ +\overline{p+qx+rxx} \text{ in } y \end{array}}{s+tx}$$

[Or]
$$\begin{array}{l} A+Bx+Cxx+Dx^3+dtx^4 \\ +\overline{E+Fx+Gxx+gtx^3} \text{ in } y \end{array} \quad [=\overline{p+qx+rxx} \text{ in } yy+\&\text{c}].$$

---

(11) Where, we need but lightly insist, the coefficients *A*, *B*, *C*, *D*, *E* are not those of the preceding equation. Newton is again interested only in the algebraic 'shape' of his result.

[And by reduction] $\quad y = \dfrac{A+Bx+Cxx+Dx^3+Px^4[+\&c]}{E+Fx+Gxx+Hx^3[+\&c]}$.

[Whence] $\quad \dfrac{k+lx+mxx+nx^3}{+p+qx+rxx} \text{ in } y = \overline{s+tx} \text{ in } \dfrac{AA * * * * +PPx^8}{EE * * * +Hx^6}$. [12]

§2.[1] THE CARTESIAN EQUATION OF THE NODAL CUBIC WHICH IS MANUALLY DESCRIBED BY A PIVOTED, SLIDING 'SECTOR'.[2]

[1]   [Put] $AF = a = FG = \tfrac{1}{2}ED$.[3]

$FR = b. \ AR = a - b = EC = GP.$

$GR = a + b = CD$.[4] $FT = x = BC.$

$AB = z$. [that is] $CT = a - z.$

[Whence]

$\quad TD$[5] $= \sqrt{2ab + bb + 2az - zz}.$

[and in consequence]

$FD = PE$

$\quad = x + \sqrt{2ab + bb + 2az - zz}.$

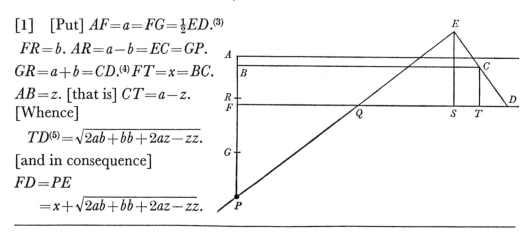

(12) As we have already remarked in note (4) above, the improved versions of this scheme of elimination between two cubics which survive on Bodleian MS New College 361.3:2^{v} have earlier been reproduced on v: 518–19.

(1) Add. 3964.8:28^{r}. In the top right-hand corner of this loose scrap Newton has scribbled a reminder to himself that 'Desunt Com[m]er Epi[s]t 15,1. 24,1. 88,1. 92,1. 207': this would seem to refer to deficiencies of some sort or other which he had noticed on pages 15, 24, 88, 92 and 207 of the proof sheets of the second edition of his collection of the *Commercium Epistolicum D. Johannis Collins et Aliorum, de Analysi Promota* which he was at this time (about the end of 1721 or the beginning of the next year) seeing through the press in London. (In his preliminary list of *errata* for the 1722 *Commercium* which is now ULC. Add. 3968.14:277^{r} he has indeed tabled an emendation of 'P. 15 l. 14' to read 'data' which was subsequently made good in the published book.) But the exact significance of the adjoined codings ',1' is not clear to us.

(2) See pages 236–7 of the parent 'Arithmetica' (v: 464–6). Here as there, the *norma PED*, right-angled at *E*, is conceived to travel smoothly along (in the plane of the paper) so that its one leg passes ever through the pivot *P* while the end-point *D* of its other leg *ED*, of fixed length, lies perpetually in the base-line *FD*: where the point *P* is at a distance vertically below *F* equal to *ED*, the mid-point *C* of *ED* traces out a *Cissois Veterum*, as Newton had already proved in 'Prob. 20' on pages 81–2 of the 'Arithmetica' (see v: 218–20), but in restating the construction on its page 237 he carelessly put the distance *PF* to be only $\tfrac{1}{2}ED$ and in his accompanying figure (see v: Plate III, opposite p. 466) he erroneously drew the ensuing 'cissoid' to have its cusp at *P*. His related determination of the more general conchoidal cubics traced out in this fashion by other points of the leg *ED* of the 'sector'—and by whose aid he there sought to

[Furthermore] $CD = a + b \cdot DE = 2a :: TD \cdot SD = \dfrac{2a}{a+b} \sqrt{2ab + bb + 2az - zz}.$

[so that] $FS^{(6)} = x + \dfrac{b-a}{a+b} \sqrt{2ab + bb [+] 2az - zz}.$

[Again] $a + b \cdot a - z :: x + \sqrt{2ab + bb + 2az - zz} \cdot FS = \dfrac{ax - zx}{a+b} + \dfrac{a-z}{a+b} \sqrt{}.^{(7)}$

[Whence]$^{(8)}$ $\dfrac{+bx + zx}{a+b} + \dfrac{b - 2a + z}{a+b} \sqrt{2ab + bb [+] 2az - zz} = 0.^{(9)}$

---

[2] [Put] $AF = FG = CD = a. \quad GP = EC = b. \quad ED = F[P] = c.^{(10)}$
[and] $FT = x = BC. \quad BF = CT = y.$ [so that] $TD = \sqrt{aa - yy}.$ [and therefore]

$$FD = x + \sqrt{aa - yy} = PE. \quad \frac{ca}{\sqrt{aa - yy}} = QD = PQ. \quad \frac{cy}{\sqrt{aa - yy}} = EQ.$$

[and so] $\dfrac{ca + cy}{\sqrt{aa - yy}} = PE.$ [Whence]$^{(11)}$

$$ca + cy = aa - yy + x\sqrt{aa - yy}. \quad [\text{or}] \quad ca - aa + cy + yy = x\sqrt{aa - yy}.$$

---

construct the cubic equation $x^3 \pm px^2 - qx + r = 0$ (see v: 466)—was, we need not emphasize, likewise mistaken. We have already mentioned (see v: 469, note (686)) that the slip passed unremarked by Whiston into the *editio princeps* in 1707, and was only caught by Newton himself after the sheets of the *Arithmetica*'s second edition were printed off, last-minute correction of the error there being made by cutting out the original pages 315/16 and pasting in a corrected sheet on the stub. The present calculation, itself faulty in its first state, is manifestly that by which he made good his lapse of forty years earlier—one which, so far as we know, no one else had in the meantime come to observe.

(3) This should be '$= CD$'. Since in the sequel the point C is $(b \neq 0)$ not the mid-point of ED, the ensuing calculation is fundamentally faulted; see the next note.

(4) This geometrical equality can be true only when $b = 0$, and so R coincides with F, since correctly $CD = GF$. In this instance C is the mid-point of the leg ED, and it traces a simple *cissois Veterum*.

(5) That is, $\sqrt{(CD^2 - CT^2)}$ where $CT = BF = a - z$.

(6) Namely $FD - SD$.

(7) The expression understood within the radical sign is of course again '$2ab + bb + 2az - zz$'.

(8) On equating the two values for FS.

(9) As we have remarked (see notes (3) and (4) above), Newton's argument is only valid when $b = 0$ and hence C is the mid-point of ED: in this instance this Cartesian defining equation reduces to be the pair of $z = 0$ and the classical cissoid $x^2 z = (2a - z)^3$ traced by the mid-point C.

(10) That is, $c = a + b$. In the sequel Newton will in fact employ only the constants a and c.

(11) On equating the two values of PE found, and then multiplying through by $\sqrt{(a^2 - y^2)}$; the latter, as Newton afterwards appreciates, introduces the unwanted factor $a + y = 0$ or $y = -a$ (corresponding to the particular solution where the locus of C is the straight line through G parallel to FD).

[and so on squaring]

$$a^4 - 2a^3c + aacc + 2accy + 2acyy + 2cy^3 + y^4 = aaxx - yyxx.^{(12)}$$
$$-2aacy \quad +cc$$
$$-2aayy$$

[Better] $\quad c = a - y + x \dfrac{\sqrt{a-y}}{\sqrt{a+y}}.$ [or] $\quad c + y - a = x \sqrt{\dfrac{a-y}{a+y}}.$ [Whence]

$$cc + 2cy - 2ca + yy - 2ay + aa = \frac{a-y}{a+y}\, xx.^{(13)}$$

---

(12) Where (compare the previous note) the two sides of the equation have the common factor $a+y$, which therefore needs to be eliminated. Newton passes to do so at source.

(13) That is, $x^2(y-a) + (y+a)(y+c-a)^2 = 0$: the required Cartesian equation of the conchoidal cubic traced by the general point C of the describing leg ED of the sector PED. If for $ED - CD = c - a$ we here substitute its equal $EC = b$, there results (as we need not elaborate) the instance where $e = 0$ (and $c \to a$) of the yet more general 'Descriptio cujusdam generis Curvarum secundi ordinis' which Newton had set down some thirty years before on the page (f. 78^r) of his Waste Book which is reproduced on VII: 658. If, alternatively, we fix the point L in GF such that $LF = PG = b$, and refer the curve (C) to the new abscissa

$$LB = y + b = X,$$

then its defining equation assumes the form $x^2(X-c) + (X+a-b)X^2 = 0$ or equivalently $X^3 + (a-b)X^2 + x^2X - cx^2 = 0$; whence the real root(s) of the general cubic equation $X^3 \pm pX^2 + qX - r = 0$ may be constructed by putting $a - b = p$ and $c\,(= a+b) = r/q$, that is, making $GF\,(=PL) = a = \frac{1}{2}(p+r/q)$ and $PG\,(=LF) = b = \frac{1}{2}(-p+r/q)$, and then determining the point(s) of the cubic traced by the point C of the leg $ED = PF$ (where, as before, $CD = GF$) which are at the distance $BC = x = \sqrt{q}$ from the axis PA. In his corrected version of this construction on the reprinted pages 315–16 of the 1722 *Arithmetica* (we have cited this in full in V: 469, note (686)) Newton has, in his endeavour to adapt this as closely as possible to his preceding analogous 'manual' description of the simple cissoid traced by the mid-point of ED (see the figure on V: 467), somewhat masked this simple scheme by transposing G to be the mid-point of PF and maintaining A as its mirror-image in FD (so that now $AF = FG = \frac{1}{2}c = \frac{1}{2}r/q$ and $GL = \frac{1}{2}(a-b) = \frac{1}{2}p$) with the consequence that the parallel through A to FD is no longer the cubic's asymptote (which is shown as the broken horizontal line through A' in our figure in the footnote on V: 468). Newton's last act at this point in his 1722 revise was, as we before remarked, to omit the too hastily dismissive sentence 'Sed in his nimius sum' (But I am too much on these matters) with which he terminated his outline of the construction in his original 'Arithmetica' (see V: 466). Let us be equally laconic in repressing wider comment as we conclude this our final footnote upon the corpus of Newton's technical mathematical papers.

PART 2

NEWTON'S VARIED EFFORTS TO SUBSTANTIATE HIS CLAIM TO CALCULUS PRIORITY

(1712–1722)

INTRODUCTION

It would leave behind something of a hollow were we to bring this edition of Newton's mathematical papers to its close without saying a little of his wrangle with the Leibnizians over first 'invention' of the principles and techniques of the calculus: that long-festering boil of antagonistic dispute which came slowly to a head over many years and broke suddenly open early in 1712 when Leibniz directly appealed to the Royal Society to sustain what he regarded as his own inviolable author's right to its discovery, and which polluted the whole European world for a decade afterwards with the corruption of its discharging pus. We have ourselves no interest in assiduously tracing the detailed cut and thrust and countering parry of the spats and wranglings which went endlessly on, in public view and behind scenes. And we shall not be blamed unduly for that. Whatever the secondary value of the bulky records of this relentless dispute over false priorities may be as case histories of the contemporary techniques of propagandist wilful mis-implication and the toe-to-toe verbal slanging match, those who have sat down without prejudice to read through their thick dossiers of scribbled manuscript and more neatly printed items have always risen bleary-eyed and shaking their heads that so many hundreds of thousands, if not millions, of words were penned to so little effect other than for nearly a century afterwards near-completely to sever fruitful contact between British and Continental mathematicians. But despite the essential sterility of so much of these out-pourings, born more often of emotional prejudice than calm, rational assessment, it is undeniable that both Newton and Leibniz remained convinced throughout of the objective, historical truth of their claim to be first 'inventor' of what the one called 'infinitesimal analysis' and the other named *calculus differentialis et summatorius*. And there is some limited value in exploring the broad outline of the claim which each presented.

Lopsided and unfair though this may seem, we leave the specialist students of his extensive manuscript *Nachlass* in Hanover to root out what further complements to Leibniz' published papers on the dispute[1] may yet still survive there.

---

(1) We mean not only his extensive correspondence with his contemporaries, but especially the long essay (unprinted in his own day) where Leibniz attempted with commendable honesty to set out, in about the autumn of 1714, what he conceived to be the true 'Historia et Origo Calculi Differentialis'. (This autobiographical essay, first published by C. I. Gerhardt in 1846 'aus den Handschriften der Königlichen Bibliothek zu Hannover' where its original still rests, and reprinted by him in *Leibnizens Mathematische Schriften*, 5 (Halle, 1858): 392–410, is also to be read in a not entirely accurate Englished version by J. M. Child in his gathering of *The Early Mathematical Manuscripts of Leibniz translated from the Latin texts published by . . . Gerhardt* (Chicago/London, 1920): 22–57, packed out with a now badly dated—and never very percipient—stream of would-be 'critical and historical [foot]notes'.) Of the extensive secondary

We have ourselves already, in the first volume,[2] raided the equally rich store of Newton's private papers in Cambridge University Library[3] to stitch together a patchwork of his retrospects from old age back over his earliest researches in calculus; and we have elsewhere not been shy in citing other snippets from the anti-Leibnizian writings of his last years to fill in the otherwise often thinly known background to his later papers on infinitesimal analysis. It will be a perhaps not entirely unfitting termination to our reproduction of the known corpus of his mathematical work if we adjoin the full texts of the autographs from which we previously selected our chosen snippets, and also present in introduction a *précis* of the dispute over calculus priority which will set in context their nuances of reference and allusion. And above this, we believe, there is point in reading the uninterrupted flow of Newton's words as he sets out his own case, historically ignorant and often wilfully distorted from the past truth though its detail was and is.

Newton's reluctance in his later life to make public the detail of his early researches into fluxional increase—whether it was through apprehension at how they might be received by a new generation of mathematicians who had outgrown the primitive algorithms in which he couched his methods, or merely because of a personal disinclination to go to the hard effort necessary in reshaping their texts for the press, is not here important to decide—made a head-on clash sooner or later with the Continental proponents of the primacy and superiority of Leibniz' rival calculus of infinitesimals virtually inevitable. After the middle 1690's, certainly, when both Newton and Leibniz retired from active investigation of the mathematical unknown, each of them already deeply convinced that he it was who had 'first invented' the mode and basic technique of the new analysis, only the time and place of the coming confrontation was any longer in doubt. Meanwhile each basked contentedly in the warm glow of being

---

literature which now chronicles and passes informed judgement upon Leibniz' independent path to his discovery of calculus let us mention only, among nineteenth-century monographs which it is still useful to read, Gerhardt's *Die Entdeckung der Differentialrechnung durch Leibniz, mit Benutzung der Leibnizischen Manuskripte*...(Halle, 1858; English version in Child (1920): 59–158) and H. Sloman's much harsher examination of *The Claims of Leibniz to the Invention of the Differential Calculus* (Cambridge, 1860); and, of more recent accounts and assessments, J. E. Hofmann's authoritative study of *Die Entwicklungsgeschichte der Leibnizschen Mathematik während des Aufenthaltes in Paris (1672–1676)* (Munich, 1949; translated into English and extensively revised as *Leibniz in Paris, 1672–1676: his Growth to Mathematical Maturity*, Cambridge 1974).

(2) See I: 148–52.

(3) Notably Add[itional MS] 3868, the so-called 'fluxion priority box' (now split into two separate parts for ease of handling). All manuscripts whose present location is not specified in the sequel are to be understood as being in this main corpus of Newton's scientific papers in the University Library, Cambridge.

regarded, in his own country, as founding father-figure of the 'new' mathematics. Any spark of reaction there might have been in England to Leibniz' several provocative assertions abroad of his 'droit d'Inventeur' of the 'nouveau calcul'[4] was speedily quenched by Newton's wariness and lassitude before it could be fanned into open flame. When, in particular, Nicolas Fatio de Duillier in 1699 set in print[5] his own claim to belated but independent discovery of the fundaments and principal algorithms of the infinitesimal calculus, and added in codicil that he was 'compelled by the very evidence of the facts' (he went on to specify his 'letters and other manuscript texts') to acknowledge Newton as 'first inventor, and the oldest one by several years' despite Leibniz' show of 'everywhere attributing the invention of this calculus to himself',[6] Newton himself was doubtless well pleased with this unsolicited testimonial to his priority, but also at the same time not a little apprehensive that he might be called upon to confirm the truth of Fatio's statement; at any event he took precautions to remind himself of the sequence of his earliest mathematical discoveries.[7] Leibniz, on the other hand, must have been furious to read Fatio's strong hint that he, the 'second inventor', might have 'borrowed' something

---

(4) We have already (see II: 262–3, note (17)) cited the self-important article in the *Nouvelles de la Republique des Lettres*, **27**, 1706: 521–8 where Leibniz used this phrase in correcting a careless preceding reference by its editor, Bernard, to the Bernoulli brothers' extensive prior publication of its 'Système' of basic algorithms and techniques.

(5) In his short pamphlet displaying as its primary function the *Lineæ Brevissimi Descensus Investigatio Geometrica Duplex* (on whose argument see **1**, 1, Appendix 2: note (1) preceding).

(6) To quote the original Latin words in which Fatio answered his self-posed question: 'Quæret forsan Cl. Leibnitius, unde mihi cognitus sit iste Calculus, quo utor?' (*Lineæ Brevissimi Descensus Investigatio...*: 18), he went on to respond that 'Ejus...Fundamenta universa, ac plerasque Regulas, proprio Marte, Anno 1687, circa Mensem Aprilem & sequentes, aliis deinceps Annis, inveni; quo tempore neminem eo Calculi genere, præter me ipsum, uti putabam. Nec mihi minus cognitus foret, si nondum natus esset Leibnitius....Newtonum tamen primum, ac pluribus Annis vetustissimum, hujus Calculi Inventorem, ipsa rerum evidentia coactus, agnosco: a quo utrum quicquam mutuatus sit Leibnitius, secundus ejus Inventor, malo eorum, quam meum, sit Judicium, quibus visæ fuerint Newtoni Litteræ, aliiique ejusdem Manuscripti Codices. Neque modestioris Newtoni Silentium, aut prona Leibnitii Sedulitas, Inventionem hujus Calculi sibi passim tribuentis, ullis imponet, qui ea pertractarint, quæ ipse evolvi, Instrumenta'. (Slightly variant English translations of this passage are given by L. T. More in his *Isaac Newton: A Biography* (New York, 1934): 574–5 and by A. R. Hall in *Correspondence of Newton*, **5**, 1975: 98, note (3).) We may remind that Fatio had (see VII: 12) been shown by its author the greater part of the original full-length treatise 'De quadratura Curvarum' during Newton's stay in London in the early winter of 1691/2; and that he may well have seen further items from Newton's private hoard of earlier papers in the course of his own visit to Cambridge the following autumn.

(7) For surely it is not mere coincidence that it was just at this time—'July 4th 1699' by his manuscript dating at its head—that Newton set down in the little notebook (now ULC. Add. 4000) where he had penned many of his earliest mathematical notes the oft-quoted autobiographical passage (*ibid.*: 14v) whose text is reproduced on I: 7–8.

of his own differential calculus from Newton's writings, though at the time he did not speak out publicly against this false imputation.[8] Indeed it would seem that Fatio's slur still rankled five years afterwards when Leibniz came to pass his public opinion upon Newton's newly appeared 'Tractatus de Quadratura Curvarum'. In his anonymous review of this fluxions tract in the Leipzig *Acta* in January 1705 he thought fit to make a comparison in itself not unjust, but in the overtones of its context demeaning to Newton's mathematical originality by all but implying that the latter had done no more in essence than transmute Leibniz' prior discovery of calculus: just as Fabry more than thirty years before had replaced Cavalieri's 'indivisibles of *continua*' by equivalent continuously generated 'flowing motions', so now (claimed Leibniz in his unsigned comment) 'in place of Leibnizian differences Mr Newton employs fluxions, and has ever employed them'.[9]

It is easy, two and a half centuries afterwards, to be wise and tolerant, to urge that when six years later Newton and his most steadfast supporter John Keill convinced one another that Leibniz had intended his statement to reflect his own sure priority in the discovery of the calculus they read what was not there. The conviction of being prime 'inventor' of the essential method of relating 'indivisible' limit-increases was strong in Leibniz, and, whatever his particular statement of it may have been intended to mean, any attempt to attack its validity was to be taken as an impugning of his rights of priority. The phrase which he used in anonymous recension of Newton's first published piece on his algorithm of fluxions certainly took no notice of the latter's own firm statement in his introduction to the 'De Quadratura' that he had fallen on the 'Method

(8) Writing privately to John Wallis on 4 August 1699 (N.S.) he complained with some vehemence, as one 'qui inter veteranos Collegas [Societatis Regiæ] me esse glorior', about Fatio's 'in me immerita et inexpectata incursatio', inquiring plaintively 'Quid tantum commeritus sim, ut ita læderer publice [?]'. Wallis diplomatically passed the complaint on to the Royal Society's secretary, Hans Sloane, and meanwhile he smoothed Leibniz' ruffled feathers with a suitably mollifying reply on 29 August. (See *Leibnizens Mathematische Schriften*, 4 (Halle, 1889): 70–3.) In Leibniz' anonymous review of Fatio's tract the following autumn (in *Acta Eruditorum* (November 1699): 510–13) the offending passage is not mentioned; but in his 'G. G. Leibnitii Responsio ad Dn. Nic. Fatii Duillerii Imputationes' which appeared six months later (*Acta* (May 1700): 198–206 [= *Mathematische Schriften*, 5 (Halle, 1858): 340–9]) he did remark with sour sarcasm (*ibid.*: 202) that 'Neque...mirum est odisse pronam quam vocat... sedulitatem meam, ...quemadmodum quidam Veterum dicebat: *pereant qui ante nos nostra dixere*....Pauci ad tantum virtutem perveniunt, ut noxiam sibi virtutem alterius amare possint; quanto minus, si (ut ipse de me) suspiciones sibi fingant, ...non recta via sed obliquis artibus alium ad laudem esse grassatum'.

(9) *Acta Eruditorum* (January 1705, especially 35: 'Pro differentiis...Leibnitianis Dn. Newtonus adhibet semperque adhibuit fluxiones..., quemadmodum & Honoratus Fabri in sua *Synopsi Geometrica* motuum progressus Cavallerianæ Methodo substituit'. We have already dwelt at some length upon the justice of this observation in our introduction to Part 1 (see pages 26–30 above).

of Fluxions' which he there used 'in the years 1665 & 1666'[10]—a decade, almost, before Leibniz had contrived his *calculus differentialis*. Once it had been uttered, it lay there ready to blow sky-high at the touch of a light all future good relations between Newton and Leibniz until the latter's death in November 1716, and long years beyond.

For some little further while, however, all remained outwardly quiet. If Newton ever glanced through the text of this only considerable review which his 'De Quadratura' received when it first appeared, the event left no memory with him.[11] But in 1708 John Keill (a sometime *protégé* of David Gregory's who had followed him south from Edinburgh to Oxford in the 1690's, and subsequently made local reputation both there and in London for the zest and vigour of his teaching rather than for the depth and originality of his talent) added to an open letter to his friend Edmond Halley advocating and expounding the employment of the tangential polar in giving fluxional solutions of problems of central-force orbits along lines pioneered by Newton and de Moivre[12] a short paragraph where he stated in aside:

All these things follow from the nowadays very celebrated arithmetic of fluxions, whose first inventor was beyond all doubt Mr Newton, as will readily be agreed by anyone who reads the letters published by Dr Wallis;[13] the same arithmetic was, however, afterwards published with changes in name and notation by Mr Leibniz in the *Acta Eruditorum*.[14]

This reissue of Fatio's earlier insinuation of Leibniz' debt to Newton for his

---

(10) 'has motuum vel incrementorum velocitates nominando *Fluxiones* & quantitates genitas nominando *Fluentes*, incidi paulatim *Annis* 1665 & 1666 in Methodum Fluxionum qua hic usus sum in Quadratura Curvarum' (*Opticks* (London, ₁1704): ₂166). Newton's original autograph of this claim-staking affirmation is reproduced on page 122 above in its context of the 'Introductio' to the printed *Tractatus de Quadratura Curvarum*.

(11) To Sloane (was it?) he drafted in April (?) 1711 a letter declaring of the closing pages of Leibniz' 1705 *Acta* review that he had 'not seen those passages before' (see his *Correspondence*, 5, 1975: 117).

(12) 'Jo. Keill ex Æde Christi Oxoniensis, A. M.₍,₎ Epistola ad Clarissimum Virum Edmundum Halleium Geometriæ Professorem Savilianum, de Legibus Virium Centripetarum', *Philosophical Transactions*, 26, No. 317 [for September/October 1708]: §II: 174–88. On the technical background to Keill's article and his acknowledged debt therein to Newton (by way of de Moivre) see VI: 548–9, note (25).

(13) Understand in the 'Epistolarum Collectio' appended by him to his *Opera Mathematica*, 3 (Oxford, 1699): 617–95. Keill refers particularly, of course, to Wallis' first full printing of Newton's *epistolæ prior et posterior* to Leibniz in 1676 (*ibid.*: 622–9, 634–45 respectively; compare VII: xxvi).

(14) 'Hæc omnia sequuntur ex celebratissimâ nunc dierum Fluxionum Arithmeticâ, quam sine omni dubio Primus Invenit Dominus Newtonus, ut cuilibet ejus Epistolas à Wallisio editas legenti, facile constabit[;] eadem tamen Arithmetica postea mutatis nomine & notationis modo, à Domino Leibnitio in *Actis Eruditorum* edita est' (*Philosophical Transactions*, 26: 185).

insights into the foundations of infinitesimal calculus neatly (if maybe not intentionally) twisted the former's 1705 simile to make Leibniz a Fabry to Newton's Cavalieri, of course, and from then on it would have taken an angel such as none of the participants were in order not to be stung into direct response. The issue of the *Philosophical Transactions* making public Keill's 'Epistola ad Halleium' did not appear, it would seem, until 1709 and the bound volume containing it not for another year, and when a copy of this arrived in Hanover (from the Royal Society's secretary Hans Sloane) Leibniz chanced to be away in Berlin on court business, and he did not see it till his return. But his anger blazed out when he at last read Keill's offending paragraph in February 1711, and he made haste to answer it. Because, Leibniz irately and scornfully wrote back to Sloane, Keill's renewal of Fatio's 'most silly accusation' of his having published Newton's 'calculus of fluxions' in changed form and notation—a charge whose falsity 'no one knew better than Newton himself' and for which (Leibniz claimed at least) Fatio himself had been 'taught better'—might even so 'be often repeated by the wicked or the foolhardy', he was driven to seek remedy for the libel from the Royal Society itself: he had not even heard speak of the name of fluxions or seen Newton's name for them before the account of them published by Wallis in his *Opera*, and Keill must now in fairness 'publicly testify that he had no mind to impute to me, what his words seem to insinuate, that I learnt my discovery, such as it was, from another and ascribed it to myself'.[15]

Such a formal complaint laid by one (longstanding, if foreign) Fellow of the Society against another[16] placed Newton, both (as Keill would have it) the plagiarised party and also the Society's President called upon by Leibniz to redress (as the latter saw it) a gross calumny, in an exceedingly delicate midway position. Not that his private confidence in his own absolute priority in dis-

---

(15) Leibniz to Sloane, 4 March 1711 (N.S.): 'Olim Nicolaus Fatius Duillierius me pupugerat in publico scripto, tanquam alienum inventum mihi attribuissem. Ego eum... meliora docui....Et tamen D. Keillius...renovare ineptissimam accusationem visus est.... Id quidem quam falsum sit nemo melius ipso Newtono novit: certe Ego nec nomen Calculi fluxionum fando audivi, antequam in Wallisianis *Operibus* [*viz.* 2 (Oxford, 1693): 392–6 (= VII: 172–80)] prodiere...: quomodo ergo aliena mutata edidi quæ ignorabam?...[At] quia verendum est ne [accusatio] sæpe vel ab improbis vel ab imprudentibus repetatur; cogor remedium ab inclyta Vestra Societate Regia petere. Nempe æquum esse Vos ipsi credo judicabitis, ut D. Keillius testetur publice, non fuisse sibi animum imputandi mihi quod verba insinuare videntur, quasi ab alio hoc quicquid est inventi didicerim et mihi attribuerim' (*Correspondence of Newton*, 5, 1975: 96). Slightly to blunt the hammer-blows of Leibniz' stern rhetoric, we may recall that Newton had briefly referred to the 'velocitates incrementorum vel decrementorum, quas etiam...fluxiones quantitatum nominare licet' in Lemma II of Book 2 of his *Principia* ($_1$1687: 251; compare IV: 522).

(16) Leibniz had been admitted F.R.S. in April 1673 in the wake of his first visit to London two months before; Keill's election exactly twenty-eight years later in April 1701 followed by a few months his appointment as the Society's clerk.

covering the calculus in his preferred fluxional form of it was at all shaken by Leibniz' outburst when it was received in London, but sheerly because he did not want to have to take sides, he seems to have hedged for a while. And so when Leibniz' letter was read out to the Society in late March it was noncommittally resolved by those present, no doubt with Newton's tactful acquiescence if not at his positive prompting, that Sloane be 'ordered to write an Answer'[17] without any direction as to how the reply should be couched. But Keill, it is clear, was not prepared meekly to fall in with any mollifying response. Having (we may presume) in the interim briefly talked the matter over with him, on 3 April he sent Newton a copy of the anonymous *Acta* review of his book, desiring him to read the offending passages in it 'from pag. 3[4]...to the end'.[18] What Newton saw there was enough to bring him heavily down from his presidential fence, deciding him not only to let Keill have his head, but to support him behind-scenes in proving his 'libel' of Leibniz true. He straightaway dashed off a brisk note to Sloane, informing him that

Upon speaking wth Mr Keil about ye complaint of Mr Leibnitz concerning what he had inserted into the *Ph. Transactions*, he represented to me that what he said there was to obviate the usage which I & my friends met wth in the *Acta Leipsica*, & shewed me some passages in those *Acta*, to justify what he said....upon reading them I found that I have more reason to complain of the collectors of ye mathematical papers in those *Acta* [who] every where insinuate to their readers that ye method of fluxions is the differential method of Mr Leibnitz & in such a manner as if he was the true author & I had taken it from him, & give such an account of the Booke of Quadratures as if it was nothing else then an improvement of what had been found out before by Mr Leibnitz Dr Sheen [Cheyne] & Mr Craig.[19]

Two days afterwards on 5 April, when Keill similarly insisted to the assembled Royal Society that 'in the Lipsick *Acta Eruditorum* for ye year 1705, there is an unfair Account given of Sir Isaac Newton's Discourse of Quadratures, asserting ye Method of Demonstration by him made use of to Mr Leibnitz', the minutes of the meeting record that he was supported by 'the President [who] gave a short account of that matter, with ye particular time of his first mentioning or discovering his Invention, referring to some Letters published by Dr Wallis. Upon which Mr Keill was desired to draw up an account of ye Matter in dispute,

(17) So it was minuted in the Royal Society's Journal Book, xi: 209.
(18) See *Correspondence*, **5**: 115.
(19) From the more finished of the two drafts on Add. 3968.30: 440^v; the text of this is reproduced in full in *Correspondence*, **5**: 117. On the works by Cheyne (*Fluxionum Methodus Inversa*, London, 1703) and Craige (*Tractatus de Figurarum Curvilinearum Quadraturis*, London, 1693) to which Leibniz chiefly referred in his review in mentioning the 'tractatus de Quadraturis Dn. Cheynæi, Medici Scotici Londini degentis [et] Dn. Craigii Scoti' (*Acta* (January 1705): 35–6) see notes (47) and (48) to the Introduction to Part 1 of the present volume.

and set it in a just light, and also to vindicate himself. . .'.[20] And a week later when those minutes were read out at the Society's next meeting, giving 'occasion to further discussion of yᵉ Matter mentioned in yᵉ Leipsick *Acta*', Newton from his presidential chair was in addition 'pleased to mention his Letters many years ago to Mʳ Collins about his method of treating Curves &c'[21] and 'Mʳ Keill being present was again desired to draw up a Paper to assert the President's right in this matter'.[22]

How narrowly Newton was involved in framing the response to Leibniz which Keill drew up during the next six weeks is not known. While the surviving draft of its first portion[23] shows no physical traces of his guiding mentor's pen, Newton must at the least have given its author a clear general notion of what to write, and he was certainly asked to approve the final version[24]—making some

(20) Royal Society Journal Book, xi: 210. Newton clearly benefited from Keill's stimulus in framing his preferred 'account of the matter', for in its prior draft on Add. 3968.30: 440ᵛ (compare the previous note) Newton set down in his letter to Sloane: 'Mʳ Keil. . .represents that by the letter of Mʳ L. to Mʳ [Oldenburg] dated [12 July 1675] & published by Dʳ Wallis [in his *Opera*, 3, 1699: 646] he was satisfied that Mʳ Leibnitz did not at that time use the differential method & that in my two letters [the *epistolæ prior et posterior*] sent the next year by Mʳ Oldenburg to Mʳ Leibnitz. . .I sufficiently described the nature uses & elements of that method[,] of wᶜʰ I there represented that I had written a treatise five years before & that the method of fluxions published in the book of *Quadratures* is that very method. . .'. And 'since in those [passages in the] *Acta*', Newton went on, 'the method in dispute is taken from me & given to Mʳ Leibnits & yet I never could learn that he himself ever pretended to be the first author, I desire that you would send him this'.

(21) Those, we may presume, of 10 December 1672 where (see *Correspondence*, 1: 247–8) he had worked an instance of his general rule for drawing tangents to algebraic curves (using the universal method for Cartesian curves which he had set down the year before in his 1671 tract; see iii: 120–2), and of 8 November 1676 where he briefly announced his method for the quadrature of curves 'exprest by any [algebraic] æquation of three terms' (see *Correspondence*, 2: 179–80; and compare iii: 373–85). In Keill's Latin translation the former was—at Newton's suggestion beyond a doubt—put to be the main set-piece of his ensuing reply to Leibniz (on which see note (23) following); the pertinent extract from the latter, in its original English and in a Latin rendering, was made public on the last page of the 'Excerpta ex Epistolis D. Newtoni ad Methodum Fluxionum et Serierum Infinitarum' which William Jones included in his collection of Newton's short mathematical pieces which had come out at the beginning of the year (*Analysis per Quantitatum Series, Fluxiones et Differentias. . .*, London, 1711: 23–38).

(22) Royal Society Journal Book xi: 213.

(23) Now ULC. Res. 1894, Packet 11 (2nd MS). This roughly written autograph embraces the essence of the first six paragraphs of Keill's finished letter (≈ *Correspondence*, 5: 133–5) along with a last lame assertion (compare *ibid*.: 137) that 'Hæc'—the method underlying Newton's *regula generalis* for constructing tangents as instanced (see note (21) preceding) in his letter of December 1672 to Collins—'est Fluxionum methodus quam Dⁿᵘˢ Newtonus invenit circa annum 1665 vel 1666, hoc est decem[!] annos antequam calculum suum differentialem Leibnitius in *Actis Lipsiensibus* publicavit'.

(24) The original is now Add. 3968.22: 333ʳ–338ʳ, whence it is printed in *Correspondence*, 5: 133–41. Its text was first published by Newton in his edition of the *Commercium Epistolicum D. Johannis Collins et aliorum. . .* the next year (pages 110–18).

ameliorations in its phrasing and one considerable addition to it[25]—of the text of this formal 'Letter from Mr John Keill to Dr Sloane...relating to the Dispute concerning the Priority of the Invention of the Arithmetick of Fluxions, &c: between Sir Isaac Newton and Mr Libnitz'. Of this 'A Copy...was order'd to be sent to Mr Leibnitz, and Dr Sloane desired to draw up a Letter to accompany it, before it was made publique in y^e *Transactions*, which should not be 'till after ye receipt of Mr Leibnitz's Answer'.[26]

To make good its primary claim that Newton was indeed 'first inventor of the Arithmetick of Fluxions, or Differential Calculus' and to make plausible its secondary suggestion that Leibniz might have 'drawn the principles' from the 'obvious enough indications' there given, the letter itself leaned heavily upon Newton's two mathematical *epistolæ* to Oldenburg in 1676 which were (with some intervening delay) transmitted by the latter to Leibniz,[27] combining extensive citations of points in these with direct quotation of the passage in Newton's letter to Collins on 10 December 1672 'found among Collins' papers' (and surely furnished by Newton himself?) which sets out his general rule for constructing tangents to algebraic curves,[28] and adjoining a brief remark that the corresponding rule for quadratures is 'had demonstrated from the ratio of the nascent augments of its abscissa and ordinate' at the end of his 1669 treatise 'de Analysi per æquationes infinitas' also found among Collins' papers and 'just now published by Mr Jones'.[29] Above all, perhaps, the 'prime' theorem on quadrature by series which Newton had presented in his 1676 *epistola posterior*[30] was now made the basis of a train of argument, historically shaky in its inferences, which Newton was in years to come to repeat over and over again with variations:

---

(25) Namely, a couple of lines (set within square brackets on *Correspondence*, **5**: 141) added in his hand in the manuscript to the fourth paragraph from the end: this has the effect of softening Keill's harsh preceding query 'Quis est jam qui...non videbit Indicia & Exempla Newtoni satis a Leibnitio perspecta fuisse' by adjoining 'saltem quoad differentias primas', but with the tail-sting 'Nam quoad differentias secundas, Leibnitius methodum Newtonianam tardius intellexisse videtur, quod brevi forsan clarius monstrabo'. (The bulk of Newton's other significant amendments are cited in *ibid.*: 151–2, notes (23), (35) and (39).)

(26) Royal Society Journal Book, xi: 224.

(27) '...innuebam D^{num} Newtonum fuisse primum inventorem Arithmeticæ Fluxionum, seu Calculi Differentialis; [et] in duabus ad Oldenburgum scriptis Epistolis, & ab illo ad Leibnitium transmissis, indicia dedisse perspicacissimi ingenij viro, satis obvia, unde Leibnitius principia istius Calculi...haurire potuit' (see *Correspondence*, **5**: 133).

(28) Again see note (21) above, and compare note (23).

(29) 'Anno 1669 misit Newtonus ad D^{num} Collinium Tractatum de Analysi per æquationes infinitas; quam inter schedas Collinij repertum [see ii: 206–7, note (2)] D^{nus} Jones nuper edidit. Sub hujus fine habetur [see ii: 242–4] demonstratio Regulæ pro quadraturis curvarum nata ex proportione augmentorum nascentium abscissæ & ordinatæ' (*Correspondence*, **5**: 136).

(30) See *Correspondence*, **2**, 1960: 115–17 [=iv: 621–3]; and compare vii: 48–54.

That theorem is, however, the 5th proposition in the treatise *de Quadratura*, and its 6th proposition also has the same regard but extends to more involved cases. The third and fourth propositions are lemmas premissed to their demonstration.[31] The 2nd proposition in the *De Quadratura*, however, is already in the treatise *de Analysi per æquationes infinitas*,[32] while the first proposition is the very one which in the aforesaid *epistola* he calls the foundation of these operations, and was then of a will to conceal in transposed letters.[33] Newton also writes that he has some while ago listed in a catalogue certain theorems which shall serve for the comparison of curves with conic sections, and in the same letter writes down the ordinates of curves comparable according to that criterion— ones patently the same, to be sure, as those which he presents in the second table in the scholium after Proposition 10 in his treatise on quadratures;[34] whence it is plain enough that Newton found that table and [!] Propositions 7, 8, 9 and 10 in the treatise on quadratures (on which that table depends) some while before the year 1676 in which that letter was written. But since he says in his *epistola prior* to Oldenburg that he had left off studies of this sort during the five years past,[35] it can hence clearly enough be gathered that the propositions in the treatise on quadratures were found by Newton at least five years before those letters were written to Oldenburg. And the whole doctrine of fluxions was before that time further advanced by Newton than has right to this very day been done by anyone under the name of differential calculus.[36]

---

(31) See VII: 520, 528 and 518 respectively. Proposition V of the (revised 1693) 'Tractatus de Quadratura Curvarum' does indeed elaborate the *primum Theorema* of the 1676 *epistola posterior*; in the parent 1691 treatise the latter had been set in pride of place as Proposition I (see VII: 24–6, 50–4).

(32) Again see II: 242–4. This general problem 'Invenire Curvas quæ quadrari possunt' (see VII: 516 for its 1693 restatement) derives in fact, by way of 'Prob. 7' of the 1671 fluxions treatise (III: 194–6), from the still earlier tract of October 1666 (see I: 430).

(33) See *Correspondence*, 2: 115; and compare VII: 25, note (6).

(34) See *Correspondence*, 2: 119 and VII: 550. This table of 'Curvæ simpliciores quæ cum Ellipsi & Hyperbola comparari possunt' (as Newton retitled it in 1704; see page 132 above) in fact repeats with only small rearrangement the equivalent 'Catalogus Curvarum aliquot ad Conicas Sectiones relatarum' which Newton had previously adjoined to Problem 9 of his 1671 tract (see III: 244–54).

(35) Newton's phrase in his June 1676 letter was the slightly vaguer 'jam per quinque fere annos' in fact (see *Correspondence*, 2: 29), and that took no note of his not inconsiderable mathematical correspondence during the period late 1671–spring 1676 (on which see III: 567–71; IV: 659–66). We may add that Newton almost certainly computed the main portion of the 'Catalogus posterior' of curvilinear integrals which he set out in his 1671 tract freshly for the occasion, using any and every technique of evaluation which came pertinently to hand (compare our outline restoration of the various manners of resolution on III: 256–7). It inverts the historical sequence—however convenient a 'proof' of Newton's calculus priority it might seem to afford—to make the derivation of these particular tabulated integrals succeed in time the systematisation of their modes of reduction which Newton afterwards spelled out at length in Propositions V–X of the 'De Quadratura' (see V: 520–46; the first versions of these, *ibid.*: 36–46, 68–70, 138–46, were drafted only in late 1691).

(36) In Keill's original Latin (see *Correspondence*, 5: 136–7): 'Est autem Theorema illud propositio 5ta in Tractatu de *Quadraturis*, eodem etiam spectat ejusdem prop. 6ta sed ad Casus magis implicatos se extendit. Propositiones Tertia & Quarta sunt Lemmata Theor[ematibus]

Such arguments supporting Newton's priority in discovering and elaborating the techniques and algorithms of infinitesimal calculus were not, of course, without their core of truth, and they might, had they been put more judiciously and circumspectly to Leibniz, have served as a healthy corrective to his unduly inflated notion of his lone rôle as 'prime' inventor. But such was not to be. It was Keill's overriding purpose not merely to knock Leibniz off his pedestal, but thoroughly to discredit his pretensions. Let us not single out the several 'indications' adduced in the latter half of Keill's letter—did they come from Newton himself? we cannot know—as to how Leibniz might have drawn on Newton's 1670's letters as the fount of his own 'indivisibles' techniques of determining tangents to curves and the quadrature of their areas. The underlying premiss that Leibniz was grossly culpable of plagiarism is belied by the historical facts as we may now solidly document them; but these were then hidden from sight in Leibniz' private papers at Hanover, and we ought not perhaps too strongly to condemn Keill (and Newton) for drawing an over-partial picture of the past truth from the scanty evidence then available. The time was ripe for wilful misunderstanding.

When Keill's letter was finally put in the post, in about May 1711, it took a long time to reach its recipient: during the following summer and autumn Leibniz was away in Berlin on court matters—meeting at one point the 'great' Czar Peter of Russia, beside whom even Newton cut a lowly temporal figure—and he seems to have read it only on his return to Hanover at the beginning of December.[37] For nearly a month more, certainly, he pondered what to reply. He could surely not have failed to wonder if Newton himself was solidly behind this (as it seemed to him) disparaging attack upon his intellectual probity and

hisce demonstrandis præmissa, 2ᵈᵃ autem in *Quadraturis* propositio extat in Tractatu de *Analysi per æquationes infinitas* & prima propositio est ea ipsa, quam in dicta epistola fundamentum operationum vocat, & transpositis literis celari tunc voluit. Scribit etiam Newtonus se dudum Theoremata quædam quæ comparationi curvarum cum sectionibus Conicis inserviant, in catalogum retulisse, & ordinatas curvarum quæ ad eam normam comparari possunt, in eadem Epistola describit; quæ profecto eadem plane sunt cum ijs quas Tabula 2ᵈᵃ ad scholium propositionis 10 in Tractatu de *Quadraturis* exhibet; unde satis liquet Tabulam illam & propositiones 7, 8, 9 & decimam quæ sunt in Tractatu de *Quadraturis*, (a quibus Tabula pendet) Newtonum dudum invenisse ante annum 1676, quo scripta est Epistola illa posterior; Cum vero in prima ad Oldenburgum Epistola dicit se ab ejusmodi studijs per Quinquennium abstinuisse, hinc satis clare colligi potest, Propositiones in Tractatu de Quadraturis a Dⁿᵒ Newtono inventas fuisse Quinquennio saltem antequam Epistolæ illæ ad Oldenburgum scriptæ essent. Totamque illam de Fluxionibus Doctrinam, ante illud Tempus ulterius a Newtono provectam esse; quam ad hunc usque diem a quoquam alio factum est, sub nomine calculi differentialis'.

(37) He began his letter to Johann Bernoulli on 3 December (N.S.): 'Redux Torgavia, ubi Nuptiarum Czarigenæ solemnia spectavi et ipsi magno Russorum Czari collocutus sum...' (*Mathematische Schriften*, 3. 2: 877). What a come-down to return from this hob-nobbing with royalty to find his 'right' as first inventor of the calculus in dispute!

mathematical originality, but if such thoughts persisted with him he suppressed them from public view. In a short, painstakingly honed and polished response to Sloane in late December he pointedly forewent any attempt to defend himself against this yet 'more open onslaught' upon his 'sincerity' by a man 'learned, but newly risen', one with 'too little experience of what had been done in times past' and 'lacking a mandate to act from the person concerned' in putting his case; rather, he wrote to Sloane, he would 'entrust to your [sense of] fairness' as to whether or not to 'curb empty and unjust bawlings which offend, I think, Newton himself' and which he 'will, I am confident, readily decry'.[38]

Fine and stirring phrases indeed, but with more panache than underlying good sense. Just how ill-advised this ploy was of calling jointly upon the Secretary and President of the Royal Society—and so in effect the Society itself —to sit in tribunal over Keill in his challenge to Leibniz' priority of discovery we may only now, a quarter of a millennium after the event, begin fully to encompass. After Leibniz' reply was read out to the Society on 31 January 1711/12 it was 'delivr'd to the President to consider of the Contents thereof'[39] and Newton accordingly began a few days later to draft an official statement upon it, first as a formal letter to Sloane[40] which he then reshaped to be a direct address to the 'Gentlemen' of the Society.[41] The latter he seems to have delivered the next month on 6 March, when 'Upon account of Mr Leibnitz's Letter to Dr Sloane concerning the Dispute...between him and Mr Keill, a Committee was appointed by the Society to inspect the Letters and Papers relating thereto'.[42] To those—'Dr Arbuthnot, Mr [Abraham] Hill, Dr Halley, Mr [William] Jones, Mr Machin, and Mr [William] Burnet'—originally appointed to serve on it, Francis Robartes was added on 20 March, the Prussian ambassador Bonet was coopted the next week evidently to give a semblance of 'Germanic' balance to the committee, and on 17 April de Moivre, Francis Aston and Brook Taylor (the last had been elected F.R.S. but a fortnight before)

---

(38) Leibniz to Sloane, 29 December 1711 (N.S.): 'Quæ...Keilius nuper ad Te scripsit, candorem meum apertius quam ante, oppugnant: quem ut ego hac ætate post tot documenta vitæ, Apologia defendam, et cum homine docto, sed novo, et parum perito rerum anteactarum cognitore, nec mandatam habente ab eo cujus interest, tanquam pro Tribunali litigem: nemo prudens æquusque probabit.... Itaque vestræ æquitati committo, an non coercendæ sint vanæ et injustæ vociferationes, quas ipsi Newtono...improbari arbitror, ejusque sententiæ suæ libenter daturum indicia mihi persuadeo' (*Commercium*, ₁1712: 118[≈*Correspondence*, **5**: 207]).

(39) Royal Society Journal Book, xi: 267.

(40) Two drafts of this abortive letter are preserved in Cambridge University Library (Add. 3968.30: 438ʳ/438ᵛ→441ʳ; the former is printed in *Correspondence*, **5**: 212–14).

(41) Add. 3965.8: 79ᵛ, printed in *Correspondence*, **5**: xxiv/xxv and also in Appendix 1.1 following; this breaks off at the beginning of a sentence, but the sequel may be restored from the second of the prior drafts to Sloane (on which see the previous note).

(42) Royal Society Journal Book, xi: 275.

were also taken on to its strength. Just a week later the committee delivered in its report, predictably vindicating Newton's rights as 'first Inventor', and finding that 'Mr Keill in asserting the same has been noways injurious to Mr Leibnitz':[43] in which judgement the full gathered Society 'agreed *nemine contradicente*'.

In the sterner moral climate of a century ago[44] it seemed utterly deplorable that the powers-that-were at the Royal Society should have solidly packed its investigating committee with men so stalwartly convinced *à priori* of Newton's supremacy in the new techniques of infinitesimal analysis. (Though with the exception of John Craige and John Harris, each of whom in his own way had already made public show of his involvement and partiality, it is difficult to know which other Fellows at all knowledgeable in the recent development of the calculus might have been appointed in their place.) In our more cynical present age it will seem but justifiable *Realpolitik*, a swift and efficient means to achieve an end of whose propriety all members of the Society were already persuaded. In the historical fact as we have but recently come narrowly to know it, the

(43) Royal Society Journal Book, xi: 289.

(44) Above all in the steely-eyed gaze of Augustus De Morgan, who over a number of years took it as his puritanistic mission to ferret out the details which we summarised in the preceding paragraph; see especially his note 'On a Point connected with the Dispute between Keil and Leibnitz about the Invention of Fluxions' (*Philosophical Transactions*, **136**. 2, 1846: 107–9). De Morgan rationalised his squeamish disgust at what he found by subsequently urging that the committee's ensuing report was an '*ex-parte* statement' of 'agents and counsel' for Newton; see note † to his article 'On the Additions made to the Second Edition of the *Commercium Epistolicum*' (in *Philosophical Magazine*, (3) **32**, No. 217 [for June 1848]: 446–56). Four years later he had swung round to writing of the '*partisan* Committee' and its 'atrocious unfairness' in his 'Short Account of some Recent Discoveries...relative to the Controversy on the Invention of Fluxions' (*The Companion to the Almanac...for 1852* (London, 1852): 5–20 [= (ed. P. E. B. Jourdain) *Essays on the Life and Work of Newton* (Chicago/London, 1914): 67–101, especially 68, 75]).

Let us remark that in after years Newton himself tried hard to oppose the impression that the 1712 committee had been prejudiced in its findings or too narrowly British in its constitution. In particular in his anonymous 'Account of the Book entituled *Commercium Epistolicum*...' (*Philosophical Transactions*, **29**, No. 342 [for January/February 1714/15]: 173–224; the largest part of this is reproduced from Newton's autograph drafts of its text in Appendix 2 below) he insisted disingenuously that its members 'were not appointed to examin Mr Leibnitz or Mr Keill, but only to report what they found in the ancient Letters & Papers: and he that compares their Report therewith will find it just. The Committee was numerous & skilfull & composed of Gentlemen of several Nations, & the Society are satisfied in their fidelity in examining the hands & other circumstances, and in printing what they found in the ancient Letters & Papers so examined, without adding omitting or altering any thing in favour of either party' (*ibid.*: 221, with the orthography changed to be that of Newton's original on Add. 3968.41: 112^v). In his comment upon this passage in his ensuing letter to Conti on 9 April 1716 (N.S.) Leibniz frostily observed that 'Je n'ay pas eu connoissance du *numerous Committee of Gentlemen of several Nations*..., & je ne say pas encore presentement les Noms de tous ces Commissaires, & particulierement de ceux qui ne sont pas des Isles Britaniques' (*Correspondence of Newton*, **6**, 1976: 305).

committee was less an active partisan group than a mere passive mouthpiece for the slashing riposte which Newton could not himself be seen openly to deliver from his impartial President's chair. During its seven weeks' existence the committee's members may well have made some conscientious effort to read through the extensive corpus of letters and documents in the Royal Society's archives and among John Collins' private papers (then mostly in the possession of one of their number, Jones) which they claimed in their subsequent report to have consulted. But it really mattered not at all. Newton was not one to permit any amateurish defence of his lawful rights, and he at once put himself to a professional hunt for documents which would establish his priority with all legal watertightness. He deftly assembled pertinent 'Extracts of Letters found among the papers of Mr John Collins'[45] (making good use of the ready *entrée* to these which their owner allowed) and interlarded these with related slices of other 'Letters in ye hand of Mr Leibnitz' which he found in the Royal Society's Letter Books,[46] and the pastiche he put to the investigating committee as the documentary basis on which it should come to its findings. In the report which the committee members duly delivered to the assembled Society on 24 April these extracts from 'Letters...in the Custody of the R. Society, and those found among the Papers of Mr John Collins dated between the years 1669 and 1677' were accepted as 'genuine and authentic', and the recommendation was made that they should be 'made publick' together with 'what is Extant to the same purpose in Dr Wallis's 3d Volume [of works]'.[47] It should not shock us that the four itemised conclusions from this evidence to which the reporting committee then set its name were likewise drafted by Newton without any essential help from any of its members.[48] And we will be equally unsurprised to learn that the

(45) The first half of a tentative title for the joint 'epistolary commerce'—it continues '...relating to the collection of mathematical Letters published by Dr Wallis in the third volume of his Works'—jotted down by Newton on Add. 3968.19: 289r, where it follows the initial essay of his at introducing the 'Extracts of ye MS Papers of Mr John Collins concerning some late improvements of Algebra' whose drafts we reproduce in Appendix 1.2/3 below.

(46) See Add. 3968.19: 289v–290v. Newton there lists all the letters known to him which passed between Leibniz and the Society's secretary Henry Oldenburg during February 1673–December 1675, tagging each entry with its corresponding page in the Letter Books 'No. 6' and 'No. 7', and then under the head which we here cite makes (f. 280r) a briefer note of earlier correspondence between the two which is to be found in Letter Books IV and V, beginning with Leibniz' letter from 'Mentz 8/18 Septemb 1670'.

(47) In its gathered 'Epistolarum Collectio' of course (on which compare note (13) above). We quote here from Royal Society Journal Book, XI: 287–8. This compendium of extracts furnished by Newton for the committee's nominal inspection and approval could not have differed greatly from that assembled by him in Add. 3968.19: [in correct sequence] 265r/270r/ 266r/269r/267r/268r, which collects 'Extracts of [36] Letters found in the custody of the R. Society among the papers of Mr John Collins, relating to the Collection of Mathematical Letters published by Dr Wallis in the third volume of his works' (as his title on f. 265r puts it).

(48) The text of Newton's autograph draft of the committee's report—which does not

Society unanimously accepted its findings, ordering that these, along with 'the whole Matter, from the Beginning, with the Extracts of all the Letters relating thereto, and Mr Keill's and Mr Leibnitz's Letters', be 'published with all convenient speed'; nor any the more astonished to hear that, even though Halley, Jones and Machin were at the same meeting 'desired to take care of the said Impression (which they promised)' and Jones was additionally ordered to 'make an Estimate of the Charges',[49] it was in fact Newton who, virtually single-handedly, edited and annotated the volume of letters which was not quite a year later printed off and privately circulated in would-be complete substantiation of his priority of invention of infinitesimal calculus.

Intermittently over the ensuing late spring, summer and autumn of 1712, in fact, Newton spent a considerable amount of time and trouble in reshaping and elaborating the 'Extract of Papers & Letters' which was put to the Society at the end of April, first adjoining to it a companion 'Brief History of the Method of Series drawn from Ancient Memoirs'[50] and then combining the whole to be a comprehensive 'Epistolary Commerce of Mr John Collins and others on Advanced Analysis'.[51] Without halting minutely to substantiate our statement in the fulsome detail which his surviving store of papers relating to the fluxions dispute affords, let us here say plainly that not only did Newton select all the extracts of the letters and papers which were duly printed—many for the first time—in this *Commercium Epistolicum*, but he also penned each last nuance of the editorial fill-ins and numerous explicative footnotes which mould and condition the uncommitted reader's response in a multiplicity of subtle ways of which Leibniz afterwards rightly complained.[52] And Newton it was, moreover, who drafted each of the known preliminary versions of the book's opening address

diverge in its substance from the finished set of findings recorded in Royal Society Journal Book, XI: 288–9 on 24 April 1712 (for which see *Correspondence*, 5: xxvi–xxvii) and afterwards printed at the end of the *Commercium Epistolicum* (London, $_1$1712: 120–2)—is reproduced (from the original in private possession; a preliminary tentative in his hand survives on Add. 3968.10: 140^r) in Appendix 1.4 below.

(49) Royal Society Journal Book, XI: 289.

(50) 'Historia brevis Methodi serierum ex Monumentis antiquis desumpta' as Newton titled this Latin piece (Add. 3968.10: 123^r–125^r; preliminary drafts exist on *ibid.*: 120^r–122^v) on its first recto.

(51) 'Commercium Epistolicum D. Johannis Collins et aliorum de Analysi promota' as the book's final Latin title put it. We will remind that Newton was during this same period much preoccupied in answering Cotes' queries upon the later pages of the second edition of his *Principia* as it passed through press at Cambridge, and for several weeks in the autumn of 1712 in particular his attention was taken up by the very different problems involved in correcting (see **1**, 6 preceding: *passim*) the erroneous Proposition X of its second book, and thereafter in drafting and revising the lengthy, newly added terminal 'Scholium Generale' (→*Principia*, $_2$1713: 481–4; see also the suppressed final portion, now Add. 3970.3: 427^r–428^v, which is newly printed for the first time in *Correspondence of Newton*, 5: 362–5).

(52) Notably in his letter to Conti on 9 April 1716 (N.S.) where *inter alia* he affirmed that

'Ad Lectorem' which yet further stressed the absolute unchallengeability of his claim to priority of discovery over Leibniz by many years.[53] Such lavish overkill of an opponent was in no way unusual with Newton, but rather (as Hooke, Flamsteed and many another more minor figure before Leibniz could have witnessed) the usual reaction of this infinitely careful and cautious man; nor, after two and a half centuries, ought we to be unreasonably shocked at the covert way in which he here applied it. At length, after an unexplained delay at the printer's in the last months of 1712,[54] all was ready to make its public bow.

'dans la publication du *Commercium Epistolicum* on a supprimé des endroits qui pouvoient être au désavantage de M. N[ewton], au lieu qu'on[!] n'y a rien omis de ce qu'on croyoit pouvoir tourner contre moi par des gloses forcées' (see *Correspondence of Newton*, **6**, 1976: 306). Newton's rough draftings of virtually all the editorial interpolations and footnotes in the published *Commercium*—and also of much else which he had a mind to put in, but never did so (or severely summarised there)—are now to be found scattered in the still largely unordered bulk of his writings during 1712–22 on the dispute both in Cambridge University Library (where see especially Add. 3965.12: 220^r, 366^r/.18: 730^r; 3968.3: 10^r–13^r/.4: 15^r–16^v/.12: 158^v/.14: 243^r–246^v/.19: [eight unpaginated sheets]/.41: 19^r/19^v, 58^r–59^v, 108^r, 119^r/119^v, 132^v, 137^v, 146^r/146^v; 3977.8: 8^r–9^v) and also in private possession (compare for instance **1**, Appendix 2: note (16) preceding). From Newton's manner of citation in these autographs it is abundantly clear that many were added to an already existing printer's page proof (now lost) of the primary text of the 'epistolary commerce' itself; several of these, we have already remarked (see **1**, 6, §1.6: note (76); §2, Appendix 2: notes (16) and (43) above) were penned by him about the beginning of October on the same sheets as those where he was at the time labouring to mend the erroneous Proposition X of Book 2 of his *Principia* and refine its corrected version for the book's second edition. With knowledge of the real reason for their inclusion in the *Commercium* we may smile at the disingenuity of Newton's excuse in the last paragraph of his editor's preface to the volume that these notes were 'added to ye Letters'—and indeed also to the text of the 'De Analysi' which Newton reprinted, as the prime documentary evidence of the quality of his earliest researches in infinitesimal analysis, on its pages 3–20 let us add— 'to enable such Readers as want leasure, to compare them with more ease & see the sense of them at one reading' (so he wrote in his original English version, here reproduced in full in Appendix 1.6 below; his Latin rendering in the 'Ad Lectorem' to the published book notices equivalently that 'Subjunctæ sunt Epistolis Annotationes quædam; quò Lectores, quibus minus est otii, & Epistolas inter se faciliùs conferre, & semel perlectas intelligere queant').

(53) As David Brewster long ago observed in his *Memoirs of the Life, Writings and Discoveries of Sir Isaac Newton*, **2** (Edinburgh, 1855): 75, note 3, half a dozen separate English drafts by Newton of this preface survive (mostly in Add. 3968.37, but see also 3968.41: 36^r, 96^r–97^v, 105^r–106^r, 109^r, 115^r/115^v, 127^r and 138^r). Of these the longest one (Add 3968.37: 551^r–553^v +554^v) and the considerably shorter final version (*ibid.*: 549^r/549^v)—rendered into Latin, the latter does not differ in its essence from the printed 'Ad Lectorem' (*Commercium Epistolicum*, $_1$1712: signatures A2^r/A2^v)—are reproduced in Appendix 1.5/6 respectively. As with the footnotes to the ensuing 'Commerce of Letters' (see note (52) preceding) the time at which Newton composed these several preliminary tentatives, and hence the Latin 'Ad Lectorem' too we need not doubt, can narrowly be pinpointed to the early autumn of 1712 inasmuch as a number of preparatory draftings for them occur on worksheets where he was also engaged in setting the erroneous argument of Proposition X of Book 2 of the *Principia* to rights for its impending second edition (compare **1**, 6, §2: notes (45) and (51), and note (77) of Appendix 2.10 thereto).

(54) Writing to Johann Bernoulli on 8 October, when he came to mention the 'recueil de

On 8 January following (still 1712 in the old calendar) the Royal Society's Journal Book records that 'Some copies of a book entitled *Commercium Episto-licum* &c being brought, the President [*sc*. Newton] ordered one to be delivered to each person of the Committee appointed for that purpose, to examine it before its publication'. Manifestly no objection was lodged by anyone—as who would have dared!—and within a month some hundreds of copies of the printed book were on their way, as gifts of the Society, to institutions and selected interested individuals in Britain and on the Continent.[55]

Here is not the place to examine detail by detail the lengthy deposition of evidence which, under the ægis of the Society's *imprimatur*, Newton presents as pertinent to the case of Leibniz *v*. Keill being tried before his Presidential bench; or to point item by item to the frequent partiality and indeed hostility of the asides to the jury of readers of this 'Commerce of Letters' which he presents in footnote and interpolated running commentary as anonymous 'friend of the court'. Nor do we need to. While the *editio princeps* of the *Commercium* is a book not easily found in libraries outside of London, Cambridge and Oxford,[56] its text was reprinted ten years afterwards in 1722 with (to De

quelques lettres écrites autrefois entre Messieurs Leibnitz, Newton, Wallis, Gregory, Olden-burg, Collins...qu'on imprime par ordre de la Société...accompagnées d'un petit[!] nombre de notes marginales' de Moivre added that 'j'aurai l'honneur de vous les envoyer aussi tôt qu'elles seront imprimées: je crois que cela ne passera pas une quinzaine de jours' (see K. Wollenschläger, 'Der mathematische Briefwechsel zwischen Johann I Bernoulli und Abraham de Moivre' (= *Verhandlungen der Naturforschenden Gesellschaft in Basel*, **43**, 1931–2: 151–317): 273–4); but in his next letter on 17 December he had to make apology for still not being able to send along a copy of the *Commercium*, which 'ne sera publié que la semaine prochaine' (*ibid.*: 280). De Moivre accurately foresaw that 'ces lettres là pourront bien quelque jour exciter bien des tempêtes parmi les sçavants' (*ibid.*: 274), but Bernoulli was too wrapped up in the problem of resisted motion—and in particular in Newton's reaction to being informed that the argument of Proposition X, Book 2 in the *Principia*'s first edition must contain some fundamental error—to be interested in de Moivre's news, sight unseen of the *Commercium* itself.

(55) We may infer that the full number of copies of the book had been received at the Society by 29 January 1712/13 when the Council Minutes record that the Treasurer was to 'pay the charges of printing' (see Joseph Edleston, *Correspondence of Sir Isaac Newton and Professor Cotes* (London, 1850): lxxiii, note (148)). How, and exactly to whom, examples of the printed book were distributed is not known with clarity. On 13 February following (see Edleston, *Correspondence*: 222) Roger Cotes wrote to William Jones, acknowledging receipt of a parcel of four copies of the *Commercium* to be distributed in Cambridge, and stating that he had kept one for himself, while of the remaining three he had 'delivered one Copy to the University Library Keeper[,] another to the Library-keeper of Our [Trinity] College, and the third to Mr [Nicholas] Sanderson [the newly appointed Lucasian Professor] as from the Society'. But this was hardly a typical arrangement, and most of the other presentations were probably more formally made.

(56) We have ourselves most often used the Cambridge University Library copy, press-marked Bb.7.18*, which was (see the previous note) delivered to the 'Keeper' in early February 1713. The copy (now Trinity College, Cambridge. NQ.16.76²) which Newton kept on his own library shelf is clean of annotation.

Morgan's subsequent puritanical wrath) a few minor, unacknowledged alterations by Newton[57] and is yet more widely accessible in Lefort's *variorum* edition of a century ago.[58] A modern re-edition—or even minimally annotated photofacsimile—of any or all of these is, unfortunately, one of the many lacunas in Newtonian scholarship which yet await their turn to be filled. Suffice it to stress that the historical truth, as we can now much more precisely than ever before establish it from the still growing bulk of the known documentary record, supports none of Newton's tentative charges against Leibniz of having borrowed from his own prior-existing methods of series and fluxions in creating the equivalent techniques of the *calculus differentialis*. It is true that Leibniz fruitfully corresponded, by way of Oldenburg, with both Collins and Newton during his prime formative period in the early/middle 1670's, and in particular while on his second visit to London in October 1676 he was shown by Collins copies of a number of Newton's mathematical manuscripts, not least the 'De Analysi',[59]

---

(57) *Commercium Epistolicum D. Johannis Collins, et Aliorum, de Analysi Promota, jussu Societatis Regiæ in lucem editum: et jam Unà cum ejusdem Recensione præmissa, & Judicio primarii, ut ferebatur, Mathematici subjuncto, iterum impressum* (London, 1722; reissued with a new title-page in 1725). The main portion of this (pp. 61–244, of which pp. 145–76 [=sheets L/M] are misnumbered 129–60) reprints the 1712 text substantially as it is (title-page and 'Ad Lectorem' included); it is itself introduced by a new 6-page foreword 'Ad Lectorem'—of Newton's own composition, we need not say—and prefaced (pp. 1–58) by a Latin rendering of his own lengthy anonymous 'Account' of the *Commercium* in the *Philosophical Transactions* in 1715 (see Appendix 2 below) which once more conceals its author's identity; and it is concluded by an Appendix (pp. 245–6) in which Newton sets out the 'Judicium' delivered by Johann Bernoulli privately to Leibniz in early June 1713 and broadcast by the last six weeks later in a flysheet (see below), and then passes (pp. 247–50) to demolish its criticisms item by item in a tight-packed, ever-anonymous 'Annotatio' thereon. De Morgan has assiduously identified the greatest part of the small (and fairly innocuous) departures here made from the 1712 *princeps* text in his article (already cited in note (44)) 'On the Additions made'—he does not seem to suspect that the agent was Newton—'to the Second Edition of the *Commercium Epistolicum*'.

(58) *Commercium Epistolicum J. Collins et Aliorum de Analysi promota, etc., ou Correspondance de J. Collins et d'autres savants célèbres du XVIIᵉ siècle, relative à l'analyse supérieure, réimprimée sur l'édition originale de 1712 avec l'indication des variantes de l'édition de 1722, complétée par une collection de pièces justificatives et de documents* . . . (Paris, 1856). Though the title goes on to add *et publiée par J.-B. Biot et F. Lefort*, Biot in his 'Avertissement' to the volume generously makes it clear that 'je n'. . .ai de part que pour le projet [de cette publication]. L'exécution appartient tout entière à M. Lefort. . .'. There are not a few misreadings in the reprinted texts, and the 1712 *editio princeps* is a little confusingly run together with the 1722 re-edition to give a 'complete' work, but the main inconvenience of Lefort's *variorum* text for the Newtonian student working with contemporary documents is that he does not indicate the 1712 or 1722 page-divisions.

(59) The text of the 'Excerpta ex tractatu Newtoni Mso De Analysi per æquationes numero terminorum infinitas' which Leibniz then made is reproduced on II: 248–59; with one or two emendations this is now also printed in J. E. Hofmann's standard edition of his 'mathematischer Briefwechsel. . .1672–1676' in *Leibniz: Sämtliche Schriften und Briefe. Dritte Reihe*, 1 (Berlin, 1976): 664–77, together with (*ibid.*: 677–81; compare I: 490, note (6)) the extracts taken by him on the same occasion from Newton's letter to Collins on 20 August 1672.

but we have no ground to disbelieve his firm denial in later years that he then saw any paper in which Newton gave a systematic exposition of the foundation of his method of fluxions.[60] That Newton in the *Commercium* appealed so strongly to the secondary evidence of letters to establish the existence and extent of his own earliest researches in fluxional calculus, resting his case otherwise on the flimsy base of the short tract 'De Analysi' (which, we have argued,[61] he had prepared to support the very different end of substantiating his claim to supremacy over Nicolaus Mercator in analysis by means of infinite series), was, we may say, a lawyer's tactic which invited no more positive response from Leibniz than the arid, verbal counter-affirmation of priority of 'invention' which it forthwith provoked, and an inevitable consequent degeneration of the dispute into empty catcalling and legalistic cross-challenge. If only Newton could have seen that the only effective way of substantiating his youthful pre-eminence and achievement in the field of infinitesimal analysis was to publish to the world the essence of what he had discovered and elaborated half a century before! Except, however, for the briefest of excerpts he allowed none of his 1660's papers on fluxions to be made known to the outside world, and when the text of his fullest exposition of his basic methods—that in his suppressed 1671 treatise—at length appeared in print (in Colson's English translation) in 1736,[62] nine years after his death, it was two decades too late to have its proper impact, and it did little to raise his mathematical reputation other than among the already converted. The very exactness with which in the 1712 *Commercium*

---

(60) In a celebrated baring of his heart to Conti in a letter of 9 April 1716 (N.S.) he declared: 'Je n'ay jamais nié qu'à mon second Voyage en Angleterre j'aye vû quelques Lettres de M. N[ewton] chez Monsieur Collins, mais je n'en ay jamais vû ou M. N[ewton] ait expliqué sa Méthode des Fluxions' (*Correspondence of Newton*, 6: 307). The historical fact, as we may now know it, is a little less clean-cut than this. When Collins first met Newton seven years earlier 'somewhat late upon a Saturday night'—this must (compare II: x, note (6)) have been 27 November 1669—at the London inn where he was briefly staying, the latter had shown him the mathematical pocketbook which is now ULC. Add. 4000, and (either then or during the next few days) permitted him to make the extensive excerpts from it which (compare I: 47, note (1); 97, note (1); 560, note (2)) survive in his hand in private possession. It may be (see I: 400, note (1)) that Collins also then acquired the complete copy (in Wickins' hand) of the October 1666 fluxions tract which has been with these excerpts from his own pen since at least the early eighteenth century. In October 1676, accordingly, when Collins made Leibniz welcome at his London home, he had it in his hands to give him at least a fleeting glimpse of Newton's earliest fluxional techniques of ten years before, and must surely have been sorely tempted at least broadly to hint of their existence. Whatever suggestion was (or was not) made, it clearly did not take root in Leibniz' mind. Let us leave it as one of the all-but-might-have-beens of history that he sat briefly within a few feet of a transcript of Newton's earliest mathematical compendium, and was seemingly denied a sight of it (and maybe yet other things Newtonian) by its owner.

(61) See II: 165–7.

(62) See III: 13, note (32).

he sought to present a watertight claim in 'law' to be initial discoverer (and hence sole 'inventor') of the fundamental concepts and techniques of the calculus was to hamstring it in the eyes of those who knew that Newton had no unique right to be so regarded, and to attract the anger and derision of those who supported Leibniz in his rival contentions to the prime founder's seat of honour. And in directing that the dispute should be argued out in the courtroom of public debate, rather than deepened and enriched by the introduction of material mathematical evidence from his youthful pen, Newton more than anyone (even including Leibniz) determined that its future level of discussion would all too rarely rise above that of the sterile repetition of fixed, unyielding statements of entrenched position coupled with the equally rigid expression of unseeing prejudice.

Further to lessen the now slim chance that the two disputants might settle upon some gainful compromise, almost a year before the *Commercium* was publicly distributed Leibniz had started a secondary hare. In the *Acta Eruditorum* for February 1712,[63] namely, he had—anonymously as always—delivered himself of a review of William Jones' 1711 *Analysis* in which Newton's 'De Analysi' was first printed, there declaring that 'the method of first and last ratios preferred by the editor[64] to the method of infinitely small quantities varies from this merely in the manner of speech and according to the rigour of its demonstration', and that, unlike Leibniz' *calculus differentialis* (where the 'nullescent' *d* is the sole differential 'affection' introduced), in its fundamental appeal to separate related vanishingly small fluxional 'moments', one for each variable, Newton's equivalent method unnecessarily 'multiplied quantities'.[65] This new offending passage came to Newton's attention too late to be castigated by him in an editorial interjection in the *Commercium*, but nevertheless in the early winter of 1712–13 privately inspired a deep and complicated reaction on his part. In first impulse to complain directly (so it would appear) to the editors of the *Acta* he drafted an initial statement riposting that[66]

whereas Mr Jones... preferr'd his [Newton's] method of *rationes primæ & ultimæ* to ye method of infinitely littles[,] & the Author of ye said Account replies that ye difference is only in the mode of speaking[,] methods proceeding alike by quantities infinitely little: I

(63) Pages 74–7; the main portion of this Latin recension has been reproduced in II: 259–62.

(64) In Jones' introductory 'Præfatio Editoris', that is; the pertinent passage is quoted *in extenso* in II: 261, note (13).

(65) We part translate, part paraphrase Leibniz' Latin statement on *Acta Eruditorum* (1712): 76–7 [=II: 261–2].

(66) Public Record Office. Mint Papers II: 88ᵛ, first printed in *Correspondence*, 5: 383. On the same manuscript sheet there are two drafts by Newton of the 'Scholium Generale' to the *Principia*'s second edition, whose finished version he sent to Cotes on 2 March 1712/13 (see *ibid.*: 384).

beg leave to acquaint you that the difference between the methods is real. In the method of *rationes primæ & ultimæ* quantities are never considered as infinitely little. The whole computation is done in finite quantities by the Geometry of Euclid untill you come to an Equation[,] & then by reducing the equation & making one of the quantities vanish you obtain the equation desired.

He then went on to compile a more elaborate set of Latin notes on the *Acta* review,[67] whose comment on 'P 77 Lin 2'[68] became, in its turn, the opening paragraph of a paper giving 'A general description of the method of fluxions',[69] this last itself to be superseded by the several English drafts of 'Observations' on Leibniz' unsigned 'synopsis' of Jones' work whose principal portion has been set out in the second volume in appendix to our reproduction of the text of the 'De Analysi'.[70] In these remarks—were they, too, intended to be cast into Latin for onwards communication to the editors of the *Acta Eruditorum*? or to appear, perhaps, in their untranslated original in the *Philosophical Transactions*? —Newton on the whole sought to make fair (if at times somewhat sneering) response to the charges brought by his critic, whom he all but names to be Leibniz in his final version.[71] In particular he emphasises the basic distinction which in his view must, in any comparison of his and Leibniz' modes of analysis, be made between the finite instantaneous 'speed' of a quantity which is its

(67) 'Notæ ad *Acta eruditorum* mensis Feb. 1712. p. 75 [et seq.]' (ULC. Add. 3968.4: 17ʳ/18ʳ; there is a considerably variant preliminary version on f. 19 following). Since the first of these annotations—that on '[pag] 75 Lin 2' [=ɪɪ: 260, ll. 2/3], where Leibniz had written of the desirability of making such a richer excerpt from the 'epistolary commerce' of Collins of which Jones had (see ɪɪ: 259, note (5)) made a brief mention in his *Analysis*—states that 'Uberiorum excerptorum publicatio facta est in *Commercio epistolico* nuper edito', we will only be days out in assigning the date of their composition as early February 1713.

(68) Where (see ɪɪ: 262, ll. 2/3) Leibniz shakily affirmed that 'Fermatius [!] aliique' had earlier employed Newton's symbol *o* for the vanishingly small increment of a quantity.

(69) 'Descriptio generalis methodi fluxionum' (see Add. 3968.10: 131ᵛ, where it follows immediately upon the cancelled opening of one last(?) tentative draft of a direct letter of complaint to the *Acta*'s editors). The content of this 'Descriptio' was a few months afterwards absorbed into a much broader set of 'resolutions' of queries framed by Newton in regard to points raised by his own newly edited Collinsian 'epistolary commerce' (see note (94) below), this in turn to become the basis of the full-blown 'Account' of the *Commercium* which he was anonymously to publish early in 1715.

(70) Add. 3968.32: 463ʳ/462ᵛ→460ʳ–461ᵛ→461ᵛ–462ᵛ (=ɪɪ: 263–5/265–70/270–3). Since f. 464 (on the same folded sheet as f. 463) carries a preliminary draft of part of the 'Scholium Generale' to the 1713 *Principia*, these most finished versions of Newton's objections to Leibniz' *Acta* review were, like their preliminaries (see note (67) preceding), almost certainly composed in February 1713.

(71) See ɪɪ: 271, note (31). We would now, however, withdraw our too-hasty conjecture there that Newton had already seen Leibniz' *Charta volans* of July 1713 when he wrote these countering observations upon Leibniz' *Acta* critique. The increasing impatience and abruptness of his tone as the drafts go on is readily explained by mere personal irascibility and tartness.

fluxion (given some related uniformly 'flowing' measure of time) and the Leibnizian infinitesimal *differentia* corresponding to his own notion of its 'momentaneous' increment or decrement.[72] But events were already moving in the spring of 1713 to bypass such mild appeals to reason, and come the early summer the courteous 'phoney war' of the first months of open dispute over priority in calculus invention yielded to the grimmer realities of a no longer gentlemanly total conflict between Newton and Leibniz and their warrior champions in which the blindness of hot passion and the furtiveness of conscious deceit too often nullified all good sense and rational argument. The student of human frailty and unseeing prejudice has as good a claim as the mathematical historian rightfully to assess the flood of emotionally charged, often crudely propagandist writings which was to come pouring from the pens of the principal combatants—that of Newton no less than anyone else—during the next decade. Let us try to pick a meaningful path through the quagmire of empty, angry wranglings.

When Johann Bernoulli initially passed on from William Burnet in the summer of 1712 the news that 'People are busy just now at the [Royal] Society in proving by original letters that the method of fluxions was known by Newton more than seven years before Mr Leibniz published anything of it, and that Mr Leibniz could have seen the elements of it at the house of one Mr Collins who had them at London during the time Mr Leibniz was there, and that he subsequently asked by letter for enlightenments which showed he did not yet understand the matter, five years after Mᵣ Newton showed it complete to his friends', transmitting from the same source the opinion that 'The controversy has been caused by the Leipzig men, whose criticism of Mᵣ Newton's book on quadratures was ill to the purpose',[73] Leibniz' reaction had been sternly to ridicule the 'vain endeavours of certain Englishmen' unspecified—for 'they will never show what they boast of, unless they either corrupt my words or interpret them per-

---

(72) Compare II: 265, 266, 271.

(73) Johann Bernoulli to Leibniz, 13 August 1712 (N.S.): 'In literis novissimis Burneti sequentia inveni ad Te spectantia. "L'on est occupé présentement à la Societé à demontrer par des Lettres originales, que la Methode des fluxions a été connüe de Newton plus de 7 ans, avant que Monsieur Leibnits n'en ait rien publié, & que Mᵣ Leibnits en pouvoit avoir vu les principes chez un Monsieur Collins, qui les avoit à Londres dans le tems que Monsieur Leibnits y a été, & qu'ensuite, par des Lettres, il a demandé des éclaircissemens, qui montroient qu'il n'entendoit pas encore la matiére, cinq ans après que Monsieur Newton l'a fait voir complette à ses amis. Cette controverse a été causée par les Messieurs de Leipsic, qui ont critiqué mal à propos le Livre de Monsieur Newton sur les quadratures...". Tuum est videre, quomodo ridiculam hanc Anglorum ambitionem retundas' (*Commercium Philosophicum et Mathematicum*, **2** (Lausanne/Geneva, 1745): 283 [= *Leibnizens Mathematische Schriften*, **3**, 2 (Halle, 1856): 892–3]). As we have stated above, Burnet—himself without any known proclivities to mathematics—was one of the Fellows initially appointed, on the previous 6 March, to serve on the Royal Society's investigating committee.

versely'[74]—and then to ignore them. And when Bernoulli, primed this time by de Moivre, wrote six months later to inform him of the imminent publication of 'Collins' History of the Differential Calculus',[75] whatever sudden chill the title may have struck in his heart Leibniz could still quietly reply that 'We will see what the Collinsian letters shall give us'.[76] Bernoulli's sight of a copy of the *Commercium* (one brought back from Paris in the spring of 1713 by his nephew Niklaus) was quickly, through the report on it which he promptly delivered to Leibniz, to whip that surface calm into storm. Having, in a long letter dated 7 June (N.S.) to him, rightly complained that the form of procedure in the *Commercium* was to charge Leibniz before 'a tribunal consisting, it would seem, of the very principals and witnesses', then adduce evidence of his guilt and pass sentence upon him, he went on to give some more positive impression of what new light was cast on the mathematical past by the letters now published. 'It is probable', he wrote, that Newton in his initial mathematical studies 'put all his effort into developing infinite series, nor at that time, I believe, did he even dream of his calculus of fluxions and fluents, or its reduction to general analytical operations'; and 'the strongest evidence for this conjecture of mine' was that 'nowhere in all those epistles, nor even in the *Principia* where he should have had so frequent an occasion for applying his calculus of fluxions, shall you find any slightest trace of the [pointed] letters which he now employs in place of differentials', while 'further witness that the calculus of fluxions was not born before the differential calculus' is had from '*Principia* [₁1687], page 263' where Newton 'had, it is patent, not yet come to know the true method of taking the fluxions of fluxions, that is, of differentiating differentials'.[77] Leibniz had still

(74) Leibniz to Johann Bernoulli, 18 September 1712 (N.S.): 'Anglorum quorumdam vanos conatus rideo; nunquam illi monstrabunt quæ jactant, nisi aut verba mea corrumpant, aut perverse interpretentur' (*Commercium*, **2**: 286 [= *Math. Schriften*, **3**: 894]). He went on to add a revealingly tart comment on the broad intellectual, political and moral decline which he saw to prevail in England: 'Valde mihi declinare videtur doctrina in Anglia; nec tot egregiis Viris defunctis pares, aut vel longo proximi intervallo successere. Scilicet nunc ingenia politicis nugis, aut theologicis controversiis destinentur: Successioni legibus stabilitæ merito timetur, quando illi apud quos est potestas nihil agunt quod ad eam firmandam facere possit, multa faciunt aut patiuntur quæ ad infirmandam....' He was nonetheless considerably put out when the newly created George I did not take him with him to London from Hanover when the latter was made king of this 'decadent' country two years later.

(75) Bernoulli to Leibniz, 28 Febrary 1713 (N.S.): 'Collinsii Historia Calculi Differentialis, cum mihi scriberet Moyvræus novissime'—in his letter of 8 October preceding (see note (54))— 'nondum erat evulgata, eam autem sequenti hebdomada, post literarum suarum scriptionem, dicebat certo prodituram' (*Commercium*, **2**: 299 [= *Math. Schriften*, **3**: 903]).

(76) Leibniz to Bernoulli, 25 March 1713 (N.S.): 'Videbimus quid daturæ sint nobis literæ Collinsianæ, in quibus quidam Angli reperisse putant, non quod Pueri in Faba [*sc.* clamitant]' (*Commercium*, **2**: 299 [= *Math. Schriften*, **3**: 904]); and 'not a bean' was doubtless his expectation of enlightenment from the *Commercium*. Compare Plautus, *Aulularia*, 5, 11.

(77) 'Newtonus...in iis [infinitis seriebus] excolendis, ut verisimile est, ab initio omne

not seen the *Commercium* for himself when he responded to this three weeks later, but he showed himself very eager to accept Bernoulli's basic conclusion that Newton 'did not come to know the correct method of differentiating differentials till a long time after it had become familiar to us'.[78] And more than eager. 'Those bungling reasonings', he disdainfully wrote back, 'which are, I infer from your letter, now adduced deserve to smart under the sting of ridicule.... It appears that Newton no more knew our calculus than Apollonius knew the algebraic analysis of Viète and Descartes. He knew fluxions, but not the calculus of fluxions—that, as you rightly adjudge, he got up only later when our own was already published'.[79] That countering whiplash he began at once himself to weave, though without telling Bernoulli till some time afterwards.[80] During the next month he carefully honed and polished the script of an anonymous flysheet in which he dismissed with sarcastic sneer Newton's every pretension to priority in 'inventing' such a calculus (and *à fortiori* rejected all charge that he himself might have learned his own superior *calculus differentialis* from him), setting Bernoulli's 'judgement' thereon in his letter of 7 June in pride of place as that of an un-named 'mathematician of first rank'. And at the end of July he circulated his printed *charta volans* widely among interested parties in Europe.[81] By its

suum studium unice posuit, nec, credo, tunc temporis vel somniavit adhuc de Calculo suo fluxionum et fluentium, vel de reductione ejus ad generales operationes Analyticas....Cujus meæ conjecturæ validissimum indicium est, quod de literis [punctatis] quas pro differentialibus ...nunc adhibet, in omnibus istis Epistolis nec volam nec vestigium invenias; imo nequidem in *Principiis Philos. Natural.* ubi Calculo suo fluxionum utendi tam frequentem habuisset occasionem.... Alterum indicium, quo conjicere licet, Calculum fluxionum non fuisse natum ante Calculum differentialem...est quod veram rationem fluxiones fluxionum capiendi, hoc est differentiandi differentialia...Newtonus nondum cognitam habuerit, quod patet ex ipsis *Princ. Phil. Nat.* [Lib II. Prop. X] pag. 263' (*Commercium,* **2**: 309–10 [=*Math. Schriften,* **3**: 910–11]). The inadequacy of this last piece of 'evidence', whose validity Bernoulli went on to seek to support by a strained—and, one might in hindview at least call it, unworthy—interpretation of what is (compare **1**, **2**, §3: note (46) preceding) no more than verbal clumsiness on Newton's part in the final scholium to his 1704 treatise *De Quadratura Curvarum,* is dwelt on by us at length in **1**, 6, Appendix 1: note (14) above.

(78) 'Saltem constat, Newtono rectam Methodum differentiandi differentialia non innotuisse, longo tempore postquam nobis fuisset familiaris' (*Commercium,* **2**: 311 [=*Math. Schriften,* **3**: 912]).

(79) 'Merentur illæ insulsæ rationes, quas afferri ex Tuis conjicio, sale satyrico defricari.... [A]pparet non magis [Newtonum] cognovisse Calculum nostrum, quam Apollonius cognovit calculum Vietæ et Cartesii speciosum. Fluxiones cognovit, non Calculum fluxionum, quem demum, ut recte judicas, nostro jam edito conflavit' were his Latin words to Bernoulli on 28 June 1713 (N.S.) (*Commercium,* **2**: 313 [=*Math. Schriften,* **3**: 913]).

(80) In his letter to Bernoulli of 19 August 1713 (N.S.), at whose end he airily remarked—disingenuously employing a future tense even then—that 'ego exigua, credo, scheda efficiam[!] ut pœniteat eos [*sc.* Newtoni adulatores] nugarum' (*Commercium,* **2**: 321 [=*Math. Schriften,* **3**: 919]).

(81) The text of this *Charta volans* is now conveniently to be read in *Correspondence,* **6**: 15–17, along with its translation into English (*ibid.*: 18–19). Newton himself in due course received

immediate *succès de scandale* Leibniz' 'chart' was, perhaps more decisively than ever he intended, to drive a broad wedge between his adherents and those of Newton; and the cleavage was not to be closed, despite several optimistic efforts to join it, even after his death in November 1716.

It has long been known how deeply Newton involved himself in the ensuing ructions and altercations, cautiously and often reluctantly in his own name, but much more freely and not a deal intemperately under that of his most fearlessly outspoken and enthusiastic disciple John Keill, long passages of whose subsequent blasts against both Leibniz and Bernoulli he himself in fact wrote. The precise part which he played as fencing-master in lending his force and skill to these Keillian thrusts and parries we need not here specify and analyse in detail. In the spring of 1713 Keill had written to the editor of the *Journal Literaire*—without Newton's connivance, as far as we can see—a letter summarising the light which the newly published *Commercium* shed on the origins of the calculus, and an unsigned Frenchified extract from this 'London letter' was published in that journal in June.[82] Leibniz countered equally anonymously in the same place a few months afterwards, fortifying his 'remarks' making stern correction of its 'several misinformations' with a French translation of the *Charta volans*.[83] And to this Keill, now considerably assisted by Newton, returned a lengthy response[84]

---

several uncut examples of the original broadsheet (splittable by bisection into twin copies)—one of these is now with the main store of Newton's papers on the fluxion priority dispute (loose in ULC. Add. 3968.34), and another was sold at the 1936 Sotheby's sale (item 278) but soon afterwards divided into two by its purchaser—but he has also transcribed its text (before he had a personal copy? or as an *aide-mémoire*?) word for word on a separate sheet, now Add. 3968.34: 478^r–479^r. Eight years later, to anticipate, under new provocation (as he saw it) from Bernoulli he excerpted the 'Judicium primarii Mathematici' from Leibniz' 1713 flysheet and reprinted it separately (see note (57) above) at the end of the second edition of his *Commercium*, along with a refutation item by item of its assertions and conjectures; an unpublished preliminary English version of these 'Observations', intended to be adjoined to Desmaizeaux' 1720 *Recueil*, is printed in Appendix 9.

(82) 'Extrait d'une Lettre de Londres', *Journal Literaire*, **1**. 1 (May/June 1713): 206–10; there followed by (*ibid.*: 210–13) a French rendering of the 'Raport des Membres de la Societé Royale, commis pour examiner le different entre M. *Leibnitz* & M. *Keil*' and also (*ibid.*: 213–14) the like Frenchified 'Extrait d'une Lettre de M. *Newton* à M. *Collins* le 10. de Décembre 1672'. Newton was no doubt told beforehand by Keill of his survey of the prehistory of his dispute with Leibniz as the *Commercium* now supported his heavily pro-Newtonian stance, but we know no evidence which indicates that Newton himself was directly involved in preparing it.

(83) 'Remarques sur le different entre M. de *Leibnitz*, & M. *Newton*', *Journal Literaire*, **2**. 2 (November/December 1713): 445–8 (=*Correspondence of Newton*, **6**: 30–2); 'Traduction de la Piéce Latine, où se trouve [le] jugement [d'un illustre Mathematicien]', *ibid.*: 448–53. Newton in fact also saw the somewhat longer unpublished version of Leibniz' 'Remarques' which is now Add. 3968.35: 487^r–488^v.

(84) 'Réponse de M. *Keill*, M.D. Professeur d'Astronomie *Savilien*, aux Auteurs des Remarques sur le Différent entre M. de *Leibnitz* & M. *Newton*', *Journal Literaire*, **4**. 2 (July/August 1714): 318–58. This is preceded by an 'Extrait' from Keill's covering letter to the

in the *Journal* the next summer, there *inter alia* rescuing the 1687 version of Proposition X of the *Principia*'s second book from Bernoulli's spurious assignation of its error as lying in second differences[85] and, in a tit-for-tat, pointing a Newtonian finger at an analogous lapse by Leibniz in his derivation in 1689 of the radial acceleration generated in Keplerian elliptical orbit round a focus.[86]

editor, Thomas Johnson, which affirms: 'Ayant vû dans le *Journal Literaire* de Novembre et Decembre 1713. des Ecrits de trois différens Auteurs Anonimes [*viz.* Leibniz, Bernoulli and his *alter ego* of the 'primarius Mathematicus' in the *Charta volans*] contre le *Commercium Epistolicum*, j'ai cru y devoir faire quelque Réponse, & prendre une methode toute différente de celle de ces Messieurs. Car au lieu qu'ils se tiennent dans des généralitez, j'ai cru qu'il étoit nécessaire d'entrer dans le détail, afin de donner des preuves claires & solides de tout ce qui pourroit être de quelque poids dans cette dispute' (*ibid.*: 319).

(85) On this see **1**, Introduction: note (196) preceding.

(86) Namely in his 'Tentamen de Motuum Cœlestium Causis' (*Acta Eruditorum* (February 1689): 82–96); on which see the articles by E. J. Aiton cited in VI: 11, note (32). The care which Newton here took in briefing Keill on what to say may be gathered from his letters to him on 20 April and 11/15 May 1714 (see *Correspondence*, **6**: 107–10, 128–30 and 136–7 respectively; preliminary drafts of the second letter of 11 May may be found on ULC. Add. 3968.4: 14^r and .41: 3^v–2^v/124^r). His correction of Leibniz' mistake on 'p. 89 1 20' of his 1689 *Acta* article states (in his prior draft on Add. 3968.4: 14^r) '*lege*. Igitur congrua erunt triangula $_1MNG$ et $_3M_2DG$ et erit $_1MG$ æqu. G_3M, et NG æqu. G_2D', whence 'Ib. 1. 24. *lege* Jam. P_2M æqu. $PN+N_2M$ æqu. $PN+NG-_2MG = PN+G_2D-_2MG$; et $_2T_2M$ æqu. $_2MG+G_2D-_2D_2T$. Ergo $P_2M-_2T_2M = PN-2,_2MG+_2D_2T = 2$ in $_2D_2T-_2MG$' and so 'Ib. 1 31. Est autem $_2D_2T$ vel NP conatus centrifugus circulationis in circulo, quippe sinus versus... & $_2MG$ seu $_3ML$ est conatus centrifugus circulationis in Orbe [adeoჳ] solicitatio gravitatis.... Itaჳ elementum velocitatis paracentricæ æquatur duplo differentiæ inter conatum centrifugum in circulo & conatum centrifugum in Orbe alio'; these emendations Newton soon afterwards introduced into a long passage regarding §15 of the 'Tentamen' in a set of more generally critical 'Notæ in *Acta Eruditorum* an. 89 p. 84 & sequ.', but deleted this in afterthought (see *Correspondence*, **6**: 122, note (10); the 'Notæ' themselves are printed on *ibid.*: 116–17). His checking computation that in Leibniz' Keplerian ellipse the second-order increment of the radius vector r, as this travels round a focus, is in time $\theta(=dt)$ indeed (as Leibniz states) '$ddr = \dfrac{bb-2qr}{b^2r^3}a^2\theta^2$', where a^2/r^3 is the *conatus centrifugus*, and so the *solicitatio gravitatis* $(2a^2q/b^2)r^2$ does consequently vary as the inverse-square power of r, is preserved on Add. 3968.41: 124^v. (We may notice in passing that Newton's attempts in 1714 to find fault with the 'Tentamen' had their origin two years earlier in an extended critique by him of this and two companion pieces by Leibniz in the 1689 *Acta*—on resisted projectile motion (pages 38–47; see VI: 8, note (23) and 70–1, note (109)) and, much more briefly, on diacaustics (pages 36–8 preceding) —which was published, from the original now in ULC. Res. 1894, Packet 8, only a century and a half later by Edleston at the end of his edition of the *Correspondence of Sir Isaac Newton and Professor Cotes* (London, 1850): 308–14; preparatory drafts of this survive in Add. 3968.12: 176^r/176^v, 201^v and .41: 145^r/145^v. By way of an intermediary revise, now in private possession (the text is reproduced by S. P. Rigaud on pages 66–8 of the Appendix to his *Historical Essay on Sir Isaac Newton's Principia*, Oxford, 1838), this was *via* Add. 3968.5: 15^r–16^v→17^r–19^r subsumed in much abbreviated form into Newton's editorial interpolation on page 97 of the *Commercium Epistolicum*. At the head of his original critique Newton subsequently inserted the title 'Ex Epistola cujusdam ad Amicum'—Keill, one is tempted to suppose—but there is no indication that he ever went further in any design he might have had to publish this separately. To say more would be to pile conjecture upon conjecture....)

We have already sketchily outlined how Keill in a further article in the *Journal Literaire* two years later went on, again spoon-fed at crucial points by Newton, publicly to deflate Bernoulli's overdone previous criticisms (in the Paris *Memoires* for 1710 and 1711, neither published till three years after their titular date) of the *Principia*'s inadequacies in determining the totality of orbits possible in a given inverse-square or inverse-cube central force-field.[87] Like the anonymous Latin letter in which Bernoulli later the same year, 1716, endeavoured crushingly to lambast him in the *Acta Eruditorum*[88] this contained a deal more of emotional invective than rational counter-argument, and we will say no more of either. Let us equally speedily pass over the arid and wearisomely repetitive response which Johann Kruse, doing a Keill in his own master's behalf, likewise made on behalf of Bernoulli in the *Acta* two years later;[89] and also the replies to both Bernoulli and Kruse which Keill was in consequence stimulated to compose, as ever with Newton as his mentor, and set in print in 1719 and 1720[90] shortly before his death (in late August 1721) put an end to

(87) 'Défense du Chevalier Newton. Dans laquelle on répond aux Remarques des Messieurs *Jean & Nicolas Bernoully*, insérées dans les Memoires de l'Académie Royale des Sciences pour les Années 1710. & 1711. par Jean Keill, Professeur d'Astronomie à Oxford', *Journal Literaire*, **8**. 1, 1716: 418–33. (De Moivre's original rendering of its author's English original into French, bearing corrections in Keill's own hand and entitled somewhat differently 'Apologie pour le Chevalier Newton...', is preserved in ULC. Res. 1893, Packet 5.) That Newton was more than superficially involved in the composition of this *apologia* for his technical prowess in calculus and dynamics is shown most clearly by the fact that his original English draft of a crucial passage in it dealing with inverse-square and inverse-cube orbits (*Journal*, **8**: 427–8; on its import see VI: 349–50, note (209)) is roughed-out by him—a slightly more finished version is found on a separate sheet now in private possession—on the back of Keill's letter to him of 29 October 1715 (Add. 3985.14, printed—Keill's letter only—in *Correspondence*, **6**: 245–6).

(88) 'Epistola pro Eminente Mathematico Dn. Johanne Bernoullio, contra quendam ex Anglia antagonistam scripta', *Acta Eruditorum* (July 1716): 296–315. Bernoulli's authorship of this piece was, as is well known, laid bare through an omission by the *Acta*'s editor Christian Wolf to suppress a give-away 'meam' in its printed text (see the excerpt from the 'Epistola' quoted on *Correspondence*, **6**: 302, and note (4) thereto). Keill in particular was quick to pounce on the slip, drawing Newton's notice to this place where Bernoulli 'foregott himself' in his letter of 17 May 1717 (see *Correspondence*, **6**: 385–6).

(89) 'M. *Jo. Henr. Crusii* Responsio ad Cl. Viri *Johannis Keil*...Defensionem pro Nobilissimo Viro *Is. Newton* in *Diario Literario Hagiensi* A. 1716', *Acta Eruditorum* (October 1718): 454–66.

(90) 'Lettre de Monsieur *Jean Keyll*...à Monsieur *Jean Bernoulli*...', *Journal Literaire*, **10**. 2, 1719 [1720]: 261–87; *Joannis Keill... Epistola ad Virum Clarissimum Joannem Bernoulli* (London, 1720). Both these tirelessly—and tiresomely—polemical pieces had their origin in a yet unpublished 'Second Letter of John Keill...to Mr John Bernoulli' (the autograph is preserved in ULC. Res. 1893, Packet 5) which was, in its own proud words, 'a vindication of Mr Newton and himself from the Objections and Aspersions of one Mr Crusius, who has published them in the *Acta Eruditorum Lipsiæ* for the year 1718: with some observations on the new Method of writing Invectives in an Index [*sc.* to *Acta*, 1718 where (p. 960) it is 'twice assert[ed] without any *Perhaps* that I have need of Mr Newton to help me...'] first invented by the Leipsick Journalists', Keill had Halley read the central portion of this parent English

this increasingly thin secondary caterwaul to the main spat of priority dispute.[91]

We may now also know from his private papers that it had been Newton's initial impulse to publish a refutation directly from his own pen (if not over his signature) to Leibniz' accusatory 'Remarks' and appended Frenchified text of his flysheet in the 1713 *Journal Literaire*: 'pieces...full of assertions without any proof & without the name of the Author & so...of no credit or authority', he wrote.[92] Not to take his ultimate omission to do so as an indication that, upon maturer reflection, he opted merely to play a subordinate adjutant's rôle to Keill. It was rather a temporary withdrawal in order to mount a more broadly based future attack on Leibniz' pretensions which would (so he purposed) slash them utterly into shreds. Already in the late autumn of 1713 he had, his unpublished manuscripts of the period reveal, a mind to make a series of detailed historical annotations on the Collinsian 'epistolary commerce' as he had

---

piece out for him at the Royal Society on 28 May 1719, and then four weeks later passed its full text to Newton for comment (see his letter to him of 24 June, printed in *Correspondence*, **7**, 1977: 48); Newton evidently encouraged him to expand it into the fuller letter which (in de Moivre's French) Keill soon after sent to the *Journal*'s editor, Johnson. (Keill's original fair copy of this enlarged version exists in Packet 5 of the Lucasian Papers in Cambridge University Library, and Newton's own elaborate suggestions for improving its content and expression—virtually all these were incorporated in the printed 'Lettre'—are preserved on Add. 3968.33: 364^r–367^r, with preliminary castings on .41: 55^v/135^r.) Two extended preparatory drafts of the 1720 Latin booklet are also to be found in Packet 5 of the Lucasian Papers, along with an unbound printer's proof of the *Epistola* itself (an 8-page *Additamentum* was adjoined in the published text). Newton had (see note (156) below) thought briefly of making a direct reply to Bernoulli's 'squabbling Paper', but quickly saw the wisdom of withdrawing to be Keill's ghost writer in a terrain where only the mud of insult accrued to the front-runners.

(91) In **1**, Introduction: note (198), however, we cited one mildly interesting exception to one's over-riding impression of sterile repetitiveness in this side-wrangle: namely, where Keill in a neat display of one-upmanship drew attention in his 1720 *Epistola* to an article in the Paris *Memoires* for 1706 in which Bernoulli had himself slipped badly, much like Leibniz in his 1689 'Tentamen' (see note (86) above), in evaluating the second-order *differentia* of two infinitesimally differing related geometrical quantities. Would that more of the surrounding gush of heated, emotive prose held interest for the mathematical reader!

(92) We quote this judgement of Newton's from the opening paragraph of the fullest surviving draft of his intended counterblast on Add. 3968.35: 496^r–497^v. (A considerable number of drafts of this, both partial and full, survive on *ibid*.: 489^r→492^r–493^r→502^r–503^r →500^r–501^v/504^r→498^r–499^r→494^r–495^v (+3968.41: 114^v/Mint Papers II: 335^v/340^v).) He went so far, to be sure, as to have de Moivre render this into French for him (the translation is now Add. 3985.20) before deciding to let Keill send a 'Response' in his place. (Compare *Correspondence*, **6**: 80–90, which reproduces the text of ff. 494^r–495^v, interlarded with additional phrases from ff. 496^r–497^v and from de Moivre's French version.) It is worthy of mention that Newton based his riposte not on the text of Leibniz 'Remarques' as published in the *Journal Literaire* in 1713 (see note (84) above), but on a somewhat larger and different version (now Add. 3968.35: 487^r–488^v) sent to him—it would appear—privately by Johnson: the last twenty lines of this, in particular, do not appear anywhere in the printed version.

brought its text out just a few months before.[93] But soon he broke off to ask—and answer—a series of wider questions to which 'the old manuscripts and letters printed in the *Commercium Epistolicum* provide a grip',[94] namely:

1. Who was the first author of the series for the area of the hyperbola which is everywhere attributed to Mr N. Mercator by Mr Leibniz?

2. Was Mr N. Mercator the first discoverer of an infinite series [found] through an unending division?

3. Did Leibniz do right in denying to all men that Newton had a general reduction of fractions to infinite series and the quadrature of figures by these series?

4. Was Mr Leibniz right in wanting to be co-discoverer of Mouton's method of differences?

5. Who was first discoverer of the series for the arc of a circle in terms of its tangent given?

6. Was Mr Leibniz co-discoverer of this series?

7. What manner of series did Mr Leibniz have in the year 1674 for the area of a circle?

---

(93) See Add. 3968.41: (9^r→) 71^r/9 –14^v, 54^r/54^v, 76^r/76^v, 134^r and .14: 244^r–245 *bis*^{r} →236^r–238^v (where, on f. 236^r, the title of 'Annotationes in Commercium Epistolicum' first appears) →239^r/240^r (the most finished version). In confirmation of our dating of these (mostly roughly scrawled) sequential drafts we may remark that Add. 3968.41: 54^v carries the draft (another is on 3965.13: 472 *bis*^{v}) of a receipt on 'September 22th 1713' for 'fifty Medals of fine Gold' struck to celebrate the Peace of Utrecht; *ibid.*: 136^v is penned upon a formal business invitation sent to Newton on 6 October following; and *ibid.*: 9^v bears the missive directed to him by John Thorpe on 9 November which is printed in *Correspondence*, 6: 36. The long note on *ibid.*: 76^r/76^v spelling out the import of his *epistola posterior* to Leibniz in 1676—and finding room to ram home yet again (compare notes (86) preceding and (97) below) that 'Planetas circum Solem in Ellipsibus revolvi ac radijs ad Solem ductis areas describere temporibus proportionales, Newtonus ante annum 1677[!] demonstrare didicit.... Prodijt hæc demonstratio in lucem anno 1687, et Leibnitius anno 1689 Theorema Newtonianum calculo diff[er]entiali sed errante ad incudem revocavit ut suum faceret' (f. 76^v)—follows a lone first-person declaration by Newton in English of some interest: 'In the winter between the years 1664 & 1665 upon reading Dr Wallis's *Arithmetica Infinitorum* & trying to interpole his progressions for squaring the circle, I found out first an infinite series for squaring the circle, & then another infinite series for squaring the Hyperbola & soon after. This latter series was the same with that published by Mr Mercator about 3 or 4 years after [*viz.* in his *Logarithmotechnia* in 1668] & two years before he published it I had a general method of squaring curves by such series by the help of division & extraction of roots' (f. 76^r). (See *Correspondence*, 2: 111–14 for the corresponding passage in Newton's 1676 *epistola*. The annotations upon Wallis' *Arithmetica* to which he here makes reference in retrospect, half a century after he recorded these first mathematical discoveries of his, are reproduced in ɪ: 96–134; compare also D. T. Whiteside, 'Newton's Discovery of the General Binomial Theorem' (*Mathematical Gazette*, 45, 1961: 175–80): 176.)

(94) Add. 3968.10: 130^r–131^r/133^v (with preparatory roughings-out on *ibid.*: 132^r–133^v/112^r–120^r); the fourteen 'quæstiones' which follow (in our English version)—ones to which, in Newton's Latin words, 'MSS^{ti} veteres et Epistolæ in *Commercio Epistolico* impressi ansam præbēt'—are found on the opening f. 130^r. (On f. 131^v is drafted the opening paragraph of the unfinished 'Descriptio generalis methodi fluxionum' referred to in note (69) above.)

8. Had he discovered that series or received it from another?

9. Did Mr Leibniz discover the proposition that a body revolving in an ellipse, and describing about its lower focus areas proportional to the time, shall be drawn to that centre by some force which is reciprocally as the square of the distance from the centre?

10. How happened it that, although Newton's *Principia* was sent to the Royal Society in 1686, published in 1687 and described in the *Acta Eruditorum* in 1688, Mr Leibniz in 1689 took pains to print Newton's chief propositions under the titles of 'A letter on optical lines', 'A scheme on the resistance of a medium, and the motion of heavy projectiles in a resisting medium' and 'An essay on the causes of celestial motions'?

11. What is Newton's universal analysis?

12. When was it invented?

13. What may Leibniz' differential method have been, and when was it invented?

14. In what ways do the two methods differ from one another?...[96]

---

(95) In *Acta Eruditorum* (January 1689): 36–8, 38–47 and (February 1689): 82–96 respectively. On the somewhat confused and opaque 'Schediasma de resistentia Medii, & Motu projectorum gravium in medio resistente' see VI: 70–1, note (109) and also 1, 6, §2. 3: note (50) above; for Newton's criticisms of Leibniz' 'errors' in the much more original and profound 'Tentamen de motuum cœlestium causis', see note (86) preceding and compare note (97) below. The very brief opening essay 'de lineis opticis' merely specifies the *constructio simplicissima* (p. 38) of a bifocal catacaustic ('Acamptos' is Leibniz' invented name for it) together with that of its involute, and hints that 'eadem & dioptricis applicari possunt' much as in the final section of Book 1 of Newton's *Principia* (compare III: 549–52). (In a draft of his 'Account' of the *Commercium* the next year, in the autumn of 1714, Newton here 'caustically' commented that what he 'had said about refractions [Mr Leibnitz] needed nothing more to apply it to reflexions then to put ye sine of refraction equal to ye sine of incidence & to change its signe'; see Add. 3968.41:80^v) In an introductory statement embracing all three essays Leibniz specified (p. 37) that he had as yet seen only the review of Newton's work which had appeared in the *Acta* the previous June, '& ex relatione *Actorum* video, cum multa prorsus nova, magni sane momenti, tum quædam etiam ibi tradi, a me nonnihil tractata; nam præter motuum cœlestium causas, etiam lineas catoptricas vel dioptricas, & resistentiam medii explicare aggressus est', adding that '*Lineas* illas *Opticas* habuit, sed celavit.... Eas postea ab Hugenio... inventas intelligo. Etiam mihi, sed per diversam, ut arbitror, viam innotuere'. We may guess that Leibniz in fact took some care not to read the *Principia* itself till his own 'efforts' were in print.

(96) Newton's original Latin (on Add. 3968.10:130^r) reads:

1. Quis fuerit primus author seriei pro area Hyperbolæ quæ D. N. Mercatori a D. Leibnitio passim tribuitur.

2. An D. N. Mercator primus fuerit inventor seriei infinitæ per divisionem perpetuam.

3. An recte fecerit Leibnitius qui totis viribus generalem reductionem fractionum in series infinitas & quadraturam figurarum per has series Newtono denegat.

4. An recte D. Leibnitius se coinventorem esse voluit methodi differentialis Moutoni.

5. Quis fuerit primus inventor seriei pro arcu circuli ex data Tangente.

6. An D. Leibnitius fuerit hujus seriei coinventor.

7. Quamnam seriem D. Leibnitius habuerit anno 1674 pro area circuli.

8. Utrum seriem illam D. Leibnitius invenerat vel aliunde acceperat.

9. An D. Leibnitius Propositionem invenit quod corpus in Ellipsi revolvens & areas tempori proportionales circum focum inferiorem describens, attrahatur in centrū vi aliqua quæ sit reciproce ut quadratum distantiæ a centro.

An imposing list of queries indeed, some readily answered but most not. The 'resolutions' to them which Newton proceeded to rough out—we need not specify the details—bleakly allowed Leibniz independent creation of the basic modes and techniques of 'universal' mathematical 'analysis', but in contriving the 'diverse method' whereby the latter claimed to have 'opened up new geometrical ways' of treating the principal themes of his *Principia* he all but openly charged him with outright plagiarism.[97] And we cannot find that he ever forgave Leibniz for his maladroit attempt to claim to be 'co-inventor' of its principal propositions.

Small wonder that half-hearted attempts by John Chamberlayne and Willem 'sGravesande to effect a reconciliation between the two in the early months of 1714[98] came to nothing. Not only, in fact, did Newton rebuff Chamberlayne

---

10. Unde factum est quod cum Newtoni *Principia Mathematica* anno 1686 ad Societatem Regiam mitterentur, anno 1687 ederentur & anno 1688 describerentur in *Actis Eruditorum*, D. Leibnitius anno 1689 præcipuas Newtoni Propositiones [imprimi curavit] sub titulis Epistolæ de Lineis Opticis, Schediasmatis de resistentia Medij et motu Projectorum gravium in Medio resistente & Tentaminis de motuum cœlestium causis.

11. Quænam sit Analysis Univ[ers]alis Newtoni.

12. Quando fuit hæc Analysis inventa.

13. Quænam fuerit Methodus differentialis L[eibnitij] et quando fuerit inventa.

14. Quomodo differunt hæ duæ Methodi ab invicem.

In sequel Newton began a further query '15 Quibus argumentis', but there broke off to start to pen in its place the answering 'Quæstionum Resolutiones' as he drew these 'ex ijsdem MSS & Epistolis': in these the familiar form of the 'Account' of his *Commercium* which he elaborated and published a year afterwards (see Appendix 2 below) already takes firm shape.

(97) 'Cum Newtoni *Principia Mathematica*...anno 1687 ederentur & anno 1688 in *Actis Eruditorum* in epitomen redigerentur, & D. Leibnitius...Epistolam de Lineis Opticis, Schediasma de Resistentia Medij...& Tentamen de motuum cœlestium causis componeret et in *Actis Lipsicis* anno 1689 imprimi curaret quasi ipse quoⳓ præcipuas Newtoni de his rebus Propositiones invenisset idⳓ diversa methodo qua vias novas Geometricas aperuisset; et librum Newtoni tamen nondum vidisset: quis fuerit harum Propositionum inventor primus censendus?' (Add. 3968.10: 130ᵛ). Compare the equivalent passage on Add. 3968.41: 76ᵛ already cited in note (93) above.

(98) To Chamberlayne's overture to him in late February 1714 (see *Correspondence*, 6, 71) Leibniz replied in April (*ibid.*: 103–4) that he was the injured party, and that it was therefore up to the Royal Society to denounce the 'Jugement prétendu' rendered in its name by their committee and retract what in the *Commercium* 'Mr. Newton...a fait publier dans le Monde... exprès pour me décréditer'. When Newton was shown Leibniz' unbudging letter early in May his own reaction (*ibid.*: 126–7) was as icily phrased as it was tartly dismissive. And when in time Chamberlayne reported back (*ibid.*: 152–3) his lack of success in softening Newton's own rigid stance, Leibniz' reaction (*ibid.*: 173) was stiffly to make it known that, 'quand je serai de retour à Hanover', he also was minded to prepare a *Commercium Epistolicum* of his own 'qui pourra servir à l'Histoire Littéraire'. As for 'sGravesande—one of Johnson's editorial advisers on the *Journal Literaire*—though he wrote to Newton on 8 June 1714 (N.S.) *inter alia* to urge 'avec quel soin je rechercherai toujours les occasions de vous faire voir l'estime que j'ai pour vous' (*Correspondence*, 6: 145), and indeed thereafter showed himself a good friend to Newton's cause in several ways, he baulked the next year at putting into print in the *Journal* 'as of his

and continue to help Keill prepare 'his' response to the *Charta volans* for publication in the *Journal Literaire* later that year, but he began privately to compose 'An Account of the Method of Fluxions until the year 1676 inclusively'[99] wherein he appealed to the warrant of his *epistola posterior* to Leibniz and of his 'De Analysi' seven years earlier in 1669 to establish the oldness of the fluxional analysis which he had delivered to the world in his 1704 'De Quadratura';[100] and also to rough out the opening to a considerably more polemical 'Account of the Differential Method from the year 1677[!] inclusively'[101] where, it was

---

own head' a submitted French translation of Newton's *Phil. Trans.* 'Account' of the *Commercium* (so Halley wrote to Keill on 3 October 1715; see *ibid.*: 242), and no further test of his pledge unsparingly to serve Newton 'dans toutes les occasions que je pourai vous estre de quelque utilité' seems to have been made.

(99) Now Add. 3968.12: 167^r/167^v (+168^r)/170^r/170^v. Newton's original title (on f. 167^r) concluded more condemningly: '... until Mr Leibnitz had notice of it in the year 1676. (We may add that what appears to be part of a preliminary revise is found at Add. 3968.8: 87^r–90^r.) The authoritative statement by him on his method of fluxions which was reproduced (*ex* Add. 3968.41: 83^r) is an earnest of the opening paragraph of his present 'Account', which (transposed to an anonymous third-person narrative) declares:

> The method of fluxions is this. Mr Newton considers two or more quantities as growing in magnitude [a tentative replacement of this phrase by 'flowing' is deleted] by continual increase in the same time, & the velocities of their increase he calls their fluxions, & their parts generated in moments of time he calls their moments, the names of fluxions & moments being taken from the fluxion & moments of time. For the flowing quantities or fluents he puts any symbols as z, y, x, & for their fluxions he puts any other symbols or even the same symbols distinguished by their magnitude or form or by any mark as $\dot{z}$, $\dot{y}$, $\dot{x}$. But one of the fluents he usually considers as flowing uniformly or in proportion to time, & usually puts an unit for its fluxion & the letter o for its moment. And for the moments of the other fluents he puts the letter o multiplied by their fluxions. For as 1 the fluxion of time (or of its exponent) is to $\dot{z}$ the fluxion of z, so is o the moment of time to $\dot{z}o$ the moment of z. When he is demonstrating any Proposition he takes the moment o in the sense of the vulgar for an indefinitely (not infinitely) small part of time, & performs the whole operation in finite figures by the Geometry of the Ancients without any approximation, & when the calculation is finished & the Equation reduced, he supposes that ye moment o decreases in infinitum & vanishes & thereby makes all the terms of the equation vanish wch are affected with the moment o, & from the remaining terms draws his conclusion. But when he is only investigating a truth, he supposes the moment o to be infinitely little, & for making dispatch neglects to write it down, & works in figures infinitely small by all manner of approximations wch he conceives will make no error in the conclusion. The first way is sure & exact[,] the second expedite.

'& both are natural', Newton initially concluded his paragraph before striking out this somewhat meaningless final phrase.

(100) He ended his 'Account', we may notice, with the vigorous affirmation (Add. 3968.12: 170^v) that 'when Mr Newton wrote that *Analysis* he had observed the great relation which the method of converging series & that of fluxions have to one another. For he every where applys the method of series to ye solution of Problems by the method of fluxions & by the method of series [there] demonstrates the first Rule wch is the foundation of the method of fluxions, & by reason of this relation between the methods he had then composed one general method of them both [*viz.* in his 1671 tract]'.

(101) Add. 3968.11: 145^r/145^v.

affirmed, Leibniz in fashioning his variant *calculus differentialis* 'began where Dr Barrow left off' in framing the 'more simple' version of Fermat's 'method of Tangents' which 'adopted a proper calculation to it, but was still wanting, *viz* to exclude fractions & radicals', and yet in publishing his calculus in the *Acta Eruditorum* 'he never made any acknowledgem^t of any advantage wch he had received either from Dr Barrow or from Mr Newton or from M^r [James] Gregory or from Mr Collins or Mr Oldenburg or any body else in England, unless where he could not avoid it'. And he went straightaway to subsume both these pre-liminary descriptive sketches of what he could make out to be the past truth— we see no cause here to challenge Newton's honesty in compiling these surveys as best as his heavily blinkered vision would permit—in a fuller and broader 'Account of the *Commercium Epistolicum*'[102] which over the summer and autumn of 1714 grew through draft after draft to be at length the elaborate 'author's review' of the book which appeared anonymously in the *Philosophical Transactions* early the next year.[103] Since we reproduce the greatest portion of the text

---

(102) Such, *tout court*, is the title of the first draft (now Add. 3968.8: 67^r–72^r). This was at once superseded by 'An Account of y^e *Commercium Epistolicum D. Joannis Collinij* [*sic*] *& aliorum De Analysi promota*' (*ibid.*: 81^r–82^v), with which he quickly conjoined a related 'Account of the *Analysis per Quantitatum Series, Fluxiones ac Differentias*...published by Mr Jones (396.12: [154^r/154^v→] 150^r/150^v + .41: 80^r/80^v); and the composite went rapidly through a succession of further full and partial redraftings (3968.39: 579^r–580^r.(+ .41:151^r/151^v)→.8: 91^r–92^v + .12: 149^v/149^r, 157^r–158^r) to emerge as 'An Account of the Book entitled *Commercium Epistolicum*...' scheduled to be comprised 'under these four Heads. 1 Of the method of con-verging Series. 2 Of the method of fluxions & moments. 3 Of the Differential method. 4 Of the three Papers [*sc.* by Leibniz; see notes (86) and (95) preceding] entitled *Epistola de Lineis Opticis, Schediasma de resistentia Medij*, & *Tentamen de motuum cœlestium causis*' (see Add. 3968.8: 94^r), but of these sections only the first two have survived (in .8: 94^r–96^v and .12: [157^r/157^v→] 148^r–149^r/.8: 73^r–80^r respectively) and no more, it seems to us, were written.

(103) 'An Account of the Book entitled *Commercium Epistolicum Collinii & aliorum, De Analysi promota*; published by order of the *Royal-Society*, in relation to the Dispute between Mr. *Leibnitz* and Dr. *Keill*, about the Right of Invention of the Method of *Fluxions*, by some call'd the Differential Method' (*Philosophical Transactions*, **29**, No. 342 [for January/February 1714/15]: 173–224); since the preliminary version of its page 205 to be found on Add. 3968.41: 79^r is there entered below a draft of Newton's letter to Popple on 13 April 1714 (whose version as sent, only trivially different, is printed in *Correspondence*, **6**: 99) we have a firm date *ante quem non* to the composition of the finished piece. This public text—readily consultable in the recently issued photofacsimile (New York, 1968) of the Royal Society's holding of the early volumes of the *Phil. Trans.*—is marred, unfortunately, by a variety of obtrusive printer's emendations and clumsy would-be typographical 'improvements' (not least a veritable splatter of uselessly inserted hyphens and commas). In making reproduction of the core of his text in Appendix 2 below we have preferred to restore the orthography and punctuation used by Newton from his surviving autograph drafts, kneading and pasting these together with the minimum of editorial guile necessary to achieve a smooth end-product verbally indistinguishable from the 'Account' as printed. (For a detailed collation of the folios, both in Cambridge University Library and in private possession, which went to make up this pastiche see note (3)

of this in the second of our following appendixes, there is no need to detail the several 'proofs' of his priority in calculus discovery, and of Leibniz' arrogance and deceit (so it seemed to him) in 'illegally' advancing his own claim, which Newton there built up point by point out of the documents printed in his Collinsian 'Epistolary Commerce' (and also, let us notice, on cited extracts from two then as yet unpublished letters to him, from Leibniz and from Wallis,[104] which he there spoke of as 'upon a late Occasion communicated to the Royal Society'[105] so as nominally to be in the public domain). Still less is there necessity one by one to probe and demolish the lengthily argued demonstrations by which he there seeks to substantiate his preconceived opinion of Leibniz' extensive plagiarism of the 'inventions' of himself and other precursors in the field of universal analysis by infinite series and methods of 'tangents' and 'quadratures': the validity, and even in most instances the very plausibility of these charges has not been evinced by any test which the documentary record, as it has survived and as we have more sharply and fully come to know it in the two and a half centuries since, can provide. Newton's 'Account' has always been a manifesto which convinces the uncritical, unreasoning believer in his mathematical supremacy, but a recital of thoroughly biased and fanciful reconstruction of the might-have-been which has ever failed to sway the agnostic onlooker. Its core of good sense in fairly reconstructing the sequence and import of events already, in 1714, in their main part more than forty years old (and thoroughly confused in the memory of the very few still living who had witnessed any part of them) is there submerged beneath a frothy, verbal dross of rationalisation and overstatement as grossly prejudiced as anything which Leibniz and Johann Bernoulli put into print on the dispute. One can well appreciate Newton's reluctance to set his name to so outspoken an interpretation of the significance of the letters (and his 'De Analysi') in the 1713 *Commercium*, and his wider examination of the demerits of Leibniz' claim to priority in 'inventing' the calculus.

The identity of the author of the 'Account' must have been known from the

---

of Appendix 2. In subsequent footnotes thereto, let us add, we also take pertinent cognizance of the private 'Corrigenda' to the published text listed by Newton shortly afterwards on loose sheets which are now Add. 3968.13: 173^r/223^r, along with certain more substantial *addenda* on Add. 3968.42: 612^v which he had it in mind to incorporate some while later, in about the late autumn of 1719.)

(104) Of 17 March 1693 (N.S.) and 10 April 1695 respectively (see *Correspondence*, **3**: 257–8 and **4**: 100–1). Fuller publication of the texts of both letters was to be made by Newton two years afterwards in the Appendix (pages 119–21) of his 1717 reissue of Raphson's *History of Fluxions*; see note (145) below.

(105) So Newton described Leibniz' 1693 letter on page 198 of the 'Account'; Wallis' letter of two years later is similarly introduced by him on the next page as 'lately communicated to the Royal Society'.

first within the narrow circle of Newton's confidants, and was doubtless an open secret to many more in London, Oxford and perhaps Cambridge. In particular, Edmond Halley (one of the two Secretaries of the Royal Society at this time)[106] wrote knowingly to Keill in regard to 'a French translation of ye account of the *Commercium* given in the *Transactions*' which 'we have printed in order to send it abroad' that

> Sr Isaac is desirous it should be publisht in the *Journal Literaire*...[but] is unwilling to appear in [ye controversy between you and Mr Leibnitz] himself, for reasons I need not tell you, and therfore has ordered me to write to you about it, who has been his avowed Champion in this quarrel; and...proposes that you would signifie to Mr Johnson at the Hague, by a letter enclosed either to Sr Isaac or me, that you are desirous that the said French paper be inserted in his *Journal*, as containing the whole state of ye controversy between you and Mr Leibnitz.[107]

Keill, we may remark, dutifully did as he was asked, and the French version of the 'Account' (made by de Moivre, we must presume) was printed by Johnson in his *Journal* soon after, in train to a suitable covering letter from Keill.[108] In later years in England, certainly, it became widely accepted that it was Newton who had written the review—in 1761 James Wilson could make numerous citations of it as being Newton's without any air other than of stating a truism[109

---

(106) He had been appointed to the post in 1713 upon Hans Sloane's resignation, and was joined in the twin post in 1714 by Brook Taylor, another of the committee appointed by the Society in 1712 to pass judgement upon Leibniz' complaint against Keill (see page 480 above).

(107) Edleston, *Correspondence of Newton and Cotes* (note (86)): 184 (=*Correspondence*, **6**, 1975: 242).

(108) See *Journal Literaire*, **7**. 1/2 (October/December 1715): §vi/§xv: 114–58/344–65: 'Extrait[!] du Livre intitulé *Commercium Epistolicum Collinii & aliorum, de Analysi promota;* publié par ordre de la *Societé Royale,* à l'Occasion de la Dispute élevée entre M. *Leibnitz,* & le Dr. *Keill,* sur le Droit d'invention à la Methode des *Fluxions,* par quelques-uns appellée, *Methode Differentielle'*; the preceding 'Extrait d'une Lettre de M. le Docteur *Keill...*' (*ibid.*: 113) begins: 'Nous avons fait imprimer dans nos *Transactions Philosophiques* une piéce curieuse qui contient un détail de la dispute sur l'Invention de la *Methode des Fluxions,* ou *Methode differentielle,* & qui en même tems fait l'Histoire des progrès & des découvertes qu'on a faites depuis cinquante ans dans la Geometrie & dans l'Algebre. De plus on y fait voir clairement que tout ce que M. de *Leibnitz* a écrit sur cette matiére est plutôt de l'Invention des autres que de la sienne...'.

(109) In the Appendix (pages 307–76) to the second volume of his edition of *Mathematical Tracts of the late Benjamin Robins* (London, 1761)—by far, in our opinion, the most informed and balanced account of Newton's mathematical development composed before the full detail of the latter's private papers was opened up to scholarly study at the end of the nineteenth century. A hundred years after the 'Account' appeared, it is only fair to add, Newton's authorship of it began to be queried in England just when it had become accepted in France— notably by Biot—that he had indeed written it (and the lightly revised Latin version, the 'Recensio Libri...', which was added to the second edition of the *Commercium* in 1722; see note (57) above). David Brewster in his one-volume *Life of Sir Isaac Newton* (London, $_1$1831) went so far as to state that the 'general review of its contents' prefixed to the 1722 *Commercium* 'has been falsely ascribed to Newton' (p. 215). He was taken severely to task twenty years later by

—but the ploy of having Keill introduce its Frenchified text to non-English-speaking readers achieved its aim of diverting Continental conjecture as to who was its author away from its onlie begetter onto his willing front-man, scapegoat though in some ways he turned out to be.

Not that in the immediate outcome it mattered much. Neither of the 'prime antagonists' Leibniz and Bernoulli read the 'Account' when it came out in its original English, and when simultaneously in November 1715 Conti wrote to Leibniz and Monmort to Bernoulli's nephew Niklaus to tell of the newly appeared first part of the French translation, both Leibniz and Johann were already preoccupied in selecting a proper test of the 'pulse of the English analysts'—to wit by resurrecting Bernoulli's twenty-year-old problem of constructing orthogonals to a given family of curves.[110] To his letter to Conti on

De Morgan in an article 'On the Authorship of the Account of the *Commercium Philosophicum...* (*London, Edinburgh and Dublin Philosophical Magazine and Journal of Science* (4) **3**, No. 20 (June 1852): 440–4) which gave (*ibid.*: 442) explicit citation of the half dozen main instances in Wilson's 1761 Appendix where Newton is referred to as the writer without 'laying any stress on the assertion of authorship'. In his *Memoirs of the Life, Writings and Discoveries of Sir Isaac Newton* just three years later, having in the meantime seen 'scrolls of almost the whole of the *Recensio...*in [Newton's] own hand' among his manuscripts (then at Hurstbourne Park; see 1: xxix), Brewster did a complete about-face to state that 'the *Recensio...*was written by... Newton, a fact which Professor De Morgan had deduced from a variety of evidence' (**2**: 63, note 1; compare 75, note 3).

(110) See **1**, 7: note (1) preceding. It was Bernoulli who, having read an excerpt from Keill's 1714 'Réponse' to the *Charta volans* (see note (84) above) 'maximam partem ad me spectans' which Jakob Hermann had written out for him, and hastened to lambast him for his 'lepida excusatio' of Newton's error in Proposition X of Book 2 of the *Principia* in its first edition (see p. 345 of Keill's 'Réponse'; the 'excuse' was, of course, Newton's) that this was 'commissum...per accidens, producendo lineam in plagam non debitam, sed debitæ omnino contrariam', went on in a letter to Leibniz on 6 February 1715 (N.S.) to urge that 'Recte facies, si quædam edas, in quibus Newtono aquam hærere scis. Suppetunt haud dubie multa eorum, quæ olim inter nos agitata fuere, et quæ per communem differentialium methodum non facile obvia sunt: qualia sunt quæ de transitu ex curva in curvam habuimus, quæ peraguntur singulari quadam differentiatione adhibita' (*Commercium Philosophicum et Mathematicum*, **2**: 343–4 [=*Leibnizens Mathematische Schriften*, **3**: 936–7]). But only nine months (and three letters) later did Leibniz—his attention returned to the dispute by one John Arnold 'Anglus, Vir, ut apparet, doctus et bonus'—make direct answer to Bernoulli's suggestion. 'Keilius', he wrote at last on 4 November, 'responsione indignus est, sed rem ipsam brevi narratione [his 'Historia et Origo Calculi Differentialis'? (compare note (1) above)] complecti e re erit, et adjicere problemata quædam, unde intelligamus quid ipsi [Angli] possint. Cum illam novam differentiandi rationem considerasses,...excogitaveras inde applicationem quandam sic satis generalem, qua...problemata quædam, quæ ad differentio-differentiales descendere solent, intra differentiales primi gradus coërcentur. Ego nunc non bene memini, nec in literis antiquis quærere vacat. Te autem melius meminisse puto, itaque rogo ut si commodum est, iterum communices. Inserviet enim fortasse ad aliquod problema proponendum, cujus non statim apparebunt fontes' (*Commercium*, **2**: 358–9 [=*Math. Schriften*, **3**: 948]). Bernoulli obliged nineteen days afterwards by sending along an extract from his earlier letter to Leibniz of mid-August 1697 where he had outlined two methods 'ex nova illa differentiandi

6 December (N.S.) with which he included a 'Billet'—on a separate sheet of paper no doubt—enunciating this general problem (with a suggested instance of a group of hyperbolas which the challenged Anglo-Scots-*cum*-expatriate-Huguenot mathematicians were, Newton excepted, to take as its whole) Leibniz elected to append a long postscript vindicating his claim to priority over Newton in attaining the 'infinitesimal algorithm', and mounting a strong broader attack upon the latter's 'somewhat strange philosophy' with its appeals to 'occult scholastic qualities' and 'miracles' and its setting of 'bounds to God's wisdom and power'.[111] This was to be one disparagement of his mathematical originality—and scornful dispraise of his philosophy, natural and divine—

---

secundum parametrorum variabilitate erutas' (*Commercium*, **2**: 361 [= *Math. Schriften*, **3**: 949]), and the latter chose to put to Conti 'pour tater un peu le pouls à nos Analystes Anglois' an example of the orthogonal instance of the latter of these, that of determining curves intersecting given sequences of curves at assigned angles.

We should add that an offending phrase ('Tout le monde sait que M. Leibnitz a voulu disputer à M. Newton l'Invention du Calcul des Differences') in a brief French report upon the 'Account' which appeared in du Sauzet's weekly review, the *Nouvelles Litteraires* in mid-September 1715 drew at this time from Leibniz a sharp rebuttal (*Nouvelles*, 28 December 1715: 413–14, reprinted in *Correspondence*, **6**: 256–7) where—evidently still sight unseen of the English original—he made reference to the French 'Extrait' from it published 'dans l'Article 5 du 7 Tome du *Journal Literaire*' (*vide* note (108) above).

(111) 'Il ne paroist point que M. Newton ait eu avant moy...l'Algorithme infinitesimal, ...quoyqu'il lui auroit été fort aisé d'y parvenir s'il s'en fut avisé. ... ... Sa Philosophie me paroist un peu étrange, et je ne crois pas qu'elle puisse s'etablir. Si tout corps est grave, il faut necessairement...que la gravité soit une qualité occulte Scholastique, ou l'effect d'un miracle. ...[Mais] parcequ'on ne sait pas encor parfaitement et en detail comment se produit la gravité, ou la force...magnetique &c, on n'a pas raison pour cela d'en faire des qualités occultes Scholastiques ou des miracles; ...on a encore moins raison de donner des bornes à la sagesse et à la puissance de Dieu' (*Correspondence*, **6**, 250, 251, 253). Let us notice that Leibniz' draft of this 'P.S.' (now with the grand corpus of his manuscript writings in the Niedersächsische Landesbibliothek at Hanover, whence its text is reproduced by C. I. Gerhardt in his *Briefwechsel von G. W. Leibniz mit Mathematikern*, **1** (Berlin, 1899): 263–7, and, from this last, in *Correspondence*, **6**: 250–3) is rather longer and verbally slightly different from the secretary copy (now Add. 3968.36: 520^r–522^r) afterwards furnished by Conti to Newton (who printed this version on pages 97–9 of his 1717 Appendix to Joseph Raphson's *History of Fluxions*; on which see below). Lacking Leibniz' letter as sent to Conti in December 1715, we cannot now rigidly distinguish the last-minute changes which Leibniz effected in his draft from the alterations made by Conti (and his amanuensis) in the copy which Newton saw. Of two additions to the first paragraph of this 'grande Apostille' which Leibniz asked Nicholas Remond, its transmitter, to add before passing his letter on to Conti (see Pierre Desmaizeaux, *Recueil de Diverses Pieces*..., **2** (Amsterdam, $_1$1720): 114–15), only the first is found in Newton's version. Newton certainly never saw the concluding portion of Leibniz' draft (Gerhardt, *Briefwechsel*: 266, ll. 28 *et seq.* [= *Correspondence*, **6**, 253, ll. 8–21]), which named 'M. le Chevalier Wren, de qui M. Newton[!] et beaucoup d'autres ont appris quand il étoit jeune' and 'un François en Angleterre, nommé M. Moyvre, dont j'estime les connoissances Mathematiques' among 'autres habiles gens...qui ne font point de bruit' who might restrain the 'Sectateurs de Newton' from raucously parading their master's 'nouvelle Philosophie'.

which Newton was to be compelled to meet head on in his own name, for Conti made the content of Leibniz' *Apostille* known not only to the small inner circle of 'English analysts' but also to a wider circle of ministers and diplomats at Court in London.

To such people, without technical training or narrow scientific interest, old mathematical letters and papers meant little or nothing of course. And when a group of foreign ambassadors assembled by Conti at Crane Court proved singularly unimpressed by those in the Royal Society's archives which (so it was intended) should have convinced them of the validity of Newton's claim to originality and priority of discovery, considerable pressure was put upon him from 'above'—from the Princess, Caroline, of Wales and maybe even from King George himself (who certainly read and approved the reply subsequently made)[112]—to make direct answer to Leibniz' charges and insinuations. Newton could do nothing but acquiesce, and he set conscientiously to. Only his private papers reveal just how painstakingly he approached the task: a succession of preliminary notes[113] were energetically worked up by him into several drafts of an initial response[114] which ultimately, because (so he would have it) 'the more I consider the Postscript of Mr Leibnitz the less I think it deserves an answer. For it is nothing but a piece of railery from the beginning to the end',[115] he curtailed to be the letter finally sent off to Conti on 26 February 1715/16.[116]

Over its stern injunctions that Leibniz should accept as unchallengeable 'matter of fact' the 'ancient Letters & Papers' published in the *Commercium*, and either renounce the 'Answer of the Mathematician (or pretended Mathematician) dated 7 June 1713' which had been adduced so approvingly in the

(112) See Conti's letter to Brook Taylor of 22 May 1721 (N.S.) (*Correspondence of Newton*, 7: 137–9, especially 137–8). Before his recent coronation, of course, George had been Elector of Hanover, there employing Leibniz as Court librarian and Royal historian.

(113) These are now Add. 3968.28: 414^r(+415^v)–415^r/412^r–413^r (+.41: 23^r)→[in sequence] 416^r/419^r/417^r/418^r.

(114) Namely Add. 3968.38: (559^v+) 558^r/558^v→(.41: 95^v–94^v/94^r/94^v→) .38: 566^r–567^v (+436^r/436^v+437^v)→568^r/568^v + Trinity College, Cambridge. R.16.38: 432A^v)→569^v→ (571^r–572^r→) 573^r/573^v+(.39: 587^r/.41: 90^r/90^v→Trinity R.16.38: 432A^r→) 588^r/591^r (+587^v/ Lehigh University MS 731)/592^r. The text of much of the non-mathematical portion of these drafts is printed in I. B. Cohen and A. Koyré, 'Newton and the Leibniz–Clarke Correspondence' (*Archives Internationales d'Histoire des Sciences*, 15, 1962: 63–126): 73–5/104–15. The preliminary draft on Lehigh MS 731 was first brought to notice by A. R. and M. B. Hall in *Isis*, 52, 1961: 583–5.

(115) His dismissive statement in the unpublished transitional draft now Add. 3968.38: (571^r→) 573^r.

(116) Add. 3968.38: 560^r(+561^v)–561^r→564^r–565^r [initial state]→562^r/562^v→564^r(+.41: 91^r)–565^r [revised/augmented state], whence Newton published its text the next year on pages 100–3 of his Appendix to the 1717 reissue of Raphson's *History of Fluxions* (on which see below). (This definitive draft on ff. 564^r–565^r is more accurately reproduced on *Correspondence*, 6, 285–8.)

Charta volans or prove such 'an accusation wch amounts to plagiary... on pain of being accounted guilty of calumny' we may swiftly pass. The written and hearsay evidence upon which Newton seeks to rely in demonstrating that what is 'called by me the method of fluxions, & by him the differential method' was 'invented by me' many years before it was known to Leibniz—'ten years before [1676] or above' he follows Wallis' 1695 preface to his *Opera* in stating—contains no surprises. Nor, after Conti duly forwarded Newton's letter to him, was Leibniz in the least persuaded by its arguments or put off by its censures, but in the long reply which he returned a month later[117] calmly (and on the whole judiciously) parried them all—though he could not resist a sour opening stab at the 'forlorn boys' Keill and Cotes whom Newton had earlier 'detached against me' and against whom 'I had no wish at all to enter the lists'.[118] The letters published in the *Commercium Epistolicum* contained, he insisted, 'not a word which can call into question my invention of the calculus of differences'; but he went on to stress in his postscript that there was, on the other hand, 'not the slightest trace or shadow of the calculus of fluxions in all the ancient letters of Mr Newton which I have seen, except in the one he wrote on 24 October 1676

(117) On 9 April 1716 (N.S.). The original is now lost, but the text of the copy (now Add. 3968.36: [in sequence] 531^r/536^r/523^r–534^r) which Newton took of it—the first ten paragraphs in his own hand, and the remainder in secretary transcript—was subsequently published by him on pages 103–11 of his 1717 Appendix to Raphson's *History of Fluxions*, and from this *editio princeps* all later printings (by Desmaizeaux, Gerhardt and, most recently, by the editors of Newton's *Correspondence*) derive. Leibniz' accompanying 'P.S.' (Add. 3968.36: 5: 537^r–538^r) remained in manuscript till 1720 when Desmaizeaux set it in print—under the variant subhead of 'Apostille' and with other differences in spelling and punctuation which suggest that he worked from a variant copy—on pages 67–71 of the second volume of his *Recueil de Diverses Pieces*.... (Two years before, he had—see note (160) below—sent Newton a further secretary transcript of it which is Add. 3968.36: 527^r–529^r.) Newton's reluctance to publish this postscript in 1717 is readily explained: at the end of his 'P.S.' Leibniz had indicated to Conti that 'l'Apostille est pour vous, Monsieur: & la Lettre est plûtôt pour M. Newton' (see *Correspondence*, 6: 312). When sensitive issues of personal integrity were at issue, Newton could have had no wish for it publicly to be suspected that he had seen more of Leibniz' letter than the latter wished him to.

(118) 'Je n'ay point voulu entrer en lice avec des enfans perdus, qu'il [M. Newton] avoit detachés contre moy; soit qu'on entende celuy qui a fait l'Accusateur sur le fondement du *Commercium Epistolicum*, soit qu'on regarde la Preface pleine d'aigreur qu'un autre a mise devant la nouvelle Edition de ses *Principes*' (see *Correspondence*, 6: 304). Newton, we may note, never publicly countered this charge that he had loosed against Leibniz a detachment of desperadoes to attack him, be it anonymously, in the 1715 *Phil. Trans.* 'Account' of the *Commercium* or in the 'spiteful' editor's preface to the 1713 revised edition of the *Principia*, though he certainly had it in mind more than once to make clear that Cotes was 'none of his forlorn hope' (his words in an unfinished letter to Bernoulli in 1721; see note (208) below). 'The Præface of the Editor præfixed to the second Edition of my Book of *Principles* at Cambridge', he specified in one draft of his ensuing 'Observations' upon Leibniz' two letters to Conti, 'I did not see till the Book came abroad' (Add. 3968.36: 506^r, quoted *hors de contexte* by I. B. Cohen in his *Introduction to Newton's 'Principia'* (Cambridge, 1971): 240).

where he spoke of it only in an anagram, the solution of which he did not give till ten years afterwards'.[119] Huygens, not Newton or indeed Barrow, had been his guide into the intricacies of higher mathematics at Paris, leading him on to find 'my arithmetical quadrature of the circle and soon my general method by arbitrary series' and finally to gain entrance into the differential calculus; 'however, if someone profited from Mr Barrow it will rather be Mr Newton, who studied under him, than I, who never read his books with attention'.[120] But Leibniz spoilt this general show of open-handed reasonableness by maintaining that the words of his anonymous 1705 *Acta* review of Newton's *De Quadratura Curvarum* (whose authorship he here, as ever, did not take to himself) meant, in stating that 'In place of Leibnizian differences Mr Newton employs, has always employed, fluxions...',[121] in no way to insinuate that Newton had come to employ fluxions only after he had had a sight of his own method of differences—'and I defy who it may be to give another reasonable intention to these words *semperque adhibuit*. It is hence necessary to conclude either that Mr Newton has let himself be deceived by a man who has poisoned these words of the *Acta* [review], supposed not to have been published without my knowledge, and has come to imagine that he is charged with being a plagiarist; or that he was well content to find a pretext to attribute the new calculus to himself, or have it privately so attributed (since he noticed its success and the noise it was making in the world...)'.[122]

---

(119) 'Lors que j'eus enfin le *Commercium Epistolicum*, je vis...que les Lettres qu'on publioit ne contenoient pas un mot qui pût faire revoquer en doute mon invention du Calcul des differences.....Il n'y a pas la moindre trace ny ombre du calcul des... fluxions dans toutes les anciennes Lettres de M. Newton que j'ay vües excepté dans celle qu'il a écrite le 24 d'Octobre 1676, ou il n'en a parlé que par enigme (*Commerce* p: 56.) Et la solution de cet Enigme il n'a donnée que dix ans après...' (Add. 3968.36: 536ʳ, 537ʳ [=*Correspondence*, 6: 305–6, 312]).

(120) 'Il est bon de savoir qu' à mon premier Voyage d'Angleterre en 1673 je n'avois pas la moindre connoissance des Series infinies, ...ny d'autres Matieres de la Geometrie, avancées par les dernieres methodes....Ce fut [après] peu à peu que Mʳ Hugens me fit entrer en ces matieres, quand je le pratiquois à Paris, et cela...me fit trouver environ vers la fin de l'an 1673 ma Quadrature Arithmetique du Cercle, [et] bientôt ma methode generale par des series arbitraires, et j'entray enfin dans mon Calcul des differences.......Mʳ N[ewton]...pretend que ce que j'ay écrit pour luy à Mʳ. Oldenbourg en 1677 est un deguisement de la methode de Mʳ. Barrow. Mais...une infinité de gens liront le Livre de Mʳ Barrow sans y trouver notre calcul....Cependant si quelqu'un a profité de Mʳ Barrow, ce sera plus tôt Mʳ N[ewton] qui a etudié sous luy que moy qui...n'ay...jamais lûs [ses Livres] avec attention...' (Add. 3968.36: 532ʳ–532ᵛ, 533ᵛ [=*Correspondence*, 6: 308, 310]).

(121) 'Pro differentijs Leibnitianis D. Newtonus adhibet semperque adhibuit fluxiones...' (*Acta Eruditorum* (January 1705): 35); see page 26 above.

(122) '...c'est une interpretation maligne d'un homme qui cherchoit noise...insinuer que [c'est] apres la veue de mes differences [que Mʳ Newton] s'est servi de fluxions. Et je défie qui que ce soit de donner un autre but raisonnable à ces paroles *semperque adhibuit*....D'ou il faut

When Newton read these words he blew up. 'S^r', he thundered in a first intended reply$^{(123)}$ to their intermediary Conti,

> Mr Leibnitz Letter...is like all the rest of his Letters & like the Libel of [29 July 1713] published in Germany without the name of y^e Author or Printer or place where it was printed, vizt full of assertions accusations & railing reflexions without proving any thing. He...complains that the Passage in the *Acta Lipsiensia* of January 1715 [*sic*] has been poisoned by a malicious interpretation of a man who would pick a quarrel. In the Introduction to the Book of *Quadratures*$^{(124)}$ I affirmed that I found the Method of fluxions gradually in the years 1665 & 1666. And in opposition to this the said *Acta* in giving an Account of this book represent that for the better understanding the said Introduction we must have recourse to the Differential Method....

The seething outrage beneath these rough, unpolished phrases is evident, and we will never approach nearer to a stark expression by Newton of his true feelings. In succeeding paragraphs, however, as the hot flush of his anger subsided and calmness of mind reasserted itself, he for the first time declared his readiness to bring forward in irrefutable witness of his priority of discovery the texts of his unpublished early papers on calculus to which he had before, in the introduction to his *De Quadratura Curvarum* but fleetingly alluded. And in the meanwhile he gave a foretaste of what was in them. 'Mr Leibnitz', he subsequently went on,$^{(125)}$

> saith that he found the Arithmetical Quadrature of the Circle towards the end of y^e year 1673 & the general method for arbitrary series in the year 1676 & soon after the differential calculus in the same year.... And am not I as good an evidence that I found the method of fluxions in the year 1665 & improved it in the year 1666. And if I should add that before the end of the year 1666 I wrote a small tract on this subject wch was the grownd of that larger tract wch I wrote in the year 1671, & in this smaller tract, tho I generally put Letters for differences, yet in giving a general Rule for finding the curvature of curves I put the letter ‭X‬ with one prick for first fluxions & with two pricks for second fluxions:$^{(126)}$...especially since both these Tracts are still in being & ready to be produced upon occasion.

---

conclure ou que M. N[ewton] s'est laissé tromper par un homme qui a empoisonné ces Paroles des *Actes*, qu'on supposoit n'avoir pas été publiées sans ma connoissance, et s'est imaginé qu'on l'accusoit d'être plagiaire; ou bien qu'il a été bien aise de trouver un pretexte de s'attribuer ou faire attribuer privément l'invention du nouveau Calcul (depuis qu'il en remarquoit le succès, et le bruit qu'il faisoit dans le monde...)' (Add. 3968.36: 531^r [=*Correspondence*, **6**, 305]).

(123) Add. 3968.29: 420^r–421^v. We cite the opening sentences on f. 420^r.

(124) This is reproduced in **1**, **2**, §3.1 preceding; see especially page 122.

(125) Add. 3968.29: 421^r.

(126) On the rule for curvature in the October 1666 tract to which Newton here refers, see 1: 421–2. When he came to incorporate this sentence in the 'Observations' upon Leibniz' 'Epistle' which he afterwards appended to his printing of its text in his 1717 reissue of Raphson's *History of Fluxions*, he was more careful to specify that 'I put the letter ‭X‬ with one prick for

Newton's succeeding drafts[127] of his planned letter of riposte to Conti elaborate and embellish the detail of this first wrathful explosion in ways one would mostly expect, though at one point he introduced a memory of thirty years before whose authenticity we see no reason to dispute: 'In the year 1684 when Mr Leibnitz first printed the Elements of his method Mr Craige who was then at Cambridge brought me the *Acta Eruditorum* & desired me to explain. Whereupon I explained it to him & I told him that the method was mine in a new dress. . . . Mr Craig is still alive & remembers this'.[128] Above all he was adamant that 'The Royal Society. . . acted by good authority & in a regular manner. . . when [Mr Leibnitz] put them upon the necessity of appointing a Committee to examin old Letters & Papers. . . & report their opinion thereupon. . . & the Society ordered the letters & papers with the Report to be published'; Leibniz, on the other hand, had operated with 'tricking, double-faced, maskerading management' in electing to appeal 'from them to a Mathematician of his own chusing'—one who 'in his Answer dated 7 Ju[ne] 1713. . .published in a clandestine back-biting manner (as defamatory Libels use to be). . .cited John Bernoulli as a man different from himself [but whom] Mr Leibnitz lately. . . tells us was Mr Bernoulli himself'[129]—and, although the *Commercium* had been published early in 1713, 'in all the three years & four months wch are since elapsed has not yet been able to produce any further proof [of his priority] against Dr Keill then what was then before them'.[130]

Strong words, but Conti was never in fact to read them. As Newton's pen

---

first fluxions drawn into their fluents & with two pricks for second fluxions drawn into the square of their fluents' (we cite Newton's draft on Add. 3968.31: 448^r; in the version printed (*History*: 116; compare page 515 below) he replaced 'the letter ϽϹ' by the anonymous 'a Letter', leaving his readers with the impression that in 1666 he had set his 'pricks' upon his fluents to denote their first- and second-order fluxions in the standard way made public by him in his 1704 *De Quadratura*).

(127) Add. 3968.41: 156^r→.29: 422^r–423^v→.29: 424^r–425^r→.31: 449^r–450^v. A rough preliminary version of ff. 422^r (bottom)-422^v (top) is found on Add. 3970.3: 294^r amid contemporary draftings by him of additions to the terminal queries of his *Opticks* for its English re-edition (London, ₂1717).

(128) Add. 3968.29: 422^v. The draft on .41: 156^r states slightly differently that 'In the year 1684 Mr Craig desired me to explain to him the elements of the differential calculus then newly printed in the *Acta Eruditorum*. I did so & told him that it was my method in another dress. . .'. In his revise on .29:424^r, let us add, Newton altered the year to be, much more plausibly, '1685' (compare vii: 3–4) and so too it is put in the final, otherwise minimally differing version on .31: 449^r.

(129) Namely 'in the *Novelles Litterairs* [*sic*] for 28 Decem. 1[71]5' (so Newton specifies on .29: 423^r), where the writer of the letter is named to be 'M. Jean Bernoulli, l'Auteur le plus profond dans ces matières' (see *Correspondence*, 6: 257) and—Newton went on—'the citation . . .whereby the author distinguishes himself from [the *eminens Mathematicus*] Bernoulli is omitted'.

(130) We conjoin phrases on Add. 3968.29: 423^r and .31: 449^v.

raced over the paper in the early summer of 1716 coining its rich variety of slashing epithet, his mood swung towards making a more public refutation of the unfounded pretensions and provocations of his 'aggressor'.[131] And, when he looked around for means and opportunity to do so, his eye alighted on a book by his late acquaintance Joseph Raphson, posthumously published the year before, which sought to recount *The History of Fluxions* by *Shewing in a Compendious Manner the first Rise of, and various Improvements made in, that Incomparable Method*.[132] By his own later testimony Newton had delayed the publication of this brief historical account (whose 96 quarto pages in their greatest part gave mere direct quotation or minimally variant paraphrase of Leibniz' 'Nova Methodus pro Maximis et Minimis itemque Tangentibus' in the 1684 *Acta*, his own *De*

---

(131) In a draft of a letter to Desmaizeaux two years afterwards he not a little disingenuously affirmed that he regarded addressing his reply to Leibniz through Conti's hands—'that he might have neutral & intelligent witnesses of what passed between us'—as 'an indirect practise', and so 'forbore writing an Answer in the form of a Letter to be sent to him, & only wrote some Observations upon his Letter to satisfy my friends here that it was easy to have answered him had I thought fit to let him go on with his politicks' (Add. 3968.27: 401ʳ, quoted by Brewster in his *Memoirs*, **2**, 1855: 288–9 and cited more fully and accurately on *Correspondence*, **6**: 458).

(132) Such is the title of the original English version of the book published by Richard Mount at London in 1715; the Latin translation issued simultaneously with it (or perhaps a little afterwards?) asserts equivalently that the work is a *Historia Fluxionum, sive Tractatus Originem & Progressum Peregregiæ Istius Methodi Brevissimo Compendio (et quasi Synopticè) Exhibens*. We have earlier (in III: 11, note (29)) innocently pointed to one instance, on the book's second page, where the latter deviates significantly from its English parent—and ought then to have known that the divergence was no accidental straying on the translator's part from the literal rendering which he almost everywhere else gives. It is now clear that Mount submitted to Newton for his prior inspection and comment at least the Latin text of the opening portion of the book (where Raphson in his English original gives a blurred summary of Newton's entrance upon the 'incomparable' method of fluxions 'about the Years 1665. & 1666'). This no longer survives, so far as we know, but Newton's refinings of its wordings and sharpenings of its detail into the form ultimately printed in 1715 are preserved on Add. 3968.10: 139ᵛ→134ʳ →135ʳ. It would not seem that he saw before its publication the Latin of the book's Chapter XIII (pages 92–6) where, in Raphson's original phrasing, there was presented 'A Synopsis or Review of the Inventions...of the Method of Fluxions, or the Differential Calculus' summarising 'the whole Controversy' as 'delivered in the Original Letters between Sir *Isaac Newton*, Mr. *Leibnitz*, and others' which are found 'fully and clearly laid open' in the *Commercium Epistolicum*—already become 'a very scarce Book, but few printed, and none to be met with amongst the Booksellers' according to the 'Author's Preface' (signature A2ᵛ)—together with a *verbatim* citation of the 1712 'Sentence of the *Royal Society*, annex'd to the end of the *Epistolical Commerce*; to [*sc.* against] which...Mr. *Leibnitz* had Appeal'd, upon the occasion of Mr. *Keil*...his Vindicating and asserting the Invention to its proper Author'. When, however, he received his copy (now Trinity College, Cambridge NQ.16.173) of the printed 1715 *Historia*, he made in its margins a number of corrections and emendations to the text of this concluding chapter, and the more significant of these were gathered (by way of Add. 3968 [.10: 144ʳ+134ʳ→] .40: 593ʳ) as the printed *Emendanda...in Historia Fluxionum* added at the bottom of the last page of the Appendix in the 1717 reissue.

Quadratura Curvarum and George Cheyne's *Fluxionum Methodus Inversa*[133] of the previous year, 1703) 'for four years together & could stop it no longer without paying for the edition of it'[134]—from which we may surmise that he was not greatly keen to have Raphson stand in for him as recorder of the past truth of calculus discovery. But with the book appeared—and its author-*cum*-editor then more than a year dead[135]—Newton towards the latter half of 1716 felt free to arrange with its publisher, Richard Mount, to attach to its unsold copies (in both English and Latin versions) an appendix bringing it up to date, *viz.* by printing the texts of Leibniz' two letters to Conti along with his own intervening response to the first, and by adjoining to the latter an extended, further reworked version of the riposte which he had before planned to make directly and more privately, now restyling it (and toning down its more stinging outspokennesses) to be a series of 'Observations upon the foregoing Epistle'.[136]

---

(133) On this work see **1**, Introduction: note (47) above. Raphson also cites single periodical articles by Johann Bernoulli (*Acta Eruditorum* (March 1697): 125 ff.) on 'Principia Calculi Exponentialium...'—with his Leibnizian *differentiæ dx*, *dy*, *dz* replaced by (so Raphson follows de Moivre in taking them to be) 'equivalent' Newtonian fluxions $\dot{x}$, $\dot{y}$, $\dot{z}$, and such phrases as 'Ad differentiandam quantitatem...' likewise recast to be 'Ad inveniandam fluxionem quantitatis...'!—and by John Craige (*Philosophical Transactions*, **17**, No. 232 (September 1697): 708 ff) on 'The Quadrature of Figures Geometrically Irrational'; see *History*: ₁41–2/₂42–4 and 51–6 (=*Historia*: 39–44 and 51–6) respectively.

(134) So he stated in the draft (Add. 3968.42: 601ʳ, now printed in *Correspondence*, **7**, 1978: 82) of a letter—probably never sent—to (we assume) Fontenelle. But his further statement there that 'Ralp[h]sons book was written & in the Press before I knew of it' can (see note (132) above) refer truthfully at best only to the original English text. It was of course to Newton's best interest to disclaim all advance knowledge of the content of Raphson's *History*, but he could not but have known from the first that the book was being prepared for publication. Roger Cotes, certainly no bosom friend of Raphson's, had been acquainted by him of his 'design of writing yᵉ History of yᵉ Method of Fluxions' already some 'Year & an half ago' when he wrote to William Jones on 15 February 1710/11 (see Edleston, *Correspondence of...Cotes* (London, 1850): 206), and had in the mean time been led to understand that Jones had given Raphson a 'sight' of 'some papers of Sr Isaacs in Yʳ hands which were communicated long ago to Mʳ Collins' but that 'he thought 'em not to be for his purpose, which I do now very much wonder at, if his intention was to do justice to Sʳ Isaac' (*ibid.*).

(135) The title of the 1715 *History of Fluxions* records that it is 'By (the late) Mr. Joseph Raphson...' (and a Greek parenthesis to that of the Latin *Historia* likewise speaks of him as τὸν νῦν μακαρίτην), while the *Emendanda* at the end of the 1717 Appendix has (p. 123) the parting phrase: 'Cætera, cùm Author emortuus sit, & Liber [understand the printed sheets of the English *History*] annis abhinc plus tribus impressus fuerit, emendet Lector', but the precise date of Raphson's death we have not been able to discover.

(136) The title which Newton set upon these comments in the draft which is now Add. 3968.31: 447ʳ–448ᵛ. This was subsequently revised (and considerably amplified)—by way of [in sequence of their incorporation] .41: 116ᵛ–117ᵛ/(.38: 506ʳ/506ᵛ→) 28ʳ/40ᵛ(+ .10: 144ᵛ)/(38ᵛ→) 38ʳ+28ʳ—to be the 'Observations upon the preceding Epistle' which were eventually published on pages 111–19 of the 1717 Appendix (and whose text is reprinted by Samuel Horsley in his *Isaaci Newtoni Opera quæ exstant Omnia*, **4** (London, 1782): 607–14; by C. I.

As ever, not to attempt to recount the individual thrusts of Newton's blade or summarise its defensive parries in their detail. Whatever sharpness there might have been in such a charge as that

In the year 1684 Mr Leibnitz published only the Elements of the Calculus differentialis & applied them to questions about Tangents & Maxima & Minima as Fermat Gregory & Barrow had done before, & shewed how to proceed in these Questions without taking away surds, but proceeded not to the higher Problemes. The *Principia Mathematica* gave the first instances made publick of applying the Calculus to the higher Problemes. [137]

is dulled for us, if not for Newton's contemporary reader, by its blurred over-statement and wearisome repetition. 'Both methods [of infinite series and fluxions] are', he continued strongly to insist, 'but two branches of one general method' which he 'joyned' in his *De Analysi* in 1669 and 'interwove in the Tract which I wrote in the year 1671'; and, if Leibniz seriously wished to challenge his own prior claim to its discovery, then he must establish the right he himself had to it as a whole, instead of 'tearing this general method in pieces & taking from me first one part & then another part whereby the rest is maimed'. [138] But what is unprecedented is the precise account which he proceeded to give of the 'several mathematical papers written in the years 1664, 1665 & 1666 [w^ch] I still have in my custody'—'The original Manuscripts have been seen by Mathematicians many years ago'—as providing the best 'witness that I invented the methods of series & fluxions in the year 1665 & improved them in the year 1666'. [139] The testimony of these earliest papers of his on calculus Newton initially planned to present in a separate following 'P.S.' illustrating the 'progress by which the method was invented' with lavish *verbatim* extracts from his papers of 13 November 1665 and 16 May 1666 and from 'a little Tract written in October 1666'; [140] but he broke off after he had,

Gerhardt in *Briefwechsel von G. W. Leibniz mit Mathematikern*, **1** (Berlin, 1899): 285–94; and most recently in *Correspondence of Newton*, **6**: 341–8).

(137) Add. 3968.41:448^r (= 1717 Appendix: 117). We may be sure that the 'Aggressor' was not in sequel allowed to escape without again (compare page 494 above) being faced with the accusation (*ibid.*: 448^r/448^v) that 'In the year 1689 Mr Leibnitz published the principal Propositions of this Book as his own in three papers [in the Leipzig *Acta*]..., pretending that he had found them all before that book came abroad. And to make the principal Proposition his own [he] adapted to it an erroneous demonstration'—for a fuller account of which see note (86) preceding. (In the printed version Newton rammed his point home by adding 'and thereby discovered that he did not yet understand how to work in second Differences'.)

(138) Add. 3968.31: 447^r/447^v (= 1717 Appendix: 112, 113).

(139) We excerpt these phrases from Newton's preliminary revise, now Add. 3968.41:38^r, of his intended postscript to his Leibnizian 'Observations'.

(140) 'where', he went on, 'the same method [of resolving Problems by motion] is set down in the same seven Propositions [as in my] paper dated 16 May 1666...but the seventh'—'the same with that conteined in the Paper of 13 Novem. 1665 tho exprest in other words'—'is enlarged & an eighth added to them'.

with manifestly fast increasing impatience, got so far as to transcribe the two examples instancing the application of the eighth of the rules introductory to the last.[141] In a humbler succeeding revision of his over-optimistic original scheme—now filling the postscript with the verbal witness of Leibniz' letter to him of 7/17 March 1693 and an extract from Wallis' to him of 10 April 1695, 'both wch upon a fresh occasion have been produced & left in the Archives of the R. Society'[142]—he transposed the enunciation of the central 'Probleme' of his November 1665 paper, along with the bare statement of its construction, to the main text of his 'Observations' and then went on to comment[143] that

this resolution is there illustrated with examples & demonstrated & applied to Problems about Tangents & the Curvature of Curves. And...in another Paper dated 16 May 1666 a general method of resolving Problems by motion is set down in seven Propositions the last of wch is the same with the Problem conteined in the aforesaid paper of 13 Nov. 1665.[144] And...in a small Tract written in November[!] 1666, the same seven Propositions are set down again & the seventh is improved by shewing how to proceed

---

(141) See Add. 3968.41: 39^r/39^v. Having there transcribed virtually *in toto* both the 'enlargement' upon Proposition 7 in the October 1666 tract (see 1: 411–12) and the statement of Proposition 8 together with its two illustrative examples (see 1: 403–4), Newton proceeded to strike out their text, leaving in his 'P.S.' only its introductory words: 'When the Committee of the R.S. published the *Commercium Epistolicum* the Papers in my custody were not produced.... Among my papers written in the years 1664, 1665 & 1666 some happen to be dated: amongst wch is that which follows.

> 13 Novem. 1665
> *Probleme.*

An Equation being given &c.'
On a related stray manuscript sheet, in Shirburn Castle, which was published over a century ago by S. P. Rigaud under the title of an 'Early Notice of Fluxions' (*Historical Essay on ... Newton's 'Principia'* (Oxford, 1838): Appendix: 20–4) Newton about this time set out a fuller *précis*—nowadays often quoted—of his early 'Waste Book' papers on fluxional analysis, not only those of 13 November 1665 (there, too, transcribed *in extenso*) and May 1666 but also that of 20 May 1665 (see 1: 272–80) where 'the same method is set down in other words, and fluxions applied to [*sc.* multiplied by] their fluents are represented by prickt letters'; and then passed to outline the variety of problems resolved in the 'small tract dated in October 1666', notably the rule (see 1: 421–2) for 'finding the curvature of curves without calculation' (though he broke off with his specification of it scarcely begun). None of its discarded detail, unfortunately, found its way into Newton's printed 'Observations' on Leibniz' 1716 letter to Conti.

(142) By Newton himself, of course; compare note (104) preceding. Let us remark, however, that in a transitional state of the 'P.S.' on Add. 3968.41: 29^v the text of the fundamental paper of 13 November 1665 is still set to be the 'Schediasma hereunto annexed' though he is there clearly upon the point of inserting it in the main 'Observations'.

(143) On pages 116–17 of the published Appendix to Raphson's *History*; we have, however, adapted the orthography and punctuation—crudely mangled by the copy-reader in the printed text—to agree with those of Newton's autograph draft on Add. 3968.41: (40^r +) 38^r.

(144) 'tho exprest in other words' of course, as Newton had been more careful to specify in his initial draft on .41: 39^r (see note (140)).

without sticking at fractions or surds or such quantities as are now called transcendent. And an eighth Proposition is here added conteining the inverse method of fluxions so far as I had then attained it, namely by Quadratures of Curvilinear figures & particularly by the three Rules upon wᶜʰ the *Analysis per Æquationes numero terminorum infinitas* is founded & by most of the Theorems set down in the Scholium to the Tenth Proposition of the Book of *Quadratures*. And...in this Tract when the area arising from any of the terms in the Valor of the Ordinate cannot be expressed by Vulgar Analysis I represent it by prefixing the symbol □ to yᵉ term....And...in the same Tract I sometimes used a Letter with one prick for quantities involving first fluxions & the same Letter with two pricks for quantities involving second fluxions. And...a larger Tract wᶜʰ I wrote in the year 1671 & mentioned in my Letter of 24 Octob. 1676 was founded upon this smaller Tract & began with the Reduction of finite quantities to converging Series & with the solution of these two Problems. 1 *Relatione quantitatum fluentium inter se data fluxionum relationem determinare.* 2 *Exposita Æquatione fluxiones quantitatum involvente invenire relationem quantitatum inter se.* And...when I wrote this Tract I had made my Analysis composed of the methods of series & fluxions together so universal as to reach to almost all sorts of Problemes as I mentioned in my Letter dated 13 June 1676....⁽¹⁴⁵⁾

---

(145) In preliminary draft on .41: 38ʳ Newton adjoined: 'These things do not rely meerely upon my testimony, but the ancient Manuscripts themselves are ready to be produced....The original Manuscripts...Mʳ Fatio mentioned...at the end of his book entituled *Fruit walls improved*, which was published in the year 1699'. He subsequently decided to attach a full *verbatim* extract of the pertinent passage—we have ourselves quoted it in note (6) above— 'Out of the Book of Mʳ Nicolas Fatio de Duillier intituled *Investigatio Geometrica Solidi rotundi in quod minima fiat resistentia*, & published in the year 1699' (see .41: (155ʳ→) 143ʳ), and so it appears on page 123 of the published Appendix to Raphson's *History* with the 'N.B.' that 'Mr *Fatio* wrote this...as a Witnesse. He related what he had seen, and his testimony is the stronger because it was against himself'. In the printed 'P.S.', as we have already said, pride of place was given to the texts of the letters sent to him by Leibniz on 7/17 March 1693 (pages 119–20) and from Wallis on 10 April 1695 (pages 120–1), extracts from both which he had inserted in his 1715 'Account' of the *Commercium* (compare note (104) preceding). In their train in the printed Appendix he also set Wallis' 'short Intimation' of Newton's invention of fluxions as it had appeared twenty-one years before in preface to the first volume of his *Opera Mathematica* at short notice 'while the Press stay'd' (page 121), and the related passage in the anonymous review of Wallis' book—by Leibniz, we now know—the next year, June 1696, in the *Acta Eruditorum* (*ibid.*: 121–2). Newton had also initially intended to cite 'Part of a Letter of Dr Wallis to Mʳ Leibnitz of July 30. 1697' (see .41: (155ʳ→155ᵛ→) 93ᵛ/93ʳ; the letter had been printed in full by Wallis in the 'Epistolarum Collectio' inserted in his *Opera*, 3 (Oxford, 1699): 681–5), but he afterwards replaced this by his survey of parallel affirmations by Leibniz in his preceding letters to Wallis of 19/29 March and 18/28 May 1697 (*ex* Wallis, *Opera*, 3: 672–4 and 678–80 respectively) in the lengthy 'N.B.' on page 122 of the 1717 Appendix (Newton's drafts of which survive on .41: 93ʳ). His initial thoughts of quoting allied extracts 'Out of the Answer of Mʳ Leibnitz [to Fatio] published in the *Acta Eruditorum* for May 1700 p. 202' (see .41: (92ᵛ→) 155ᵛ + 93ʳ) and 'Out of the Introduction of Mʳ James Bernoulli to the third part of the Positions *de Seriebus Infinitis* maintained by Mʳ James Herman & published at Basil in the year 1696' (see .41: 143ʳ; the passage cited is reprinted in Jakob Bernoulli, *Opera*, 2 (Lausanne/Geneva, 1744): 748, ll. 2–19) were stillborn. On .10: 128 *bisʳ-cum*-142ʳ there are, we may add, to be found similar excerpts penned by Newton at this time

Brief and imprecise though this authorised account of his creation of his single 'universal' method of analysis is, Newton did not improve publicly[146] upon it during his remaining ten years of life; and with the sequestering of his private papers after his death, first in the tight grip of John Conduitt and then all but forgotten in the archives of the Portsmouths,[147] for all but the privileged few who gained access to that unsorted hoard of manuscripts (or the fewer still who owned the rare copies of Newton's major unpublished writings which were to be had) it had to suffice as the well-nigh unique *locus classicus* for authentic information about the early stages of his development of fluxional calculus.[148] What reaction Leibniz himself, however, might have had to these autobiographical revelations we will never know—though we may hazard a shrewd guess!—for, such was fate, the 'aggressor' against whom Newton primarily

'Out of a Paper of Mʳ James Bernoulli printed in the *Acta Eruditorum* for January 1691 [pag. 14]' and 'Out of the Preface to the *Analysis des Infinitement Petits* [*sic*] published A.C. 1696', which respectively urge the merits of Barrow's 'calculus' of tangents, and affirm that Newton's *Principia* 'est presque tout de ce calcul [différentiel]'; and also one 'Out of a Paper of Mʳ Leibnitz published in the *Journal des Sçavans* du 30 Aoust 1694' which more timidly allows that Newton 'a eu quelque chose de semblable', with a subjoined 'N.B.' suggesting that when Leibniz wrote this he had not yet seen the preliminary exposition of Newton's notation 'by prickt letters' for fluxions—'more convenient then differentials tho not necessary to the Method'—published the previous year by Wallis in the second volume of his *Opera* (see VII: 174 ff.).

(146) He subsequently had it strongly in mind to attach an elaborate Latin description of the prime propositions of his *chartæ antiquæ* to an intended reissue, around 1719, of the *editio secunda* of his *Principia*, incorporating it in an elaborate 'Enarratio plenior' (the finished text is reproduced in Appendix 5 below) of the scholium to Lemma II of its second book; but Pemberton was given no instruction to incorporate it in the *editio tertia* of 1726, and this 'fuller enarration' disappeared into the limbo of Newton's private papers like so many other products of his pen. In preliminary to this he with his usual painstaking efficiency wrote out (on Add. 3968.1: 1ʳ/2ʳ) a dress rehearsal for what, in the 'Enarratio plenior' (see .36: 506ʳ) he was to title 'Apographum Schediasmatis a Newtono olim scripti cui tempus scribendi forte appositum fuit, vizᵗ 13 Novem 1665'; and elsewhere (on .19: 263ʳ/264ᵛ) sketched out an analogous scheme in which, *inter alia*, copious excerpts were to be adduced 'Ex Tractatu parvo sub finem anni 1666 conscripta' and 'Ex Tractatu de methodis serierum et fluxionum anno 1671 composito' (this last we cited in full on III: 14–15, there slightly—perhaps?—mistaking its context).

(147) On this see I: xvii ff.

(148) It—together with the related manuscript published by Rigaud in 1838 (see note (141) preceding)—still filled this rôle for J. E. Hofmann when he came in the early 1940's to write his elaborate 'Studien zur Vorgeschichte des Prioritätstreites zwischen Leibniz und Newton.... I. Materialien zur ersten mathematischen Schaffensperiode Newtons 1665–1675' (*Preussische Akademie der Wissenschaften (1943). Math.-naturw. Klasse*, No. 2); and it led him in particular, like P. E. B. Jourdain (see his edition of Augustus De Morgan's *Essays on the Life and Work of Newton* (Chicago/London, 1914): 108–9) and others before him, erroneously to distinguish (page 116) the October 1666 fluxions tract from its ghost of 'November 1666' which Newton's carelessness—his mind still thinking of his paper of 13 November of the previous year?—summed up in his present 'Observations' (when he copied Add. 3968.41: 39ʳ onto *ibid.*: 40ʳ) and failed instantaneously to exorcise.

directed his swingeing comments died suddenly in mid-November 1716 while the Appendix to the *History* was still in press and he was never to see it.[149]

In any generous view Leibniz' abrupt removal from the scene ought to have taken the heat out of the dispute over priority of 'invention', if not at once to have ended its fracas as Conti expected in briskly breaking the news of his death to Newton a few weeks later.[150] But such was not to be. Fearful of an 'imperfect' publication in France of Leibniz' two letters to Conti and his own intervening first reply, Newton hastened to make his own edition of them with his 'Observations' thereon public, and copies of Raphson's *History* (and Latin *Historia*) augmented with its new 27-page Appendix appeared in the bookshops early the next year, 1717.[151] Then for a while all was quiet. But when Newton

(149) Nor indeed did Bernoulli (who had a minimal command of English) read Newton's 'Observations' till three years afterwards when Desmaizeaux' reprinting of their French version in the second volume of his *Recueil* first came his way; see page 524 below.

(150) 'M. Leibniz est mort; et la dispute est finie' was the theatrical phrase, now so often quoted, which he used in his letter to Newton on 10 December 1716 (N.S.); see *Correspondence*, **6**: 376. Newton's reaction to this news was doubtless tinged with regret that his 'foe' could not have lived long enough publicly to admit defeat in his fight over priority of discovery of analytical calculus.

(151) So we gather from a number of Newton's subsequent statements. In particular, in the draft of an intended letter to Bernoulli early in 1720 he wrote of 'Mr Ralphson['s augmented] book' as having been 'published three years ago' (Add. 3968.42: 617^v, printed in *Correspondence*, **7**: 80), while a year and a half or so before he had affirmed to Desmaizeaux—in immediate continuation to the passage from this August 1718 letter which we quoted in note (132) above—that 'As soon as I heard that he [Leibniz] was dead I caused the Letters & Observations to be printed least they should come abroad imperfectly in France' (.27: 401^r; see *Correspondence*, **6**: 458). Newton's papers otherwise confirm both the haste with which the Appendix to Raphson's *History* was pushed through the press, and his genuine fear of being anticipated by a French edition of Leibniz' letters to Conti which would leave his own answers unheard. His draft (.36: (506^r→) 505^r) of the half-title to the Appendix (on its opening page 97) which introduces the 'Epistolæ sequentes a D. Leibnitio cum amicis suis in Gallia et alibi communicatæ' carries a scrawled note in the publisher's hand: 'This to be at the Top of the first page of ye French Coppy. I pray loose noe Time. You shall have more Coppy Tomorrow. R. Mount'. The 'French Coppy' here referred to (a pair of proofs of its text survive in Add. 4005.3) was a separately numbered 15-page 'Traduction Francoise des Lettres'—only Newton's to Conti of 26 February 1715/16 needed to be translated of course—'et des Remarques imprimées cy-dessus en Anglois', evidently meant to be attached to the Appendix for the sake of non-English-speaking Continental readers; but we know of no published copy of the augmented Raphson *History* where this was done, and we suppose that the plan was dropped before it was implemented. (The excellent translation of Newton's 1716 letter to Conti and of his ensuing 'Observations' on Leibniz' reply in April following—the work, no doubt, of de Moivre yet again—was not wasted, however, but fed to Desmaizeaux for publication in his 1720 *Recueil de Diverses Pieces*...; see note (159) below.) Newton's autograph drafts, now Add. 3968.10: 144^r→ .40: 593^r (+ .10: 134^r) of the concluding 'Emendanda in Epistolis [et] in Historia Fluxionum' printed at the end of the Appendix (after the 'FINIS' on page 123) include, we may lastly notice, an unused correction (for which, anyway, there would have been no space enough left to take it) of 'Pag 92 l. 17' in Raphson's *Historia*, namely: '*lege* ante annos

came to read the 'Elogium' of Leibniz in the Leipzig *Acta*[152]—this by its editor Christian Wolf—and the parallel 'Éloge' of him in the *Histoire* of the Paris *Académie*—this by its *Secrétaire perpétuel* Fontenelle[153]—his mood had not mellowed towards his defunct German 'rival'. In 1718, indeed, after he had studied the detail of Fontenelle's elegantly turned memorial tribute he privately reacted by making stern correction of its many faults (as he saw them to be) in a long itemised list where, among other things, he roundly declared once more that 'the Principles upon w^ch [Mr Leibnitz] founded his Dynamique is erroneous' and went on to castigate Wolf's 'pretence [in his *Acta* 'Elogium'] that Mr Leibnitz intended to write a *Commercium* [*sc.* of his own]' as 'a sham'.[154] Above all he spoke out angrily[155] against 'the charge of Plagiary published in the *Acta Lips.* 1705' and the 'Paper...dated 29 Julij 1713' in which

> Mr Leibnitz...then at Vienna [whither] Copies of the *Commercium* were sent to him by several hands from England...appealed from the judgm^t of the [Royal Society's] Committee to that of Mr J. Bernoulli, desiring him to examin the Book, & pretending that he had not yet seen it; & [with his own inserted] Mr Bernoulli's answer dated 7 June 1713...& caused them to be published without their names, as if written by other persons unconcerned.

For this 'defamatory Libell'[156] with its implicit accusation that he had based his calculus of fluxions upon Leibniz' prior differential method—for its 'designe',

---

44 vel 45 hanc methodum invenisse ac dignitates et dignitatum latera in speci[e]bus designasse hoc modo, vizt per $\overline{x+o}|^{\frac{m}{n}}$ in seriem $x^{\frac{m}{n}}+\ldots+$&c conversam, et ejus fluxiones omnes ad infinitum, primas, secundas &c per earum analogiam cum terminis hujus seriei designasse, uti in prima Newtoni Epistola ad Oldenburg. et fluxionum Scholio sub calcem *Quadraturarum* annexo videre est....' (.40: 593^r + .10: 134^r).

(152) *Acta Eruditorum* (July 1717): 332–6.

(153) See *Histoire de l'Académie Royale des Sciences pour l'Année M.D.CCXVI* (Paris, ₁1718): 24–62, especially 38–57. Newton also read Fontenelle's memorial tribute in the collection of *Éloges Historiques de Tous les Académiciens morts depuis* [*1699*], **2** (Paris, 1717): 274–33 which the latter simultaneously published, and the well 'dog-eared' presentation copy of which 'Pour Monsieur Newton' is now Trinity College, Cambridge. NQ. 10. [131/]132.

(154) We cite these comments by Newton on 'P. 38 l. 14' and 'P. 49 l. 19' of Fontenelle's *éloge* in the Paris *Histoire* (see the previous note) from his initial draft, now Add. 3968.41: 46^r. These observations were subsequently incorporated by him into an intended response (.26: 381^r/381^v, printed in *Correspondence*, **7**: 17–19) to Pierre Varignon, who had sent him copies of both printed versions of Fontenelle's memorial tribute; this was in turn quickly converted (.26: 372^r–373^v→377^r–378^v→379^r–380^v) to become a set of full-blown 'Historical Annotations on the Elogium of Mr Leibnitz' (.26: (375^r/375^v→) 374^r/374^v; compare Brewster, *Memoirs of ...Newton*, **2**, 1855: 290) which were almost certainly never passed on to Varignon in any form and remain yet unpublished.

(155) In his initial comment on 'P. 49. l. 10' of Fontenelle's *éloge* (.41: 46^r; see the previous note).

(156) We take this epithet from the draft (Add. 3968.34: (473^v→) 476^r/476^v) of a reply which Newton intended to make around the beginning of 1717 to Bernoulli's 'squabbling Paper published by Dr Menkenius in the *Acta Eruditorum* for July 1716 [see note (88)] without

as he elsewhere stated it,[157] 'to propagate an opinion that [my] Book of Quadratures...is a piece of plagiary & that Mr Leibnitz was the inventor of the direct method of Fluxions & Mr Bernoully the inventor of the inverse method'—he never forgave Leibniz dead nor Bernoulli still living. And with the shade of the one and the very much alive person of the other (grown to be an equal to Newton in cantankerousness no less than in mathematical power) there was to be continual warring for four years more till Varignon at last parleyed an uneasy outward truce.

---

the name of the Author... & yet naming Mr Bernoulli's formula of an Equation *meam formulam*'. (On this see note (90) above. Newton's Latin version on .34: 474r speaks with still more cutting asperity of a 'Charta quædam rixis & accusationibus plena'; though his revised English draft on (.34: 477r→) .9: 100v/100r(+.41: 107r) opens much more quietly by referring just to 'a Paper published in the *Acta*....) As so often, once his initial impulse to slam straight back at Bernoulli for his veiled slangings was no more, Newton slid behind the scene where he much preferred to be, letting his 'voice' John Keill make public counterblast to the *Acta* article on stage in the *Journal Literaire* in 1719 (again see note (90) above).

(157) His words (.9: 100v) in the most finished version of his projected 'Advertisement' in direct riposte to Bernoulli's article in the *Acta* for July 1716 (see the previous note). He there went on to thrust back that 'The first Proposition of the Book *de Quadratura Curvarum* with its Solution & examples in first & second fluxions was published a[l]most verbatim by Dr Wallis in the second Volume of his Works A.C. 1693, (pag. 391, 392, 393 &c [= vii: 174–6]) being sent to ye Dr...the year before; & therefore this Book was...in MS before the differential method began to be spread abroad by Mr John Bernoulli & his brother'; and furthermore that 'In this Book are many things wch had they been proposed as Problems to be solved by others, might have puzzled all the Mathematicians in Europe. As for instance, to reduce the integration [!] of the following equations to the quadrature of the conic sections.

$$\frac{d\dot{z}z^{2\eta-1}}{e+fz^\eta+gz^{2\eta}} = \dot{y}. \quad d\dot{z}z^{\frac{1}{2}\eta-1} = e\dot{y}+f\dot{y}z^\eta+g\dot{y}z^{2\eta}. \quad d\dot{z}z^{\frac{3}{2}\eta-1} = e\dot{y}+f\dot{y}z^\eta+g\dot{y}z^{2\eta}.$$

$$d\dot{z}\sqrt{e+fz^\eta+gz^{2\eta}} = z\dot{y}. \quad d\dot{z}\sqrt{\frac{e+fz^\eta}{g+hz^\eta}} = z\dot{y}.$$

Or to reduce the integration of the following equations to the simplest cases of quadratures.

$$a\dot{z}^q z^m + b\dot{z}^{q-p} z^n \dot{y}^p = c\dot{y}^q.$$

But the pedantry of proposing Problems to be solved by others is not in fashion in England'. (The former group of instanced equations are of course but minimal fluxional restatements of *Formæ* 5.2, 6.1, 6.2, 7.1 and 11.1 respectively of the latter 'Tabula Curvarum simpliciorum quæ cum Ellipsi & Hyperbola comparari possunt' appended in scholium to Proposition X of the 1704 *De Quadratura*; see pages 140–6 above. The latter general trinomial equation received its fullest squaring by Newton in Proposition III of his 1691 treatise 'De quadratura Curvarum'; see vii: 56–62. Both have a Newtonian pedigree reaching back to the early 1670's, namely, in—the original *Ordines* 3.2, 4.1, 4.2, 6.1 and 11.1 (see iii: 246–54) of—the parent 'Catalogus Curvarum aliquot ad Conicas sectiones relatarum' of the 1671 fluxions tract, and in his contemporary study (see iii: 380–2) of the 'Quadrature of all Curves whose æquations consist of but three terms. But notice—as Newton himself does not appear to do—how old-fashioned such 'Problemes' are become in the new Bernoullian age where differential equations incapable of exact, algebraic solution are increasingly the rule rather than the exception.)

We may be brisk in reporting the ensuing external event. Newton's anticipation of Conti's intention to arrange a separate edition of the letters sent through
him as intermediary was not misgauged, and their appearance in the Appendix
to Raphson's *History* forestalled their printing by only a few months, though
Newton was to be able to delay publication till 1720. Conti in fact early in 1718
furnished to the émigré French *entrepreneur* Pierre Desmaizeaux—who for
two years past had been planning a comprehensive 'gathering' of Leibniz'
letters on natural philosophy of which the mainpiece was to be his exchanges
during 1715–16 with Samuel Clarke [158]—transcripts of his full correspondence
with Leibniz during the winter and early spring of 1715/16 and also of related
letters which had passed between Leibniz and Nicolas Rémond, the Baroness
von Kilmansegge and Count Bothmar, each of which in its own small way
contributed minor *éclaircissements* on Leibniz' attitude to the question of who had
discovered the calculus. We need not say that arrangement was speedily made
to set these in a second volume of Desmaizeaux' forthcoming *Recueil de Diverses
Pieces sur la Philosophie... &c.* In turn Desmaizeaux approached Newton for
permission to include with this collection of Leibniz' epistolary 'remains' the
text of a French version of his Raphsonian 'Observations' on Leibniz' April
1716 letter to Conti which Newton had had printed off at the same time[159] but
never made public. And of course as soon as Desmaizeaux received proof-sheets
of the printed volume he in all innocence passed a set on to Newton for him to
run his eyes over[160]—to receive a sharp lesson in the realities of behind-

---

(158) And with which—contrary to entrenched belief—Newton had seemingly little to do,
advising Clarke on how to frame his answers in only one or two small points. (The single
solidly documented instance of his intervention—a short passage on 'computing...the
Quantity of impulsive Force in an Ascending [or falling] Body' in Clarke's Fifth Reply—is
discussed at length in Appendix Two (pp. 116–22) to I. B. Cohen and A. Koyré, 'Newton &
the Leibniz–Clarke Correspondence', *Archives Internationales d'Histoire des Sciences*, **15**, 1962:
63–126.) The publisher of Desmaizeaux' forthcoming *Recueil* was to be the Amsterdam book-
seller Henri Du Sauzet; see *Correspondence*, **6**: 441, note (1).

(159) See note (151) preceding. When Desmaizeaux—an *émigré* French Huguenot who had
come first to London in 1699, henceforth to lead a precarious existence there and in Holland
as a literary *entrepreneur*—first met Newton is not known to us, but he had Montagu (Halifax)
as a patron and his fellow expatriate de Moivre as a friend, and the introduction could well
have been effected through one or other of these. The texts of both Newton's letter to Conti on
26 February 1715/16 and of his ensuing 'Observations' on Leibniz' reply as rendered in the
suppressed 'Traduction Françoise' were, to anticipate, followed word for word by Desmaizeaux
in the 1718 pre-printing of his *Recueil* (pages 16–25 and 71–93 respectively); on which see the
next note.

(160) The first four sheets (pages 1–88 = signatures A: 1ʳ–8ᵛ/B, C, D: each 1ʳ–12ᵛ) of this
set of proofs are preserved in Cambridge University Library (shelfmarked Adv.d.39.2), but
by Newton's own subsequent back-reference to a page 93[= E2ʳ] in them (see Add. 3968.29:
432ᵛ) it is clear that there was originally also a fifth sheet [E: 1ʳ–(?) 12ᵛ]—and indeed, as it
now survives, the set breaks off in the middle of Newton's Frenchified 'Observations'. The

scenes infighting. Though he later smoothly told Varignon[161] that he had known nothing of Desmaizeaux's 'design of printing...the Letters sent to Madam Kilmansegg & Count Bothmar...till I saw them in print', leading his correspondent to infer that he had no hand in its execution, Newton at once drew up a schedule for reordering the letters to be printed in the *Recueil*'s second volume into a better reading sequence.[162] Except for one small instance where

opening recto has, we may notice, the same half-title 'Lettres de M. Leibniz et de M. le Chevalier Newton, Sur l'invention des Fluxions & du Calcul Differentiel' as was afterwards to precede the second tome of the published *Recueil*. There follow the texts of (pages 3–11) Leibniz' December 1715 letter to Conti, (pages 16–25) the 'Traduction Françoise' (see the previous note) of Newton's response on 26 February 1715/16 and (pages 48–66) Leibniz' reply to Conti on 9 April 1716 (N.S.)—but not yet its 'P.S.' (which Desmaizeaux included, however, in a 'M.S.' secretary copy, now .36: 527^r–529^r, for Newton's inspection)—along with (pages 71–[?93]) Newton's Frenchified 'Remarques' thereon; and sandwiched between these (here printed for the first time) are (pages 12–15) Conti's covering letter to Leibniz transmitting a copy of Newton's February riposte, and also Leibniz' letters (pages 26–8) to Conti on 3/14 April 1716, (pages 29–41) to Baroness von Kilmansegge on 7/18 April, (pages 42–7) to Count Bothmar at about the same time and—this last badly out of chronological sequence—(pages 67–70) to Nicholas Remond in about late December 1715 communicating two slight additions to his postscript to Conti a few days before.

Along with these printed sheets Desmaizeaux also provided Newton with a 'M.S.' (furnished doubtless by Conti) of Leibniz' letter to Remond on 9 April 1716 (N.S.). (Newton's copy of this, now .31: 445^r, is partially reproduced in *Correspondence*, 6: 314; Desmaizeaux afterwards printed the full text in his *Recueil*, 2 (Amsterdam, $_1$1720): 72–4.) He further sent along, it would appear, the 4-page 'Excerptum ex epistolâ D^{ni} Monmort ad D. N. Bernoullium, datâ Aug: 20, 1713, & impressâ in secundâ editione *Analysis* suæ in *ludos aleatorios*' which is now preserved in .33: 465^r–466^v together with, *ibid.*: 467^r, some outline 'Hints of an Answer' by Newton to 'passages' in it 'wch deserve to be rectified'. (He had already four years before received a gift copy, now Trinity College, Cambridge. NQ.10.26, of this second edition (Paris, 1714) of Monmort's *Essai d'Analyse sur les Jeux de Hazard* from its author; in his till recently unpublished draft letter of thanks for the presentation Newton wrote—on the back (.41: 99^v) of a page destined for his 'Account' of the *Commercium* in autumn 1714—that 'I...cannot but applaud it much. I wish heartily that France may always flourish with men who improve these sciences'.) Newton afterwards elaborated these 'Hints' to be a lengthy set of Latin 'Animadversiones' on Monmort's letter (.33: 469^r–470^r), of which Desmaizeaux made a fair copy (*ibid.*: 471^r–472^v), returning this to Newton to have its neat script splattered with further interlineations by the latter before he grew tired of lambasting Leibniz' pretensions in this private, out-of-the-way place. Nothing of this critique—which seems to have escaped scholarly attention for all that it devotes itself in the main to non-mathematical aspects of attractive forces and their physical manifestation in terrestrial and celestial motion—has ever been published, but we would stray too far from our present purpose to say more about it here.

(161) In the rough English draft of a letter to him about the beginning of 1721 (Add. 3968.42: 616^r; see *Correspondence*, 7: 124).

(162) This not without several changes of mind *en route*. The succession of Newton's adjustments of his preferred reordering of the 'Table des Pieces selon l'ordre qu'elles doivent être lûes & qu'elles auroient dû etre imprimees' survive on .36: 511^r→516^r→514^r→512^r→515^r. On the last folio he initially set out the letters in strict chronological sequence of composition, but then yet again altered his mind, delaying the Leibniz letter to Conti of '9 d'Avril 1716' to be immediately before his 'Remarques' thereon, now for the first time dated to be of '29 de

he (doubtless to avoid the cost of resetting type to no purpose) maintained the original order[163] Desmaizeaux accepted all of Newton's recommended changes into the text finally printed, but that was only to be the beginning of it. Newton proceeded to pen a set of '*Corrigenda*'[164] to the submitted proof-sheets which rapidly mushroomed from being minor rectifications of dates and typographical slips to urge in a page-length addition to 'Pag. 88. lin. 2'[165] that

In using the method of fluxions I commonly put letters without pricks for fluxions where I considered only first fluxions, but where I considered also second third & fourth fluxions &c (as for instance in extracting fluents out of equations involving fluxions) I distinguished them by letters with one two or more pricks, a Notation which I reccon more convenient then that of Mr Leibnitz tho not necessary to the method. As I found the methods of Series & Fluxions in the year 1665 so I found the Theory of Colours in the beginning of the next year & in the year 1671 was upon a design of publishing them all; but for a reason mentioned in my Letter of 24 Octob. 1676 laid aside that designe till the year 1704. However, when Mr Leibnitz by his Letter of 12 May 1676 had put me upon resuming the consideration of the methods of Series & Fluxions I wrote the book of Quadratures[166] (excepting the Introduction & Conclusion) extracting most of it out of old Papers; & when I had newly finished the tenth Proposition with its Corollaries & they were fresh in memory I wrote upon them that Letter to Mr Collins wch was dated 8 Novem. 1676 & published by Mr Jones. The Tables at the end of that Proposition for squaring of some Curves & comparing others with the Conic Sections were invented by the Inverse Method of fluxions before the year 1671 as may be understood by my Letter of 24 Octob. 1676. And in the same Letter, where I represented that the general Theoremes there mentioned for squaring of Curves were founded on the Method of fluxions, I meant the method described in the first six Propositions of the Book of Quadratures. For I know of no other Method by which those Theorems could be invented. By the inverse method of fluxions I found in the year 1677[!] the Demonstration of Kepler's Astronomical Proposition wch is the eleventh Proposition of the first Book of *Principles*, & in the year 1683[!] at the importunity of Dr Halley I resumed the consideration thereof & added some more Propositions about the motions of the heavenly bodies wch were by him communicated to the R. Society & entred in their Books the winter following, & upon their request that these things might be published,

May 1716' as in the printed *Recueil* (but how can that be right?). On *ibid.*: 508 Desmaizeaux has copied out the 'Table des Pieces' as Newton would finally have them read.

(163) In the case, namely of Leibniz' letter to Remond in late December 1715, which is retained in the printed *Recueil* in its incongruous place at the end of the other letters, as in the 1718 preprint (see note (160)).

(164) This is Newton's title for the draft corrections which survive on Add. 3968.29: (432^r–433^r→) 434^r–435^r; see also .36: 516^v.

(165) That is, at the end of the passage in his 'Remarques' (=*Recueil*, 2, $_1$1720: 92, l. −4) —we quote Newton's original English of this paragraph in his 'Observations' on pages 514–15 above—where brief description is given of his earliest papers on fluxional 'calculus'.

(166) On this often repeated but unsupportable claim—which Newton must himself have come sincerely to believe against all the evidence of his private papers? —see VII: 15–16.

I wrote the Book of *Principles* in the years 1684, 1685, 1686, & in writing it made much use of the method of fluxions direct & inverse, but did not set down the calculations in the Book it self because the Book was written by the Method of Composition as all Geometry ought to be.

'And ever since I wrote that Book', he added with a deeper truth than perhaps he had in mind, 'I have been forgetting the Methods by which I wrote it.'[167]

In drafting his covering letter to Desmaizeaux he likewise passed from a mild initial remark that '[having] run my eye over the printed Letters...I think it is right to let them come abroad. I have observed some faults wch may be put into ye errata',[168] to delivering a stern declaration that 'In all this controversy the true Question has been whether Mr Leibnitz or I were the first inventor [&] not who invented this or that method' and following his stark ensuing statement 'Mr Leibnitz has constantly avoided it' with a familiar parade of evidence from the *Commercium* to support its concluding 'judgm^t' that he himself was the first Inventor by many years.[169] And in like vein, seeking in particular to determine from Leibniz' letter to Oldenburg of 17/27 August 1676 what was then (on the eve of his second visit to London) 'the top of his skill in Analysis', he continued through several further versions[170] of his intended letter before putting these

---

(167) See Add. 3968.29: (434^r→) 435^r. We have omitted from the preceding quotation a sentence which Newton indicates in his manuscript (f. 434^r) should follow on immediately after, namely: 'At the request of Dr Wallis, when he was printing the second Volume of his Works, I sent to him in two Letters dated 27 Aug & 17 Sept. 1692 the first Proposition of the Book of *Quadratures* copied almost verbatim from the Book & the Method of extracting Fluents out of Equations involving fluxions mentioned in my Letter of Octob. 24. 1676 & copied from my older Paper [of 1671]: & the Doctor printed them both the same year (vizt A.C. 1692) in that Volume of his works, pag. 391, 392, 393, 394, 395, 396 [see VII: 170–80], together with an explication of the method of fluxions direct & inverse...'. With this in revision—now scheduling the whole to be added 'At the end of the Remarks after the words *du même genre de celles que nous venons de marquer*'—Newton conjoined the decidedly untrustworthy witness of 'Mr Collins in his Letter to Mr [Strode] dated 26 July 1672' (the pertinent portion of this, rendered into Latin from the original in Royal Society MS LXXXI, No. 24, had been printed in *Commercium Epistolicum*, $_1$1713: 28–9) that 'By the testimony of Dr Barrow grounded upon papers communicated to him from time to time...it appeared that I had the Method conteined in the *Analysis per æquationes numero terminorum infinitas* some years before the Doctor [in 1669] sent this *Anaysis* to Mr Collins. A part of this Method is to square figures accurately when it may be done or else by perpetual approximation: & this requires skill in the method of fluxions, so far at least as it is contained in the first six Propositions of the Book of *Quadratures* as was said above' (see f. 428^r).

(168) Add. 3968.27: 383^r; the *corrigenda* themselves follow on *ibid.*: 383^r/383^v. In first draft on .29: 431^v he had written: 'Upon reading the Letters wch you are printing I see no need of saying any thing further about that matter....'!

(169) We cite these phrases from Newton's interim draft on Add. 3968.27: 407^r.

(170) Add. 3968.27: 407^v/407^r→406^r/406^v→385^r/385^v→(385^v+) 387^r/388^r+.12:164^v. The opening paragraph (on f. 387^r) of this last draft is, we may add, quoted in *Correspondence*, 6: 459–60, note (7); while the opening paragraph and two others of the middle version on ff. 385^r/385^v are also there cited (*ibid.*: 462/459, note (1)).

aside, to begin in their stead a separate, still more elaborate 'History of the Differential Method, written by Sr Isaac Newton'.[171] This 'History of the Method of Moments, called Differences by Mr Leibnitz' as he retitled its more finished version[172] was, we have no doubt, composed to reinforce and complement the 'Observations' appended to Raphson's *History* the previous year—suitably rendered, like these, into French—in their re-edition by Desmaizeaux in his coming *Recueil*: with what success the reader may judge for himself in Appendix 3.1 where we reproduce its text. It was never made public, but cannibalised for its meat in a further flurry of more private outpourings to Desmaizeaux[173] which came equally to naught. From this abortive revised batch of letter drafts in turn Newton drew a second more formal statement of his amended claim to calculus priority, remarshalling their principal individual paragraphs and interlarding these to produce the carefully penned (though not fully completed) *addendum* to his earlier 'Observations' which is printed in Appendix 3.2 below.[174] And then, in yet one more turnabout, this intended 'Supplement to the Remarks' fathered a third series of draftings of a response to Desmaizeaux' inquiry which, too, were left unfinished in their much cancelled and reworked surviving manuscript roughings-out.[175] (Afterwards, let us anti-

(171) This is the title of the first full draft on Add. 3968.12: 171^r, where—in intended preliminary to a preceding revised account of Leibniz' introduction to series and techniques of calculus (*ibid*.: 163^r/(163^v + 166^v→) 164^r)—Newton draws upon his 1669 *De Analysi* and his 'larger Tract on this subject' of two years later to make clear 'the state of the method [of fluxions and series] in the year 1671' when he 'left of these studies'.

(172) Namely Add. 3968.12: 159^r–161^r/(165^v +) 166^r; the preparatory draft on .12: 165^r–166^r, taken together with the amplification of its text on .5: 21^r/21^v, is not greatly different.

(173) See Add. 3968.29: 426^r–427^r/.31: 455^r→(.29: 433^v/.27: 384^v–383^v→).27: 404^r/405^r→401^r/402^r. The opening of the last draft (on f. 401^r) is printed in *Correspondence*, 6: 458, and its two final paragraphs—which Newton has cancelled in the manuscript—are quoted by I. B. Cohen in his *Introduction to Newton's 'Principia'* (Cambridge, 1971): 295–6; while the first version of the penultimate paragraph on f. 405^r is reproduced in *Correspondence*, 6: 458–9.

(174) From the original in Add. 3968.27: 453^r–458^r; a preliminary casting by Newton of this 'Supplement to the Remarks' is found on the preceding ff. 451/2, while minor intermediary drafts exist on .12:164^v, .27: 386^r and .41: 140^r/140^v.

(175) Add. 3968.27: 391^r–392^r→397^r–398^r(+398^v)→398^r–390^r/(.12:164^r + .41:85^r/85^v→) 390^v→393^r–395^r; which last(?) version—an uncompleted reworking of the preceding one which may not be as full as the letter ultimately sent to Desmaizeaux—is printed in *Correspondence*, 6: 454–7, together with (*ibid*.: 460–2, notes (8), (11) and (15)) two snippets from the preceding ff. 389^r/390^r and a lengthy excerpt, omitting the first three paragraphs there, from (the reworked text on) ff. 389^v/390^v. The tentative addition (afterwards crossed through) on .41: 85^r has, ever since the nineteenth-century editors of Newton's 'remains' first published it, inaccurately and without indication of its context, in the preface (p. xviii) to their 1888 *Catalogue* of the Portsmouth Collection of his books and papers, become deservedly known to all the world as an 'authentic' memorandum in which he surveys (in a number of eminently quotable phrases) the 'days' when he was in the 'prime' of his 'age for invention & minded Mathematicks & [natural] Philosophy', even despite the manifest impossibility of several of its datings of his discoveries. (We cited a pertinent snatch from this chronicle on 1: 152, in the

cipate, Newton further redressed portions of his unutilised 'Supplement' to support a yet more bitter third-person invective[176] against Johann Bernoulli—happily for future peace it was never publicly delivered—for permitting his 'judgment of an eminent Mathematician' to be anonymously inserted into Leibniz' 'defamatory Letter dated July 29 [1713] published in a flying paper without the names of the Authors or printer or city where it was printed[,] & dispersed over all the western parts of Europe: a backbiting infamous way of proceeding wch in England is punishable by the civil Magistrate'.)

The precise phrasing of the answer which Desmaizeaux finally had from Newton in the late summer of 1718 we must needs, for lack of the text of the letter as sent, surmise from the emendations and additions made in the *Recueil* of the 'Letters of Mr Leibniz and Sir [Isaac] Newton on the invention of fluxions and the differential calculus' when it appeared two years later.[177] Desmaizeaux certainly used what new material was there supplied to him with a deal of caution and circumspection, setting out amid the general preface to his published 'collection of pieces'[178] an intendedly neutral, factual narrative

pastiche there made from his autobiographical writings setting out his 'inventions' in mathematical analysis in the mid-1660's; a more critical discussion of its historical verities will be found in our 'Newton's Marvellous Year: 1666 and all that', *Notes and Records of the Royal Society of London*, **21**, 1966: 32–41. A fuller and more exact—still not quite perfect—transcription of the text of Newton's autobiographical retrospect will be found in Cohen's 1971 *Introduction to Newton's 'Principia'*: 291–2.)

(176) See Add. 3968.9: 97^r(+99^v)–99^r; the following quotation is from f. 97^v.

(177) Newton himself was the prime cause of the delay in publication, it would seem, being certainly unwilling to let the letters which passed between himself and Leibniz by way of Chamberlayne in 1714 (see note (98) above) appear in the *Recueil*; in November 1719, in particular, he successfully offered the book's publisher Du Sauzet twelve guineas further to hold back publication 'till Lady [Day] next'—that is, 25 March 1720 (O.S.)—but he could do nothing thereafter to prevent the correspondence being published in the *Recueil*'s second volume (₁1720: 116–24) 'à la fin de la dispute sur le Calcul différentiel'. (See the draft of Newton's covering letter to Desmaizeaux, printed—from the original now Add. 3968.38: 570^r—in *Correspondence*, **7**: 73, and also the extracts from Du Sauzet's related letters to Desmaizeaux of 17 October, 1 November and 1 December (N.S.) which are cited in *ibid.*: 73–4, note (2).)

(178) See *Recueil de Diverses Pièces*, **1** (₁1720): xi–lxvi. He justified himself at the end of his précis by observing: 'On n'a dans les Pays étrangers [*sc.* outside England and Germany] que des idées très confuses de cette Dispute, &...il m'a paru qu'un Narré simple & historique de ce qui s'est dit de part & d'autre étoit nécessaire pour l'intelligence des Pièces que je publie' (*ibid.*: lxvi). To Conti in September 1720 he privately wrote: 'Je me suis borné à la simple qualité de Rapporteur; n'ayant aucun interêt à prendre parti dans cette dispute' (see *Correspondence*, **7**: 99). Even so, he found that at times his position as fence-straddling neutral reporter invited fire from both sides simultaneously: Newton could have been no more enamoured of his opening statement on the 'invention de la Methode des Fluxions, ou du Calcul differentiel' that 'c'est une même Methode d'Analyse, sous deux noms differens' (*Recueil*, **1**: xi), than Bernoulli was taken with the hesitant presentation (*ibid.*: xlviii) of his disclaimer of Leibniz' public attribution to him of authorship of the 'Judicium primarii Mathematici' in the 1713 *Charta volans*.

account of the antecedents of the dispute over priority of discovery of the calculus and of its squabbling confrontations which, for all its length, trod but lightly and inoffensively on anyone's toes; nor in the body of his compendium of letters did he, so far as we can discern, yield to any pressure from Newton to add (in the style of his own *Commercium*) provocative 'correction' of the Leibnizian case, in interpretative footnote or attached comment.[179] Which inevitably satisfied no one, even on the Continent (where Desmaizeaux's soft-pedalling went widely unappreciated).[180] Such is always the lot of the would-be neutral observer where prejudice is rife and the downright lie (however motivated) at a premium.

When, during the course of the next year, it became increasingly clear to Newton that the *Recueil* yet in press was not going to carry his preferred 'true' account of what was at issue in the priority dispute, he was doubtless disappointed and not a little angry with Desmaizeaux. But not unduly dismayed. His pen, thus stimulated to rectify and refute the falsities and calumnies (as he saw them) which Leibniz' lieutenant Bernoulli and his pupils began now

---

(179) The editorial footnotes to Newton's Frenchified 'Observations' (*Recueil*, **2**: 77, 78, 83 and 96) were, we may notice, already inserted by Desmaizeaux in the proof-sheets of these 'Remarques' (ULC. Adv. d. 39.2; see note (160) above) which he passed to Newton in the summer of 1718. He did, however, permit the latter—then (see below) under gentle pressure from Varignon to make his peace with Leibniz' lieutenant—to spell out in a terminal 'Erratum' (*Recueil*, **2**: 125–6) that he had not meant to 'disparage the skill of Mr Bernoulli' in calling the author of the 1713 'Judicium' a 'Mathematician, or pretended Mathematician' in his letter to Conti in February 1716 (as he had stated in his 'Observations'; see (Add. 3968.31: 447^r→) Raphson, *History*, ₂1717: 112); rather 'J'ai seulement voulu faire entendre, que le *Mathematicien*...ayant cité M. *Bernoulli* (& même avec éloge l'appellant *excellent Mathematicien*) il s'est representé comme une personne differente de M. *Bernoulli*; & que d'un autre côté, M. *Leibniz* ayant fait réimprimer depuis peu cette même Lettre sous le nom de M. *Bernoulli* & supprimé la citation'—a footnote hereto cites 'Dans les *Nouvelles Litteraires* du 28. Decembre 1715. page 413 & *suiv.*' (*vide* note (129) preceding)—'a marqué que M. *Bernoulli* en étoit l'Auteur: de sorte que je ne savois lequel des deux on en devoit croire, ou du *Mathematicien* qui *prétendoit* n'être pas M. *Bernoulli*, ou de M. *Leibniz* qui assuroit que c'étoit M. *Bernoulli* lui-même'. In one last letter to him early in 1723 (see note (225) below) Bernoulli was to set right—for what it mattered—Newton's mistaken presumption here that Leibniz' inserted parenthesis 'quemadmodum ab eminente quodam Mathematico dudum notatum est' on page 3 of the *Charta volans*, supporting Bernoulli's criticism that 'N[ewtonus]... regulam [fluxiones capiendi] circa gradus ulteriores falsam', was to be read back half a dozen lines to signify that this 'eminent mathematician' was of necessity the 'primarius Mathematicus' who erected this to be an 'Alterum indicium, quo coniicere licet Calculum fluxionum non fuisse natum ante Calculum differentialem'. Newton, praise be, did not respond.

(180) The review received by the second volume of the *Recueil* in *Acta Eruditorum* (February 1721): 82–94 (from its editor Wolf?) gave, as one would have expected, prime notice to Leibniz' two letters to Conti (*ibid.*: 90–1) and his intervening one to Nicolas Remond (*ibid.*: 91–4) without even mentioning Newton's replies to the two first of these other than to say in opening remark that 'liter[æ] *Newtonianæ*...idiomate Anglicano exaratæ...subjunctæ fuerunt *Historiæ fluxionum*, quam parum nec a studio partium alienam Londini A[nno] 1715 ediderat *Josephus Raphson*'.

increasingly to churn out,[181] quickly found an alternative outlet for its insistent message that 'second inventors have no right' in the joint edition of his *Principia* with his *De Quadratura Curvarum* which he was planning at this time to issue[182]— though, as with so many other of his projects, nothing in fact came of it except a sheaf of papers bearing drafts of its prefaces and one or two intended additions to the body of the text. To be here very brief, in Appendixes 4 and 5 we set out the main augmentations which Newton drafted at this time to the scholium to Lemma II of the *Principia*'s second book, purposing to bring its 'mind' out more clearly and to trace its historical context with some care; while in Appendix 6 is printed the English roughing-out of a projected new preface to the work as a whole, where he recurs to his shaky thesis[183] that the *Principia*'s propositions were initially 'invented by Analysis: but... composed to make it Geometrically authentic & fit for the publick' and then passes, inevitably, to frame one more statement of his claim to priority over Leibniz in discovering the elements of that self-same universal analytical calculus. The three draft prefaces to the appended *De Quadratura* which are displayed in Appendix 7 yet again parade the half-truth that (so Newton would have it) 'in writing the Book of *Principles* I made very much use of the Book of *Quadratures*' and go on to document his concomitant affirmation in introduction to the latter—one, he asserts, 'called in question'— that he had 'found the method of fluxions gradually in the years 1665 & 1666'; and the same message is drummed out in his succeeding Latin 'Ad Lectorem' (Appendix 8.1) and 'Annotationes' (Appendix 8.2) to his intended re-issue of the *De Quadratura*. And when that planned joint edition fell through, he began to think of coupling his book on quadrature with a reprint of his 1669 'De Analysi', recounting their antecedents—that these 'Tracts may be the better understood'—in a premised 'History of the Method of Series, Fluxions & Moments' which revamped and amplified his 1715 *Phil. Trans.* 'Account' of his *Commercium*....[184] All very repetitive and dubious, to say the least, in its recon-

---

(181) See notes (88) and (89) above.

(182) This, as ever (compare **1**, Introduction: note (102)), so that the modes of algebraic integration set out in the *De Quadratura*—and its long lists of evaluated curve-areas in particular—should serve to fill the many appeals in the *Principia* to be 'granted the quadrature of curves'. Already in late August 1718, in the draft of a letter to Varignon, Newton had declared his intention to 'write a short Preface to the Book...to shew that [it] was in MS before [1676]' (.42: 613^r [=*Correspondence*, **7**, 1978: 3]).

(183) This recurrent theme Newton most fully expounds in his draft prefaces to the accompanying *Principia* which are reproduced in **1**, 8; see also the preceding pages 68–70 setting their background.

(184) The page-corrections and more elaborate emendations which Newton made at this time to his 1715 review are preserved on the worksheets now collected in Add. 3968.12: 146^r– 147^v/.13: 173^r–185^r/.39: 575^r–577^r and also on a couple of other stray related folios in private possession. The title of this 'History' as we cite it is found on .12: 146^r (where Newton omits the first two paragraphs of his printed 'Account', plunging straight into its text at the top of

struction of the issues and motives underlying the events of the recent mathematical past, and the reader may be forgiven his yawn as he ploughs through
the successive draftings which we juxtapose below. At the same time, what is
common to their texts does represent Newton's ultimate opinion of when and
how he 'invented' the joint method of analysis by infinite series and fluxions
of which he was so intensely proud, and also how what he had discovered
related to similar findings by Leibniz and others. And let us not look too harshly
upon the distortions and myopic blurrings in his view of what he too often saw
with blinkered and purblind eyes. That we now are unable to accept so much
of the detail of his retrospective survey of what had happened in analytical
mathematics in his lifetime—a sincere expression of what he there saw, we
are sure—is due in greatest part to the clarity of our own modern hindsight,
furnished as we are with near-complete access to past records held private
in Newton's day.

One last matter, and we will be done with his pursuance of his quarrel with
Leibniz into and beyond the grave. When the leading French mathematical
physicist Pierre Varignon began in late 1718 to try to make peace between
Newton and Johann Bernoulli[185]—a thankless self-appointed task on which he
was to spend much of the last four years to his life—he was to find from the start
a near-insurmountable obstacle in the tactless and in great part sheerly wrong-
headed words criticising the several 'inferiorities' in Newton's grasp of differential calculus which Bernoulli had put privately to Leibniz in the early
summer of 1713, to have the latter at once publish these in his *Charta volans* as the

---

page 174, but going on more fully to inform that 'In the winter between the years 1664 &
1665 Mr Newton (now Sr Isaac Newton) by trying to interpole a series of Dr Wallis for
squaring the circle found out the [general summatory] series for squaring the circle & Hyperbola & their segments.... And at the same time he found out also the reduction of the dignities
of Binomials into converging series...'); while the phrase set in parenthesis is extracted from
the revised preface 'To the Reader' on .39: 575ʳ ff (which breaks off at a point towards the
bottom of page 180 of the printed 'Account', but not without first interpolating from his
Raphsonian 'Observations' a brief notice, on ff. 576ʳ/577ʳ, that 'Mr Newton in his Answer to
a Letter of Mr Leibnitz dated 9 Apr 1716, has told us that he hath still in his custody several
Mathematical Papers written in the years 1664, 1665 & 1666 some of wᶜʰ happen to be dated;
& that in one of these dated 13ᵗʰ of Novem 1665 the direct Method of fluxions is set down in
these words. PROB. An Equation being given, expressing - - - - [*vide* Raphson, *History*, ₂1717:
116, ll. 4–13] - - - - gives the relation of *p*, *q*, *r*, &c....').

(185) As an opening offering Varignon passed on to Bernoulli a presentation copy sent to
him by its author of the second (1717) English edition of the *Opticks*, 'tuo nomine...ut
generosum ipsi cor tuum exhiberem' he wrote to Newton on 17 November (N.S.), adding that
'rectè judicaveris me innocentem omnino esse factorum contrà te in controversiâ Leibnitianâ:
tantum abest ut eorum quantulumcunque particeps fuerim, quin potius de illâ controversiâ
semper silui in meis ad D.D. Leibnitium & Bernoullum litteris, mecum & intùs solummodò
dolens eâ tantos viros agitari, quos, si penes me fuisset, in pristinam concordiam reduxissem'
(King's College, Cambridge. Keynes MS 142(C), printed in *Correspondence*, 7: 15).

dictum of an un-named 'primary mathematician'.[186] The Latin letter in which Newton replied to Varignon has not survived, but in first version of his preparatory English draft he wrote formally stating his obligation 'for your Letter in order principally to bring Mr John Bernoulli & me to a better understanding' and then went curtly on to state[187] that

if that Letter [of June 7th 1713 inserted into a flying Paper] be his he cites himself by the title of *Eminens quidam Mathematicus* & took upon him to act as judge between Mr Leibnitz & the Committee of the R. Society & in that sentence [*sc.* the anonymous 'Judicium primarii Mathematici'] denied (in effect) that when I wrote the Book of *Quadratures*, (wch I can assure was above 40 yeares ago, except the Introduction & Conclusion) I...underst[oo]d the first Proposition of the Book.

Eventually, at Varignon's gentle insistence, Bernoulli wrote to Newton in the early summer of the next year, 1719, an appeasing letter in the course of which he gave a poker-faced assurance that 'they are unquestionably mistaken who have reported me to you as the author of certain of those flysheets in which not honourable mention enough, perhaps, is made of you'.[188] But when Newton responded (as ever through Varignon) by taking these words at their face value —or being prepared to do so to outward appearance, at least—as a categorical denial that he was the 'primary' mathematician who had passed judgement in Leibniz' 1713 *Charta*, and he on that basis pledged his future friendship,[189]

---

(186) See page 492 above.

(187) Add. 3968.26: 381^r (=*Correspondence*, 7: 17).

(188) In the original Latin words of Bernoulli in his epistle of 5 July 1719 (N.S.): 'Fallunt haud dubie, qui me Tibi detulerunt tanquam Auctorem quarundum ex schedis istis volantibus, in quibus forsan non satis honorifica Tui fit mentio' (see *Correspondence*, 7: 44)—to which he adjoined in a fine semblance of outraged innocence 'sed obsecro...atque per omnia humanitatis sacra obtestor, ut Tibi certo persuadeas, quicquid hoc modo sine nomine in lucem prodierit, id mihi falso imputari. Non enim mihi est in more positum, talia protrudere anonyma quæ pro meis agnoscere nec vellem nec auderem'.

(189) As he wrote in an enclosure for Bernoulli with his letter to Varignon on 29 September 1719 (see *Correspondence*, 7: 62–4): 'Cum primum Literas tuas ad me mediante D^no Abbate Varignone missas acceperam, et ex ijs intellexeram te non esse authorem Epistolæ cujusdam ad D. Leibnitiū 7 Junij 1713 datæ, in animum statim induxi me non tantum lites mathematicas nuper commotas negligere velle (id enim prius feceram [!]) sed etiam amicitiam tuam colere et ob ingentia tua in rem mathematicam merita magni æstimare.... Studijs Mathematicis jam senex minime delector, neçz opinionibus per orbem propagandis operam dedi, sed caveo potius ne earum gratia disputationibus involvar. Nam Lites semper odi' (from Newton's draft, now Add. 3968.42: 614^r, printed in *Correspondence*, 7: 69; he there went on to adjoin 'et conabor etiam ut amici mei'—he means Keill of course, even then seeing through press his 'Epistola ad...Bernoulli' (*vide* note (90) above)—'a litibus commotis abstineant'). His draft on .42: 615^v, may we add, was much more elaborate, again insisting on the now familiar documentary published 'proofs' that he had long had the elements of his method of fluxions, and in particular a 'Regula capiendi fluxiones primas secundas tertias & sequentes verissima': this was doubtless deleted by him from his final letter because of its possible 'provocation'

Bernoulli uneasily added in December that he could not, 'because I do not keep copies of all the letters I have written', be 'entirely' sure that he had not written *some* letter to Leibniz on the day of 7 June 1713, and, if he had, what its content was.[190] That lame and shifty proviso to his previous firm disclaimer of authorship of (by implication) any portion of any flysheet directed against Newton satisfied the latter not at all, of course, and when he came to re-read Bernoulli's evasive letter his anger grew apace.[191] After some hesitation he decided to make no direct answer to him[192] but instead let it be publicly known to friends that, while he was 'far from making Mr Leibnitz a witness against Mr J. Bernoulli' and indeed had 'openly declared the contrary', the letter which he had most recently received was 'full of such misunderstandings as may require some further testimony then my own to sett them right'.[193] And more privily, so we may now know from his papers, he began to prepare a set of elaborate observations upon Bernoulli's anonymous 1713 'Judgement' of his own 'defective' method of calculus, the most finished version of which[194] is reproduced in Appendix 9 below. This, with its declaration that 'Mr J. Bernoulli ...in a Letter to Mr Newton hath positively'—in first draft Newton wrote

---

to Bernoulli (who in his anonymous *alter ego* as author of the 1713 'Judicium' had, we need not say, emphatically denied that he could have had such a rule when he composed his *Principia*).

(190) Bernoulli to Newton, 21 December 1719 (N.S.): 'Qualis sit epistola illa, de qua dicis quod sit 7 Junij 1713 data ad D. Leibnitium, mihi non constat: Non memini ad illum eo die me scripsisse, non tamen omnino negaverim, quandoquidem non omnium epistolarum a me scriptarum apographa retinui. Quodsi fortassis inter innumeras quas Ipsi exaraveram una reperiretur, quæ dictum diem et annum præ se ferrent, pro certo asseverare ausim, nihil in ea contineri quod probitatis nomen Tuum ullo modo convellat...' (see *Correspondence*, **7**: 75).

(191) Compare de Moivre's eye-witness report, quoted by Varignon to Bernoulli in July the next year (see *Correspondence*, **7**: 80, note (1)), that 'Quoy que M. Newton voye avec plaisir que les choses s'acheminent à la paix, je vous diray en confidence qu'à la seconde lecture qu'il fist de la derniere lettre de M. Bernoulli, il trouva quelques petits traits qui lui firent de la peine'.

(192) Though he did go so far as to rough-out the beginning of a reply (Add. 3968.42: 617^v, printed on *Correspondence*, **7**: 80) in which he affirmed that 'I do not admitt Mr Leibnitz a witness against you, but have told my friends that I admit the author of that Letter [dated 7 June 1713], Mr Leibnitz, & you to be three witnesses against Mr Leibnitz'.

(193) We quote from the unpublished draft of a letter on .42: 617^v (immediately preceding that to Bernoulli cited in the previous note) to some acquaintance in Paris—Fontenelle, we must presume—in which he went on to declare, as two-facedly as ever Bernoulli in his two 1719 letters to him, that it was his 'desire to be at rest'—'from squabbles' is deleted in immediate sequel—'& to let Mr Leibnitz rest in his grave'. (The revised versions of this draft on .42: 616^r→607^r/607^v are printed in full, as from 'Newton to (?) De Moivre', in *Correspondence*, **7**: 83–5 and 81–3 respectively.)

(194) Add. 3968.34: 480^v/481^r, keyed to the text of the 1713 'Judicium' itself (as it had appeared six years before in the *Charta volans*) on f. 480^r preceding. Preliminary draftings of these English 'Observations' are found on .34: (485^r→) 483^r/484^r.

'solemnly'—'declared that he was not the author' and its ensuing trapping of the quarry with the *argumentum ad hominem* 'And if he had been the author thereof he would not have said the Eminent Mathematician charged Mr Newton wth a false Rule...in the Solution of Prob. III Lib. II. *Princip. Philos.*',[195] may well have been intended to be submitted to Desmaizeaux for publication in his *Recueil*. Be this so or no—and whether Desmaizeaux, with his publisher Du Sauzet breathing down his neck, could have made room for any further addition to his 'collection of pieces' when now, at the beginning of 1720, all its sheets had been printed off and were being bound—Newton soon laid aside these till now unpublished 'Observations upon the foregoing Letter [of 7 June 1713] to Mr Leibnitz'. When, however, some time afterwards he wrote to Varignon to report that he had shown Bernoulli's denial of authorship to John Keill, we need not stress that he spoke for himself when he affirmed that the latter 'could not be led to believe it'.[196] And that was far from being the last of it.

An uneasy truce ensued on the matter during 1720, with Bernoulli's attention taken by Keill's two most recent outpourings against him,[197] but the calm was shattered early the next year when the former opened a newly received copy of Desmaizeaux' *Recueil* to read for the first time the text (only slightly sugared in its French version) of Newton's appended Raphsonian 'Observations' on Leibniz' April 1716 letter to Conti. As soon as his eyes lit upon Newton's terse rebuff that he, Bernoulli, was no less a 'new man, too little learned in things gone before' than Keill (against whom Leibniz had originally laid the deprecating remark long years before[198]) and above all upon the sarcasm that, whereas in 1713 he had done battle for Leibniz concealed under the mask of a 'great Mathematician' and 'impartial Judge', he was two years later prepared to champion his master with open helm by throwing down a challenger's gauntlet to the English in the mathematical lists—'as if', Newton had written, 'a Duell...were a fitter way to decide the truth then an appeal to ancient & authentick writings, & Mathematicks must henceforward be filled with atchievements in Knight Errantry instead of reasons & demonstrations'[199]—

(195) See .34: 480^v (=page 679 below).

(196) Newton to Varignon, 19 January 1720/1: 'ostendi Epistolam illam D[omino] Keill ut illi suaderem D. Bernoullium non esse auctorem. Ille autem...adduci non po[terat] ut crederet' (Add. 3968.42: 604^r=*Correspondence*, 7: 120).

(197) Namely, his 'Lettre' in the 1719 *Journal Literaire* and his ensuing *Epistola ad... Joannem Bernoullium* (London, 1720); on both which see note (90) above.

(198) In his letter to Sloane of 18/29 December 1711, where (see *Commercium Epistolicum*, $_1$1713: 118 [=*Correspondence*, 5: 207]) he had bitterly complained that any 'wise and right-minded person' should think he 'at my age and after so many proofs of my [way of] life' needed to defend his candour against a 'man learned but new-fledged, one with too little experience and awareness of things done before'. (Leibniz' original Latin words have already been cited in note (38) above.)

(199) 'And so', Newton had written in his original English (Raphson, *History*, $_2$1717:

Bernoulli's anger erupted. And Varignon again it was who, standing between the two crusty pugilists, tried to cool its heat, endeavouring earnestly over the late winter, spring and early summer of 1721 to persuade Newton, by way of de Moivre, to make some public retraction of his worst-offending phrases.[200] The unpublished draft of a reply which Newton had it in mind to make directly to Varignon about the end of July reveals just how little concession he was prepared to make. 'I understand', he wrote,[201] 'by your letter to Mr Moiver[202] that Mr Bernoulli complains of me for calling him a Mathematician or pretended Mathematician[203] & *homo novus & rerum anteactarum parum peritus* & a Knight errant.[204] And yet', he went on,

---

Appendix: 114) of the offending passage whose Frenchified version Bernoulli now read, 'the ancient Letters & Papers must be laid aside, & the Question must be run off into a squabble..., & the great Mathematician, who in his Letter to Mr Leibnitz dated 7 June 1713, concealed his name that he might pass for an impartial Judge, must now pull off his mask'— 'vizzard' was put by Newton in his preliminary draft on Add. 3968.10: 144^v (to whose orthography and punctuation we here cleave)—'& become a partiman in this squabble & send a challenge by Mr Leibnitz to the Mathematicians in England'—'as if', he continued in his draft, 'a Duell was a fitter way to decide the Question then an appeal to antiquity, & Mathematicks it self must henceforward become the subject of Knight Errantry, & the Differential Method be the Du[e]llinea of the first Knight'. The French translation, we may notice, expanded this to include the phrase (drafted on .40: 573^r) added in the *Errata* to the 1717 Raphson Appendix: '[...a Duell] or perhaps a Battel with what he calls my forlorn Hope & the Army of Disciples in which he boasts himself happy [were a fitter way to decide the truth...]'.

(200) See the *précis* given in *Correspondence*, **7**: 132, note (2) which Varignon wrote to both Bernoulli and de Moivre during February–August 1721.

(201) See Add. 3968.27: 411^v.

(202) This is lost, but it must have been written towards the middle of June; again see *Correspondence*, **7**: 132, note (2).

(203) This in his letter to Conti of 26 February 1715/16 (see page 506 above), though as Newton proceeds to emphasise he had there used this phrase strictly only of the author of the 1713 'Judicium'—'La Reponse du Mathematicien, ou prétendu Mathematicien, datée du 7. *Juin* 1713....inserée dans une Lettre diffamatoire datée du 29. *Juillet* de la même année, & publiée en *Allemagne*' Bernoulli read it in the letter's French version in Desmaizeaux' *Recueil de Diverses Pieces...*, **2**, ₁1720: 17.

(204) These on pages 112 and 114 respectively of his 1717 Raphsonian 'Observations' on Leibniz' letter of 9 April 1716 (N.S.) to Conti, seen by Bernoulli on pages 80 and 84–5 of the Frenchified 'Remarques' in the second volume of Desmaizeaux' *Recueil*. On Newton's latter complaint that 'le grand Mathematicien' who had authored the letter to Leibniz of 7 June 1713 'se demasque à present, prene partie dans la dispute, &...envoye un deffi à tous les Mathematiciens d'*Angleterre* par l'entremise de Monsieur *Leibniz*: comme si un Duel...étoit une voye plus propre pour decider qui a raison dans cette Dispute, qu'un appel fait à d'anciens & authentiques Manuscrits; & qu'il fallut qu'à l'avenir les Mathematiques fussent remplies des prouesses d'une Chevalerie Errante...', see note (199) preceding. At the former place in his 1717 Appendix to Raphson's *History* Newton had declared (we cite his draft on Add. 3968.41: 28^r): 'Mr Bernoulli had the differential Method from Mr Leibnitz & was the chief of his disciples, & gave his opinion for his Master in the *Acta Leipsica* before he saw the *Commercium Epistolicum*; at wch time he was *homo novus & rerum anteactarum parum peritus*, as [see note

by the first I meant only to call him the author or pretended author of the judgment published in the flying paper & said to be written by a great mathematician. And this was in his favour: For he has since declared that he was not the author. If he was the Judge then the Judge was in a conspiracy wth M^r L[eibnitz] to be carried on ag^t Dr Keill & the Com$^{e[e]}$ of y^e R[oyal] S[ociety] by concealing the name of the judge for two years together. And by calling Mr Bernoulli the pretended Judge I absolved him from being in such a conspiracy. And by the second I meant [*sc.* only] that he was risen up since the days of Mr Oldenburg & unacquainted wth the transactions of those days, as Mr Leibnitz objected against Dr Keill.$^{(205)}$ And by the third I meant only to correct Mr Leibnitz for referring the matter in question to a duell or battel instead of deciding it by reason & old records, & making Mr Bernoulli at one & the same time an impartial judge between the two parties & a partiman in whose name he sent a challenge to y^e English Mathematicians.$^{(206)}$ Mr Bernoulli's Problem was published in [y^e] Leipsic *Acts* for May 1697 p. 211. Mr Leibnitz in the year 1700 challenged Mr Fatio to solve this Probleme.$^{(207)}$ It reflects upon Mr Leibnitz (not upon Mr Bernoulli) for appealing from the evidence wch Mr Fatio offered to a Mathematical Duell, & thereby treating Mr Bernoulli as a Hero or Don Quixot[!] in Mathematicks. And the same thing Mr Leibnitz repeated [in 1715] by challenging the English Mathematicians to solve the same Probleme.'$^{(208)}$

---

(38) above] Mr Leibnitz objected against Dr Keil...' which came out no better in the equivalent passage of the 'Remarques' in Desmaizeaux' *Recueil*, where Bernoulli was put to be 'le chef' of Leibniz' followers and to have 'pris le parti de son Maître...'.

(205) Again see note (38) above, and compare note (198).

(206) Bernoulli had indeed had, we may now know, a far from passive rôle in deciding his 'master' Leibniz to communicate his problem of orthogonals to the British mathematicians in 1715 as a feeler of their 'pulse'; see note (110) above.

(207) Namely, in his 'Responsio ad Dn. Nic. Fatii Duillerii imputationes' (*Acta Eruditorum* (May 1700): 198–214) where (page 204) Leibniz had briefly referred to Bernoulli's particular problem of constructing the synchrone (at right angles to the family of cycloids having the same vertex and their bases in the horizontal through this) and its generalisation—see **1**, 7: note (25) preceding— 'Quale etiam est a Domino Joh. Bernoullio propositum in *Actis Erud.* Lips. Maji 1697, p. 211, invenire curvam...quæ curvas etiam transcendentes ordinatim datas secet ad angulos rectos. Nam si ea tantum [Dominus Duillierius] producit quorum jam datæ sunt a nobis methodi, satis intelligit quantum proprio Marte consequi potuerit.' We know nothing to suggest that Fatio made any headway in rising to Leibniz' challenge to solve either the particular or general problems of determining orthogonal curves which Bernoulli had outlined in the *Acta* three years earlier.

(208) Newton subsequently inserted the gist of this draft reply—but not the comparison of Bernoulli with Cervantes' windmill-tilting knight—in a Latin version of his response (King's College, Cambridge. Keynes MS 142(T), printed in *Correspondence*, **7**: 141) to Varignon, but no mention of the matter could have been made in the letter ultimately dispatched, for the latter subsequently told Bernoulli (once more see *Correspondence*, **7**: 132, note (2)) that he received word of Newton's unyielding reaction only by way of de Moivre on 19 August. Newton also had a brief-lived impulse to write a snappy riposte straight to Bernoulli himself, where (Add. 3968.42: 603^r) he curtly began: 'Sr The words *Atchievements in knight errantry instead of reasons & demonstrations* wch you meet with in the Remarks upon Mr Leibnitz Letters to Abbe Conti relate to Mr Leibnitz his declining to answer the *Commercium Epistolicum* & instead thereof

In the slightly more amiable Latin letter which he did send to Bernoulli's faithful Parisian Sancho a few weeks later, for onward transmission to Basel, Newton without more than partly concealing his own iron fist in velvet glove again put the blame for the current *contretemps* upon the dead and unprotesting Leibniz, as being himself 'the knight errant who...provoked me to this duel for the favour of the infinitesimal method, that beauteous maiden for whom our knight did battle':[209] which minimal concession to Bernoulli's pride—one which, Newton added, was not to be made public—Varignon duly passed on in October.[210] And in this way Bernoulli's ruffled feathers were largely smoothed.

As so often, however, what Newton let escape to Varignon was only the outward flare of a deeper, fiercer fire which did not so easily die down. All the while over this same summer and autumn of 1721 his pen was privately racing over the pages, preparing and honing a fuller retaliation to Bernoulli's new act of incitation. In the first instance, if we read the sequence of his mass of surviving writings from this last energetic period of his life aright, he dusted off the set of 'Observations' on Bernoulli's 1713 'Judicium' which he had held back nearly two years before[211] and amplified these to be a comprehensive batch of Latin notes on the full text of the *Charta volans*.[212] And thence he was led to end this 'squabbling humour', continued by his lieutenant 'tho Mr Leibnitz is dead',

---

proposing new Questions to be disputed & problems to be resolved, as if the first Question were to be decided in things of another kind, as men appeal to a Duell for deciding the truth of what they fall out about. Those words are in opposition to the first Paragraph of the Letter of Mr Leibnitz to Abbe Conti dated 9th April 1716.... The Preface [to the new edition of the *Principia*] here mentioned I did not see till the book...was published & a copy presented to me: & so the Author thereof [*viz.* Cotes] was none of my forlorn hope'—meaning the stormtrooping squad of 'enfans perdus' which, so Leibniz wrote (we have cited the pertinent passage in note (118) above), Newton had detached against him.

(209) Newton to Varignon, 26 September 1721: 'Leibnitius fuit eques ille erraticus qui vice argumentorum ex veteribus et authenticis scriptis desumendorum, introduxit alias disputationes quas ipse contulit cum duello. Ad hoc duellum ille me provocavit Methodi infinitesimalis gratia. Hæc methodus erat virgo'—a less innocent 'fœmina' was hastily replaced!— 'pulchra pro qua eques noster pugnabat' (Add. 3968.42: 599v [=*Correspondence*, 7: 162]).

(210) See Varignon's reply to Newton on 9 December 1721 (N.S.) (*Correspondence*, 7: 178). For future peace he had elected not to pass on to Bernoulli the last paragraph of Newton's September letter (see *ibid.*: 162) which made it a condition for Bernoulli to have a requested portrait of him that the latter should publicly acknowledge that he had had his 'methodus fluxionum et momentorum' in 1672, one which embraced a 'Regula vera differen[tian]di differentialia'—as Newton put it in his prior English draft of the letter (now Add. 3968.42: 597r/597v) at this point: 'Mr Bernoulli desires my picture[,] but should I give him my picture while the world beleives that he was the Author of the Letter inserted into the flying paper, they would think that I confess my self guilty of what I am there charged with'. May the Freudian student of Newton's *psyche* make what he will of the taboo which he here places on letting his knightly foe have the totem of his painted likeness.

(211) See page 531 above, and Appendix 9 below.

(212) These 'Annotationes in Libellum a D. Leibnitio 29 Julij 1713 scriptum' (the title set

by republishing the text of his nine-year-old *Commercium Epistolicum D. Johannis Collins et Aliorum de Analysi promota* 'together wth the Account given thereof in the *Philosophical Transactions* for January [and February] $171\frac{4}{5}$' and 'so leave it to posterity to judge of this matter by the ancient Records'.[213] And in preface to this re-edition of this documentary evidence and his (ever anonymously presented) preferred interpretation of its significances, to tell those readers—there could have been few!—who were ignorant of the outward boxing and less scrupulous infighting of the past decade, he also started to prepare in preface 'An Account of what has been done since the publishing of the *Commercium*'.[214] Over the phrasing and styling of this scene-setting address 'To The Reader' Newton spent, we need not say, enormous care and labour, for ever refining and polishing a verbal commentary in which Leibniz took *d'outre tombe* the main fury of his onslaught upon unsustainable 'pretences', but not neglecting to aim a glancing shot or two at Bernoulli whether masked as Leibniz's 'impartial Judge' or visored for combat as his mathematical 'parfit Knight'.[215] For its several interests and novelties—not least an anachronistic remoulding of his general rule for drawing tangents to curves 'as it is in Newton's letter to Collins of 10 Dec. 1672' which has never been published—we reproduce the text of the fullest manuscript version of this proem[216] in Appendix 10, there also indicating

---

by Newton on a preliminary recasting now in private possession) are successively elaborated on Add. 3968.41: $20^r/20^v(+68^r)\rightarrow$.35: (490^r+) 491^v–490^v.

(213) We draw these phrases from the opening paragraph of an English draft, on .41: 21^r–22^r, of Newton's preface to this planned re-edition. 'It was', he there began (*ibid.*: 21^r), 'hoped that a few copies of the following *Commercium Epistolicum* printed off & sent to Mathematicians who were able to judge of these things, might have silenced the complaint that Mr Newtons book of *Quadratures* was a peice of plagiare: but Mr Leibnitz & his friends have declined answering it, & endeavoured to run the dispute into a squabble about occult qualities, miracles, gravity, attraction, sensoriums, the perfection of the world, a vacuum$_{[,]}$ atoms, the solving of problems & the like: all [!] wch are nothing to the purpose.'

(214) This is the title of the initial English draft of the preface on .41: $22^v/22^r$ (subsequently altered and recast into the fuller version, on ff. 21^r–22^r now preceding, whose opening sentences we quoted in the previous note).

(215) Related scraps (such as .41: 130^v and $144^r/144^v$) aside, there survive no less than six successive versions of this retitled 'Historia Methodi Infinitesimalis ex Epistolis antiquis eruta', *viz.* on (.14: 235^r/.13: $187^v\rightarrow$).14: 227^r–228^r ($+$ .9: 111^r); .14: 224^r–226^r/(.13: 187^r–$188^r\rightarrow$ .5: $31^v/31^r\rightarrow$) .41: 30^r–31^v; .14: $230^r/231^r$/(Adv. b. 39.2: *ad pag.* $226\rightarrow$) 232^r–234^r; .13: 189^r–195^r; .13: $199^r/201^r/(202^r\rightarrow)$ $204^r/205^r$–208^r; and lastly .13: 209^r–216^r+ (.13: 186^v–$185^v\rightarrow$) .41: $57^v/56^r$–57^r. Since the preparatory drafting on .9: 111^r there precedes one of the last paragraphs of Appendix 8.1 below (see our note (15) thereto), the earlier of these may date from 1719 rather than 1721.

(216) Now set (on .2: $(4^r\rightarrow)$ 9^r; there are a dozen other preliminary variations of the title on .13: $187^r/187^v$; *vide* Brewster's 1855 *Memoirs*, **2**: 77) to be but the 'Præfatio' to a grand 'HISTORIA METHODI ANALYSEOS/ Quam Newtonus Methodum Fluxionum/ Leibnitius Methodum Differentialem/ vocavit; / in Commercio Epistolico Collinij et aliorum, / et Recen-

how it was condensed and abridged to be the 'Ad Lectorem' which introduced the new edition of the *Commercium* when this made its public bow in the summer of 1722.[217] To the 'Papers & Records' there reprinted it had the previous autumn briefly been his design to adjoin an appendix yet further underlining the sound justice of the 1712 'Committee's' conclusion that Newton was 'first inventor' of the 'method of moments & differences' by many years over its secondary discoverer,[218] but he quickly left this off to take up anew, as we just now mentioned, a systematic critique of Leibniz' 1713 flysheet. Through this *Charta volans* he went line by line, making new notes which he incorporated in a yet more lavish set of Latin observations upon its detail,[219] and then encapsulated these in a more pithy set of annotations confined once more to its included 'Judicium primarii Mathematici' as Leibniz had extracted it out of Bernoulli's letter to him of 7 June 1713 (N.S.).[220] And at the very end of 1721, with the sheets of main text of the reissued *Commercium* before him, he further reshaped these notes upon Bernoulli's anonymously published judgement in the single, continuously written 'Annotation' upon it which is annexed to the printed volume.[221]

sione Commercij/contenta:/Quorum prius iussu Regiæ Societatis ex antiquis Literis/collectum fuit et editum/Anno 1712/Altera in Actis Philosophicis ejusdem Societatis/Anno 1715/(anno et aliquot mensibus ante obitum Leibnitij)/lucem vidit' which Newton has broken up as we indicate and carefully centred line by line.

(217) See note (57) above.

(218) The sole surviving draft of this broader intended 'Appendix' (a roughly cast preliminary English one which breaks off abruptly in mid-sentence) is found on Add. 3968.15: 253^r/253^v. 'By the foregoing', he there commenced as he had stated so many times previously, 'it may be understood that Mr Leibnitz did not understand the higher Geometry when he was the first time in England, wch was in February 1673.... [In] June...of the same year...he began to be instructed in the higher Geometry by Mr Hygens, beginning wth his *Horologium Oscillatorium* published about a month before. The next year in July he resumed his correspondence wth Mr Oldenburg & began to boast of his skill in [it].... And these things are plane matter of fact.... As for his *Methodus differentialis* it doth [not] appear that he understood any thing of it before the year 1677. He was [yet] learning the higher Geometry in the year 1674....'

(219) See .14: 253^r/253^v→ .41: 62^r/(.13: 220^r→) 63^r/64^r+65^r, which comprises sixteen separate annotations, lettered *a* to *r* (*j* is omitted), on the text of the 1713 *Charta*; the unfinished further elaboration on .41: (61^r→) 60^r/67^r interposes three further notes in this set.

(220) These revised 'Annotationes in...sententiam [primarij Mathematici]' are found on .13: 219^r/222^r. We may notice that Newton here leaves out his firm statement in his fuller previous Latin notes (in *m* on .41: 63^r; compare the equivalent note 7 in the English original printed in Appendix 9 below) that the 'Eminens Mathematicus'—as Leibniz had called him in a parenthesis to the 1713 'Judicium' in making it public in the *Charta volans* (see note (179) above)—'est Johannes Bernoullius'.

(221) This final version of Newton's appended comments on the 1713 'Judicium' began its existence as a simple ' NB ' entered on the verso (.14: 252^v) of a letter from John Arnold to him on 5 November [1721] (this is epitomised in *Correspondence*, 7: 374) and was thereafter honed by him over several more rewritings (see .14: (248^r→) 250^r–251^r+248^v–249^v/251^v) to be the 'Annotatio' ultimately printed (*Commercium Epistolicum*, $_2$1722: 247–50).

And so the new edition of the 'Epistolary Commerce of Mr John Collins and others on Advanced Analysis...now imprinted a second time, together with a premised Recension of the same, and the subjoined Judgement of a Mathematician of, so it was held, primary rank'[222] came in the late summer of 1722—after the acceptance of a last-minute plea from Varignon to recast one small verbal phrase in the 'Ad Lectorem' so as not to offend Bernoulli[223]—to be put to a world which by and large had had more than enough of increasingly barren squabbling over vain priorities in the discovery and development of calculus. To a letter from Bernoulli early in 1723,[224] written ostensibly to convey thanks for gift copies of the newly appeared second French edition of Newton's *Opticks*, but quickly passing to plead his innocence in the matter of the epithet of 'eminent mathematician' which Leibniz in the 1713 *Charta volans* had set in parenthesis upon the author of its included 'Judicium',[225] Newton himself (by now a sickly man feeling very much the weight of his eighty years) made no reply. And with that his part in the ongoing but now much muted dispute between 'English' and 'Germans' over priorities of 'invention' was at an end—and so too is our present interest in it.

---

(222) So we render the book's Latin title, which we have earlier cited in full in note (57). (For what it matters, in the initial draft of this on .2: 5^r the '...*ut ferebatur*' was originally '*ut fingebatur*'—in Newton's preferred translation, 'so it was pretended'.) The 'Corrigenda' to the volume which Newton systematically compiled (on .14: 247^r/247^v) as he found them were, we may add, in their earlier portion corrected in proof in the text itself before this was printed off, and for the rest gathered in the book's terminal list of *Errata* (*Commercium Epistolicum*, $_2$1722: signature S2^r [facing page 250]).

(223) See Varignon to Newton, 4 August 1722 (N.S.) (King's College. Keynes MS 142(L), printed in *Correspondence*, 7: 206–7): 'in præfatione [ad lectorem] pag. 5. ...lin. 12 & 13 legitur, *Jam velo sublato, ut militem in hac rixâ pro se inducere*: mallem simpliciter, *Jam in hac rixâ pro se inducere*, ne quis sub illo *velo* prius latitantem putet D^{um} Bernoullum, cui D^{us} Leibnitius Epistolam prædictam ascripsit. Adde quod *ut militem*, vilior est denominatio quam ut eundem D^{um} Bernoullum non offendat'. Newton compromised in the published 'Ad Lectorem' (*Commercium*, $_2$1722: signature A4^r) by rounding out Varignon's preferred curtailment with an 'ut advoca[t]um'. He paid no heed to a second mollifying suggestion from Varignon that, at line 29 on the same page of his printed preface, he should—as would have been a bare-faced lie—add to his report that 'Quæ de Questionibus Philosophicis disputata sunt [*sc.* in annis 1715, 1716, 1717] D. *Des Maizeaux*...in lucem edidit' the declaration that this was done 'me'—read 'Newtono'!—'non consulto'.

(224) Bernoulli to Newton, 6 February 1723 (N.S.), printed in *Correspondence*, 7: 218–20.

(225) As Hartsoeker had (see *Correspondence*, 7: 222, note (5)) recently suggested in a set of 'Remarques' on a thesis of Bernoulli's in his *Recueil de Plusieurs Pieces de Physique* published the previous year. To Newton (see *ibid.*: 219) Bernoulli now over-elaborately wrote: 'Licet indignus sit [Hartsoekerus] cui ego respondeam, unum tamen est quod me magnopere urit; scilicet ut me omnium risui exponat...me mihimet ipsi tribuisse titulum *excellentis mathematici*, et ut calumniæ crimen a se amoliatur Te...ejus Auctorem facit, dum locum citat ex Tomo 2. *Collectaneorum* D^i Desmaizeaux...p. 125. l. 32. [et seq.]'—we have quoted this *addendum* to Newton's 'Remarques' in the *Recueil* in note (179) above—'ubi loqueris de epistola illa 7 Junij

In conclusion, let us pass no retrospective censure upon the morality of any or all of those involved on the two sides of the squabble. The fullness and wisdom of hindsight should lead us less to condemn all that underhandedly went on, or headshakingly to despair that so much bickering continued so long to so mean and futile an end, than to be humble in our awareness that even the greatest of intellects are but fraily human, and above all—as seekers after past truth— grateful that our modern documentary knowledge allows us so much sharper and broader a picture than to them was possible of the complex skein of discovery of the algorithmic mode of analysis by limit-infinitesimals which Newton knew from the middle 1660's as the method of fluxions and which Leibniz a decade later christened by its now standard name of differential calculus. When, as both grew themselves more and more to realise (though neither would have publicly admitted so), each of these independently framed variant modes of analysis of the infinitesimally small and the instantaneously moving drew so heavily on the insights of so many who had gone before, the priority in time of creation of his fluxional method which Newton indubitably has must seem of minimal significance. The rest, as they say, is 'history'.

1713...in qua prout erat impressa in scheda...volante 29 Julij 1713 elogium illud, sed quod parenthesi includebatur, mihi erat adscriptum...'. And so the barren discord went unendingly on.

APPENDIX 1. PAVING THE WAY TO THE 'COMMERCIUM EPISTOLICUM' (1712).[1]

From originals in the University Library, Cambridge and in private possession

[1][2]   Gentlemen

The Letter of Mr Leibnitz[3] wch was read before you when I was last here[4]

---

(1) On pages 469–80 above we summarised the train of events, beginning with Leibniz' needlessly offensive unsigned review in 1705 of his treatise *De Quadratura Curvarum*, which led Newton at the beginning of 1712 to start to assemble the set of extracts from old papers and letters which ultimately issued from the printer in January the following year as a booklet of some hundred and twenty small quarto pages under the title of *Commercium Epistolicum D. Johannis Collins, et Aliorum de Analysi promota: jussu Societatis Regiæ in lucem editum*. There is no room in an edition of Newton's mathematical papers to fill a long-standing scholarly want by reprinting the text of these carefully chosen and elaborately annotated *pièces justificatives*, designed though they were to provide unshakeably documented support for the judgement of the Committee set up by the Royal Society to report on the matter which at their end set down its reckoning that Newton was 'first inventor' of the method of fluxional analysis which Leibniz afterwards published in equivalent form as his *calculus differentialis* and for whose discovery he claimed priority. But to give body and colour to our sketchy preceding outline of the prehistory and compass of this Collinsian 'Epistolary Commerce', every last word of which (as their surviving autograph drafts amply confirm) came from Newton's pen as it wrote, wrote, wrote away over the summer and autumn of 1712, we here adjoin—from his extensive *Nachlass* of papers on the dispute, notably those now gathered (in massive disorder) in ULC. Add. 3968—a miscellany of mostly unpublished ancillary pieces: these range from the speech ([1] following) in which he first declared to the Royal Society his support for Keill against Leibniz, and (see [2] and [3]) two initially intended openings to his selected 'Extracts of ye MS Papers of Mr John Collins'; to his preparatory drafting (reproduced in [4]) of the Committee's report on the 'controversy' between Leibniz and Keill, and (in [5] and [6]) the fullest along with the most finished of Newton's English originals of the Latin preface which he set to the *Commercium Epistolicum* in its final printed form.

(2) ULC. Add. 3965.8: 79^v. This draft by Newton of the speech to the 'Gentlemen' of the Royal Society in which (see page 480 above) he delivered his consideration of Leibniz' letter of late December 1711 to its secretary Hans Sloane, answering Keill's charges against him the previous summer, was first printed in *The Correspondence of Isaac Newton*, 5: xxiv/xxv. We diverge slightly, however, from A. R. Hall in his surmise there (*ibid.*: xxiv) that the finished version of this rough preliminary casting was read out to the gathered Fellows on 14 February 1711/12, preferring to postpone the date of its delivery by Newton till the meeting of three Thursdays later on 6 March when, the Society's Journal Book records, 'account'—from the President's chair, we assume—was given 'of Mr Leibnitz's Letter to Dr Sloane concerning the Dispute...between him and Mr Keill': on which occasion 'a Committee was appointed by the Society to inspect the Letters and Papers relating thereto' which was 'to make their Report to the Society' (Journal Book, xi: 275).

(3) That to Sloane of 29 December 1711 (N.S.), which was read out to the assembled Royal Society on 31 January 1711/12 and then 'deliver'd to the President to consider of the Contents thereof' (Journal Book, xi: 267).

(4) Newton initially specified 'this day fortnight [*sc.* past]', meaning (see the previous note) 31 January. It is, however, our belief (see note (2)) that he subsequently delayed making his speech for another three weeks.

relating to me as well as to Mr Keil I have considered it, & can[5] acquaint you that I did not see the papers in the *Acta Leipsica* till the last summer[6] & therefore had no hand in beginning this controversy. [7]The controversy is between the author of those papers & Mr Keil. And I have as much reason to complain of that author for questioning my candor & to desire that Mr Leibnitz would set the matter right without engaging me in a dispute wth that author as Mr Leibnitz has to complain of Mr Keil for questioning his candor & to desire that I would set the matter right without engaging him in a controversy wth Mr Keil. For if that author in giving an account of my book of *Quadratures* gave every man his own, as Mr Leibnitz affirms, he has taxed me with borrowing from other men & thereby opposed my candor as much as Mr Keil has opposed the candor of Mr Leibnitz.[8] Mr Leibnitz & his friends allow that I was the inventor of the method of fluxions: [9]& claim that he was the inventor[10] of the differential method. Both may be true because the same thing is often invented by several men. For the two methods are one & ye same method variously explained & no man could invent the method of fluxions w^thout knowing first how to work in the augmenta momentanea of fluent quantities[,] wch augmenta Mr Leibnitz calls differences. [11]Dr Barrow & Mr Gregory drew tangents by the differential

---

(5) Originally, with a slight change in nuance of meaning, 'beg leave to' was inserted.

(6) Read 'spring', since Keill had drawn Newton's attention to Leibniz' offending anonymous recension of the *De Quadratura* (in the *Acta Eruditorum* for January 1705) in the first days of the preceding April; see page 475 above.

(7) In the revised draft of a private letter to Hans Sloane at this time which rehearses much of Newton's present, more public speech—a first version, now Add. 3968.30: 438^r/438^v, is printed in *Correspondence*, **5**: 212–14 and there are two further rough preparatory draftings on .30: 440^v—he here initially went on to make his comparison a little differently but no less cogently: 'Mr Leibnitz thinks that one of his age & reputation should not enter into a dispute wth Mr Keil, & [I] have as much reason to forbeare disputing wth the author of those papers' (.30: 441^r).

(8) '& gave occasion to Mr Keil to retaliate' is deleted in sequel in the manuscript.

(9) Newton first went on to write: '& its impossible to invent that method without considering the proportion of y^e *augmenta momentanea* of quantities wch he calls differences'.

(10) 'author' is replaced.

(11) In an elaboration of the two following sentences entered in a blank space on Add. 3968.30: 441^r below the revised draft (on which see note (7) above) of his private letter to Sloane at this time Newton recollected that 'I had the hint of this method from Fermats[!] way of drawing Tangents & by applying it to abstracted Æquations directly & invertedly I made it general. Mr Gregory & Dr Barrow used & improved the same method in drawing of tangents. A paper of mine [*sc.* his 'De Analysi' in 1669] gave occasion to Dr Barrow to shew'—Newton initially wrote 'Upon my communicating some things of this kind to Dr Barrow he shewed'—'me his method of Tangents before he inserted it into his 10th Ge[o]metrical Lecture. For I am that friend wch he there mentions'. We pooh-poohed in 1: 149, note (5) the possibility that Newton did in historical fact draw inspiration from either of the two minimal accounts, by Hérigone and Schooten, of Fermat's rule for tangents which were available in print before 1667; certainly, nearly half a century later, he has forgotten that in his own

method before the year 1669. I applied it to abstracted æquations before that year & thereby made it general. Mr Leibnitz might do the like about the same time; but I heard nothing of his having the method before the year 1677. When & how he found it must come from himself. By putting the fluxions of quantities to be in the first ratios of ye augmenta momentanea I demonstrated the method & thence called it the method of fluxions:[12] Mr Leibnitz uses it w^thout a Demonstration.[13]

---

[2][14]   *Extracts of ye MS Papers of Mr John Collins,*[15]
concerning some late improvements of Algebra.

By the proportion of the increase of lines Archimedes drew tangents to Spirals & Mr Fermat applied the method to Equations for determining maxima &

---

earliest papers on the topic he owed a massive debt to the second Latin edition of Descartes *Geometrie*, and in particular to Florimond de Beaune's 'notes on Cartes pag 131' (see 1: 280 and 281, note (33)). In citing these two sentences in 1: xv, note (1)—where, we regret, we partly mistook their context—we gave full reference to the passage in *Lectio* X of Barrow's 1670 *Lectiones Geometricæ* where a there unnamed 'amicus' is (p. 80) thanked for his advice that Barrow should, in a work otherwise devoted to expounding techniques of geometrical investigation, set out his improved version of the method of algebraically determining tangents to curves which James Gregory had displayed in Proposition 7 of his *Geometriæ Pars Universalis* two years before.

(12) 'The Demonstration', Newton added in his preliminary drafting to Sloane on Add. 3968.30: 441^r (see note (7) above), 'you have in the end of my *Analysis per æquationes infinitas* & in the first Proposi[ti]on of my Treatise of *Quadratures*.'

(13) A lone 'I'—pronoun? or first letter?—ensues in the manuscript. The reader's guess as to how Newton had it in mind to continue is as good as the editor's.

(14) ULC. Add. 3968.19: 289^r; we have earlier, on 1: 148–9, reproduced a substantial portion of the initial version of this which is found on *ibid.*: 290^v.

(15) The collection of *Collinsiana* which is now in the Royal Society's 'Commercium Epistolicum' volume (MS LXXXI), and which their then owner, William Jones, placed at Newton's disposal in or shortly after March 1712 in order to provide documented epistolary evidence supporting his claim to priority over Leibniz in the creation of a universal method of analysis *per æquationes infinitas et augmenta momentanea*: one whose prehistory Newton here—all but ignoring his principal mentor Descartes—somewhat fancifully (see the next note) traces back through Fermat's method *de maximis et minimis* to Archimedes, before passing to inform his untutored reader that he had expounded its elements in a paper on the topic which he in 1669 communicated (through Barrow) to Collins, who forthwith over the next couple of years outlined the content of this tract 'De Analysi', directly and through Oldenburg, to a wide range of contemporaries (none except Leibniz, unfortunately, in 1712 still living). In a subsequent preparatory draft for his ensuing *Phil. Trans.* 'Account' of the *Commercium Epistolicum* Newton drew on the detail of his following historical survey to assert a deal more trenchantly (on the verso of a loose scrap now Add. 3968.41: 97) that 'Archimedes began the method of squaring curves & drawing tangents to them by con[si]dering the infinitesimals of quantity. Cavallerius & Fermat applied this method to Equations. Fermats method was first published by Herigon & Schooten. Gregory, Barrow & Slusius improved it for Tangents.

minima[,] & thereby drew tangents to curve lines.[16] The method of Fermat was published by Herigon in his *Cursus* A.C. 1631,[17] & an instance of it was inserted by Schooten into his Commentary on the 2ᵈ Book of Cartes's *Geometry*. Mʳ Newton[18] met wᵗʰ this method upon reading that *Geometry* wᵗʰ the Commentators[19] thereon & by applying it to abstracted equations found the proportions of the increases of indeterminate quantities in the manner described in the first Proposition of his book of *Quadratures*. Those increases or *augmenta momentanea* he called moments[,] as may be seen in his *Analysis per æquationes infinitas*[20] & in his *Principia Philosophiæ* Lib. II Lem. II & Prop. VIII, IX, X, & in his Book of *Quadratures*: &[21] the velocities by wᶜʰ the quantities increased he called motions, velocities of increase & fluxions.[22] Mʳ James Gregory &

Mʳ Newton made it extend to the solution of all sorts of difficulter Problems & notice of his having made it general was sent to Collins Gregory Slusius & others in the year 1669.' We will spare the reader the full detail of the sequel where Newton declared: 'Mʳ Leibnits having notice thereof...conceals what he learnt by correspondence with Mʳ Oldenburg & Mʳ Collins, & will not allow young men [*viz.* Keill] to be capable of understanding the letters & papers then written, & yet appeals to yᵉ judgemᵗ of yᵉ R. Society amongst whom there is not a man now to be found besides himself who is now privy to it'.

(16) Not without some considerable difficulty, as Newton probably himself only vaguely knew (see the next note) and so we need not here emphasise, but merely refer in passing to Fermat's protracted *querelle* with Descartes on the best method of drawing tangents to curves, one or two of the subtler points of which are still to be brought out in any published secondary account.

(17) Read 'in the *Supplementum* to his *Cursus* A.C. 1642' as we observed in 1: 149, note (5) at the equivalent place in the initial version of the present text (Add. 3968.19: 290ᵛ), there drawing the only possible inference that Newton 'had never seen Hérigone's work'. The source of this misinformation was almost certainly a long critique of Descartes' mathematical achievements which Collins drew up in May 1676, and is (in its somewhat disordered draft) still preserved among 'yᵉ MS Papers of Mʳ Collins' in the Royal Society (MS LXXXI, No. 39ᴬ, whence it is now printed in J. E. Hofmann's 'normalised' version in *Gottfried Wilhelm Leibniz: Sämtliche Schriften und Briefe*, (3) **1** (Berlin, 1976): 383–407). In a section on page 13 of this ragbag of Cartesian truths, half-truths and falsities where he held forth 'About his Method of Tangents' he communicated the story, 'as Mʳ [William?] Joyner and others assert', of how 'Robervall basted Deschartes in a publique Assembly—we know from other reports of this confrontation that it took place at Paris in the summer of 1648—'accusing him of taking his Algebra out of Harriot and his method of Tangents out of Herigon published in 1631 as the Inventum of Fermat...' (see *Sämtliche Schriften*, (3) **1**: 390–1, where however 'basted' is mistranscribed as 'bafled'). If we remind that Harriot's 'Algebra' was edited posthumously at London in 1631 (as his *Artis Analyticæ Praxis*) Collins' slip is readily explained.

(18) 'acknowledges that he first' is deleted in sequel.

(19) As we remarked in note (11) to [1] preceding, Newton in his original papers on his 'universal' method of tangents here (see 1: 280) singled out Florimond de Beaune.

(20) Whose text, as Newton passes below to state, was 'lately published by Mʳ Jones' in his 1711 compendium of Newtonian *Analysis*; see II: 207, note (2) for full details.

(21) 'the proportion[s] of these increases or' is deleted.

(22) Newton initially concluded his sentence with a further clause 'but it was three or four years before he used the name of fluxions' prior to discarding the remark— rightly, for he

Mr Barrow improved the method of Fermat for drawing of Tangents & published their improvements, vizt Mr Gregory A.C. 1668 in the 7th Proposition of his *Geometria universalis* & Mr Barrow A.C. 1670 in the end of his 10th Geometrical Lecture.[23] Mr Newton's *Analysis* aforesaid conteining his methods of series & moments together,[24] was communicated by Mr Barrow to Mr Collins A.C. 1669 & lately published by Mr Jones.[25] This *Analysis* gave occasion to several things in the Epistolary commerce between Mr Collins & others, the following extracts of wch have been taken out of Letters & papers written in the hand of Mr Collins & left to his Executors,[26] & [27]examined by some who know the hand.

. . .   .[28]

---

first employed this medieval *terminus technicus* for '[rate of] flowing' of motion in his 1671 tract (see III: 72 ff.) only some six years after he began (see I: 344) to develop methods for determining the relationship of the instantaneous 'speeds' of fluxion in contemporaneously 'fluent' quantities.

(23) 'but they applied it only to the drawing of tangents' is deleted. Newton elsewhere (see [1]: note (11) preceding) stated that it was at his 'friendly' advice that Barrow included his improved version of Gregory's Fermatian method of constructing tangents in *Lectio* X of his Lucasian *Lectiones Geometricæ*.

(24) Originally just '...his general method of endless series'.

(25) On pages 1–21 of his 1711 compendium of Newtonian mathematical tracts, the generically titled *Analysis per Quantitatum Series, Fluxiones, ac Differentias...*; again see II: 209, note (2). Newton initially went on to end his paragraph: 'On wch occasion Mr Collins communicated several things [*sc.* in his correspondence], as will appear by the following extracts out of his papers written in his own hand & lately fallen into the hands of Mr Jones: wch extracts have'—understand where (in the case of Collins' copy of his 'De Analysi', notably) this was possible—'been compared wth the originalls & found genuine'.

(26) Newton could hardly have known who these were (Collins died in obscurity in London in 1683 at a time when he himself was a virtual recluse in Cambridge): he probably meant to write only '& lately fallen into the hands of Mr Jones' as he had initially phrased it (see the previous note). In the 'Præfatio Editoris' of his 1711 *Analysis* Jones himself stated in regard to the pedigree of his Collinsian papers that 'secundus jam agitur annus ex quo scrinia D. Collinsii...meas in manus inciderunt' (signature A1^r).

(27) 'carefully' is here deleted.

(28) Initially intended to follow here, no doubt, were his penned *verbatim* 'Extracts out of [ten] Letters found among the papers of Mr John Collins'—beginning with six lines 'Out of a Letter of Mr Collins to Mr James Gregory dated the 25 of November 1669' and passing to include yet lengthier excerpts from, *inter alia*, Collins' letter to Strode of 26 July 1672 and Newton's own letter to Collins of 10 December 1672 communicating his general method of algebraic tangents—which survive on Add. 3968.19: 277^r–278^v/289^v–290^r/287^r–288^r/281^r–282^r/278^r–278^v. Immediately below in the manuscript, however, Newton has drafted the more pithy preface to his selection from Collins' 'Epistolary Commerce' which follows in [3], one which begins *in medio* without any elaborate survey of the prehistory of his analysis *per augmenta momentanea* such as he here thought to present.

WMM

[3][29] *Extracts of Letters found*[30] *among the papers of Mr John Collins relating to the Collection of mathematical Letters published by Dr Wallis in the third volume of his works.*

Dr Wallis having published in the third Volume of his works several Letters between himself Mr Oldenburg Mr Collins Mr Leibnitz & Mr Newton, & some other Letters relatin[g] to that Collection being[31] found among the papers of Mr J. Collins, it has been thought [fit] to make them publick.[32]

. . .   .

(29) This rephrased introduction by Newton to his excerpts from Collins' 'Epistolary Commerce' is drafted by him on Add. 3968.19: 289^r immediately below the more leisurely and historically slanted initial preface which is reproduced in [2] preceding. He now, it will be clear, broadens his original scope to embrace not only the *Collinsiana* in the 'Epistolarum Collectio' gathered by John Wallis in the third volume of his *Opera Mathematica* in 1699—particularly Collins' letter to Newton of 5 March 1676/7 (*ibid,*: 646–7)—but also the numerous items there printed from Leibniz' correspondence with Oldenburg between July 1674 and July 1677—notably Newton's two majestic *epistolæ prior/posterior* to Oldenburg for Leibniz here (*ibid.*: 622–9/634–45) first made public in their full Latin texts—along with Wallis' own not inconsiderable exchange of letters with Leibniz between December 1695 and January 1698/9. Newton's annotations on a number of the letters in Wallis' 'Collectio' survive in Add. 3968.3 and 3977.8; whence, having been given a final polish in Add. 3968.41: 33^v/33^r, they were ultimately to emerge to be set as footnotes at the pertinent places in the printed *Commercium Epistolicum*.

(30) Newton's augmentation of the present title on Add. 3968.19: 265^r (see note (32) below) here inserts '*in the custody of the R[oyal] Society &*'.

(31) 'lately' is deleted in sequel in the manuscript.

(32) A superfluous 'for completing that collection' is cancelled. Straightaway, in fact, Newton further broadened the purview of his original notation merely to gather selected 'Extracts of ye MS Papers of Mr John Collins' by adjoining also excerpts from 'Letters in ye hand of Mr Leibnitz' written by him to Oldenburg during 1673–5 (these he summarily and none too accurately lists at the bottom of f. 290^r) together with Oldenburg's replies as he found them copied into the Royal Society's Letter Books: his full register of fourteen Leibniz letters therein, beginning with that of 3 February 1672/3 (from 'No. 6. page. 35') 'concerning Moutons differential Method in wch Mr Leibnitz had made progress...', along with an odd one from Oldenburg to Sluse of 10 February (dated, from its draft, 29 January in *The Correspondence of Henry Oldenburg*, **9** (Madison, Wisconsin, 1973): 427–8) 1672/3 passing on Newton's account of his tangent-rule for algebraic curves as he had communicated it to Collins on the previous 10 December, can be found on the intervening ff. 289^v/290^r, in each case referred to the corresponding page of Letter Books 'No. 6' and 'No. 7'. Subsequently, on ff. [in sequence] 265^r/270^r/266^r/269^r/267^r/268^r he set out a detailed enumeration of the component items in this amplified gathering of 'Extracts of Letters'—now numbering 36 altogether—'found in the custody of the R. Society & among the papers of Mr John Collins, relating to the Collection of Mathematical Letters published by Dr Wallis...'. And so the 'Epistolary Commerce' ramified and grew: a long extract was interpolated 'Out of Mr Newton's Treatise *de Analysi per æquationes infinitas* sent by Dr Barrow to Mr Collins ye 31th of July 1669'—in the final version, of course, the text of the 'De Analysi' was reprinted *in toto* from Jones' *editio princeps* of the year before—and appropriately annotated (see Add. 3968.10: 126^r/126^v); and then the whole swelling mass of excerpts was rendered into Latin (where the originals were in English) and

[4]⁽³³⁾ We have consulted the Letters & Letterbooks in the custody of the R.S. & those found among the papers of Mr John Collins dated between the years 1669 & 1677 inclusively & shewed them to such as know the hands of Mr Barrow, Mr Collins, Mr Oldenburg, Mr Leibnitz, Mr Gregory⁽³⁴⁾ & extracted from them what relates to the matter in dispute.⁽³⁵⁾ By their Letters and papers we find

---

further honed and reworked to be (see Add. 3968.10: [120^r–122^r→] 123^r–125^r) a 'Historia brevis Methodi serierum ex Monumentis antiquis desumpta' (as it is titled on f. 123^r). From this, with the addition of the unabridged 'De Analysi' and of the letters from Leibniz and Keill to Sloane which were the *raison d'être* for the exercise, was born the *Commercium Epistolicum D. Johannis Collins et Aliorum de Analysi promota* which Newton at length put to press in about the late summer of 1712—with yet further editorial interpolations and annotations we need not say.

(33) Newton's rough draft of the 'findings' which the Royal Society Committee set up on 6 March 1711/12 to 'inspect the Letters and Papers relating [to] the Dispute between... Mr Leibnitz...and Mr Keill...concerning the Priority of the Invention of the Arithmetick of Fluxions' (see Journal Book, XI: 224, 275) brought in just seven weeks later on 24 April; compare page 482 above. The finished piece which was presented to the Society, as recorded in its Journal Book (*ibid.*: 287–9), is reproduced in *The Correspondence of Isaac Newton*, 5: xxvi–xxvii with the portions of it which are already present in this rough draft (the original of which is now in private possession) italicised. Here *vice versa* we cite in following footnotes the additions which convert Newton's autograph into the Committee's official report. We can only conclude—without passing any moral judgement on the circumstance—that the latter's members merely rubber-stamped, without any essential display of initiative or counter-opinion of their own, the pre-dictated judgement set before them by the Society's nominally impartial President upon such selection as he then chose to place before them from the *Collinsiana*, *Wallisiana* and *Oldenburgensia* which he was hard at work gathering together to back his case for priority over Leibniz. Newton's cancelled first tentative findings from his jackdaw-store that 'Mr Leibnitz was in London in Febr. 167$\frac{2}{3}$ as appears by his Letters to Mr Oldenberg dated at London 3^d & 10th of Feb. 1673 O. st. He was then upon improving a differential [method] wch Mouton had wrote of before & being suspected of borrowing from Mouton defended himself in the first of those two Letters, but made no mention of his having any other differential method at that time. [He] went to Paris before the $\frac{16}{26}$ Apr. & from that time kept a correspondence wth Mr Oldenburgh & by his means wth Mr Collins as appears by his Letters to Mr Oldenburgh... entred in the books of y^e Society....' may be found on Add. 3968.10: 140^r, where they precede (on ff. 140^r–141^r) items '24', '25', '26' and '33'—excerpts respectively 'Out of Mr Oldenbergs Answer to Mr Leibnitz dated 8 Dec. 1674', 'Out of a Letter of Mr Leibnitz to Mr Oldenburg dated from Paris 30th March 1675', 'Out of Mr Oldenburghs Answer dated at London 15th April 1675 & sent to Mr Leibnitz at Paris (I think by Mr Tschirnhause' and 'Out of Mr Leibnitz's Letter to Mr Oldenburgh dated at Paris 12 May, 1676', the last two accompanied by annotations in English—which show that the compendium of 'Extracts of Letters found in the custody of the Royal Society & among the papers of Mr John Collins' had already achieved ample proportions.

(34) In the full report this last entry is enlarged to read 'and compared those of Mr Gregory with one another and with Copies of some of them taken in the hand of Mr Collins'.

(35) Afterwards changed to be 'to the matter referred to us'. The following sentence was then recast to read: 'All which extracts herewith delivered to you we believe to be genuine and authentic, and by these Letters and Papers we find'.

1 That Mr Leibnitz was in London in ye beginning of the year 1673 & went thence to Paris where he kept a correspondence wth Mr Collins by means of Mr Oldenberg till about September 1676 & then returned by London & Amsterdam to Hannover. And that Mr Collins was free in communicating[36] what he had received from Mr N. & M^r Gr.

2 That when Mr Leibnitz was the first time in London he pretended to the differential method of Monsr Mouton, but quitted his pretension upon being told by Dr Pell that it was Moutons method.[37] And we find no mention of his having any other differential method then Moutons before his Letter of [21] June 1677[,] wch was a year after[38] Mr Newton's Letter of [10 December] 1672 had been sent to him at Paris, & above four years after Mr O. began to communicate the Letter to his correspondents[,] in wch Letter the method of Fluxions was[39] described by Mr Newton to an[40] intellig^t person.

3 That by Mr Newton's Letter of 13 June 1676 it appears that he had the method of Fluxions above five years before his writing of that Letter. And by what we find in his *Analysis per Æquationes numero terminorum infinitas* communicated by Mr Barrow to Mr Collins in July 1669 we find[41] that he invented the method of Fluxions before that time.

4 That the Differential method is one & the same method with the method of Fluxions, excepting the name & mode of notation[,] Mr Leibnits calling those quantities differences wch Mr Newton calls Fluxions & Moments & Incrementa momentanea, & marking them with the letter *d* wch is a mark not used by Mr Newton. And we take the proper question to be not who invented this or that, but who was the first author of the method,[42] & beleive that those who have reputed Mr Leibnitz the first Author of this method knew little or nothing[43] of his correspondence wth Mr Collins & Mr Oldenburg,[44] nor of Mr Newtons

---

(36) 'very free in communicating to able Mathematicians' in the final report.

(37) Afterwards changed to assert that 'when Mr Leibnitz was the first time in London he contended for the Invention of another Differential Method properly so called; and notwithstanding that he was shown by Dr Pell that it was Mouton's Method, he persisted in maintaining it to be his own Invention, by reason that he found it by himself without knowing what Mouton had done before, and had much improved it'. Here of course, ungenerous as ever, Newton cites Leibniz' letter to Oldenburg of 3 February 1672/3 against him (compare note (30) above). On Mouton's prior 'invention' of a method of subtabulation by finite differences (already known in more sophisticated form to Henry Briggs half a century earlier) see IV: 4–5.

(38) Trivial insertion of 'a Copy of' was afterwards here made.

(39) 'sufficiently' is deleted in the manuscript—to be replaced in the final report.

(40) 'any' by a minimal enlargement in the report submitted.

(41) Newton first wrote, slightly less forcibly, 'are satisfied'.

(42) Originally, in anticipation of the 'Committee's' formal conclusion, 'And we are satisfied that Mr Newton was the first author of this method'.

(43) A totally dismissive 'nothing' *tout court* was first written.

(44) This replaces a vaguer 'wth the friends of Mr Newton': in the final report a scarcely necessary 'long before' was inserted in sequel.

having that Method above 15 years before Mr Leibnitz began to publish it in the *Acta Leipsica*.[45]

For wch reasons we reccon Mr Newton the first inventor.[46]

---

[5][47]   The occasion of publishing this Collection of Letters will be understood by the Letters of Mr Leibnitz & Mr Keil in the end thereof: Mr Leibnitz taking offence at a passage in a discourse of Mr Keil published in the *Transactions*

---

(45) Minimally corrected to be '*Acta Eruditorum* of Leipsick' in the report put to the Royal Society. In his present draft Newton initially went on: 'For in ye year 1669 when he communicated the aforesaid *Analysis* to his friends he had so far advanced the method as to affirm that by the method of infinite series'—here, of course, he renders the last sentence of the central portion of his 1669 'De Analysi' (see II: 242)—'the areas & lengths of curves &c (when it could be done) were determined exactly & Geometrically; that is, by the breaking off of those series & their becoming finite. In his Letter of June 13 1676 he represents that the series for Quadratures wch break off & become finite when the Curve admits of an exact quadrature were found out by the method of Fluxions. And such series   '. At this point, doubtless becoming aware that such monotonous axe-grinding was here pointless, he abruptly broke off continuing his own 'endless' series of words.

(46) A superfluous 'of the method' is deleted in sequel in the manuscript. The final report goes on to add: 'and are of opinion that Mr Keill in asserting the same has been noways injurious to Mr Leibnitz. And we submit to the Judgement of the Society whether the Extract of Letters and Papers now presented'—this collection, whatever precisely it contained, was certainly already an ample gathering (see notes (32) and (33) preceding)—'together with what is extant to the same purpose in Dr Wallis's 3^d Volume, may not deserve to be made publick'. 'To which Report', the Journal Book (XI: 289) goes on to record for its meeting on 24 April 1712, 'the Society agreed *nemine contradicente*, and ordered that the whole Matter, from the Beginning, with the Extracts of all the Letters relating thereto, and Mr Keill's and Mr Leibnitz's Letters, be published with all convenient speed may be, together with the Report of the said Committee' with 'Dr Halley, Mr Jones and Mr Machin...desired to take care of the same Impression (which they promised)...'. We scarcely need reiterate that it was Newton himself, in fact, who at once took complete charge of the editorial preparation of the *Commercium Epistolicum*,; see page 483 above.

(47) Add. 3968.37: 551^r–553^v, with a last paragraph introduced from the predraft on 554^v; a concatenation of preliminary roughings-out and recastings survives (compare note (53) to the preceding introduction) on Add. 3968(.41: 127^r/138^r→).37: 539^r/539^v + 540^r→540^r/540^v (+ .41: 36^r/109^r/108^v)→541^r–544^v→547^r/548^r(+ .41: 115^r/115^v/105^r–106^v/96^r–97^v). When proofs of the finally styled *Commercium* | *Epistolicum* | *D. Johannis Collins,* | *et Aliorum* | *de* | *Analysi* | *promota:* | *jussu* | *Societatis Regiæ* | *In lucem editum*—whether this was Newton's own title or no, he found no slightest amendment to make in the proofsheet (now Add. 3968.2: 6) which bore these staggered words on its recto—began at length to issue from the printer in autumn 1712, the question urgently posed itself of how best the main text of this 'Epistolary Commerce' should briefly be introduced to the reader. Here, and in the shorter revise in [6] following, we reproduce Newton's principal English originals of such a prefatory 'Ad Lectorem' concisely setting the scene.

Throughout, it will be evident, Newton keys his outline of his case page by page to the *Commercium*. Since the *editio princeps* of this work 'Londini: Typis Pearsonianis. Anno M DCC XII [= 1712/13]' is not easily come by even in the largest libraries (and has never, to our knowledge, been reprinted in photofacsimile) the following broad breakdown of its

A.C. 1708,[48] wrote a Letter to the Secretary of the R. Society[49] complaining thereof as a calumny, desiring a remedy from the Society & suggesting that he

content may possibly be of help: (pages 1–2) excerpts from Barrow to Collins of 20 and 31 July and 20 August 1669, regarding (3–20) the 'De Analysi' (here reprinted *in toto* from Jones' 1711 *Analysis*); (21–6) pertinent excerpts from (Collins to) Oldenburg for Sluse, 14 September 1669, Collins to Gregory of 25 November 1669, Gregory to Collins of 20 April, 5 September, 23 November and 19 December 1670, and Collins to Gregory of 24 December 1670; (25–6) Gregory to Collins, 15 February 1670/1; (26–9) extracts from Collins' drafts of his letters to Bertet of 21 February 1670/1, to Borelli in December 1671, to Vernon of 26 December 1671, and to Strode of 26 July 1672 (all rendered *Latine*; see III: 21–3 for the English originals); (29–30) Collins to Newton of 30 July 1672, and Newton to Collins of 10 December 1672 (portion on tangent-rule only); (31–2) excerpts from Oldenburg's letter to Sluse of 29 January 1672/3, Sluse's reply of 3 May, and Oldenburg's further letter to him on 10 July 1673; (32–7) Leibniz to Oldenburg of 3 February 1672/3; (37–45) brief excerpts from their ensuing correspondence on 15 July, 26 October and 8 December 1674, 30 March, 15 April, 20 May, 24 June, 15 July, 30 September and 28 December 1675, and 12 May 1676, together with (pages 42–3) extracts from *Acta Eruditorum* (April 1691): 178 and Collins' critique of Descartes' achievement in mathematics (for Tschirnhaus in May 1676; see note (17) above); (46–7) a brief excerpt from Collins' 'Abridgement' (for Leibniz in June 1676) and his introduction to the fuller 'Historiola' (never sent) of his correspondence with James Gregory, 1668–75; (47–8) extract from Collins to David Gregory *père* on 11 August 1676 (English draft rendered *Latine*); (49–57) full text of Newton's *epistola prior* to Oldenburg for Leibniz of 13 June 1676; (58–65) Leibniz' response of 17/27 August; (66) extract from Tschirnhaus to Oldenburg of 1 September 1676; (67–86) full text of Newton's *epistola posterior* to Oldenburg for Leibniz of 24 October 1676; (87) extract from Collins to Newton of 5 March 1676/7 (transmitting Leibniz to Oldenburg of 18/28 November preceding); (88–95/96–7) Leibniz' letters to Oldenburg of 21 June and 12 July 1677 (N.S.); (98–9) extract, regarding Newton *v.* Leibniz from the anonymous review (by Leibniz) in *Acta Erud.* (June 1696): 257 ff. of Wallis' *Opera* (**2**, 1693/1, 1695); (99–106) excerpts from Wallis to Leibniz of 1 December 1696 and 6 April/ 30 July 1697, with Leibniz' intervening responses of 19/29 March and 18/28 May 1697; (107) extracts from Fatio's 1699 *Investigatio Geometrica*, p. 18 and Leibniz' counterblast in *Acta Erud.* (May 1700); (108–9) the offending final paragraphs from the anonymous review (by Leibniz) in *Acta Erud.* (January 1705) of Newton's *De Quadratura*; (109) the offensive sentences in Keill's 'Epistola' in the *Phil. Trans.* in 1708 (see the next note); (109–10/118–19, 110–18) Leibniz' letters of complaint to Sloane of 4 March and 29 December 1711 (N.S.) together with Keill's intervening riposte in May; (120–2) the four conclusions delivered by the reporting Committee (English original with adjoined Latin translation). As we need not itemise, the autograph originals of most of the letters here printed, in full or in excerpt (and in Latin version, most often, where the original is in English), are preserved in the Royal Society's 'Commercium Epistolicum' volume (MS LXXXI: individual papers are there keyed to their number and first-page in the *Commercium*'s 1722 second edition, on which see note (57) of the preceding introduction).

(48) In his 'Epistola ad...Halleium...de Legibus Virium Centripetarum', namely; see page 473 above, and note (14) thereto where the offending aside is quoted in full. We will not in sequel duplicate what we have said at length before in our preceding introduction, but restrict our attention largely to citing significant textual variants.

(49) Those of early March, mid-May and late December 1711 respectively to the Society's (only active) Secretary, Hans Sloane, which were the immediate occasion for Newton, as stage-managing President and behind-scenes onlie true begetter, to convene 'his' reporting Committee on 4 March 1712.

beleived they would judge it equal that Mr Keil should make a public acknowledgment of his fault. Mr Keil chose rather to return an answer in writing,[49] wherein he explained his meaning in that passage & defended it. Mr Leibnitz not meeting with that satisfaction he desired wrote a second Letter to the Secretary of the R. Society,[49] wherein he still complained of Mr Keil, representing him a young man not acquainted with what was done before his time nor authorized by the person concerned[,] & appealed to the equity of the Society to cheq[50] his unjust clamours.

Mr Leibnitz was in England in the beginning of the year 1673 & again in October 1676 & during the interval in France, & all that time kept a correspondence with Mr Oldenburgh & by his means with Mr John Collins, & what he learnt from the English by that correspondence is the main Question. Mr Leibnitz appeals from young men to old ones who knew what passed in those days, refuses to let any man be heard against him w^thout authority from Sr Isaac Newton & desires that Sr Isaac himself would give judgment. Sr Isaac lived then at Cambridge, & knew only his own correspondence printed by Dr Wallis. Mr Oldenburgh & Mr Collins are long since dead & so are Dr Barrow Mr Gregory & Dr Wallis who corresponded with Mr Collins, & Dr Wallis gave judgment against Mr Leibnitz in a Letter dated Apr 10th 1695 & not yet printed.[51] The R. Society therefore being twice pressed by Mr Leibnitz & having no[52] means of enquiring into this matter by living evidence, appointed a Committee to search out & examin the Letters Letter-books & papers left by Mr Oldenburgh in the hands of the Society & those found among the papers of Mr Collins relating to the matters in dispute, & ordered the Report of the Committee with the extracts of the Letters & Papers to be published.

The Question is, Who was the first author of the method called by Mr Leibnitz the infinitesimal method the Analysis of indivisibles & infinites, & the Differential & summatory method & by Mr Newton the Method of fluents fluxions & moments. Mr Leibnitz claims *inventoris jura*, & sometimes allows that Mr Newton might also find it apart. Mr Keil asserts Mr Newton to be the first inventor & [53]the Committe is of the same opinion.

---

(50) Read 'check'.

(51) Its recipient was of course Newton, who still at this time retained it in his private possession, though he was (see note (104) of the preceding introduction) a couple of years later to deposit it in the Royal Society's archives for future use as documentary evidence against Leibniz in his anonymous 1715 'Account' of the *Commercium*, and ultimately himself to publish its full text in his 1717 Appendix to Raphson's *History of Fluxions* (see *ibid.*: note (145)).

(52) 'other' is deleted in sequel. Newton regards himself, of course, as President of the 'court of justice' to which Leibniz has made his complaint, to be above the matter—and, in any case, unwilling to appear openly to give such 'living evidence' as he now decrees to be no longer possible. One must never look too closely into the rigorous logic of his arguments.

(53) An intervening nicety 'the Report of' is here deleted (perhaps because it comes too uncomfortably close to the truth?).

The Letters & Papers till the year 1676 inclusively shew that Mr Newton had a general method of solving Problemes by reducing them to equations finite or infinite, whether those equations include moments (the exponents of fluxions) or do not include them, & by deducing fluents & their moments from one another by means of those equations.

[54]The *Analysis* printed in the beginning of this Collection shews that he had such a general method in the year 1669. And by the Letters & Papers wch follow, it appears that in the year 1671, at the desire of his friends he composed a larger Treatise upon this method (p. 27. l. 10, 27 & p. 71. l. 4, 26)$_{[,]}$ that it was very general & easy without sticking at surds or mechanical Curves, & extended to the finding tangents areas lengths centers of gravity & curvatures of Curves &c[55] (p. 27, 30, 85)$_{[,]}$ that in Problems reducible to Quadratures it proceeded by the Propositions since printed in the book of Quadratures, wch Propositions are there founded upon the method of fluents (p. 72, 74, 76)$_{[,]}$ that it extended to the extracting of fluents out of æquations involving their fluxions & proceeded in difficulter cases by assuming the terms of a series & determining them by the conditions of the Probleme (p. 86)$_{[,]}$ that it determined the[56] curve by the length thereof (p. 24) & extended to inverse Problems of tangents & others more difficult, & was so general as to reach almost all Problemes except numeral ones like those of Diophantus (p. 55, 85, 86). And all this was known to Mr Newton before Mr Leibnitz understood[57] any thing of the method, as appears by the dates of their Letters.

For in the year 1673 Mr Leibnitz was upon another differential method (p. 32). In May 1676 he desired Mr Oldenburg to procure him the method of infinite series (p. 45) & in his Letter of 27 Aug 1676 he wrote that he did not beleive Mr Newton's method to be so general as Mr Newton had described it. For, said he, there are many Problemes & particularly the inverse Problems of Tangents wch cannot be reduced to æquations or Quadratures (p. 65)$_{[,]}$ wch words make it evident that Mr Leibnitz had not yet the method of differential equations. And in the year 1675 he communicated to his friends at Paris a tract written in a vulgar manner about a series wch he received from Mr Oldenburgh, & continued to polish in the year 1676 wth intention to print it :[58] but it swelling

(54) The text of the two following paragraphs has already been cited on III: 20–1 from Newton's minimal revision of these on Add. 3968.39: 583^r.

(55) Initially '... extended to Problemes of Tangents direct & inverse... & solving other more difficult Problemes'.

(56) The revision on .39: 583^r (see note (54)) here inserts 'species of the'.

(57) Minimally amended in the manuscript from 'all this was found out by Mr Newton before Mr Leibnitz knew'.

(58) Leibniz' most polished tract (of early? summer 1676) 'De quadratura arithmetica circuli, ellipseos et hyperbolæ' has never been printed, but the highly individual manner in which he derived 'his' series for the inverse-tangent, about October 1673 and before Oldenburg

in bulk he left off polishing it after other business came upon him, & afterwards finding the differential Analysis he did not think it worth publishing because written in a vulgar manner (p. 42, 45). In all these Letters & Papers there appears nothing of his knowing the Differential method before the year 1677. It is first mentioned by him in his Letter of 21 June 1677, & there he began his description of it with these words: *Hinc nominando* IN POSTERVM *d$\bar{y}$ differentiam duarum proximarum y* &c (p. 88).

Mr Newton was therefore the first inventor, & whether Mr Leibnitz invented it *proprio Marte* afterwards or not, is a question of no consequence. The first inventor is the inventor, & *inventoris jura* are due to him alone. He has the sole right till another finds it out, & then to take his right from him without his consent & share it with another would be an Act of injustice, & an endless encouragement to pretenders. But however, there are great reasons to beleive that Mr Leibnitz did not invent it *proprio marte*, but received some light from Mr Newton.

For it is to be observed that wherever Mr Newton in his Letters spake of his general method, he understands his method of series & fluents taken together as two parts of one general method. In his *Analysis* the series are applied to the solution of problemes by the method of fluents & thereby give new series, & the method of fluents is demonstrated by the method of series (p. 14, 15, 18, 19) & in the year 1671 he wrote also of both together (p. 71). The series in his book of *Quadratures* are derived from the method of fluents & were derived from it before the year 1676 (p. 72). The method of extracting fluents out of equations involving their fluxions comprehends both together, & the method of assuming the terms of a series & determining them by the conditions of the probleme proceeds by means of the method of fluents (p. 86). When Mr Newton represented that his method of series extended to the solution of almost all Problemes except numeral ones like those of Diophantus, he included inverse problemes of Tangents (p. 55, 56, 85) & those problems are not tractable wthout the method of fluents (p. 86). And sometimes series are considered as fluents & their second terms as moments. And Mr Newton sometimes derives his method of fluents from the series into wch the power of a binomium is resolved (p. 19. lin. 19, 20).

In the next place it is to be observed that Mr Newton at the request of Mr Leibnitz communicated to him freely & plainly one half of this general method, namely the method of series (p. 45, 49). Mr Gregory by the help of but one

had communicated to him James Gregory's prior version of it (as reported by Collins) in his letter of 12 April 1675, is set out at some length by J. E. Hofmann in his *Entwicklungsgeschichte der Leibnizschen Mathematik während des Aufenthaltes in Paris (1672–1676)* (Munich, 1949): 32–6 [= *Leibniz in Paris, 1672–1676: His growth to mathematical maturity* (Cambridge, 1974): 54–61]; see also IV: 531, note (13).

series, with notice that it was the result of a general method, found out the method within the space of a year (p. 22, 23, 24). Mr Leibnitz pretended to have two series in the year 1674, & had eight others sent him by Mr Oldenburg in April 1675 as the result of a general method & took a years time to consider them (p. 38, 40, 41, 42): but this Method being *altioris indaginis*[59] he could not find it out but at length requested Mr Oldenburg to procure it from Mr Collins (p. 45)[60] & at the request of Mr Oldenburg & Mr Collins, Mr Newton sent it to him. And when he had it he understood it with difficulty & desired Mr Newton to explain some things further (p. 49, 63).

And as for the other half of the method, Mr Leibnitz had a general description of it in Mr Newton's Letters of 10 Decemb. 1672, 13 June 1676 & 24 Octob. 1676, with examples in drawing of Tangents (p. 30, 47) squaring of curves (p. 42) & solving of inverse problemes of Tangents (p. 86), & understanding by the same Letters that the method of tangents printed by Slusius was a branch & corollary of Mr Newtons general method (p. 30) he set his mind upon improving this method of Tangents so as to bring it to a general method of solving problems. For in his journey home from Paris by London & Amsterdam, he was upon a project of extending it to the solution of all sorts of problemes by calculating a certain Table of Tangents as the most easy & usefull thing he could then think of for a calculator wch he wanted,[61] & wrote of this designe to Mr Oldenburg in a Letter dated at Amsterdam 28 Novemb. 1676 (pag. 87.) & therefore he had not then improved the method of Slusius into a general method of solving all sorts of problems but was endeavouring to do it.[62] Now Mr Newton had told him that his method extended to Tangents of mechanical Curves & to Quadratures centers of gravity & Curvatures of Curves in general & to inverse problemes of Tangents, & he was thereby sufficiently informed that this method was founded upon the consideration of the small particles of quantity called *augmenta momentanea* &

(59) 'of deeper investigation'; that is, one requiring more profound inquiry.

(60) In his letter of 12 May 1676 (N.S.), that is, where he wrote: 'Cum Georgius Mohr Danus...nobis attulerit communicatam sibi a Doctissimo Collinio vestro expressionem relationis inter Arcum & Sinum per infinitas Series...quæ mihi valde ingenios[æ] videntur, ...ideo rem gratam mihi feceris...si demonstrationem transmiseris'. Newton does not, of course, stress that Leibniz asks for the demonstration of but one series, and not for his general method of deriving such series; and omits to mention that the latter promised his own 'Leibniz' series in return, 'demonstratione tamen non addita quam nunc polio'—at which phrase Newton in the *Commercium* adjoined the cynical footnote: 'Opusculum prædictum de Quadratura Arithmetica D. *Leibnitius* polire [*sc.* antea] perrexit'.

(61) A wholly conjectured and highly dubious surmise '& spake to Mr Collins to help him' is rightly cancelled in sequel.

(62) Initially '& therefore he had not then invented the differential method but was freely' —subsequently 'only'—'endeavouring to find out such a general method as Mr Newton had described'.

moments by Mr Newton, & infinitesimals indivisibles & differences by Mr Leibnitz. For there is no other way of resolving any of those sorts of Problemes, then by considering these particles of quantity. This consideration therefore might make him$^{(63)}$ think upon the methods of Fermat Gregory Barrow & Slusius, who drew tangents by the proportion of the particles of lines. For he tells us that he found out the Differential method by considering how to draw Tangents by the differences of the Ordinates & how thereby to render the method of Slusius more general (p. 88) & considered that as the summs of the Ordinates gave the area so their differences gave the tangents, & thence received the first light into the differential method (p. 104). And with Slusius he gave the name of differences to the moments of dignities. (*Transact. Philosoph.* Num. 95.)$^{(64)}$ And when he had found the Method, he saw that it answered to the description wch Mr Newton had given of his method in drawing of Tangents, in rendring Problems of Quadratures more easy, & in bringing inverse Problems of Tangents to Equations & Quadratures, wch in his letter of 27th August 1676 he had represented impossible (p. 65, 88, 89, 90, 91, 93).

But when he sent his method to Mr Newton he forgot to acknowledge that he had but newly found it, & that the want of it had made him of opinion the year before that inverse problemes of Tangents & such like could not be reduced to Equations & quadratures. He forgot to acknowledge that by means of this invention he now perceived that Mr Newtons method wch extended to such Problems was much more general then he could beleive the year before. He forgot to acknowledge that in the collection of Gregories Letters & Papers wch at his own request were sent to him at Paris by Mr Oldenburg & Mr Collins, he found the copy of Mr Newtons Letter of Decem. 10th 1672, conteining his method of Tangents & representing it a branch or corollary of a general method of solving all sorts of problems & that the agreement of this method of Tangents wth that of Slusius, put him upon considering how to enlarge the method of

---

(63) 'begin to' is deleted, along with the tentative insertion at this point 'lay aside his project of a Table of Tangents &'.

(64) Newton's reference here is to Sluse's letter of 3 May 1673 (N.S.) to Oldenburg, printed straightaway by its recipient in *Philosophical Transactions*, **8**, No. 95 (for 23 June 1673): 6059, where he outlined 'meam [methodum], qua nempe tot ante annos usus sum, et cuius ope flexus curvarum contrarios ac Problematum limites ostendi...'. In essence, Sluse computes the derivative of any power x^n (in his instance $n = 3$) of a base 'fluent' x by letting the latter be augmented to be y, and then computing the quotient of the 'differentia dignitatum $[y^n - x^n]$ applicata ad differentiam laterum $[y - x]$', namely, $y^{n-1} + y^{n-2}x + ... + x^{n-1}$; whence at once his version of the Leibnitian derivative $\lim_{y \to x} [(y^n - x^n)/(y - x)]$ reduces to be

$$\lim_{y \to x} [y^{n-1} + y^{n-2}x + ... + x^{n-1}] = nx^{n-1}.$$

(See *The Correspondence of Henry Oldenburg*, **9** (Madison, Wisconsin, 1973): 617–18, and note 2 thereto; also Appendix 2: note (72) below.)

Slusius, by the Differences of the Ordinates. He forgot to acknowledge that Mr Newtons Letters of 13 June & 24 Octob. 1676, gave him any light into the method. And those things are now so far out of memory, that he has told the world that when he published the elements of his differential method he knew nothing more of Mr Newtons inventions of this sort then what Mr Newton formerly signified in his Letters, namely that he could draw tangents without taking away irrationals: wch Hugenius had signified that he could also do before he understood the infinitesimal method (p. 104, 107).

In like manner when Mr Collins had begun to make known Mr Newtons method of series to the mathematicians in London & to communicate the series even to forreigners, Mr Leibnitz about three or four years after coming to London where he conversed with the mathematicians & going thence to Paris, wrote from thence the next year to Mr Oldenburg as if he had not heard of Mr Newton's method of series, & put in for the invention next after Mercator, pretending to be the first inventor of two series for the circle both found by the same method (p. 37, 38) but wrote afterwards for the Demonstration or method of finding one of them & did not know it to be his own (p. 45). The next year he received eight series from Mr Oldenburg & Mr Collins & knew none of them to be his own$_{[,]}$ but before the end of the year forgot the receipt of them & communicated to his friends at Paris an Opusculum upon one of them[65] as his own series (p. 40, 41, 42). [66]And having now got eight or ten series, he endeavoured in May foll[ow]ing to get the method also from Mr Oldenburg & Mr Collins wthout the knowledge of Mr Newton (p. 45), tho by his own rule if he should have forgot the receipt of the method they were not to reclaim it without authority from Mr Newton (p. 118). And in recompense for the method he promised them the series of wch he had written to them some years before, representing it his own & very different from theirs (p. 45) but meaning one of the eight series wch he had received from them the year before (p. 61). For they were not to question his candor wthout authority from the Executors of Mr Gregory the author of the series. Mr Collins had Mr Newton's method composed by himself in the *Analysis* here printed but instead of sending it after the eight series, both he & Mr Oldenburg wrote earnestly to Mr Newton to send his own method himself, & Mr Newton did so (p. 49) & then Mr Leibnitz rewarded them with the said series as he had promised (p. 61), & sent them also his Transmutation method whereby that series might be found$_{[,]}$ representing it a general method to be recconed among the principal inventions of Analysis, for wch he hoped that

(65) That yielding the 'arithmetical quadrature' of the quadrant of a circle, a particular instance of the general Gregorian series for the inverse-tangent whose independent 'invention' was (compare note (58) above) his prime mathematical discovery in autumn 1673.

(66) As we have already previously remarked in **1**, 6, §2.2: note (23) (page 352 above), the following sentences are drafted on Add. 3968.41: 119^v.

they would not deny him what was notable[67] among them (p. 58, 59). And by vertu of this method he has eversince numbered himself amongst the inventors of the methods of series, wch discovers his designe in sending it. And yet this method was not generall till Mr Newton made it so & is of no use.[68] When he sent this method he desired Mr Newton['s] method of deriving reciprocal[69] series from one another (p. 63) & at the same time endeavoured to take two of Mr Newtons reciprocal series from him tho he wanted the method of finding them (p. 61, 62). And when Mr Newton sent him the method, tho he understood it with difficulty, yet he wrote back that he had found it before as he perceived by his old papers, but not meeting with[70] an elegant example of its use had neglected [it] (p. 63, 96). And when he had published the above mentioned series of Gregory in the *Acta Lipsiensia* as his own (p. 97), he had not only forgot that he had received that series from Mr Oldenburg & Mr Collins but also that the Collection of Gregories Letters had been sent him at his own request, in one of wch dated 15 Feb. 167$\frac{9}{4}$ Gregory had then sent that series to Collins (p. 25, 47). He forgot also that the series for Quadratures wch breake off in some cases & become finite, were deduced from the method of fluents before the writing of the Letters wch passed between them in the year 1676 & even before the writing of the *Analysis* (p. 18, 72, 109). & that the method of assuming the terms of a series & determining them by the conditions of the Problem was a part of Mr Newtons general method before the writing of those Letters (p. 86. *Wallisij Operū* Vol 2. p. 393[71]). In the year 1689 Mr Leibnitz began to pretend to this method (*Act. Leips.* An. 1689 p. 37[72] & An. 1693 p. 178[73]) & has been since admonished that it is

---

(67) Unable at once to seize on the *mot juste*, Newton first wrote 'eminent' and then 'valuable'.

(68) Newton means practical utility, 'there being', he initially went on, 'not one series to be found by it wch cannot be more readily found without it'.

(69) That is, inverse.

(70) Initially 'but for want of'.

(71) Here (see VII: 176) was first printed the enodation of the Latin anagram in which Newton had in his *epistola posterior* for Leibniz in October 1676 concealed his twin methods of resolving the general 'inversum de tangentibus Problema'.

(72) Among other *obiter dicta* in preliminary to this brief overview by him 'De lineis opticis' in a letter to Mencke which its recipient at once printed (in its greatest part) in *Acta Eruditorum* (January 1689): 36–8 Leibniz here touched in passing on his yet unpublished new general 'methodus serierum promovenda' whereby 'Assumo...seriem arbitrariam, eamque ex legibus problematis tractando, obtineo ejus coefficientes', innocently unaware (we see no reason to doubt his later word, tempting though it was for Newton to do so) that this is precisely the 'altera methodus [consistens] in assumptione seriei pro quantitate qualibet incognita' whose brief enunciation had been encoded in anagram in Newton's letter of 24 October 1676 to Oldenburg for him (see the previous note).

(73) We gave some bare account in VII: 182, note (26) of this article in *Acta Eruditorum* (April 1693): 178–80 where Leibniz presented and briefly instanced his 'nova Methodus generalissima' of resolving problems (reducible to differential equations, it is implied) 'per series infinitas', in continuing ignorance of Newton's prior discovery of it.

Mr Newtons method (*Wallisij Opera* Vol. 2 p. 393[74] & Vol 3 p. 645[75] & Cheynes *Fluxionum methodus inversa* p. 46[76]), but forgets to acknowledge the true author. A year or two after Mr Newton had published his *Principia Philosophiæ* Mr Leibnitz [printed] his schediasme *de Resistentia Medij*[77] conteining Mr Newtons Propositions on that subject put into another dress, & tells his Reader that he found them for the most part at Paris twelve years before & communicated some of them to the Royal Academy. And in the end of his discourse he adds: *Nobis nunc fundamenta Geometrica jecisse suffecerit in quibus maxima consistebat difficultas. Et fortassis attente consideranti vias quasdam novas vel certe satis antea impeditas aperuisse videbimur. Omnia autem respondent nostræ Analysi infinitorum, hoc est ca[l]culo summarum & differentiarum cujus elementa quædam in his Actis dedimus.* But he [78]forgot to tell us which of the Propositions he communicated to the Academy at Paris & to acknowledge that he left Paris before he understood the differential method. And Mathematicians say that in writing his *Tentamen de motuum cælestiu[m] causis*[79] he forgot to reexamin the process of finding his 19th Proposition, wch is

---

(74) Again see VII: 176. Wallis of course, in his *Opera Mathematica*, **2** (Oxford, 1693): 393–6, gives main attention to Newton's 'methodus prior' of extracting a fluent quantity 'ex æquatione simul involvente fluxionem ejus', and nowhere admonishes Leibniz for being but second discoverer of the 'methodus altera' as Newton here implies by over-energetic wishful thinking.

(75) This, in Wallis' *editio princeps* of the full text of Newton's second letter to Leibniz in his 1699 'Epistolarum Collectio' (on which see note (29) above), is its last page where the 1676 anagram is set out, now with a footnote in which its unravelled equivalent Latin sentences are adjoined.

(76) More accurately, pages 46–50 of this 1703 work of George Cheyne's expound Newton's 'methodus altera', as we earlier observed in VII: 18, note (73).

(77) In *Acta Eruditorum* (January 1689): 38–47, namely, in immediate sequel to his 'De lineis opticis, et alia...' (on which see note (72) preceding). We made comparison of the mathematical portion of this 'Schediasma de resistentia Medii, & Motu projectorum gravium in medio resistente' with Newton's treatment of the simplest cases of resisted projectile motion in his 1684 'De motu Corporum' (and in the equivalent opening propositions of the second book of the *Principia*) in VI: 70–1, note (109), there stressing the relative deficiency of Leibniz' understanding of how to compound horizontal and vertical components of motion resisted according as the square of the instantaneous speed. Notice, however, that Newton's following phrase 'put into another dress' leaves his reader to conclude that Leibniz must have had direct knowledge of the *Principia*'s pertinent propositions before composing his 'Schediasma': a surmise which, however plausible and 'evidently' true it may have seemed to Newton, is supported by no independent evidence, and which is rejected by all recent students of the matter. (See, notably, E. J. Aiton's study of 'Leibniz on Motion in a Resisting Medium', *Archive for History of Exact Sciences*, **9**, 1972: 257–74, especially 257–9.)

(78) Newton first went on to state 'had now forgot that when he wrote his Letter of 27 August 1676, wch was but a month or six weeks before he left Paris, he was of opinion that inverse Problems of tangents could not be reduced to equations or quadratures. For this is a demonstration that he had not then found the differential method'.

(79) On this paper printed in *Acta Eruditorum* (February 1689): 82–96, the third of the trio (on the others see notes (72) and (77) above, and note (95) to our preceding introduction) where Leibniz made claim to independent 'invention' of three of the principal groups of

the chief of Mr Newton's Propositions relating to the motions of the Planets, & that no man could find that Proposition by such an erroneous process as he has set down. They blame him not for committing some errors in his first essay[80] but for adapting a calculation to another man's Proposition with a designe to make himself the inventor, tho by the errors it appears that he did not invent it. And when he published an Account of Mr Newton's book of Quadratures & represented that the method of fluxions was from the first beginning thereof substituted & used by Mr Newton instead of the Differential method (p. 108)[,] he had forgot that the very first Proposition of the Book was set down in Mr Newtons Letter of Octob. 24 1676 as the foundation of the method of fluxions upon wch Mr Newton had written a treatise in the year 1671. And tho Mr Keil put him in mind of this & explained to him that the second Proposition of the book was in the *Analysis* (p. 19) & the eight following Propositions (which are all founded upon the method of fluents) were found before the writing of the said Letter (p. 113) & by consequence before Mr Leibnitz understood the differential

---

propositions in Books 1 and 2 of the *Principia* more than a year after this was published (and the review of which in the *Acta* in June 1688 he cheerfully admitted to have stimulated him to gather together his own duplicated notions), see note (86) of our introduction. We there cite Newton's original correction of an error in §15 (at 'p. 89 l. 20') of the 'Tentamen', and also his checking computation that Leibniz' calculation, in the ensuing §19 (pages 91–2), of the radial acceleration along a focal vector of the Keplerian planetary ellipse is indeed

$$d^2r/dt^2 = c^2/r^3 - k/r^2,$$

where r is the vector's length and t the time of passage. (His complaint against Leibniz is that he appeals to §15 to distinguish the two components in this total acceleration as respectively the inverse-cube *conatus centrifugus circulationis* and the wanted contrary, inverse-square *gravitatis solicitatio* acting upon the orbiting planet.) In an editorial interpolation on page 97 in the main text of the *Commercium Epistolicum* he had already waxed sarcastic on the topic of the 1688 'Epitome' of his *Principia* 'in *Actis Lipsicis* impressa...: qua lecta D. *Leibnitius* Epistolam de lineis Opticis, Schediasma de resistentia Medii & motu Projectilium gravium in Medio resistente, & Tentamen de Motuum Cœlestium causis composuit, & in *Actis Lipsicis* ineunte anno 1689 imprimi curavit, quasi ipse quoque præcipuas *Newtoni* de his rebus Propositiones invenisset, idque diversa methodo qua vias novas Geometricas aperuisset; & librum *Newtoni* tamen nondum vidisset'. In the considerably truncated final version (see [6] following) of the present 'Ad Lectorem', we may add, no mention of Leibniz' three '*Principia*' papers in the 1689 *Acta* is made, but to compensate he squeezed in a late footnote to his mild remark about them on page 97 of the *Commercium*—his surviving drafts of it on Add. 3968. 41 : 148^r/148^v→. 12 : (176^r→) 176^v show the care he took in framing its words—which states a deal more angrily, much as here: 'Hac licentia concessa authores quilibet inventis suis facile privari possunt. Viderat *Leibnitius* Epitomen Libri in *Actis Lipsicis*. Per commercium Epistolicum, quod cum Viris doctis passim habebat, cognoscere potuit Propositiones in libro illo contentas. Si librum non vidisset, videre tamen debuisset antequam suas de iisdem rebus in itinere [*viz.* in Italia] scriptas compositiones publicaret. Dicunt aliqui[!] falsas esse *Tentaminis* Propositiones 11, 12, & 15, & D. *Leibnitium* ab his per calculum suum deduxisse Propositiones 19 & 20 ejusdem *Tentaminis*. Talis autem calculus ad Propositiones priùs inventas aptari quidem potuit, non autem inventorem constituere.'

(80) That is, at his first trial attempt.

method: yet Mr Leibnitz continues to justify the Account that he has given of the book of *Quadratures*[,] representing that it has taken nothing from any man but given every man his own (p. 119).

As for those Gentlemen who have used the differential method, & particularly the Marquess de l'Hospital, Monsr Varignon, & the brothers Jacobus & Joannes Bernoullius, there is nothing in these Letters & Papers which can affect them. They were strangers to the correspondence between Mr Leibnitz & Mr Oldenburg, it was before their time[:] Mr Leibnitz handed the infinitesimal method to them, they found it very usefull & they are much to be commended for the use & improvements that they have made of it.

---

[6]$^{(81)}$ The occasion of publishing this Collection of Letters & Papers will be understood by the Letters of Mr Leibnitz & Mr Keill in the end thereof. Mr Leibnitz taking offence at a passage in a discourse of Mr Keill published in the *Transactions* A.C. 1708, wrote a Letter to y^e Secretary of the R. Society complaining thereof as a calumny, desired a remedy from the Society, & suggested that he beleived they would judge it equal that he should make a publick acknowledgement of his fault. Mr Keil chose rather to return an answer in writing, wherein he explained his meaning in that passage & defended it. Mr Leibnitz not meeting with that satisfaction he desired wrote a second Letter to the Society, wherein he still complained of Mr Keill, representing him a young man & not acquainted with things done before his time nor authorized by the person concerned, & appealed to the equity of the Society to cheq his unjust clamours.

Mr Leibnitz was in England in the beginning of the year 1673 & again in October 1676, & during ye intervall in France, & all that time kept a correspondence wth Mr Oldenburg, & by his means wth Mr John Collins & sometimes with M^r Is. Newton, & what he might learn from the English either in London or by that correspondence is the main question. Mr Oldenburg & Mr Collins are long since dead, & Mr Newton lived then at Cambridge & knew little more then his own correspondence since published by Dr Wallis. Mr Newton can be no witness for Mr Keill nor Mr Leibnitz for himself, & there appears no other

---

(81) Add. 3968.37: 549^r/549^v; a preliminary draft (not quite as complete) is found at *ibid.*: 545^r/545^v. This most finished English version of Newton's preface to the *Commercium Epistolicum* was rendered word for word—by him?—to be the Latin 'Ad Lectorem' sent to the printer in late autumn 1712, as the original proof-copy of the latter now Add. 3968.2: 7 (the second leaf of the sheet A whose first recto, f. 6^r, bears the title of the *Commercium* as we cited it in note (47) above) confirms. In a late insertion in this, however, Newton introduced the postil which we quote in the next note.

living evidence. The R. Society therefore being twice pressed by Mr Leibnitz against Mr Keill, appointed a Committee to search out & examin the Letters Letter-books & papers left by Mr Oldenburg in the hands of the Society & those found among the papers of Mr John Collins relating to the matters in dispute & report their opinion thereupon[,] & ordered the Report of the Committee with the extracts of the Letters & Papers to be published.

When Mr Newton wrote the *Analysis* printed in the beginning of this Collection, he had a method of resolving finite equations into infinite ones, & of applying both finite & infinite equations to the solution of Problemes by means of the proportions of the *augmenta momentanea* of growing or increasing quantities.[82] These *augmenta* Mr Newton calls particles & moments, & Mr Leibnitz infinitesimals indivisibles & differences. The increasing quantities Mr Newton calls fluents, & Mr Leibnitz summs, & the velocities of increase Mr Newton calls fluxions & exposes these fluxions by the moments of the flowing quantities.[83] That part of the method wch consists in resolving finite equations into infinite ones, was at the request of Mr Leibnitz communicated to him by Mr Newton in his Letters of June 13th & Octob^r 24th 1676. And Mr Newton having so far touched upon the other part as to reckon that it was become ^{a}sufficiently obvious, to secure it from being taken from him before he should have occasion to explain it, he expressed it in cyphre after the manner

^{a}pag. 72 lin.1.[84]

---

(82) At this point in the corresponding Latin 'Ad Lectorem' Newton subsequently introduced into its printer's proof a note alongside (his draft of it survives on Add. 3968.41: 137^v) referring for support of this statement to passages in the ensuing texts of his 1669 'De Analysi', in his letter to Collins of 10 December 1672 transmitting his general rule for algebraic tangents, and in his *epistolæ prior et posterior* of June/October 1676 to Leibniz. This reads: 'De hac Methodo ex Methodis Serierum & Fluentium [*sic*: 'fluxionum' was clearly meant, and so indeed Newton put it in his draft, but the slip stands uncorrected in the 1722 reprint] composita scripsit infra'—in the draft 'loquitur in sequentibus'—'*Newtonus*, pag. 14, 15, 18, 30, 55, 56, 71, 85, 86'.

(83) 'istasꝗ Fluxiones exponit per quantitatum fluentium momenta' the equivalent published Latin text here reads. In his proof-copy of the 'Ad Lectorem' (on this see note (81) preceding) Newton briefly thought to replace 'exponit' by 'determinat per analoga' (determines by analogy), and then he went on to adjoin a long passage which likewise never appeared in print: 'Fluentes & fluxiones is designat per symbola quæcunꝗ[:] tempus per quantitatem quamvis uniformiter fluentem, fluxionem temporis per unitatem, momentum temporis per *o*, momenta aliarum quantitatum per symbola fluxionum ducta in momentum temporis. Ubi aliquid investigandum est, momentum *o* est infinite parvum & compendij causa subintelligi solet. Ubi aliquid demonstrandum est, momentum *o* est particula temporis indefinite parva sed finita tamen, et semper exprimitur. Priore casu computus per approximationes pergere potest, posteriore computus fit in finitis quantitatibus per vulgarem Geometriam quam accuratissime. Hac methodo Newtonus olim usus est et usꝗ nunc utitur et per hanc methodum æquationes tam infinitas quam finitas ad [Problemata quæ]libet applicat.'

(84) 'Fundamentum harum Operationum, satis obvium quidem...' was the phrase here referred to by Newton at the pertinent place in his 1676 *epistola posterior* (=*Correspondence of Newton*, 2, 1960: 115, l. 11).

used by Galilæo and Hugenius upon other like occasions. Mr Leibnitz claims the invention of this other part, & Mr Keill asserts it to Mr Newton & is favoured in his opinion by the Report of the Committee. But there is nothing in these Papers wch can affect other persons abroad who have received the method from Mr Leibnitz. They were strangers to the correspondence between Mr Leibnitz & Mr Oldenburg. They found the method usefull & are much to be commended for the use & improvements that they have made of it.

Some Notes are added to ye Letters to enable such Readers as want leasure, to compare them with more ease & see the sense of them at one reading.

APPENDIX 2. EXTRACTS FROM NEWTON'S ANONYMOUS REVIEW OF THE 'COMMERCIUM' (LATE SUMMER/AUTUMN? 1714).[1]

From the text printed in the *Philosophical Transactions*,[2] adapted to the autograph drafts in the University Library, Cambridge[3] and in private possession

(1) On pages 493–501 above we noticed how, in rejoinder to the 'defamatory Libell' of Leibniz' anonymous *Charta volans* of July 1713 (and not least to Bernoulli's 'Judicium' therein inserted of his purported deficiencies in advanced mathematics), Newton began in the autumn of 1713 to prepare a set of elaborate annotations upon the *Commercium Epistolicum*, marshalling—and bringing out where these were implicit—plainly for all to see the massive witness of its numerous 'proofs' of his own absolute priority in discovering and developing his 'universal' method of analysis through 'converging' infinite series and the vanishingly small increments of 'contemporaneously' variable 'fluent' quantities; and we also went on to outline how over the next year these comments upon his published Collinsian 'Epistolary Commerce' grew and branched out to be separate historical 'Accounts' both setting forth his own early researches and studies in his joint method of series and fluxions 'until the year 1676' and also reconstructing (honestly, we believe, as best he could from the little evidence available, though certainly without generosity) how Leibniz 'from the year 1677' onwards began to develop the variant of his own fluxionary analysis which he published in the Leipzig *Acta* in 1684 as his *calculus differentialis*. Some time in the middle months of the next year—we can give no finer dating, though a loose *terminus ante quem non* is yielded by the fact (see note (103) of the preceding introduction) that the preliminary drafting on Add. 3968.41:79^r of what was eventually to be page 205 of the printed text is there penned below a roughing-out of his letter of 13 April 1714 to Popple—Newton combined these two complementary historical-*cum*-analytical reports to form (see note (102) of the introduction) the core of his present 'author's' review of the *Commercium*. Augmented with multifold further caustic imprecations to Leibniz 'honestly' to acknowledge and accept his position as 'second Inventor' of the calculus (without 'Right' to its discovery) and topped up with three final pages stressing that 'The Philosophy wch Mr Newton in his *Principles* & *Opticks* has pursued is experimental, &...Hypotheses have no place [in it] unless as conjectures or Questions proposed to be examined by experiments' (we cite these sentences of page 222 from their draft on Add. 3968.39: 585^r), Newton's finished script was despatched to the printer in late 1714 and made its public bow early the next year in the *Philosophical Transactions* without indication (as was then indeed normal practice) as to who its un-named author was.

One might easily fill a volume in 'fuss'noting this gospel of the birth of mathematical analysis according to Newton if one made earnest effort to chronicle the full extent of his ignorance of past event and circumstance, let alone tried in adequate detail to pinpoint, assess and correct his numerous distortions of historical vision, myopic and astigmatic, wilfully blinkered or less consciously skew-eyed as these variously were. Tempted though we are to take him to task for his lack of generosity in assigning the guiding motives which underlay Leibniz' claims and for want of insight into his own equally jealous ones, to swamp Newton's text with an extravagant deluge of corrections and qualifications would go counter to our main aim here of letting him speak for himself, sagely and informedly or ignorantly and foolishly as may be. Our running editorial commentary will therefore be sparse other than in recording verbal variants and unused prior phrasings and in giving pertinent references to primary documents and related secondary literature.

(2) Under the amplified heading '*An Account of the Book entituled* Commercium Epistolicum Collinii & aliorum, De Analysi promota; *published by order of the* Royal-Society, *in relation to the*

An Account of the Book entituled

Commercium Epistolicum D. Johannis Collinij & aliorum De Analysi promota.

‖[173] ‖ Several Accounts[4] having been published abroad of this *Commercium*, all of them very imperfect, it has been thought fit to publish the Account which follows.

---

dispute between Mr. Leibnitz *and Dr.* Keill, *about the Right of Invention of the Method of* Fluxions, *by some call'd* the Differential Method' (*Philosophical Transactions*, **29**, No. 342 [for January/ February 1714/15]: 173–224). The shorter title which we set on our following text of the review is that found on the most finished draft (now Add. 3968.8: 85[r]) of its opening pages. Newton's initial head (*ibid.*: 67[r]) was the pithy 'An Account of the *Commercium Epistolicum*', while intervening versions (*ibid.*: 81[r]→91[r]→94[r]→93[r]) severally interpolate 'Book (entituled)' and adjoin '*& aliorum De Analysi promota*'.

(3) In order, namely, to re-establish a near approximation to the orthography and punctuation used by Newton in the script which he put to the printer (this is now lost), and which the latter very evidently mangled and coarsened in setting the published 'Account' in type in accord both with house-style and (it would appear) personal idiosyncrasy. (Why else convert Newton's citation of his '*Opticks*'—see note (1) above—to be '*Optiques*' on page 222 of the printed review?) To that end we ransack Add. 3968 for Newton's original phrasings, cannibalising [in sequence] .8: 85[r] (+ .41: 99[v])–86[r]; .41: 118[r]/118[v]/100[r]/99[v]/34[r]/98[r]; .39: 581[r]– 582[r]; .41: 143[r]/143[v]/100[v]; .16: 255[r]; .13: 177[r]–178[r]; and .41: 41[v]/41[r], which (together with two related sheets now privately owned) furnish an all-but-identical autograph twin, word by word and paragraph by paragraph, to the published text. On .8: 85[r]/85[v], we may notice, Newton initially inserted marginal postils directing the reader to the sources for his factual statements—except for one instance (see the next note) where he made oblique reference to Derham's lately appeared *Physico-Theology* his citations are, as we would expect, all of pages in the parent 'Commerc. Epist.'—but he soon discontinued the practice (so permitting himself, we need not say, the greater liberties possible from not having everywhere to control his argument by the tight discipline of documenting its finer points).

(4) Principally, of course, Newton has in mind Leibniz' *Charta volans* of July 1713 and his equally anonymous 'Remarques sur le different entre M. de *Leibnitz*, & M. *Newton*' the following autumn in *Journal Literaire*, **2**.2: 445–8 (on which see page 493 of our preceding introduction, and note (83) thereto). Indeed, even as he began in the spring of 1714 to put together his present essay-review of the *Commercium* he was also engaged in helping Keill prepare the elaborate 'Réponse' to these which was published in the *Journal*, **4**.2: 319–58, and whose text the latter had previously put to him on 2 May. (See *Correspondence*, **6**: 113; we gave instance of the care which Newton took in rounding out Keill's 'Answer' in note (86) of the preceding introduction, and we may more broadly cite his letters to Keill of 20 April and 11/15 May 1714, first published by Edleston in appendix to his edition of the *Corrrespondence of Sir Isaac Newton and Professor Cotes* (London, 1850): 170–7 [=*Correspondence*, **6**: 117–19/128– 30/136–7].) Unexpectedly closer to home, however, a stray marginal citation by Newton in his draft on Add. 3968.8: 85[r] (see note (15) below) originally directed the reader in addition to 'Derham p. 308'. In footnote (13) at this point in Book V, Chapter 1 'Of the Soul of Man' in the first edition (London, 1713) of his Boyle Lectures of the year before on *Physico-Theology: Or, A Demonstration of the Being and Attributes of God, from his Works of Creation* William Derham

This *Commercium* is composed of several ancient Letters & Papers put together in order of time & either copied or translated into Latin from such Originals as are described in the title of every Letter & Paper; a numerous Committee of the R. Society being appointed to examin the sincerity of the Originals & compare therewith the copies taken from them. It relates to a general method of resolving finite æquations into infinite ones & applying these æquations both finite & infinite to the solution of Problems by the method of fluxions & moments. We will first give an account of that part of the method wch consists in resolving finite equations into infinite ones & squaring curvilinear figures thereby. By infinite Equations are ment[5] such as involve a series of terms converging or approaching the truth nearer & nearer *in infinitum* so as at length to differ from the truth less then by any given quantity[6] & if continued in infinitum, to leave no difference.

‖ D[r] Wallis in his *Opus Arithmeticum*[7] published A.C. 1657, cap. 33 Prop. 68, ‖[174]

---

discoursed at some length—inspired thereto (see note (15) below) by a 1672 letter of Collins to Richard Towneley which he had unearthed among the latter's papers 'in my Hands'—upon the prime instance of 'Man's Invention' furnished by the achievements of England's 'Mathematical Wits' since 'the middle of the last Century; when in the Year 1657 those very ingenious and great Men, Mr. *William Neile* and my Lord *Brounker*, and Sir *Christopher Wren* afterwards in the same Year, geometrically demonstrated the Equality of some Curves to a strait Line'. (On these rectifications by Neil—and in improved form by Brouncker—of the semicubical parabola and by Wren of the cycloid, all first published by John Wallis in his *Tractatus Duo. De Cycloide, et...De Cissoide...* (Oxford, 1658): 91–4 and 72–6 respectively, see our 'Patterns of Mathematical Thought in the later Seventeenth Century' (*Archive for History of Exact Sciences*, **1**, 1961: 179–388): 328–30 and 333–5.) 'Soon after which', Derham went on, 'others at home and abroad did the like in other Curves. And not long afterwards this was brought under an *Analytical Calculus*: the first Specimen whereof, that was ever publish'd, Mr. *Mercator* gave in 1668, in a Demonstration of my Lord *Brounker's* Quadrature of the *Hyperbola* by Dr. *Wallis's* Reduction of a Fraction into an Infinite Series by Division. But the penetrating Genius of Sir *Isaac Newton* had discovered a Way of attaining the Quantity of all quadrible Curves analytically by his Method of *Fluxions*, some time before...'. In his much fuller present historical sketch of recent developments in mathematical analysis Newton deals swiftly in his opening paragraphs with these precursors in the 'Calculus' of infinite series.

(5) 'meant'.

(6) The draft on .8: 85[v] adds 'w[ch] can be assigned'.

(7) In its fuller title *Mathesis Universalis: sive, Arithmeticum Opus Integrum, tum Philologice, tum Mathematice traditum....* This second work in Wallis' *Operum Mathematicorum Pars Prima* (Oxford, 1657) is reprinted in his collected *Opera Mathematica*, **1** (Oxford, 1695), 11–228; Cap. xxxiii. 'Progressio Geometrica fusius traditur' is there to be found on pages 169–80. Wallis was, we may add, by no means the first to turn the reciprocal of a general binomial into a 'converging' infinite geometrical series in this way. Archimedean precedents apart, Grégoire de Saint-Vincent had in Book 2 'De Progressionibus Geometricis' (pages 51–177) of his *Opus Geometricum Quadraturæ Circuli et Sectionum Coni decem libris comprehensum* (Antwerp, 1647) given an exhaustive discussion of the limit-summation of such a series. (Compare our 'Patterns of Mathematical Thought...': 254.)

reduced the fraction $\dfrac{A}{1-R}$ by perpetual division into the series

$$A + AR + AR^2 + AR^3 + AR^4 + \&c.$$

Vicount Brounker squared the Hyperbola by this series

$$\frac{1}{1 \times 2} + \frac{1}{3 \times 4} + \frac{1}{5 \times 6} + \frac{1}{7 \times 8} + \&c:$$

that is by this $1 - \frac{1}{2} + \frac{1}{3} - \frac{1}{4} + \frac{1}{5} - \frac{1}{6} + \frac{1}{7} - \frac{1}{8} + \&c$, conjoyning every two terms into one. And the Quadrature was published in the *Phil. Transactions* for April 1668.[8]

Mr Mercator soon after published a Demonstration of this Quadrature by the division of Dr Wallis, & soon after that Mr James Gregory published a Geometrical Demonstration thereof. And these books[9] were a few months after sent by Mr John Collins to Dr Barrow at Cambridge & by Dr Barrow communicated to Mr Newton[10] in June 1669. Whereupon Dr Barrow mutually sent to Mr Collins a Tract of Mr Newtons entituled *Analysis per æquationes numero terminorum infinitas*. And this is the first piece published in the *Commercium* & conteins a general method of doing in all figures what my Lord Brounker & Mr Mercator did in the Hyperbola alone. Mr Mercator lived above ten[11] years

---

(8) 'The Squaring of the Hyperbola by an Infinite Series of Rational Numbers...' (*Phil. Trans. 3*, No. 34 [for 13 April 1668]: 645–9). Brouncker's series is, of course, that for log 2—and had earlier been derived in greatly different manner by Pietro Mengoli in his *Geometria Speciosa* (Bologna, 1659)—but his manner of generating it by repeated dissection of the area under a rectangular hyperbola may readily, as he himself hints, be adapted to produce *mutatis mutandis* a poorly convergent general series for $\log(1+x)$. (See our 'Patterns of Mathematical Thought...': 222–5; and J. E. Hofmann, *Leibniz in Paris, 1672–1676* (Cambridge, 1974): 96–7.)

(9) Namely the *Logarithmo-Technia: sive Methodus construendi Logarithmos nova, accurata, & facilis* (London, 1668) in which (see II: 166) Nicolaus Mercator first published 'his' infinite series for $\log(1+a)$ yielding—for $a \leqslant 1$ by implication—the 'Vera Quadratura' of the rectangular hyperbola $e = 1/(1+a)$; and the *Geometriæ Pars Universalis, inserviens Quantitatum Curvarum transmutationi & mensuræ* (Padua, 1668) in whose Prop. 7 (pages 20–2) James Gregory instanced his Fermatian algorithm for resolving the 'Problema. Rectam ducere datam curvam tangentem in eius puncto dato, si modo curva sit ex earum numero, quas Cartesius appellat Geometricas' in the case of the cubic of ordinate proportional to $\sqrt[3]{[(a+b)\,b^2]}$ by supposing—independently of Newton (see I: 557, note (21))—that its abscissa b receives an instantaneous decrement of magnitude 'nihil seu serum o' (page 20). Newton's unmarked library copies of both these works are now bound together in Trinity College, Cambridge. NQ.9.48.

(10) An irrelevant, blatant reminder 'now Sir *Isaac Newton*' was inserted in a following parenthesis at this point in the printed version—not at Newton's own behest, we would like to think.

(11) '10 or 11' in initial draft on Add. 3968.41: 118^r; but, since Mercator did not die till early January 1687, Newton's later correction of this to be 'fifteen'—afterwards increased by yet a further year to 'sexdecim' in the Latinised 'Recensio' ([.41: 134^v→] *Commercium*, $_2$1722: 3])—is nearer the mark. This unpublished amendment is entered as the first of twenty

longer without proceeding further then to yᵉ single Quadrature of the Hyperbola. The progress made by Mʳ Newton shews that he wanted not Mʳ Mercators assistance. However, for avoyding disputes, he supposes that my Lᵈ Br. invented & Mʳ Mercator demonstrated the series for the Hyperbola some[12] years before they published it, & by consequence before he found his general method.

The aforesaid Treatise of *Analysis* Mʳ Newton in his Letter to Mʳ Oldenburg dated 24 Octob. 1676 mentions in the following manner. *Eo ipso tempore quo Mercatoris Logarithmotechnia prodijt, communicatum est per amicum D. Barrow (tunc Matheseos Professorem Cantab.) cum D. Collinio Compendium quoddam harum serierum, in quo significaveram Areas & Longitudines Curvarum omnium & solidorum superficies & contenta ex ‖ datis rectis; et vice versa ex his datis Rectas determinari posse: & methodum* ‖[175] *indicatam illustraveram diversis seriebus.*[13] Mʳ Collins in yᵉ years 1669, 1670, 1671 & 1672 gave notice of this Compendium to Mʳ James Gregory in Scotland,[14] Mʳ Bertet & Mʳ Vernon then at Paris, Mʳ Alphonsus Borelli in Italy, Mʳ Strode Mʳ Townsend[15] Mʳ Oldenburg Mʳ Dary & others in England[16] as appears by

such subsequent 'Corrigenda in *Philosoph. Transact.* pro Jan. et Feb. 171⅔' listed in tidy page-order on the loose leaf now .13: 223ʳ: we will cite the remainder at pertinent places in following footnotes.

(12) Originally 'three or four' in the draft on .41: 118ʳ.

(13) *Commercium*, ₁1713: 70 [*Correspondence*, **2**, 1960: 114].

(14) In the *Commercium* Newton had (pages 22 and 24 respectively) presented Latinised extracts only from Collins' letter of 25 November 1669 where he wrote that 'Mʳ Barrow hath resigned his Lecturers place to one Mʳ Newton of Cambridge..., one who (before Mercators *Logarithmotechnia* was extant) invented the same method and applyed it generally to all Curves, and diverse wayes to the Circle...' (see (ed.) H. W. Turnbull, *Gregory Tercentenary Volume* (London, 1939): 74), and from that more than a year later where, on 24 December 1670, he cited Newton's series for the sine, cosine and inverse sine and also his series quadrature of the 'ancient Quadratrix' (see *ibid.*: 154–5).

(15) So also it is put in the Latin 'Recensio' (*Commercium*, ₂1722: 3) but read 'Townley'. A marginal note 'See Derham p. 308' in Newton's draft on Add. 3968.8: 85ʳ alongside his ensuing quotation of the Latinised extract from Collins' letter to Strode of 26 July 1672 (*vide* note (22) following) which he had printed in the *Commercium* is our *verb. sap.* that it was William Derham who, so he wrote to Newton on 20 February 1712/13, while 'perusing the *Commerc. Epist.* wᶜʰ yᵉ R. S. honoured me with' remembered that 'in some of Mʳ Collins's Lʳˢ to Mr [Richard] Townley of Lanc[ashire] (now in my hands) there was something relating to that subject, & looking over Mʳ Towneleys papers I found a long Lʳ of Mʳ Collins's giving a sort of Historical account of the matter' (*Correspondence*, **5**, 1975: 379); and that Derham took quick opportunity to print the gist of the letter which he had dug up in his store of Towneley's papers, setting this virtually as he had earlier reported it privately to Newton in a footnote to the text of his recent Boyle Lectures, then in press (*Physico-Theology* (London, ₁1713): 308, note (13); we cited the preceding portion of this somewhat irrelevant *addendum* in note (4) above). The phrases Derham cites from this now seemingly lost letter of Collins—'In September 1668 Mʳ Mercator published his *Logarithmotechnia*, one of wᶜʰ he sent to Dr Wallis... & another to Dʳ Barrow, who thereupon sent him up some papers of Mʳ Newton (now his successor) by wᶜʰ, & some former Communications made thereof by the Author to yᵉ Dʳ it appears yᵗ yᵉ method was invented, some years before by yᵉ said Mr Newton, & generally applied...'—do

his Letters. And Mr Oldenburg in a letter dated 14 Sept. 1669 & entred in the Letter book of the R. Society gave notice of it to Mr Francis Slusius at Liege,[17] & cited several sentences out of it. And particularly Mr Collins in a Letter to Mr James Gregory dated 25 Novem. 1669, spake thus of the Method conteined in it.[18] *Barrovius provinciam suam publice prælegendi remisit cuidam nomine Newtono Cantabrigiensi, cujus tanquam viri acutissimo ingenio præditi, in Præfatione Prælectionum Opticarum, meminit: quippe antequam ederetur Mercatoris Logarithmotechnia, eandem methodum adinvenerat, eamꝗ ad omnes curvas generaliter, et ad Circulum diversimode applicarat.* And in a Letter to Mr David Gregory dated 11 August 1676 he mentions it in this manner.[19] *Paucos post menses quam editi sunt hi libri* (vizt Mercatoris *Logarithmotechnia & Exercitationes Geometricæ* Gregorii) *missi sunt ad Barrovium Cantabrigiæ. Ille autem responsum dedit hanc infinitarum serierum doctrinam a Newtono biennium ante excogitatam fuisse quam ederetur Mercatoris Logarithmotechnia & generaliter omnibus figuris applicatam, simulꝗ transmisit D. Newtoni opus manuscriptum.* The last[20] of the said two books came out towards the end of ye year

---

not greatly differ from those of Collins' letter to Strode in July 1672 (see note (22) below), and we presume that it was sent around the same date.

(16) Latinised excerpts from Collins' letters to Bertet of 21 February 1670/1, to Vernon on 26 December 1671 and to Alphonse Borelli around the same time are printed in the *Commercium*: 26–7, 27–8 and 27 respectively; we gave the original English of these extracts in III: 21–3 (from Collins' drafts in Royal Society MS LXXXI). Collins' letter to Oldenburg early in September 1669 giving account of the 'universall Analytical method' newly passed on by Barrow, to whom it had been imparted by 'Mr Isaac Newton his Collegiate', is now published in (ed.) A. R. and M. B. Hall, *The Correspondence of Henry Oldenburg*, 6 (Madison, Wisconsin, 1969): 226–9, especially 227–8; Oldenburg merely rendered Collins' letter word for word into Latin before passing its content on to Sluse on 14 September (see the next note).

(17) The pertinent excerpt from Oldenburg's letter to Sluse—a straight Latin rendering by him of the equivalent passage in Collins' English letter to him of a few days before (on which see the previous note)—was printed by Newton in the *Commercium*: 21. For the full text of the letter see *The Correspondence of Henry Oldenburg*, 6: 232–5.

(18) See note (14) above.

(19) Newton quotes the following passage from the Latinised version on *Commercium*: 48 of what is in fact a draft preface to his 'Historiola' of James Gregory's letters to him over the previous eight years which is penned—with several third-person references to 'Collins'—on the second leaf of a double sheet on whose first folio he roughed-out his letter to James' brother David in August 1676 (a Latinised extract from whose first paragraph appears on *Commercium*: 47; for the parent English see *Gregory Tercentenary Volume* (note (14)): 344–5). In Collins' original words (Royal Society MS LXXXI, No. 47: 2^r): 'The said Bookes being some few Months after they were published sent to Mr Barrow at Cambridge, in returne he gave answer that the said doctrine of infinite Series was invented by Mr Isaac Newton above 2 yeares before and generally applyed to all figures. With which Answer a Manuscript of Mr Newtons paines was transmitted to Collins and communicated to the Lord Brouncker President of the Royall Society.' We have earlier (III: xv, note (15)) cited his ensuing sentence 'And afterwards when Mr Barrow left his place as Mathematick Professor he recommended Mr Newton to it, who read Lectures on the said Doctrine [of infinite Series] which were put into the Publique[!] Library at Cambridge'.

(20) That is, Gregory's *Exercitationes Geometricæ*.

1668, & Dʳ Barrow sent the said Compendium to Mʳ Collins in July following as appears by three of Dʳ Barrows Letters.⁽²¹⁾ And in a Letter to Mʳ Strode dated 26 July 1672 Mʳ Collins wrote thus of it.⁽²²⁾ *Exemplar ejus (Logarithmotechniæ) misi Barrovio Cantabrigiam qui quasdam Newtoni chartas extemplo remisit: E quibus et alijs, quæ olim ab Authore cum Barrovio communicata fuerant, patet illam methodum a dicto Newtono aliquot annis antea excogitatam et modo univer‖sali applicatam fuisse: ita ut ejus* ‖[176] *ope in quavis Figura Curvilinea proposita, quæ una vel pluribus proprietatibus definitur Quadratura vel Area dictæ figuræ, accurata si possibile sit [sin] minus infinite vero propinqua, Evolutio vel longitudo Lineæ Curvæ, centrum gravitatis figuræ, solida ejus rotatione genita & eorum superficies; sine ulla radicum extractione obtineri queant. Postquam intellexerat D. Gregorius hanc methodum a D. Mercatore in Logarithmotechnia usurpatam & Hyperbolæ quadrandæ adhibitam, quamꝗ adauxerat ipse Gregorius, jam universalem redditam esse, omnibusꝗ figuris applicatam; acri studio eandem acquisivit multumꝗ in ea enodanda desudavit. Uterꝗ D. Newtonus & Gregorius in animo habet hanc methodum exornare: D. Gregorius autem D. Newtonum primum ejus inventorem anticipare haud*

---

(21) Namely, of 20 and 31 July and 20 August 1669; the pertinent excerpts from these are printed, in Latin version with their parent English texts in footnote thereto, in *Commercium*: 1–2. (Fuller reproduction of the originals is made in *Correspondence*, **1**, 1959: 13–15; see, further, F. E. Manuel's *A Portrait of Isaac Newton* (Cambridge, Massachusetts, 1968): 97.)

(22) Newton cites (with some unstated telescoping) the Latinised extract from Collins' draft of his letter to Strode which is printed in *Commercium*: 28–9. In Collins' (hitherto unpublished) original words: 'In September 1668 Mʳ Mercator published his *Logarithmotechnia* containing a Specimen of this Method in one only figure[,] to witt in the quadrature of the Hyperbola: not long after the Book came out I sent one of them to Dr Wallis at Oxford, who forthwith gave his Sense of it in the *Transactions* [**3**, No. 38 (for 17 August 1668): 753–6]; another of them I sent to Dʳ Barrow at Cambridge who forthwith sent me up some papers of Mʳ Newton who is since become yᵉ Dʳˢ'—'Barrows' was afterwards interlineated by Collins to avoid confusion with Wallis—'Successor in the Mathematicall Lecture there; by which and former Communications made thereof by the Author to the Doctor it appeares that the said Method was invented some yeares before by the said Mʳ Newton, and generally applied; so that thereby in any Curvilinear figure proposed that is determined by one or more common Properties...may be obtained the quadrature or Area of the said figure accurately when it can be done, but alwaies infinitly neare, the Evolution or Length of the [= its] Curved Line, the Center of gravity of the figure, its round Solids made by rotation and their Surfaces, and all performed without any Extractions of rootes....... After Mʳ Gregory hears that this method in Mercators *Logarithmotechnia* applied to the Hyperbola and by Mʳ Gregory commented upon was rendred generall and applyed to all figures, upon diligent study for it he attained it and hath since taken much paines therein.... ...Both Mʳ Newton and Mʳ Gregory intend to write of this method in Latin [*sc.* with scholarly detail], but Mʳ Gregory will not anticipate Mʳ Newton the first inventor thereof' (Royal Society MS LXXXI, No. 24: 2ʳ, 2ᵛ, 3ʳ). Alongside the phrase 'E quibus et alijs' in his draft on Add. 3968.8:85ʳ Newton has introduced a marginal note citing Collins' words 'By wᶜʰ & former communications made thereof by the author to the Dʳ' (as Newton spelled them) and also attaching a 'See Derham p. 308'. We have already dwelt at length in note (15) above on the (now lost?) letter of Collins to Richard Towneley of about summer 1672 whose Newtonian portion Derham quotes (from the original then in his possession) in footnote (13) on this page of the first edition of his *Physico-Theology* (London, 1713).

integrum ducit. And in another Letter written to Mr Oldenburg to be communicated to Mr Leibnitz & dated 14 June 1676, Mr Collins adds:[23] *Hujus autem methodi ea est præstantia ut cum tam late pateat ad nullam hæreat difficultatem. Gregorium autem aliosq̃ in ea fuisse opinionem arbitror ut quicquid uspiam antea de hac re innotuit, quasi dubia diluculi lux fuit si cum meridiana claritate conferatur.*

This Tract was first printed by Mr William Jones, being[24] found by him among the papers & in the hand writing of Mr Collins, & collated wth the original wch he afterwards borrowed of Mr Newton.[25] It conteins the above mentioned general method of Analysis teaching how to resolve finite equations into infinite ones[26] & how by the method of moments[27] to apply equations both finite & infinite to the solution of all Problems. It begins where Dr Wallis left off, & founds the method of Quadratures upon three Rules.

Dr Wallis published his *Arithmetica Infinitorum* in y^e year 1655,[28] & by the 59th Proposition of that Book, if the Abscissa of any Curvilinear figure be called x, & m & n be numbers, & the Ordinate erected at right angles be $x^{\frac{m}{n}}$, the area

||[177] of the figure shall be $\dfrac{n}{m+n} x^{\frac{m+n}{n}}$. And this is assumed || by Mr Newton as the first Rule upon wch he founds his Quadrature of Curves. Dr Wallis demonstrated this Proposition by steps in many particular Propositions, & then collected all

---

(23) The two following sentences are cited by Newton from the Latinisation printed on *Commercium*: 46 of the first paragraph of his 'Abridgment' by Collins (now Royal Society MS LXXXI, No. 45) of his much fuller previous 'Historiola' (MS LXXXI, No. 46) of Gregory's mathematical letters to him during 1668–75. In his original (see *Correspondence*, **2**, 1960: 47–8; the full text of the 'Abridgement' is now more recently published, in a heavily 'normalised' transcription, in (ed.) J. E. Hofmann, *Leibniz: Sämtliche Schriften und Briefe* (3) **1** (Berlin, 1976): 504–16), Collins had written of the 'Excellency' of Newton's method of analysis by means of infinite series 'which is so extensive that it seemes to sweepe away all difficulties. So that I apprehended Gregory &c to be of Opinion, that all that was knowne before it was but as dawning to Noone day.'

(24) Newton's subsequent *corrigenda* of his 'Account' (see note (11) above) list this word to be replaced by 'from a copy'.

(25) Initially 'wth the original remaining in the hand of Mr Newton' (.8: 86^r). The *corrigenda* on .13: 223^r here add the not very startling information that 'The impression was finished in Decem. 1710'. We need but lightly insist that the text of the 'De Analysi' here reproduced on II: 206–46 is taken directly from Newton's autograph, and so is free from the vaguenesses and imprecisions of Collins' copy.

(26) Originally 'into converging series or infinite equations' (.8: 86^r).

(27) Newton first wrote 'fluxions' in his draft on .8: 86^r.

(28) In pedantic correction, the date on the title-page of Wallis' *Arithmetica Infinitorum, sive Nova Methodus Inquirendi in Curvilineorum Quadraturam, aliaq̃ difficiliora Matheseos Problemata* is '1656', and indeed the work was published as the third tract of his *Operum Mathematicorum Pars Altera* (Oxford, 1656). Newton's library copy, now in Cambridge (Whipple Science Museum 1305), of the 'Volumen Primum' of Wallis' collected *Opera Mathematica* in which (**1**, Oxford, 1695: 355–478) the *Arithmetica* was reprinted forty years afterwards is unmarked by him.

the Propositions into one by a Table of the Cases:[29] Mr Newton reduced all the Cases into one by a dignity wth an indefinite index & at the end of his Compendium demonstrated it at once by his method of moments, he being the first who introduced indefinite indices of dignities into the operations of Analysis.[30]

By the 108th Proposition of the said *Arithmetica infinitorum* & by several other Propositions wch follow it;[31] if the Ordinate be composed of two or more Ordinates taken wth their signes + & −, the area shall be composed of two or more areas taken wth their signes + or − respectively. And this is assumed by Mr Newton as the second Rule upon wch he founds his method of Quadratures.

And the third Rule is to reduce fractions, Radicals & the affected roots of æquations into converging series when the Quadrature does not otherwise succeed, & by the first & second Rules to square the figures whose Ordinates are the single terms of the series. Mr Newton in his Letter to Mr Oldenburg dated 13 June 1676 & communicated to Mr Leibnitz taught how to reduce any dignity of any binomial into a converging series & how by that series to square the Curve whose Ordinate is that dignity.[32] And being desired by Mr Leibnitz to explain the original of this Theoreme, he replied in his Letter dated 24 Octob. 1676 that a little before the plague (wch raged in London in the year 1665[33])

---

(29) Namely in Propositio LIX of his *Arithmetica*. For an elaborate modern paraphrase of how Wallis 'induced' this general rule from the pattern of a succession of instances where the index was initially put to be an integer, then a rational number and lastly an irrational one see J. F. Scott, *The Mathematical Work of John Wallis...(1616–1703)* (London, 1938): 28–38.

(30) Wallis himself, in fact, had made free use of such general 'literal' indices in the very Cap. XXXIII on the summation of geometrical progressions in his *Mathesis Universalis* (see note (7) above) to which Newton refers in the third paragraph of his present 'Account'; but it was Newton who a decade later, initially in his Wallisian researches into the general binomial expansion (see 1: 133) and thereafter in a wider Cartesian context of squaring curvilinear areas (see 1: 315, 318 ff.), first made confident use of such 'indefinite indices of dignities' from the winter of 1664/5 onwards.

(31) 'wch follow therein' in the draft on .8: 86^r. Newton means Propositions CVIII–CXXXII of Wallis' *Arithmetica*.

(32) In recapitulation of his Wallisian researches of almost twelve years before, that is; on which see 1: 96–134. (The pertinent opening passage of Newton's *epistola prior* is printed in *Correspondence*, **2**: 21 ff.)

(33) In his draft on .8: 86^r Newton first wrote '1665 & 1666'. Earlier, in a letter to Wallis in about January 1693 he was insistent—evidently through a distorted memory of the past event—that 'The plague was in Cambridge in both ye years 1665 & 1666 but it was in 1666 y^t I was absent from Cambridge...' (ULC. Add. 3977; see *Correspondence*, **7**, 1978: 394); in his *editio princeps* of Newton's present second letter to Leibniz in 1676, Wallis half a dozen years afterwards cautiously dated only the 'Pestis ingruens', in a parenthesis (see his *Opera Mathematica*, **3**, 1699: 635) at this place in the *epistola posterior* (=*Correspondence*, **2**: 113, l. 8), as having occurred 'annis 1665, 1666'. Newton has now at long last convinced himself, whether by finding documentary evidence of his absence from Cambridge in 1665 or no, that the plague-year crucial to dating his earliest original discoveries in mathematical analysis was indeed not 1666 but before. (Compare our earlier comment in VII: xvii, note (37).)

upon reading the *Arithmetica infinitorum* of Dr Wallis, & considering how to interpole the series x, $x - \frac{1}{3}x^3$, $x - \frac{2}{3}x^3 + \frac{1}{5}x^5$, $x - \frac{3}{3}x^3 + \frac{3}{5}x^5 - \frac{1}{7}x^7$, &c, he found the area of a circle to be the series $x - \frac{\frac{1}{2}}{3}x^3 - \frac{\frac{1}{8}}{5}x^5 - \frac{\frac{1}{16}}{7}x^7 - \frac{\frac{5}{128}}{9}x^9 -$ &c. And by pursuing

‖[178] the method of interpolation ‖ he found the Theoreme above mentioned. And by means of this Theoreme he found the reduction of fractions & surds into converging series by division & extraction of roots, & then proceeded to the extraction of affected roots. And these Reductions are[34] his third Rule.

When Mr Newton had in this Compendium explained these three Rules & illustrated them wth various examples, he layd down the Idea of deducing the area from the Ordinate by considering the Area as a quantity growing or increasing by continual flux in proportion to the length of the Ordinate, supposing the Abscissa to increase uniformly in proportion to time. And from the moments of time he gave the names of moments to the momentaneous increases or infinitely small parts of the Abscissa & Area generated in moments of time. The moment of a line he called a point in the sense of Cavallerius tho it be not a geometrical point but an infinitely short line, & the moment of an Area or superficies he called a line in the sense of Cavallerius tho it be not a geometrical line but an infinitely narrow superficies. And when he considered the Ordinate as the moment of the Area he understood by it the rectangle under the geometrical Ordinate & a moment of the Abscissa, tho that moment be not always expressed. *Sit ABD*, saith he, *Curva quævis... ...momento 1 descripta conferre.*[35] This is his Idea of the work in squaring of curves, & how he applies this to other Problems he expresses in the next words: *Jam qua ratione*, saith he,

‖[179] *superficies ABD...* ‖ *... elicietur. Exemplo res fiet clarior.*[35] And after some examples he adds his method of Regression from the Area, Arc or solid Content to the Abscissa, & shews how the same method extends to Mechanical Curves for determining their Ordinates, tangents, areas, lengths &c. And that by assuming any equation expressing the Relation between the Area & Abscissa of a Curve you may find the Ordinate by this method. And this is the foundation of the method of fluxions & moments wch Mr Newton in his Letter dated 24 Octob. 1676 comprehended in this sentence. *Data æquatione quotcunq fluentes quantitates involvente invenire fluxiones; & vice versa.*

---

(34) In his draft on .8: 86^r Newton wrote a deal more circuitously: 'And the Reduction of the Ordinates of Curves into converging series by these operations when the Quadrature of Curves does not otherwise succeed is'.

(35) The full text of this instance of direct comparison of the fluxions of two contemporaneously 'flowing' geometrical magnitudes, understanding that these are in the ratio of their 'moments', will be found on II: 232, ll. 9–17/17–23 respectively; we see no need here to repeat it. (But we may perhaps be forgiven for noticing the minor oddity that the cutter of the wood-block of the accompanying figure omitted to reverse its configuration, so that its printed impress appears as a mirror-image of Newton's parent diagram.)

(36)In this Compendium Mr Newton represents the uniform fluxion of time or of any exponent of time, by an unit, the moment of time or of its exponent by the letter o, the fluxions of other quantities by any other symbols, the moments of those quantities by the rectangles under those symbols & the letter o, & the area of a Curve by the Ordinate inclosed in a square, the Area being put for a fluent & the Ordinate for its fluxion. When he is demonstrating any Proposition he uses the letter o for a finite moment of time or of its exponent &

---

(36) In a fuller first version (.12: 148^v/149^r) of the ensuing paragraph Newton initially went on: 'Here Mr Newton considers not quantities as composed of indivisibles but as generated by local motion, after the manner used by the Ancients. They considered rectangles as generated by drawing one side into the other,₎ that is by moving one side upon the other to describe the area of the rectangle: & in like manner Mr Newton considers the areas of curves as generated by drawing the Ordinate into the Abscissa, & all indeterminate quantities he considers as generated by continual increase. And from the flowing of time & the moments thereof, he gives the name of flowing quantities to all quantities wch increase in time, & that of fluxions to the velocities of their increase. He considers time as flowing uniformly, & exposes or represents it by any other quantity wch is considered as flowing uniformly & its fluxion by an unit. And the moments of time or of its exponent he considers as equal to one another, & represents such a moment by the Letter o or by any other mark drawn into an unit. The other flowing quantities he represents by any letters or marks & most commonly by the letters at the end of the alphabet. Their fluxions he represents by any other letter or marks, or by the same letters in a different form or magnitude or otherwise distinguished. And their moments he represents by their fluxions drawn into a moment of time. Fluxions are not moments but finite quantities of other kind. They are motions.... When Mr Newton is demonstrating any Proposition he considers the moment of time in the sense of the vulgar as indefinitely small but not infinitely small, & by that means performs the whole work in finite figures or schemes by the Geometry of Euclid & Apollonius exactly without any approximation: And when he has brought the work to an equation & reduced the equation to the simplest form, he supposes the moment [o] to decrease & vanish, & from the terms wch remain, he deduces the demonstration....But when he is only investigating any truth or the solution of any Problem he supposes the moment of time to be infinitely little in the sense of Philosophers, & uses works in figures or schemes infinitely small & uses any approximations wch he conceives will make no error in the conclusion, as by putting the arc, chord, sine and tangent equal to one another, & for the greater dispatch he neglects to write down the moment o'. Furthermore, he added in a preliminary revise on .8: 73^r, 'By this means [Mr Newton] performed all his computations in this method wthout any other infinitely small quantity then the moment o'. Many similar *ex cathedra* statements by Newton on his basic notions of the fluxional method survive in his private papers: on III: 17–18 and in note (99) to the preceding introduction we have quoted *in extenso* the related first- and third-person affirmations by him which are to be found on Add. 3968.41: 83^r and .12: 167^r respectively. To John Keill he likewise declared on 15 May 1714 (Edleston, *Correspondence of Newton and Cotes* (London, 1850): 176–7 (=*Correspondence*, **6**, 1976: 136–7]): 'Fluxions & moments are quantities of a different kind. Fluxions are finite motions, moments are infinitely little parts. I...multiply fluxions by the letter o to make them become infinitely little & the rectangles I put for moments....In demonstrating Propositions I always write down the letter o & proceed by the Geometry of Euclide & Apollonius without any approximation. In resolving Questions or investigating truths I use all sorts of approximations wch I think will create no error in the conclusion & neglect to write down the letter o, & this do for making dispatch...'.

performs the whole calculation by the Geometry of the Ancients in finite figures or schemes without any approximation; & so soon as the calculation is at an end & the Equation is reduced, he supposes that the moment o decreases in infinitum & vanishes. But when he is not demonstrating but only investigating a Proposition, for making dispatch he supposes the moment o to be infinitely little, & forbears to write it down & uses all manner of approximations wch he conceives will produce no error in the conclusion. An example of the first kind you have in the end of this Compendium in demonstrating the first of the three ||[180] Rules laid down in the beginning of the book.[37] || Examples of the second kind you have in the same Compendium in finding the lengths of Curve lines (p. 15), & in finding the Ordinates, Areas & lengths of Mechanical Curves (p. 18, 19). And he tells you that by the same method tangents may be drawn to Mechanical curves. And in his Letter of 10 Decem. 1672 he adds that Problems about the Curvature of Curves, geometrical or mechanical, are resolved by the same method. Whence its manifest that he had then extended the method to the second & third moments.[38] For when the Areas of Curves are considered as fluents (as is usual[39] in this Analysis) the Ordinates express the first fluxions, the tangents are given by the second fluxions & the curvatures by the third. And even in this *Analysis* (p. 16)[40] where Mr Newton saith *Momentum est superficies cum de solidis & linea cum de superficiebus & punctum cum de lineis agitur*, it is all one as if he had said that when solids are considered as fluents their moments are superficies, & the moments of those moments (or second moments) are lines, & the moments of those moments (or third moments) are points in the sense of Cavallerius. And in his *Principia Philosophiæ* where he frequently considers lines as fluents described by points whose velocities increase or decrease, the velocities are the first fluxions & their increase the second. And the Probleme *Data æquatione fluentes quantitates*[41] *involvente fluxiones invenire & vice versa* extends to all the fluxions as is manifest by the examples of the solution thereof published by

---

(37) In draft (on .41:100^r) Newton adjoined '& another Example you have in the beginning of his book of *Quadratures* in demonstrating the first Proposition of the book' and then deleted it. Proposition I of the *De Quadratura Curvarum* (see VII: 512–16) does not, of course, presuppose the increment o to be 'infinitely' small.

(38) Newton initially (on .41: 100^r) asserted to the same effect that he had 'then extended the method to the second fluxions, if not to the third'.

(39) The draft on .41: 100^r has, with minimal difference in sense, 'as is done'.

(40) The reference here should be to 'p. 15' of the *Commercium*, at which place in his 'De Analysi' ($=$ II: 234, ll. 1–3) Newton had, to make exact quotation of his words, written 'Sed notandum est quod unitas ista quæ pro momento ponitur est superficies cum de solidis... agitur'. The misnumbering passed unchecked in 1722 into the Latinised 'Recensio'.

(41) Newton's original 1676 anagrammatic sentence here (see II: 191, note (25), and compare VII: 173) had '*quotcunq*'; and so, indeed, he initially put it in his draft of the present portion of his 'Account' on .41: 100^r.

Dr Wallis [*Opera*] Tom. 2. p. 391, 392, 396.[42] And in Lib. II *Princip*. Prop. XIV he calls the second difference the difference of moments.

Now that you may know what kind of calculation Mr Newton used[43] in or before the year 1669 when he wrote this Compendium of his *Analysis* I will here set down his Demonstration of the first Rule above mentioned.

> *Sit* ‖ *Curvæ alicujus . . . .*   ‖[181]
>
> *Pro lubitu sumatur . . . .*
>
> *Vel generaliter . . .* ‖ *. . . erit* $\dfrac{n}{m+n} ax^{\frac{m+n}{n}} = z.$ *Q.E.D.*[44]   ‖[182]

By the same way of working the second Rule may be also demonstrated. And if any æquation whatever be assumed expressing the relation between the Abscissa & Area of a Curve the Ordinate may be found in the same manner, as is mentioned in the next words of the *Analysis*. And if this Ordinate drawn into an unit be put for the Area of a new Curve the Ordinate of this new Curve may be found by the same method. And so on perpetually. And these Ordinates represent the first second third fourth & following fluxions of the first Area.

This was Mr Newtons way of working in those days when he wrote this Compendium of his *Analysis*. And the same way of working he used in his Book of *Quadratures*, & still uses to this day.[45]

Among the examples wth wch he illustrates the method of series & moments set down in this Compendium are these . . . .[46]

Mr Collins gave Mr Gregory notice of this method in autumn 1669, & Mr Gregory by the help of one of Mr Newtons series after a years study found the method in December 1670; & two months after in a Letter dated 15 Feb. 1671 sent several Theorems found thereby to Mr Collins wth leave to communicate them freely. . . . Amongst the series were ‖ these two . . . .[47]   ‖[183]

---

(42) See our reproduction of these pages in vii: 173–4, 180.

(43) 'But that you may be fully satisfied that Mr Newton used this Method of calculat[io]n' is the phrasing of the initial draft on .41: 100^r, subsequently converted into its above final form on *ibid.*: 99^v.

(44) We again omit to cite the text of Newton's 'De Analysi'; the present passage will be found on ii: 242–4.

(45) 'And by the same way of working he demonstrated the two first Propositions in his Book of *Quadratures*: & the very same way of working is used by him to this day' the draft on .41: 100^r a little differently reads.

(46) Newton proceeds to cite from his 'De Analysi' (see ii: 232, 236) the series expansions of the mutually inverse $z = \sin^{-1}x$ and $x = \sin z$.

(47) We omit to cite *in extenso* the expansions of $r\tan^{-1}(t/r)$ and $r\tan(a/r)$ which Newton selects from the half dozen and more trigonometrical series passed by Gregory to Collins in his letter in 'requital' for those of Newton for the sine, cosine and inverse sine sent to him on 24 December 1670 (*Commercium*: 24, 25 [= *Gregory Tercentenary Volume*: 155, 170]). The single series of Newton's earlier communicated by Collins (on 24 March 1669/70; see *Gregory Volume*: 89) had been one for the zone of a circle, in modern terms the expansion of $\int_{-B}^{B} \sqrt{(R^2 - x^2)} \, . \, dx$,

...Mr Leibnitz...at London...In February 167$\frac{2}{3}$ meeting Dr Pell at Mr Boyle's, he pretended to the differential method[48] of Mouton. And notwithstanding that he was shewn by Dr Pell that it was Moutons method he persisted in maintaining it to be his own invention by reason that he had found it himself without knowing what Mouton had done before & had much improved it.

When one of Mr Newton's series was sent to Mr Gregory he tried to deduce it from his own series combined together as he mentions in his Letter dated 19 December 1670.[49] And by some such method Mr Leibnitz before he left London seems to have found the sum of a series of fractions decreasing in infinitum whose numerator is a given number and denominators are triangular or pyramidal or triangulo-triangular numbers &c. [50]See the mystery! From the series $\frac{1}{1}+\frac{1}{2}+\frac{1}{3}+\frac{1}{4}+\frac{1}{5}+$ &c subduct all the terms but the first (vizt $\frac{1}{2}+\frac{1}{3}+\frac{1}{4}+\frac{1}{5}$ &c) & there will remain

$$1 = 1-\tfrac{1}{2}+\tfrac{1}{2}-\tfrac{1}{3}+\tfrac{1}{3}-\tfrac{1}{4}+\tfrac{1}{4}-\tfrac{1}{5}+\tfrac{1}{5} \&c = \frac{1}{1\times 2}+\frac{1}{2\times 3}+\frac{1}{3\times 4}+\frac{1}{4\times 5}+\&c.$$

And from this series take all the terms but the first & there will remain

$$\frac{1}{2} = \frac{2}{1\times 2\times 3}+\frac{2}{2\times 3\times 4}+\frac{2}{3\times 4\times 5}+\frac{2}{4\times 5\times 6}+\&c.$$

And from the first series take all the terms but the two first & there will remain

$$\frac{3}{2} = \frac{2}{1\times 3}+\frac{2}{2\times 4}+\frac{2}{3\times 5}+\frac{2}{4\times 6}+\&c.^{[51]}$$

and indeed Gregory for a while failed to appreciate that this was but the term-by-term quadrature of the series equivalent of the binomial integrand; but, we now know, the general method of Taylorian expansion into series to which Gregory had attained by February 1671 owed nothing to the one series of Newton's which he saw before December 1670, and it is ungenerous of Newton here to assume that the latter was of crucial help to Gregory in attaining his massive insights.

(48) The method of interpolation and subtabulation by finite differences, that is. As we have before remarked, Newton here draws heavily for his facts upon Leibniz' letter to Oldenburg on 3 February 1672/3 (*Commercium*: 32–6).

(49) A Latin version of the pertinent first paragraph of this letter is printed in *Commercium*: 23. 'I had not taken notice', Gregory wrote in his original (see *Gregory Volume*: 148), 'that Mr Newton's series for the zone of a circle [see note (47) above]...may be a consectaries to that'—his version of the general binomial expansion, independently discovered by him— 'which I sent you.... I admire much my own dulness, that in such a considerable time [*viz.* the eight months and more since he had first received Newton's undemonstrated series] I had not taken notice of this; nevertheless that I had taken much pains to find out that series. But the truth is, I thought always (if so be it were a series) that I might fall upon it by some combination of my series for the circle, seeing I had such infinite numbers of them, not so much as once desiring any other method.'

(50) The remainder of this paragraph—a late insertion in the main text of the 'Account'— is drafted on a slip of paper now loose (grossly out of context in a surround of its author's early mathematical researches) in Add. 3958.3. We may notice, howsoever here irrelevantly, that Newton has set down on its verso the (to us) cryptic direction: 'In aire street going from Picad[illy] to golden squ. over ag^t y^e P[estle] & Moretar'—the last a tavern, we presume.

(51) In his draft (see the previous note) Newton initially went on: 'Or take all the terms

‖ In the end of February or beginning of March 167$\frac{2}{3}$ Mr Leibnitz went from ‖[184]
London to Paris & continuing his correspondence wth Mr Oldenburg &
Mr Collins$^{(52)}$ wrote in July 1674 that he had a wonderful Theoreme wch gave the
Area of a Circle or any sector thereof exactly in a series of rational numbers....
But the Demonstration of this Theoreme Mr Leibnitz wanted.$^{(53)}$...

In a Letter composed by Mr Collins & dated 15 April 1675 Mr Oldenburg

---

but the three first & there will remain $\frac{11}{6} = \frac{3}{1 \times 4} + \frac{3}{2 \times 5} + \frac{3}{3 \times 6}$ &c. Or from this series
$\frac{1}{1} + \frac{1}{3} + \frac{1}{5} + \frac{1}{7} + \frac{1}{9}$ &c take all the terms but the first & there will remain

$$1 = \frac{2}{1 \times 3} + \frac{2}{3 \times 5} + \frac{2}{5 \times 7} + \frac{2}{7 \times 9} \text{ \&c'}.$$

And on the verso of his roughing-out he added in generalisation:

$$\text{`}\frac{a}{b} + \frac{a}{b+c} + \frac{a}{b+2c} + \frac{a}{b+3c} \text{ [\&c]} - \frac{a}{b+c}\left[-\frac{a}{b+2c} - \frac{a}{b+3c} - \text{\&c}\right]$$

$$\text{[id est]} \frac{+ac}{bb+bc} + \frac{ac}{bb+3bc+2cc} + \frac{ac}{bb+5bc+6cc} + \frac{ac}{bb+7bc+3,4cc} \text{ [\&c]} = \frac{a'}{b}.$$

In fairness to Leibniz one must surely say that, until the 'evident' reduction is made of
splitting the general term $ac/b(b+c)$ of a given infinite series into the equivalent partial
fractions $a/b - a/(b+c)$, there is a genuine 'mystery' about the summations which Leibniz
achieved (at Paris in 1672, in fact, before his first visit to London the next February), if a fairly
low-level one. Leibniz reported to Oldenburg in his letter 'Londini' of 3 February 1672/3 that
'Multa alia circa hos numeros observata sunt a me, in quibus illud eminet, quod modum habeo
summam inveniendi Seriei Fractionum in infinitum decrescentium, quarum numerator Unitas,
nominatores vero numeri isti Triangulares aut Pyramidales aut Triangulo Triangulares &c'
(*Commercium*: 36). Newton surely over-reacts to the understandable tyro's pride and enthusiasm
with which Leibniz announced to the English this his first modest mathematical discovery (on
which see more fully J. E. Hofmann, *Leibniz in Paris, 1672–1676*: 14–19). He would, it seems to
us, have had much more reason to be snappily sarcastic about the more half-baked 'Harmony'
which Leibniz sought afterwards to trace between these simple series and similar compounds of
the alternating series of reciprocals of succeeding integers whose infinite sums yield the area, $\frac{1}{4}\pi$,
of the quadrant of a unit circle and that, log 2, under the related unit rectangular hyperbola,
and which he subsequently set out in his letter of 27 August 1676 (N.S.) to Oldenburg (see
Commercium: 61 [=*Correspondence*, **2**: 60]). And to be sure on another occasion he was: Newton's
unravellings of Leibniz' wanted *Harmonia numerorum* as expounded in this latter letter are
preserved on Add. 3968.14: 245^r, there ultimately gathered into a yet unpublished annotation
on 'pag. 61' of the *Commercium*: 'Hujus Harmoniæ series prima ex secunda et tertia componitur,
secunda duplicata ex hac $\frac{1}{1} - \frac{1}{3} + \frac{1}{3} - \frac{1}{5} + \frac{1}{5} - \frac{1}{7} + \frac{1}{7} - \frac{1}{9}$ [&c] conjunctis binis terminis, tertia
quadruplicata ex hac $\frac{1}{1} - \frac{1}{2} + \frac{1}{2} - \frac{1}{3} + \frac{1}{3} - \frac{1}{4} + \frac{1}{4} - \frac{1}{5}$ [&c] conjunctis binis terminis, quarta est
series Gregorij et quinta series Brounkeri'.

(52) The list of *corrigenda* on .13: 223^r (see note (11)) rightly directs the reader to 'blot out'
this last phrase '& Mr Collins': Leibniz did not meet Collins on his first visit to London in
1673, and during the next three years their only contact was at one remove by way of
Oldenburg.

(53) Newton's sole 'evidence' for this mistaken notion that Leibniz could not, before 1676,
prove the truth of his series for the inverse tangent—a gratuitous under-estimation (and, we
may now know, a wholly erroneous one) which would surely have sparked off a flash of anger

sent to Mr Leibnitz eight of Mr Newton's & Mr Gregorys series amongst wch were Mr Newton's series above mentioned for finding the Arc whose sine is given & the sine whose Arc is given, & Mr Gregory's two series above mentioned for finding the Arc whose tangent is given & the tangent whose Arc is given. And Mr Leibnitz in his answer dated 20 May 1675 acknowledged the receipt of this Letter in these words. *Literas tuas multa fruge Algebraica refertas accepi, pro quibus tibi & doctissimo Collinio gratias ago. Cum nunc præter ordinarias curas Mechanicis imprimis negotijs distrahar, non potui examinare Series quas misistis ac cum meis comparare. Ubi fecero perscribam tibi sententiam meam: nam aliquot jam anni sunt quod inveni meas via quadam sic satis singulari.*[54]

‖[185]   ‖ But yet Mr Leibnitz never took any further notice of his having received these series nor how his own differed from them. . . . And what he did with Mr Gregory's series for finding the Arc whose tangent is given he has told us. . . . By a Theoreme for transmuting of figures like those of Dr Barrow & Mr Gregory he had now found a demonstration of this series. . . . But he still wanted a demonstration of the rest; & meeting wth a pretence to ask for what he wanted, he wrote to Mr Oldenburg the following Letter dated at Paris 12 May 1676.[55] *Cum Georgius Mohr Danus nobis attulerit communicatam sibi à doctissimo Collinio vestro expressionem rationis inter arcum & sinum per infinitas series sequentes. . .*[56]*. . .& posterior imprimis series elegantiam quandam singularem habeat: ideo rem gratam mihi feceris. . .si demonstrationem transmiseris. Habebis vicissim mea ab his longe diversa circa hanc rem meditata, de quibus jam aliquot abhinc annis ad te perscripsisse credo, demonstratione tamen non addita quam nunc polio. . . .* [*Cl. Collinius*] *facile tibi materiam suppeditabit satisfaciendi desiderio meo.* Here. . .one would think that he had never seen these two

in the latter had he lived to read Newton's present 'Account'—was, he here went on to specify, his unfounded inference from Leibniz' request to Oldenburg on 12 May 1676 (N.S.) (see Newton's page 185 following) to procure from Collins the (particular) demonstration of the series for the sine and inverse sine, passed third-hand to him by Mohr, that he thereby meant 'the [general] Method by which Mr Newton had invented it'. In an additional list of *corrigenda* on .42: 612^r (the context suggests that this was compiled in 1719) he subsequently thought to amplify the preceding sentences by continuing after '. . .from London to Paris' with the declaration: 'He was hitherto unacquainted with the higher Geometry. But the *Horologium oscillatorium* of Mr Hugens coming out in April following, he began to learn it with reading that book. He continued his correspondence with Mr Oldenburgh about Arithmetical matters till June, spent the year following in studying the higher Geometry[,] & in July 1674 began to renew his correspondence & wrote to Mr Oldenburgh that he had a wonderfull Theoreme &c.' Huygens was indeed a mentor to Leibniz in mathematics during the crucial years 1673–5 of the latter's growth to intellectual maturity at Paris during 1672–6, but we need not insist that Newton gives a vastly oversimplified *précis* here of a complexly ramified mathematical self-education of whose detail he knew next to nothing.

(54) Quoted from *Commercium*: 42.

(55) N.S., that is. The ensuing extract from Leibniz' letter is that printed in *Commercium*: 45.

(56) These are—'Posito Sinu x, Arcu z, Radio 1' in Leibniz' phrase—the expansions of $z = \sin^{-1} x$ and $x = \sin z$ respectively, as we need not here spell out.

series before & that his *diversa circa hanc rem meditata* were something else then one of the series wch he had received from ‖ Oldenburgh the year before.... ‖[186]

Upon the receipt of this Letter Mr Oldenburg & Mr Collins wrote pressingly to Mr Newton, desiring that he himself would describe his own method to be communicated to Mr Leibnitz. Whereupon Mr Newton wrote his Letter dated 13 June 1676, describing therein the method of series as he had done before in the Compendium above mentioned, but with this difference: here he described at large the reduction of the Dignity of a binomial into a series & only touched upon the reduction by division & extraction of affected roots; there he described at large the reduction of fractions & radicals into series by division & extraction of roots & only set down the two first terms of the series into which the dignity of a binomial might be reduced. And among the examples in this letter there were series for finding the number whose logarithm is given & for finding the versed sine whose Arc is given....

The answer of Mr Leibnitz directed to Mr Oldenburgh & dated 27 August 1676 begins thus: *Literæ tuæ die Julij 26. datæ plura ac memorabiliora circa rem Analyticam continent quam* ‖ *multa volumina spissa de his rebus edita....... ...* ‖[187]

And tho Mr Leibnitz had now received this series [for squaring the circle] twice from Mr Oldenburgh, yet in his Letter...he sent it back to him by way of recompence for Mr Newtons method, pretending that he had communicated it to his friends at Paris three years before or above...: but it doth not appear that he had the demonstration thereof so early. When he found the demonstration, then he composed it in his *Opusculum* & communicated that also to his friends, & he himself has ‖ told us that this was in the year 1675.... ‖[188]

In the same Letter of 27 Aug. 1676 after Mr Leibnitz had described his Quadrature of the circle & equilateral Hyperbola,[57] he added: *Vicissim ex seriebus Regressum pro Hyperbola hunc inveni. Si sit numerus aliquis unitate minor 1 — m ejusq̃ logarithmus Hyperbolicus l, erit...*[58]. *Si numerus sit major unitate, ut 1 + n, tunc pro eo inveniendo mihi etiam prodijt Regula quæ in Newtoni Epistola expressa est, scilicet erit...*[58]. *....Quod regressum ex arcubus attinet incideram ego directe in Regulam, quæ ex dato arcu sinum complementi exhibet. Nempe sinus complementi = ...*[59]. *Sed postea quoq̃ deprehendi ex ea illam nobis communicatam pro inveniendo sinu recto, qui est...*[59] *posse demonstrari.* Thus Mr Leibnitz put in his claim for the coinvention of these four series, though the method of finding them was sent him at his own request &

---

(57) Or, yet more generally, of the sector of any central conic; on this see our earlier extended comment in IV: 531, note (13). The following passage in Leibniz' letter is quoted from *Commercium*: 61–2 (= *Correspondence*, **2**: 60–1).

(58) These omitted inverse logarithmic series in which (Newton specifies) there is in turn $l = -\log(1-m)$ and $l = \log(1+n)$ are those of the exponentials $m = 1-e^{-l}$ and $n = e^l - 1$ respectively.

(59) Leibniz here sets out the now familiar series for cos a and sin a respectively.

he did not yet understand it. For in this same Letter of 27 August 1676 he desired Mr Newton to explain it further. His words are. *Sed desideraverim ut Clarissimus Newtonus nonnulla quoqʒ amplius explicet: ut originem Theorematis quod initio posuit: Item modum quo quantitates p, q, r in suis operationibus invenit: Ac deniqʒ quomodo in Methodo regressuum se gerat, ut cum ex Logarithmo quærit Numerum. Neqʒ enim explicat quomodo ex methodo sua derivetur.*[60] . . .

‖[189]   ‖ . . . When Mr Newton had received this Letter, he wrote back[61] that all the said four Series had been communicated by him to Mr Leibnitz, the two first being one & the same series in which the letter *l* was put for the Logarithm wth its sign + or −, & the third being the excess of the Radius above the versed sine for which a series had been sent to him. Whereupon Mr Leibnitz desisted from his claim. . . .

·· ·   ·· ·   ·· ·

When Mr Newton in his Letter dated 13 June 1676 had explained his method of series, he added: *Ex his videre est quantum fines Analyseos per hujusmodi infinitas æquationes ampliantur: quippe quæ earum beneficio ad omnia pene dixerim problemata (si numeralia Diophanti et similia excipias) sese extendit. Non tamen omnino universalis evadit nisi per ulteriores quasdam Methodos eliciendi series infinitas. Sunt enim quædam Proble-mata in quibus non licet ad series infinitas per Divisionem vel extractionem radicum simpli-cium affectarumve pervenire. Sed quomodo in istis casibus procedendum sit jam non vacat* ‖[190] *dicere; ut neqʒ alia quædam tradere quæ circa reductionem infinitarum serierum* ‖ *in finitas ubi rei natura tulerit excogitavi. Nam parcius scribo, quod hæ speculationes diu mihi fa-stidio esse cœperunt, adeo ut ab ijsdem jam per quinqʒ fere annos abstinuerim.*[62] To this Mr Leibnitz in his Letter of 27 Aug. 1676 answered: *Quod dicere videmini plerasqʒ difficultates (exceptis problematibus Diophantæis) ad series Infinitas reduci; id mihi non videtur. Sunt enim multa usqʒ adeo mira & implexa ut neqʒ ab æquationibus pendeant neqʒ ex Quadraturis. Qualia sunt (ex multis alijs) Problemata methodi Tangentium inversæ.*[63] And Mr Newton in his Letter of 24 Octob. 1676 replied: *Ubi dixi omnia pene Problemata solubilia existere, volui de ijs præsertim intelligi circa quæ Mathematici se hactenus occuparunt vel saltem in quibus Ratiocinia Mathematica locum aliquem obtinere possunt. Nam alia sane adeo perplexis conditionibus implicata excogitare liceat, ut non satis comprehendere valeamus: & multo minus tantarum computationum onus sustinere quod ista requirerent. Attamen ne nimium dixisse videar, inversa de Tangentibus Problemata sunt in potestate aliaqʒ illis difficiliora. Ad quæ solvenda usus sum duplici methodo, una concinniori altera generaliori. Utramqʒ visum est impræsentia literis transpositis consignare, ne propter alios idem obtinentes institutum in aliquibus mutare cogerer. 5a cc d æ 10e ff h* &c.

---

(60) *Commercium*: 63 (=*Correspondence*, **2**: 62).

(61) In his *epistola posterior* of 24 October 1676 of course; see *Commercium*: 79–80 (=*Corre-spondence*, **2**: 122–4).

(62) *Commercium*: 55–6 (=*Correspondence*, **2**: 29).

(63) *Commercium*: 65 (=*Correspondence*, **2**: 64).

id est, *Una methodus consistit in extractione fluentis quantitatis ex æquatione simul invol-*
vente fluxionem ejus; altera tantum in assumptione seriei pro quantitate qualibet incognita
ex qua cætera commode derivari possunt, & in collatione terminorum homologorum
æquationis resultantis ad eruendos terminos assumptæ seriei.[64] By Mr Newtons two
Letters its certain that he had then (or rather above five years before) found
out the reduction of Problems to fluxional æquations & converging series, & by
the Answer of Mr Leibnitz to the first of those Letters its as[65] certain that he had
not then found out the reduction of problems either to differential equations
or to converging series.[66]

And the same is manifest also by what Mr Leibnitz wrote in the *Acta Erudi-*
torum Anno 1691 concerning this matter. ‖ *Jam anno 1675*, saith he, *compositum*   ‖[191]
habebam opusculum Quadraturæ Arithmeticæ ab amicis ab illo tempore lectum, sed quod
materia sub manibus crescente, limare ad editionem non vacavit, postquam aliæ occupationes
supervenere; præsertim cum nunc prolixius exponere vulgari more quæ Analysis nostra nova
paucis exhibet, non satis operæ pretium videatur.[67] This Quadrature composed *vulgari*
more he began to communicate at Paris in the year 1675. The next year he was
polishing the Demonstration thereof to send it to Mr Oldenburg in recompence
for Mr Newton's method as he wrote to him in his Letter dated 12 May 1676,
& accordingly in his Letter of 27 August 1676 he sent it composed & polished
vulgari more.[68] The winter following he returned into Germany by England &
Holland to enter upon publick business & had no longer any leisure to fit it for
the press, nor thought it afterwards worth his while to explain those things

---

(64) *Commercium*: 85–6 (=*Correspondence*, **2**: 129) + 86, note *.

(65) 'it is most' in the initial draft on Add. 3968.41: 98r.

(66) In his 1719 list of *emendanda* (see note (53) above) Newton here adjoined: 'The same
is manifest also from hence, that Mr Leibnitz in his Letter of Aug. 27. 1676 placed the top of
Analysis not in the Differential method but in Analytical Tables of Tangents & in the Combinatory Art; saying [*Commercium*: 64 (=*Correspondence*, **2**: 63)] of one of them: *Nihil est quod*
norim in TOTA *Analysi momenti majoris*; & of the other: *Ea nihil differt ab Analysi illa* SUPREMA, *ad*
cujus intima Cartesius non pervenit. Est enim ad eam constituendam opus Alphabeto Cogitationum
humanarum.'

(67) *Acta Eruditorum* (April 1691): 178. This is the opening sentence of a short article (*ibid.*:
178–82 [= (ed. C. I. Gerhardt) *Leibnizens Mathematische Schriften*, **5** (Halle, 1858): 128–32]) in
which Leibniz sketches the application of the basic trigonometrical series, pure and logarithmic,
to determine the 'Quadratura Arithmetica communis Sectionum Conicarum quæ centrum
habent, indeque ducta Trigonometria Canonica ad quantamcumque in numeris exactitudinem
a Tabularum necessitate liberata: cum usu speciali ad lineam Rhomborum Nauticam,
aptatumque illi planisphærium'.

(68) See *Commercium*: 61 (=*Correspondence*, **2**: 60). In his letter, in fact, Leibniz merely stated
the double series $t \mp \frac{1}{3}t^3 + \frac{1}{5}t^5 \mp \frac{1}{7}t^7 + \ldots$—the combined expansions, namely, of $\tan^{-1}t$ and $\tanh^{-1}t$
—which in his estimate yields 'omnium possibilium Circuli et Sectionis Conicæ centrum
habentis per series infinitas quadraturarum simplicissimam'; indeed, the manuscripts setting out
his original derivation of the inverse-tangent series (analysed by J. E. Hofmann in his *Leibniz in*
Paris, 1672–1676: 54–7/59–61) have yet to be published.

prolixly in the vulgar manner wch his new Analysis exhibited in short. He found this new Analysis therefore after his return into Germany & by consequence not before the year 1677.

The same is further manifest by the following consideration. Dr Barrow published his method of Tangents in the year 1670.^(69) Mr Newton in his Letter dated 10 Decemb. 1672 communicated his method of Tangents to Mr Collins,^(70) & then added.^(71) *Hoc est unum particulare vel Corollarium potius Methodi generalis quæ extendit se citra molestum ullum calculum non modo ad ducendum Tangentes ad quasvis Curvas sive Geometricas sive Mechanicas, vel quomodocunꝗ rectas Lineas aliasve Curvas respicientes, verum etiam ad resolvendum alia abstrusiora Problematum genera de Curvitatibus, Areis, Longitudinibus, centris gravitatis Curvarum &c. Neꝗ (quemadmodum Huddenij methodus de Maximis & Minimis) ad solas restringitur æquationes illas quæ quantitatibus surdis sunt immunes. Hanc methodum intertexui alteri isti qua*

---

(69) At the close, that is, of the tenth lecture in his published Lucasian *Lectiones Geometricæ*; see notes (11) and (23) to Appendix 1.1 above. In his primary list of *corrigenda* on Add. 3968.13: 223^r (on which see note (11) above) Newton here inserted the additional observation: 'From this method & his own Mr Gregory deduced a method of Tangents without computation, & notified it to Mr Collins in a Letter dated 5 [Sept.] 1670'. In a preliminary version of this interpolation on .27: 421^v he went on to cite the Latin version of the pertinent passage 'printed in the *Commercium Epistolicum* pag. 22'; Gregory's original words here were: 'I have discovered from Barrow his method of drawing tangents together with some of my own, a general geometrical method, without calculation, of drawing tangents to all curves, and comprehending not only Barrow's particular methods, but also his general analytical method in the end of the 10th lecture'. He there proceeded to amplify the sequel: 'And two years after upon notice that Slusius & Gregory had Methods of drawing Tangents without calculation, Mr Newton wrote back to Mr Collins 10 Decem. 1672 in these words. *Ex animo gaudeo...*'— 'I am heartily glad at the acceptance wch our Reverend Dr Barrow's *Lectures* finds wth forreign Mathematicians, & it pleased me not a little to understand that they are falln into the same method of drawing Tangents wth me. What I guess their method to be you will apprehend by this example' to cite his original English (see *Correspondence*, **1**, 1959: 247) of these sentences rendered into Latin on *Commercium*: 29—'And at the end of the example he added these words *Hoc est unum particulare...*'.

(70) In the *corrigenda* on .13: 223^r Newton here directed there to be inserted 'conjecturing that it was the same with that of Gregory & Slusius' (compare the preliminary *addendum* by him on .27: 421^v which is cited in the preceding note).

(71) Newton quotes the Latin version on *Commercium*: 30. In his original letter he wrote (see *Correspondence*, **1**: 247–8): 'This is one particular, or rather a Corollary of a Generall Method wch extends it selfe w^thout any troublesome calculation, not onely to the drawing tangents to all curve lines whether Geometrick or mechanick or how ever related to streight lines or to other curve lines but also to the resolving other abstruser kinds of Problems about the crookedness, areas, lengths, centers of gravity of curves &c. Nor is it (as Huddens method *de maximis et minimis...*) limited to æquations wch are free from surd quantities. This method I have interwoven wth that other of working in æquations by reducing them to infinite series.' The general method of drawing tangents to Cartesian curves which Newton here exemplified for Collins was, as we have already mentioned (III: 122, note (191)), that expounded and instanced at length as Mode 1 of Problem 4 (see *ibid.*: 120–32) of his general treatise on infinite series and fluxions of the previous year.

Æquationum exegesin instituo reducendo eas ad series infinitas. Mr Slusius sent his method of Tangents to Mr Oldenburg 17 Jan. 1673 [N.S.] & the same was ‖ ‖[192] soon after published in the *Transactions*.[72] It proved to be the same wth that of Mr Newton. It was founded upon three Lemmas, the first of wch was this: *Differentia duarum dignitatum ejusdem gradus applicata ad differentiam laterum, dat partes singulares gradus inferioris ex binomio laterum, ut* $\dfrac{y^3 - x^3}{y - x} = yy + yx + xx$. That is (in the notation[73] of Mr Leibnitz) $\dfrac{dy^3}{dy} = 3yy$. A copy of Mr Newton's Letter of 10 Decem. 1672 was[74] sent to Mr Leibnitz by Mr Oldenburg amongst the papers of Mr James Gregory at the same time wth Mr Newton's Letter of 13 June 1676. And Mr Newton having described in these two Letters that he had a very general Analysis consisting partly in the method of converging series, partly in another method by wch he applyed those series to the solution of almost all problems (except perhaps some numeral ones like those of Diophantus) & found the tangents, areas, lengths, solid contents, centers of gravity, & curvities of curves & curvilinear figures Geometrical or mechanical without sticking at surds, & that the method of Tangents of Slusius was but a branch or corollary of this other method: Mr Leibnitz in his returning home through Holland was meditating upon the improvement of the method of Slusius. For in a Letter to Mr Oldenburg dated from Amsterdam $\frac{18}{28}$ Novem. 1676 he wrote thus.[75] *Methodus Tangentium a Slusio publicata nondum rei fastigium tenet. Potest aliquid amplius præstari in eo genere quod maximi foret usus ad omnis generis Problemata: etiam ad meam (sine extractionibus) Æquationum ad series reductionem. Nimirum, posset brevis quædam calculari circa*

(72) With praiseworthy celerity indeed; see *Phil. Trans.* **7**, No. 90 (for 20 January 1672/3): 5143–7 (=*Correspondence of Henry Oldenburg*, **9**, 1973: 386–92). The lemmas upon which he founded his minimal variant upon the Gregory–Barrow rule for tangents Sluse laid bare, however, only in his following letter to Oldenburg on 3 May 1673 (N.S.); as we observed in note (64) to Appendix 1.5 preceding, the first of these—which Newton here proceeds to cite *via* the extract from it printed in *Phil. Trans.* **8**, No. 95: 6059—adumbrates (in the instance $n = 3$) his determination from first principles that the derivative of x^n, n some positive integer, is nx^{n-1}.

(73) 'language' in Newton's preliminary draft on Add. 3968.39: 581^v.

(74) A deleted *corrigendum* on .27: 421^v initially expanded this to read: 'Copies of Mr Gregories Letter of 5 Sept 1670'—on this see note (69) preceding—'& Mr Newton's Letter of 10 Decem 1672 were', and then subsequently directed that an equivalent reference to Gregory's letter should be added at the end of the sentence.

(75) Collins had transcribed the essential portion of Leibniz' letter in his own letter to Newton on 5 March 1676/7 following, first published by Wallis in the 'Epistolarum Collectio' gathered by him in the third volume of his *Opera Mathematica* in 1699, but Newton quotes from the excerpt from this printed in *Commercium*: 87. (The text of Collins' letter is reproduced in *Correspondence*, **2**: 198–200 from a contemporary copy by David Gregory, but there is a fuller extract 'Ex Leibnitij Epistola $\frac{18}{28}$ Novemb 1676 Amsterodami' in Collins' hand, preserved in ULC. Add. 3971.2: 62^r/62^v, which is yet unpublished.)

Tangentes Tabula, eousq̃ continuanda donec progressio Tabulæ apparet; ut eam scilicet quisq̃, quousq̃ libuerit, sine calculo continuare possit. This was the improvement of the Method of Slusius into a general method wch Mr Leibnitz was then thinking ‖[193] upon, & by his words ‖ *Potest aliquid amplius præstari in eo genere quod maximi foret usus ad omnis generis Problemata*, it seems to be the only improvement wch he had then in his mind for extending the method to all sorts of Problems.⁽⁷⁶⁾ The improvement by the differential calculus was not yet in his mind, but must be referred to the next year.

Mr Newton in his next Letter dated 24 Octob. 1676 mentioned the *Analysis* communicated by Dr Barrow to Mr Collins in the year 1669 & also another Tract written in 1671 about converging series & about the other method by wch Tangents were drawn after the method of Slusius & maxima & minima were determined & the Quadrature of Curves was made more easy &c & this without sticking at radicals, & by wch series were invented wch brake off & gave the Quadrature of Curves in finite æquations when it might be. And the foundation of these operations he comprehended in this sentence exprest enigmatically as above. *Data æquatione fluentes quotcunq̃ quantitates involvente fluxiones invenire & vice versa.* Which puts it past all dispute that he had invented the method of fluxions before that time. And if other things in that Letter be considered it will appear that he had then brought it to great perfection & made it exceeding general, the Propositions in his book of *Quadratures* & the methods of converging series & of drawing a Curve line through any number of given points being then known to him. For when the method of fluxions proceeds not in finite equations he reduces the equations into converging series by the binomial Theoreme & by the extraction of fluents out of equations involving or not involving their fluxions. And when finite Equations are wanting he deduces converging series from the conditions of the Probleme by assuming the terms of the series gradually & determining them by those conditions. And when fluents are to be derived ‖[194] from fluxions & the Law of the fluxions is wanting, he finds ‖ that Law *quam proxime* by drawing a Parabolick line through any number of given points. And by these improvements Mr Newton had in those days made his Method of fluxions much more universal then the Differential method of Mr Leibnitz is at present.

This Letter of Mr Newtons dated 24 Octob. 1676 came to the hands of Mr Leibnitz in the end of the Winter or beginning of Spring following,⁽⁷⁷⁾ & M^r

---

(76) In the preliminary draft on .39: 581^v Newton wrote a little differently ‘ . . . for making the method of Slusius general’.

(77) The *corrigenda* on .13: 223^r directs that this be reworded to read ‘was seen by Mr Leibnitz in London in November, & a copy thereof came to his hands at Hanover in the beginning of Spring following’; and equivalent emendation was made at the corresponding place in the published Latin ‘Recensio’ (*Commercium*, ₂1722: 25: ‘Hæc *Newtoni* Epistola data

Leibnitz soon after, vizt in a Letter dated 21 Junij 1677 wrote back: *Clarissimi Slusij methodum Tangentium nondum esse absolutam Celeberrimo Newtono assentior. Et jam a multo tempore rem Tangentium generalius tractavi, scilicet per differentias Ordinatarum.—Hinc nominando in posterum dy differentiam duarum proximarum y* &c.[78] Here Mr Leibnitz began first to propose his differential method: & there is not the least evidence that he knew it before........it lies upon Mr Leibnitz to prove that he found out this method before the receipt of Mr Newton's Letters....

The Marquess de L'Hospital (a person of very great candor) in the Preface to his book *De Analysi quantitatum infinite parvarum* published A.C. 1696[79] tells us

---

24 Octob. 1676, in fine mensis illius vel initio sequentis visa est *Leibnitio* Londini; ejusque Exemplar *Hanoveriæ* ei 'obtigit initio Veris insequentis'). This conjecture that Leibniz had a prior sight of Newton's *epistola posterior* in late October/early November 1676, six months before he officially received the copy despatched to him by Oldenburg the next May, is thoroughly mistaken; for Leibniz left London on 19 October, five days before Newton sent his thick mathematical epistle off from Cambridge *en route* to London. (See J. E. Hofmann's elaborate refutation in his *Leibniz in Paris, 1672–1676*: 274–5.)

(78) *Commercium*: 88 (=*Correspondence*, 2: 213).

(79) *Viz.* in its original French title *Analyse des Infiniment Petits (pour l'intelligence des lignes courbes)* (Paris, ₁1696 [→₂1726]). The passage (signatures aiii^v–aiv^v) in the 'Preface'—there is now good reason to think that this was written not by L'Hospital but by Fontenelle—which Newton proceeds to render into English reads: 'Peu de temps aprés la publication de la Méthode de M. *Descartes* pour les tangentes, M. *de Fermat* en trouva aussi une, que M. *Descartes* a enfin avoüé'—a marginal note citing 'Lett[re] 71. Tom. 3' of Clerselier's edition of the *Lettres de Mr Descartes* (₁1667: 412) here adduces his remark to de Beaune on 20 February 1639 (N.S.) that 'Ie ne croy pas qu'il soit possible de trouuer generalement la conuerse de ma regle pour les tangentes, ny de celle dont se sert Monsieur de Fermat non plus, bien que la pratique en soit en plusieurs cas plus aisée que de la mienne'—'luy-même être plus simple en bien des rencontres que la sienne. Il est pourtant vray qu'elle n'étoit pas encore aussi simple que M. *Barrow* l'a renduë depuis en considérant de plus prés la nature des polygones, qui présente naturellement à l'esprit un petit triangle fait d'une particule de la courbe, comprise entre deux appliquées infiniment proches, de la différence de ces deux appliquées, & de celle des coupées correspondantes; & ce triangle est semblable à celuy qui se doit former de la tangente, de l'appliquée, & de la soutangente: de sorte que par une simple Analogie cette derniere méthode épargne tout le calcul que demande celle de M. *Descartes*, & que cette méthode, elle-même, demandoit auparavant. M. *Barrow*'—'*Lect. geomet.* [London, ₁1670] p. 80'—'n'en demeura pas là, il inventa aussi une espéce de calcul propre à cette méthode; mais il luy falloit, aussi-bien que dans celle de M. *Descartes*, ôter les fractions, & faire évanoüir tous les signes radicaux pour s'en servir. Au défaut de ce calcul est survenu celuy'—'*Acta Erud. Lips.* an 1684. p. 467'—'du célébre M. *Leibnis*; & ce Sçavant Géometre a commencé où M. *Barrow* & les autres avoient fini. Son calcul l'a mené dans des païs jusqu'ici inconnus; & il y a fait des découvertes qui sont l'étonnement des plus habiles Mathématiciens de l'Europe.' The '&c' with which Newton concludes his following translation of this conceals a further sentence of L'Hospital's (or Fontenelle's) which he clearly found ill-suited to his personal claim to priority in discovery of the calculus to render: 'Mrs *Bernoulli* ont été les premiers qui se sont aperçus de la beauté de ce calcul: ils l'ont porté à un point qui les a mis en état de surmonter des difficultés qu'on n'auroit jamais osé tenter auparavant.'

that a little after the publication of the method of Tangents of Descartes,
Mr Fermat found also a method wch Descartes himself at length allowed to be
‖[195] for the most part more simple then ‖ his own. But it was not yet so simple as
Mr Barrow afterwards made it by considering more nearly the nature of
polygons wch offers naturally to the mind a little triangle composed of a particle
of the Curve lying between two Ordinates infinitely neare one another & of the
difference of these two Ordinates & of that of the two correspondent Abscissas.
And this triangle is like that which ought to be made by the tangent, the
Ordinate & the subtangent: so that by one simple Analogy this last Method
saves all the calculation which was requisite either in the method of Des Cartes
or in this same method before. Mr Barrow stopt not here, he invented also a
sort of calculation proper for this method. But it was necessary in this[80] as well
as in that of Des Cartes to take away fractions & radicals for making it usefull.
Upon the defect of this calculus that of the celebrated Mr Leibnitz was introduced & this learned Geometer began where Mr Barrow & others had left off.
This his calculus led into regions[80] hitherto unknown & there made discoveries
wch astonished the most able Mathematicians of Europe. &c. Thus far the
Marquess. He had not seen Mr Newtons *Analysis* nor his Letters of 10 Decem.
1672, 13 June 1676 & 24 Octob. 1676:[81] & so not knowing that Mr Newton had
done all this & signified it to Mr Leibnitz he recconed that Mr Leibnitz began
where Mr Barrow left off, & by teaching how to apply Dr Barrow's method
wthout sticking at fractions & Surds, had enlarged that method wonderfully.
[82]And Mr James Bernoulli in the *Acta Eruditorum* of January 1691 pag. 14 writes
thus: *Qui calculum Barrovianum (quem in Lectionibus suis Geometricis adumbravit Auctor,*

---

(80) Newton's draft on .39: 581^r here has the initial, more literal, cancelled renditions 'But
this was wanting in it' and 'went through countries' respectively. Whether the primary
translation from L'Hospital's French 'Preface' was Newton's own or one commissioned by
him from another's hand—one immediately thinks of de Moivre, who helped him in many
another similar context—is not clear, but the firm continuous manner in which the rest of the
draft translation on f. 581^r is written out would suggest the latter.

(81) 'none of them being printed before the year 1699'—and then only the two latter
epistolæ prior/posterior to Leibniz (on pages 622–9/634–45 of the 'Epistolarum Collectio' gathered
by Wallis in the third volume of his *Opera*)—is adjoined in the *corrigenda* on .13: 223^r (and a
similar reminder is found at this point in the Latin 'Recensio'; see *Commercium*, $_2$1722: 27).

(82) The following extract from Jakob Bernoulli's 'Specimen Calculi Differentialis in
dimensione Parabolæ helicoidis...' (*Acta Eruditorum* (January 1691): 13–23 [= *Opera*, **1**
(Geneva, 1744): 431–42]) does not appear in Newton's draft on .39: 581^r, and we presume
that he came upon this neat put-down of Leibniz' originality in 'inventing' a calculus
'founded' upon Barrow's (and 'differing from it merely in its notation for differentials and
some brevity of operation') only just before he put his 'Account' to press. Had he come upon
this opinion of Bernoulli's at an earlier stage, Newton would surely have made much more of it.
We may guess that it was Keill, rather than Newton himself, who unearthed from the dusty
pages of the *Acta* of nearly a quarter of a century before this still-born derogation of Leibniz'
mathematical merit.

cujusợ specimina sunt tota illa Propositionum inibi contentarum farrago,) intellexerit [calculum] alterum a Domino Leibnitio inventum ignorare vix poterit; utpote qui in priori illo fundatus est, & nisi forte ın Differentialium notatione & operationis aliquo compendio ab eo non differt.[83]

(83) For what it is here worth, Jakob Bernoulli renewed his criticism at the end of his ensuing 'Specimen Alterum Calculi differentialis in dimetienda Spirali Logarithmica...' (*Acta Eruditorum* (June 1691): 282–90 [= *Opera*, **1**: 442–53]: 'Cæterum in his Problematibus omnibus, quæ quis nequicquam alia tentet methodo, calculi *Leibnitiani* eximium & singularem plane usum esse comperi, ut ipsum propterea inter primaria seculi nostri inventa censendum esse æstimem. Quanquam enim, ut nuper innui, ansam huic dedisse credam calculum *Barrovii*, qualem appello, qui ab hujus viri tempore passim fere apud Geometras præstantiores invaluit... ; [at] si quæ conferenti mihi utrumque intercedere inter illos visa est affinitas, ea major non est, quam quæ faciat, ut uno intellectu ratio alterius [*sc.* Leibniz] facilius comprehendatur; dum unus [*viz.* Barrow] superfluas & mox delendas quantitates adhibet, quas alter compendio omittit: de cætero... compendium isthoc tale est, quod naturam rei prorsus mutat, facitque ut infinita per hunc præstari possint, quæ per alterum nequeunt: præterquam etiam quod ipsum hoc compendium reperisse utique non erat cujusvis, sed sublimis ingenii & quod Autorem quam maxime commendat.' Leibniz put a swift end to this incipient squabble in his 'G.G.L. De Solutionibus Problematis Catenarii... aliisque a Dn. J. B. propositis' (*Acta Eruditorum* (September 1691): 435–9 (= *Leibnizens Mathematische Schriften*, **5**: 255–8]) with the velvet-gloved rebuke (*ibid.*: 437–9): 'Negare non possum mirifice mihi placuisse, quæ celeb. *Bernoullius* cum ingeniosissimo juvene fratre suo [Johanne] fundamentis calculi novi a me jactis inædificavit, idque eo magis, quod excepto acutissimo Scoto *Joh. Craigio*'—who on pages 27–9 of his *Methodus Figurarum Lineis Rectis & Curvis comprehensarum* (London, 1685; see VII: 3–4 on the Newtonian portion of this eclectic tract) gave, within months of its publication (and using its author's *d*-notation for differentials), a clear account of the 'nova Methodus Tangentes inveniendi' set forth by 'G. G. Leebnitius [*sic*] in *Actis Eruditorum* Anni superioris [1684]'— 'nondum mihi occurrerat, qui eo fuisset usus.... Cæterum quia ipse... conjicere voluit, qua occasione, aut quorum ante me scriptorum auxiliis potissimum ad has meditationes devenerim, placet id... candide aperire. Eram ego hospes plane in interiore Geometria, cum Lutetiæ Parisiorum A[nno] 1672 *Christiani Hugenii* notitiam nactus sum, cui certe viro, post *Galilæum & Cartesium*, ... me privatim plurimum debere agnosco. Hujus cum legerem librum de *Horologio Oscillatorio*, adjungeremque *Dettonvillæi* (id est *Pascalii*) *Epistolas* & *Gregorii a S. Vincentio Opus [Geometricum]*, subito lucem hausi... inexpectatam, quod mox speciminibus datis ostendi. Ita mihi sese aperuit ingens numerus theorematum, quæ corollaria tantum erant methodi novæ, quorum partem deinde apud *Jac. Gregorium* & *Isaacum Barrovium* aliosque deprehendi. Sed animadverti fontes non satis adhuc patuisse, & restare interius aliquid, quo pars illa Geometriæ sublimior tandem aliquando ad analysin revocari posset, cujus antea incapax habebatur. Ejus elementa aliquot abhinc annis publicavi, consulens potius utilitati publicæ quam gloriæ meæ, cui fortasse magis velificari potuissem methodo suppressa. Sed mihi jucundius est, ex sparsis a me seminibus natas in aliorum quoque hortis fructus videre. Nam nec mihi ipsi integrum erat hæc satis excolere, nec deerant alia, in quibus aditus novos aperirem, quo ego semper palmarium judicavi, ac methodos potius quam specialia licet vulgo plausibiliora æstimavi.'

In our own century the thesis that Leibniz owed a massive, unacknowledged primary debt to Barrow for the foundation of his own 'new method of tangents' has been urged in an extreme form by J. M. Child in introduction, footnotes and running commentary to his English translations of C. I. Gerhardt's original German editions of a number of Leibniz' shorter papers under the collective title of *The Early Mathematical Manuscripts of Leibniz*

‖ Now Dr Barrow in his method of Tangents[84] draws two Ordinates indefinitely neare one another & puts the letter *a* for the difference of the Ordinates & the letter *e* for the difference of the Abscissas, & for drawing the tangent gives these three Rules. 1 *Inter computandum*, saith he, *omnes abjicio terminos in quibus ipsarum a vel e potestas habeatur vel in quibus ipsæ ducantur in se. Etenim isti termini nihil valebunt.* 2 *Post æquationem constitutam omnes abjicio terminos literis constantes quantitates notas seu determinatas significantibus aut in quibus non habentur a vel e. Etenim illi termini semper ad unam æquationis partem adducti nihilum adæquabunt.* 3 *Pro a Ordinatam & pro e subtangentem substituo. Hinc demum subtangentis quantitas dignoscetur.* Thus far Dr Barrow.[85]

And Mr Leibnitz in his Letter of 21 June 1677 above mentioned wherein he first began to propose his Differential method, has followed this method of Tangents exactly excepting that he has changed the letters *a* & *e* of Dr Barrow into *dy* & *dx*. For in the Example wch he there gives he draws two parallel lines & sets all the terms below the under line in w^ch *dx* & *dy* are (severally or joyntly) of more then one dimension & all the terms above the upper line in w^ch *dy* & *dx* are wanting & for the reasons given by Dr Barrow makes all these terms vanish. And by the terms in w^ch *dy* & *dx* are but of one dimension & wch he sets between the two lines he determines the proportion of ye subtangent to ye ordinate. Well therefore did the Marquess de L'Hospital observe that where Dr Barrow left off Mr Leibnitz began: for their methods of Tangents are exactly ye same.

But Mr Leibnitz adds this improvement of the method that the conclusion is

---

(Chicago/London, 1920); see its index *s.v.* 'Barrow, indebtedness of Leibniz to, suggested' and 'Barrow's *Lectiones*, anticipation of theorems admitted by Leibniz/ Characteristic Triangle probably suggested by/ suggestion as to the way in which Leibniz studied'. The several sub-species of this conjectured vital rôle played by Barrow's *Lectiones Geometricæ* in the evolution of Leibniz' concepts and algorithm of calculus are each adequately refuted, from the solidly documented fact of his maturing mathematical development, by J. E. Hofmann (see especially his *Leibniz in Paris, 1672–1676*: 74–8).

(84) Made public by Barrow in his *Lectiones Geometricæ: In quibus (præsertim) Generalia Curvarum Linearum Symptomata declarantur* (London, ₁1670): Lectio X: 80–4; the three 'regulæ' for pursuing the 'calculus' of the 'methodus tangentium', which Newton proceeds to quote, are found in *ibid.*: 80–1.

(85) Barrow goes on, we may add, to apply his rules *mutatis mutandis*—determining the pertinent derivatives of algebraic powers and of the tangent function *ad hoc* in the several cases (compare J. M. Child's analysis of 'Exemp. V' in his *The Geometrical Lectures of Isaac Barrow* (Chicago/London, 1916): 122–3)—to draw tangents to curves in the five worked instances (*Lectiones*: 81–4) of the Gutschoven quartic $\sqrt{(x^2+y^2)} = ay/x$ (he had earlier in Lectio VIII §18 found its osculating hyperbola; see III: 269, note (602)); the simple conchoidal cubic $x^3+y^3 = a^3$; the Cartesian folium $x^3+y^3 = axy$ ('La Galande' Barrow christens it in his margin alongside after Descartes' disparaging comparison of Roberval's kaleidoscoping of its 'leaf' with a 'galand'—a bow of ribbon then *à la mode* at Paris—in his letter to Mersenne on 27 July 1638 (N.S.), newly printed by Clerselier in his *Lettres de Mr Descartes*, 3, ₁1667: 376); the quadratrix $y = x \tan(\frac{1}{2}\pi y/a)$; and the 'curve of tangents' $y = a \tan(x/a)$.

coincident wth the Rule of Slusius & shews how that Rule presently occurs to any one who understands this method. For Mr Newton had represented in his Letters that this Rule was a corollary of his general method.[86]

|| And whereas Mr Newton had said that his method in drawing of Tangents & ||[197] determining maxima & minima &c proceeded wthout sticking at surds, Mr Leibnitz in the next place shews how this method of tangents may be improved so as[87] not to stick at surds or fractions. And then adds: *Arbitror quæ celare voluit Newtonus de tangentibus ducendis, ab his non abludere. Quod addit, Ex hoc eodem fundamento quadraturas quoq̃ reddi faciliores, me in hac sententia confirmat; nimirum semper figuræ illæ sunt Quadrabiles quæ sunt ad æquationem differentialem.*[88] By wch words its manifest that Mr Leibnitz at this time understood that Mr Newton had a method wch would do all these things & had been examining whether Dr Barrows differential method of Tangents might not be extended to the same performances.[89]

[90]In November 1684 Mr Leibnitz published the Elements of his differential method in the *Acta Eruditorum*[91] & illustrated it wth Examples of drawing Tangents & determining maxima & minima, & then added: *Et hæc quidem initia sunt tantum Geometriæ cujusdam multo sublimioris ad difficillima et pulcherrima quæq̃ etiam mistæ Matheseos Problemata pertingentis, quæ sine calculo nostro differentiali* AVT SIMILI *non temere quisquam pari facilitate tractabit.* The words AVT SIMILI plainly relate to Mr Newtons method. And the whole sentence conteins nothing more

---

(86) In a tentative *addendum* on Add. 3968.27: 421^v (compare note (53) above) Newton subsequently had it in mind to adjoin: '& Mr Gregory that he had deduced it from y^e Method of Dr Barrow. And thus far Mr Leibnitz described nothing more then what he might learn from Dr Barrow & Mr Gregory.'

(87) In the draft on .39: 582^r 'may be enlarged so as to proceed in more unknown quantities then two &'.

(88) *Commercium*: 90 (=*Correspondence*, **2**: 215).

(89) '...& that his method was either the same with Dr Barrows method of Tangents improved & made general or another like it' he concluded in his draft (.39: 582^r).

(90) In draft Newton made a less jerky transition from his preceding paragraph by here inserting the linking phrase 'At length, vizt'. It is an all too typical sloppiness of his over temporal detail that he retards by a month the true publication date—*recte* 'In October 1684' (see the next note)—of the founding paper of Leibniz' on differential calculus which he now passes skimpily to epitomise, without mentioning its novelties of *d*-ist notation or its clear opening enunciation of the basic algorithms (never before set forth in print) for determining the derivative of the sum/difference, product and quotient of two quantities. (Had Newton forgotten the heavy weather which he himself had once made in framing equivalent rules in his own 'calculus' of the Cartesian subnormal? On which see 1: 322–41.)

(91) 'Nova methodus pro maximis & minimis, itemque tangentibus, quæ nec fractas, nec irrationales quantitates moratur, & singulare pro illis calculi genus, per G.G.L.' (*Acta Eruditorum* (October 1684): 467–73 [=*Leibnizens Mathematische Schriften*, **5** (Halle, 1858): 220–6]); the sentence which Newton cites in sequel is found on pages 472–3. An annotated English translation of this first publication of Leibniz' on differential calculus is given by D. J. Struik in his *A Source Book in Mathematics, 1200–1800* (Cambridge, Massachusetts, 1968): 272–80.

then what Mr Newton had affirmed of his general method in his Letters of 1672 and 1676.

And in the *Acta Eruditorum* of June 1686 pag 297^(92) Mr Leibnitz added: *Malo autem dx & similia adhibere quam literas pro illis quia istud dx est modificatio quædam ipsius x* &c. He knew that in this method he might have used letters with Dr Barrow but chose rather to use the new symbols *dx* & *dy*, tho there is nothing wch can be done by these symbols but may be done by single letters wth more brevity.

||[198] || The next year Mr Newtons *Principia Philosophiæ* came abroad, a book full of such Problemes as Mr Leibnitz had called *difficillima & pulcherrima etiam mistæ Matheseos*....^(93) And the Marquess de L'Hospital has represented this book *presque tout de ce calcul*,^(94) composed almost wholly of this calculus. And Mr Leibnitz himself in a Letter to Newton dated from Hannover $\frac{7}{17}$ March 1693 & still extant in his own hand writing & upon a late occasion communicated to the Royal Society,^(95) acknowledged the same thing in these words: *Mirifice ampliaveras Geometriam tuis seriebus, sed edito Principiorum opere ostendisti patere tibi etiam quæ analysi receptæ non subsunt. Conatus sum ego quoq notis commodis adhibitis quæ differentias & summas exhibeant, Geometriam illam quam Transcendentem appello Analysi quodammodo subjicere, nec res male processit.*^(96)... In the second Lemma of the second book of these *Principles* the Elements of this calculus are demonstrated synthetically & at the end of the Lemma there is a Scholium in these words. *In Literis quæ mihi cum Geometra peritissimo G. G. Leibnitio annis abhinc decem inter-*

---

(92) In an essay-review of Craige's *Methodus Figurarum...Quadraturas Determinandi* of the previous year (see note (83) above) under the title 'G.G.L. De geometria recondita et analysi indivisibilium atque infinitorum...' (*Acta Eruditorum* (June 1686): 292–300 [=*Math. Schriften*, 5: 226–33]). Newton chooses not to record that Leibniz' present sentence follows immediately upon his first employment in print of the 'long-*s*' sign for integration ('...nobis summæ & differentiæ seu ∫ & *d*, reciprocæ sunt' in his classic definition of this operation as inverse-differentiation); see Struik's English translation of Leibniz' complete paragraph in his *Source Book...*: 282.

(93) In his article in the October 1684 *Acta*; see Newton's previous paragraph but one.

(94) This (Newton fights shy of mentioning) there set as grounds for rendering him the due 'justice' that, in addition to Leibniz, 'il avoit aussi trouvé quelque chose de semblable au Calcul différentiel, comme il paroît...' (*Analyse des Infiniment Petits* (Paris, 1696): signature éii^v/éiii^r); we made full quotation of this passage in L'Hospital's 'Preface'—which Newton elsewhere used to support his broader claim (as he put it on Add. 3968.41: 22^v) that 'the Analysis by which it [the *Principia*] was invented shines through the composition'—in **1**, 8.1: note (15) preceding. See also note (137) below.

(95) By Newton himself, understand, simply so that he here might quote from it without appearing publicly to enter the dispute over calculus priority as a 'parti-man' furnishing private documents in defence of his claim. He himself was anonymously to publish the full text of Leibniz' letter to him in the 1717 Appendix to Raphson's *History of Fluxions* (see note (145) to our preceding introduction).

(96) Raphson, *History*, ₂1717: 119 (=*Correspondence*, **3**, 1961: 257).

cedebant, cum significarem me compotem esse methodi determinandi Maximas & Minimas,
ducendi Tangentes & similia peragendi quæ in terminis surdis æque ac in rationalibus
procederet, & literis transpositis hanc sententiam involventibus (Data æquatione quotcunq
fluentes quantitates involvente, fluxiones invenire & vice versa) eandem celarem: rescripsit
Vir clarissimus se quoq in ejusmodi methodum incidisse & methodum suam communicavit
a mea vix abludentem præterquam in verborum & notarum formulis. Utriusq fundamentum
continetur in hoc Lemmate.[97] In those Letters & in another dated 10 Decem. 1672,
a copy of wch at that time [98] was sent to Mr Leibnitz by Mr Oldenburgh as is
mentioned above. Mr Newton had so far explained his method that it was not
difficult for ‖ Mr Leibnitz by the help of Dr Barrows method of Tangents to   ‖[199]
collect it from those Letters....

Dr Wallis had received copies of Mr Newtons two Letters of 13 June &
24 Octob. 1676 from Mr Oldenburgh[99] & published several[100] things out of
them in his *Algebra* printed in English 1683 & in Latin 1693.[101] And soon after
had intimation from Holland to print ye Letters intire because Mr Newtons

---

(97) *Principia*, ₁1687: 253–4. The scholium passed unaltered into the second edition (₂1713:
226–7) except for the addition of the phrase '& Idea generationis quantitatum' to its penultimate sentence; but with the onset of the open and guerrilla warfare over calculus priority in
the middle and late 1710's Newton projected a number of major recastings of its text—see
Appendix 4 below, and especially the 'fuller recounting' of the 'purpose' of the scholium
reproduced in Appendix 5 following—to make it clear to his reader that he did not here allow
Leibniz an equal claim to independent discovery, and in the brand-new scholium which
replaces the present one in the *Principia*'s *editio ultima* (₃1726: 246; the text is quoted in the final
footnote to Appendix 4) direct citation of the Latinised words of Newton's letter to Collins on
10 December 1672 (see page 580 above) alone is made, without reference to his *epistola posterior*
to Leibniz in 1676, or to the latter's reply in June 1677.

(98) The list of *corrigenda* on Add. 3968.13: 223^r (see note (11)) here directs '*for* at that time
read four years after', and so it was rendered *Latine* in the 'Recensio' (*Commercium*, ₂1722: 31).

(99) Though not, it would appear, with Newton's own approval or prior permission; see
IV: 671–2, note (54) (and compare II: xiii–xiv).

(100) Newton's draft on Add. 3968.41: 143^r has 'many', which indeed is very much an
exaggeration (see the next note).

(101) '& printed in Latin in the second volume of his Works A.C. 1693' reads the slightly
fuller draft on .41: 143^r; the preceding '1683' advances by two years the date of publication
of the parent *Treatise of Algebra, Both Historical and Practical* (London, 1685), but that of Wallis'
revised and augmented *De Algebra Tractatus; Historicus & Practicus* [=*Johannis Wallis...*
Operum Mathematicorum Volumen alterum: 1–482] is correct. In fact in Chapters LXXXV and
XCI–XCV of his *Algebra* (₁1685: 318–20, 330–47 [= ₂1693: 357–9, 368–77/381–90]) Wallis
cited out of Newton's *epistolæ prior et posterior* only the passages where he presented and applied
to series-quadrature of curves his general binomial expansion, and where he expounded his
broader technique for extracting the root of an 'affected' numerical or algebraic equation
term by term as a 'converging' infinite sequence. As we mentioned in note (81) above, Newton
in an added *corrigendum* (on .13: 223^r) to his preceding 'Pag 195' subsequently thought to
inform his future reader that the full texts of his two 1676 letters were 'none of them...
printed before the year 1699' when Wallis set them among the 'Epistolarum Collectio' in the
third volume of his *Opera Mathematica*.

notions of fluxions passed there wth applause by the name of the Differential Method of Mr Leibnitz, & thereupon he took notice of this matter in the Preface to the first Volume of his Works published A.C. 1695. And in a Letter to Mr Leibnitz dated 1 Decemb. 1696^(102) he gave account of it.... And in a Letter dated 10th April 1695 & lately communicated to the Royal Society^(103) he wrote ||[200] thus....|| ...

The short Intimation of this matter wch Dr Wallis inserted into y^e said Preface

(102) In the only footnote to the printed 'Account', Newton added the information at this point (switching to Latin for no reason evident to us): 'Extat hæc Epistola in tertio volumine operum *Wallisii*' (*viz.* pages 653–5 [= *Leibnizens Mathematische Schriften*, **4** (Halle, 1859) : 4–10]). He here went on to cite (with some minimal adaptation) a short passage in Wallis' letter (*Opera*, **3**: 654 [= *Math. Schriften*, **4**: 7]) where the latter had written: 'Neque Calculi differentialis vel nomen audiveram, nisi postquam utrumque volumen absolverant operæ'—the second volume (containing the Latin *Algebra* as its principal item) in fact appeared first, in spring 1693, while the *volumen primum* (collecting Wallis' earlier pieces) lingered 'in press' for almost two years more—'eratque Præfationis (præfigendæ [*sc.* ad volumen primum]) postremum folium sub prelo, ejusque typos jam posuerant typothetæ. Quippe tum me monuit amicorum quidam (harum rerum gnarus) qui peregre fuerat'—we presume this was Wallis' fellow Savilian professor David Gregory, who had paid an extended visit in late spring/early summer 1693 to the Netherlands, there meeting a number of Dutch mathematicians with whom he exchanged a wide variety of such gossip—'tum talem methodum in Belgio prædicari, tum illam cum Newtoni methodo Fluxionum coincidere. Quod fecit ut (transmotis typis jam positis) id monitum interseruerim.'

(103) By its recipient Newton—is his omission here to name himself as receiver intended? —so that, as with Leibniz' letter to him in March 1693 (see note (95) preceding), he may quote from its text 'from the original in the Society's archives' without publicly compromising his Presidential impartiality. In sequel he cited, as two years later in his 1717 Appendix: 120–1 to Raphson's *History of Fluxions* (*vide* note (145) of our above introduction), only the presently pertinent sentences in its second paragraph where Wallis had delivered the mild rebuke: 'I wish you would also print the two large Letters of June and August'—'he means October', Newton wrote in a parenthesis at this point in his transcription in the printed 'Account' (page 199, lines 23–4; Wallis made the same error in his 'intimation' in preface to his *Opera*, but Newton silently corrects the slip in his following quotation of this)—'1676. I had intimation from Holland, as desired there by your friends, that somewhat of that kind were done; because your Notions (of *Fluxions*) pass there with great applause, by the name of Leibnitz's *Calculus Differentialis*. I had this intimation'—from David Gregory (see the previous note)—'when all but (part of) the Preface to this Volume was Printed-off; so that I could onely insert (while the Press stay'd) that short intimation thereof which you there find. You are not so kind to your Reputation (& that of the Nation) as you might be, when you let things of worth ly by you so long, till others carry away the Reputation that is due to you. I have endeavoured to do you justice in that point; and am now sorry that I did not print those two letters *verbatim*'. (Wallis' full letter was to be published only much more than a century afterwards—by Joseph Edleston in appendix to his edition of *The Correspondence of Sir Isaac Newton and Professor Cotes* (London, 1850): 300–1 [= *Correspondence of Newton*, **4**, 1967: 100–1]—when it had ceased to have any but historical interest. We have earlier observed, on VII: xxvi, that Newton's reaction in spring 1695 to those stirring words was, whether through indecision or a reluctance to trust his two finest mathematical letters to Wallis' not always delicate editorial hands, to dither and procrastinate.)

was in these words. *In secundo Volumine (inter alia) habetur* Newtoni *Methodus de Fluxionibus (ut ille loquitur)*[104] *consimilis naturæ cum* Leibnitii *(ut hic loquitur) Calculo Differentiali (quod qui utramꝗ methodum contulerit satis animadvertat, utut sub loquendi formulis diversis) quam ego descripsi* (Algebræ *Cap. 91 &c. præsertim Cap 95*) *ex binis* Newtoni *Literis...Junii 13 & Octob.*[105] *24. 1676 ad* Oldenburgium *datis cum* Leibnitio *communicandis (iisdem fere verbis, saltem leviter mutatis, quæ in illis literis habentur) ubi* METHODUM HANC LEIBNITIO EXPONIT, *tum ante* DECEM ANNOS *nedum plures* (id est, anno *1666* vel *1665*) *ab ipso excogitatam. Quod moneo, nequis causetur de hoc Calculo Differentiali nihil à nobis dictum esse.*[106]

Hereupon the Editors of the *Acta Lipsiensia* the next year in June, in the style of Mr Leibnitz,[107] in giving an account of these two first Volumes of Dr Wallis took notice of this clause of the Doctors Preface & complained not of his saying that Mr Newton in his two Letters above mentioned explained to Mr Leibnitz the method of fluxions found by him ten years before or above, but that while the Doctor mentioned the Differential calculus & said that he did it *ne quis causetur de calculo differentiali nihil ab ipso dictum fuisse*, he did not tell the reader that Mr Leibnitz had this calculus at that time when those Letters passed between him & Mr Newton by means of Mr Oldenburg. And in several Letters wch followed hereupon between Mr Leibnitz & Dr Wallis concerning this matter, Mr Leibnitz denied not that Mr Newton had the method ten years before the writing of those Letters as Dr Wallis had affirmed, pretended not that he himself had the method so early, brought no proof that he had it before the year ‖ ‖[201] 1677, no other proof besides the concession of Mr Newton that he had it so early, affirmed not that he had it earlier, commended Mr Newton for his candor in this matter, allowed that the methods agreed in the main & said that he therefore used to call them by the common name of his[108] Infinitesimal Analysis, repre-

---

(104) The pertinent pages 390–6 from the second volume of Wallis' collected *Opera Mathematica* are reproduced on VII: 170–80.

(105) Silently corrected by Newton from Wallis' mistaken '*Augusti*' (on which slip see also note (103) preceding).

(106) *Ad Lectorem Præfatio*, p. [iii] (= *Opera*, 1, 1695: signature a4^r), lines 1–10. The small capitals here employed by Newton in emphasis are, we need not say, his own, and the italicisations/roman 'italics' elsewhere do not rigorously follow Wallis' original.

(107) This lengthy review in *Acta Eruditorum* (June 1696): 249–59 was, we may now know, indeed by Leibniz. His (exceedingly polite) 'complaint' that Wallis did not in his preface make any detailed mention of his own *Calculus differentialis*, making it clear that it was an 'Analyticum...summæ reciprocum' and stressing its superiorities over both classical Apollonian and even recent Cartesian modes of geometrical analysis in being able to attack 'multo magis abstrusa', will be found on *ibid.*: 258 (reprinted in *Commercium*: 98–9).

(108) The list of *corrigenda* on Add. 3968.13: 223^r (regarding which see note (11) above) here justly orders '*for* his *write* the'. Leibniz' original words in his letter to Wallis on 28 May 1697 (N.S.) (first published by Wallis in his *Opera*, 3, 1699: 678–80 [= *Leibnizens Mathematische Schriften*, 4, 1859: 23–9 in its fullest version as sent]; the central portion is reprinted in *Com-*

sented that as the methods of Vieta & Cartes were called by the common name of Analysis speciosa & yet differed in some things so perhaps the methods of Mr Newton & himself might differ in some things, & challenged to himself only those things wherein as he conceived they might differ; naming the notation, differential equations & Exponential Equations. But in his Letter of 21 June 1677 he recconed differential equations common to Mr Newton & himself.

This was the state of the dispute between Dr Wallis & Mr Leibnitz at that time. And four years after, when Mr Fatio suggested[109] that Mr Leibnitz the second inventor of this calculus might borrow something from Mr Newton the oldest inventor by many years: Mr Leibnitz in his Answer published in the *Acta*

---

mercium: 104–6) were: '*Methodum Fluxionum* profundissimi Newtoni cognatam esse *Methodo meæ Differentiali*…monui…. Itaque communi nomine designare soleo *Analyseos Infinitesimalis*…. Interim, quemadmodum & Vietæa & Cartesiana methodus *Analyseos Speciosæ* nomine venit, discrimina tamen nonnulla supersunt: ita fortasse & Newtoniana & Mea differunt in nonnullis…. Et licet fatear, quemadmodum rem ipsam in æquationibus curvarum facilioribus…tenebant Veteres [ut et calculo suo Cartesius], ita rem ipsam meis æquationibus facilioribus expressam, non potuisse Tibi…esse ignotam: non ideo minus tamen puto & Cartesium & Me aliquid utile præstitisse. Nam antequam talia ad constantes quosdam characteres calculi analytici reducuntur, tantumque omnia vi mentis et imaginationis sunt peragenda, non licent in magis composita abditaque penetrare, quæ tamen, calculo semel constituto, lusus…jocusque videntur'—which, to say the least, minimises the difficulties of solution not a little. As instances of the 'Transcendentales quantitates' which he could now bring within the scope of his general analytical method Leibniz went on to cite the 'Ordinata Cycloidis [cujus circulus generans habet diametrum *a*] methodo mea expr[essa] per Æquationem [differentialem] $y = \int \dfrac{x\,dx}{\sqrt{ay-yy}}$'—in the *Commercium* at this point Newton rightly observed '*Legendum* $y = \int \dfrac{x\,dx}{\sqrt{ax-xx}}$', nor did he lose the opportunity to notice that, in the notation of his own *De Quadratura* (see VII: 510), 'Idem sic designari potest $y = \dfrac{x\dot{x}}{\sqrt{ax-xx}}$, vel sic $\dot{y} = \dfrac{x\dot{x}}{\sqrt{ax-xx}}$'—and also the 'incognita *x* [quæ] exprimitur per hanc Æquationem [exponentialem] $x^x + x = 1$'. Newton was, let us add, something less than happy about the way in which, in this (originally) private letter to Wallis, 'Mr Leibnit[z] compared Mr Newton & himself to Vieta & Descartes in respect of what was common to theire methods' (so he chose to phrase it in his draft, on .41: 100ᵛ, in a reiterated sentence subsequently discarded by him from his following paragraph); but he did not afterwards vigorously pursue the points in which this tentative juxtaposition of himself *vis-à-vis* Leibniz as representing the classical *v.* the modern in mathematical analysis cannot be sustained.

(109) On page 18 of his *Lineæ Brevissimi Descensus Investigatio Geometrica Duplex*. We have quoted the pertinent passage at length in note (6) to the preceding intoduction, but Newton here appeals only to Fatio's central sentence (reprinted on *Commercium*: 107): 'Newtonum… *primum*, ac pluribus Annis vetustissimum, hujus Calculi Inventorem, ipsa rerum evidentia coactus, agnosco: a quo utrum quicquam mutuatus sit Leibnitius, secundus ejus Inventor, malo eorum, quam meum, sit Judicium, quibus visæ fuerint Newtoni litteræ, aliique ejusdem Manuscripti Codices'.

Lipsiensia in May 1700,[110] did not deny that Mʳ Newton was the oldest inventor by many years nor asserted any thing more to himself then that he had found the method apart or without the assistance of Mʳ Newton & pretended that when he first published it he knew not that Mʳ Newton had found any thing more of it then the method of tangents. And in making this defence he added:[111] *Quam* (methodum) *ante Dominum Newtonum et Me nullus quod sciam Geometra habuit; uti ante hunc maximi nominis Geometram* NEMO *specimine publice dato se habere probavit, ante Dominos Bernoullios et Me nullus communicavit.* Hitherto therefore Mʳ Leibnitz did not pretend to be the first inventor. He did not begin to put in such a claim till after the death of Dr Wallis, the last of the old men[112] who were acquainted with what had passed between the English & Mʳ Leibnits ‖ forty[112] years ago. The Doctor died in October A.C. 1703 & Mʳ Leibnitz began not to put in this new claim before January 1705.   ‖[202]

Mʳ Newton published his Treatise of *Quadratures* in the year 1704. This Treatise was written long before, many things being cited out of it in his Letters of 24 Octob. & 8 Novemb. 1676. It relates to the method of fluxions & that it might not be taken for a new piece Mʳ Newton repeated what Dʳ Wallis had published nine years before without being then contradicted, namely that this method was invented by degrees in the years 1665 & 1666. Hereupon the Editors of the *Acta Lipsiensia* in January 1705, in the style of Mʳ Leibnitz,[113] in giving an Account of this book represented that Mʳ Leibnitz was the first inventor of the method & that Mʳ Newton had substituted fluxions for differences. And this accusation gave a beginning to this present controversy.

---

(110) 'G. G. Leibnitii Responsio ad Dn. Nic. Fatii Duillerii Imputationes', *Acta Eruditorum* (May 1700): 198–206 (compare page 472, note (8) above). Newton in sequel paraphrases the first extract (*ibid.*: 203) from Leibniz' general concluding observations on the recent development of mathematical analysis which he had earlier reprinted on *Commercium*: 107.

(111) The list of *corrigenda* on Add. 3968.13: 223ʳ (see note (11)) here directs there to be inserted 'concerning a branch of this method by wᶜʰ Mʳ Newton A.C. 1686'—as he then summarily recorded in the Scholium to Prop. XXXV of the second book of his *Principia* (₁1687: 326–7; see VI: 460–4)— 'found the *solidum minimæ resistentiæ*'. Here, as in the parent excerpt printed on *Commercium*: 107, the emphasis on 'nemo' in the ensuing extract, from page 206 of Leibniz' 'Responsio' to Fatio, is here set by Newton and not to be found in the original.

(112) 'old mathematicians' and '40 or 50' respectively in Newton's rough draft on .41:100ᵛ.

(113) And of course this review in *Acta Eruditorum* (January 1705): 34–6 of Newton's *De Quadratura Curvarum*—the two final pages of a joint recension of the 'Tractatus Duo, de Speciebus & Magnitudine Figurarum Curvilinearum' appended to the 1704 *Opticks* which also embraced the prior *Enumeratio* (*ibid.*: 30–6)—was indeed from Leibniz' hand. We find it surprising that Newton does not in sequel give exact quotation of the 'accusation' in the *Acta* review (*ibid.*: 35) that 'Pro differentiis Leibnitianis Dn. Newtonus adhibet semperque adhibuit fluxiones...', but perhaps he considered that its simple, forceful translation into English brought out its sense beyond all possible shading of doubt? The interested reader could, we need scarcely add, at once backtrack to consult the extended excerpt from the *Acta* review printed in *Commercium*: 108–9.

For Mr Keil in an Epistle published in the *Philosophical Transactions* for Sept. &
Octob. 1708[114] retorted the accusation, saying: *Fluxionum Arithmeticam sine
omni dubio primus invenit D. Newtonus, ut cuilibet ejus Epistolas a Wallisio editas
legenti facile constabit. Eadem tamen Arithmetica postea mutatis nomine & notationis
modo a Domino Leibnitio in Actis Eruditorum edita est.*

Before Mr Newton saw what had been published in the *Acta Leipsica* he
expressed himself offended at the printing of this Paragraph of Mr Keill's Letter
least it should create a controversy. And Mr Leibnitz understanding this in a
stronger sense then Mr Keill intended it, complained of it as a calumny[115] in
a Letter to Dr Sloan dated 4 March *st. n.* 1711, & moved that the Royal
Society would cause Mr Keill to make a publick recantation. Mr Keill chose
rather to explain & defend what he had written, & Mr Newton upon being
shewed the accusation in the *Acta Leipsica*[116] gave him leave to do so. And
Mr Leibnitz in a second Letter to Dr Sloan dated 29 Decem. 1711 instead of
‖[203] making good his accu‖sation as he was bound to do that it might not be deemed
a calumny, insisted only upon his own candor as if it would be injustice to
question it, & refused to tell how he came by the method, & said that the *Acta
Lipsica* [117]had given every man his due, & that he had concealed the invention
above nine years[118] (he should have said seven years) that no body might

---

(114) 'Epistola ad Edmundum Halleium...de Legibus Virium Centripetarum', *Phil. Trans.*
26, No. 317: 174–88. The short paragraph on page 185 in which Keill set down his provocative aside which Newton proceeds to quote (from *Commercium*: 109, but with the omission
of its opening phrase) is rendered by us in English on page 473 above.

(115) 'gave his reasons against it' Newton wrote much more weakly at this point in his
draft on Add. 3968.16: 255^r. He carefully refrains in either case from stating these 'reasons'
why, in his letter of complaint to Sloane (printed in full in *Commercium*: 109–10 [=*Correspondence*, 5, 1975: 96]), Leibniz found 'calumny'—unmaliciously intended as it might have
been—in Keill's words.

(116) At the beginning of April, namely; see page 475 above. What broad advice Newton
may then have given on how best to refute Leibniz' complaint is not known, but Keill surely did
not proceed to seek to demonstrate the truth of what he had before written without some prior
briefing. Some weeks later, certainly, Newton added a number of finishing touches (on which
see note (25) of the preceding introduction) to the final text of the long reply which Keill
subsequently, in late May, returned through Sloane (*Commercium*: 110–18 [=*Correspondence*, 5:
133–41]).

(117) In his ensuing list of *corrigenda* on Add. 3968.13: 223^r (see note (11)) Newton here
commands there to be inserted 'had not in the least detracted from any body but' so as fully
to render Leibniz' declaration in his December 1711 letter that 'in illis [*Actis Lipsiensibus*
Januarii 1705] circa hanc rem quicquam cuiquam detractum non reperio, sed potius passim
suum cuique tributum' (*Commercium*: 119 [=*Correspondence*, 5: 207]).

(118) Before, understand, he made publication to the world of his 'Nova Methodus pro
Tangentibus...' in the *Acta Eruditorum* in 1684 (see note (91)). Since the mid-nineteenth
century the published witness of Leibniz' major early papers on the topic, many dated, has
amply supported his many assertions in his lifetime that he had already fashioned the basic
algorithms and notation of his *calculus differentialis* almost a decade earlier; but this wealth of

pretend...to have been before him in it, & called Mr Keill a novice[119] unacquainted with things past & one that acted without authority from Mr Newton & a clamorous man who deserved to be silenced, & desired that Mr Newton himself would give his opinion in the matter. He knew that Mr Keill affirmed nothing more then what Dr Wallis had published thirteen years before without being then contradicted. He knew that Mr Newton had[120] given his opinion in the Introduction to his book of *Quadratures* published before this controversy began, but Dr Wallis was dead & the mathematicians which remained in England were novices. Mr Leibnitz may question any man's candor without injustice & Mr Newton must now retract what he had published or not be quiet.

The Royal Society therefore having as much authority over Mr Leibnitz as over Mr Keil & being now twice pressed by Mr Leibnitz to interpose & seeing no reason to condemn or censure Mr Keil without inquiring into the matter, & that neither Mr Newton nor Mr Leibnitz (the only persons alive who knew & remembred any thing of what had passed in these matters forty years ago) could be witnesses for or against Mr Keill, appointed a numerous Committee to search old Letters & Papers & report their opinion upon what they found, & ordered the Letters & Papers with the Report of their Committee to be published. And by these Letters & Papers it appeared to them that Mr Newton had the Method[121] in or before the year 1669 & it did not appear to them that Mr Leibnitz had it before the year 1677.

‖ For making himself the first inventor of the differential method he has ‖[204] represented that Mr Newton at first used the letter o in the vulgar manner for the given increment of x, wch destroys the advantages of the differential method; but after the writing of his *Principles*, changed o into $\dot{x}$, substituting $\dot{x}$ for dx. It lies

---

evidence lay obscurely tucked away in his private manuscripts throughout Newton's lifetime, and indeed for more than a century afterwards. In 1714, when the latter wrote his present 'Account' of the *Commercium*, Leibniz' first documented reference to his 'new method for tangents' remained (see Newton's page 194 above) the sketch of it which he set out two years afterwards in his letter of 21 June 1677 (N.S.) to Oldenburg; and when Newton printed Leibniz' December 1711 letter to Sloane, it was not mere lack of generosity but a demand for better demonstration than just verbal statement unsupported by independent evidence which led him to attach to Leibniz' declaration that by 1684 'inventum plusquam nonum in annum pressi' a laconic rejoinder 'Probandum est' (*Commercium*: 119).

(119) '*for* novice', orders Newton's list of *corrigenda* on .13: 223^r (see note (11)), '*write* new-man', thus more precisely rendering Leibniz' own distinction between the total beginner and the trained but inexperienced graduate in dubbing Keill a 'homo doctus, sed novus, & parum peritus rerum anteactarum cognitor' (see *Commercium*: 118). In a less successful attempt to express the nuance, the 'Recensio' retranslates Leibniz' 'homo novus' back as 'homo juvenis' (see *Commercium*, ₂1722: 35).

(120) Newton's preparatory draft on .16: 255^r adds 'already'.

(121) 'of fluxions & moments' is deleted in sequel on .16: 255^r.

upon him to prove that Mʳ Newton ever changed o into $\dot{x}$ or used $\dot{x}$ for dx, or left off the use of the letter o. Mʳ Newton used the letter o in his *Analysis* written in or before yᵉ year 1669, & also in his book of *Quadratures*, & in his *Principia Philo-sophiæ*, & still uses it in the very same sense as at first. In his book of *Quadratures* he used it in conjunction with the symbol $\dot{x}$, & therefore did not use that symbol in its room. These symbols o and $\dot{x}$ are put for things of a different kind. The one is a moment, the other a fluxion or velocity as has been explained above.[122] When the letter x is put for a quantity wᶜʰ flows uniformly, the symbol $\dot{x}$ is an unit & the letter o a moment & $\dot{x}o$ & dx signify the same moment. Prickt letters never signify moments unless when they are multiplied by the moment o either exprest or understood to make them infinitely little, & then yᵉ rectangles are put for moments.

Mʳ Newton doth not place his method in forms of symbols nor confine himself to any particular sort of symbols for fluents or fluxions. Where he puts the Areas of Curves for fluents he frequently[123] puts the Ordinates for fluxions & denotes the fluxions by the symbols of the Ordinates, as in his *Analysis*. Where he puts lines for fluents he puts any symbols for the velocities of the points wᶜʰ describe the lines, that is, for the first fluxions, & any other symbols for the increase of those velocities, that is, for the second fluxions, as is frequently[124] done in his *Principia Philosophiæ*. And where he puts the letters x, y, z for fluents, he denotes their fluxions either by other letters as p, q, r, or by the same letters ‖[205] in other forms as X, Y, Z[125] or $\dot{x}$, $\dot{y}$, $\dot{z}$, or by ‖ any lines as DE, FG, HI[126] con-sidered as their exponents. And this is evident by his book of *Quadratures* where he represents fluxions by prickt letters in the first Proposition, by Ordinates of Curves in the last Proposition & by other symbols in explaining the method &

---

(122) On Newton's page 180.

(123) Newton wrote 'usually' in his draft on .16: 256ʳ: an entirely accurate emphasis on his frequency of usage, it seems to us.

(124) This 'frequently', on the other hand, merits an exclamation mark! Newton in the *Principia* widely makes use of the notion of accelerative/decelerative fluxional increase in speed, but while for example in Proposition **XXXIX** of Book 1 he denotes a general *velocitas* by V and its *incrementum* (in vanishingly small time) by I, and elsewhere in Proposition X of Book 2— which on page 207 in his present 'Account' he insists uses but the 'terms of a converging series for solving of [its] Probleme'—he sets out the Taylor expansion of an incremented ordinate where (in its initial 1687 version; see **1**, 6, Appendix 1 preceding) the base variable is time, representing its successive coefficients by Q, R, S, ..., we know no instance in Newton's book where he (in any edition of it) gave an explicit symbol to the rate of change of a speed, or indeed to the 'force' engendering this second fluxion of a fluent line.

(125) This claimed usage by Newton of upper-case fluxions corresponding to lower-case fluents comes as news to us, and we must presume that he misremembers his occasional employment (compare ɪᴠ: 430 and ᴠɪɪ: 448) of related lower-case letters to denote *vice versa* the contemporaneous increments of upper-case fluents.

(126) The related increments, understand, of (say) $AD = x$, $BF = y$ and $CH = z$ and hence, where the increment o of time comes to be zero, in proportion as $o\dot{x} : o\dot{y} : o\dot{z} = \dot{x} : \dot{y} : \dot{z}$.

illustrating it with examples in the Introduction.[127] Mr Leibnitz hath no symbols of fluxions in his method, & therefore Mr Newtons symbols of fluxions are the oldest in the kind. Mr Leibnitz began to use the symbols of moments or differences *dx, dy, dz* in the year 1677: Mr Newton represented moments by the rectangles under the fluxions & the moment *o* when he wrote his *Analysis*, which was at least forty six[128] years ago. Mr Leibnitz has used the symbols $\int x, \int y, \int z$ for the summs of Ordinates ever since the year 1686:[129] Mr Newton represented the same thing in his *Analysis* by inscribing the Ordinate in a square or rectangle.[130] All Mr Newtons symbols are the oldest in their several kinds by many years.[131]

And whereas it has been represented[132] that the use of the letter *o* is vulgar & destroys the advantages of the Differential Method: on the contrary the method of fluxions as used by Mr Newton has all the advantages of the differential & some others. It is more elegant because in his calculus there is but one infinitely little quantity represented by a symbol, the symbol *o*. We have no ideas of infinitely little quantities & therefore Mr Newton introduced fluxions into his method that it might proceed by finite quantities as much as possible. It is more natural & geometrical because founded upon the *primæ quantitatum nascentium rationes* wch have a being in Geometry, whilst *indivisibles* upon which the Differential method is founded have no being either in Geometry or in nature. There are *rationes primæ quantitatum nascentium*, but not *quantitates primæ nascentes*.

---

(127) See vII: 512–14/550–2 and pages 122–30 above.

(128) On subtracting (July) 1669 from (January/February) 1715.

(129) When he briefly introduced it in his essay-review of Craige's *Methodus Figurarum*... in the *Acta Eruditorum*; see note (92).

(130) The *corrigenda* list on Add. 3968.13: 223ʳ directs here to add 'for there he uses the symbol $\boxed{\dfrac{aa}{64x}}$ in the same sense in wch Mr Leibnitz uses the symbol $\int \dfrac{aa}{64x}$'. (See II: 226.)

(131) Not so in the case of Newton's □ for 'square'. We have already observed in I: 147, note (6) that Pietro Mengoli in his *Geometria Speciosa* (Bologna, 1659) used the letter O to sum *omnes ordinatæ* under a curve in exactly the same 'indivisibles' sense some half a dozen years before Newton himself felt the need to devise his own equivalent symbol for integration.

(132) In his draft on .16: 256ᵛ Newton wrote explicitly that 'Mr Leibnitz has represented'. And indeed—though he cannot yet, in late 1714, 'dare' to say so openly—it was Leibniz who in his *alter ego* as author of the July 1713 *Charta volans* had interpolated in Johann Bernoulli's included 'Alterum indicium, quo conjicere licet, Calculum fluxionum non fuisse natum ante Calculum differentialem'—'hoc est, quod veram rationem fluxiones fluxionum capiendi... Newtonus nondum [in 1687] cognitam habuerit, quod patet ex ipsis *Principiis Phil. Math.* [sc. Lib. II, Prop. X] ubi...incrementum constans ipsius *x*, quod nunc notaret per *x* punctatum uno puncto [!], designat per *o*'—a parenthesis stating that this denotation of the increment of the base variable by a separate lower-case letter 'more vulgari' (that is, in the fashion of Descartes, Fermat and especially—see note (9) above—James Gregory) was one 'qui calculi differentialis commoda destruit'. (see *Correspondence*, **6**, 1976: 16; and compare the corresponding portion of Bernoulli's letter to Leibniz on 7 June 1713 (N.S.), conveniently reprinted *ibid.* : 2 from Gerhardt's *Leibnizens Mathematische Schriften*, **3**.2: 911.)

Nature generates quantities by continual flux or increase, & the ancient Geometers admitted such a generation of areas & solids when they drew one
||[206] line into another by local motion || to generate an area & the area into a line by local motion to generate a solid. But the summing up of indivisibles to compose an area or solid was never yet admitted into Geometry. Mᵣ Newtons method is also of a greater use & certainty[133] being adapted either to the ready finding out of a Proposition by such approximations as will create no error in the conclusion or to the Demonstrating it exactly: Mᵣ Leibnitz's is only for finding it out. When the work succeeds not in finite equations Mᵣ Newton has recourse to converging series & thereby his method becomes incomparably more universal then that of Mᵣ Leibnitz wᶜʰ is confined to finite equations: [134]for he has no share in the method of infinite series. Some years after the method of series was invented Mᵣ Leibnitz invented a Proposition for transmuting curvilinear figures into other curvilinear figures of equal areas[135] in order to square them by converging series, but the methods of squaring those other figures by such series were not his.[136] By the help of this new Analysis Mᵣ Newton found out most of the

---

(133) 'extent' was first written in the draft on .16: 256ᵛ.

(134) In first draft on .13: 177ᵛ/178ʳ Newton initially concluded his paragraph: 'And when the law of the fluxions is not known but the fluxions are had only in a few particular cases, Mᵣ Newton finds that law *quamproxime* by drawing a Curve line through any number of given points, & thence deduces the solution of the Probleme. And to this degree of perfection Mᵣ Newton had brought his method sometime before the year 1676, as appears by his *Analysis* & his Letters of 10 Decem. 1672, & 13 June & 24 Octob. 1676. And when he wrote his *Principia Philosophiæ*, he had recourse upon all occasions to this method & most frequently to that part of it wᶜʰ is contained in his book of *Quadratures*. And because by the law of the Ancients Propositions invented by Analysis were not admitted into Geometry till they were demonstrated synthetically, the Propositions which he found out by his Analysis, he composed, which makes it now difficult for unskilful men to see the Analysis by wᶜʰ they were found out. But this Book will be a lasting Demonstration'—'to all posterity' is deleted—'that Mᵣ Newton when he wrote it was master of an Analysis wᶜʰ carried him through many such difficulties as Mᵣ Leibnitz would have stuck at'. (A 'compared with the *Tentamen de motuum cœlestium causis*' subsequently interlineated in the last sentence after 'Book' shows which of Leibniz' published pieces Newton here meant.) On this preference for the synthetically 'composed' see 1, 8: *passim*.

(135) In his letter to Oldenburg on 21 June 1677 (N.S.) Leibniz had in fact sketched, in the instances of the circle and general central conic respectively, two separate types of area-preserving transformation by which reduction from an irrational integrand to a rational one may be made (see IV: 530–1, note (13)). He had not gone on to claim any originality for the further reduction of the ensuing integrands—by expansion of the inverse cube of a binomial and by 'Mercator' division respectively—into an equivalent infinite sequence of terms which can then be separately integrated to yield the 'converging' series sought.

(136) A related paragraph on Newton's ensuing page 212 (not here reproduced) makes it clear—when at least we there insert two small *éclaircissements* from the list of *corrigenda* on .13: 223ʳ (see note (11))—that he is thinking above all of his most general method of deriving a required series expansion by first assuming it in an appropriately undetermined form, entering this in the equation (algebraic or differential) given, and then finding its coefficients piecemeal by collating homologous terms: one which 'Mr Leibnitz in the *Acta Eruditorum* for

Propositions in his *Principia Philosophiæ*. But because the Ancients for making things certain admitted nothing into Geometry before it was demonstrated synthetically, he demonstrated the Propositions synthetically that the systeme of the heavens might be founded upon good Geometry. And this makes it now difficult for unskillful men to see the Analysis by wch those Propositions were found out.[137]

It has been represented[138] that Mr Newton in the Scholium at the end of his book of *Quadratures* has put the third fourth & fift terms of a converging series respectively equall to the second third & fourth differences of ye first Term, & therefore did not then understand the method of second third & fourth differences. But in the first Proposition of that Book[139] he shewed how to find the first second third & following fluxions *in infinitum* & therefore when he wrote that Book, wch was before the year 1676,[140] he did understand the method of

A[pril] 1693 [pag. 179 *et seq.*] published...as his own, claiming to himself the first invention thereof & calling it *Methodus universalissima pro seriebus*'. Leibniz' title for his 2-page paper was in fact 'Nova Methodus generalissima per series' (see VII: 182, where we briefly survey its content) and it has only broad connection with Newton's method, of which first public account was not given till afterwards in Wallis' *Opera*, **2**: 393–6 (=VII: 176–80). Short of supposing that Leibniz did indeed perform the impossible and unscramble the anagram of Newton's cryptic enunciation as it was put to him in the latter's 1676 *epistola posterior* (see *Correspondence*, **2**, 1960: 129 and note (72) thereto; and compare II: 191, note (25))—which to us is incredible—one must, exasperating though it be to Newton's cherished belief in the sanctity of 'first invention', allow Leibniz here as elsewhere his proper recognition as independent discoverer (and, equally importantly, first publisher).

(137) So Newton asserts here. On other occasions (compare note (94) above) he could equally claim that 'the Analysis by which [the *Principia*] was invented shines through the composition' (.41: 22^v). In footnoting the papers, reproduced in 1, 8 preceding, in which Newton elaborated his present thesis that 'most' of the *Principia*'s propositions were discovered by a prior application of the 'new [*sc.* infinitesimal] Analysis' we have found little to support its truth; on the contrary, it seems to us that, except for a number of opening propositions in its second book, the *Principia* was in its greatest part 'invented' in the highly geometrical form in which it was 'composed' for the press. Newton in his old age was, we need but lightly remind, a frequent sufferer from that convenient amnesia which forgets the complexities of past circumstance and erects a factitious, simplified version of what 'ought' in retrospect to have happened in its place.

(138) By the Bernoullis, uncle and nephew: by Johann in the scholium (pages 93–5) to the first instalment of his two-part article 'De Motu Corporum Gravium...' in the *Acta Eruditorum* (February/March 1713): 77–95/115–32, and by Niklaus in an 'Addition' (pages 54–6) to Johann's more recently published 'Lettre...écrite de Basle le 10. Janvier 1711 touchant la maniere de trouver les forces centrales dans des milieux resistans...' (*Memoires de l'Académie Royale des Sciences. Année MDCCXI* (Paris, 1714): 47–54). On both which see 1, 2, §3.2: note (46) preceding; and compare 1, 6, Appendix 1: note (14).

(139) In anticipation of his equivalent remark in the next paragraph but two following, Newton's list of *corrigenda* on .13: 223^r (see note (11)) here directs there to be added 'published by Dr Wallis in the year 1693 ([*Opera*] Vol. 2 pag. 392)'.

(140) We have dwelt at length upon this piece of wishful thinking in VII: 15–16.

‖[207] all the fluxions & by con‖sequence of all the Differences. And if he did not understand it when he added that scholium to the end of the book, wᶜʰ was in the year 1704, it must have been because he had then forgot it. And so the Question is only whether Mʳ Newton had forgot the method of second & third differences before the year 1704.

In the tenth Proposition of the second Book of his *Principia Philosophiæ* in describing some of the uses of the terms of a converging series for solving of Problems he tells us that if the first term of the series represents the Ordinate *BC* of any Curve line *ACG*(141) & *CBDI* be a parallelogram(142) infinitely narrow whose side *DI* cuts the Curve in *G* & its tangent *CF* in *F*: the second term of the series will represent the line *IF*, & the third term(143) the line *FG*. Now the line *FG* is but half the second difference of the Ordinate: & therefore Mʳ Newton when he wrote

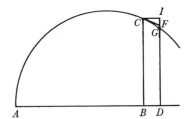

(141) Newton's accompanying figure is abstracted from the one set to illustrate the partially mistaken argument of this proposition in its first published version (*Principia*, ₁1687: 260–9, reproduced in **1**, 6, Appendix 1 preceding)—properly so, the more forcefully to bring home the absurdity of the joint attempt by the Bernoullis Johann and Niklaus (see note (138) above) to locate and 'correct' its error. In the *Principia*'s figure, which has to do double duty both for the main proposition and its prime 'Exempl. 1' of circular motion, the curve $\overset{\frown}{ACG}$ is drawn as a semi-circle (meeting the base *ABD* in a second diametral end-point *K*). In his manuscript draft of the present page of his 'Account' Newton roughly sketches (.41: 41ᵛ) an arc which over most of its length is somewhat more flattened and elliptical but which to the right of *G* shoots over the paper's edge in a wobbly straight line (through some careless spasm in his guiding hand we presume) and he also inserts in his figure a second, unwanted ordinate to the left of *BC* (corresponding, it would seem, to *il* in the original figure; *vide* page 374 above). The printed 'Account' has a crude wood-cut in which $\overset{\frown}{ACG}$ is drawn accurately as a semi-circle, extended beyond *G* to meet the base in an unlettered diametral point. We ourselves seek to have our cake and eat it, drawing $\overset{\frown}{ACG}$ as a circular arc of diameter *ABD*, but (as Newton in his surviving sketch, intentionally or no) omitting its rightmost portion the better to convey that it is any specified curve.

(142) In the case of Newton's *Principia* proposition understand 'rectangle' (as it is depicted); but for the purposes of his present rejection of the Bernoullis' would-be criticism of the erroneous first (1687) version of its argument—that he mistakenly took *FG* to be the full 'second difference' of the ordinate *BC* ($= f(AB)$ say) when the base *AB* receives the infinitesimal increment *BD*—it is of course unnecessary that the parallel ordinates *BC* and *DG* or

$$f(AB+BD) = (DI \text{ or}) \, BC + (IF \text{ or}) \, f'(AB).BD + (FG \text{ or}) \, \tfrac{1}{2}f''(AB).BD^2 + \dots$$

be restricted to be at right angles to the base. In a discarded initial phrasing in the draft on .41: 41ᵛ he originally wrote directly to the point '& *DG* be another Ordinate infinitely nearer to the former'.

(143) That is, $\tfrac{1}{2}f''(AB).BD^2$ in our preceding note. Strictly, however, *FG* is the value of 'the third and succeeding terms', and in the parent Proposition X of the *Principia*'s second book it had been very necessary to evaluate the fourth term $\tfrac{1}{6}f'''(AB).BD^3$ in the Taylorian expansion of the incremented ordinate.

his *Principia*, put the third term of the series equall to half the second difference of the first term & by consequence had not then forgotten the method of second differences.

In writing that book he had frequent occasion to consider the increase or decrease of the velocities wth wch quantities are generated, & argues right about it. That increase or decrease is the second fluxion of the quantity, & therefore he had not then forgotten the method of second fluxions.

In the year 1692 Mr Newton at the request of Dr Wallis sent to him a copy of the first Proposition of the book of *Quadratures*, with examples thereof in first second & third fluxions, as you may see in the second Volume of the Doctors works pag. 391, 392, 393 & 396.[144] And therefore he had not then forgotten the method of second fluxions.

‖ Nor is it likely that in the year 1704 when he added the aforesaid Scholium ‖[208] to the end of the book of *Quadratures* he had forgotten not only the first Proposition of that book but also the last Proposition upon wch that Scholium was written. If ye word (*ut*) wch in that Scholium may have been accidentally omitted between the words (*erit*) & (*ejus*) be restored, that Scholium will agree with the two Propositions & the rest of his writings, & the Objection will vanish.[145]

---

(144) See vii: 174–7, 180. The list of *corrigenda* on .13: 223^r here adjoins: 'This was the first Rule made publick for finding second third & fourth differences &c, & is the best.' In his draft of this sentence on .41: 41^v Newton originally reminded his reader that Proposition I of the *De Quadratura* 'explicated' how 'Data æquatione fluentes quotcuncq quantitates [involvente, fluxiones] invenire'.

(145) In the copy of Jones' 1711 *Analysis* which he handed to Niklaus Bernoulli in October 1712 as a present to his uncle Johann, Newton had indeed (as we have already remarked in **1**, **2**, § 3.2: note (46)) previously entered this *corrigendum* in the *De Quadratura*'s terminal scholium—only to have Johann take this not very adroit emendation of a sloppy phrasing as evidence, so he reported to Leibniz in his celebrated Letter of 7 June 1713 (N.S.), that Newton had not noticed the error (in 'second differences' Bernoulli took it to be) 'nisi brevi ante, et forte nonnisi post adventum Agnati mei in Anglia' (*Leibnizens Mathematische Schriften*, **3**: 912 [=*Correspondence*, **6**: 3]). A stray sheet in private possession, yet unprinted, shows the elaborate care which Newton took behind-scenes to insist that 'the first Paragraph of the Scholium disagrees with the next paragraph & also with what has been shewed above'—'in the Introduction' where (see page 152 above) it is stated that 'if any indeterminate quantity x increase or flow uniformly & its fluxion or velocity of increasing be represented by an unit & its moment or increase generated in a moment of time be called o, ... the Dignity $\overline{x+o}|^n$ whose index is n...resolved into a series will be $x^n + nox^{n-1} + n \times \dfrac{n-1}{2} \times oox^{n-2} + n \times \dfrac{n-1}{2} \times \dfrac{n-2}{3} \times o^3x^{n-3}$ &c'— 'unless the word *ut* be twice inserted, vizt in y^e 11th & 13th line of the Scholium'. In sequel to which he put the questions:

'Quære 1. Whether the word *ut* was omitted by accident or designe [?]

Qu. 2. Whether there be any thing in the *Principia* by wch it can be gathered that Mr Newton took the 2^d 3^d & 4th terms of the series for the 1st 2^d & 3^d moments of the first term of the series [?]'

And then passed on to elaborate his responses in predictable fashion.

Thus much concerning the nature & history of these methods, it will not be amiss to make some Observations thereupon.

. . .   .(146)

---

(146) We omit this concluding portion (pages 208–24) of his 'Account' of the *Commercium* in which Newton, largely on the basis of the documents which he has previously adduced, reiterates that because (page 218) 'second Inventors have no right' Leibniz must either 'prove' his case to priority of discovery in calculus or 'quit his claim'. Carrying the attack to Leibniz, he dwells at some length (pages 208–9) on errors made by the latter in developing the argument of his twin 1689 *Acta Eruditorum* articles on resisted projectile and on centrally enforced planetary motion—on these see notes (86) and (95) of our preceding introduction—and of course urges that this is a demonstration that 'when Mʳ Leibnitz wrote these three Tracts he did not well understand the ways of working in second Differences'. And he waxes sarcastic that Leibniz should have 'tried to adapt' a proof to his own prior propositions in this way. 'It seems therefore', he went on, 'that as he learnt the Differential Method by Mʳ Newton's aforesaid three letters [*viz.* of 10 December 1672 and 13 June/24 October 1676] compared with Dʳ Barrow's method of Tangents, so ten years after when Mʳ Newton's *Principia Philosophiæ* came abroad, he improved his knowledge in these matters by trying to extend this method to the principal Propositions in that book.... For the Propositions contained in [his three tracts] (errors and trifles excepted) are Mʳ Newton's (or easy corollaries from them) being published by him in other forms of words before. And yet Mʳ Leibnitz published them as invented by himself.' And so the catfight continued....

Let us say no more of the ensuing pages 210–21 than that in them Newton presents, with pertinent anti-Leibniz asides and blanket interpretations, his prosecutor's concluding survey of the asserted evidence presented: amid such an unyieldingly stern address to the juryman reader the omissions and inadequacies of Leibniz' case receive of course short shrift, but the sheer general repetitiveness of Newton's charges and imprecations quickly bores. (He or she who is interested in pursuing the dubious 'fuller' riches of Newton's preparatory drafts in Add. 3968 may here look to .41: 125ᵛ/120ʳ–120ᵛ (jotted in the blank spaces of Katther[i]n Rastall's begging letter dated 'Oct. 19. 1714', for whose original text see *Correspondence*, **6**: 183] /25ʳ/103ʳ– 103ᵛ/142ᵛ–142ʳ/26ᵛ/112ᵛ(+68ᵛ); yet other related sheets survive in private possession.) The three concluding pages of the 'Account' in which Newton digresses from the straight fight over priorities of mathematical 'invention' to defend the 'experimental Philosophy' pursued by him in his *Principia* and *Opticks*—where he took it to be his 'business' not to 'teach the causes of things any further then they can be proved by Experiments' and where 'Hypotheses have no place unless as conjectures or Questions proposed to be examined by experiments'—lie wholly outside our purview. (We take our quotations from the opening recto of Add. 3968.39: 585ʳ–586ʳ, where Newton carefully drafted the last three pages 222–4 of the printed 'Account' effectively in their final form; rougher preliminary tentatives may be found on .41: 24ʳ–27ᵛ, 44ʳ/44ᵛ and 125ʳ.)

The full text of the 'Account' as it was printed in the *Philosophical Transactions* in January/ February 1715 (see note (2) above) is, we may lastly adjoin, now set in appendix to A. R. Hall's newly appeared *Philosophers at War* (Cambridge, 1980).

APPENDIX 3. TWO VERSIONS OF A STATEMENT BY NEWTON, INTENDED FOR PUBLICATION BY DESMAIZEAUX IN HIS *RECUEIL*, ON THE SEQUENCE OF 'INVENTION' OF THE CALCULUS AND ON THE RIGHTS OF PRIORITY THEREIN.[1]

[summer 1718][2]

From the autograph originals in the University Library, Cambridge

(1) In our introduction (see pages 517ff. above) we recounted how Leibniz' sudden death in November 1716, far from checking Newton's impassioned pursuing of his 'rights' as 'first inventor' of the basic modes and techniques of the calculus, in fact spurred him on to renew it and reinforce it with new evidence, documented and personally attested. Within weeks, in the event, he had hastened into print, in appendix to his reissue of Raphson's *History of Fluxions*, an elaborate set of 'Observations' upon Leibniz' criticisms (to Conti) of his own claim to priority of discovery in particular and of the competence and achievement of English mathematicians in general: these pulled no punches in their countering disparagements. And his anger against the dead Leibniz still smouldered on a year and a half later when Pierre Desmaizeaux sent him proof-sheets of the second volume of his forthcoming *Recueil de Diverses Pieces* where de Moivre's Frenchification of those remarks was given prominent place, doubtless expecting back a few minor corrections of detail and the odd amelioration of a verbal phrase. In that summer of 1718, however, Newton's unmellowed pen in response began to race afresh over the page, filling dozens of surviving folios with rearranged 'proofs' of his absolute priority of invention which Desmaizeaux might introduce into his editorial 'Preface' to the *Recueil*, and also in between times refining his author's witness as to how he had 'gradually' found out and developed the method of fluxions from the middle 1660's onwards: this, so the title of its most finished version (that given in [2] below) specifically announces, in 'Supplement' to his previous remarks upon Leibniz' April 1716 letter. In a familiar pattern, however, once the initial flurry of activity had subsided Newton's enduring wariness to make public avowals under his own name once again asserted itself. We do not have the covering letter which he finally returned to Desmaizeaux, but it was probably some very near approximation to the relatively subdued final draft of it which is preserved in Add. 3968.27: 393^r–395^r (whence it is reproduced in *Correspondence*, 6, 1976: 454–7 along with, in ancillary notes *ibid.*: 458–62, lavish excerpts from a selection of its widely variant preliminary castings). Newton's accompanying grand testimony regarding his steps to discovery in fluxional calculus was thereafter engulfed among his mass of other equally stillborn writings on the topic, unglimpsed by its intended public; and in that limbo it has remained during the quarter of a millennium since, known only to the world at large by the few jejune snippets quoted from its text by Brewster in his 1855 *Memoirs* (see note (27) below). It seems an undeserving fate that what is perhaps his definitive personal statement of priority over Leibniz in finding out the 'method of moments' (and the conjoint mode of analysis *per series infinitas*) should for ever go unprinted. Here we repair the omission, reproducing the texts of the two principal versions of this Newtonian 'Historia et Origo Calculi Differentialis' with some running editorial comment (in footnotes at the pertinent places) upon its intermix of basic historical truth with an overlay of self-deceiving fiction.

(2) In reporting upon his small rôle in the preparation of the second volume of Desmaizeaux' *Recueil*, Newton stated in a first English draft of a letter to Varignon in January 1721 that it was 'in August 1718' that Desmaizeaux had 'shewed me' its first printed sheets (see the verso of the slip of paper stuck on to the recto of Add. 3968.27: 409^r) though he straightaway amended

[1]⁽³⁾   THE HISTORY OF THE METHOD OF MOMENTS, CALLED
DIFFERENCES BY Mʳ LEIBNITZ.

In the Introduction to [the] Book of *Quadratures* published A.C. 1704 I wrote that I⁽⁴⁾ found the Method of fluxions gradually in the years 1665 & 1666, & tho this was not so much as Dʳ Wallis said nine years before in the Preface to the first volume of his works⁽⁵⁾ without being then contradicted, yet in the *Acta Eruditorum* for January 1705, in giving an Account of this Book Mʳ Leibnitz is called the *Inventor*; & from thence is deduced this conclusion:⁽⁶⁾ *Pro differentijs igitur Leibnitianis Newtonus adhibet semperɋ* (pro ijsdem) *adhibuit fluxiones—ijsɋ tum in Principijs Naturæ Mathematicis tum in alijs postea editis* (pro differentijs illis) *eleganter est usus, quemadmodum et Honoratus Fabrius in sua Synopsi Geometrica motuum progressus Cavallerianæ methodo substituit.* Dʳ Wallis was *homo vetus*⁽⁷⁾ & informed himself of these matters from the beginning, being very inquisitive in Mathematicall affairs, & having received from Mʳ Oldenburg copies of my two Letters of 13 June & 24 Octob. 1676 when they were newly written.⁽⁸⁾ And in the said

the time to be more loosely 'about a year before I received from you Mr J. Bernoulli's Letter of 5 July 1719 [N.S.]'. We will not insist rigidly upon which late summer month in 1718 it was when Newton received from Desmaizeaux the five proof-sheets of the latter's *Recueil* (those now ULC. Adv. d. 39.2) which led him in indirect reaction to compose, in succession, the two pieces which we here reproduce. In the most finished draft of his letter to Varignon (.42: 610ʳ/610ᵛ; see *Correspondence*, **7**, 1978: 123–4) Newton in his twin datings there of having received the proof-sheets from Desmaizeaux 'in   1718' each time left the space for the month blank, clearly unwilling to trust his unaided memory of the time of year when these had in fact come to him. 'Summer 1718' is more than accurate enough for our purposes.

(3) Add. 3968.12: 159ʳ. 161ʳ/166ʳ(+165ᵛ); preliminary drafts are found on (.5: 21ʳ→).12: 165ᵛ–166ʳ.

(4) Initially 'Mr Newton wrote that he'. Most of the opening paragraph was likewise written in the cloaking anonymity of a third-person narrative, and Newton only firmly settled for the more forceful 'I' in the sentence beginning 'And how I found them I explained...', going back at that point to replace all preceding occurrences of 'Mʳ Newton' and 'he' in similar fashion (and a lone 'Mʳ Newton's' by 'my'). Since he had delivered his parent 'Observations' upon Leibniz' 'Epistle' of 9 April 1716 (N.S.) to Conti (in his 1717 Appendix to Raphson's *History of Fluxions*) forthrightly from his own mouth, there could have been no real concealment possible of his authorship of the present piece had he chosen to go on trying to hide the fact. In his revision in [2] below he abandons all cloak firmly to title his *apologia* 'A supplement to the Remarks'.

(5) *Opera*, **1**, 1695: *Ad Lectorem*, p. [iii], ll. 1–10; quoted on page 200 of Newton's 1715 'Account' of the *Commercium* (=page 591 preceding).

(6) See pages 26–7 above.

(7) '*& rerum anteactarum peritissimus*' Newton initially went on to add before rendering its sense into English in his ensuing phrase. The apposition is of course with Leibniz' labelling of the 'upstart' John Keill as 'homo novus, & parum peritus rerum anteactarum cognitor' in his letter claiming redress before the bar of the Royal Society in late December 1711 (see *Commercium*: 118 [=*Correspondence*, **5**: 207]).

(8) Though not immediately so; indeed Wallis may not have had access to the full text of the *epistola posterior* for some considerable time (see ɪᴠ: 671–2, note (54)).

Preface he said that in those Letters I had explained to Mʳ Leibnitz the Method
found by me ten years before or above; meaning that I had found the Method
above ten years before Mʳ Leibnitz, & that I had so far discovered it to him by
those Letters, as to leave it easy to find out the rest. And even before Mʳ Leibnitz
had the Method, I said in one of those Letters (that of 24 Octob. 1676) that the
foundation of the Method was obvious, & therefore, since I had not then leasure
to describe it at large, I would conceale it in an Ænigma. This I did, not to make
a mystery of it, but to prevent its being taken from me because it was obvious.
And in that Ænigma I set down the first Proposition of the book of *Quadratures*
in the very words of the Proposition; & therefore had this Proposition with the
Method founded upon it when I wrote that Enigma: or rather, because the very
words of the Proposition are copied in the Ænigma, it argues that the Book of
Quadratures was then before me, & so was written before Mʳ Leibnitz had the
Method. This Book was⁽⁹⁾ extracted out of older papers. In the said Letter of
24 Octob. 1676 I set down a series for squaring of figures wᶜʰ in some cases
breaks off & becomes finite₍,₎ & illustrated it with examples & said that I found
this & some other Theorems of the same kind by the method whose foundation
was comprehended in that Ænigma, that is, by the method of fluxions. And how
I found them I explained in the first six Propositions of the Book of *Quadratures*,
& do not know any other method by which they could be found: & therefore
when I wrote that Letter I had the Method of fluxions so far as it is contemed
in those six Propositions. After I had finished the Book & while the 7ᵗʰ 8ᵗʰ 9ᵗʰ &
10ᵗʰ Propositions were fresh in memory⁽¹⁰⁾ I wrote upon them to Mʳ J. Collins
that Letter wᶜʰ was dated 8 Novem 1676 & being found amongst his Papers was
published by Mʳ Jones.⁽¹¹⁾ The Theoremes at the end of the tenth Proposition

(9) A needlessly indecisive 'is said to have been' is replaced.

(10) Newton mistakenly antedates the composition of these Propositions VII–X of the
published *De Quadratura* by some fifteen years: such are the compelling freshnesses of memory
of events which one wants to have been but which never were, outside of wilful misrecollection.
We have already (on VII: 15–16) briefly traced the stages by which Newton came erroneously
to conclude that it was, as he not long afterwards stated in the superseded first roughing-out
(on .41: 86ʳ) of a new preface to a reissue of the *De Quadratura* whose reworked version we
reproduce in Appendix 7.3 below, 'in the year 1676 in summer' that he had 'extracted' his
'Book of Quadratures' out of his 1671 fluxions treatise '& from other early papers'.

(11) In his 'Excerpta ex Epistolis D. *Newtoni* Ad Methodum Fluxionum et Serierum
Infinitarum spectantibus' (=*Analysis per Quantitatum Series, Fluxiones ac Differentias:*...
(London, 1711): 23–38):38 William Jones in fact published only the concluding portion of
Newton's letter where he had declared to Collins—in a corrected transcript of the original
(compare *Correspondence*, 2: 179–80)—that 'there is no curve line exprest by any æquation of
three terms...but I can in less then half a quarter of an hower tell whether it may be squared
or what are yᵉ simplest figures it may be compared wᵗʰ, be those figures Conic sections or
others. And then by a direct & short way...compare them. And so if any two figures exprest
by such æquations be propounded I can by yᵉ same rule compare them if they may be compared....The same method extends to æquations of four terms & others also but not so

for comparing curvilinear figures with the Conic sections are mentioned in my said Letter of 24 Octob 1676, & all the Ordinates of the figures in the second part of the Table[12] are there copied from the Book. And therefore the Book was then before me. To understand the two Letters of Octob 24 & Novem 8, 1676 & how to perform the things & find the Theorems mentioned in them requires skill in the Method of fluxions so far as it is comprehended in all the first ten Propositions of the Book. And the eleventh & last Proposition depends upon a series of first second third & fourth fluxions.

The first Proposition of this Book & its solution illustrated with examples in first & second fluxions &c was at the request of Dr Wallis sent to him almost

---

generally'. As we insisted on III: 19 in introducing the hitherto unpublished papers on the squaring of all trinomials and certain types of quadrinomial and higher polynomial curve (see III: 373–82/383–5) which Newton did write about the late summer of 1676, full coverage of this somewhat restricted and specialised topic was subsequently given by him in an undemonstrated set of *regulæ* for attaining such particular quadrature which he introduced in late 1691 into the preliminary versions (see VII: [30–6→] 56–62) of his comprehensive general treatise 'De quadratura Curvarum', but these complicated rules were omitted from the abridged text (see VII: 508 ff.) which he prepared two years afterwards as his 'Geometriæ Liber Secundus' and passed ultimately into print as the core of the lightly reshaped and amplified *Tractatus de Quadratura Curvarum* (see 1, 2 preceding) which first appeared in annexe to his *Opticks* in 1704.

(12) Understand the table of 'Curvæ simpliciores quæ per quadraturam Sectionum Conicarum quadrari possunt' which is found adjoined in Newton's published *De Quadratura* (pages 132–46 above)—as in its 1693 parent 'Geometriæ Liber Secundus' (see VII: 550)—in general scholium to its principal Propositions I–X. But the 'Theoremes' cited by him in his *epistola posterior* to Leibniz derived not from this, but from the 'Catalogus Curvarum aliquot ad Conicas Sectiones relatarum' which he had (see III: 244–54) set to illustrate Problem 9 of his 1671 tract on series and fluxions, and which he afterwards slavishly copied in his 1693 *Tabula* except for changing the sequence of certain of its component entries and verbally amending its listings. There is, as we showed in our analysis of his original 1671 'Catalogus' (see III: 256–9, notes (550)–(569)), much to bear out his here ensuing claim that 'To understand...how to... find the Theorems mentioned in [my said Letter of 24 Octob 1676] requires skill in the Method of fluxions so far as it is comprehended in all the first ten Propositions of the Book [of Quadratures]'. But it is disingenuous of Newton to conflate the 'Book' on fluxional analysis which he wrote in 1671 with that, dealing more narrowly with the squaring of algebraic curves, which he began to put together only twenty years later; and thence to infer by a series of plausible concatenations of would-be *prima facie* indisputable evidence and personal remembrances—'rationally' reconstructed where direct recollection of the past had grown to be dim and confused—that the fine web of propositions set out in the published text of his *De Quadratura* was already physically (as well as potentially) 'before' him in the middle 1670's in a richness of texture far superior to any parallel mode of 'summatory calculus' which Leibniz was then able to contrive. To have hinted that this polished final treatise was not, in its published version, put together till the early 1690's would of course virtually have destroyed its weight of authority as prime exhibit of Newton's grasp and expertise in calculus a decade and a half earlier. Nearly three centuries afterward we may acknowledge the resounding success of Newton's subterfuge, but let us also murmur our dismay that his distortion of the historical record has endured even down to our present age so as thoroughly to mislead even the most percipient students of his mathematical development regarding the past existence of a chimerical '1676' treatise 'De Quadratura Curvarum'.

verbatim in a Letter dated 27 Aug. 1692, & printed by him that year in the second Volume of his works,[13] which came abroad the next year, A.C. 1693. And thus the Rule for finding second third & fourth fluxions there set down was published some years before the Rule[14] for finding second third & fourth differences[,] & was at least seventeen[15] years in manuscript before it was published. In the Introduction to this Book the method of fluxions is taught without the use of prickt letters; for I seldom used prickt letters when I considered only first fluxions: but when I considered also second third & fourth fluxions I distinguished them by the number of pricks. And this notation is not only the oldest but is also the most expedite, tho it was not known to the Marquess de l'Hospital when he recommended the differential Notation.

In my *Analysis per æquationes numero terminorum infinitas* communicated by Dr Barrow to Mr Collins in July 1669, I said that my Method by series gave the areas of Curvilinear figures exactly when it might be, that is, by the Series breaking off & becoming finite: & thence it appears that when I wrote that *Analysis*, I had the Method of fluxions so far, at the least, as it is conteined in the first six Propositions of the Book of *Quadratures*; tho those Propositions were not then drawn up in the very words of the Book. [16]In that Tract of *Analysis* I represented time by a line increasing or flowing uniformly & a moment of time by a particle of the line generated in the moment of time, & thence I called the particle a moment of the line; & the particles of all other quantities generated in the same moment of time I called the moments of those quantities; & the fact[17] under the rectangular Ordinate & a moment of the Abscissa I considered as the moment of the area (rectilinear or curvilinear) described by that Ordinate moving uniformly upon the Abscissa. For a moment of time I put the letter *o*, & thence computed the moments of the other quantities generated in the moment of time, & for those moments put any other symbols. And for the Area of a figure I sometimes put the Ordinate included in a square. And by considering how to deduce moments from increasing quantities & quantities from their moments, I deduced Ordinates of figures from their areas & the Areas from the Ordinates: w[ch] is the same thing with deducing fluxions from fluents & fluents from fluxions. And in the end of the Book I demonstrated by this sort of

---

(13) *Opera Mathematica*, **2**, 1693: 392–3 (=VII: 174–7).

(14) Understand 'of M[r] Leibnitz' (in his 'Nova methodus pro...tangentibus, ... & singulare pro illis calculi genus' in the *Acta Eruditorum* in October 1684; on which see Appendix 2: note (91) preceding).

(15) That is, 1676–1693. Newton initially computed 'eighteen' in his draft on .12: 165[r].

(16) A new paragraph begins at this point in the preliminary drafts on .12: (167[r]→) 166[r]. As we suggested in note (99) of the introduction, the original template for what follows is probably the authoritative statement by Newton now found in lonely grandeur on .41: 83[r] which we reproduced in III: 17–18.

(17) 'product'.

calculus the first of the three Rules set down in the beginning thereof. And in this Rule for the index of a Dignity I put an indefinite quantity affirmative or negative, integer or fract, & thereby introduced indefinite indices of Dignities into Analysis. And applying this Method of Moments not only to finite equations but also to equations involving converging series I gave this Tract the name of *Analysis per æquationes numero terminorum infinitas*. And shewing also how thereby to find the Ordinates & Areas of Mechanical Curves &c I said:[18] *Nec quicquam hujusmodi scio ad quod hæc Methodus idɋ varijs modis sese non extendit. Imo...& quicquid vulgaris Analysis per æquationes ex finito terminorum numero constantes (quando id sit possibile) perficit, hæc per æquationes infinitas semper perficiat.* And in my Letter of 13 June 1676 I said of this Analysis: *Ex his videre est quantum fines Analyseos per hujusmodi infinitas æquationes ampliantur: Quippe quæ earum beneficio ad omnia pene dixerim Problemata (si numeralia Diophanti et similia excipias) sese extendit.*[19] And Mr Leibnitz in his Letter of 27 Aug. 1676 replied that he did not beleive that my method was so general, the inverse Problem of Tang^ts & many others not being reducible to Equations or Quadratures. And placed the perfection of Analysis not in this Method but in a Method composed of Analyticall Tables of Tangents & the Combinatory Art, saying of the one; *Nihil est quod norim in tota Analysi momenti majoris*:[20] & of the other; *Ea vero nihil differt ab Analysi illa suprema, ad cujus intima, quantum judicare possum, Cartesius non pervenit. Est enim ad eam constituendam opus Alphabeto cogi[ta]tionum humanarum.*[20] This was the top of his Analytical skill at that time.

Mr Collins in a Letter to Mr Strode dated 26 July 1672 gave this account[21] of these Methods. *Mense Septembri 1668 Mercator Logarithmotechniam edidit suam... idɋ non obstantibus radicalibus, quæ eam non morantur, obtineri queunt. Postquam intellexerat D. Gregorius...D. Newtonum primum ejus inventorem anticipare haud integrum ducit.* So then by the testimony of Dr Barrow grounded upon papers wch I had communicated to him from time to time[22] I had the method here described some years before the Doctor sent my *Analysis* to Mr Collins, that is,

---

(18) See II: 240, 242. Newton has silently omitted several lines after the first sentence (between '...extendit.' and 'Imo...'). We see no need to follow him in making full quotation once again of the passage on II:242.

(19) See *Correspondence*, **2**, 1960: 29, ll. 14–16.

(20) See *Correspondence*, **2**: 63, ll. 26 and 30–2 respectively.

(21) Newton quotes from his Latin rendering of Collins' letter printed in *Commercium Epistolicum*: 28–9. We have ourselves cited the original English of Collins' draft (in Royal Society MS LXXXI, No. 24) in note (22) of Appendix 2 preceding.

(22) So far as we have been able to discover, the 'De Analysi' which Newton submitted to Barrow some little while before the end of July 1669—'towards the end of June' was our conjecture on II: 167, though Barrow's phrase 'the other day' in his letter to Collins on 20 July informing him of the event (see *Correspondence*, **1**, 1959: 13) perhaps suggests a date in early July—was the only mathematical paper of his which Newton ever allowed his 'mentor' to see.

some years before July 1669. And this is sufficient to justify what I said in the Introduction to the Book of *Quadratures*,[23] vizt that I found the Method gradually in the years 1665 & 1666. Dr Barrow then read his Lectures about Motion & that might put me upon taking these things into consideration.[24] If it be asked why I did not publish this method sooner, it was for the same reason that I did not publish the Theory of colours sooner. I found the Method of Series in the beginning of the year 1665 & the method of fluents & moments soon after & the Theory of colours in the beginning of the year 1666, & in the year 1671 was about to publish them: but for a reason mentioned in my Letter of 24 Octob. 1676, I desisted till the year 1704,[25] excepting that some of my Letters were published before by Dr Wallis & Mr Oldenburg.[26]

---

(23) *Tractatus de Quadratura Curvarum* (= *Opticks*, $_1$1704: $_2$165–211): $_2$166: '...& has motuum vel incrementorum velocitates nominando *Fluxiones*...incidi paulatim Annis 1665 & 1666 in Methodum Fluxionum qua hic usus sum...'. Newton's manuscript drafts of this passage—at the end of the second paragraph of his 'Introductio'—are reproduced in **1**, **2**, (§2.1→) §3.1 preceding (see pages 106, 122).

(24) 'tho I not now remember it' Newton was in honesty afterwards here (on .41: 86^v) to adjoin in the reworking of the present piece which he later had in mind to set before the *De Quadratura* in a planned reissue of its text jointly with an *editio tertia* of the *Principia*; see Appendix 7.3 below (page 668, *lin. ult.*). We have already cited on **1**: 150 the main part of a concluding passage on .5: 21^r where he initially went on to elaborate: 'In the winter between the years 1664 & 1665 I had a method of Tangents like that of Hudden Gregory & Slusius & a method of finding the crokedness of Curves at any given point, & by considering how to interpole certain series of Dr Wallis I found the Rule set down in my Letter of 13 June 1676 for reducing any power or dignity of any Binomium into an approximating Series, & in the following Spring, before the Plague, wch invaded us that summer, forced me from Cambridge, I found how to do the same thing by continual division & extraction of roots, as I mentioned in my Letter of 24 Octob. 1676. And soon after I extended the Method to the extraction of the roots of affected Equations in species. And from all this I learnt how to deduce the Ordinates or Abscissas of Curvilinear figures from their Areas or A[r]cs, as well as the Areas & Arcs from the Abscissas & Ordinates. Thus far I proceeded before the plague forced me from Cambridge. And in a paper dated 13 Nov. 1665 I find the direct Method of first fluxions set down with examples & a Demonstration.' There is no need here to give precise references to the texts of Newton's early papers themselves (as these are set out in the first volume of this edition) or to the related passages in his *epistolæ prior/posterior* to Leibniz in 1676 (which are readily consultable in their now standard texts in *Correspondence*, **2**).

(25) Namely 'out of a desire to be quiet'—so he will phrase it in the last sentence of [2] following—and free of squabble.

(26) *Viz.* in the 'Epistolarum Collectio' added by Wallis to his *Opera*, **3**: 617–708 (see especially 617 [= Collins to Newton, 18 June 1673 *Latine reddita*], 622–9/634–45 [= Newton's *epistolæ prior/posterior* to Leibniz on 13 June/24 October 1676] and 646–7 [= Collins to Newton, 5 March 1676/7 *Latine reddita*]); and in the pages of the *Philosophical Transactions* where the Royal Society's secretary Henry Oldenburg printed major items from the correspondence on optics which Newton maintained through him in the four years from 1676 (these are now conveniently gathered in photofacsimile in (ed.) I. B. Cohen, *Isaac Newton's Papers and Letters on Natural Philosophy*...(Cambridge, $_1$1958 = $_2$1978): 47–176).

[2]⁽²⁷⁾   A Supplement to the Remarks.⁽²⁸⁾

Mr Leibnitz by telling his own story in his Letter of 9 April 1716⁽²⁹⁾ put me upon doing the like in my Remarks upon it. And since in his Letters to the

---

(27) Add. 3968.31:453^r(+452^v/451^v)/458^r/454^r/457^r/455^r/455^v/456^r (Newton has paginated these seven carefully written pages 1, 2, ..., 7 at their upper, outer corner), together with the concluding portion of the preliminary draft on .12:164^v which completes the text of this finished fair copy of the main text. It will be evident that this 'Supplement' to the 'Observations' on Leibniz' letter to Conti which he had previously published (see the next note) is written in the form of an open letter to the editor, Desmaizeaux, of the *Recueil de Diverses Pieces* in which, so it was his intention, Frenchified versions of both should appear. As we noticed in our introduction (see page 524 above) this *addendum* to his earlier remarks was never, as such, sent on to Desmaizeaux, who there reprinted de Moivre's French rendering of Newton's parent 'Observations'—prepared in 1717 for a Continental issue of Raphson's augmented *History of Fluxions* (on this see note (151) of the introduction) which never publicly appeared—without any major embellishment. No more is heard of the 'Supplement' during Newton's lifetime, and after his death it lingered in limbo along with his other forgotten private writings on priorities of calculus invention till David Brewster brought its fund of personal declaration and 'authoritative' comment to scholarly attention more than a century later in the chapters (xiv/xv) of his *Memoirs of the Life, Writings and Discoveries of Sir Isaac Newton* (Edinburgh, 1855) in which he dealt at length with the prehistory and justices of the squabble over first discovery; the lavish citations which he there (**2**: 30, 34, 41–2) made of its juicier passages will be pin-pointed in following footnotes.

We have underlined in the introduction that Newton's present intended 'Supplement' was but one superseded stage in his complicated reaction to reading the proofs of the second tome of the *Recueil* which its editor sent him in summer 1718. There seems no need, accordingly, to make minute record of the small divergences in phrasing and ordering which are found in the variety of surviving partial preparatory drafts now located at .27:386^v/386^r→.31:452^r/452^v/ (.27:384^r→)/451^r/459^r; or similarly to enumerate every one of the subsequent changes in its text which Newton began straightaway to enter on .41:140^r/140^v in prelude to reshaping its statements and remarshalling its arguments in a renewed series of draftings of a more private response to Desmaizeaux. Let us notice in passing, however, that his first use of .27:384^r was to draft his letter to the Treasury dated 'Mint Office/6 May 1718' which is printed in *Correpondence*, **6**, 1976:444. This affords, we need not say, a firm *terminus ante quem non* for the composition of the 'Supplement' which fits in well with Newton's recollection to Varignon three years afterwards (see note (2) preceding) that it was 'in August 1718' that Desmaizeaux sent him the advance proof-sheets of his forthcoming *Recueil* which stimulated Newton to write this *addendum* to it.

(28) Understand de Moivre's 'Traduction Françoise' of Newton's original English 'Observations' (again see note (151) of the introduction for details of how this 'French Coppy' came to be) which Desmaizeaux included in the second volume of his *Recueil de Diverses Pieces* (→**2**, ₁1720:75–98) under the title of 'Remarques de M. le Chevalier Newton sur la Lettre de M. Leibniz à M. l'Abbé Conti [*sc.* écrite d'Hanover le 9 d'avril 1716]'. Newton initially, we may notice, addressed himself 'To the Reader' (on .27:386^v) in words which he evidently expected Desmaizeaux to print in general preface to his reissue of the 1717 Appendix to Raphson's *History* in the second volume of the *Recueil*: 'Towards the end of the year 1715 (I think about November) Mr Leibnitz writing to M^r l'Abbe Conti then at London..., added a large Postscript about the invention of the Differential Method. And the Postscript was sent also to M^r [Nicolas] Remond at Paris. The Postscript not being written to me I did not think my self concerned to meddle with it. But Mr Leibnitz declining to answer the *Commercium*

Comtesse of Kilmansegger[30] & the Comte de Bothmer[30] he has told it at large,
I will tell the rest of my story & leave you to beleive what you please.

In my *Analysis per series numero terminorum infinitas* communicated by Dr Barrow
to Mr Collins in July 1669 I represented that I had the Method of finding the
areas & lengths of curves exactly[31] when it might be done, that is by series w^{ch}
in those cases break off & become finite equations. And this proves that I had

---

Epistolicum or to meddle with any body but me: M^r l'Abbé le Conti at length pressed me to
write an Answer to the Postscript that they might both be shewed to the King. And when my
Answer was sent [on 26 February 1715/16] to Mr Leibnitz, he wrote an Answer & sent it...
open to Mr Remond at Paris & the Answer was sent by Mr Remond to M^r l'Abbe Conti. This
tricking management compared with the beginning of his Answer made me sensible that he
was upon a design of setting aside the ancient Records published in the *Commercium Epistolicum*'—'...the Committee of the R. Society' was first written— '& stripping me of my
friends, & attaquing me alone with all his posse of disciples & all his interest. Whereupon
I refused to write an Answer'—'in the form of a Letter' is deleted—'to be sent to him[,] & yet
to satisfy my friends that it was easy to have answered him had I thought fit to let him go
on wth his p[o]litiques, I wrote an Answer in the form of Observations & shewed them
privately to some of my friends. And as soon as I heard that Mr Leibnitz was dead I caused
the Letters & Observations to be printed least they should at any time hereafter come abroad
imperfectly in France. There were some other Letters writ by Mr Leibnitz, M^r l'Abbé Conti &
Mr Remond, & by M^r l'Abbé Conti left in the hands of Mr de M. [*sc.* Desmaizeaux] to be
published together with what I had caused to be published before. And a few days since a copy
of the same was brought to me'—on this set of advance proofs of the opening of the second
tome of the *Recueil* see note (160) of the preceding introduction—'printed off except the last
sheet [*viz.* E]. But they being put together I have caused them to be reprinted in due order of
time, together with a Paper'—'scandal[o]us libel' was first written—'published about the same
time in the *Acta Lipsiensia* by a nameless author who called a solution of Mr John Bernoulli
solutionem ME$\bar{\text{A}}$.' (When it eventually made its public bow in 1720 the *Recueil* did not, we may
notice, include any reprint, in whole or in part, of the anonymous self-excusing 'Epistola pro
Eminente Mathematico Dn. Johanne Bernoullio, contra quendam ex Anglia antagonistam
[*sc.* Keillium] scripta' published in *Acta Eruditorum* (July 1716): 296–315 in which Bernoulli
failed to suppress the give-away 'meam'; on which see more fully note (88) of the preceding
introduction.)

(29) New Style, that is. This had (compare Newton's intended preface 'To the Reader'
quoted in the previous note) first been made public by Newton himself on pages 103–10 of his
1717 Appendix to Raphson's *History*, along with (pages 111–19) his ensuing 'Observations'
upon it and its 'Apostille', and was now to be reprinted by Desmaizeaux in his *Recueil* ($\rightarrow$**2**,
$_1$1720: 48–71). We have summarised its content in the preceding introduction (see pages 507–8
above).

(30) Desmaizeaux was to publish Leibniz' letter of 18 April 1716 (N.S.) to the powerfully
placed King's sister, Sophia Charlotte, Baroness von Kilmansegge, in full in his forthcoming
Recueil (**2**, $_1$1720: 29–41 [$\rightarrow$*Correspondence*, **6**, 1976: 324–9]), along with (*ibid.*: 42–7 [compare
Correspondence, **6**: 329, note (1)/2nd paragraph]) the pertinent 'Apostille' of his letter of around
the same time to Baron Bothmar where he outlined in a more summary fashion how he had
found his 'nouveau Calcul Mathematique' in the middle 1670's, and how, after he had
published its elements in 1684, it was quickly 'introduit par tout, & appliqué utilement à cent
Questions difficiles', while it was only 'par Enigme' that Newton 'avoit donné il y avoit
longtems quelque chose de cette nature'.

(31) '& geometrically' is added in the draft on .31: 452^r.

at that time the method of fluxions direct & inverse so far as it is explained in the first six Propositions of the Book of *Quadratures*, there being no other way of finding such series.

In the end of the year 1669[32] M[r] Collins sent notice to M[r] James Gregory that I had a general method of Series & M[r] Gregory by this Notice & one of my Series being put upon searching after this method found it after a years study. But tho he found it *proprio Marte* yet because he knew that I had it before him, he never claimed a right to it.

In the beginning of the year 1666 I found the Theory of colours & in the year 1671 I was upon a designe of publishing it with the methods of series & fluxions[,] & for that end I wrote a Tract that year upon the Method of series & fluxions together, but did not finish it & for a reason mentioned in my Letter of 24 Octob. 1676[33] I laid aside my designe of publishing them till the year 1704.

After the year 1671 I intermitted these studies almost five years, that is till June 1676 (as I mentioned in my Letters of 13 June & 24 Octob. 1676), but before I left of I had made my Analytical Method so general as is described in those Letters & particularly in the Letter of 13 June 1676, where I represented that it extends to almost all Problems except perhaps some numeral ones like those of Diopha[n]tus.[34] And this is that Method w[ch] I described in my Letter of 10 Decem 1672.[35] This method consisted in reducing Problemes to Equations finite or infinite & applying the method of fluxions either to the Equations or to any other conditions of the Problemes.

(32) In the Julian calendar, understood. It was in fact only on the very last day of the year, 24 March 1699/70 (=3 April 1670 N.S.) that Collins passed on to Gregory any example of Newton's method of series, and then but a single undemonstrated expansion for the zone of a circle; see our previous comment (in note (47) thereto) at the equivalent place on pages 182–3 of Newton's 1715 *Phil. Trans.* 'Account' of his *Commercium*, in Appendix 2 preceding.

(33) Newton there pointed to the 'crebræ interpellationes' occasioned by his having to respond to the numerous objections raised to the outline of his notions on the nature of light which he had sent to Oldenburg in February 1672 'ex occasione Telescopij catadioptrici'; see *Correspondence*, **2**, 1960: 14. Whether or not these wrangles about his 'New Theory of Light and Colors' were his sole reason for laying his 1671 fluxions treatise aside unfinished—he would certainly have found it difficult to have the completed tract printed at that time, short of paying all costs himself (compare III: 6–9)—after the early summer of 1672 he made no great effort further to implement his original scheme for the work (on which see III: 28–30). The treatise on series and fluxions eventually published in 1704 was the very different *Tractatus de Quadratura Curvarum* (**1**, **2** preceding) which was put together only twenty years afterwards in the early 1690's.

(34) 'Ex his videre est quantum fines Analyseos per hujusmodi infinitas æquationes ampliantur: quippe quæ earum beneficio, ad omnia, pene dixerim, problemata (si numeralia Diophanti et similia excipias) sese extendit' (*Correspondence*, **2**: 29; compare page 578 above).

(35) Where (see *Correspondence*, **1**, 1959: 247–8) he communicated to Collins the prime 'particular, or rather...Corollary' of his 'Generall Method of tangents' which he had the year before set out as Mode 1 of Problem 4 of his 1671 tract (see III: 120–2).

Mr Leibnitz was in London in the beginning of the year 1673, & going from thence to Paris in February, corresponded with Mr Oldenburg about Arithmetical matters till June following, being not yet acquainted with the higher Geometry. In April 1673 the *Horologium oscillatorium* of Mr Huygens came abroad, & this was the first book wch he studied in learning the higher Geometry, Mr Huygens introducing him.[36] In the year 1674 Mr Leibnitz after a years intermission renewed his correspondence with Mr Oldenburg & began to write to him about series for finding the Area or Circumference of a circle or any Arc whose sine[37] was given. And in the year 1675 Mr Oldenburg sent from Mr Collins to Mr Leibnitz several of mine & Gregories series for the same purpose; & Gregory dying in the end of the year, Mr Collins at the request of Mr Leibnitz collected the Letters wch he had received from Gregory, & Mr Oldenburg in June 1676 sent the Collection[38] to Paris to be perused & returned, & it is now in the Archives of the R. Society. In this Collection were a copy of Mr Gregories Letter of 5 Sept. 1670 & a Copy of my Letter of 10 Decem 1672 to Mr Collins: and by these Letters Mr Leibnitz had notice that Dr Barrow's method of

---

(36) '...& this was the first book wch he began to read of that sort of Geometry Mr Huygens himself introducing him' in the draft on .27: 386^v. Newton's source here would seem to be Leibniz' letter of 18 April 1716 (N.S.) to the Baroness von Kilmansegge (*Recueil*, **2**, $_1$1720: 29–41, just now shown to Newton in its printer's proof by Desmaizeaux) where the latter wrote that on returning to Paris in spring 1673 from his first visit to London 'Comme j'y pratiquai M. *Huygens* de Zulichem, ...grand Geometre, je commençai à prendre goût aux meditations Geometriques. J'y avançai en peu de temps...' (*ibid.*: 31 [=*Correspondence*, **6**: 325]).

(37) Read 'tangent' in historical truth: in 1674 Leibniz was (compare Appendix 2: note (53)) concerned only to enquire whether the 'English' knew of his own 'Gregory–Leibniz' series for the inverse-tangent, and did not become interested in the corresponding expansion of the inverse-sine till nearly two years later after Georg Mohr had passed its Newtonian statement—but not its proof—on to him at third hand. (Newton's series for both the sine and its inverse, and also James Gregory's for the tangent and its inverse, had already previously been sent to him by Oldenburg in April 1675 among a whole bunch of trigonometrical series garnered for him by Collins out of Newton's and Gregory's papers and letters; compare J. E. Hofmann, *Leibniz in Paris, 1672–1676* (Cambridge, 1974): 131–5.)

(38) Or rather the 'Abridgement' of this *Historiola* of Gregory's mathematical correspondence whose draft is now Royal Society MS LXXXI, No. 45 [=*Leibniz: Sämtliche Schriften und Briefe* (3) **1**, Berlin, 1976: 504–16]; compare Hofmann, *Leibniz in Paris*: 214–15, and see also IV: 666, note (30). Leibniz saw the full gathering of Collins' preparatory extracts only in mid-October 1676 during his second visit to London; the 'Excerpta ex commercio epistolico inter Collinium et Gregorium' which he then made are now printed by Hofmann in *Sämtliche Schriften*, (3) **1**: 485–503, and include (see *ibid.*: 495) his annotation of the essence of Newton's letter of 10 December 1672 to Collins which was, so Newton in his old age firmly believed, Leibniz' source of knowledge of the 'universal' method of Cartesian tangents. (Let us add, however, that Collins' *excerptum* from Newton's letter exists no longer with the main portion of the *Historiola* preserved as Royal Society MS LXXXI, No. 46 [=*Sämtliche Schriften*, (3) **1**: 433–84] but is now with the dozen pages or so retained by its then owner William Jones when he deposited its greatest part in the Society's archives in 1712.)

Tangents was capable of improvement so as to give the Method of Tangents of Gregory, & that the method of tangents of Gregory & Slusius was capable of improvement so as to give my general Analysis mentioned in my said Letter & that this Analysis proceeded without sticking at surds, & that I had interwoven it with the Method of Series, vizt in my *Analysis per series* abovementioned & in [the] Tract wch I wrote upon them in the year 1671. Mr Leibnitz wrote also to Mr Oldenburg[39] for the demonstration of some of my series, that is, for the method of finding them, & promised him a Reward, & told him that Mr Collins could help him to it; & therefore he had heard that Mr Collins had my method of series, that is, my *Analysis per series* above mentioned. For I had sent my method of series to Mr Collins in no other Paper then that.[40] But Mr Collins instead of sending a copy of that Tract joyned with Mr Oldenburg in solliciting me to write an Answer to Mr Leibnitz's Letter. And thereupon I wrote my Letter of 13 June 1676, wch was sent to Mr Leibnitz at the same time with the aforesaid Collection. And Mr Leibnitz in his Answer dated 27 Aug. 1676, replied that he did not believe that my Analysis was so general as I represented, there being many Problemes, & particularly the inverse Problemes of Tangents, not reducible to equations or quadratures. And [in] the same Letter he placed the perfection of Analysis not in the differential calculus as he did after he found it, but in another method founded on Analytical Tables of Tangents & the Combinatory Art. *Nihil est*, said he, *quod norim in tota Analysi momenti majoris.* And a little after: *Ea verò non differt ab Analysi illa* SVPREMA *ad cujus intima Cartesius non pervenit. Est enim ad eam constituendam opus Alphabeto cogitationum humanarum.*[41] Mr Leibnitz never pretended to have found the differential Analysis before this year,[42] & these circumstances satisfy me that he did not find it till after the writing of this Letter.

In October following he came again to London, & there met wth Dr Barrows Lectures & saw my Letter of 24 Octob. 1676 & therein had fresh notice of the said method & of [the] Compendium of series sent by Dr Barrow to Mr Collins

---

(39) On 12 May 1676 (N.S.). The pertinent portion of the letter is set out on *Correspondence of Newton*, **2**: 3–4; and its full text is now given with full and elaborate editorial comment in *Leibniz: Sämtliche Schriften und Briefe*, (3) **1**, 1976: 375–8.

(40) This is untrue. Newton would seem to have forgotten that on or soon after meeting Collins for the first time, at London in late November 1669, he had let him make lavish excerpts from the student pocketbook (now ULC. Add. 4000) where he had set out his earliest researches into the binomial expansion, establishing its general truth by a succession of Wallisian 'inductions' from its simplest particular cases, and thereafter applying it to square the area under a variety of algebraic curves (notably the circle and rectangular hyperbola) through the 'converging' series which arise from integrating such expansions of binomial ordinates term by term; see our fuller comment in note (60) of the introduction (page 487 above).

(41) As at the equivalent place in [1] preceding (*vide* note (20) thereto), again consult *Correspondence*, **2**: 63 for the Leibnizian surround to these two excerpted sentences.

(42) 1676, namely.

in the year 1669 under the title of *Analysis per series* &c & consulting Mr Collins saw in his hands several of mine & Gregories Letters, especially those relating to series; & in his way home from London was meditating how to improve the method of Tangents of Slusius as appears by his Letter to Mr Oldenburgh dated from Amsterdam $\frac{18}{28}$ Novem 1676.[43] And the next year upon his arrival at Hannover a copy of my Letter of 24 Octob. 1676 was sent after him, & in a Letter to Mr Oldenburg dated 21 June 1677 he sent us his new Method with this Introduction.[44] *Clarissimi Slusij methodum tangentium nondum esse absolutam cele-berrimo Newtono assentior.* And in describing this method he abbreviated Dr Barrow's method of Tangents, & shewed how it might be improved so as to give the Method of Slusius & to proceed in æquations involving surds, & then subjoyned:[45] *Arbitror quæ celare voluit Newtonus de tangentibus ducendis ab his non abludere: Quod addit, ex eodem fundamento quadraturas reddi faciliores me in sententia hac confirmat.* And [46]after seven years, vizt in October 1684[1] he published the elements of this method as his own without mentioning the correspondence which he had formerly had with the English about these matters. He mentioned indeed a *Methodus similis*, but whose that method was & what he knew of it, he did not say, as he should have done. And this his silence put me upon a necessity of writing the Scholium upon the second Lemma of the second Book of *Principles* least it should be thought that I borrowed that Lemma from Mr Leibnitz. In my Letter of 24 Octob. 1676 when I had been speaking of the method of Fluxions I added:[47] *Fundamentum harum operationum, satis obvium quidem, quoniam non possum explicationem ejus prosequi sic potius celavi 6a cc d æ 13e f f 7i 3l 9n 4o 4q rr 4s 9t 12v x.* And in the said Scholium I opened this ænigma, saying that it conteined the sentence *Data æquatione quotcunqp fluentes quantitates involvente, fluxiones invenire et vice versa*; & was written in the year 1676. For I looked upon this as a sufficient security without entring into a wrangle: but Mr Leibnitz was of another opinion.

Mr James Bernoulli in the *Acta Eruditorum* for December[48] 1691 pag. 14, said that the Calculus of Mr Leibnitz was founded on that of Dr Barrow & differed not from it except in the notation of Differentials & some compendium of

(43) John Collins had passed the central portion of this letter to Newton four months after-wards in his letter of 5 March 1676/7 (on which see *Correspondence*, **2**: 198–200).

(44) See *Correspondence*, **2**: 213; Leibniz' elaboration of the simple Gregory–Barrow method of tangents ensues on *ibid.*: 213–15.

(45) *Correspondence*, **2**: 215.

(46) The remainder of this paragraph was (compare note (27) above) reproduced *in extenso* by Brewster in 1855 in his *Memoirs*, **2**: 30–1.

(47) See *Correspondence*, **2**: 115; and compare VII: 24, 48–50.

(48) Read 'January'. Newton had earlier given full quotation of this passage in Jakob Bernoulli's 'Specimen Calculi Differentialis in dimensione Parabolæ helicoidis...' at the bottom of page 195 of his 1715 'Account' of the *Commercium* (Appendix 2 preceding; compare notes (82) and (83) thereto).

operation. And the Marquess de l'Hospital in the Preface to his *Analysis of infinite petits* published A.C. 1696 represented[49] that where Dr Barrow left off Mr Leibnitz proceeded, & that the improvement wch he made to Dr Barrow's Analysis consisted in excluding fractions & surds: but he did not then know that Mr Leibnitz had notice of this improvemt from me by the two Letters above mentioned dated 10 Decem 1672 & 24 Octob. 1676. After he had notice that such an improvement was to be made, he might find it *proprio Marte*, but by that notice knew that I had it before him.

I wrote the Book of *Quadratures* before Mr Leibnitz understood the Differential Analysis. For I wrote it in the year 1676, except the Introduction & Conclusion, extracting most of it out of old Papers. And when I had finished it & the 7th 8th 9th & 10th Propositions with their Corollaries were fresh in memory, I wrote upon them to Mr Collins that Letter wch was dated 8 Novem. 1676 & published by Mr Jones.[50] The Tables at the end of the tenth Proposition for squaring of some Curves & comparing others with the Conic Sections were invented by the inverse Method of fluxions before the year 1671 as may be understood by my Letter of 24 Octob. 1676, where the Ordinates of the Curves are set down. And all the ten first Propositions of the Book of *Quadratures* except the fift & sixt are in the Tract wch I wrote in the year 1671 tho not in the same words & some of them not in words but in equations,[51] & most of them are in a little Tract wch I wrote in autumn A.C. 1666.[52]

In my *Analysis* above mentioned I said of this new Method:[53] *Ad Analyticam merito pertinere censeatur cujus beneficio Curvarum areæ et longitudines &c, id modo fiat, exacte et Geometrice determinantur: sed ista narrandi non est locus.* And Mr Collins in his Letter to Mr Strode [dated 26 July 1672][54] said that by this Tract of *Analysis &*

---

(49) We have cited the original French text of this passage (signatures aiiiv–aivv) from the 'Preface' to L'Hospital's *Analyse des Infiniment Petits* (Paris, ₁1696) in our footnote (79) to Appendix 2 preceding (page 583 above).

(50) In his 1711 *Analysis*: 38; see our fuller comment in note (11) to [1] preceding. We earlier cited the two most recent sentences in III: 19, note (60), there slightly mistaking their context.

(51) Newton again aims disingenuously to conflate the text of his (yet unpublished) 1671 treatise of infinite series and fluxions with the phantom of the non-existent precursor to his later 1690's tract *De Quadratura Curvarum* which he seeks to conjure up out of the 'evidence' of his *epistola posterior* to Leibniz in 1676. Compare our note (12) to [1] preceding.

(52) The text of this October 1666 fluxions tract—no portion of which was ever made public in Newton's own lifetime—is set out on I: 400–48.

(53) Again see II: 242.

(54) We provide an explicit citation in place of a mistaken 'above mentioned' in the manuscript; Newton makes no prior reference to this letter of Collins' in his present 'Supplement', though at the corresponding place in his immediately preceding 'History of the Method of Moments...' he had indeed (see page 608 above) quoted at length the pertinent passage in the Latin version of the letter which is printed in *Commercium Epistolicum*, ₁1713: 28–9. (Let us again remind that the original English words of Collins' draft, now Royal Society MS LXXXI, No. 24, are given in Appendix 2: note (22); see page 567.)

other things communicated before to Dr Barrow, it appeared that I knew this method some years before the *Logarithmotechnia* of M^r Me[r]cator came abroad so as to find the area of any figure accurately if it may be or at least by approximation *in infinitum*. And in my Letter of 24 Octob. 1676 I represented that the Quadrature of Curves was improved by the method of fluxions[,] & that by that method I had found some general Theoremes for that end & there set down one of those Theoremes & illustrated it with examples. And in the six first Propositions of the Book of *Quadratures* I shewed how such Theoremes were to be found by that Method. And therefore that Method so far as it is conteined in the first six Propositions of the Book of *Quadratures* was known to me before I wrote the said Letter, & even before I wrote the said *Analysis* & before Mercators *Logarithmotechnia* came abroad, there being no other method then that conteined in those six Propositions by which such Theoremes could be found.

By the inverse Method of fluxions I found in the year 1677[55] the demonstration of Keplers Astronomical Proposition vizt that the Planets move in Ellipses about the lower focus with an angular velocity reciprocally proportional to their distance from the focus: & in the year 1683[56] at the importunity of Dr Halley

(55) Read 'in the winter between 1679 and 1680', as his short-lived correspondence with Robert Hooke between November 1679 and January 1680 on the topic of celestial dynamics establishes beyond a doubt. If '1677' had any significance in the prehistory of the squabble over priority in discovery of calculus, it was the year when (as he had put it in the *Commercium Epistolicum*: 97) 'mors *Oldenburgi* huic literarum Commercio [cum Leibnitio annis 1674, 1675, 1676, 1677] finem imposuit'. Newton made a similar slip in the draft (Add. 3968.41: 85^r) of an ensuing letter to Desmaizeaux, first printed in the preface (p. xviii) to *A Catalogue of the Portsmouth Collection of Books and Papers written by or belonging to Sir Isaac Newton* (Cambridge, 1888) and since widely quoted *hors de contexte* in its own right as an independent 'memorandum', where he affirmed: '...At length in the winter between the years 1676'—'1666' was first written!—'& 1677 I found the Proposition that by a centrifugal force reciprocally as the square of the distance a Planet must revolve in an Ellipsis about the center of the force placed in the lower umbilicus... & with a radius drawn to that center describe areas proportional to the times'. (See also I. B. Cohen's *Introduction to Newton's 'Principia'* (Cambridge, 1971): 291–2.) Let us but lightly insist that Newton's memory for dates—and the niceties of past fact in general—was none too trustworthy in his old age, and all too prone to be 'rationally' reconstructed (through laziness or more wilful blinkering) even where he had documents to hand which would have accurately reinforced it.

(56) Here read '1684', as is likewise beyond question. (Compare our full parade of the supporting evidence in VI: 16–17.) In his ensuing 'memorandum' to Desmaizeaux in this summer of 1718—see the previous note—Newton went on similarly to adjoin: 'And in the winter between the years 1683 & 1684 this Proposition wth the Demonstration was entered in the Register Book of the R. Society....'. This false recollection that it was 'Anno 1683 ad finem vergente' that 'Propositiones principales earum quæ in *Philosophiæ Principijs Mathematicis* habentur Londinum misit' was born in fact in autumn 1712 when he set this as his opening phrase in an open 'Epistola' to an un-named 'Amicus' (ULC. Res. 1893.8, first printed by Joseph Edleston on pages 308–14 of the Appendix to his edition of the *Correspondence of Sir Isaac Newton and Professor Cotes* (London, 1850); see note (86) of the introduction for its provenance); there 'Newton first of all clearly wrote 1684, then altered the 4 to a 3, afterwards

I resumed the consideration thereof, & added some other Propositions about the heavenly bodies[57] & sent them to him that year in autumn, & they were by him communicated to the R. Society & by their order entered in their Letter Book. And all this was done before Mr Leibnitz published the elements of the Differential calculus.

In writing the Book of *Principles* I made much use of the method of fluxions direct & inverse but did not set down the calculations in the Book itself because the Book was written by the method of composition, as all Geometry ought to be.[58] And this Book was the first specimen made publick of the use of this method in the difficulter Problemes. The Marquess de l'Hospital said that this Book was *presque tout de ce calcul*.[59] And Mr Leibnitz in a Letter to me dated $\frac{7}{17}$ Mart. 1693: *Mirifice ampliaveras Geometriam tuis seriebus, sed edito Principiorum opere ostendisti patere tibi etiam quæ Analysi receptæ non subsunt. Conatus sum ego quoφ, notis commodis adhibitis quæ differentias & summas exhibent, Geometriam illam quam transcendentem appello, Analysi quodammodo subjicere.*[60] And in the *Acta Eruditorum* for May 1700 pag. 203.[61] *Certe cum elementa calculi mea edidi anno 1684, ne constabat quidem mihi aliud de inventis ejus* (*Newtoni*) *in hoc genere quam quod ipse olim significaverat in litteris, posse se tangentes invenire non sublatis irrationalibus &c sed majora multi consecutum Newtonum, viso demum libro Principiorum ejus satis intellexi.* And pag 206 speaking of the method by wch they solved the Problem of the *linea celerrimi*

crossed all the figures out and wrote distinctly 1683' (Edleston: 308). The phrase was later introduced without change by him into the linking editorial passage on page 97 of the *Commercium Epistolicum*.

(57) The text of the resulting tract 'De motu corporum in gyrum' is reconstructed on VI: 30–90 from the much corrected and overwritten original draft which is now ULC. Add. 3965.7: 55^r–62 *bis^r*.

(58) In commenting upon his like claim on page 206 of his 1715 'Account' of the *Commercium* that he had 'found out most of the propositions' in his *Principia* by his 'new analysis' of fluxional increase, but had then given them synthetic demonstration in the published book 'that the systeme of the heavens might be founded upon good Geometry', we have (Appendix 2: note (137)) indicated that we can find little to support its veracity. Just a few months later, in the drafts whose mathematical portions are reproduced in 1, 8 preceding, Newton was to elaborate the contention in perhaps its extreme form in preface to an intended re-edition of the *Principia*. Let us not here linger to repeat the variety of caveats and corrections of detail which need to be stated before each and any of these several affidavits can be allowed to be an accurate presentation of the past reality of the way in which he had come to discover the *Principia*'s propositions and then set his preparatory analyses out on paper for others to read.

(59) In preface to his *Analyse des Infiniment Petits* in 1696; see our fuller comment in Appendix 2: note (94) preceding.

(60) Raphson, *History of Fluxions*: Appendix: 119 (=*Correspondence*, **3**, 1961: 257). All this goes much as on page 198 of Newton's 1715 'Account' of the *Commercium* (see page 588 above).

(61) In his 'Responsio ad Dn. Nic. Fatii Duillerii Imputationes' (*Acta Eruditorum* (May 1700): 198–206 [=*Leibnizens Mathematische Schriften*, **5** (Halle, 1858): 340–8]), countering Fatio's judgement the year before, 'rerum evidentia coactus', that he was merely 'Calculi... secundus Inventor'; see notes (6) and (8) of the preceding introduction.

descensus & I found the solid of least resistance, he calls it *methodum summi momenti valdeq diffusam—quam ante Dominum Newtonum et me nullus quod sciam Geometra habuit*; *uti* [*ante*] *hunc maximi nominis Geometram nemo specimine publice dato se habere probavit.*

The Book of *Principles* came abroad in spring[62] 1687 & in the *Acta eruditorum* for January 1689[63] M[r] Leibnitz published a *Schediasma de resistentia Medij & motu Projectorum gravium in Medio resistente.* This Tract was writ in words at length without any calculations, & in the end of it Mr Leibnitz added: *Et fortassis attente consideranti vias quasdam novas vel certe satis antea impeditas aperuisse videbimur. Omnia autem respondent nostræ Analysi infinitorum, hoc est, calculo summarum et differentiarum (cujus elementa quædam in his Actis dedimus) communibus quoad licuit verbis hic expresso.* And this was the second specimen made public of the use of this method in the difficulter Problems. And yet it was nothing else then the two first Sections of the second book of *Principles* reduced into another order & form of words, & enlarged by an errone[o]us Proposition. To find the Curve described in a Medium where the resistance was in a duplicate ratio of the velocity he composed the horizontal & perpen[di]cular motions of the projectile.[64]

In the same *Acta Eruditorum* for February[65] he published another Paper entitled *Tentamen de motuum cœlestium causis* & therin he endeavoured to demonstrate Keplers Proposition above mentioned, by the Differential Method. But the Demonstration proved an erroneous one.[66] And this was the third specimen made publick of the use of this Method in the difficulter Problems.

---

(62) Read 'summer'. It was only on 5 July 1687 (see *Correspondence*, **2**: 481) that Halley was able to announce to Newton that he had 'at length brought your Book to an end'.

(63) Pages 38–47 (= *Mathematische Schriften*, **6**: 1860: 135–43); on this see **1**, 6, §2.3: note (50), and also note (95) of the introduction to the present section.

(64) We have already quoted this last sentence in VI: 71, note (109), where we discuss the point of Newton's criticism in some detail.

(65) Pages 82–96 (= *Mathematische Schriften*, **6**: 144–61).

(66) Newton had dwelt upon this error of Leibniz' in second, geometrical differences in (§15 of) his 1689 *Acta* article at considerable length more than four years before when he furnished this to John Keill as one of the prime 'preuves claires et solides' of Leibniz' inferior grasp of the subtleties of calculus which the latter desired to publish in the *Journal Literaire* in 1714 in 'Réponse' to the more generalised counter-assertions of Newton's inadequacy in the realm of infinitesimal analysis. We cited in note (86) of the preceding introduction (from the draft on Add. 3968.4: 14[r]) the detailed corrections to page 89 of the 'Tentamen' which Newton then adjudged to be necessary; these emendations he subsequently subsumed into a broader set of 'Notæ in *Acta Eruditorum* an. 89 p. 84 & sequ.' whose full text has only recently been published (in *Correspondence*, **6**, 1976: 116–17; compare also E. J. Aiton, 'The celestial mechanics of Leibniz in the light of Newtonian criticism', *Annals of Science*, **18**, 1962: 31–41). In his 'Tentamen' Leibniz deftly—and probably (in the first instance) quite unperceivingly—absorbed the consequences of his mathematical slip by a far from evident countering allowance in his dynamical presuppositions regarding the measure of centrifugal force, but eventually made proper estimate of the quantity of such *vis paracentrica* after a vigorous exchange of letters

[67]At the request of Dr Wallis I sent to him in two Letters dated 27 Aug. & 17 Septem 1692 the first Proposition of the Book of *Quadratures* copied almost *verbatim* from the Book & also the method of extracting fluents out of equations invol[v]ing fluxions mentioned in my Letter of 24 Octob 1676 & copied from an older Paper,[68] & an explication of the Method of fluxions direct & inverse comprehended in the sentence *Data æquatione quotcunꝗ fluentes quantitates involvente, invenire fluxiones & vice versa*: & the Doctor printed them all the same year (viz *anno* 1692) in the second volume of his works pag. 391, 392, 393, 394, 395, 396,[69] this Volume being then in the press & coming abroad the next year, two years before the first Volume was printed off. And this is the first time that the use of letters with pricks & a Rule for finding second third & fourth fluxions were published, tho they were long before in Manuscript.[70] When I considered only first fluxions I seldome used letters with a prick: but when I considered also second third & fourth fluxions &c I distinguished them by Letters with one two or more pricks: & for fluents I put the fluxion either included within a

---

during 1704–6 with Pierre Varignon brought him to see the light, and he straightaway published an 'Excerptum' from this correspondence in *Acta Eruditorum* (October 1706): 446–51 where due correction of his blunder in the 1689 'Tentamen' was made. (Compare Aiton's fundamental secondary account of 'The celestial mechanics of Leibniz' in *Annals of Science*, **16**, 1960: 65–82, especially 77–82; and see also his ensuing 'New interpretation'—which we ourselves cannot everywhere accept—in *Annals*, **20**, 1964: 111–23.)

In his draft of the present passage on .31: 452ʳ, where he stated much as here (compare notes (55) and (56) above) that 'In the year 1677[!] I found the Demonstration of Keplers Proposition that the Planets revolve in Ellipses about the inferior focus of their Orbs & communicated it to the mathematicians in London A.C. 1683[!] & published it in the Book of *Principles* A.C. 1687', Newton had passed a little more forthrightly to declare that 'Mr Leibnits in the *Acta Eruditorum* for Febr. 1689 published a Demonstration of the same Proposition, pretending that he had found it *proprio Marte*, but the Demonstration for want of skill in the differential method is an erroneous one'. The soundness of this inference, in itself not unreasonable if (as ever) lacking in charity, he went on to spoil, however, by adjoining his entirely prejudiced conjecture—we need not stress how wildly it conflicts with the documentable historical reality—that 'After this year Mr Leibnitz grew better acquainted with the Method, the two Bernoullis coming in to his assistance, & in the years 1691, 1692 & 1693 the Method began to be celebrated'.

(67) The paragraph which follows is cited *in toto* by Brewster in his *Memoirs* (see note (27) above): 34, note 1. In its content it is but a minimal elaboration of the equivalent portion of [1] preceding (see pages 606–7 above).

(68) In context Newton evidently means Problem 2 of his 1671 tract (see III: 82–116) rather than his subsequent refinements (see IV: 564–88 and especially VII: 70–98) of the techniques there adumbrated for extracting the roots of fluxional equations.

(69) See VII: 174–80.

(70) This is not true, as we have already spelled out in VII: 64–5, note (35). Although Newton had briefly employed a superscript 'double-dot' notation for first-order fluxions in autumn 1665 (see I: 363), his standard use of 'prickt' letters in this signification was born in late 1691 when he began to reshape the first version of his 'De quadratura Curvarum' (see VII: [131→] 64 ff.).

square (as in the aforesaid *Analysis*) or with a square prefixed (as in some other papers[71]) or with an oblique line upon it.[72] And these notations by pricks & oblique lines are the most compendious yet used, but were not known to the Marquess de l'Hospital when he recommended the diff[er]ential notation, nor are necessary to the method.

In the end[73] of the year [1691] M[r] James Bernoulli spake contemptib[l]y of the Differential Method as above. In the years 1692, 1693 & 1694 it grew into reputation. And in spring 1695 D[r] Wallis hearing that my Method of fluxions began to be celebrated in Holland under the name of the Differential Method of M[r] Leibnitz, wrote in the Preface to the first Volume of his works that the two methods were the same & that in my Letters of 13 June & 24 October 1676 I explained to M[r] Leibnitz this method found by me ten years before or above. D[r] Wallis was not *homo novus & rerum anteactarum parum peritus* as M[r] Leibnitz objected against D[r] Keill.[74] He was *homo vetus & rerum anteactarum peritus*, having received copies of my said two Letters from M[r] Oldenburgh in the very year 1676 when they were newly written, & having had sufficient opportunity in those days to inform himself about this matter. By my explaining the Method to M[r] Leibnitz, I suppose he meant no more then that in those two Letters I had said so much of it as to make it easy to find the rest.

In the year 1696 the Marquess de l'Hospital published his *Analysis de infinitement petits* & this I think was the first time that a Rule for finding second third & fourth differences was published. This Book being an Introduction to the differential Method made it spread much more then before. But in the year 1699 M[r] Fatio published that I was the oldest inventor by many years[75] & D[r] Wallis published my Letters in the third Volume of his works. M[r] Leibnitz returned a reply to M[r] Fatio in the year 1700[76] & M[r] Fatio wrote an Answer & sent it to the Editors of the *Acta Eruditorum* A.C. 1701, but they refused to print it, pretending an aversion to controversies. See the *Acta* for March [1701] page. 134.

In...the year 1704 [I] published the Theory of colours & the Book of *Quadratures* together.[77] And writing an introduction to the Book of *Quadratures*

---

(71) Notably in his October 1666 tract (see I: 404 ff.).

(72) This usage first appears in the 1693 revision of the 'De quadratura Curvarum'; see our note (9) on vii: 510–11.

(73) Read 'beginning'. (Newton repeats his earlier slip in assigning the date of Jakob Bernoulli's article; see note (48).)

(74) In his letter to Sloane of 18/29 December 1711; we quoted Leibniz' full sentence of complaint against Keill in note (38) of the preceding introduction.

(75) On page 18 of his *Lineæ Brevissimi Descensus Investigatio Geometrica Duplex*; the essential core of this aside of Fatio's on the source of his knowledge of the 'Calculus quo utor' is given in note (6) to the preceding introduction.

(76) *Acta Eruditorum* (May 1700): 198–206. Two passages from this lengthy response of Leibniz to Fatio have here been cited by Newton four paragraphs earlier; see pages 618–19.

(77) Namely in his *Opticks: Or, A Treatise of the Reflexions, Refractions, Inflexions and Colours of*

with relation to the method of fluxions upon wch that Book depended: I said in that Introduction^(78) that I found the Method gradually in the years 1665 & 1666; this being not so much as Dr Wallis had said nine years before without being then contradicted.^(79) And thus the Method was claimed from time to time from Mr Leibnitz till this year so that he never was in quiet possession.^(80)

But ^(81) in the *Acta Eruditorum* for January 1705 an account of the Introduction to the Book of *Quadratures* was published in these words.^(82) *Quæ* (Isagoge) *ut* MELIVS *intelligatur, sciendum est cum magnitudo aliqua continue crescit, veluti linea (exempli gratia) crescit fluxu puncti quod eam describit, incrementa illa momentanea appellari* DIFFERENTIAS, *nempe inter magnitudinem quæ antea erat & quæ per mutationem*

Light, in whose *princeps* edition the 'Tractatus de Quadratura Curvarum' is placed (₁1704: ₂163–211) as the latter of the adjoined *Two Treatises of the Species and Magnitude of Curvilinear Figures* heralded in this composite work's main title.

(78) See page 123, note (7) above (= *Opticks*, ₁1704: ₂166).

(79) In the preface, that is, to the first volume of his collected *Opera*; Newton had quoted the pertinent passage (*Ad Lectorem Præfatio*: [iii]) at length on page 200 of his 1715 *Phil. Trans.* 'Account' of the *Commercium Epistolicum* (see page 591 above). In his further reshaping of the present 'Supplement' to be an open letter to Desmaizeaux, Newton also appealed to the evidence of John Collins in his early 1670's letters—in turn calling upon 'the testimony of Dr Barrow grounded upon what I had communicated to him from time to time before [1669]'—as a like 'ancient, knowing and credible witness' whose variety of statements 'may suffice to excuse me for saying…that I found the Method by degrees in the years 1665 & 1666': 'I was then', he added, 'in the prime of my age for invention & most intent upon mathematicks & philosophy & found out in those two years the methods of series & fluxions & the theory of colours & began [to think of gravity extending to the orb of the Moon…]'. (See Add. 3968.27: 390^v; a full citation of the passge is made in *Correspondence*, **6**, 1976: 461–2, note (15).) The further elaboration of this would-be authoritative pronouncement upon the chronology of his earliest discoveries in exact science to declare that 'In the beginning of the year 1665 I found the Method of approximating series & the'—'general' is delected—'Rule for reducing any dignity of any Binomial into such a series. The same year in May I found the method of Tangents of Gregory & Slusius, & in November had the direct method of fluxions & the next year…in May…I had entrance into ye inverse method…. All this was in the two plague years of 1665 & 1666. For in those days I was in the prime of my age for invention & minded mathematicks & philosophy more then at any time since…' (.41: 85^r) has, ever since it was printed on page xviii of the 1888 *Catalogue of the Portsmouth Collection* (see note (55) above; a more complete transcription of the draft text is given by I. B. Cohen in his *Introduction to Newton's 'Principia'*: 291–2), been ubiquitously quoted as gospel truth. For some of the qualifications which must be made in this somewhat oversimplified account of the birth of fluxional calculus we may cite our own 'Newton's marvellous year: 1666 and all that' (*Notes and Records of the Royal Society of London*, **21**, 1966: 32–41).

(80) In his draft on .12: 164^r Newton here adjoined 'abroad, nor I out of possession in England'.

(81) The remainder of this paragraph is quoted in full by Brewster in his 1855 *Memoirs*, **2**: 41–2 (compare note (27) preceding).

(82) *Acta* (1705): 34–5. We have given our English version of this portion of Leibniz' anonymous *critique* of the *De Quadratura* in introducing the latter text on pages 26–7 above. (As ever, of course, the small capitals in the following quotation are Newton's preferred emphases, and do not appear in the original.)

momentaneam est producta; atɋ hinc natum esse Calculum Differentialem, eiɋ reciprocum
summatorium; cujus elementa ab INVENTORE *D. Godofredo Guillielmo Leibnitio in his*
Actis sunt tradita varijɋ usus tum ab ipso, tum a DD. Fratribus Bernoullijs tum a
D. Marchione Hospitalio sunt ostensi. Pro Differentijs IGITVR *Leibnitianis D. Newtonus*
adhibet, semperɋ (pro ijsdem) adhibuit, fluxiones,—ijsɋ tum in suis Principijs Naturæ
Mathematicis, tum in alijs postea editis (pro Differentijs Leibnitianis) *eleganter est*
usus, QVEMADMODVM *et Honoratus Fabrius in sua Synopsi Geometrica, motuum progressus*
Cavallerianæ methodo SVBSTITVIT. And all this is as much as to say that I did not
invent the method of fluxions in the years 1665 & 1666 as I affirmed in this
Introduction, but after Mr Leibnitz in his Letter of 21 June 1677 had sent me
his differential method, instead of that method I began to use & have ever since
used the method of fluxions.[83]

In the *Philosophical Transaction[s]* for September & October 1708 Dr Keill
published to the contrary: *Fluxionum Arithmeticam sine omni dubio primus invenit*
D. Newtonus ut cuilibet ejus epistolas a Wallisio editas legenti facile constabit. Eadem
tamen Arithmetica postea mutatis Nomine & Notationis modo a Domino Leibnitio in
Actis Eruditorum edita est.[84] And Mr Leibnitz in a Letter to Sr Hans Sloane dated
4 March 1711 complained of this.... And Dr Keill in an Epistle read before the
R. Society 24 May 1711 replied: *Agnosco me dixisse fluxionum Arithmeticam a*
D. Newtono inventam fuisse, quæ mutato nomine & notationis modo a Leibnitio edita fuit:
sed nollem hæc verba ita accipi, quasi aut nomen quod Methodo suæ imposuit Newtonus aut
Notationis formam quam adhibuit, D. Leibnitio innotuisse contenderem; sed hoc solum
innuebam, D. Newtonum fuisse primum inventorem Arithmeticæ fluxionum seu Calculi
Differentialis; eum autem in duabus ad Oldenburgum scriptis Epistolis & ab illo ad
Leibnitium transmissis, indicia dedisse perspicacissimi ingenij Viro satis obvia; unde
Leibnitius principia illius calculi hausit, vel saltem haurire potuit. At cum loquendi et
Notandi formulas quibus usus est Newtonus, ratiocinando assequi nequiret Vir illustris, suas
imposuit....

Afterwards Mr Leibnitz in his Letter to Dr Sloan of 29 Decem st.n. 1711
declined answering Dr Keill & pressed me to declare my opinion.[85] In a Paper

---

(83) 'For', Newton originally went on to add before suddenly striking the sentence out
(doubtless when he realised the unfortunate sense which might be put upon his sarcastic words),
'the way to understand the method of fluxions is to learn the differential method first.'

(84) For our English rendering of this paragraph at the end (pag 185) of Keill's paper 'de
Legibus Virium Centripetarum' (*Phil. Trans.*, **26**: 174–88)—the Sarajevo incident, one might
say, which soon after its delayed publication (in 1710) led to the outbreak of open war in the
long-gathering fight over priorities of calculus discovery—see page 473 above. We there in
sequel present a considerably more detailed overview of Newton's evolving reaction to Leibniz'
pressing insistence in his two letters in 1711 to the Royal Society's secretary, Hans Sloane,
that his own over-riding claim to 'first invention' of the *calculus differentialis* should publicly be
sustained.

(85) Which, we need not say, Newton eventually did under the cloaking anonymity of the
Society committee appointed in March 1712 (see pages 480–3 above) to report on the

dated 29 July 1713 & written by one who used the phrase *illaudibili laudis amore*[86] & knew what Mr Leibnitz did at Paris 40 years before, I was singled out & treated very reproachfully. And this was to make me appear.[87] And in his Postscript to M^r l'Abbé Conti written in Autumn 1715[88] I was again singled out & treated in a very provoking manner to make me appear.[87] And when I was prevailed with to return an Answer to this Postscript he declared in his Letter of Apr. 9 1716 to M^r l'Abbé Conti that he would by no means enter the lists with my forlorn hope[89] (meaning Mr Keill & Mr Cotes &c) but since I was willing to appear my self[87] he would give me satisfaction. Nothing could content him but to make me appear & declare my opinion, & now I have declared it, I leave every body to his own opinion, out of a desire to be quiet.[90]

---

justice and historical truth of Leibniz' claim to priority. It is curious that he here makes no mention of the *Commercium Epistolicum* in which the 'Committee's' findings against Leibniz (*ibid.*: 120–2) were given the permanence of print.

(86) 'Phrasis est Leibnitiana' he later still more explicitly declared in a set of yet unpublished Latin 'Annotationes' (now in private possession) on the 1713 *Charta volans*. For our fuller account of the background to this anonymously edited 'flying Paper' of Leibniz' (and, less willingly so, of Johann Bernoulli's) composition, see pages 490–2 preceding.

(87) That is, publicly to speak out in his own defence.

(88) On this 'Apostille'—which Conti lost no time in passing on to Newton—and the flurry of correspondence ensuing in its train in the early months of 1716, see pages 504–10 above.

(89) 'Je n'ay point voulu entrer en lice avec des enfans perdus...detachés contre moy' were Leibniz' exact words in his letter (see note (118) of the preceding introduction). He did not, however, go on to name any individual members of the stormtrooping *verloren hoop* of 'desperate lads' which (so he would have it) Newton had detached against him.

(90) In draft on .41: 140^v Newton added a final sentence which well merits preservation from oblivion: 'The inverse Method of fluxions is capable of great improvements, & the improvements are his who finds them out'.

APPENDIX 4. A PREFACE TO AN INTENDED JOINT EDITION OF THE *PRINCIPIA* AND THE *DE QUADRATURA*, AND ITS SUBSEQUENT INCORPORATION IN THE FORMER'S FLUXIONS SCHOLIUM.[1]

[*c.* mid-1719][2]
From the original drafts in the University Library, Cambridge

(1) Already in the early summer of 1712, when the second edition of his *Principia* was nearing the end of its long passage through the press, Newton had planned to add to it an amplified version of his published tract *De Quadratura Curvarum* (the surviving drafts of this are set out in **1**, 5 preceding) which would—with the help of additional 'Propositions taken from my letters already published for reducing quantities into converging series & thereby squaring the figures' (so he phrases it in the last paragraph of the piece reproduced in Appendix 7.1 below)—provide a full account of the 'Analysis per quantitates fluentes et earum momenta' used by him, so he claimed, in composing his masterwork and also a listing of the basic quadratures of curves to which, in particular instances and in general assumptions *concessis curvilinearum quadraturis*, he there appealed as prior givens. Although he never did put into print any such augmented 'Book of Quadratures', he returned to think seriously of doing so in the late 1710's when the need for a yet further corrected and rewritten *editio ultima* of the parent *Principia* began to make itself urgently felt. And, with passions still running high in England and on the Continent over 'first invention' of the methods and algorithms of calculus long months after the prime antagonist Leibniz had (so the optimistic phrase is) been laid to rest in his grave, we will not be surprised that Newton in no wise reluctantly seized the opportunity to capitalise upon the projected publication by advancing anew his own claim to absolute priority of discovery. In late August 1718 he was minded to write to Pierre Varignon that, though 'while the *Commercium Epistolicum* remains unshaken I see no need of my medling with that Controversy any further', yet 'perhaps I should write a short Preface to the Book *De Quadratura Curvarum* to shew that that Book (which has been accused of Plagiary) was in MS before Mr Leibnitz knew anything of the Differential Method' (Add. 3968.42: 613r, printed in *Correspondence*, **7**, 1978: 3). In this passing comment—at once deleted in the manuscript— there is, to be sure, no indication that this new preface was to introduce a reissue of the *De Quadratura* in annexe to a coming re-edition of the *Principia*. But he made his intention clear the next summer in a memorandum (now in private possession) jotted on the back of a preliminary English draft—on this see *Correspondence*, **7**: 68, note (11)—of his response to an ensuing letter to him from Varignon on 26 July 1719 (N.S.), where he wrote:

'In the book of *Principles* in the end of the Schol. upon Lem. 2 Lib. 2 explain how that Lemma conteins a solution of the Proposition *Data æquatione fluentes quotcunꝗ Quantitates involvente invenire fluxiones*, demonstrated synthetically. & how the same solution may be deduced from my Letter [to Mr Collins] of 10 Decem 1672.

'To the book of *Quadratures*, prefix a Preface to ye Reader confirming what I said in ye Introduction, viz. that I found the method of fluxions in year[s] 1665 & 1666'— 'or do it in a Note upon the Introduction' he subsequently here inserted—'& add Notes upon ye Book, Prop. II, V, & Corol 2 Prop. X, & Tab. II. part. ult.'

The extracts set out in the next five Appendixes 4–8 will reveal the zeal and diligence with which Newton went on to implement this scheme. In the present Appendix 4 we print the major pieces in which he drafted and elaborated his historical 'Schol.' upon Lemma II of the *Principia*'s second book (whose original 1687 text is quoted in IV: 524, note (11))—the first of these is a rough draft of a tentative 'Præfatio' to the joint edition whose content is in greatest

[1]⁽³⁾   *Præfatio.*

Analysin per series & momenta a me scriptam Barrovius noster anno 1669 ad Collinium misit. Methodos ibi expositas in alio tractatu plenius explicui anno 1671. In epistola 10 Decem 1672 ad Collinium data methodum momentorum exemplo tangentium more Slusiano ducendarum illustravi, dixicʒ eandem etiam ad abstrusiora Problematum genera de curvitatibus & longitudinibus Curvarum decʒ areis & centris gravitatis &c sese extendere et esse generalem & ad quantitates surdas non hærere: et Collinius exemplar hujus Epistolæ ad D. Leibnitium tunc in Gallia agentem misit idcʒ mense Junio anni 1676. Et hoc anno sub autumno ex tractatu prædicto quem scripseram anno 1671, tractatum de quadratura curvilinearum⁽⁴⁾ extraxi, eodemcʒ plurimum usus sum in componendo libro præcedente de *Philosophiæ naturalis principijs mathematicis.* Et in lib III de *Principijs Philosophiæ* Lem. V, specimen dedi methodi cujusdam differentialis.⁽⁵⁾ Quapropter tractatum prædictum de quadratura figurarum subjung[er]e visum est, ut et tractatum de *methodo* illa *differentiali.*⁽⁶⁾

---

part subsumed in what comes after—and in Appendix 5 is set out the unpublished text of an intended *Scholium secundum* where Newton gives yet fuller account of the 'mind' of his printed fluxions scholium. In Appendix 6 is reproduced the largest part of an English draft of a preface to the coming reissue of the *Principia* where 'Lem. 2 Lib. 2' and its scholium is the centre of attention. In Appendix 7 there follow three successive English drafts of a new preface to the *De Quadratura* variously stressing in now familiar terms the unchallengeableness of his claim in its original 'Introductio' to have 'found the method of fluxions gradually in the years 1665 & 1666' and to have given an extended exposition of it in 'the Tract wch I wrote in the year 1671'. In Appendix 8, lastly, we set out the finished Latin text of Newton's address 'Ad Lectorem', together with that of the additional 'Notes upon ye Book' which he had promised himself to attach to the *De Quadratura*, setting forth the dated pedigree of its main propositions.

(2) This approximate dating, that (see the previous note) of his scheme for adding supplementary historical material to his projected joint edition of the *Principia* and *De Quadratura*—along with, he initially planned (see note (6) following), the *Methodus Differentialis*—may be a few months out. Newton's handwriting in the manuscripts agrees well with their having been penned in 1719–20, but this can afford only a loose confirmation of the suggested date of composition.

(3) Add. 3968.21: 221^v; a first draft is found overleaf on the recto side of the manuscript page. The greater part (the three opening sentences) of this initial version of a preface to the planned joint edition of the *Principia* with *De Quadratura* was subsequently absorbed by Newton into the re-done Scholium to Lemma II of the former's Book 2 whose text is reproduced in [5] below. The remainder is now printed, with English translation alongside, on page 348 of I. B. Cohen's *Introduction to Newton's 'Principia'* (Cambridge, 1971).

(4) This, so Newton implies (but see VII: 15–16), is the *Tractatus de Quadratura Curvarum* as printed in 1704 (pages 122–58 above +VII: 508–52); certainly, it was his intention to reprint this text (in the minimally revised form set out by William Jones on pages 41–66 of his 1711 compendium of Newtonian *Analysis per Quantitatum Series, Fluxiones ac Differentias*) as the 'Book of Quadratures' now to be annexed to the *Principia*.

(5) *Principia,* ₁1687: 481–3 [=IV: 70–3].

(6) On this tract setting forth the elements of interpolation by the *Methodus Differentialis* (mocked up by Newton in late 1710 from his 1670's researches into finite differences, and

[2]⁽⁷⁾ In literis quæ mihi cum Geometra peritissimo G. G. Leibnitio annis abhinc decem⁽⁸⁾ intercedebant, cum significarem me compotem esse methodi determinandi maximas et minimas, ducendi tangentes, quadrandi figuras curvilineas⁽⁹⁾ & similia peragendi quæ in terminis surdis æque ac in rationalibus procederet methodumcɟ exemplis illustrarem⁽⁹⁾ sed fundamentum ejus literis transpositis hanc sententiam involventibus (*Data æquatione quotcuncɟ fluentes quantitates involvente, fluxiones invenire, et vice versa*) celarem:⁽¹⁰⁾ rescripsit Vir Clarissimus anno proximo,⁽¹¹⁾ se quocɟ in ejusmodi methodum incidisse, & methodum suam communicavit a mea vix abludentem præterquam in

published by Jones in his *Analysis* just a few months after it was written) see **1**, 5 preceding. In the early summer of 1712 Newton had (see page 47 above) fleetingly thought to adjoin it to the second edition of the *Principia*, then in press, 'cum in exponenda Cometarum Theoria' —and nowhere else, we may interject—'usus sim methodo mea differentiali' (Add. 3968.32: 464ᵛ; the full text of this discarded 'Ad Lectorem' is given, again with English translation, on page 349 of Cohen's 1971 *Introduction*). In a draft (.5: 29ʳ; the pertinent portion is cited by Cohen on page 348 of *Introduction*) of the last paragraph of the English preface to *De Quadratura* reproduced in Appendix 7.1 below Newton was still minded to write that 'since by the help of this Method [of Fluxions] I wrote the Book of *Principles* I have therefore subjoyned the Book of *Quadratures* to the end of it. And for the same reason I have subjoyned also the Differential Method.' But the latter sentence was omitted from the finished version of the piece (Appendix 7.2), and Newton never afterwards gave any notice to adjoin the *Methodus Differentialis* to his coming re-edition of the *Principia* (for whose cometary propositions Lemma V of its third book is perfectly adequate; see ɪᴠ: 73, note (5) and ᴠɪɪ: 682–6).

(7) Add. 3968.5: 20ʳ. This initial recasting of the scholium to the *Principia*'s Lemm. II, Lib. 2 (₁1687: 253–4 [=ɪᴠ: 524, note (11)]→₂1713: 226–7 with the single addition of the four-word phrase to the penultimate sentence cited in note (12)) is quoted by David Brewster *in extenso* in his *Memoirs of the Life, Writings and Discoveries of Sir Isaac Newton*, **2** (Edinburgh, 1855): 33, note 1 with the divergences from the 1687 printed text (here signalled in notes (9), (10) and (11) following) italicised. All but the last of these are separately itemised at the end of his ensuing Appendix I (*ibid.*: 425–6), where he adjoins some ancillary comment—in particular he aptly cites from the *Principia* preface here reproduced as Appendix 6 Newton's declaration (.9: 105ʳ) that 'because Mʳ Leibnitz had published th[e] elements [of the Analytic Method] a year & some months before [I set them down in the second Lemma of the second Book of the *Principles*] without making any mention of the Correspondence wᶜʰ I had with him by means of Mʳ Oldenburg ten years before that time [*sc.* in 1676], I added a Scholium not to give away the Lemma, but to put him in mind of that Correspondence in order to his making a public acknowledgemᵗ thereof before he proceeded to claim that Lemma from me'—and then passes to give the text of the fuller draft scholium whose latter part is here set out in [4] below.

(8) So still in 1719, as in *Principia*, ₂1713: 226! Newton means, of course, ten years back from the time, early 1686, when he put the first two books of the *Principia* initially to press.

(9) These two three-word phrases are not found in the original printed scholium (*Principia*, ₁1687: 253–4 = ₂1713: 226).

(10) In both parent printed versions Newton's phrasing here is '& literis transpositis hanc sententiam involventibus (*Data...versa*) eandem celarem'.

(11) In 1677, namely, in his letter to Oldenburg for Newton of 21 June; this insistence upon the long intervening interval of time—in fact Leibniz answered Newton's *epistola posterior* of October 1676 within hours of receiving the copy of it despatched by Oldenburg only a month before in mid-May (see J. E. Hofmann, *Leibniz in Paris, 1672–1676* (Cambridge, 1974): 274)— is not found in the original scholium (*Principia*, ₁1687: 254 = ₂1713: 227).

verborum & notarum formulis.[12] Utriusꝗ fundamentum continetur in hoc Lemmate.

---

[3][13] [...] et literis transpositis hanc sententiam involventibus (*Data æquatione quotcunꝗ fluentes quantitates involvente, fluxiones invenire & vice versa*) fundamentum ejusdem celarem: rescripsit vir clarissimus mense Junio anni sequentis[14] se quoꝗ in ejusmodi methodum incidisse & methodum suam a Barroviana tangentium methodo cum Epistolis meis collata derivatam communicavit, a mea vix abludentem præterquam in verborum et notarum formulis; deinde agnovit se ex epistolis meis didicisse me methodum similem tunc habuisse. Ejus verba erant:[15] *Arbitror quæ celare voluit Newtonus de tangentibus ducendis ab his non abludere. Quod addit ex hoc eodem fundamento quadraturas quoꝗ reddi faciliores, me in sententia hac confirmat.* Hæc ille. Methodi utriusꝗ fundamentum continetur in hoc Lemmate.

Scholium secundum.[16]

Mense Julio anni 1669 Barrovius noster *Analysin* meam *per Æquationes numero terminorum infinitas* ad Collinium misit.[17] Hæc Analysis ex methodo serierum et methodo fluxionum conjunctis constabat. De ijsdem methodis anno 1671

---

(12) So the *Principia*'s first edition reads; the *editio secunda* goes on to adjoin '& Idea generationis quantitatum', as we have already observed in Appendix 2: note (97). Leibniz himself had stated his belief that 'quæ celare voluit Neutonius de tangentibus ducendis ab his [*sc.* meis methodis] non abludere'; but the listing of the points in which the two methods of tangents differ is Newton's gloss. Leibniz would surely have agreed regarding the discrepancy in verbal form and notation, but his writings *post* 1713 show that he was something less than willing to acknowledge the basic difference in concept of generation of quantities posited by Newton in his *addendum*.

(13) Add. 3965.10: 145ᵛ/146ʳ. This recasting of [2], now amplified to include a 'Scholium secundum' where brief mention is made of Newton's 1669 'De Analysi' and the fuller treatise of 1671 where he expounded his conjoint methods of series and fluxions at greater length, opens abruptly *in medio*: understand that this revise begins with the unchanged text of the parent printed scholium, *viz.* 'In literis quæ mihi cum Geometra peritissimo *G. G. Leibnitio* annis abhinc decem intercedebant, cum significarem me compotem esse methodi determinandi Maximas & Minimas, ducendi Tangentes, & similia peragendi, quæ in terminis surdis æque ac in rationalibus procederet,...'. Newton subsequently entered the complete double scholium into his annotated copy of the *Principia*'s second edition (now ULC. Adv. b.39.2) as a replacement 'Ad pag. 226', there combining the *Scholium secundum* with its parent (see notes (16) and (18) following) and lightly reworking the three final sentences (see note (24)): this amended 'fair copy' is printed in full, with some confusion between what Newton in the manuscript original intended to be cancelled and what he meant to be retained, on pages 794–6 of A. Koyré and I. B. Cohen, *Isaac Newton's 'Philosophiæ Naturalis Principia Mathematica': the third edition (1726) with variant readings* (Cambridge, 1972).

(14) That is, 1677; see note (11) to [2] preceding.

(15) *Commercium Epistolicum*: 90 [=*Correspondence*, 2: 215].

(16) In the amended text entered by Newton into his interleaved second-edition *Principia* (see note (13)), he subsumed the first two sentences here following into the added paragraph of the preceding main scholium which is cited in note (18), suppressed the third one and inserted its gist in a clause '& me Tractatum de eadem anno 1671 scripsisse' after '...æque

Tractatum ampliorem scripsi. Anno 1670 methodus Analytica Barrovij ducendi Tangentes lucem vidit.[18] Ab hoc methodo D. Jacobus Gregorius methodum ducendi tangentes abscꝗ calculo meo deduxit, ut ipse ad Collinium 5 Sept 1670 scripsit.[19] Et Slusius anno 1672 ad Oldenburgum[20] scripsit se methodum expeditam ducendi tangentes invenisse. Et Collini[o] hæc mihi significanti rescripsi 10 Decem 1672, in verba sequentia.[21] *Ex animo gaudeo D. Barrovij… quæ quantitatibus surdis sunt immunes… reducendo eas in series infinitas.* Anno 1675 ad finem vergente Gregorius emortuus est. Et mense Junio anni sequentis Excerpta Epistolarum ejus[22] ad Leibnitium missa sunt. In his Excerptis erant exemplaria Epistolæ meæ jam citatæ[23] ut et Epistolæ prædictæ Gregorij.[24]

---

ac in rationalibus procederet' in the first sentence, and reordered the fourth—now beginning —one to read 'Methodus Analytica Barrovij ducendi Tangentes lucem vidit anno 1670'. As an afterthought, lastly, he ran his new additional paragraph into the present 'Scholium secundum', understanding this subhead to be cancelled.

(17) Newton began to continue this sentence with 'eandem laudando' but there broke off with his intended meaning yet unclear.

(18) In Lectio X of his *Lectiones Geometricæ*, namely; see Appendix 1.1: note (11) and compare Appendix 2: note (69). In revise (see note (16)) this became, with a trivial recordering of its words, the opening sentence of the paragraph, there following an inserted passage which in its final form—for the preliminary roughings-out of it see page 795 of the Koyré/Cohen *variorum* edition of Newton's *Principia* (note (13) preceding)—reads: 'Sed et in *Analysi* mea *per series numero terminorum infinitas* quam Barrovius noster mense Julio anni 1669 ad Collinium misit, eandem Methodum descripsi idcꝗ ponendo literas quascuncꝗ pro fluentibus & literas alias pro fluxionibus et rectangula sub fluxionibus et momento temporis *o* pro momentis fluentium, ac designando etiam fluentia per fluxiones rectangulis inscriptas, et sub finem Tractatus illius exhibendo specimen calculi.'

(19) In Gregory's original words to Collins in his letter: 'I have read over Mr Barrow's lectures with much pleasure and attention, wherein I find him to have infinitely transcended all that ever wrote before him. I have discovered from Barrow his method of drawing tangents together with some of my own, a general geometrical method, without calculation, of drawing tangents to all curves, and comprehending not only Barrow's particular methods, but also his general analytical method in the end of the 10th lecture' (H. W. Turnbull, *James Gregory Tercentenary Memorial Volume* (London, 1939): 103); a Latin translation of this passage is printed in *Commercium Epistolicum*: 22. In forming his own variant upon the general rule for drawing tangents to 'geometrical' (algebraic) curves Gregory does not here, of course, admit a unique debt to Barrow's *Lectiones Geometricæ*, and Newton—ever ungenerous to him (compare Appendix 2: note (47) preceding)—is wrong to say so.

(20) See Appendix 2: note (72).

(21) Newton cites the Latin version of his 1672 letter to Collins as he had published it in *Commercium Epistolicum*: 29–30; for his original English words see *Correspondence*, **1**, 1959: 247, l. 11–248, l. 2.

(22) Newton means the 'Historiola' (Royal Society MS LXXXI, No. 46, together with a dozen stray pages now in private possession); in fact, only a greatly abridged version of this collection by Collins of excerpts from Gregory's mathematical correspondence was passed on to Leibniz (see Appendix 3.2: note (38) preceding).

(23) Collins to Newton, 10 December 1672; see note (21).

(24) The revised version inserted 'Ad pag. 226' in Newton's interleaved copy of the second

[4]⁽²⁵⁾   *Scholium.*

In literis quæ mihi cum Geometra peritissimo G. G. Leibnitio anno 1676 intercedebant, cum significarem me compotem esse methodi Analyticæ determinandi Maximas et Minimas, ducendi tangentes, quadrandi Figuras curvilineas, conferendi easdem inter se, & similia peragendi, quæ in terminis surdis æque ac in rationalibus procederet; et Tractatus du[o]s de hujsmodi rebus scripsisse, alterum quem Barrovius⁽²⁶⁾ anno 1669 ad Collinium misit, & alterum anno 1671 in quo hanc methodum fusius⁽²⁷⁾ exposueram; cumcɞ fundamentum hujus methodi literis transpositis hanc sententiam involventibus (*Data æquatione quotcuncɞ fluentes quantitates involvente, Fluxiones invenire, et vice versa*) celarem, specimen vero ejusdem in Curvis quadrandis subjungerem & exemplis illustrarem; et cum Collinius Epistolam 10 Decem 1672 datam a me accepisset, in qua methodum hanc descripseram et exemplo Tangentium more Slusiano ducendarum⁽²⁸⁾ illustraveram, & hujus Epistolæ exemplar mense Junio anni 1676 in Galliam ad D. Leibnitium misisset; & Vir clarissimus sub finem mensis Octobris, in reditu suo e Gallia per Angliam in Germaniam, epistolas meas in manu Collinij insuper consuluisset: incidit is tandem⁽²⁹⁾ in methodum similem sub diversis verborum et notarum formulis, et mense Junio sequente specimen ejusdem in Tangentibus more Slusiano ducendis ad me misit, & subjunxit se credere methodum meam a sua non abludere, præsertim cum quadraturæ curvarum per utrācɞ methodum faciliores redderentur. Methodi vero utriuscɞ fundamentum continetur in hoc Lemmate.

---

edition of his *Principia* (see note (13) above) lightly recasts these three final sentences to read, following on the quoted passage from his 1672 letter to Collins: 'Methodus de qua hic locutus sum ea est cujus fundamentum literis transpositis celabam ut supra & cujus elementa continentur in Lemmate præcedente. Et in excerptis Epistolarum D. Gregorij mense Junio anni 167[6] ad Leibnitium missis, habebatur exemplar hujus Epistolæ ut et Epistolæ prædictæ Gregorij. Ejusdem anni mense Octobri, D. Leibnitius e Gallia per Angliam domum rediens vidit Epistolam meam die 24 ejusdem mensis ad Oldenburgum datam in qua methodum eandem generalem [? descripsi]'. We have earlier stated that Leibniz had left London on his way home to Germany five days before Newton sent his *epistola posterior* to him off from Cambridge (see Appendix 2: note (77) preceding); and that Oldenburg did not transmit a copy of it to Hanover till late the next spring (see [2]: note (11) above).

   (25) Add. 3968.5: 22^r; preliminary draftings are found on *ibid.*: 35^r→32^r/32^v. This further recasting of the *Principia*'s fluxions scholium was first published by **Brewster** in Appendix I to his 1855 *Memoirs* (see [2]: note (7) above): 425–6, following the initial text on the manuscript sheet, with indication on *ibid.*: 426 of the ensuing changes which Newton there made. We here reproduce the finished scholium as it appears when these minor alterations in afterthought are incorporated.

   (26) 'noster' is added in the initial version in the manuscript.
   (27) 'plenius' is here replaced by Newton in his preliminary drafting.
   (28) See Newton's *précis* of this in his 1715 'Account' (page 581 above).
   (29) 'finito itinere' was written in a cancelled preparatory version of this sentence.

[5]$^{(30)}$ *Pag. 226.*

Scholium.

Analysin per series et momenta a me scriptam Barrovius noster anno 1669 ad Collinium misit. Methodos ibi expositas in alio Tractatu plenius explicui anno 1671, et inde Tractatum *de Quadratura Curvilinearum* anno 1676 extraxi. In Epistola 10 Decem 1672 ad Collinium data, methodum momentorum exemplo tangentium more Slusiano ducendarum illustravi, dixiϙ eandem etiam ad quæstiones de curvitatibus et longitudinibus Curvarum ac de areis & centris gravitatis &c sese extendere, & esse generalem, et ad quantitates surdas non hærere, & methodo serierum in scriptis meis intertextam esse: et Collinius exemplar hujus Epistolæ mense Junio anni 1676 ad D. Leibnitium tunc in Gallia agentem misit.

In literis insuper quæ mihi cum D. Leibnitio anno 1676 intercedebant, cum verba facerem de Tractatibus prædictis & significarem me compotem esse methodi Analyticæ determinandi maximas et minimas, ducendi tangentes, quadrandi figuras curvilineas, conferendi easdem inter se, et similia peragendi quæ in terminis surdis æque ac in rationalibus procederet; et literis transpositis hanc sententiam involventibus (*Data æquatione quotcunϙ fluentes quantitates involvente, fluxiones invenire, et vice versa*) fundamentum hujus methodi celarem, specimen vero ejusdem in curvilineis per seriem quadrandis subjungerem et exemplis illustrarem, sed me quietis gratia hæc in lucem non edere dicerem; et cum D. Leibnitius eodem anno in reditu suo de Gallia per Angliam in Germaniam, sub finem mensis Octobris, Commercium meum in manu Collinij etiam consuleret: incidit is$^{(31)}$ in methodum momentorum sub diversis verborum et notarum formulis, et mense Junio anni sequentis specimen ejusdem in Tangentibus more Slusiano ducendis ad me misit, et subjunxit methodum suam ad irrationales non hærere & se credere methodum meam a sua non abludere, præsertim cum quadraturæ per methodum utramϙ faciliores redderentur. *Fluxiones* utiϙ sunt velocitates quibus *fluentium momenta* generantur, et *momenta* a D. Leibnitio vocantur *differentiæ*: et inde Methodus fluentium et Methodus fluxionum & Methodus momentorum eandem methodum significare possunt.

---

(30) Add. 3968.5: 36^r; predrafts are found on *ibid.*: 23^v/23^r→34^r. In this most finished version (so it would appear) of these 1719 recastings of the fluxions scholium 'Ad pag. 226' in the *Principia*'s second edition, the only considerable novelties are that Newton takes the first three sentences of [1] above, inserting these as an opening paragraph leading into the published scholium, and also in his penultimate, newly introduced sentence insists, in counter to Leibniz and Johann Bernoulli, that fluxions are (finite) 'speeds' whereas Leibnizian *differentiæ* are the (infinitesimal) 'moments' of these fluxions which are 'instantaneously' generated.

(31) 'non multo post'—itself replacing 'tandem' at this point in [4] preceding—is deleted in sequel in the manuscript.

Methodi hujus fundamentum continetur in Lemmate præcedente, et in prædicto de Quadraturis tractatu fusius exponitur.[32]

(32) How little of this rewritten scholium and its predecessors was, even provisionally, entered by Newton into his private interleaved and annotated correction-copies (ULC. Adv. b.39.2 and Trinity College, Cambridge. NQ.16.196) of the *Principia*'s second edition may be seen conveniently listed (under the *sigla* E_2i and E_2a respectively) on page 369 of Koyré and Cohen's 1972 *variorum* text. Four years later, about the summer of 1723, when Henry Pemberton took on the hard grind of seeing the *editio tertia* through press, Newton began a further series of revampings of his fluxions scholium where he adjoined a *précis* of his manner of computing the fluxion of a fluent quantity from first principles and then applied the ensuing algorithm to elaborate his general rule for drawing tangents to curves as he had first communicated it to John Collins half a century before in December 1672; since this further spasm of revisions has close affinity with the latter portion of the intended 'Præfatio' to the 1722 re-edition of the *Commercium Epistolicum* which we reproduce in our final Appendix 10 (see pages 691–5 below) we will here say no more of it. Ultimately, in the *Principia*'s third edition as it was at length published in 1726 he was to settle for replacing the scholium as printed in 1687(→1713 with the minor addition cited in note (12) above) by a relatively inoffensive brace of sentences: 'In epistola quadam ad D. *J. Collinium* nostratem 10 Decem. 1672, cum descripsissem methodum tangentium quam suspicabar'—'conjiciebam' was first written—'eandem esse cum methodo *Slusij* tum nondum communicata; subjunxi: *Hoc est unum particulare vel corollarium potius Methodi generalis, quæ extendit se, citra molestum ullum calculum, non modo ad ducendum tangentes ad quasvis curvas sive Geometricas sive Mechanicas vel quomodocunꝗ rectas lineas aliasve curvas respicientes; verum etiam ad resolvendum alia abstrusiora problematum genera de curvitatibus, Areis, longitudinibus, centris gravitatis curvarum &c. Neque (quemadmodum Huddenij methodus de Maximis et Minimis) ad solas restringitur æquationes illas quæ quantitatibus surdis sunt immunes. Hanc methodum intertexui alteri isti, qua æquationum exegesin instituo, reducendo eas ad series infinitas.* Hæc ultima verba spectant ad Tractatum quem anno 1671'—'tunc ante annum' is replaced—'de his rebus scripseram. Methodi vero hujus generalis fundamentum continetur in Lemmate præcedente'. (We cleave to the orthography and punctuation of Newton's essentially identical draft on Add. 3968.5: 33^r, whence we also notice two minor preparatory variants. A deleted final clause there initially adjoined 'per quod utiꝗ resolvitur pars prior hujus problematis, *Data æquatione fluentes quotcunꝗ quantitates involvente, invenire fluxiones; & vice versa*'.)

APPENDIX 5. THE 'MIND' OF THE
PRINCIPIA'S FLUXIONS SCHOLIUM.[1]

[*c.* mid-1719]
From the original[2] in the University Library, Cambridge

ENARRATIO PLENIOR[3] SCHOLIJ PRÆCEDENTIS.

Brevitate verborum effectum est ut Scholium præcedens a nonnullis perperam expositum fuerit et in disputationes tractum, et propterea rem totam fusius enarrabo.

Analysin per æquationes numero terminorum infinitas a me scriptam Barrovius noster misit ad Collinium mense Julio anni 1669. Continet autem hic Tractatus Analysin qua Problemata per methodos serierum & momentorum[4] conjunctas

---

(1) While for want of any precise knowledge of it we are unable to say anything about the immediate background to this fuller recital by Newton of what forty years before he had had in mind to point out in the scholium to Lemma II of the second book of his *Principia*, his broad purpose of furnishing the reader of its new edition with a fuller account of what had now become long-past history will be evident. For much of this lengthy second scholium he unfolds a tale to us familiar—one certainly too winding to be adjoined, in a work upon the dynamics of celestial and terrestrial motion, as a historical aside upon a mere ancillary mathematical technique (as he doubtless perceived in deciding ultimately to discard this 'Enarratio plenior' from his planned *editio tertia*). What, however, raises this largely repetitive statement of his case to be acknowledged as 'first inventor' of the calculus above the humdrum level of most of his other assertions in old age of his priority and his originality is the section where, elaborating upon the briefer similar passage on pages 115–17 of his 'Observations' in the 1717 Appendix to Raphson's *History of Fluxions* (cited on pages 514–15 above), he summarises and quotes Latinised extracts from his earliest writings on fluxional analysis in the 1660's, and also pithily details the twin opening problems of the comprehensive treatise on infinite series and fluxions wherein in 1671 he digested the main portion of his previous findings.

(2) Add. 3968.6: 38ʳ–46ʳ. A somewhat different preliminary version of this carefully penned piece is to be found at .7: (59ʳ–60ʳ→) 57ʳ–58ʳ, further revised (by way of additions roughed out on .41: 7ʳ/7ᵛ and .13: 374ʳ) to be .7: 49ʳ–56ʳ. Our suggested date of around (or shortly after) summer 1719 for the preparation by Newton of this lengthy 'Enarratio plenior' of his path to discovery in calculus is, of course, that assigned by us to his prior recastings of the *Principia* scholium itself which are printed in Appendix 4 preceding. This assumption that the two were composed at about the same time is well borne out by an inspection of Newton's autograph hand, which throughout the manuscripts is substantially the same in shape and slope and other chacteristics. A loose *terminus ante quem non* is, we may not very usefully add, furnished by the fact that .41: 7 is the cover 'For Sʳ Isaac Newton/ at his house in Sᵗ Martins Street/ nigh Leister Fields/ These' of William Newton's letter to him of 7 August 1716 (for whose text see *Correspondence*, **6**, 1976: 364).

(3) In the manuscript this more elaborately descriptive titling replaces a simple 'MENS'. This heading 'MENS SCHOLIJ PRÆCEDENTIS' is also found on the immediately preceding draft on .7: 49ʳ, but the preparatory castings of the piece on .7: (59ʳ→) 57ʳ are noncommittally named 'SCHOLIUM SECUNDUM'.

(4) 'fluxionum' is replaced in the manuscript, to be more aptly reintroduced in the next sentence.

tractantur. Problemata utiꝗ ubi vulgaris Analysis non sufficit, deducuntur ad æquationes per methodum momentorum, quæ et methodus fluxionum dicitur; et æquationes finitæ per methodos in hoc Tractatu descriptas convertuntur (ubi visum est) in series perpetuo convergentes; et series nonnunquam redeunt in Æquationes finitas; et ubi symbolum aliquod pro serie tota ponitur, Series inter operandum pro symbolo illo nonnunquam substituitur; et ex fluxionibus per Regulas tres initio hujus Tractatus positas eruuntur fluentes; et ex fluentibus vicissim eruuntur fluxiones per easdem Regulas inversas; et methodum ad quantitates surdas non hærere ostenditur; et Curvas Mechanicas hujus ope ad æquationes reduci, et Problemata omnia quæ in Curvis Analyticis[5] tractari solebant, tractari etiam in Curvis Mechanicis per hanc Methodum[;] et quicquid Analysis vulgaris per æquationes ex finito terminorum numero constantes (quando id sit possibile) perficiat, hanc Methodum per æquationes numero terminorum infinitas semper perficere. Quibus de causis nomen Analyseos huic methodo a me recte impositum fuisse disputatur.

In hoc Tractatu pro fluxione vel Temporis vel quantitatis cujuscunꝗ uniformiter fluentis qua tempus exponitur, usurpatur unitas, & pro fluxionibus aliarum quantitatum ponuntur alia quæcunꝗ symbola, et pro fluentium momentis (seu particulis momento temporis genitis,) ponuntur rectangula sub fluxionibus & momento dato o. Et siquando symbolum fluxionis pro momento ponitur subintelligitur coefficiens o. Nam fluxio non est momentum sed fluendi velocitas. Fluxio vero rectangulo inclusa fluentem designat. Et symbola pro lubitu variantur cum methodus in forma symbolorum minime consistat. Et specimen calculi exhibetur sub finem Tractatus.

Collinius autem ex hoc Tractatu Series cum amicis mox communicare cœpit & methodum tanquam generalem celebrare. Et D. Jacobus Gregorius Abredonensis Scotus de his admonitus methodum inveniendi seriem quandam a Collinio ad ipsum missam aliquandiu quæsivit, & sub finem anni 1670 invenit, & mox per Epistolam ad Collinium 15 Feb. 167$\frac{0}{1}$ datam misit series plures per eandem methodum inventas, e quarum numero erat hæc.[6] In circulo sit Radius r, Arcus a, et Tangens t, et erit

$$a = t - \frac{t^3}{3r^2} + \frac{t^5}{5r^4} - \frac{t^7}{7r^6} + \frac{t^9}{9r^8} - \&c.$$

Exemplar autem hujus Epistolæ ad D. Leibnitium missum fuit mense Junio anni 1676, cum series ipsa ad illum prius missa fuisset, nempe mense Aprili anni 1675.

---

(5) That is, 'geometricis' in Cartesian parlance.

(6) *Commercium Epistolicum*, ₁1712: 25 (= *Correspondence*, **2**, 1960: 62). We need not yet again summarise Newton's ungenerous (and historically invalid) circumstantial argument as to how this following series-expansion of $r \tan^{-1}(t/r)$ was sent across the Channel to its independent discoverer in the mid-1670's.

Interea Gregorius anno 1670, Sept 5, scripsit ad Collinium se ex methodo Tangentium Barovij (hoc anno edita) et suis methodum generalem invenisse ducendi Tangentes absq̃ calculo; et anno 1672 sub autumno, Slusius scripsit se methodum Tangentium expeditam habere: et subinde Collinius postulavit ut methodum meam communicarem. Qua occasione sequentem Epistolam 10 Decem 1672 ad ipsum scripsi.[7] *Ex animo gaudeo...ad me transmittere ne grave ducas.*

Missum est autem exemplar hujus Epistolæ meæ ab Oldenburgo ad Slusium 29 Jan. 167⅔, et a Collinio ad Tschirnhausium mense Maio 1675, et ad Leibnitium mense Junio 1676. Et eodem mense Junio missum est etiam exemplar Epistolæ jam dictæ[8] Gregorij ad Leibnitium. Et ex his Epistolis innotescere potuit quod hæcce Tangentium Methodus et Corollarium esset Methodi generalis de qua hic locutus sum, et consequeretur etiam ex Methodo Tangentium Barrovij.

Ineunte anno 1673 D. Leibnitius Londinum venit et cum Pellio nostro de rebus Arithmeticis sermones habuit, in Geometria sublimiore nondum instructus. At a Pellio (qui Collinium noverat) de serie Mercatoris &c admonitus, *Logarithmotechniam* ejus emit & secum Lutetiam tulit. Scripsit vero ad Oldenburgum de numeris usq̃ ad mensem Junium hujus anni,[9] dein siluit per annum integrum & interea Geometriam sublimiorem Hugenio magistro didicit, et anni 1674 mense Julio silentium abrumpens, de seriebus ad Oldenburgum scribere cœpit, dicendo[10] se *Theorema invenisse cujus ope area circuli vel sectoris ejus dati exacte exprimi posset per seriem numerorum rationalium continue productam in infinitum.* Et mense Octobri ejusdem anni exposuit quale esset hoc Theorema....[11]

· · ·   · · ·   · · ·

---

(7) Newton passes to cite in full the Latinised version of the main portion of his letter to Collins on 10 December 1672 communicating his general rule for tangents (compare *Correspondence*, **1**: 247, l. 10–248, l. 5/252 ll. 19–21), of which he had given first publication in *Commercium*: 29–30.

(8) To Collins on 15 February 1670/1; see note (6) above.

(9) Namely 1673.

(10) In his letter to Oldenburg of 15 July 1674 (N.S.). Newton had cited a slightly fuller extract from this letter in *Commercium Epistolicum*: 37.

(11) In his letter to Oldenburg of 26 October 1674 (N.S.). We see no need to weary the reader with yet one more citation, by way of Newton, of Leibniz' affirmation there that 'Eadem methodo etiam arcus cujuslibet cujus sinus daretur, geometrice exhiberi per ejusmodi seriem valor posset...' (first published in *Commercium*: 38). In sequel, likewise, we omit to reproduce Newton's supporting quotations of similar remarks of Leibniz regarding his further inventions of infinite series 'via quadam sic satis singulari' in his ensuing letters to Oldenburg over the next eighteen months.

Hac occasione,[12] sollicitantibus Oldenburgo et Collinio, scripsi Epistolam ad Oldenburgum 13 Junij proxime sequentis datam, in qua methodum serierum descripsi et addidi, *Analysin per easdem ad omnia pene Problemata (si numeralia quædam Diophantæis similia excipiantur) sese extendere; non tamen omnino universalem evadere nisi per ulteriores quasdam methodos eliciendi series infinitas quas non vacabat describere, cum hæ speculationes diu mihi fastidio esse cœpissent adeo ut ab ijsdem tum per quinq̃ fere annos abstinuissem.*[13] Hæc Analysis per Series & ulteriores quasdem methodos procedens, illa ipsa est cujus specimen sub titulo *Analyseos per æquationes numero terminorum infinitas* Barrovius noster anno 1669 ad Collinium misit.

His respondit D. Leibnitius 27 Aug. 1676 in hæc verba.[14] *Quod dicere videmini... ...Problemata methodi Tangentium inversæ.* Et addidit se Curvam cujus subtangens datur certa Analysi statim invenisse, sed quicquid in hoc genere desiderari potest nondum consecutum. Nempe si Abscissa crescit in Progressione Arithmetica Ordinata crescet vel decrescet in Progressione Geometrica; ideoq̃ si Abscissa sit ut Logarithmus Ordinata erit ut Numerus. Nulla alia Analysis ad inventionem hujus Curvæ requiritur. In eadem Epistola[15] D. Leibnitius misit etiam Quadraturam suam Arithmeticam per hanc seriem $\frac{t}{1} - \frac{t^3}{3} + \frac{t^5}{5} - \frac{t^7}{7}$ &c eandemq̃ triennio ante et ultra a se amicis communicatam esse dixit, id est, ubi primum venit ab Anglia in Galliam.

Ipse vero in Epistola 24 Octob. 1676 ad Oldenburgum data et a D. Leibnitio (qui tum Londinum secunda vice venerat) statim lecta,[16] rescripsi quod *eo circiter*[17] *tempore quo Mercatoris Logarithmotechnia prodijt, communicatum fuit per*

---

(12) In his immediately preceding paragraph, here omitted (see the previous note), Newton had quoted at length from Leibniz' letter to Oldenburg on 12 May 1676 (N.S.) the paragraph in which he reported having received from Georg Mohr the 'series valde ingeniosa' for the inverse sine, and went on to ask for its demonstration; see *Commercium*: 45 (=*Correspondence*, **2**: 3).

(13) Newton sets a familiar passage from his 1676 *epistola prior* to Leibniz (*Commercium*: 55–6 [=*Correspondence*, **2**: 29]) into *oratio obliqua*.

(14) We again omit to follow Newton in making full quotation of the following first sentence and a half of the penultimate passage of Leibniz' response (*Commercium*: 65 [=*Correspondence*, **2**: 64]).

(15) See *Commercium*: 63 (=*Correspondence*, **2**: 60).

(16) Not true, as we observed in Appendix 2: note (77); Leibniz had his first sight of Newton's *epistola posterior* only in the secretary copy transmitted to him by Oldenburg in May 1677, long months after his return to Hanover. We will not in sequel give precise reference for every detail of the citations and paraphrases from his 1676 letter (*Commercium*: 67–86 [=*Correspondence*, **2**: 110–29]) which Newton passes to make.

(17) In his original letter, in fact, Newton had written more sharply 'eo ipso' (*Commercium*: 70 [=*Correspondence*, **2**: 114]). The implication that Mercator's *Logarithmotechnia* did not narrowly provoke Newton to set down his own more far-reaching results on infinite series in his 'De Analysi' is not in tune with our reconstruction from the meagre evidence (see II: 165–7)

amicum D. Barrow (tunc Matheseos Professorem Cantabrigiensem) cum D. Collinio
Compendium quoddam methodi harum serierum in quo significaveram areas et longitudines
Curvarum omnium & solidorum superficies et contenta ex datis Rectis, et vice versa ex his
datis Rectas determinari posse; et Methodum ibi indicatam illustraveram diversis seriebus.
Hoc Compendium est *Analysis* illa per series et momenta et motuum progressus
a Barrovio mense Julio anni 1669 ad Collinium missa[,] ut supra. Et his
admonitus D. Leibnitius Compendium hocce in manu Collinij videre potuit.
Is enim Collinium de Commercio Gregorij et meo ad Series maxime spectante
consuluit, & partem Literarum nostrarum in ejus manu vidit, et ab Oldenburgo
paulo ante postulaverat ut is Demonstrationem mearum serierum a Collinio
procuraret et ad se mitteret: quæ Demonstratio extabat in hoc Compendio.

In eadem Epistola subjunxi[18] quod Collinius subinde *non destitit suggerere ut*
hæc publici juris facerem, et quod *ante annos quinqȝ* (anno scilicet 1671) *cum suadentibus*
amicis con[s]ilium c[e]peram edendi Tractatum de Refractione Lucis et Coloribus quem
tunc in promptu habebam, cœpi de his seriebus iterum cogitare, et Tractatum de ijs etiam
conscripsi ut utrumqȝ simul ederem. Sed antequam Tractatum absolvissem[,] lites
de coloribus subortæ me quietis amantem a consilio deterruerunt. In eo autem
Tractatu fundamentum me aliquatenus posuisse dixi[19] solvendi Problemata

---

that Mercator's book directly stimulated him to compose his own compendium on analysis
by series and to show it to Barrow as confirmation that he himself had gone a long way
beyond just independently discovering the logarithmic expansion. We have on II: 170 already
cited John Collins' firm declaration to Oldenburg in April 1675, for onward transmission to
Leibniz, that 'as soone as [Mercators Logarithmotechnia] came foorth I sent [it] to Dr Barrow'
(Royal Society MS LXXXI, No. 36: 2r; the full text of this letter of 10 April is now printed in
(ed. J. E. Hofmann) *Leibniz: Sämtliche Schriften und Briefe* (3) **1** (Berlin, 1976): 217–30 and also in
(ed. A. R. and M. B. Hall) *The Correspondence of Henry Oldenburg*, **11** (London, 1977): 253–63).

(18) *Commercium*: 70–1 (=*Correspondence*, **2**: 114). On Add. 3968.6: 47r Newton, evidently
understanding the preceding paragraph to be deleted *in toto*, began to expand the opening to
the present one before abruptly breaking off, there writing: 'In Epistola mea 6 Febr. 16$\frac{71}{72}$ ad
D. Oldenburgum scripta, et in *Transactionibus Philosophicis* eodem Mense [**6**, No. 80 (for
19 February 1671/2): 3075–87 (=*Correspondence*, **1**, 1959: 92–102)] impressa, scripsi me initio
anni 1666 in Theoriam novam lucis et colorum incidisse. Et eandem mihi tunc plene innotuisse
ex ipsa Epistola constare potest. Ut hæc Philosophia eodem fere tempore cum methodo
fluxionum inventa fuit[,] sic etiam eodem tempore utramqȝ in lucem edere olim in animum
induxeram & eodem destiti a consilio easdem edendi. Etenim in Epistola mea 24 Octob. 1676
ad Oldenburgum data[,] ubi dixeram Compendiũ meum serierum a D. Barrow ad D. Collins
missum fuisse, statim subjunxi: *Suborta deinde inter nos Epistolari consuetudine, D. Collinius Vir in*
rem Mathematicam promovendam natus, non destitit suggerere ut hæc publi[ci] juris facerem. Et ante annos
quinqȝ (1671)...*cœpi de his Seriebus iterum cogitare, & Tractatum de ijs etiam conscripsi ut utrumqȝ simul*
ederem. Sed ex occasione [*Telescopij Catadioptrici...*]'. Presumably it was his intention, before
interrupting this quotation from his *epistola posterior*, to cite his now famous ensuing statement
that 'subortæ statim per diversorum epistolas objectionibus alijsqȝ refertas crebræ interpellationes me prorsus a consilio deterruerunt, et effecerunt ut me arguerem imprudentiæ quod
umbram captando eatenus perdideram quietem meam, rem prorsus substantialem'.

(19) Newton's citations, direct and paraphrased, in the next several sentences are from
Commercium: 71 (=*Correspondence*, **2**: 115).

quæ ad quadraturas reduci nequeunt, et quomodo methodus Slusiana ducendi Tangentes ex hoc fundamento statim prodiret et quod hic non hæreretur *ad æquationes radicalibus unam vel utramᵽ indefinitam quantitatem involventibus utcunᵽ affectas sed absᵽ aliqua talium æquationum reductione* (*quæ opus plerumᵽ redderet immensum*) *Tangens confestim duceretur.* Et quod *eodem modo se res haberet in quæstionibus de Maximis et Minimis alijsᵽ quibusdam.* Et *fundamentum harum operationum satis obvium* esse dixi, sed cum explicationem ejus prosequi non vacaret, id celavi hac sententia ænigmatice posita: *Data æquatione quotcunᵽ fluentes quantitates involvente fluxiones invenire et vice versa.* Et *hoc fundamento* dixi me etiam conatum esse *reddere speculationes de Quadratura Curvarum simpliciores* et pervenisse ad *Theoremata quædam generalia,* et Theorema primum ibi posui et exemplis illustravi. Addidi etiam⁽²⁰⁾ quod *alia* haberem *Theoremata pro comparatione Figurarum* cum Conicis Sectionibus, alijsᵽ figuris quibuscum possent comparari, meque hujusmodi Theoremata aliqua in Catalogum dudum retulisse; quæ Theoremata *vix per transmutationem figurarum quibus Jacobus Gregorius et alij usi sunt absᵽ ulteriore fundamento* (nempe fundamento meo prædicto) *inveniri posse* putarem. Addidi deniᵽ⁽²¹⁾ quod *inversa de Tangentibus Problemata essent in potestate, aliaᵽ illis difficiliora* ad quæ solvenda usus essem *duplici methodo, una concinniori, altera generaliori*; & utramᵽ literis transpositis consignavi hanc sententiam involventibus: *Una Methodus consistit in extractione fluentis quantitatis ex æquatione simul involvente fluxionem ejus: altera tantum in assumptione seriei pro quantitate qualibet incognita ex qua cætera commode derivari possunt et in collatione terminorum homologorum æquationis resultantis ad eruendos terminos assumptæ seriei.* Et methodum ex his omnibus compositum ibi vocabam meam *methodum generalem.*⁽²²⁾

His abunde satis patet me anno 1676 & annis minimum quinᵽ vel septem prioribus⁽²³⁾ methodum generalem habuisse reducendi Problemata ad æquationes fluxionales & series convergentes, & nomen Analyseos eidem imposuisse; & huic affines fuisse methodos Tangentium Barrovij et Gregorij, propterea quod methodum Tangentium Slusij similiter producerent, sed ab Authoribus in Analysin generalem minime perfectas; et Barrovium Analysin meam ad Collinium ut methodum novam misisse; et in hac Analysi me exemplo ostendisse quid faciendum sit ubi occurrunt irrationales.

(20) *Commercium*: 74, 76–7 (=*Correspondence*, **2**: 117, 119–20).

(21) *Commercium*: 86 (=*Correspondence*, **2**: 129).

(22) Newton had not, to be precise, specified that the totality of his 'methodus generalis' was comprised in the twin methods for solving 'inversa de tangentibus Problemata...aliaᵽ illis difficiliora' whose statement he had set down for Leibniz in anagram, though of course the reader would naturally conclude that this was so. If, however, Newton had in the meantime come to develop a third general method of analysis, he would no doubt have embraced that also within the vague terms of his 1676 letter, throwing back (as was so easy for him to do) the date of its 'discovery' to a time when its flower was but a seed in his mind.

(23) From the time, that is, of his 1671 fluxions treatise and 1669 'De Analysi' respectively.

Lecta hac Epistola D. Leibnitius Londino mox discessit, in Hollandiam navigans, et ubi nunciatum est ipsum tandem Hannoveram pervenisse, D. Oldenburgus exemplar ejusdem ad ipsum misit. Hoc fecit Mense Martio anni 1677, et D. Leibnitius Literis 21 Junij 1677 respondit in hæc verba.[24] *Clarissimi Slusij methodum Tangentium nondum esse absolutam Celeberrimo Newtono assentior. Et jam a multo tempore rem Tangentium longe generalius tractavi, scilicet per differentias Ordina[ta]rum.* Et subinde descripsit methodum Tangentium Barrovij symbolis mutatis, & ostendit quod Methodus Slusiana statim occurreret hanc methodum intelligenti, & quomodo irrationales eam nullo morarentur modo; deinde subjunxit:[25] *Arbitror quæ celare voluit Newtonus ab his non abludere. Quod addit ex hoc eodem fundamento Quadraturas quoꝗ reddi faciliores me in sententia hac confirmat. Nimirum semper figuræ illæ sunt quadrabiles quæ sunt ad æquationem differentialem.* Et his verbis agnovit me methodum similem anno 1676 et annis minimum quinꝗ præcedentibus habuisse.

Cum vero D. Leibnitius anno 1684 Elementa quædam hujus Analyseos in lucem emitteret[26] et silentio præteriret ea omnia ad hanc methodum spectantia quæ vel ab Oldenburgo acceperat vel in manibus Collinij viderat vel in hac Epistola sua agnoverat: posui Scholium superius[27] ut inde constaret me primum de hac methodo scripsisse, D. Leibnitium eandem tardius intellexisse, Lemma superius[27] ab editis Leibnitianis non fuisse desumptum, et Propositiones difficiliores in hocce *Principiorum* Libro synthetice demonstratas, vi hujus Analyseos inventas fuisse.

[23]Chartas habeo his omnibus antiquiores, annis scilicet 1665 & 1666 scriptas, in quibus tempora scribendi nonnunquam notantur. Et in Schediasmate

---

(24) *Commercium*: 88 (=*Correspondence*, **2**: 213).

(25) *Commercium*: 90–1 (=*Correspondence*, **2**: 215).

(26) In his pioneering sketch of his 'Nova methodus pro maximis & minimis, itemꝗ tangentibus...& singulare pro illis calculi genus' (*Acta Eruditorum* (October 1684): 467–73); for a fuller reference to this see Appendix 2: note (91) above.

(27) The primary scholium to the parent preceding Lemma II of the *Principia*'s second book, of course. We presume that this was to be reprinted in the lightly augmented form of the second edition (on which see Appendix 4.2: note (7) preceding).

(28) The following section elaborates the passage in Newton's 1717 'Observations' upon Leibniz' letter to Conti of 9 April 1716 (N.S.) where (Raphson, *History of Fluxions*, ₂1717: 115–16; compare note (141) of our preceding introduction) he passed, in substantiation of his own absolute priority of invention in calculus, to declare: 'And am not I as good a witness that I invented the methods of series & fluxions in the year 1665 & improved them in the year 1666, & that I still have in my custody several mathematical papers written in the years 1664, 1665 & 1666 some of wch happen to be dated, & that in one of them dated 13 Novem 1665 the direct method of fluxions is set down in these words: PROB. *An equation being given expressing the Relation of two or more lines x, y, z &c...*'. (We follow the orthography of Newton's draft on Add. 3968.41: 38^r.)

13 Novemb. 1665 conscripto extat Lemma superius verbis sequentibus propositum ac demonstratum.[29]

Novemb. 13. 1665.

Data æquatione mutuam designante relationem duarum vel plurium linearum x, y, z &c a duobus vel pluribus corporibus A, B, C &c simul descriptarum: invenire relationem velocitatum p, q, r, &c quibus corpora lineas illas describunt.

Resolutio.

...   .[30]

Demonstratio

...   .[30]

Hic observandum venit primo quod termini illi semper evanescunt in quibus o non extat, propterea quod ex hypothesi sunt nihilo æquales; deinde quod in æquatione residua per o divisa termini in quibus o adhuc manet semper evanescunt cum sint infinite parvi: ac deniq quod termini residui semper habebunt formam illam quam per præcedentem Regulam habere debent.

Hæc Regula eodem modo demonstratur ubi tres vel plures habentur quantitates indeterminatæ x, y, z &c.

Hactenus Manuscriptum illud vetus. In alio Manuscripto 16 Maij 1666 composito,[31] methodum solvendi Problemata per motum complexus sum

---

(29) In sequel Newton sets out a—surely his own?—Latin version of the paper 'To find ye velocitys of bodies by ye lines they describe' whose original English text is reproduced on 1: 382–6. Here and in his ensuing excerpts from his early papers, we may notice, he works to a scheme outlined on Add. 3968.19: 263^r in prelude to drafting the parallel passage in his 1717 'Observations' whose opening we cited in the previous note; on this see also note (146) of our preceding introduction. Let us add that on .1: 1^r/2^r, in evident preliminary to rendering it into the present Latinised 'Apographum Schediasmatis a Newtono olim scripti cui forte appositum fuit scribendi tempus 13 Novem. 1665' (so he framed its title on .36: 506^r, a trivial subsequent reordering of the words in its subordinate clause apart), he found point in copying out the original English of this extracted piece in full.

(30) We see no need to reproduce Newton's Latin of his 'Resolution' and 'Demonstration' of this his 1665 'Probleme' (for whose original English text see 1: 383–4, 385).

(31) See 1: 392–9. On a slip of paper now loose in Add. 3958.3 Newton began to pen a full Latin translation of this set of what propositions 'Ad resolutionem Problematum per motum... requiruntur et sufficiunt', but left off making this *verbatim* presentation of its content after completing the enunciation of its 'Prop. 1'. In the initial version of this page (.6: 44^r) of the manuscript scholium which is preserved as .9: 113^r it was not, we may add, in Newton's mind to make any reference at all to this paper of 16 May 1666 or to the ensuing fluxions tracts of October 1666 and of 1671, but he there went straightaway on after 'Hactenus Manuscriptum illud vetus' to write: 'Inde vero hæc descripsi ut vera Lemmatis hujus origo pateret & quale esset methodi meæ fundamentum illud quod anno 1676 literis transpositis celavi...'. Interpolation of the next following paragraphs which here intercede is made in the preparatory recasting on .9: 115^r(+116^v)→115^v(+116^r).

Propositionibus septem, quarum ultima est Regula jam descripta eliciendi veloci[ta]tes crescendi vel decrescendi ex æquatione quantitates crescentes vel decrescentes involvente. Et in alio Manuscripto quod mense Octobri ejusdem anni[32] composui, descripsi easdem septem Propositiones, et octavam addidi; septimam vero sequentibus[33] adauxi.

Si in æquatione quavis occurrat quantitas aliqua vel fracta, vel surda, vel mechanica.... ...et habebitur æquatio quam invenire oportuit.

Exempl. 1. ...   .
Exempl. 2. ...   .
Exempl. 3. ...   .

In octava Propositione[34] docebam vicissim quomodo ex Æquatione velocitates crescendi vel decrescendi[35] involvente quantitates crescentes vel decrescentes deduci possent, idꝗ regrediendo vel reducendo Problema ad quadraturam Curvarum et quadrando Curvam per tres illas Regulas quas etiam postea descripsi in principio Tractatus *de Analysi per Æquationes numero terminorum infinitas*,[36] ut et per Catalogum Curvarum quæ vel quadrari possunt vel cum Conicis Sectionibus comparari[37] et quarum Ordinatas posui postea in Epistola mea ad Oldenburgum 24 Octob. 1676 data. Et hæc est methodus a Leibnitio summatoria, a me inversa methodus fluxionum et momentorum dicta.

Hanc methodum annis 1665 et 1666 a me inventam fuisse Barrovius noster per ea tempora Lucasianus Matheseos apud Cantabrigienses Professor, idoneus est testis; et ejus testimonium Collinius noster in Epistola sua ad D. Strode 26 Julij 1672 data sic protulit.[38] *Mense Septembri 1668 Mercator Logarithmotechniam*

---

(32) Namely 1666. The original English text of this small but densely compact fluxions tract is reproduced on 1: 400–48.

(33) Newton in sequel sets out his Latin version of what is printed by us on 1: 411 l. 6–412. As ever, we see no need here to repeat this secondary translation (of no historical significance), but we may remark that it is drafted by him on .30: 439ᵛ/440ʳ around and below his certificate (*ibid.*: 439ᵛ, reproduced in full in *Correspondence*, 7, 1978: 370) that 'Edward Carter is Waterman & Servant to the Mint & one of that corporation & on that account is exempted from all Parrochial Services...' and so may serve to date the latter as *c.* late 1719.

(34) See 1: 403–4.

(35) 'velocitates augmentorum vel decrementorum' is replaced.

(36) See 1: 404–5, and compare 11: 206–12.

(37) See 1: 405–10. Since this first catalogue of curves was to be much extended and restructured into the twin tables of Problem 9 of his 1671 tract (see 111: 236–54), and instances of ordinates in the latter's 'Catalogus [Arearum] Curvarum aliquot ad Conicas Sectiones relatarum...' (*ibid.*: 244 ff.) were the ones quoted by Newton in his *epistola posterior* to Leibniz in 1676 (see *Commercium*: 76 [=*Correspondence*, 2: 119]), the following statement must not be taken for literal truth.

(38) Newton's following quotation is of the second paragraph of the Latinised excerpt from Collins' letter printed in *Commercium*: 28–9. We have cited the original English of this in Appendix 2: note (22) above, let us again add.

edidit suam.... ...sine ulla radicum extractione obtineri queant. Hactenus Collinius. Cum vero hæc spectent ad Tractatum *de Analysi per series* quem Barrovius ad Collinium miserat, & Barrovius hunc Tractatum legerat et intellexerat, et methodus in hoc Tractatu tradita pergat per Series & momenta[39] conjunctim, et ejus ope area figuræ *accurata si possibile sit, sin minus infinite vero propinqua* prodire dicatur,[40] et series quarum ope hoc fit inventæ fuerunt per methodum fluxionum ut in Epistola mea 24 Octob. 1676 ad Oldenburgum data traditur: inde discas Analysin in Tractatu illo expositam quæ ex methodis serierum et momentorum componitur, a me annis aliquot antequam Tractatus ille ad Collinium mitteretur inventam et generalem redditam fuisse.

In Tractatu quem anno 1671 conscripsi, primam docui reductionem quantitatum in series convergentes & extractiones radicum tam affectarum quam simplicium. Et his præmissis, methodum fluxionum exposui docendo solutionem plurium Problematum, quorum duo prima erant hæc[41]

Prob. 1. Relatione quantitatum fluentium inter se data, fluxionum relationem determinare.

Prob. 2. Exposita æquatione fluxiones quantitatum involvente, invenire relationem quantitatum inter se.

Hæc omnia ex veteribus Manuscriptis protuli ut vera Lemmatis hujus origo pateret, et quale esset methodi meæ fundamentum illud quod anno 1676 literis transpositis celavi, sententiam in Scholio præcedente expositam involventibus, id est, sententiam: *Data æquatione quotcunqꝫ fluentes quantitates involvente, fluxiones invenire, et vice versa.*[42]

---

(39) 'Fluxiones' is rightly replaced.

(40) By Collins in his preceding letter to Strode, as set in Latin in *Commercium*: 29, ll. 2/3—'accurately when it can be done, but alwaies infinitly neare' in his original English words (on which again see Appendix 2: note (22) above). Newton himself in his 'De Analysi' (II: 226 ff.) makes no such explicit statement, though in conclusion (*ibid.*: 240–2) he is at pains to stress that 'quicquid Vulgaris analysis per æquationes ex finito terminorum numero constantes (quando id sit possibile) perficit, hæc per æquationes infinitas'—presupposing (compare *ibid.*: 226) that the 'valor areæ tanto magis veritati accedat' the more terms of its approximating series one takes into the reckoning—'semper perficiat...(id modò fiat) exactè & Geometricè' (in Descartes' analytical sense in the *Geometrie*).

(41) Newton here contents himself with merely enunciating the two primary problems of his 1671 tract (see III: 74, 82), though, as we observed in III: 15, note (43), in first draft on .9: 115ʳ he initially began to adjoin his 'Solutio' to 'Prob. 1'.

(42) In his preliminary recasting on .9: 115ᵛ (see note (31)) Newton here went on to elaborate: 'et quod Scholium illud verbis fusioribus enarratum ita sonabit. In literis quæ mihi cum Geometra peritissimo G. G. Leibnitio anno 1676 intercedebant, cum significarem me compotem esse Methodi determinandi Maximas et Minimas, ducendi Tangentes, quadrandi figuras curvilineas, solvendi inversa de Tangentibus Problemata, quadrandi areas Curvarum in Seriebus quæ finitæ evadunt ubi area per finitam æquationem exhiberi potest, comparandi areas curvarum cum areis Sectionum Conicarum aliarumve figurarum simplicissimarum cum quibus comparari possunt, & omnia fere Problemata solvendi si forte numeralia quædam

Si fluxiones pro fluentibus habeantur, operatione repetita prodibunt earum fluxiones, id est fluentium primarum fluxiones secundæ, et sic deinceps in infinitum. Fluxionibus autem secundis et momentis secundis in hisce *Principiorum* Libris nonnunquam usus sum.[43] In Lib. II Prop. X Exempl. 1 fluxionem secundam Curvaturæ vocavi variationem variationis ejus, et in ejusdem Libri Prop. XIV Cas. 3, momentum secundum Areæ vocavi dif-

---

Diophantæis similia excipiantur, et hanc methodum in terminis surdis æque ac in rationalibus procedere, et methodum tangentium Slusij ex eadem primo intuitu profluere[,] et me anno 1671 de hac methodo Tractatum scripsisse, et fundamentum ejus satis obvium esse, sed quoniam explicare non vacaret, me literis transpositis hanc sententiam involventibus *Data æquatione quotcunqɜ fluentes quantitates involvente fluxiones invenire, et vice versa* fundamentum celasse [;] Cum insuper ex literis Gregorij et meis datis 5 Sept. 1670 & 10 Decem 1672 & ad ipsum mense Junio Anni 1676 missis, methodum Tangentium Slusij tam ex Methodo Tangentium Barrovij quam ex mea tanquam Corollarium profluere didicisset & hanc methodum esse particul[ar]e quoddam vel potius Corollarium methodi meæ generalis, et quomodo difficultates quantitatum surdarum inter computandum effugerem in Analysi mea vidisset: Vir clarissimus mense Junio anni sequentis, postquam in methodum similem ope præcedentium incidisset vel saltem incidere potuisset, rescripsit se quoqɜ in ejusmodi methodum incidisse et methodum suam communicavit a mea vix abludentem præterquam in verborum et notarum formulis, et ex ijs quæ de methodo mea didicisset affinitatem inter methodos simul agnovit.' When he came to make the fair copy on .6: 46ʳ which we reproduce Newton evidently at once realised the incongruity of attaching in a second scholium an improved version of its primary one while still retaining this latter—unchanged, we presume, from the lightly augmented version in the *Principia*'s second edition (compare note (27) above)—for he there did not even begin to enter any portion of this intended *addendum*.

In sequel in his initial draft on .9: 113ʳ, we may further notice, he adjoined a following paragraph whose first sentence he took over into his recasting on *ibid.*: 115ʳ but there crossed out, namely: 'In Epistolis meis 10 Decem. 1672 & 24 Octob. 1676 datis, dixi quantitates surdas methodum meam non morari, et hanc rem exemplo explicui in *Analysi* mea prædicta. Substituatur utiqɜ in æquatione pro quantitate radicali symbolum quodvis, tractetur symbolum ut quantitas fluens & completo opere pro symbolo et ejus fluxione scribatur quantitas radicalis et ejus fluxio.'

(43) In his initial draft on .9: 113ʳ Newton went on at this point more shortly and somewhat differently to write: 'ut videre licet in Lib. II, Prop. XIV, Cas. 3. Earum subsidio inveni tum demonstrationem illam Propositionis Keplerianæ'—'anno 1677' is inserted as an interlineation—'quam in Lib. I, Prop. XI descripsi, tum Curvaturam Curvarum multo ante, de qua utiqɜ locutus sum in Epistola mea 10 Decemb. 1672 ad Collinium data. Sed et in chartis vetustioribus determinando Linearum Curvaturam, nunc literis punctatis nunc alijs symbolis usus sum'. On Add. 3965.13: 374ʳ the first exemplification was expanded to be '[ut...in] Lib. II Prop. XIV Cas. 3 et Prop. X Exempl. 1 ubi fluxio fluxionis Curvaturæ dicitur variatio variationis & momentum secundum dicitur differentia momentorum', and then the whole was elaborated to be substantially as it here reads; the complete paragraph, less the last sentence which is a late drafting on .9: 116ᵛ (see note (49) following), was then minimally reshaped on .9: 114ʳ and marked to replace the original on the preceding folio. (We have in 1, Introduction: note (221) outlined the gist of Newton's somewhat confused calculations on Add. 3965.13: 374ʳ, around which his present reshaping of .9: 113ʳ was written in order to ascertain—or, more fairly, refresh his memory of—the value of a conic's *latus rectum* in terms of the distance of a focus from the nearer vertex and the length of its semi-axis, first 'in Ellipsi' and thence, in the limit as this axis comes to be infinite, 'in Parabola'.)

ferentiam momentorum ejus.[44] Momentorum secundorum subsidio Demonstrationem illam Propositionis Keplerianæ quæ in Lib. I Prop. XI habetur, inveni anno 1677,[45] et multo ante eorum subsidio inveni Curvaturam Curvarum, de qua utiꞓ locutus sum in Epistola mea 10 Decem 1672 ad Collinium data, ut et variationem Curvaturæ de qua egi in Tractatu quem anno 1671 composui,[46] et curvaturam maximam vel minimam de qua egi in eodem Tractatu ut et in prædicto Manuscripto quod mense Octobri anni 1666 composui,[47] in quo etiam literis punctatis[48] nonnunquam usus sum. Sed et considerando momenta prima ut quantitates fluentes inveni solidum resistentiæ minimæ cujus memini in Scholio ad Prop. XXXIV[49] Lib. II.

---

(44) In **1**, 8: note (44) we have dwelt at length on this mistaken notion of Newton's, now become an *idée fixe*, that the *differentia momentorum* computed in this proposition of his *Principia* was a second-order infinitesimal *momentum secundum*.

(45) On this antedating by two years—a late insertion in the original draft on .9: 113^r— of his first proof that planetary bodies may orbit in an ellipse under an inverse-square 'pull' towards a focus see Appendix 3: note (55) (page 617 above).

(46) In its Problem 5, namely (see III: 150–76). Newton there passes briefly to discuss the 'point of straightness' and cusp where a curve's curvature is respectively infinite and zero in the second and third of his attached 'Quæstiones quædam cognatæ' (*ibid.*: 178–80).

(47) '...ut et in Tractatu quodam minore quem scripsi sub finem anni 1666' in the initial amplification on Add. 3965.13: 374^r of his anonymous citation 'in chartis vetustioribus' of his 1666 and 1671 tracts in his original draft on .9: 113^r (for which see note (43) above). Newton here refers, of course, to Problem 4 of the former (compare I: 425–6). We may notice that on the worksheet, .41: 61^v, where he proceeded to draft his next paragraph he briefly jotted down a neat fluxional restatement of his construction in the preceding 'Prob 2^d' (see I: 419–20) of the radius of curvature—or more strictly its vertical projection—at a general point of a curve, *viz.* (with some editorial interpolation to bring out the underlying argument):

'[Ponendo] $ab = x.$ $bc = y.$ [erit] $\dot{x} = nb = p.$ $\dot{y} = cb[= q].$ $\overline{ac} = cn.$ $cg = nd = \dot{x} + bd.$

[Dic] $bd\left[= \dfrac{\dot{y}\ddot{y}}{\dot{x}} \right] = z.$ [erit] $\dot{x} + \dot{z} = \overline{ad} = dk.$ $\dot{x} + \dfrac{\dot{y}\ddot{y}}{\dot{x}} = cg.$

[Unde quia] $cg - dk = \dfrac{\dot{y}\ddot{y}}{\dot{x}} [-]\dot{z}.y :: cg.ck.$ [prodit] $ck = \dfrac{\dot{x}y + \dfrac{\dot{y}\ddot{y}y}{\dot{x}}}{\dfrac{\dot{y}\ddot{y}}{\dot{x}}[-]\dot{z}} = \dfrac{\dot{x}\dot{x}y + \dot{y}\ddot{y}y}{\dot{y}\ddot{y}[-]z\dot{x}},$

that is, since $\dot{z}\dot{x} = \overline{\dot{y}\ddot{y}} = \ddot{y}^2 + y\dddot{y}$, there is $ck = -(\dot{x}^2 + \dot{y}^2)/\ddot{y}$. The figure understood is that on I: 419; we would again remind that Newton has, not a little confusingly, designated two separate points in it by the letter k.)

(48) These 'pricked' letters in his 1666 tract (see I: 421–7), like their parents in his paper of 21 May 1665 (see I: 280–94), denote partial derivatives in modern equivalent and except in the loosest sense are not the progenitors of Newton's later standard 'dotted' full fluxions—to be introduced by him (compare VII: 64, note (35)) only in December 1691 in his first treatise 'De quadratura Curvarum'—with which he here, without directly stating so, suggests they may properly be compared.

(49) In the second edition of the *Principia* ($_2$1713: 299–300), that is; in the first edition the proposition is numbered XXXV (see $_1$1687: 324–6; Newton's draft of the scholium on *ibid.*: 326–7 is reproduced on VI: 460–4 with our accompanying technical comment). In sequel to his initial version of the present sentence on .9: 116^v Newton adjoined 'Et Leibnitius ipse

Computationes[50] per fluentium momenta sæpe contrahuntur resolvendo fluentem uno temporis momento fluendo auctam in seriem convergentem, ut fit in Scholio ad Prop. XCIII Lib. I. Nam termini seriei proportionales sunt fluxionibus et momentis, secundus terminus fluxioni primæ et momento primo, tertius fluxioni secundæ et momento secundo, et sic deinceps; et multiplicati respective per terminos hujus seriei $1.$ $1\times 2.$ $1\times 2\times 3.$ $1\times 2\times 3\times 4.$ &c, vertuntur in momenta; deinde divisi per terminos hujus $o.$ $oo.$ $o^3.$ o^4 &c vertuntur in fluxiones.[51] Et ob hanc methodorum affinitatem eædem conjungi merebantur.[52]

---

fatetur hoc fuisse specimen primum hujus methodi', there deleting a final phrase 'in Lucem editum' which might seem to detract from his claim to absolute priority in discovery as well as first publication; but nothing of this is found in his present revise (on .6:46ʳ). In his 'Responsio ad Dn. Nic. Fatii Duillerii imputationes...' in the *Acta Eruditorum* (May 1700): 198–214) (on the context to this stinging rebuke see VI: 466–8, note (25)) Leibniz had in fact (page 206) lauded the 'Methodi de maximis & minimis pars sublimior' as an instance of the 'nova inventa' which Newton had given to the world in his *Principia* in 1687, but he had also earlier in the same article (pages 201–2) accepted Fatio's own criticism of the 'Newtoniana constructio solidi minimum medio resistendi' as 'perplexa' and one shedding no light on its prior analysis —'quæ solutionem quærenti lucem nullam præferat' he had put it in his anonymous review the previous autumn (*Acta* (November 1699): 510–13, especially 512) of Fatio's *Investigatio Geometrica Solidi Rotundi, in quod Minima fiat Resistentia.*

(50) 'Operationes' in the draft of the following paragraph on .41: 61ᵛ.

(51) Newton initially wrote in his draft on .41: 61ᵛ: 'Nam termi[ni] seriei multiplicati per correspondentes terminos hujus progressionis $1.1\times 2.1\times 2\times 3.1\times 2\times 3\times 4.$ &c convertuntur in momentum primum, secundum, tertium, quartum &c. Sint A & A^n fluentes duæ sintq $A+o$ & $\overline{A+o}|^n$ fluentes eædem post momentum temporis & resolvatur $\overline{A+o}|^n$ in seriem

$$A^n + noA^{n-1} + \frac{nn-n}{2}\, ooA^{n-2} + \frac{n^3-3nn+2n}{6}\, o^3A^{n-3} + \frac{n^4-6n^3+11nn-6n}{24}\, o^4A^{n-4}\ [\&c].$$

Et hujus termini multiplicati per terminos correspondentes seriei $1.1\times 2.1\times 2\times 3.1\times 2\times 3\times 4$ [&c] evadent fluentis A^n momentum primum noA^{n-1}, secundum $\overline{nn-n}\times ooA^{n-2}$, tertium $\overline{n^3-3nn+2n}\, o^3A^{n-3}$, quartum $[\overline{n^4-6n^3+11nn-6n}\, o^4A^{n-4}\ \&c]$'; and then he shortened this to read simply 'termini seriei erunt ut fluxiones & momenta, secundus terminus ut fluxio prima et momentum primum, tertius ut fluxio secunda et momentum [secundum &c]'. Although the latter version carefully interlards the 'ut' with which he defended himself, in his anonymous *Phil. Trans.* 'Account' of the *Commercium Epistolicum* and elsewhere (see Appendix 2: note (145) preceding), against the wilful misunderstanding by Johann Bernoulli of the corresponding first paragraph of the *De Quadratura*'s terminal scholium, we will forgive Newton that in his present final phrasing of his sentence he returns to stating explicitly the factors of proportionality in each case between the terms of the incremented series and the related fluxions/moments of its principal term.

(52) The draft on .41: 61ᵛ has in place of this last sentence: 'Magna est igitur inter methodum fluxionum & methodum serierum affinitas & harmonia, et propterea utramq conjunxi ab initio et ex utraq Analysin meam' —'methodum unam' was initially written— 'generalem conflavi'. Little of this key declaration by Newton of the foundation of his unified method of mathematical analysis survives in his recast version. Why he chose to tone down its message is not clear to us.

Ad eandem Analysin pertinet etiam artificium ducendi Curvam per puncta quotcunq data, et ea ratione interpolandi Series quascunq. Nam si, verbi gratia, series aliqua vel fluentium vel fluxionum habeatur, sed fluentes vel fluxiones in intermedijs seriei locis non habeantur, per interpolationem seriei habebuntur eædem in locis quibuscunq. Deinde ex lege fluentium sic inventa prodibit lex fluxionum per methodum nostram, et contra. Artificij autem describendi curvam per puncta data memini in Epistola prædicta 24 Octobris 1676 ad Oldenburgum data.[53]

Atq hactenus de Analysi qua usus sum in investigatione rerum quas in hisce *Principiorum* Libris composui.

---

(53) In the briefest of terms: ' . . . aliam nondum communicatam Methodum habeo, qua pro lubitu acceditur ad quæsitum. Ejus fundamentum est commoda, expedita et generalis solutio hujus Problematis, *Curvam geometricam describere, quæ per data quotcunq puncta transibit*' (*Commercium*, ₁1712: 75–6 [= *Correspondence*, **2**: 119]). It had (see IV: 22–34) been Newton's original intention in the *epistola posterior* to afford Leibniz some more detailed glimpse of his 'Rule of Differences' in operation, but none of the seven undemonstrated 'regulæ in usum Calculatorum' which he initially set down (*ibid.*: 28–34) for Oldenburg to communicate— let alone the general 'Regulæ Differentiarum' by advancing and central differences on which these particular rules for interpolation were founded (see *ibid.*: 36–50 and 52–68 respectively; the latter was, more than thirty years afterwards, to form the main corpus of his printed *Methodus Differentialis* [= **1**, 5 preceding] of course)—in fact left Newton's hands in 1676.

APPENDIX 6. DRAFT PREFACE TO AN INTENDED REISSUE OF THE *PRINCIPIA*.[1]

[*c.* mid-1719]
From the original[2] in the University Library, Cambridge

In Mathematical Sciences the Ancients had two Methods which they called Synthesis & Analysis or Composition & Resolution. By the Method of Analysis they found their inventions & by the Method of Synthesis they composed them for the publick. The Mathematicians of the last age have very much improved Analysis, but stop there & think they have solved a Problem when they have only resolved it, & by this means the method of Synthesis is almost laid aside. The Propositions in the following Book[3] were invented by Analysis: but considering that the Ancients (so far as I can find) admitted nothing into Geometry before it was (for the greater certainty) demonstrated by composition, I com-

---

(1) That in whose train (compare Appendix 4: note (1) preceding) he intended at this time, in or a little after the summer of 1719, to adjoin a lightly augmented version of his tract *De Quadratura Curvarum* (on which see Appendix 8.2 following) as an ancillary companion-piece supporting his several assumptions in the *Principia* itself of particular squarings of curves or his frequent more generalised appeals there to be granted *curvilinearum quadraturæ*. In this tentative author's preface Newton parades yet again his contention (whose justice we examined at length in **1**, 8) that the *Principia*'s propositions were—in their greatest part at least—'invented by Analysis' but afterwards 'demonstrated by composition' to make them 'Geometrically authentic'; then quickly passes on to air once more his reasons for attaching the short historical scholium which he placed after Lemma II of the second book, where he had set out the 'Elements' of his 'Analytic Method'. The two final paragraphs where he responds (without naming him) to criticisms by Johann Bernoulli of inadequacies and errors in particular propositions of the *Principia* do not directly relate to the issue of calculus priority, but their mathematical significance will be evident and we need make no apology for here retaining them. No equivalent Latin version of this multiply drafted piece is known to us, and, in default, we take it that Newton never came round to composing one, though on Add. 3968.9: 112ᵛ/112ʳ there survives his rough jotting of what was evidently meant as a portion of a common preface to the present joint edition of the *Principia* with the *De Quadratura*; among other things he there affirmed that 'anno 1685 et parte anni 1686 beneficio hujus methodi [*sc.* analysis per series et fluxiones] & subsidio libri *de Quadraturis* scripsi libros duos primos *Principiorum mathematicorum Philosophiæ*. Et propterea Librum *de Quadraturis* subjunxi Libro *Principiorum*'. (The full text is printed in I. B. Cohen, *Introduction to Newton's 'Principia'* (Cambridge, 1971): 347). On the dubious general validity of this claim see Appendix 1, 8.4 preceding.

(2) Add. 3968.9: 105ʳ–106ʳ with the penultimate (cancelled) paragraph inserted from *ibid.*: 102ᵛ; the last paragraph was first printed by David Brewster in his *Memoirs of the Life, Writings and Discoveries of Sir Isaac Newton*, 1 (Edinburgh, 1855): 471, but a more correct transcription of it will be found on page 295 of Cohen's 1971 *Introduction*. Three successive prior draftings of this preface, the first two run (and partially intertwined) together and all manifestly written at one go, exist on .9: 101ʳ→101ʳ/101ᵛ→101ᵛ–102ᵛ; the unseparated texts of the former, along with the opening paragraph (alone) of the last, are reproduced by Cohen in his *Introduction*: 293–4.

(3) The whole *Principia* of course, not just its opening book.

posed what I invented by Analysis, to make it Geometrically authentic & fit for the publick.[4] And this is the reason why this Book was written in words at length after the manner of the Ancients without Analytical Symbols & Calculations. But if any man who understands Analysis, will reduce the Demonstrations of the Propositions from their composition back into Analysis (wch is easy[5] to be done) he will see by what method of Analysis the Propositions were invented. And by this means the Marquess de l'Hospital was able to affirm that this Book was (*presque tout de ce calcule*) almost wholly of the infinitesimal Analysis.[6]

---

(4) That is, proper to be published—'good Geometry' Newton had called it on page 206 of his anonymous 1715 'Account' of the *Commercium Epistolicum* (Appendix 2 above). In initial draft on .9: 101^v of the present late-inserted paragraph Newton first wrote 'more fit...' before deleting the attenuating adverb.

(5) Tentatively amended by interlineation to be 'very easy' in the draft on .9: 101^v, where in sequel Newton first passed to write 'he will esily see...'. Whether this be so the reader may himself decide from our translations back(?) into analytical equivalent, in the sixth volume *passim*, of many of the *Principia*'s propositions. (See also our related discussion of 'The Mathematical Principles underlying Newton's *Principia Mathematica*' in *Journal for the History of Astronomy*, **1**, 1970: 116–38, especially 117–21.) Newton elaborated his highly questionable thesis in fullest and most detailed form, we have seen, in the drafts for a somewhat earlier *Principia* preface—one devoid of references to calculus priorities—whose mathematically interesting portions are set out in **1**, 8 preceding.

(6) Full citation of the pertinent place in the 'Preface'—by Fontenelle, it would appear—to L'Hospital's *Analyse des Infiniment Petits* (Paris, 1696) was made in **1**, 8.1: note (15). As in the similar passage on page 198 of his anonymous 1715 'Account' of the *Commercium Epistolicum* (see Appendix 2: note (94) above) Newton here, too, first wrote 'Calculus' in literal rendering of *calcul* before replacing it by 'Analysis': the former, we may guess, now held for him an impermissible Leibnizian overtone of a 'Calcul différentiel'.

This first paragraph echoes, we may notice, the like opening (drafted on .13: 185^r) of a parallel 'Ad Lectorem', written by Newton (in the third person) in general introductory—so it would appear—to his planned joint edition of the *Principia* and *De Quadratura*, where he declared: 'The ancients had their method of Analysis, but admitted nothing into Geometry before it was demonstrated by Composition. The moderns'—'neglect Composition &' is cancelled in first sequel—'are intent only upon Analysis, & admit analytical inventions into Geometry before they are demonstrated by Composition. Synthetical Demonstrations are easier to be read, render Propositions more certain, & convey them better to posterity. For the symbols used in Analysis are apt to be changed from time to time. Mr Newton for these reasons & that the Propositions in his *Principia Philosophiæ* might be fit to be received into Geometry, after he had invented them by Analysis, demonstrated them by Composition. But the Analysis is so conspicuous through the composition that the Marquess de l'Hospital said that this Book was *almost* wholly of the Infinitesimal calculus'; and then added: '& Mr Leibnitz in a Letter to Mr Newton dated 7 Mart. 1693 wrote thus of it.'—see Raphson, *History of Fluxions*, ₂1717: 119 (=*Correspondence*, **3**, 1961: 257)—'*Mirifice amplicaveras Geometriam tuis seriebus, sed edito Principiorum opere ostendisti patere tibi etiam quæ Analysi receptæ non subsunt. Conatus sum Ego quoqʒ notis commodis adhibitis quæ differentias & summas exhibent, Geometriam illam quam transcendentem appello, Analysi quodammodo subjicere, nec res male successit.* The Elements of this method are contained in the first Proposition of the Book of *Quadratures* & also in the second Lemma of the second Book of *Principles* with the Scholium thereupon. The Proposition is in the Scholium & the Solution theoreof in the Lemma. The same Elements are also conteined in

In the second Lemma of the second Book of these *Principles*, I set down the Elements of this Analytic Method & demonstrated the Lemma by Composition in order to make use of it in the Demonstration of some following Propositions.[7] And because Mr Leibnitz had published those elements a year & some[8] months before without making any mention of the Correspondence wch I had with him by means of Mr Oldenburg ten years before that time, I added a Scholium not to give away the Lemma, but to put him in mind of that Correspondence in order to his making a publick acknowledgm^t thereof before he proceeded to claim that Lemma from me. [9] For in my Letter dated June 13th 1676 I said that the method of converging series in conjunction with some other methods (meaning the Methods of Fluxions & Arbitrary Series) extended to almost all Questions except perhaps some numeral ones like those of Diophantus. And in his Answer dated Aug. 27 1676, he replied that he did not believe that my methods were so general, there being many Problemes wch could not be reduced to Equations or Quadratures. And in mine dated 24 Octob. 1676 I represented

Mr Newton's Letter to Mr Collins dated 10 Decem. 1672: a copy of wch was sent to Mr Leibnitz by Mr Oldenburg among the extracts of Gregories Letters 26 June [!] 1676...'. In the draft of his present paragraph on .9: 101^v Newton here, too, similarly first went on: 'And Mr Leibnitz in a letter to me dated [7] Mar. 1693 st. vet. [wrote] *Mirifice ampliaveras Geometriam...* ... *Analysi quodammodo subjicere.* And in the *Acta Eruditorum* for May 1700'—see Appendix 5: note (49) preceding—'he allowed that I was the first who by a specimen made publick (meaning in the Scholium upon the 35th Proposition of ye second Book) had proved that I had the method of maxima & minima in infinitesimals. And when he himself had '—in his 'Schediasma de Resistentia Medii, & Motu Projectorum Gravium in Medio Resistente' (*Acta Eruditorum* (January 1689): 38–47)—'composed the first & second Sections of the second Book of my *Principles* in another form of words without calculations'—'Analysis' is replaced—he concluded [pag. 46]: *Multa ex his deduci possent praxi accommodata, sed nobis nunc fundamenta Geometrica jecisse suffecerit, in quibus maxima consistebat difficultas. Et fortassis attente consideranti vias quasdam novas vel certe satis antea impeditas aperuisse videbimur. Omnia autem respondent nostræ Analysi infinitorum* —Newton began to adjoin Leibniz' ensuing phrase '*hoc est calculo summarum & differentiarum*' but broke off after three words (without amending the gender of the respondent '*expresso*' in sequel)—(*cujus elementa quædam in his Actis [sc. Octob. 1684] dedimus) communibus quoad lic uitverbis hic expresso*'.

(7) Specifically (in amplification of our earlier note (1) on IV: 521) in Propositions VIII, IX, XI, XIV and XXIX of the *Principia*'s second book. The eminently citable ensuing sentence is, we may again notice, quoted by Brewster in his *Memoirs*, 2: 425.

(8) Initially numbered to be 'about eighteen' in the draft on .9: 101^v.

(9) In his original version (.9: 101^v) Newton inverted the order of his reference in sequel to his *epistolæ prior et posterior* to Leibniz in 1676, there first continuing: 'For in my Letter of 24 Octob. 1676 I mentioned an Analysis wch proceeded without stopping at surds & gave me the method of tangents & faciliated Quadratures of Curves & gave me the Quadrature of Curves by converging series wch brake off & became finite equations when the curve could be squared by such finite equations[,] & I said that I had wrote a Treatise upon this method & the method of Series together five years before, that is in the year 1671, wch was two years before Mr Leibnitz began to study the higher Geometry. And the Problem upon which this method was grounded I set down enigmatically. And in a former Letter dated 13 June 1676 I said that...'.

that my Analysis proceeded without stopping at surds, & readily gave the method of Tangents of Slusius & faciliated Quadratures & extended also to Problems which could not be reduced to Quadratures & gave me converging series for squaring of Curves which become finite whenever the Curve can be squared by a finite equation. And I said also that I had written a Treatise on this subject five years before that time, that is in the year 1671: wch was two years before Mr Leibnitz began to study the higher Geometry. And the foundation of this method I said was obvious, & wrote it down enigmatically in this sentence: *Data æquatione fluentes quotcunq̃ quantitates involvente invenire fluxiones, & vice versa.* And in both my Letters I said that I had then absteined from this subject five years, being tyred with it before. When this Letter of 24 Octob. arrived at London Mr Leibnitz was there the second time & saw it, & procured Dr Barrows Lectures, wherein was his method of Tangents invented above 12[10] years before, & Mr Leibnitz in his way to Hanover was considering how to make the Method of Tangents of Slusius general by a Table of Tangents, as I find by a Letter of his to Mr Oldenburg dated at Amsterdam Novem 18 1676 *st. vet.* And when Mr Oldenburg heard that Mr Leibnitz was arrived at Hannover, wch was in March following, he sent to him a copy of my last Letter. And Mr Leibnitz in his Answer dated 21 June 1677 sent back Dr Barrows Method of Tangents under the differential notation & how this method might be improved so as to give the method of Tangents of Slusius; & then how it might be further improved so as to proceed without taking away fractions & surds; & then added.[11] *Arbitror quæ celare voluit Newtonus, ab his non abludere. Quod addit, ex hoc eodem fundamento Quadraturas quoq̃ reddi faciliores me in sententia hac confirmat, nimirum semper figuræ illæ sunt quadrabiles, quæ sunt ad æquationem differentialem.* And to put Mr Leibnitz in mind of all this was the meaning of that Scholium above mentioned.[12]

At the same time that Mr Oldenburg sent my Letter of 13 June 1676 to Mr Leibnitz (wch was the 26th day of the same June[13]) he sent also (at the request of Mr Leibnitz) a collection of extracts of the Letters & papers of Mr James

(10) Presumably Newton meant to write '6' or '7'. Barrow set out his (improvement of Gregory's 1668) method of algebraic tangents only in his *Lectiones Geometricæ* of 1670 'in the end of his 10th...Lecture', as Newton had phrased it seven years before (see Appendix 1.2 above, and compare note (11) to Appendix 1.1 preceding). He had made no mention of the problem of tangents in his first series of Cambridge *Lectiones Mathematicæ* which he delivered from his Lucasian chair during 1664–6.

(11) *Commercium Epistolicum,* ₁1713: 90 (=*Correspondence,* **2**: 215).

(12) That to Lemma II of Book 2 of the *Principia*, of course.

(13) Read 'of July'. Newton perpetuates, by way of his head to this 'Epistola prior D. *Isaaci Newton*' on *Commercium*: 90, a typesetter's error of 'Junii' in the editorial preliminary to Wallis' first complete publication of the latter in his *Opera Mathematica,* **3** (Oxford, 1699): 622 ff.—one caught by Wallis himself in his late list of *Errata Emendanda* on signature a4^v in an entry itself keyed by printer's slip to 'p. 662'.

Gregory then deceased. And amongst these extracts was a Letter of M[r] Gregory to M[r] Collins dated 5 Sept. 1670[14] in w[ch] M[r] Gregory represented that he had improved the Method of Tangents beyond what Dr Barrow had done, so as to draw Tangents to all Curves without Calculation. There was also a Copy of a Letter written by me to Mr Collins 10 Decem. 1672 in which I wrote that the methods of Tangents of Gregory & Slusius were only *Corollarium Methodi generalis quæ extendit se citra molestum ullum calculum non modo ad ducendum Tangentes ad quasvis Curvas sive Geometricas sive Mechanicas, vel quomodocunꝗ rectas lineas aliasve curvas respicientes; verum etiam ad resolvendum alia abstrusiora Problematum genera de Curvitatibus, Areis, Longitudinibus, Centris gravitatum Curvarum &c. Neꝗ (quemadmodum Huddenij Methodus de Maximis et Minimis) ad solas restringitur æquationes illas quæ quantitatibus surdis sunt immunes. Hanc methodum intertexui alteri isti, qua Æquationum Exegesin instituo, reducendo eas a[d] Series infinitas.*[15] These Letters M[r] Leibnitz received in July 1676. And in October following he procured D[r] Barrows Lectures at London as above & saw my Letter of Octob. 24. And all this was enough to put him upon considering how to improve the Method of Tangents of D[r] Barrow, as Gregory had done before[16] so as to draw Tangents without calculation; and how to improve the Method of Tangents of Gregory & Slusius so as to make it proceed without stopping at fractions & surds, & to extend it not only to Tangents & Maxima & Minima, but also to Quadratures &[17] other sorts of Problemes, so as to become such a general Method as I described in my Letters of 10 Decem. 1672, 13th June 1676 & 24 Octob. 1676. And after all this light received from England (besides what he saw in the hands of M[r] Collins when he was last in London) I had great reason by the Scholium above mentioned to put him in mind of his correspondence wth M[r] Oldenburgh & M[r] Collins.[18]

···   ···   ···[19]

---

(14) We have cited the original English (Anglo-Scottish!) of the pertinent passage from this letter of Gregory's in note (19) of Appendix 4.3 above.

(15) *Commercium Epistolicum*: 30. Newton's original English words (earlier partially cited by us in III: 122, note (191) at the point—Mode 1 of Problem 4—of the 1671 tract where he set out his solution to the general problem of tangents in Cartesian coordinates as he expounded it to Collins the next year) are given on *Correspondence*, 1, 1959: 247–8.

(16) In fact (again see Appendix 1.2: note (23) preceding) it was Barrow who in 'Lectio X' of his 1670 *Lectiones Geometricæ* improved upon the method adumbrated by James Gregory two years earlier in Proposition 7 of his *Geometriæ Pars Universalis*.

(17) 'all' is rightly here deleted.

(18) This last sentence replaces a separate paragraph in the draft on .9: 102[r]/102[v] where Newton went on at much greater length to state: 'In this same year [1676] in a Letter dated May 12 Mr Leibnitz desired Mr Oldenburg to procure from M[r] Collins the Demonstration of two of my series for finding the Arc of a circle whose sine is given & the sine whose Arc is given; that is, the method of finding them. And in October following coming to London he consulted M[r] Collins to see the Mathematical Letters & Papers w[ch] M[r] Collins had received from

[20]The Demonstration of the first Corollary of the 13th Proposition[21] of the first Book being very obvious, I omitted it in the first Edition of this Book [of *Principles*] & contented my self with determining in the 17th Proposition what will be the Conic Section described in all cases by a body going from any place with any velocity in any determination,[22] if attracted by a force reciprocally as the square of the distance.[23]

Mr James Gregory & me. And no doubt he would desire to see the demonstration of the two series wch he wanted, that is, the *Analysis per series numero terminorum infinitas*, wch Dr Barrow had sent from me to Mr Collins in July 1669, wch Analysis consisted in reducing quantities to converging series & applying those series to the solution of Problems by the Method of Moments & Fluxions. For the direct method of fluxions described in the four first Propositions of the Book of *Quadratures* is there described under this Title: *Inventio Curvarum quæ quadrari possunt*. And there are examples of the inverse method in finding the Areas & lengths of some Curves & demo[n]strating the first Proposition of the Book. And the universality of the Method is described in these words. *Quicquid vulgaris Analysis per æquationes ex finito terminorum numero constantes (quando id sit possibile) perficiat, hæc per æquationes infinitas semper perficit*: .... This last is comprehended in the fift & sixt Propositions of the Book of *Quadratures*. And therefore the Method comprehended the first six Propositions of the Book of *Quadratures* in the year 1669. And all this may suffice to justify me in publishing the second Lemma of the second Book of *Principles* as my own'.

(19) We here omit a paragraph, relating to Proposition XXXVI—in the second edition; this is XXXVII in the first—of the *Principia*'s second book (for the background to which see *Correspondence*, **5**, 1975: 66–160, *passim*), where Newton went on to observe: 'By measuring the quantity of water wch runs out of a vessel in a given time through a given round hole in the bottom of the vessel I found that the velocity of the water in the hole was that which a body would acquire in falling half the height of the water stagnating in the vessel. And by other experiments I found afterwards that the water accelerated after it was out of the vessel untill it arrived at a distance from the vessel equal to the diameter of the hole, & by accelerating acquired a velocity equal to that which a body would acquire in falling the whole height of the water stagnating in the vessel or thereabouts. ...'

(20) The following paragraph does not appear in the final version of the present piece (on f. 106r), but because of its especial mathematical interest we here interpolate it, as a 'cancellation in the manuscript' in its proper place, from the immediately preceding draft on .9: 102v (where Newton has struck its text through). We have earlier in vi: 147, note (124) partially cited its first draft on .9: 101v, which in full reads: 'The Demonstration of the first Corollary of the 11th 12th & 13th Propositions being very obvious, I omitted it in the first edition & contented my self with adding the 17th Proposition whereby it is proved that a body in going from any place with any velocity will in all cases describe a conic Section: wch is that very Corollary'. (See Cohen's *Introduction to Newton's 'Principia'*: 294, ll. 16–20 for a yet fuller transcription recording insignificant interim deletions and minor recastings.)

(21) More accurately, this corollary is to the three preceding '11th 12th & 13th Propositions' jointly, as Newton rightly put it in his first draft (see the previous note).

(22) That is, direction of motion.

(23) Which of course may always be done, given any point of the path, the 'pull' of force towards the focus there, and the speed and direction of motion therefrom. But even the great Johann Bernoulli, like many a more recent reader of the *Principia*, did not see—chose not to see?—as at all 'obvious' this unstated inference from the involved geometrical construction of Proposition XVII that, because (by Propositions XI/XII/XIII respectively) motion in an ellipse/hyperbola/parabola may always be generated under an instantaneous inverse-square

In the tenth Proposition of the second Book there was a mistake in the first edition by drawing the Tangent of the Arch *GH*[24] from the wrong end of the Arch, which caused an error in the conclusion:[25] but in the second Edition

---

deviation to a focus, no other species of motion—if we exclude the degenerate cases (*Principia*, Book I, Propositions XXXII–XXXVII) where the 'determination' is through the centre of force and the ensuing conic paths collapse into their main, rectilinear axes—is possible in an inverse-square force-field. (See our fuller comment in VI: 146–9, note (124).) By the late summer of 1719 Newton had already introduced into an annotated copy of the second edition, at the end of Proposition XVII, the sentence by which he sought to bring out the cogency of this implication, *viz.* 'si corpus in his casibus'—as we would put it, of initial speed respectively less than, equal to, or greater than 'escape' velocity (the species of the conic orbit is independent of the angle of projection; see VI: 56–7, note (73))—'revolvatur in Conica Sectione sic inventa, demonstratum est in Prop. XI, XII, XIII quod vis centripeta erit ut quadratum distantiæ corporis a centro virium...reciprocè' (*Principia*, $_3$1726: 64, but we maintain Newton's original orthography); for in the letter which he wrote to Pierre Varignon on 29 September of that year, in response (compare page 529 above) to the latter's attempt to promote a concord with Bernoulli after so much earlier wrangling over calculus priority, he noted that he had made this addition 'in fine Prop. XVII Lib. I' in preparation for a forthcoming *editio tertia* of his book 'in Exemplari [secundæ editionis]'—that now ULC. Adv. b.39.2, we presume, where the new sentence is inserted on an interleaved sheet—'quod hunc in finem correxi', but now offered to suppress its publication should Varignon think this wise. (See the Latin draft of Newton's letter printed in *Correspondence*, **7**, 1978: 62–4, especially 63— essentially that sent, one assumes, though the latter has not been found—and the preliminary English draft, partially quoted in *ibid.*: 68, note (11) where he wrote more explicitly that 'I can strike [these words] out if you think it better to omit them'.) At the same time Newton took pains in an accompanying letter to Bernoulli himself, duly passed on by Varignon in November along with lavish quotation of the first, to assert that 'In editione secunda Libri mei *Principiorum* postulabat D. Cotes ut Corol. 1 Prop. XIII Lib I demonstratione munirem'— the begetter was Newton himself in fact (see his letter to Cotes of 11 October 1709 in *Correspondence*, **5**, 1975: 5) but he now understandably hides behind the coat of his dead editor in white lie—'et ea occasione Corollarium illud verbis nonnullis auxi: SED hoc factum est antequam hæ lites cœperunt [*sc.* anno 1713]....Et hoc annoto ut intelligas me animo candido Corollarium illud auxisse et hactenus nullas tecum lites agitasse' (*Correspondence*, **7**: 70). In his answer, directly to Newton, on 21 December (N.S.) Bernoulli was only in part prepared to be assuaged: 'Lubens credo quod ais de aucto Corollario 1, Prop. XIII Lib. 1 Operis Tui incomparabilis *Princip. Phil.*..., neque dubitavi unquam, Tibi esse demonstrationem propositionis inversæ quam nude asserueras in prima Operis editione[:] aliquid dicebam tantum contra formam illius asserti, atque optabam, ut quis analysin daret'—as he himself had done in 1710! (again see VI: 148, note (124))—'qua inversæ veritatem inveniret a priori, ac non supposita directa jam cognita. Hoc vero, quod Te non invito dixerim, a me primo præstitum esse puto, quantum saltem hactenus mihi constat' (*Correspondence*, **7**: 76). Newton does not refer to this in the unfinished surviving fragment of an English draft of his reply (*ibid.*: 80) and we may assume that he was content to let the matter rest there, honours equally divided.

(24) That is, the 'first moment of the Arch *FH*'. Newton understands the relettered accompanying figure in the *Principia*'s second edition (see page 356 above).

(25) '...wch made some'—meaning Johann Bernoulli and his nephew Niklaus (see note (175) of the introduction to Part 1 of this volume)—'think that there was a error in second fluxions' was written in first draft (on .9: 101^v). As we observed in **1**, 6, Appendix 3: note (22) preceding, the error in the first-edition version of *Principia*, **2**, X had neither at bottom to do

I rectified the mistake.[26] And there may have been some other mistakes occasioned by the shortness of the time in which the book was written & by its being copied by an Emanuensis who understood not what he copied;[27] besides the press-faults. For I wrote it in 17 or 18 months, beginning in the end of December 1684 & sending it to ye R. Society in May 1686: excepting that about ten or twelve of the Propositions were composed before, vizt the 1st & 11th in December 1679, the 6th 7th 8th 9th 10th 12th 13th 17th Lib. I & the 1, 2, 3 & 4 Lib. II in June & July 1684.[28]

---

with drawing the tangent from the wrong end (*sc. H*) of the infinitesimal arc *GH*, nor was its mending through a simple adjustment of the flawed argument at all difficult or complicated (even though Newton failed to see how this might be).

(26) Namely, by recasting the argument of the proposition into a chain of reasoning which invoked the length of the tangent *GL* drawn from the opposite end of $\overset{\frown}{GH}$; see (**1, 6**, §2.3→) *Principia*, $_2$1713: 232 ff.

(27) 'who understood not Mathematicks' Newton more fairly summarised the limitations of his former secretary, Humphrey Newton, in first draft on f. 101^v.

(28) This last date is a late replacement for an initial final phrase 'composed in the summer time of the year 1684'. Had he remembered what had happened so long before more exactly—or looked up Halley's letter to him of 29 June 1686, where an accurate chronology of dynamical events in that crowded twelve months was set down in writing only two years after their occurrence (see *Correspondence*, **2**, 1960: 442 and also VI: 16–18)—he would surely have put the date to be 'August' stretching on through September and October till early November, when Paget bore the completed demonstrations of his tract 'De motu Corporum' to Halley in London. (We take it that the 'ten Propositions composed in the summer time of 1684' were merely this celebrated treatise in its first version (VI: 30–74); on pages 69–76 of his *Introduction to Newton's 'Principia'* I. B. Cohen has, taking Newton's present concordance with propositions in the published *Principia* stringently, developed a tenuous argument according to which the equivalents in the 'De motu' to Propositions XV and XXXII of Book 1 'would have been discovered only after Halley's [first] visit [to Cambridge] in...August'.) In preliminary draft on .9: 102^v of this concluding sentence, let us adjoin, Newton affirmed somewhat differently that 'after I had found tenn or twelve'—'eight or tenn' is replaced—'of the Propositions [*viz.* in the *Principia*] relating to the heavens, they were communicated to the R.S. in December 1684, & at their request that the Propositions might be printed I set upon composing this Book & sent it to the R.S. in May 1686 as I find entred in their Journal Books Decem 10 1684 & May 19th 1686'; and went on to assert that 'I was enabled to make the greater dispatch by means of the Book of *Quadratures* composed some years before'—prior to 1684, that is—'& annexed to this Edition'. (The last phrase he subsequently crossed through: do we take it that it was at this point that he abandoned his plan to publish the *De Quadratura* as an ancillary to the *Principia* in its forthcoming re-edition?)

As with the penultimate paragraph (see note (23) preceding) this final one, too, is designed to forestall further objection by Bernoulli to Newton's author's competence in composing the first (1687) edition of the *Principia*. We will not be surprised that he spoke of Proposition X of the second book in much the same terms as here when he came to write to Varignon in late September 1719. 'When', he affirmed in English draft (see *Correspondence*, **7**: 68, note (11)), 'Mr N. Bernoulli was in London (wch was in Autumn 1712) & told me that there was a fault in the resolution of Prob. III [= Prop. X] Lib. II. & that he took it [to] be a mistak in second

differences; I answered that I would examine it & toold him…within a day or two that…
there was a mistake, but it lay in drawing the tangent of the Arch *lC*'—understand the figure
of the first edition (page 374 above)—'from the wrong end…. For the tangents of both arches
lC & *CG* should have been drawn the same way with the motion of the body because they
represent the moments of the motion. I gave him also the scheme set right & the calculation
suited thereunto, & added that I would cause that sheet to be reprinted. The Book of *Principles*
was writ in about 17 or 18 months, whereof about two months were taken up with journeys,
& the MS was sent to yᵉ R.S. in spring 1686; & the shortness of the time in which I wrote it,
makes me not ashamed of having committ[ed] some faults & omitt[ed] some things…'.
The Latin version passed to Varignon in due course stated much more abruptly that 'mense
Octobri anni 1712 ubi Liber totus'—Cotes' revised edition of the *Principia*, then at an
advanced state in press—'usᵩ ad pag. 456 inclusive impressa esset, D. Nicolaus Bernoulli me
monuit quod error aliquis admissus fuisset in resolutione Prop. X Lib. II Edit. 1, et resolutionem subinde examinavi et correxi & correctam ei ostendi, & imprimi curavi non subdole
sed eo cognoscente' (*Correspondence*, **7**: 63; a prior Latin draft in the possession of the Earl of
Macclesfield was printed by S. P. Rigaud in his *Correspondence of Scientific Men of the Seventeenth
Century*, **2** (Oxford, 1841): 437, but its recipient was first identified by Brewster in his *Memoirs*,
2: 300, note 1).

APPENDIX 7. THREE PRELIMINARY ENGLISH DRAFTS OF A PREFACE TO THE *DE QUADRATURA CURVARUM* IN ITS INTENDED JOINT EDITION WITH THE *PRINCIPIA*.[1]

[c. mid-1719]
From the originals in the University Library, Cambridge

[1][2]   In writing the Book of *Principles* I made very much use of the following Book of *Quadratures* & therefore have subjoyned it. This Book is founded upon the method of fluxions, & in the Introduction to it I said that I found the method of fluxions gradually in the years 1665 & 1666, & this has been called in question. But the accusation is not yet proved as it ought to have been, nor hath any answer been given to the collection of ancient Letters & Papers published by order of the R. Society under the title of *Commercium Epistolicum* for setting this matter in a fair light.[3] However, upon publishing this Book anew,

---

(1) Again see note (1) to Appendix 4 above for the little information we have been able to glean regarding this planned re-edition, about the late summer of 1719 or shortly after, of the *Principia* with the *De Quadratura* in its train as technical adjunct. These preparatory English earnests of the more finished 'Ad Lectorem' and historical annotations to the latter which are reproduced in Appendix 8 following largely repeat one another and also much of what has been churned out *ad nauseam* already in foregoing pages. They merit publication, however, for Newton's few novelties of comment—notably his reluctant confession in the next to last sentence of [3] that 'its probable' that his attendance at Barrow's professorial lectures at Cambridge in 1665 'might put me upon considering the generation of figures by motion, tho I not now remember...' (see note (64) below)—and for their revelations (if further be needed) of the endless pains which he was tirelessly yet ready to take, three years after Leibniz' death, in cutting up and repasting and everywhere further polishing his multi-faceted claim to absolute priority of calculus 'invention' by the creation and development from the middle 1660's onwards of his fluxional method, and by its varied application to problems of theoretical physics and astronomy in his *Principia*.

(2) A composite of Newton's main drafting on Add. 3968 (.41: 16ᵛ/16ʳ→) .5: 28ʳ–29ᵛ (+ .28: 403ʳ) and the partial revise on .5: 24ʳ–25ʳ; we follow the latter text in the main, suitably eking it out from its preparatory versions.

(3) 'But instead of doing this', Newton initially went on to write on f. 24ʳ, 'the accusers have been at work to set up judeges of their own₍₎ freeing themselves from calumny'; and thereafter to assert: 'On the contrary, the Accusers have been endeavouring to avoid answering it by telling us that we shall not have the pleasure to see them answer & that to answer it would require a book as big as the *Commercium*, & by appealing to Judges of their own setting up, & by chall[eng]ing the English Mathematicians to solve their Problems, & by framing many new occasions to make a squabble, & writing Letters to Ladies & other persons unacquainted with these matters for making an interest'. This outburst against the now long-dead Leibniz was straightaway deleted, but we may glimpse in it the anger and resentment against his old foe which still burnt away deep within Newton.

it may be expected that I say some thing in justification of what I said in the Preface of the former edition.[4] . . .   . . .

Mr James Gregory in a Letter to Mr Collins dated 5 Sept 1670[5] wrote that by comparing Dr Barrows methods of drawing Tangents with his own he had found a method of drawing Tangents to all Curves without calculation. And upon notice thereof[6] & that Mr Slusius had such another Method wch he intended to communicate to Mr Oldenburg I wrote to Mr Collins the following Letter dated 10 Decem 1672.[7] *Ex animo gaudeo . . . reducendo eas ad series infinitas.* In this Letter you have an example of the Method of Fluxions in drawing of Tangents & a description of the large extent of it in this & other more difficult Problems without stopping at surds; & the last words of the Letter related to a Tract wch I wrote the year before & in which the method here described was interwoven with another Method wherein æquations are reduced to converging series, both wch methods together constitute the general Method described in my Letters dated 13 June & 24 Octob. 1676[,] which Letters were published by Dr Wallis in the third volume of his works.[8]

For from the Tract wch I wrote in the year 1671, I extracted in the year 1676 the following Book of *Quadratures* & in my aforesd Letter to Mr Oldenburg of Octob 24. 1676 I set down the first Proposition thereof with its inverse verbatim[9] in this sentence *Data æquatione quotcunqɜ fluentes quantitates involvente invenire fluxiones & vice versa* & represented it the foundation of the method then spoken of, & then I added that *Hoc fundamento conatus sum etiam reddere speculationes de quadratura*

---

(4) So the final version on f. 24r; the draft on f. 28r (which lacks the preceding sentence) reads more shortly ' . . . it will not be amiss to justify what I said'.

(5) Gregory's exact words are quoted in note (19) of Appendix 4.3 preceding.

(6) In Collins' lost letter to Newton early in December 1672, to which the latter made the reply which he proceeds to cite *Latinè* (see the next note).

(7) Newton quotes from the Latinisation of (part of) his letter in *Commercium Epistolicum*, ₁1713: 28–9; the original English text of the following extract is, let us notice yet once more, to be found in *Correspondence*, **1**: 247–8.

(8) As items VI and VIII respectively in the 'Epistolarum Collectio' thereto appended (*Opera Mathematica*, **3** (Oxford, 1699): 622–9/634–45).

(9) A curious description of Newton's 1676 anagram, separating as it does the ensuing Latin sentence into its component letters and displaying these by counted number in alphabetical sequence. The phrase 'with its inverse' we here (f. 24v) insert from .28: 403r, along with the ensuing amplification (drafted on .5: 28r/28v) of his laconic sequel on f. 24v: '& copied without any alteration the two [!] Tables set down in the Scholium upon the tenth Proposition'. In the prior version of the opening to the present paragraph on f. 28r, we may add, Newton passed along to his reader the further information from his October 1676 letter (compare III: 9, note (24)) that the treatise 'concerning the method of converging Series & Fluxions joyntly' which he wrote in 1671 he 'did not finish . . . that part of it being wanting in wch I intended to explain the manner of solving such problems as can not be reduced to quadratures, as I mentioned'— the original Latin (see *Correspondence*, **2**: 114) here reads: 'Deerat . . . pars illa qua decreveram explicare nodum solvendi Problemata quæ ad Quadraturas reduci nequeunt, licet aliquid de fundamento ejus posuissem'—'in my Letter of 24 Octob. 1676 aforesaid'.

Curvarum simpliciores; perveniᵹ ad Theoremata quædam generalia. And then I set down a Theorem for squaring Curves whose ordinates were dignities of⁽¹⁰⁾ Binomials & illustrated it wᵗʰ examples & then added that I had other such like Theorems for Trinomials. And all this implies the knowledge of the method of fluxions so far as it is conteined in the first six Propositions of the Book of *Quadratures*. In the same Letter [of 24 Octob.] I said also: *At quando hujusmodi Curva aliqua non potest Geometrice quadrari; sunt ad manus alia Theoremata pro comparatione ejus cum Conicis Sectionibus; vel saltem cum alijs Figuris simplicissimis quibuscum potest comparari—et Theoremata quædam pro comparatione Curvarum cum Conicis Sectionibus in Catalogum dudum retuli.*⁽¹¹⁾ And then I set down the Ordinates of the Curves in the latter part of the⁽¹²⁾ Table for comparing Curves wᵗʰ the Conic Sections set down in the 10ᵗʰ Proposition of this book, wᶜʰ is a proof that I had that Table some years before I wrote that Letter.⁽¹³⁾ And in my Letter to Mʳ Collins dated 8 Novem. 1676 & published by Mʳ Jones, I had relation to the 10ᵗʰ Proposition of this Book in saying:⁽¹⁴⁾ *Nulla extat Curva cujus Æquatio ex tribus constat terminis, in qua, licet quantitates incognitæ se mutuo afficiant & indices dignitatum sint surdæ quantitates . . . nulla inquam hujusmodi est Curva de qua an quadrari possit, necne, vel quænam sint Figuræ simplicissimæ quibuscum comparari possit, sive sint Conicæ Sectiones sive aliæ magis complicatæ, intra horæ octantem respondere non possim. Deinde methodo directa & brevi, imo methodorum omnium generalium brevissima* (de qua vide Coroll. 2 Prop. 10 Libri sequentis) *eas modo comparari possint, comparo. Affirmatio quidem videri potest temeraria, propterea quod perdifficile sit dictu an Figura quadrari vel cum alia comparari possit, necne; mihi autem manifestum est, ex eo unde deduxi fonte, quanquam id alijs demonstrare in me suscipere nollem. Eadem methodus Æquationes quatuor terminorum aliasᵹ complectitur, haud tamen adeo generaliter.* All this relates to the method of squaring figures set down in the Book of *Quadratures*, & chiefly to the tenth Proposition of that

---

(10) 'any' is deleted (.28: 403ʳ).

(11) In here quoting from his *epistola posterior* Newton slightly adapts the original text (for which see *Correspondence*, 2: 117, 119).

(12) 'second' is deleted. Newton has no wish to suggest that there might be two such catalogues of curves whose quadrature is related to that of conics, from which he extracted the instances he named in his 1676 letter.

(13) But not of course that this 'Catalogus posterior' in Problem 9 of his 1671 treatise (see III: 244–56) had by 1676 already been adjoined to the later *De Quadratura*: a work indeed (we now know) whose preliminary version Newton was not in fact to compose for yet another fifteen years. In first draft on .5: 28ʳ he had here—as several times elsewhere at this time (see VII: 16)—cast historical accuracy wholly to the wind by claiming that in this phantom 'Book of Quadratures' which he 'extracted in the year 1676' from his earlier treatise he had 'therein copied from the former Tract without any alteration the two Tables set down in the Scholium upon the tenth Proposition, the one for squaring some Curves, & the other for comparing others with the Conick Sections'.

(14) Newton quotes the Latinised 'fragmentum' from his letter published, along with the original English text (= *Correspondence*, 2: 179–80), by William Jones in his edition of Newton's several tracts and pieces on *Analysis*. . . (London, 1711): 38.

Book:[15] & the *Methodus directa & brevis* mentioned in this Letter, by wch I compare such trinomial figures as may be compared$_{[,]}$ is that mentioned[16] in Corol. 2 Prop. 10; & this Proposition with its Corollaries is deduced from the 5^t 6th 7th 8th & 9th Propositions of this Book$_{[,]}$ & these are deduced from the 1st 2^d 3^d & 4th Propositions of the same Book: & therfore the method of Quadratures so far as it is conteined in the ten first Propositions of this Book was known to me in the year 1676 when I wrote this Letter.

[17]About three years after the writing of these Letters[18] by the help of this method of Quadratures I found the Demonstration of Keplers Propositions that the Planets revolve in Ellipses describing with a Radius drawn to the sun in the lower focus of the Ellipsis, areas proportional to the times. And in the year 1686 I set down the Elements of the Method of Fluxions & Moments, & demonstrated them synthetically in the second Lemma of the second Book of *Principles* in order to make use of the Lemma in demonstrating some[19] following Propositions. And because Mr Leibnitz had published those Elements in the *Acta Eruditorum* two years before[20] without taking any notice of the correspondence wch had passed between him & me by means of Mr Oldenburg ten years

---

(15) Not so. As we have many times before remarked, Newton in his letter of 8 November 1676 harks back to his recent separate researches into the quadrature of curves defined by a trinomial Cartesian equation; see III: 373–85. Newton is of course correct to state that these were afterwards (but not till 1693) subsumed into the various component corollaries to Proposition IX, the general strategy for applying which is outlined in Corollary 2 to the ensuing Proposition X, of the *De Quadratura* (on which see VII: 536–42, 546–8).

(16) Newton has rightly replaced an initial 'described' here. The 'direct & short way' for reducing the squaring of 'trinomial figures' to quadrable form upon which he reported to Collins in November 1676 is only sketchily outlined in Corollary 2 to Proposition X of the later *De Quadratura*; see the previous note.

(17) The next four sentences are introduced from the recast opening to the ensuing paragraph which Newton afterwards penned on .28: 403^r. In initial draft on .5: 28^v—the first sentence is printed, *hors de contexte* (but with full notice taken of a verbal stumbling and a cancelled dittograph which we ignore), by I. B. Cohen in his *Introduction to Newton's 'Principia'* (Cambridge, 1971): 348—he had gone on a deal more shortly to affirm: 'The Book I made use of in the year 1679, when I found the demonstration of Keplers Proposition that the Planets move in Ellipses, & again in the years 1684, 1685 & 1686 when I wrote the Book of *Mathematical Principles of Philosophy*, & for that reason I have now subjoyned it to that Book. In the year 1691 it was in the hands of Mr Ralpson & Dr Halley as one of them attested in print before his death & the other still attests....'

(18) That is, in late 1679. The stimulus to apply himself anew to planetary dynamics was, we need not say, his short-lived but exceedingly fruitful exchange of letters with Robert Hooke during November and December of that year (on which see VI: 9–14).

(19) 'five' we enumerated in note (7) to Appendix 6 preceding.

(20) On .5: 25^r—in further revise or preliminary caution subsequently discarded?—Newton toned down this bleak statement, appearing (as he did not outrightly intend) to charge Leibniz with plagiarising his yet unpublished *Principia*, by adjoining the escape-clause 'in another form'.

before:[21] I added a Scholium to that Lemma not to give away the Lemma but to put Mr Leibnitz in mind of making a candid & publick acknowledgement of that correspondence. And in demonstrating the XIV Prop. of that second Book I considered the *differentia momentorum arearum* or [*ar*]*earum momentum differentiæ*, & therefore when I wrote that Book I was no stranger to second moments or differences.[22] After this, in the year 1691 the Book of *Quadratures* was in the hands of Mr Ralp[h]son & Dr Halley[23] as one of them attested in print before his death[24] & the other still attests. And the next year at the request of Dr Wallis that I would explain how I found the Theorems set down in my Letter of 24 Octob. 1676 for squaring Curvilinea[r area]s by series wch break off & become finite equations when the Curves can be squared by finite equations, I sent to him in a Letter dated Aug 27 the fift Proposition of the Book of *Quadratures* as a more general Theoreme wch comprehended the Theorems mentioned by him & at the same time I sent to him also the first Proposition of this book illustrated with examples in finding first & second fluxions. And these two Propositions were printed before the end of the year in the second Volume of his works pag. 391, 392, 393[25] which Volume came abroad in August 1693. And this was the first time that any Rule was published for finding second third fourth & other fluxions or differences, & [it] is still the shortest, the clearest & the best. In March 1695 Dr Wallis upon notice from Holland that the Method of fluxions was celebrated there by the name of the Differential Method of Mr Leibnitz, inserted into the Preface of the first Volume of his works (wch came abroad after the second Volume) the following Paragraph.[26] *Quæ in secundo Volumine habentur, in Præfatione eidem præfixa dicitur. Ubi (inter alia) habetur Newtoni Methodus de Fluxionibus (ut ille loquitur) consimilis naturæ cum Leibnitij (ut hic loquitur) Calculo Differentiali, (quod qui utramq̃ methodum contulerit, satis animadvertat, utut sub loquendi formulis diversis,) quam ego descripsi (Algebræ cap. 91 &c. præsertim cap. 95) ex binis*

---

(21) Namely in (1686 − 10 =) 1676; .5: 25^r reads 'eight years before' equivalently, counting back from Leibniz' publication of his 'Nova Methodus pro Maximis et Minimis itemque Tangentibus...' in the *Acta* in October 1684.

(22) Newton yet again repeats his curious misapprehension in overhasty retrospect that the *differentia* of moments of area which he had employed in the argument of *Principia*, **2**, XIV was their infinitesimal second difference; once more see on this our note (44) to **1**, 8 preceding.

(23) Originally 'Dr Hally & Mr Raphson' (.5: 24^v), to which .28: 403^r adjoins '...at Cambridge', for what it matters.

(24) By Raphson on page 2 of his account of *The History of Fluxions* (published in 1715 posthumously); the essence of the passage (already quoted on III: 11) is given in note (55) below.

(25) These pages from the second volume (first to be published) of Wallis' *Opera Mathematica* have been reprinted on VII: 171–6. Newton's following firm citation of 'August 1693' is doubtless in line with the date 'Aug. 28. 1693', of the *Imprimatur* of the Vice-Chancellor of Oxford (then Henry Aldrich) set on the verso of the volume's title-page. Bound copies cannot have 'come abroad' for some weeks after.

(26) *Opera*, **1** (Oxford, 1695): signature a4^r; compare Appendix 2: note (106) above.

Newtoni literis. . .Junij 13 & Octob. 24, 1676 ad Oldenburgum datis, cum Leibnitio communicandis. . .ubi methodum hanc Leibnitio exponit tum ante decem annos, nedum plures, ab ipso excogitatam. And as soon as the Volume was printed off [he] wrote to me the following Letter dated Apr. 10th 1695.[27]. . . . . . And the next year the Editor of the *Acta Eruditorum* in g[i]ving an Account of this Volume cited some words out of this Paragraph, & Dr Wallis in a Letter dated 1 Decem. 1696 gave notice to Mr Leibnitz of the same Paragraph.[28] And yet neither the Editor of the *Acta* nor Mr Leibnitz denied what Dr Wallis had affirmed. On the contrary the Marquess de L'Hospital in the Preface to his book *de Infinite parvis* published this year allowed that the Book of *Principles* was almost wholy of the Differential calculus,[29] & Mr Leibnitz himself. . .in the *Acta Eruditorum* for May 1700[30]. . .acknowledged that I was the first who had manifested by a specimen made publick that I had the method of maxima & minima in infinitesimals or moments, meaning the specimen in the Scholium upon the 34th Proposition of the [second] Book of *Math. Principles.* And in general this Book of *Principles* is the first specimen made publick of applying the Method of Fluxions & Moments to the difficulter Problems. . . . . . But the Book of *Quadratures* continued in MS till the year 1704. . . . . . .

At the request of Mr Leibnitz I described in my said Letter of 24 Oct. 1676, how, before the plague wch raged in London in the years 1665 & 1666, by interpoling the series of Dr Wallis, I found the method of converging series together with the Rule for resolving the dignities of Binomials into such series. I there

---

(27) A gap follows in the manuscript (at the bottom of .5: 24ᵛ). We assume that Newton intended to cite in sequel the extract from Wallis' letter which he had published two years earlier on pages 120–1 of his 1717 appendix to Raphson's *History of Fluxions*, namely: 'I wish you would print the two large Letters of *June* and *October* 1676. I had Intimation from *Holland*, as desired there by your Friends, that somewhat of that kind were done, because your Notions of Fluxions pass there with great Applause by the Name of *Leibnitz's Calculus Differentialis.* I had this Intimation when all but Part of the *Preface* to this Volume was printed off; so that I could only insert (while the Press stay'd) that short Intimation thereof, which you there find. You are not so kind to your Reputation (and that of the Nation) as you might be, when you let things of worth lie by you so long, till others carry away the Reputation which is due to you. I have endeavoured to do you Justice in that, and am now sorry that I did not print those two Letters *verbatim*'. (For Wallis' original orthography see Joseph Edleston's full printing of the letter in appendix to his edition of *The Correspondence of Sir Isaac Newton and Professor Cotes* (London, 1850): 300 [=*Correspondence of Newton*, **4**, 1967: 100].)

(28) This anonymous review of the first two volumes of Wallis' *Opera Mathematica* in *Acta Eruditorum* (June 1696): 249–60 was—we may now know for sure, and doubtless as Newton (if not Wallis) suspected—by Leibniz himself.

(29) See **1**, 8.1: note (15), and compare Appendix 6: note (6) preceding. The original phrase in the preface to L'Hospital's *Analyse des Infiniment Petits* (Paris, 1696) was of course 'presque tout de ce calcul [*sc.* différentiel]' (signature éiiiʳ).

(30) 'pag. 206'. We have quoted Leibniz' lauding of Newton's 'Methodi de maximis & minimis pars sublimior', as evidenced in his brief exposition of the solid of revolution of least resistance in the *Principia*'s second book, in Appendix 5: note (49) above.

mentioned also that upon the publication of Mercator's *Logarithmotechnia* Dr Barrow sent [31]to Mr Collins a Compendium of these series. A copy of this Compendium in the handwriting of Mr Collins was found by Mr Jones & published after it had been collated wth the original wch Mr Jones borrowed of me for that purpose. The title thereof was *Analysis per series numero terminorum infinitas*. And in this Tract the method of series is interwoven with that of fluxions. For after I had found the method of series it quickly led me into the method of fluxions, & their affinity made me write of them both together as composing one general method of Analysis. In this Tract I affirmed that this method extends to all Problemes[32] & that *ejus beneficio Curvarum areæ & longitudines &c (id modo fiat) exacte et Geometrice determinantur*. Which I could not have said without understanding at that time so much of the method as is conteined in the first five or six Propositions of the Book of *Quadratures*. . . .

And all this may suffice to justify my saying in the Introduction to the following Book of *Quadratures*, that I found the method gradually in the years 1665 & 1666. However, the method is capable of improvements & the improvements are theirs who make them.[33]

[34]Now since by the help of this Method I wrote the Book of *Principles* I have therefore subjoined the Book of *Quadratures* to the end of it.[35]. . . And because

(31) Newton here (on .5: 28^v) first wrote 'abroad'—meaning just 'out of Cambridge'?—before quickly discarding it for the phrase which follows. It was of course Collins who, after receiving it from Barrow in the summer of 1669, passed on to Continental mathematicians both by direct correspondence and *via* Oldenburg a summary account of what was contained in the 'De Analysi'.

(32) Newton paraphrases the opening sentence of the *Conclusio* to his 'De Analysi', *viz.* 'Nec quicquam hujusmodi scio ad quod hæc methodus idqȝ varijs modis, sese non extendit' (see II: 240; the phrase next quoted ensues on *ibid.*: 242).

(33) We have minimally altered Newton's original paragraph (on .5: 29^r) to accord with his light recasting of it on the manuscript's verso.

(34) A page-long square bracket in the left-hand margins on .5: 29^r/29^v is Newton's mark that the remainder of the present draft is to be discarded; but we reproduce the first paragraph of this final portion, even so, for its explicit statement by him of what the reprinted *De Quadratura* was to contain in amplification of the published text. (The opening sentences are, let us add, cited also by I. B. Cohen on page 348 of his 1971 *Introduction to Newton's 'Principia'*, along with the corresponding section of Newton's immediately preceding first rough casting of it.) It will be evident that Newton here repeats the first words of his piece, and doubtless initially intended to strike these out, passing straightaway to declare: 'The following Book of *Quadratures* is founded upon the method of fluxions, & in the Introduction to it I said . . .'.

(35) We omit an intervening sentence, little to the present point (and not in the first casting of this paragraph immediately before it on f. 29^r) where Newton added 'And for the same reason I have subjoyned also the Differential Method'. This fleeting thought of annexing the central-difference formulas of his 1711 *Methodus Differentialis* to the forthcoming re-edition of the *Principia*—in complement to the advancing-difference theorems of Lemma V of its third book, we must assume—endured into the Latin preface to the joint edition set out in Appendix 4.1 (see note (6) thereto), but would essentially have been a publication for publication's sake. It did not survive past the preliminary revise printed in [2] below (see note (51)).

several Problems are proposed in the Book of *Principles* to be solved *concessis Figurarum Quadraturis*, I have added to the end of the Book of *Quadratures* some Propositions taken from my Letters already published for reducing quantities into converging series & thereby squaring the figures.[36] For Quadratures by such series have the same place in Arithmetick & Algebra with operations in decimal numbers.[37] . . . . . . [38]

---

(36) In his immediately preceding prior draft Newton wrote a little differently: 'And because the methods of series & fluxions are nearly related to one another & were invented in the same year (the year 1665), & were conjoyned by me in the Tracts wch I wrote in those days'—'above 49 years ago' (in 1669/71) is replaced—'& jointly compose one very general method of Analysis, I have here added to the end of this Book out of my Letters formerly published, some Propositions for reducing quantities into series, & for resolving Problems by the help of'—'such series *quamproxime*' he initially went on—'both methods together'. This augmented 'Book of Quadratures' would evidently have diverged little in its substance from the similarly amplified 'Analysis per Quantitates Fluentes et earum Momenta' (**1**, 5 above) which he had briefly, seven years earlier in the summer of 1712, planned likewise to adjoin to the *Principia*'s second edition. Subsequently (see Appendix 8.2 following) Newton was to settle for a much more concise annotation of the published text of the *De Quadratura*.

(37) '& the extraction of roots in simple & affected [æquations]' Newton began to add before changing his mind; on this compare II: 214–26→III: 40–66. In his preceding first draft he had more fully spelled out that 'For tho the resolution of Problems'—'into series' is deleted—'by continual approximation be not Geometrical, yet it is allowed in Arithmetick, & by consequence also in Algebra; as is manifest by continual approximations in decimal numbers & by perpetual divisions & extractions of square & cube roots & the roots of affected æquations allowed in Arithmetick, & it may sometimes lead to exact resolutions of Problems by finite equations'.

(38) In the final cancelled paragraph which ensues in the manuscript (.5: 29^r/29^v) Newton drifted away from his professed intention of supporting his own claim in the preface of the *De Quadratura* to have 'found the method of fluxions gradually in the years 1665 & 1666', once more tediously to attack Leibniz for publishing the 'series of Mr Gregory' for the inverse-tangent 'as his own' in the *Acta Eruditorum* in 1682 'without taking any notice of the correspondence between him & Mr Oldenburgh by w^ch [in April 1675] he had received it from London', and more broadly to query his mathematical originality and competence. 'In the *Analysis per series* abovementioned', he began, 'were several instances [see II: 232–40] of squaring the Circle & Conic Sections & other Figures & of finding the lengths of Curve lines, & on the contrary of finding the Abscissas & Ordinates of Figures whose Areas or lengths of the Curve lines are given[,] & Mr Collins was very free in communicating to Mathematicians the series there set down. Mr Leibnitz was in London in the beginning of ye year 1673 & conversed wth Dr Pell &c about numerical series & carried with him to Paris Mercator's *Logarithmotechnia*, but did not yet understand the higher Geometry. The next year he studied this Geometry & in two Letters dated 15 July & 26 Octob. he wrote to Mr Oldenburg that he had found the circumference of a circle in a series of rational numbers. . . . The next year Mr Oldenburg in a Letter dated 15 Apr. 1675 sent to Mr Leibnitz these two series [for the inverse-sine and sine] wch Mr Collins had received from me. . . And these two [for the inverse-tangent and tangent] wch Mr Collins had received from Mr James Gregory. . .'. We need quote no more of this attempt to take from Leibniz his independent discovery of the inverse-tangent series and the particular series of reciprocals of successive odd numbers, alternating in sign,

[2]⁽³⁹⁾   In composing these Books [of *Principles*] I was much assisted by the Book of *Quadratures*. At the request of Dᴿ Wallis I sent to him the first Proposition of this in the year 1692 Aug. 27, & he printed it in the second Volume of his works before the end of the year & the Book came abroad in Spring⁽⁴⁰⁾ 1693. & this was the first time that any rule for finding second third & fourth fluxions was published.⁽⁴¹⁾ This is a proof that the Book of *Quadratures* was then in Manuscript. And Dᴿ Halley & Mᴿ Ralp[h]son saw it in my hands at Cambridge⁽⁴²⁾ in summer 1691, as Mᴿ Ralphson has left attested in print, & Dᴿ Halley a living evidence still attests. And therefore it was in MS in yᵉ year 1691, & continued in MS thirteen years at the least before it was published. In my Letter of 24 Octob. 1676 I cited many things out of it, particularly I cited (in an Ænigma) the very words of the first Proposition *Data æquatione fluentes quotcunꝗ quantitates involvente invenire fluxiones*. I mentioned also the substance of the fift & sixt Propositions⁽⁴³⁾ & gave a solution of the fift wᵗʰ some examples. These two are the inverse of the third & fourth Propositions & these two are examples of the second & all of them are deduced from the first, & therefore I was in those days no stranger to the first six Propositions of this Book.

In the same Letter of 24 Octob. 1676, I wrote thus.⁽⁴⁴⁾ *Seriei a D. Leibnitio pro Quadratura Conicarum Sectionum propositæ, affinia sunt Theoremata quædam quæ pro comparatione Curvarum cum Conicis Sectionibus in Catalogum dudum retuli.* The series for squaring the Conic Sections⁽⁴⁵⁾ Mᴿ Leibnitz had twice from Mᴿ Oldenburg. The Theorems for comparing other Curves with the Conic Sections, I reduced into a Catalogue in [a] Tract wᶜʰ I wrote in the year 1671, & thenc[e] I copied it into the Book of *Quadratures*, & the Ordinates of the Curves in the more intricate part of the Table I set down in the very same order⁽⁴⁶⁾ & in the very same letters & symbols in wᶜʰ you will now find them in the Scholium upon the

---

for the quadrant of a circle wherein he detected a 'harmony' which the more practical, down-to-earth Newton could not—or at least would not—see. (Compare ɪᴠ: 531, note (13).)

(39) Add. 3968.5: 26ʳ/26ᵛ, a first recasting of the previous initial English version.

(40) Read 'late Summer' or even 'Autumn'; in [1] (see note (25) above) Newton had specified 'August'.

(41) An ill-considered concluding phrase ' & the Rule is the best' is wisely deleted in sequel.

(42) Originally 'borrowed [it] of me...& carried it with them to London'. Newton once more appeals to a stray remark by Raphson in his posthumous *History of Fluxions*, counting on Halley verbally to support its truth should the need arise; compare note (24) to [1] preceding.

(43) Only that of the 'fift', in fact. Proposition VI of the *De Quadratura* is referred to obliquely at best as one of the unnamed 'Theoremata quædam generalia' merely mentioned in passing by Newton in his *epistola posterior* to Leibniz in 1676 (see the next paragraph).

(44) *Commercium Epistolicum*: 76 (=*Correspondence*, **2**: 117).

(45) Understand those for the sine and the tangent and especially their inverses, sent by Oldenburg to Leibniz in April 1675 and again—this time only the sine and its inverse—in June 1676 when he transmitted Newton's *epistola prior*.

(46) An exaggeration, as we need not here insist.

10th Proposition of the Book of *Quadratures*. And therefore that Table was composed before I wrote that Letter; & the 7th 8th 9th & 10th Propositions upon wch it depends were then known to me.

Between the years 1671 & 1676 I medled not wth these studies[,] being tyred with them before: but in the year 1676 I extracted the Book of *Quadratures* from the Tract wch I wrote in [ye] year 1671 & from other older papers.[47] And soon after I had finished it I wrote to Mr Collins a Letter dated Novem. 8. 1676[,] a part of which is here set down.[48] *Nulla extat Curva, ... modo comparari possint, comparo.* To do all this is the inverse method of fluxions so far as that method is carried on in the Book of *Quadratures*. And by the help of this method I composed the Book of *Principles*, & therefore in this Edition have added the Book of *Quadratures* to the end of it.

I made use also of the method of maxima & minima in Infinitesimals, & by the confession of Mr Leibnitz was the first who shewed by a specimen made publick[49] that I had this method. I made use also of the method by me called the differential Method,[50] & for that reason have annexed this Method[51] to the Book of *Quadratures*. In my Letter of Octob. 24 1676 I said that the Tract wch I was writing in the year 1671 I never finished, & that that part of it was wanting in wch I intended to teach the manner of resolving Problems wch cannot be reduced to Quadratures.[52] But what I then intended to write is now gone out of my mind through long disuse of these methods.[53]

---

(47) In an editorial comment upon the following letter of Collins to Newton on 8 November 1676, whose full text he reproduced in *Correspondence*, 2: 179–80, H. W. Turnbull unwisely cited this sentence and the first part of the ensuing one as evidence 'according to...Newton' that he had written the *De Quadratura* shortly before. Let us yet again recur to vii: 15–16 for the historical—certainly far different—truth as we there sought to ferret it out.

(48) From the Latinised extract printed by Jones in his 1711 *Analysis*; see [1]: note (14).

(49) In the scholium to Proposition XXXV (in the first edition) of the *Principia*'s second book, of course; see vi: 460–4. Newton again (see Appendix 5: note (49), and compare [1]: note (30)) appeals to Leibniz' something less than laudatory aside upon the *methodus de maximis & minimis in infinitesimalibus* which there underlay his bare geometrical definition of the curve of the solid of revolution of least resistance.

(50) Understand the method of finite differences.

(51) As contained, that is, in Newton's 1711 tract on this *Methodus Differentialis*. This cancelled paragraph would seem to be the last occasion on which he announced his intention thus somewhat uselessly (see Appendix 4.1: note (6) and also note (35) to [1] above) to pad out the *Principia*'s text.

(52) 'Ipse autem Tractatum meum non penitus absolveram, ubi destiti a proposito.... Deerat quippe pars ea qua decreveram explicare modum solvendi Problemata, quæ ad Quadraturas reduci nequeunt...' (*Commercium Epistolicum*: 71 [=*Correspondence*, 2: 114]). See our fuller comment upon this point on iii: 9.

(53) Having in an unguarded moment let slip this scarcely politic home truth—if he could not remember what he had planned to write in completion of his 1671 tract, how could he recall after equal 'long disuse' of them his other early achievements in methods of calculus?— Newton here (at the top of f. 26v) pulls himself up short, cancelling the whole of this last

[3]⁽⁵⁴⁾   In the Introduction to the following Book I said that I invented the Method of fluxions gradually in the years 1665 & 1666. The first Proposition thereof at the request of Dr Wallis I sent to him in [August] 1692 & it was printed that year in the second Volume of his works, & came abroad in Spring⁽⁴⁰⁾ following, & this was the first time that any rule for finding second third & fourth fluxions came abroad.⁽⁵⁵⁾ In the Preface to the first Volume of his works Dr Wallis affirmed that in my Letters of 13 June & 24 Octob. 1676 I explained to M^r L[eibnitz] the Method of fluxions found by me ten years before that time or above.⁽⁵⁶⁾ And in the second Lemma of the second Book of *Principles* I set down & illustrated the Elements of this Method & mentioned in a Scholium the correspondence between me & Mr L. ten years before about these things to put him in mind of making such an acknowledgement thereof publickly as he had made privately in his Letter of 21 June 1677 wherein he sent me a specimen of his improvement of D^r B[arrow]s method of Tang^{ts}.⁽⁵⁷⁾

---

paragraph, and then passes straightaway to pen (on ff. 26^v/27^r immediately after) a first draft of the yet further recast preface whose text we give in [3] next following.

(54) Add. 3968.41: 86^r–87^r; a first version of this revise is found on .5: 26^v/27^r, in sequel to [2] (see the previous note).

(55) In sequel Newton first took over from [1], or rather its initial draft on .5: 28^v (see note (17) above), the supporting 'evidence' that 'The MS of the Book of *Quadratures* Mr Raphson & Dr Halley had in their hands in the year 1691 as one of them has left attested & the other still attests'. The reader may choose to believe that, in now deleting it, Newton at last recalled that the manuscript which Raphson and Halley in 1691 'had in our hands at *Cambridge*, in order to bring it up [to London] to be printed' was of 'that Treatise...which in the Year 1671 [Sir Isaac] had prepar'd for the Press' (see *History of Fluxions* (London, 1715): 2, and our comment in III: 12, note (29)).

(56) Much as in the corresponding passages of [1] (see pages 659, 662 above) Newton here also went on initially to declare: 'By this Method I invented the Demonstration of Keplers Proposition [that the Planets move in Ellipses] in the year 1679 & almost all the rest of the Difficulter Propositions of the Book of *Principles* in the years 1684, 1685 & in the beginning of the year 1686. And in Demonstrating the XIVth Prop. of the second Book I made use of second moments by the name of *differentia momentorū* & *momentum differentiæ*'.

(57) On VII: 16 we have already quoted the opening sentence of a further cancelled following passage where Newton made the now familiar, distorted declaration: 'In the year 1676 in summer I composed the Book of *Quadratures*, & then in my Letters of Octob. 24 & Novem. 8 1676 published by Dr Wallis & Mr Jones, I cited severall things out of it,'—'& other Papers from whence I had extracted it' he justly inserted in afterthought—'particularly the very words of the first Proposition; the Theorems for squaring figures by converging series wch break of & become finite equations when the Curve can be squared by such equations; the ordinates of Curves wch I could compare with the Conic Sections; & the squaring of all Curves exprest by equations of three terms, or comparing them wth the simplest Curves [wth] wch they can be compared: each of wch are a sufficient proof that I had the method of fluxions when I wrote that Letter. And in that Letter I represented that five years before, that is, in the year 1671 I wrote a Tract of that Method & the method of series together, wth a designe to have published it together with another Tract wch I had written about the refractions & colours of light. But finding that by what I had already communicated I began to be entangled in

From Dr Barrows method of Tangents compared with his own Methods Mr James Gregory deduced a method of drawing Tangents without calculation..., & upon notice from Mr Collins that Mr Gregory & Mr Slusius had such Methods I wrote to Mr Collins the following Letter dated 10 Decem 1672.[58] *Ex animo gaudeo D. Barrovij amici nostri reverendi Lectiones Mathematicas exteris adeo placuisse... ne grave ducas.* Thus far my Letter, in wch the method of fluxions wth its large extent is sufficiently described, & illustrated wth an example of drawing Tangents. And by saying: *Hanc methodum intertexui alteri isti qua Æquationum Exegesin instituo reducendo eas ad series infinitas* I refer to the method described in the Tract wch I wrote in the year 1671, in wch I joyned the methods of series & fluxions together, as I mentioned in my Letter of 24 Octob. 1676.

In the same Letter I mentioned also that when Mr Mercators *Logarithmotechnia* came abroad Dr Barrow communicated to Mr Collins a Compendium of my method of series. And this is the Tract entituled *Analysis per series numero terminorum infinitas.* The *Logarithmotechnia* came abroad in September 1668. Mr Collins a few months after sent a copy to Dr Barrow, who replied that the Method of Series was invented & made general by me about two years before the publication of Mercators *Logarithmotechnia* & at the same time sent back to Mr Collins the said Tract of *Analysis per Series*. This was in July 1669. In a letter to Mr Bertet dated 21 Feb 167$\frac{0}{1}$[59] Mr Collins represented that about four years before that time I had invented a general method of Analysis, meaning the method described in the *Analysis per series*.[60] And in a Letter to Mr Strode dated 26 July 1672[61] he said that after he had sent a copy of the *Logarithmotechnia* to Dr Barrow at Cambridge, the D^r *quasdam Newtoni chartas extemplo remisit: e quibus et alijs quæ prius ab Auctore cum Barrovio de eadem methodo communicata fuerant, patet*

---

disputes, I forbore to publish them for the sake of quiet; & as soon as I could get rid of the disputes I was engaged in I forbore to publish the method of Fluxions & Quadratures & the Theory of refractions & colours till the year 1704 wch was above thirty years together'.

(58) Newton goes on to quote at full length (as we here, following the manuscript, only summarily indicate) the Latinised rendering in *Commercium Epistolicum*: 29–30 of the central portion of his original English letter (for whose text see *Correspondence*, 1: 247–8) where he gave account to Collins in December 1672 of his general solution of the problem of drawing tangents to algebraic curves referred to Cartesian coordinates (originally set out in Mode 1 of Problem 4 of his 1671 tract the previous year; see III: 122 ff.).

(59) See the Latinised excerpt from Collins' draft of this (now Royal Society MS LXXXI, No. 21) which Newton had published in *Commercium*: 26–7. (The corresponding portion of the original English text is cited on III: 21–2.)

(60) 'For', Newton initially went on to add in sequel, 'by the title of *Analysis* I understood not merely the reduction of quantities into converging series but much more the method of solving all sorts of Problems by working in æquations wch involve converging series when they cannot be solved by finite equations. And this is the method of moments'.

(61) Quotation of the original English text of the Latinised excerpt from its draft which Newton proceeds to cite (from *Commercium*: 28–9) is made in note (22) to Appendix 2 preceding.

illam Methodum a dicto Newtoro aliquot annis antea excogitatam & modo universali applicatam fuisse: ita ut ejus ope in quavis figura curvilinea proposita quæ una vel pluribus proprietatibus definitur, Quadratura vel Area dictæ figuræ ACCURATA SI POSSIBILE SIT, *sin minus infinite vero propinqua, Evolutio vel longitudo lineæ curvæ, Centrum gravitatis figuræ, Solida ejus rotatione genita, et eorum superficies, obtineri queant.* Here the words *accurata si possibile sit* relate to the words in the *Analysis*[:] *Cujus* (methodi) *beneficio Curvarum areæ et longitudines &c* (id modo fiat) *exacte & Geometrice determinentur. Sed ista narrandi non est locus.* How this is done is described in the first six Propositions of the Book of *Quadratures*. And without the method described in those Propositions it cannot be done. And thus it appears by the testimony of Dr Barrow & Mr Collins grounded upon this & former papers communicated to Dr Barrow that I had the method of moments & fluxions & made it general some years before July 1669 or (as Mr Collins explains himself in his Letter to Mr David Gregory [of 11 August 1676][62]) about two years before the publication of Mr Mercators *Logarithmotechnia*, that is, in the year 1666. And all this...may suffice to justify my saying in the Introduction to the Book of *Quadratures* that I found this Method gradually in the years 1665 & 1666.

The method of converging series I found first by interpolation & later by division & extraction of roots in the[63] year 1665 as at the request of M^r L[eibnitz] I described in my Letter to Mr Oldenburg 24 Octob. 1676. But Dr Wallis had before that given an instance of reducing fractions into converging series by Division. The same year I got some light into the methods of moments & fluctions. And its probable that Dr Barrows Lectures might put me upon considering the generation of figures by motion, tho I not now remember it.[64]

---

(62) See the Latinised excerpt from this letter to David Gregory *père* printed in *Commercium Epistolicum*: 48. In his original letter (the surviving draft of which is now Royal Society MS LXXXI, No. 47) Collins wrote (f. 2^r): 'As to the doctrine of infinite series...The said Bookes [*viz.* Mercator's *Logarithmotechnia* and James Gregory's *Exercitationes Geometricæ*] being some few Months after they were published sent to Mr Barrow at Cambridge, in returne he gave answer that the said doctrine of infinite Series was invented by Mr Isaac Newton above 2 yeares before and generally applyed to all figures. With which Answer a Manuscript of Mr Newtons paines [*sc.* the 'De Analysi'] was transmitted to Collins and communicated to the Lord Brouncker President of the Royall Society'....

(63) 'beginning of the' is deleted here.

(64) A slightly more positive variation upon a similar remark that Barrow's lectures 'might put me upon taking these things into consideration' in Newton's preceding tentative 'History of the Method of Moments...' (Appendix 3.1; compare note (24) thereto) which we have earlier quoted in 1: 344, note (4) in discussing whether or not he in fact owed a real debt to Barrow's lectures on mathematics at Cambridge in the middle 1660's for directing his own youthful attention to the fruitful notions of fluxional increase and limit-speed. (See also 1: 10 and note (26) thereto for our doubts about the extent of personal contact between the two while Newton was still, before 1667, *in statu pupillari*.)

In the beginning of the year 1666 I found out the different refrangibility of the rays of light & the Theory of colours depending upon it as is mentioned in the *Philosophical Transactions* for [19 Feb.] A.C. 1672.[65]

---

(65) In a final cancelled passage on f. 87ʳ which we do not adjoin, Newton returned *inter alia* briefly to pick up the theme of his last (undeleted) paragraph in draft [1] above, urging that 'The direct method is sufficiently perfect. The Inverse Method is not yet perfected nor perhaps ever will be. The first step therein is equipollent to the Quadrature of Curves & this is the subject of the following book. And I write of it here no further then it is conteined in the following Tract & the methods of converging series, & leave the further improvements of it to others who have made or shall make any'. A generous note on which to finish, indeed. It is typical of Newton that he should ultimately cross it through.

APPENDIX 8. A LATIN PREFACE TO THE 'DE QUADRATURA', AND A SET OF TEXTUAL FOOTNOTES WHICH FILL ITS PLACE.[1]

[late 1719?]
From autograph drafts in private possession and in the University Library, Cambridge

[1]   Tractatui *de Quadratura Curvarum* præmittatur hæc admonitio.

AD LECTOREM.

In Introductione ad hunc Librum affirmatur Newtonum[2] incidisse paulatim Annis 1665 & 1666 in Methodum fluxionum qua hic usus est in Quadratura Curvarum. Hoc idem Wallisius noster prius affirmavit in Præfatione ad Volumen primum *Operum* suorum.[3] Et idem sic ostenditur. Literis ad Wallisium datis Aug. 27 & Sept. 17 1692, Newtonus mittebat duas hujus Libri Propositiones primam et quintam una cum Propositione extrahendi radicem ex æquatione fluxionem radicis involvente. Et Wallisius hæc edidit anno 1693 in secundo Volumine operum suorum,[4] et ibi notavit has tres Propositiones in Epistola Newtoni ad Oldenburgum 24 Octob. 1676 data, describi. In eadem Epistola Propositio prima hujus Libri dicitur esse fundamentum Methodi generalis de qua Newtonus Anno 1671 Tractatum scripserat.[5] Propositio secunda hujus Libri extat in *Analysi*

---

(1) In Newton's planned reissue of the *De Quadratura* in annexe to his coming re-edition of his *Principia*; on which again see note (1) to Appendix 4 above. Here he begins (in [1]) to cast into Latin fair copy the finished preface at which he had earlier arrived in the English drafts reproduced in Appendix 7 preceding, and then breaks off to replace this by the equivalent but terser annotations to the text of the *De Quadratura* itself (in William Jones' 1711 printing of it) which follow in [2]. While it would seem natural to assume, as we do, that he wrote these final clarifications of the historical background to his composition of the tract only a short time after he penned the parent English passages upon which he draws so heavily, it is beyond question that the preparatory Latin draft on Add. 3968.9: 111ʳ of the main portion of the last paragraph of [1] is (see note (15) below) entered after a first revision of the opening to the initial version of Newton's 'Historia Methodi Infinitesimalis ex Epistolis antiquis eruta' which we tentatively dated, like its maturer recastings, to 1721 in our general introduction to the calculus priority squabble (see page 535 above, especially note (215) thereon) and it may be that we advance the time of writing by more than a year. Since Newton never made publicly known his intention to implement his new edition of the *Principia* with his 'Book of Quadratures', let alone did so when he called on Pemberton to bring the *editio tertia* out in the mid-1720's, it is of small importance whether the true date of composition be late 1719 or early 1721.

(2) 'me' was first written. Newton's decision here to withdraw behind the cloak of editorial anonymity was evidently a last-minute one.

(3) *Opera Mathematica*, **1** (Oxford, 1695): signature a4ʳ; again see page 200 of the 1715 'Account' (Appendix 2 above) for Newton's quotation of Wallis' words themselves.

(4) *Opera*, **2** (1693): 390–6 [=vii: 170–80].

(5) In his 1676 *epistola posterior* Newton had, we need scarcely remind, adduced this key proposition 'Data æquatione quotcunque fluentes quantitates involvente invenire fluxiones'— enunciated *mutatis mutandis* as 'Prob: 1' of his 1671 tract (iii: 74 ff.) and not till afterwards

per series sub finem,[6] et pendet a Propositione prima: ideoꝗ Propositiones duæ hujus Libri Newtono innotuere Anno 1669, quo utiꝗ Barrovius hanc Analysin ad Collinium misit. Sed et Propositio quinta eodem anno Newtono innotuit. Nam in *Analysi* illa dicitur,[7] quod illius *beneficio curvarum areæ et longitudines &c* (*id modo fiat*) *exactè et Geometricè determinantur.* Et hoc fit per Propositionem illam quintam. Propositio autem tertia et quarta sunt tantum exempla Propositionis secundæ, ut in hoc libro dicitur. Ideoꝗ Methodus Fluxionum quatenus habetur in Propositionibus quinꝗ primis Libri hujus *de Quadraturis*, Newtono innotuit anno 1669.

Porro Collinius in Epistola ad Tho. Strode 26 Julij 1672 data scripsit[8] quod *Mense Septembri anni 1668 Mercator Logarithmotechniam suam edidit* et quod ex *Analysi per series* et ex *alijs chartis quæ olim a Newtono cum Barrovio communicatæ fuerant*, pateret *illam Methodum a dicto Newtono aliquot annis antea excogitatam et modo universali applicatam fuisse: ita ut ejus ope in quavis figura curvilinea proposita quæ una vel pluribus proprietatibus definitur, Quadratura vel Area dictæ figuræ, accurata si possibile sit, sin minus infinite vero propinqua—obtineri queat.* Hoc fit per Propositionem illam Quintam ut et in casibus difficilioribus per sextam & alias in serie in infinitum pergente. Ideoꝗ methodus fluxionum quatenus habetur in Propositionibus quinꝗ vel sex primis libri *de Quadraturis* Newtono[9] innotuit annis aliquot antequam prodiret Mercatoris *Logarithmotechnia*, testibus Barrovio et Collinio, id est anno 1666 aut antea ut Wallisius etiam affirmavit.[10]

D. Fatio de Duillier incidit in hanc methodum A. 1687, sed visis postea Newtoni MSS antiquis in anno 1699 in Tractatu de solido minimæ resistentiæ testimonium pro Newtono perhibuit his verbis.[11] *Newtonum tamen primum ac pluribus annis vetustissimum hujus calculi inventorem ipse rerum evidentia coactus agnosco. A quo utrum quicquam mutuatus sit Leibnitius secundus ejus inventor, malo eorum quam meum sit judicium quibus visæ fuerint Newtoni Literæ alijꝗ ejusdem manuscripti codices.*

Corollarium secundum Propositionis decimæ habetur in Epistola Newtoni ad

---

included in his *De Quadratura* as its like 'Propositio I'—only in the scrambled anagram of its component letters ordered in alphabetical sequence. See *Commercium Epistolicum*, ₁1713: 71–2 (=*Correspondence*, **2**, 1960: 114–15).

(6) See II: 242–4.

(7) See II: 242, ll. 7–9.

(8) Newton yet again quotes from the Latinised extract of Collins' letter in *Commercium*: 28–9; the corresponding words of the parent English draft (now Royal Society MS LXXXI, No. 24) have been cited in Appendix 2: note (22).

(9) Originally 'mihi'; compare note (2).

(10) In an editorial interpolation 'quæ contigit annis 1665, 1666' after the fourth word of the paragraph beginning 'Eo tempore Pestis ingruens coegit me hinc fugere...' in his *princeps* edition of the *epistola posterior* in his 1699 'Epistolarum Collectio' (*Opera Mathematica*, **3**: 634–45, especially 635 [=*Correspondence*, **2**: 113, ll. 8 ff.]).

(11) *Lineæ Brevissimi Descensus Investigatio Geometrica Duplex* (London, 1699): 18. Fuller citation of Fatio's words has been made in note (6) to our preceding introduction (page 471 above).

Collinium Nov 8, 1676 data, et a Jonesio in *Analysi per Quantitatum Series, Fluxiones ac Differentias* edita[12] in hæc verba. *Nulla extat Curva cujus Æquatio... haud tamen adeo generaliter.*

Ordinatæ Curvarum quæ in Tabularum ultima in Scholio ad Prop. X habentur, recitantur eodem ordine & ijsdem literis in Epistola Newtoni[13] ad Oldenburgum Octob. 24 1676; et ibi Tabula illa, Catalogus Theorematum vocatur pro comparatione Curvarum cum Conicis sectionibus dudum conditis, id est diu ante annum 1676, et propterea anno 1671 aut antea. Inde vero colligitur Propositionem illam decimam (& propterea etiam nonam & octavam) Newtono[13] innotuisse anno 1671.

Extractus utiqǂ fuit hic Liber circa annum 1676 ex Libro antiqu[i]ore quem Newtonus scripsit[14] anno 1671. [15]In scribendis *Philosophiæ Principijs Mathematicis* Newtonus hoc Libro plurimum est usus, ideoqǂ eundem Libro *Principiorum* subjungi jam voluit. Investigavit itaqǂ Propositiones per Analysin, investigatas demonstravit per Synthesin pro lege Veterum qui Propositiones non prius admittebant in Geometriam quam demonstratæ essent synthetice. Analysis hodierna nihil aliud est quam *Arithmetica* in speciebus. Hæc *Arithmetica* ad res *Geometricas* applicari potest, et Propositiones sic inventæ sunt *Arithmeticè* inventæ. Demonstrari debent syntheticè more veterum et tum demum pro *Geometricis* haberi.[16]

---

(12) See Appendix 7.1: note (14).

(13) These citations by Newton of himself in the third person are late replacements for a first-person 'mea' and 'mihi' respectively; again see note (2).

(14) Here, likewise, Newton first wrote 'quem [ego] scripsi'.

(15) A preliminary, minimally variant casting of the rest of the paragraph on Add. 3968.9: 111ʳ is printed by I. B. Cohen in his *Introduction to Newton's 'Principia'* (Cambridge, 1971): 347. It there follows immediately after a preparatory touching-up of the opening of the first version of Newton's contemporary 'Historia Methodi Infinitesimalis ex Epistolis antiquis eruta' (see note (215) to the preceding introduction), which after further honing and remoulding came ultimately to be the unpublished 'Præfatio' to the second (1722) edition of the *Commercium Epistolicum* which is set out as our final Appendix 10. What Newton proceeds to state of his *Principia*—that he made prior use of a generalised 'arithmetical' (that is, algebraic-*cum*-fluxional) analysis to discover a great many of its propositions, in prelude to composing their demonstrations synthetically in proper geometrical form in the finished work—is argued by him at greater length in the draft prefaces to the forthcoming re-edition of the book whose mathematically pertinent portions are reproduced in **1**,8 preceding. Let us again (see our previous criticisms in footnote thereto) urge caution upon anyone who is tempted to accept this distinctly dubious contention at its face value solely because Newton so firmly presses it.

(16) Newton here suddenly leaves off writing out this general introductory 'admonition' to the reader of his *De Quadratura* (at the top of a new page in the manuscript which he otherwise left blank) to begin to draft in its place separate annotations upon individual points in the text itself which collectively carry its equivalent message.

[2]⁽¹⁷⁾ *p. 41.*⁽¹⁸⁾   In *Analysi per series numero terminorum infinitas* quam Barrovius mense Julio anni 1669 ad Collinium misit, dixi⁽¹⁹⁾ quod methodi ibi compositæ

---

(17) Add. 3968.32: 464ʳ, augmented by two related 'NB in Prop. II' and 'NB in Prop. V' penned by Newton on the verso of the second folio of the manuscript sheet (in private possession) whose first leaf carries the now superseded prefatory 'Ad Lectorem' to his intended reissue of the *De Quadratura* which is printed in [1]. In this most finished set of historical notes upon his 'Book of Quadratures'—these are keyed to the pages of the lightly amended *editio secunda* of it included by William Jones in his 1711 compendium of Newtonian *Analysis per Quantitatum Series, Fluxiones, ac Differentias*: 41–66 (on which see page 31 above)—he draws narrowly, in ways which we need not precisely specify, upon his fulsome previous arguments where in a plethora of repetition he had sought likewise to substantiate his claim that the main propositions in the published *De Quadratura* were already known to him by late 1676 at the latest, and that he had discovered the essential content of several of these half a dozen years earlier still when he wrote his 'De Analysi'. A number of other equivalent draftings of such *annotationes* survive, for instance on .9: 112ᵛ/112ʳ—two paragraphs from this preparatory worksheet are printed by Cohen in his 1971 *Introduction to Newton's 'Principia'*: 347—and on .10: 137ᵛ/136ʳ. On .41: 88ʳ(+89ᵛ)→.14: 234ʳ there is a more elaborate and continuously written 'NB', in whose initial statement Newton declared *inter alia*: 'Hunc Librum in M.S. nostrates Halleius et Ralphsonus in manibus suis habuerunt'—this was at once changed to 'triverunt' (.9: 88ʳ) and then 'tractarunt' (.9: 88ᵛ→.14: 234ʳ)—anno 1691, uti Ralphsonus publice testatus reliquit & Halleius adhuc testatur. Propositionem primam exemplis [in fluxionibus primis & secundis inveniendis] illustratam D. Wallisius anno 1692 [a me accepit &] in secundo *Operum* suorum Volumine...sub initio anni sequentis in lucem emisit. Et hæc fuit Regula omnium [hujus generis] prima...quæ lucem vidit, estcʒ Regula verissima brevissima et optima. Eandem Propositionem ijsdem verbis, tanquam fundamentum methodi fluxionum, e Libro quem anno 1671 conscripseram desumptum, posui in Epistola mea 24 Octob. 1676 ad Oldenburgum data, et a Wallisio [in tertio *Operum* suorum volumine] edita. In eadem Epistola posui Propositionem quintam hujus Libri pro Quadratura Curvilinearum quarum Ordinatæ sunt dignitates binomiorum, eandemcʒ exemplis aliquot illustravi, dixicʒ Regulas hujusmodi ad Trinomia et alia magis composita se extendere, et has Regulas Quadraturam accuratam dare quoties fieri potuit. Sed et *in Analysi per series numero terminorum infinitas* mense Julio anni 1669 a D. Barrovio ad Collinium missa, descripsi methodum momentorum & quomodo Problemata per eandem ad series convergentes deduci possint, et quod hujus *beneficio Curvarum areæ et longitudines &c (id modo fiat) exacte et Geometrice determinantur.* Ideocʒ Propositio quinta Libri hujus de Quadraturis tunc mihi innotuit. Et propterea etiam methodum fluxionum et momentorum quatenus in Propositionibus quincʒ primis habetur, tunc intellexeram. Nam Propositio quinta a quatuor primis dependet'. (We interpolate in square brackets a few clarifying phrases from the equivalent recasting on .14: 234ʳ.) 'Est autem', Newton went on in his revised version, 'Theorema primum pro Trinomijs idem cum Propositione sexta [hujus] libri....Sed et Propositio septima et octava ejusdem sunt generis. Nona autem et decima requiruntur ad solutionem Problematis quod posui in Epistola ad D. Collinium Novem. 8 1676 data & a D. Jonesio edita verbis sic Latine redditis *Nulla extat Figura curvilinea cujus Æquatio - - - haud tamen adeo generaliter*'; whence, he shakily concluded, 'Anno igitur 1676 Methodum fluxionum intellexeram quatenus in Libro hocce de Quadraturis exponitur' (.14: 234ʳ).

(18) Understand the bottom lines where, at the end of the second paragraph of the 'Introductio ad Quadraturam Curvarum', Newton delivers his unsupported affirmation '...& has motuum vel incrementorum velocitates nominando *Fluxiones* & quantitates genitas nominando *Fluentes*, incidi paulatim Annis 1665 & 1666 in Methodum Fluxionum qua hic usus sum in Quadratura Curvarum'.

beneficio, Curvarum areæ & longitudines &c (id modo fiat) exacte & Geometrice determi-nantur: sed ista narrandi non esse locum. Et Collinius in Epistola ad Thomam Strode, 26 Julij anno 1672 data, scripsit[20] quod ex hac *Analysi* et *alijs quæ olim a me cum Barrovio communicata fuerant,* pateret *illam methodum* a me *aliquot annis antea* (i.e. ante mensem Julium anni 1669) *excogitatam & methodo universali appli-catam fuisse: ita ut ejus ope in quavis figura Curvilinea proposita quæ una vel pluribus proprietatibus definitur, Quadratura vel Area dictæ figuræ, accurata si possibile sit, sin minus, infinite vero propinqua—obtineri queat.* Et in Epistola mea ad Oldenburgum Octob. 24. 1676 data posui fundamentum harum operationum Propositione sequente: *Data Æquatione quotcunſp fluentes quantitates involvente fluxiones invenire; & vice versa.* deinde addidi Theorema primum inde derivatum quo Curvæ geometrice quadrantur ubi fieri potest. Et hoc idem fit per Proposit[ionem] quintam hujus Libri, quæ Propositio pendet a Propositionibus quatuor primis: ideoſp Methodus fluxionum quatenus exponitur in Propositionibus quinſp primis hujus libri, mihi innotuit annis aliquot ante Septembrem anni 1668 quo Mercatoris *Logarithmotechnia* prodijt.

Pag. 45.[21] In *Analysi per series numero terminorum infinitas,* pro fluxionibus posui literas quascunſp ut *z* vel *y*; pro momentis literas easdem multiplicatas per literam *o*; & pro fluentibus vel literas alias quascunſp ut *v* et *x*, vel fluxiones in quadrato inclusas ut $\boxed{z}$ et $\boxed{y}$. Et sub finem Tractatus illius specimen dabam calculi demon trando Propositionem primam illius Tractatus.

Pag. 46. [*Prop. I.*] Propositionis hujus solutionem, cum exemplis in fluxionibus primis et secundis, Wallisius noster in lucem edidit anno 1693. Et hæc fuit Regula omnium prima quæ lucem vidit pro fluxionibus secundis, tertijs & alijs omnibus inveniendis.

Pag. 48. [*Prop. II.*] Extat hæc Propositio et ejus solutio in *Analysi per Æqua-tiones numero terminorum infinitas* prope finem.[22] Et ejus solutio eadem est cum solutione Propositionis primæ.

[23][*Pag. 49.*] *Prop. V.* Hæc Propositio a Wallisio edita [est anno 1693 in secundo Volumine *Operum* suorum pag. 392.]

Pag. 52. Prop. VI. Hanc Propositionem anno 1671 mihi innotuisse patet ex Epistola mea prædicta ad D. Oldenburgum 24 Octob. [1676] data ubi dicitur *Pro trinomijs etiam et alijs quibusdam Regulas quasdam concinnavi.*[24]

---

(19) Again see II: 242 (compare note (7) of [1]).

(20) See [1]: note (8).

(21) The opening to the main text of the *De Quadratura* in which (compare pages 128–30 preceding) Newton sets out his notation for fluxions and fluents.

(22) Again see II: 242–4, where Newton adjoins his 'Demonstratio' of this the *regula prima* of his 1669 'De Analysi'.

(23) This and the next (cancelled) annotation are here interpolated from the verso of the sheet bearing the text of [1]; compare note (17) above.

(24) *Commercium Epistolicum:* 74 (=*Correspondence,* **2**: 117). In his *epistola posterior,* however,

Pag. 61. Prop. X. Corol. II. Ad hoc Corollarium spectabat Epistola mea Novem. 8. 1676 ad Collinium scripta his verbis.[25] *Nulla extat Curva cujus Æquatio —haud tamen adeo generaliter.*

Pag. 62. [*Prop. X. Scholium.*] Hanc Tabulam [posteriorem Curvarum simpliciorum quæ cum Ellipsi & Hyperbola comparari possunt] diu ante annum 1676 compositam fuisse[26] patet per ordinatas Curvarum in Epistola mea prædicta Octob. 24 1676 ad Oldenburgum data positas.

Newton does not claim—as he could not in truth—that these rules for squaring trinomials (on which see III: 373–85) were set out five years previously in his 1671 treatise, and he rightly deletes his present false inference from this mistaken premiss to replace it by what comes next.

(25) See Appendix 7.1: note (14), and compare note (15) there following.

(26) 'proindeqȝ anno 1671 mihi innotuisse' is cancelled in sequel. This latter table of curves whose quadrature is reducible to the squaring of related conic-areas is, some superficial reordering apart, indeed identical with the corresponding one (see III: 242–54) which Newton had earlier adjoined to his—in 1719 still to be printed—1671 treatise, but in his *epistola posterior* he merely made mention of 'Theoremata quædam, quæ pro Comparatione Curvarum cum Conicis Sectionibus in Catalogum dudum retuli', passing to instance seven of the orders of curve listed therein. (See *Commercium*: 76 [=*Correspondence*, **2**: 119], in an anonymous editorial footnote in which Newton emphasised to his reader that 'Ex his patet Propositiones *Newtoni* de Quadratura Curvarum diu ante annum 1676 inventas fuisse'.)

APPENDIX 9. OBSERVATIONS UPON BERNOULLI'S 1713 LETTER TO LEIBNIZ.[1]

[early 1720?]
From the original[2] in the University Library, Cambridge

(1) As is so often the case, nothing is known for sure of the immediate circumstances which stimulated Newton to compile these elaborate annotations upon the central section of Leibniz' 1713 *Charta volans*, where the latter, to support his own anonymous counter there to the *Commercium*'s recent public vindication of Newton's priority in calculus invention, recurred (without naming its giver) to the 'judicium primarii Mathematici, & harum rerum peritissimi' which Johann Bernoulli had passed privately to him a month or so before in his letter of 7 June 1713 (N.S.). We have (see page 530 above) suggested—without having any firm supporting evidence, let us add—that these 'observations' were put together shortly after he received, by way of Varignon and de Moivre, Bernoulli's letter of 21 December 1719 (N.S.) with its less than honest evasion that 'Non memini ad [Leibnitium] eo die [*viz.* 7 Junij 1713] me scripsisse, non tamen negaverim, quandoquidem non omnium epistolarum a me scriptarum apographa retinui' (see *Correspondence*, 7, 1978: 75). From the editorial 'we' of Newton's title and his reference to himself throughout the ensuing text always in the third person, together with his citations by page of Wallis' preface to the first volume of his collected *Opera* (see note (4) below) and, at the very end, of Leibniz' letter of 9 April 1716 (N.S.) to Conti 'printed above' (see our final note), it would seem that he intended to submit the piece—with its English rendered (by de Moivre?) into French, we assume—to Pierre Desmaizeaux for insertion after the 'Lettres de M. Leibniz et de M. le Chevalier Newton, Sur l'invention des Fluxions & du Calcul Differentiel' with which the latter opened the second volume (pages 3–124 [+125–6]) of the *Recueil de Diverses Pieces, Sur la Philosophie, ...les Mathematiques, &c. Par Mrs. Leibniz, Clarke, Newton, & autres Autheurs célèbres* then about to be published by Du Sauzet in Amsterdam. Such, however, was certainly not to be, and the *Recueil* (whose passage through the press the preceding autumn Newton had already tried to delay for a similar purpose; see the draft of his letter then to Desmaizeaux which is printed in *Correspondence*, 7: 73 and note (2) thereto) appeared in the summer of 1720 without any annexe of this sort. Undeterred, Newton went on to cast his present 'Observations' upon Bernoulli's 1713 letter into Latin, and in much the same form (but with many divergences in detail) they were made public by him in 1722 in the second edition of the *Commercium Epistolicum*, where they are appended as an equivalent 'Annotatio' (pages 247–50) upon the 'Judicium Mathematici' (*ibid.*: 245–6); compare note (221) of the preceding introduction. His remarks there repeat item by item, we do not have to say, what is stated over and over again in the pieces here set out in previous Appendixes, but we need not apologise for reproducing in its totality the English text in which Newton made his most considered and rounded answer to Bernoulli's charges against him in 1713, under a cloak of anonymity which he here respects in the letter, of lack of mathematical originality and understanding and even bare technical competence.

(2) Add. 3968.34: 480^r–481^r. Sketchier preliminary draftings are found on *ibid.*: 485^r →483^r–484^r; the former of these may have been intended as a direct letter to Desmaizeaux about this intended addition to the latter's *Recueil* (see the previous note), for its opening sentence—in which Newton at once deleted the 'I have'—reads: 'Since the printing of the foregoing Letters I have the following Observations upon a flying paper dated [29 July 1713]'. Subsequent recastings of this, in an intermixture of English and Latin, prior to subsuming it to be the 'Annotatio' upon Bernoulli's 1713 'Judicium' appended two years later to the *Commercium*'s re-edition exist widely both in ULC. Add. 3968 (for instance at .35: 490^r/ (.41: 68^r→20^r/20^v→) 490^v–489^v) and in private possession. (We have listed the more finished

The following Letter to Mr Leibnitz was written originally in Latin,
& we have met with the ensuing Observations upon it.

Videtur N...us occasionem nactus serierum opus multum promovisse per Extractiones Radicum, quas primus in usum adhibuit, et quidem in ijs excolendis ut verisimile est ab initio omne suum studium posuit, nec credo tunc temporis [1]vel somniavit adhuc de calculo suo fluxionum et fluentium, vel de reductione ejus ad generales operationes Analyticas ad instar Algorithmi vel Regularum Arithmeticarum aut Algebraicarum. Ejuscp meæ conjecturæ (*primum*)[3] validissimum *indicium* est, [2]quod de literis x vel y punctatis uno, duobus, tribus &c punctis superpositis, quas pro dx, ddx, d^3x, dy, ddy, &c nunc adhibet, in omnibus istis Epistolis (*Commercij Epistolici Collinsiani*, unde argumenta ducere volunt) nec volam, nec vestigium invenias. Imo ne quidem in *Principijs Naturæ Mathematicis* N....i, [3]ubi calculo suo fluxionum utendi tam frequentem habuit occasionem, ejus vel verbulo fit mentio, aut notam hujusmodi unicam cernere licet, sed omnia fere per lineas figurarum sine certa Analysi ibi peraguntur more non ipsi tantum, sed et Hugenio, imo jam antea (in nonnullis) dudum Torricellio, Robervallio, Cavallerio, alijs, usitato Prima[4] vice hæ literæ punctatæ comparuerunt in tertio Volumine *Operum* Wallisij, multis annis postquam Calculus differentialis jam ubicp locorum invaluisset. *Alterum indicium*, quo conjicere licet Calculum fluxionum non fuisse natum ante Calculum differentialem, hoc est, [5]quod veram rationem fluxiones fluxionum capiendi, hoc est differentiandi differentialia, N....us nondum cognitam habuerit, quod patet ex ipsis *Principijs Phil. Math.* ubi [6]non tantum incrementum constans ipsius x, quod nunc notaret per x punctatum uno puncto, designat per o (more vulgari, qui calculi differentialis commoda destruit) sed etiam regulam circa gradus ulteriores falsam dedit (quemadmodum [7]ab eminente quodam Mathematico notatum est)......... Saltem apparet N....o rectam methodum differentiandi differentialia non innotuisse longo tempore, postquam alijs fuisset familiaris &c.

---

drafts of this 'Annotatio' in note (221) to the preceding introduction.) Though Newton possessed at least two copies of the uncut twin printing of the original 1713 *Charta volans*—one is now found loose in Add. 3968.34—let us add that on .34: 478r–479r he has (as an *aide-mémoire* perhaps) penned out its text word for word without comment or interjection of any kind. The reader without access to Leibniz' broadsheet itself will find its text conveniently reproduced *in toto* in *Correspondence*, **6**, 1976: 15–17.

(3) This and following like-bracketed phrases are Leibniz' editorial insertions in the excerpt from Bernoulli's original letter of 7 June 1713 (N.S.) to him which he published seven weeks afterwards in his *Charta* as the 'Judicium primarii Mathematici'.

Observations upon the foregoing Letter.

Mr Leibniz appealed from the *Commercium Epistolicum* to the judgment of the anonymous Author of this Letter; but [**1**] Dr Wallis an older & abler Mathematician has attested in the Preface to the first Volume of his works[4] that Mr Newton explained this Method to Mr L....z in his Letters of 13 June & 24 Octob. 1676 & invented it ten years before or above. [**2,3**] In the second Lemma of the second Book of *Mathematical Principles* the Elements of this Method are taught & demonstrated, & in the Introduction to the Book of *Quadratures* the method it self is expresly taught & illustrated by examples, & in Mr Newtons Letter of 10 Decem 1672 the Method is plainly described & illustrated with an example of Drawing Tangents thereby: & all this is done without the use of pricked letters. The Book of *Principles* was writ by Composition & therefore there was no occasion of using the calculus of fluxions in it. [**4**] Prickt letters appeared in the second Volume of the works of Dr Wallis,[5] & this Volume was almost all printed in the year 1692 & came abroad the next spring before the Differential method began to make a noise. The Manuscript of the Book of *Quadratures* was in the hands of Dr Halley & Mr Ralphson in the year 1691 as both of them hath attested.[6] And this Book is sufficiently described in Mr Newtons Letters of Octob. 24 & Novem 8 1676, & the Quadratures there cited out of it are not to be attained without the Method in dispute.

[**5**] The Only Rule wch Mr Newton has given for finding first second third fourth & other fluxions is conteined in the first Proposition of the Book of *Quadratures* & is a very true one & was published with examples in the second Volume of the works of Dr Wallis before any other Rule for finding second third

fourth differences came abroad. And the very words of the Proposition were set down in the Scholium upon the second Lemma of the second Book of *Principles*, & in Mr Newtons Letter of 24 Octob. 1676, as the foundation of the method of fluxions. And the inverse of this Rule is the first Rule in the *Analysis*

(4) 'cited above pag. 105, 106' Newton added in a first cancelled 'Obs. 1' on f. 480r. In fact, the pertinent extract from Wallis' preface to the first volume of his *Opera* (on which see Appendix 2: note (106) above) is set out—as one item in Desmaizeaux' reprinting of Newton's postscript to his 1717 'Observations' upon Leibniz' letter to Conti of 9 April 1716 (N.S.) in appendix to Raphson's *History of Fluxions*: 119–23 (on which see note (145) of the preceding introduction)—on pages 104–5 of the second volume of the *Recueil de Diverses Pieces* in which (see note (1)) Newton planned to adjoin his present annotations upon Bernoulli's 1713 'Judicium'.

(5) Yet once more see VII: 171–6, where the pertinent pages of Wallis' *Opera Mathematica*, **2** (1693): 390–6 are reproduced. On the mistaken following date of publication as in 'spring'— which should be 'late summer' (or perhaps 'autumn')—see note (25) to Appendix 7.1 above.

(6) Again let us adjoin (compare notes (24), (42) and (55) to Appendix 7) that Newton depends solely on the witness of the (now half-dozen years dead) Raphson on page 2 of his *History of Fluxions* (see III: 11), expecting Halley to corroborate this testimony should the need arise.

per series numero terminorum infinitas communicated by Dr Barrow to Mr Collins in July 1669, & the Rule is demonstrated by the method of fluxions in the end of that *Analysis*. And without this Rule the Series for Quadratures wch break off & become finite when the Curve can be squared by a finite equation, & wch are mentioned in the said *Analysis*, are not to be attained.

[**6**] The constant fluxion of *x* Mr Newton denotes by $\dot{x}$ wth a point above it; but the constant increment or moment of *x* Mr Newton denotes not by $\dot{x}$ but by $o_{[,]}$ & still uses this notation as convenient: Mr Leibnitz hath no Notation for fluxions, & therein his method is defective.

[**7**] In translating this Letter of 7 June 1713 into French & printing it in the *Novelles Literairs* Decem. 28 1715, we are told[7] that the Author of this Letter was Mr John Bernoulli, & to make this credible, the citation of the Eminent Mathematician is omitted in the body of the Letter. For if Mr J. Bernoulli be the Eminent Mathematician there cited by the author of the Letter he cannot be the Author himself. And Mr J. Bernoulli in a Letter to Mr Newton[8] hath positively declared that he was not the author thereof. And if he had been the author thereof he would not have said that the Eminent Mathematician charged Mr Newton wth a false Rule.

That Mathematician noted only that there was an error in the Solution of Prob. III Lib. II *Princip. Philos.* & suspected that it lay in second differences. Mr Nicholas Bernoulli told Mr Newton in his Unkles name in autumn 1712 that there was an error in the conclusion of the solution.... Mr Newton corrected the error himself, shewed him the correction & told him that the Proposition should be reprinted in the new Edition wch was then coming abroad.[9] The Tangents of the Arcs *GH* & *HI* are first moments of the Arcs *FG* & *FH*[10]

---

(7) The writer of the anonymous letter to the publisher, Du Sauzet, of the *Nouvelles Litteraires* to which Newton here refers—and where Bernoulli was indeed squarely named as the author of the 'Lettre...de Bâle, du 7. de Juin 1713' from which the 'judicium primarii Mathematici' printed a month later in the *Charta volans* was taken—was in fact Leibniz. (See *Correspondence*, **6**: 257 and note (1) thereto; and compare note (129) to our own preceding introduction.)

(8) That of 5 July 1719 (N.S.). Bernoulli's words were, however, something less than a positive declaration, merely affirming to Newton that 'Fallunt haud dubie, qui me Tibi detulerunt tanquam Auctorem quarundam ex schedis istis volantibus, in quibus forsan non satis honorifica Tui fit mentio' (see *Correspondence*, **7**: 44).

(9) The essential portion of the surviving sheets in which Newton came to detect the mistake in his working-out of this problem on motion under resistance has been reproduced in **1**, 6 (pages 312–68 above). In almost the same (Latin) phrasing he reported at this time to Varignon in his letter of 29 September 1719 that 'mense Octobri anni 1712...D. Nicolaus Bernoulli me monuit quod error aliquis admissus fuisset in resolutione Prop. X Lib. II Edit. 1... Resolutionem subinde examinavi et correxi & correctam ei ostendi, & imprimi curavi...' (*Correspondence*, **7**: 63); we have made fuller quotation of this on page 655.

(10) Compare Appendix 6: note (24) preceding. Newton again understands the redrawn figure of the *Principia*'s second edition (see page 356 above).

& should have been drawn the same way with the motion describing those arcs, whereas through inadvertency one of them had been drawn the contrary way, & this occasioned the error in the conclusion.[11]

There is an error of much greater consequence committed in second Differences by Mr L....z A.C. 1689 in his *Tentamen de motuum cœlestium causis* sect. 15: which tho often complained of, is not yet corrected nor so much as acknowledged.[12] Nor doth it appear by any instance that Mr L....z knew how to work in second differences before Mr Newtons Book of *Principles* came abroad: whereas it plainly appears by Prop. XIV Lib II *Princip.* that w^n he wrote that Book knew how to work in second differences wch he there calls *Differentia Momentorum, id est Momentum differentiæ.*[13] And Mr Leibnitz himself in the *Acta Eruditorum* for May 1700 has acknowledged that M^r N[ewton] was the first who shewed by a specimen made publick,[14] that he had the method of maxima & minima in infinitesimals, & there calls it a method of the highest moment & greatest extent.[15] And in the *Analysis per series numero terminorū infinitas* communicated by Dr Barrow to Mr Collins in the year 1669 Mr Newton mentions that by that method of Analysis *Curvarum areæ & longitudines &c (id modo fiat) exacte et Geometrice determinantur.*[16] And how this [is] done by the Method of Fl[uxions] is explained in the first six Propositions of the Book of *Quadratures.* And without that Method it cannot be done.

And Mr Collins in a Letter to Mr Tho. Strode dated 26 July 1672 & printed from the Original in the *Commercium Epistolicum*; mentioning the Papers wch he had received from Dr Barrow in July 1669, subjoyns: *Ex quibus (chartis) et alijs quæ olim ab auctore cum Barrovio communicata...obtineri queant.*[17] Thus by the

---

(11) In a fundamental sense this is untrue. See **1**, 6, Appendix 3: note (22); and compare note (25) to our present Appendix (page 477 above).

(12) 'And', Newton went on in a first casting of this sentence at the bottom of f. 480^v, 'his friends will not see clearly to pull motes out of other mens eyes till they have pulled the beam out of their own'—a Biblical flourish rightly discarded as out of place in the present terse annotation. For our detailed comment upon the rights and wrongs of the criticism of this key proposition in Leibniz' essay on the dynamics of celestial motion see note (86) to the preceding introduction.

(13) Yet again see **1**, 8.4: note (44) for our exegesis of this *idée fixe* of Newton in his old age, thoroughly mistaking the nature of the 'differentia momentorum' which he had, more than thirty years earlier, used in developing the argument of Proposition XIV of his *Principia*'s second book.

(14) Namely, the determining of the solid of revolution of least resistance to motion along its axis—that whose undemonstrated defining differential condition Newton had set out in the scholium to Proposition XXXV of the second book of his 1687 *Principia* (see vi: 462–4).

(15) 'Methodi de maximis & minimis pars sublimior' in Leibniz' Latin words (*Acta* (May 1700): 206)—but see our codicil in Appendix 5: note (49) above.

(16) Once more see ii: 242.

(17) Newton yet again understands the full Latinised excerpt from Collins' letter to Strode as printed in his *Commercium*: 28–9 (the original English text of which is cited in Appendix 2: note (22) above).

testimony of Dr Barrow & Mr Collins, as well as by that of Dr Wallis, Mr Newton had the method described in the first six Propositions of the Book of *Quadratures*, some years before July 1669. And these three ancient & able Mathematicians knew what they wrote: but the author of the Letter of [7] June 1713 above mentioned, wrote only by conjecture. His words are *Ejusꝗ conjecturæ meæ primum validissimum indicium est* &c. He accused Mr Newton of plagiary without any better proof then conjecture & therefore is guilty of calumny, even by the concession of Mr Leibnitz in the end of his Letter of 9 April 1716 to Mr Conti printed above pag [66].[18]

(18) We fill a blank in the manuscript with the pertinent page of the second tome of Desmaizeaux' *Recueil*. (See note (1) above for our reasons why this should be so.) In the next to last paragraph of his main letter Leibniz had indeed allowed: 'je conviens avec luy [Newton], que la malice de celuy qui intente une telle accusation sans la prouver, le rend coupable de calomnie' (compare *Correspondence*, **6**: 311).

APPENDIX 10. AN INTENDED PREFACE
TO THE 1722 'COMMERCIUM'.[1]

[late 1721]
From the original in the University Library, Cambridge[2]

PRÆFATIO.

[3]Consessus Arbitrorum a Regia Societate constitutus *Commercij* subsequentis

---

(1) The purpose of this new edition by Newton of the *Commercium Epistolicum D. Johannis Collins et Aliorum de Analysi promota* which he had (cloaking in anonymity his own, begetter's rôle) compiled 'at the command of the Royal Society' to lay bare and document the primacy of his contribution to the 'invention' of the calculus of the infinitely small was, so we have urged in our preceding introduction (see page 534 above), once for all to still the 'squabbling humour' maintained to the contrary, five years after Leibniz' death, by his lieutenant Johann Bernoulli and 'so leave it to posterity to judge of this matter by the ancient Records' now here for the first time laid before the general public. (The original 1713 edition of the *Commercium* had been printed solely for private distribution to interested parties.) The elaborate preface, here given in its final form, by which Newton initially planned to introduce this reissue—now augmented by the Latinised texts of his equally anonymous 1715 'Account' of its first edition (see Appendix 2 preceding) and of his unpublished annotations upon Bernoulli's 'Judicium' in Leibniz' 1713 *Charta volans* (see Appendix 9) to his reader, summarising the complicated chain of events of the past decade, is the last of a long sequence of drafts of what he originally titled a 'Historia Methodi Infinitesimalis ex Epistolis antiquis eruta' which was in the first instance perhaps meant to appear as a separate pamphlet. We see no point in tracing minutely through so many reworkings the steps by which the first roughings-out of this 'Historia' passed, word by word, sentence by sentence, paragraph by paragraph, to be the finished text which we reproduce in this the last piece in our edition of Newton's mathematical papers. Let us adjoin that the 'Ad Lectorem' finally set in preamble (signatures A2r–A4v) to the 1722 *Commercium* takes over only the seventh, eighth, ninth and (most of the) tenth paragraphs of this discarded 'Præfatio', severely compressing its six opening ones into a few lines (see note (3) following), and lopping off its most novel feature—Newton's anachronistic remodelling of the rule for algebraic tangents which he had communicated to Collins half a century before in December 1672—in favour of the generalised *dicta* which we quote in our final note below. May our modern reader think the resurrection of it worthwhile.

(2) Add. 3968.13: 209r–216r, partially amended in line with Newton's corrections on .41: 56r–57r, and with a last paragraph rounding the text off borrowed from the end (.13: 208r) of the immediately prior draft of this 'Præfatio' (*ibid.*: 199r–208r)—there (compare the previous note) initially under the head of 'Historia Methodi Fluxionum et Methodi Differentialis ex Epistolis antiquis eruta'. (The main preparatory draftings of this 'Historia', and Newton's variant castings of its title, are listed in notes (215) and (216) of our preceding introduction. We will not here discuss their differences.)

(3) In the printed 'Ad Lectorem' the first six paragraphs which ensue were drastically encapsulated to read (*Commercium Epistolicum*, ₂1722: signature A2r; we do not maintain the printer's italicisings): 'Cum primum *Commercium Epistolicum* lucem vidit, D. Leibnitius Viennæ agens, ut librum sine responso dimitteret, prætendit per biennium se eundem non vidisse, sed ad judicium primarii Mathematici & harum rerum peritissimi & a partium studio alieni se provocasse. Et sententiam ejus 7 Jun. 1713 datam, schedulæ volanti 2[9] Julii datæ inclusam, per orbem sparsit, sine nomine vel Judicis vel Impressoris vel Urbis in qua impressa fuit. Et sub finem anni 1715 in Literis quæ ad Abbatem de Comitibus tunc Londini agentem scripsit,

Epistolici exemplaria tantum pauca Anno 1712 imprimi curavit, et ad Mathematicos mitti qui soli de his rebus judicare possent. Cum vero D. Leibnitius huic Libro minime responderet, sed Quæstionem de primo Inventore desereret & ad Quæstiones Metaphysicas aliasqʒ ad hanc rem nihil spectantes & sine fine tractandas, id est ad rixas confugeret, et ejus amici quidam[4] adhuc rixentur: visum est hunc Librum una cum ejus Recensione quæ in *Transactionibus Philosophicis*[5] ac *Diario Literario* Anno 1715[6] (anno et septem vel octo mensibus ante obitum D. Leibnitij) impressa fuit, in lucem iterum mittere, ut Historia vera ex antiquis monumentis deducta ad posteros absqʒ rixis[7] perveniat, et sic finis imponatur huic controversiæ. Nam depulso plagij crimine res non digna est de qua ulterius disputetur.

Epistolas ad Oldenburgium Leibnitius scripsit 3 Feb. 20 Feb. 30 Mar. 26 Apr. 24 Maij & 8 Jun. 1673; 15 Jul. & 26 Octob. 1674; 30 Mar. 20 Maij, 12 Jul. & 28 Decem. 1675; 12 Maij, 27 Aug. & 18 Novem. 1676; 21 Jun. & 12 Jul. 1677. Et harum omnium Autographa adhuc asservantur[8] si duas tantum excipias 27 Aug. & 18 Novem. 1676 datas et in tertio *Operum* Wallisij Volumine impressas: ubi etiam eæ 15 Jul. & 26 Octob. 1674; 12 Jul. & 28 Decem. 1675; 21 Jun. & 12 Jul. 1677 leguntur. Harum etiam septendecim Epistolarum Apographa (si tertiam et ultimas quinqʒ excipias) extant in Libris antiquis Epistolicis Regiæ Societatis Num. 6, pag. 35, 34, *, 101, 115, 137, & No. 7, pag. 93, 110, 213, 235, 149, 189. Et hæ omnes Leibnitij Epistolæ una cum Epistolis mutuis Oldenburgij perpetuum inter eos constituunt per Epistolas commercium a die 3 Feb. 1673 ad usqʒ mortem Oldenburgij, præter Epistolam qua Leibnitius

---

confugit ad Quæstiones novas de Qualitatibus occultis, gravitate universali, Miraculis, Organis & Sensorio Dei, spatio, Tempore, Vacuo, Atomis, Perfectione mundi, & Intelligentia supramundana; & Problema ex *Actis Eruditorum* desumptum proposuit ab Analystis Anglis solvendum. Quæ omnia ad rem nil spectant. Sed & Consessum a Regia Societate constitutum, qui *Commercium* ex antiquis monumentis ediderant, accusavit quasi partibus studuissent, & Literas antiquas edendo omisissent omnia quæ vel pro ipso vel contra Newtonum facerent.' Thereafter he passed to tie this new opening into his seventh paragraph below by writing: 'Et ut hoc probaret, scripsit is in prima sua ad Abbatem epistola, quod *in secundo* suo *in Angliam itinere* Collinius ostenderit ipsi *partem Commercii sui in qua Newtonus agnoscebat ignorantiam suam in pluribus*...' (page 687, ll. 10 ensuing).

(4) Meaning Johann Bernoulli above all, of course.

(5) The greater part of this unsigned English 'Account' by Newton of his 1713 *Commercium* in *Philosophical Transactions*, **29**, No. 342 (for January/February 1714/15): 173–224 has been reproduced in Appendix 2 above.

(6) *Journal Literaire*, **7**.1, §vɪ/.2, §xv: 114–58/344–65; see page 503: note (108) above.

(7) 'ac disputationibus ad rem nil spectantibus' is deleted in sequel.

(8) In the archives of the Royal Society, collected especially in the manuscript volume (now LXXXI) which still, after some losses over the centuries, contains the greatest part of the 'antique' letters and papers whose text is printed in the *Commercium*. Newton passes to list the pages of Royal Society Letter Books 6 and 7 where copies of items in this correspondence of Leibniz with Oldenburg may still be found.

postulabat excerpta ex epistolis Gregorij ad se mitti & Epistolam qua Oldenburgius excerpta illa misit. Epistolæ Leibnitij versabantur circa numeros ad
usq̃ 8 Junij 1673: dein Leibnitio Geometriam addiscente, Commercium
aliquamdiu intermissum est, et 15 Julij 1674 renovatum est a Leibnitio sic
scribente: *Diu est quod nullas a me habuisti literas.* Et ab hoc tempore Commercium
quod Leibnitius cum Oldenburgio Collinio et Newtono habuit, circa series et
altiorem Geometriam versabatur, et hic integrum (quoad hanc disputationem)
imprimitur, præter dictas duas epistolas quæ interciderunt. Nam Collinius &
Newtonus nullum cum Leibnitio Commercium habuerunt præterquam per
Oldenburgium.[9] De fide epistolarum minime dubitatur, certe non apud Anglos.

Ubi primum *Commercium Epistolicum* lucem vidit, D. Leibnitius Viennæ agens,
ne libro responderet, causabatur per biennium se librum non vidisse, sed ad
judicium primarij et a partium studio alieni provocasse, cum ipse per occupationes diversas rem tunc non satis discutere posset. Et Judicium *mirabile*
nomine hujus Mathematici 7 Junij 1713 datum in schedula contumelijs referta
die 29 Julij data describi, & utrumq̃ per Europam spargi curavit, sine nomine
vel Mathematici vel Impressoris vel Urbis in qua impressa fuit; adjuvante ni
fallor Menkenio. Author schedulæ utitur voce *illaudibili* quæ Leibnitio fere
propria fuit, et narrat quæ inter Hugenium & Leibnitium Parisijs ante annos
37 vel 38 privatim gesta fuerant; et quæ de serie Gregorij habet, de Commer[c]io
Epistolico desumpta sunt. Annon Leibnitius hæc scripserit, & Commercium
viderit?[10] Sententiam in schedula descriptam voco *mirabilem* quia Judex
methodum collocat in characteristica, in methodo synthetica desiderat symbola
analytica, contra Newtonum disputat ex usu literæ *o*, literas punctis notatas
lucem primo vidisse ait in tertio volumine *Operum* Wallisij, id est anno 1699, et
Newtonum Regulam falsam dedisse pro differentijs secundis, et rectam
methodum differentiandi differentialia non cognovisse nisi longo tempore
postquam alijs fuisset familiaris: cum tamen characteristica mutari possit non
mutata methodo, in methodo synthetica nulla sit occasio utendi symbolis
analyticis, Newtonus adhuc utatur litera *o* eodem sensu quo prius, & literæ
punctis superimpositis notatæ methodusq̃ Newtoni differentiandi diff[er]en-
tialia lucem viderint in secundo *Operum* Wallisij volumine, pag. 392,[11] id

---

(9) Understand before Oldenburg's death in September 1677. A decade and a half later,
in the late winter of 1692/3, Leibniz wrote a single letter to Newton without using any
intermediary (see *Correspondence*, **3**, 1961: 257–8), and received an equally direct if belated
response the following autumn (*ibid.*: 285–6).

(10) Leibniz had not in fact himself seen a copy of the 1713 *Commercium* before, on hearing
of its appearance from Johann Bernoulli in June, he hastened to set in print the latter's private
assessment of Newton's mathematical (un)originality and (in)competence, sandwiching this
'Judicium' in his July flysheet between equally damning but more generalised comments of his
own (and of course not naming the author of either); see page 492 above.

(11) See vii: 174.

est anno 1693, annis tribus antequam Marchio Hospitalius quæ a Bernoullio didicerat in lucem edidit, & methodus illa verissima sit et optima.

Mathematicus ille Judex in scripto prædicto latine edito Bernoullium citabat tanquam a se diversum: Leibnitius vero sub finem anni 1715 in ejusdem versione Gallica, citationem illam (nescio qua fide) delevit et Mathematicum esse Bernoullium ipsum scripsit, et literis ad Abbatem de Comitibus datis Problemata Bernoullij Analystis Anglis solvenda proposuit, & chartam illam volantem denuo dispersit, et ad sententiam Bernoullij appellans, amicos suos rerum mathematicarum inscios in Newtonum totis viribus per literas impellere conatus est: cum tamen Bernoullius judex constitui jure nullo posset nisi ipse jure omni in methodum differentialem prius renunciasset; et Wallisius longe antea de hac re judicium contrarium tulisset, Leibnitio et Menkenio per ea tempora non mussitantibus. [12]Sub idem tempus Abbas de Comitibus ad Leibnitium de systemate Nigrisoli scripsit[13]& Leibnitius respondendo[14] Postscriptum addidit, in quo dixit Bernoullium tulisse judicium illud, & se *Commercio* non esse responsurum, & Newtonum modis omnibus aggressus est, proponendo quæstiones de gravitate & attractione, qualitati[bu]s occultis, & miraculis, & sensorio Dei, et vacuo et atomis et spatio et tempore & perfectione mundi, & prædicando fœlicitatem suam in discipulis, & proponendo Problema ab Analystis Anglis solvendum. Quæ omnia ad rem nil spectant.

In charta illa volante Leibnitius epistolam 1[2] Apr. 1675 scriptam[15] (qua Oldenburgius series aliquot ad Leibnitium miserat, et inter alias seriem Gregorij quam Leibnitius postea ut suam edidit) suspectam reddere conatus est; *Tale quiddam*, inquiens, *Gregorium habuisse ipsi Angli et Scoti, Wallisius, Hookius, et Newtonus & junior Gregorius ultra triginta sex annos ignoraverunt et Leibnitij esse inventum crediderunt.* Verum hæc Epistola in Libro Epistolico Regiæ Societatis asservata, ut et Epistola autographa Leibnitij se series missas accepisse agnoscentis, cum ijsdem epistolis in *Commercio* editis, coram Comite de Kilmansegg, Abbate de Comitibus, Ministris aliquot publicis exterorum Principum, & alijs

---

(12) The rest of this paragraph is taken from Newton's *addenda* on .41: 57ʳ.

(13) This letter from Conti to Leibniz, in the early spring of 1715, is printed by C. I. Gerhardt in *Der Briefwechsel von Gottfried Wilhelm Leibniz mit Mathematikern*, **1** (Berlin, 1899): 258–62. Its two paragraphs (*ibid.*: 259, 262) of main Newtonian concern are cited by A. R. Hall in *Correspondence*, **6**, 1976: 215.

(14) At the end of the year, on 6 December (N.S.); the lengthy offending 'P.S.' to this (and the text of its included 'Billet', on which see page 505 above) may be found in Gerhardt's *Briefwechsel von Leibniz*: 263–7 (=*Correspondence*, **6**: 250–3).

(15) Now most recently and accurately printed by J. E. Hofmann in his edition of Leibniz' 'Mathematischer, Naturwissenschaftlicher und Technischer Briefwechsel, 1672–1676' in *Leibniz: Sämtliche Schriften und Briefe*, (3) **1** (Berlin, 1976): 230–45; and also (*ex* Gerhardt's 1899 *Briefwechsel*: 113–22) by A. R. and M. B. Hall in their edition of *The Correspondence of Henry Oldenburg*, **11** (London, 1977): 265–73. Newton mistakenly here dates this letter '15 Apr.'.

exteris non paucis, Anno 1715,[16] in domo Regiæ Societatis collatæ sunt, et impressionis fides probata. Sed et Leibnitius ipse anno proximo in Epistola sua ad Cometissam de Kilmanseg 18 Apr. data, idem agnovit dum narrat, ut cum ipse de serie quam pro circulo invenerat, ad Oldenburgium scriberet (vizt per epistolas 15 Julij & 26 Octob. 1674;) Oldenburgius responderit (8 Decem. 1674) Newtonum quendam Cantabrigiensem jam antea similia dedisse, non solum pro circulo sed etiam pro omni figurarum aliarum genere, et ipsi miserit †*des Essais*[17] serierum †specimina. His verbis Leibnitius agnoscit se epistolam Oldenburgij 15 Apr. 1675 datam accepisse. Nam specimina illa erant in hac epistola. Et in eadem erat series Gregorij, ut in *Commercio epistolico*[18] videre licet. Sed pergit Leibnitius: *Hoc non obstante*, ait, *series mea satis laudata fuit per Newtonum ipsum. Postea inventum est Gregorium quendam eandem etiam seriem invenisse: se hoc didici tarde.* Hæc Leibnitius. *Literas* utiꝗ *multa fruge Algebraica refertas* acceperat, sed tunc *præter ordinarias curas Mechanicis imprimis negotijs distractus* non potuit *examinare series quas Oldenburgius miserat ac cum suis comparare* (ut ipse tunc rescripsit[19]) neque unquam comparavit, sed priusquam Epistolam 12 Maij 1676 datam ad Oldenburgium scripsit, oblitus est se series anno superiore missas accepisse, et seriem quam anno 1682 pro sua edidit, Gregorium quendam invenisse didicit tarde. Newtonus autem et Wallisius et junior Gregorius hanc seriem a Gregorio seniore ad Collinium et ab Oldenburgio ad Leibnitium fuisse missam, per ea tempora multo magis ignorarunt. Leibnitius itaꝗ epistolam Oldenburgij seriebus refertam accepit, sed series illas, si fas est credere, nunquam contulit cum suis.

Contra fidem epistolarum in *Commercio* editarum scripsit insuper Leibnitius, Arbitrorum Consessum a Regia Societate constitutum omnia edisse quæ contra ipsum facerent, omnia omisisse quæ contra Newtonum. Et præterea per epistolam 25 Aug. 1714 ad D. Chamberlain datam postulavit ille ut Societas Regia

---

(16) See page 506 above.

(17) The original French of what Newton renders in the text alongside as 'specimina', quoting from Desmaizeaux' edition of Leibniz' letter of 18 April 1716 (N.S.) to the Baroness von Kilmansegge in his *Recueil de Diverses Pieces*...(Amsterdam, ₁1720): 29–41 (→₂1740: 33–46=*Correspondence*, **6**: 324–9). In full citation of his words, Leibniz there wrote that 'M. Oldenbourg m'écrivit qu'un M. Newton à Cambridge avoit déja donné des choses semblables, non-seulement sur le Cercle, mais encore sur toutes sortes d'autres figures, & m'en envoya des essais', adding thereafter (to anticipate Newton's present sequel) that 'Cependant le mien fut assez aplaudi par M. Newton même. Il s'est trouvé par après, qu'un nommé M. Gregory avoit trouvé justement la même *series* que moi. Mais c'est ce que j'appris tard' (*ibid.*: 31–2, with italicisations of proper names ignored).

(18) *Commercium*, ₁1713: 39–41 (=₂1722: 118–21). This, however, but runs together the 2nd–8th and 9th–12th opening paragraphs of Oldenburg's lengthy full letter (on which see note (15) above).

(19) On 20 May 1675 (N.S.). Newton here partially cites the paragraph of Leibniz' letter which he had published in *Commercium*, ₁1713: 42 (=₂1722: 121–2).

epistolas nondum editas ad ipsum mitterent. *Nam cum Hanoveram,* inquit, *rediero, possum etiam in lucem mittere Commercium aliud Epistolicum quod historiæ Literariæ inservire possit; et literas quæ contra me allegari possunt non minus publici juris faciam quàm quæ pro me faciunt.*[20] Hæc Leibnitius. Sed omnes inter ipsum et Oldenburgium epistolæ, quatenus ad hanc rem spectant, continua serie jam antea in *Commercio* edito impressæ sunt præter duas quæ non extant (uti jam dictum est[21]) et nullius esse momenti videntur.

Attamen ut accusationem suam confirmaret, scripsit Leibnitius sub finem anni 1715, in Epistola sua prima ad Abbatem de Comitibus per Galliam missa, quod *in secundo ejus in Angliam itinere* Collinius ostenderit ipsi *partem Commercij sui in qua Newtonus agnoscebat ignorantiam suam in pluribus, dicebatǭ (inter alia) quod nihil invenisset circa dimensiones Curvilinearum quæ celebrantur, præter dimensionem Cissoidis; sed Corsessus hoc totum suppressit.*[22] Et Newtonus in epistola sua ad dictum Abbatem 26 Feb. 171$\frac{5}{6}$, respondit, hoc non fuisse omissum sed extare in epistola sua ad Oldenburgum 24 Octob. [1676] missa, & impressum fuisse in *Commercio Epistolico* pag. 74. lin. 10, 11.[23] Et subinde Leibnitius in Epistola sua proxima ad Abbatem de Comitibus Apr. 9. 1716 agnovit se errasse. *Sed,* inquit, *exemplum dabo aliud. Newtonus in una Epistolarum ejus ad Collinium agnovit se non posse invenire magnitudinem sectionum secundarum (vel segmentorum secundorum) sphæroidum & corporum similium, sed Consessus hunc locum vel hanc Epistolam minime edidit.*[24]

---

(20) 'Car quand je serai de retour à Hanover, je pourrai publier aussi un *Commercium Epistolicum,* qui pourra servir à l'Histoire Litteraire. Je serai disposé de ne pas publier moins les Lettres qu'on peut alleguer contre moi, que celles qui me favorisent...' in Leibniz' original French words (Desmaizeaux, *Recueil,* ₁1720: 124 [→₂1740: 128–9 = *Correspondence,* **6**: 173]).

(21) In the second paragraph above, page 683.

(22) Newton slightly paraphrases the sentences in the opening paragraph of Leibniz' 'P.S.' to Conti on 6 December 1715 (N.S.): '...à mon second voyage [d'Angleterre] M. Collins me fit voir une partie de son commerce'—'épistolique', that is—'& j'y remarquay que M. Newton avoua aussi son ignorance sur plusieurs choses, et dit entre autres qu'il n'avoit rien trouvé sur la dimension des Curvilignes celebres que la dimension de la Cissoide. Mais on [!] a supprimé tout cela' (see Raphson, *History of Fluxions,* ₂1717: Appendix: 98 [= *Correspondence,* **6**: 251]). It was of course in his *epistola posterior* to Leibniz in 1676 that Newton had modestly introduced his rectification of the cissoid's general arc—that set out by him five years earlier in 'Exempl: 5' of Problem 11 of his 1671 tract (see III: 320)—with the throw-away sentence 'Sed in simplicioribus vulgoǭ celebratis figuris vix aliquid relatu dignum reperi quod evasit aliorum conatus nisi fortè longitudo Cissoidis ejusmodi censeatur' (see *Correspondence,* **2**, 1960: 117). It is hard to see how Leibniz could seriously take Newton's diffidence at this place in his October 1676 letter as an acknowledgement of any ignorance on his part of the 'dimension of curved lines'; nor, since Newton's phrase is not omitted from the text of the *epistola* printed in full in the *Commercium Epistolicum* (₁1713: 67–86, especially 74), how he could possibly sustain his claim that 'one' had 'suppressed all that'.

(23) Of the 1713 edition, that is—'& I am not ashamed of it' he stoutly adjoined in his letter to Conti (Raphson, *History of Fluxions,* ₂1717: 101 = *Correspondence,* **6**: 286).

(24) 'Comme je n'ay pas daigné lire le *Commercium Epistolicum* avec beaucoup d'attention, je me suis trompé dans l'Exemple que j'ay cité, n'ayant pas pris garde, ou ayant oublié qu'il

Newtonus autem in Observationibus quas in hanc Leibnitij Epistolam scripsit, respondit:[25] *Si Consessus hoc omisisset, recte omnino omissum fuisse, cum hujusmodi cavillationes ad Quæstionem de qua agitur nil spectent; sed Consessum hoc minime omisisse. Collinius in Epistola ad D. Gregorium 24 Decem. 1670, et in altera ad D. Bertet 21 Feb. 1671 (utrisq̃ impressis in Commercio p. 24, 26) scripsit quod methodus Newtoni se extenderet ad secunda solidorum segmenta quæ per rotationem generantur. Et Oldenburgius idem scripsit ad Leibnitium ipsum 8 Decem. 1674, ut videre est in Commercio, pag. [3]9.* Leibnitius igitur accusationem finxit.[26] *Nam et Abbas de Comitibus per horas aliquot inspexit Epistolas antiquas in Archivis Reg. Societatis asservatas ut et Societatis Libros MSS in quibus Apographa sunt Epistolarum, ut [aliquod] invenret quæ pro Leibnitio vel contra Newtonum facerent, et omissa fuissent in Commercio Epistolico Collinij et aliorum: et ejus generis nihil invenire potuit, ut in Transactionibus Philosophicis pro Jan. & Feb. 1718 habetur.*[27]

---

s'y trouvoit; mais j'en citeray un autre: M. N[ewton] avouoit dans un de ses Lettres à M. Collins'—that of 20 August 1672 (see *Correspondence*, **1**, 1959: 229)—'qu'il ne pouvoit point venir à bout des Sections secondes (ou Segments seconds) de Spheroides ou corps semblables: mais on n'a point inseré ce Passage ou cette Lettre dans le *Commercium Epistolicum*...' (Raphson, *History*, ₂1717: 105=*Correspondence*, **6**: 306). Collins had shown Newton's letter to Leibniz during the latter's second visit to London in October 1676, and the rough notes which he then made upon it (now Hanover. Niedersächsische Landesbibliothek, **35**, VIII, 19: 2ʳ) are reproduced in photocopy by J. E. Hofmann at the end of his 'Studien zur Vorgeschichte des Prioritätstreites zwischen Leibniz und Newton.... I: Materialien zur ersten Schaffensperiode Newtons (1665–1675)' (*Preussische Akademie der Wissenschaften*/(1943). *Math.-naturw. Klasse*, No. 2) together with a transcription of their text on *ibid.*: 80. (Leibniz' Latin notes may now also be found set out in parallel column alongside the corresponding passages of Newton's 1672 English letter in *Leibniz: Sämtliche Schrifte und Briefe*, (3) **1**, 1976: 677–80.) This new instance of (so Leibniz would have it) suppression of evidence in the 1713 *Commercium* again badly backfires. Not only, as he here hastens to drum home, had Gregory's rule for 'second segments' (as sent to Collins in December 1670) itself been printed in the *Commercium*, but in there making no mention of his August 1672 letter he had, along with his admission that he could find no approximating series for the purpose 'more simple' than Gregory's, omitted to make public a rule for 'guageing' a 'Parabolick spindle'—one which Leibniz passed by in his 1676 notes and doubtless forty years later had long forgotten ever existed—whose subtlety we explored in III: 568–9. (The analysis of this rule by H. W. Turnbull in *Correspondence*, **1**: 232–3 is, we may add, a deal deficient and in error.)

(25) See Raphson, *History*, ₂1717: 113 (=*Correspondence*, **6**: 343) for Newton's original English phrasing in his 'Observations' upon Leibniz' letter to Conti.

(26) The 1722 'Ad Lectorem' here has in diminution '...iterum erravit'. Newton's words in his 1717 'Observations' were yet stronger: 'So you see that Mr Leibnitz hath accused the Committee of the R. Society without knowing the truth of his accusation & therefore is guilty of a misdemeanour'. (We cleave to Newton's orthography in his autograph draft on Add. 3968.41: 447ᵛ.)

(27) (*Philosophical Transactions*, **30**, No. 359: 'pag. 925'. There, in an unsigned 'N.B.' to Conti's letter to Leibniz in March 1716 (*ibid.*: 923–5 [=*Correspondence*, **6**: 295–6]) here now first published, Newton stressed that 'Mr. l'Abbé Conti spent some Hours...in looking over the old Letters and Letter Books kept in the Archives of the Royal Society, to see if he could find any thing which made either for Mr. Leibnitz, or against Mr. Newton, and had been

Cæterum Leibnitius[28] in prima sua ad Abbatem de Comitibus epistola dixit, *eos qui contra ipsum scripsissent* (id est Consessum a Regia Societate constitutum) *candorem ejus aggressos esse per interpretationes duras et male fundatas, & voluptatem non habituros esse videndi Responsa ejus ad pusillas rationes eorum qui ijs tam male utuntur.*[29] Interpretationes illæ nullius quidem sunt autoritatis nisi quam ab Epistolis derivant; at male fundatas esse Leibnitius nunquam ostendit.

Subinde vero Newtonus[30] in prima sua ad Abbatem Epistola 26 Feb. 171$\frac{5}{6}$ ita rescripsit. *D. Leibnitius hactenus respondere recusavit, bene intelligens impossibile esse res factas refutare. Silentium suum hac in re excusat, prætexens se librum nondum vidisse, & otium illi non esse ad examinandum, sed se orasse Mathematicum celebrem ut hoc negotium in se susciperet.——Utitur et novo prætextu ne respondeat, dicens quod Angli voluptatem non habebunt videndi responsum ejus ad pusillas eorum rationes; & proponens disputationes novas philosophicas ineundas, & Problemata solvenda: quæ duo ad rem nil spectant.*[31]

D. Leibnitius autem in proxima sua ad Abbatem Epistola 9 Apr. 1716 data, et per Galliam in Angliam missa,[32] pergebat se excusare ne respondeat. *Ut operi,* inquit, *contra me edito sigillatim respondeam, opus erit alio opere non minore quam hoc est; percurrendum erit corpus magnum minutorum ante annos 30 vel 40 præteritorum quorum perparvum reminiscor; examinandæ erunt veteres epistolæ quarum plurima sunt perditæ, præterquam quod maxima ex parte non conservavi minuta mearum, et reliquæ sepultæ sunt in maximo chartarum acervo quem non possum sine tempore et patientia discutere. Sed otium minime mihi suppetit, alijs negotijs alterius prorsus generis occupato.*[33]

---

omitted in the *Commercium Epistolicum*...: but could find nothing of that kind'. (We ignore the printer's italicisations.) The sentence, rendered into Latin, is quoted in full in the printed 'Ad Lectorem' (*Commercium*, ₂1722: signature A2ᵛ, bottom; Newton's draft is on .41: 56ʳ).

(28) 'Insuper D. Leibnitius, ut *Commercium Epistolicum* sine responso dimitteret' in the 1722 'Ad Lectorem' (signature A3ʳ, ll. 1–2).

(29) In the original French words used by Leibniz in the 'P.S.' of his first letter to Conti on 6 December 1715 (N.S.): 'Ceux qui ont ecrit contre moy n'ayant pas fait difficulté d'attaquer ma candeur par des interpretations forcées et mal fondées, ils n'auront point le plaisir de me voir répondre à de petites raisons de gens qui en usent si mal...' (see *Correspondence*, **6**: 250).

(30) 'qui ægre adductus est ut scriberet' Newton adjoined in the printed 'Ad Lectorem'.

(31) In rendering them into Latin, Newton somewhat paraphrases his assertions in his original letter to Conti that 'Mʳ Leibnitz has hitherto avoided returning an Answer to the [*Commercium Epistolicum*]; for the Book is matter of fact & uncapable of an Answer. To avoid answering it he pretended the first year'—after its publication, namely 1713—'that he had not seen this Book nor had leasure to examin it, but had desired an eminent Mathematician to examin it....And now he avoids it by telling you that the English shall not have the pleasure to see him return an Answer to their slender reasonings (as he calls them) & by endeavouring to engage me in dispute about Philosophy & about solving of Problems, both which are nothing to the Question' (see *Correspondence*, **6**: 285).

(32) This superfluous phrase is omitted from the printed 'Ad Lectorem'.

(33) 'Pour repondre donc de point en point à l'Ouvrage publié contre moi, il falloit un autre Ouvrage aussi grand pour le moins que celuy-là, il falloit entrer dans un grand detail de quantité de Minuties passées il y a 30 ou 40 Ans dont je ne me souvenois gueres; il me

Et paulo post. *Epistolæ non erant truncandæ. Nam parvum est quod restat inter meas chartas vel cujus mihi relinquuntur minuta. Sic omnibus perpensis, videns tanta malignitatis & fraudis indicia,*[34] *cre[di]di indignum esse me rem discutere cum hominum genere qui se tam male gerunt. Sentio quod in ijs refutandis difficile fuerit ab opprobrijs et expressionibus asperis abstinere, qualibus eorum acta*[35] *merentur, & tale spectaculum dare publico non cupio,*[36] *in animo habens tempus meum melius impendere, quod mihi pretiosum esse debet, & satis contemptui habens*[36] *judicium eorum qui super tale opus (*Commercium *scilicet) contra me sententiam dicere voluerunt: præsertim cum Societas Regia hoc facere noluerit, uti didici per Extractum ex eorum Registris.*[37] Hæc Leibnitius.[38]

Attamen post ejus mortem (quæ contigit proximo mense Septembri[39]) in Elogio ejus quod in *Actis Eruditorum* pro mense Julio anni 1717 impressum fuit, amici ejus[40] scripserunt eum *Commercio Epistolico Anglorum aliud quoddam suum idemq amplius opponere decrevisse; et paucis ante obitum diebus Cl. Wolfio significasse se Anglos famam ipsius lacessentes reipsa refutaturum: quamprimum enim a laboribus historicis vacaturus sit, daturum se aliquid in Analysi prorsus inexpectatum, et cum inventis quæ hactenus in publicum prostant, sive Newtoni sive aliorum nihil quicquam affine habens.* Hæc illi. Verum ex jam dictis patet illum aliud nullum cum Oldenburgio[41] Commercium epistolicum habuisse [quod ederet]. Et inventum novum his

---

falloit chercher mes vielles Lettres, dont plusieurs se sont perdues, outre que le plus souvent je n'ay pas gardé les Minutes des miennes; & les autres sont ensevelies dans un grand tas de Papiers, que je ne pouvois debrouiller qu'avec du Temps & de la Patience. Mais je n'en avois gueres le loisir, êtant chargé presentement d'Occupations d'une toute autre Nature' (*Correspondence*, **6**: 306).

(34) 'tantas...notas' in the printed 'Ad Lectorem' (A3ᵛ), after which Newton wrote in slightly different Latin rendering 'credidi indignum esse me ingredi discussionem cum...'.

(35) 'facta' in the printed version.

(36) Lightly amended to read '& non cupio hujusmodi spectaculum exhibere publico' and '& contemnens' *tout court* respectively in the printed 'Ad Lectorem'.

(37) '...il auroit êté plus sincere...particulierement de ne pas tronquer les Lettres, car il y en a peu parmi mes Papiers, ou dont il me reste des Minutes. Ainsi tout consideré, voyant tant de marques de malignité & de chicane, je crûs indigne de moi d'entrer en discussion avec des gens qui en usoient si mal. Je voyois qu'en les refutant on auroit de la peine à éviter des Reproches, & des Expressions fortes, telles que meritoit leur Procedé; & je n'avois point envie de donner ce Spectacle au Public, ayant dessein de mieux employer mon temps, qui me doit être precieux, & meprisant assez le Jugement de ceux qui sur un tel Ouvrage voudroient prononcer contre moi, d'autant que la Societé Royale même ne la point voulu faire; comme je l'ay appris par un extrait de ses Registres'. Newton's rendering of Leibniz' last phrase is omitted in the 'Ad Lectorem' to the reissued *Commercium*.

(38) 'Quæstionem primam deserit rixando, & Quæstiones novas proponit' Newton adjoined in the ensuing printed 'Ad Lectorem'.

(39) Rightly corrected to be 'Novembris' (*sic*) in the printed preface.

(40) The *Acta*'s editor Christian Wolf would seem to have been the author of this Latin *éloge*.

(41) This unnecessarily restrictive specification of Leibniz' possible partner in an epistolary commerce with England in the 1670's is omitted in the printed 'Ad Lectorem'.

nihil affine habens, ad rem nihil spectat. Missis ægrorum somnijs Quæstio tota ad Epistolas antiquas referri debet. [42]Et hæc Quæstio est (secundum Compilatores *Actorum* Lipsiensium): Utrum Leibnitius sit inventor methodi de qua disputatur, et *pro Differentijs igitur Leibnitianis Newtonus adhibet semperǫ* (ex quo usus est hac methodo) *adhibuit Fluxiones, quemadmodum* [*et*] *Honoratus Fabrius motuum progressus Cavallerianæ methodo substituit.*[43] Quæritur, non quis methodum totam invenit (nam tota nondum inventa est) sed quis Methodum invenit quatenus in scriptis a Newtono editis habetur. Quæstiones aliæ omnes dimittendæ sunt & hæc sola discutienda.

Ad hanc Quæstionem spectat quod Leibnitius differentias & methodum differentialem vocat quas Newtonus momenta & methodum momentorum, et quod methodus momentorum et methodus fluxionum una et eadem sit methodus. Momenta sunt partes quas Leibnitius differentias vocat, fluxiones sunt velocitates quibus partes generantur. In methodo Leibnitij considerantur partes, in ea Newtoni considerantur etiam velocitates. Newtoni methodus est amplior & Leibnitij methodum complectitur. Sed Newtonus methodum inversam reliquit imperfectam, et quæritur quid alij addiderint.

Eodem spectat quod D. Wallisius Propositionem primam Libri *de Quadraturis*, exemplis inveniendi fluxiones primas et secundas illustratam edidit anno 1693 in Volumine secundo *Operum* suarum, pag. 392,[44] ut supradictum est. Et ejusdem Propositionis solutionem Newtonus demonstravit synthetice in Lem. 2. Lib. 2 *Principiorum* anno 1686.[45] Sunto quantitates datæ a, b, c, fluentes x, y, z, fluxiones p, q, r, & momenta op, oq, or. Et proponatur æquatio quævis fluentes involvens, puta $x^4 - axyy + by^3 - z^4 + bbcc = 0$. Et per Lemma prædictum, si sola fluat x, momentum totius erit $4x^3op - ayyop$; si sola fluat y, momentum totius erit $-2axyoq + 3byyoq$; si sola fluat z, momentum totius erit $-4z^3or$; si fluant omnes, momentum totius erit

$$4x^3op - ayyop - 2axyoq + 3byyoq - 4z^3or.$$

Et quoniam totum semper est æquale nihilo, momentum totius erit æquale nihilo. Hæc est æquatio involvens fluentium momenta. Si eadem dividatur per o, habebitur æquatio involvens fluxiones. Per hoc Lemma igitur solvitur Propositio: *Data æquatione fluentes quotcunǫ quantitates involvente, fluxiones invenire.*

---

(42) The remainder of this paragraph—urging that the only question to be asked regarding priority of invention of calculus is 'Who [*sc.* first] found the method (as it is set out in Newton's writings)?' and setting aside Leibniz' one of 'Who found the whole method (as it is now known)?'—is omitted from the printed 'Ad Lectorem', and the sequel discarded in favour of the shorter, greatly different generalised sally at Leibniz' provocations and prevarications which is quoted *in extenso* in our final footnote.

(43) Leibniz' words, we surely need not say, on page 35 of his January 1705 *Acta* review (unsigned by him) of Newton's then newly published *De Quadratura Curvarum*.

(44) See VII: 174–6.   (45) See IV: 521–5.

Et in hujus Solutione fundatur methodus fluxionum uti dictum est in Scholio quod eidem Propositioni subjungitur. Eadem Propositio extat in Epistola Newtoni ad Oldenburgium 24 Octob. 1676,⁽⁴⁶⁾ et ibi dicitur esse fundamentum methodi generalis de qua Newtonus Tractatum scripserat tum ante annos quincȝ, id est anno 1671. Hujus autem Solutio exhibet *Algorithmum* seu calculum Arithmeticum Methodi ejusdem, ideocȝ *Algorithmus* ille Newtono innotuit anno 1671.⁽⁴⁷⁾

Eodem spectat quod Propositio secunda libri *de Quadraturis* extat soluta in *Analysi per series* quam Barrovius anno 1669 ad Collinium misit, [ut videre est in *Commercio*] pag. 19, ubi docetur Inventio Curvarum quæ quadrari possunt. Nam haec Propositio secunda pendet a Propositione prima Libri ejusdem; ideocȝ Propositiones duæ primæ Libri *de Quadraturis* Newtono innotuere anno 1669. Propositio autem tertia et quarta sunt exempla tantum Propositionis secundæ, ut ibi dicitur. Et propterea Methodus fluxionum quatenus in Propositionibus quatuor primis Libri *de Quadraturis* habetur, Newtono innotuit anno 1669.

Eodem spectat quod Propositionem quintam Libri *de Quadraturis* Wallisius edidit anno 1693 in secundo *Operum* suorum volumine pag. 391. Hac Propositione quadrantur figuræ accuratè et Geometricè si fieri potest. Et hoc artificium Newtono innotuit anno 1676, uti patet per Epistolam ejus 24 Octob. ejusdem anni ad Oldenburgium missam pag [72].⁽⁴⁸⁾ Ut et anno 1669 uti affirmatur in *Analysi per series* quam Barrovius eo anno ad Collinium misit, pag. [18].⁽⁴⁸⁾ Imò et annis aliquot antequam Mercatoris *Logarithmotechnia* prodijt, testibus Barrovio & Collinio in Epistola Collinij ad D. Strode [26 Julij 1672 data], pag. [28].⁽⁴⁸⁾ Atqui Propositio illa quinta pendet a Propositionibus quatuor prioribus. Ideocȝ methodus fluxionum quatenus continetur in Propositionibus quincȝ primis Libri *de Quadraturis* Newtono innotuit annis aliquot antequam prodiret Mercatoris *Logarithmotechnia*, id est anno 1666 aut antea,

(46) Stated merely in the alphabetically ordered anagram of its enunciation, of course. (See *Correspondence*, **2**: 115.)

(47) In a late addition on .41: 57ʳ Newton here subjoined: 'Sed et Leibnitius in P.S. sub Epistola [secunda 9 Apr. 1716] ad Abbé Conti sic scripsit. *Nulla sunt vestigia necȝ umbræ*'—'Il n'y a pas la moindre trace, ni ombre' was Leibniz' French phrase (see *Correspondence*, **6**: 312)—'*calculi differentiarum aut fluxionum in antiquis omnibus Newtoni Epistolis præterquam in illa quam scripsit 24 Octob. 1676. ubi ille non loquitur nisi per ænigma. Et hujus ænigmatis solutio quam ille dedit post decennium*'—in the scholium to Lemma II of Book 2 of the 1687 *Principia*, that is—'*aliquid loquitur, sed non dicit illud omne quod quis desiderare possit. Hoc non obstante, dignatus sum alibi de ea loqui quasi diceret fere totum, & alij post me idem de ea locuti sunt. Hæc* Leibnitius'—who adjoined, however, that 'Mon honneté a été mal reconnuë'. Newton left unfinished a last sentence where he began to comment: 'Et qui hoc concedunt, certe concedunt Newtonum non substituisse fluxiones pro differentijs sed method[um de novo invenisse?]'.

(48) Following his practice elsewhere in this 'Præfatio', we fill the blanks which Newton has here left in the manuscript by citations of the pertinent page-numbers in the first (1713) edition of the *Commercium*.

testibus Barrovio et Collinio. Id quod testatus est etiam Wallisius in Præfatione ad *Operum* suorum Volumen primum.

Ad eandem Quæstionem spectat quod in Libro de *Analysi per Series Fluxiones ac Differentias*, anno 1711 a Jonesio edito, pag. 38, extet Fragmentum Epistolæ Newtoni ad Collinium Nov. 8 [1676] datæ, his verbis: *Nulla extat Curva cujus Æquatio ex tribus constat terminis, in qua, licet quantitates incognitæ se mutuo afficiant, & indices dignitatum sint surdæ quantitates (v.g.* $ax^\lambda + bx^\mu y^\sigma + cy^\tau = 0$: *ubi x designat basin; y ordinatam;* $\lambda, \mu, \sigma, \tau$ *indices dignitatum ipsarum x & y; & a, b, c quantitates cognitas, una cum signis suis* $+$ *&* $-$) *nulla, in quam, hujusmodi est Curva de qua an quadrari possit necne, vel quænam sint figuræ simplicissimæ quibuscum comparari possit, sive sint Coricæ Sectiones, sive aliæ magis complicatæ, intra horæ Octantem respondere non possim. Deinde methodo directa et brevi, imo methodorum omnium generalium brevissima, eas comparare queo. Quinetiam si duæ quævis figuræ per hujusmodi æquationes expressæ proponantur, per eandem Regulam eas modo comparari possint, comparo.——Eadem methodus æquationes quatuor terminorum, aliasq complectitur, haud tamen adeo generaliter.*[49] Hactenus Newtonus. Hæc autem absq methodo fluxionum fieri non possunt; indicant vero Methodum quadrandi Curvilineas in Libro *de Quadraturis* expositam, et methodum fluxionum in qua methodus altera fundatur, eousq promotas fuisse ante 8 Novem. 1676.

Eodem spectat etiam quod in Epistola Newtoni ad Oldenburgium 24 Octob. 1676 data, descriptæ habentur Ordinatæ Curvilinearum, quarum collationes cum Conicis sectionibus Newtonus in Catalogum tunc *olim* retulerat, id est anno 1671 aut antea. Nam anno 1676 Newtonus annos quinq ab hac methodo promovenda abstinuerat, ut ipse ibidem refert. Earundem Curvilinearum, et eodem ordine et modo ijsdemq literis descriptarum collationes cum Conicis sectionibus ponuntur in Tabula posteriore duarum quæ in Scholio ad Propositionem decimam Libri *de Quadraturis* habentur, ideoq Tabula illa composita fuit et methodus quadrandi Curvilineas eousq producta annis minimum quinq ante annum 1676. Id quod absq methodo Fluxionum fieri non potuit. Jam vero Propositio decima Libri *de Quadraturis* pendet a Propositionibus novem primis ejusdem Libri: ideoq Propositiones decem primæ hujus Libri Newtono innotuere anno 1676, vel potius anno 1671.[50]

---

(49) As Newton says, this is William Jones' Latinisation of the main portion of this November 1676 letter to Collins, as printed by him on page 38 of his 1711 compendium of Newtonian mathematical *Analysis*. For the original English text (also given by Jones *ibid.*) see *Correspondence*, 2: 179–80. Newton's following assertion that the rules for squaring trinomial curves (see III: 373–85) to which he referred in his letter 'cannot be derived without the method of fluxions' and the 'method of squaring curves expounded in the *De Quadratura*' goes against the historical fact, of course; compare Appendix 7.1: note (15) above.

(50) Newton repeats the dubious chain of reasoning which he had earlier refined and set down, honed to a nicety, in the 'admonition' to the reader of his *De Quadratura* (in his intended

Ad eandem Quæstionem spectat quod Newtonus in Epistola sua prædicta ad Oldenburgium 24 Octob. 1676 data, ubi Problematum genera quædam nominasset quæ per Methodum suam solverentur, et methodum Tangentium Slusij inde fluere, idçз absçз æquationum Reductione dixisset; subjungit: *Fundamentum harum operationum satis* OBVIUM *quidem, quoniam jam non possum explicationem ejus prosequi, sic potius celavi. 6acc dæ* &c.[51] Celavit igitur ut *obvium*, ne subriperetur. Quam vero fuit *obvium* et quam facile subripi potuit sic patebit.

Gregorius scripsit ad Collinium 5 Sept. 1670 se ex Barrovij Methodis Tangentes ducendi invenisse methodum generalem et Geometricam ducendi Tangentes ad omnes Curvas sine Calculo. Slusius se ejusmodi methodum Tangentium habere scripsit ad Oldenburgium mense Octobri vel Novembri 1672. Et Newtonus 10 Decem. 1672 scripsit ad Collinium in hæc verba: *Ex animo gaudeo D. Barrovij amici nostri Reverendi Lectiones Mathematicas exteris adeo placuisse, neçз parum me juvat intelligere eos* (Slusium et Gregorium) *in eandem mecum incidisse ducendi Tangentes methodum* &c.[52] Et subinde Newtonus in eadem Epistola methodum suam ducendi Tangentes descripsit, et addidit hanc methodum esse partem vel Corollarium potius methodi suæ generalis solvendi abstrusiora Problemata, et non hærere ad quantitates surdas. Epistolas totas Gregorij et Newtoni habes infra in *Commercio Epistolico*, et earum Apographa Oldenburgius 26 Junij 1676 misit ad Leibnitium inter excerpta ex Gregorij epistolis, et Leibnitius incidit in Prælectiones Barrovij in Anglia mense Octobri anni 1676, ut ipse scripsit in Epistola ad Abbatem de Comitibus 9 Apr. 1716.[53] Sunto jam, ut in Epistola Newtoni quantitates datæ *a, b, c*, [&c], Abscissa $AB = x$, Ordinata $BC = y$, & Linea curva ACF; et proponatur Æquatio quævis quantitates illas duas fluentes *x* et *y* involvens, puta

$$x^3 - 2xxy + bxx - bb[x] + byy - y^3 = 0,$$

ut in eadem Epistola: et ducenda sit recta CD quæ Curvam tangat in C et Abscissam utrinçз productam secet in D. Multiplicetur omnis Æquationis terminus per indicem

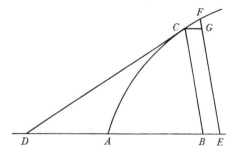

reissue of it in annexe to the *Principia*) whose text is given in Appendix 7.1 above; see pages 658–9. We need not here, yet again, be elaborate in our comment upon its weaknesses.

(51) See note (46) preceding.

(52) 'I am heartily glad at the acceptance wch our Reverend friend Dr Barrow's Lectures finds wth forreign Mathematicians, & it pleased me not a little to understand that they are falln into the same method of drawing Tangents wth me...' in Newton's original words (see *Correspondence*, **1**: 247). Here he quotes the Latin rendering on *Commercium*, ₁1713:29 (parenthesis and all).

(53) This does not adequately convey the tenor of Leibniz' own careful statement to Conti (see *Correspondence*, **6**: 310) that 'si quelqu'un a profité de M. Barrow, ce sera plus tôt M.

dignitatis x et productum divisum per x (videlicet $3xx - 4xy + 2bx - bb$) vocetur R. Multiplicetur omnis æquationis terminus per indicem dignitatis y, et productum divisum per y (videlicet $-2xx + 2by - 3yy$) vocetur S. Et per Regulam in Epistola illa Newtoni traditam, erit subtangens $BD = \dfrac{Sy}{R}$, vel potius $= -\dfrac{Sy}{R}$, propterea quod AB et BD ducantur ad partes contrarias. Et hæc est Regula ducendi Tangentes quam Newtonus in Epistola illa posuit ut partem aliquam vel specimen vel Corollarium Methodi suæ generalis.[54] Methodus vero tota ex hac ejus parte, et Propositio generalis ex hoc ejus Corollario sic deducitur.

Agatur secundum methodum Tangentium Barrovij & Gregorij, Ordinata nova EF priori BC proxima, et compleatur parallelogrammum $BCGE$; et pro differentijs vel momentis BE et FG, scribantur p, et q. Et erit FG ad GC ut est CB ad BD, id est, q ad p ut est y ad BD, seu $\dfrac{py}{q} = BD = -\dfrac{Sy}{R}$, et facta reductione, $Rp + Sq = 0$. Hæc Æquatio ubi duæ tantum sunt fluentes involvit earum differentias. Et ubi plures sunt fluentes, operatio similis ad omnes applicata dabit æquationem involventem omnium Differentias. Et Theorema hocce quod sic ex Newtoni Epistola facillime colligitur, illud omne comprehendit quod Leibnitius ad Newtonum Anno 1677 rescripsit, ut et illud omne quod in *Actis Eruditorum* Anno 1684 in lucem edidit.[55] Nam solutionem comprehendit Propositionis primæ Libri *de Quadraturis*.

Eodem deniqʒ spectat quod D. Leibnitius per Literas 12 Maij 1676 datas, peteret ab Oldenburgio, ut is demonstrationem serierum duarum Newtoni, id est methodum easdem inveniendi in *Analysi* per series descriptam, postularet a Collinio et ad se mitteret;[56] et quod sub finem mensis Octobris ejusdem anni

---

N[ewton] qui a étudié sous luy que moi qui (autant que je puis m'en souvenir,) n'ay veu les Livres de M. Barrow qu'à mon second Voyage d'Angleterre [en Octobre 1676], & ne les ay jamais lûs avec attention, parce qu'en voyant le Livre je m'apperçus que par la consideration du Triangle Characteristique...j'étois venu comme en me jouant aux Quadratures, Surfaces & Solides dont M. Barrow avoit remply un Chapitre des plus considerables de ses *Leçons*...'.

(54) In essence, yes. Notice, however, that Newton here sets down an anachronistic generalisation of the 'one particular'—that of the Cartesian curve

$$x^3 - 2x^2y + bx^2 - b^2x + by^2 - y^3 = 0$$

(compare III: 122–4)—with which he had half a century earlier in his letter to Collins in December 1672 (see *Correspondence*, **1**: 247) instanced what he 'guessed' to be the method of tangents employed by the 'foreigners' James Gregory and Sluse. There are those who believe that notation matters all.

(55) In his 'Nova methodus pro maximis & minimis, itemque tangentibus, ...& singulare pro illis calculi genus' (*Acta* (October 1684): 467–73) namely; see our fuller bibliographical citation of this in Appendix 2: note (91) preceding.

(56) See *Correspondence*, **2**: 3. Newton could never be persuaded to see Leibniz' innocent request for enlightenment as anything but deceitfully two-faced.

Leibnitius videret in manibus Collinij epistolas plures Newtoni Gregorij et aliorum quæ præcipue de seriebus scriptæ erant, et inter alias Epistolam Newtoni ad Oldenburgium 24 Octobris ejusdem datam, ubi Newtonus *Analysin* illam per series *Compendium* methodi serierum vocat. Et nondum probatum fuit quod Leibnitius eo tempore non viderit hoc *Compendium* vel hanc *Analysin* per series; ubi Methodus fluxionum describitur, & symbola o, ov, oy, $\boxed{\dfrac{aa}{64x}}$, idem significant cum symbolis dz, dy, dx, $\int \dfrac{aa}{64x}$, & similibus, quæ D. Leibnitius postea adhibuit.

Et his præmissis legatur jam Recensio *Commercij Epistolici*, et consulatur *Commercium* ipsum, sicubi de factis dubitatur.[57]

---

(57) As stated in note (2) above, we borrow this last paragraph from Newton's preceding draft on .13: 208^r, supposing that he had it in mind to round off this final preliminary version of his preface to the 1722 *Commercium* in much the same manner had he gone on to finish it. The printed 'Ad Lectorem' (signatures A3^v–A4^v) discards the present last eleven paragraphs (compare note (42) above) and concludes in their place:

'Initio secundæ ad Abbatem de Comitibus Epistolæ [*viz.* 9 Aprilis 1716] D. Leibnitius primam Newtoni Epistolam vocavit *speciem chartæ provocatoriæ* [une espece de cartel] *ex parte Newtoni*, dein'—see Raphson, *History of Fluxions*, ₂1717: 103 (=*Correspondence*, 6: 304) for Leibniz' original French remarks about his unwillingness 'entrer en lice' 'with the enfans perdus' that Newton had detached against him—'addidit; *In arenam descendere nolui contra ejus milites emissarios, sive intelligas Accusatorem supra fundamentum Commercii Epistolici, sive Præfationem spectes acrimoniæ plenam, quam alius quidam novæ Principiorum [e]ditioni præmisit: Sed cum is per se jam lubens apparebit, paratus sum ipsi satisfactionem dare.* Et Newtonus'—in his adjoined 'Observations' upon Leibniz' letter in Raphson, *History*: 114 (=*Correspondence*, 6: 344)—respondit, *D. Leibnitium literas & chartas antiquas seponere, & ad Quæstiones circa philosophiam & res alias confugere. Et magnum illum Mathematicum, cui sine nomine ut Judici Epistolam 7 Jun. 1713 datam attribuerat, jam ut advoca[t]um*'—Newton here initially described the transparently disguised 'Party-man' Bernoulli as '*militem*', now ready to do battle '*velo sublato*', but was persuaded by Varignon in August 1722 to retract the phrase in the interest of future peace between them (see note (223) of our preceding introduction)—'*in hac rixa pro se inducere, mathematicos in Anglia provocantem* (uti fingitur) *ad problemata solvenda; quasi Duellum* (cum Leibnitio scilicet) *vel forte prælium cum exercitu discipulorum ejus* (quos jactat) *methodus esset magis idonea ad veritatem dirimendam, quam discussio veterum & authenticorum scriptorum, & Mathesis factis heroicis vice rationum ac demonstrationum abhinc implenda esset.* Hic rationes ac demonstrationes alludunt ad argumenta e scriptis veteribus desumpta, & facta heroica [*Atchievements in Knight-errantry*] ad contentiones philosophicas & problematicas ad rem nil spectantes, ad quas D. Leibnitius a prioribus aufugit.

Quæ novæ *Principiorum* editioni præmissa sunt Newtonus non vidit antequam Liber in lucem prodiit. Quæ de Quæstionibus Philosophicis disputata sunt D. Des Maizeaux a D. Leibnitio & aliis accepit & in lucem edidit. (Vide Epistolas D. Leibnitii ad D. Des Maizeaux 21 Aug. 1716, & D. Des Maizeaux ad Abbatem de Comitibus 21 Aug. 1718, in *Collectionum* [*Recueil*] Tomo secundo, pag. 356, & 362 impressas.) Solutiones Problematum'—those, stemming from Bernoulli, which were sent by Leibniz to 'test the pulse' of English mathematicians in the winter/spring of 1715/16 (see pages 62–5, 66–7 above)—'maxima ex parte lucem viderunt in *Actis Eruditorum*'.

And there Newton closed his unsigned address 'Ad Lectorem', much as in the first paragraph of his preliminary 'Præfatio' above, by adducing his immediate reason for reissuing the text of the 1713 *Commercium* in amplified form:

> '*Commercii Epistolici* exempla tantum pauca impressa fuerunt, & ad Mathematicos missa qui de his rebus judicare possent, neque prostant venalia. Ideoque hunc Librum, ut & ejus Recensionem quæ in *Transactionibus Philosophicis* ac *Diario Literario*, anno 1715 (anno & septem vel octo mensibus ante obitum Leibnitii) impressa sunt, iterum imprimere visum est, ut historia vera ex antiquis monumentis deducta, missis disputationibus quæ ad rem nil spectant, ad posteros perveniat, & sic finis imponatur huic controversiæ. Nam. D. Leibnitius a Quæstione desciscens emortuus est, & judicium posteris relinquitur. Denique Judicium primarii Mathematici'—on this see Appendix 9 preceding—'subjunctum est, unà cum Notis quibus pateat, eidem in Recensione prædicta, vivente Leibnitio, responsum esse, & scopum ejus fuisse tantum, ut Commercium Epistolicum sine Responso dimitteretur'.

At which point we may fitly ourselves write *finis* to erecting Newton's mathematical monument.

INDEX OF NAMES